Engineering Design Communication

...n through Graphics

Shawna D. Lockhart
Montana State University

Cindy M. Johnson

Prentice Hall
Upper Saddle River, NJ 07458

Library of Congress Cataloging-in-Publication Data

Lockhart, Shawna D.,
Engineering design communication : conveying design through graphics / Shawna D. Lockhart, Cindy M. Johnson.
p. cm.
Includes bibliographical references (p.).
ISBN 0-201-33151-9
1. Engineering design—Graphic methods. 2. Engineering graphics.
I. Johnson, Cindy M. II. Title.
TA337.L63 2000
620'.0042—dc21 99-41992
CIP

Editor-in-Chief: Marcia Horton
Acquisitions Editor: Eric Svendsen
Editorial Assistant: Kristen Blanco
Associate Editor: Joe Russo
Marketing Manager: Danny Hoyt
Production Editor: Kerry Reardon
Executive Managing Editor: Vince O'Brien
Managing Editor: David A. George
Art Director: Gus Vibal
Cover Art Director: Jayne Conte
Artists: Marita Froimson, Marla Goodman, and Mike Ohar
Page Layout and Composition: Greg Johnson, Art Directions
Cover Art: Courtesy of Wayne Fehres and Carl Fehres and IRI International, Barrett Technology, Tektronix, Inc., Albert Brown, Jr., and Park City Entertainment.
Cover design: Bruce Kenselaar
Manufacturing Buyer: Beth Sturla
Assistant Vice President of Production and Manufacturing: David W. Riccardi

The authors and publisher of this book have used their best efforts in preparing this book. These efforts include the development, research, and testing of the theories to determine their effectiveness.

Printed in the United States of America

10 9 8 7 6 5 4 3 2 1

ISBN 0-201-33151-9

Prentice-Hall International (UK) Limited, *London*
Prentice-Hall of Australia Pty. Limited, *Sydney*
Prentice-Hall Canada Inc., *Toronto*
Prentice-Hall Hispanoamericana, S.A., *Mexico*
Prentice-Hall of India Private Limited, *New Delhi*
Prentice-Hall of Japan, Inc., *Tokyo*
Pearson Education Asia Pte. Ltd., *Singapore*
Editora Prentice-Hall do Brasil, Ltda., *Rio de Janeiro*

Contents

DESIGNER'S NOTEBOOK 2

CHAPTER 1 ENGINEERING GRAPHICS AND THE DESIGN PROCESS 3

Objectives 3
What Is Engineering Design? 5
The Role of Graphics 6
Visualization 6
Communication 7
Documentation 8
A Model of the Design Process 9
Graphics Tools in Action 11
Problem Identification 11
Graphics Tools/Skills: Reading Existing Drawings 11
Tektronix 12
Ideation 12
Graphics Tools/Skills: Freehand Sketching 12
Design Selection 14
Graphics Tools/Skills: Interpreting Charts and Sketches 14
Refinement 15
Graphics Tools/Skills: 3D and Parametric Models 15
Analysis 17
Graphic Tools/Skills: Exporting Model Geometry for Analysis 17
Design Selection 18
Implementation 19
Graphics Tools/Skills: Manufacturing and Assembly Drawings 19
Documentation 20
Graphics Tools/Skills: Rendered Views, Exploded Views 20
Types of Design Processes 21
Design Tasks 21
Concurrent Engineering 23
The Concurrent Design Process 24
Concurrent Engineering and the CAD Database 27
Design Teams 28
Who's on a Design Team? 29
Teamwork and the CAD Database 30
Communication for Design 32
The Designer's Notebook 32
Exercises 35

DESIGNER'S NOTEBOOK 38

CHAPTER 2 IDEATION AND VISUALIZATION 39

Objectives 39
The Design Process: Problem Identification 40
Design Constraints 41
Ideation 44
Where Do You Get Your Ideas? 44
Literature Search 45
Consumer Surveys 47
Competition Reviews 47
Brainstorming 47
Good Survey Question Design 48
Individual Brainstorming 49
Listing 50
The Design Process 50
Product Review/Design Review Meetings 51
Product Review Stages 52
Design Review 53
The Role of Visualization in Ideation 54
Building Visual Ability 57
Tools 57
Practice 58
Sketching Basic Shapes 58
Forget the Symbols You've Learned 58
Optical Illusions 58
Estimating Proportions 59
Viewpoint 60
Shading 60
Positive and Negative Space 60
Visual Perception 61
The Physiology and Psychology of Seeing 62
Construction Lines/Sketching Aid 63
Schematic Drawings 64
Standards 64
Sketching a Block Diagram 66
Drop Shadows 66
Sketching an Electronic Schematic 67
Sketching Thread and Fasteners 68
Detailed Representation 70
Simplified Representation 70
Schematic Thread Representation 70
Thread Notes 71
Bolt Heads 71
Springs 72
INDUSTRY CASE: SKETCHING FOR IDEATION: DOWN YOUR ARM AND OUT YOUR PENCIL JOHN HOULIHAN, TIMEX 74
Skills Summary 78
Exercises 78

DESIGNER'S NOTEBOOK 82

CHAPTER 3 PICTORIAL SKETCHING FOR DESIGN VISUALIZATION 83

Objectives 83
Geometry of Solid Objects 85
Pictorial Sketching 86
Euler's Formula 86
Isometric Sketches 87
The True Size of an Angle 88
Isometric Grid Paper 89
Orientations of Isometric Views 89
Creating an Isometric Sketch Using Isometric Grid Paper 90
Sketching Isometrics without Grid Paper 90
Nonisometric Lines 91
Isometric Ellipses 91
Creating a Scale Isometric Piping Layout 92
Oblique Sketches 94
Perspective Sketches 95
Creating a Perspective Sketch 97
Lettering 99
General and Specific Notes 101
Sketching Assemblies 102
Sketching in the Design Process 103
INDUSTRY CASE: SKETCHING TECHNIQUES: ONE THING LEADS TO ANOTHER
ALBERT BROWN, AFFYMAX RESEARCH INSTITUTE 104
Skills Summary 108
Exercises 108

DESIGNER'S NOTEBOOK 114

CHAPTER 4 MULTIVIEW SKETCHING 115

Objectives 115
Multiview Sketches 116
The Geometry of Projection 117
Orthographic Projection 117
Standard Arrangement of Views 118
Spacing between Views 120
Relating Views to a Cartesian Coordinate System 120
Systems of Projection 122
The Glass Box 123
First-Angle and Third-Angle Projection 124
Using Projection Lines 125
Reference Surfaces (Coordinate System) 126
Sketching Orthographic Views 128
Choosing the Front View 128
Interpreting Multiview Drawings 130
Lines and Points in a Multiview Drawing 130
Normal, Inclined, and Oblique Surfaces 131
Normal, Inclined, and Oblique Edges 132

Cylinders, Spheres, and Cones in a Drawing 132
Sketching Cylinders, Spheres, and Cones 133
Intersections and Tangencies 134
Projecting Curves 137
Fillets and Rounds 138
Scale 138
Sketching Dimensions 138
Visualizing the 3D Object 140
Interpreting Views by Removed Portions 141
Building a Physical Model 142
Making a Clay Model 142
Interpreting Linetypes 143
Hidden Lines 144
Centerlines 144
Sketching Hex Head Bolts and Nuts 145
Section Views 146
Auxiliary Views 148
Selecting Views 148
INDUSTRY CASE: VISUALIZATION: FROM 2D TO 3D AND BACK AGAIN
KAREN MARKUS, CRONUS INTEGRATED MICROSYSTEMS, INC. 150
Skills Summary 157
Exercises 157

DESIGNER'S NOTEBOOK 164

CHAPTER 5 USING GEOMETRY FOR MODELING AND DESIGN 165

Objectives 165
Drawing Geometry 166
Working with Coordinates for 3D CAD Modeling 167
The First Coordinate System 167
Specifying Location 169
Absolute Coordinates 169
Relative Coordinates 169
Polar Coordinates 170
Using Existing Geometry to Specify Location 170
Geometric Entities 171
Points 171
Lines 172
Planes 173
Circles 173
Arcs 174
Ellipses 174
Formulas for Circles and Arcs 175
The Perimeter of an Ellipse 176
Spline Curves 176
Geometric Relationships 178
Solid Primitives 180
Making Complex Shapes with Boolean Operations 181
Recognizing Symmetry 182
Right- and Left-Hand Parts 183

Parting Line Symmetry 183
Extruded Forms 184
Swept Shapes 184
Revolved Forms 185
Irregular Surfaces 186
User Coordinate Systems 186
Transformations 187
Geometric Transformations 188
Viewing Transformations 189
INDUSTRY CASE: THE GEOMETRY OF 3D MODELING: USE THE SYMMETRY
MARTY ALBINI, STRATEGIX ID **195**
Skills Summary 196
Exercises 196

DESIGNER'S NOTEBOOK 202

CHAPTER 6 MODELING FOR REFINEMENT 203

Objectives 203
Refining the Design Idea through Modeling 204
What Is a Model? 206
Kinds of Models 206
Model Qualities 208
2D Models 210
Paper Drawings 210
2D CAD Models 212
3D Models 214
Physical Models 214
Computer-Generated 3D Models 217
Virtual Reality 217
Types of 3D Models 218
Wireframe Modeling 218
Wireframe Modeler versus Wireframe Display 221
Surface Models 223
Surface Information in the Database 223
Extruded and Revolved Surfaces 224
Meshes 224
NURBS-Based Surfaces 225
Reverse Engineering 225
Complex Surfaces/Combining Surfaces 227
Editing Surfaces 228
Surface Model Accuracy 229
Using Surface Models 230
Solid Models 231
CSG Method and Boolean Operations 232
BREP Method and Boolean Operations 232
Hybrid Systems 233
Solid Model Accuracy 233
Modeling Kernel 235
Using 3D Solid Models 236
Parametric Models/Intelligent Models 236

Choosing the Right Modeling Method 237
INDUSTRY CASE: 3D MODELING METHODS: NOT JUST A PRETTY PICTURE
MARK GERISCH, M&L AUTO SPECIALISTS, INC. 241
Skills Summary 244
Exercises 244

DESIGNER'S NOTEBOOK 246

CHAPTER 7 PARAMETRIC MODELING AND DESIGN 247

Objectives 247
Advantages of Parametric Modeling 248
Parameters Control the Geometry 250
Feature-Based Modeling 251
Planning Parts for Design Flexibility 254
Sketch Constraints 256
Overconstrained and Underconstrained Sketches 258
Setting the Base Point 259
The Base Feature 260
Adding Features to the Model 261
Parent-Child Relationships 262
Datum Planes and Surfaces 263
Parametric Dimensions 265
Formulas in Dimensions 266
Editing the Model 268
Standard Features 270
Working with Built-in Features 270
Complex Shapes 272
Parametric Modeling Modes 272
Assembly Mode 272
Drawing Mode 273
Bidirectional Associativity 273
Nonparametric Dimensions 274
INDUSTRY CASE: PARAMETRIC MODELING: CAPTURING DESIGN INTENT
BRANDON LAROCQUE, BARRETT TECHNOLOGY 275
Skills Summary 280
Exercises 280

DESIGNER'S NOTEBOOK 288

CHAPTER 8 MODELING FOR MANUFACTURE AND ASSEMBLY 289

Objectives 289
Combining Parts in an Assembly Model 290
External References 291
Parametric Assemblies 292
Using Dynamic Assemblies 293
Managing Assembly Files 295
Assembly Constraints 296
Intelligent Assemblies 296
Layout Drawings 298
Skeleton Models 299

Global Parameters 301
Seed Parts 302
Parametric Drawing Elements 302
Building a Useful Database: Bringing It All Together 303
Standard Parts 303
Adding Static Parts to a Parametric Database 304
Threads and Fasteners 305
Thread Forms 305
Double and Triple Thread 306
Right- and Left-Hand Thread 306
Springs 307
Modeling Springs 308
Fillets and Rounds 308
Using Your 3D Model to Determine Fits 309
Interference Checking 309
Tolerances 310
Fit between Mating Parts 311
Nominal Size and Basic Size 312
Tolerances per Manufacturing Process 313
Measurement and Inspection 314
Accessibility Checking 316
Design for Manufacturing 317
Modeling Injection Molded Plastic Parts 318
Cast Parts 321
Modeling Machined Parts 321
Modeling Sheet Metal Parts 322
INDUSTRY CASE: MODELING SHEET METAL PARTS: THINK FLAT
STAN MCLEAN, VKI TECHNOLOGIES 325
Skills Summary 328
Exercises 328

DESIGNER'S NOTEBOOK 336

CHAPTER 9 DRAWING CONTROL AND DATA MANAGEMENT 337

Objectives 337
Requirements for Engineering Documentation 339
Drawing Approval and Release 340
Change Orders 342
Revision Block 342
A Drawing as a Snapshot in Time 343
Good Practices for Electronic Drawing Storage 344
Storing Electronic Files 344
Organized Directory Structures 344
File Naming Conventions 346
Drawing Standards 347
Permission and Ownership 348
Backing Up Drawing Files 350
Storage Media 350
Using the 3D Design Database in Concurrent Engineering 351
Quality Management 352
ISO 9000 352

Product Data Management 354
Organized by Product 355
A Relational Database 356
Managing Work Flow 359
Archiving Work History 360
Individual Productivity 361
Interface Tools 361
Data Management and the Web 363
CAD Files on the Web 364
INDUSTRY CASE: MANAGING ENGINEERING DRAWINGS: A LIBRARY METAPHOR
CARL FEHRES, IRI INTERNATIONAL **366**
Skills Summary 371
Exercises 371

DESIGNER'S NOTEBOOK 374

CHAPTER 10 USING THE MODEL FOR ANALYSIS AND PROTOTYPING 375

Objectives 375
Determining Mass Properties 376
Factors of Safety 376
Mass Property Calculations 378
Understanding Mass Property Calculations 380
Verifying Accuracy 380
Units and Assumptions 381
Materials 381
Pounds Mass versus Pounds Force 382
Specific Gravity 383
Exporting Data from the Database 383
File Formats 384
Common Formats for Export 385
Standard for the Exchange of Product Model Data 386
Vector versus Raster Data 388
Translating Data 388
Downstream Applications 389
Spreadsheets 389
Equation Solvers 390
Graphs 392
Finite Element Analysis 395
Simulation Software 399
Human Factors 402
Integrated Modeling and Design Software 403
Prototyping Your Design 405
Rapid Prototyping 406
Translating the Model 407
Current Rapid Prototyping Systems 407
3D Printing 411
Rapid Tooling 412
Cores and Cavities 413
INDUSTRY CASE: MASS PROPERTIES: FROM I-DEAS TO EXCEL
SCOTT ADAMS, BALL AEROSPACE **414**

Skills Summary 416
Exercises 417

DESIGNER'S NOTEBOOK 422

CHAPTER 11 IMPLEMENTATION 423

Objectives 423
Manufacturing Processes Overview 425
Manufacturing Processes 426
Machining: A Manual Example 428
Designing the Impact Tower Fixture 429
Machining the Fixture 430
Milling Machine 431
Drill Press 433
Lathe 434
Inspection 435
Computer-Aided Manufacturing (CAM) 436
Communicating from CAD to Machine Code 439
Generic Numerical Control Codes (NCI) 442
NC Code 442
G-Codes and M-Codes 443
Tooling 443
Jigs 444
Fixtures 444
Metal Forming 445
Sand Castings 445
Investment Casting 446
Centrifugal Casting 446
Die Casting 447
Forging 448
Sheet Metal Manufacturing 448
Sheet Metal Fasteners 448
Pattern Layout 449
Developments 449
Bend Allowance 450
Stamping 452
Packaging 452
Plastics Manufacturing 453
Injection Molding 454
Extrusion Molding 454
Injection and Extrusion Blow Molding 455
Compression Molding 457
Vacuum Molding 457
INDUSTRY CASE: MODELING FOR MANUFACTURE: NO GLUES, NO SCREWS
HUGO HASELHUHN, C&K SYSTEMS **458**
Skills Summary 463
Exercises 463

DESIGNER'S NOTEBOOK 466

CHAPTER 12 DOCUMENTATION GRAPHICS 467

Objectives 467
The CAD Database as Design Documentation 469
Standards 470
ANSI Drawing Standards 471
Working Drawing 472
Assembly Drawings 472
Part Drawings 476
Preparing Working Drawings 483
Standard Sheet Sizes 483
Scale 484
Developing Views from 3D Models 485
Placing the Views 487
Isometric Views 489
Auxiliary Views 490
Section Views 492
Types of Sections 494
Linetypes 497
Phantom Lines 498
Title and Revision Blocks 499
Dimensioning 500
Units 501
Advantages of Metric Units 501
Standard Dimension Appearances 501
Dimension Lines 502
Extension Lines 502
Dimension Value 502
Orientation 504
Placement of Dimensions 504
Choosing the Best Dimensions to Show 505
Dimensioning Arcs and Circles 509
Coordinate and Grid Dimensioning 511
Standard Symbols for Dimensioning Common Features 511
Dimensioning Chamfer 512
Dimensioning Knurling 512
Dimensioning Keyseats and Keyways 513
Overdimensioning 514
Tolerancing 514
General Tolerance Notes 515
Limit Tolerances 516
Plus/Minus Tolerances 516
Tolerance Stacking 516
Baseline Dimensioning 517
Fit 518
The Basic Hole System 520
The Basic Shaft System 520
Allowance 520
Maximum Material Condition 521
Perfect Form 522

Modeling for Tolerance Studies 522
Metric Fit Tables and Designations 523
Geometric Dimensioning and Tolerancing 524
Geometric Characteristic Symbols 525
The Feature Control Frame 525
Form Tolerances for Individual Features 526
Datum Surfaces and Features 531
Tolerances for Related Features 532
Datum Identifiers 532
Tolerances for Related Features 532
Runout 534
Basic Dimensions 534
Positional Tolerance 536
Tolerancing Angles 538
Supplementary Symbols and Modifiers 539
Regradless of Feature Size 540
Surface Control 540
Specifications 545
Patent Drawings 546
INDUSTRY CASE: DESIGN DOCUMENTATION: YOU CAN BUILD *ONE* OF ANYTHING
JAE ELLERS, TEKTRONIX **548**
Skills Summary 554
Exercises 554

DESIGNER'S NOTEBOOK 560

CHAPTER 13 ANIMATION AND PRESENTATION GRAPHICS 561

Objectives 561
Effective Presentations 562
Text 565
Charts and Graphs 566
Pie Charts 567
Bar Charts 567
Line Charts 568
Area Charts 568
Presentation Software 569
Formatting 570
Organizing 570
Importing Graphics 571
Graphic Formats for Reports and Presentations 572
Resolution 572
Color Depth 573
Graphic File Formats 573
File Formats for 2D Images 574
Object Linking and Embedding 574
Rendering 575
Lighting 575
Placing Lights 578
Materials 579
Rendering Options 580

Choosing Resolution and Color Depth 580
Antialiasing 582
Rendering Algorithms 582
Flat Shading 582
Gouraud Shading 583
Phong Shading 583
Ray Tracing 584
Radiosity 585
Animation 585
Planning Your Animation 587
Storyboarding 587
Importing Geometry 588
Keyframing Methods 589
Transformations 589
Inverse Kinematics 590
Rendering the Animation 590
VRML 592
INDUSTRY CASE: ANIMATING YOUR MODELS: IF A PICTURE IS WORTH A THOUSAND WORDS . . .
ROBERT REISINGER, MOUNTAIN CYCLE **596**
Skills Summary 603
Exercises 603

APPENDIX CONTENTS 607

GLOSSARY 689

INDEX 703

Preface

Rationale

Engineering Design Communication is a new approach to the traditional foundation course in engineering graphics. We believe that the role of the CAD database as the hub of the concurrent engineering environment has implications for the way engineering graphics is taught. In the concurrent environment, engineers are expected to communicate effectively with customers as well as other technical personnel and to prepare the CAD database so it will be useful for analysis, manufacturing, and presentation purposes.

New software tools, especially solid and parametric modeling software, have changed the way engineers create drawings. These packages automate much of the work of creating drawing views and offer functions that make it easier to modify and refine a solid model. They also offer many options for viewing the model and exporting it for use in other applications. These software developments have shifted the rationale for CAD modeling from the creation of documentation drawings to the creation of multipurpose models for design visualization and communication. At the same time, the ability of parametric models to capture the design intent for the part or assembly in its geometric relationships puts an added burden on the beginning student to grasp the bigger picture of the design process in order to use these tools effectively.

Approach

Engineering Design Communication addresses this shift in two ways. First, it presents the CAD file as a database of design information that is developed over time. The use of solid and parametric modeling software makes it possible to follow the design process through the course—students first learn to sketch their ideas, to transform 2D sketches into 3D models, to refine the models and use them for analysis, and finally to use the models to document the design. This organization more closely reflects industry practice and gives students a strong framework for understanding why they should learn to sketch, when it is appropriate to use different kinds of models, and what they need to discover in order to prepare a model for manufacture.

Second, it introduces topics that will help students learn to reflect design intent in their models. Building effective models that will update correctly requires some consideration of design goals. For beginning students unfamiliar with design issues, *Engineering Design Communication* anchors modeling techniques in the context of design to help them build an understanding of design intent as they learn to model. Case histories in the text show how design goals influence the way models are made. The text encourages students to think about the broader context for their models so they are more receptive to rules of thumb being taught in the CAD lab. By presenting the big picture of the many uses of the CAD database—and the power of automatic model updates—the text motivates students to practice building design intent into their models.

Engineering Design Communication is designed to be a true foundation course in graphic communication. From sketching to modeling to documentation drawings, the text emphasizes the nature of visual information. Because the concurrent engineering environment expects engineers who can present ideas and work with nontechnical personnel, *EDC* presents the wide variety of software and presentation tools that an engineer will use to visualize design. Students will learn what a technique communicates well and when it might be useful. Each technique is illustrated with real examples. Understanding when to sketch and when to model—and when to use a different medium—will help students use these tools effectively in future courses and the workplace. It will also prepare them to evaluate and adopt new graphics tools as they are developed.

Organization

Engineering Design Communication is designed to be used with a tutorial guide for a specific computer-aided design package. Step-by-step tutorial guides that follow the organization of the text are being prepared for popular parametric modeling packages. As students learn the specifics of CAD modeling in the lab, the text presents the sketching and modeling techniques as they are used in the design process.

Design Framework

Engineering Design Communication introduces students to the process of design in Chapter 1, then uses the design process as an organizing framework for the rest of the book. Each chapter opens with an introduction that explains how the graphic skills presented in the chapter relate to and are used in engineering practice. Numerous examples and illustrations from industry motivate students with real examples of engineering practice and reinforce the connection between the technique and its application.

Industry Cases

Case histories at the end of each chapter reinforce key topics by showing students how a practicing engineer applies them. These case histories present specific instances of general principles presented in the text, giving students a clear idea of the contemporary software tools and techniques used to create design. Although focused on the chapter content, most industry cases detail the use of specific graphics tools to solve a design problem.

Sketching

Sketching is the focus of the early chapters (2 through 4) of the text, which correspond to the ideation and visualization stages of the design process. Many topics traditionally taught in the context of documentation drawings are presented in these chapters so that students learn to use a full range of drawing views and projections in their sketches. Actual sketches are used to illustrate these early chapters so that students may compare their efforts with other sketches, not instrument or CAD drawings.

Designer's Notebook

The role of sketches in planning and visualization is reinforced in each chapter through the Designer's Notebook. Each chapter opens with sketches and notes from a practicing engineer's notebook. The excerpt previews the information to be conveyed through graphics in that chapter and gives students real examples of how they might use their own design notebook. (Chapter 1 encourages them to begin and use such a notebook throughout the course.)

Modeling

Modeling—and creating the design database—is the focus of the middle chapters of the book (5 through 10), where it is linked to the process of design refinement and analysis. These chapters begin with an introduction to geometry that takes students from the 2D sketching environment to the 3D environment that they will work with on screen. After a survey of different modeling methods, separate chapters on parametric modeling and assembly modeling introduce modeling considerations unique to 3D solid models. Each general modeling principle is illustrated with real models and examples showing the application of key ideas. A chapter (10) on exporting and using the model data with downstream applications, including rapid prototyping, presents additional considerations for creating a useful design database.

Documentation

Implementation and documentation are the focus of the latter chapters of the book (11 through 13). Because many students lack an understanding of basic manufacturing processes, an overview of these processes presents key questions to be asked and answered before a design is ready for manufacture. As in the modeling chapters, documentation and presentation techniques are presented in the context of the 3D model that students will use to generate drawings, rendered views, animations, and files for numerically controlled machinery.

Flexibility

Engineering Design Communication is designed to be used in a one- or two-semester sequence in engineering graphics. An instructor teaching a one-semester sequence may choose material from or omit chapters 2, 9, 10, 11, or 13 to match the emphasis in the course.

Each chapter opener signals to the student which stage of the design process corresponds to the chapter content. This device makes it possible to cover chapters earlier in the text without losing the link to the design process. A comprehensive glossary is included as an aid to students who choose to read chapters out of order.

The implementation chapter (11) is a stand-alone chapter that may be covered at any point before the documentation chapter. Because manufacturing issues are addressed in the context of modeling in Chapter 8, some instructors may want to assign 11 before 8.

For instructors who wish to ask students to start making drawings from their solid models earlier, the documentation chapter (12) may be covered any time after Chapter 4.

The drawing control chapter (9) may be covered at any point in the text. For courses with more emphasis on rendering and animation, the presentation graphics chapter (13) may be covered any time after Chapter 6.

Annotated Overview

Chapter 1

This introductory chapter presents the many ways that graphics are used in the engineering design process and introduces key themes of the text, such as concurrent engineering, design teams, and the CAD model as a design database. It uses a comprehensive case study to illustrate the role of graphics in the design process from start to finish.

Chapter 2

This chapter introduces sketching as a tool in the ideation process. Needs assessment and design constraints provide a context for sketching as a means to refining the problem definition. Students begin to develop their sketching ability with contour sketches and schematic drawings methods and with exercises that help them "see" objects as they truly appear. For courses with less emphasis on design, the latter half of this chapter may be assigned.

Chapter 3

In this chapter, students learn to create freehand pictorial sketches in the context of design visualization. The chapter introduces 2D projections of 3D objects and the kinds of 3D sketches that can be used to convey design information.

Chapter 4

Techniques of multiview sketching and orthographic projection are presented in the context of design communication and visualization. The first half of the chapter illustrates the techniques of creating multiview sketches, while the latter half helps students interpret multiview drawings and their conventions.

Chapter 5

The geometry of modeling presents key geometric entities, relationships (such as symmetry), and transformations as they apply to both sketching and modeling. The chapter links the entities in a sketch to those used by the CAD modeling software and introduces topics such as coordinate systems, extrusion, revolution, and Boolean operations, which are common in working with geometry in most 3D modeling packages.

Chapter 6

This broad introduction to modeling defines, illustrates, and compares the many kinds of models that may be used to represent and test a design. The chapter presents strengths and weaknesses of each kind of model, including 2D and 3D CAD modeling systems, and builds a context for understanding parametric modeling.

Chapter 7

This chapter defines and explains the terms and techniques used in creating parametric models. It shows the steps in planning and adding features to create a model, explains how the information is stored, and demonstrates the role of parameters in updating the model.

Chapter 8

Bringing individual parts together into a complete assembly and adding details required by the manufacturing process are the focus of this chapter. Students are introduced to dynamic assemblies, how the model can be used to check for fit and interference, and ways to augment the 3D database with standard and static parts. They also learn about model modifications related to key manufacturing processes (such as draft) that are added after issues of functionality have been resolved.

Chapter 9

This chapter discusses the engineer's responsibility for managing the design database and introduces students to drawing control and file-sharing issues in a networked engineering environment.

Chapter 10

Chapter 10 familiarizes students with exporting data from the model to be used for testing and analysis. Common analysis applications as well as rapid prototyping systems are introduced.

Chapter 11

This overview of common manufacturing processes gives students an appreciation for issues addressed in the implementation phase. They also learn about the use of the 3D model to generate NC machine code and how this may affect the way design information is transmitted.

Chapter 12

This chapter shows students how drawings are generated from the 3D model and the role of the engineer in applying dimensioning and tolerancing to the model, then documenting it in the drawing.

Chapter 13

The final chapter introduces good presentation practice and media—from charts and graphs to rendered views to animated sequences—that are commonly used to present engineering information. The pages of the textbook are a "flip book" that previews the animation coverage in Chapter 13; see page 55 of the text for a description of this feature.

Associated Materials

Web Resources

Many of the examples and case histories in the text can be explored further through World Wide Web sites that include additional images, animations, and details about the design. These links and additional course materials are available on the EDC Web site at www.prenhall.com/lockhart.

We would like to thank Todd Amundson of Lower Columbia College for developing and providing the content for the *EDC* Web site.

Presentation Slides

A series of PowerPoint slides prepared by the authors are available for adopter use. Visit the EDC Web site for more information.

Acknowledgments

There are many individuals we would like to thank for contributing time, energy, and expertise to the development of this text. First, our advisory board of reviewers, who provided valuable feedback and advice:

- Robert Chin, East Carolina University
- Barry Crittenden, Virginia Polytechnic Institute and State University
- Larry Goss, University of Southern Indiana
- Steve Kaminaka, California Polytechnic State University, San Luis Obispo
- Bill Law, Tektronix, Inc.
- Dennis Lieu, University of California, Berkeley
- Fritz D. Meyers, Ohio State University
- James Shahan, Iowa State University.

We are especially grateful to Addison Wesley Longman for publishing a preliminary edition that could be reviewed and tested prior to this first edition. We would like to thank the following preliminary edition reviewers, whose feedback was instrumental in strengthening the text for this first edition:

- R. Glenn Allen, Southern Polytechnic State University
- Todd Amundson, Lower Columbia College
- Radha Bala, Northern Illinois University
- Samir B. Billatos, University of Texas, Brownsville
- John Cherng, University of Michigan, Dearborn
- Robert A. Chin, East Carolina University
- Clark Cory, Purdue University
- Frank Gerlitz, Washtenaw College
- Randy J. Gu, Oakland University
- Kevin Henry, Columbia College, Chicago
- Craig Johnson, Central Washington University
- Steve Kaminaka, California Polytechnic State University, San Luis Obispo
- Hamid A. Khan, East Carolina University
- Kathleen Kitto, Western Washington University
- Steve Lamm, San Antonio College
- Dee Lauritzen, Eastern Arizona College
- Chris Merritt, CADMAX Consulting
- Ahad Nasab, Middle Tennessee State University
- John G. Nee, Central Michigan University
- Ben Nwoke, Virginia State University
- Patrick O'Connor, Texas State Technical College
- Ken Perry, Indiana University–Purdue University, Fort Wayne

- S. Manian Ramkumar, Rochester Institute of Technology
- Virgil Seaman, California State University, Los Angeles
- Arijit Sengupta, New Jersey Institute of Technology
- Randy Shih, Oregon Institute of Technology
- Nancy E. Study, Southwest Missouri State University
- Lee Waite , Rose-Hulman Institute
- Jianping Yue, Essex Community College

We owe a special thank-you to Albert W. Brown, Jr., of Affymax Research Institute, for contributing so many of his notebook pages and sketches, and for creating original sketches for his case study; to Bill Law, Tektronix, Inc., who not only served as a reviewer and advisor, but also facilitated two different case studies; and to Kent Johnson, Boeing Commercial Aircraft Group, for his advisory support and assistance with many examples and illustrations.

For the many case studies and the models and drawings that illustrate them, we owe a huge debt of gratitude to:

- Scott Ketterer, Paul Chadwick, and Dilys Waters, MBD Division, Tektronix, Inc.
- Jae Ellers and Bruce Weaver, Color Printing and Imaging Division, Tektronix, Inc.
- John Houlihan, Timex Corporation
- Marty Albini and Bill Clem, Strategix ID
- Mark Gerisch and Erik Nielsen, M&L Auto Specialists, Inc.
- Brandon Larocque, Barrett Technology, Inc.
- Stan Mclean, VKI Technologies
- Carl Fehres, IRI International
- Scott Adams, Ball Aerospace & Technologies Corp.
- Hugo Haselhuhn, C&K Systems
- Karen Markus, Cronos Integrated Microsystems, Inc.
- Robert Reisinger, Mountain Cycle
- Greg Smith, Immersive Design
- Tony Prosser, Real Time Visualisation

Thank you also to Robb Larson, Montana State University, for assistance with the exercises, and to Will Callahan and Jerry Riekenberg at Implemax Equipment Co., Inc., for the many drawings and sketches provided. We also wish to thank Shawn Alstad, Boeing; Dana Rockel, Timex Corporation; Mark Huppenthal, ACT Technologies; Rob Mesaros, Strategix ID; Mark Soares and Neal Story, J.E. Soares; Alan Hooker, Powerhouse Technologies, Inc.; Dale Pickard, Radiant Engineering; Dan Rooney, Smith Equipment Co.; Darryl Jensen, Tektronix, Inc.; Bruce Olson, Dynojet Research, Inc.; Jerry Baker, CTA Architects; Jeffrey O. Pentecost, Oregon Health Sciences University; Brad Wright, Salient Technologies, Inc.; Mark Wheeler, Tri-Cycle Product Design, Inc.; Darin Graf and Jay Conant, Montana State University; Dr. Daniel Gaffney; Ron Hunter; Trent Wetherbee; Chris Moler; Aaron Morrow; Mark Perkins; and Marla Goodman.

We would like to thank Bob Woodbury, Denise Olson, Phoebe Ling, Pat Unubun, and Karen Wernholm for their professional approach to publishing the preliminary edition of this text. Thanks also to Eric Svendsen and the team at Prentice Hall for the enthusiastic support and publishing craftsmanship that was applied to this first edition.

Contacting the Authors

Our goal has been and will continue to be a textbook that presents a contemporary view of the engineering graphics concepts and skills needed for a strong foundation. We welcome your comments and encourage you to join us in continuing to shape a text that will meet the needs of tomorrow's engineers. Please send mail to us at lockhart@me.montana.edu or cjohnson@seacoast.com, or via our Web site, www.prenhall.com/lockhart.

About the Cover Art

You will read more about the models, sketches, and drawings on the cover in the case studies and examples in the text. The rendered view of the drilling rig is used courtesy of Wayne Fehres and Carl Fehres and IRI International (see Chapter 13 and the case study in Chapter 9). The BarrettHand is featured in the case study in Chapter 7 and used with permission from Barrett Technology. The background drawing is featured in the WFM 90 case study in Chapter 1 and reprinted with the permission of Tektronix, Inc. On the back cover, the assembly sketch is used courtesy of Albert W. Brown, Jr., who is featured in the case study in Chapter 3. The flywheel assembly model is the subject of the case study in Chapter 5 and used with the permission of Strategix ID and Park City Entertainment.

Engineering Design Communication

Conveying Design through Graphics

DESIGNER'S NOTEBOOK

Capturing ideas in graphical form is an efficient and effective way to communicate and record design information. This page is taken from one of the design notebooks kept by Albert Brown for the Affymax Research Institute in Palo Alto, California. Note the hand sketches, meeting notes, and dated entries that relate to his work on this design problem. (Courtesy of Albert W. Brown, Jr., Affymax Research Institute.)

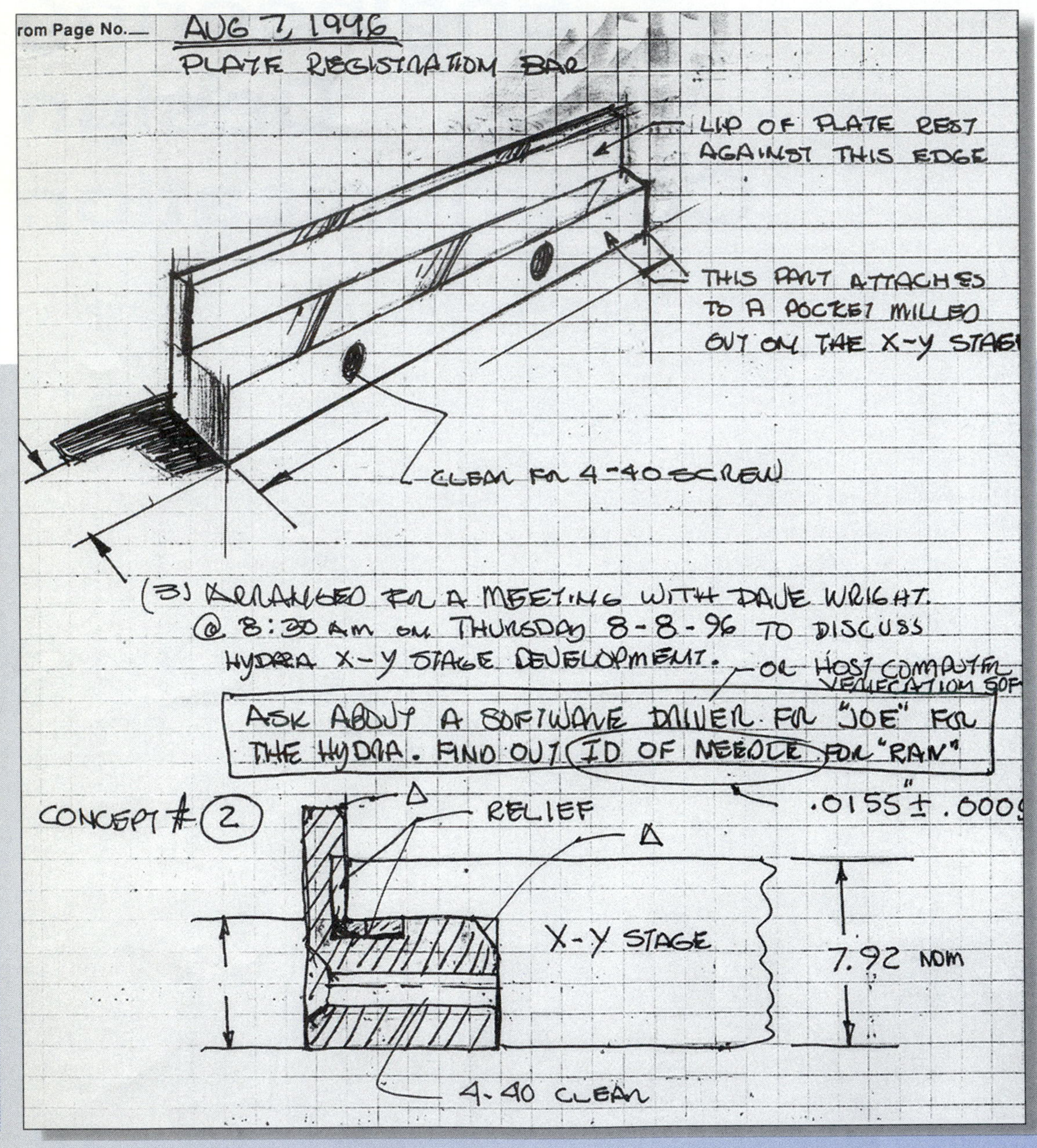

CHAPTER 1

Engineering Graphics and the Design Process

Objectives

When you have completed this chapter you will be able to

- describe the large and varied role that graphical information plays in engineering design;
- visualize a model of engineering design;
- list different types of design and design processes;
- identify the role of graphics in and the benefits of concurrent engineering;
- explain how communication of all kinds is critical to the team approach to design.

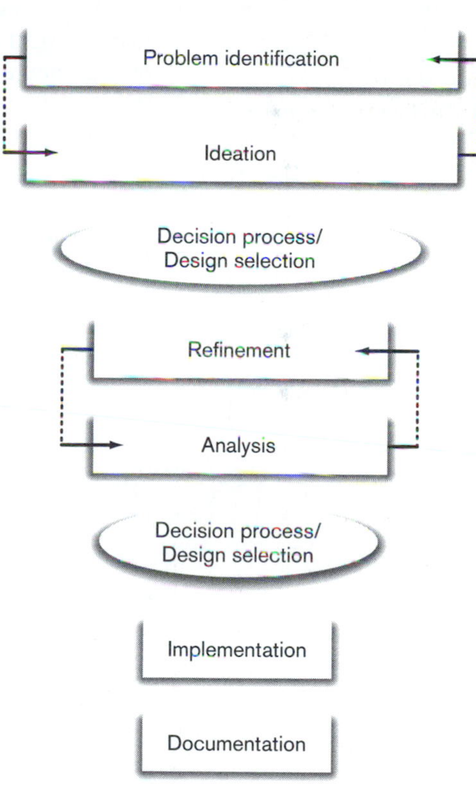

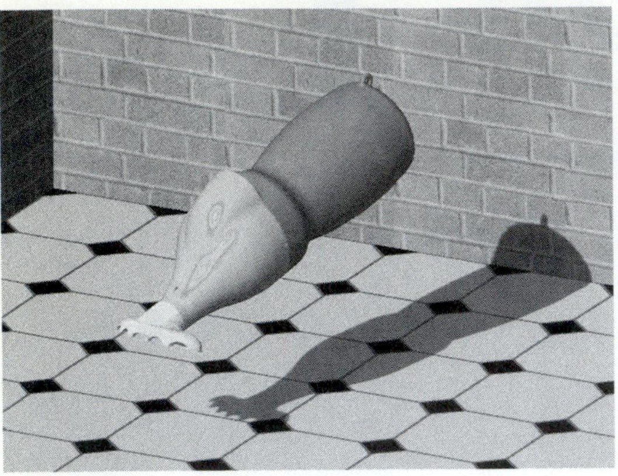

FIGURE 1.1
This view of the hot tub vacuum was created by rendering the solid model. In a rendered view, lights and background can be added to add realism to the image. (Courtesy of SWS Corporation and Brad Wright, Salient Technologies, Inc.)

Powerful desktop computers and new software tools are transforming the way engineers use graphics. Figures 1.1–1.5 were all derived from a single solid model created in SolidWorks, a parametric solid modeling package. When engineers "draw" a model of a part in a solid modeling package, the software creates a database of information that describes the model in three dimensions as if it were a physical entity. This database makes it possible for engineers to present their work in ways that are useful to non-engineers, such as marketing specialists and even the customer. An accurate database also makes it possible to perform analysis, check for interferences, and directly manufacture parts from the model.

Today's graphics tools are being used to accomplish the same tasks as traditional engineering drawing tools, plus many additional tasks. They are more aptly called *computer-aided design* tools because they can be used in many ways throughout the design process. By supporting more design tasks, they are enabling companies to work in teams more effectively, reduce new product costs, and get products to market faster—even to change the design process itself.

Boeing's 777 was the first airplane ever produced for which a full-scale model was not built. Because all of the parts (except for some fasteners) were designed and modeled electronically, a *virtual* prototype of the solid models was used in place of a real one to check

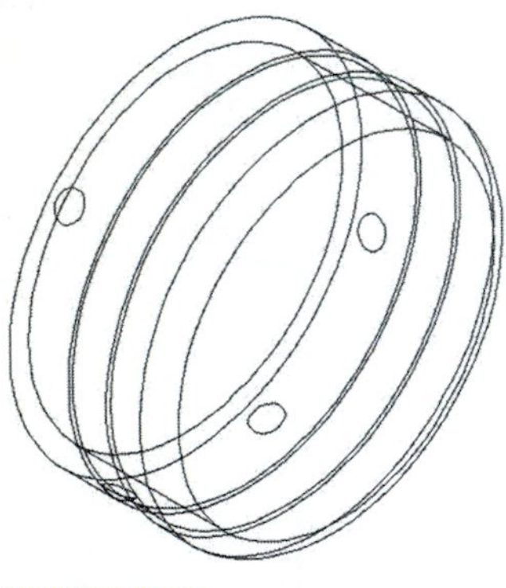

FIGURE 1.2
Wireframe views of the model from any angle can be generated automatically from the 3D model. (Courtesy of SWS Corporation and Brad Wright, Salient Technologies, Inc.)

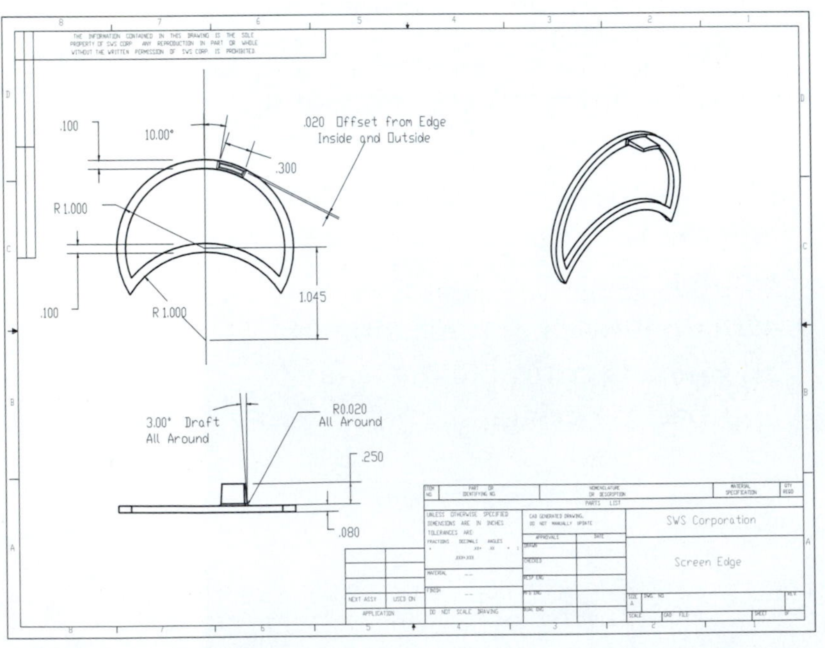

FIGURE 1.3
Detailed dimensioned drawings of each individual part can be created from the information in the solid model. (Courtesy of SWS Corporation and Brad Wright, Salient Technologies, Inc.)

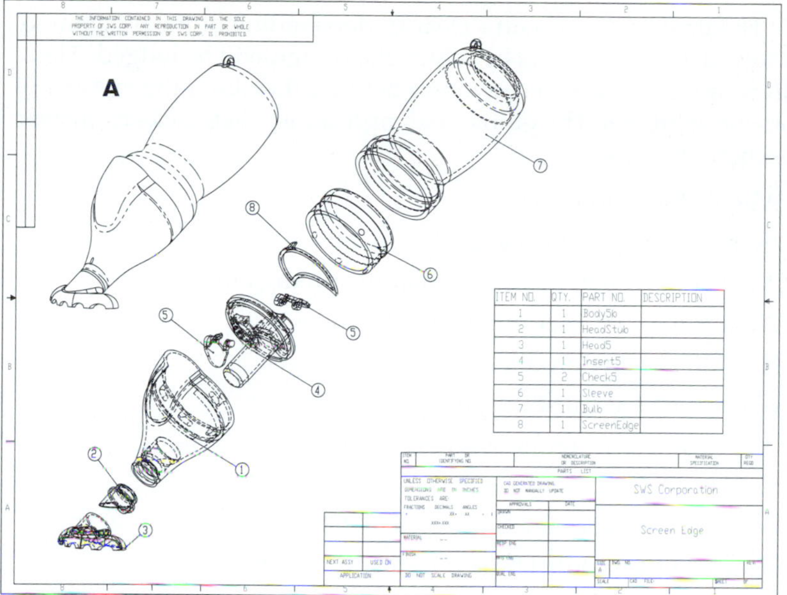

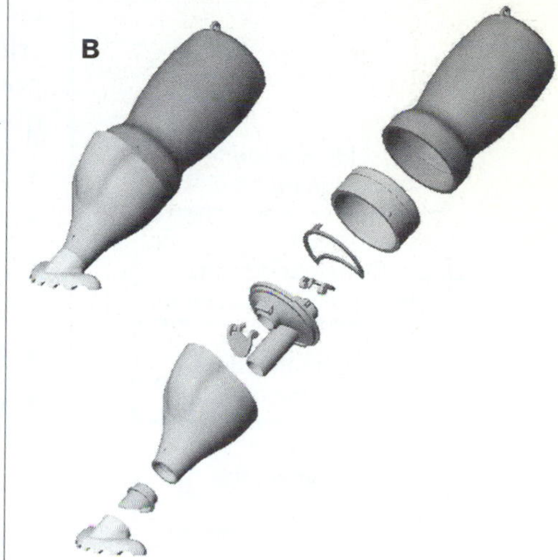

FIGURE 1.4
Information associated with the individual parts in the model can be shown in a drawing to document the part for manufacture and assembly. An exploded view generated from the assembled solid model makes it easy to see how the parts fits together. Exploded views can be wireframe, as in (a) and shaded, as in (b). (Courtesy of SWS Corporation and Brad Wright, Salient Technologies, Inc.)

FIGURE 1.5
Finite element analysis software can use the solid model to analyze physical properties of the design and display the results graphically. (Courtesy of SWS Corporation and Brad Wright, Salient Technologies, Inc.)

the design, fit, and manufacturability of each part. The software tools developed by and for Boeing—which you'll read more about later—made this transformation possible, but the shift also was influenced by the company's desire to work smarter by integrating the teams of designers into a design/build organization. The graphics tools they use work together to support team interaction.

Despite the changing nature of the design process, however, engineers use graphics in design in three basic ways: to **visualize** the design idea, to **communicate** the idea so others can evaluate it, and to **document** the design so the product or process can be reliably reproduced and maintained. In this book, you will learn to draw and use computer-generated graphics in the context of the design process. Understanding the role of graphics in design will help you use the tools more efficiently and will equip you to evaluate new technologies for design as they are developed.

What Is Engineering Design?

Engineering design is the act of creating the specifications for a product or process that best satisfies the design criteria. Every design task is undertaken to solve a problem: to make a faster race car, to build a larger airplane, to design a more reliable disk drive. And every design task ends with specifications and drawings that can be used to make or implement the design.

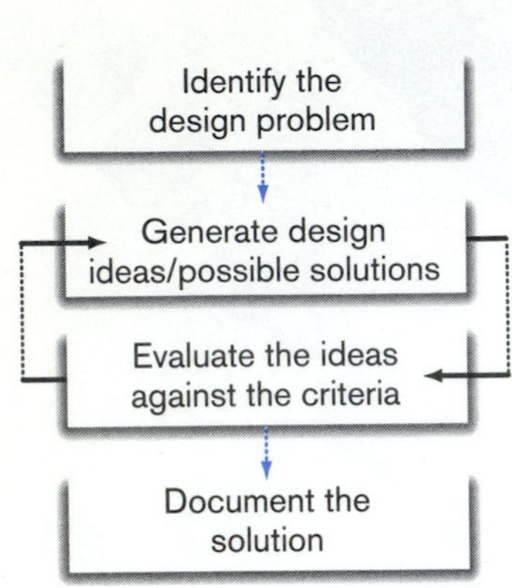

FIGURE 1.6
Simplified design model

The process starts with a clear understanding of the problem to be solved and the criteria by which the design will be judged. The criteria are called *design constraints* because they limit the number of possible solutions. The problem identification stage seeks to answer questions such as:

- Who is the customer/user?
- How will the product be used?
- What are the customer needs that the product must satisfy?
- What is the customer problem to be solved?
- How do existing products meet those needs?
- What functions must the product perform?
- What limits exist for its size and shape?
- How is it to be manufactured?
- What performance goals must be met?
- How will the product be serviced and repaired?

Effective design requires creativity, or the ability to generate many alternate design solutions for a given situation. It also requires the ability to evaluate your design to see how well it satisfies the design criteria. Design problems have many possible solutions. The heart of the design process is an *iterative* loop: generate a design idea, evaluate it against the criteria, refine the idea, test it again, and so on until the design idea becomes the design solution.

Once a solution has been determined, the design process ends with documentation that can be used to reliably manufacture or reproduce it.

A simple model of the design process is shown in Figure 1.6.

The Role of Graphics

This simple model is enough to illustrate the three key roles that graphics play in engineering design: visualization, communication, and documentation.

Visualization

Learning to draw helps you communicate your design ideas, but it also develops your ability to visualize design solutions—making you a better designer. As you identify the design problem, graphics can help you "see" the problem and possible solutions. Good designers capture their design ideas as hand-drawn sketches and use these sketches to show the ideas to others for feedback (see Fig. 1.7).

For every design problem, there may be several viable solutions. As you generate design ideas, you will want to capture a variety of approaches to the problem before you select one to pursue. Investing too much time in a fully detailed drawing of a design idea at this stage may be more than is needed for others to react to the

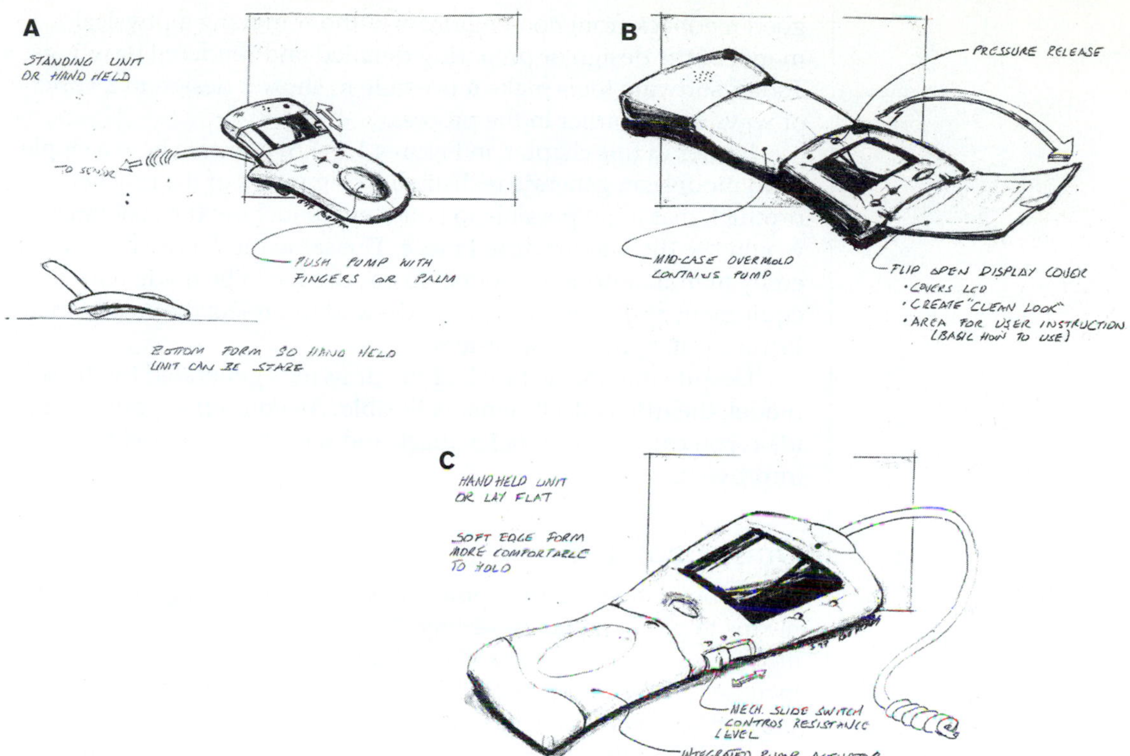

FIGURE 1.7
These sketches depict three alternative solutions to the challenge of designing a hand-held exercise trainer that would be compact and easy to operate. (Courtesy of Robert Mesaros, Strategix ID.)

idea—and it may narrow your vision of the solution prematurely. Learning to sketch your ideas quickly and effectively will allow you to capture all of your ideas (before you forget them) so you can share them or return to them in the future.

Few design problems will come to you fully defined. As you generate design ideas and ask others to evaluate them, you will learn more about the design constraints and how different ideas satisfy them. The ability to sketch your ideas by hand is a powerful tool for letting others "see" your ideas so they can understand and evaluate them.

Communication

As you refine your design idea, you will add more and more detail to your drawing of the solution. At the same time, the drawing must remain flexible so details can be added and changed without too much effort. A computer-assisted drawing or model of the part will help you add needed detail while still allowing you to make changes easily.

During the evaluation and refinement stages, designers need to be able to speak to each member of the product team—as well as customers—about the design idea. In the past, it was difficult to get good feedback from non-engineers without making a physical model of the design or preparing detailed and rendered drawings. Today, software tools make it possible to show a design in a variety of ways much earlier in the process.

Earlier in this chapter, in Figures 1.1–1.5, you saw how a single solid model can generate realistic-looking views of the finished product, making it possible to better evaluate how the customer would see the manufactured piece. The same model can be evaluated by manufacturing personnel to be sure it can be made with the equipment and materials specified—and to project the cost of manufacturing the product or system.

Despite the finished look of the drawings generated by the solid model, the model itself remains flexible. As you refine your design idea, you can incorporate feedback and modify the model to further improve it.

Documentation

Once you've reached the optimal design solution, graphics are the means to permanently recording that solution. *Documentation drawing* is the term used to describe a drawing or set of drawings that includes all the information necessary to unambiguously describe the part or process so it can be made.

Traditionally, paper drawings have been used to document the design. All the information needed is captured on paper and interpreted by the machine operators who manufacture the part (see Fig. 1.8). With today's software tools, more and more designs are

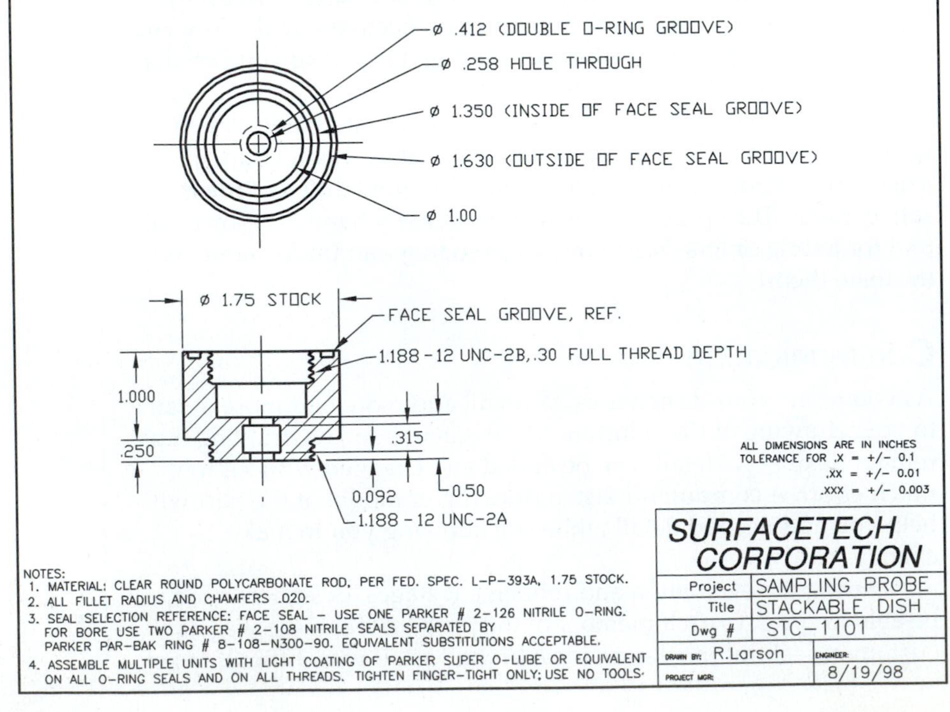

FIGURE 1.8
This fully dimensioned drawing defines the shape of the part, the dimensions of each feature, and the allowable tolerances (the range within which parts will be considered acceptable). It also includes notes that explain materials and other information needed to manufacture the part. (Courtesy of Robb Larson.)

captured as solid models that interface directly with computer-aided manufacturing software. Instead of providing the manufacturer with paper drawings only, companies transmit electronic files that can be used to generate the code that controls the manufacturing machinery. In both cases, however, the finished parts must be inspected to see if they are acceptable renditions of the design. Documentation drawings serve as a "contract" with the manufacturer that says, "If you make the parts as described here, we will pay you for them."

Learning to document your designs involves using drawing conventions that will be understood by the manufacturer. You will learn how designs are checked after manufacture and how to specify what the criteria will be for whether a part is acceptable.

Documentation in a CAD/CAM environment also means preparing a model that will be useful for manufacturing. Some models may look good and generate acceptable paper drawings but may not operate correctly when interpreted by CAM software. Structuring a solid model correctly is part of good documentation (as well as an important facet of being able to easily modify the model).

Documentation also serves legal and archival purposes. To protect your design, you need to be able to capture it for patent applications and other legal uses. Although the Boeing 777 was modeled and prototyped electronically, complete records of each of its parts are stored as documentation drawings as well as electronic models. FAA acceptance of a new aircraft is a detailed process that requires a significant amount of documentation. Documentation is a "snapshot" of the model that is fixed at a point in time—either as hard copy (paper or mylar prints, for example) or as a controlled electronic document.

Good file storage habits and an understanding of the documentation requirements in your discipline are part of your responsibilities as an engineer. You must learn good habits for backing up files and for documenting changes to existing files.

Engineers use a wide variety of tools to help them visualize, communicate, and document their work. From pencil and paper to modeling software to rendering and animation tools, each technique can be used at various points in the design process. To better understand the many tools and techniques you will learn about in this book, take a look at them in the context of a more detailed model of the engineering design process.

A Model of the Design Process

The simplified design process in Figure 1.6 is a good illustration of the iterative nature of design, but its simplicity makes it less useful as a guide for the efficient design of complex products.

To be competitive, most firms want to encourage their designers' creativity and allow them to examine a wealth of possible solutions to a design problem. At the same time, these firms wish to control the cost of refining and developing each of these ideas. One way to

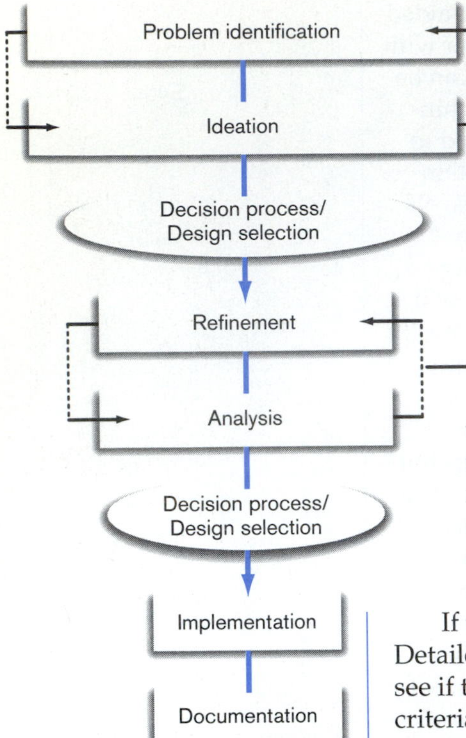

FIGURE 1.9
The design process

balance the importance of generating a wealth of possible solutions with the cost of refining them is to use a design process that breaks development into two stages.

The design process model shown in Figure 1.9 includes three instances of the simple loop presented in Figure 1.6: one loop for problem identification and ideation, one for refinement and analysis, and a third showing that there may be some iteration between the two. In fact, iteration can occur at all points in the process if circumstances demand it.

In the problem identification/ideation loop, the goal is to define the design problem and to identify all the possible ways the problem may be solved. In this stage, the design team strives to translate customer needs into engineering specifications for the product. A written needs statement or problem statement can help keep the project on track. Good design solves the problem that was identified, not some other related problem.

If the customer needs a faster airplane, how fast is fast enough? Detailed design criteria make it possible to evaluate design ideas to see if they can meet specified performance benchmarks and other criteria. Because design problems are frequently loosely defined at the outset, however, it can be difficult to determine all of the requirements ahead of time. A needs statement or problem definition may be very general at the start. As design ideas are generated and evaluated they serve to further define the problem. If a concept for a faster airplane with reduced cargo space is unacceptable, the engineering specification for minimum cargo space should be defined. New ideas are generated to satisfy the problem statement as it becomes better defined until a set of viable concepts is developed.

At this point, firms employ some process for deciding which design ideas to pursue. This may be an informal process or it may consist of formal sign-offs from several departments that will be responsible for making, marketing, and selling the product. The funnel shape of the process in Figure 1.9 shows how the ideas generated in the first cycle of problem identification and ideation are narrowed to a smaller number that pass muster and proceed to the refinement and analysis stages.

In the refinement/analysis loop, also called the product development stage, viable ideas are fleshed out to include more detail and to allow for different types of analysis. In this stage, models are created to test the design against the engineering specifications. The results of these tests feed back into the refinement process to further improve the design. As detail is added, new features may be added that require the team to revisit concept development for that particular part of the project. Modifications to the design may also result from the manufacturer's requirements or to eliminate assembly difficulties or address product safety.

Before the design is manufactured, another round of design approval is required. The decision-makers at this stage may differ from those at the first stage, and considerations such as cost, assembly, and packaging issues may be foremost.

By the time a design has reached this stage, it is possible that the design criteria have changed. A competitor may have introduced a new product, a manufacturing process may no longer be cost-effective, or the customers for the product may have changed. When this happens, a company may decide to return to the ideation stage to generate alternative designs, as indicated by the third iterative loop in the model.

Finally, any remaining details relating to implementation are clarified and the design is documented for manufacturing. Changes may be made to the design to incorporate the use of purchased or stock parts, the assembly of the product is clarified, and detail drawings with allowable tolerances are prepared.

Because design is a means to an end, the process itself is a flexible one. Different combinations of design teams, software tools, and problem types will generate slightly different versions of this basic model. In the next section, you'll learn about the various tools available to accomplish each stage's goals.

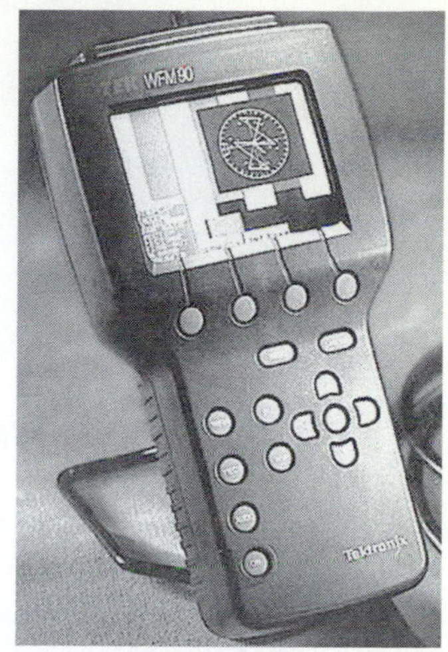

FIGURE 1.10
The Tektronix WFM 90 is a hand-held multifunction monitor that allows a video technician to set up, time, and troubleshoot complex video and audio systems in remote locations. The WFM 90 combines the power of four separate instruments into a single portable device. First, the 4-inch/102-mm color LCD display provides a picture of the incoming video signal. Alternately, the waveform, vector, or audio displays may be shown full-screen, or any one may be cut into any corner of the display. The WFM 90 monitor can be powered with a NiCad battery pack, six alkaline C cells, or an AC adapter. It has an optional viewing hood for operating in bright light, an optional desk stand for benchtop use, and a carrying case with a shoulder strap. (Copyright 1998 Tektronix, Inc. All rights reserved. Reproduced by permission.)

Graphics Tools in Action

Engineers use graphics for visualization, communication, and documentation throughout the design process. A case study from Tektronix will illustrate the tools that its MBD division used in the development of the WFM 90, a hand-held instrument for video technicians, shown in Figure 1.10.

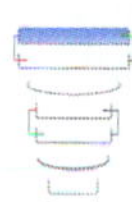

Problem Identification

Graphics Tool/Skill: Reading Existing Drawings

The impetus for the WFM 90 came from two sources. The first was from the marketing department. Based on customer feedback about instrumentation used to monitor and calibrate video signals, marketing generated a list of criteria for a new portable device that would replace three rack-mounted instruments. The needs statement generated by marketing included the following characteristics:

- portability, so it would be easy to use outside the studio and in the field;
- compact size, so the user could carry it in one hand;
- screen display that would allow the user to calibrate and monitor signals from video cameras;
- full range of tools for checking video signals and diagnosing problems;
- cost that would allow competitive pricing.

The project leader provided the second stimulus. His knowledge of forthcoming LCD displays encouraged him to think that the new technology could better meet the needs of the video technician. He

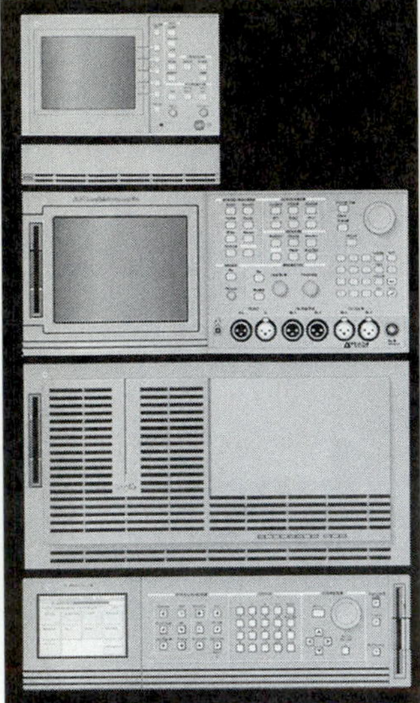

FIGURE 1.11
This illustration depicts four rack-mounted instruments used for video setup and calibration that the WFM 90 was designed to replace for field operations. (Copyright 1998 Tektronix, Inc. All rights reserved. Reproduced by permission.)

Tektronix

Tektronix manufactures more than 3000 products within three broad groups: measurement, color printing, and video and networking. The Measurement Business Division, headquartered in Beaverton, Oregon, started producing oscilloscopes–used to measure and display electrical signals–as the demand for television and FM radio boomed after World War II. Its test equipment line grew as it was applied to the development of computers and the delivery of high-quality color television signals. Today, it is the largest Tektronix division, producing a broad range of instruments for measurement applications. The Color Printing and Imaging Division (featured in a case study in Chapter 12) is the second-largest division, and is headquartered in Wilsonville, Oregon, where it develops and markets color printers.

brought together a team from marketing, mechanical engineering, and industrial design to work toward a product concept.

The team members had a wealth of information to help them define the problem: they were familiar with the design drawings, performance specifications, and manufacturing costs of the rack-mounted instruments already being produced by Tektronix (see Fig. 1.11).

For this instrument, the challenge was to get the circuitry and power supply needed to support all the functions into a device small enough to be handheld. A key element of the design challenge was shrinking the circuit boards so they would fit into the size available. Being able to read and interpret documentation drawings of existing designs were important skills at this stage.

The team also made use of additional information in defining the problem. Data about the performance, cost, size, and operating parameters (such as the amount of heat generated) for the forthcoming LCD panels was gathered. Mechanical engineering was able to estimate the number of batteries the unit would require and how many square inches of circuit board would be needed to accommodate the electronics.

At the problem identification stage, the team worked to enlarge the definition of the criteria by which the solution would be judged. Some information gathered at this stage is obvious; more will be identified during the ideation stage.

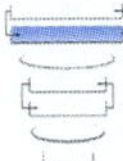

Ideation

Graphics Tool/Skill: Freehand Sketching

Having studied the problem, the Tektronix design team met weekly as a group (and more frequently informally) to discuss possible solutions. Because of the iterative nature of problem identification and ideation, these stages at Tektronix are rolled into a single stage called *definition*. At each meeting, ideas were evaluated and new information shared.

At the ideation stage, being able to draw a readable freehand sketch is vital to "selling" your idea to your design team colleagues. Fitting all the components into a portable size and shape was the key design challenge for the WFM 90, so the process started with Scott Ketterer, the industrial designer on the team, bringing pencil sketches of various ways in which the parts could be combined. Scott presented sketches of devices with vertical and horizontal orientations (see Fig. 1.12), and one that opened like a clam shell.

This stage is called *universal possibilities* because the group seeks to consider every possible solution. At this stage, groups might hold brainstorming sessions to identify more design ideas around the power supply, the shape of the case, and any other component that might be redesigned to better suit the purpose. This type of exercise generally spawns new questions about the design problem which can be used to further define the problem, too. For the WFM 90, the size of the circuit boards was a barrier. It was overcome early in the process when the team decided to design a custom silicon chip to do what several larger circuits did in the rack-mounted instruments.

Once the team decided on a vertical, calculator-like orientation, it worked to define the product more. At this point, Scott Ketterer began to generate drawings on the computer, using Adobe Illustrator, a 2D drawing program. This software allows free-form drawing, much like hand sketching, but it also provides grid and measurement tools so drawings can be prepared to size. Mechanical engineering exported basic shapes of components from Pro/Engineer, the 3D modeling software used at Tektronix, so Scott could use the full-size 2D images to test fit the components into the vertically oriented case he was designing.

A key challenge was making the WFM 90 narrow enough to fit the human hand. As the team members got more details about the LCD panel for the display, they discovered it would be wider than they had anticipated. Working with actual-size drawings of the various components, Scott modified the shape of the WFM 90 to allow for the width of the LCD panel by creating a "handle" section that would be narrow enough to be hand held (see Fig. 1.13).

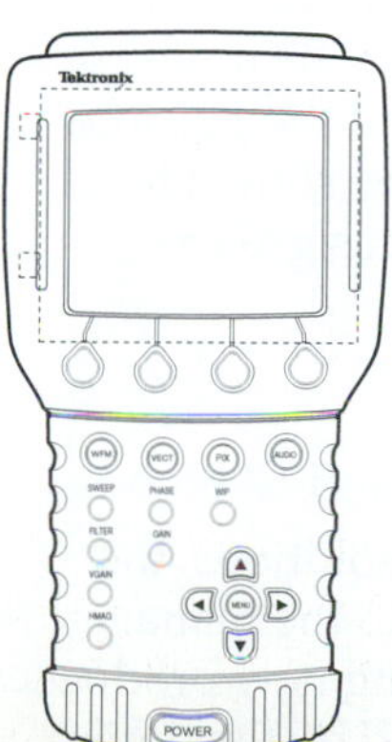

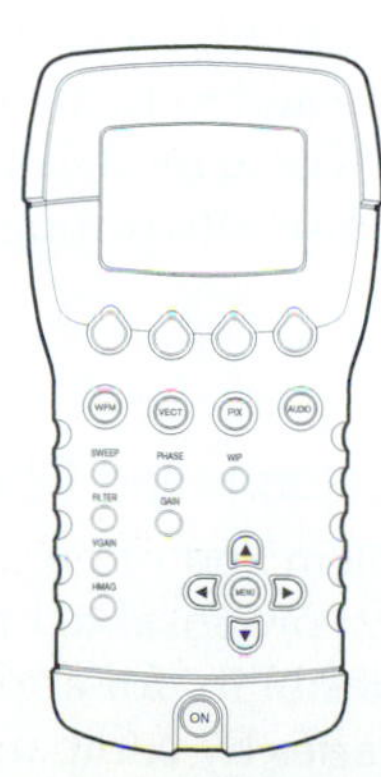

FIGURE 1.13
The original small-screen concept on the right was modified to accommodate a larger LCD display yet keep the handle narrow enough for the human hand. The dotted lines in the concept drawing on the left represent the LCD panel.

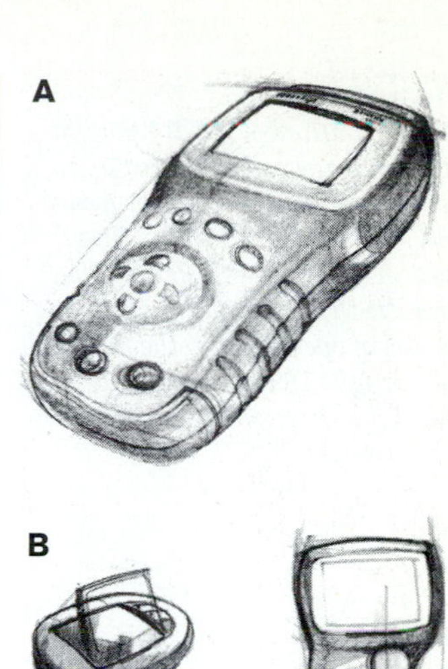

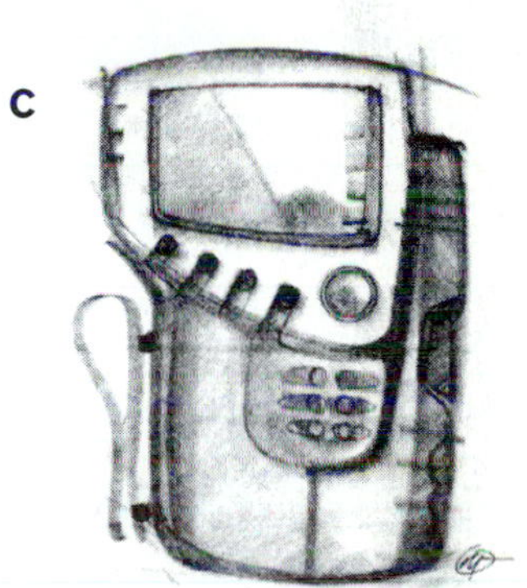

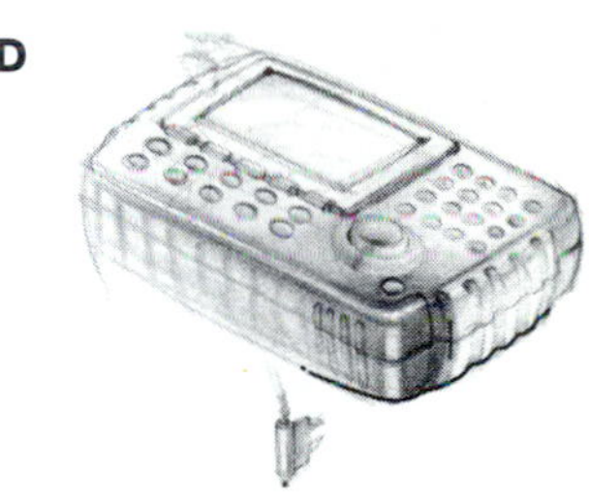

D

FIGURE 1.12
Each of these concept sketches for the WFM 90 includes the LCD monitor and interface buttons in a shape that can be gripped in one hand. Note that (b) and (c) also include the attachment for a shoulder strap.

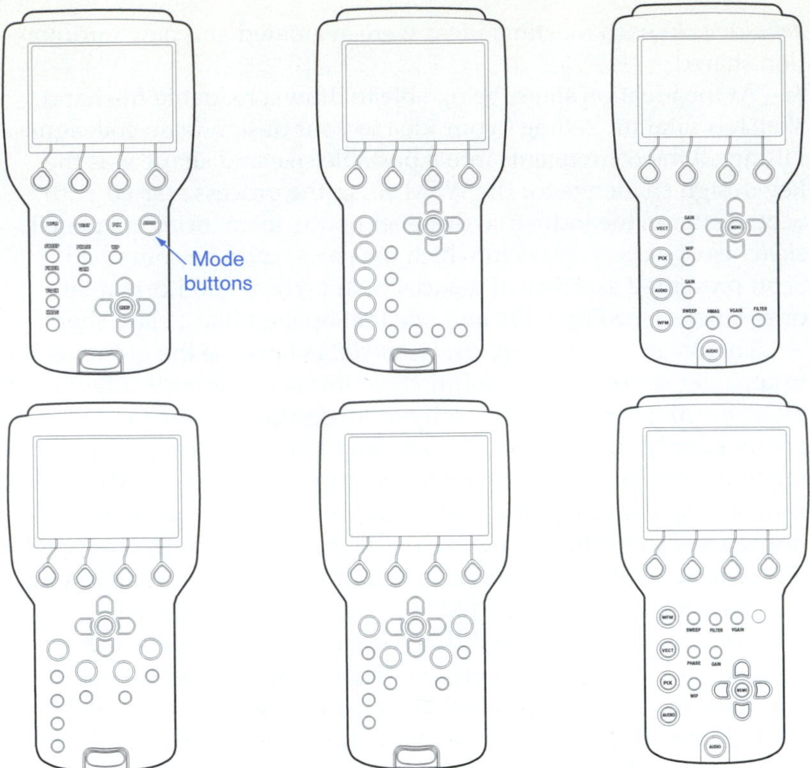

FIGURE 1.14
The 2D outline of the drawing in Illustrator was used to create several variations on the keyboard interface for the WFM 90. The buttons labeled Vect, Pix, Audio, and WFM correspond to the four modes of operation for the monitor. (Copyright 1998 Tektronix, Inc. All rights reserved. Reproduced by permission.)

The design team also considered various options for the placement of the buttons that control the video display, some of which are illustrated in Figure 1.14. Buttons were needed for four "modes" of operation (waveform, vector, audio, and the raw video display), plus controls for functions within each of these modes. To reduce the number of buttons on the keyboard, the team decided to use the LCD panel for some of the controls. A single button could then control several functions, depending on the "mode" the unit was operating in. Moving some of the controls to the screen simplified the final keyboard and reduced the number of buttons that had to be fit into the handle of the device.

Even at this early stage, drawings and models can be used to elicit customer feedback. At this stage, the team uses models and drawings as needed to communicate relevant features to the user. If the interface operation is important, for example, a 3D model or a computer simulation may be created to further detail the portions that the user will interact with. These models may be shown to customers informally, or more formally through focus groups.

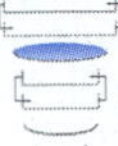

Design Selection

Graphics Tools/Skills: Interpreting Charts and Sketches

After further study of the problem and of possible solutions, the design team prepared a product proposal to bring to the management group that had the responsibility for approving new products. The design selection process varies by company and by the design challenge. It may be very informal and conducted by a single individual, or it may be a highly formal process in which different teams compete for development dollars for new products.

The design selection group for the WFM 90 consisted of managers from engineering, manufacturing, and finance, as well as executive managers. Each member received a product plan, sales estimates, and cost figures for developing the product, and an estimate of the profit to be generated by the expenditures. The members' role was to decide among several product proposals to identify those that best fit with the company's goals and promised the best return on the investment over five years. The ability of the team to communicate effectively through drawings, spreadsheets, written communication, and oral presentation all played a role in the product's acceptance for development.

2D preliminary drawings were prepared to fully represent the product concept. These drawings are similar to multiview drawings used for engineering documentation: drawings showing the top, front, and side views of the device, plus an isometric view that simulates 3D viewing (see Fig. 1.15). Relevant dimensions were added to some of the drawings so the design team could assess critical dimensions, such as the width of the handle. These dimensioned drawings also provided a starting point for the engineers who would model the part in the next phase of the project.

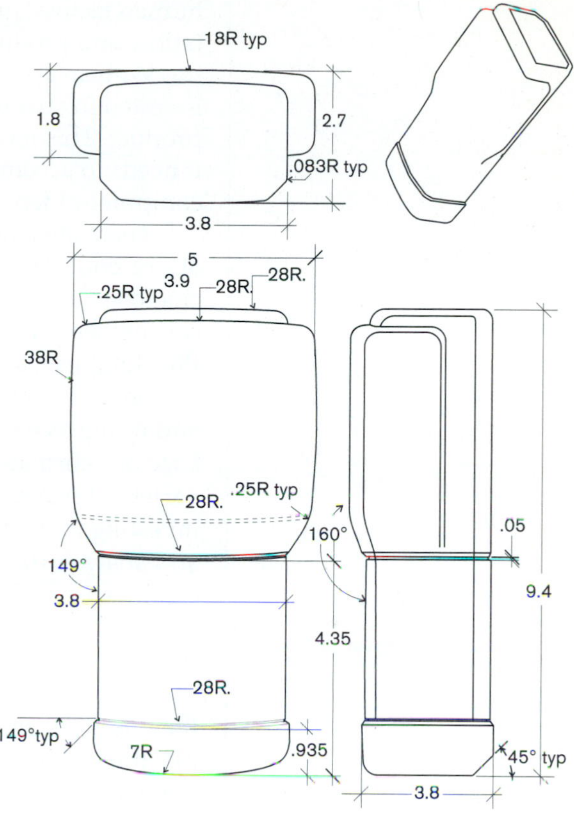

FIGURE 1.15
Dimensioned drawings created at the concept stage show front, top, right-side, and isometric views of the shape of the WFM 90. (Copyright 1998 Tektronix, Inc. All rights reserved. Reproduced by permission.)

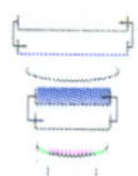

Refinement

Graphics Tools/Skills: 3D and Parametric Models

When the WFM 90 was approved for further development, the team moved into the refinement and analysis phases of the project. Because these two phases form another "loop" in the design process, they are combined into a single stage at Tektronix called the *development phase*.

During the refinement stage, the product concept is solidified into an accurate plan for making the product. Different options are evaluated in terms of how much they will cost, how long they will take to make, how well they satisfy the customer's requirements, how durable they will be, how difficult they are to assemble and service, etc. Adding members to the design team can provide expertise in these different areas and help finalize the design.

Project leadership for the WFM 90 passed at this point to a hardware designer, Bob Prebstell. Given the pivotal role of the custom chip being designed for the WFM 90, it made sense for Prebstell to head the project. The team also gained a second mechanical engineer, an electrical engineer, and members who would address human factors, product service, manufacturing, technical documentation, and product safety.

With an approved concept in hand, time and money are invested in creating a model that will accurately represent the final product. This model will be used to analyze and test the plan, so it needs to accommodate detail without sacrificing flexibility. A computer-aided drawing package is essential at this stage.

There are a host of modeling packages that create models in 2D, 3D, or both. Their fitness as a modeling tool depends on the design challenge and the needs of the design team. At Tektronix, mechanical engineer Paul Chadwick modeled the WFM 90 using Pro/Engineer, a parametric solid modeling package.

Solid modeling creates a computer model that can be analyzed and tested as if it were the real thing. The model is actually a database of information that can be used to simulate the product in a variety of ways. The ability to simulate the final product without investing in its manufacture allows the team to perform more testing and analysis during the refinement cycle of the design while changes are still relatively low cost. The completeness of the solid model database makes it the "hub" that communicates the plan to the various members of the team.

Once the mechanical engineers began to model the WFM 90, they used the model to establish parameters for the team. For example: the shape, size, height, and clearances for the circuit boards were defined in the Pro/Engineer model, then passed to the electrical engineers who would design the boards. If the board size or height had to change, the model would be changed to reflect the new parameters.

At Tektronix, their solid models are also *parametric* models. Parametric models capture *relationships* between part features and the sizes of the features. When a part changes, any related parts then update automatically. For example, the casing for the WFM 90 was formed in two parts and snapped together with pins on one side of the casing that fit into matching holes on the other side of the casing (see Fig. 1.16). The locations of these pins and holes were somewhat

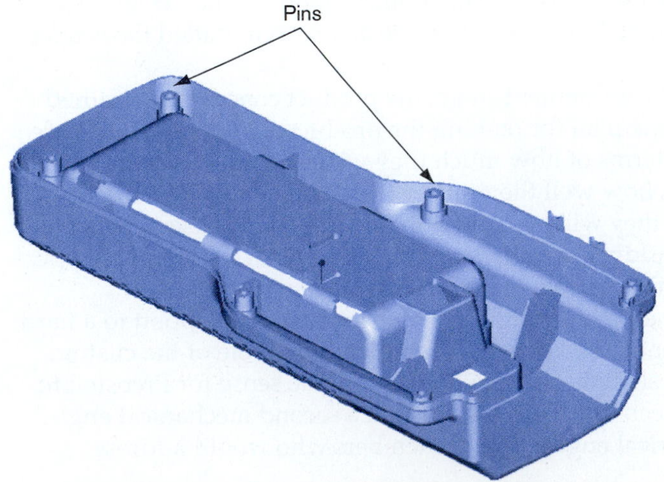

FIGURE 1.16
The parametric model for the bottom half of the case shows the pins that were designed to fit into holes in the top half of the case. (Copyright 1998 Tektronix, Inc. All rights reserved. Reproduced by permission.)

arbitrary, but each pair had to match up exactly. The location of each pair was specified in the model by a formula that put them half the width of the case from the midline of the case. Anytime the width of the case changed, all of the pins and holes automatically shifted to their new locations.

During refinement, design changes are made as more is learned about actual components and manufacturing processes. The initial power supply estimates for the WFM 90 were not large enough, so the decision was made to go to larger, C-cell batteries. Once the case was modeled and allowances made for the required wall thicknesses, the interior width would not accommodate the circuit boards. Both necessitated making the case wider and deeper.

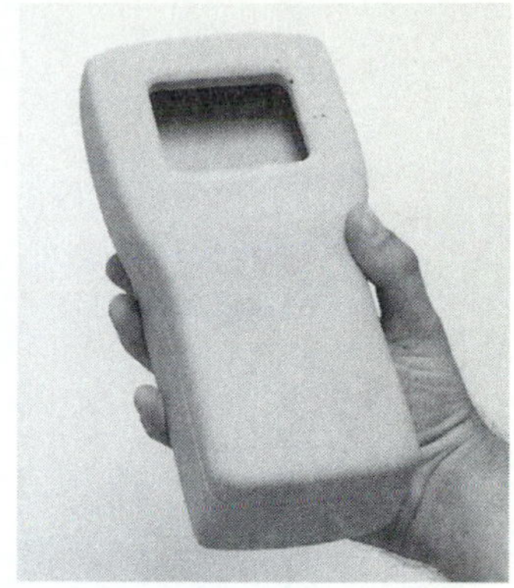

FIGURE 1.17
Because a key design challenge was making a monitor that would be small enough to be hand held, this foam model made it possible for the team to test the fit with different individuals—each with different-sized hands—to ensure that they would be able to hold the device. (Copyright 1998 Tektronix, Inc. All rights reserved. Reproduced by permission.)

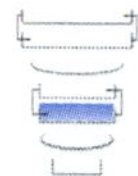

Analysis

Graphics Tools/Skills: Exporting Model Geometry for Analysis

The analysis phase tests the design and feeds back information to further refine the design, forming the second key iteration or "loop" in the design process. Crucial here are the performance criteria that the design must meet. For the WFM 90, size, ruggedness, and electro-magnetic emission control were key criteria.

After the size of the case changed during refinement to accommodate the larger batteries, the wider and deeper case was tested in the human hand. Analysis of human factors tested a foam model of the case for large and small hands, gloved hands, and use in hot and cold weather (see Fig. 1.17). The conclusion was that the handle was too wide to fit comfortably in the hand. The case design was refined further by adding 45° cuts to the back sides of the case to allow the fingers to wrap around the device (see Fig. 1.18). Although the handle was still as wide overall, this refinement met the criteria for a comfortable grip.

Mechanical engineering did structural analysis of the case's strength to ensure that it would survive a fall from 3 feet. The plastic case was tested to determine the rate at which it would dissipate the heat generated by the display. Electrical engineering tested the circuit board design with circuit modeling software, and components such as the AC adapter were checked for fit and location on the circuit board. Open circuit boards were used to control emissions, but a coating was added to the inside of the case for insurance.

When the WFM 90 was designed, many of the software packages available today for analyzing and prototyping 3D models were

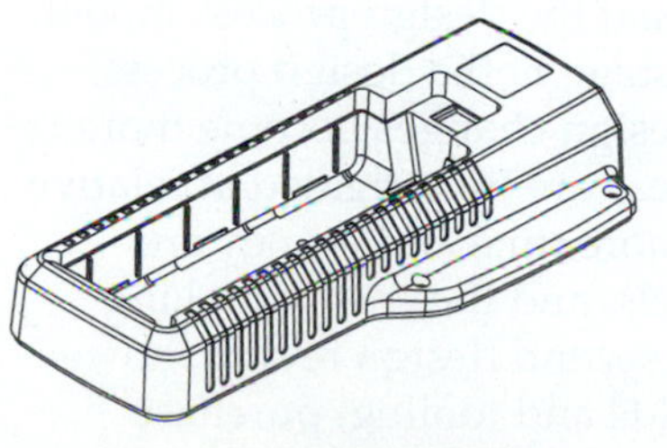

Isometric View

(.030)
(.153)
(.960)
2×45°

Top View

FIGURE 1.18
These views of the bottom half of the case show the bevel cuts added to reduce the overall circumference of the handle to make it easier to grip. (Copyright 1998 Tektronix, Inc. All rights reserved. Reproduced by permission.)

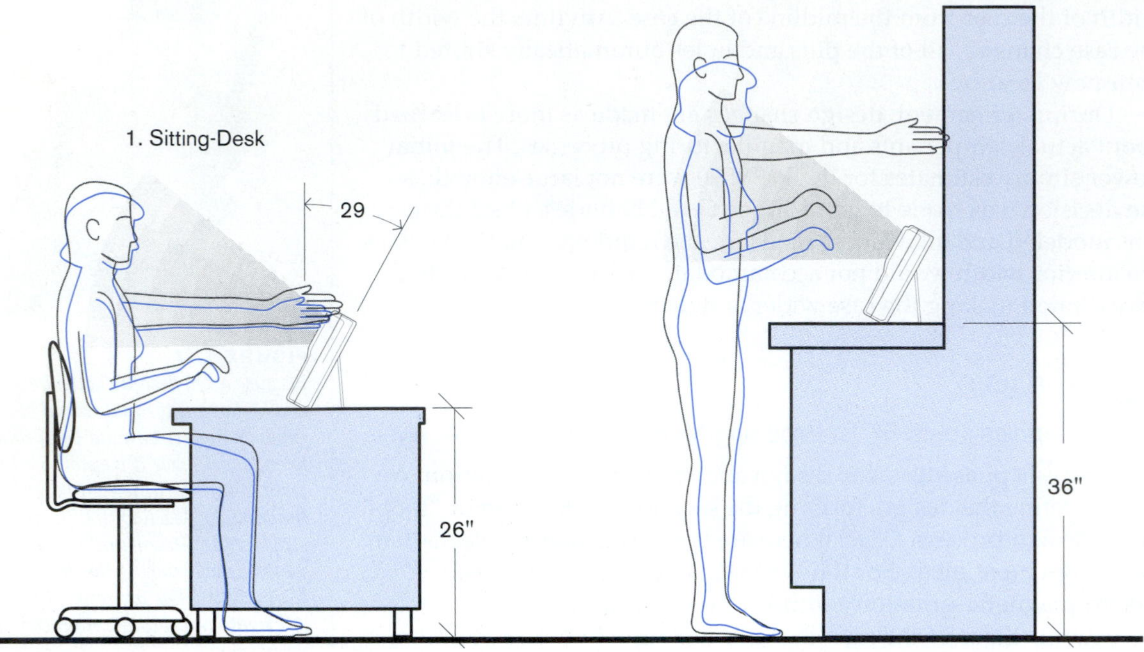

FIGURE 1.19
The WFM 90 monitor comes with a stand so it can also be used at a desk or counter. These illustrations were part of human factors analysis done for the angle of the monitor on the stand. Note that the analysis considered the difference in average height between men and women. (Copyright 1998 Tektronix, Inc. All rights reserved. Reproduced by permission.)

not yet in wide use. As more companies use the solid model as the core of the design process, analysis tools are being developed to work with solid modeling packages. Motion analysis for moving parts, for example, can be done with a 3D modeler such as Pro/Engineer, but the model's parts can also be exported to a kinematics package such as Working Model and tested under a wide range of physical conditions (even zero gravity!). Other analysis tools work with the modeling package to make the changes needed to optimize the model for specific performance criteria.

Whether the analysis software recommends a change or simply reports results, it is up to the design engineer to interpret the tests and take steps to further refine the design and continue to test until the desired results are achieved (see Fig. 1.19).

Design Selection

Once a design is ready to manufacture, another design review may be required. Remember that the design process model is somewhat funnel-shaped. At each stage of the design process, more is invested in the design, and design changes become more costly. At the ideation stage, many ideas are investigated at relatively low cost. In the refinement stage, more time and money are devoted to building and testing models, and the cost of making major design changes rises. After the second design review, however, the company will commission molds and tooling, purchase materials, and commit other resources to manufacturing the product. After this point, major design changes become very costly and can threaten a product's success. The second design selection process is a filter that allows only the products with the greatest chance of success to proceed.

The selection team and review process may be very similar to the one that precedes refinement, or it may involve a completely different set of criteria. As a member of the design team, it will be important for you to understand what the criteria are. Understanding the review criteria can help you make not only an effective presentation, but also help you make better design decisions in the context of your company's goals.

For the WFM 90, a key criteria was time to market. Could the team deliver the product in twelve months? Its plan to do so opened the gate to the implementation phase.

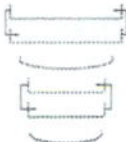

Implementation

Graphics Tools/Skills: Manufacturing and Assembly Drawings

During the implementation phase, the design is communicated to those responsible for manufacturing, assembling, and distributing the product. Up to now, drawings and models have served a variety of purposes. In the implementation phase, they become contracts with suppliers. The model and the accompanying drawings (if any) must clearly specify what will constitute an acceptable part.

As in refinement, the 3D solid model serves as the information database. Although fully dimensioned multiview drawings may be generated from the solid model, more and more processes can be directed from the model itself. In these cases, critical dimensions marked on a plot of the model may be sufficient.

For the WFM 90, the tooling for the plastic injection-molded case was controlled by information from the Pro/Engineer solid model. The model itself had been created to require no more than standard tolerances, and the experienced tool builder had worked with Tektronix in the past. Paul Chadwick noted critical dimensions on a plot of the model for the case, then provided the Pro/Engineer model database to define the other dimensions. With a less experienced tool builder, or for a part that had to fit more closely, more specifications would have been spelled out so that the definition of an acceptable part would have been clear and unambiguous—as a contract should be.

Although the design has been finalized, there still may be changes during the implementation phase (as well as later when the product is in production). Unavailable parts or materials can force a substitution, a supplier can recommend a cheaper alternative, standard parts may be substituted for newly designed ones, or unanticipated problems can crop up during assembly. With the proliferation of rapid prototyping processes, it is easier and more cost effective to anticipate these issues in the refinement stage, but it is not uncommon for fine-tuning to occur during implementation.

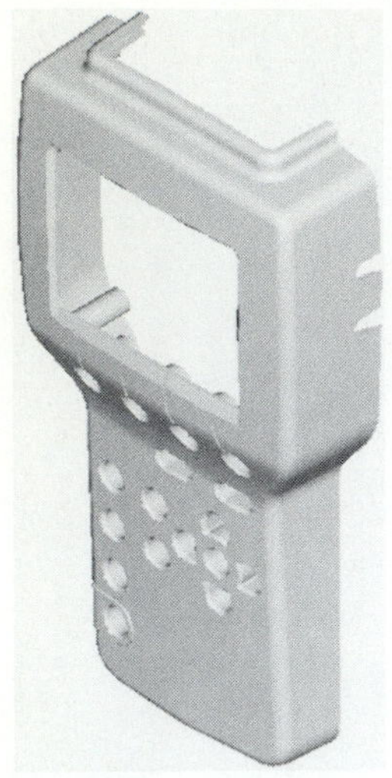

FIGURE 1.20
This 3D view of the top of the WFM 90's case is smoother and lacks the gripper indentations present in the earlier concept views of the monitor. These were preserved on the bottom of the case but removed from the top. (Copyright 1998 Tektronix, Inc. All rights reserved. Reproduced by permission.)

The accelerated production schedule for the WFM 90 is a case in point. After the solid model had been delivered for tooling in January (tooling was to take twelve to sixteen weeks), the model was also sent to another firm, Plynetics, which would generate a rapid prototype, use it to make a urethane mold, then produce epoxy versions of the cases. These cases would take five weeks to make, cost about $200 each, and be ready in time to be assembled with circuit boards and LCD panels and shown at a major trade show in March. Plynetics looked at the solid model for the case and made some suggestions for simplifying the exterior detail on the case to reduce the cost of making the mold (see Fig. 1.20). Tektronix liked the changes, and the model was revised. The revised copy of the electronic file was subsequently used for the tooling of the final steel mold. Given the long lead time required by the tool maker, the changes were incorporated without delaying the project.

Assembly drawings are generated to document the parts included in an assembly and to illustrate how they fit together. The parts list, or bill of materials, is an important facet of inventory planning and assembly. At Tektronix, the solid model is used to generate plots of individual parts to be catalogued in the company-wide parts database. Parts in the database can be reused or shared by other products, and standard parts inventories can be tracked and monitored.

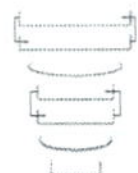

Documentation

Graphics Tools/Skills: Rendered Views, Exploded Views

The documentation phase of the design process captures the final design and freezes it. Each company has its own procedures for managing models and controlling changes to the model database, but it must also create permanent records for archive purposes, patent applications, and other legal uses.

For the WFM 90, dimensioned drawings were prepared after the fact to document the final product as manufactured and assembled (see Fig. 1.21). Measurements were taken from the parts and the solid model modified to match them, then dimensioned drawings were prepared and stored in the archive. (This is the reverse of the traditional model, in which the documentation drawings were prepared first to transmit the design to manufacturing.)

Fully dimensioned engineering drawings provide a permanent record of the design at a point in time. The solid model may also be archived for use in product redesign, but its data may not be acceptable as a permanent record. Because files stored in electronic format can be altered so that they no longer represent the actual product design, the information in the solid model must be controlled or captured in another medium to be a permanent record.

The solid model can also be used to generate a wide range of graphics used to document a product or process. Views of the model are imported into step-by-step illustrated guides for assembly line personnel, and the same guides are sent to service personnel who need to disassemble and reassemble the products after sale.

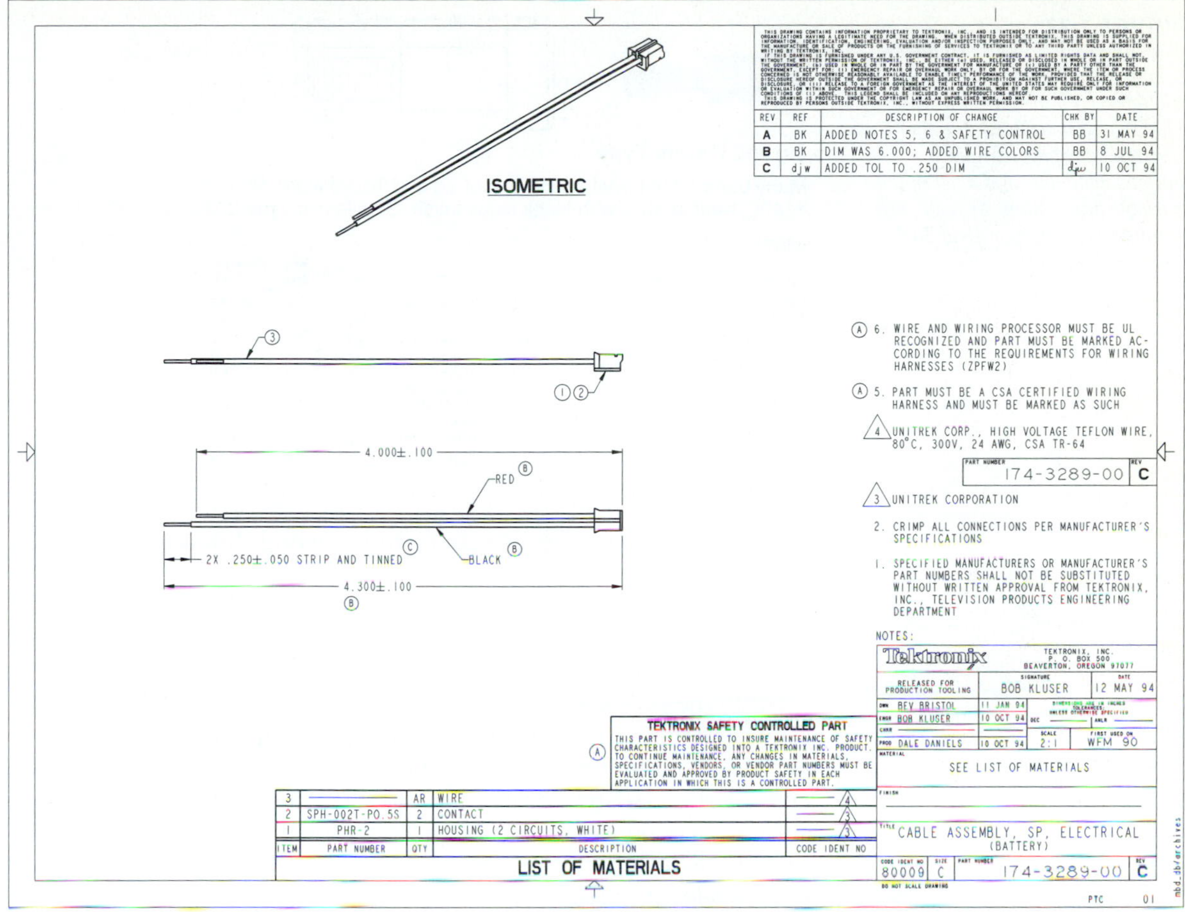

FIGURE 1.21
Dimensioned part drawings for each component of the WFM 90 are part of the Tektronix archive and referenced by number to the parts database. This cable assembly drawing documents the part, who made it, the materials and specifications to be used to make the assembly, and a listing of any changes to the drawing since it was first released for manufacture. (Copyright 1998 Tektronix, Inc. All rights reserved. Reproduced by permission.)

Types of Design Processes

The Tektronix WFM 90 case study illustrates the many ways graphics supports the design process, but it is just one example of the way design tasks can be organized and carried out. The tools the Tektronix team used and the process they followed were designed to fit the needs of their organization and product.

In large companies, the design process may be well defined and coordinated. In smaller companies, or if you're working alone, it may be up to you to decide how best to organize the design effort. The first step in designing a product is planning the process itself. Planning up front for the tasks to be accomplished, the interaction among team members, and the decision points in the process will help you select the right graphics tools and use graphics effectively throughout the process.

Design Tasks

The tasks outlined in the design process model may or may not be required for a design problem. There are several different kinds of design problems, each with a different set of tasks and objectives.

FIGURE 1.22
The catalog listing for shoulder screws shown here is typical: it includes a 3D view (in this case, a photo), a dimensioned drawing to illustrate the fastener and the options available, and a list of the variable dimensions for each part number. (Courtesy of Reid Tool Supply Company.)

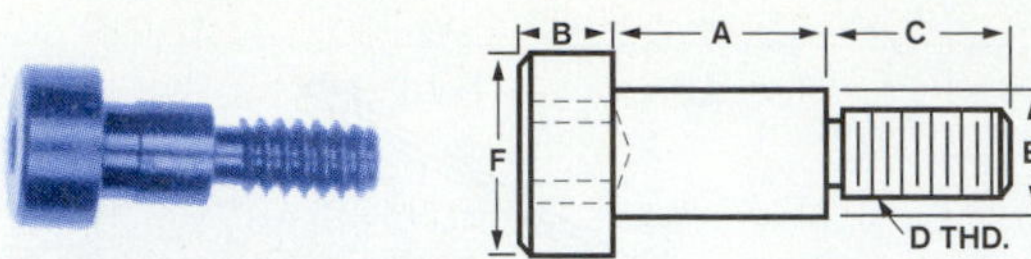

Jig and Fixture Type

Many uses in the construction of tool, jig and fixture work. Steel is 4140L, heat treated with black oxide finish. Stainless is type 303.

Steel

Cat. No.	A	B	C	D	E	F	Key Size	Price Each 1-11
SS-00	.250	3/16	3/8	10-24	.2480/.2460	3/8	1/8	**$1.68**
SS-7	.250	7/32	7/16	1/4-20	.3105/.3085	7/16	5/32	**1.87**
SS-0	.312	7/32	7/16	1/4-20	.3105/.3085	7/16	5/32	**1.98**
SS-1	.250	3/16	3/8	1/4-20	.3730/.3710	5/8	3/16	**2.13**
SS-2	.312	3/16	3/8	1/4-20	.3730/.3710	5/8	3/16	**2.08**
SS-3	.375	1/4	1/2	5/16-18	.4980/.4960	3/4	1/4	**2.97**
SS-4	.500	1/4	5/8	3/8-16	.4980/.4960	3/4	1/4	**3.33**
SS-5	.625	1/4	5/8	3/8-16	.4980/.4960	3/4	1/4	**3.38**
SS-6	.750	1/4	5/8	3/8-16	.4980/.4960	3/4	1/4	**3.42**
SS-8	1.000	3/8	3/4	1/2-13	.6230/.6210	7/8	5/16	**4.34**

Complex design processes may include several different design problems.

For example, a *selection* design problem could be as simple as choosing the appropriate fastener from a catalog (see Fig. 1.22). This design problem would be well served by a more simplified model: identify the problem and engineering specifications for the part, select a part, analyze it to see if it works, and document the design decision. In this simple example you would need the ability to read and interpret existing drawings (and written specifications) to be sure the fastener satisfies the criteria for the part. You may also need to edit the design database to incorporate a reference to the new part and its relevant features.

Configuration design is another kind of design problem in which the challenge is to arrange a set of components into a redefined space or shape. Fitting the power supply, LCD panel, circuit boards, and connectors into the casing for the WFM 90 is an example of configuration design within the larger product design process (see Fig. 1.23). Your ability to visualize and sketch various configurations,

FIGURE 1.23
The dotted lines in this concept drawing of the WFM 90 represent the size and shape of the components to be contained in the plastic case. The two side views illustrate the impact of going from size AA to size C batteries. (Copyright 1998 Tektronix, Inc. All rights reserved. Reproduced by permission.)

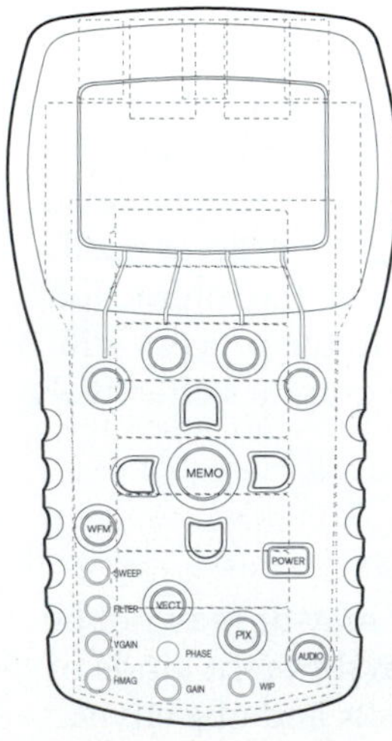

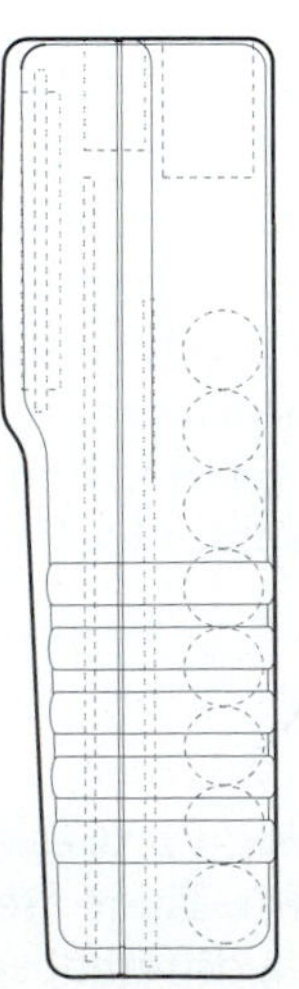

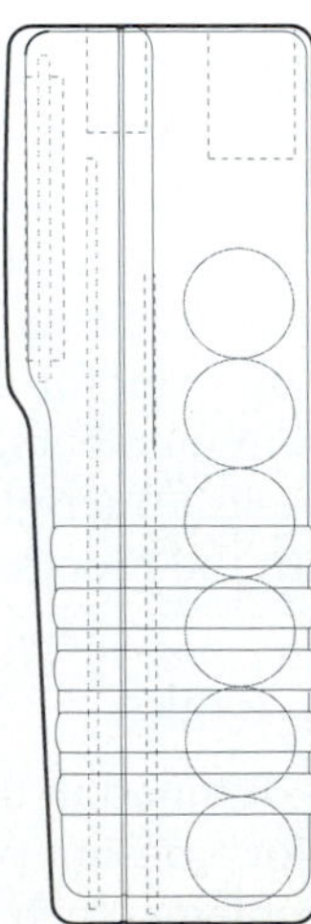

then use more accurate computer models (either 2D or 3D) of the various components to test the configuration will help you solve this type of design problem.

Redesign involves making improvements to an existing part or product. This type of design problem can range from redesigning a component to completely redesigning an entire product. Many times, these design challenges address components that cause problems after a product is released. Customer complaints, service problems, assembly difficulties, and changes in manufacturing options are all reasons why a part might be redesigned. For example, the printer part shown in Figure 1.24(a) was originally designed as a steel assembly in six parts plus a fastener. The redesigned part, shown in Figure 1.24(c), is a molded plastic piece that requires no assembly yet provides the strength and flexibility required. Redesign requires you to be able to work with models and drawings for parts that have already been designed and produced. Reading documentation drawings and building a computer model from existing specifications are two skills common to this type of design. Your knowledge of manufacturing processes—or the ability to gather this information from manufacturing engineers—also can be very important to your success with redesign.

Original or new product design was illustrated by the WFM 90 case in this chapter. It encompasses the full range of graphics and communications skills, including sketching to support ideation, 3D modeling, prototypes for user input, presentations to users and managers, and documentation and assembly drawings.

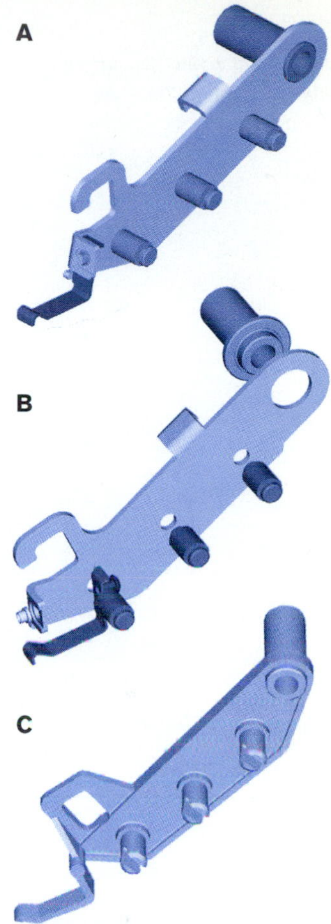

FIGURE 1.24
The exploded view in (b) shows the different pieces that had to be assembled for this part before it was redesigned into a single plastic part, shown in (c). (Copyright 1998 Tektronix, Inc. All rights reserved. Reproduced by permission.)

Concurrent Engineering

The design process used at Tektronix is a contemporary one—not just because of the graphics tools the company uses, but because it practices *concurrent engineering*. Concurrent engineering recognizes the many influences on a successful product and applies the knowledge of different departments early in the design process. Instead of a linear process that moves from marketing (which asks for the product) to engineering (which designs it) to manufacturing (which makes it) and to the warehouse (which assembles and ships it), concurrent engineering describes a process in which teams of individuals from all of these departments are engaged in defining the problem, evaluating concepts, and developing the product design (see Fig. 1.25).

Concurrent engineering recognizes the knowledge base of different individuals and puts it to work. As the design team works to clarify the design problem and draft specifications for a project, it may be working with a vague description of the customer's needs. Marketing's involvement throughout this stage of the process can ensure that the team understands the customer's desires—and can help to identify additional criteria that may not have been spelled out. Sketches, rendered views, and 3D models can play an important role in communicating design ideas to non-engineers—for the

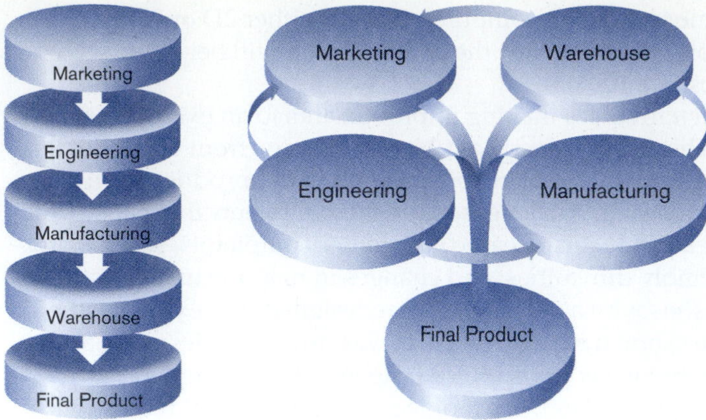

FIGURE 1.25
Traditional versus concurrent engineering design

simple reason that "a picture is worth a thousand words." Concurrent engineering adds to the design criteria by adding consideration of all facets of the product's manufacture, assembly, distribution, repair, and, increasingly, disposal to the problem definition.

For example, service engineers can help a design team anticipate and avoid choices that will make it difficult to service and repair a product. For the WFM 90, service personnel were on the team to steer the design away from using glue to join the two halves of the molded case. Although glue would be an effective way to seal the case, it would make it difficult to service the components inside and increase the risk of cases being damaged during service. Planning for a different type of closure helped avoid after-sale replacement costs.

As a new engineer, you will bring a broad base of engineering knowledge to your first job. But you may know very little about manufacturing—and even less about more specialized processes important to your employer. The manufacturing engineer on the design team is a resource of information about what *can* be built and the options for building it. Just because you can draw a design—or even create a 3D model of it—doesn't mean that it can be made with today's technology. Understanding common manufacturing processes will make you a better designer. Many manufacturing companies are happy to work with designers to ensure a design that can be built easily and will be cost-efficient, and the World Wide Web contains many excellent resources for manufacturing information. But it may not be the best use of a company's talent to train all of its engineers to also be manufacturing engineers. Concurrent engineering recognizes the value of this specialized information and allows for designers and manufacturing engineers to interact with and learn from each other early in the process.

Similarly, assembly and distribution processes can dramatically affect the final costs of getting a product to the customer. Assembly engineers evaluate designs to be sure they can be assembled. A fastener in a difficult-to-reach position may add to assembly costs and require that a special tool be made. Identifying it early allows for redesign that can eliminate it or replace it with a different feature.

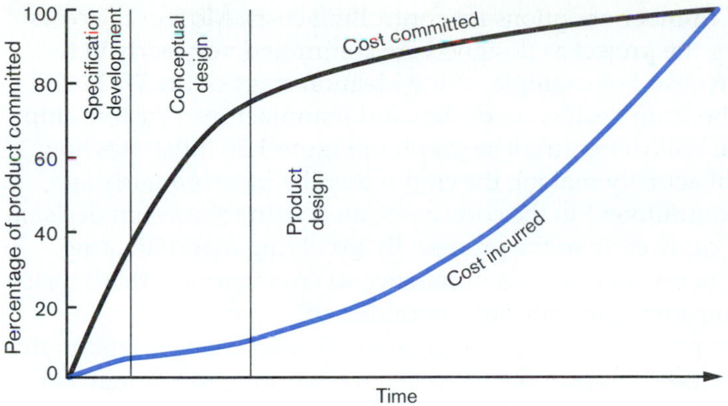

FIGURE 1.26
Early in the design process a large percentage of the life cycle cost of the product is committed. Late in the process, when costs are actually incurred, it may be impossible to reduce costs without major changes to the design. In order to produce products that are viable on the world market, you will want to consider as many alternative designs as possible early in the design process. (Reprinted with permission of The McGraw-Hill Companies from The Mechanical Design Process, Second Edition, *by David G. Ullman, 1997.)*

Computer-Integrated Manufacturing

While concurrent engineering describes the effect of integrated systems on the product design process, *computer-integrated manufacturing* (CIM) describes the use of technology to integrate data common to engineering and product design, manufacturing planning, and automated factory processes. As microcomputers became more affordable and numerically controlled machines automated manufacturing processes, these technologies made it possible to streamline operations for greater competitiveness. Today's concurrent engineering is supported and made possible by the degree to which design tools have improved the collaboration between manufacturing and engineering. This trend continues today, as even more functional areas in the organization share information in a common database.

The Concurrent Design Process

In the traditional design process, the stages of the design model corresponded to different engineering groups that each performed their part of the process more or less independently. Other company groups would not be represented in the model, except perhaps as groups that provided input or handled output. Marketing would toss the design problem "over the wall" to engineering, and engineering would, in turn, toss it "over the wall" to manufacturing. As you might imagine, the opportunity for miscommunication in a traditional model is great, and difficult problems could cause the project to be canceled or returned to an earlier stage of the design process. The design process allows for a certain amount of iteration during ideation/visualization and refinement/analysis, but today's companies operate in competitive environments where speed to market is the difference between success and failure. Concurrent engineering allows today's companies to get products to market faster at less cost.

Most of the costs of producing, assembling, and shipping a product are determined by decisions made early in the design process. Once a product has been released to manufacturing, there is a limited number of options for controlling costs. Most of the costs of making the project as designed are committed very early in the design process. For example, in the ideation stage of the WFM 90 design, the team decided to design and manufacture a custom chip to replace bulky circuitry. The graph in Figure 1.26 illustrates how the cost of actually making the chip would be incurred fairly late, but the commitment to the costs associated with this design decision occurred quite early in the process. By involving manufacturing early in the process—as it is in concurrent engineering—these costs will be considered as they are committed to.

When problems are not anticipated, making changes later in the process causes delays and adds costs in reworking the design. For complex products, any change to a subassembly or system can cause ripple effects through several other systems. Boeing's shift to a concurrent "design/build" model was driven in part by its need to reduce the amount of rework required when the many components of a jet airplane came together to be produced. The Boeing 777, which was the first airplane designed and built under this concurrent model, experienced significant reductions in the cost of design changes, as illustrated in Figure 1.27.

A concurrent engineering process is sometimes depicted as overlapping circles representing the functional areas, such as assembly and manufacturing, that are involved in the design process sooner. Boeing's concurrent model, shown in Figure 1.28, is one way to illustrate the involvement of different groups in the development phase of new product design. Note the functional groups included in the design/build process circle.

Depending on the design problem and the size of the organization, the degree to which functions such as marketing, manufacturing, service, and sales interact with design engineers will vary considerably. In each case, however, they will participate in and contribute to the generation, testing, and selection of alternative approaches that will result in a final product or process design. The overall funnel shape of the design model that we began with in Figure 1.9 is the container that integrates the ideas and expertise of the different members of the organization and the design team, as shown in Figure 1.29.

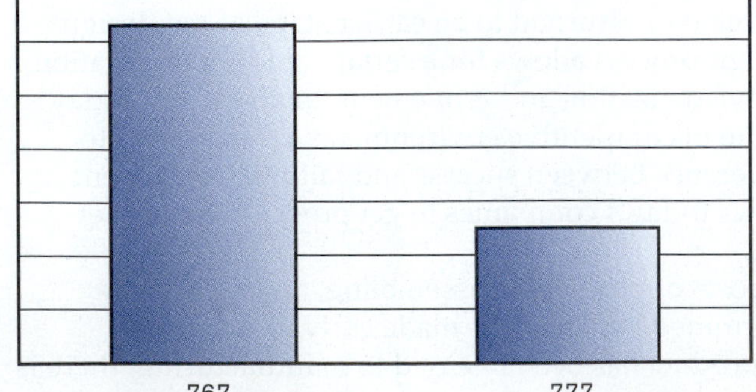

FIGURE 1.27
Boeing tracked and compared the costs incurred for different kinds of changes during the development of the 767 and 777 aircraft and graphed them in the chart shown here. Notice the reduction in change activity with the shift to concurrent engineering and solid modeling. (Courtesy of The Boeing Company.)

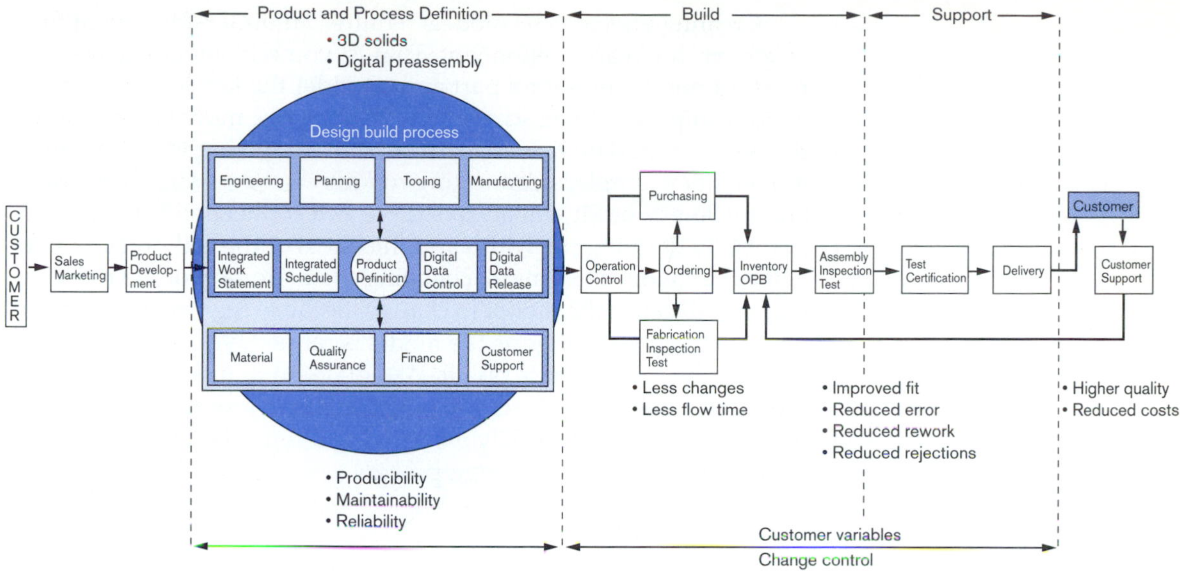

FIGURE 1.28
Boeing's Preferred Business Process illustrates the concurrent nature of the design/build model at the heart of its design process. (Courtesy of The Boeing Company.)

Concurrent Engineering and the CAD Database

The evolution of CAD systems to 3D solid modelers fueled, even enabled, the move to concurrent engineering. 3D solid models offer four advantages over traditional 2D engineering drawings by allowing design teams to work together more effectively and to test and refine ideas more quickly:

- 3D models offer a wide range of ways to view the model, such as animated and rendered views, in addition to traditional 2D views.
- The completeness and accuracy of the design definition in the CAD database make the models suitable for use in analysis.
- The CAD database can drive prototyping and manufacturing machinery.
- The 3D modeling software can automatically update related parts of the model when changes are made.

You've already seen how the CAD database supports graphical communication, allows for analysis earlier in the design process, and supports lower-cost prototypes that can be used to test and refine the model. Underlying all of these benefits, however, is the ability of the model to be updated and changed. The time and expense involved in creating a 3D solid model are balanced by its greater utility. If a model cannot be modified or updated, this equation no longer balances.

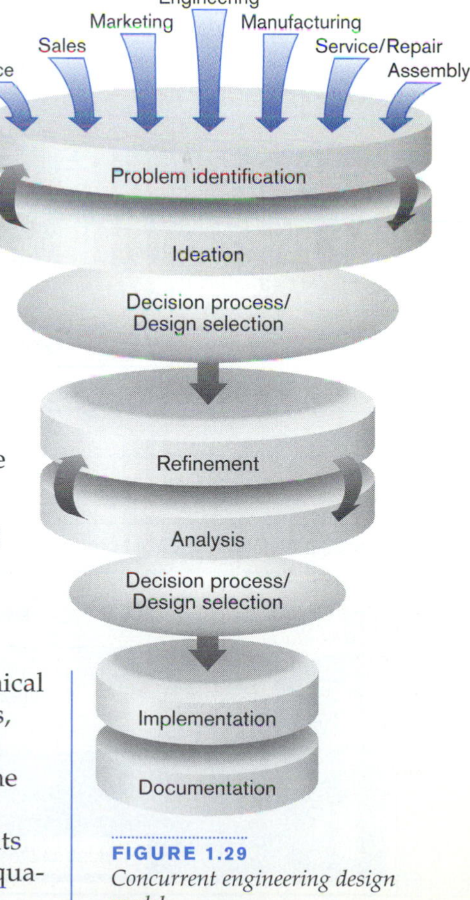

FIGURE 1.29
Concurrent engineering design model

Creating effective 3D models requires an understanding of the *design intent* of each component. That is, you will anticipate the range of possibilities for a part and consider the key attributes and relationships. By doing so, you will define your model in such a way that you can update related features without having to redraw the model. For example, a connecting rod may need to vary in its width and thickness, but it will always serve as a connection between two points. By defining the width and thickness of the rod in relation to a line that defines its length, changing the definition of the rod's length will cause the model to change. Because the width and thickness are likely to change as the model is refined, you would not want to use them as base points. As you learn to use the graphics tools described in this text, you will also learn how to prepare your models so they can be easily modified and reused and are appropriate for each stage of the design process.

Design Teams

Practicing concurrent engineering means building design teams. Through the WFM 90 example, you saw how the design team changed as the product moved through the process. To use teams effectively, it is important to recognize the varying levels of involvement of different functions through the process. Figure 1.30 illustrates one example of how these activity levels can change over time.

It would not be a good use of anyone's time to build teams that required members from every functional area to participate at the same level throughout the design process. Effective teams match involvement levels to the design process to use each member's time wisely. Deciding *who* should be involved and *when* is a key part of setting up the design process for a given project.

FIGURE 1.30
Industrial design, engineering, and manufacturing are involved throughout the design process in the MBD division at Tektronix, but their activity levels vary over time. The numbered milestones correspond to key events in the typical product introduction process.

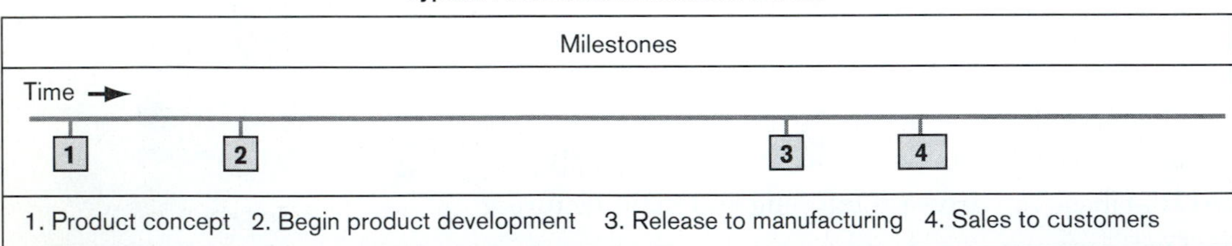

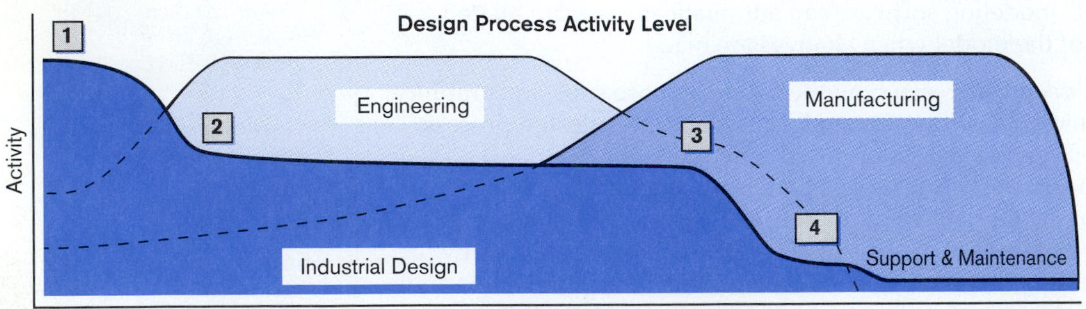

Who's on a Design Team?

Tektronix employs individuals in a variety of positions that may serve on a product team. The MBD Division project teams consist of representatives from all functional areas involved in a project, including engineering, marketing, manufacturing, service, documentation, industrial design, quality and safety, and purchasing. This "core team" is formed at the product concept stage and carries the product through completion. Member responsibilities include:

Program Manager: The program manager has overall program responsibility from conception through introduction and is multidisciplined–usually with a strong technical background and experience in engineering, manufacturing, and marketing.

Mechanical Engineer: Mechanical engineers are responsible for the mechanical and electromechanical product development. They work closely with the electrical engineers, industrial designers, manufacturing, and marketing to satisfy the hardware requirements (cooling, ruggedness, aesthetics, component locations, controls, safety, etc.), the look/feel desired by marketing and industrial design, and the manufacturability goals. The mechanical engineer will work closely with purchasing and component vendors as well, and seek early involvement with the component vendor and tool designers. On large projects, mechanical design efforts are divided up in logical groupings. For example, one engineer may be responsible for the exterior plastic package, another for interior sheet metal, another for front panels and graphics.

Electrical Engineers: Electrical engineers are responsible for the electronic functions in a product. This may include front panel controls, power supply, processor system, acquisition or signal conditioning, the calibration/diagnostics algorithm, and integrated circuit design. Responsibilities are divided up functionally among the team. Electrical engineers are responsible for insuring the product meets required safety and regulatory requirements, and they work closely with the mechanical engineers to insure the products meets ruggedness requirements of the customer. This includes areas such as operating temperature range, shake, shock, and vibration requirements.

Software Engineers: Almost all products today contain some amount of embedded software. The software engineer works closely with the electrical engineers to develop the code to control the hardware. This may include user interface, power-on diagnostics, internal control, and drivers for remote interfaces.

Quality and Safety Engineers: Responsible for testing and insuring the product meets applicable safety requirements and product quality expectations, safety engineers are versed in the safety requirements of countries around the world and insure the products are certified to these standards. Quality engineers work with the design team to predict product reliability based on components used, operating environment, etc., and develop test plans to validate the product reliability.

Industrial Design Engineer: The industrial designers work with the design team to insure the product "look and feel" meets customer needs and is consistent across similar products. The industrial design engineer is involved with everything from color choices to knob spacing and text font and size.

Industrial Engineers: Industrial engineers are responsible for the overall design of a manufacturing line to support a new product. They will work with the design team to integrate the product into an existing manufacturing line, modify an existing manufacturing line, or develop a new manufacturing line as appropriate. They will develop time estimates for the entire manufacturing process, recommend purchasing strategies, develop models for manufacturing capacity requirements, and develop new test and assembly fixtures as needed.

Manufacturing Test Engineers: Test engineers develop the automated tests necessary to calibrate and test products in manufacturing. Often they will work with the service group to modify or develop test procedures for periodic calibration or repair of products.

Documentation: Technical writers work with the design team to compose the manuals and other customer documentation for a product. This can include installation/setup information, operator information, use examples, calibration/performance verification, diagnostics, repair, and a programmer's manual for remote operation and control.

Once the teams are defined, they need structure. *How* team members work together, share information, coordinate their efforts, and make decisions will determine the team's effectiveness. The team structure may vary over time, but it needs to be defined.

Teamwork and the CAD Database

Modeling tools and the CAD database are a vital part of the structure of today's design teams and how they collaborate. For small projects, one person may be responsible for modeling the entire assembly. Weekly team meetings could be a viable mechanism for reviewing progress and providing input to the person building the model. But the complexity of a product such as a laser printer or jet airplane—with its many subsystems and integrated mechanical/electronic functions—requires that the effort of many individuals be coordinated into a cohesive design process. Today's modeling tools can be used to facilitate this coordination.

The printer division at Tektronix uses parametric modeling software to distribute and integrate its design efforts. As refinement and modeling begin, the design team decomposes the printer into its subassemblies and creates a "skeleton" model that defines the relationships among these parts. Each subassembly can then be modeled independently within the parameters established by the skeleton, or master, model. The master model's defined relationships make it possible for subassemblies to be uploaded and tested with other subassemblies during refinement. Subdividing the project in this way accelerates product development because it allows more than one person to work on the model. (It also reduces modeling time and the computer power needed by each designer because he or she can work with just his or her own subassembly.)

FIGURE 1.31
Boeing's 3D visualization tools allow engineers to view parts of the model in relationship to other systems being modeled at the same time. (Courtesy of The Boeing Company.)

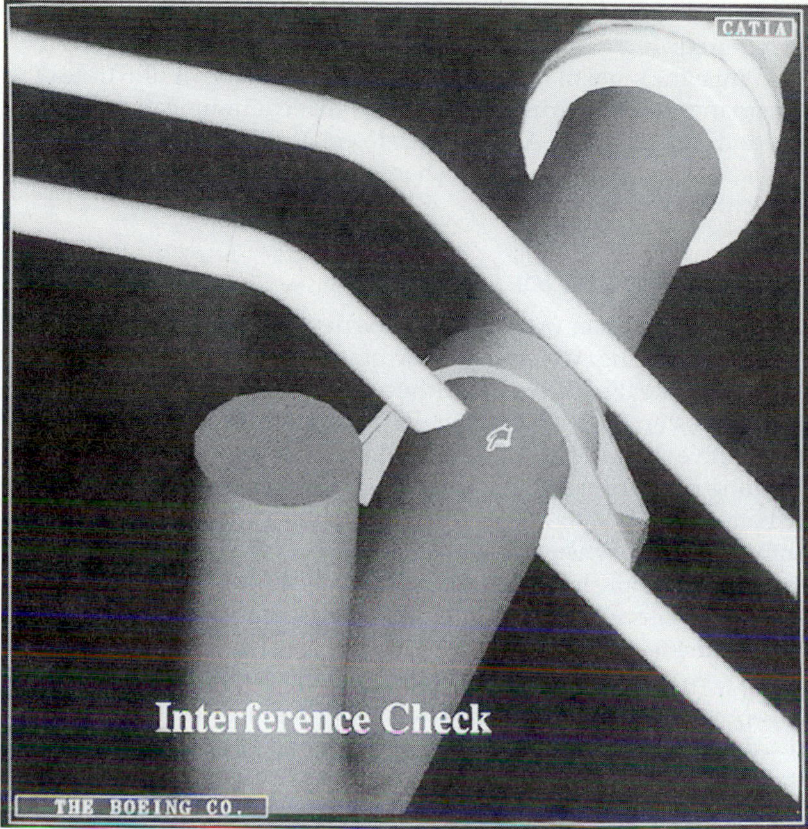

FIGURE 1.32
This close-up view shows part interference that the design teams will have to resolve. (Courtesy of The Boeing Company.)

Boeing developed its own proprietary system for coordinating the design/build process it first used with the 777 aircraft. To enjoy the benefits of concurrent engineering and to reduce the amount of time needed to produce an aircraft, Boeing needed to find a way to enable groups to work in parallel yet communicate effectively about design decisions as they were made. Because of the massive scale of aircraft design, communication among the many different engineering groups had to be structured so that it did not take more time than what would be saved by working in parallel. (With 9700 CATIA workstations in operation in their Commercial Airplane Group, the importance of structure to the team cannot be overstated.) Because the 777 was designed 100% digitally, Boeing used the models themselves as the tool to integrate the project teams. Using CATIA as the primary design tool, engineers were responsible for uploading their models to Boeing's solids database. This database organized each model into a virtual prototype—a "digital preassembly"—of the airplane that could be used by other software tools that would check for interference and allow an engineer to "walk though" a 3D visualization of the assembled airplane (see Fig. 1.31). Each design team was responsible for identifying problems. If, for example, the interference check found that an air duct was positioned in the same space as wiring conduit, as in Figure 1.32, the engineers responsible for the two parts would work together to resolve the conflict. If no interference was found, no contact would be required.

While the scale of the Boeing example is unique, it illustrates the importance of structuring teamwork so that communication is effective and cost-effective. As you learn to create 3D solid models, you will learn ways to manage and share the CAD database to facilitate efficient teamwork.

Communication for Design

Much of this chapter (and this book) is devoted to graphical means of communicating. We've seen how graphics tools were used in the design of the WFM 90 and how they support the concurrent engineering design process. But you will need more than graphics to be an effective designer and team member. By developing your understanding of the process and different members' roles in it, you will be better equipped to determine what you need to communicate with whom and when. Sometimes this process with be fairly informal and accomplished through regular meetings, but there are several reasons why you will also have to document your work and team decisions more formally.

Where spoken communication can be misheard, misunderstood, or forgotten, creating a written record can capture decisions more clearly and communicate them to a wider audience. A written need statement serves as a reference point for a team's ideation. Specifications and performance criteria are benchmarks against which to measure refinement efforts.

At the same time, you need to be able to work with tools that allow you to manage and update information shared by the team. Dated memoranda can give members the most up-to-date version of a needs statement, for example, but the same information could be stored on an intranet Web page where any team member can go to find the most current version. Shared databases may be needed to facilitate sharing of sketches, models, performance data, finite element analyses, and team contact information that is developed over the course of the project. There are several options for "publishing" your information that can make your design team more effective.

In addition to communicating for effectiveness, you will also find that you need to document your work for legal purposes. With product liability being what it is today, you may need to document your design process to show that products have been analyzed properly with customer safety in mind. Each company will have its own documentation requirements that you will be expected to adhere to.

The Designer's Notebook

Most companies require their design engineers to keep a notebook as a record of their work. For the engineer, the notebook serves as a personal "diary" of design ideas, project information, and things that need to be done. This notebook captures ideas before they fade from memory and information that is not ready to be more formally documented. Legally, it is a record of design ideas generated in the course of an engineer's employment that others may not claim as their own. It can also help to document procedures and decisions that may come into question at a later date (but that were not formally documented elsewhere).

The page at the beginning of this chapter—and each chapter throughout the book—was taken from a design notebook kept by a practicing engineer. Each notebook reflects the individual's sketching technique and method for recording ideas and events but also illustrates what your notebook might look like as you take on some of the activities discussed in the chapter. (See Figs. 1.33–1.34.)

As you work through this book, we encourage you to keep your own designer's notebook (see Fig. 1.35). Any size bound notebook (not loose-leaf) with grid paper is suitable. Date each entry as you use this notebook to record your design ideas, assignments, sketches, and more. For team projects, use the notebook to capture notes from your team meetings. Use the notebook for all of your sketching: at the end of the term, you will have a record of your course and your personal growth.

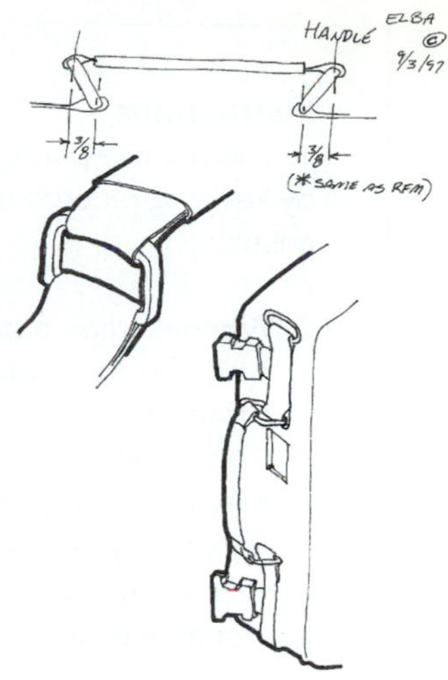

FIGURE 1.33
Notebook sketches for a carrying case (Courtesy of Scott Ketterer, Tektronix, Inc.)

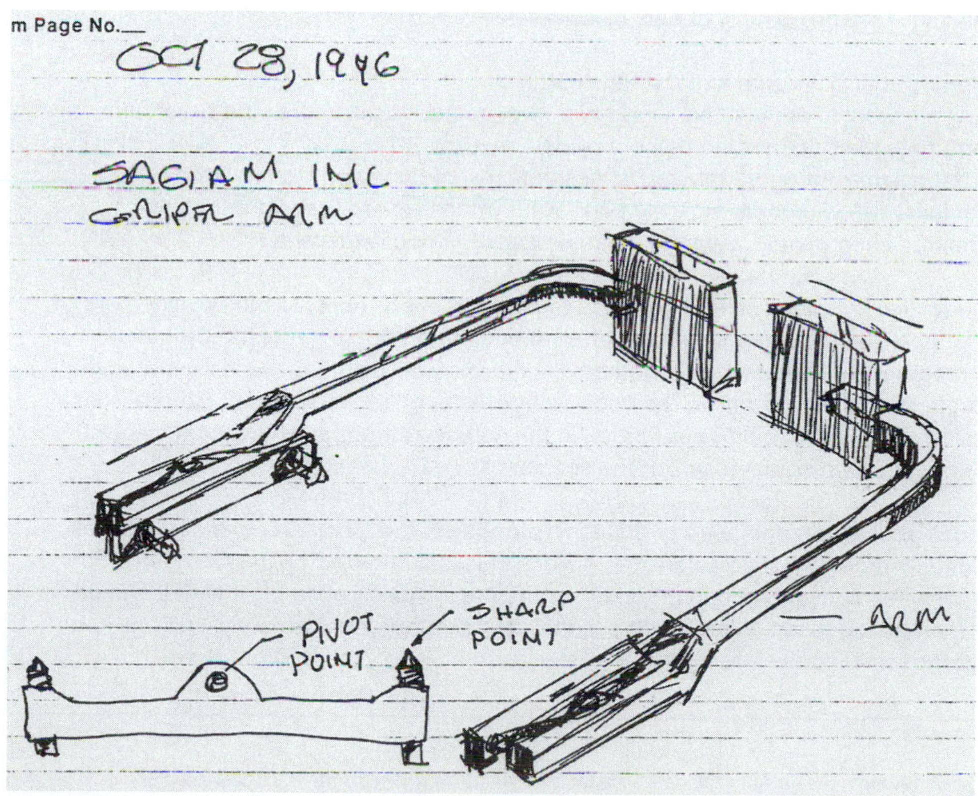

FIGURE 1.34
Gripper arm sketches from a designer's notebook (Courtesy of Albert W. Brown, Jr., Affymax Research Institute.)

Instructions

1. The primary purpose of this notebook is to protect your and the Company's Patent-Rights by keeping records of all original work in a form acceptable as evidence if any legal conflict arises.

2. When starting a page, enter the title, project number, and book number.
   - Use ink for permanence–avoid pencil.
   - Record your work as you progress, including any spur-of-the-moment ideas which may be developed later.
   - Avoid making notes on loose paper to be recopied.
   - Record your work in such a manner that a co-worker can continue from where you stop. You might become ill and to protect your priority it could be urgent that the work continue while you are absent.

3. Give a complete account of your experiments and the results, both positive and negative, including your observations.
   - Record all diagrams, layouts, plans, procedures, new ideas, or anything pertinent to your work including the details of any discussions with suppliers, or other people outside the Company.
   - Do not try to erase any incorrect entries; draw lines deleting them, note the corrections, sign and date the changes. This extra care is worthwhile because of the necessity of original data to prove priority of new discoveries.

4. After entering your data, sign and date the entries.
   - Explain your work to at least two witnesses who are not co-inventors, and have them sign and date the pages in the place provided.
   - Record the names of operators and witnesses present during any demonstration and have at least two witnesses sign the page. If no witnesses are present during an experiment of importance, repeat it in the presence of two witnesses.

5. Since computer programs can be patented these instructions apply to the development of computer software. In this case a description of structure and operation of the program should be recorded in the notebook together with a basic flow diagram which illustrates the essential features of the program. In the course of developing the code, the number of lines of code written each day should be recorded in the notebook together with a statement of the portion of the flow diagram to which the section of code is directed.

This notebook and its contents are the exclusive property of the Company. It is confidential and the contents are not to be disclosed to anyone unless authorized by the Company. You must return it when completed, upon request, or upon termination of employment. It should be kept in a protected place. If loss occurs, notify your supervisor immediately, and make a written report describing the circumstance of the loss.

FIGURE 1.35
Many companies spell out the rationale for, legal issues surrounding, and requirements for an engineer's notebook. This excerpt is taken from the printed instructions for keeping a design notebook used by Affymax Research Institute. (Reprinted with permission from Harold Gallup, Scientific Notebook Co.)

FIGURE 1.36
This cartoon is a humorous reminder of the many facets of being a successful engineer. (DILBERT reprinted by permission of United Feature Syndicate, Inc.)

Writing well, interpreting numerical data, understanding statistical data, reading and creating engineering drawings, sketching, listening, and making presentations are all communications skills you will need to be a successful engineer (see Fig. 1.36). In this text, you will learn about engineering graphics and how to apply the many software tools available for engineering design graphics. The more you practice these skills in the context of real problem solving, the more you will develop your communications abilities in other areas.

EXERCISES

1. Think about a vacation spot you are familiar with. Write a needs statement for a resort to be located in that area.
2. Consider the recent developments in the disposable cameras that are available everywhere. Put yourself in the place of the original designers of these cameras and re-create the written statement of need that might have guided the project.
3. Ergonomics, or human factors engineering, is an important consideration when designing control handles or other interfaces between a tool and the human user. It has been shown that one's productivity improves when more functional interfaces are present. With this in mind, write a needs statement for the ergonomic redesign of the following items:
   - **a.** toothbrush
   - **b.** bicycle seat
   - **c.** fishing rod
   - **d.** ski pole
   - **e.** kitchen knife

4. Design the process you would use to design the items for which you have written needs statements (or use your partner's needs statement). Consider the team of people you would need to work with, when each of them would be involved, and how and when you would present ideas to the team for its approval and input.
5. List five attributes of a successful product design team. What skills are important to its success?
6. You are a member of a three-person design team chosen to develop a pocket-sized flashlight that will be sold through high-quality sporting goods/camping equipment catalogs. The basic idea passed on to your team is that a small, flattened, palm-sized flashlight would be desirable.
   a. Write a needs statement for this product.
   b. Outline the design process that will be used to go from the first step of identifying a need to production of the part.
   c. Create a simple sketch of three versions of this device to communicate your initial ideas to the other members of your team.
7. Consider the Tektronix case study in this chapter for the WFM 90. What if customer feedback presented a need for double the battery life? How would you address this need if it arose early in the design process? What if it arose after the product was introduced? Consider the impact of the change at both stages and list and explain the items that would have to be changed to meet the customer's needs.
8. Computer applications play a critical role in the modern product design field. List at least four ways that a design team developing a new product could use various types of computer-generated information. What types of software programs do you think a mechanical design engineer should be skilled with in this environment?
9. Design Solutions Corporation provides engineering design and analysis consulting services for large industrial customers: company policy is to strictly follow the steps of the design process as shown schematically in Figure 1.9. Explain how a customer would benefit from this philosophy, and give three examples of what might happen if steps were skipped.
10. Interview a design engineer who works for a company in your area to better understand the process and graphics tools used. Compare and contrast the process with that shown in Figure 1.9.
11. Write a set of specifications for packaging that would protect an egg when it is dropped from 30 feet. Develop a process for designing, testing, and refining the device that does not use more than two eggs. Follow your procedure to design the device and use your notebook to record your all notes, sketches, ideas, and results.

12. Visit the two Web sites below and locate the product names they market for the following stages of the design process:
    a. Problem identification/ideation
    b. Refinement/analysis
    c. Implementation/documentation

    Parametric Technology Corporation (www.ptc.com)
    Dassault Systemes (www.dsweb.com)

DESIGNER'S NOTEBOOK

These freehand sketches from Albert Brown's design notebook show four different concepts for a coaxial cable clamp that would fit into a movable disk drive test arm. These cross-sectional sketches show four different ways the cable holder (shaded) could be fit into and around the test arm. The model in the lower right was developed further. (Courtesy of Albert W. Brown, Jr., Affymax Research Institute.)

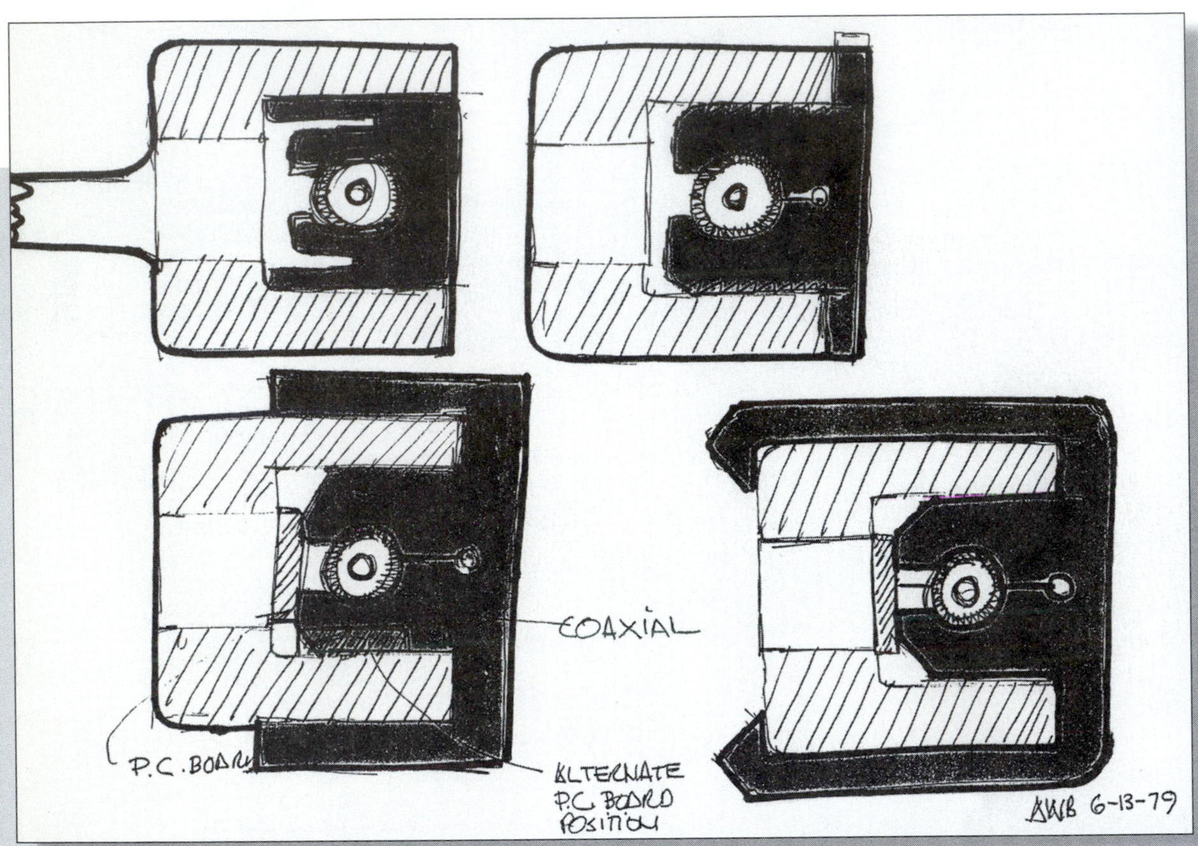

CHAPTER

2

Ideation and Visualization

Objectives

When you have completed this chapter you will be able to

- list four types of factors that constrain design problems;
- describe techniques for generating design ideas;
- discuss the role of visualization in ideation and problem identification;
- use a block diagram to represent a system or device;
- create freehand contour sketches for use in ideation;
- recognize graphic/visual symbols and schematics;
- locate standard drawing symbols;
- sketch graphic symbols for threads, fasteners, and springs.

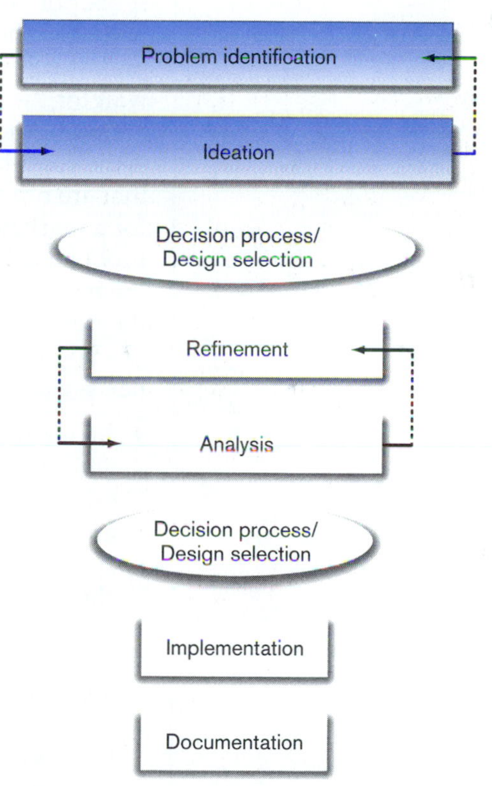

The Design Process: Problem Identification

Design is a problem-solving process. The design challenge is the "problem" that the engineer must "solve." Engineers work together with other design team members of a variety of backgrounds to solve problems that include issues such as the economic feasibility of the design, safety, reliability, aesthetics, ethics, and even the social impact of the project. Diversity among team members can be a valuable asset to the design team. Carefully defining the problem is the first step toward a solution.

A problem definition containing a needs statement helps provide direction for the design team. The problem definition helps to make sure that the problem solved is the one the design team has set out to solve. While there are examples of products and scientific discoveries resulting from accidental discoveries, such as vulcanized rubber or penicillin, to be competitive in today's marketplace you will need the skills to solve problems while under time and financial pressure. It is unlikely on any job that you will have the leisure to wait for the solution to come to you. Much more common than the announcement of a wonderful new accidental discovery or product is the complaint that engineers are adding features and gadgetry that are not desired while neglecting to provide the features needed for a marketable product or system. A clear understanding of the problem and the development of design specifications that will meet the basic objectives set out in the problem statement will help keep the design effort moving in the right direction.

Some examples of problem definitions are listed below, with the needs statement italicized:

- An environmental science company is developing a process for decontaminating polluted ground water on-site using bacteria (bioremediation). To demonstrate the process in the laboratory, it grows bacteria in a corrosive soup in five pressurized and sealed cylindrical glass, teflon, and stainless steel containers (called chemostats or reactors) with dimensions 6 inches in diameter by 18 inches in height. The chemostats are maintained at a constant temperature in a water bath. *A means is needed to keep the bacteria from adhering to the interior surfaces (e.g., the walls) and from settling to the bottom of the container*. It is also necessary to stir the soup so it stays completely mixed. No perforation of the top of the reactor can occur.
- Consider the advancements in dry-fire laser technology to *develop a device useful for marksmanship training and for target spotting*. The device should mount to rifle barrels of various sizes.
- Removing and disposing of large amounts of snow from parking lots are a major expense for the city of West Yellowstone. Currently, collected snow must be transported several hundred miles to a location where it can be dumped. *A local facility is needed to allow disposal of excess snow removed from streets, parking lots, and other areas*. Research and design an energy-efficient and cost-effective snow holding and disposal system.

- A therapy center has a patient who is an avid hunter and would like to be able to hunt independently once again. He uses a powered wheelchair full-time. *This patient needs an adapted wheelchair for hunting*. He finds it easiest to operate devices that are located directly in front of him in the mid-line position. He would like to be able to store the gun on the wheelchair when he is not shooting. The mount needs to be accessible to him on the wheelchair. It may be possible to use the battery system on the wheelchair to power the mount.

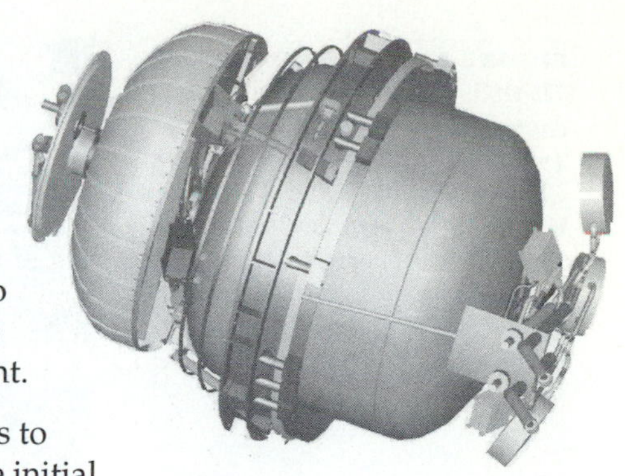

FIGURE 2.1
This cryostat (modeled in I-DEAS) was designed to keep the infrared instruments of the SIRTF space telescope at temperatures as low as 1.4° Kelvin. The outer vacuum shell, which has been removed to show detail, shields the cryostat from external temperature extremes. (Courtesy of NASA and Ball Aerospace & Technologies Corp.)

The designer must first define the problem he or she wishes to solve. However, if the criteria are too rigidly defined during the initial stages of the design process, they may limit the designer's thinking and creativity in providing a solution. If the criteria do not reflect the actual need or problem to be solved, then the design may be a wonderful and creative solution, but not for the problem at hand.

A general needs statement or problem definition may initiate the design process. As some preliminary research is performed, more general design constraints or criteria can be developed.

Design Constraints

Designing an acceptable solution requires the engineer to consider a number of *design constraints*. A constraint limits the range of options that are acceptable. Design constraints can be functional, aesthetic, economic, legal, time-related, or all of the above.

Functional design constraints relate to how the product or process will work. What are the limits within which the product or process must operate? A designer needs to know how much load a robot arm must be able to bear, or how much fluid needs to flow through a system. If the design must fit into an existing space or attach to other equipment, the size may be limited to a certain range. Under what conditions will the product be used? Will it need to operate at extreme temperatures or pressures? (See Fig. 2.1.) The failure of the *Challenger* space shuttle was caused by the fact that the material for the O-ring seal couldn't expand quickly enough in cold weather to make a tight seal under extreme pressure.

Ergonomics is the term used to describe the study of human interaction, or how well a product fits the human body (see Fig. 2.2). A designer needs to consider who will use the product or process. Visibility may be an issue if an operator needs to see portions of the equipment or beyond it in order to operate it properly. Designing a handle shape to fit most comfortably in the human hand requires ergonomic testing—and a decision whether to design for left-handed people, too.

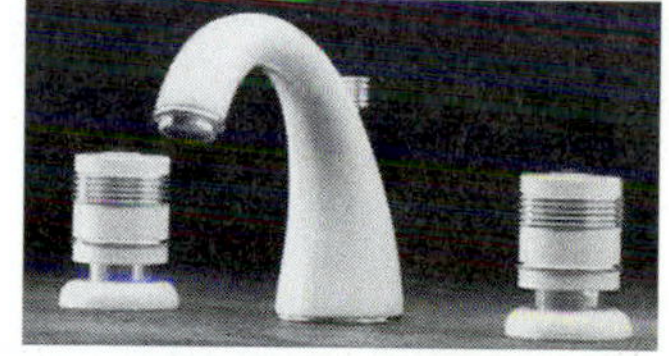

FIGURE 2.2
These faucet handles look nice, but the designer failed to consider the conditions under which they would be used. When your hands are wet and soapy, it is very difficult to get a grip on these smooth, round handles. (For more examples of bad design, visit the Web site at www.baddesigns.com.)

Aesthetic constraints express what is known about taste and appeal in a certain market (see Fig. 2.3). What "look" will speak to the consumer or customer? Automobile design over the last forty years illustrates how the public's sense of what is contemporary and fashionable has changed: tail fins are as unappealing now as black bumpers are "in." Black & Decker revitalized its small appliance line

FIGURE 2.3
The 1959 Cadillac in (a) looked as contemporary in its day as the 1998 Cadillac in (b) does today. Can you identify elements of each that characterize the aesthetics of the period? (Copyright General Motors Corporation, 1978 and 1997. Used with permission from GM Media Archives.)

in the early 1990s in part through greater sensitivity to the aesthetic aspects of product design. Today, the Black & Decker Ergo line of products, shown in Figure 2.4, reflects the aesthetic appeal of ergonomic features to today's consumer.

Economic constraints are another major kind of design constraint. How much can be spent to produce a product? How much should it cost to operate it? Finding a less expensive material that satisfies other design goals can make a product marketable. At the same time, a complex design that requires plant retooling or that cannot be manufactured by more than one or two companies in the world may make the product too expensive to be viable.

Legal issues can be another type of design constraint. Governmental safety and environmental regulations may restrict certain design options. Many industries are regulated through various codes and standards that must be followed and that limit the broadest range of possibilities for the design. For example, the American Society of Mechanical Engineers publishes the *ASME Boiler and Pressure Vessel Code, an American National Standard.* All commercial, institutional, and governmental design and manufacture of pressure vessels in the United States must follow this code. Section VIII, Part UG, Item UG-5 of this code describes the use of plate. It states,

> "Plate used in the construction of pressure parts of pressure vessels shall conform to one of the specifications in Section II for which allowable stress values are given in Subsection C, except as otherwise provided in (b) and UG-4, UG-10, and UG-11. Material not identified in accordance with UG-93 may be used for nonpressure parts such as skirts, baffles, and supporting lugs. A double-welded butt-joint test plate shall be made from each lot of all such material over 3/8 in. thick that is to be welded. Guided bend specimens made from the test plates shall pass the tests specified in QW-451 of Section IX..."

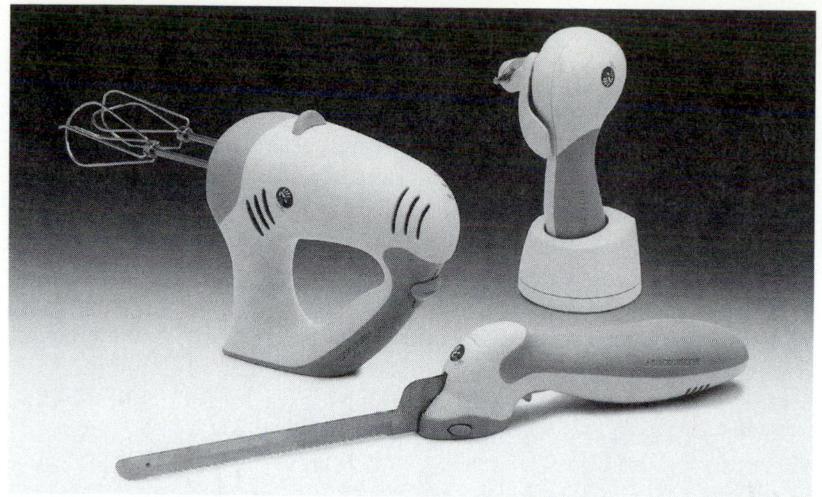

FIGURE 2.4
The new Black & Decker Ergo line is an example of contemporary styling that reflects today's greater sensitivity to ergonomics. The new appliances are designed to provide maximum accessibility and ease of use for a wide range of individual users. (Courtesy of Household Products, Inc., exclusive licensees of Black & Decker household products.)

Finally, *time* is an important design constraint. If the product is delayed for six months, what other products that fill this need will already be on the market? What new materials or processes will be available that could make this design obsolete? Is the aesthetic appeal rapidly changing in this area so that the product will look dated before it is ever released?

Today, worldwide competition makes it difficult to succeed if all of these major areas are not reflected in the design process. While engineers may be most comfortable in the realm of functional design, they will undoubtedly work in an environment where customer feedback and manufacturing concerns are integrated into the design process. At the same time, high-speed global communication makes time to market an equally important factor in the ultimate success of the design. While a certain amount of iteration is necessary (and desirable) in the design process, going back to the drawing board too many times can add costs and allow a competitor to solve the problem before you do.

A *specification* is a measurable statement of the objectives the design should achieve. Developing written specifications for the design provides feedback so that you know when the target design has been achieved. As new considerations come into play in the design, various trade-offs may have to be considered. For example, a slightly higher target cost may have to be considered in order to meet the time schedule. Or one of the performance specifications may have to be lessened in order to meet the target price.

Engineering design today is occurring in ways that incorporate a wide range of design considerations into the early stages of the process. As an engineer, you will be required to work with other engineers, but also representatives from marketing, sales, manufacturing, and other departments who will be involved with the product. In addition, you may be asked to interact directly with key customers. As companies strive to better meet the needs of the market, you as an engineer need the skills to communicate effectively with a variety of individuals, not just to present your work, but to find the information you need to design an acceptable (superior) solution.

Luckily, technology today makes it possible to communicate better and faster and in a wider variety of ways than ever before. Distance communication methods (e-mail, video conferencing, Internet, and intranet sites) present opportunities for working as a team that transcend the limits of traditional face-to-face meetings. Simulation and analysis tools (spreadsheets, finite element analysis software, circuit modeling software, and so on) make it possible to give team members the information they need to evaluate a design. Visualization tools (3D models, animations, and so on) make it easier to address aesthetic and functional concerns and to create more realistic representations of the design necessary for interaction with customers or other nontechnical members of the team.

FIGURE 2.5
The tenaciousness of a cocklebur inspired the invention of Velcro®. The design of the Velcro gripper spines shown in (a) mimics the shape of the hooked ends of the cocklebur, shown in (b). The hooks attach to the loops on the fuzzy side of Velcro strip in the same way that cockleburs hook onto fabric or hair. (Cocklebur courtesy of the Brousseau Project, http://elib.cs.berkeley.edu/flowers.)

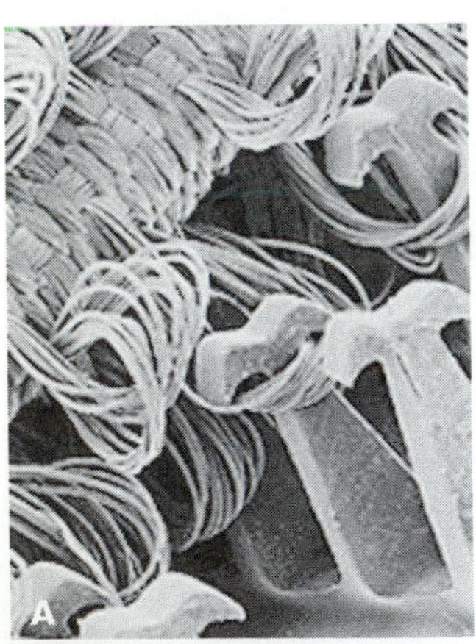

Ideation

Once the problem has been defined, the design team generates ideas for solving it. Have you ever thought of a great idea for a new product only to find out later that someone else is already making it? When you are designing a device or system, the first thing to check is what designs may already be in use. A thorough literature search for similar products and background information is a good starting place. It can keep you from "reinventing the wheel" and provide inspiration that improves the quality of the ideas you generate.

In the ideation phase of design, the emphasis is on creativity and new ways of looking at the problem. Look at the problem from different angles. Don't lock into one approach too quickly. Consider several alternative avenues to investigate further.

As ideas are generated, questions may come up that were overlooked in the problem identification phase. When they do, the design team needs to add the information to the problem definition, briefly returning to the first step, before continuing to generate ideas.

After the design solutions have been considered, one may rise to the top as the best. More frequently, a smaller group of design ideas fit many criteria, and the information gained is used to generate a second round of options. These options are then refined and evaluated until a satisfactory solution is achieved.

Where Do You Get Your Ideas?

Creativity is a much discussed notion in design. Where do original ideas come from? Many designs and patents are adaptations of other existing devices or ideas. Even ideas that may seem revolutionary started from somewhere. Swiss inventor Georg de Mestral came up with the revolutionary idea for Velcro while walking his dog. The combination of examining the cockleburs stuck in the dog's fur and trying to solve his wife's problem with a stuck zipper inspired him to invent the new product. But it took years of work before the idea became a marketable product (see Fig. 2.5).

Literature Search

Many people argue that all creative problem solving starts with an in-depth knowledge of the subject area and that it requires only ordinary types of thinking skills. Or that creative ideas are just new combinations of existing information or other ideas. This is a good reason for starting out your search for inspiration by researching and collecting data about the subject of your design. Some of the places to look for information are

- patent searches;
- the World Wide Web;
- consumer surveys;
- competition reviews.

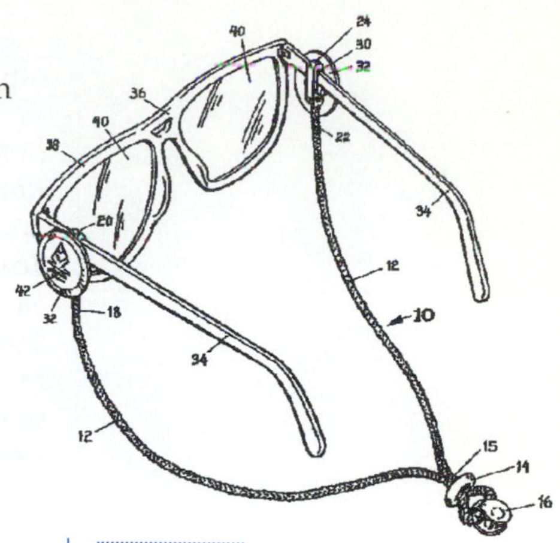

FIGURE 2.6
This figure is typical of the drawings used to document patent applications. Compare the ease with which you can visualize the product from the drawing to the effort required to do so with the prose description at left. (United States Patent 5,654,787)

An easy way to search the U.S. patent office is by using its Web site at www.uspto.gov. You can search for keywords or browse through different categories. A search on the keywords *eyeglasses* and *retainer*, for example, turned up these results: The word *eyeglasses* had 331 occurrences in 163 patents. The word *retainer* had 2881 occurrences in 1121 patents for a combination of 11 patents that used both the words *eyeglasses* and *retainer*. Selecting one of the patents resulted in the following information, which is depicted in Figure 2.6.

> United States Patent 5,654,787
> Eyewear and information holder
> Inventors: Barison; Joseph I
> (P.O. Box 9787, Denver, CO 80209).
> Appl. No.: 549,127
> Filed: Oct. 27, 1995
> Primary Examiner: Dang; Hung X.
> Attorney, Agent or Firm: Crabtree, Edwin H.; Pizarro, Ramon L.; Margolis, Donald W.
>
> Abstract: Eyewear and information holder for holding eyewear and information next to a user of the holder. The holder includes an elongated strap having a first end and a second end. The first end is attached to one end of a first flexible retainer. The second end is attached to one end of a second flexible retainer. The first and second retainers include a stud opening and a temple member opening. When the holder is used for holding eyewear, the temple member openings are adapted for receiving a first and a second temple member there through of a pair of eyeglasses. A first and a second advertising piece with an attachment stud on the back of each piece are secured next to the sides of the retainers with the attachment stud received through the stud openings. When the holder is used for holding the various types of information, the temple member openings are adapted for receiving a first and a second engagement arm of a connector device. The connector device includes a foldable attachment arm with tube member thereon and a post extending outwardly from the connector device. The post is adapted for insertion through an opening in a frame used for holding and displaying different types of information. The attachment arm is folded over and the tube member is inserted around the post securing the frame to the connector device. The strap of the holder is then placed around the neck and both the information and the advertising pieces are displayed next to the user of the holder.
>
> 12 Claims, 9 Drawing Figures

In addition to searching the U.S. patent office, there are many other excellent resources for engineering information on the World Wide Web. Engineering organizations such as the American Society of Mechanical Engineers (ASME); the American Society of Heating, Refrigeration, and Air Conditioning Engineers (ASHRAE); the Society of Manufacturing Engineers (SME); the Industrial Designers Society of America (IDSA); the American Society for Testing and Materials (ASTM); the Institute of Electrical and Electronics Engineers (IEEE); the American Institute of Chemical Engineers (AICE); and the American Society of Civil Engineers (ASCE) all maintain Web sites providing guidelines, focus groups, useful engineering information, and contacts. Many of these organizations also publish engineering references and handbooks that are invaluable in engineering design work. Vendors and manufacturers are another great place to go for information. They often have design guidelines for the products and processes they sell. An appendix at the back of this book contains a list of many useful organizations for engineering and their Web addresses and contact information.

The Thomas Register maintains a list of manufacturing companies. You can use its Web site at www2.thomasregister.com to search for information about companies, general products, or brand names. For example: if you were considering a design solution that used extruded aluminum, you could search the Thomas Register for companies producing aluminum extrusions as shown in Figure 2.7.

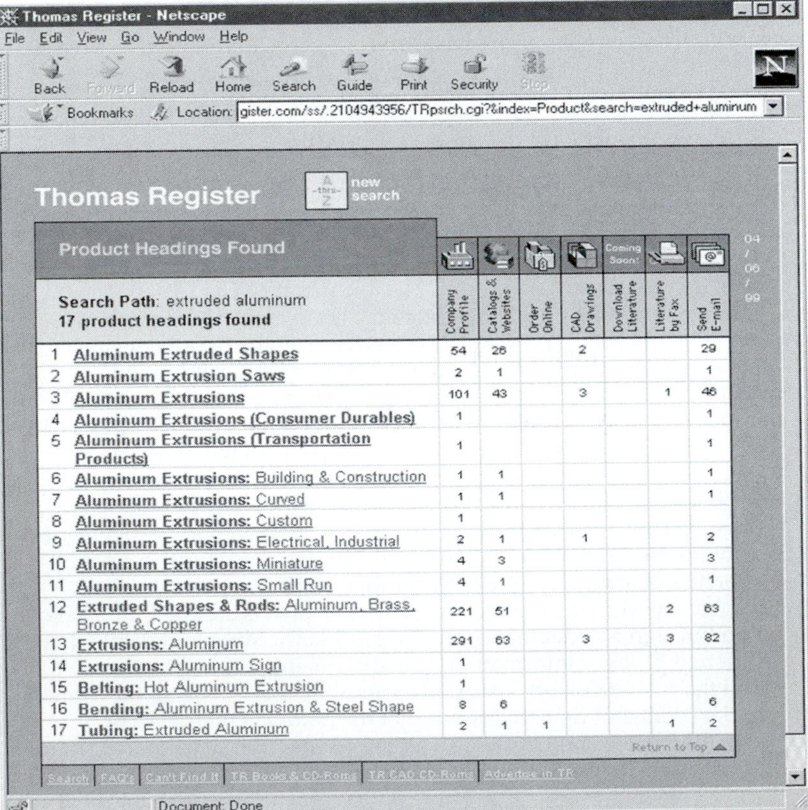

Search Path: extruded aluminum
17 product headings found

	Product Heading	Company Profile	Catalogs & Websites	Order Online	CAD Drawings	Download Literature	Literature by Fax	Send E-mail
1	**Aluminum Extruded Shapes**	54	26		2			29
2	**Aluminum Extrusion Saws**	2	1					1
3	**Aluminum Extrusions**	101	43		3		1	46
4	**Aluminum Extrusions (Consumer Durables)**	1						1
5	**Aluminum Extrusions (Transportation Products)**	1						1
6	**Aluminum Extrusions:** Building & Construction	1	1					1
7	**Aluminum Extrusions:** Curved	1	1					1
8	**Aluminum Extrusions:** Custom	1						
9	**Aluminum Extrusions:** Electrical, Industrial	2	1		1			2
10	**Aluminum Extrusions:** Miniature	4	3					3
11	**Aluminum Extrusions:** Small Run	4	1					1
12	**Extruded Shapes & Rods:** Aluminum, Brass, Bronze & Copper	221	51				2	63
13	**Extrusions:** Aluminum	291	63		3		3	82
14	**Extrusions:** Aluminum Sign	1						
15	**Belting:** Hot Aluminum Extrusion	1						
16	**Bending:** Aluminum Extrusion & Steel Shape	8	6					6
17	**Tubing:** Extruded Aluminum	2	1	1			1	2

FIGURE 2.7
Searching the Thomas Register for aluminum extrusions produced this list of product headings and information about the companies that produce them.

You could then contact specific companies to find out more particulars about pricing, specifications, and manufacturing times. Some companies have Web sites that you can link to directly from the Thomas Register site. Many companies can provide some general design guidelines containing manufacturing tolerances, material specifications, and other useful information.

Consumer Surveys

Another place to get information is from the consumer. Marketing surveys and consumer questionnaires provide important feedback. Think about all of those little product registration cards that you have filled out. Why does the manufacturer need all of that information? Probably so that it can have some idea of who buys the product, how to price it, what other features the consumer would pay for, and any problems they could solve or improvements the manufacturer could make. However, you should take consumer surveys with a grain of salt. Sometimes consumers will not state their preferences accurately—or the information may be incomplete. For example, the customer may rate a particular feature of utmost importance but is willing to pay only a few cents for it. If you are designing your own survey to test customer opinion, you should review the principles of good survey design, taught in most statistics courses, to be sure the data you get back is useful to you.

Competition Reviews

Don't forget to look at the competition. If you are going to design a better exercise bike, look at the ones that are already on the market. What are the features of the top-selling exercise bikes on the market? Are certain features critical to sales? What about their price? Styling? What areas can you improve on?

Sometimes seemingly unrelated products can yield design ideas, too. What other products are commonly used by the same group of people or businesses that will be using yours? Which features are consumers accustomed to and likely to expect from your product, too?

Brainstorming

Another popular method for generating ideas is *brainstorming*, an approach defined and popularized by Alex Osborne. Brainstorming works in two ways: First, it gets as many ideas out of the group as possible. Second, these ideas spark more ideas and generate a larger pool of ideas to work from. The key to success in using brainstorming is what Osborne calls *deferred judgment*. Deferring judgment—not being critical of the ideas at the time they are presented—may help to generate more ideas. One reason is that each person is free to list wild ideas that may not be workable but may spark an idea that is less wild and just might work. Deferring judgment lets you use the creative side of your mind freely by postponing the need to be analytical or saying which ideas may not work. It also may allow people to feel comfortable in presenting their ideas to the group because they won't be criticized. It therefore lessens the chance that good ideas will not be brought up because of an individual's fear of

Good Survey Question Design

A survey is a "conversation" with the consumer. Although your questions are going out on paper, over the Web, or via a professional researcher, they will be answered by a person with time constraints, other interests, and, many times, less of a commitment to the topic than you have. To avoid surveys that are too long or too difficult, put yourself in the place of the respondent and consider the many ways questions could be answered. Some common pitfalls to avoid are as follows:

Leading questions that subtly tell the respondent what you think about a topic.

Poor	Better
Do you think the quality of engineering drawing is as high today as it was ten years ago? ❍ Yes ❍ No ❍ Undecided Do you think the quality of engineering drawing is higher today than it was ten years ago? ❍ Yes ❍ No ❍ Undecided	Which of the following best describes the way you feel about the quality of engineering drawing today? ❍ Higher than it was ten years ago ❍ No change in quality over ten years ❍ Lower than it was ten years ago ❍ No opinion

Double-barreled questions that pose two questions but present only one set of responses. The respondent doesn't know which question to answer. In the question below, the reader who has a degree in engineering technology or civil engineering has to decide whether you really want to know when the degree was earned or if you want to know only about mechanical engineering degrees.

Poor	Better
When did you get your degree in mechanical engineering?	What year did you get your engineering degree? In which engineering discipline?

Questions with incomplete options or options that are not mutually exclusive. The respondent may not be able to find a response that fits and then has to decide whether to skip the question, add an option, or choose one that is close. In interpreting the results, you will have the same problem–deciding how the respondent handled it.

Poor	Better
In which engineering discipline did you get your degree? ❍ Mechanical ❍ Civil ❍ Electrical	In which engineering discipline did you get your degree? ❍ Mechanical ❍ Civil ❍ Electrical ❍ Other (please note): ______________

Inconsistent frame of reference: one person's "occasionally" is another person's "almost never." Choose a frame of reference that allows you to compare responses across questions and respondents. Using a scale (such as ranking from 1 to 5) or objective criteria ensures that the respondent will see your choices as a scale and not as a set of random options.

Poor	Better
How often do you use your food processor? ❍ Frequently ❍ Sometimes ❍ Almost Never ❍ Never	How often do you use your food processor? ❍ Less than once a week ❍ 2–10 times a week ❍ More than 10 times a week

criticism or of appearing silly. Osborne gives four rules for brainstorming sessions:

1. *No criticism.* Wait until you have finished generating ideas before you rule any out or begin to criticize. This way all of the ideas can be brought out.
2. *Wild ideas are welcome.* Don't limit yourself to ideas that may work. Thinking outside the sometimes closed seeming statement of the problem to come up with novel solutions is encouraged. If the wild idea doesn't work, it might spark another idea that will work.
3. *More is better.* Keep at it until you have generated a long list of ideas. These can be winnowed down to form a shorter list after you have finished brainstorming. This is basically the old adage that two heads are better than one. If you have a lot of ideas to pick from, some of them should be good ones. It is a common tendency of people to quit brainstorming before they have really come up with creative solutions. Often the first fifty solutions listed are common and not creative solutions. After that, people have to start stretching their minds for creative solutions.
4. *Work off other ideas.* Each suggestion does not have to be entirely original. If your idea is sparked by another one that was given, don't worry. Your special twist might be just the thing that is needed. Feel free to combine other ideas into a new idea or otherwise change the previous suggestions.

Brainstorming works best with groups of six to ten people working for about one hour. Including representatives from marketing, manufacturing, and consumers can increase the variety of ideas generated. The process is usually headed by a moderator who writes down the ideas, encourages the participants, and keeps any critical analysis of the ideas shut down.

Individual Brainstorming

In his book, *Planning and Creating Successful Engineering Designs*, Sydney F. Love describes another brainstorming technique called *individual brainstorming*, which was developed by William J. Osborne and George Muller. Using this idea generation method, participants are given 120 seconds to silently record key words relating their ideas. The moderator reminds them of the time remaining in 30- to 15-second intervals. (Pressure is supposed to help the flow of ideas.) Once the first list is generated, the moderator then calls on each participant one at a time in a clockwise progression to rapidly present his or her first list of ideas. During this time the other participants cross off the duplicated ideas from their list and start a second list of ideas triggered by hearing the speaker. The next participant reads only the ideas that were not crossed off his or her first list. If the participants are slow to generate a second list, the moderator may call for the speaker to present the third idea on his or her second list. The process continues one more time, proceeding counterclockwise with the participants reading the ideas from their second list and generating new ideas in a third list. This process is

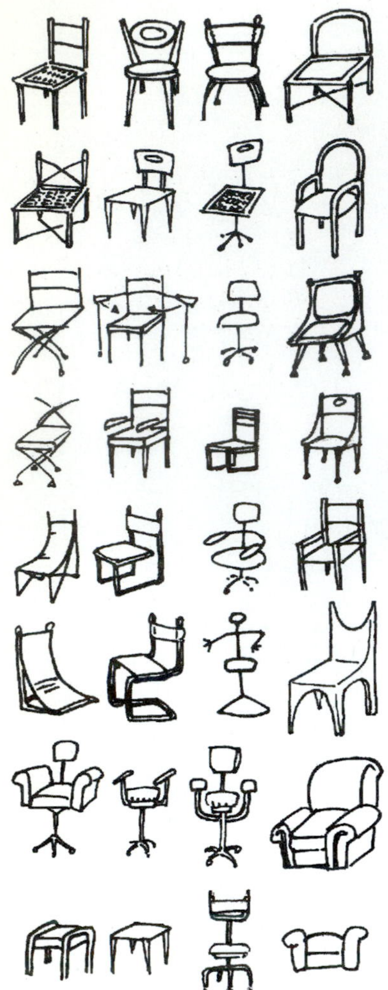

FIGURE 2.8
Even a relatively short list of alternatives can be combined into a large number of different designs, as illustrated by these chair design sketches.

supposed to help generate ideas because the second and third lists of ideas are ones that are new and not the ones that were already at the back of someone's mind. Also, the pressure from having to cross off duplicated ideas forces people to think more originally and in different ways. Or, if the ideas were not duplicated, the originator may be encouraged and will produce even more ideas for the second and third lists.

In both kinds of brainstorming sessions, it is also important how the question to be solved is posed. A good needs statement can provide needed direction but should not be too narrow. Sometimes a broader statement will generate more creative solutions.

Listing

Another method for generating ideas, sometimes called *listing*, or *morphological analysis*, is to break down the system or device into its individual components, then consider every possibility for that individual component. The individual components then are combined to come up with new possibilities that may not have been previously considered. For example, to design a chair you might break it down into the categories of seat, support base, back, and arms. For each of these components you could list every possibility that comes to mind.

Seat	Support	Back	Arms
padded	integral	padded	no arms
wood	star-type	wood	padded arms
plastic	casters	metal	plastic arms
woven	legs	contoured	separate arms
cane	floor mount	straight	attached arms
metal	pedestal	open	metal arms
		fabric sling	sling support

Once you have the list you can combine the different options and consider the alternatives. For example, one chair design may mount to the floor with a padded seat and a fabric sling back and arms. Figure 2.8 shows sketches of some of the listed concepts.

The Decision Process

After a number of possible solutions to the design problem have been generated, one idea (or sometimes more than one) is selected to develop further. There is often a certain amount of subjectivity in the selection of the "best" design concept to pursue further. The decision can be based on economic considerations often evaluated by comparing the potential cost-to-profit ratio for each design. Marketing surveys or focus groups may also be used to help select among design alternatives. Often a decision matrix is used to help rank the alternatives in a systematic way. There are two steps in producing a decision matrix. The first is to identify all of the design constraints and rank them in order of importance. The ranking may be done by assigning a certain number of points to each design consideration. Often the weighting number is based on 10 or 100

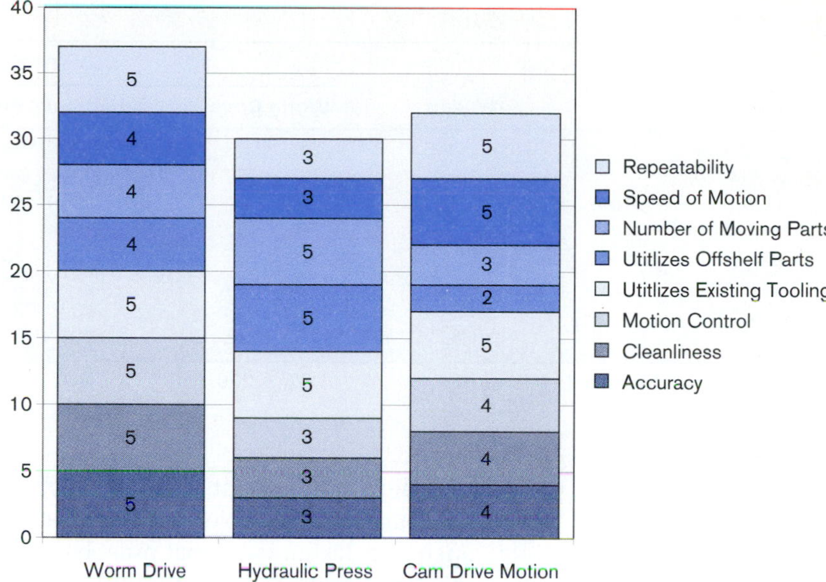

FIGURE 2.9
A decision matrix was used to evaluate design alternatives for a press brake. The worm drive, hydraulic press, and cam drive motion were rated from 1 to 5 according to the criteria on the right. The ratings were then summed (as shown in the stacked bar chart for each alternative) to determine the option with the highest overall rating. (Courtesy of Trent Wetherbee, Andrew Beddoe, and Ezra Miller.)

because most people have some sense of how they would rank based on this number of points. The criteria can also be divided into categories of must have (needs) or would like to have (wants). This makes it easier to determine trade-offs in the wants category. Any design that will be considered for further evaluation will meet all of the needs criteria but can vary among the wants criteria.

Once the criteria are fully identified and ranked, the designs are rated according to each of the criteria. The results are typically listed in a chart or decision matrix showing each criteria in one direction and each design alternative in the other direction (see Fig. 2.9). The team as a whole can discuss each item and rank each design and record the values. If the team cannot meet together to produce the rankings, the individual members can rank the items, and the results can be averaged. This averaging process, however, can tend to make the alternatives rank toward the middle of the range if individuals perceive the value of the designs differently.

For a weighted decision matrix, the weight for each criteria is multiplied by the rank before the values are summed. The criteria in Figure 2.9 were weighted and a spreadsheet was used to calculate the ranking for each design alternative, as shown in Table 2.1.

For designs that will be divided into need and want categories, the chart can be divided into two portions, the upper portion for the constraints (or needs), and the lower portion for the design criteria (or wants). Many times the need constraint items do not need to be ranked, just checked off. If an item does not meet all of the constraints, it does not have to be evaluated further.

Product Review/Design Review Meetings

Product review and design review meetings are helpful milestones during the development of the design. The purpose of these reviews is to acknowledge and minimize risk in proceeding with the project

TABLE 2.1

Criteria	Weight	Score/Weighted Score					
		Worm Drive		Hydraulic Press		Cam Drive Motion	
Repeatability	10	5	50	3	30	5	50
Speed of Motion	8	4	32	3	24	5	40
Number of Moving Parts	8	4	32	5	40	3	24
Utilizes Offshelf Parts	8	4	32	5	40	2	16
Utilizes Existing Tooling	7	5	35	5	35	5	35
Motion Control	7	5	35	3	21	4	28
Cleanliness	7	5	35	3	21	4	28
Accuracy	9	5	45	3	27	4	36
Total Score Raw/Weighted		37	296	30	238	32	257

or design at that point. The risk may have to do with the funds committed to the project or it may be related to the project itself—such as a design issue that must be resolved at a particular point or the threat that the customer will call an end to the project. A typical series of product review meetings might come at these milestones in the project:

- product proposal
- design proposal
- development plan
- engineering release
- product release

Product Review Stages

The review at the *product proposal* stage may be driven by marketing and used largely to determine whether there is consumer demand for the product or to identify what consumers need. At this point, the company begins the process of archiving data for the project so that all of the useful project information is captured. Having this information may help in the development of other projects and in providing complete documentation for the current one.

At the *design proposal* review, a plan to solve a problem or meet a need (usually one that will produce revenue for the company) is presented. The purpose of the review is to get a commitment to go ahead with the proposal as defined at that stage. A considerable amount of monetary resources and people are usually required to move ahead from this point, so the proposal includes enough detail to allow for the review committee to make an informed decision. The proposal should indicate that the product definition, specifications, conceptual design, and identification of the consumer interface are sufficiently developed. It should also show that the product's market potential has been evaluated and present the results of that evaluation. A plan for the timely completion of the project and an estimate of the manpower it will require should be part of the proposal, as should plans for future design review. The design proposal meeting may also include a review of the product's architecture (how the component systems will fit and work together); fastening, joining, and assembly plans; industrial design review; and the business plan for the project.

After the commitment has been made to go ahead with the project as defined in the proposal, many questions still have to be answered. For example, "What power supply will we use?" "How can we cool the device if a certain power requirement is necessary?" "What is the preferred placement for the user controls?" Answers to these questions should be in hand before the *development plan* review. The development plan includes all of the groups involved with the product. The participants in this level of review usually include representatives from engineering, marketing, manufacturing, service, customer groups, and management. Involving all of the groups helps ensure that key concerns are not overlooked in the plan. For example, including members of the service department gives them the opportunity to voice concerns about parts that have historically been problematic or difficult to service.

The next phase of the project is engineering and detailed design. In this phase, the design of each part is completed, CAD geometry is finalized, and prototypes of the design are created. The *engineering release* is a milestone ascertaining that satisfactory design reviews have been completed for all components and modules and that all outstanding questions have been answered. Questions such as "Can service provide field support?" "Has the product passed a review for safety? "Is all of the project CAD data created and archived?" may arise at this meeting. After the engineering release of the project, it makes a transition to being manufactured.

During the manufacturing phase, each part produced or acquired is tested to satisfy that it will function in the product as designed and produced in a process called *qualifying the production*. The final product goes through a similar series of qualifying tests. When everything is tested and in place the next milestone is the *product release*. Approval at this stage means that the product is now ready for use by the customer.

Design Review

As an entry-level engineer, you will probably have a more active role in design review meetings than in product level reviews as described above. A design review may be a *product review* (meaning that it is for the entire system or device) or a *part review* for single parts or systems that function at the lowest level of the whole.

Typically at this level the review is by more experienced individuals with a background in the same discipline you are working in—mechanical engineers reviewing the work of mechanical part designers, for example. These reviews tend to include more technical details and are often led by the team project leader. For this type of meeting, you need to provide the documentation for any major design decisions such as the fastener selections and choices you made along with justification for your choices (such as costs, company standard parts, industry codes). You may have to present calculations for cooling data, dynamic analysis, loads, or other information showing how the part as you have designed it will meet required performance specifications. You may use charts and graphs to show thermal curves with accepted engineering margins indicated. You may present sketches or drawings of the part or show a prototype model.

The sketches, graphs, drawings, CAD models, and other graphical work that you produce is not just for your use, but for the company as a whole. Make your sketches, drawings, calculations, graphs, and other documentation neat and legible. Your direct supervisor almost certainly will be included in the design review meetings where you present information, so do your best. As an engineer you may want to initiate a design review yourself to help you judge risk and design issues for your own work, especially when it is still in a conceptual phase. Getting the input of more experienced engineers or of a cross-disciplinary team can improve your designs, save time later in the process, and contribute to your learning about pertinent processes.

The Role of Visualization in Ideation

What happens when you get an idea? Do you picture something in your head? Many people when they have ideas about new designs, problem solutions, or other creative things say that they "see" the answer. They form a mental picture of the solution. There is a lot of debate about whether there can be any thought without language, but many creative inventors say that they see pictures in their mind even before they can express these ideas in words. They communicate these picture ideas to others through sketches. Albert Einstein described this phenomenon in a letter to Jacques Hadamard:

> "The words or the language, as they are written or spoken, do not seem to play any role in my mechanism of thought. The psychical entities which seem to serve as elements in thought are certain signs and more or less clear images which can be 'voluntarily' reproduced and combined—this combinatory play seems to be the essential feature in productive thought before there is any connection with logical construction in words or other kinds of signs which can be communicated to others."

Certainly the opposite process is true: when we hear expressive language we form a mental image. For example, when you read the bumper sticker "Visualize Whirled Peas," you form a much different mental image than when you read "Visualize World Peace." (See Fig. 2.10.)

FIGURE 2.10
Words conjure mental images.

The process of visualization is important in forming and expressing your ideas. As an engineer, you will need the ability to visualize things. You will need to visualize how things work, how problems might be solved, what a device drawn on paper would look like if it were real, mental models for complex systems, how parts will fit together, the magnitude of various objects from given dimensions, and many other sorts of things. The ability to apply visualization to solving problems and generating ideas will be a useful addition to the verbal and mathematical problem solving skills that you will be learning throughout your engineering career.

Some people seem to be born with excellent visual abilities, others must learn it. Chances are you enrolled in engineering because you are already pretty good at it, but practice and understanding basic principles will make you better at it.

You will also need to be able to communicate your ideas to members of the design team and others. One of the most direct means of communicating your visual ideas is sketching. Unlike a verbal description, hand sketches don't require verbal processing of the image but seem to flow directly from the mind to the hand. This basic method is one way of "thinking out loud" about mental images. As you sketch, you capture the visual ideas for your own future reference as well as for others. Sketching your ideas can also help you develop your ability to visualize.

Making a visual representation of information can help you incorporate key information into your thinking about the problem and can help others see the relationships, too. Relationships and comparisons are often best communicated through a visual image. For example, the graph shown in Figure 2.11 makes it easy to quickly see the positive linear relationship between amperage and the dynamic head for the pump. Can you see the relationship as clearly from the data in the table? (Graphs and their uses in presenting design information will be covered more in Chapter 13.)

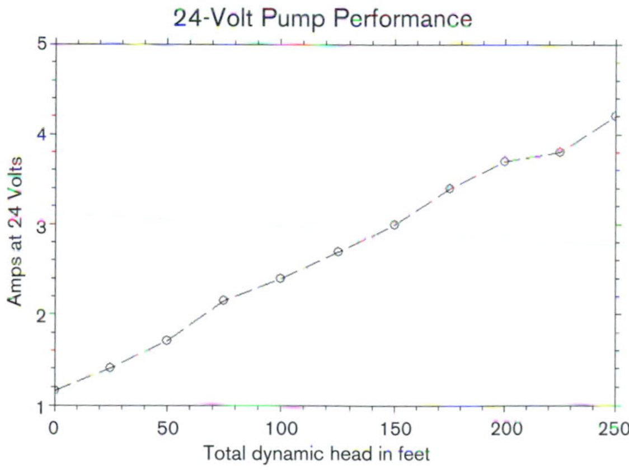

FIGURE 2.11
The simple X-Y plot shown here shows the relationship between the Y variable (amps at 24 volts) and the X variable (total dynamic head in feet) by plotting the data points from Table 2.2.

TABLE 2.2

24-Volt Pump Performance

Total Dynamic Head in Feet	Amps at 24 Volts
0	1.15
25	1.4
50	1.7
75	2.15
100	2.4
125	2.7
150	3.0
175	3.4
200	3.7
225	3.8
250	4.2

ANIMATION AND VISUALIZATION

Flip the pages of your textbook to see the sketch of the screw-type jack develop into a 3D model that demonstrates the motion of the jack. (Hold your book so your right thumb is on the right-hand edge of the pages and the rest of your fingers are underneath the book. Press up with your fingers to flex the pages of the book slightly so that the pages slip out from under your thumb and flip past your view.)

"Moving pictures" use the brain's ability to blend a sequence of images into a single moving image. If you flip the pages quickly enough, you should see the model of the jack emerge from the sketch, then move up and down. Animating your 3D models is a powerful way to visualize and demonstrate the motion built into a design. As you work through this text, you will learn to sketch, make 3D models, and use those models in various ways to visualize design. The techniques of animation used to create this flip book are presented in Chapter 13.

FIGURE 2.12
Role-playing games use 3D modeling to create environments that a player can explore and interact with. Notice how the texture applied to the wall and the clouds in the sky contribute to this scene from Sierra's Lighthouse. (Courtesy of Sierra On-Line, Inc.)

Words and mathematical symbols are important in communicating about design, too, but their pathways through the processing portions of the brain take a less direct route than a visual image. The reader must interpret your verbal or mathematical description and rebuild the visual image in order to see your design. It is not just coincidence that we say, "I see" when we mean, "I understand." However, written notes to accompany a drawing have an important role in making it easier for someone to understand how to manufacture the device or system you are designing.

The power of visuals is perhaps best exemplified in computer-generated graphics (see Fig. 2.12). Think back to the last computer game you played. How were the graphics? Would you have kept on playing it if it were a text-based game? Were the graphics in 3D? If they were not, did you wish they were?

3D computer models that are shaded and rendered are a powerful way to present a very realistic view of your designs. Today's computer models enable designers to create "virtual worlds" that are programmed to let the consumer interact with the model as if it were real and to see the model respond to events and actions.

In some ways, this rapidly developing computer-aided ability to show things using very clear and realistic graphics creates even more of a demand for it in the engineering professions. Using a virtual reality interface with 3D computer models of the design, you can even "touch" the design as well as "see" it as though it were right in front of you. Using as many of your senses as possible can add to your abilities to solve problems.

Building Visual Ability

To communicate visually you will need a set of skills. Some of these will be physical skills, such as the ability to sketch clear crisp lines that others can interpret. Because sketch paper is only a 2D media, you will also need a system to show 3D objects on a flat sheet of paper. You will also need the ability to look at drawings someone else has created and build a mental picture for yourself of the object. You can build your visual abilities through practice.

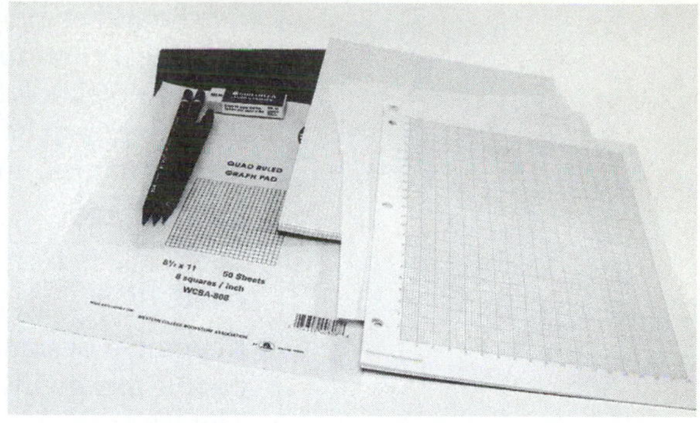

FIGURE 2.13
Grid paper of various types will help you sketch shapes and graphs more accurately.

Tools

The tools you need to begin sketching are simple—just paper, pencil, and eraser. Later you may want to add engineering markers to your sketching tools, as engineers are often required to sketch in pen or marker to provide a permanent record of their designs. Until you become more proficient at sketching, you will probably prefer to sketch in pencil so you can erase some lines. Modeling clay is also an invaluable tool, especially as you are beginning to visualize and sketch objects. Figure 2.13 shows examples of typical sketching supplies.

Pencils are available in different grades of lead (See Fig. 2.14). The pencil grade indicates the pencil's hardness or softness, which relates to how light or dark a line it will draw. Hard leads make light lines, soft leads make darker lines. Pencil grades range from the hard leads at 9H through 4H; to medium leads 3H, 2H, H, F, HB, B; then to soft leads 2B through 7B. For most engineering sketches you will need only an HB pencil, or one near that portion of the range. You may also want to use a hard pencil such as 6H to make light lines for blocking in your sketches. When you are more practiced, you can probably switch to using only HB lead, remembering to block in shapes with it very lightly at first.

It is often helpful to sketch on grid paper with divisions of either 5 mm, 10 mm, 1/8 inch, or 1/4 inch (see Fig. 2.13). The grid lines can help you develop your ability to sketch straight lines. But spend some of your practice time drawing on plain white paper in order to develop your ability to sketch quickly and not become dependent on the grid lines.

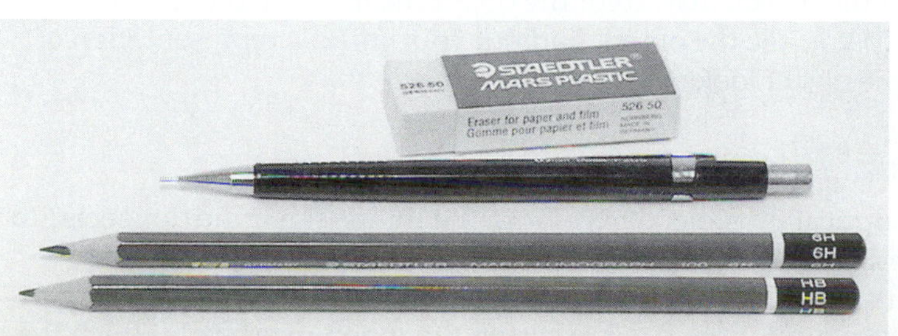

FIGURE 2.14
Different kinds of pencils and quality erasers will help you make thick and thin lines and keep your sketches neat.

Bound engineering notebooks are also frequently used to provide a continuous documented record of the design. When a bound notebook is used, it is clear if pages have been removed. Also, it's convenient because the sketches, ideas, phone numbers, notes, contact names, and other information for the design are all available in one place.

Practice

Keeping a design notebook is a good way to start practicing. Many people feel awkward jotting down their ideas and making sketches. The tendency to be overly critical of your ability to sketch can prevent some from wanting to try it at all. But you need to practice this in order to develop the ability in the first place. Start using your notebook to record ideas as sketches. Make sketches of interesting things you see and don't be too critical of your drawing ability. As you sketch, it will improve and become a skill that you can use to communicate your ideas to others.

Sketching Basic Shapes

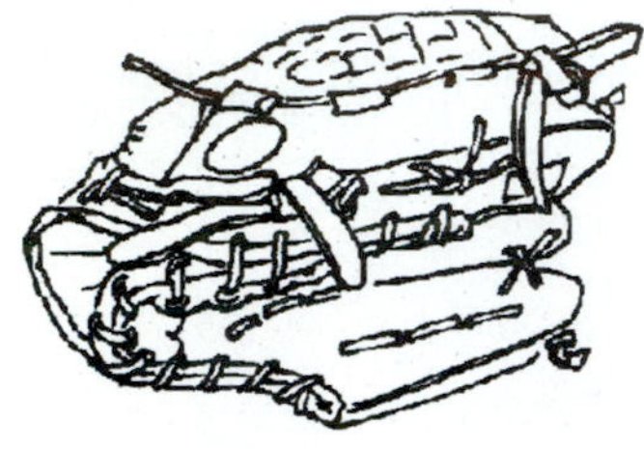

FIGURE 2.15
This freehand contour sketch of a baseball glove shows its shape and features using only lines.

One way to begin freehand sketching is to draw an object's *contour* or outline. An example is shown in Figure 2.15. Contour drawing is really the process of noting the relationship between the object you are trying to draw and the space surrounding it. You can think of it as similar to graphing the shape of the object on a sheet of paper. To draw successfully, you need to represent the object's spatial relationships accurately. Which is the highest point of the object from your point of view? Which is the lowest point? How big are the various features?

Forget the Symbols You've Learned

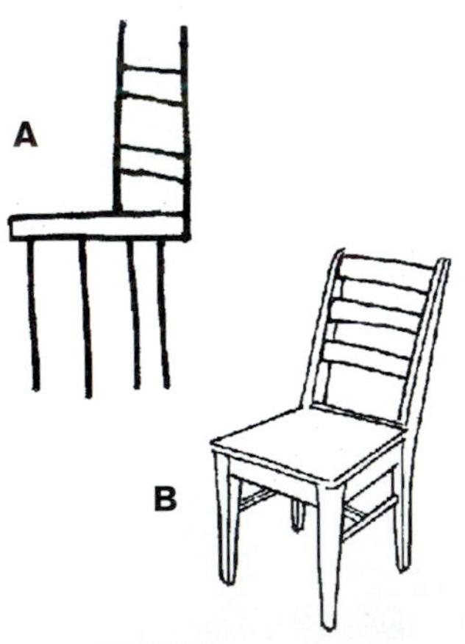

FIGURE 2.16
Compare the chair shown as a symbol in (a) to the more realistic contour sketch of the same chair shown in (b). Both are easily recognizable as chairs, but the contour sketch provides more information about the actual shape.

As children, we learn a way of drawing common objects that doesn't necessarily describe the object's shape accurately. For example, when you draw a picture of a chair, you may draw a flat seat with legs, as shown in Figure 2.16(a), even though a more realistic view of the chair would show more of the seat and the four legs. These drawings serve more as "symbols" for objects than as accurate representations of their appearance. You can probably easily sketch several of the symbols that appear in elementary school art to depict a cat, pig, house, tree, and so on.

To draw an object's shape as it actually appears, you have to examine it without any preexisting notion of what that object should look like. Drawing an object upside down and drawing the space around the object, instead of the object itself, are two exercises that can help you see the object, and not your mind's representation of what the object looks like.

Optical Illusions

Many optical illusions derive from the things the mind "expects" to see. Did you ever wonder why the rising moon on the horizon appears so much larger than the same moon higher in the sky? Because we are more used to seeing the moon overhead (which the mind interprets as closer because there are no other visual references between you and the moon overhead), it appears smaller overhead than when it is on the horizon (which seems to be farther away because you have the context of a large number of objects between you and the moon on the horizon). Actually, it occupies the same amount of our field of vision when it is overhead or on the horizon. Try holding a quarter at arm's length when the moon is on the horizon and appears large. It should just about block out the moon. Then wait until it has risen and appears smaller. Try the same thing: judge the size of the moon overhead using a quarter. You should see that they are the same size in relationship to the quarter.

Examine the optical illusion shown in Figure 2.17. Lay your book flat on the table and look at the figure. Now tip the book so that the figure is vertical in space and look again. Do the lines appear closer together when it is lying down or when it is standing up? You should see that the lines appear closer when the figure is lying flat (horizontally).

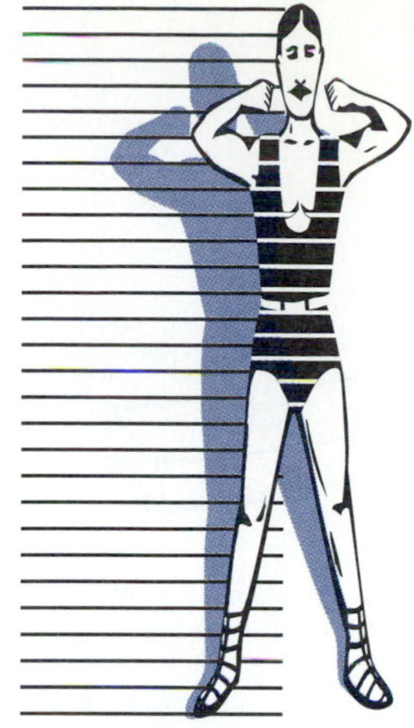

FIGURE 2.17
The appearance of the strong man figure in this illustration shifts as your viewing angle does. Does his face appear to be as thin when you lay the book flat on the desk as it does when you hold it in a vertical position? Do the horizontal lines appear closer together?

The same illusory effect can happen when you are sketching on a sheet of paper lying flat on your desk, leading you to overcompensate in your drawing. It is sometimes helpful to hold your paper vertically and get a different view of your sketch from time to time.

Try creating a contour drawing of the rubber stamp shown in Figure 2.18. Use a sheet of plain white paper attached to a clipboard, if possible, so that you can hold the paper up and look at the sketch straight on. When the paper is laid horizontally on a flat surface like your desktop, the lines you are sketching can appear foreshortened, causing you to make a distorted sketch.

FIGURE 2.18
Rubber stamp

Estimating Proportions

Start out your sketch by noting the relationships of the stamp's features. The overall shape, including the handle, is slightly wider than it is tall. The center handle sticks up above the back edge of the base about half of the height. The lowest point is in the lower right. The highest point is on the handle. Imagine tracing the outline of the stamp in the air with your pencil. Next very lightly sketch the outline of the stamp on your paper. Try to keep it in proportion as shown in Figure 2.19. After you have sketched the basic outline of the stamp, add the details representing the edges of the object. Add the curved contour for the bottom of the handle where it joins the stamp base and the flat top surface of the handle.

Showing the features of an object in correct proportion to one another is key when sketching. To help you determine the proportions of the object you are trying to sketch, use your pencil or light marks on the edge of your paper to "measure" the object's features. You have probably seen artists on TV or in cartoons holding their paint brush in front of the object they are painting (see Fig. 2.20). Try it yourself by holding your pencil up at arm's length and measuring how many pencil lengths (or what portion of the pencil) make up the vertical height of a door. Holding your arm at the same distance, measure the width. Is the height about three times the width? You can use this to make rudimentary measurements to keep your sketch in proportion.

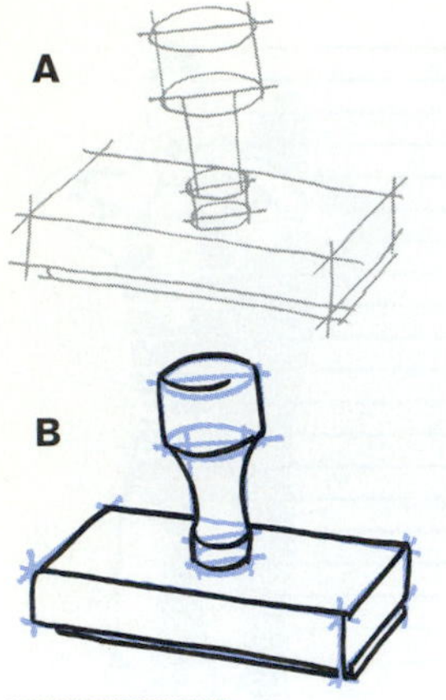

FIGURE 2.19
(a) The blocked-in shape of the rubber stamp shows the overall width, height, and height of handle over back edge. (b) The darkened final sketch of the rubber stamp is drawn over the lightly sketched shapes.

FIGURE 2.20
This artist is using his paintbrush to measure the proportions of the castle.

FIGURE 2.21
Contour sketch with hatching

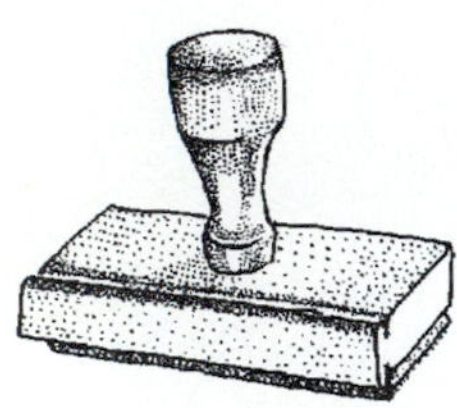

FIGURE 2.22
Contour sketch with stippling

Another technique is to hold your paper vertically in front of the object and use the edge of the paper to help you determine whether the object's edges are horizontal, vertical, or at some other angle. Visually judge the size of the angle in relationship to the edge of the paper.

Viewpoint

Keep in mind that as you are sketching objects, you want to maintain a consistent viewpoint. This can be easy when you are sketching a picture from a book because you can't move around the object. But when you can move around an object, you will see a different view of it depending on where you stand. Sometimes people have difficulty sketching because they show parts of the object that cannot really be seen from the viewpoint. For example, knowing that the handle of the rubber stamp appears circular from the top, you may want to show it round in other views, even though it appears elliptical.

Shading

The edges of an object are visible because one surface reflects more light than another. The angle of the light illuminating an object determines how different surfaces appear. Adding shading to your sketch can give it a more realistic appearance because it represents the way the actual object would reflect the light.

Hatch lines, shown in Figure 2.21, and stippling, shown in Figure 2.22, are two different ways to add shading to your sketch. These methods are used rather than continuous pencil shading because they reproduce better on copies. Add hatching lines or stippling to add shading to the stamp's surfaces where they are darker. After you do, darken the outline of the stamp more so that it defines the shape clearly and boldly. Remember that when you are communicating by using a sketch, its subject should be clear, just as it should be when you are writing a sentence. To make the subject—in this case, the stamp—clear, make it stand out with thick bold lines.

Positive and Negative Space

Another way to think about freehand contour sketching is the contrast between the positive and negative space. Figure 2.23(a) shows a drawing of the rubber stamp with the positive space shaded in. Does Figure 2.23(b), showing the negative space, help you see the shape of the contour better?

Some people sketch more accurately when they try to draw the negative space that surrounds the object. Try sketching the negative spaces that define the shape of the chair shown in Figure 2.24. What is the shape of the space between the legs? What is the shape of the space between the rungs and the seat? Make a sketch of the chair, paying careful attention to sketching the negative spaces of the chair as they really appear. Of course, the positive and negative spaces should add up to define the chair, as shown in Figure 2.25. If you have difficulty, make corrections to your sketch by defining the positive shapes and then check to see if the negative shapes match.

An 8.5-by-11 sheet of Plexiglas (available at most glass stores) is an excellent tool for developing your sketching ability. Using a dry erase marker, hold the Plexiglas up in front of an object and sketch it on the Plexiglas. The outline should match the object's outline exactly if you don't move. Lower the Plexiglas and look at the orientation of the lines. Are they what you expected? Try looking at the object and drawing the sketch with the Plexiglas laying on your desktop or knees. Then raise it up and see if your drawing matches the object. Everyday objects work great for developing sketching ability—try your toaster, printer, lamp, as well as exterior and interior views of buildings and equipment.

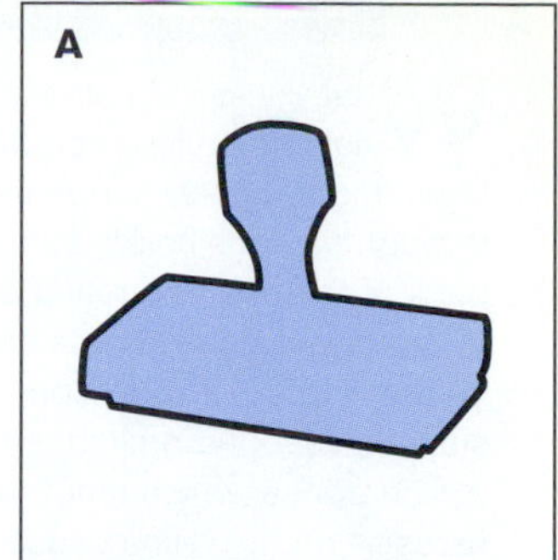

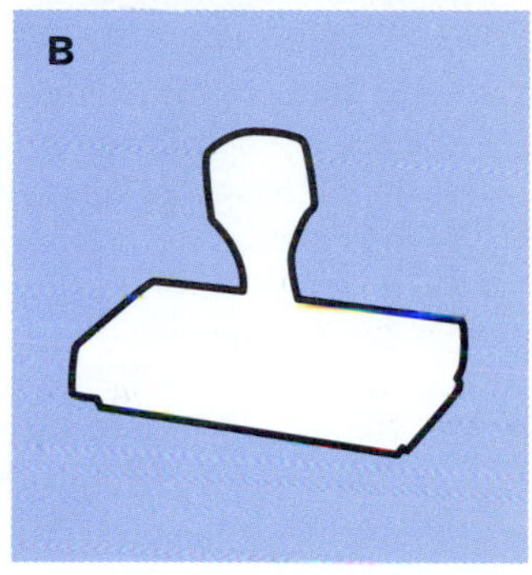

FIGURE 2.23
Rubber stamp contour drawings; (a) shows positive space shaded; (b) shows negative space shaded.

Visual Perception

Optical illusions and magic tricks are both based on the fact that what you see is actually your mind's interpretation of the data coming in through the optic nerve. The more you know about visual effects, the better you will be at re-creating them for another viewer. Learning to sketch is not only an effective means of capturing your design ideas, it can also change the way you "see" the world around you.

One way to practicing "seeing" is to look hard at an image for a few seconds, then close your eyes and try to visualize the scene. Try to re-create as much of the detail as possible, then compare the image you built with the actual scene. Notice what you could "see" in your mind's eye and what you couldn't.

People's interests and experiences cause them to remember different aspects of a given scene. Did you notice colors? The relative size of items? Texture? You can build your visual ability by practicing different ways of "seeing" and noticing how lighting changes, movement, and distance alter how your eye perceives different objects.

Imagining how objects and scenes look from different viewpoints is another way to build your visualization skills. One technique that Albert Brown, design engineer at Affymax Research Institute, uses is to select a bird on a rooftop or telephone pole and consider what the buildings, cars, and objects in the area would look like from that viewpoint. What would be visible and what would be

FIGURE 2.24
Photograph of the chair

FIGURE 2.25
Improve your sketches by showing the negative space accurately.

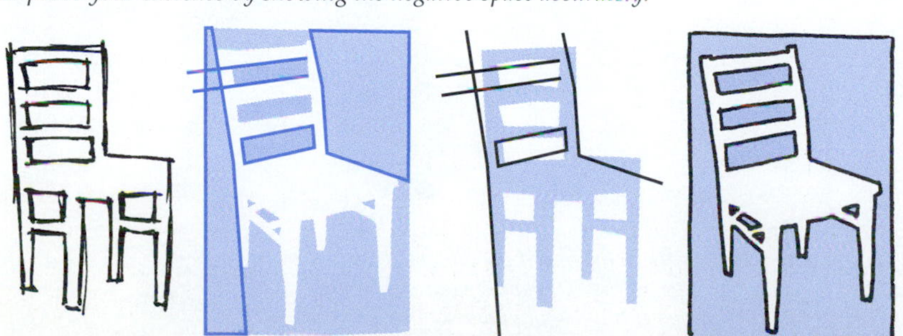

The Physiology and Psychology of Seeing

What we see is actually an interpretation of nerve impulses sent from the retina to the brain. Light reflected from an object enters the eye through the lens and is focused on the retina at the back of the eye. The retina contains light-sensitive receptor cells called rods and cones. Each eye contains about 125 million rods, which are responsible for recognizing light and dark, and about 7 million cones, which react to colored light. Light focused on the retina causes rods and cones to send impulses to the brain. Three kinds of cones–for red, blue, and green–send impulses that are mixed by the brain to create the wide range of colors we "see." Cones are concentrated in the center of the retina, where vision is sharpest, whereas rods are more numerous around the outside of the retina. Peripheral vision–or what you see out of the corner of your eye–is therefore primarily black and white.

Because humans have two eyes, two different sets of impulses are created–each dependent on the way the light enters that eye. The impulses from both eyes are combined by the brain(see Fig. 2.26). If you close one eye and then the other, you can see how the object you are looking at seems to shift side to side. The object you see with both eyes is actually a composite made of the signals from each eye.

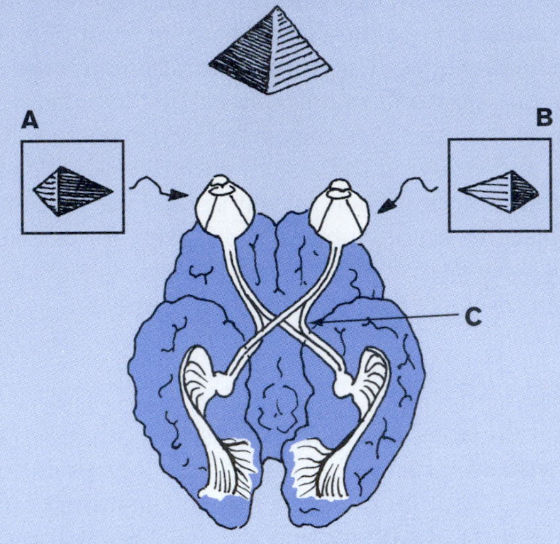

FIGURE 2.26
The light rays from the object strike the retina in the patterns shown in (a) and (b). The impulses cross over the optic chiasma, travel through the brain along the pathways illustrated in (c), and result in "seeing" the object in three dimensions.

But seeing is more than a simple mapping of impulses onto a "screen" in the mind. Current brain research has demonstrated that the impulses are combined and recombined as they pass through different parts of the brain. Impulses from each eye cross over into the opposite hemisphere of the brain at the optic chiasma, for example, but some do not. This incomplete crossing over means each half of the brain gets signals from both eyes, which scientists believe is responsible for the stereoscopic vision that allows us to see the depth of objects.

The processing of the visual impulses also explains why we are unaware of the "blind spot" in each of our eyes. The optic nerve, where the nerve fibers from the rods and cones come together to exit the eye, has no cells receptive to light, even though it is located on the back of the eye and surrounded by receptor cells. The fact that we don't see the spot is evidence of the way the brain "fills in" missing information as it interprets visual impulses (see Fig. 2.27).

As the visual impulses travel through the brain, different areas of the cerebral cortex affect how the impulses are translated into meaning. Memories, experiences, and visual cues affect how we interpret the impulses–and how we sometimes "see" things that are not there, simply because they were expected. This interpretive seeing is why it can be difficult to "see" an object from a single viewpoint–or to see the actual shapes of objects as they appear from a given viewpoint: the brain "sees" the missing information and adds it to the impression created by the visual stimuli. To improve engineering visualization and drawing skill, it can help to be aware of interpretive seeing and to practice seeing the true geometric shape of objects.

FIGURE 2.27
To "notice" your blind spot, hold this page with the drawing of the + and the dot about 16 inches in front of you. Close your left eye and focus your right eye on the +. Keeping your eye focused on the +, move the paper slowly until the dot moves into your blind spot and disappears.

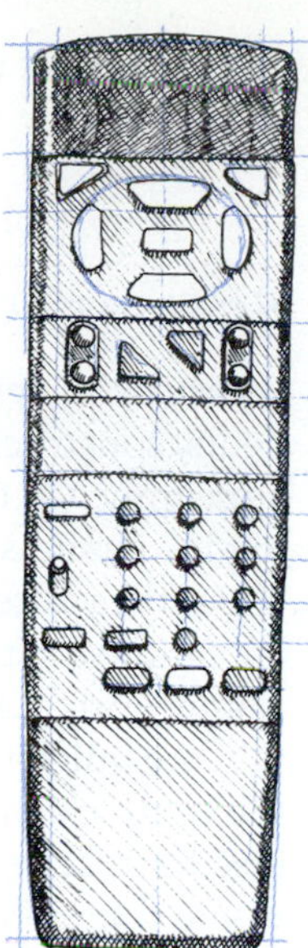

FIGURE 2.28
Blocking in the object's shape and using construction lines to arrange features, as was done in this sketch of a remote control, can help you orient features that need to line up.

obscured by other objects? What shape would a building or parking lot have from a view from the top? The ability to see objects from different vantage points is a key skill in creating and using engineering drawings, as you will see in later chapters.

Construction Lines/Sketching Aids

Contour sketches help you develop your ability to "see" an object as it actually appears so you can re-create that appearance on paper. Another technique is to sketch the basic shape of the object and use that as a guide for the object outline you want to create. *Construction lines* are guides that can help you preserve the basic shape as you add the details of the object.

Before you begin to draw the outline for an object, think about its overall shape and the relationships between its parts. Construction lines can also help you preserve the overall dimensions of the object as you sketch. The drawings in Figure 2.28 show how the sketch of the remote control started with construction lines for a rectangular box. These lines helped preserve the overall shape as the rounded corners of the outline were added. Additional construction lines parallel to the edges of the control were added to make it easy to add the shapes of the buttons.

Try this technique with objects that are generally square, round, or elliptical. Use the overall shape of the object to help you determine the relative size and placement of features on the object. Think about breaking down more complex objects into their simpler geometric shapes as shown in Figure 2.29. Sketch these shapes showing their relationships to one another accurately.

Look for the essential shapes of objects. If you were to make a clay model of an object, what would you start with? A ball? A box? Squint your eyes and look at familiar objects. Do you see their basic

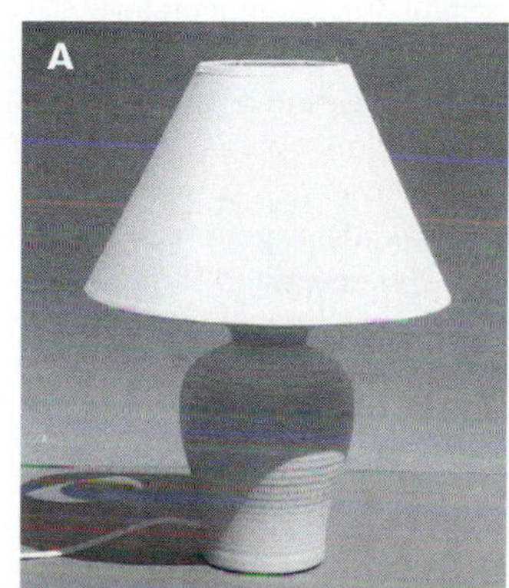

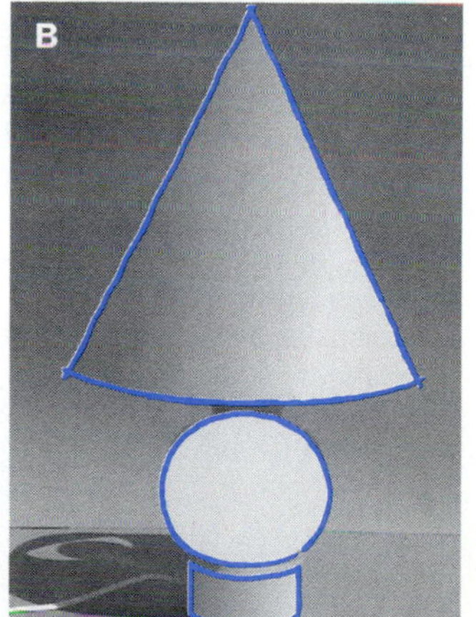

FIGURE 2.29
The essential shapes of the common household lamp in (a) are superimposed on the lamp in the illustration in (b).

shapes better? What else do you notice when you look at them in this way?

Schematic Drawings

Schematic sketches or drawings use symbols to represent components, constructs, or relationships, instead of showing their actual appearance in the drawing. Think back to the discussion of sketching the chair as a symbol. Children do this without really thinking about the process. For example, a cat has a round head and triangular ears. We can recognize this shape easily and identify it as a cat. Engineering schematic drawings use the same principle. Symbols that are quick to draw and that reproduce well are used to replace the complex shapes of the real objects. For example, electrical and electronic schematics represent electrical power systems and electronic components, such as capacitors and resistors, using symbols such as those shown in Figure 2.30.

Schematic drawings use standard symbols in order to make them easy to interpret. These standard symbols and their sizes are published as *standards* by the American National Standards Institute (ANSI) so they will be used consistently and their meaning will be clear. Standard symbols are published for several different disciplines; some of the most useful ones appear in an appendix at the end of this book.

Similarly, *freebody diagrams* use vectors, masses, and simplified objects to depict the forces in a mechanical system or device. *Block diagrams* show the relationships between equipment used in a system, such as the layout of tanks and piping in a fluid system, or the layout of computer equipment in a network. Figure 2.31 shows examples of freebody, block, and piping schematics.

Standards

ANSI is the American National Standards Institute, a governing body that establishes and publishes standards, or rules, for technical drawings and other engineering practices. The institute publishes standards for engineering drawings that govern drawing sheet sizes, linetypes, multiview drawings, and representations for screws and gears, among others. You may be familiar with the term "ANSI standard" from computer programming language courses, too.

ISO is the International Organization for Standardization, the governing body for international standards, which is headquartered in Geneva, Switzerland. The group name is always abbreviated ISO regardless of the language it is translated from. Nearly 100 member countries cooperate to develop these international standards.

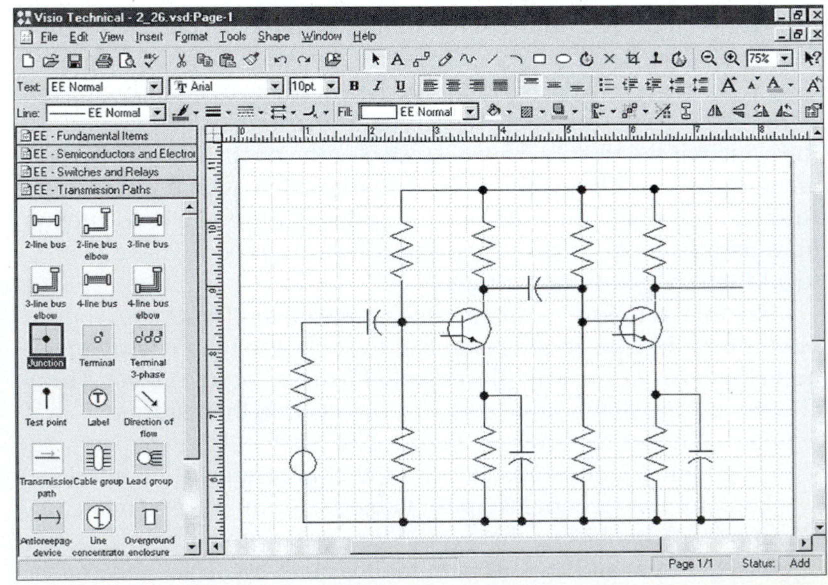

FIGURE 2.30
Visio Technical is a schematic drawing program whose drag-and-drop interface makes it easy to produce electrical, block, mapping, layout, and many other types of schematic drawings. This electrical schematic can be completed by dragging symbols onto the drawing from the buttons on the left side of the screen.

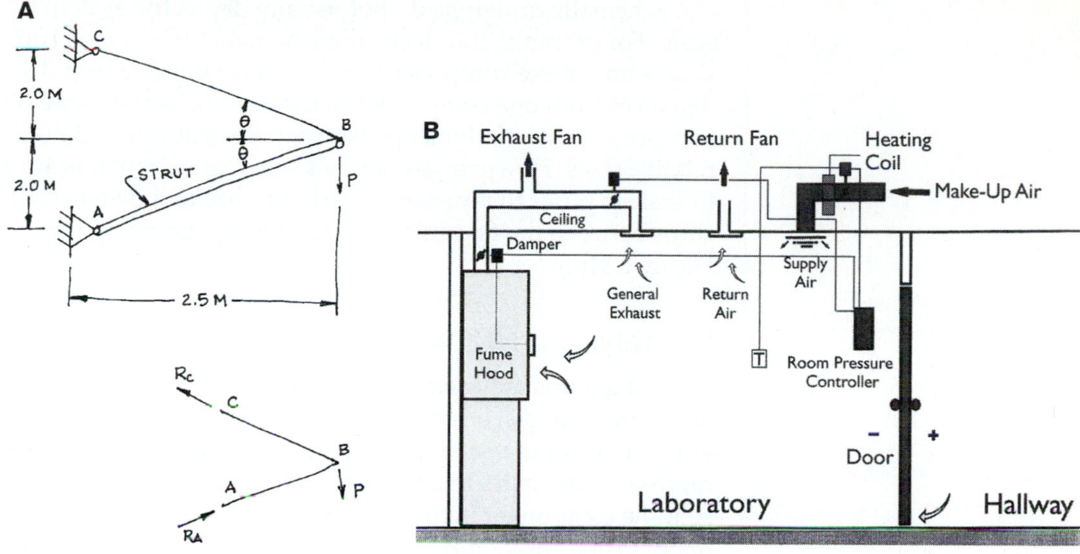

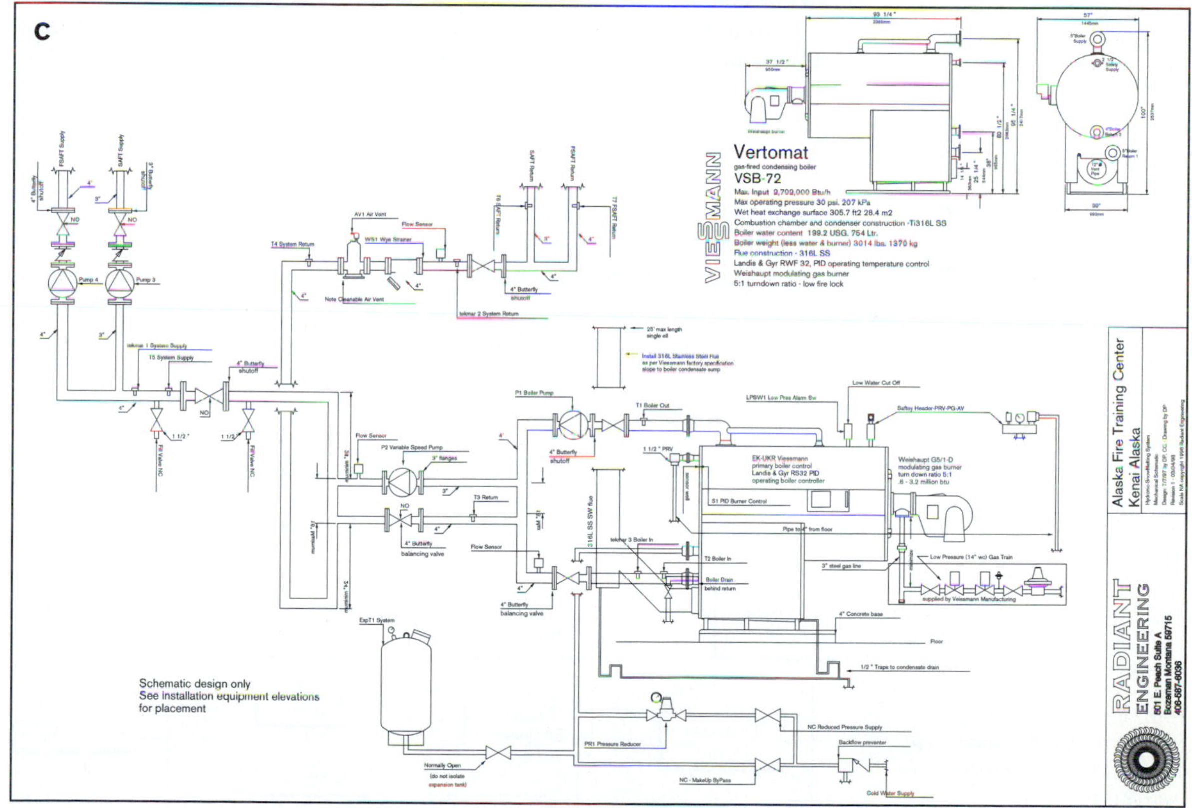

FIGURE 2.31

The freebody diagram in (a), laboratory heating and cooling system block diagram in (b), and piping control diagram in (c) are all examples of schematic drawings that use simple shapes in lieu of realistic images to convey information. (Block diagram courtesy of Jerry Baker, CTA Architects; piping diagram courtesy of Robert Knebel, Radiant Engineering.)

Schematic drawings do not usually depict the system to any scale. For example, the electronic schematic shown in Figure 2.30 shows how these components will connect but does not describe the distances from one component to the next. Schematics are often used to represent the relationships between the features and not their relative sizes. However, sometimes schematic drawings are shown to scale in order to represent clearly the size and distance relationships between the features, as in the piping diagram shown in Figure 2.31(c).

Sketching a Block Diagram

Block diagrams are used to show the relationship between components, individuals, or groups. Because they can be sketched quickly, you may want to use block diagrams early in the design process to represent how parts in a system will relate. Boxes are used to represent the component items, and lines are used to indicate how they relate to one another. Figure 2.32 shows a block diagram for a corporate structure. Each block indicates a position within the company and the lines indicate a supervision structure between the positions.

In the next steps you will see how to sketch a block diagram for the design of a squirt gun (see Fig. 2.33). Each block will represent a part that needs to be designed. The lines will indicate an assembly relationship between the parts.

Drop Shadows

Drop shadows on block diagrams and other sketches are an effective way to make the shapes stand out boldly on the page. Figure 2.34 shows how a dropped shadow can easily be created by thickening the lines along the bottom and (usually) right-hand edges. A nice way to make a dropped shadow is to use a marker that is two to

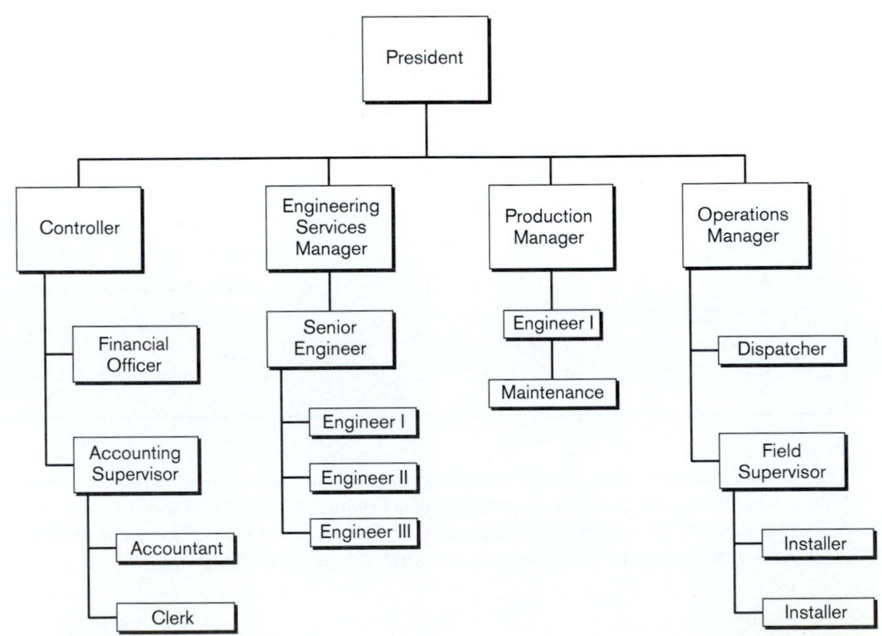

FIGURE 2.32
In the block diagram shown here, the layering of the blocks indicates levels in the corporate hierarchy. Lines represent the reporting structure of the company. For example, the line from the Accounting Supervisor to the Accountant indicates that the Accounting Supervisor is the Accountant's manager. Horizontal or vertical lines group several individuals who all report to the same person. In matrix organizations, in which people work in teams and report to more than one person, these relationships may be represented by a dotted line.

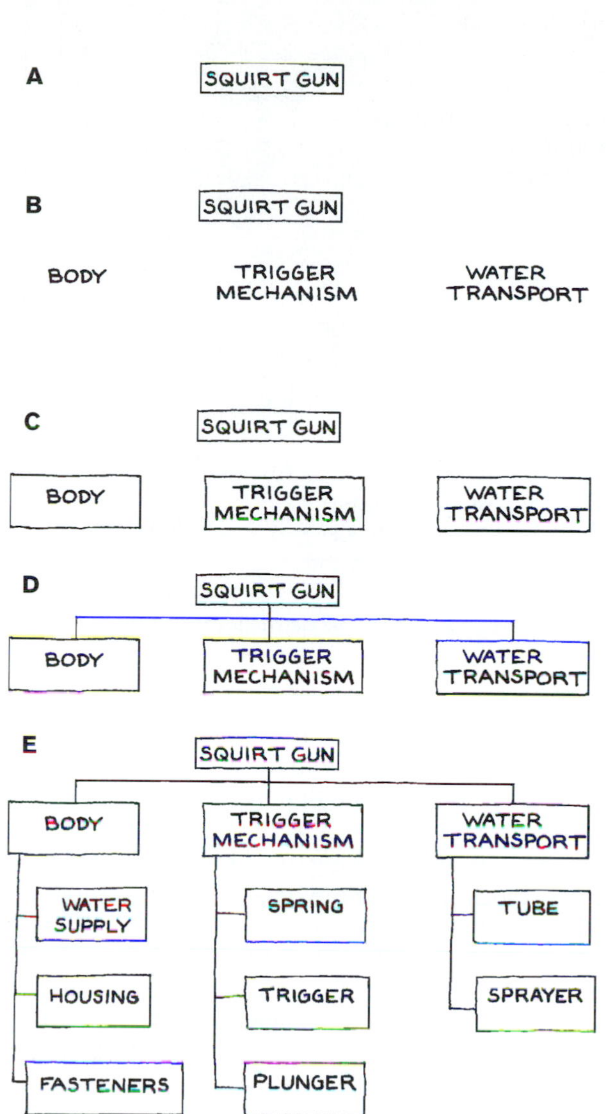

FIGURE 2.33
A block diagram

(a) Start by identifying the top level of the gun. Letter the text SQUIRT GUN neatly at the top center of the drawing and draw a box around it so that the text is roughly centered in the box.

(b) The next step is to divide the gun into its major functional areas. Into what main functional portions can you divide it? For this example, it will have three major subdivisions: the squirt gun body, the trigger mechanism, and the water transport system. Represent each of these below the main squirt gun block. Again, it is easier if you letter the text first and draw the box second.

(c) Keep this second row of boxes aligned with each other to indicate that they are all at the same level.

(d) Next draw a line halfway between the top row and the second row of boxes. Use vertical lines to connect the boxes to this "pipeline" to show that they connect.

(e) Add the subcategories below each major functional area next. Below BODY, add WATER SUPPLY, HOUSING, and FASTENERS. Below TRIGGER MECHANISM, add SPRING, TRIGGER, and PLUNGER. Below WATER TRANPORT, add TUBE and SPRAYER. Because these will not fit well if you make another horizontal row, list these vertically and add a vertical pipeline to group them together and connect them to their functional area.

three times as thick as the pen used for the outline and to make a single stroke along each of the bottom and right-hand edges. If you are adding dropped shadows, be consistent. Always add them on the same sides. This makes it look as though light is striking the object from one direction, producing the shadow. In general, use thick lines to represent the important parts of the drawing so they can be clearly seen.

SQUIRT GUN

FIGURE 2.34
Drop shadow

Sketching an Electronic Schematic

When you are sketching a schematic, refer to a current standard to learn how to draw the symbols you will use. Figure 2.35 shows some standard electronic symbols. You will become familiar with the standard symbols for the components that you use frequently so that

(a) Lightly block in major components or those that have many connections near the center of the drawing.

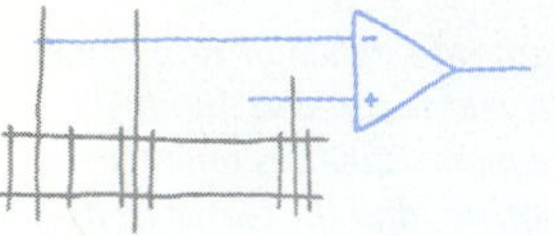

(b) Align components in rows for a neat looking sketch.

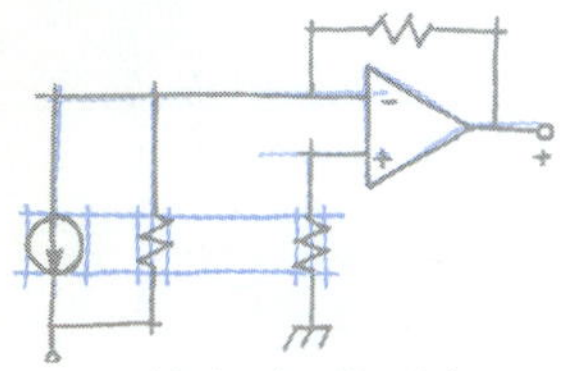

(c) Add the details of the symbols.

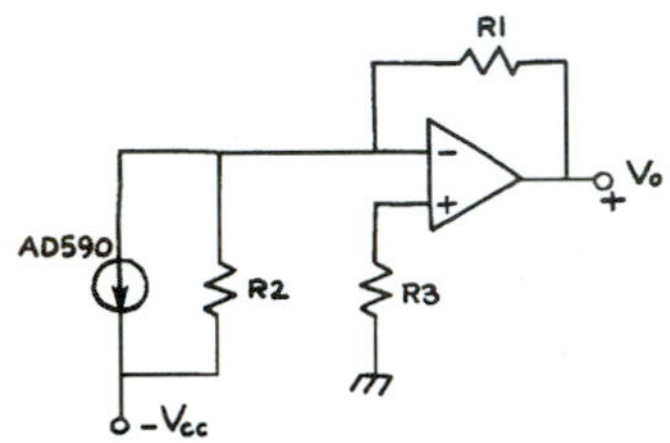

(d) Darken the final sketch and add labels and notes.

FIGURE 2.36
Sketching a Circuit
The operational amplifier circuit shown in this sketch converts a temperature to a voltage that can be displayed on a panel meter or voltmeter. The AD590 is an electronic device that acts like a current source. Its current is proportional to temperature in Kelvin. The circuit allows the temperature to be converted to a voltage where $V_o = (I + V_{cc}/R2)^*R1$. This relationship is useful for converting a current proportional to the temperature in Kelvin to a voltage proportional to the temperature in Centigrade or Fahrenheit.

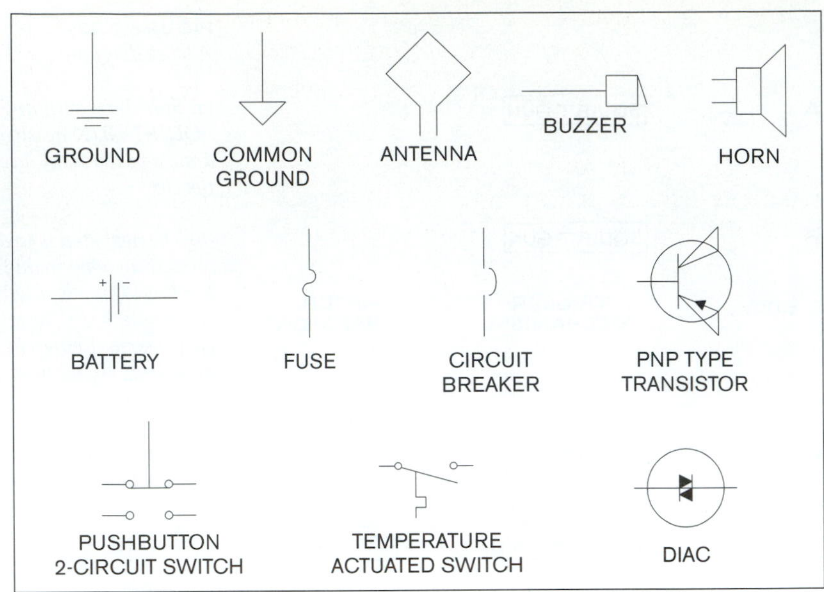

FIGURE 2.35
These are a few of the standard symbols for representing electronic components.

you can sketch them readily. Use the ANSI-IEEE standard to look up other symbols and the proportions for how to draw them. Do not make up your own new symbols, as this will make your drawing difficult for others to interpret.

Figure 2.36 shows the steps in sketching an operational amplifier circuit. Start by lightly blocking in major components or components that have many connections. Place these near the center of the sheet, so you will have room around them to add other components and connections. Try to align a number of components in a row to create a neat legible drawing. Make all of the symbols a similar size, as specified in the standard, or close to it. After you have blocked in the general locations, add the details of the symbols, then darken in the final sketch. Electronic schematics use a dot to indicate an electrical connection. Lines that cross without a dot are not considered connecting. Finally, add necessary text and labels neatly and legibly.

Sketching Thread and Fasteners

Another use for schematic drawings is to represent standard threaded fasteners and springs. The actual shape of a thread or a spring is a helix, which can be difficult to represent. By sketching thread symbolically, you can show fasteners quickly. A note can be used to indicate exactly which fastener to use. *Thread* is the term for the helical shape, similar to the striping on a candy cane, that you commonly see on screws, bolts, and other devices. Thread has three typical uses:

- to hold parts together, as do the threads on fasteners such as screws or bolts;

- to provide adjustment between parts, as the thread on the adjusting screw on your carburetor does;
- to transmit power, as the driving screw in a screw-type jack does (see Fig. 2.37).

Figure 2.38 shows a typical threaded shaft with labels identifying the major features of the thread. You should be familiar with the following thread terminology:

- *Major diameter:* the largest diameter of a threaded hole or shaft
- *Minor diameter:* the smallest diameter of a threaded hole or shaft
- *Thread angle:* the angle between the sides of the thread
- *Thread axis:* the centerline of the screw where it appears rectangular
- *Crest:* the top surface of a thread where the two sides join
- *Root:* the bottom surface of a thread where the two sides join
- *Side:* the surface of the thread connecting the crest and root
- *Pitch:* the distance from a point on a thread to the corresponding point on the next thread
- *Lead:* the distance the thread advances when turned one complete turn (single thread advances one pitch distance in one complete turn)
- *Thread depth:* the distance from the crest to the root of the thread measured perpendicular to the thread axis

You will often sketch threaded fasteners when you are sketching your ideas for the design of a device or system in which thread is used to hold parts together in assembly, provide for adjustment, or transmit power.

In sketches and drawings the helical shape of the thread is not represented exactly, but schematically or in a simplified fashion, which saves time. The three methods used to represent thread are detailed, simplified, and schematic.

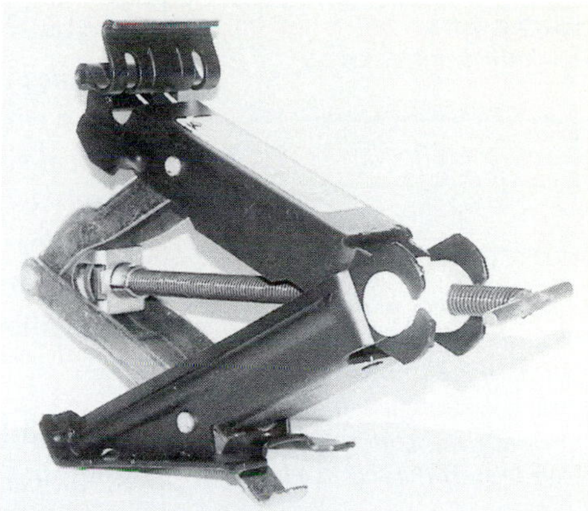

FIGURE 2.37
This screw-type jack uses thread to transmit power to raise the car. Smaller thread is used on the screws that hold its parts together.

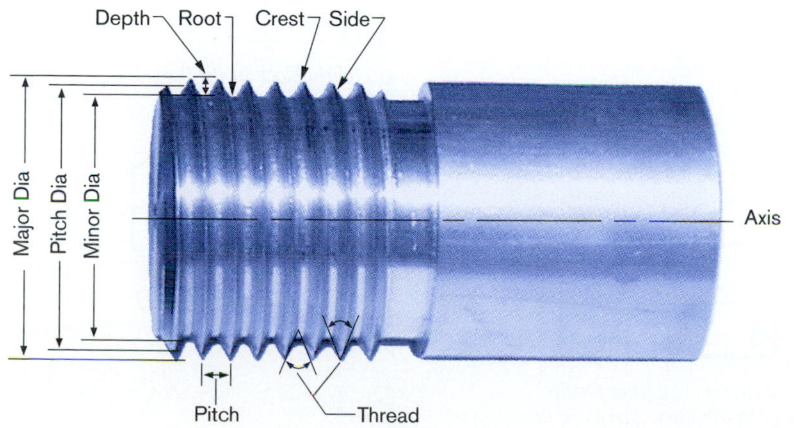

FIGURE 2.38
Thread features

FIGURE 2.39
Sketching detailed thread.

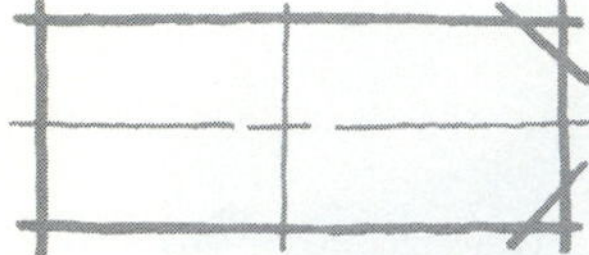

(a) Lightly block in the shape of the threaded shaft and draw a light vertical line defining the threaded portion. Add angled lines at 45° to represent the chamfer that makes it easier to start the bolt in a hole.

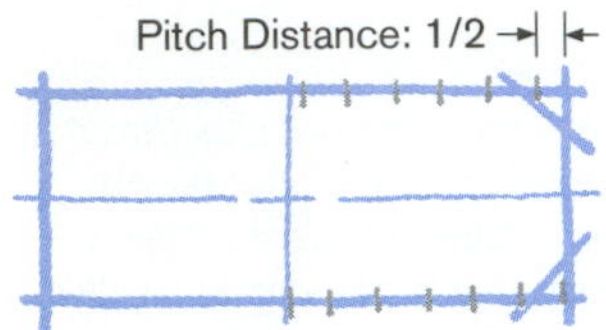

(b) Make light ticks to locate pitch distances as shown.

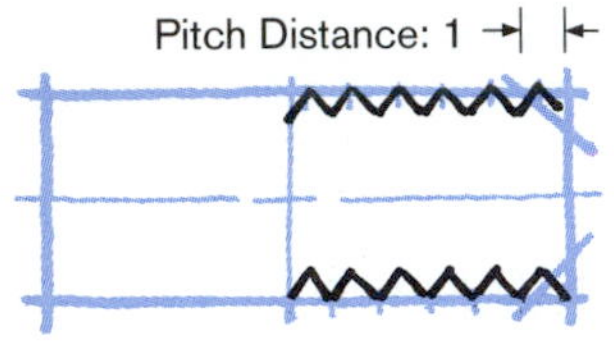

(c) Sketch 60° V's representing the thread form.

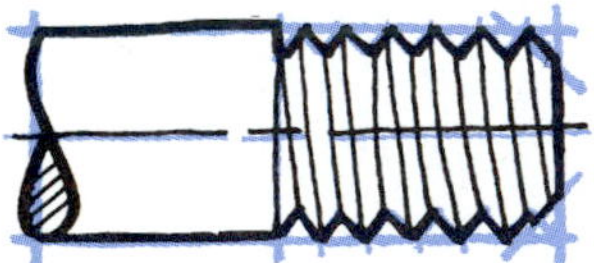

(d) Connect crest to crest and root to root and darken in the final lines.

Detailed Representation

Detailed representation is the most realistic looking, although it is still a symbolic way of representing thread and does not show the helical shape accurately (see Fig. 2.39). Detailed representation is rarely ever sketched because it is time-consuming to draw the large number of lines making up its shape. Detailed representation does not reproduce well when the thread diameter on the sheet of paper is smaller than 1 inch or 25 mm.

Simplified Representation

Simplified thread representation shows a hidden line representing the depth of the thread. Because the thread depth is different for different pitches or thread series, and can sometimes be quite small, this hidden line is often shown at 1/16 inch or 2 mm in order for it to be easily identified on a sketched or printed drawing. Figure 2.40 shows examples of simplified thread representation. Notice the thin lines that represent the chamfer and the end of the threaded length.

Schematic Thread Representation

Schematic thread representation uses shorter thick parallel lines to represent the roots of the thread and longer thin lines to represent the crests. The ends of screws and bolts are shown chamfered as they are in detailed and simplified representation. The distances from crest to crest or root to root are not always shown accurately; instead they are often shown about 1/8 inch or 3 mm apart in order to give the appearance of thread. Figure 2.41 shows the steps in sketching schematic representation for thread.

FIGURE 2.40
Sketching simplified thread.

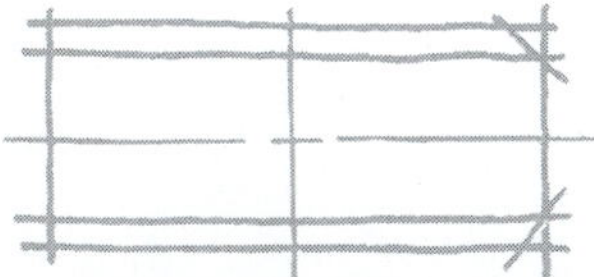

(a) Lightly block in the shape of the threaded shaft, making a light vertical line to define the threaded portion of the bolt. Use a light line to indicate the thread depth.

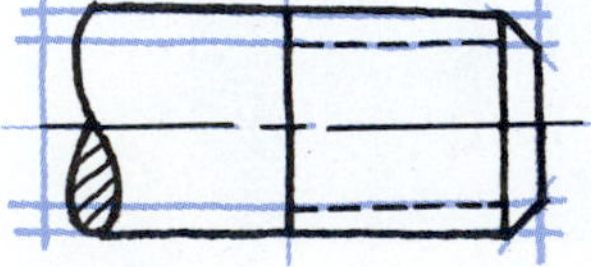

(b) Sketch short dashes for the thread depth, then darken in the final lines.

FIGURE 2.41
Sketching schematic thread.

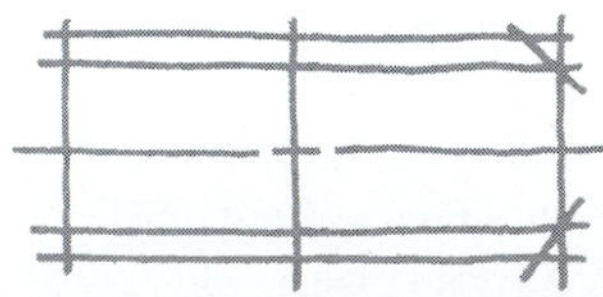

(a) Lightly block in the threaded shaft.

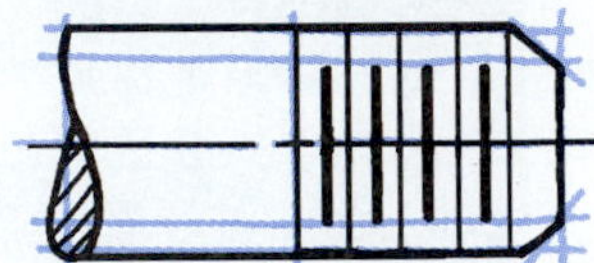

(b) Use short thick lines to represent the roots and long thin lines to represent the crests.

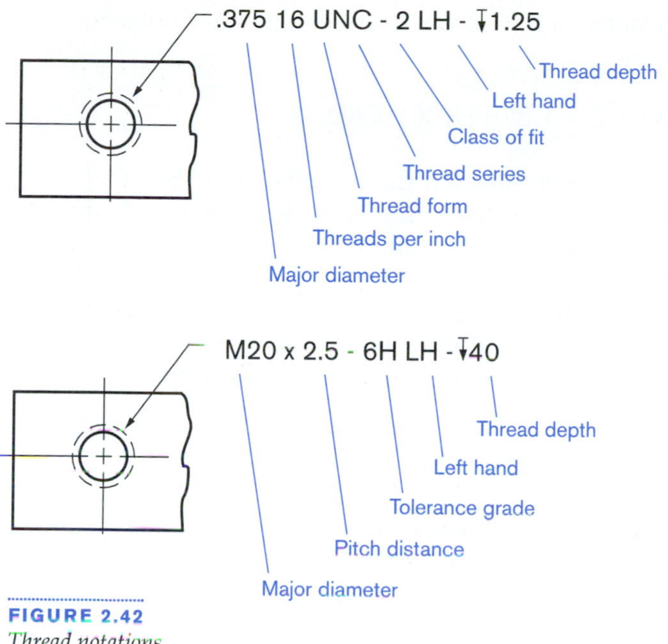

FIGURE 2.42
Thread notations

Thread Notes

Thread notes give all the information needed to purchase or manufacture the thread fastener or part indicated (see Fig. 2.42). Instead of dimensioning thread with many separate notes, a standard notation provides this information in the simplest fashion. The thread note is usually placed where a threaded hole appears round or where a threaded shaft appears rectangular.

Bolt Heads

Bolt heads are in proportion to the diameter of the threaded shaft. Figure 2.43 shows the proportions for a typical hex head bolt and nut. When sketching bolt heads, represent the shape so that it looks in proportion, but don't worry about getting the dimensions exact. Often in a quick sketch, as shown in Figure 2.44, all bolt heads are shown as hex type heads even if another type may be used.

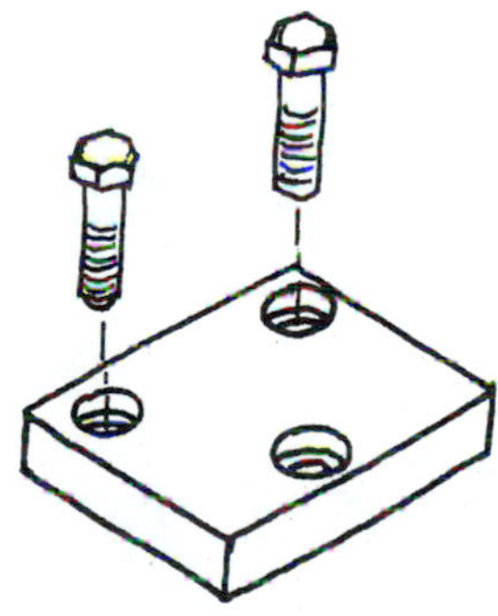

FIGURE 2.44
These bolts are sketched as hex heads, but another bolt may be called out in a note.

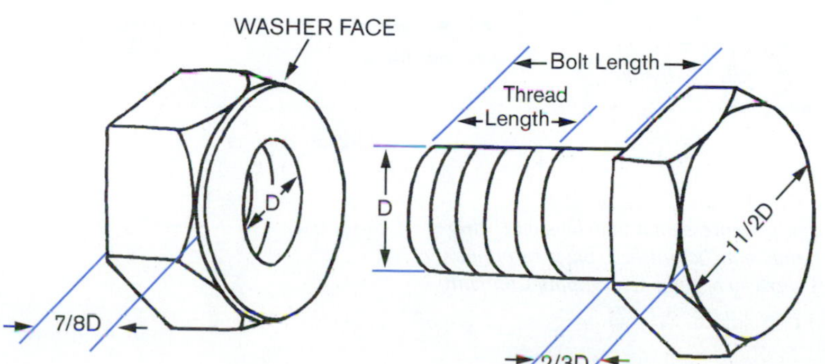

FIGURE 2.43
Bolt and nut proportions

The shape of the bolt head can be called out in the note. For example, the note

.375 - 16 UNC x 1.875 HEX HEAD

indicates a.375 diameter threaded shaft on a hex head bolt that is 1.875 inches long. The length for most bolts does not include the head portion of the bolt. Countersunk machine screws do include the head in the specified length. Refer to the appendices at the back of this book for the proportions and dimensions for typical bolt heads and for other useful information such as thread tables for standard series, bolt proportions, and typical proportions for other fasteners.

Springs

Springs are designed so that they can be either extended or compressed and still return to their original size after the load is removed. A spring designed for the purpose of being compressed is called a *compression spring*. *Extension springs* are designed to be extended. Both are usually formed in the shape of a helical coil that can be compressed or extended and still return to its original shape. A Slinky™ is a familiar example of the action of an extension spring—the Slinky stretches and returns to its original shape unless the force applied to it exceeds the limits of its material properties, in which case it no longer springs back. The spring used in many retractable ball point pens to return the clicker button to its normal position is an example of a compression spring, as are springs used in many shock absorbers.

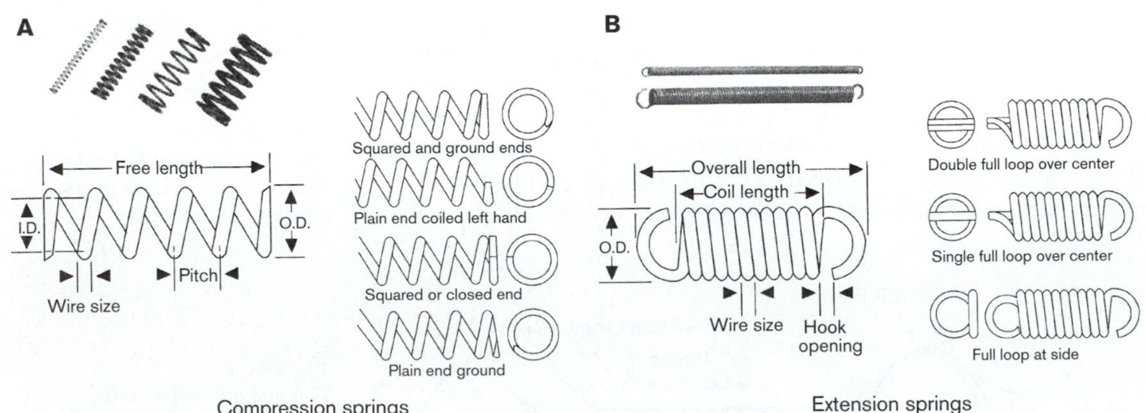

FIGURE 2.45
(a) Compression springs; (b) Extension springs. Compression and extension springs differ in how they are specified and how their ends may be finished. Basic terminology and common end types are illustrated here. (Courtesy of Reid Tool Supply Company.)

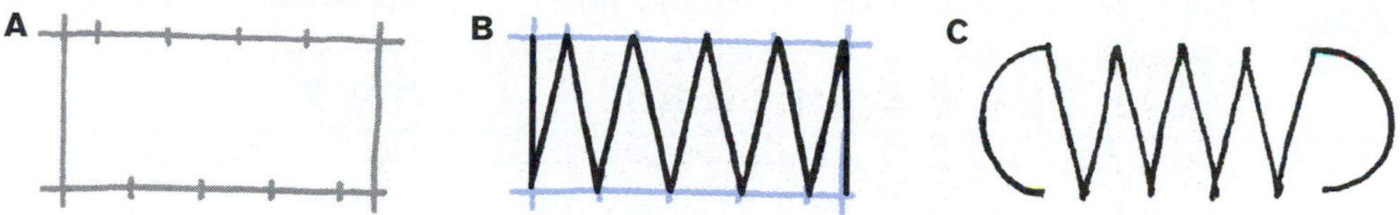

FIGURE 2.46
Sketching a schematic representation: (a) Lightly block in the major diameter of the spring; mark off the pitch distance; (b) Connect the points for the pitch with uniform Vs. Add straight segments at the ends for compression springs or; (c) Add arcs to the ends for extension springs.

Springs usually have a round or square cross-sectional shape but can have other shapes, such as elliptical or rectangular. The thickness of the spring may also be referred to as the wire size; it is the width of the unwound coil. The *major diameter* is the outside diameter of the spring's coil. The *free length* of a spring is the length of the spring when it is not compressed or extended. For an extension spring, you would specify the coil length and the *overall length,* which is the length of the coil plus the hooks at each end used to attach the spring. The *pitch* of the spring is the distance between the coils, just as pitch is the distance between the crests on a threaded item. Like thread, the helical coil of a spring can be wound right- or left-handed. Finally, the ends of the spring can be finished in several ways. Figure 2.45 illustrates several of these options.

Sketching Springs

Springs can be sketched or drawn using schematic or simplified representation similar to the way you sketch thread. Figures 2.46 and 2.47 illustrate methods for sketching schematic and simplified representations.

A

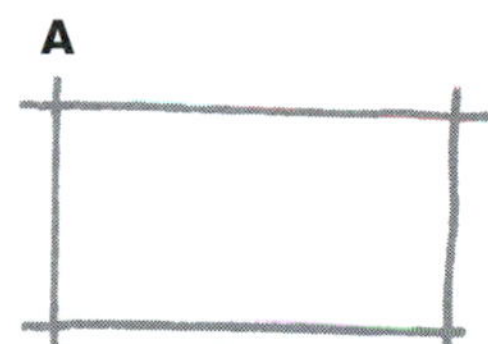

B

SPRING DATA:
MATERIAL: STAINLESS STEEL
WIRE DIAMETER: .08"
FREELENGTH: .75"
OD (OUTSIDE DIAMETER) .42"
ID (INSIDE DIAMETER) .26"
NUMBER OF COILS: 7
TYPE OF ENDS: CLOSED GROUND
MAX LOAD: 37 POUNDS
RATE: 263 LBS/IN.

FIGURE 2.47
Sketching a simplified representation: (a) Lightly block in the rectangular shape of the major diameter of the spring and its free length; (b) Sketch the diagonals through the box; (c) Darken the final lines and provide a note with the spring data.

INDUSTRY CASE

John Houlihan, Timex

Sketching for Ideation: Down Your Arm and Out Your Pencil

The concept for the Timex Turn and Pull® twist ring alarm answered a long-standing need: how to make an analog alarm that would be accurate to the minute. Digital watches popularized the wrist alarm, but making an accurate alarm for an analog watch was more difficult. Existing analog alarm watches—set with a single "hand" to indicate the desired time—vary as much as ten minutes either side of the desired time. The concept for the Turn and Pull® solved this problem by letting the user set the alarm for 8 A.M., say, as she would a clock: by setting the hour indicator to 8 and the minute indicator to 12.

This engineering concept required a completely new watch movement (the inner mechanism of the watch), which can take up to two years to design and perfect. Once it was clear that the movement design was a success, the design of the watch's exterior could begin.

In the case of the Turn and Pull®, the watch's mechanism constrained the design in a key way: to set the alarm, the user would twist rings around the watch face to line up the hour and minute indicators with the desired time. The look of the watch needed to incorporate these rings and to address concerns that the way they were stacked would make them sit up too high above the face of the watch. The design team set out to minimize the depth of the rings and maximize the size of the face as they created a salable look for the watch. Because the depth of the rings and the length of the watch hands were constrained by their functions, the industrial designers worked closely with the engineering team (which was in France and Germany) to be sure they did not compromise the watch's movement.

Quick freehand sketches are what the industrial designers use to start the ideation process. Houlihan has yet to find the software that can handle the intuitive, "idea generating, down your arm, out your pencil, onto a piece of paper" sketching process. "The design process is mostly about thinking and drawing at the same time. It's the interaction between what you're seeing on the page and what you think you want. The images are very illusive; they come and they go. You start sketching, you see what you're sketching and you know that isn't it, so you sketch right over it. It's like a language—fluency means speaking without thinking about it. If you have to think about the sketch, it detracts from your ability to capture ideas."

Once the idea is on paper, however, much of the load of communicating and refining the image is done with the computer. Houlihan estimates that hand sketching makes up about 5% of the industrial design group's work, the rest being done on the computer. A large number of quick sketches were used to determine what the case design for the Turn and Pull® alarm would look like (see

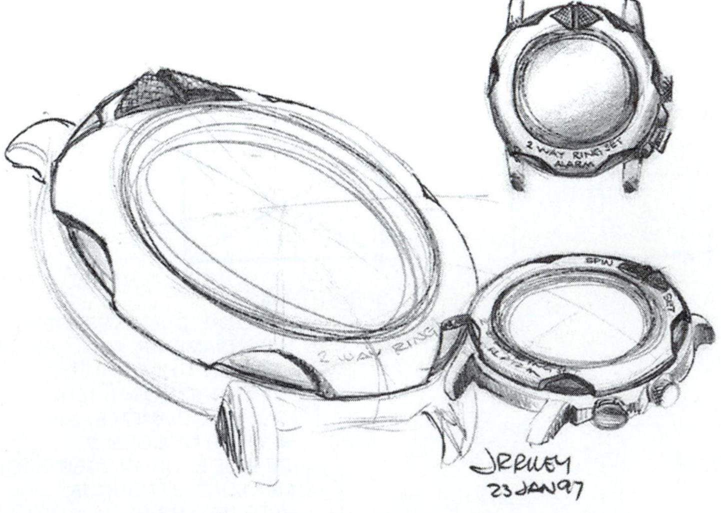

FIGURE 2.48
Hand sketches of several design concepts were evaluated for the Turn and Pull® twist ring alarm.

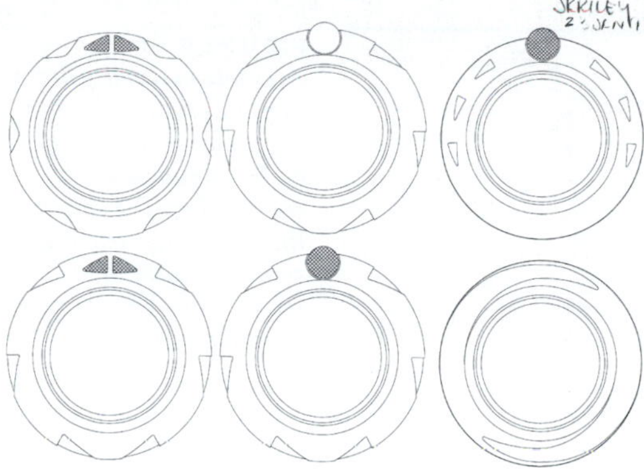

FIGURE 2.49
Concepts for the alarm's design were further visualized using drawings created in Vellum.

Fig. 2.48). Once the designers knew where they wanted to go with it, they began to model the case in Vellum, an intuitive, easy-to-learn CAD package for the Macintosh (the ID department works on Macintosh computers), in which they created multiview 2D drawings of the form (see Fig. 2.49).

In the past, the designer used hand sketches to communicate a vision of the object, not to accurately define it. If he or she adjusted a view to better match his or her concept, the other views might not be adjusted accordingly. When engineering translated the concept into final specifications, it had to resolve the discrepancies in the views. This led to what Houlihan refers to as PMD, or "post-mock-up depression," when the first mock-up of the product was made and the designer could see how the design had been interpreted from the drawings. Today, the industrial design team avoids PMD by using a variety of applications that require more discipline up-front than the pencil and paper approach. The drawings created in Vellum were exported as an IGES file to Unigraphics, the parametric solid modeler used by Timex for all product definitions (see Fig. 2.50). Conversely, they used Cad-Mover (a translation utility for CAD data) to import a 3D IGES file from the Unigraphics system used to model the watch (see Fig. 2.51). With it, the designers had an accurate 3-view drawing of the mechanism that they could design the case around. Houlihan feels the 2D accuracy of Vellum forces the designers to consider relationships that will be important later in the process and helps them learn to think in 3D early on. Viewing the shaded solid model in Unigraphics to check the surfaces and proportional relationships helps to eliminate PMD and multiple mock-ups, too. By the time the finished mock-up is made, most of the mystery regarding its final appearance is gone.

After drawing the case in Vellum, the design team used Cad-Mover again to import the design into Freehand, a 2D illustration application (see Fig. 2.52). In the past, presentation sketches were always done by hand, using magic marker, pencil, or chalk. Today, the designers use Freehand to make a colored and shaded rendering with the highlights and surface textures that create a 3D appearance. This

(continued on p. 76)

FIGURE 2.50
This Unigraphics solid model of the Turn and Pull® alarm's movement was exported as a 3D outline and imported into Vellum where it was used to visualize and evaluate how the case design would interface with the mechanism.

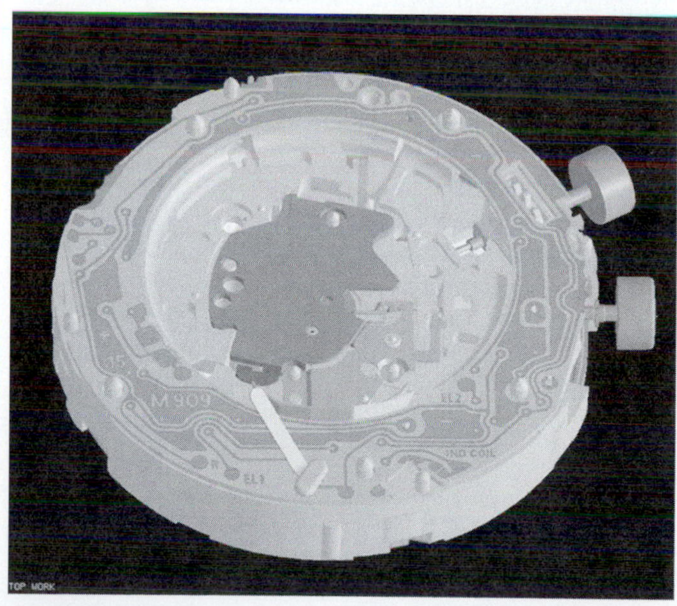

INDUSTRY CASE

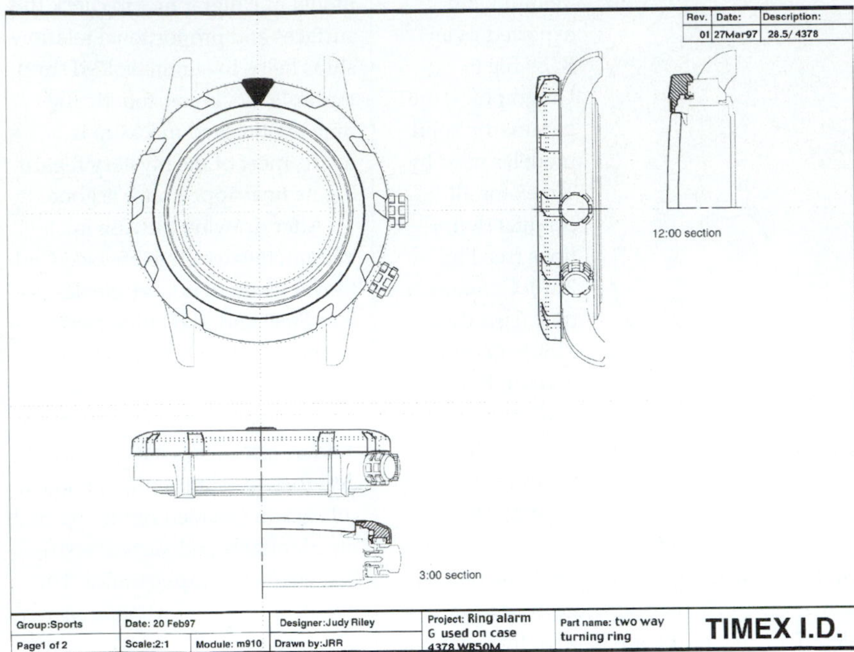

FIGURE 2.51
Accurate multiview drawings are created in Vellum to visualize the case design and communicate it to engineering. The dotted lines in the figure represent the outline of the mechanism exported from the solid model.

rendering is used to communicate the design to others in the product development team. The designers also use Freehand to design the graphics that are part of the product. The Freehand file is used in production for the numbers printed on the dial or the lettering on a digital watch lens.

For projects with complex surfaces, the Vellum files can be used by engineering to translate the case design into Unigraphics. In other cases, engineering builds the 3D solid model from the drawing views derived from the Vellum files. This transition is one that Houlihan would like to see improved. If his designers could work in the same package as the engineers who build the final solid model, they could avoid errors in "interpretation" as the surface of the object is translated into a solid model in another package. The engineers evaluate the solid model to look for gaps between surfaces and what Houlihan calls "conquagulations"—places where surfaces come together in a vortex-like knot that the NC machining software will have difficulty with. But they may not be as sensitive to the flow of the surface and how it reflects light. By using the same

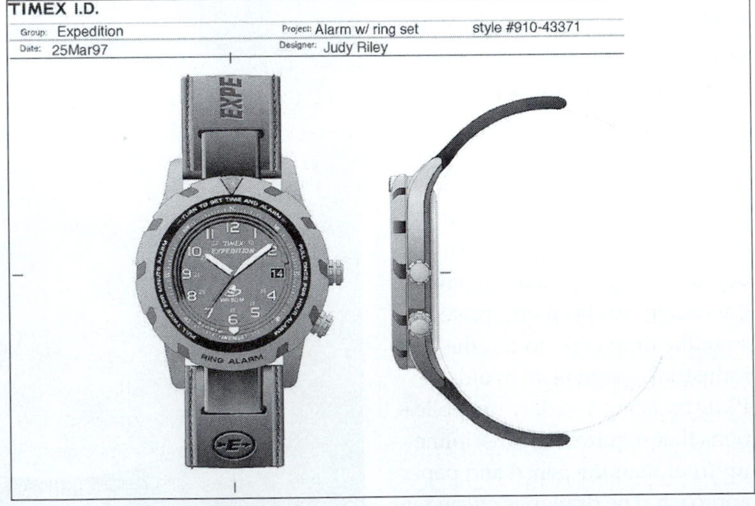

FIGURE 2.52
Views of the case design from Vellum were embellished in Freehand to create this realistic view of the Turn and Pull® alarm.

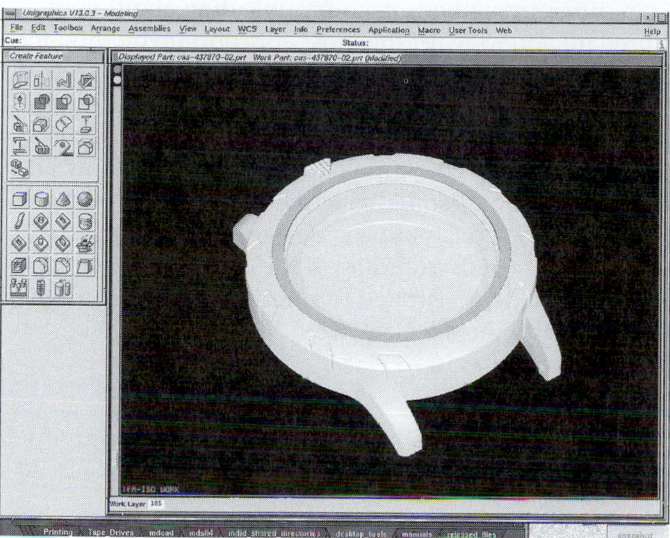

FIGURE 2.53
The Unigraphics solid model of the watch can be used to create rapid prototypes for form studies and to create tooling for part manufacture. The many uses of the model support John Houlihan's contention that the computer has created "the fourth state of matter"—a model with enough definition and information that it can be used as if it were a manufactured part for visualization, testing, and analysis.

modeler, each group would be able to contribute to the same database of information. As it is now, the design teams work together to ensure that changes to the 3D model do not digress too much from the visual appearance planned for the watch.

An added plus would be the built-in rendering capabilities of the solid modeler, which would eliminate the need to go to Freehand for presentation sketches. They could be generated almost automatically from the 3D solid model. For approximately the same time as it takes to make a Vellum model and render it in Freehand, Houlihan feels they could benefit from the immediate what-if capability of a 3D solid modeler.

Having said that, however, Houlihan is quick to reiterate that sketching is the best way to facilitate communicating the visual ideas in a designer's head. Once an idea is captured, however, fluency with a 3D modeler, like sketching fluency, is a powerful language for capturing design ideas (see Fig. 2.53). Houlihan points out how author Marshall McLuhan's maxim "the medium is the message" is also relevant to engineering design. From his own experience, he knows that if you are uncomfortable drawing a shape, or don't have the tools to create a model of a shape, it will affect your ability to represent ideas. For example, when he was required to draw and dimension a surface by hand, he found that he would avoid designing surfaces with amorphous shapes as they would be difficult to define in 2D. With today's modeling tools, amorphous shapes are common to many products because they can be modeled and represented in 3D. Whatever the medium, skill with the technique or the software is the key to being able to communicate the design you have in your head.

All illustrations are used with the permission of Timex Corporation. Thanks to Judy Riley for Figures 2.48, 2.49, 2.51 and 2.52; to Dave Quinlan and Herbert Schwartz for Figure 2.50; and to Dana Rockel for Figure 2.53.

John Houlihan is director of Industrial Design at Timex, Inc., in Middlebury, Connecticut. Houlihan graduated from the University of Notre Dame, began his design career at General Motors, then moved to consumer electronics at General Electric before joining Timex in 1979. He has experienced the transition from hand sketching tools and models made of illustration board and foam-core to the use of computer-based tools in his career and has actively investigated new software tools to better achieve the goal of communicating design. He is currently learning to use Unigraphics, the 3D modeling package used at Timex, to see if the software has become intuitive enough to be used by his team of industrial designers.

Skills Summary

Being skilled in different methods of communicating about your designs will prepare you to select the most appropriate method for a given task. In addition to being able to represent your ideas so that you can communicate about them with others, you also need to be able to interpret information available in drawings and sketches early in the design process. From this chapter, you should now be familiar with one way of conceptualizing the design process, the use of a needs statement, and constraints on designs. Answer Exercises 1–7 to demonstrate your understanding of these topics. Sketching for visualization can be demonstrated through Exercises 8–10. Schematic sketching and threads, fasteners, and springs can be demonstrated by doing Exercises 11–15.

EXERCISES

1. What are four types of factors that constrain design problems? Give examples of each.
2. Respond to this statement from an engineer: "My main concern in designing products is their function. If my boss is worried about money she should talk to the accounting department."
3. Many upper-division engineering design courses emphasize student participation in "open-ended" design problems, meaning there may be more than one correct answer to the problem. How many possible solutions are generated depends on the technique used. Describe idea-generation methods from this chapter that would lead to many alternative solutions.
4. You are a member of a team designing an automatic door-opening device for wheelchair users to enter and exit buildings on campus. Many design constraints exist, including those associated with the abilities and limitations of the users, the physical environment, security and safety concerns, etc. What specific design constraints can you list? Explain five of these constraints, including their origin and their possible impact on your design.
5. Create a survey questionnaire to determine:
   - **a.** what computer options are desirable to engineering students.
   - **b.** what options should be included on a 4 x 4 diesel pick-up truck.
   - **c.** what features are desired in a new snowboard binding design.
   - **d.** what features are desired in a hand-held cordless mixer.
6. Create a decision matrix comparing:
   - **a.** 3 brands of mechanical pencils.
   - **b.** 3 brands of cereal.

c. 3 brands of computers.

d. 3 brands of calculators.

7. Use Web search tools to find the names and contact information for three companies that manufacture ball-type universal joints. Record the information you find in your notebook. Write a letter to one of them requesting specifications and pricing.

8. Create a simple, proportionally accurate, single-view sketch of a bicycle, from memory. Now look carefully at a bike, noticing frame geometry, crank length, wheel size. Repeat the sketch, and compare both with the actual bike. How did you do?

9. Imagine that your roommate has asked you to design a car to protect the teenager in the household who is learning to drive. You have probably seen the Nerf® foam balls that you can throw in the house without breaking things. The car you design will be entirely made of or coated with a similar material. It must protect the teenager and others from accidents. What will it look like? Will your car start from scratch or be remodeled from the 1976 Dodge Dart that is already sitting in the driveway? What color will it be? Other accessories?

10. Create contour sketches of the pictured objects.

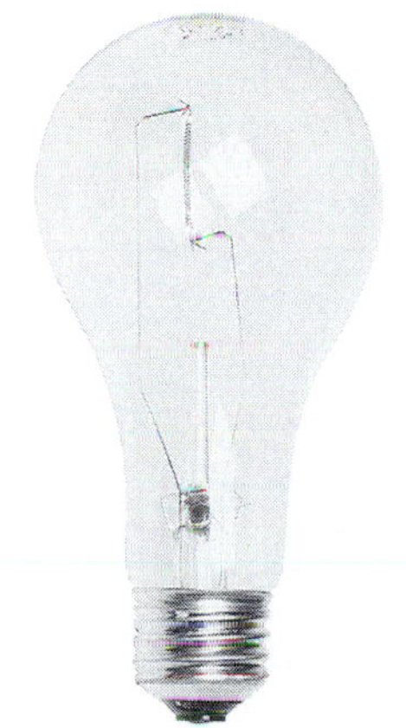

a. lightbulb

b. electrical outlet

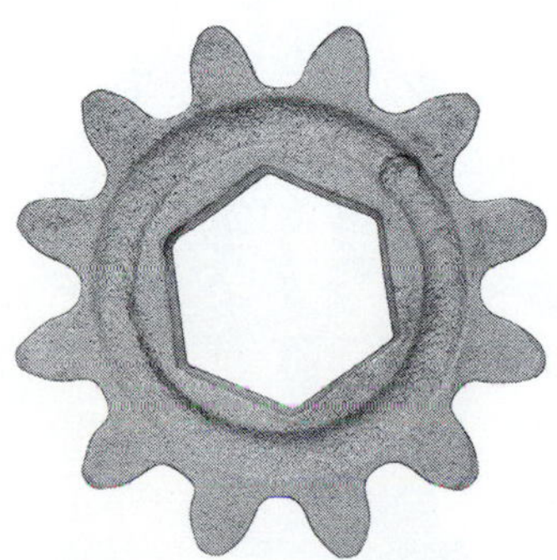

c. gear

d. wagon

e. hex bolt and screw

f. telephone

g. cup and saucer

h. suitcase

i. hat

j. baseball

k. chair

l. birdhouse

11. Sketch a block diagram for the design of a children's wagon. Divide the design to the furthest part level for your understanding of the components.
12. Sketch a block diagram showing the relationships among your family members for the last three generations.
13. Sketch a schematic or simplified representation for the fasteners listed below:
    - a. .50 – 13 UNC
    - b. .375 – 16 UNC
    - c. M10 x 1.5
14. Sketch a schematic representation of
    - a. an extension spring.
    - b. a compression spring.
15. Sketch a simplified representation of a .375" diameter extension spring with a free length of 1.25".

DESIGNER'S NOTEBOOK

In designing the lateral motion gripper, Albert Brown used a pictorial (or 3D) sketch to illustrate how the parts of the gripper would appear when assembled. This page from his notebook includes notes about the assembly and how the different parts of the mechanism will interact. (Courtesy of Albert W. Brown, Jr., Affymax Research Institute.)

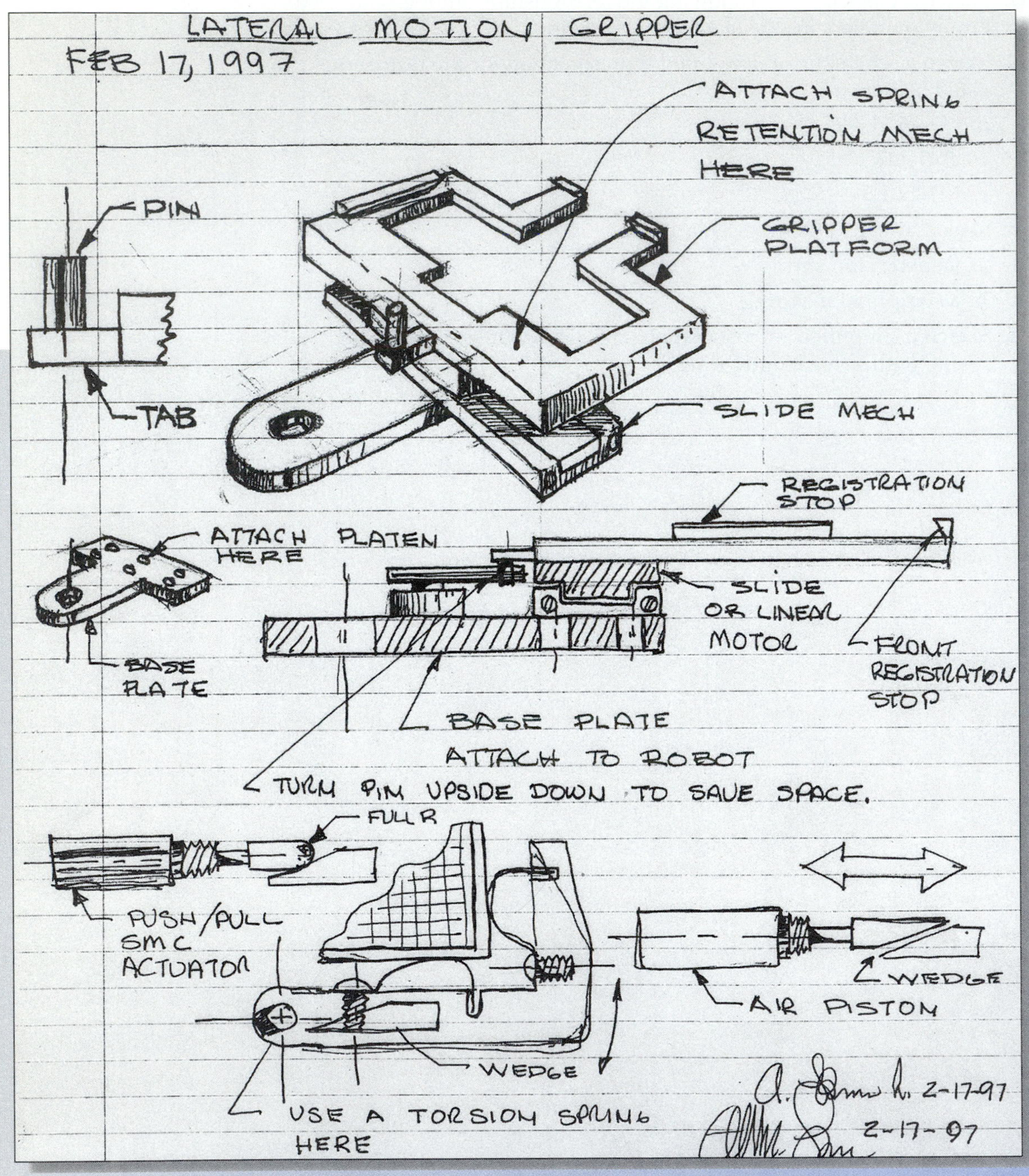

CHAPTER 3

Pictorial Sketching for Design Visualization

Objectives

When you have completed this chapter you will be able to

- create freehand isometric pictorial sketches;
- create freehand oblique pictorial sketches;
- create freehand one-point perspective sketches for ideation;
- create an isometric sketch showing multiple parts in assembly;
- add neatly lettered notes and labels to a sketch.

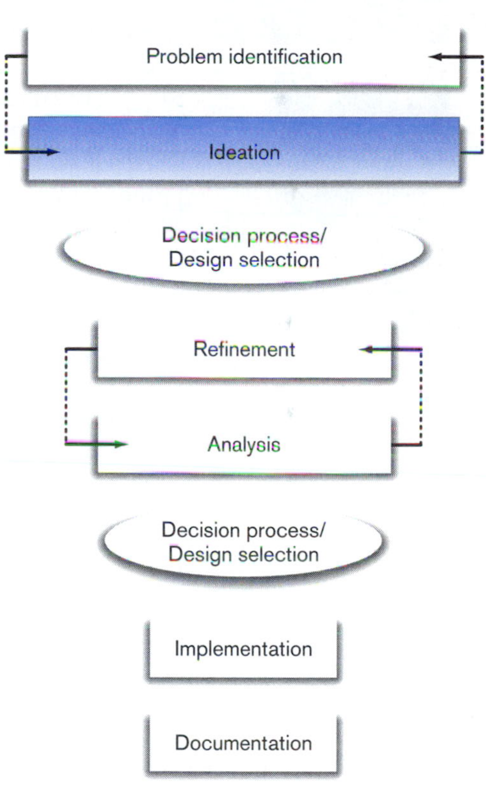

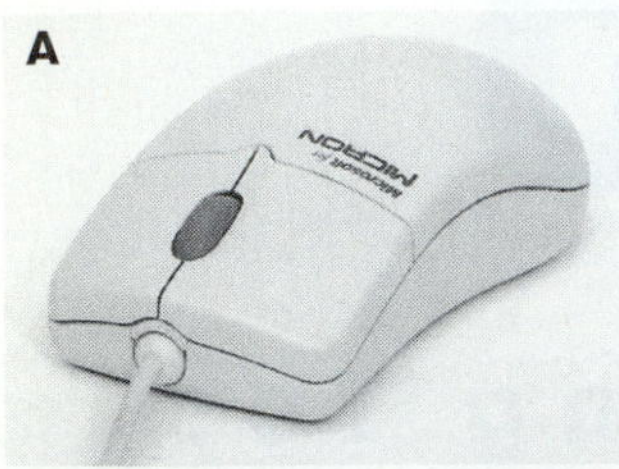

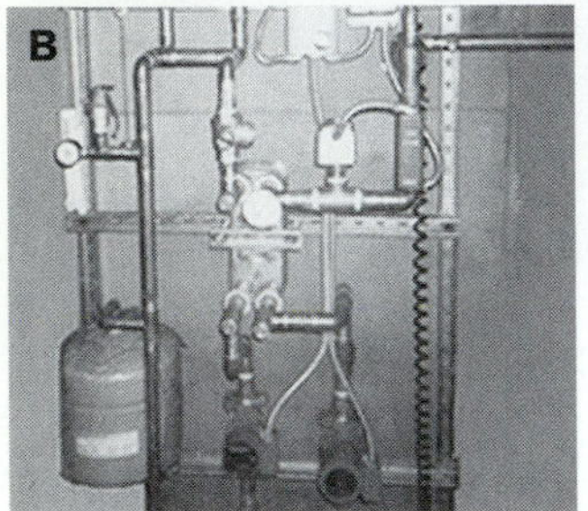

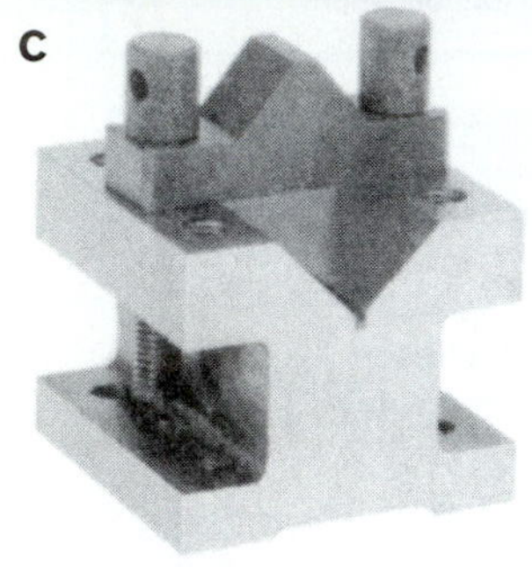

FIGURE 3.1
(a) An ergonomic mouse; (b) A piping system (Courtesy of Radiant Technology); (c) A V-block

The ability to sketch your ideas is key to your ability to capture and evaluate many different design options during the ideation phase, but it is also a skill that you will use throughout the design process to communicate quickly and effectively.

In the previous chapter, you learned how to "see" objects as they actually appear so that you can better capture their appearance on paper. The sketches you created captured all three dimensions of the object as they appeared from a particular point of view. For any given object, an unlimited number of views could be generated that would all be accurate representations of the object as "seen." Some views present a more complete view of the object than others—and some convey the size and shape of the object better than others. Others might be difficult to interpret and provide little information to the reader of the sketch. The "best" sketch would depend on your goals for the drawing in a particular context. As you are developing your sketching skills, keep in mind that sketches are used to communicate your design ideas to others. They should be neatly drawn and clearly show the information you are trying to convey.

For concept sketches, it is important to be able to quickly sketch the object so that it shows the relationship between its parts. A variety of sketching techniques have been developed that use a consistent viewpoint and follow rules for transferring 3D shapes to 2D paper so that the resulting sketches are easy to create and interpret.

For example, in designing an ergonomic mouse, such as the one shown in Figure 3.1(a), you may want to convey the appearance of the exterior of the mouse in a sketch such as that shown in Figure 3.2(a).

When designing a piping system, such as that in Figure 3.1(b), you might create a sketch that shows the relationship and orientation of the piping, valves, and tanks, such as you see in Figure 3.2(b). For a mechanical part like the V-block shown in Figure 3.1(c), the oblique sketch shown in Figure 3.2(c) may be the best means of conveying the design information quickly.

Each of these types of engineering sketches can be useful tools in the process of engineering a system or product. In this chapter, you will learn how to create the most common kinds of pictorial sketches used in engineering.

FIGURE 3.2
(a) A freehand sketch of the ergonomic mouse in Figure 3.1; (b) An isometric sketch of the piping system; (c) An oblique sketch of the V-block.

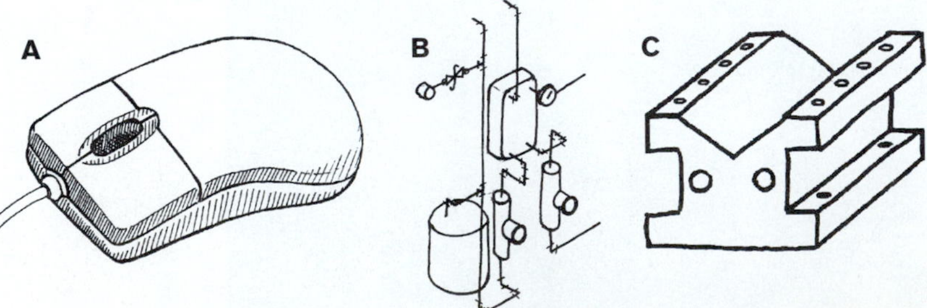

Geometry of Solid Objects

Sketching in engineering is largely used to communicate or record ideas about the shapes of 3D objects. Before you can create accurate engineering sketches, you need to understand just what it is that you are trying to show in your drawings. What is represented in drawing a 3D object?

Solid objects are bounded by the *surfaces* that contain them. These surfaces can be *planar (flat surfaces), single-curved* surfaces (a cylinder or a cone, for example), *double-curved* surfaces (such as a sphere, torus, or ellipsoid), or *warped* surfaces (like the flowing curved surfaces that you would see on a snowmobile hood, car, or other object). Figure 3.3 shows some common solid shapes and identifies their surfaces.

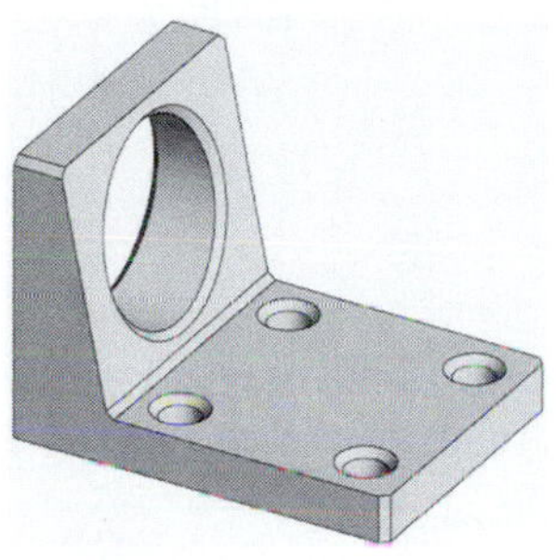

FIGURE 3.4
The inner edges of the holes in this bracket are curved surfaces, as are the bevel cuts (chamfers) on the edges of the holes. All other surfaces are planar, including the chamfers on the edges of the part.

Many common solids are bounded by a combination of only plane surfaces and single-curved surfaces. For example, the bracket shown in Figure 3.4 is commonly found in mechanical engineering. All of its features are either made up of plane or single-curved surfaces.

The intersection of different surfaces on the object results in an *edge*. Figure 3.5 shows some typical polyhedra (solids that have only plane surfaces as their boundaries) with the edges highlighted. Notice that all of the edges on a polyhedron are straight lines, because they represent the intersection of two planar surfaces, but edges can also be curved.

The term *vertex* is used to describe the location where three or more plane surfaces intersect to form a point. A *face* is the term for a planar surface on the object bounded by edges where it intersects other surfaces. You can often see the different faces on an object clearly by the way they are lighted. Figure 3.6 shows a shaded solid, first with its faces and then its vertices labeled. (Notice the different light and dark shading for the faces.)

When you sketch solid objects, you will not sketch the surfaces of the object but the edges where they intersect other surfaces. Each edge begins and ends with a vertex, where three or more plane surfaces come together. Locating the vertices where edges start and

FIGURE 3.5
A series of regular polyhedra

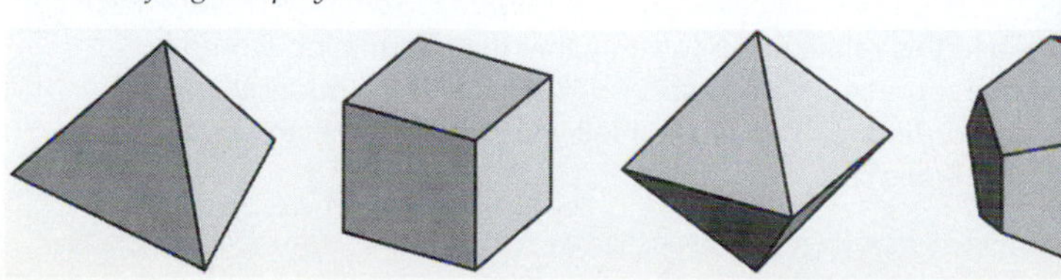

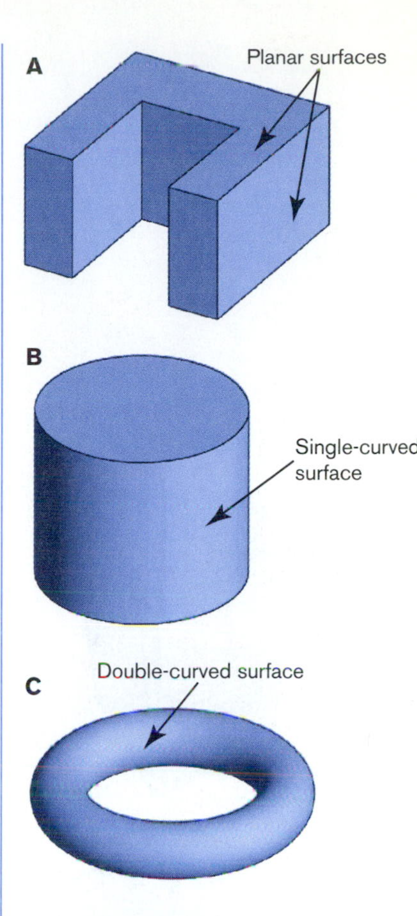

FIGURE 3.3
The objects shown here contain all of the basic kinds of surfaces. The U-shaped block in (a) is comprised of planar surfaces only. The cylinder in (b) has two planar surfaces (the circular shapes of its top and bottom) and a single-curved surface forming the rounded portion. The torus in (c) has just one surface—a double-curved surface. The irregular shape in (d) contains a warped surface. A mesh of lines has been overlain on the surface to make it easier to identify the irregular shape. Notice that the edge of this surface is not a regular geometric shape such as a line or arc.

FIGURE 3.6
Faces and vertices

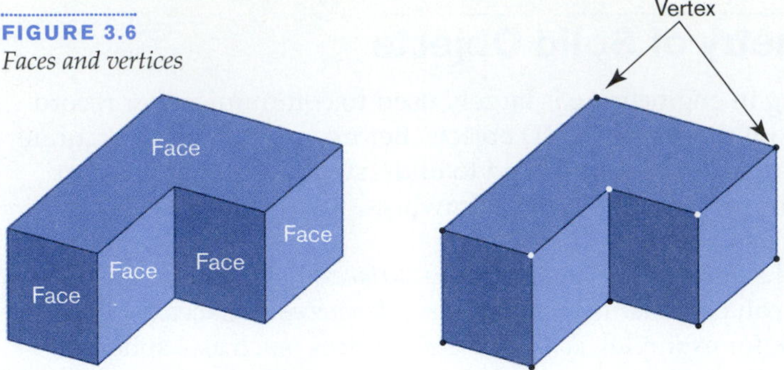

stop will help you sketch edges accurately in 3D space. In the rest of this chapter, you will learn to create sketches by drawing these features of solid objects.

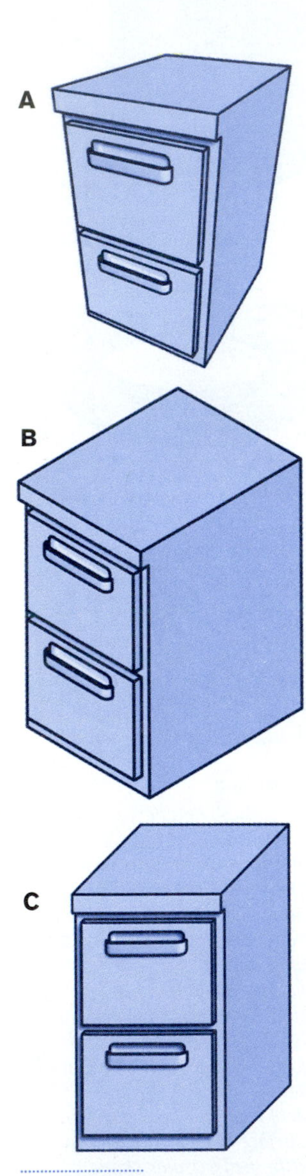

FIGURE 3.7
(a) Perspective; (b) Isometric; (c) Oblique

Pictorial Sketching

A *pictorial sketch* represents a 3D object on a sheet of 2D paper by orienting the object so you can see its width, height, and depth in a single view.

Pictorial sketches are used frequently during the ideation phase of engineering design to quickly record ideas and communicate them to others. Their similarity to how the object is viewed in the world around us makes them useful for communicating engineering designs to non-engineers. Later in the design process, pictorial drawings are also often used to show how parts fit together in an assembly and in part catalogs and manuals to make it easy to identify the objects.

There are three common methods used to sketch pictorials: isometric sketching, oblique sketching, and perspective sketching. Figure 3.7 shows perspective, isometric, and oblique sketches of a filing cabinet.

Each of the pictorial methods differs in the way points on the object are located on the 2D viewing plane (the piece of paper) as shown in Figure 3.8.

Euler's Formula

For all polyhedra, there is a regular relationship between the number of faces, edges, and vertices. The number of vertices (V) minus the number of edges (E) plus the number of faces (F) equals 2. This relationship, $V - E + F = 2$, is called *Euler's Formula*. Some CAD software packages use this relationship to verify that the information stored in the database represents a possible solid object. Think about this relationship for polyhedra. What does it tell you about sketching edges, vertices, and faces?

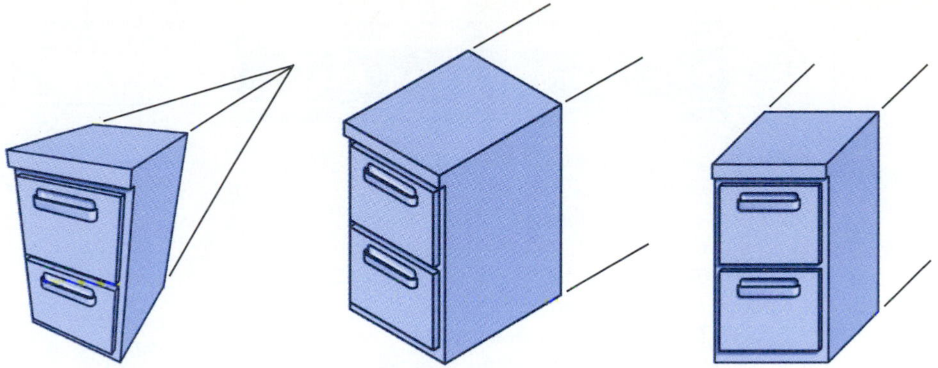

FIGURE 3.8
The file cabinet illustration with the projectors used to create the 3D view in two dimensions.

A perspective sketch presents the most realistic looking view. It shows the object much as it would appear in a photograph—portions of the object that are farther from the viewer appear smaller, and lines recede into the distance.

An isometric sketch is drawn so that lines do not recede into the distance, but remain parallel. This makes isometric views easy to sketch but takes away somewhat from the realistic appearance.

An oblique sketch shows the front surface of the object looking straight on and is easy to create, but it presents the least realistic representation, as the depth of the object appears to be out of proportion.

You will learn how to create pictorial freehand sketches using each of these methods.

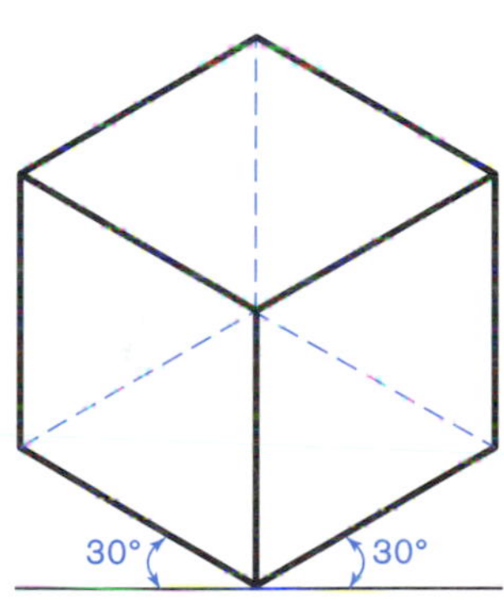

FIGURE 3.9
Isometric view of a cube

Isometric Sketches

An isometric sketch shows the object rotated and tipped so that you are looking onto a corner where the major axis directions—width, height, and depth—meet. Imagine a cube rotated 45°, so that you are looking onto one of its edges, and tipped toward you so that you are looking through its body diagonal, as shown in Figure 3.9. In this isometric view, the bottom edges of the cube are angled at 30° from horizontal.

If you think of the edges of this cube as defining a Cartesian coordinate system, the isometric view of the cube defines the isometric coordinate system used to create isometric sketches. Figure 3.10 shows an isometric pictorial view of a set of Cartesian coordinate axes. In the isometric view, the angles between the axes all appear as 120°, but they represent the actual 90° angles between the axes in the Cartesian system.

Because the top, front, and side edges of the cube are all *foreshortened* (or shown smaller than their actual size) the same amount to produce an isometric view, you can make measurements along these axis lines.

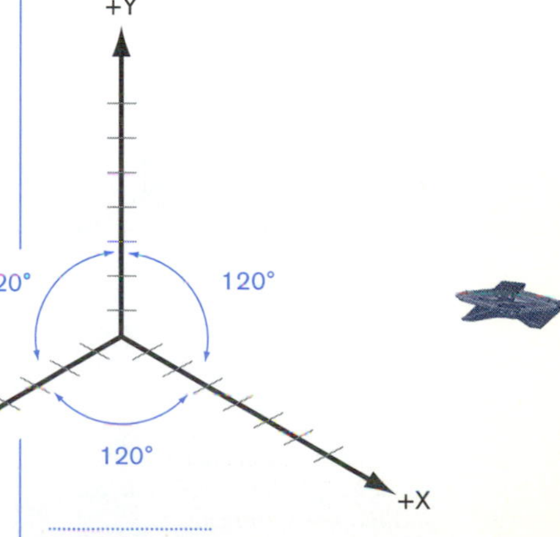

FIGURE 3.10
A set of isometric axes

The True Size of an Angle

Angles do not always appear true size in a view. Sometimes the angle will appear larger than its actual size. Try this: Take a 45°, 45°, 90° triangle and hold it so that the long edge is along the bottom and parallel to your view. Then tip the triangle away from yourself. The 90° angle on the triangle will appear larger than its actual size. Now hold the triangle so that one of the short legs is horizontal and parallel to your view. Tip the triangle away from yourself. Notice that in this orientation the 90° angle always appears as 90°. This is because a 90° angle always appears true size whenever one of its legs appears true length (parallel to your view).

Unlike angles, edges and surfaces never appear larger than their actual size in a view. They appear only true size or shorter than their actual size (foreshortened). To visualize this, take a book and tip it away from you. You will find that there is no direction you can tip it so that its edges appear larger than their actual size.

45°
45°
Angle doesn't appear 90° when neither leg is seen true length.
45°
45°
45°
45°
45°
45°
Angle appears as 90° even though triangle is tipped away from view because bottom leg remains true length.

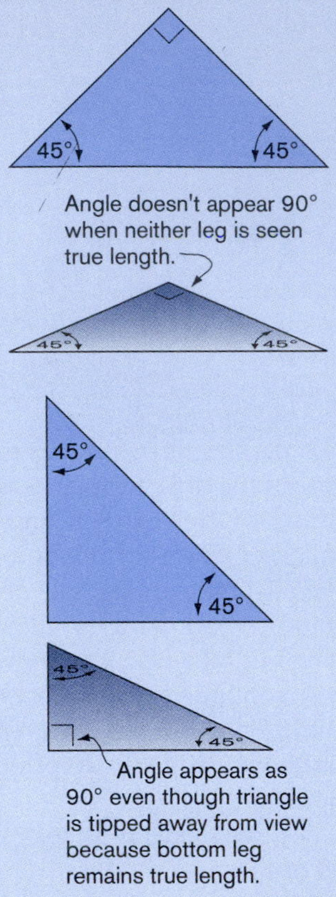

FIGURE 3.11

In fact, isometric means *equal measure*. In an isometric sketch, lines that are parallel to the axes (called *isometric lines*) are foreshortened equally along the isometric axis lines. An example is shown in Figure 3.12. Because each measurement along the axis lines is equally foreshortened, a unit of length is the same length on each of the isometric axes. Because each measurement on the object would be foreshortened equally, isometric sketches are generally drawn without depicting this foreshortening. The resulting sketch is larger than the isometric view that would be produced using CAD or by projection onto a plane, but it is proportionally correct and easier to draw.

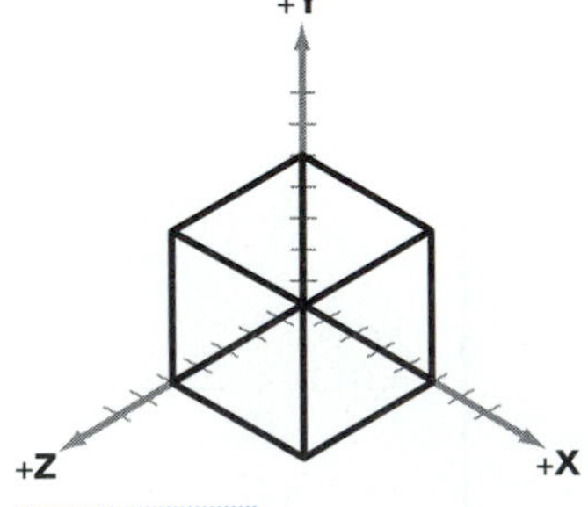

FIGURE 3.12
Each edge of the cube shown here is parallel to one of the isometric axes, so they all are equally foreshortened. Equal-length edges along each axis are depicted as equal length in the isometric view, in this case 5 units.

To sketch an isometric view, you plot the Cartesian coordinates of points on the object on the isometric axes. Typically the X axis on the Cartesian coordinate system is used to represent the horizontal location of an object relative to the *origin* or 0,0,0 point; the width of the part can be plotted along the X axis. The Y axis of the Cartesian coordinate system is used to represent vertical measurements relative to the origin of the coordinate system. In your isometric sketch, use the Y axis to represent the height of the object. The Z axis of the Cartesian coordinate system is perpendicular to the X-Y plane and can be used to represent the depth of the part. The choice of which dimension of the object to plot along which axis line produces different isometric orientations for the part. Overall you should choose to orient the part so that it shows the object clearly. Likewise it does not really matter which axis—X, Y, or Z—is labeled with which orientation as long as you are consistent within the sketch. You will learn more about viewing coordinate systems in different orientations in Chapter 5.

If you know the dimensions of the object, you can plot the width, height, and depth on the isometric axes, as shown in Figure 3.13.

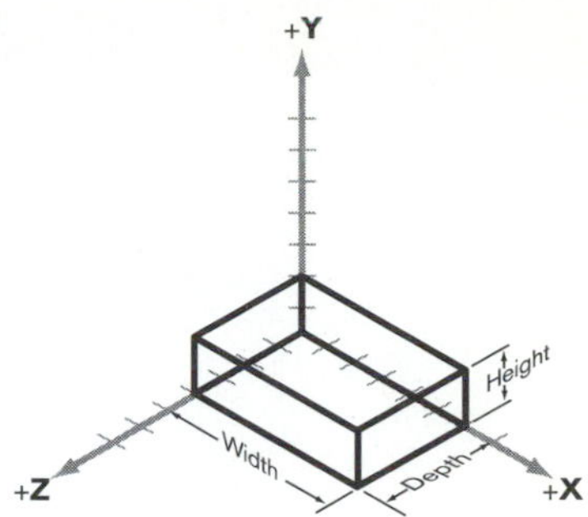

FIGURE 3.13
The block shown here is 6 units wide, 2 units high, and 4 units deep. Parallel edges on the block are represented by parallel lines in the sketch.

Isometric Grid Paper

Isometric grid paper, shown in Figure 3.14, makes it easy to plot points on the isometric axes and sketch isometric views. An isometric grid is composed of lines that are angled at 30° from horizontal in two directions. You can think of these lines as representing two of the axis directions (or the bottom edges of the cube). Vertical lines through points where the 30° lines intersect represent the vertical axis (or the vertical edges of the cube). Each isometric grid unit on the paper corresponds to one unit on the object you are drawing, making it easy to locate points and keep your drawing in proportion.

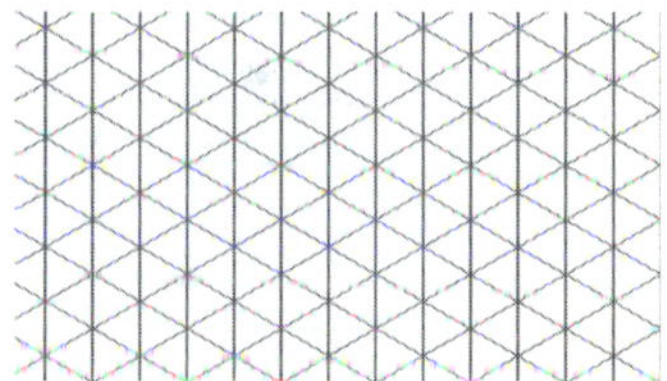

FIGURE 3.14
Isometric grid paper is available with various size grid spacing, such as four, five, or eight divisions per inch, or with separations of 5 or 10 mm between the lines.

Orientations of Isometric Views

The same object can be sketched with different orientations in isometric views (depending on which diagonal of the box you are looking at). Figure 3.15 shows an object sketched in all of the possible isometric orientations. Which one shows the object best?

Select an orientation for your isometric sketch that shows the object clearly so that the person reading the drawing can look at the surfaces shown to infer what the back corners of the object must be like. If the object is complex, you may need to show it in more than one orientation or give a note explaining the depth of a hole, for example. In Chapter 4 you will learn to use a line pattern to represent features that are hidden from view. (In general you do not show hidden features in isometric sketches.)

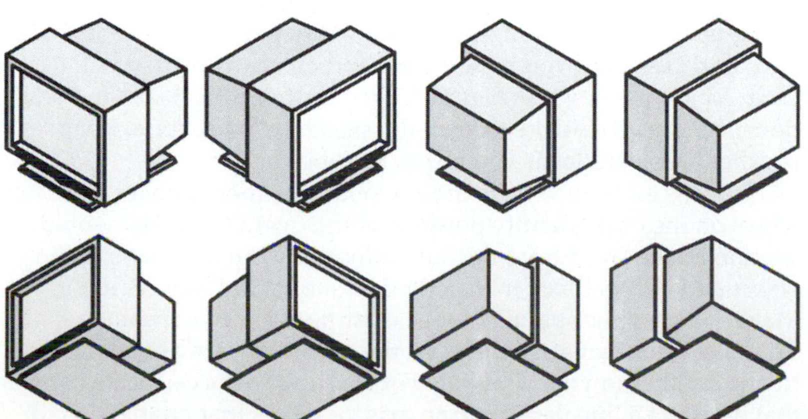

FIGURE 3.15
All of the views shown here are different possible isometric orientations of a monitor. Which one shows the monitor most clearly? Plan your isometric sketches to produce an orientation that shows the object clearly.

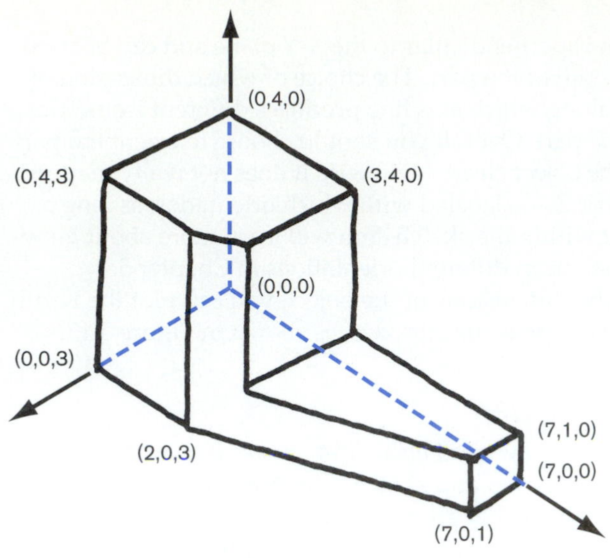

FIGURE 3.16
These overall dimensions will help you locate additional features on the sketch by identifying the location of all of the outside edges.

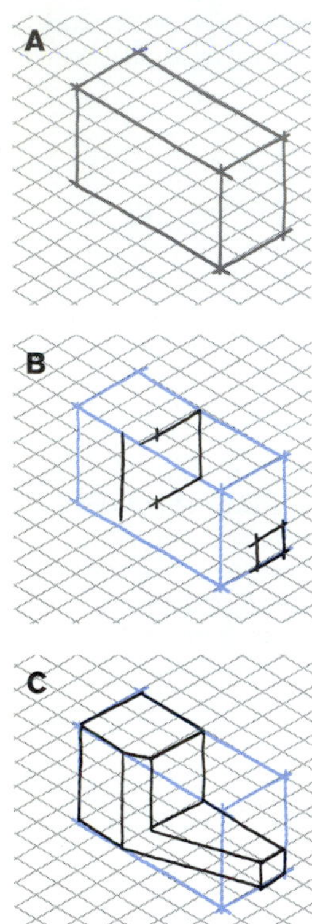

FIGURE 3.17
(a) Grid paper makes it easy to block in the overall dimensions of the object. (b) The vertices along the edges of the object can be plotted on the isometric axes as they would be on Cartesian coordinates. With the origin at a corner of the part, distance along an edge is equal to units on the axes.

Creating an Isometric Sketch Using Isometric Grid Paper

Start your isometric sketch by selecting the orientation for your view. Figure 3.16 shows an isometric sketch with the vertices labeled with their X, Y, and Z coordinates. Decide which isometric axis will show the width, which will show the height, which will show the depth. Then block in the overall dimensions of the object to form a box that would fit the object you will draw, as shown in Figure 3.17(a). Keep in mind the proportions. If the height is twice the depth, make it appear so in your drawing.

Next determine the measurements for features in your sketch. Using isometric grid paper, you can plot each corner of the object using its coordinates in the X, Y, and Z directions, and connect the vertices to show the object's edges as shown in Figure 3.17.

Sketching Isometrics Without Grid Paper

To create an isometric sketch when you are not using grid paper, you need to create a set of axes to use as a guide. Sketch light lines at 30° from the horizontal, then add a vertical line through their intersection to establish the direction and orientation for the part as shown in Figure 3.18(a). Block in the overall width, height, and depth of the part along the lines you have drawn and create a box that will fit the entire object as shown in Figure 3.18(b). Estimate the sizes and relationships of each feature to keep the sketch in proportion. Keep parallel lines parallel as you box in the object.

Use the edges you have drawn to locate other surfaces and edges on the part. Identify portions of the overall box that would need to be removed from the outer block to form the shape of the object and lightly block in the edges as shown in Figure 3.18(c). Make the new lines parallel to the existing lines. In an isometric drawing, lines that are parallel on the object will always be parallel to one another in your drawing. For slanting surfaces, locate corners that intersect a line parallel to an axis, or edges that are parallel to

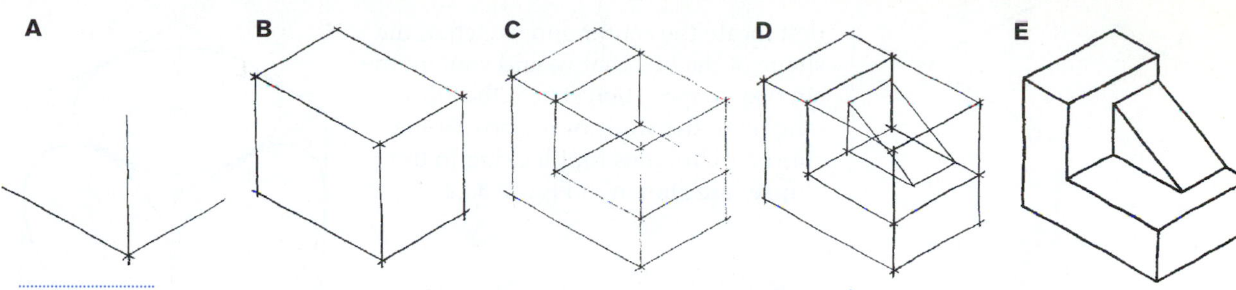

FIGURE 3.18
Creating an isometric sketch: (a) Lightly sketch a set of isometric axes on which to locate edges of the object. (b) Sketch the overall width, height, and depth of the part. (c) Use the lines drawn to plot the location of other surfaces and edges in the drawing. (d) For lines that are not parallel to the axes, locate endpoints in relation to the axes and connect the points to define the surface. (e) Darken the object lines to complete the sketch.

the existing lines. Then sketch the slanting lines between the points you have located, as shown in Figure 3.18(d).

Nonisometric Lines

In a full-size isometric sketch, lines that are parallel to the isometric axis lines are sketched to represent their actual lengths on the object (or proportionately smaller if the object is too large to fit on the paper full size). Lines that are not in the same direction as the isometric axis lines will appear other than their actual lengths because they are not foreshortened equally. In order to draw these *nonisometric* lines, locate their endpoints by measuring along isometric lines as shown in Figure 3.19.

Isometric Ellipses

As a circular shape is tipped away from you (as in an isometric view), it appears elliptical. Imagine looking straight onto a circular shape, such as the top of the can shown in Figure 3.20. As the can is tipped away from your view, it appears as a flattened, elliptical shape. For this reason, circular shapes will appear elliptical in an isometric sketch. These ellipses appear differently depending on which isometric plane they are drawn in—top, front, or side—as shown in Figure 3.21. To draw an isometric ellipse representing a circle,

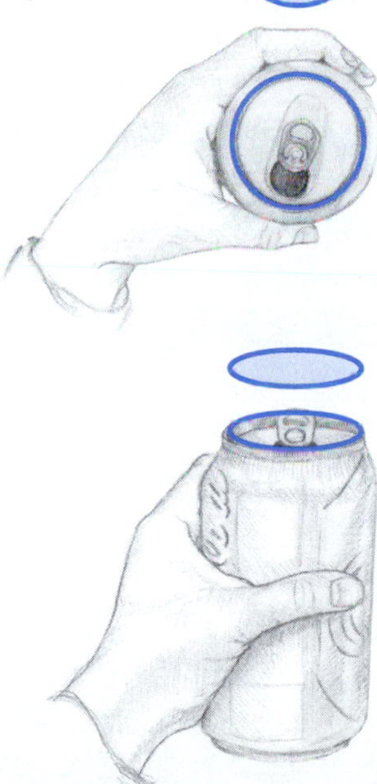

FIGURE 3.20
Starting from an end view of a can and tipping it away from you is a good illustration of how circular shapes are foreshortened into ellipses in an isometric sketch.

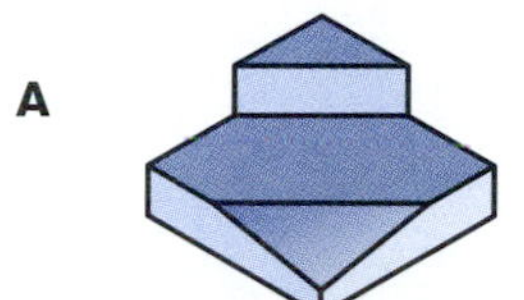

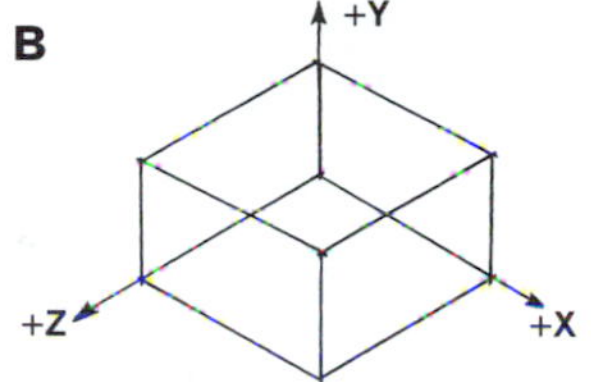

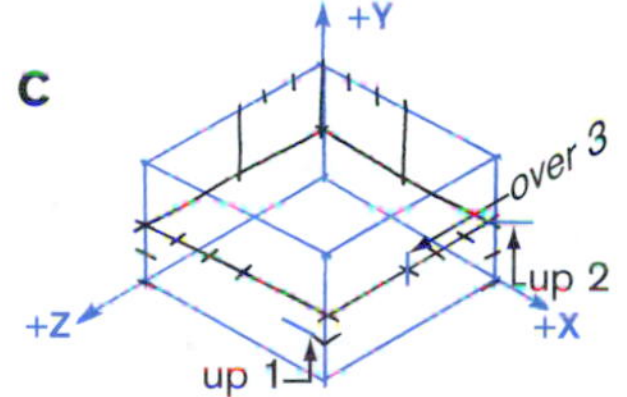

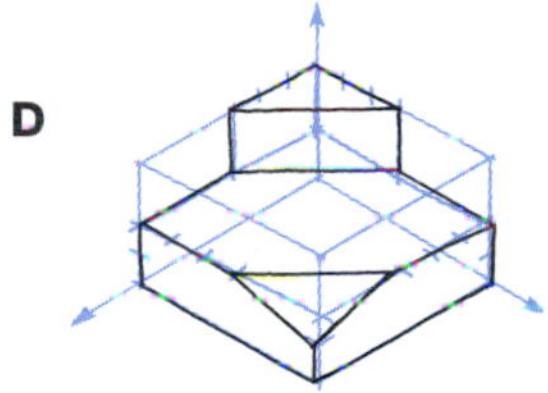

FIGURE 3.19
Sketching an oblique surface

(a) The surface on the front corner is oblique, bounded by three non-isometric lines.

(b) First block in the overall dimensions of the object on the isometric axes.

(c) Use the grid paper or tick marks on the axes to locate the object's vertices along the axis lines. Locate the vertices for the oblique surface by measuring along the isometric lines on the object.

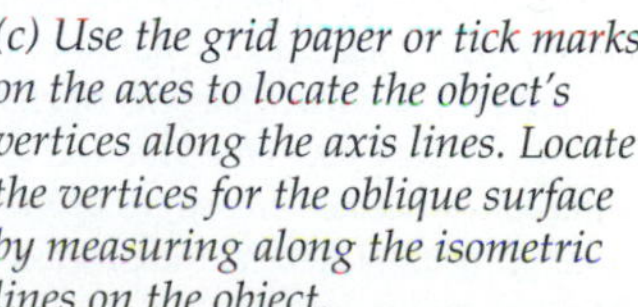

(d) Connect the vertices to define the edges of the object.

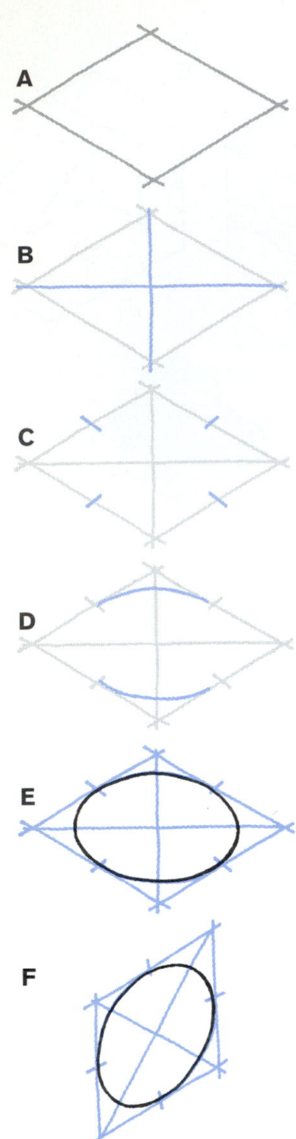

FIGURE 3.22
Sketching an isometric ellipse: (a) Locate the center of the ellipse (the circle's centerpoint) in relation to other features in the sketch, then block in the overall dimensions of the ellipse with lines parallel to the isometric axes for the view. The right-side view is shown here. (b) Identify the center of the ellipse with construction lines as shown. (c) Locate the midpoints of the edges; these locate as the major and minor axes of the ellipse. (d) Sketch arcs from each pair of midpoints as shown. (e) Darken the edges of the final ellipse. (f) Repeat the process in other views as needed. An ellipse in the right plane would appear as shown here.

first locate the center and sketch in the shape of the box that would contain the circular shape. Then sketch the ellipse tangent to the walls of the box. Some simple techniques for blocking in the ellipses are shown in Figure 3.22.

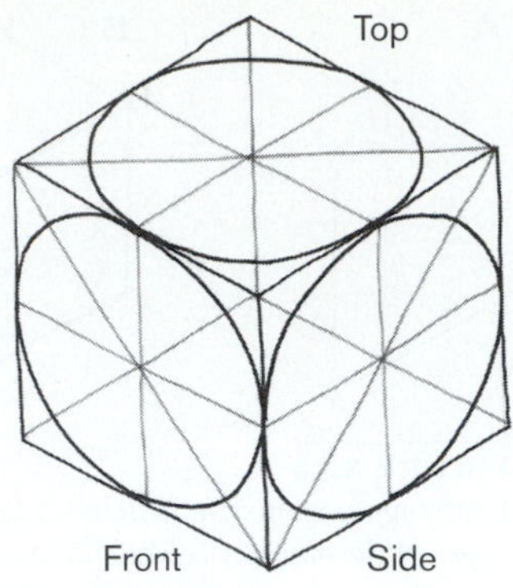

FIGURE 3.21
Circles appear as ellipses in all three isometric planes.

Creating a Scale Isometric Piping Layout

Isometric sketches are often used to show the arrangement of pipes, pumps, and valves in process and flow systems, such as the hydronic snow melting system shown in Figure 3.23. The photograph shows the system components that will be represented in the isometric sketch. The components are represented schematically to make them easy to sketch quickly, but the distances are often (but not always) represented to scale to provide the information needed to lay out the equipment.

To create the isometric piping layout, begin by using grid paper and determining the approximate scale you will use. It will be useful during the process of designing this system to see how the

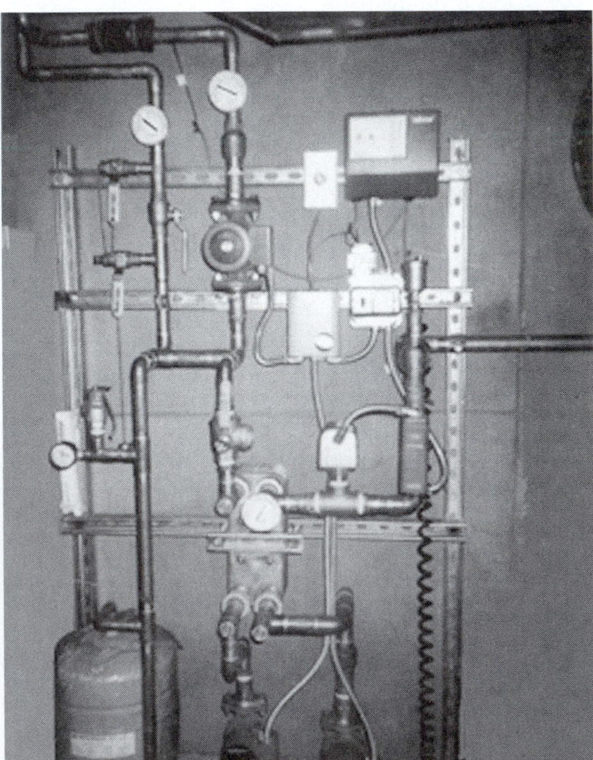

FIGURE 3.23
This hydronic snow melting system was designed for the Virginia City Courthouse in Virginia City, Montana, where heated antifreeze is pumped through tubing laid in the concrete steps. Slips and falls on icy steps prompted the city managers to put in this system, which is projected to pay for itself in two years through reduced snow removal and lawsuit costs. (Courtesy of Radiant Engineering, Bozeman, Montana.)

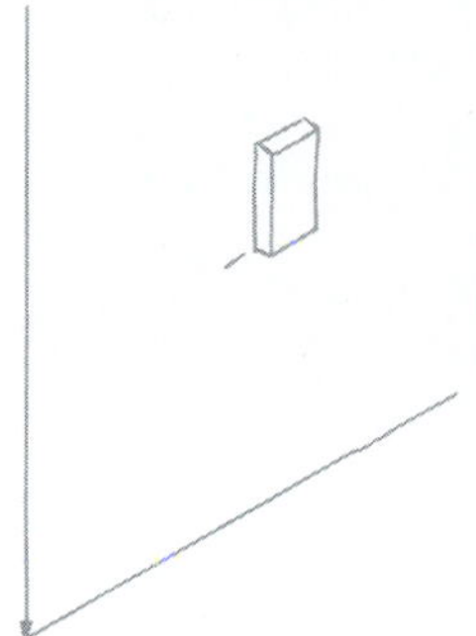

FIGURE 3.24
The overall dimension of the height of the back wall (8 feet) and the height of the central heat exchanger (30 inches) are drawn parallel to the Y isometric axis; the bottom of the wall serves as the Z axis for the sketch. Then the overall dimensions of the heat exchanger are sketched.

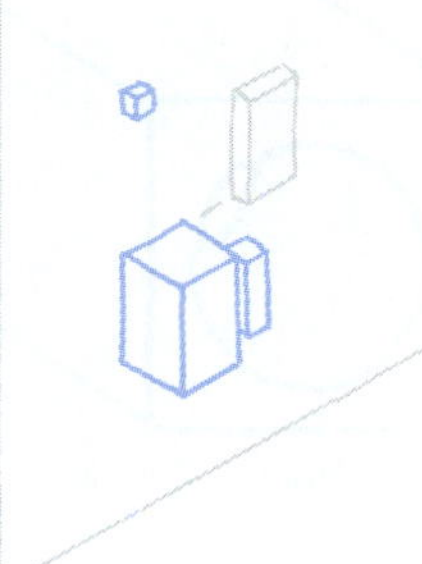

FIGURE 3.25
Other parts of the system are sketched in proportion and relative to the wall and central heat exchanger.

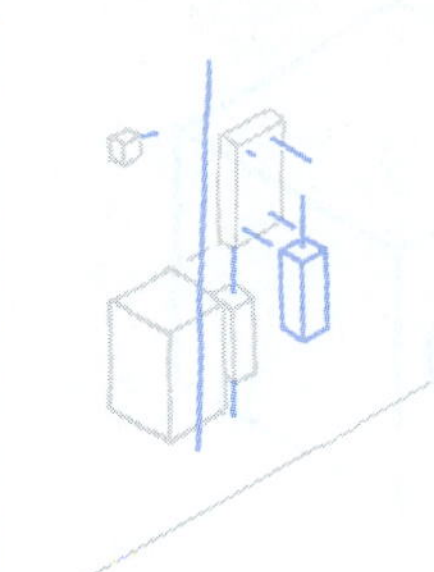

FIGURE 3.26
Straight line segments represent the centers of the pipes coming off the components. The isometric grid serves as a reference so that the segments are drawn to the correct length.

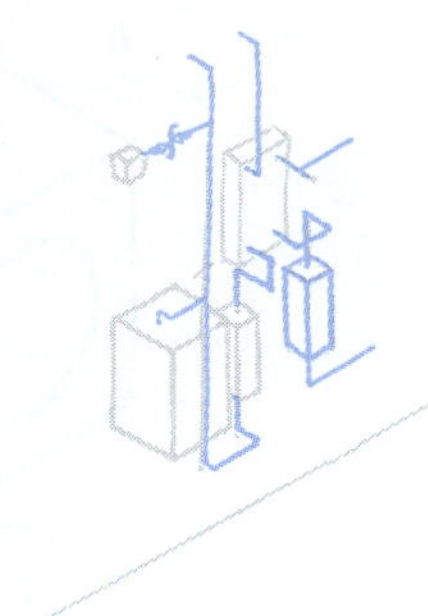

FIGURE 3.27
Additional segments of pipe are drawn as centerlines. The valve in the sketch is represented by a standard piping symbol. Notice that the vertical edges of the symbol align with the Y axis.

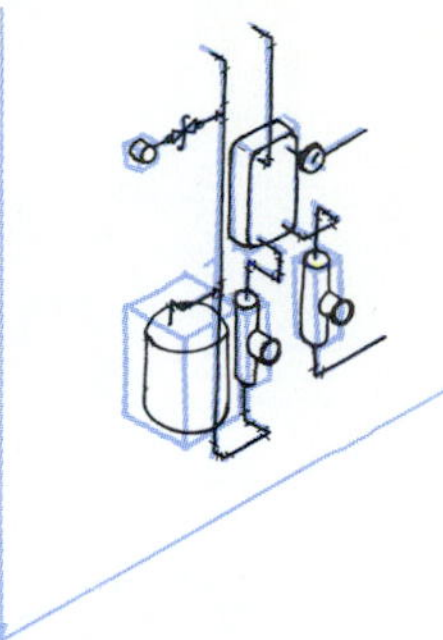

FIGURE 3.28
The actual shape of the components are added to the sketch, using the boxes with the overall dimensions as guidelines. Darkening the edges of the final sketch (and adding shading as desired) completes the sketch.

equipment will fit and mount in the mechanical room, so showing the piping to scale will be beneficial. You will not be able to sketch this piping full size on your paper, so you will need to show it to a smaller scale. Let each quarter-inch division on the paper equal 1 foot on the actual objects. This will give your isometric sketch a scale of 1/4 inch:1 foot. Start by laying out some of the overall dimensions of the back wall of the mechanical room along the Y and Z axis directions in your sketch as shown in Figure 3.24.

Locate the central plate heat exchanger on the wall and sketch a box representing its overall height, width, and depth as shown in Figure 3.25. The direction of flow through this heat exchanger will determine where other components such as pumps are placed in the system. (The flow is in an upward direction on the right side of the heat exchanger and in a downward direction on the left side.)

Next, box in the overall dimensions of the pumps, tanks, and boilers in their proper locations with respect to the back wall and the heat exchanger as shown in Figure 3.26.

Next draw the centerline of the pipe where it connects to the tank. How many inches does it go straight out before it turns? Represent this proportion of a foot along the isometric axis direction. Draw vertical lines to represent vertical pipe as shown in Figure 3.27.

Show the valves using standard piping symbols. (You can look these up in the appendix at the back of this text.) When the symbols are shown in an isometric view, lines that are horizontal or vertical in the symbol are drawn parallel to the axis directions in the isometric sketch. The finished sketch should look similar to Figure 3.28.

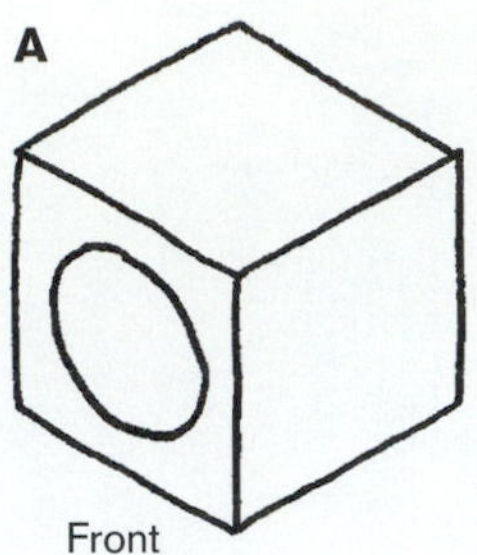

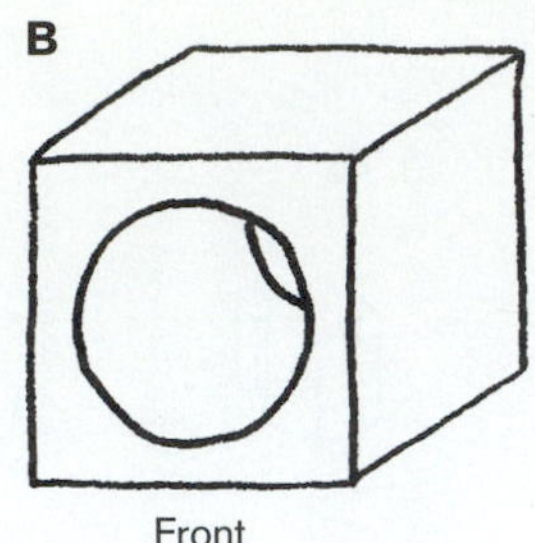

FIGURE 3.29
The two cube sketches illustrate one key difference between an isometric and oblique sketch. In an isometric sketch (a), circles appear as ellipses in all views. In an oblique sketch (b), features in the front view are drawn full size, so circles appear round.

Oblique Sketches

Oblique sketches are another type of pictorial drawing. Unlike the isometric sketch, however, the oblique sketch starts with a sketch of the front surface of the object, as if you were looking straight onto it (see Fig. 3.29). The depth of the object is represented by parallel lines extending back from the front surface. You have probably been creating oblique sketches since elementary school, as they are an easy way to add depth to a drawing. They are not very realistic, however, because you would never be able to look straight onto the front of an object and see its depth clearly at the same time.

Oblique sketches have the advantage that circular shapes in the front view appear circular (not elliptical as in isometric sketches) and the front surface of the object appears true size. For this reason, oblique pictorials can be useful when it is important to provide a true view of an irregular or curved surface that would be difficult to sketch quickly in an isometric view.

Objects shown in oblique sketches are usually oriented so that their most characteristic shape is shown as the front in the sketch. The depth of the object is shown receding from the front view. The angle of the receding lines, showing the object's depth, is arbitrary, but it is often 30° from horizontal because it gives a fairly realistic appearance and is easy to sketch. When the depth is shown full scale (actual length) the drawing is called a *cavalier projection*. When the depth is shown half size it is called a *cabinet projection*. Showing the depth full size contributes to the distorted appearance of the object, as you can see in Figure 3.30, so oblique pictorials generally show depth less than full size.

If the back surface of the object is the same size and shape as the front surface, a quick way to create an oblique sketch is to draw the front surface, then sketch it again (to represent the back surface) offset up and to the right. Sketch the receding lines by connecting the corners on the front and back surfaces as shown in Figure 3.31. Darken the visible lines to complete the sketch.

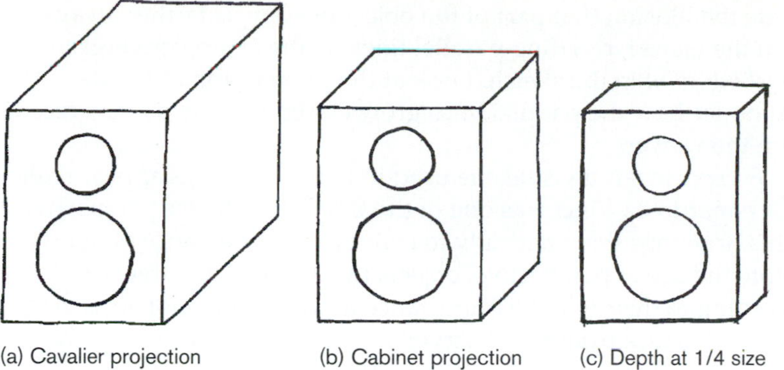

FIGURE 3.30
These three oblique sketches illustrate the effect of changing the scale at which the depth of the object is shown: (a) shows depth at full size (cavalier projection); (b) shows depth at half size (cabinet projection); (c) shows depth at quarter size.

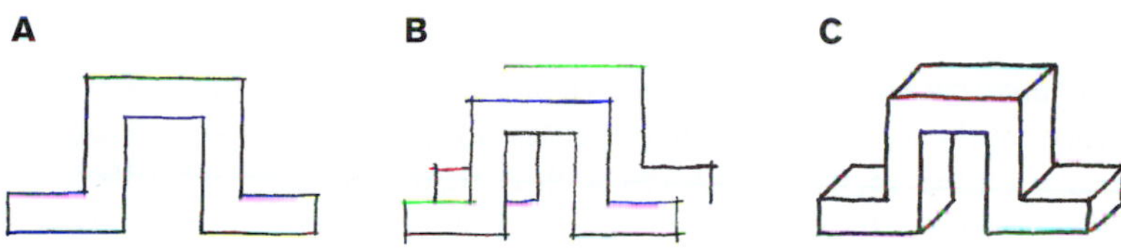

FIGURE 3.31
Three-step oblique. *When the back surface of the object is the same shape and size as the front: (a) Sketch the front surface true size. (b) Sketch the edges of the front surface again, offsetting the lines by an equal amount up and to the right. (c) Connect the corners to draw the receding edges.*

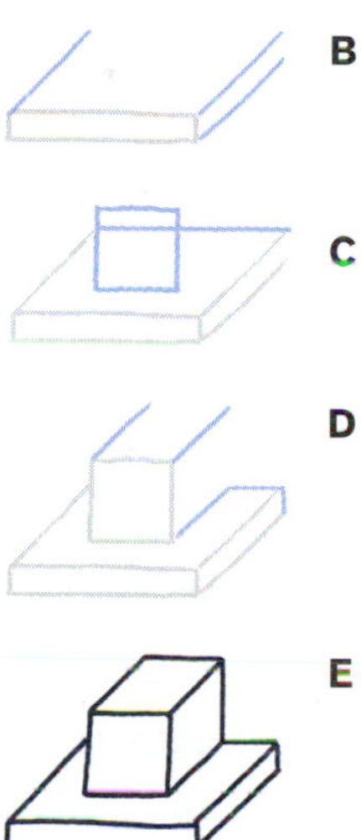

FIGURE 3.32
To draw an oblique sketch of a more complex object:

(a) Sketch the edges of the front surface.

(b) Sketch the edges that recede from the corners of the front surface, keeping parallel edges parallel.

(c) Sketch the next surface that is parallel to the front surface at the approximate depth.

(d) Draw the receding lines from the second surface, repeating steps (c) and (d) as needed.

(e) Connect the corners of the receding lines at the appropriate depth to indicate the back edges of the part.

To create an oblique sketch of a more complex object, begin by selecting the most irregular or characteristic surface of the object. Avoid selecting a surface that will result in the longest dimension of the object being shown as receding depth lines, as this will make the sketch seem more distorted. Sketch this surface true size (using grid paper, if desired, or estimating units to keep the drawing in proportion).

Sketch in the edges that recede from this front surface. Use the same angle for all receding lines that are perpendicular to the front surface so that parallel receding edges appear as parallel lines.

Estimate the depth of the object or the length along the receding lines to the next surface. Draw the next surface that is parallel to the first surface true size, then draw the receding lines from its corners as you did for the first surface, and so on (see Fig. 3.32).

Perspective Sketches

Perspective sketches are a third kind of pictorial sketch. They differ from isometric and oblique sketches in that receding lines that are parallel on the object are not shown parallel in the drawing. To

FIGURE 3.33
The railroad tracks in this photo appear to converge in the distance, even though they remain parallel. When you look at the tracks, you don't "see" them converging, because this is a cue that the mind interprets as an effect of increasing distance. Perspective drawings use this technique to trick the viewer into seeing depth in a flat image.

create the illusion that part of the object or scene is farther away from the viewer, receding parallel lines on the object converge at the *vanishing point* in the sketch. Look at the photograph in Figure 3.33 and notice how the parallel lines grow closer together as they recede from the viewer.

Perspective is a visual cue used to make drawings appear realistic. Leonardo da Vinci was one of the first artists to use perspective in his drawings systematically to convey the appearance of depth. Before the use of perspective, objects in a scene were often sized according to their relative importance to the painter, not according to their distance from the observer. A perspective sketch uses the vanishing point to simulate the way the eye perceives distance, which makes the objects in the sketch more realistic.

The appearance of a perspective sketch depends on your viewpoint in relation to the object. Select some reachable object in the room and move so that you are looking at it from above and really notice its shape. Now gradually move so that you are looking at it from below. Notice how the change of viewpoint changes the appearance of its surfaces—which ones are visible and their relative size.

The *horizon line* in a perspective sketch is a horizontal line that represents the eye level of the observer. Locating the sketched object below the horizon line produces a view from above (or a bird's eye view). Locating the sketched object above the horizon line produces a view from below (or a worm's eye view). Figure 3.34 illustrates the horizon line in a drawing and the effect of placing the object above or below the horizon line.

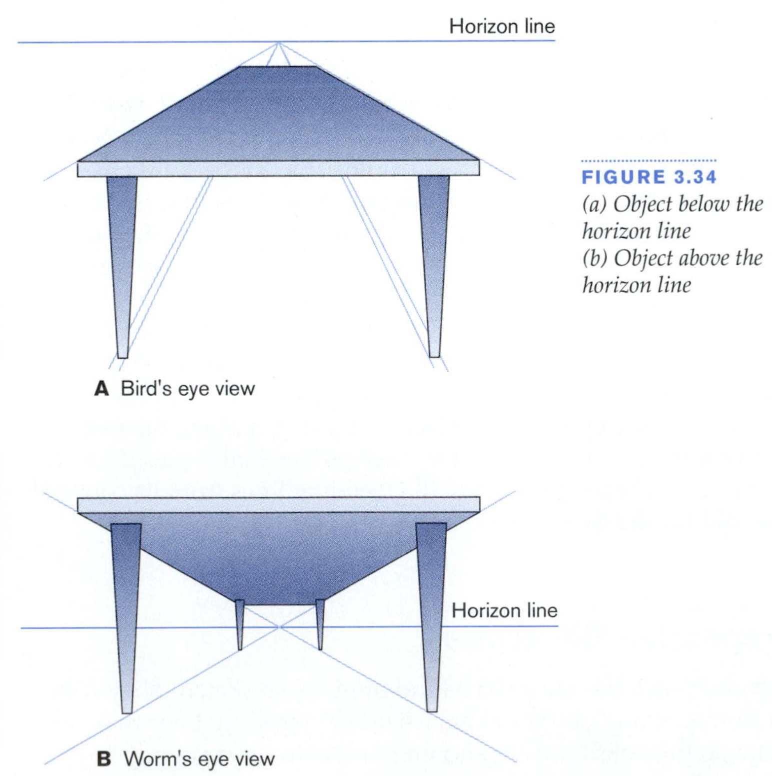

FIGURE 3.34
(a) Object below the horizon line
(b) Object above the horizon line

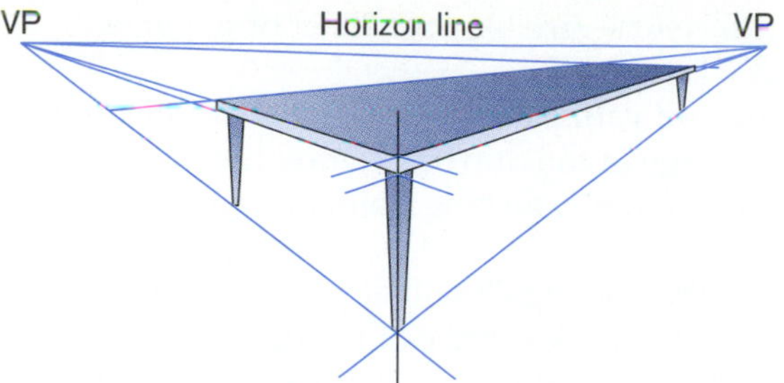

FIGURE 3.35
Receding lines converge at the horizon line at one or two vanishing points (VP).

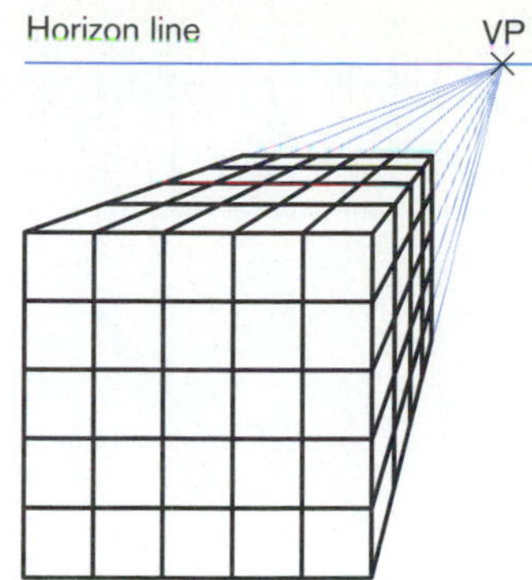

FIGURE 3.36
The dimensions of the rectangular box shown here are indicated by the unit squares in the sketch. The width of the box along the back surface remains 4 units, although it appears smaller in the sketch. The ratio of width to height, 4:4, also remains the same. The projectors from the front surface were used to sketch the back surface and keep the dimensions in proportion.

The *vanishing point* in a perspective sketch is a point on the horizon line where receding lines from the object converge, as illustrated in Figure 3.35.

To create the illusion of depth in a sketch, use guide lines, called *projectors*, that run from the corners of the object to the vanishing point. Figure 3.36 illustrates how these guidelines match up to the way the eye perceives distance. In a perspective sketch, the projectors help you keep the back surfaces of the object in proportion. The distance along the projector represents the distance that a feature is from the viewer. Each projector is scaled equally as the object recedes, so the ratios of width to height in the object remain the same.

Perspective sketches can have multiple vanishing points, as shown in Figure 3.37. Two-point perspective sketches have two vanishing points; there are three vanishing points in a three-point perspective. In each case projectors from the object to the vanishing points serve the same purpose as guidelines for drawing receding lines to create the illusion of distance.

Creating a Perspective Sketch

To create a one-point perspective sketch, orient the object you are sketching as you would for an oblique sketch so that you are looking directly onto the front surface of the object, as shown in Figure 3.38.

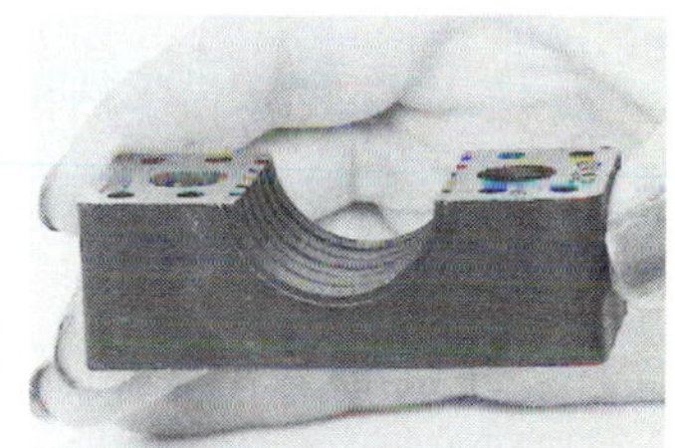

FIGURE 3.38
Orient the object so you can see the front view clearly.

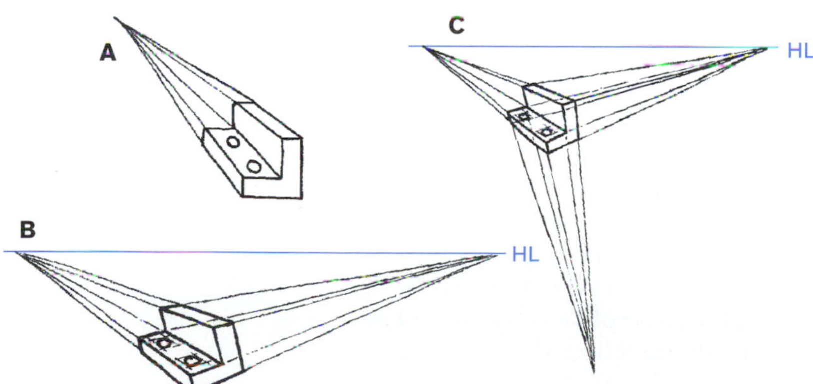

FIGURE 3.37
The light lines in each perspective sketch serve as guidelines for creating the perspective view. One-point perspective, shown in (a), has one vanishing point. Two-point perspective has two vanishing points located on the horizon line, as shown in (b). Three-point perspective adds a third vanishing point to indicate whether the view is from above or below the object. The sketch in (c) is drawn from a viewpoint above the object.

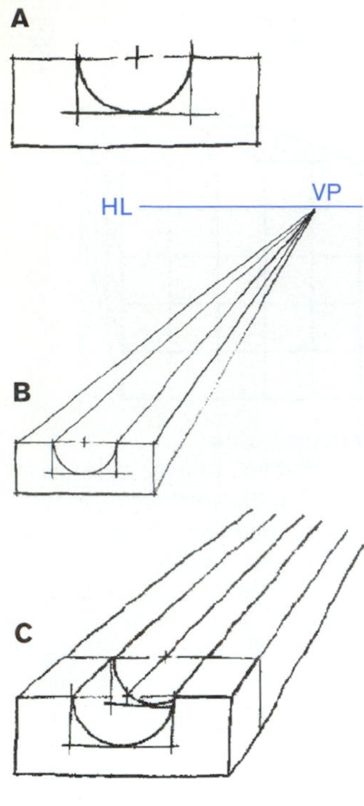

FIGURE 3.39

(a) The front surface of the object is sketched first;

(b) The vanishing point is added and projectors drawn from the corners of the front surface to the vanishing point;

(c) The back surface is sketched using the projectors to locate its vertices in the correct relationship to points on the front surface;

(d) The object lines are darkened in the final sketch and some shading is added to the surfaces.

Sketch this front surface true size on your sheet of paper, as shown in Figure 3.39(a).

Add the horizon line and vanishing point. This is the location where the projectors converge in the drawing. Draw the projectors from the front of the object to the vanishing point as shown in Figure 3.39(b).

Estimate the depth of the object as it will appear along the receding lines and lightly sketch the back surface of the object between the projectors. It should appear smaller than the front surface, but in proportion, as shown in Figure 3.39(c). If the back surface of the object were not the same size or shape as the front surface, you would use the projectors to calculate its location. That is, if the distance between the projectors is 8 units, and the back surface is 10 units wide, you can use the size of the unit at a given depth to locate the back edges. You can also add projectors from the front surface to the vanishing point to locate features along the receding surfaces.

Connect the corners of the front and back surfaces to create the edges of the object that recede into the distance. To complete the sketch, darken the visible lines in the final sketch and add shading as shown in Figure 3.39(d). Be sure to use a consistent direction for the light and shadow when you shade your sketch.

You can use these same steps to sketch any object in perspective. To produce a good perspective sketch, you first want to consider the orientation for the object so that it will be shown to advantage in the perspective sketch. Imagine a cigar-shaped space vehicle. You may want to orient the long direction of the part in the direction of the vanishing point in order to call attention to its size. You may want to start by orienting its long axis toward the vanishing point, as shown in Figure 3.40.

Although perspective projection produces the most realistic pictorial view, it can be more time consuming to sketch. In perspective, the size of the object shown depends on the distance from the viewer, and the foreshortening in the picture makes it difficult to assess whether objects are the same size or simply farther away.

Perspective views are frequently used in architectural applications, where the large scale of the objects makes the effect of distance so obvious that it must be reflected in the view. For smaller objects and mechanical parts, it is generally as effective to use an isometric or oblique pictorial, which are easier to sketch (although many CAD packages will generate a perspective view automatically).

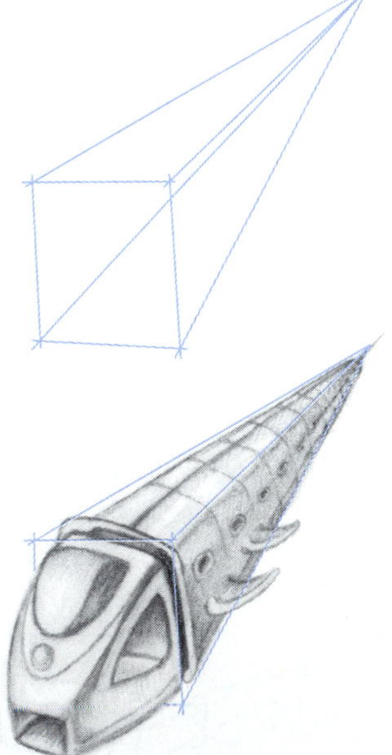

FIGURE 3.40

Using a perspective sketch for the concept sketch of this space vehicle helps to convey its size, as it would be difficult to view such a large object without some perspective effect.

Figure 3.41 illustrates the steps in sketching the three different pictorial views of a common object.

Lettering

The shape of an object can be described by drawing multiview projections of it, but other information needs to be neatly lettered on the drawing. Engineering drawings are usually lettered using 1/8-inch upper-case *gothic style* letters. Gothic style letters have no *serifs*, which is the name for the ending accent strokes or bars on letters. These letters are easy to make and read. Figure 3.42 shows the letter and number shapes in the gothic style.

The arrows on the letter and number shapes indicate pencil strokes that will result in well-formed shapes. The strokes used in cursive letter forms are designed for a flowing script and are less effective in producing balanced and evenly rounded letters and numbers. Practice with the strokes indicated in the figure as you letter your sketches until it becomes a habit.

You can use a lettering guide to help you make uniform lettering on your drawings. The lettering guide template shown in Figure 3.43 is very useful for making quick lettering guidelines for your sketches. Guidelines are thin horizontal and vertical lines that help you keep lines of text straight and letters evenly spaced. Figure 3.44 shows an example of neat freehand lettering used in a drawing.

All lettering should be drawn horizontally so that it reads from the bottom of the sheet; that is, the letters are oriented as if the bottom of the sheet were the bottom of a page of text. Sometimes lettering is aligned with other objects in the drawing, however. When it is, it should read from the bottom or right side of the drawing.

FIGURE 3.42
Use the pencil strokes indicated by the arrows to letter the alphabet and numbers 1 through 10 in the gothic style.

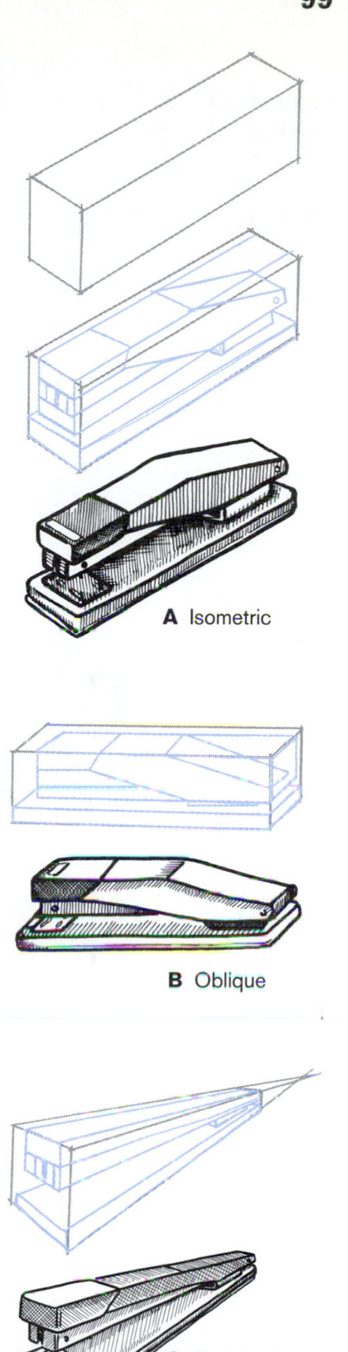

FIGURE 3.41

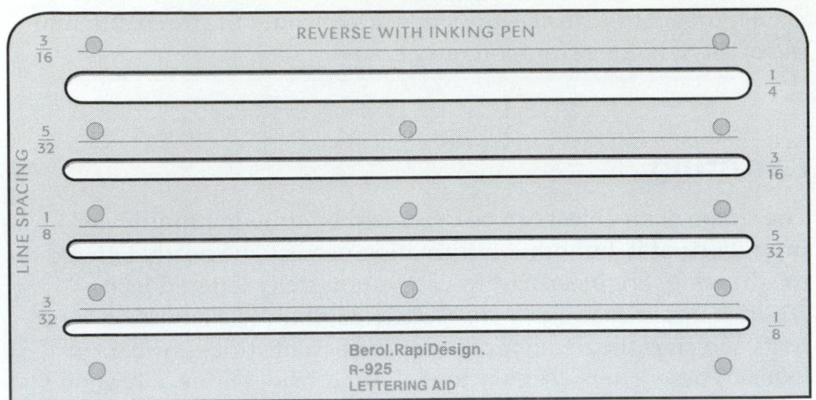

FIGURE 3.43
Lettering guide template

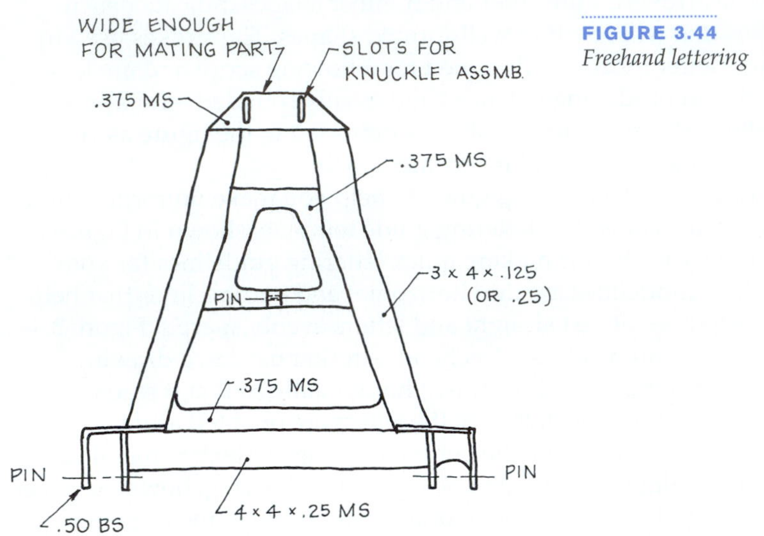

FIGURE 3.44
Freehand lettering

It is very frustrating to read large drawings that have lettering running different directions, and costly mistakes may result. For example, a contracting company received a drawing in which a critical dimension had been lettered upside down. The dimension, 1.11, was read as 11.1. Because of the number shapes involved, the contractor couldn't tell that the dimension was upside down. Relying on the misleading dimension in the drawing, the contractor allowed 11.1 inches for the glass in the steel around a rough window opening. The mistake cost about $10,000 in changes and accommodations to this wrongly sized feature. Remember that the drawing is a legal contract between the designer and the manufacturer of the design. Making all dimensions read from the bottom may seem insignificant, but it helps avoid costly mistakes.

Lettering sizes in a drawing are based on the sheet size of the final drawing. When you are adding text to a CAD drawing, you need to think about the final plotted drawing so that you size your text correctly. Titles and important headings are usually larger than the standard size for notes, dimensions, and other text. Standards for lettering sizes are reproduced in Table 3.1.

TABLE 3.1

Minimum Letter Size Standards

Use	Minimum Letter Heights: Freehand	Minimum Letter Heights: Instrumental	Drawing Size
Drawing number in title block	.312" (5/16), 7 mm	.290", 7 mm	Larger than 17" x 22"
	.250" (1/4), 7 mm	.240", 7 mm	Up to and including 17" x 22"
Drawing title	.250" (1/4), 7 mm	.240", 7 mm	All
Section and tabulation letters	.250" (1/4), 7 mm	.240", 7 mm	All
Zone letters and numerals in borders	.188" (3/16), 7 mm	.175", 5 mm	All
Dimensions, tolerances, limits, notes, subtitles for special views, tables, revisions, and zone letters for the body of the drawing	.125" (1/8), 3.5 mm	.120", 3.5 mm	Up to and including 17" x 22"
	.156" (5/32), 5 mm	.140", 5 mm	Larger than 17" x 22"

General and Specific Notes

Often you must add notes to your sketch to explain and identify items. General notes are notes that apply to the entire object, design, or process. They are usually placed at the right side of the sheet, preferably near the lower right. General notes may be placed anywhere on the sketch so long as they are neatly lettered and organized. Group all of the general notes in one place. Don't scatter them in various places on the sheet. Figure 3.45 shows an example of general notes.

Specific notes apply just to a particular part or feature. To show which part or feature a note refers to, a *leader* is drawn from the location to the note. The leader should be drawn with a horizontal tail about 1/8" long, and an angled line ending at the feature. Usually an arrowhead is used to point to the edge of the part, and a dot is

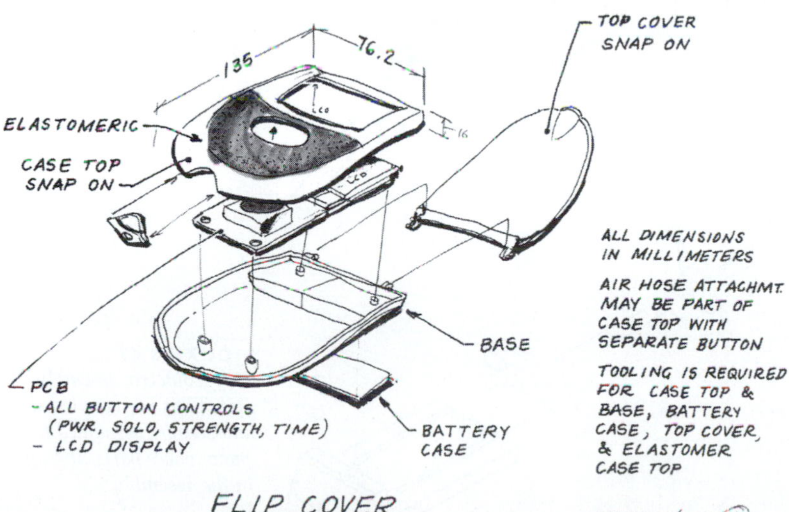

FIGURE 3.45
A specific note requires a leader line to indicate which part of the sketch the note applies to. Notice the use of dots and arrows with the leader lines in this drawing. (Courtesy of Robert Mesaros, Strategix ID.)

used if the feature is inside the object's outline. Figure 3.45 shows an example of specific notes and leader use.

Sketching Assemblies

Assembly drawings are used to show how parts assemble together. Because they do not need to provide all of the information to make individual parts, often isometric sketches or drawings are used that show only the exterior view of the assembled parts. You can use assembly sketches as you design to identify areas that must fit together, have common dimensions, or maintain critical distances, to ensure that the device will function as expected. Figure 3.46 shows an isometric sketch of an assembly.

Sketched assembly drawings are useful in documenting your ideas in a rougher state. An exploded isometric assembly drawing shows the individual parts moved apart from one another along the isometric axis line directions. A centerline pattern is used to show how the parts relate and fit together. A centerline pattern is a thin dark line that starts with a long line, then has a short dash of about 3 mm (1/8 inch), and ends with a long line. The length of the long portion of the line is adjusted to fit what you are drawing, but it is typically in the range of 40 to 80 mm (1-1/2 to 3 inches)

When creating an exploded assembly sketch, keep the parts lined up with the other parts where they assemble. If possible, move any single parts along only one axis direction. Figure 3.47 shows an

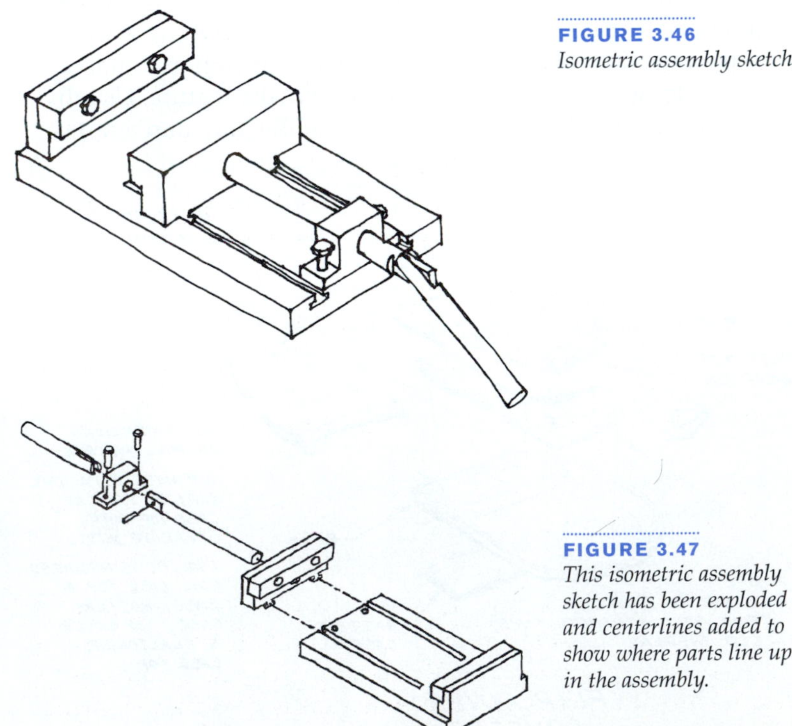

FIGURE 3.46
Isometric assembly sketch

FIGURE 3.47
This isometric assembly sketch has been exploded and centerlines added to show where parts line up in the assembly.

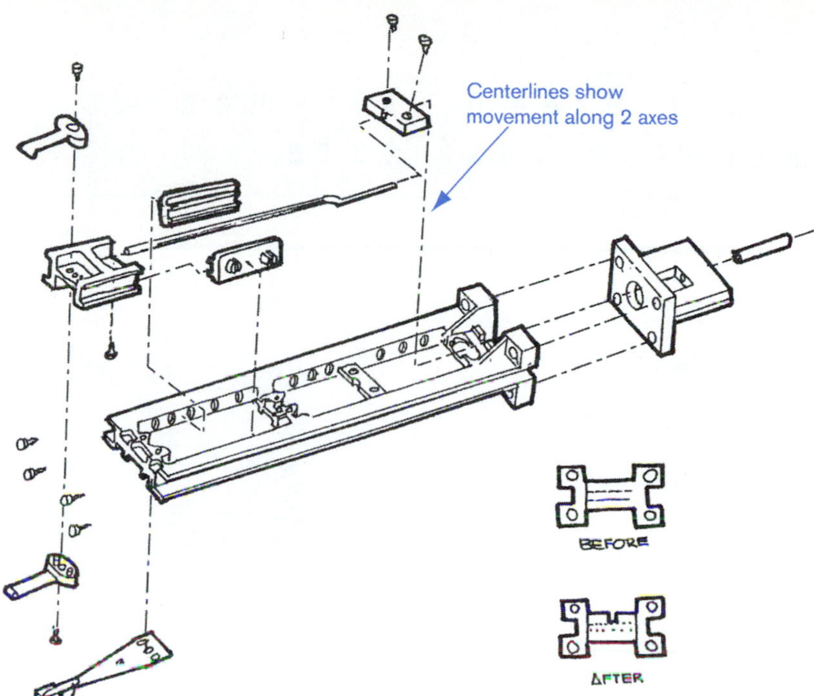

FIGURE 3.48
In this isometric assembly, two parts have been exploded in two directions so they do not overlap other parts in the assembly. The centerlines indicate the path the part must follow to assemble correctly. (Courtesy of Albert W. Brown, Jr.)

exploded isometric assembly in which all of the parts are moved along only one axis direction. Notice the centerline pattern showing how the parts align. If you must move a part in two different directions so that it can be seen clearly, be sure to add a centerline showing how the part must move to get to its proper location in the assembly. Figure 3.48 shows an exploded assembly with a part moved in two axis directions. You can see the jog in the centerline indicating how the part has been moved.

Finished assembly drawings that will be given to manufacturing workers to show them how to produce a device also contain a *parts list* that identifies each part in the assembly, the quantity necessary, the material of each part, and a number identifying which item it is on the drawing. A parts list can also be included on a sketched assembly drawing. Sometimes it is helpful to give a quick assembly sketch to manufacturing along with part drawings, especially if they are parts you need to have manufactured quickly from a sketch. Seeing how the parts fit together in assembly will help the manufacturer understand the entire device and identify potential problems. You will learn more about assemblies in Chapter 8.

Sketching in the Design Process

The techniques you have learned in this chapter will add to your repertoire of skills used to visualize and communicate your designs. Effective sketching and visualization require practice. In order to develop these skills, keep a notebook and make drawings showing your ideas and reflecting objects you see around you.

INDUSTRY CASE

Albert Brown, Affymax Research Institute

Sketching Techniques: One Thing Leads to Another

When sketching for ideation, Albert Brown starts in pencil—using either a No. 2 or a mechanical pencil—and rarely erases. "Once I get something down on paper, I usually leave it, even if it is off the wall," he explains. "When I was a new mechanical engineer, my manager emphasized the creation of many thumbnail sketches to conceptualize a solution. He told me that showing him a lot of sketches showed him how I was trying to solve the problem [see Fig. 3.49]. He and the other designers and I would meet as a group and share sketches and come up with different ideas for solving each other's design problems. There's a certain amount of synergy that happens when you show someone an idea. It stimulates their thought processes so they can elaborate on what you did—or come up with something entirely new that was triggered by what you showed them."

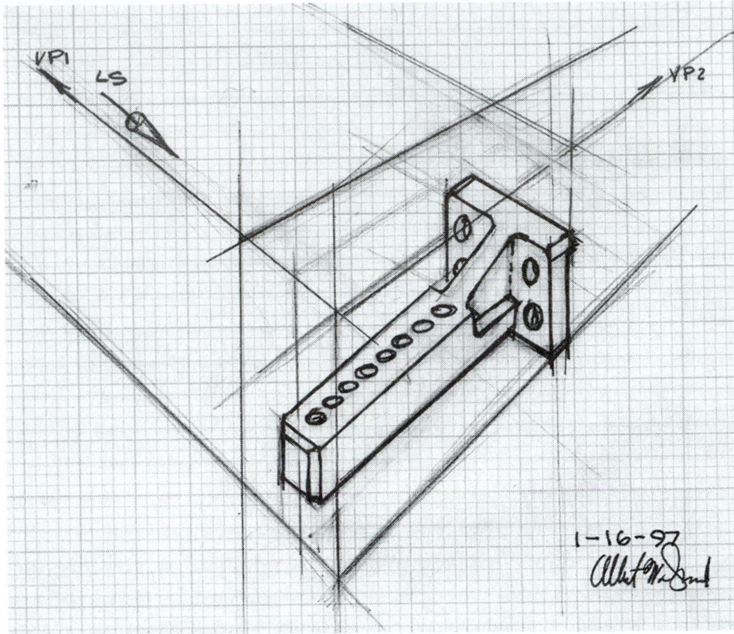

FIGURE 3.50
The construction lines to the vanishing points (labeled VP1 and VP2 in the sketch) serve as a guide for the 3D sketch of the array holder.

To capture his design ideas, Brown uses a notebook of 1/4-inch grid paper. The grid makes it easier to keep scale in mind and gives him a better feel for the proportions of the parts of the sketch. He usually sketches 2D views of the object first: profiles, front, top, and side views, but not necessarily all three views for any one concept. His sketches (six to eight sketches for a concept) are designed to illustrate different views of the same

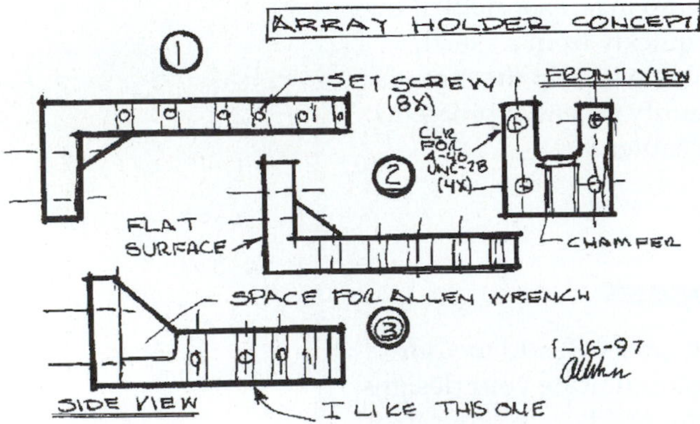

FIGURE 3.49
In the series of concept sketches for the array holder shown here, the shape of the attachment end of the holder appears most clearly in the side view. Three variations on this aspect of the design are sketched as side views.

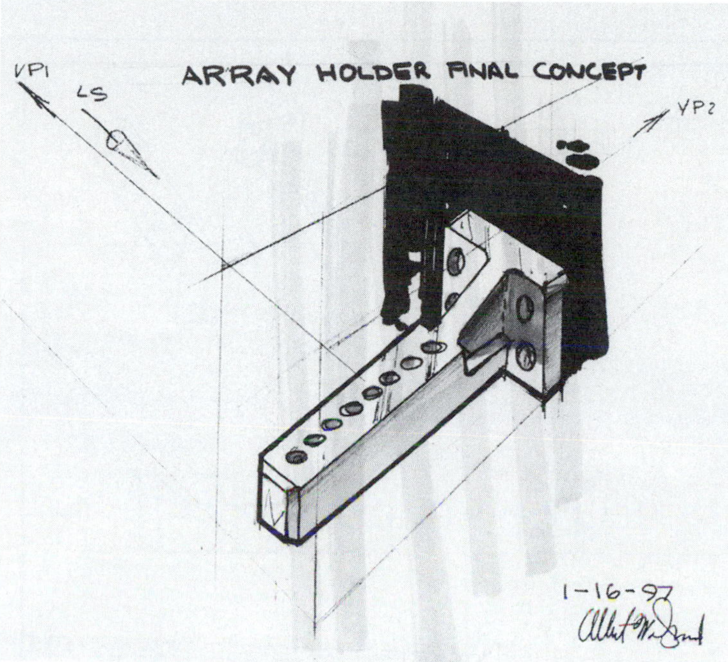

FIGURE 3.51
Notice the arrow in the top left of the final concept sketch that indicates the direction of the light source for the drawing. This arrow is an aid to visualizing the way light will hit the object and which surfaces will be in shadow.

concept, or several different variations on solving the problem. "If I like a concept, I do a front, right, and top view, but for the thumbnails, several might be front or side."

From the thumbnails, Brown will pick one or two that look most promising and sketch them in a 3D, usually isometric, view (see Fig. 3.50). These views are easier to draw on isometric grid paper, but with practice, it is easy to do on quadrille paper. He first uses a No. 1-1/2 pencil to lightly sketch an isometric cube. The cube helps him block in the overall shape of the part and keep parallel lines parallel in his sketch. He then uses a No. 2 pencil to sketch the part shape on the sides of the box.

Once again, the role of the sketch in visualizing his design is uppermost in Brown's mind. When he selects a viewing direction for the 3D sketch, he puts himself in the user's place and captures the view the user will see when he approaches the machine or device. "You usually see an object first, then formulate in your mind how you will touch it. Drawing the object from this perspective helps me imagine how a user will interact with it. I ask myself, 'What tools would I need to assemble or disassemble the part?' 'Where would you expect the handle to be?'"

After sketching the 3D view, Brown adds shadows to give it more of a 3D feel. He selects a direction for the light source in the drawing and represents it with a perspective arrow (shaped like a cone with a tail as shown in Fig. 3.51). He locates this at any angle above the drawing, then imagines how the light will come off it and strike the object. He likes to choose a direction for the light that will generate high contrast shadows so the image looks like it is coming up out of the paper. "One thing that helped me with some of the mechanics of drawing was a basic art course," said Brown. "Sketching in itself is an art form, and a basic course can teach you a lot about using light, perspective, color, and texture in your sketches."

When the sketch is done, Brown takes a fine felt-tip pen and goes over all of the object lines in his sketch to emphasize all of the important features. He also uses a medium or broad felt-tip pen to outline just the outside of the sketch. The heavy outline characteristic of his sketches is a technique he developed to make his sketches hold up for presentation purposes. By using dark lines and shadows, his sketches photocopy well and result in clear, crisp images. In addition, the darker lines hold up better when the sketch is scanned into the computer.

The same technique also helps when he needs to make a model look like a sketch (see *Designer's Notebook* in Chapter 11). Brown finds that he rarely creates sectional views of assemblies by hand, for example, as it is too difficult to sketch them well at the concept level. If he needs a section sketch of an assembly, he is more likely to

(continued on p. 106)

INDUSTRY CASE

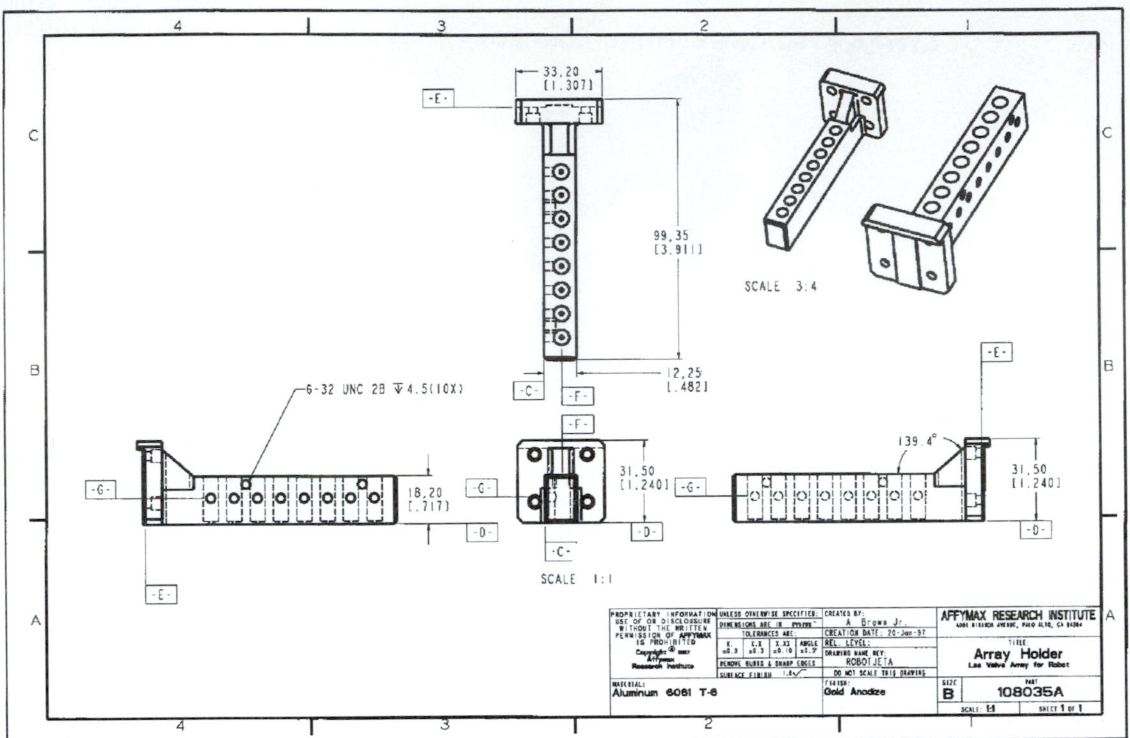

FIGURE 3.52
This documentation drawing was generated from the 3D model of the array holder. Notice that a pictorial view of the holder is included in the drawing as an aid to visualization, and is very similar to the freehand pictorial sketch of the final concept.

create the parts in his CAD package, assemble them, and use the computer to slice the model and print a section view. When presenting such a sketch early in the design process, Brown will often add the dark outline by hand so it looks like the sketch was hand drawn. "You don't want to give the impression that the design is cast in stone," Brown explains. "If it looks like a sketch, it gives others an opportunity to participate in the design."

If a product has a lot of parts, Brown will spend a fair amount of time animating it in his mind before he begins to model it. He imagines an exploded view, how the parts will be assembled, where screws will go, how much clearance will be needed for the head of a socket wrench used to tighten a bolt, and so on. He doesn't begin to model the parts until he has looked at the assembly in his mind and passed the ideas by others to see what they have to say about it. Once he's fully considered the design, he begins to model the parts in his parametric modeling package (he now uses SolidWorks) so he can be sure that the parts don't interfere and that they will work in the context of the space he has available. He finds that the sketching features built into newer parametric modeling software make it easy to use the CAD package sooner for conceptual work.

Nonetheless, Brown's advice to new engineers is simple: "Learn to generate sketches of as many ideas as you can as fast as you can. The number you generate has a lot to do with how well you can solve a problem—because one concept can lead to another if you can get them down on paper."

Illustrations are courtesy of Albert W. Brown, Jr. and Affymax Research Institute.

Albert W. Brown, Jr. is Staff Mechanical Design Engineer for Affymax Research Institute, a drug discovery and development company in Palo Alto, California. As a student at the University of Missouri, Brown won the Fisher Body automobile design contest before he graduated with a degree in mechanical engineering. From there, he went on to work for IBM, where he designed parts of the first automated teller machine in the United States. In 1976, Brown earned a Master of Science degree in product design engineering from Stanford University. While working for Grid Systems, he designed a digital camera in partnership with Kodak. Since 1996, he has been designing robotic systems to be used in drug discovery and development for Affymax, a wholly owned subsidiary of Glaxo Welcome. In his current position, he uses his sketching skills frequently and is a proficient user of parametric modeling software and other computer-based tools for evaluating his designs.

Sketching techniques are generally used early in the design process to generate ideas and show them to other members of the design team. The portability of pencil and paper, however, makes the ability to sketch an important means of capturing your ideas, working out relationships between parts, and recording your work.

As more details of the design are finalized, you will use 2D CAD drawings or 3D models to further define your designs, but many of the techniques you learned for sketching 3D shapes will also help you define them in the CAD environment. Typically if you cannot sketch the design, you do not have the details sufficiently worked out to begin modeling in CAD. In the next chapter you will learn about techniques for creating orthographic views to accurately show the true sizes of features in your drawings.

Skills Summary

Now that you have finished this chapter you should be able to create three types of pictorial drawings used in engineering and the lettering shapes that are used to add notes to engineering drawings. From this point on you should always letter your drawing notes neatly using engineering letter shapes. Practice the letter shapes by adding titles to your sketches in the following exercises. Gain additional practice by using engineering letter shapes when writing out your homework problems for other classes. Practice them while talking on the phone.

Use Exercises 1–5 and 24 to practice your skill in perspective sketching. Make several assembly sketches in either fully assembled or exploded isometric views. Show the centerlines correctly for exploded assemblies.

Do Exercises 6–24 to practice your skill in isometric and oblique sketching. Create some of your isometric and oblique sketches using grid paper. Do many others on plain white paper to gain experience in quickly capturing the shape of an object or idea. Practice some sketches using engineering markers as well as pencil. If you have access to a whiteboard or chalkboard, try this sketching surface also.

EXERCISES

1. Sketch a set of single-point perspective projectors and try creating a drawing of the houses on your street from the viewpoint of someone standing in the middle of the road and looking down the street.
2. Work with a partner on this exercise. Choose one of the objects in Exercises 6–23. One of you will describe the part while the other sketches it. Try it two ways: one in which the describer can not see the sketch, the other in which he or she can see it and give feedback while it is being drawn. Then trade roles. Try it again with a third partner: one of you describes, one sketches, one gives encouragement and feedback.

3. Sketch a freehand perspective view of the oarlock, the thread tap, and the phone receiver shown below. Scales are not provided; try to draw all features of each object to the same relative scale.

   a. Oarlock

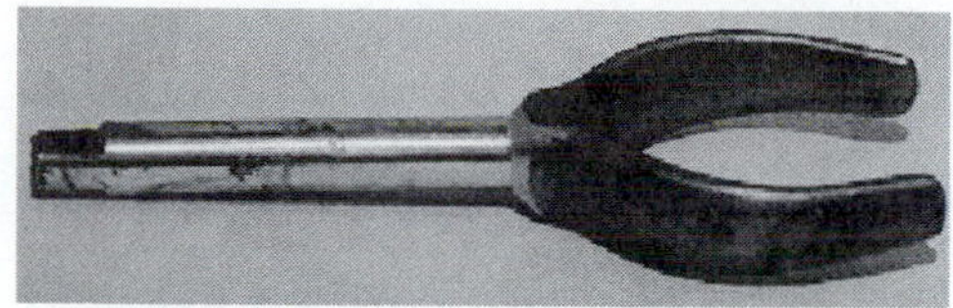

   b. Thread tap

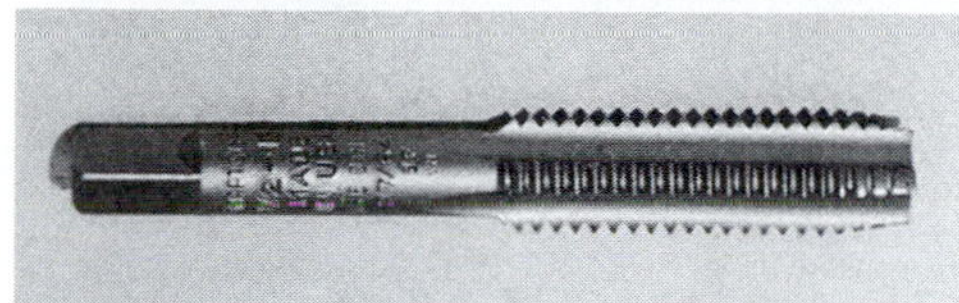

   c. Phone receiver

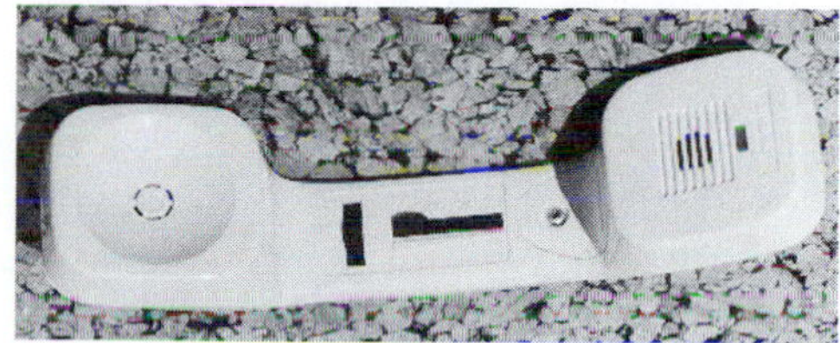

4. Create a freehand isometric sketch of the stepper motor shown below. Ignore the electrical wiring and connector if you find it easier to visualize that way. Try drawing a perspective sketch of the object. In what ways do the sketches differ?

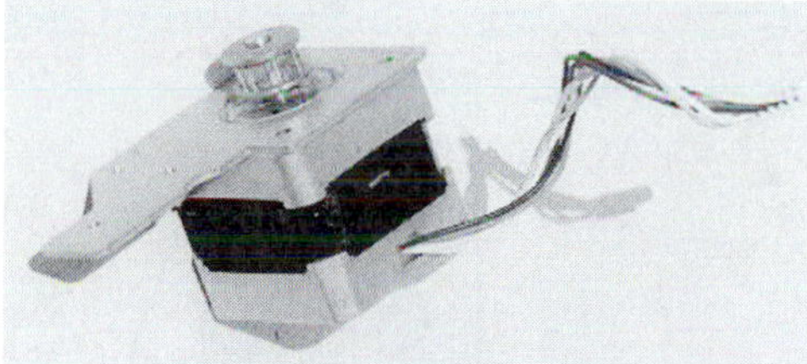

5. The CAD model shown below is an isometric view of a tool stop. The model contains (or omits) some information that would usually be omitted (or present) in different drawing representations. Redraw the tool stop as

   a. an isometric sketch **b.** a perspective sketch

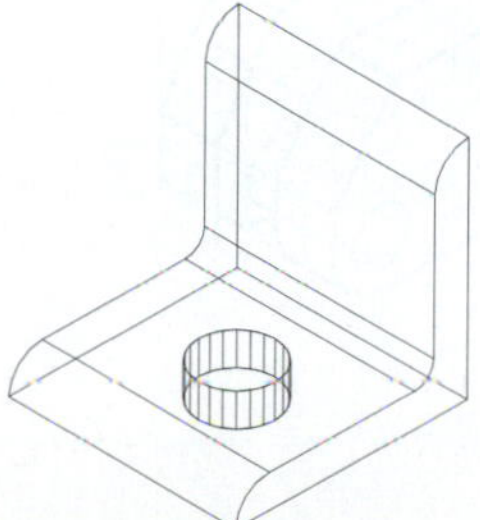

Exercises 6–23: Pictorial Sketching Problems. Create isometric and oblique sketches of the objects shown. Use grid paper to draw the objects to size, or scale them consistently. Assume that each grid square equals 1/2 inch, or that each square equals 10 mm.

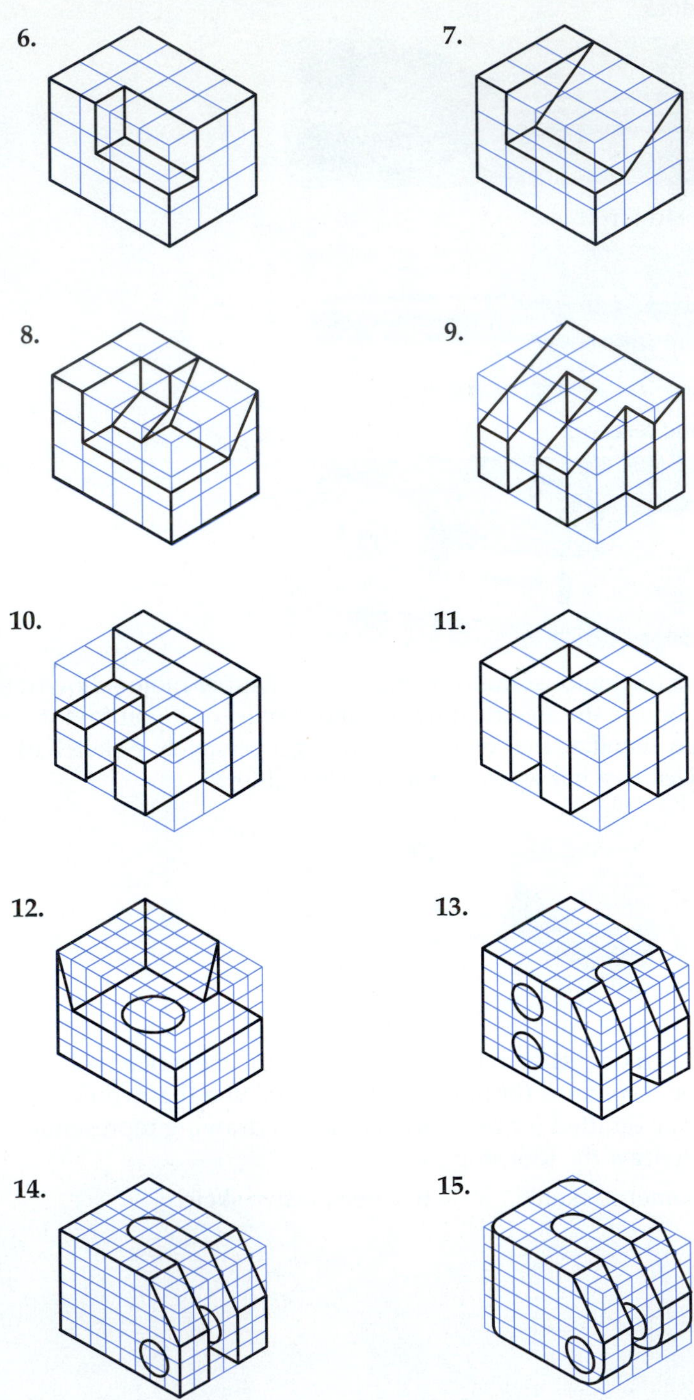

16.

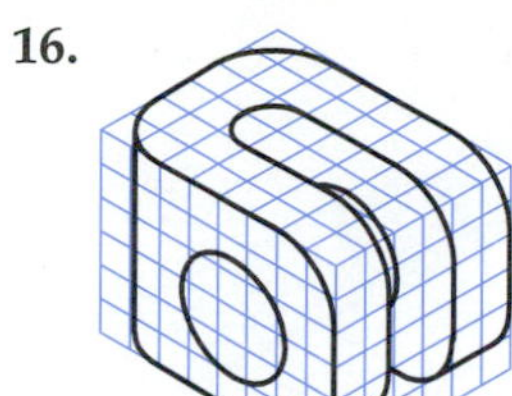

17.

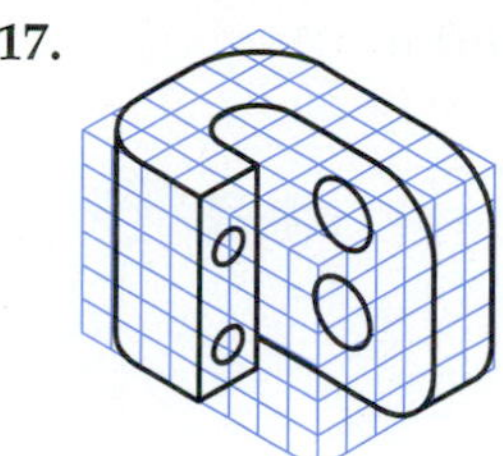

18.

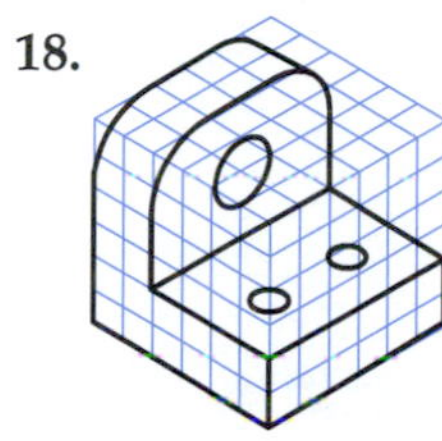

19.

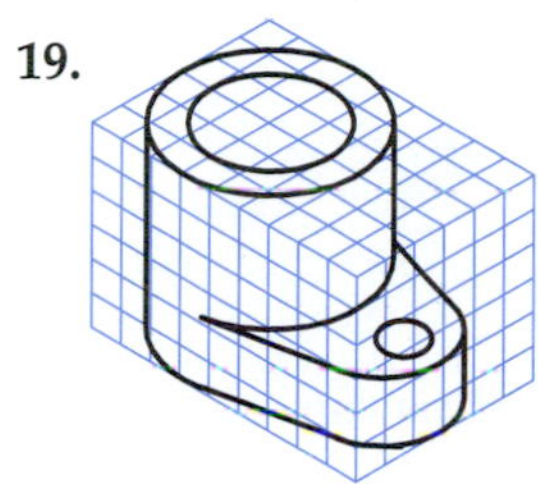

20.

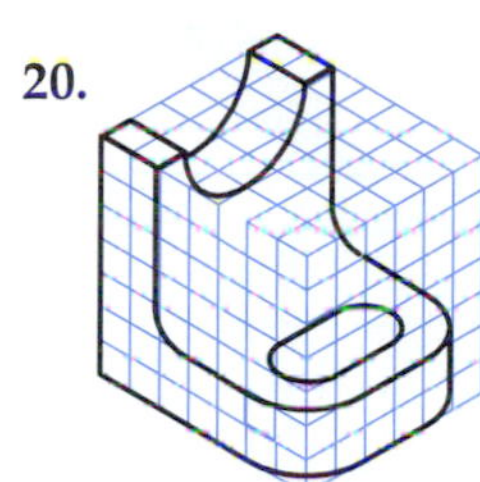

21.

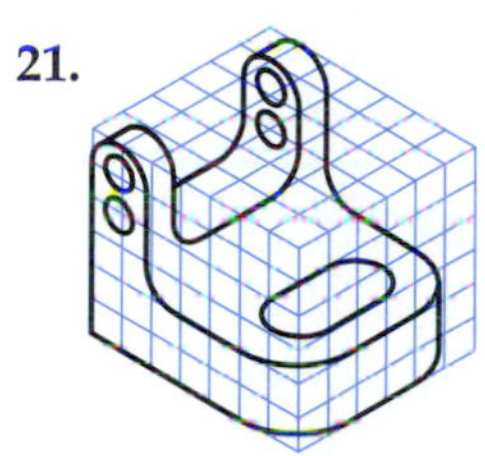

22.

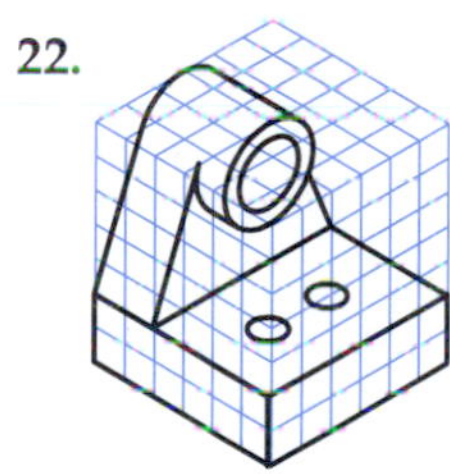

23.

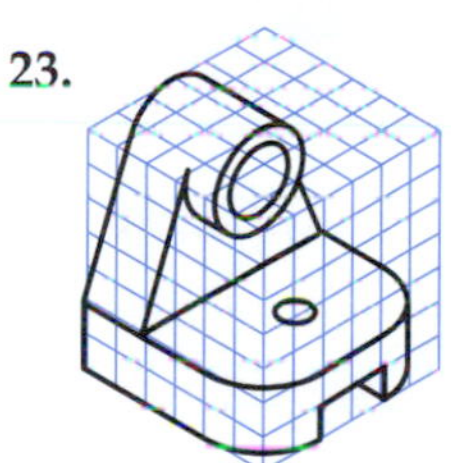

24. Sketch isometric, oblique, and perspective pictorials of the objects shown in the photographs.

a. Monitor

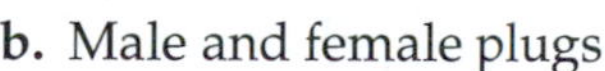

b. Male and female plugs

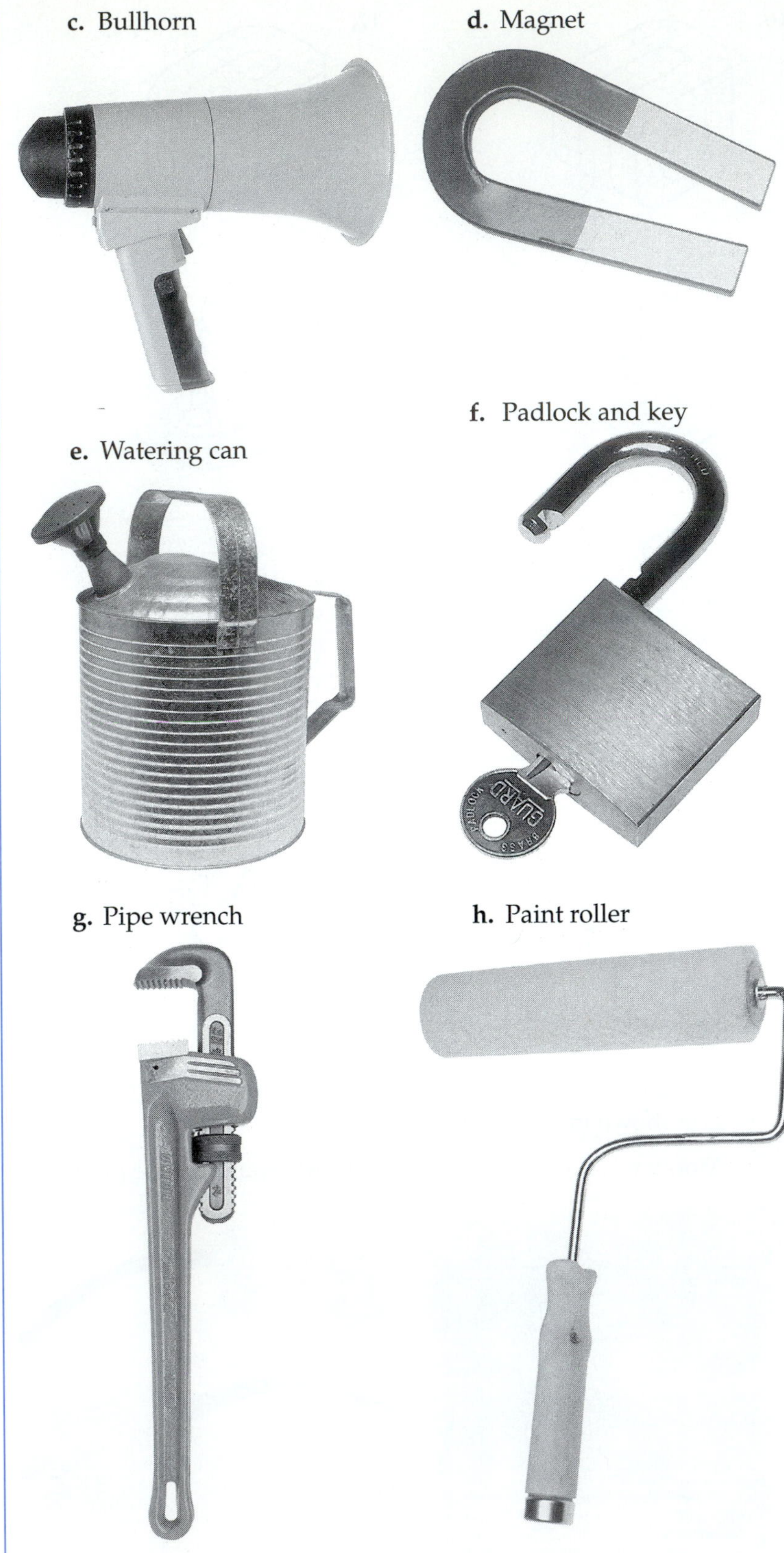

c. Bullhorn

d. Magnet

e. Watering can

f. Padlock and key

g. Pipe wrench

h. Paint roller

i. Rocket

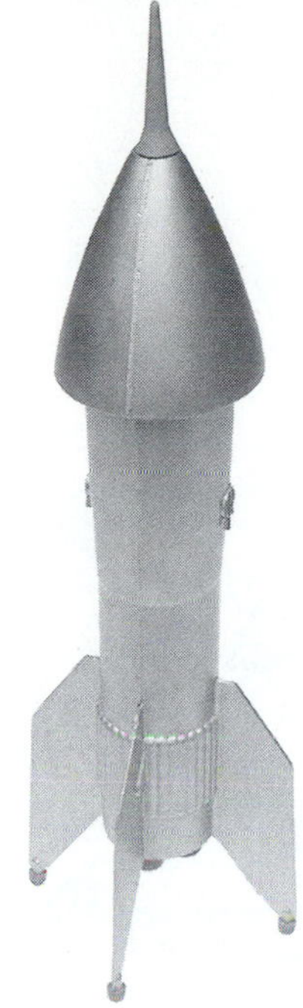

j. Chair

DESIGNER'S NOTEBOOK

The 3D view of the coaxial cable clamp on this page from Albert Brown's notebook shows its relationship to the cast arm it will clamp onto, but the sketches of one-piece and two-piece construction, and the alternate shape for the clamp's snap, are all 2D views that show the true shape of the clamp more accurately. (Courtesy of Albert W. Brown, Jr., Affymax Research Institute.)

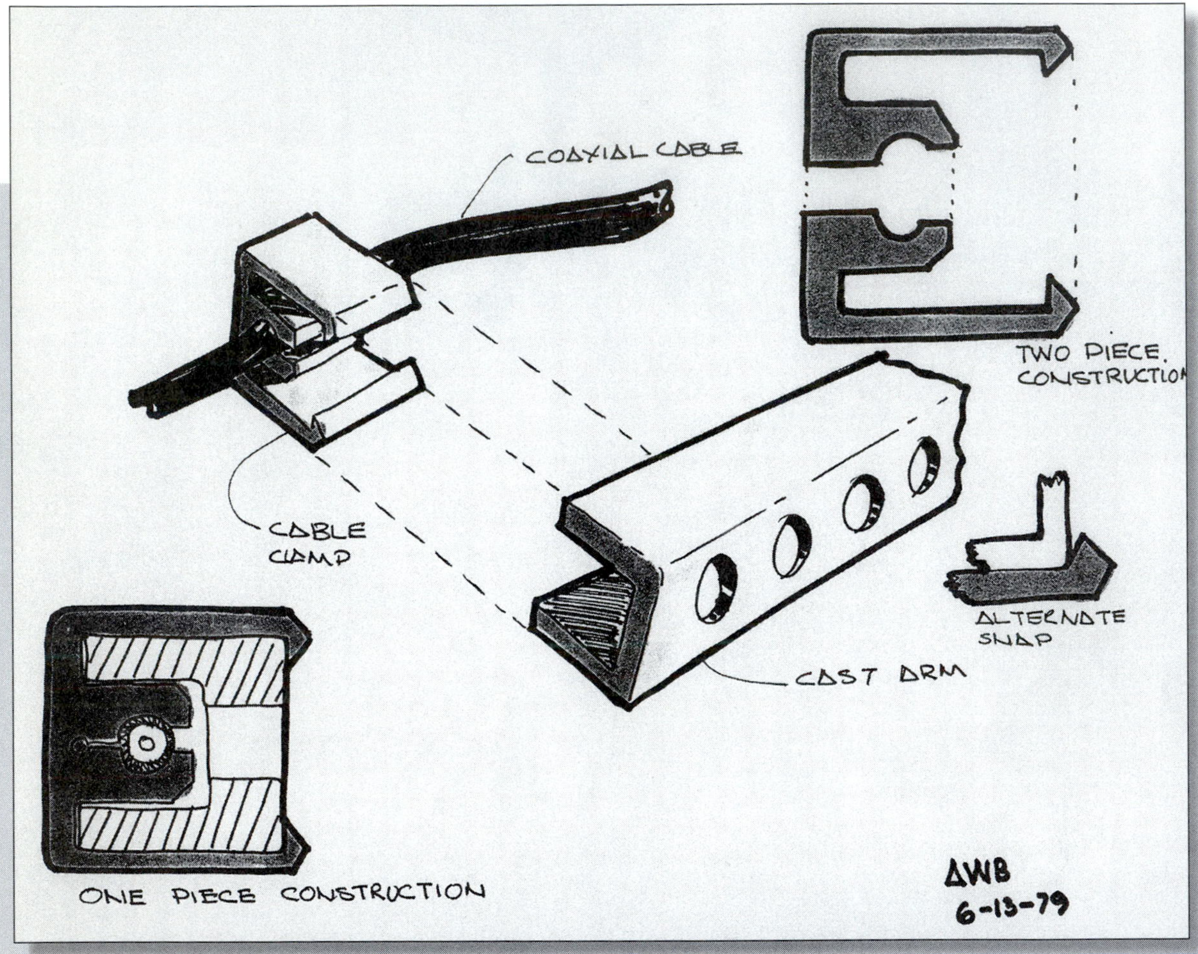

CHAPTER 4

Multiview Sketching

Objectives

When you have completed this chapter you will be able to

- project 2D views from a 3D object;
- use projection lines to align 2D views;
- sketch multiview drawings in the standard arrangement;
- visualize 3D shapes from 2D drawings and make sketches or models to illustrate them;
- interpret engineering linetypes and use them in your sketches;
- sketch simple section and auxiliary views.

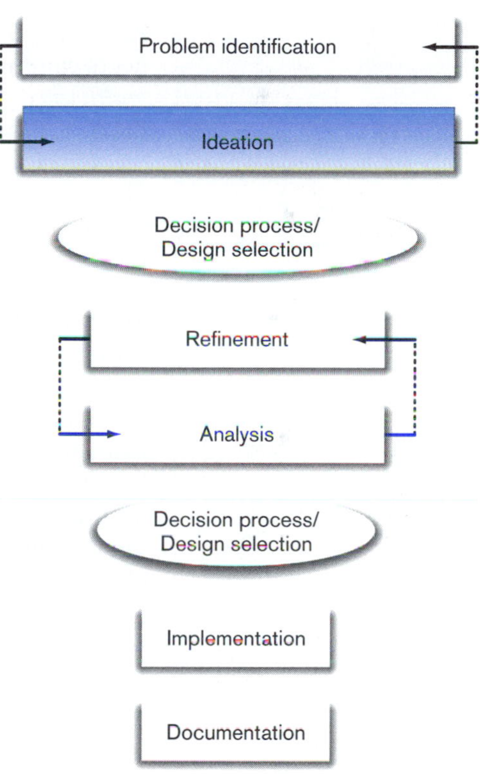

Pictorial sketches are good for getting your ideas down on paper. They also have the advantage that they make it easy for other people to visualize the shape of the object. However, pictorial drawings show the object tipped in a way that the surfaces are foreshortened, making it difficult to interpret the exact sizes of features by looking at them. A realistic appearance is often less important than a system that accurately reflects the size and shape of an object and the relationship between its parts. Orthographic projection (or multiview drawing) is a method that has been developed for transferring 3D shapes to 2D paper so that the resulting sketches can show object features true size, making it easier to define the size and relationships between features as well as the shape.

Multiview Sketches

As well as using pictorial sketches, more frequently perhaps, engineers communicate visually with a special type of drawing called a *multiview drawing* or *orthographic projection*. A multiview sketch and the object that it depicts are shown in Figure 4.1.

Consider the photograph of the V-block shown in Figure 4.1(a). The view in the photograph shows the object in perspective (as the eye sees it in the world around us). As you have learned, portions of the object that are further away from the viewer appear smaller when viewed in perspective. In this case, the back surfaces do not appear much smaller, because the V-block in the photograph is close to the camera and its surfaces are too short to see much perspective effect.

In engineering drawings, it is important to be able to accurately specify all of the information needed to make the object. Perspective drawings are not very useful for this because the back surfaces of the object appear smaller than the front surfaces of the object. In addition, some parts of the object are hidden from sight in the single view. Multiview drawings overcome these limitations to show the object's size and shape accurately and completely. They use multiple views of the object to show features on the object true size, as you will learn to do in the next section.

FIGURE 4.1
This V-block clamp is shown here in (a) a catalog photograph and in (b) a standard multiview sketch.

A

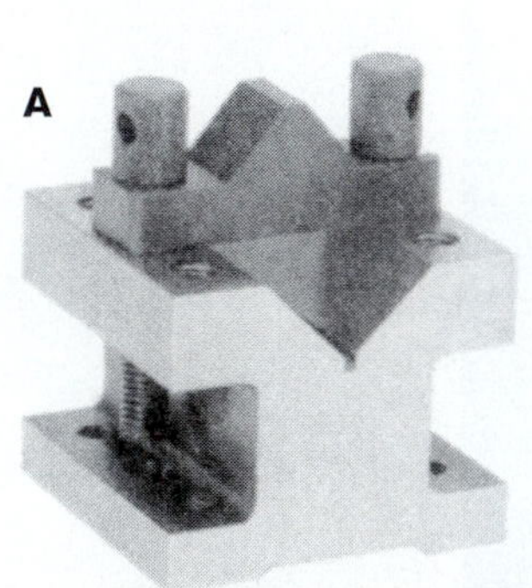

B

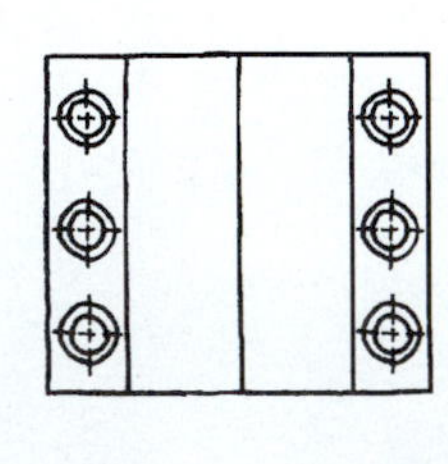

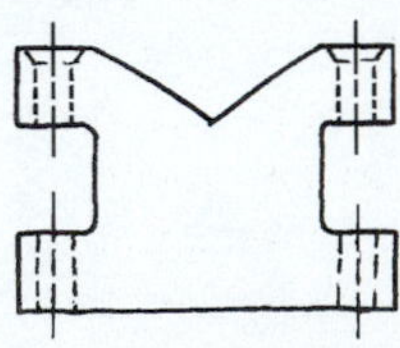

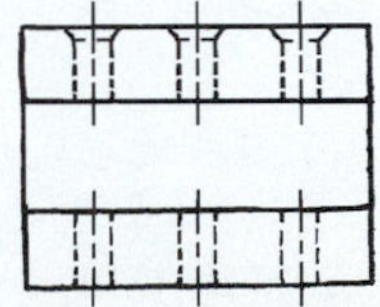

The Geometry of Projection

Gaspard Monge, born in France in 1746, is considered to be the father of orthographic projection. Prior to his time, there was no systematic approach to engineering drawings. As a professor at the Polytechnic School in France, Monge developed his ideas for orthographic projection. His method for describing a structure or device so it could be built reliably was so important that it was kept a military secret until Monge published *La Geometrie Descriptive* in 1795. Shortly thereafter, the basic principles of descriptive geometry and the projection of multiple views became a standard part of engineering education.

Even though many companies are using CAD to create 3D models and engineering drawings, existing multiview drawings outnumber them by far. For the last two hundred years, multiview drawings have been used by engineers, while 3D CAD is relatively new. Understanding how to interpret multiview drawings is an important skill for any engineer.

You will also find that the ability to create orthographic multiview sketches will aid you in visualizing and presenting your designs to others. Many items are easier to sketch orthographically than pictorially. Multiview sketching is another technique you can add to the set of skills you use to convey information about your engineering designs.

Orthographic Projection

Multiview drawings are made up of views of the object created by looking straight onto (perpendicular to) each point on the object from a particular viewing direction. Of course this is not how you see objects in real life. It is a construct that allows features on the object to be shown true size when the view is perpendicular to them.

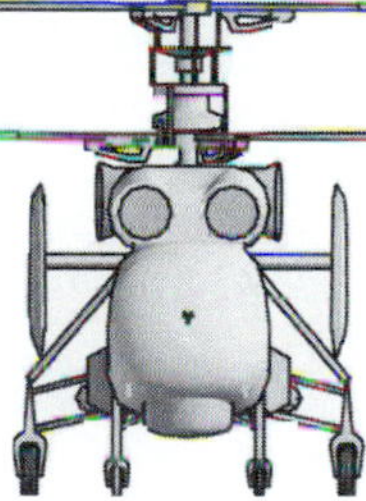

FIGURE 4.2
This view of the helicopter shows only its width and height. It is not possible to see how long (deep) it is from this viewpoint.

The drawings are called *multiview drawings* because more than one view is necessary to completely define a 3D object; each view shows only two of the three principal dimensions. Unlike the techniques used in pictorial sketches to show all three dimensions in a single 2D view, each 2D orthographic view maps to only two of the three dimensions of the object. No single orthographic view can show all three *principal dimensions*: height, width, and depth. Consider the single view of the helicopter in Figure 4.2. In this view, you can see only the helicopter's width and height. To see the depth of the helicopter, you need another view.

In multiview drawings, the principal dimensions are associated with certain views of the object, as shown in Figure 4.3. The horizontal distance in the front view is always defined as the *width* whether or not it is the widest dimension on the object. The vertical dimension in the front view is defined as the *height*. The *depth* is shown in

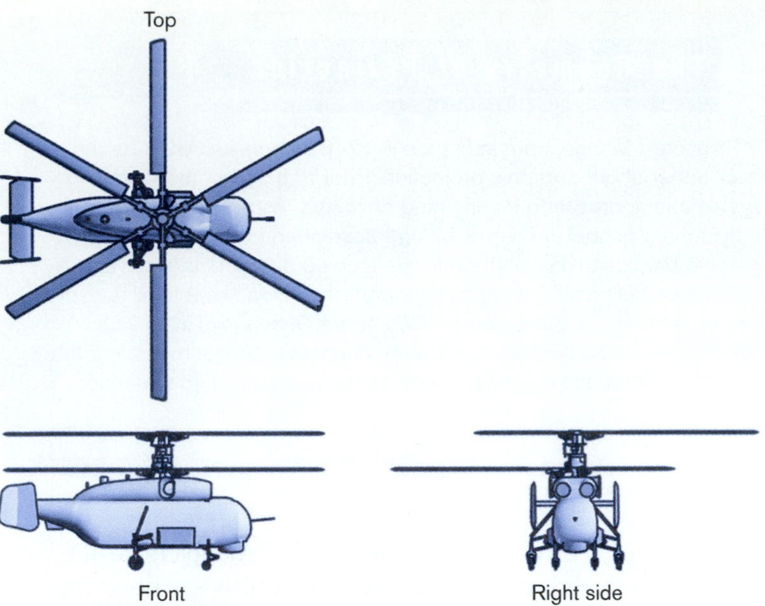

FIGURE 4.3
Each of the three 2D views in a multiview drawing shows only two of the three principal dimensions of the object. Look at the views of the helicopter. Which views show the height? Width? Depth?

any view from a viewpoint that is 90° from that of the front view. For example, both the side and top views would show the depth of the object. Associating the dimensions with the views of the object allows the terms *height*, *width*, and *depth* to be used consistently from view to view.

Standard Arrangement of Views

Multiple views are shown in a standard arrangement on the drawing sheet in order to make them easy to interpret. The front view is usually central and shows the object's shape most clearly. In the United States, Canada, and some other countries, the top view is always aligned above the front view, and the right-side view is placed directly to the right of the front view. (You will learn more about how drawing views can be arranged on the sheet later in this chapter.)

Six views of the object can be produced by looking straight onto the object from the front, back, right side, left side, top, and bottom. The techniques of orthographic projection depend on the views being mutually perpendicular to each other, so each view is produced by changing the viewing direction by 90°. The six possible views that can be produced by moving in 90° increments around the object are shown in Figure 4.4.

The helicopter's shape is most clearly shown in the front view. In this case, what you would ordinarily think of as the side of the helicopter shows its most characteristic shape, so it is depicted in the front view.

Figure 4.5 shows the same six views of the V-block. Notice that the front, top, rear, and bottom views all show the width of the object. The front, rear, right-side, and left-side views all show the object's height. The top, right-side, left-side, and bottom views all show the

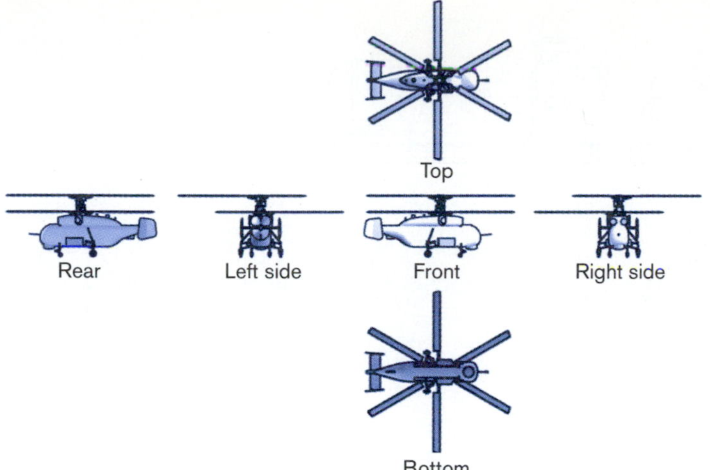

Figure 4.4
The viewpoint for each of the six views of the helicopter is rotated 90° from the view adjacent to it.

depth of the object. This relationship between the views is the key to interpreting multiview drawings. By locating surfaces, edges, vertices, and other features of the object in more than one view, you can form a mental picture of what the object looks like in real 3D space.

Notice that pairs of the six views show essentially the same information. The top and bottom views; right and left side views; and front and rear views all show similar information, just viewed from exactly opposite directions. For this reason, typically the front, top, and right-side views are shown and the redundant views are left out.

Some objects may not require even these three views to communicate the object's shape clearly. When you are creating sketches and drawings, keep in mind that your purpose is to provide the necessary information to the reader. Do not spend time drawing unnecessary views of the object. Do, however, add additional views as needed to clearly present the object to the reader of the drawing.

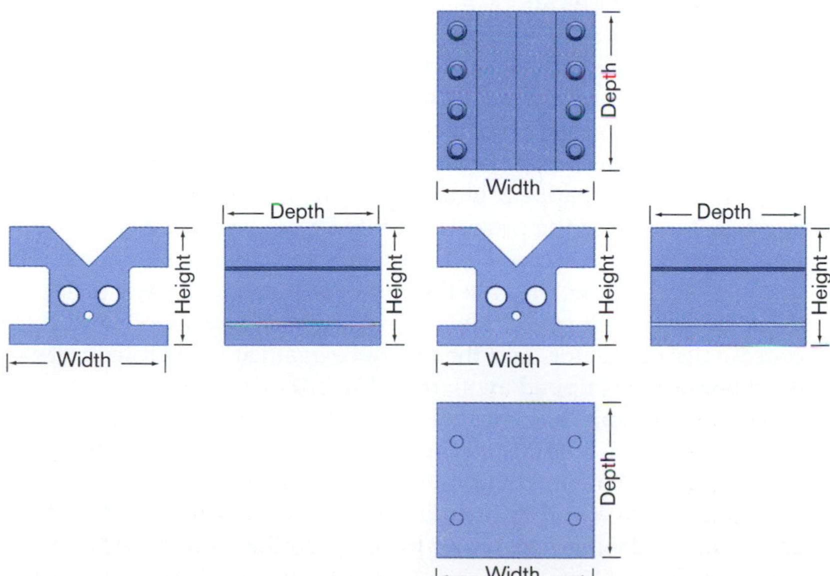

Figure 4.5
Six views of the V-block shown in Figure 4.1

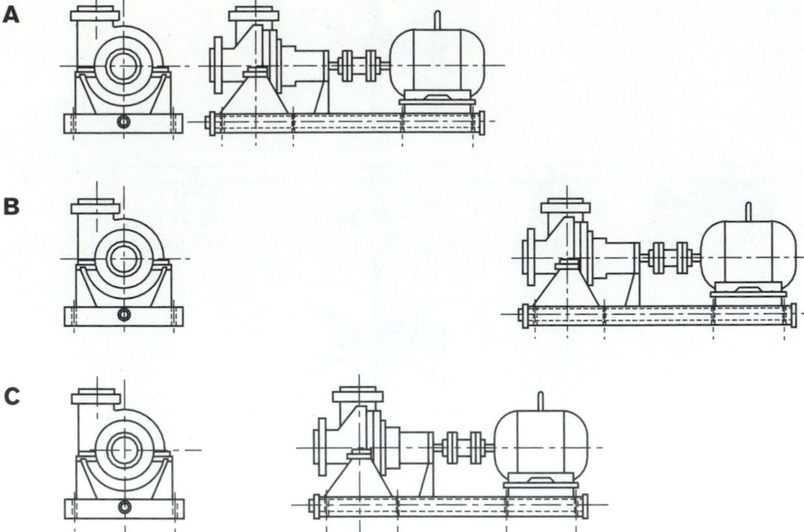

FIGURE 4.6
(a) Views are too close together; (b) Views are too far apart; (c) Spacing of the views is about right.

Spacing Between Views

The spacing between views is arbitrary, but you should use enough space so that the views are not crowded. You will want to add notes and dimensions to your sketch, so leave room between the views for this type of information. But do not space the views so far apart that they appear unrelated (see Fig. 4.6).

Relating Views to a Cartesian Coordinate System

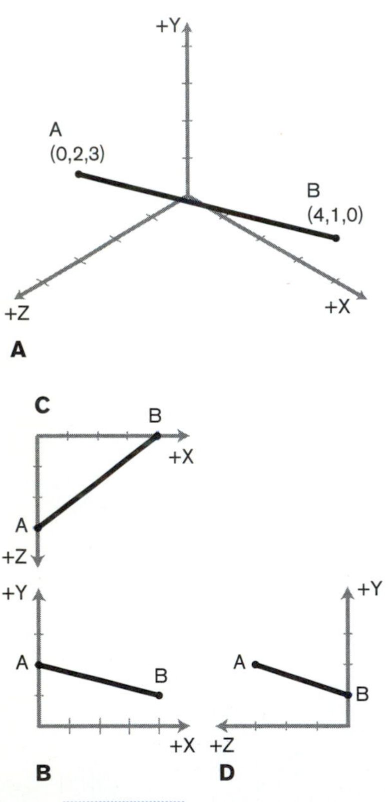

FIGURE 4.7
The line in (a) is plotted from A to B on isometric axes to show its appearance in 3D space. It is plotted again in front (b), top (c), and right-side (d) views that show the line and the appearance of the axes in each view.

To understand how the 2D views work together to fully define a 3D object, consider the angled line shown in 3D space as defined by the Cartesian coordinate system shown in Figure 4.7(a). Its vertices are identified by their locations relative to the origin in the X, Y, and Z directions.

The same line is shown in three orthographic views: top, front, and right-side. Each orthographic view shows only two of the three coordinates. The top view in this case shows the X and Z coordinates but not the Y coordinates. Because the line is angled, the true length of the line does not appear in the top view; instead it is foreshortened. It shows only the projection of its length onto the X–Z plane. In the front view you can see the X and Y coordinates of the line. The line appears foreshortened in the front view also because it is tipped away from the X–Y plane. The right-side view shows the Y and Z coordinates of the line, and the line once again appears foreshortened because it is tipped away from the Y–Z plane.

Use your pencil as a visual aid to help you picture the orientation of this line in space from the information in the three orthographic views. Imagine that your pencil is the line shown, where A is the eraser end and B is the point. The top view tells you that the eraser end of the pencil is closer to you than the pointed end. The front view tells you that the eraser end of the pencil is higher than the pointed end. The side view confirms that the eraser end is higher and closer to you than the pointed end. Figure 4.8 shows a pictorial representation of a pencil oriented similarly to the way the line is oriented in space.

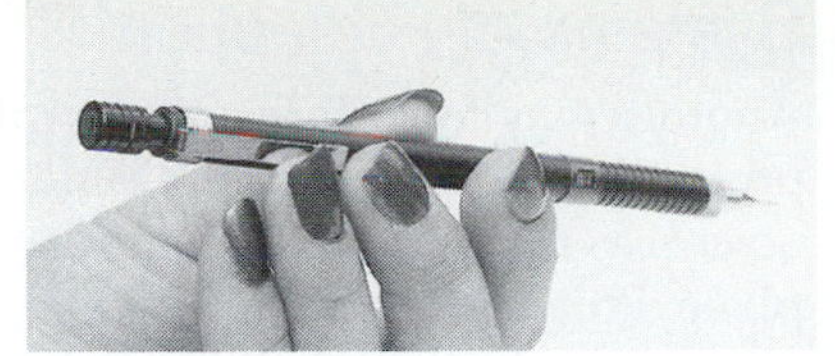

FIGURE 4.8
Did you orient the pencil this way from the drawings above?

Each orthographic view shows only two of the three Cartesian coordinates for the location of the endpoints, but by interpreting the views together you can produce an orientation for the lines of the drawing in 3D space. Figure 4. 9 shows an isometric view of an angled block. Its vertices are labeled with their X, Y, and Z coordinates.

Figure 4. 10 shows the same block drawn using three orthographic views: top, front, and side. Notice that the top view shows only the X axis direction and the Z axis direction. The Y axis is pointing out of the paper and appears as a dot at the origin. The drawing in the top view would appear the same regardless of the value for the Y coordinate for each point.

The front view shows only the X and Y coordinates for the vertices, and the Z axis is pointing toward you coming out of the paper. The front view would appear the same regardless of what the Z coordinate of each point was.

The side view shows the Y and Z plane, and the X axis direction is pointing out of the paper toward you. In the side view, the view would appear the same regardless of what the X coordinate of each point was (see Fig. 4.11).

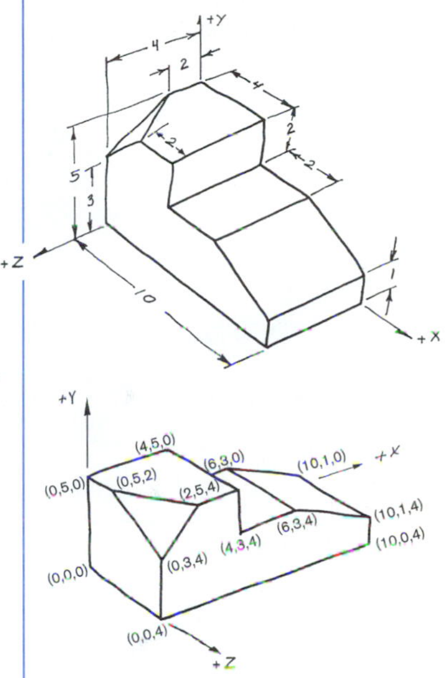

FIGURE 4.9
An isometric sketch of the angled block

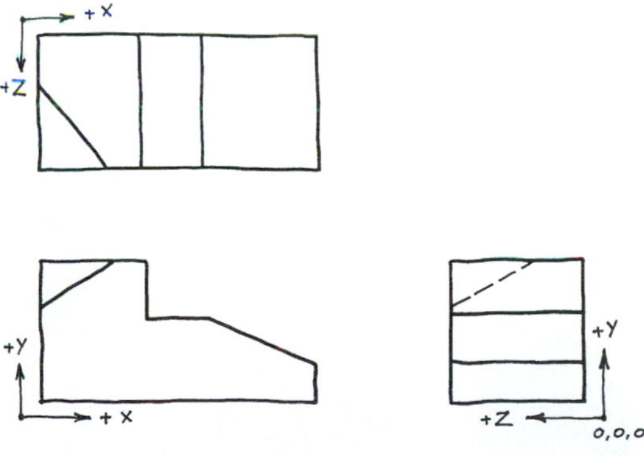

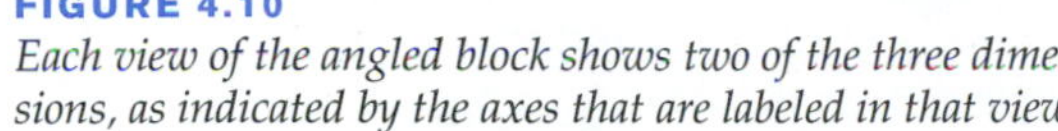

FIGURE 4.10
Each view of the angled block shows two of the three dimensions, as indicated by the axes that are labeled in that view.

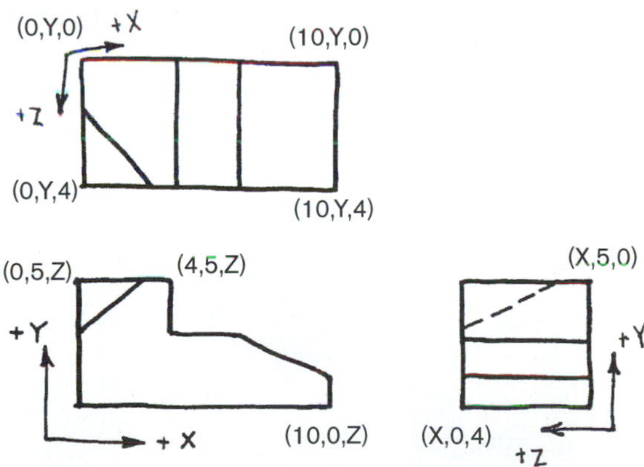

FIGURE 4.11
Dimensions along the axes shown in Figure 4.10 are visible in each view. The coordinates are labeled to show that the coordinate on one of the axes is not visible in each view. That is, (10,Y,0) indicates that the X and Z coordinates are visible, but all Y coordinate locations (like the Y axis) appear as a point.

Systems of Projection

Another way to picture orthographic views is to consider them as projections. A system of projection is a method of showing 3D objects as they would appear when projected onto a 2D viewing plane. Projection systems involve four basic parts: the object, the plane of projection, the viewpoint or direction of sight, and the projectors, as shown in Figure 4.12. To create a projection, projectors are extended from every point on edges of the object (and other significant features) to the plane of projection toward the viewpoint, tracing the appearance of the object onto the viewing plane.

FIGURE 4.12
A parallel, orthographic projection of the V-block onto a viewing plane aligned with the front surface

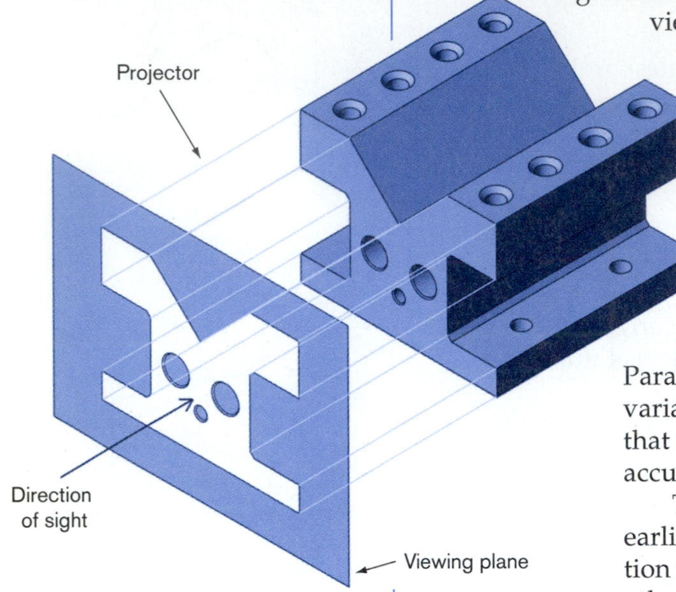

There are two major types of projection systems: parallel and perspective. You have already seen how perspective projectors converge to a point, while parallel projectors remain parallel to each other. The key difference between them is that in perspective projection, the distance from the object to the plane of projection affects the size of the object shown. Parallel projection eliminates distance as a variable and makes it easier to draw objects so that their true shapes and sizes are represented accurately.

The oblique sketches you learned to create earlier in this chapter are a type of parallel projection in which the projectors are parallel to each other but strike the viewing plane at some other angle, not 90°. In oblique projection systems, surfaces parallel to the viewing plane show their true size and shape, but surfaces that are perpendicular to the viewing plane show their foreshortened shape (shown smaller than their actual size).

FIGURE 4.13
Compare the way the object is oriented to the viewing plane and the way the projectors strike the plane. In each case, the viewpoint is perpendicular to the viewing plane.

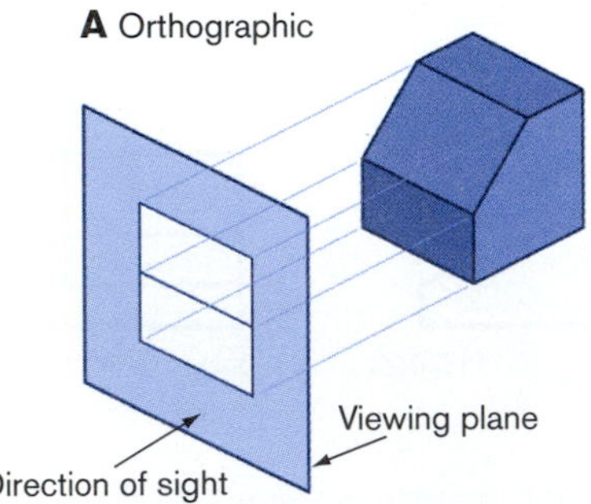

(a) Orthographic:
- *Viewing plane is aligned with a surface on the object.*
- *Projectors are perpendicular to the viewing plane.*

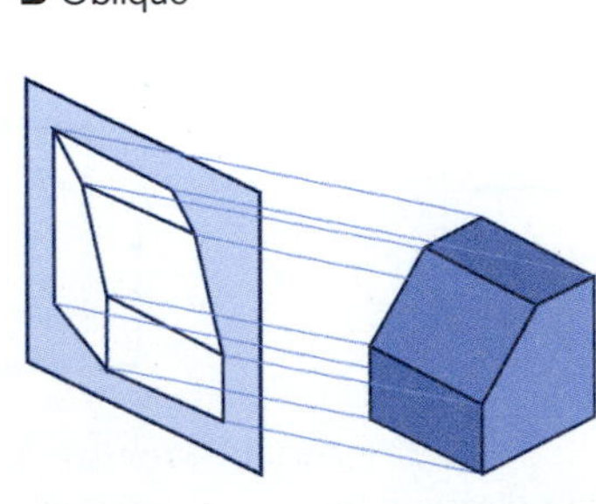

(b) Oblique:
- *Viewing plane is aligned with a surface of the object.*
- *Projectors are angled other than 90° to the viewing plane.*

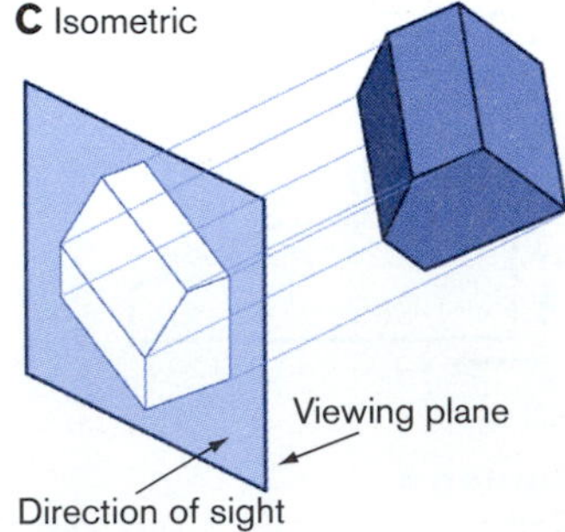

(c) Isometric:
- *Viewing plane does not align with a major surface of the object.*
- *Projectors are perpendicular to the viewing plane.*

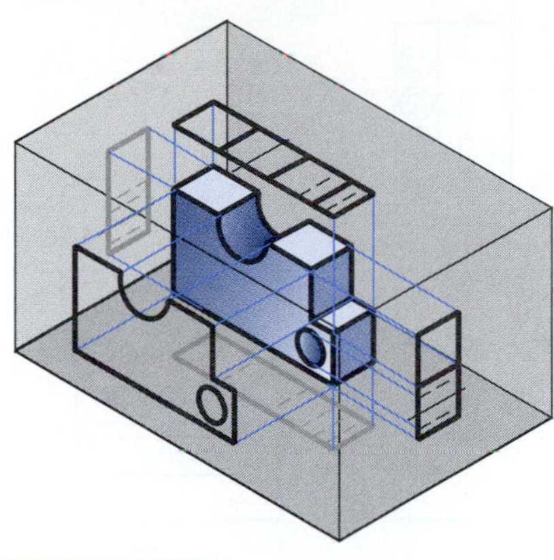

FIGURE 4.14
The walls of the glass box are six mutually perpendicular planes of projection that produce the six orthographic views.

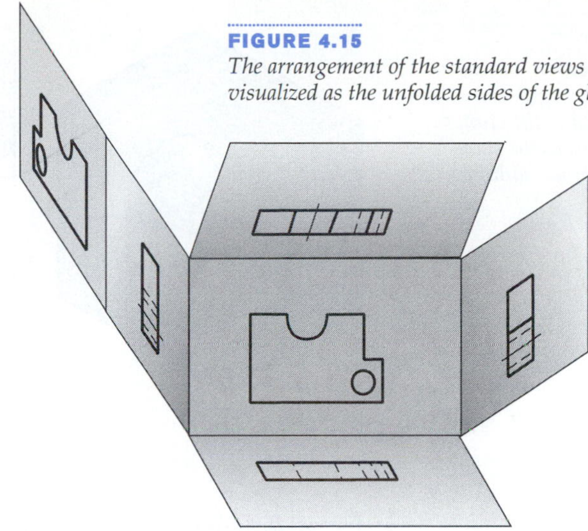

FIGURE 4.15
The arrangement of the standard views can be visualized as the unfolded sides of the glass box.

In both orthographic and isometric projection, the line of sight and parallel projectors are perpendicular to the viewing plane. The key difference between them is that the object in an isometric view is rotated and tipped so that the three coordinate axes appear in the view. In an orthographic view, the viewing plane is aligned with one of the surfaces on the object.

Figure 4.13 contrasts an orthographic projection, in which the projectors are perpendicular to the viewing plane (and parallel to the direction of sight) and the object is oriented so that its surface is parallel to the viewing plane, with oblique and isometric projection.

The techniques of orthographic projection are used to depict the surfaces of an object so they appear true size. Projecting the points on the object onto one of the six mutually perpendicular 2D viewing planes is comparable to plotting them on the X-Y, Y-Z, or X-Y plane of a Cartesian coordinate system.

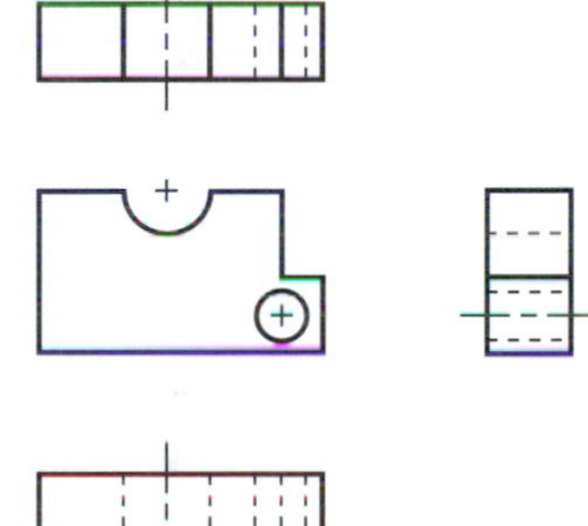

FIGURE 4.16
Multiview engineering drawings align the views as they would be aligned on the sides of the glass box, but do not show the fold lines between the views.

The Glass Box

You can think of the standard arrangement of orthographic views as views of an object being projected onto the walls of a glass box. The walls of the glass box are six mutually perpendicular planes of projection, representing the top, front, right-side, left-side, rear, and bottom views of the object. Figure 4.14 depicts the projection of an object onto the walls of the glass box. Unfolding the box, as shown in Figure 4.15, produces the arrangement of views used in the United States (called third-angle projection). The lines representing the edges of the glass box are not shown in engineering drawings. Figure 4.16 shows the final multiview drawing of the object.

FIGURE 4.17
Third-angle projection, in which the top view is arranged above the front view, is used in the United States and Canada as the standard arrangement for multiview drawings.

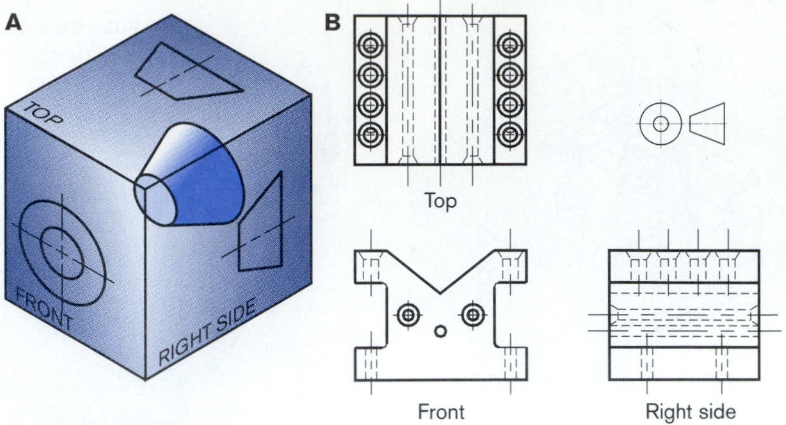

First-Angle and Third-Angle Projection

The arrangement of views shown in Figure 4.17 is called *third-angle projection* and will be used for all multiview drawings in this book. In Europe, Asia, and several other countries, views are arranged differently, using what is called *first-angle projection*. Because many industries provide products and services all over the world, you should be familiar with the differences between third-angle and first-angle projection. In third-angle projection, the viewing plane for the front view (or frontal viewing plane) is placed in front of the object and the views are projected onto it (see Fig. 4.17(a)). The viewing plane for the right-side view (or profile viewing plane) is placed to the right of the object and the view projected onto it. The viewing plane for the top view (or horizontal viewing plane) is placed above the object and the view is projected onto it. The views are then unfolded into the standard arrangement for third-angle projection.

Both first- and third-angle projection arrange the views around the front view. First-angle projection projects the front view onto a plane of projection behind the front view. The right-side view is

FIGURE 4.18
First-angle projection, in which the top view is arranged below the front view, is used in Europe and Asia as the standard arrangement for multiview drawings.

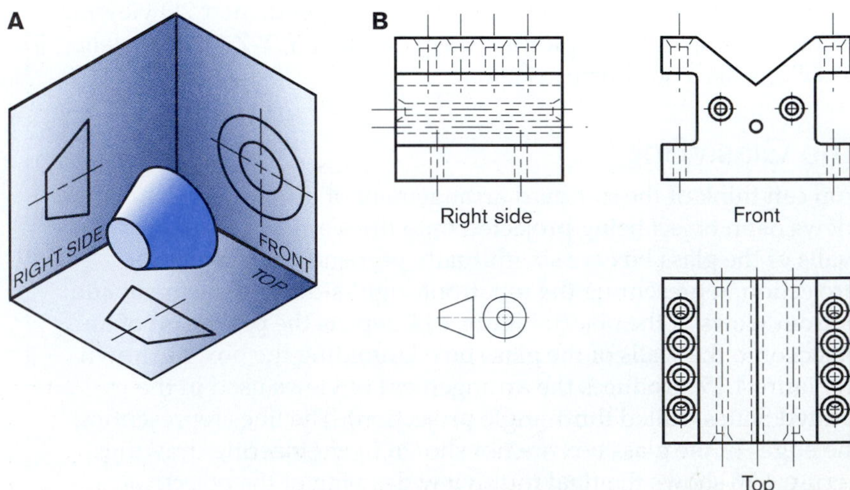

projected onto a viewing plane to the left of the object. A first-angle projection is shown projected onto viewing planes in Figure 4.18(a), then with the viewing planes unfolded into their standard view arrangement in (b).

Note the symbols that show two views of a truncated cone on the multiview drawings. The symbol in Figure 4.19(a) is a signal to the reader that third-angle projection was used to create the drawing. The symbol shows the arrangement of the front and right-side views of the cone as they would appear in third-angle projection. A similar symbol in Figure 4.19(b) indicates first-angle projection, where the right-side view is located to the left of the front view.

When you are preparing drawings for international use you should always include the symbol for the projection system used. Within the United States, drawings are assumed to be third-angle projection unless otherwise specified.

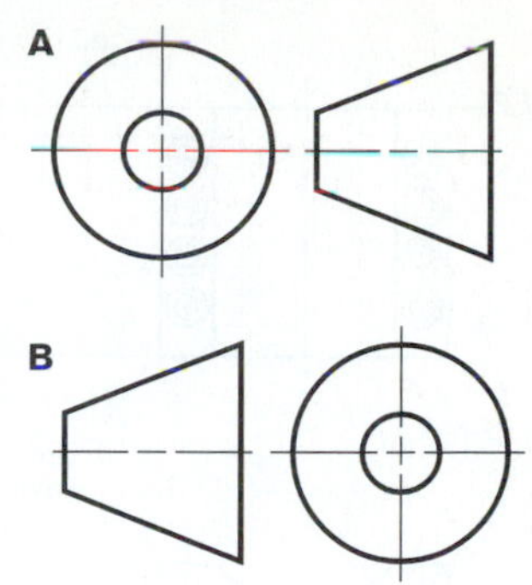

FIGURE 4.19
(a) Third-angle projection symbol
(b) First-angle projection symbol

Using Projection Lines

The unfolded box used to show the arrangement of views in the previous section is commonly used to illustrate how projection lines can be used to create multiview sketches.

Notice how the views of the V-block in Figure 4.20 line up with each other. Any point on the object shown in the front view lines up along a straight line, called a *projection line*, with that same point on the object in the top view. Notice that points in the top and front views have their X coordinates in common. A point seen in the side view aligns with that same point shown in the front view, and vice versa, because the Y coordinate of the point is the same in the front and side views. If you think of the object in a box that has been unfolded to produce the views, it is easy to see why the points line up.

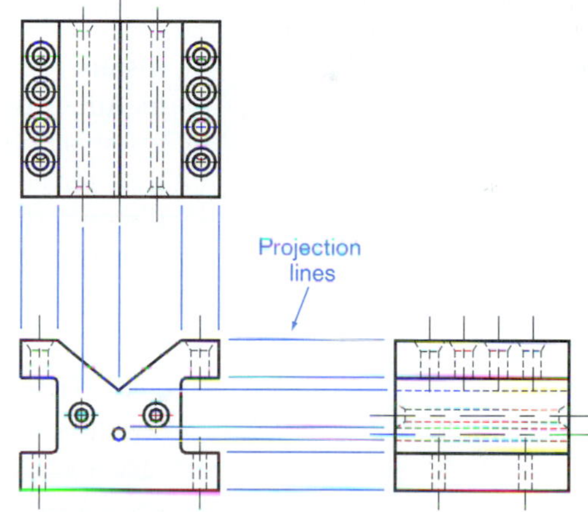

FIGURE 4.20
Features line up along projection lines between views.

You can think of these projection lines as the set of all of the points on the coordinate system that share a coordinate value. For example, the projection line and the points it runs between in the front and side views shown in Figure 4.21 all have the same Y coordinate, or height, on the part. To draw the side view, you can draw the shape in the front view, then project a line from each point in the front view to locate its Y coordinate in the side view. For example, the lower end of the "vee shape" must have the same Y coordinate in the front and side views, so a projection line into the side view can help you locate its height in the side view.

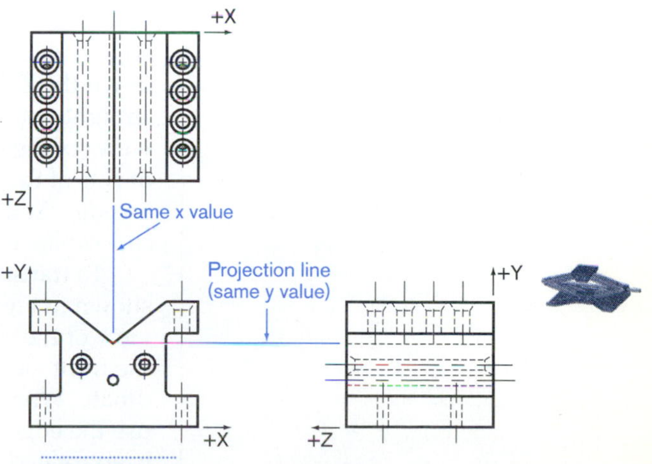

FIGURE 4.21
You can think of a projection line as showing all of the points that have a common coordinate value.

When two views sharing a common dimension are placed next to each other so that the dimension is aligned in both views, the views are referred to as *adjacent views*. The front and side view share the

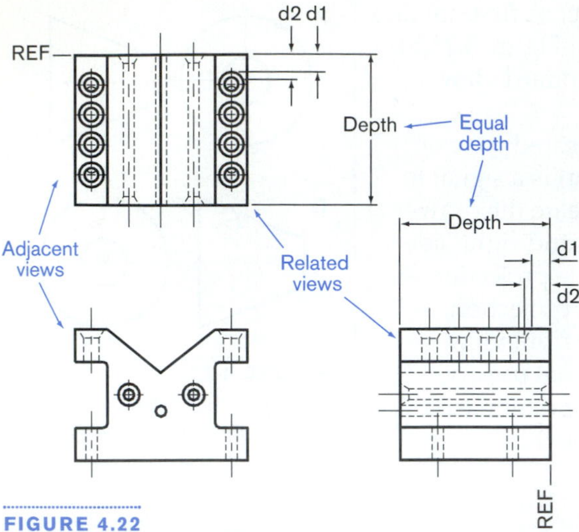

FIGURE 4.22
A reference surface can be used to locate points from one view to the next.

height dimension and are aligned so that points with the same height dimension line up along the same projection line.

To fully locate the point in the side view, you also need to locate the depth for the feature, or its Z coordinate. The techniques of orthographic projection and arrangement of views make it possible to transfer drawing information from one view to another. Since the position of the object in space does not change when it is projected to top, bottom, and side views, this depth measurement will be the same in any view that shows the object's depth. Notice that the top and side views of any point on the object show the same depth or Z coordinates.

The side and top views are referred to as *related views*. Related views are adjacent to the same view (in this case, the front view), and the distance that is measured perpendicular to projection lines between features appearing in both views is the same. If the location of the point is defined in the top view, its Z coordinate (or depth) can be transferred to the side view by measuring the perpendicular distance from a common reference surface, then transferring that distance to the side view, as illustrated in Figure 4.22.

Reference Surfaces (Coordinate System)

Another way to look at reference surfaces is to think back to the object as located on a Cartesian coordinate system. Figure 4.23(a) shows an isometric view of the Cartesian coordinate system. Think back to an orthographic view of the object as looking perpendicular to all of the points in the view along the direction of one coordinate system axis. When you look straight down the X axis, you see the X-Y and X-Z planes on edge and look straight onto the Y-Z plane. When you look straight down the Y axis, you see the X-Y and Y-Z planes on edge and look straight onto the X-Z plane. When you look straight down the Z axis, you see the X-Z and Y-Z planes on edge and look straight onto the X-Y plane. Figure 4.23(b) shows the orthographic views of the same Cartesian coordinate system. Notice that each set of two of these views show a common plane of the coordinate system on edge. You can use this edge view of the common plane as a *reference surface* to transfer measurements from view to view.

To transfer a dimension from one view to an adjacent view as shown in Figure 4.24, you can think of measuring from the edge view of the plane showing on edge in both views. A projection line in the front view is the set of all of the points that have the same Y coordinate. When transferring a location from a related view you can still use the edge view of their common coordinate plane. It will have been rotated to preserve its orientation. In the top and side views, the X-Y coordinate plane is on edge. You can locate the Z coordinate of vertices by measuring from this plane.

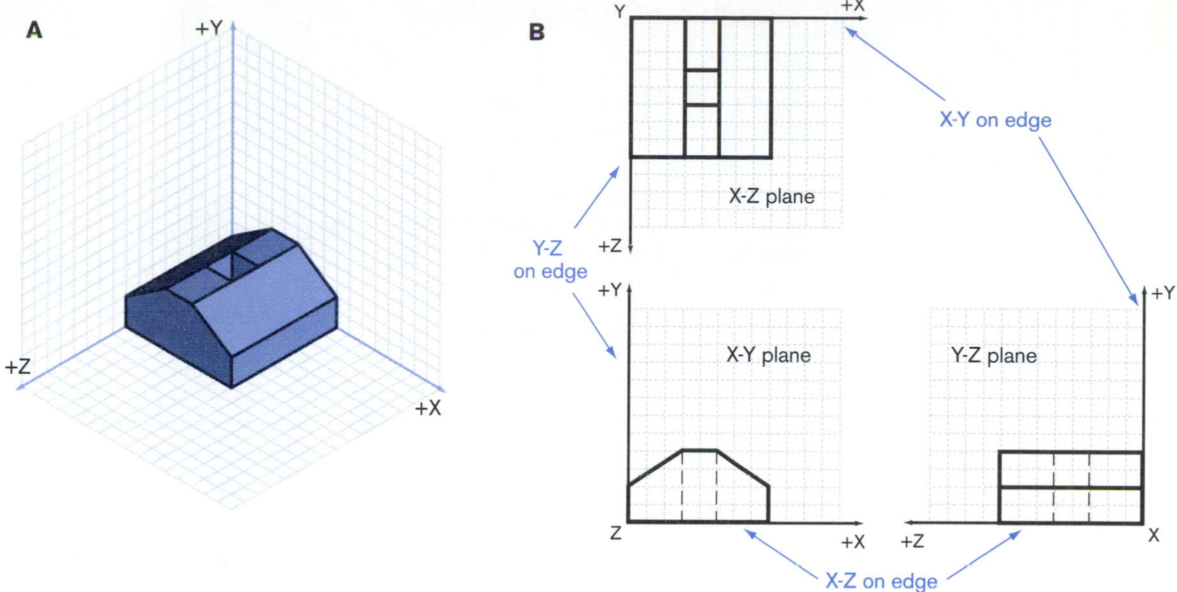

FIGURE 4.23
(a) The planes formed by the coordinate axes are visible in this isometric view. The intersection of the axes forms three planes. Each axis in a 2D view also represents a plane formed with the axis that is not visible; (b) this plane is on edge in the view so it appears as the axis line.

Using projections lines to align features in your sketch is an important technique to master. When you are creating multiview sketches, you should develop the views together, keeping them in the correct alignment.

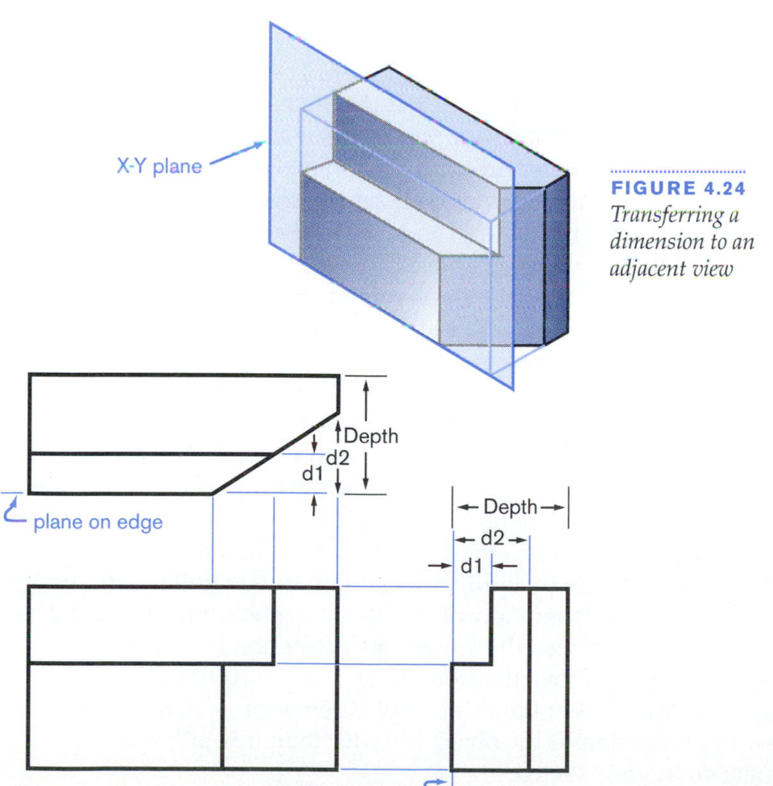

FIGURE 4.24
Transferring a dimension to an adjacent view

Sketching Orthographic Views

Figure 4.25 shows an isometric view of a block with the end cut off at an angle. In order to get started sketching multiple orthographic views of this object, first you should determine which view of the object you will orient as the front view.

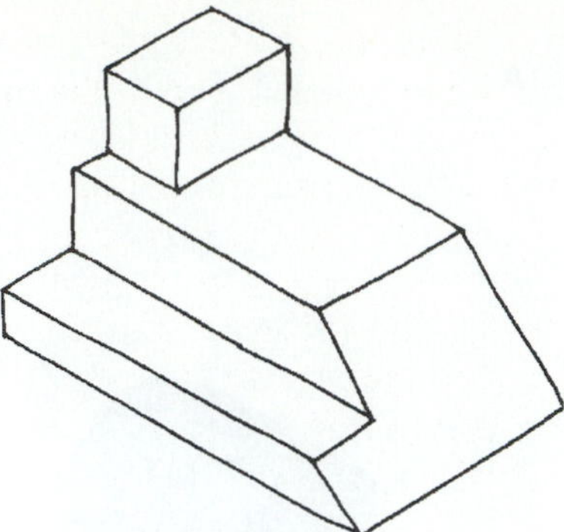

FIGURE 4.25
To begin sketching this angled block, you should decide how you will orient the block to produce the front view.

Choosing the Front View

There are some general rules to follow for determining which view of the object should be shown as the front view.

The front view in a drawing should:

- show the shape of the object clearly;
- show the object in a usual operating position;
- have large flat surfaces parallel to the viewing direction.

Think back to the six views of the helicopter. Refer to Figure 4.26. If you were sketching the helicopter, its side would be the best choice for the front view as it shows the distinctive shape most clearly. Also, you would tend to orient the helicopter in your sketch so that the top rotor is toward the top of your drawing sheet. It would look odd for it to be shown upside down. The helicopter does not have many large flat surfaces, but it is oriented so that it is essentially parallel to the view and not tipped at some angle.

After you have selected the direction of sight you will use to show the object in the front view, it is preferred to show the top view (aligned above the front view) and the right side view (aligned to the right of the front view), unless other views are clearer. If the left-side view is clearer than the right-side view, you may want to consider reorienting the object in the front view (by 180°) in order to produce a clear right-side view.

Keep in mind that sketches and drawings do not always have to show three views. If you are drawing an object that has a uniform thickness, you can show a single view and provide the thickness in a drawing note. The time you save by not drawing unnecessary views can be productively spent on other tasks.

Figure 4.27 illustrates the steps used to sketch the block shown in Figure 4.25. Begin by lightly blocking in the overall height, width, and depth for the front view of the block, as shown in Figure 4.27(a). Leaving space between the views, use projection lines to keep the top view aligned above the front view, and the right-side view aligned to the right of the front view. Keep your sketch of the front view in proportion. If the object is wider than it is tall, make it appear so in your sketch.

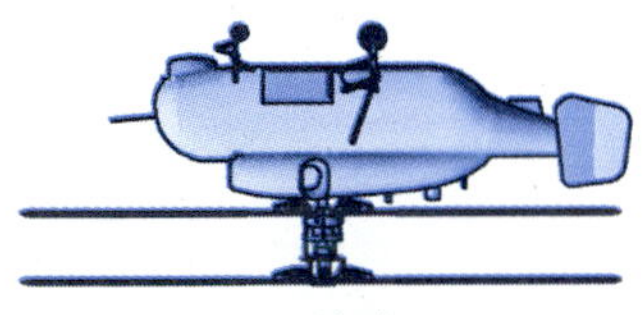

poor choice

poor choice

good choice of front view

FIGURE 4.26
Choosing the best front view

Next lightly block in the features that will be visible in the front view, paying attention to the proportions as shown in (b). At this point, you may want to slightly darken the object lines so that the shape in the front view begins to stand out clearly as shown in (c). Otherwise you may become confused by so many lines and project incorrect features into the other views.

Use projection lines to locate the features shown in the front view in the side and top views as shown in steps (d) and (e). Darken the final lines of the sketch so that the object stands out clearly. Object lines should be thick and dark as shown in (f).

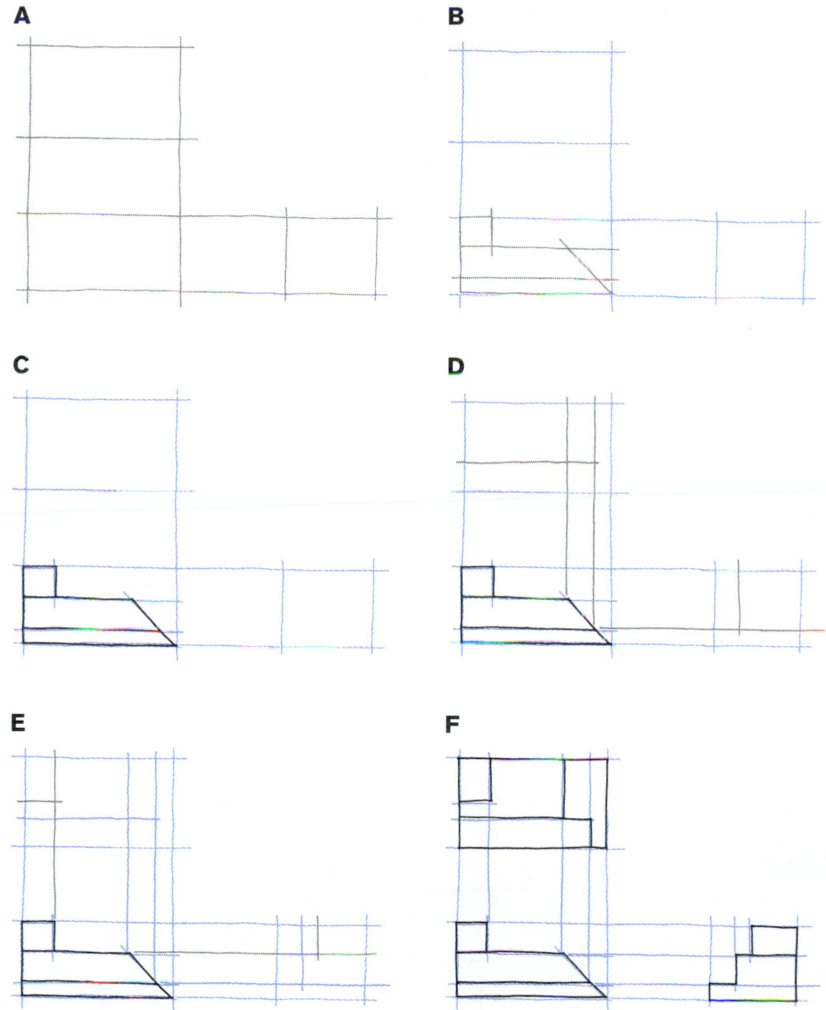

FIGURE 4.27

Sketching orthographic views: (a) Lightly block in the overall dimensions; (b) Use light lines to locate features in the front view, which should show the overall shape of the object and represent features in proportion; (c) Slightly darken the lines of the object as needed to help you visualize the shape; (d) Project points from corners and intersections in the front view to locate features in the top and side views; (e) Transfer depth dimensions; (f) Darken final object lines to produce a clear sketch.

FIGURE 4.28
Frank Zappa used this single-view illustration on the cover of his album, "Ship Arriving Too Late to Save a Drowning Witch." (Title and cover art used by permission of The Zappa Family Trust. Original illustration by Roger Price © 1953, 1981. All rights reserved.)

Interpreting Multiview Drawings

Can you interpret the single-view drawing in the cartoon shown in Figure 4.28? If you did not know that the drawing was titled "Ship Arriving Too Late to Save a Drowning Witch," would you be able to "see" the familiar objects and their relationship in 3D? The cartoonist uses a single view that one would not easily identify and surprises you with a humorous title. In engineering, less familiar objects are depicted in multiple views to provide more information, but you still need to interpret those views to "see" what the object looks like. Understanding how to interpret multiview engineering drawings will help you learn to use the same techniques to create 2D views of 3D objects yourself.

Lines and Points in a Multiview Drawing

Lines in a multiview sketch or drawing represent where *surfaces* on the object meet to form an edge. Most lines in a drawing represent an intersection where two flat (or planar) surfaces join together on the object. Lines can also represent a flat surface that is viewed on edge (when the surface is perpendicular to the view). An edge view of a curved surface would appear as an arc or irregular curve, depending on the shape of the surface where it intersects the other surface.

A straight line can also represent the extreme edge of a curved surface, called the *limiting element* of the contour. Examples are shown in Figure 4.29.

Points in a drawing always represent where three or more surfaces join together to form a vertex on the object. A point represents a location in space and appears in a drawing as a corner or the endpoint of a line. An example of a point in a drawing and on the object is shown in Figure 4.30. When labeling points in a drawing or

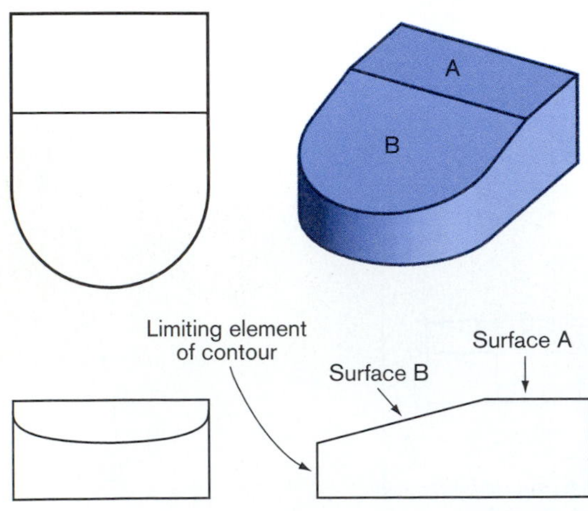

FIGURE 4.29
Most of the lines in the views of this model represent the intersection of planar surfaces, or surfaces viewed edgewise. One edge in the side view is actually the limiting element of the curved surface on the model.

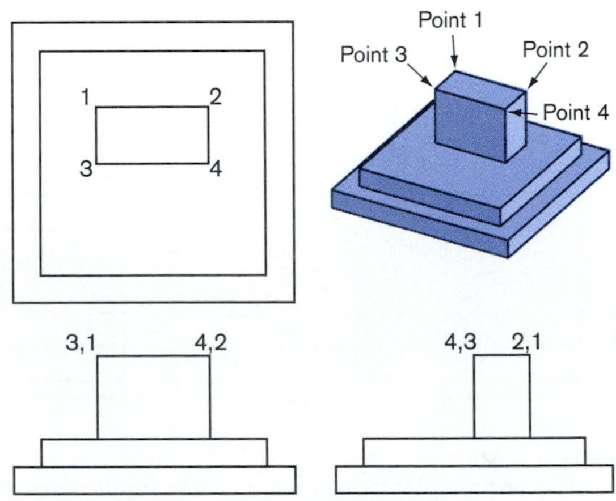

FIGURE 4.30
Points represent vertices and appear as corners.

sketch, list the closest point first when two points coincide, as in the point view of a line.

Any point on the object shown in a view must align along a projection line with that same point in any other view as shown in Figure 4.31.

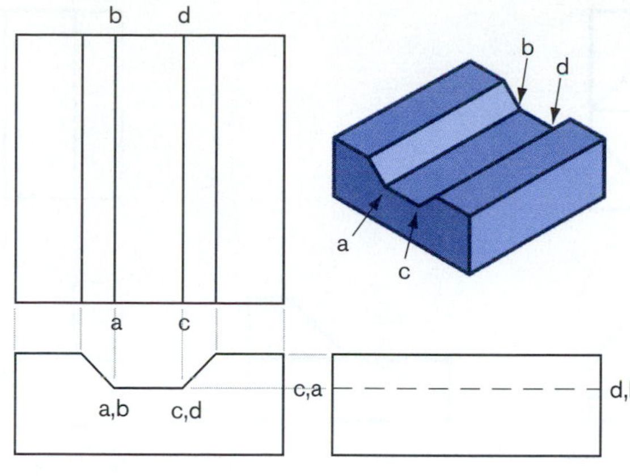

FIGURE 4.31
Points in any view line up with the same point in any other view.

Normal, Inclined, and Oblique Surfaces

Understanding the orientations that planes (flat surfaces) can have with respect to the viewing plane can help you interpret the object. There are three possible orientations that a surface can have with respect to the viewing plane: normal, inclined, or oblique. When you understand these three typical orientations, you will be able to recognize these surfaces in the standard views.

A *normal* surface is parallel to one viewing plane (or perpendicular to the line of sight for the view) and shows true size in that view. It shows on edge in the two other standard views, where it appears as a line (see Fig. 4.32).

An *inclined* surface is perpendicular to one of the viewing planes, but not true size in any view (see Fig. 4.33). It appears on edge as an angled line in the view to which it is perpendicular. In the two other standard views, its shape appears foreshortened.

An *oblique* surface is not parallel *or* perpendicular to any of the standard views (see Fig. 4.34). Its shape is foreshortened in all three standard views; it doesn't appear on edge in any view.

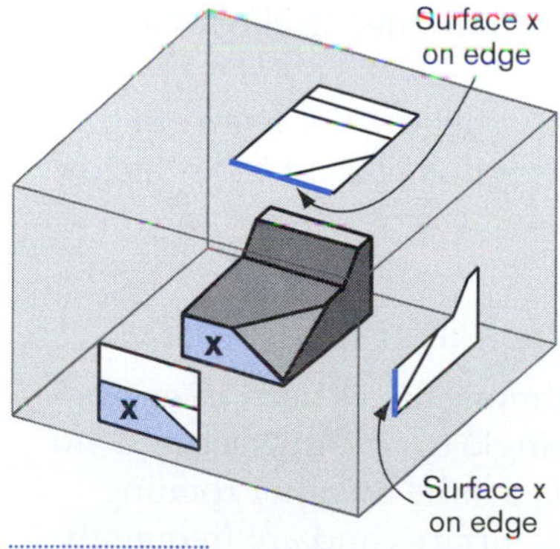

FIGURE 4.32
The highlighted surface on the object is normal to the viewing plane, where it appears true size. It is highlighted as it appears on each viewing plane, too. Notice that the normal surface appears on edge in the other two views.

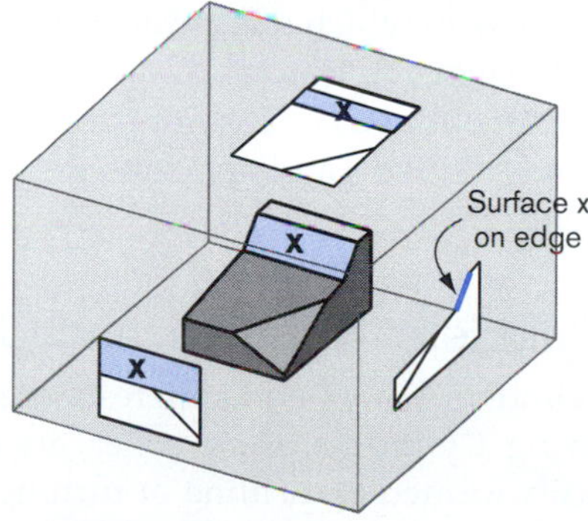

FIGURE 4.33
The highlighted surface on the object is an inclined surface. Its appearance in the three views is highlighted on the three viewing planes. Notice that the inclined surface appears as an angled line in one view, and foreshortened in the other two. It appears true size in none, as it is parallel to none.

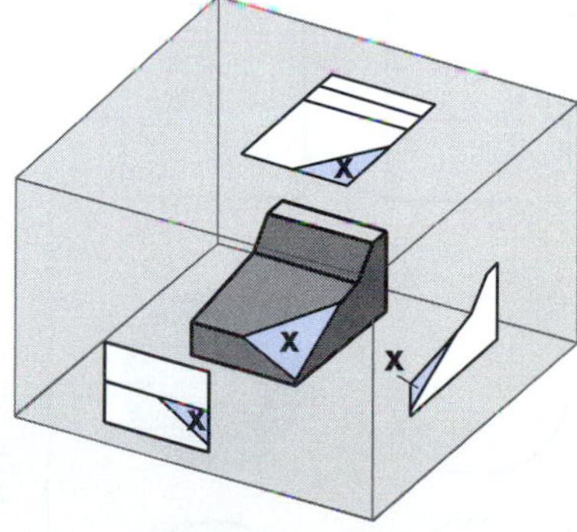

FIGURE 4.34
The highlighted surface on the object is an oblique surface. Its appearance in the three views is highlighted on the three viewing planes. Notice that the inclined surface appears foreshortened in each view. It appears true size in none, as it is parallel to none. It appears on edge in none, as it is perpendicular to none.

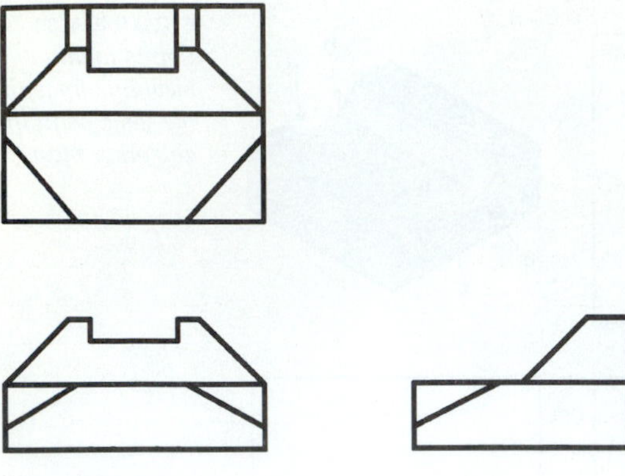

FIGURE 4.35
After you've identified each surface as normal, oblique, or inclined, check the answer on page 134.

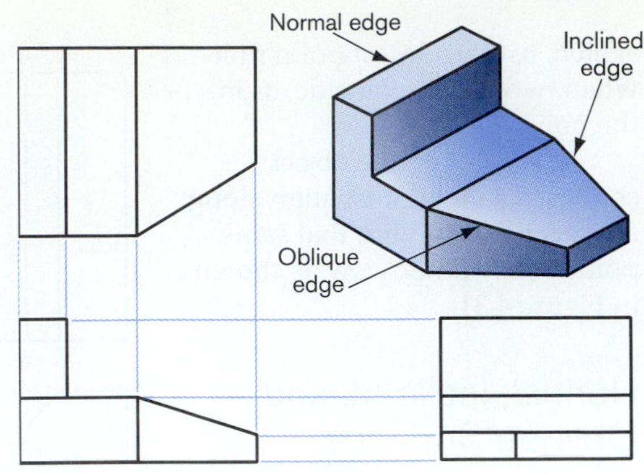

FIGURE 4.36
Normal, oblique, and inclined edges

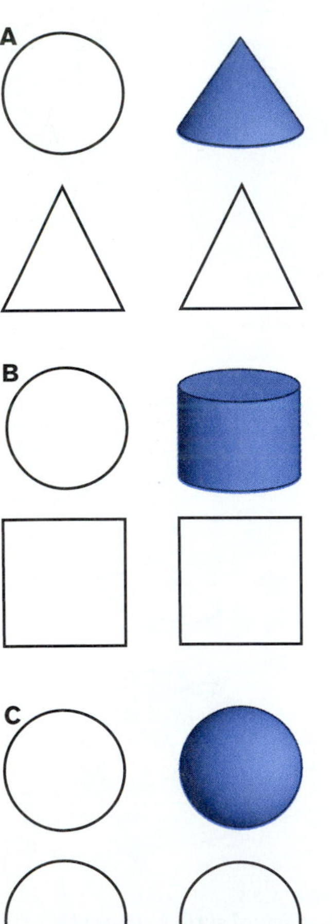

FIGURE 4.37
Front, top, side, and pictorial views of a (a) Cone, (b) Cylinder, and (c) Sphere.

Figure 4.35 shows normal, inclined, and oblique surfaces in a multiview drawing. Can you identify which is which from the preceding definitions without looking at the key?

Normal, Inclined, and Oblique Edges

The terms *normal*, *inclined*, and *oblique* can also be applied to a single edge on an object. Figure 4.36 shows examples of normal, inclined, and oblique edges.

A normal edge, similar to a normal surface, is parallel to two of the principal views and shows true length in those views. It is perpendicular to one of the principal viewing planes and in that view the endpoints line up to produce a point view.

An inclined edge is parallel to one of the principal viewing planes and tipped toward the other two. It shows true length in the view to which it is parallel, and foreshortened in the other two views.

An oblique edge is not parallel or perpendicular to any of the principal views and appears foreshortened in each of the principal views.

Cylinders, Spheres, and Cones in a Drawing

Cylinders, cones, and spheres are common shapes used in engineering. Cylinders, as are cones, are a single curved surface and are easily formed by drilling or turning on a lathe by other rotating machinery. Spheres are double curved surfaces and are frequently used as bearings, knobs, and in other features. You should become familiar with the appearance of these shapes in orthographic views. A cylinder appears as a circle in one view and as a rectangle in the adjacent view. Spheres appear as circles in two adjacent views. Cones appear triangular in one view and generally circular in the adjacent view. Use your knowledge of these primitive shapes to help you identify these types of features on the object. Figure 4.37 shows cylinders, cones, and spheres both as they would appear in a multiview drawing and as shaded 3D models.

Sketching Cylinders, Spheres, and Cones

Practice sketching cylinders, cones, and spheres in both orthographic and pictorial views so you can easily sketch these shapes when they are features of other objects. Use projection lines to help align features between orthographic views. Recall the appearance of a cylinder in two views. In one view the cylinder appears circular and in the adjacent view it appears rectangular. A cone appears triangular in one orthographic view and circular in the adjacent view. Spheres appear circular in all orthographic views. Figure 4.38 illustrates the steps in sketching orthographic views of objects with cylindrical, conical, and spherical features. Refer to Figure 4.39 for steps in sketching and shading pictorial views of these features.

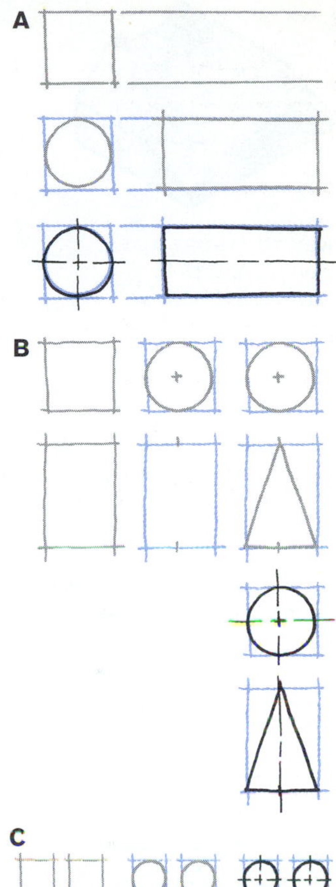

FIGURE 4.38
Sketching orthographic views of objects with cylindrical, conical, and spherical features

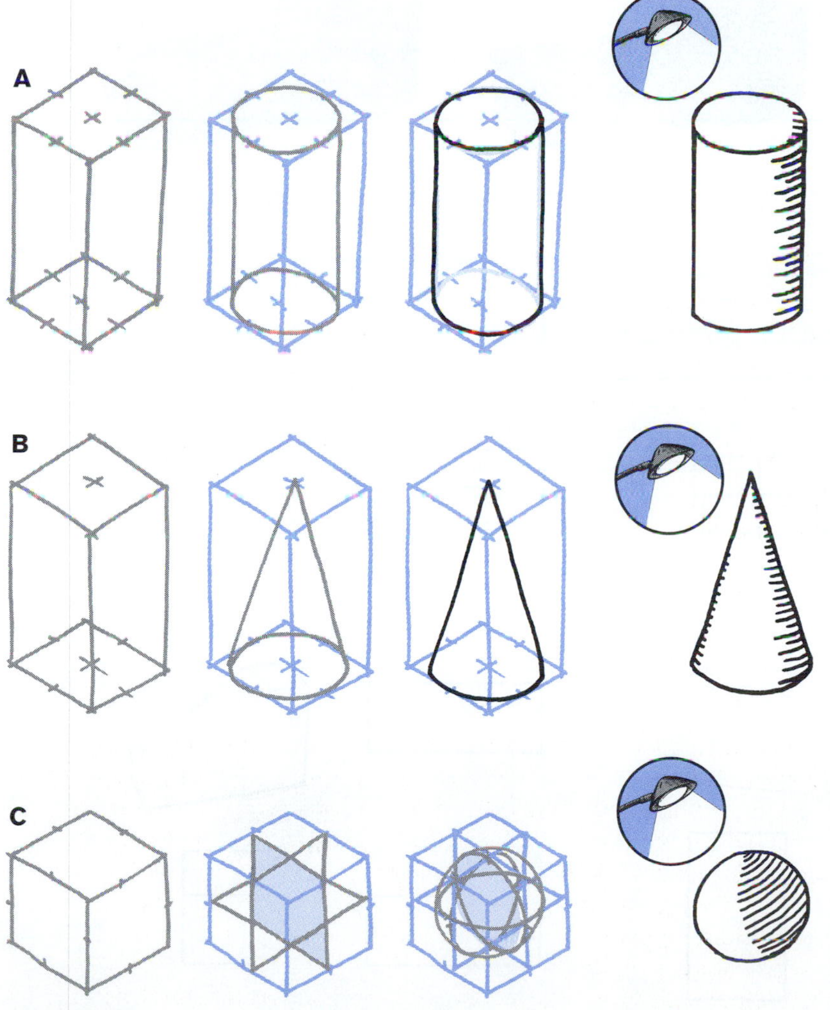

FIGURE 4.39
Sketching pictorial views of cylinders, cones, and spheres.

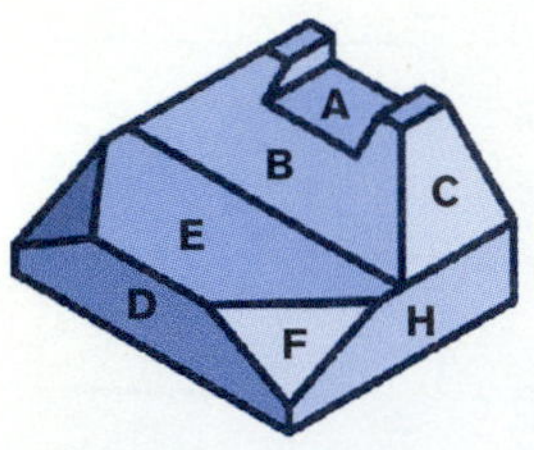

KEY TO FIGURE 4.35
Normal: A, E, D, H.
Inclined: B, C. Oblique: F.

Intersections and Tangencies

To represent complex objects, multiview drawings also have conventions for depicting the way planar and curved surfaces meet. A plane surface can intersect or be tangent to a contoured surface, as shown in Figure 4.40. When the surfaces are tangent, no line is drawn to indicate a change between the plane and contoured surface. When the surfaces are intersecting, an edge is formed that is shown as a line in the drawing. Figure 4.41 shows examples of intersecting and tangent surfaces.

Use projection to help you draw intersections between surfaces in orthographic views. Figure 4.42 shows some typical intersections between different diameters of pipe. When you are sketching these in orthographic views, identify points on the intersection and project those points into the remaining views as shown in Figure 4.43.

FIGURE 4.40
(a) Intersecting and (b) tangent planes in pictorial views

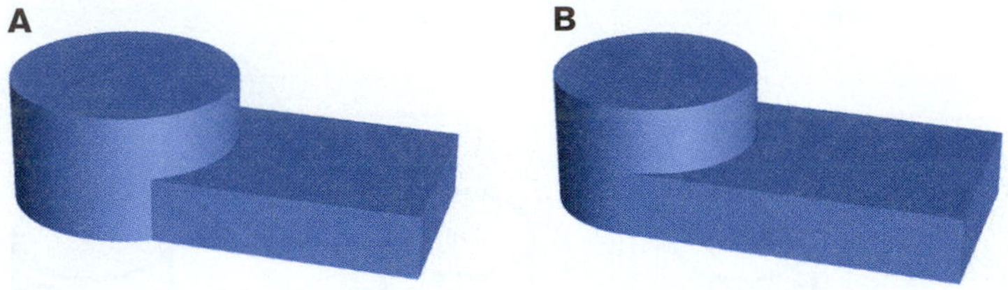

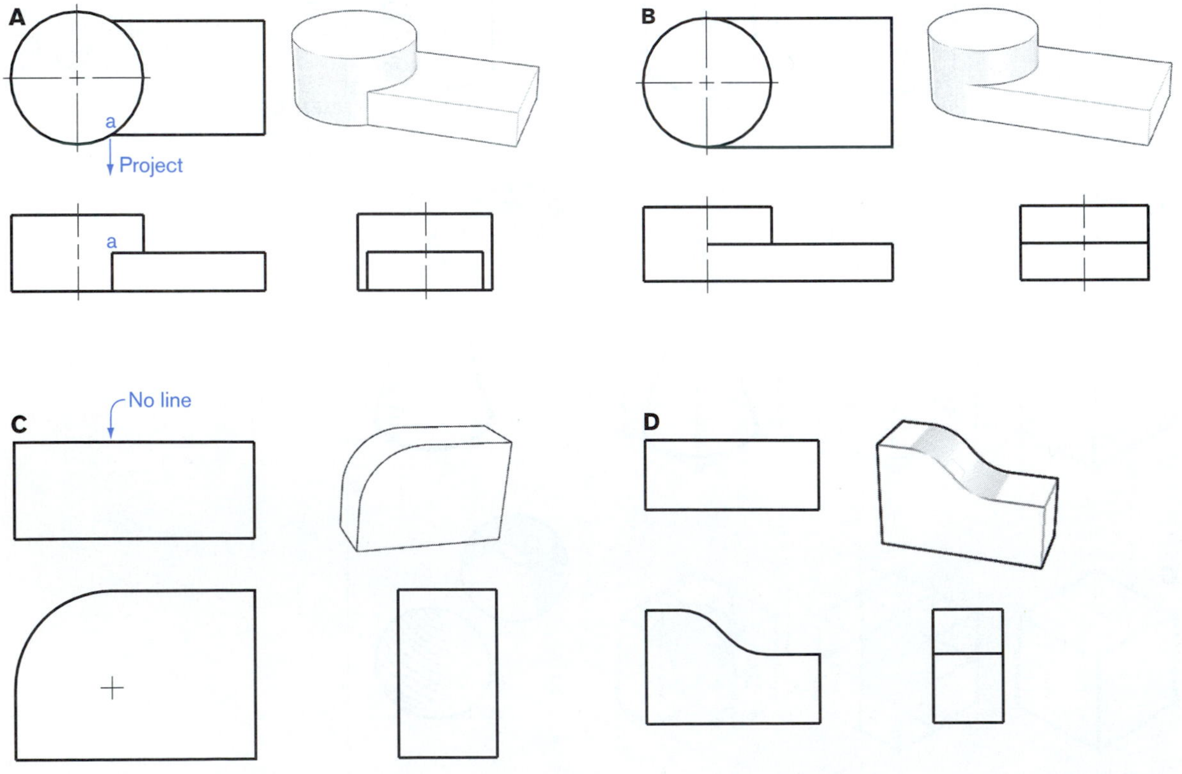

FIGURE 4.41
Examples of intersecting and tangent planes in orthographic and pictorial views.

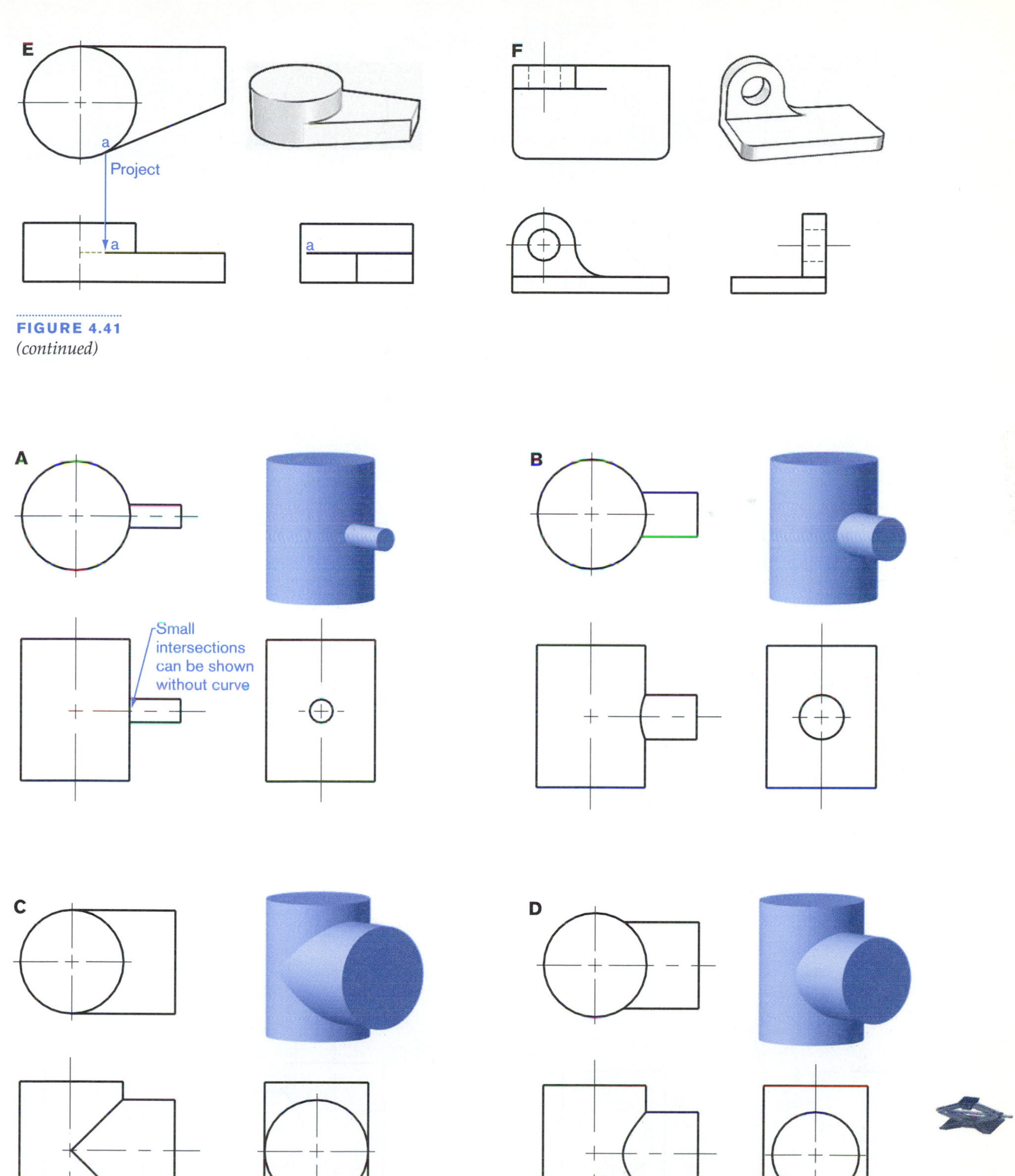

FIGURE 4.41
(continued)

FIGURE 4.42
Intersections of different diameter cylinders

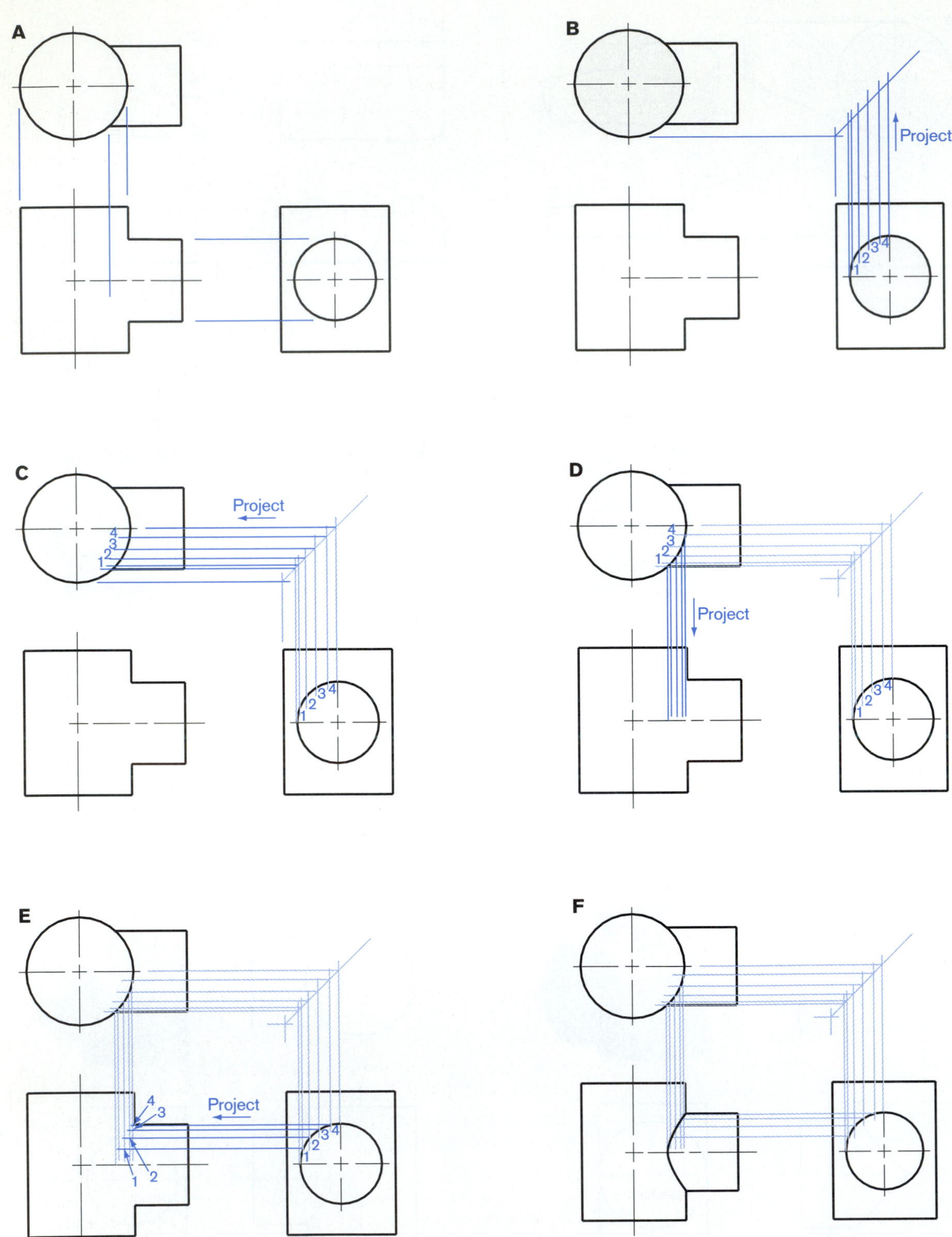

FIGURE 4.43
Projecting an intersection

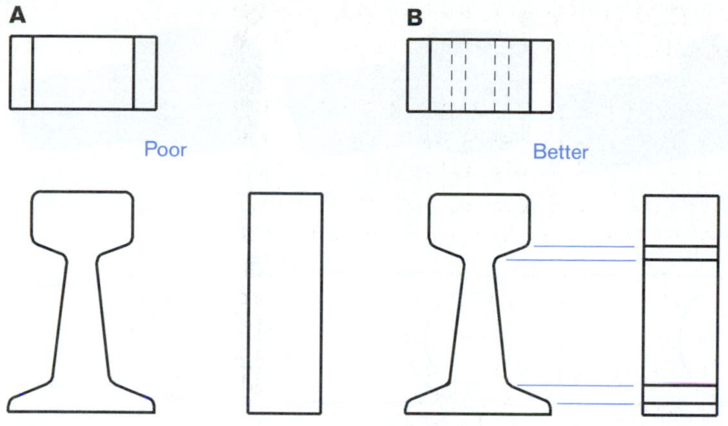

FIGURE 4.44
Conventional lines are used to show changes in direction between two surfaces when leaving lines out might make the view misleading.

Sometimes a drawing may appear misleading if it contains planar surfaces that change direction through tangency with a contoured surface. Because the surfaces are tangent, no line is drawn where they meet, so the change in direction is not depicted. In such cases, a line may be drawn to indicate the change of direction of the surface even though tangent surfaces do not form a visible edge. Figure 4.44 shows an example where the projection might be misleading without the addition of such lines.

Projecting Curves

Use projection techniques to locate irregular curves in orthographic views. Figure 4.45 shows the steps in projecting a curve.

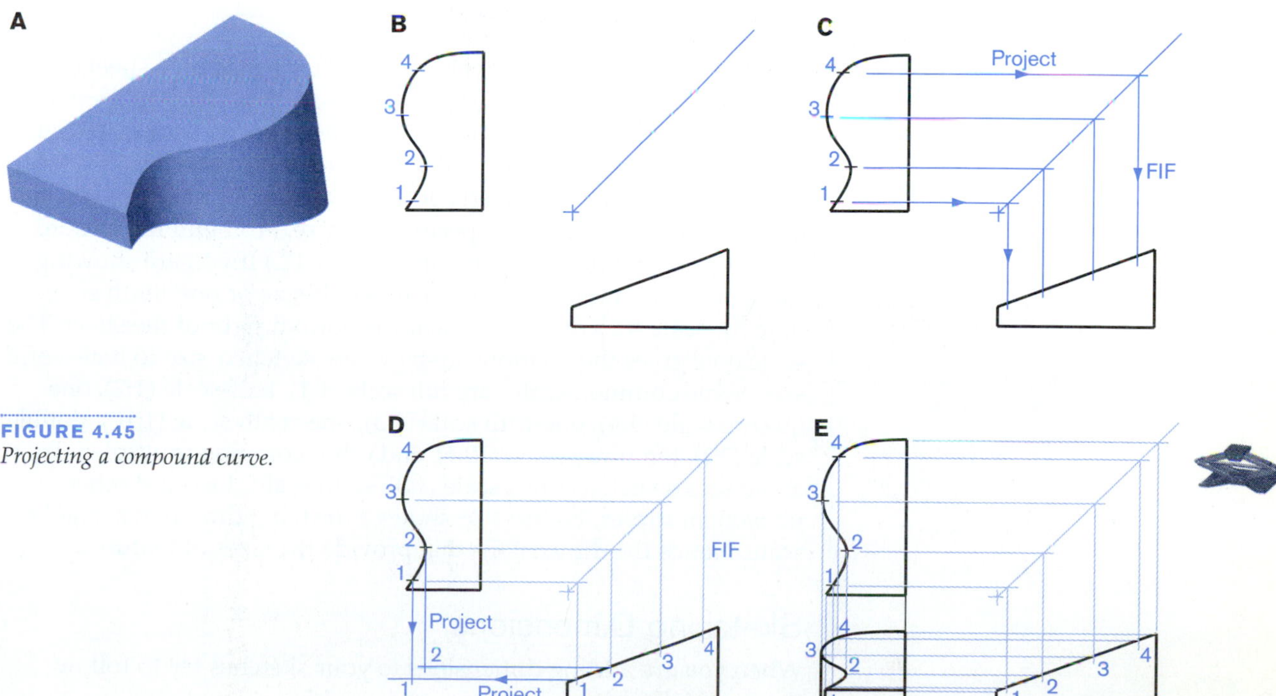

FIGURE 4.45
Projecting a compound curve.

FIGURE 4.46
Examples of fillets and rounds in pictorial views.

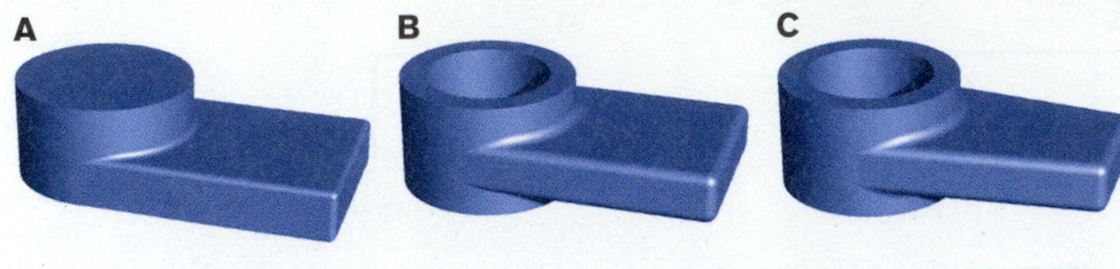

FIGURE 4.47
Drawing fillets and rounds in orthographic views.

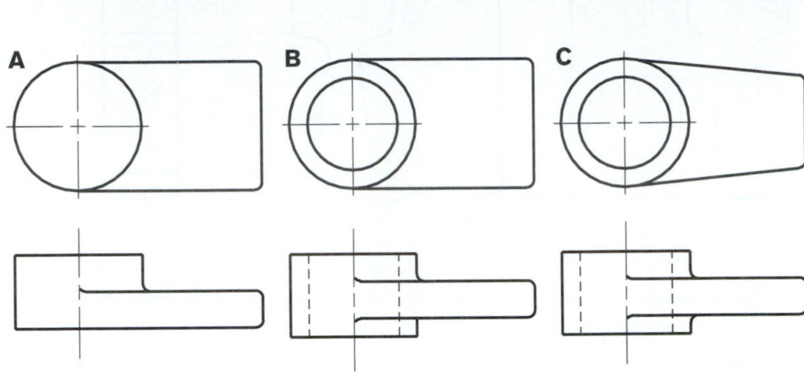

Fillets and Rounds

A *fillet* is an interior rounded blend between two surfaces. A *round* is an exterior rounded blend. Both are illustrated in Figure 4.46. You should draw the fillets and rounds on an object in the views where you see their curved shape. In views where you do not see their rounded shapes, ignore the fillets and rounds and draw the views as if there was no fillet or round present, as shown in Figure 4.47.

Scale

Sketched views of parts will not always fit full size on a sheet of paper. A reduced scale is used to represent a large part on a small sheet of paper. When sketching you often do not indicate the scale; rather, you show the object so that it fits on the paper and its features appear proportioned correctly. Sizes of features are labeled by sketched dimensions to specify their size. (You will learn more about scales and dimensioning in Chapter 12.) If you are showing the sketch to a common scale (such as half-size or one-tenth size), note the scale for your sketch near the bottom right of the sheet. The scale note gives the relationship between sketched size to real-world size. Some common scales are full scale (1:1), half-scale (1:2), one-quarter scale (1:4), one-fifth scale (1:5), one-tenth scale (1:10), double scale (2:1), ten times scale (10:1), and other common multiples. Try to avoid scales such as third scale (1:3), sixth scale (1:6), and other uneven multiples. Figure 4.48 shows a drawing drawn to a smaller scale. Notice the dimensions that provide the sizes of features.

Sketching Dimensions

When you are adding dimensions to your sketches try to follow these guidelines:

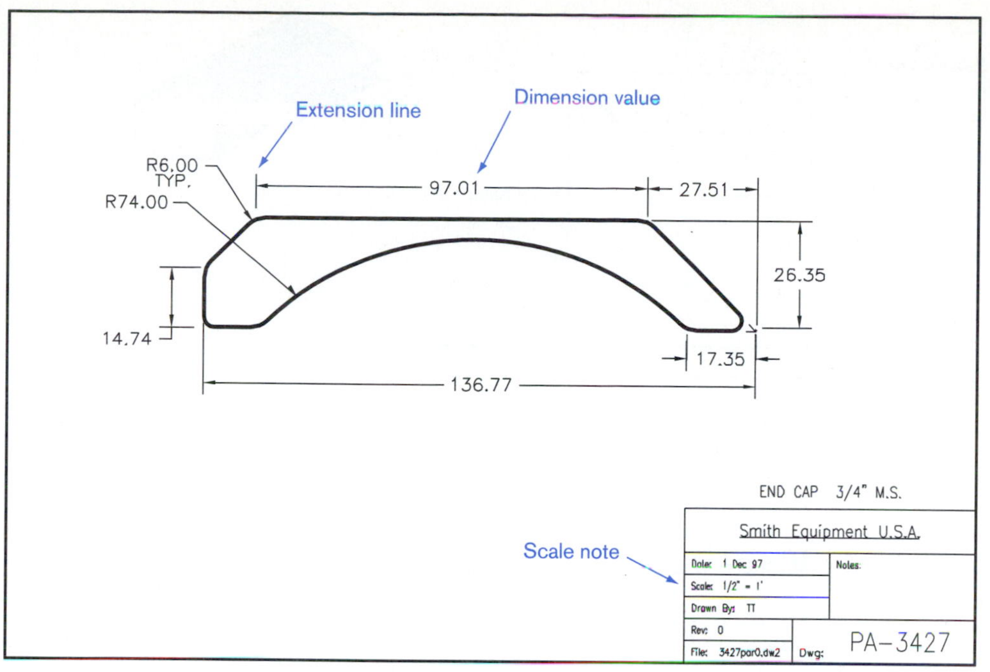

FIGURE 4.48 *This large part was drawn at a scale of 1/2"= 1′. Note the scale indicated in the title block near the lower-right corner of the drawing. (Courtesy of Smith Equipment USA.)*

- Place dimensions where the feature shows its shape clearly and true size.
- Use thin dark lines for dimensions and their text.
- Locate the dimension to the feature using an extension line.
- Place dimensions off the drawing views.
- Do not duplicate dimensions from one view to the next.
- Make arrowheads, text, and other symbols about 3 mm (1/8 inch) tall.

  Space dimensions away from the object by about 9 mm (3/8 inch).

  Group dimensions together to make them easy to read.

Figure 4.49 shows examples of good practice for sketching dimensions.

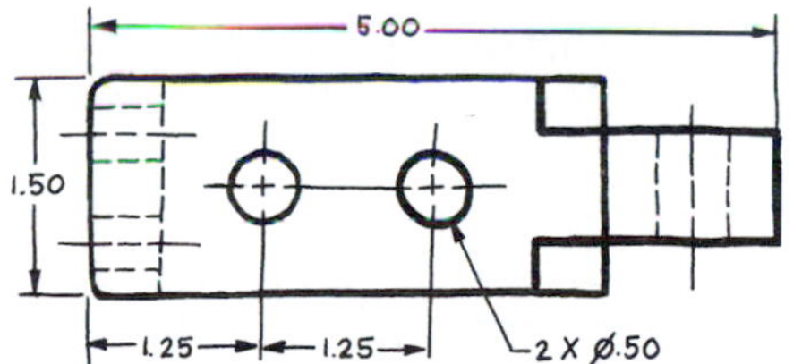

Sketched dimensions.

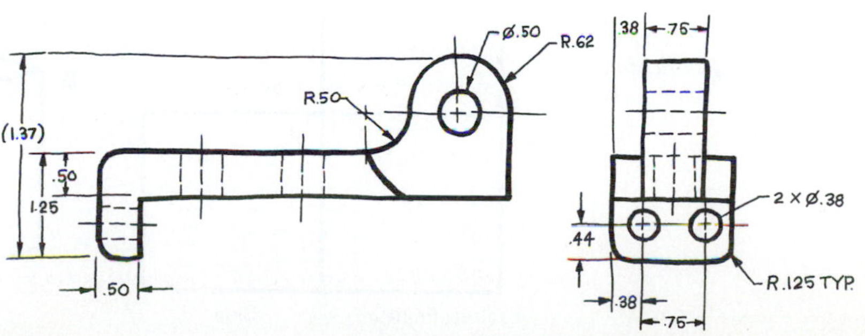

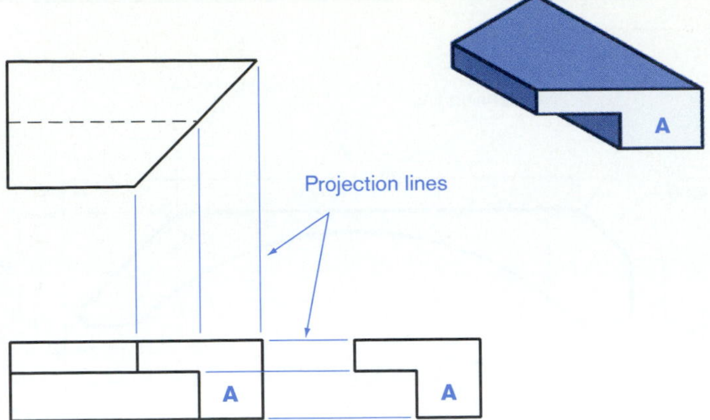

FIGURE 4.50
Surface A on the model is the same in each view. The projection lines from one view to the next help you locate the surface in each view.

Visualizing the 3D Object

When you interpret a multiview drawing, you combine the views into a 3D image of the object in your mind. The first step is to identify surfaces that are clearly shown in a view. For example, surface A in Figure 4.50 shows its shape in the front- and right-side views.

Keep in mind that a surface has the same basic shape no matter how it is viewed. It always has the same number of vertices, although these vertices may be hidden or appear differently in some views. When a surface is perpendicular to a view, for example, all of the vertices line up to produce a straight line, as they do in the front view of the highlighted surface in Figure 4.50.

Because of the standard arrangement of orthographic views, surfaces are always in alignment from one view to the next. Visualizing projection lines is one way to help identify the same surface in more than one view.

If you can identify the same point in two views, you know its location in space. Think back to the viewing planes (or the object's normal surfaces) as defining the axes of a 3D Cartesian coordinate system. If you know the location of a point in any two adjacent

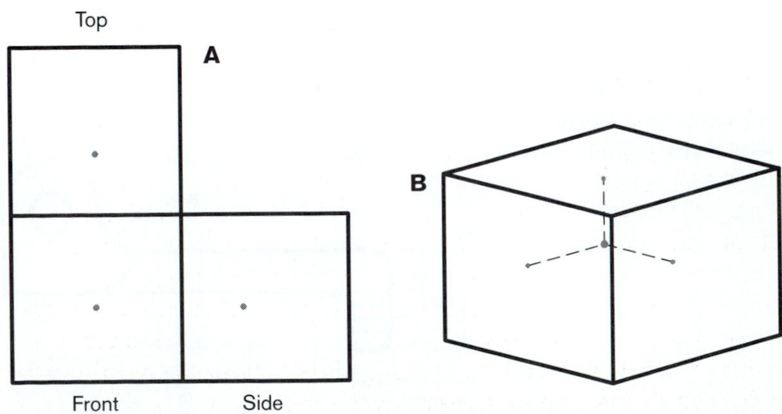

FIGURE 4.51
(a) A point as it appears in each view is defined by two of the three dimensions; (b) Any two views can provide the missing third dimension and locate the point in space.

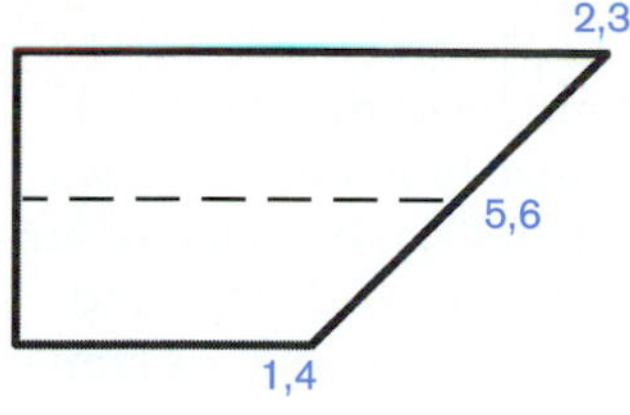

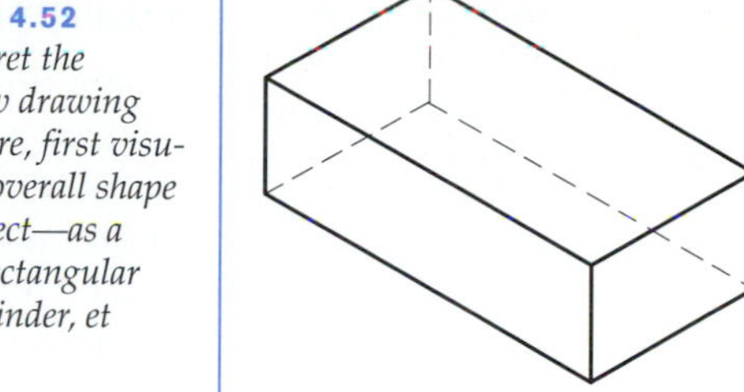

FIGURE 4.52
To interpret the multiview drawing shown here, first visualize the overall shape of the object—as a square, rectangular block, cylinder, et cetera.

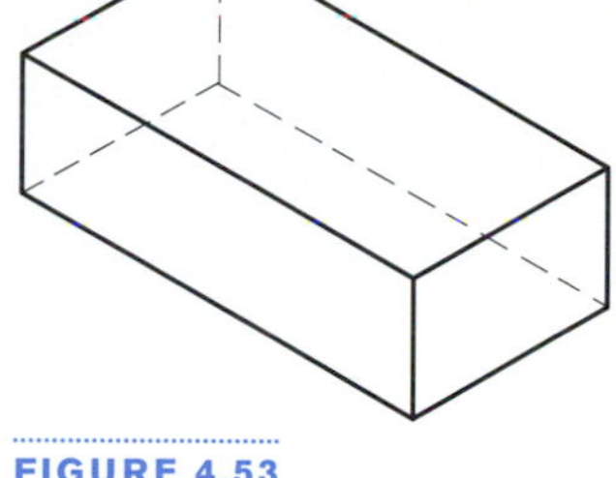

FIGURE 4.53
Sketch this isometric block so that it has the same overall proportions as the object in the views.

views, you have all of the three dimensions needed to plot its location in 3D space as shown in Figure 4. 51. To interpret views of an object, you combine the location information from the views in a mental process that constructs a 3D image of the features on the object.

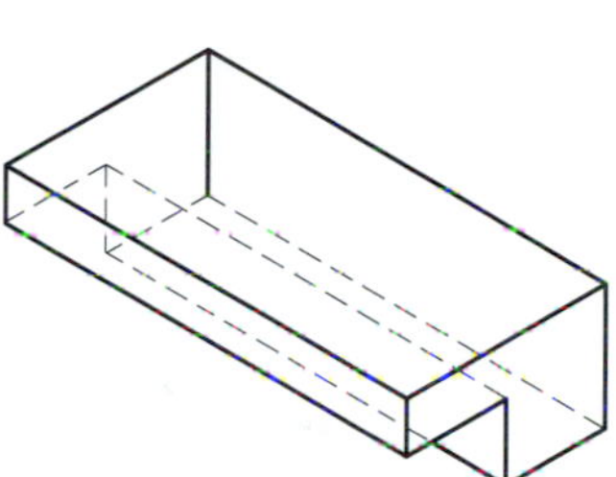

FIGURE 4.54
Identify places where material would have to be removed from the block to form the shape and mark it on the sketch.

Interpreting Views by Removed Portions

Another way to help you visualize an object from 2D views is to start out by thinking of a rectangular block (prism) that would just contain the object. Then examine the views and look for areas that are cut off.

Figure 4.52 shows three views of an object. Start by imagining a block of clay, foam, or some other material with the same general proportions as the views, and sketch an isometric block that has the same overall height, width, and depth, as shown in Figure 4.53.

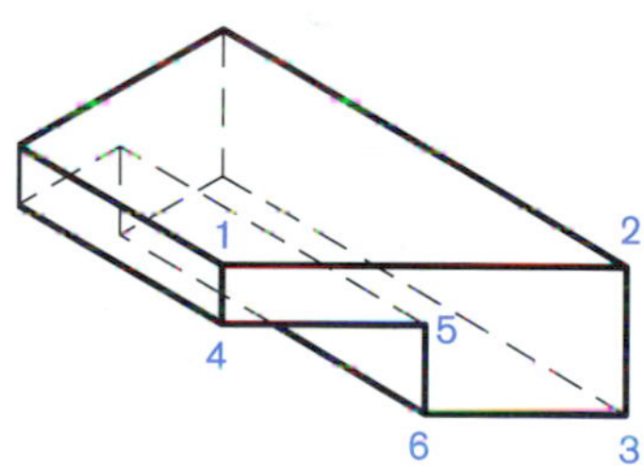

FIGURE 4.55
Using numbers to match up points in each view helps you visualize their location in 3D. Marking them on the sketch locates them so they can be connected to locate the surfaces they define.

Now examine the views to see if you can identify any inclined or normal surfaces on edge in one of the views. Can you imagine material removed from the block where the normal surface shows on edge in the side view? Mark the ends of the surfaces in your isometric sketch, then lightly block in the portion that is removed. Figure 4.54 shows the isometric view that results.

Now look for other surfaces that show you where material is removed. Oblique and inclined surfaces are usually the hardest for people to visualize. To picture them it usually helps to assign point numbers to each vertex and use the numbers to keep track of the location of the vertex in each view. Look at Figure 4.52, where the vertices of the inclined surface have been labeled. Notice that point 1 is left of and in front of point 2. Point 3 is directly below point 2. To imagine this surface cut off, think about locating each vertex on the block (as shown in Fig. 4.55) and think of cutting through the object with a broad knife that passes through each vertex. Mark these points by locating them relative to the edges in your isometric sketch, then sketch straight lines between the points.

Building a Physical Model

If you find it difficult to construct a mental model of the 3D object, making a physical model can help you visualize how the 2D views work to describe the 3D object. A physical model can be made of any solid material that is easy to cut, but clay is easy to work with and can be reused. The steps below for creating the clay model are similar to the mental process used to visualize the object.

Making a Clay Model

1. Shape a block of clay into the same overall proportions as the drawing views. Figure 4.56 shows a block formed to match the overall width, height, and depth shown in the principal views. Use a paring knife or a jackknife to square up your block of clay. Using a knife helps to make crisp cuts in the clay so you can clearly see the edges where surfaces join.
2. Lightly scribe the lines in the drawing views on the appropriate surfaces of the clay block. Remember that the edges of the block correspond to the overall width, height, and depth of the object and therefore describe edges of the object.
3. Look at the outlines of the views to identify portions of the block that are missing from each view. Cut these away first.
4. Identify the remaining surfaces on the object and turn the block as necessary to see how they are defined in the views. Make the necessary cuts to finish the model.

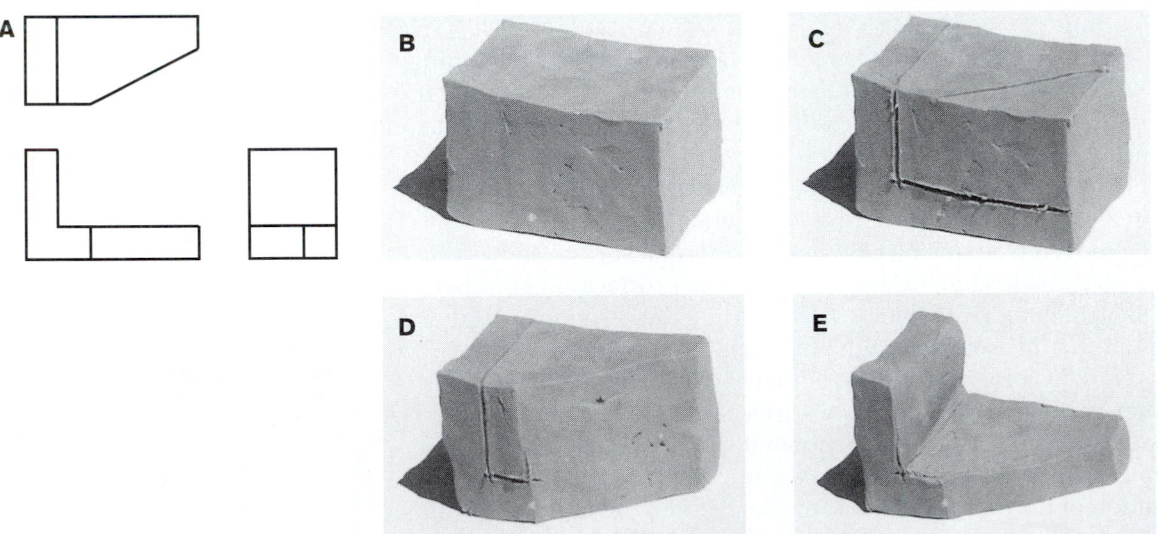

(a) Clay models can help you visualize the object shown in a multiview drawing; (b) First, square up your clay so that it has the same overall height, width, and depth shown in the drawing; (c) Scribe the lines of the drawing on the corresponding surfaces of the block of clay; (d) Cut away areas that do not show in a view. Here the angled surface was created by cutting along the angled line shown in the top view; (e) The L-shape was created by removing the clay from areas that are not shown in the front view.

Interpreting Linetypes

To more fully describe an object in a 2D drawing, engineering drawings use different kinds of lines, or *linetypes*, to add information to the depiction. Remember that each view of the object shows the entire object *as viewed from that direction*. The object's edges are depicted with thick lines to clearly outline the object. These lines are referred to as *visible* lines, as they are what you would see when viewing the object from that direction.

There are several other kinds of lines used to represent features on the object or to indicate how views were constructed. You should be familiar with the standard linetypes in Table 4.1.

TABLE 4.1

Standard Linetypes

Linetype Name	Description	Illustration
Visible	Thick, dark, bold lines (0.6 mm)	
Hidden	Thin, dark, dashed lines (dashes about 1/8 inch, spaces about 1/16 inch) (0.3 mm)	
Center	Thin, dark, long-short-long-dash-patterned line, starting with a long dash (about 1.5 inches up to several inches), short dash (about 1/8 inch), and ending with a long dash (about 1.5 inch to several inches). Alternate the pattern as necessary for the length of line you want to draw, always beginning and ending with a long dash. (0.3 mm)	
Phantom	Thin, dark, long-short-short-long-patterned line, similar to centerline with two dashes (0.3 mm)	
Cutting plane	Thick, dark, long-dash-patterned line (0.6 mm), or Thick, dark, long-short-short-long pattern (0.6 mm)	
Projection	Light lines used to locate features between views	
Extension	Thin, dark lines extending from a feature locating the extent of a dimension (0.3 mm)	
Dimension	Thin, dark lines locating a dimension value between extension lines (0.3 mm)	20.5
Long break	Thin, dark line with break mark indicating a large portion of the object not shown in view	
Short break	Thick, jagged line representing broken edge of object	
Hatch	Thin, dark lines representing cut surfaces in cross section	

Hidden Lines

When edges or other features are not directly visible in a view, a special *hidden* linetype is used to represent them. Hidden lines allow you to see edges on the back, bottom, left, or interior of the object in the standard views. Hidden lines are thin lines made of short dashes.

To visualize hidden lines, imagine an object made of clear plastic or glass, such as the one shown in Figure 4.57. From the front, you can look right through the object to see the edges where surfaces join together inside and in the back of the object. These edges would appear as hidden lines in the drawing. In the front view, shown in Figure 4.58, you can see how these edges are differentiated from the thick visible lines in the same view.

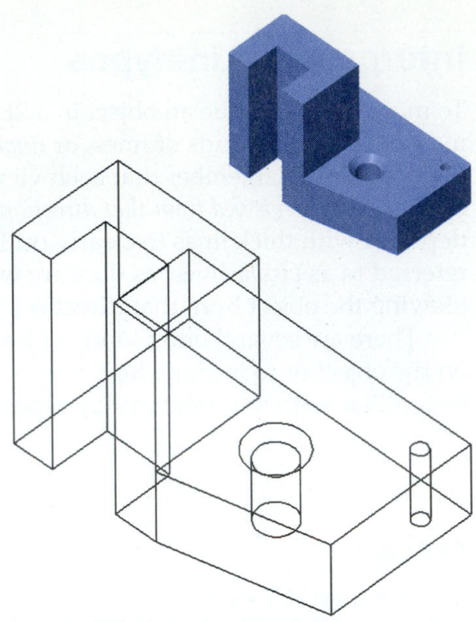

FIGURE 4.57
A transparent model makes it easy to see edges inside and on the back of the model that would appear as hidden lines in a sketch or drawing.

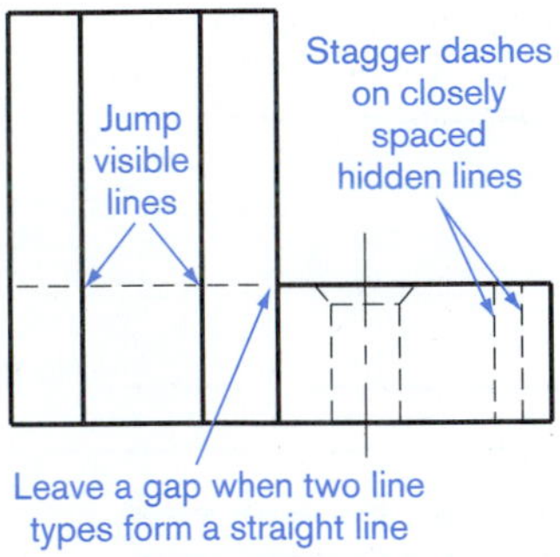

Jump visible lines

Stagger dashes on closely spaced hidden lines

Leave a gap when two line types form a straight line

FIGURE 4.58
Proper hidden line technique makes hidden lines unambiguous to the reader.

Standard hidden lines conform to drawing standards to eliminate ambiguity for the reader, as illustrated in Figure 4.58. Hidden lines should do the following:

- Form neat intersections with the edge of the object and other hidden lines.
- Jump across lines (edges) that they do not intersect on the object itself. (This is a convention observed in sketches, but may not be implemented in CAD drawings.)
- Not extend another line of a different linetype. Any two lines of different linetypes should never join to form a single straight line. Leave a visible gap in your sketch (about 1/16 inch) anytime two different lines would join to form a straight line.

Centerlines

Centerlines are used in drawings to indicate the axis of symmetry for a part or feature. Centerlines are also used to show the symmetrical alignment of a pattern of holes (often called a bolt circle or bolt pattern) and to show the path of motion for moving parts in an assembly. Figure 4.59 shows several examples of correct centerline practices.

Note that thickness as well as pattern is used to distinguish linetypes. Object lines or visible lines are drawn thick so that the subject

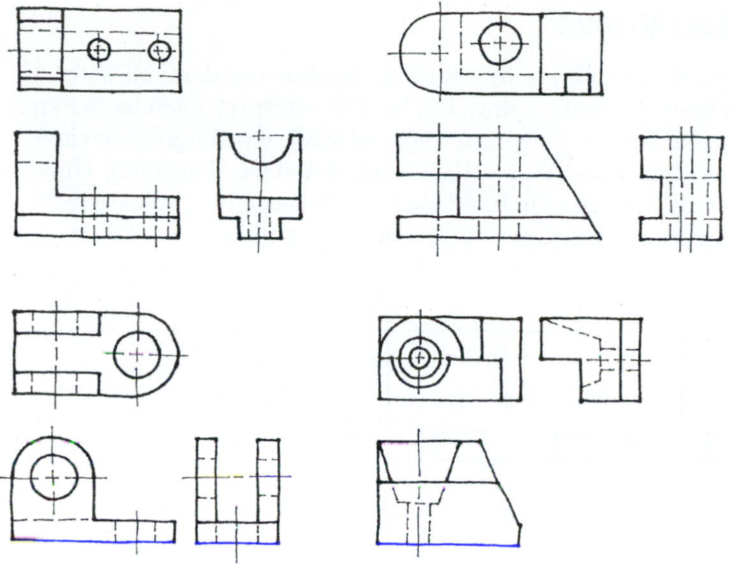

FIGURE 4.59
Proper centerline practice.

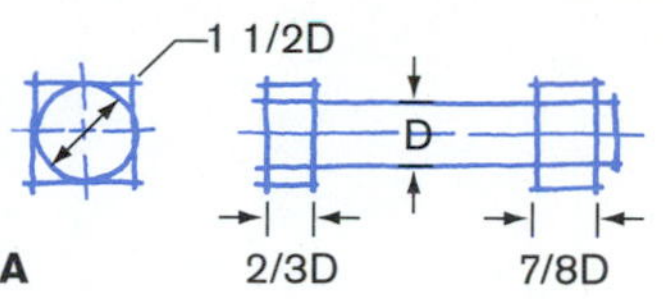

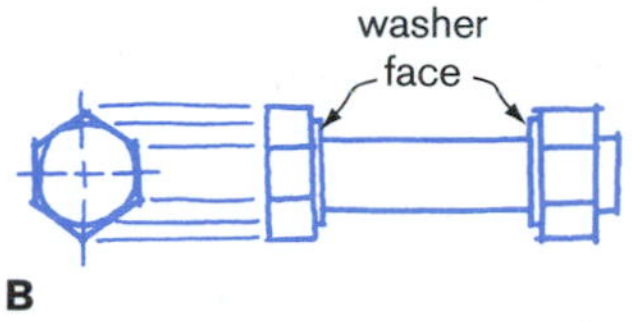

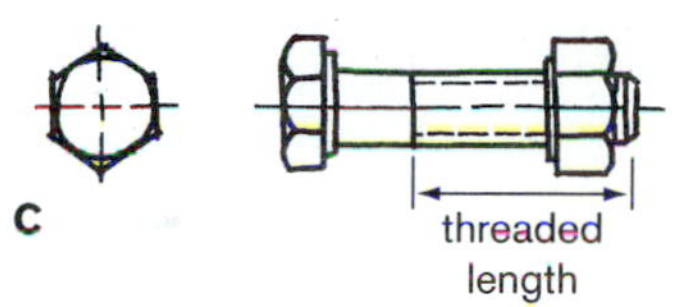

FIGURE 4.61
Sketching orthographic views of hex head bolts and nuts: (a) Block in the overall proportions based on the major diameter; (b) Show the hexhead across corners and project into the adjacent view; (c) Add arcs representing chamfer and darken final lines.

of the drawing will stand out clearly in your sketch. Hidden lines and centerlines are drawn thin. Practice making vertical, horizontal, and angled lines with each of the linetypes illustrated in the table so you can use them in your sketches.

Sketching Hex Head Bolts and Nuts

Hex head bolts and nuts are typically shown across the corners in orthographic views (see Fig. 4.60), so that they are not confused with square head bolts and nuts. Figure 4.61 illustrates the steps in sketching orthographic views of hex head nuts and bolts.

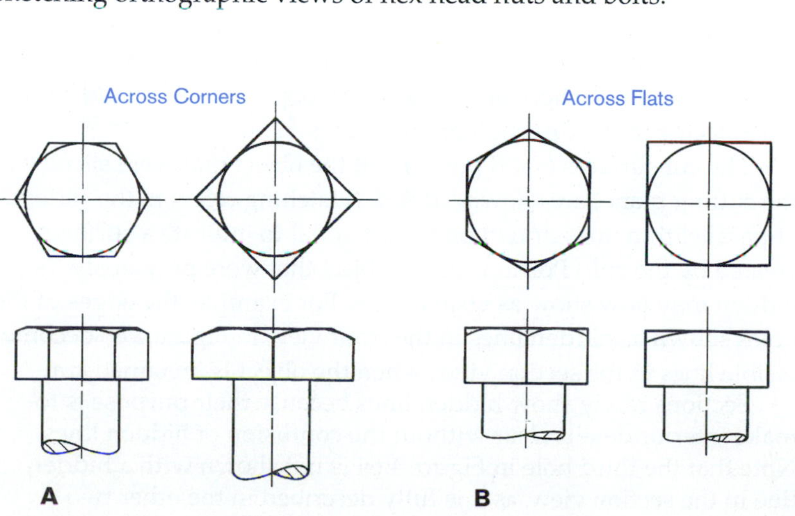

FIGURE 4.60
Hex head bolts and square head bolts across flats and across corners.

Section Views

When parts have a lot of interior details, showing them all with hidden lines can make a drawing hard to interpret, even to an experienced eye. Section views are a special kind of orthographic view that is used to show an object's interior detail (see Fig. 4.62). They are especially effective in hand sketches because they so clearly reveal the inner structure of a part or an assembly.

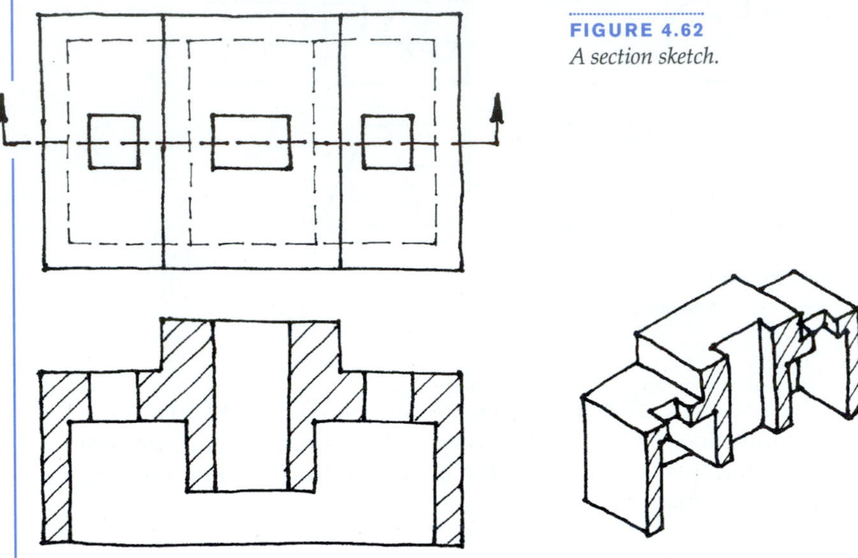

FIGURE 4.62
A section sketch.

Like a cross section, a section view shows what the object would look like if it were cut along a plane, called the *cutting plane*. The portion of the object in front of the plane is removed and the section view shows the surfaces revealed by the cut (see Fig. 4.63).

The cutting plane linetype is used in a single standard view showing the edge view of the cutting plane to indicate where the object is cut. The arrows on the cutting plane line point indicate the line of sight; the section view depicts the object as you would see it when facing in the direction of the arrows.

The cut surfaces (solid portions of the object that were sliced by the cutting plane) are shown *hatched*. (Hatching refers to the series of parallel thin lines drawn on the diagonal to indicate a surface created by the cut.) Portions of the object that were previously hidden may now show as visible lines. For example, the edges of the holes shown as hidden lines in the front view in Figure 4.63 become visible lines in the section view, when the object is imagined cut.

Sections rarely show hidden lines because their purpose is to make interior details clear without the confusion of hidden lines. Note that the third hole in Figure 4.64 is not shown with a hidden line in the section view, as it is fully described in the other two

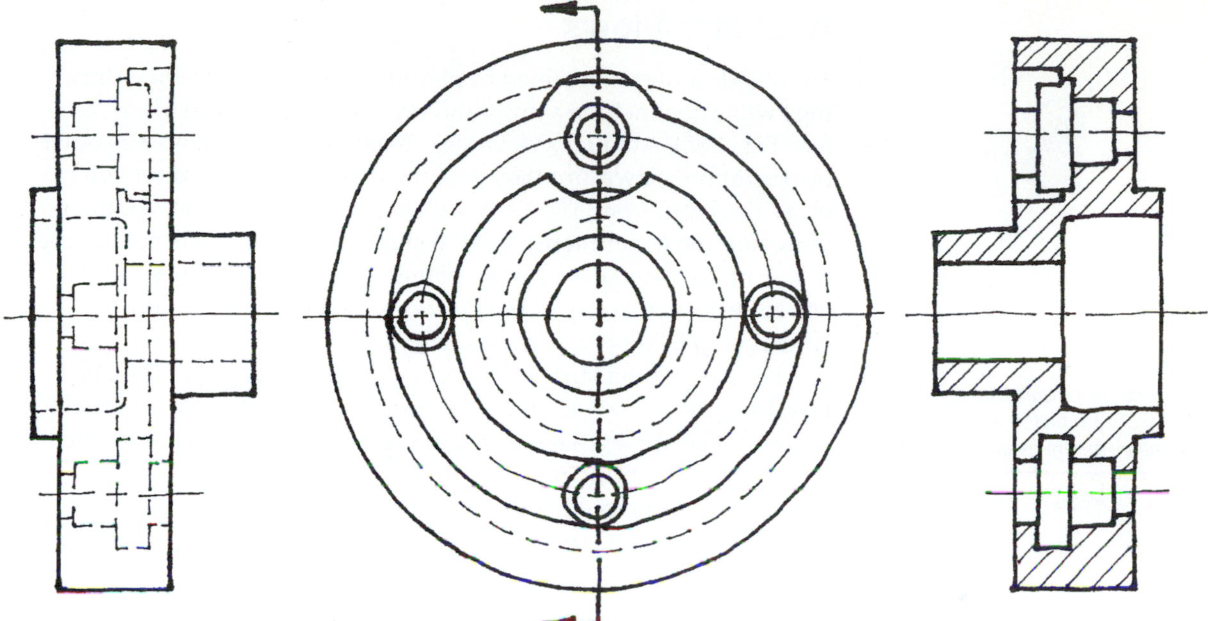

FIGURE 4.63
The hidden lines in the front and side views appear as object lines in the section view because they would be visible if the object were cut along the cutting plane line.

views. If it were not shown elsewhere, hidden lines could be used to show it in the section view, or an offset cutting plane could be used to show the hole. (An *offset cutting plane* bends 90° to pass through objects that are not all in the same plane.)

Section views often replace one of the standard views in the drawing, and there are a variety of ways in which sectional views can be depicted. You will learn more about creating sectional views from 3D models in Chapter 12.

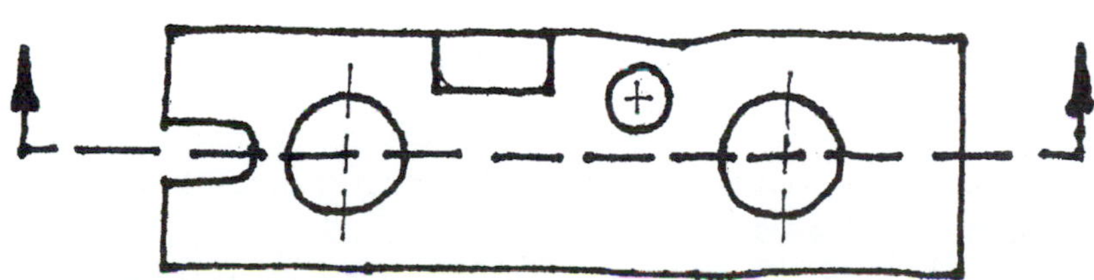

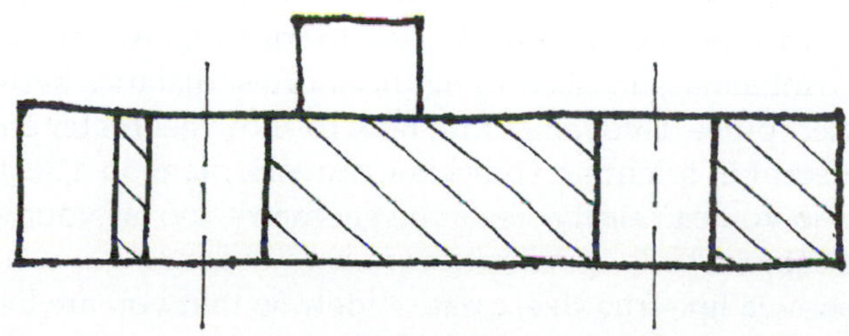

FIGURE 4.64
Hidden lines are not used for the hole not revealed by the section view, because it is fully described in other views.

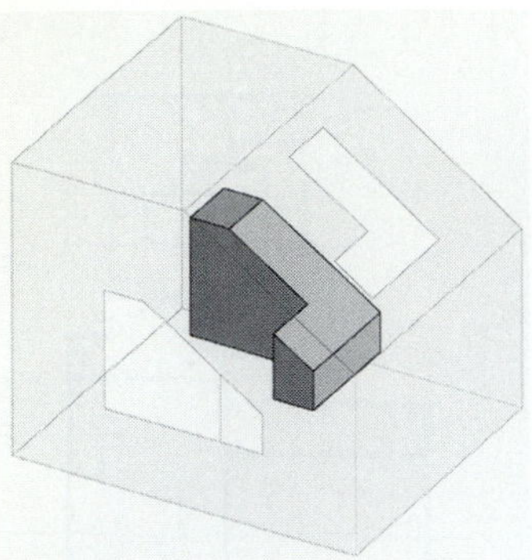

FIGURE 4.65
An auxiliary view of an inclined surface is parallel to the surface to be shown and perpendicular to the principal view where the inclined surface appears on edge.

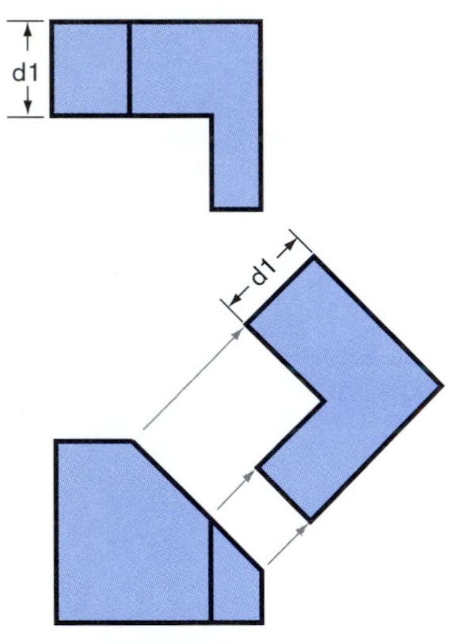

FIGURE 4.66
An auxiliary view of an inclined surface is projected from the front view.

Auxiliary Views

To fully describe an inclined or oblique surface, engineering drawings will often include a nonstandard view called an *auxiliary view* (see Fig. 4.65). Auxiliary views are created in the same way as other orthographic views, but they depict a view from some direction other than top, front, right-side, left-side, bottom, or rear. Auxiliary views are most commonly used to show a surface that is not parallel to any of the principal viewing planes (and therefore does not appear true size in any of the views); auxiliary views are often constructed to show a surface's true size.

For an inclined surface, think back to the example of the glass box. Imagine a new viewing plane added to it that is parallel to the surface to be shown. The new viewing plane would be perpendicular to one of the principal views (where the inclined surface appears on edge), but not to all of them.

Notice that the auxiliary view shown in Figure 4.66 is projected from the front view and has the same depth as the top view. The auxiliary viewing plane is hinged so that it is perpendicular to the front view. When the views are unfolded as shown in Figure 4.66, projection lines can be drawn from the front view to the auxiliary view to transfer the height locations of the object's vertices. The depth locations can be transferred by measuring from a reference plane parallel to the front viewing plane.

For oblique surfaces, a second auxiliary view is necessary, as the oblique surface is perpendicular to none of the principal views.

Auxiliary views are also frequently used to show the true angle between two planes, called the *dihedral angle.* A dihedral angle is shown true size in a view where the line of intersection between the two planes is shown as a point. In the past, the true dihedral angle would be calculated or projected in an auxiliary view and then measured. Today complex shapes that have many inclined and oblique surfaces can often be created fairly easily using 3D CAD, allowing you to make the measurements directly from your model.

Selecting Views

Even using 3D CAD, it is still important to understand which views show surfaces, lines, and angles true size. You should understand how to show a line true length, how to show a line as a point view, how to show a surface on edge, and how to show a surface true size. These four types of views can be used to solve many other engineering problems, such as finding the shortest distance between two angled pipes. Understanding how to show geometry true size will be useful in orienting your view or work plane in a 3D CAD package so you can easily create the geometry, too, as you will see in the next chapter.

To show a line true size, create a view so that you are looking perpendicular to the line (see Fig. 4.67).

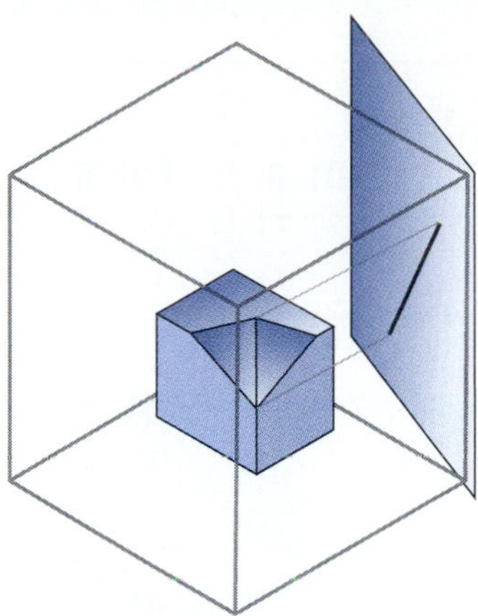

FIGURE 4.67
True-length view of line

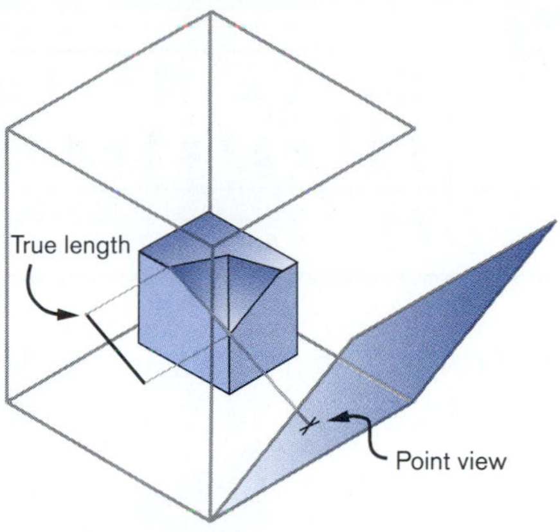

FIGURE 4.68
A line viewed as a point

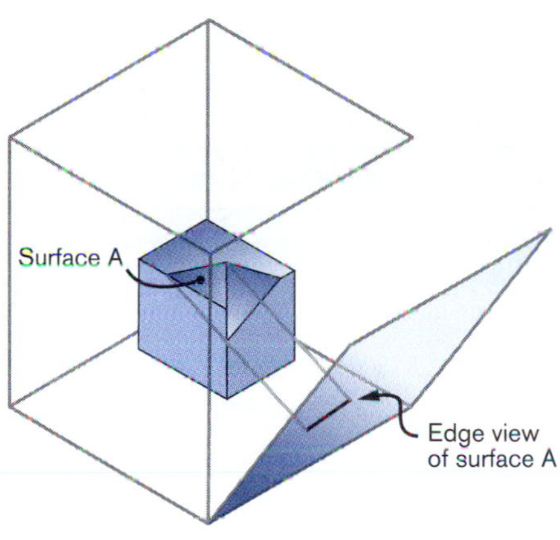

FIGURE 4.69
A surface on edge

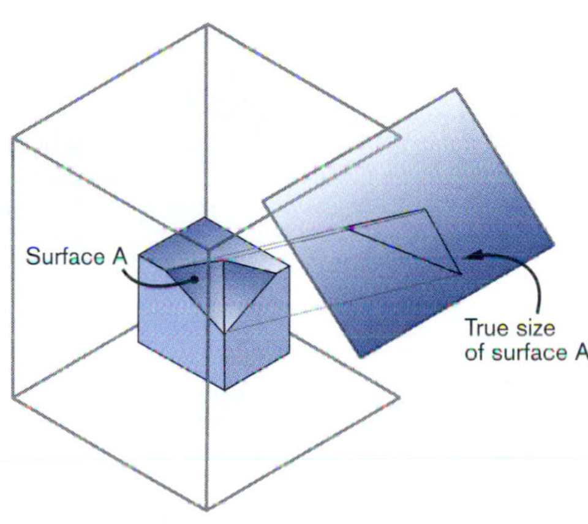

FIGURE 4.70
A surface shown true size

To show a line as a point view, create a view so that you are looking parallel to the line (perpendicular to the view where you see it true length as shown in Fig. 4.68).

To show a surface on edge, show any line in the surface as a point view (see Fig. 4.69).

To show a surface true size, your line of sight should be perpendicular to the surface (or perpendicular to the edge view of the surface as shown in Fig. 4.70).

Remember that just because you are sketching freehand does not mean that a sketch may be sloppy or illegible. The sketches that you will make as an engineer are not just for your own use. You must be able to sketch your ideas during meetings to get your point across to others, to record your ideas for the design record, to exchange your ideas with other members of the design team, and to give quick instructions to manufacturing personnel and others. Your sketches and lettering need to be neat and legible.

INDUSTRY CASE

Karen Markus, Cronos Integrated Microsystems, Inc.

Visualization: From 2D to 3D and Back Again

Designing microelectromechanical structures (MEMS) presents a visualization challenge. "MEMS use a technology similar to that used to create integrated circuits (computer chips) to create *physical, mechanical* structures," explains Karen W. Markus, VP and Chief Technical Officer of Cronos Integrated Microsystems, Inc.. "Most integrated circuits (ICs) are designed for binary logic: they signal on or off, or store a 0 or 1, which is what is needed for computation. IC processing deposits, patterns, and etches layers of material to create transistors made of silicon (and the interconnection pathways in metal layers on top of the transistors). MEMS use a similar manufacturing process to make machines—switches, relays, optical devices, sensors, accelerometers, and other devices that have moving parts—and that are built at the scale of a computer chip."

FIGURE 4.71
The size of this electrostatic motor is measured in microns like an integrated circuit, but its movable parts make it a microelectromechanical structure—a MEMS device. (Courtesy of MEMScaP S.A.)

Motion is at the heart of a MEMS device, whether that motion is an end product (as in the motor in Fig. 4.71) or a measurement function. For example, the airbag in your car is triggered by a MEMS sensor similar to the lateral resonator shown in Figure 4.72. The sensing device is basically two sets of interdigitated comb fingers—one set is fixed and one set is suspended so it can move. The fingers themselves are made of conductive material (polysilicon) and arranged so that the differential capacitance between the fixed and suspended sets can be measured. "When you slam on the brakes, the suspended set will move (through the force of inertia), and change the differential capacitance between the comb fingers from the neutral position," explains Markus. "The logic in the system looks at this change in capacitance and, if the magnitude and duration of the shift is large enough, sends a signal to trigger the CO_2 charge to inflate the airbag."

MEMS is an emerging area that reflects the growing integration of electrical and mechanical systems in many designs as well as the trend toward miniaturization. Miniaturization of an existing function is a primary motivation for many MEMS designs. Before MEMS were used, the airbag accelerometer was a ball and tube device. A conductive ball was positioned at one end of a metal tube. If the forces were large enough, the conductive ball would roll down the tube and close a switch at the other end, triggering the airbag. In a fraction of the space and weight, the MEMS device provides the same function but with greatly increased intelligence. For weight and size-sensitive applications, such as satellite design, MEMS devices offer new possibilities—a satellite the size of a can of tennis balls, for example.

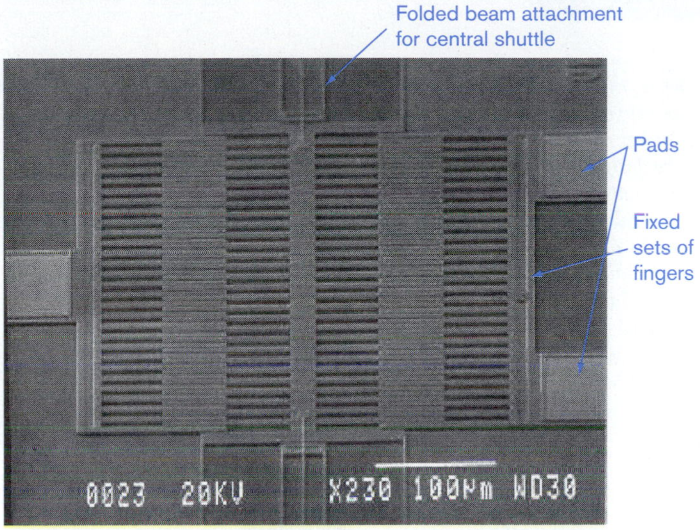

Folded beam attachment for central shuttle

Pads

Fixed sets of fingers

0023 20KV X230 100µm WD30

FIGURE 4.72
The MEMS sensor used to trigger your car's airbag is similar to the lateral resonator shown in this scanning electron micrograph. The small scale of these devices is evident in the white bar in the lower right, which represents 100 microns in this 230X enlargement. The larger squares to the far left and right are the pads anchoring the fixed sets of fingers. The narrow folded beams at the top and bottom attach the central shuttle and are designed to flex and allow the shuttle fingers to move.

Because they integrate electrical and mechanical properties into their designs, MEMS devices are also much more easily integrated into a system of logic components that will use signals from the MEMS sensor or send signals to a MEMS actuator device.

Like computer chips, MEMS are manufactured by a process that lays down successive layers of material—conductive and non-conductive—to form devices on a wafer. For each layer, a film of the material is applied to the entire surface of the wafer, the desired shape of the structure is masked, and unwanted material is removed. Then the next layer is applied. For each layer, the designer creates a 2D pattern for the material in that layer. This pattern is essentially a top view of the shapes to be formed on that layer.

For integrated circuits, the 2D representation is sufficient. Designers work with a 2D layout editor and software tools that allow them to simulate and analyze the circuit represented there. Because MEMS devices are machines and other physical structures, however, the designer needs to visualize how the material deposited in each layer will result in a 3D structure that moves.

(continued on p. 152)

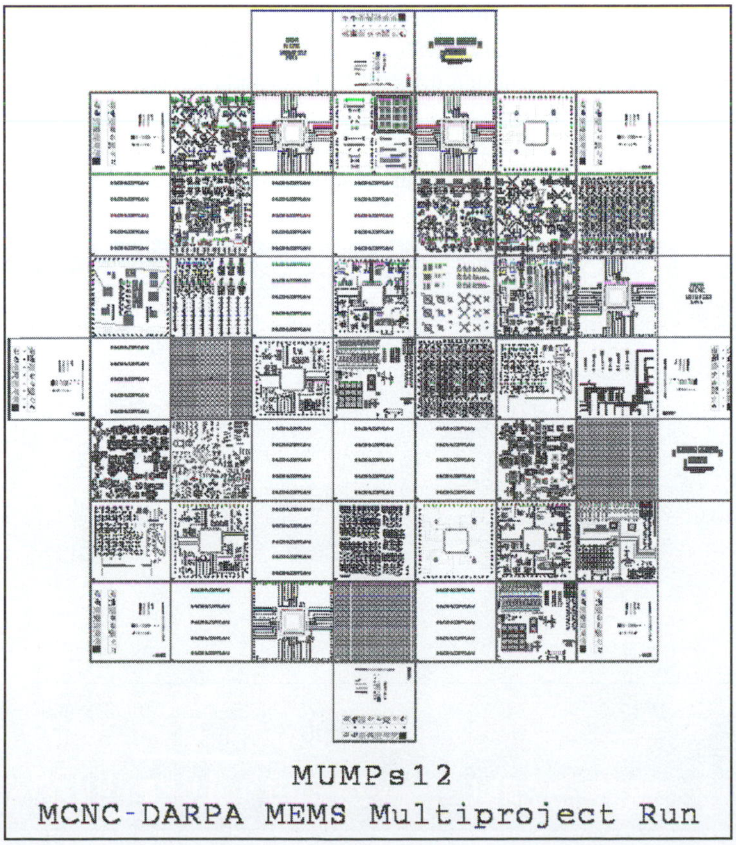

MUMPs12
MCNC-DARPA MEMS Multiproject Run

FIGURE 4.73
Layouts for several MEMS designs are imported by the Cronos engineering group into a single MUMPs™ mask layout for fabrication. This batch processing is useful for prototyping MEMS designs so they can be tested and refined.

INDUSTRY CASE

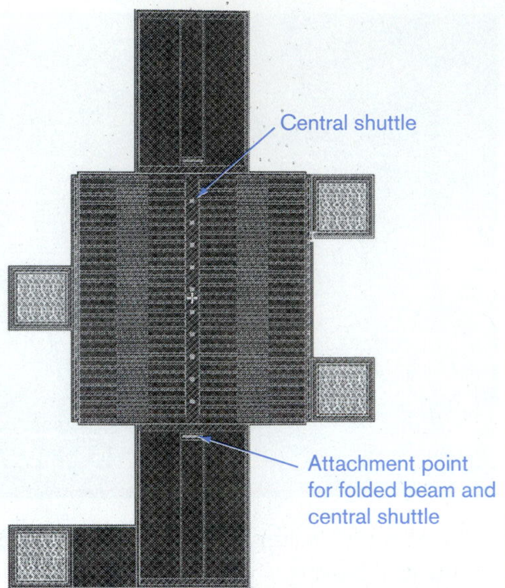

FIGURE 4.74
This top view of a layout for the lateral resonator shows the design of the folded-beam attachment for the central shuttle more clearly. Colors in the layout indicate different material layers. The interdigitated fingers are all on the same layer. (Courtesy of MEMScaP S.A.)

Sectional views from the front or side of the chip are necessary to see how shapes on different layers come together to form a 3D device.

The blending of electrical and mechanical design processes means that neither set of tools supports MEMS design visualization. 2D layout software used for integrated circuits lacks the 3D visualization and mechanical simulation capabilities of solid modelers. Solid modelers do not yet simulate the MEMS manufacturing process to enable a seamless transition from the 2D layouts. In order to help engineers access MEMS technology and begin developing their own MEMS devices, Cronos Integrated Microsystems offers MUMPs™, a surface micromachining prototyping process that reduces the barriers to MEMS design. The MUMPs process spells out the design rules for a MEMS chip that can have up to three layers of polysilicon. Users who follow these rules can submit their designs to be grouped with other chips in the fabrication process, creating a mosaic of many chips on the surface of a single wafer (see Figure 4.73). The designs are translated to a set of masks that are used to image the 2D pattern required for each layer. Cronos gathers the layouts into a large job that is then fabricated by a mask-house, such as DuPont. Because the design and analysis tools are still developing for MEMS, easy prototyping makes it possible for individuals with little knowledge of IC manufacturing processes and who lack the necessary clean room manufacturing facilities to design and test MEMS devices.

"Working with MUMPs is like making a layer cake," Markus explains. "Each layer of cake is 1-inch tall and each layer of frosting is 1/2-inch tall. You can have a layer in any shape—10 inches square, 8 inches round, or 11 x 17

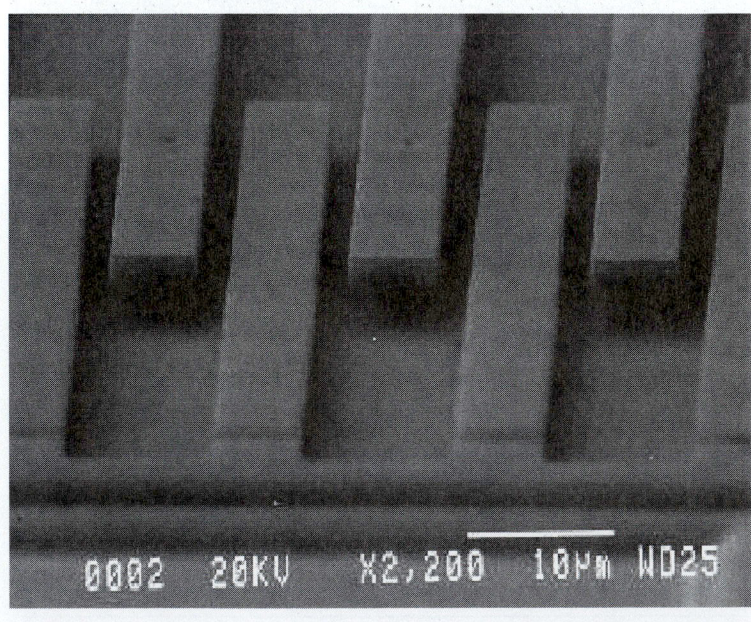

FIGURE 4.75
The interlaced fingers of the lateral resonator are shown in this scanning electron micrograph at an enlargement of 2200X. The gaps between the fingers are created by the layout for that layer; the air space beneath the fingers was created by a layer of oxide that was removed after fabrication of all of the layers.

sheet—but it will always be 1-inch tall. You can skip a layer of cake so you have two frosting layers, but you can't change the thickness of the layers." For this reason, most of the design rules pertain to the 2D shapes and their layout on the plane of the layer. For example, no polysilicon structure can be less than 2 microns wide. Gaps between polysilicon structures must be at least 4 microns wide to eliminate shorts or electrical interference between them.

The MUMPs process starts with a layer of silicon nitride as an insulating material. Each layer of polysilicon, the conductive material, is preceded by a layer of silicon dioxide. The oxide layers are sacrificial—they will be washed away at the end of the process to create air pockets and gaps between the polysilicon structures and the wafer. An orthographic view of the structures formed by a particular layer is drawn for each layer of material. The layering feature of programs such as AutoCAD could be used to separate the outlines for each layer of material onto a layer in the document, yet keep them aligned properly on top of each other.

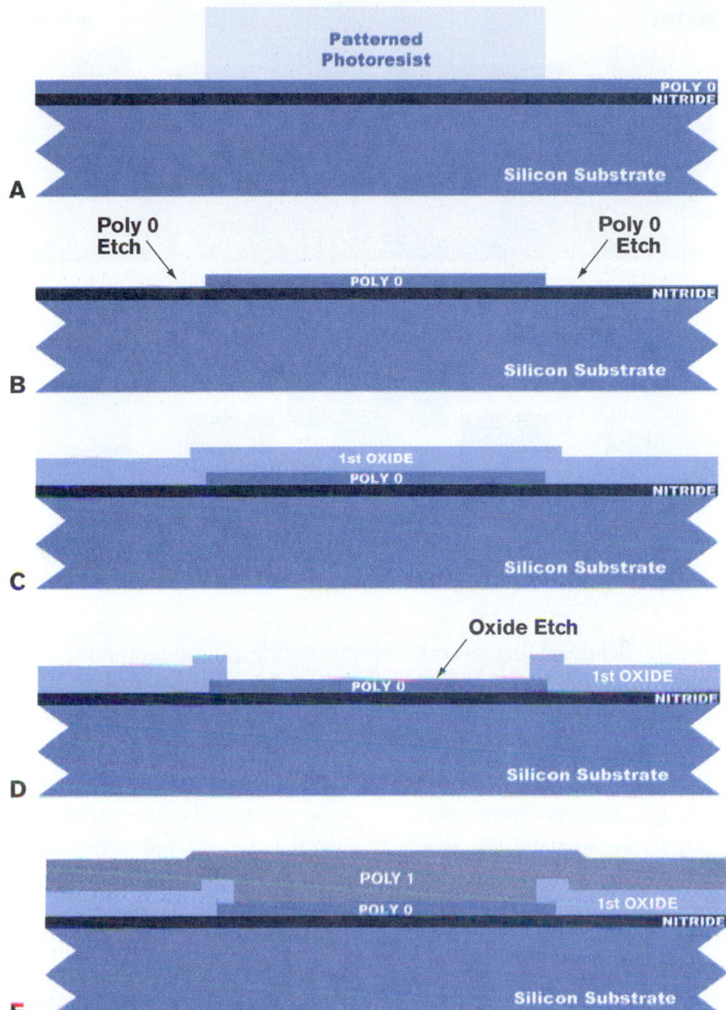

FIGURE 4.76
The basic elements of the MEMS surface micromachining manufacturing process are illustrated in this sequence of section views. In (a), a layer of nitride has been applied, followed by the first layer of polysilicon. In (b), the desired portions of the Poly 0 layer have been masked and the unwanted portions removed. The next layer is the first sacrificial oxide layer; in (c) it is applied to the surface of the chip and conforms to the shape of the structures on the previous layers. In (d), the oxide is masked and etched to expose Poly 0 so the next polysilicon layer will attach to it. The rest of the oxide remains to form a gap between the nitride and the structures in the next polysilicon layer, added in (e). The masking and etching process is repeated for subsequent layers.

"In making the lateral comb resonator, the design goal was two sets of polysilicon fingers—one fixed and one suspended set—separated by an air space across which the capacitance can be measured. The fixed set must be anchored firmly to the chip, while the suspended set must be designed to allow movement. The center structure of the device shown in Figure 4.74 is a shuttle of fingers with a folded beam suspension at either end of it. This suspension structure is attached to the substrate by anchor points, but the arrangement of the legs allows the shuttle to move when force is applied to it. Each comb structure

(continued on p. 154)

INDUSTRY CASE

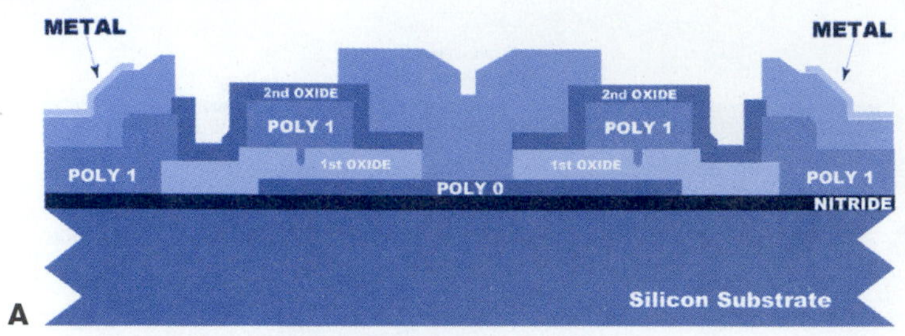

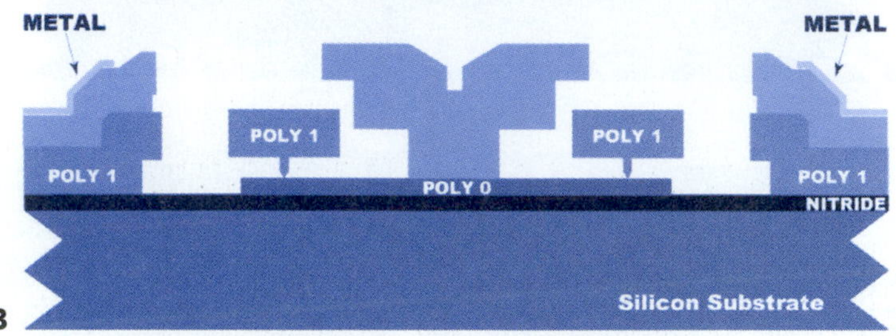

FIGURE 4.77
The MUMPs process uses seven layers of material, of which two are polysilicon material used to form structures; one is polysilicon used for electrical functions, and two are sacrificial material (oxide) that is removed in an acid bath after fabrication. The section view in (a) shows the layers and how each conforms to the topography of the structures formed before it. In (b), the cross section shows the device after the oxide has been removed.

has fixed (anchored) fingers and suspended fingers (attached to the shuttle) to provide the ability to sense and drive the resonator."

To create the resonator, both the shuttle and the fixed fingers need to be in the same plane, but an air space is needed below the shuttle so it can move (see Fig. 4.75). The structures on this layer, however, also need to be attached by anchors. The section views in Figure 4.76 illustrate how the resonator would be built up from the different layers. The first layer of polysilicon, Poly 0, forms pads that can be used as electrodes and can also be where the shuttle and fixed fingers attach. The plane of the fingers is formed by the second layer of polysilicon, Poly 1. The air gap under the plane of the fingers would be formed by the oxide layer, which would be removed after manufacture.

The challenge of designing MEMS devices is anticipating and visualizing the topographical effects of the manufacturing process and planning the layers accordingly. That is, each layer is a film that covers the entire surface of the wafer, and each subsequent layer flows over whatever structures are in place on the wafer. Each "flat" layer takes on the surface shape, or topography, of the wafer at that stage of the process. Successive sectional views are vital to visualizing the effect of each layer (see Fig. 4.77).

Structures are formed by a masking and etching process that removes unwanted material and protects the structures you want to keep. For example, after the Poly 0 layer covers the wafer, the 2D shape of the pads prepared for that layer is imaged on the wafer using a mask, ultraviolet light, and a photosensitive polymer called photoresist. Any polysilicon outside of the masked areas is then removed in an etching process.

This creates islands of polysilicon, as shown in Figure 4.76(b). The next layer is oxide, which will form the air pockets. Notice in the section view in Figure 4.76(c) that this layer covers the entire chip and flows over and around the Poly 0 structures. Because the next layer of polysilicon needs to attach to the Poly 0 layer, the oxide is removed from the places that will serve as anchor points, as shown in Figure 4.76(d). The remaining oxide separates the next layer of polysilicon, Poly 1, from the substrate. The Poly 1 layer is then deposited, patterned, and excess material removed to form the interlaced fingers.

To document a MEMS device, nothing more than top orthographic views made of basic geometric shapes are needed to convey the layout. In planning a MEMS device, however, sectional views and a knowledge of the manufacturing process are needed to anticipate the geometry of the layers as they are applied. Multiple section views, at each stage of the process and reflecting different cutting plane locations,

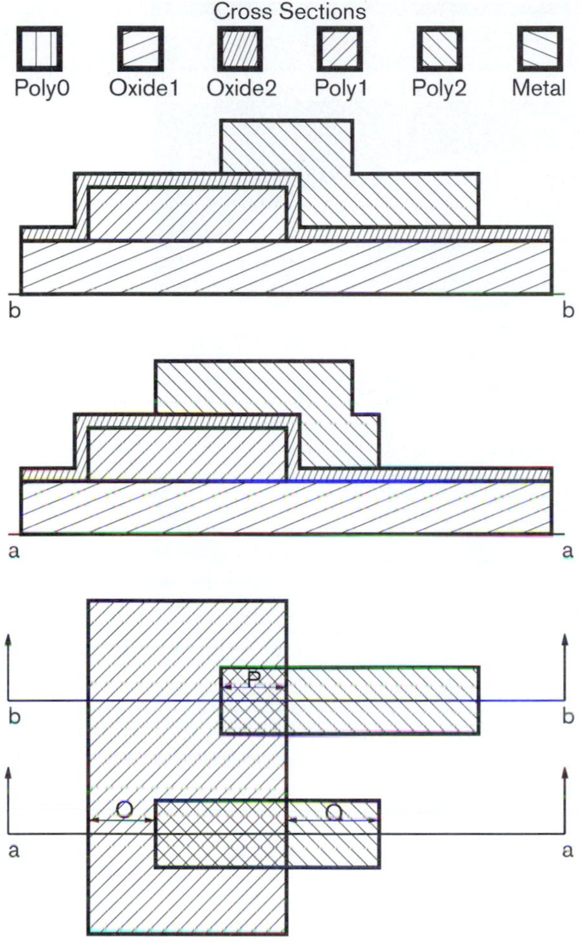

FIGURE 4.78
The interplay of top views and section views allows the designer to visualize the structures formed by the layers and ensure that all design rules are met. The outlines in the bottom figure are layouts for the masks to be applied to the Poly 1 and Poly 2 layers. The cutting plane lines on the outlines indicate the locations of the section views a-a and b-b, which show the resulting structures up to that point of the process. The dimensions O, P, and Q on the layout illustrate three of the 21 MUMPs design rules imposed by the tolerances inherent in the fabrication process and properties of the materials used. O indicates that the edge of Poly 2 must be at least 4.0 mm from the edge of Poly 1 to ensure that it does not overlap the Poly 1 edge. P is the minimum (5.0 mm) that a Poly 2 structure must overlap a Poly 1 structure to ensure overlap. Q is the minimum distance that Poly 2 must extend beyond the Poly 1 edge (4.0 mm) to ensure complete edge overlap.

may be required to visualize some devices. Although the section views may be hand sketches, consistent use of hatch patterns and careful selection of cutting plane location are important. The hatch patterns for different layers are vital to identifying each layer as it is applied, and differentiating between the oxide layers, which will be removed, and the conductive polysilicon layers. By creating sectional views of the effect of each stage of the process, the designer can be certain that the air spaces and connections needed for movable structures are drawn correctly in the plan view (see Fig. 4.78).

Design tools for MEMS devices are being developed to address the hybrid nature of MEMS design. Layout editors used for integrated circuit design are being modified to incorporate the 3D visualization tools needed for MEMS design. At the same time, these tools aim to bridge the gap to the kinds of mechanical analysis needed for movable structures—for stress, fluid flow, thermal properties, and others in addition to the electrical simulation commonly performed on IC chips. The model in Figure 4.79 was generated by MEMScaP, a design environment for MEMS that includes 3D visualization and integrated analysis functions. As these tools develop, however, the design rules and prototyping opportunities provided by Cronos make MEMS device design available to anyone capable of producing accurate 2D views of the various layers.

Figures are used with the permission of Cronos Integrated Microsystems, Inc., except where otherwise noted.

INDUSTRY CASE

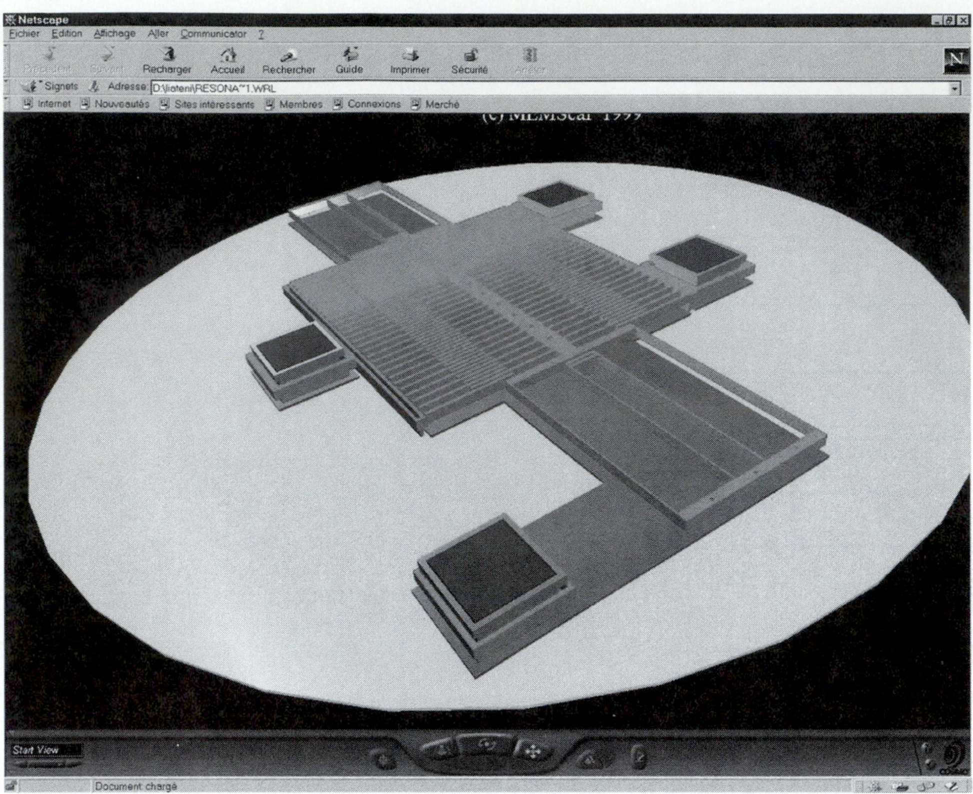

FIGURE 4.79
The lateral resonator layout in Figure 4.74 was translated into a solid model of the MEMs device by Kanaga, a MEMs design environment from MEMScaP. The solid model is "fabricated" from the layout layer by layer so that structures retain the color of the layer they are formed from. MEMScaP and other design environments use solid modeling tools or VRML to generate 3D representations of the process specified in the layouts for each layer. The model shown here is VRML. (Courtesy of MEMScaP S.A.)

Karen Markus is VP and Chief Technical Officer of Cronos Integrated Microsystems, Inc., in Research Triangle Park, North Carolina. With background in material science and engineering, physics, and a degree in electrical engineering, Markus was a Staff Engineer working in VLSI circuit process development at TRW's Microelectronics Center in California before starting the MEMS program in 1992 at MCNC in North Carolina. She enjoys working at the "bleeding edge" of engineering, where no one has really gone before, and the excitement and energy around the tiny machines that are MEMS devices. She is part of the team that has launched Cronos Integrated Microsystems, Inc., the MEMS company that will commercialize the products and processes that were developed at MCNC. For a virtual tour of the MEMS fabrication process and more about MEMS design, visit the MEMS web site at www.memsrus.com.

Skills Summary

In this chapter you have seen how to create multiview orthographic sketches and how to visualize the parts shown in an orthographic drawing. To practice multiview sketching, complete Exercises 1–24. Use correctly shown line patterns and make sure to show views in alignment. It is a serious error for adjacent drawing views not to line up in projection. Make your sketches neat and legible. To practice visualizing a shape from a multiview drawing, create pictorial sketches from the multiview drawings shown in Exercises 25–34. Practice your visualization skills further by matching the orthographic views in Exercises 35–46 with the correct pictorial view.

EXERCISES

1. Create your own cartoon that shows a single view of some event in a way that makes it visually simple and comical.
2. Create a three-view orthographic projection sketch showing front, side, and top views of a standard computer monitor. Do not be too concerned with the control knobs or switches. Use a scale of 1:4.
3. New developments in flat-screen video monitor technology permit redesign of computer screens so that depth is about 4 inches from screen face to the rear of the case. Sketch a three-view orthographic projection of this new configuration. Use a scale of 1:4.
4. Sketch a three-view orthographic projection of your telephone. Create an isometric sketch of the same phone. What advantages or disadvantages does each sketch have in communicating ideas?
5. Sketch a freehand three-view orthographic projection of the oarlock, the thread tap, and the phone receiver shown below. Scales are not provided; try to draw all features of each object to the same relative scale.

   a. Oarlock

   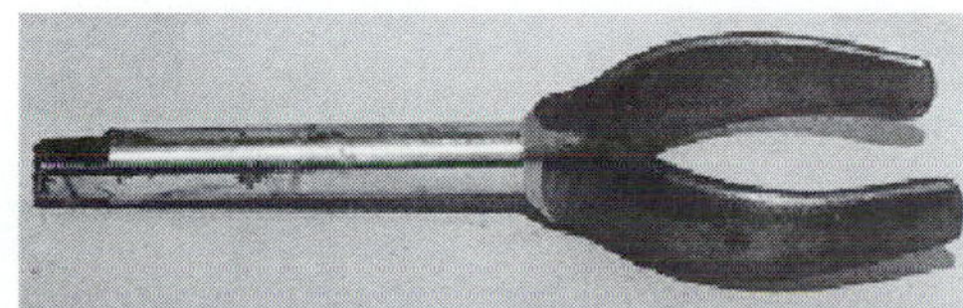

   b. Thread tap

   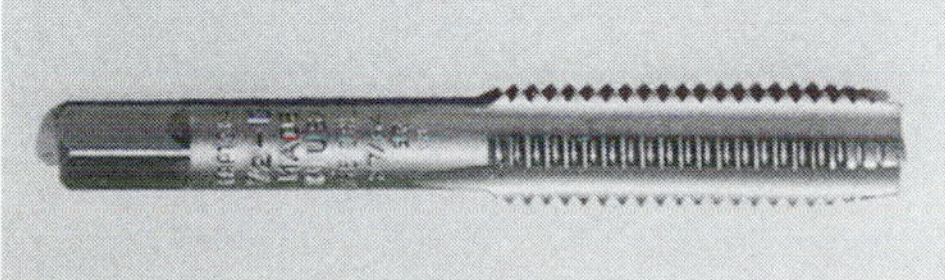

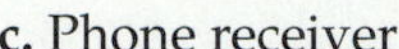

c. Phone receiver

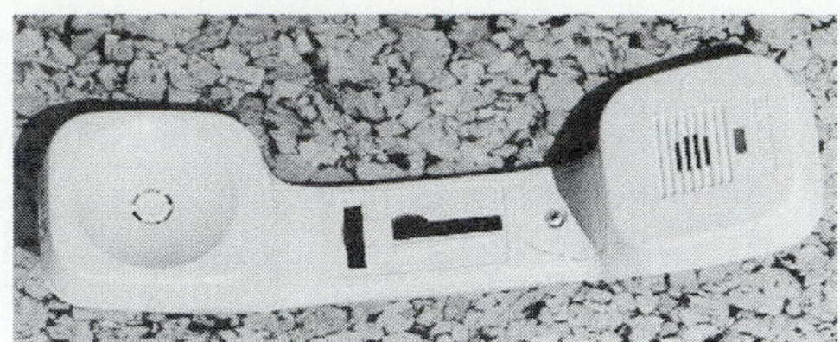

6. This CAD model is an isometric view of a tool stop. The model contains (or omits) some information that would usually be omitted (or present) in different drawing representations. Redraw the tool stop as a three-view orthographic projection drawing.

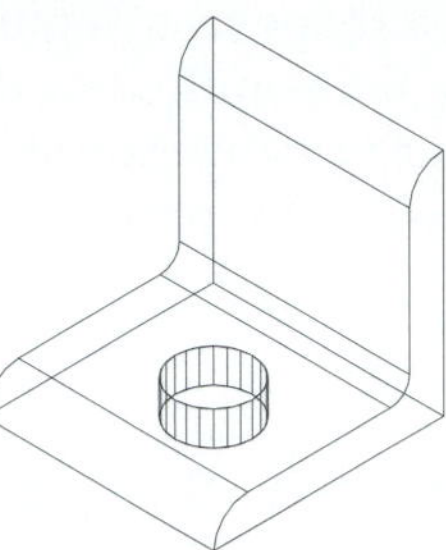

Exercises 7–12: Multiview Sketching Problems. Sketch the orthographic projection views necessary to fully define the object. How many views are needed in each case? Assume that each grid square equals 1/2 inch, or that each square equals 10 mm.

7.

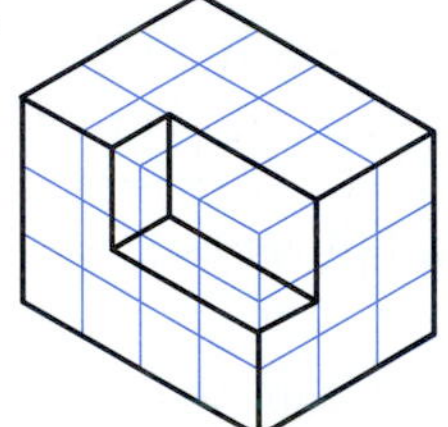

8.

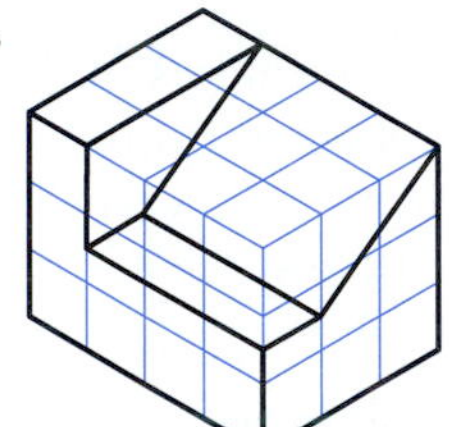

9.

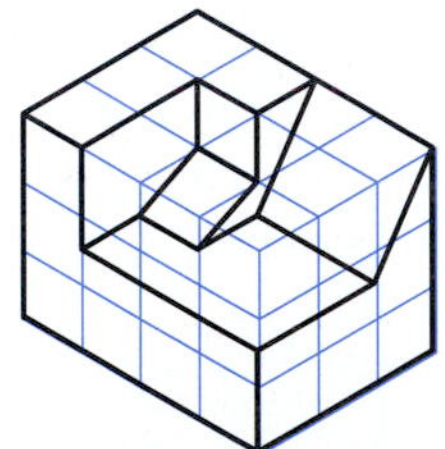

10.

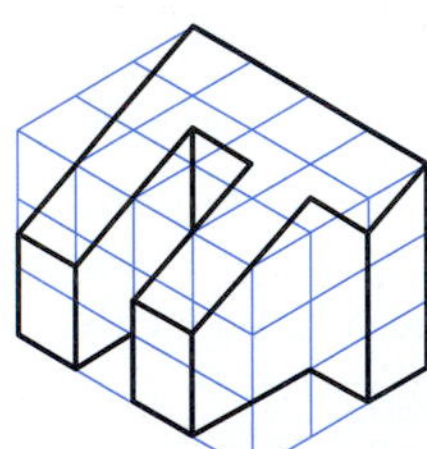

11.

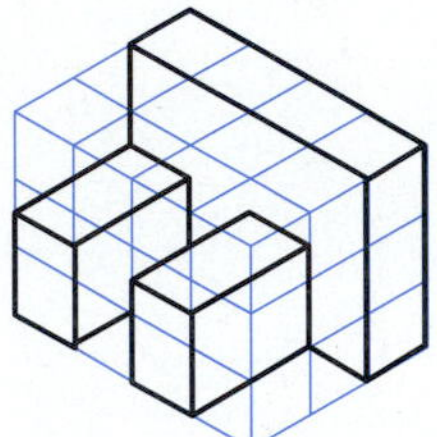

12.

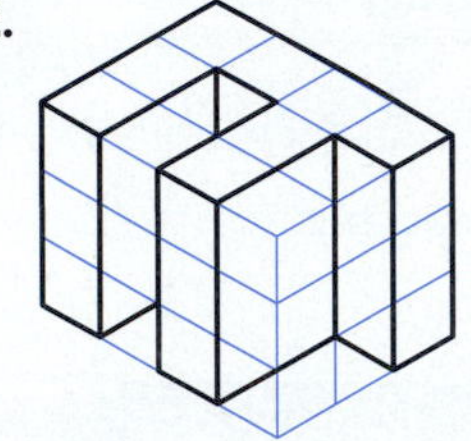

Exercises 13–18: Multiview Sketching Problems. Sketch the orthographic projection views necessary to fully define the object. Assume either that each grid square equals 1/4 inch, or that each square equals 5 mm.

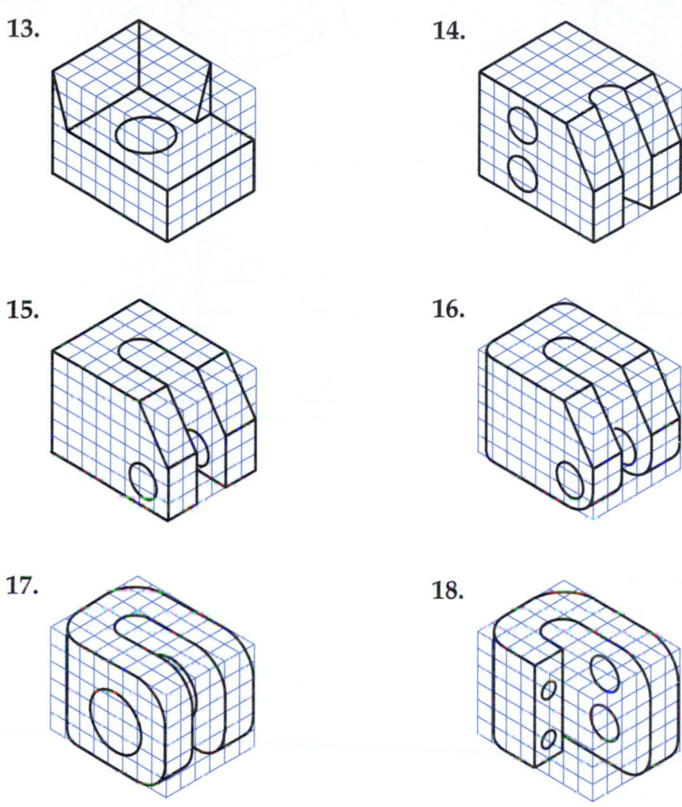

Exercises 19–24: Multiview Sketching Problems. Sketch the orthographic projection views necessary to fully define the object. Assume either that each grid square equals 1/4 inch, or that each square equals 5 mm.

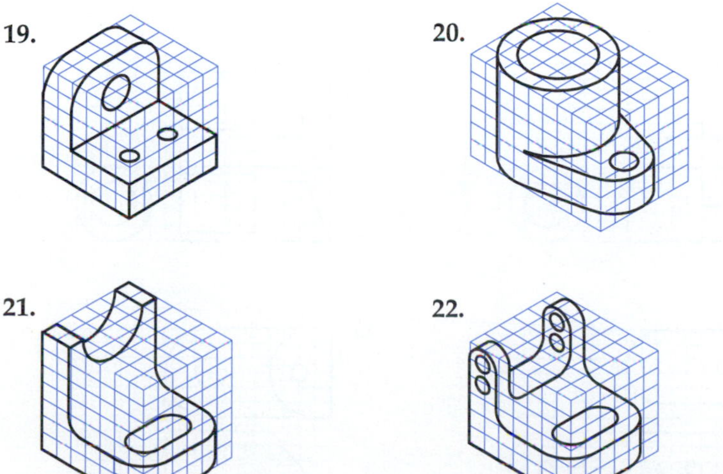

23. 

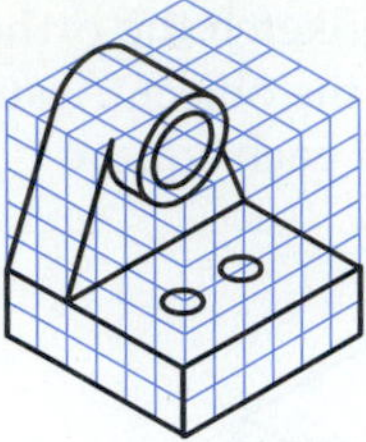

24. 

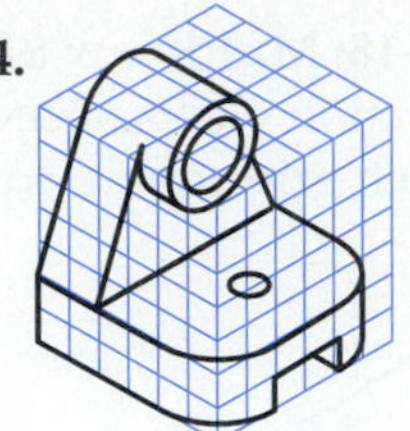

Exercises 25–34: Given the following orthographic views, create an isometric sketch of each object. Assume that each grid square equals either 1/2 inch, or 10 mm.

25.

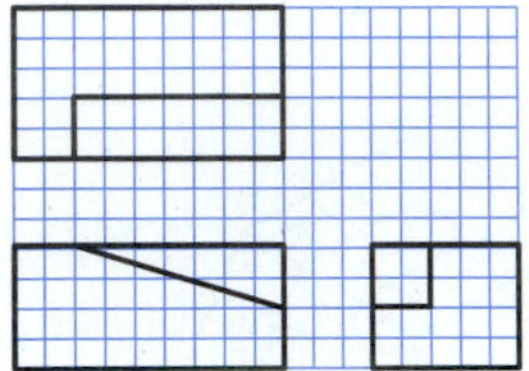

26.

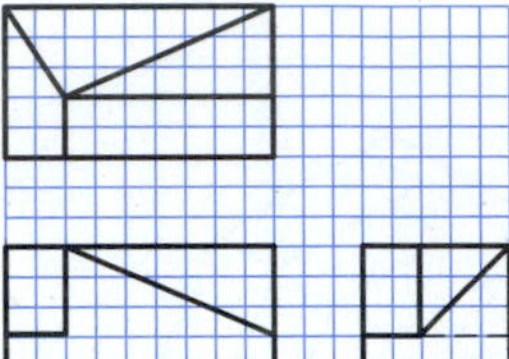

27.

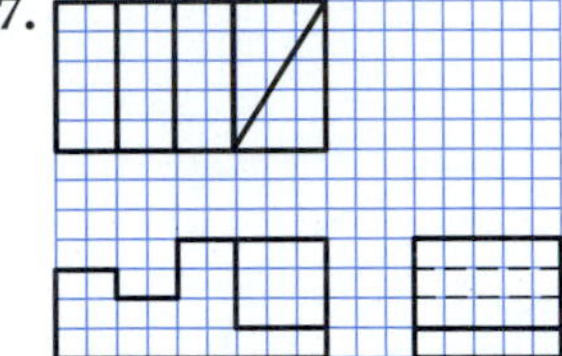

28.

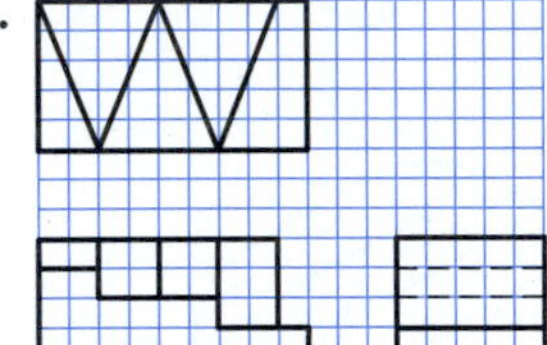

29.

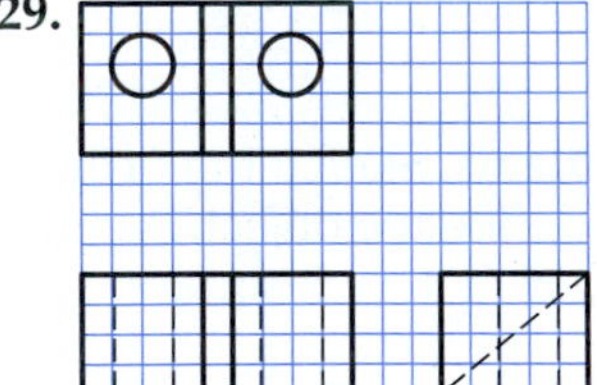

30.

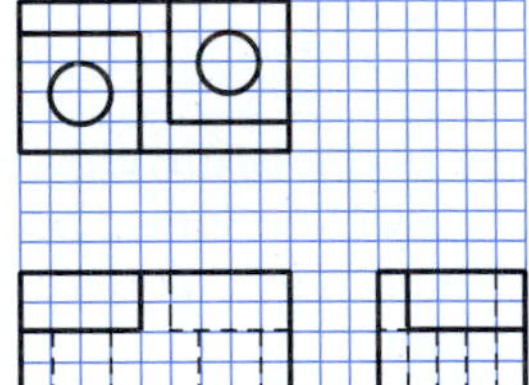

31.

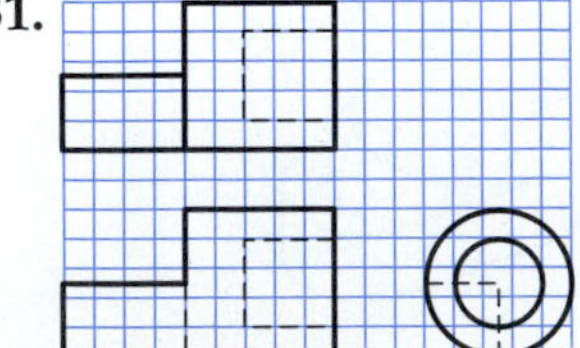

32.

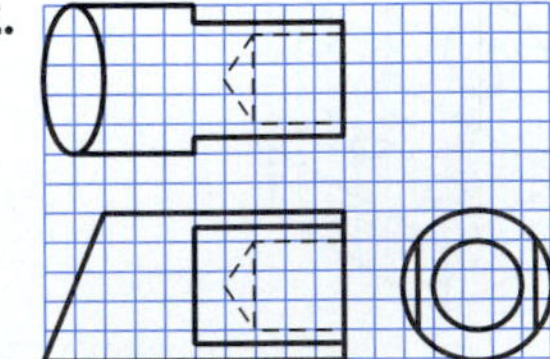

33. 

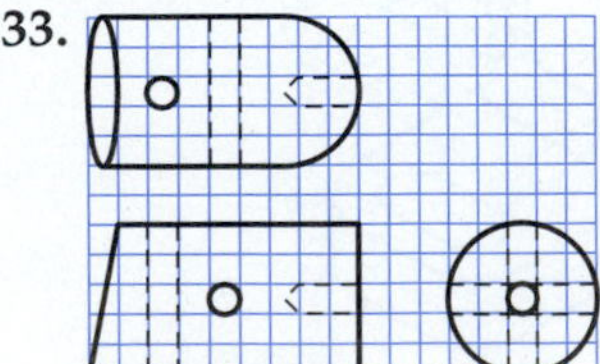

34.

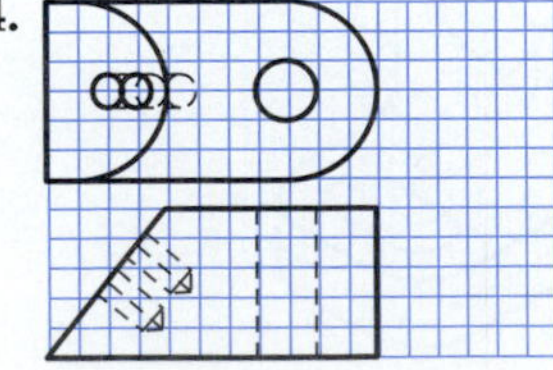

Exercises 35–46: Match each set of orthographic views with the pictorial view that it could represent.

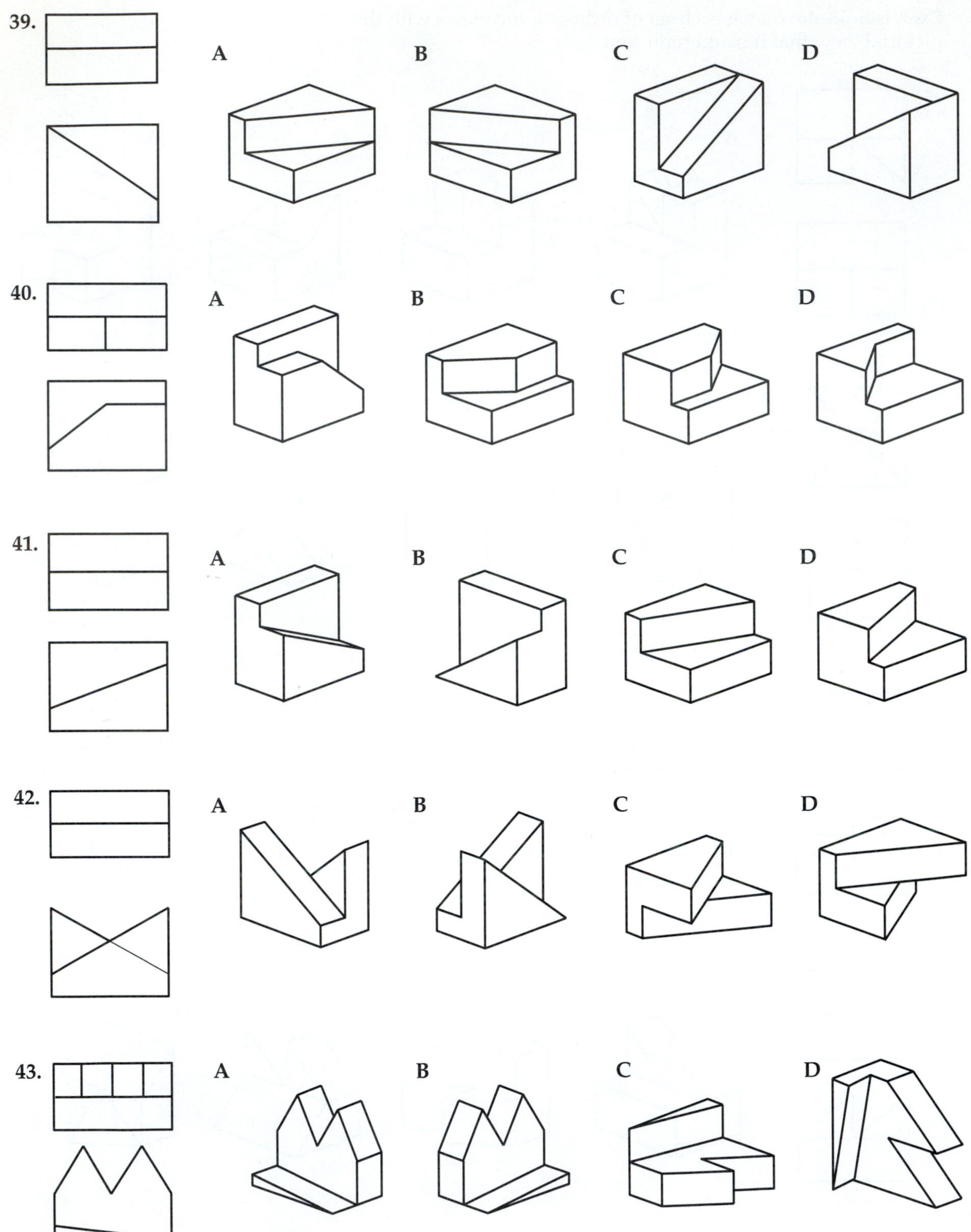
39.
A
B
C
D
40.
A
B
C
D
41.
A
B
C
D
42.
A
B
C
D
43.
A
B
C
D

44. A 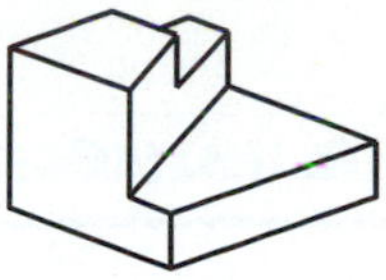 B 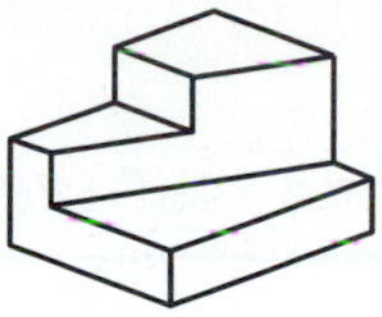 C 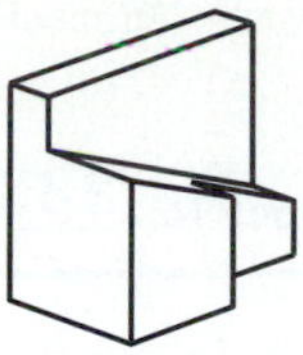 D

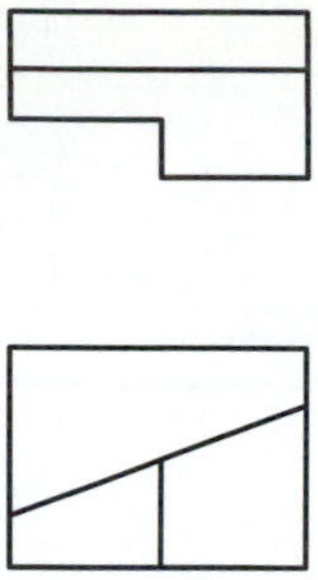

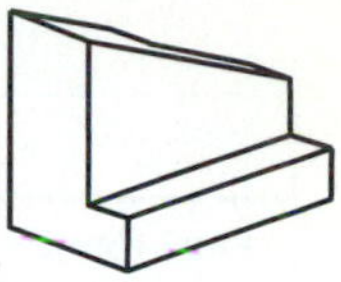

45. 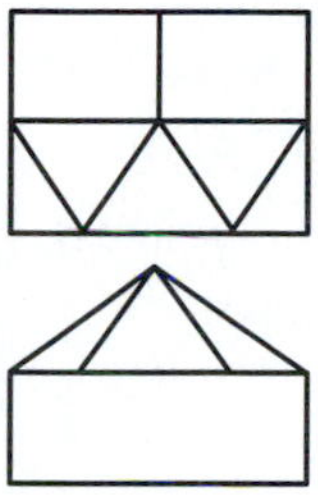 A 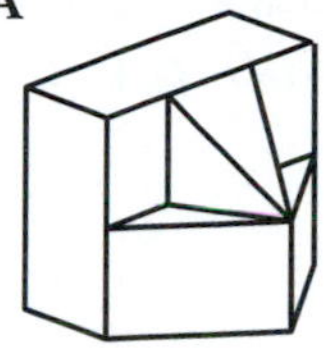 B 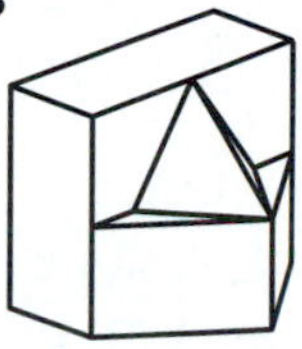 C 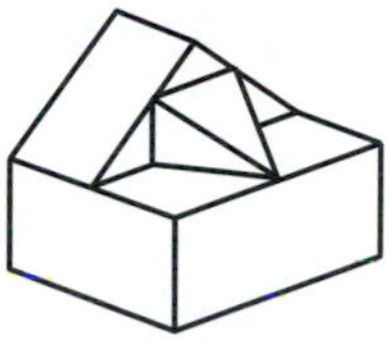 D

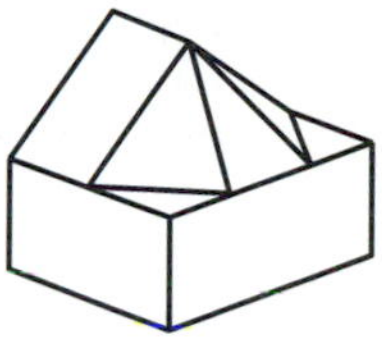

46. 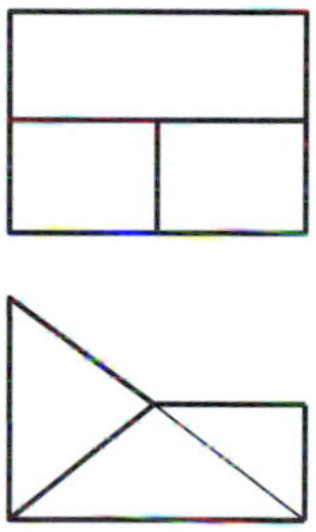 A 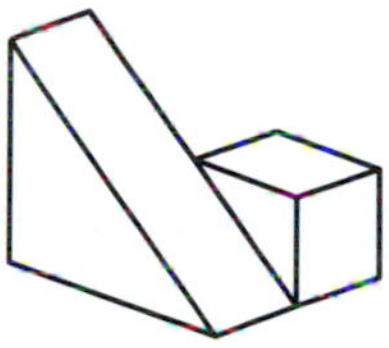 B 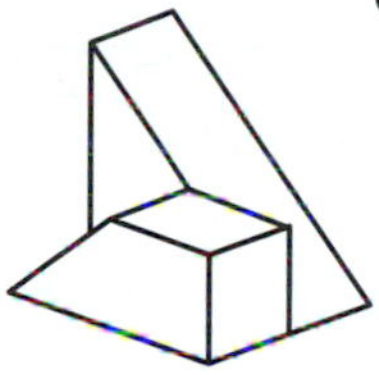 C 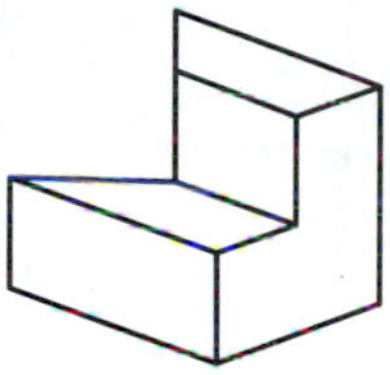 D

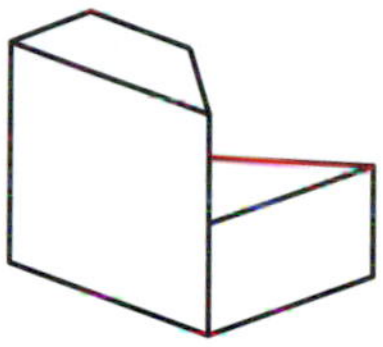

DESIGNER'S NOTEBOOK

Simple geometric shapes are often an important basis for modeling and design. These sketches of a gripper alignment tool from Albert Brown's notebook show a circular pattern of mounting holes. Note the abbreviation ID for interior diameter (OD is used to indicate outside diameter). (Courtesy of Albert W. Brown, Jr., Affymax Research Institute.)

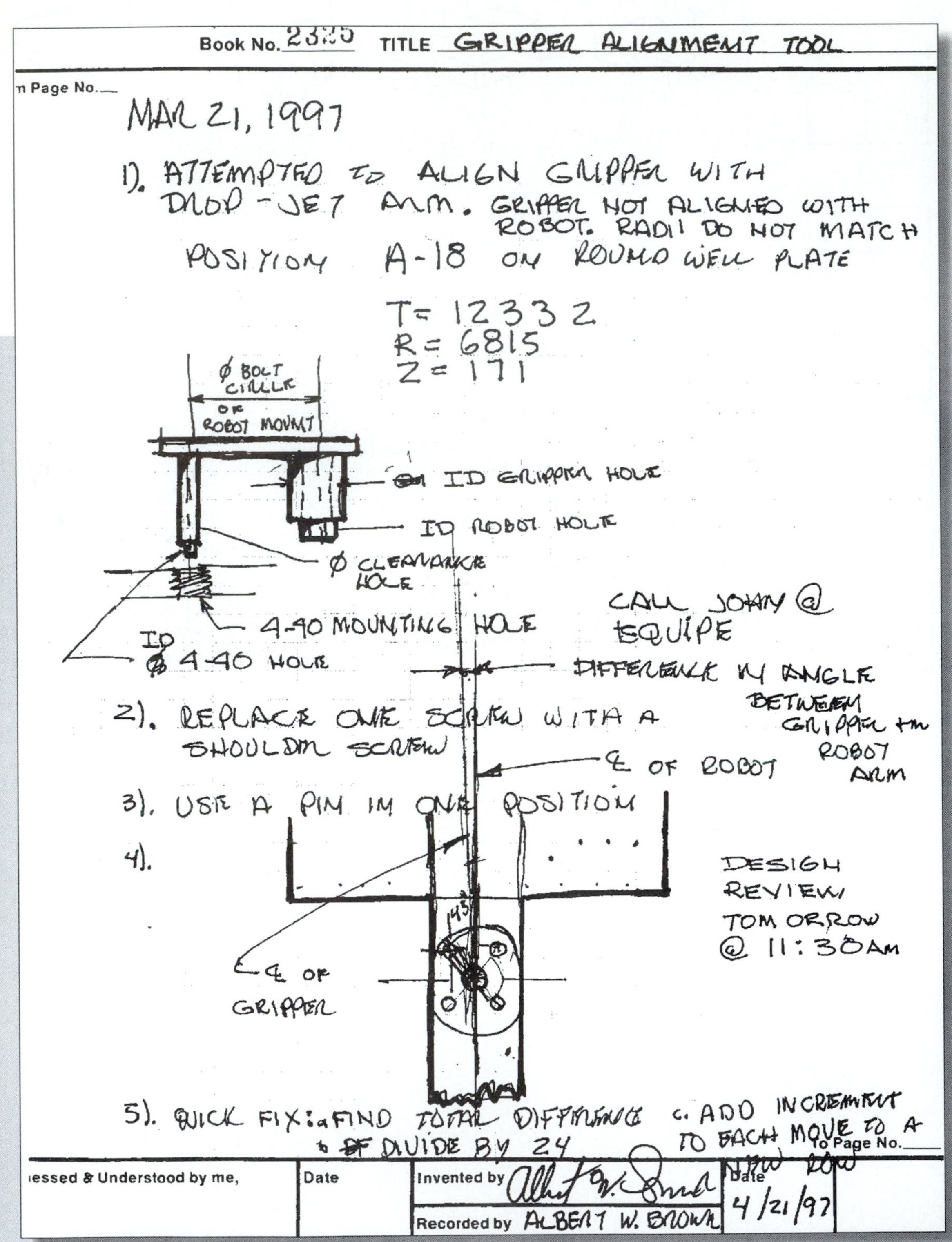

CHAPTER 5

Using Geometry for Modeling and Design

Objectives

When you have completed this chapter you will be able to

- identify and specify basic geometric elements and primitive shapes;
- select a 2D profile that best describes the shape of an object;
- identify mirrored shapes and sketch their lines of symmetry;
- identify shapes that could be formed by extrusion and sketch their cross sections;
- identify shapes that could be formed by revolution techniques and sketch their profiles;
- define Boolean operations;
- specify the Boolean operations to use to combine primitive shapes into a complex shape;
- work with Cartesian coordinates and user coordinate systems in a CAD system;
- identify the transformations common to CAD systems.

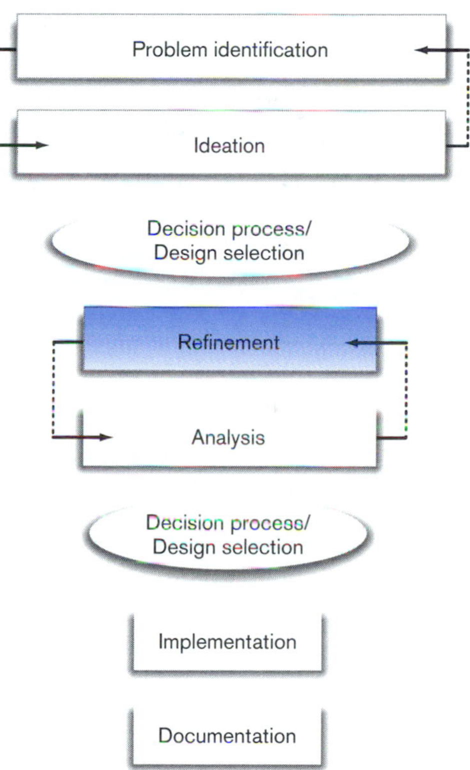

As computer-based tools, especially 3D CAD, become increasingly popular as visualization media in engineering, computer-generated models are playing more of a role in the ideation phase of the design process. Like sketching, effective modeling depends on your ability to "see" the faces, vertices, edges, and basic solid shapes that make up an object.

For many people, the link between the hand, eye, and brain is so intuitive that they are able to sketch basic shapes without specifically identifying the geometric elements (such as circles, arcs, and ellipses) that make up the sketch. To create accurate, unambiguous engineering sketches, however, it is important to be aware of the geometric entities and relationships that define a shape. This awareness is also the foundation for effective computer-based modeling, in which you must define these entities and relationships so the computer can "draw" them.

You have already seen how orthographic sketching depends on the ability to show the shape of 3D objects in a given 2D viewing plane. This ability is also the foundation for creating these shapes in a 3D CAD system. Usually, 2D outlines are the starting point for the 3D shapes that you will use to create computer-based models. Understanding which view orientation will show a surface true size in an orthographic (or auxiliary) view will also help you select an orientation that will display the surface full-size in the CAD environment.

In this chapter, you will expand your understanding of the geometry of objects and learn techniques for visualizing and modeling objects in 3D.

Drawing Geometry

You might have overheard your instructor or others in the CAD lab talking about "creating some geometry." Engineering drawings and CAD packages use basic geometric shapes and relationships to define complex objects and create accurate CAD models. Reviewing the basic geometry of the simple elements that make up CAD models will help you better understand how to combine these elements into a complex model—and also how to sketch and interpret the shapes you see around you.

All 2D drawings are composed of simple entities such as points, lines, arcs, and circles, as well as more complex entities such as ellipses and curves. These basic entities form the basis of most drawings. You use your knowledge of these geometric entities to create shapes and objects in CAD-based drawings as well as in your freehand sketches.

When you draw by hand, you don't "tell" your pencil the beginning and ending points that will make up the line. In fact, the line you draw freehand is only an approximation of the geometric definition of a straight line. When you tell the computer to draw a line, however, it will execute a true straight line to a high degree of accuracy—much greater than what you can discern on your computer monitor. Understanding the geometric properties and definitions of these entities will help you specify them accurately and appropriately in your models.

Working With Coordinates for 3D CAD Modeling

When you are using the computer to draw in either 2D or in 3D, the drawing entities are stored in relationship to a coordinate system. To create drawing geometry well, you should understand 2D and 3D Cartesian coordinates and the various ways you might enter and store geometric data based on a coordinate system. You must also understand how to orient user coordinate systems to easily create model geometry. No matter what CAD software system you will be using, you should understand some basic similarities of working with coordinates.

Most CAD systems use the right-hand rule for coordinate systems; if you point the thumb of your right hand in the positive direction for the X axis and your index finger in the positive direction for the Y axis, your remaining fingers curl in the positive direction for the Z axis (shown in Fig. 5.1).

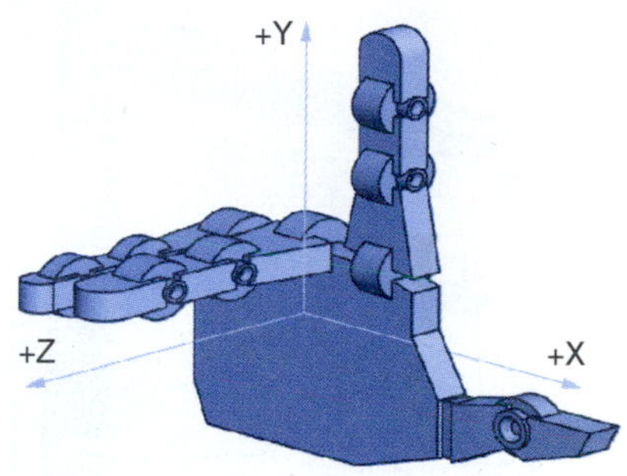

FIGURE 5.1
Right-hand rule

The right-hand rule is also used to determine the direction of rotation in most CAD systems. When thinking of the right-hand rule in the context of rotation, point your thumb in the positive direction along the axis of rotation. Your fingers will curl in the positive

The First Coordinate System

Rene Descartes (1596–1650) was the French philosopher and mathematician for whom the Cartesian coordinate system is named. Descartes was the first mathematician to link algebra and geometry and classify curves by the equations that describe them. His coordinate system remains the most commonly used coordinate system today for identifying points. A 2D coordinate system consists of a pair of lines, called the X and Y axes, drawn on a plane so that they intersect at right angles. The point of intersection is called the origin. A 3D coordinate system adds a third axis, referred to as the Z axis, that is perpendicular to both of the other axes. Each point in space can be described by numbers, called *coordinates*, that represent its distance from this set of axes. The Cartesian coordinate system made it possible to represent geometric entities by numerical and algebraic expressions. For example, a straight line can always be represented by a linear equation in the form $ax + by + c = 0$, where the x and y variables represent the X and Y coordinates for each point on the line. Descartes's work laid the foundation for the problem-solving methods of analytic geometry and was the first significant advance in geometry since the ancient Greeks.

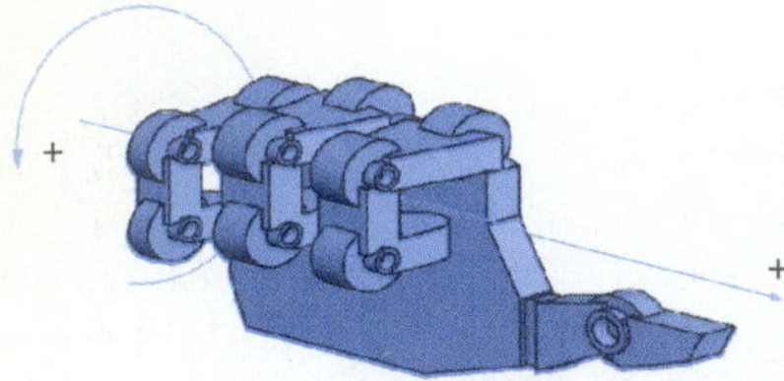

FIGURE 5.2
The curl of the fingers indicates the positive direction along the axis of rotation.

direction for the rotation, as shown in Figure 5.2. When the face of your monitor is the X-Y plane, the Z axis is pointing toward you (see Fig. 5.3).

Some CAD systems use a left-hand rule. In this case, the curl of the fingers on your *left* hand gives you the positive direction for the Z axis. In this case, when the face of your computer monitor is the X-Y plane, the positive direction for the Z axis would extend into your computer monitor, not toward you.

A 2D CAD system uses only the X and Y coordinates of the Cartesian coordinate system. Systems that are 3D use X, Y, and Z. To show a 2D representation in a 3D CAD system, the view is straight down the Z axis. Figure 5.4 shows a drawing created using only the X and Y values, leaving the Z coordinates set to 0 to produce a 2D drawing. Think back to what you learned in Chapter 4 about creating orthographic views. Recall how each orthographic view showed only two of the three coordinates because your viewing direction was looking straight down one axis. You can think of a 2D drawing on your computer monitor the same way: it shows only the X and Y coordinates because you are looking straight down the Z axis.

When the X-Y plane is aligned with the screen in a CAD system, the Z axis represents a horizontal axis, but it is commonly used to depict the vertical axis on devices and in machining applications. In both cases, the coordinate axes remain mutually perpendicular and oriented according to the right-hand or left-hand rule. Because the coordinate system can be rotated so the view is straight down any axis, understanding how to use coordinates in the model is more important than how you visualize the axes.

The technique you used to plot isometric sketches in Chapter 3 is similar to the way objects are defined in a CAD system. The 3D block shown in Figure 5.5 has the coordinates for its vertices identified as they would be described on the coordinate system (represented by the intersection of three lines). Look at the block and identify its vertices on the coordinate system. Often it is useful when modeling

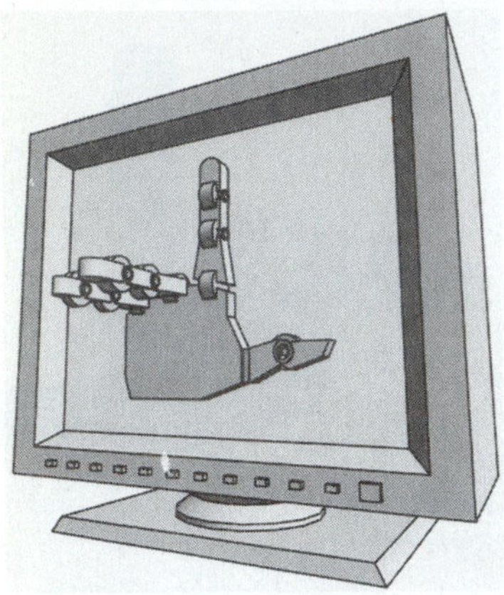

FIGURE 5.3
In systems that use the right-hand rule, the positive Z axis points toward you when the face of the monitor is parallel to the X-Y plane.

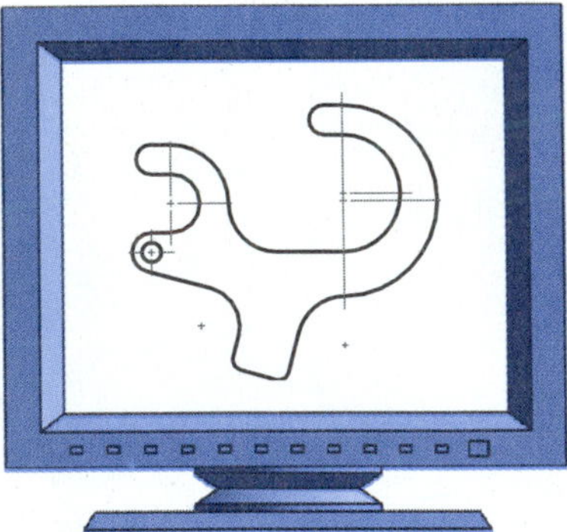

FIGURE 5.4
This drawing was created on the X-Y plane in the CAD system. It appears true size because the viewing direction is perpendicular to the plane—straight down the Z axis.

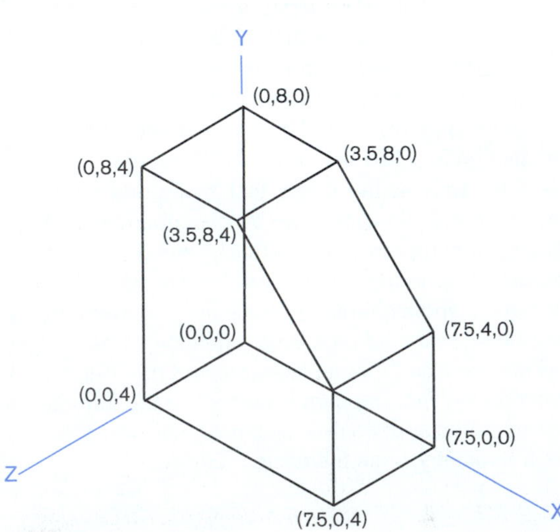

FIGURE 5.5
3D coordinates are used to locate vertices in the CAD system.

parts to have the origin of the coordinate system at the lower left of the part, as shown in Figure 5.5, because that is a useful location for the (0,0,0) point on a part when it is being machined. (Some older numerically controlled machinery will not interpret a file correctly if it has negative lengths or coordinates. Keep in mind that often your CAD models will be exported to another system for manufacturing the part, so try to create them in a way that will make them useful.)

Specifying Location

Many CAD programs, even though the model is ultimately stored in a typical Cartesian coordinate system, allow you to specify the location of objects using other types of coordinates as well. The most typical of these are relative, polar, cylindrical, and spherical coordinates. Understanding these coordinate formats and how they can be used to specify locations on your 3D model can help you to define your CAD drawing geometry.

Absolute Coordinates

Absolute coordinates are the coordinates used to store the location of points in your CAD database. These coordinates specify location in terms of distance from the origin in each of the three axis directions of the Cartesian coordinate system. Think of giving someone directions to your house (or to a house in an area where the streets are laid out in nice rectangular blocks). One way to describe how to get to your house would be to tell the person how many blocks over and how many blocks up it is from two main streets (and how many floors up in the building, for 3D). The two main streets are like the X and Y axes of the Cartesian coordinate system, with the intersection the origin. Think of this as absolute coordinates. Figure 5.6 shows how you might locate a house with Cartesian coordinates.

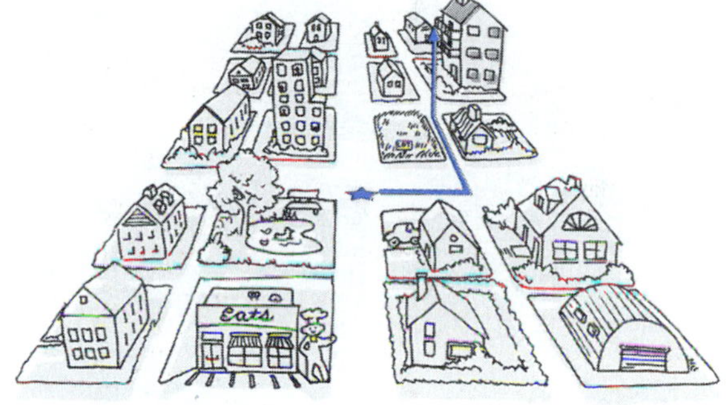

FIGURE 5.6
The directions to an apartment as described by absolute coordinates define its location in terms of the origin (0,0,0), shown here as a star. These directions are useful to anyone and will not change unless the origin changes.

Relative Coordinates

Instead of having to specify each location from the origin, using *relative coordinates* allows you specify a 3D location by giving the number of units from a previous location. In other words, the location is defined relative to your previous location. To understand relative coordinates, think about giving someone directions from his or her current position, not from two main streets. Figure 5.7 shows the same map again, but this time with the location of the house relative to the location of the person receiving directions.

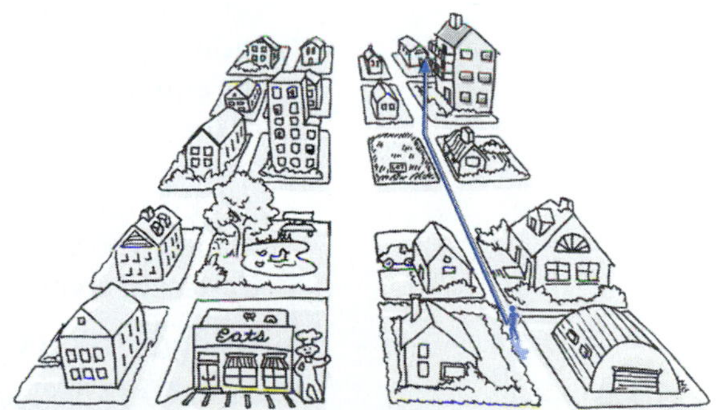

FIGURE 5.7
Directions to the same apartment using relative coordinates describe the location in terms of where the visitor is starting from. Relative coordinates for the apartment differ according to the location of the person getting them.

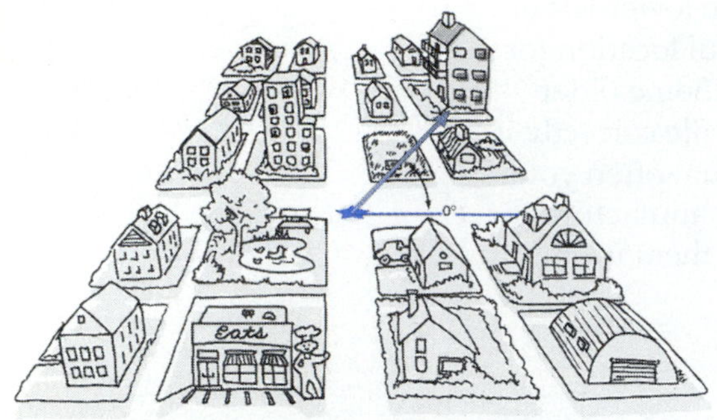

FIGURE 5.8
Absolute polar coordinates describe the location of the apartment building in terms of the origin but combine motion in two directions into a single angular measure.

Polar Coordinates

Polar coordinates are used to locate an object by giving an angle (from the X axis) and a distance. Polar coordinates can either be absolute, giving the angle and distance from the origin, or they can be relative, giving the angle and distance from the current location. Picture the same situation of having to give directions. You could tell the person to walk at a specified angle from the crossing of the two main streets, and how far to walk. Figure 5.8 shows the angle and direction for the shortcut across the empty lot using absolute polar coordinates. Figure 5.9 shows the angle and direction for the shortcut using relative polar coordinates.

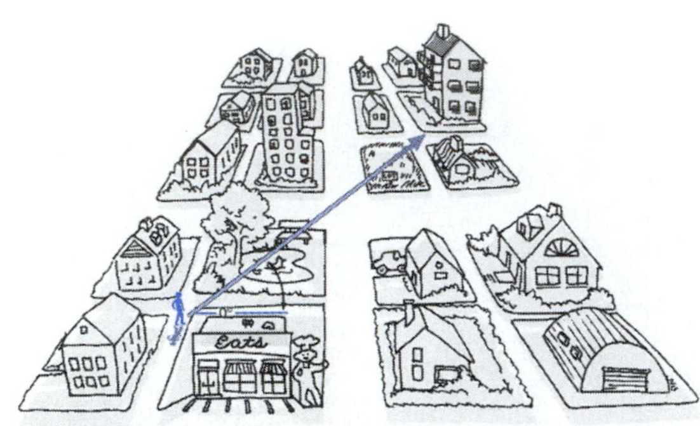

FIGURE 5.9
Relative polar coordinates describe the distance and angle of motion relative to the starting location.

Cylindrical and Spherical Coordinates

Cylindrical and *spherical coordinates* are similar to polar coordinates except that they let you specify a 3D location instead of one on a single flat plane (such as a map).

Cylindrical coordinates allow you to specify a 3D location based on a radius, angle, and distance (usually in the Z axis direction). It may be helpful to think about this as giving a location as though it were on the edge of a cylinder. The radius tells how far the point is from the center (or origin); the angle is the angle from the X axis along which the point is located; and the distance gives you the height where the point is located on the cylinder. Cylindrical coordinates are similar to polar coordinates, but they add distance in the Z direction. Figure 5.10 shows how cylindrical coordinates specify a location.

Spherical coordinates allow you to specify a 3D location by giving the radius, the angle from the X axis, and the angle from the X-Y plane. It is helpful to think of locating a point on a sphere, where the origin of the coordinate system is at the center of the sphere. The radius gives the size of the sphere, the first angle gives a location on the equator, and the second angle gives the location from the plane of the equator to the point on the sphere in line with the location specified on the equator.

Even though you may use these different systems to enter information into your 3D drawings, the end result is stored using Cartesian coordinates.

Using Existing Geometry to Specify Location

Most CAD packages offer a means of specifying location by specifying a point's relationship to existing objects in the model or drawing. For example, AutoCAD's "object snap" feature lets you enter a location by telling the software that you want to "snap" to the endpoint of a line, the center of a circle, the intersection of two lines, and so on. Using existing geometry to locate new entities can be faster than

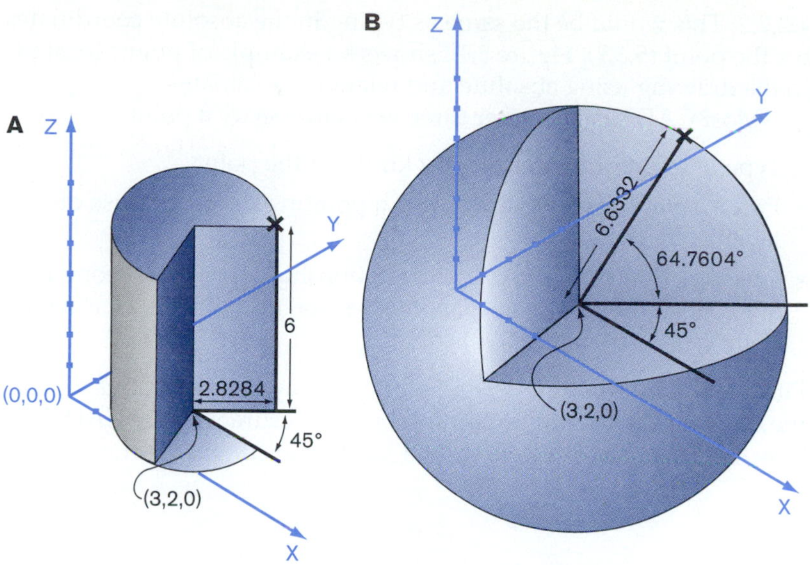

FIGURE 5.10
The target points in (a) and (b) are described by relative coordinates from the starting point, (3,2,0). In (a), the relative cylindrical coordinates 2.8284<45,6 indicate movement of 2.8284 units from the starting point along a radius that is 45° from the X axis, then a rise of 6 units (6 units in the positive Z direction) to reach the target location. In (b), the relative spherical coordinates 6.6332<45<64.7604 describe movement of 6.6332 units from the starting point along a radius that is 45° from the X axis and 64.7604 degrees from the X-Y plane. While the paths to the point differ, the resulting endpoint is the same.

entering coordinates, and can ensure that the desired location is entered exactly. This feature also allows you to capture geometric relationships between objects without calculating the exact location of a point. For example, you can "snap" to the midpoint of a line or the nearest point of tangency on a circle and the software will calculate the desired location and select it for you. Using existing geometry to specify location is an important tool in creating accurate models efficiently.

Geometric Entities

Points

Points are geometrical constructs. Points are considered to have no width, height, or depth. They are used to indicate locations in space. When you represent a point in a sketch, the convention is to make a small cross, or a bar if it is along a line, to indicate the location of the point (see Fig. 5.11). In CAD drawings, a point is located by its coordinates and usually shown with some sort of marker like a cross, circle, or other representation. Many CAD systems allow you to choose the style and size of the mark that is used to represent points.

In most CAD systems you can specify a point by typing its absolute coordinates: you select a command to create a point, then type in 3,4,7. This locates a point entity at X=3, Y=4, Z=7, relative to the origin. Often you can also locate a point relative to your last point or previous location. AutoCAD and some other software use an @ symbol to indicate that the entry to follow is relative to the last location. You might use relative entry for a point if you already entered a point at (3,4,7) and wanted to enter the next point 2 units away in each positive direction. Using AutoCAD's relative coordinate entry, you could select the command for a point, then type

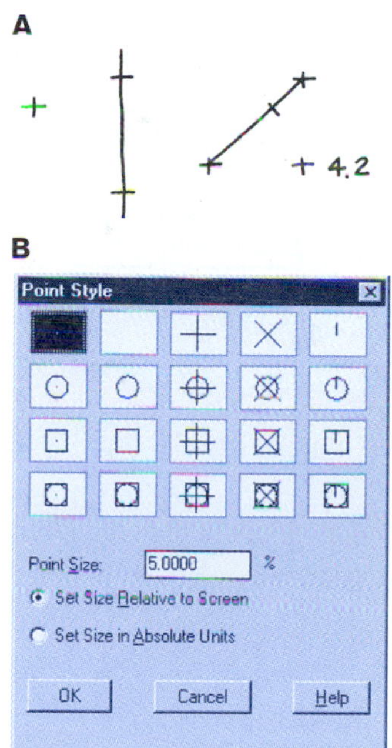

FIGURE 5.11
The crosses and bars in (a) represent points as they would appear in a sketch. In a CAD system, you can choose from a variety of ways to display points using a dialog box such as the one shown in (b).

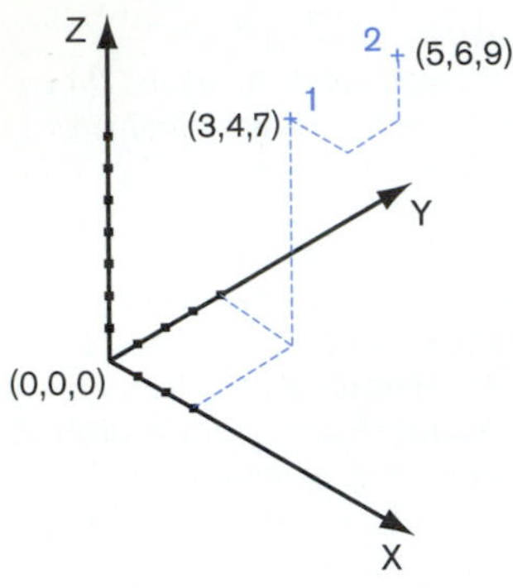

FIGURE 5.12
Point 1 was added to the drawing by typing the absolute coordinates, 3,4,7. Point 2 was added relative to Point 1 with the relative coordinates @2,2,2.

@2,2,2. This would be the same as typing in the absolute coordinates for the point (5,6,9). Figure 5.12 shows an example of points located in the drawing using absolute and relative coordinates.

Most CAD systems offer three ways to specify a point:

- Type in the coordinates (of any kind) for the point.
- Pick a point from the screen with a pointing device (mouse or tablet).
- Specify a point's location by its relationship to existing geometry (e.g., an endpoint of a line, an intersection of two lines, or a centerpoint).

Picking a point from the screen is a quick way to enter points when the exact location is not important, but the accuracy of the CAD database makes it impossible to enter a location accurately in this way.

Lines

Straight lines are defined as the shortest distance between two points. Geometrically, they are considered to have length, but no other dimension such as width or thickness. Lines are used in a drawing to represent the edge view of a surface, the limiting element of a contoured surface, or the edge formed where two surfaces on an object join. In a CAD database, lines are typically stored by the coordinates of their endpoints. For example, if you were using absolute coordinates, you would select the command for line, then type the coordinates for the endpoints based on their distance from the origin.

The endpoints of a line can be entered using a CAD system in all the ways enumerated for entering points. Using coordinates, you may specify:

- the absolute coordinates of the endpoints relative to the origin;
- relative coordinates for the endpoints relative to the last point entered (the current location);
- polar coordinates to specify the length and angle of the line; or
- spherical or cylindrical coordinates to locate the endpoints of a 3D line (if this is supported by your CAD system).

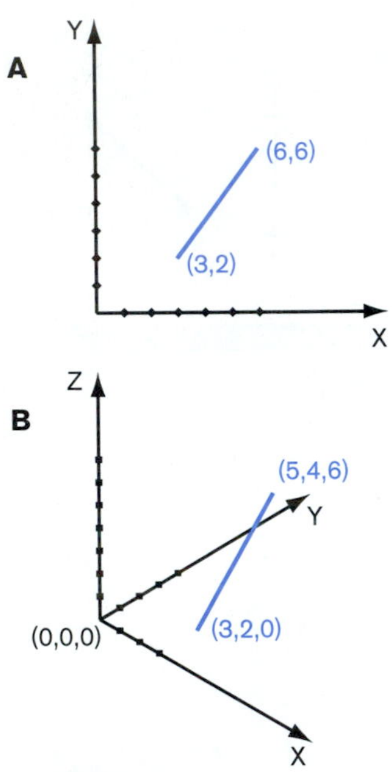

FIGURE 5.13
(a) This 2D line was drawn from endpoint (3,2) to (6,6). (b) This 3D line was drawn from endpoint (3,2,0) to (5,4,6).

For the lines shown in Figure 5.13 (a) and (b), the table shows how you can specify the second endpoint for a particular type of coordinate entry. (For either or both endpoints, you can also "snap" to existing geometry without entering any coordinates.)

	(A) Second endpoint for 2D line	**(B) Second endpoint for 3D line**
Absolute	6,6	5,4,6
Relative	@3,4	@2,2,6
Relative Polar	@5<53.13	n/a
Relative Cylindrical	n/a	@2.8284<45,6
Relative Spherical	n/a	@6.6332<45<64.7606

Planes

Planes are defined by (see Fig. 5.14):

- Three points not lying in a straight line
- Two parallel lines
- Two intersecting lines
- A point and a line

The last three ways to define a plane are all special cases of the more general case—three points not in a straight line. Knowing what can determine a plane can help you understand the geometry of solid objects—and use the geometry to work in CAD.

For example, a face on an object can be thought of as a plane that extends between the vertices and edges of the surface. Most CAD programs allow you to align new entities with an existing plane. You can use any face on the object—whether it is normal, inclined, or oblique—to define a plane that can then be used to align the new entity. You can specify that plane by using the entities defined in the list above for defining a plane.

The same plane can also serve as a 2D viewing plane that shows the surface true size. In the last chapter, you learned how oblique and inclined surfaces are not shown true size in standard orthographic views. In 3D CAD, you can easily create an auxiliary view parallel to any surface by defining the viewing plane to be parallel to the planar surface of the object (see Fig. 5.15).

Defining planes on the object or in 3D space is an important tool for working in 3D CAD. You will learn more about specifying planes to orient a user coordinate system to make it easy to create CAD geometry later in this chapter.

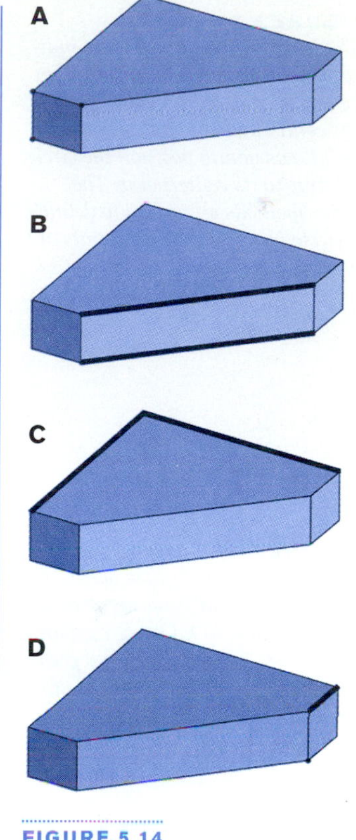

FIGURE 5.14
The highlighted entities in each image define a plane: (a) Three points not lying in a straight line; (b) Two parallel lines; (c) Two intersecting lines; (d) A point and a line.

Circles

A circle is a set of points that are equidistant from a center point. The distance from the center to one of the points is the *radius*. The distance across the center to any two points on opposite sides is the *diameter*. The circumference of a circle contains 360° of arc. In a CAD file, a circle is often stored as a center point and radius.

Most CAD systems allow you to define circles by specifying:

- the center and a diameter;
- the center and a radius;
- two points on the diameter;
- three points on the circle;
- a radius and two entities to which the circle is tangent; or
- three entities to which the circle is tangent.

These methods are illustrated in Figure 5.16. As with *any* points, the points defining a circle can be entered with absolute, relative, polar, cylindrical, or spherical coordinates; by picking points from the screen; or by specifying existing geometry.

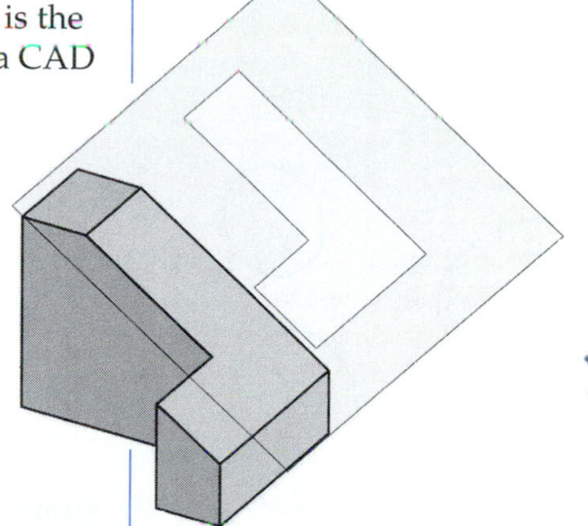

FIGURE 5.15
The viewing plane is aligned with the inclined surface and shows it true size.

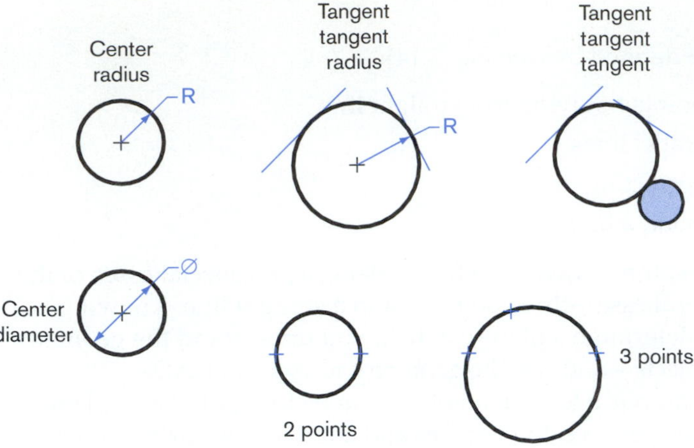

FIGURE 5.16
The labels on the figures here indicate the different ways a circle can be specified in CAD. Specifying the center and the radius or diameter allows you to position the circle relative to its centerpoint. The other methods allow you to define a circle in terms of other points or entities that it must be tangent to.

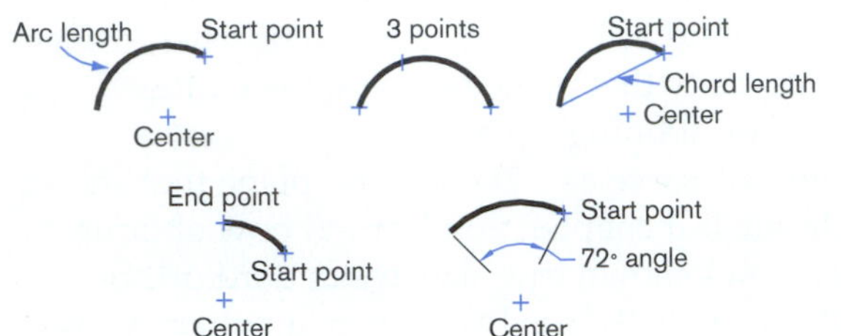

FIGURE 5.17
The labels on the figures here indicate the different ways arcs can be specified in CAD. Like circles, arcs may be located from a center point or an endpoint, making it easy to locate them relative to other entities in the model.

Arcs

An arc is a portion of a circle. An arc can be defined by specifying (see Fig. 5.17):

- a center, radius, and angle measure (sometimes called the *included angle* or delta angle);
- a center, radius, and chord length;
- a center, radius, and arc length;
- the endpoints and a radius;
- the endpoints and a chord length;
- the endpoints and arc length;
- the endpoints and one other point on the arc (3 points).

Ellipses

An ellipse can be defined by its major and minor axis distances. The major axis is the longer axis of the ellipse, the minor axis is the shorter axis. Some ellipses are shown and labeled in Figure 5.20.

An ellipse is created by a point moving along a path where the sum of its distances from two points, each called a *focus (foci* is the plural form), is equal to the major diameter. To help understand the shape of an ellipse, imagine pinning either end of a string in the location of the foci, then sliding a pencil along inside the string, keeping it tightly stretched as shown in Figure 5.21. Even though

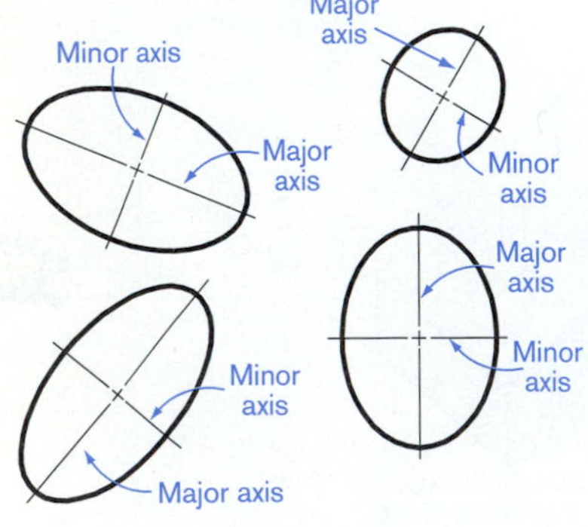

FIGURE 5.20
Major and minor axes of some ellipses

Formulas for Circles and Arcs

r = radius
C = circumference
π = pi $\cong$ 3.14159
a = arc length
A = area
L = chord length
θ (theta) = included angle

rad (radian) = the included angle of an arc length such that the arc length is equal to the radius

$C = 2\pi r$, the curved distance around a circle

$A = \pi r^2$, the area of a circle

$a = 2\pi r^*\theta/360$, so the arc length = $0.01745r\theta$ when you know its radius, r, and the included angle, θ, in degrees.

$a = r^*\theta$ (when the included angle is measured in radians)

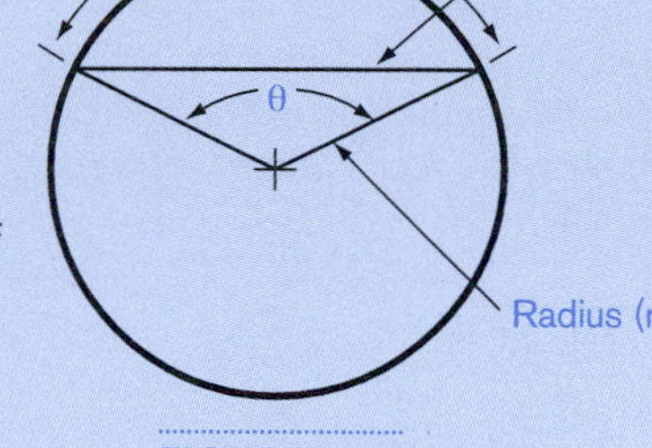

FIGURE 5.18

Bolt-Hole Circle Chord Lengths
To determine the distance between centers for equally spaced holes on a bolt-hole circle
n = 180 / number of holes in pattern
L = sin n * bolt circle diameter

Example: 8-hole pattern on a 10.00-diameter circle:
180/8 = 22.5
The sin of 22.5 is .383
.383 x 10 = 3.83 (the chord)

For more useful formulas, see Appendix 50 at the back of this book.

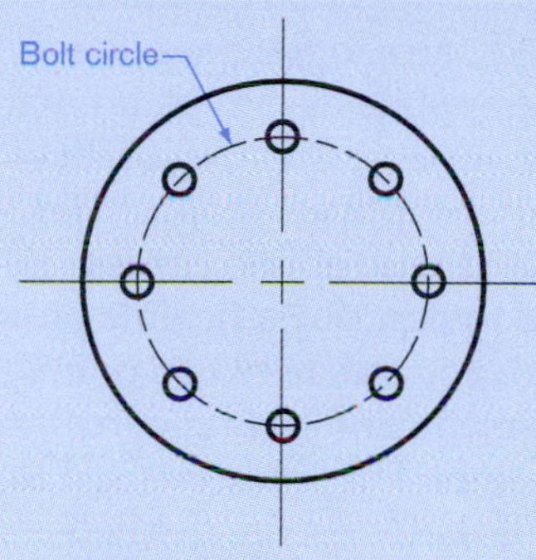

FIGURE 5.19

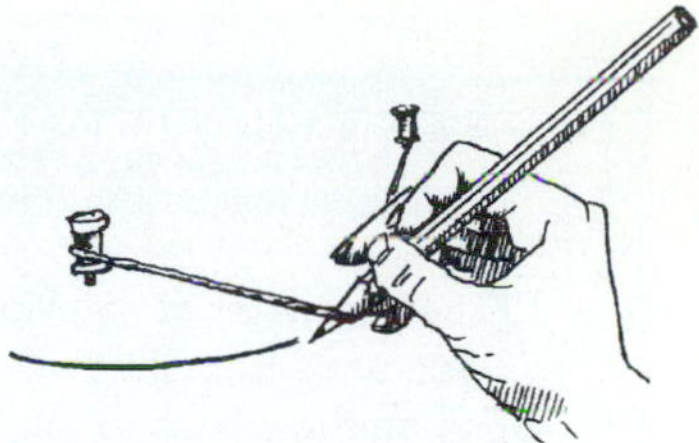

FIGURE 5.21
When creating an ellipse with the pencil-and-string method, the length of the string between the foci is equal to the length of the major axis of the ellipse. Any point that can be reached by a pencil inside the string when it is pulled taut meets the condition that its distance from the two foci sum to the length of the major diameter.

you would not use this technique when sketching (you would just block in the major and minor axes and the enclosing parallelogram, then sketch the ellipse tangent to it), it serves as a good illustration of the definition of an ellipse.

Figure 5.22 illustrates how to locate the foci of an ellipse. Most CAD packages allow you to specify an ellipse by either specifying the major and minor axes or by specifying the foci. Some older CAD systems, such as AutoCAD prior to Release 12, do not generate true ellipses, but rather an approximation of an ellipse using four arc segments. Although this is a close approximation, the information stored in the CAD database for these approximate ellipses is the centers and radii of the four arcs, not the foci or center of the true ellipse. Using this method, accurate information about the ellipse cannot be extracted.

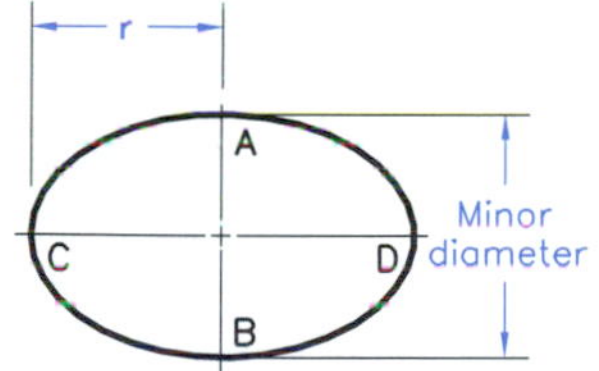

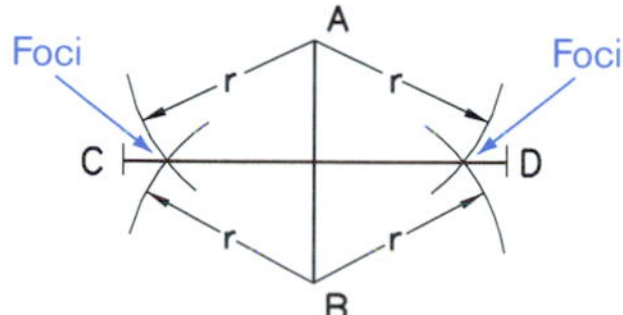

FIGURE 5.22
To locate the foci of an ellipse, draw arcs with their centers at the ends of the minor axis and their radii equal to half of the major axis. The intersection of each pair of arcs is a focus of the ellipse.

The Perimeter of an Ellipse

The perimeter of an ellipse is a set of points defined by their distance from the two foci. The sum of the distances from any point on the ellipse to the two foci must be equal to the length of the major diameter (see Fig. 5.23). The perimeter of an ellipse may be approximated in different ways. Many CAD packages use infinite series to most closely approximate the perimeter. The mathematical relationship of each point on the ellipse to the major and minor axes may be seen in the following approximation:

$$P = 2\pi\sqrt{\frac{x^2 + y^2}{2}}$$

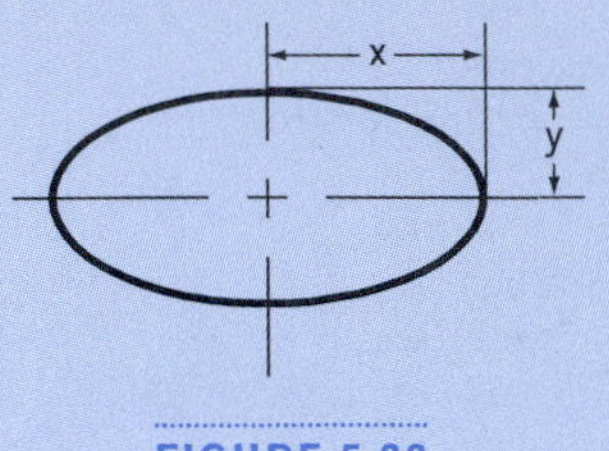

FIGURE 5.23

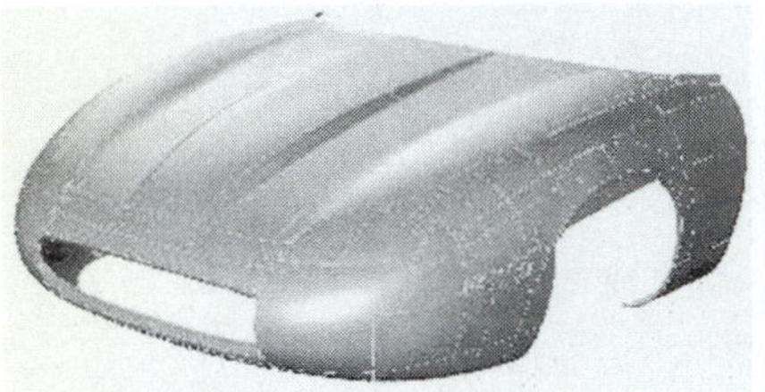

FIGURE 5.24
Spline curves were used for the front end of the Jaguar's body to create shapes that cannot be defined by regular curves such as arcs or ellipses. (Courtesy of ICEM Technologies.)

Spline Curves

Splines are used to describe complex, or *freeform*, curves. Many products that you might design include surfaces that cannot be defined with simple curves such as circles, arcs, or ellipses. For example, the flowing curves used in automobile design blend many different curves into a smooth, irregular surface. Creating lifelike shapes and aerodynamic forms may require splined shapes. The surfaces shown in Figure 5.24 were created with spline curves.

The word *spline* was originally used to describe a flexible piece of plastic or rubber used to draw irregular curves between points. The flexible shape helps smooth the segments into a single freeform curve that passes through each of the points. The spline is used to *interpolate* the curve; that is, it adds points between the defined vertices. To create such a curve on the computer, mathematical methods are used to generate the points on the curve.

One way to create an irregular curve is to draw curves between each set of points. The points and the tangencies at each point are used in a polynomial equation that determines the shape of the curve. This type of curve is useful in the design of a ship's hull or an aircraft wing. With these forms it is easy to define a few points that will define the curve. Because this kind of irregular curve passes through all of the points used to define the curve, it is sometimes called an *interpolated spline*. Because the lowest-order polynomial that can define a nonplanar curve is a cubic curve, this kind of spline is also called a *cubic spline*. An example and its vertices are shown in Figure 5.25.

Other spline curves are *approximated:* they are defined by a set of vertices, but the resulting curve doesn't pass through all the vertices. Rather, the vertices serve to "pull" the curve in the direction of the vertex. Complex curves can be created with relatively few vertices using approximation methods.

To define this kind of curve with a computer, a mathematical relationship is defined between the X and Y (and Z for a 3D shape) coordinates and a parameter, generally referred to as u. A poly-

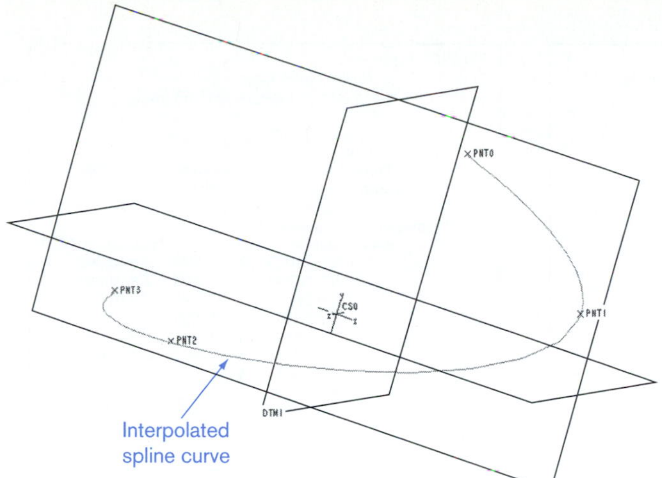

FIGURE 5.25
An interpolated spline curve passes through all of the points used to define the curve.

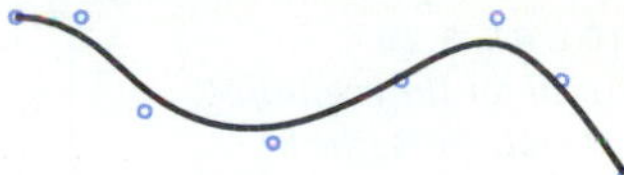

FIGURE 5.26
A Bezier curve passes through the first and last vertex but uses the other vertices as control points to generate a blended curve.

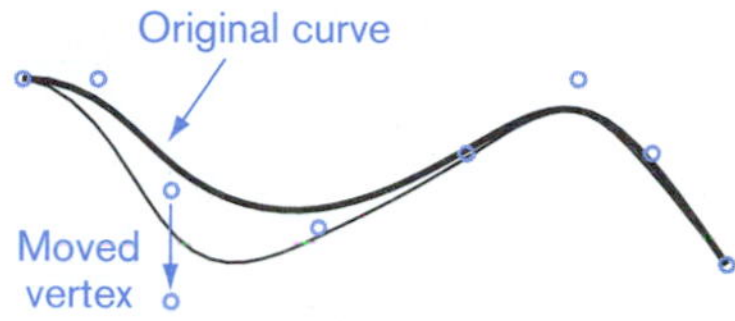

FIGURE 5.27
Every vertex contributes to the shape of a Bezier curve. Changing a single vertex's location redraws the entire curve.

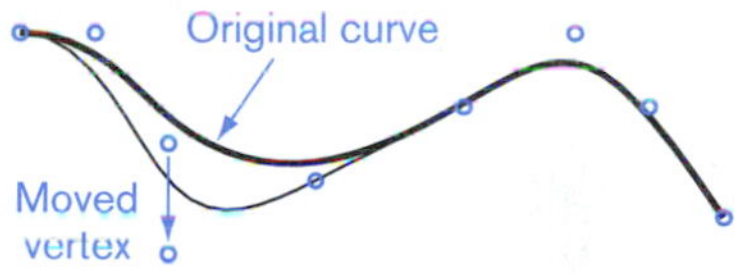

FIGURE 5.28
The B-spline is constructed piecewise so changing a vertex affects the shape of the curve near only that vertex and its neighbors.

nomial equation is used to generate functions in u for each point used to specify the curve. The resulting functions are then blended to generate a curve that is influenced by each point specified but not necessarily coincident with any of them.

The *Bezier curve* was the first CAD method that used spline approximation to create flowing curves. The first and last vertices are on the curve, but the rest of the vertices contribute to a blended curve between them. The Bezier method uses a polynomial curve to approximate the shape of a polygon formed by the specified vertices. The order of the polynomial is 1 degree less than the number of vertices in the polygon (see Fig. 5.26).

The Bezier method is named for Paul Bezier, a pioneer in computer-generated surface modeling at Renault, the French automobile manufacturer. Bezier sought an easier way of controlling complex curves, such as those defined in automobile surfaces. His technique allowed designers to shape natural-looking curves more easily than they could by specifying points that had to lie on the resulting curve, yet the technique also provided control over the shape of the curve. Changing the slope of each line segment defined by a set of vertices adjusts the slope of the resulting curve (see Fig. 5.27). Today, Bezier curves are used in many drawing programs such as Adobe Illustrator and some surface modeling programs to create freeform curves. One disadvantage of the Bezier formula, however, is that the polynomial curve is defined by the combined influence of every vertex: a change to any vertex redraws the entire curve between the start point and endpoint.

A *B-spline* approximation is a special case of the Bezier curve that is more commonly used in engineering packages because it gives the designer more control when editing the curve. A B-spline is defined as a blended piecewise polynomial curve passing near a set of control points. The spline is referred to as *piecewise* because the blending functions used to combine the polynomial curves can vary over the range of the curve. That means that when a control point changes, only the piece of the curve defined by the new point and the vertices near it change, not the whole curve (see Fig. 5.28).

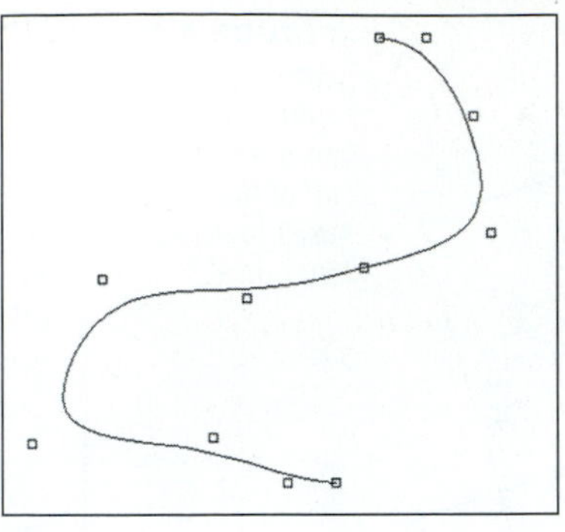

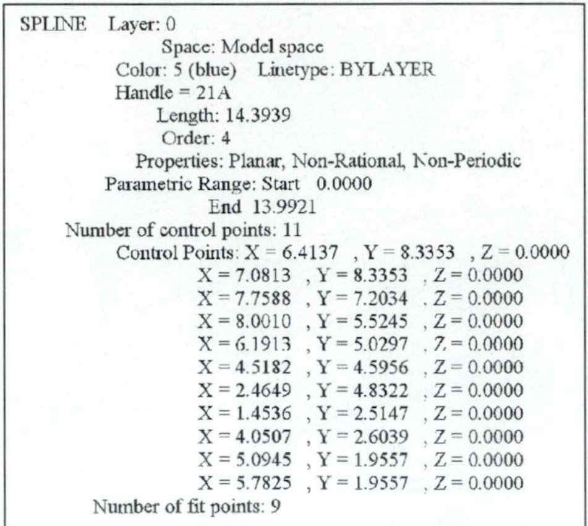

```
SPLINE   Layer: 0
            Space: Model space
         Color: 5 (blue)   Linetype: BYLAYER
         Handle = 21A
            Length: 14.3939
            Order: 4
          Properties: Planar, Non-Rational, Non-Periodic
        Parametric Range: Start   0.0000
                  End  13.9921
     Number of control points: 11
         Control Points: X = 6.4137  , Y = 8.3353  , Z = 0.0000
                  X = 7.0813  , Y = 8.3353  , Z = 0.0000
                  X = 7.7588  , Y = 7.2034  . Z = 0.0000
                  X = 8.0010  , Y = 5.5245  , Z = 0.0000
                  X = 6.1913  , Y = 5.0297  . Z = 0.0000
                  X = 4.5182  , Y = 4.5956  , Z = 0.0000
                  X = 2.4649  , Y = 4.8322  , Z = 0.0000
                  X = 1.4536  , Y = 2.5147  , Z = 0.0000
                  X = 4.0507  , Y = 2.6039  , Z = 0.0000
                  X = 5.0945  , Y = 1.9557  , Z = 0.0000
                  X = 5.7825  , Y = 1.9557  , Z = 0.0000
       Number of fit points: 9
```

FIGURE 5.29
Except for the beginning and endpoints, the fit points for the spline curve stored in the database don't all lie on the curve, but are used to derive the curve mathematically.

B-splines may or may not pass through the first and last points in the vertex set. Another difference in the B-spline is that the order of the polynomial can be set independently of the number of vertices, or control points, used to define the curve.

In addition to being able to locally modify the curve, many modelers today add to that control by allowing sets of vertices to be weighted differently. The weighting, sometimes called *tolerance,* determines how closely the curve should fit the set of vertices defining it. This measure creates a range from a curve that fits all the points (the same as an interpolated curve) to one that is only loosely controlled by the vertices. This type of curve is called a nonuniform rational B-spline, or *NURBS* curve. A rational curve (or surface) has a weight associated with each control point.

Figure 5.29 shows a 3D spline curve oriented with respect to the Cartesian coordinate system along with the list of its vertices.

Geometric Relationships

Object models are constructed by specifying relationships between the basic geometric entities that were defined in the last section. Many CAD packages have drawing aids that allow you to specify these properties as you draw an entity. When you are sketching, you often imply the relationship by the appearance of the lines or through notes or dimensions.

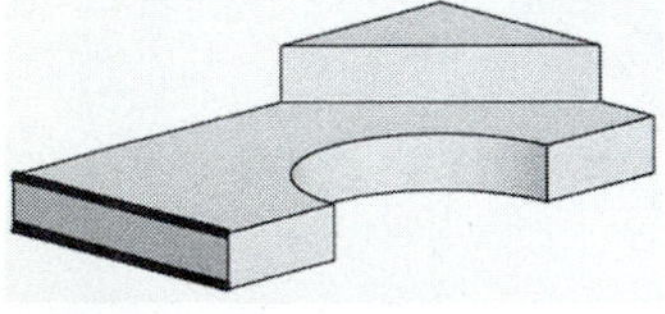

FIGURE 5.30
The highlighted lines are parallel.

Two lines or planes are *parallel* when they are an equal distance apart at every point. Parallel entities never intersect, even if extended to infinity. Figure 5.30 shows an example of parallel lines.

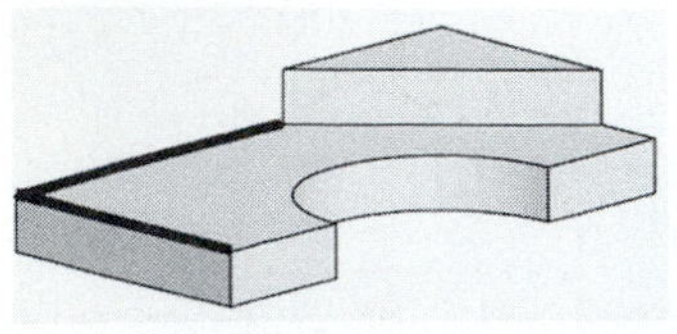

FIGURE 5.31
The highlighted lines are perpendicular.

Two lines or planes are *perpendicular* when they intersect at right angles (or when the intersection that would be formed if they were extended would be a right angle) as in Figure 5.31.

Two entities *intersect* if they have at least one point in common. Two straight lines can intersect at a single point only. A circle and a straight line that intersect will have two points in common, as shown in Figure 5.32.

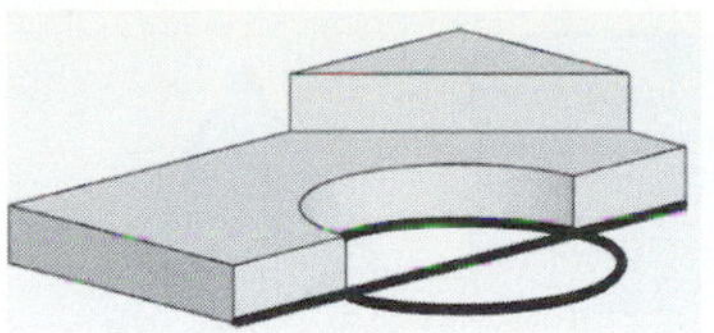

FIGURE 5.32
The highlighted circle intersects the highlighted line at two different points.

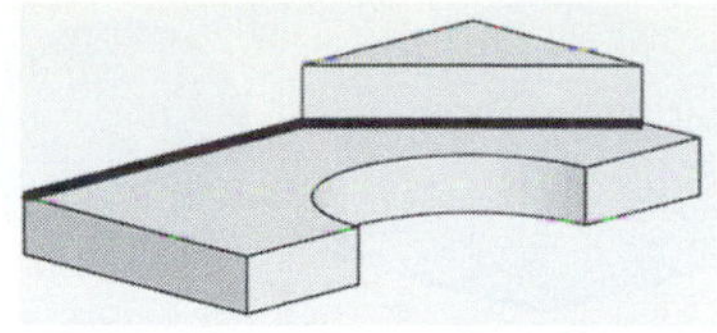

FIGURE 5.33
An angle is defined by the space between two lines (such as those highlighted here) or planes that intersect.

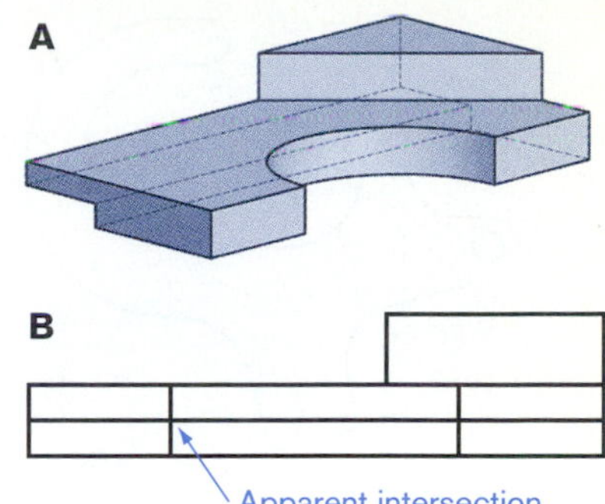

FIGURE 5.34
From the shaded view of this model in (a), it is clear that the back lines do not intersect the half-circle cutaway, even though the dotted lines do intersect the half circle in this view. If you are viewing the model as a wireframe, however, as shown in the front view in (b), the lines appear to intersect.

When two lines meet or intersect, they define an *angle,* or the measure of the space between the two lines, as shown in Figure 5.33.

The term *apparent intersection* is used to refer to lines that appear to intersect in a 2D view or on your computer monitor, but actually do not, as shown in Figure 5.34. When you look at a wireframe view of a model such as that in Figure 5.34, the 2D view may show lines crossing each other when, in fact, the lines do not intersect in 3D space. Changing the view of the model can help you determine whether an intersection is actual or apparent.

Two entities are *tangent* if they touch each other but do not intersect, even if extended to infinity, as shown in Figure 5.35. A line that is tangent to a circle will have only one point in common with the circle.

When a line is tangent to a circle, a radial line from the center of the circle is perpendicular at the point of tangency, as shown in Figure 5.36. Knowing this can be useful in creating sketches and models.

The regular geometry of points, lines, circles, arcs, and ellipses are the foundation for many CAD drawings that are created from these types of entities alone. Figure 5.37 shows a 2D CAD drawing that uses only lines, circles, and arcs to create the shapes shown.

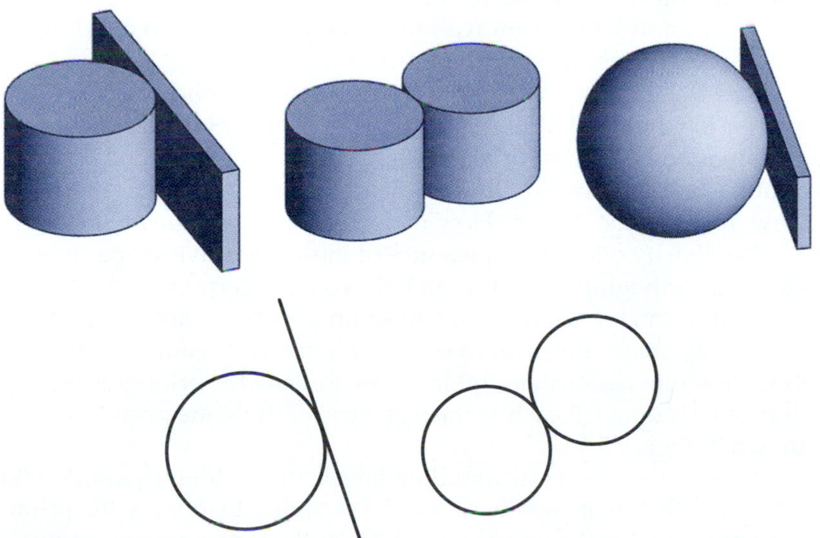

FIGURE 5.35
Lines that are tangent to an entity have one point in common but will never intersect. 3D objects may be tangent at a single point or along a line.

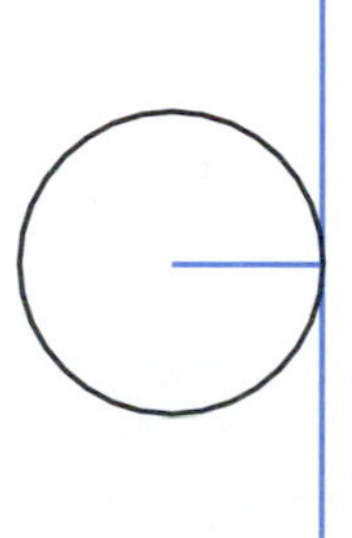

FIGURE 5.36
A radial line from the point where a line is tangent to a circle will always be perpendicular to that line.

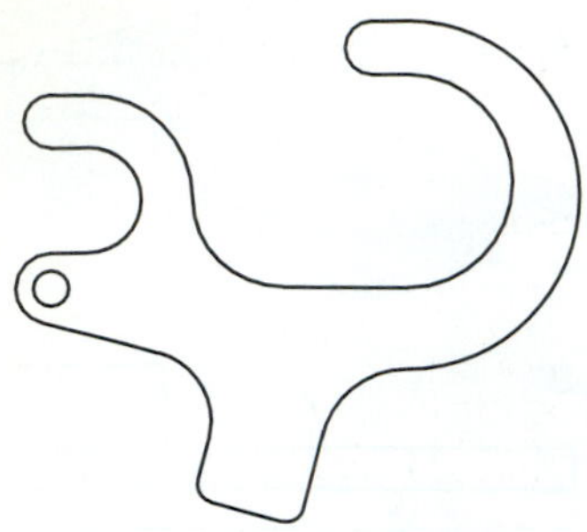

FIGURE 5.37
A 2D drawing made of only lines, circles, and arcs

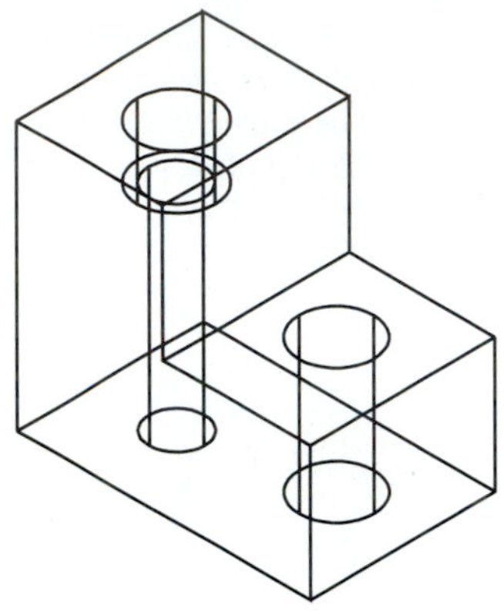

FIGURE 5.38
A 3D model made of only lines, circles, and arcs

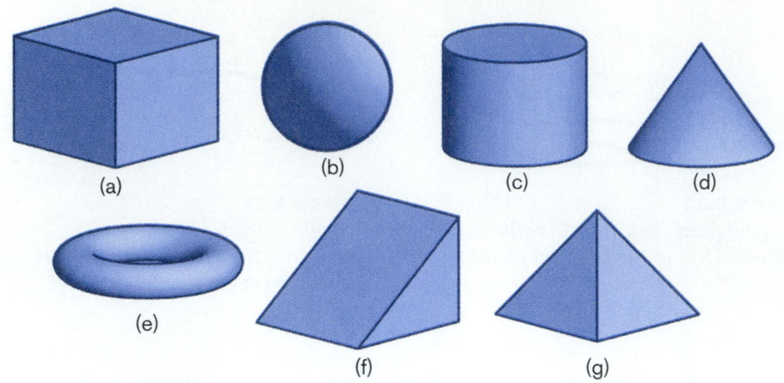

FIGURE 5.39
The most common solid primitives are (a) box, (b) sphere, (c) cylinder, (d) cone, (e) torus, (f) wedge, and (g) pyramid.

Figure 5.38 shows a 3D wireframe model that is also made entirely of lines, circles, and arcs. Many complex-looking 2D and 3D images are made solely from combinations of these shapes. Recognizing these shapes and understanding the many ways you can specify them in the CAD environment are key modeling skills.

Solid Primitives

Many 3D objects can be visualized, sketched, and modeled in a CAD system by combining simple 3D shapes or *primitives*. Solid primitives are the building blocks for many solid objects. You should become familiar with these common primitive shapes and their geometry. The same primitives that helped you understand how to sketch objects can also help you create 3D models of them using your computer.

A common set of primitive solids that you can use to build more complex objects is shown in Figure 5.39.

Recall the information about the geometry of solid objects you learned in Chapter 3. Which of these objects are polyhedra? Which are bounded by single-curved surfaces? Which are bounded by double-curved surfaces? How many vertices do you see on the cone? How many on the wedge? How many edges do you see on the box?

Familiarity with the appearance of these primitive shapes when shown in orthographic views can help you in interpreting drawings and in recognizing features that make up the objects around you. Figure 5.40 shows the primitives from Figure 5.39 again, only this time shown in two orthographic views. Review the orthographic drawings and match each to the isometric of the same primitive shown in Figure 5.39.

Look at the objects surrounding you. Can you identify some solid primitives that make up their shape? The ability to identify the primitive shapes that make up objects similar to those you are designing can help you model features of the objects in a CAD system (see Figs. 5.41 and 5.42). Also, knowing how primitive shapes appear in orthographic views can help you sketch these features correctly.

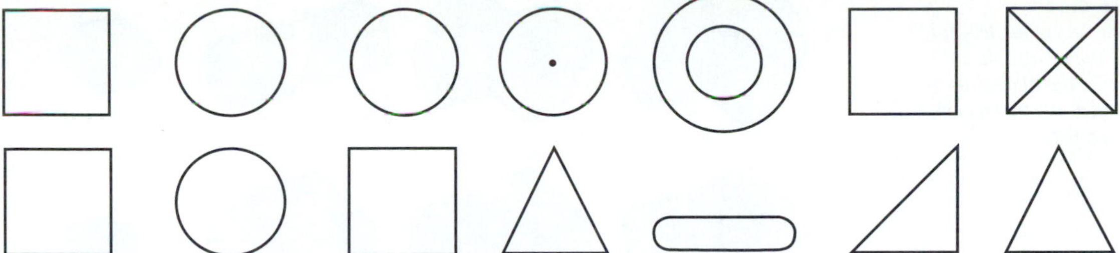

FIGURE 5.40
Can you match the top and front views shown here with the primitives shown in Figure 5.39?

Making Complex Shapes With Boolean Operations

Boolean operations, common to most 3D modelers, allow you to join, subtract, and intersect solids. Boolean operations are named for the English mathematician Charles Boole, who developed them to describe how sets can be combined. Applied to solid modeling, Boolean operations describe how volumes can be combined to create new solids.

The three Boolean operations, defined in Table 5.1, are

- union (addition)
- difference (subtraction)
- intersection

Figure 5.43 illustrates the result of the Boolean operations in pictorial views of two solid models.

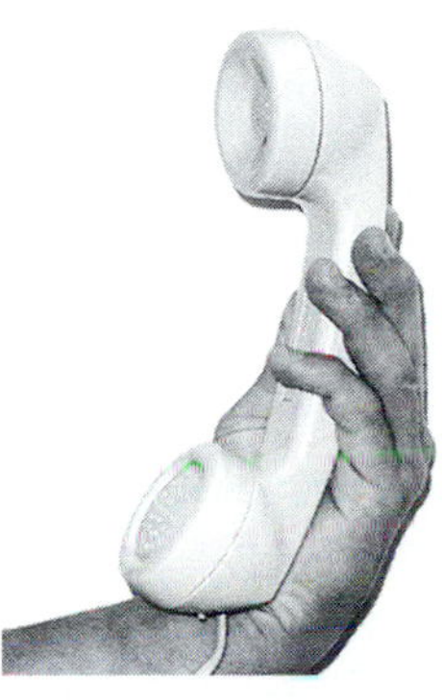

FIGURE 5.41

FIGURE 5.42
Everyday objects include shapes reminiscent of primitive solids. The 3D solid primitives in this illustration show the basic shapes that reside in the common telephone.

TABLE 5.1

Boolean Operations

Name	Definition	Venn Diagram
Union	The volume in both sets is combined or added. Overlap is eliminated. Order does not matter: A union B is same as B union A.	
Difference	The volume from one set is subtracted or eliminated from the volume in another set. The eliminated set is completely eliminated—even the portion that doesn't overlap the other volume. The order of the sets selected when using difference *does* matter (see Fig. 5.44). A subtract B is not the same as B subtract A.	
Intersection	The volume common to both sets is retained. Order does not matter: B intersect A is the same as A intersect B.	

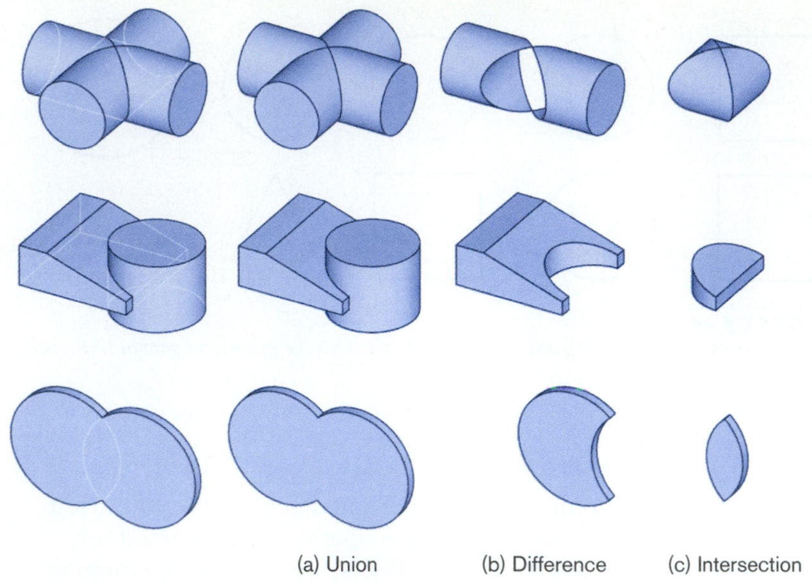

FIGURE 5.43
The three sets of models at left produce the results shown to the right when the two solids are (a) unioned, (b) subtracted, and (c) intersected.

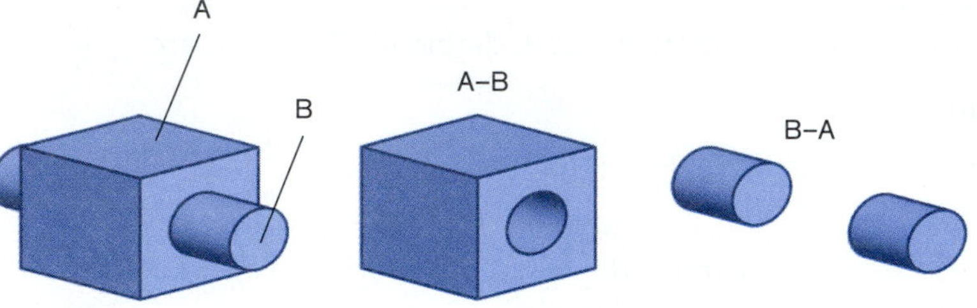

FIGURE 5.44
Order matters in subtraction. The models here illustrate how A – B differs significantly from B – A.

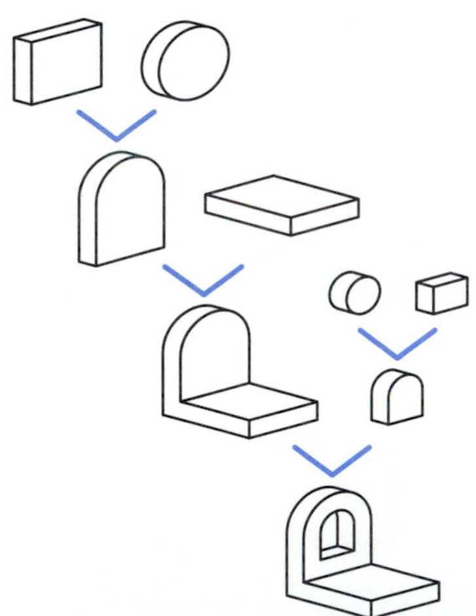

FIGURE 5.45
This diagram shows how basic shapes were combined to make a bookend. The box and cylinder at the top were unioned, then the resulting end piece and another box were unioned. To form the cutout in the end piece, another cylinder and box were unioned, then the resulting shape was subtracted from the end piece.

Look at some everyday objects around you and make a list of the Boolean operations and primitive solid shapes needed to make them. Figure 5.45 shows a bookend and a list of the primitives available in the CAD system used to create it, along with the Boolean operations used to make the part.

Recognizing Symmetry

An object is *symmetrical* when it has the same exact shape on opposite sides of a dividing line or plane or about a center or axis. Recognizing the symmetry of objects can help you in your design work and when you are sketching or using CAD to represent an object. Figure 5.46 shows a shape that is symmetrical about several axes of symmetry (of which two are shown) as well as about the circles' center point.

Mirrored shapes have symmetry so that on opposite sides of the dividing line (or mirror line) the other is the exact opposite (or mirror image). For a 2D mirrored shape, the axis of symmetry is the mirror line. For a 3D mirrored shape, the dividing line represents a plane. Examples of 3D mirrored shapes are shown in Figure 5.47.

To simplify sketching, you need to show only half of the object if it is symmetrical. When you are sketching or creating drawings, ask yourself, "How much of the object do I have to define to make the rest?" CAD systems have commands that allow you to copy or mirror a portion of the object. Figure 5.48 shows an orthographic sketch of a symmetrical shape. Notice that only half of the object is sketched. A centerline line pattern provides a visual reference for the mirror line on the part.

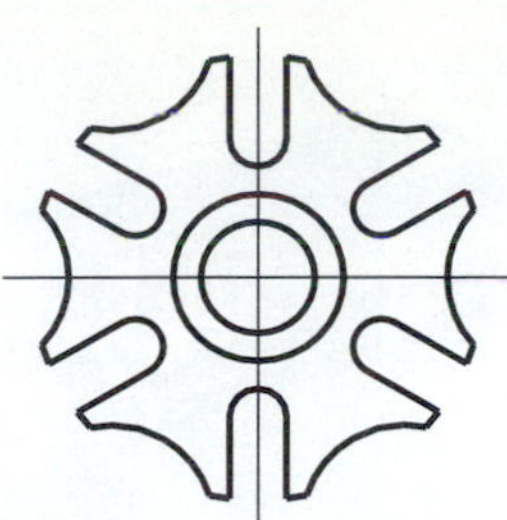

FIGURE 5.46
Symmetrical parts can have symmetry about a line or point.

Most CAD systems have a command available to mirror existing features to create new features. You can save a lot of time when creating CAD models by noticing the symmetry of the object and copying or mirroring the existing geometry and features to create new ones.

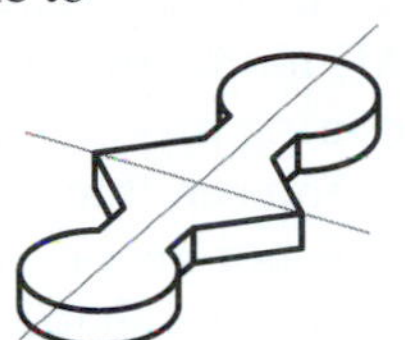

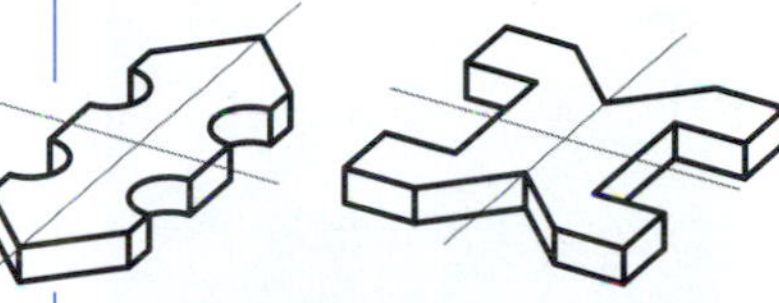

FIGURE 5.47
These symmetrical shapes each have two mirror lines, indicated by the thin axis lines. To create one of these parts, you could model one quarter of it, mirror it across one of the mirror lines, then mirror the resulting half across the perpendicular mirror line.

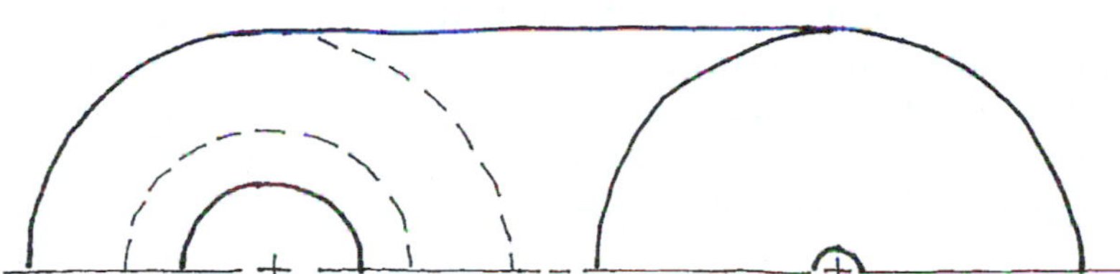

FIGURE 5.48
Orthographic sketches of symmetrical parts may show only half of the object.

Right- and Left-Hand Parts

Many parts that you will design function in pairs for the right and left sides of a device. A brake handle for the left side of a mountain bike is similar to, but actually a mirror image of, the brake handle for the right side of the bike (shown in Fig. 5.49). When using a CAD system you can create the part for the left side by mirroring the entire part. On sketches you can indicate a note on the drawing that states RIGHT-HAND PART IS SHOWN. LEFT-HAND PART IS OPPOSITE. Right-hand and left-hand are often abbreviated as RH and LH in drawing notes.

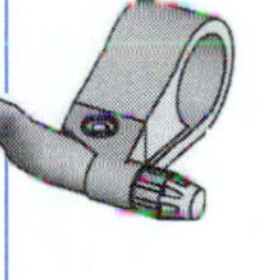

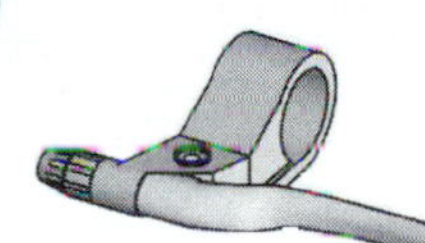

FIGURE 5.49
These brake levers are mirror images of each other and an example of right- and left-hand parts.

Parting Line Symmetry

Molded symmetrical parts are often made using a mold with two equal halves, one on each side of the axis of symmetry. The axis or line where two mold parts join is called a *parting line*. When items are removed from a mold, sometimes a small ridge of material is left on the object. See if you can notice a parting line on a molded object such as your toothbrush or a screwdriver handle like the one shown in Figure 5.50. Does the parting line define a plane about which the object is symmetrical? Can you determine why that plane was chosen? Does it make it easier to remove the part from the mold?

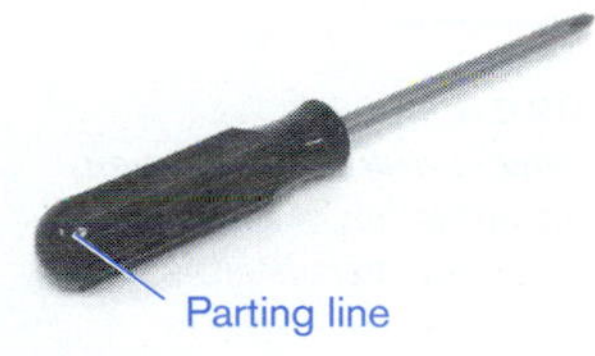

FIGURE 5.50
The parting line on a molded part is often visible as a ridge of material.

You will learn more about modeling molded parts in Chapter 8, but as you are developing your sketching skills, think about the axis of symmetry for parts and how it could affect their manufacture.

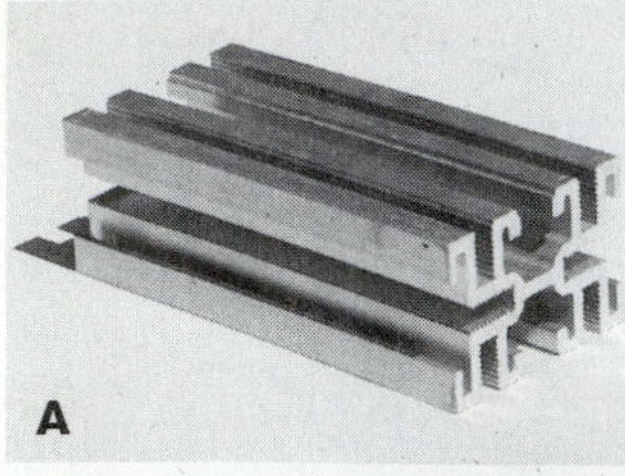

FIGURE 5.51
The aluminum part in (a) was created by forcing aluminum through an opening with the shape of its cross section. The extruded length can then be cut into a number of individual lengths. The solid model for the part in (b) was created by defining the 2D cross-sectional shape and specifying a length for the extrusion. (Courtesy of Dale Pickard, Radiant Engineering, Inc.)

Extruded Forms

In addition to mirroring, two other operations can be used to transform basic shapes into more complex ones: extrusion and revolution. *Extrusion* is the manufacturing process of forcing material through a shaped opening. Figure 5.51 shows an extruded aluminum part used to hold glazing for custom designed greenhouses. Extrusion in CAD modeling creates a 3D object by a process similar to the manufacturing process of extrusion.

You can think of the shapes of some solids as extrusions, even if that is not the method that will be used to manufacture them. You can model a shape as an extruded form if you can define the shape by sketching a 2D outline of the object's basic shape, then specifying a length in the third dimension. Many CAD systems provide an Extrude command that creates 3D shapes by mimicking the process of extrusion. Some will also allow you to specify a taper (or draft) angle so that the extruded shape narrows over its length. Some examples are shown in Figure 5.52.

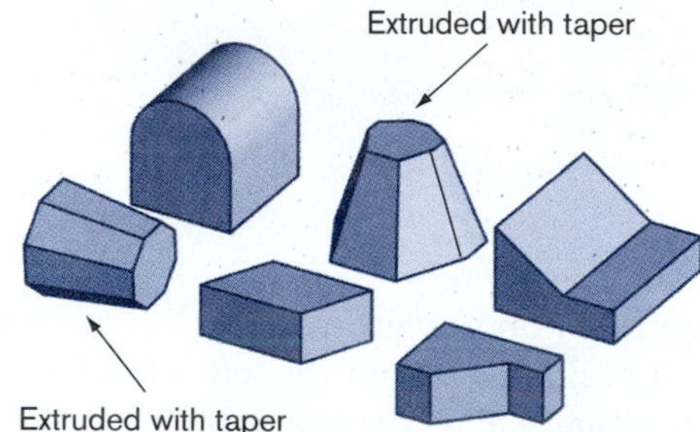

FIGURE 5.52
These CAD models were formed by extruding a 2D outline. Two of the models were extruded with a taper.

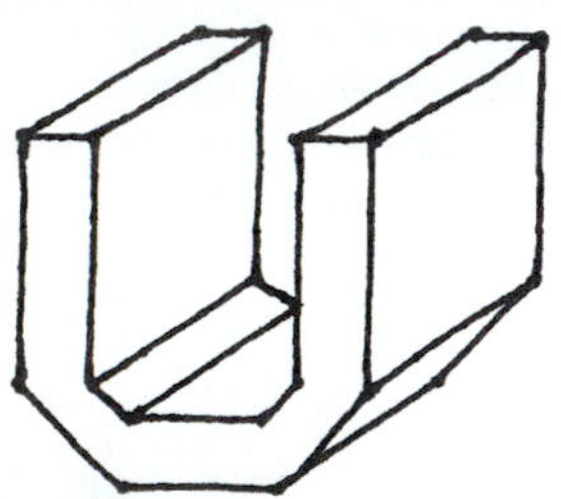

FIGURE 5.53
If a shape is easy to sketch as an oblique projection, it may be easy to create using extrusion.

Shapes that can be created using extrusion are often easily sketched as oblique projections (which you learned to create in Chapter 3). To sketch extruded shapes, show the shape (or profile) that will be extruded parallel to the front viewing plane in the sketch. Copy this same shape over and up in the sketch based on the angle and distance you want to use to represent the depth. Then sketch in the lines that represent the edges visible from the corners of the part as shown in Figure 5.53.

Swept Shapes

A swept form is a special case of an extruded form. *Sweeping* is the term used to describe extruding a shape along a curved path. To sweep a shape in CAD, you need to create the 2D shape to be swept as well as a 2D or 3D curve to serve as the path. Some objects that could be created using CAD by sweeping a planar shape along a curved path are shown in Figure 5.54.

Revolved Forms

Revolution is the second CAD operation commonly used to convert basic shapes into 3D forms. Sometimes called solids of revolution, shapes formed in this way are created by revolving a shape that lies in a single plane (sometimes called a profile) around an axis in the same plane in order to create a closed solid object. To create a revolved solid, you must first create the 2D shape to be revolved, specify an axis about which to revolve it, then indicate the number of degrees of revolution. Figure 5.55 shows some shapes created by revolution.

As these operations show, you will usually start from some sort of 2D sketch when using 3D CAD to create your engineering designs. Look at the examples shown in Figure 5.56 and match them to the 2D profile that you would use to create the feature or part. For each part, list whether you would use extrusion, revolution, or sweeping to create it.

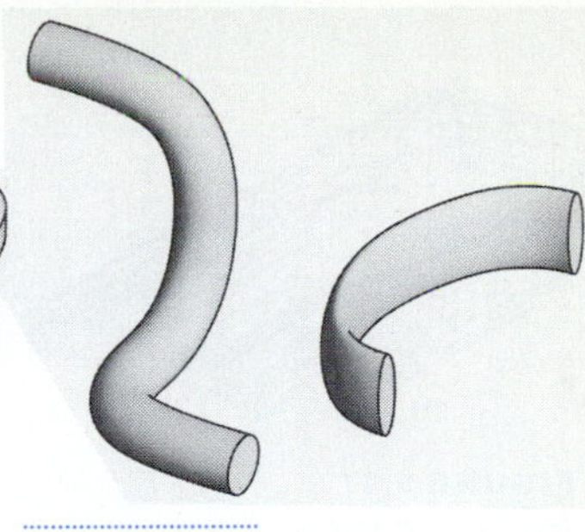

FIGURE 5.54
These shapes started as an octagon, a circle, and an ellipse, then were swept along a curved path.

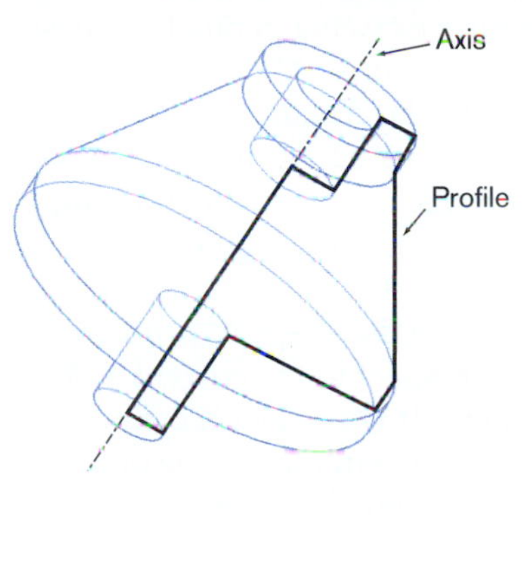

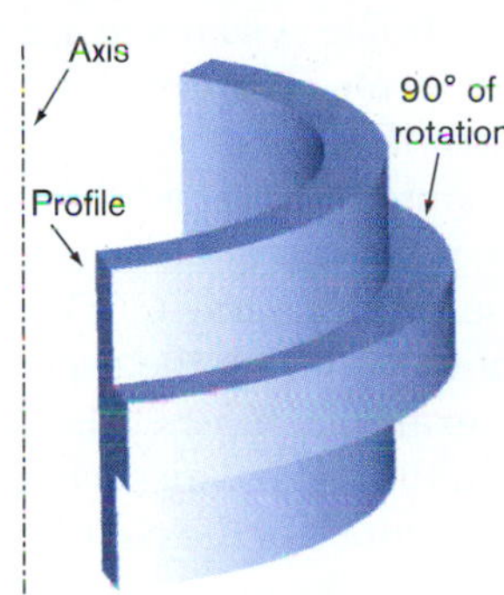

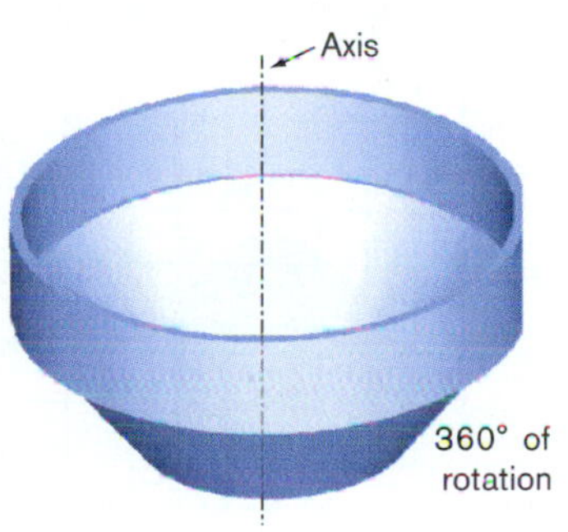

FIGURE 5.55
Each of the solids shown here were created by rotating a 2D shape around the axis.

A

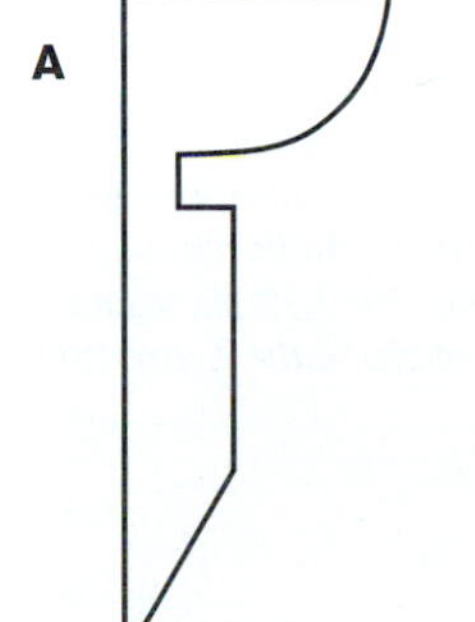

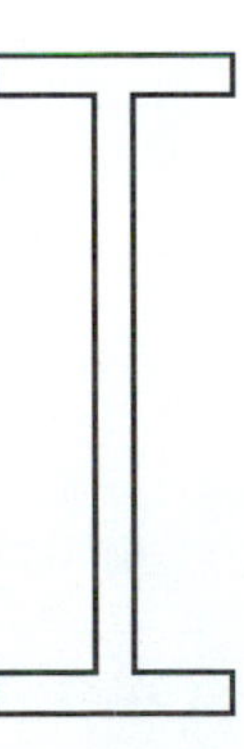

FIGURE 5.56
What operation would you choose to transform the shapes in (a) into the models in (b)?

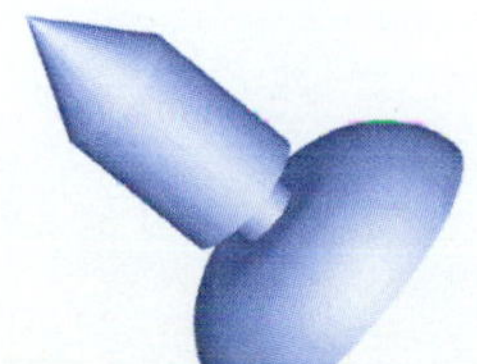

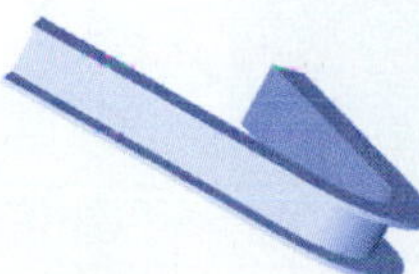

B

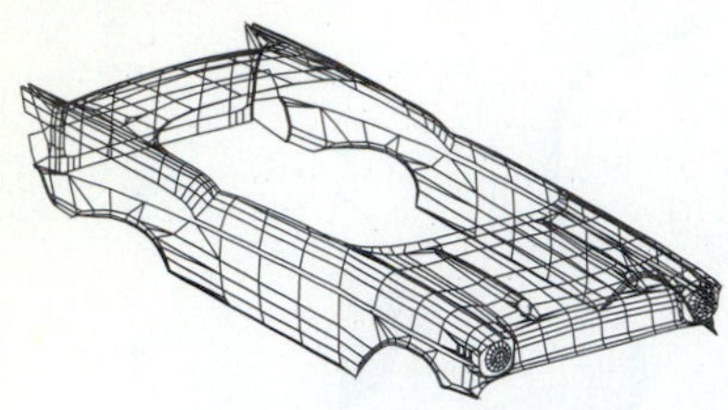

FIGURE 5.57
Irregular surfaces

Irregular Surfaces

Not every object can be modeled using the basic geometric shapes you've explored in this chapter. Irregular surfaces are those that cannot be unfolded or unrolled to lie in a flat plane. Solids that have irregular or warped surfaces cannot be created either by extrusion or revolution. These irregular surfaces are created using surface modeling techniques. Spline curves are frequently the building blocks of the irregular surfaces found on car and snowmobile bodies, molded exterior parts, aircraft, and other (usually exterior) surfaces of common objects, such as an ergonomic mouse. An example of an irregular surface is shown in Figure 5.57. You will learn more about modeling irregular surfaces in Chapter 6.

User Coordinate Systems

Many solid and surface modeling systems allow you to create your own coordinate systems to aid in creating drawing geometry. These are often termed *user coordinate systems* (in AutoCAD, for example) or *local coordinate systems*, in contrast with the *default coordinate system* (sometimes called the World Coordinate System or Absolute Coordinate System) that is used to store the model in the drawing database. To use many CAD commands effectively, you need to know how to change the user coordinate system.

For example, most CAD systems create the shape of primitive objects in the same way each time with respect to the current X, Y, and Z directions of the coordinate system. If you create a cylinder using a primitive in many CAD systems, the circular shape of the cylinder will always be oriented in the current X-Y plane, as shown in Figure 5.58.

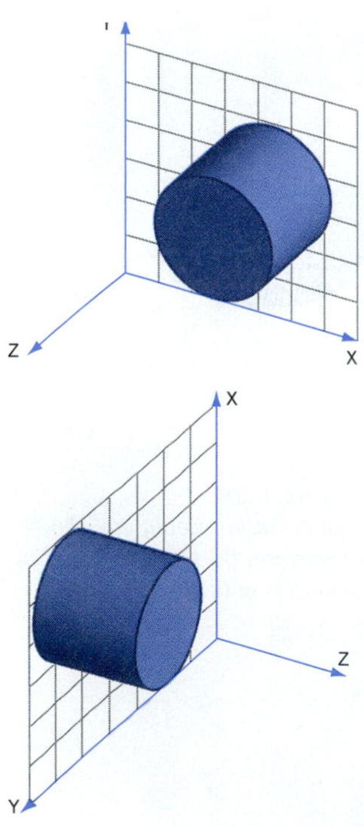

FIGURE 5.59
These cylinder primitives were created after the user coordinate system was changed to redefine the X-Y plane. Each primitive is situated by the CAD command in the same way relative to the X and Y axes, but the X and Y axes are no longer the same as the default (world) coordinate system.

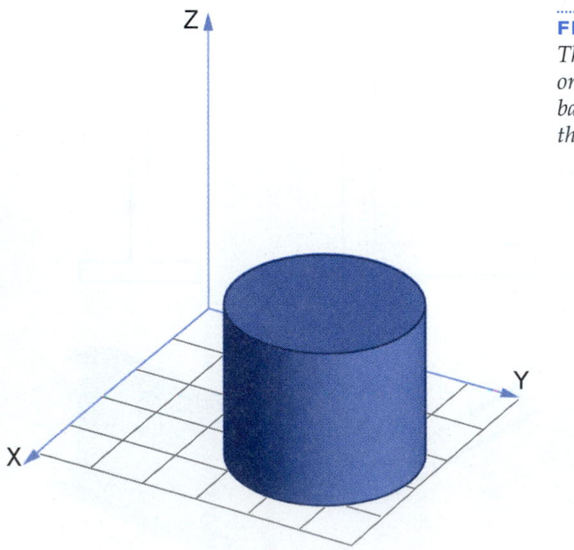

FIGURE 5.58
The cylinder primitive is oriented with the circular base on the X-Y plane and the height in the Z direction.

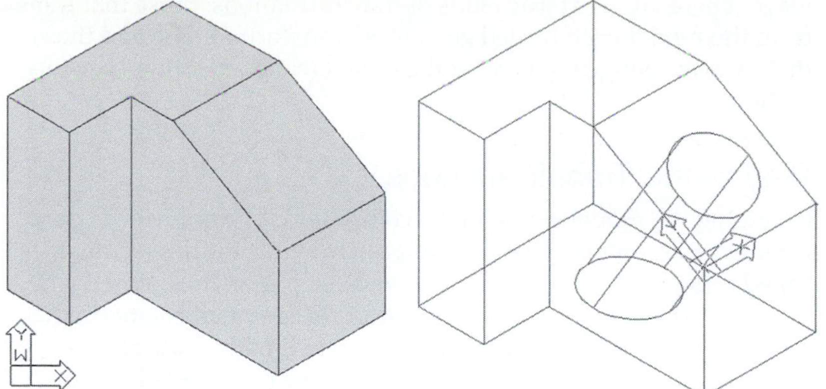

FIGURE 5.60
A user or local coordinate system can be defined relative to any object to make it easy to place geometry and effectively use CAD commands that work from the current X-Y plane.

To create a cylinder that is oriented differently, as shown in Figure 5.59, you would need to orient a new coordinate system.

For example, to create the hole perpendicular to the oblique surface shown in Figure 5.60 using 3D CAD, you could create a new local coordinate system aligned with the oblique surface. You could then use that coordinate system to create a cylinder perpendicular to the oblique surface (the new X-Y plane) to form the hole. The CAD software translates the location of the hole to the world (default) coordinate system after you have specified its location using the local coordinate system, which provided an easier reference.

A new coordinate system can be defined in any relationship to the default coordinate system. Recall from the definition of a plane that it can be defined by three points. Many CAD systems have a command to define the plane for a user coordinate system by specifying three points. This is often an easy way to create a new coordinate system that has the orientation you desire—especially when it needs to align with an oblique or inclined surface. Other solid modeling systems anticipate the user's need to reorient the coordinate system when in certain modes and make it as easy as selecting a surface on an object, such as the oblique surface in Figure 5.60. This is analogous to setting the X-Y plane of the user coordinate system to coincide with the selected surface.

Remember that the user or local coordinate system is a means to more effectively create geometry in your CAD model. Changing the local coordinate system does not change the default coordinate system where your model data is stored.

Transformations

A 3D CAD software package uses the default Cartesian coordinate system to store various amounts of information about a model of an object. The model may be stored inside the computer as a matrix (rows and columns of numbers) representing the vertices of the object. Once the object is defined, the CAD software can use mathematical methods to transform the matrix (and the object) in various

FIGURE 5.61
The location of the model shown in Figure 5.63 after being moved 2 units in the X direction and 3 units in the Y direction.

ways. There are two basic kinds of transformations: those that transform the model itself (called geometric transformations) and those that merely change the view of the model (called viewing transformations).

Geometric Transformations

Changing the model stored in the computer is accomplished using three basic *transformations* (or changes): moving entities (sometimes called *translation*), rotating entities, and scaling entities. When you select a CAD command that uses one of these transformations, the CAD data stored in your model is converted mathematically to produce the result. Commands such as move (or translate), rotate, or scale transform the object on the coordinate system and change the coordinates stored in the 3D model database.

Figure 5.61 shows a part after translation. The model was moved over 2 units in the X direction and 3 units in the Y direction. The corner of the object is no longer located at the origin of the coordinate system.

Figure 5.62 illustrates the effect of rotation and scaling. The object on the left remains the same as the object on the right is first scaled, then rotated. The coordinates for the object on the right have been transformed to define the modified object you see. The scaled object now encloses a

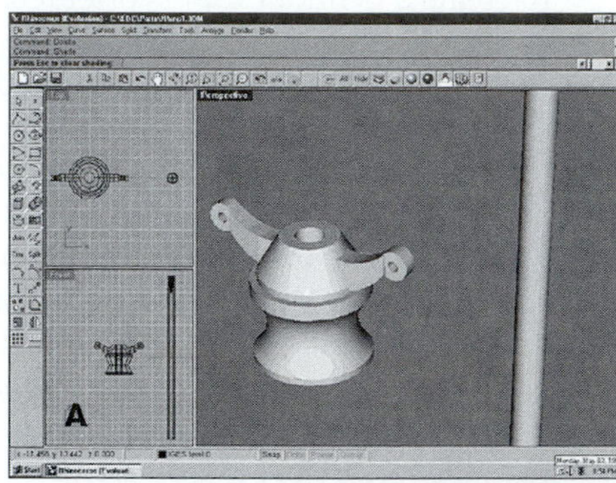

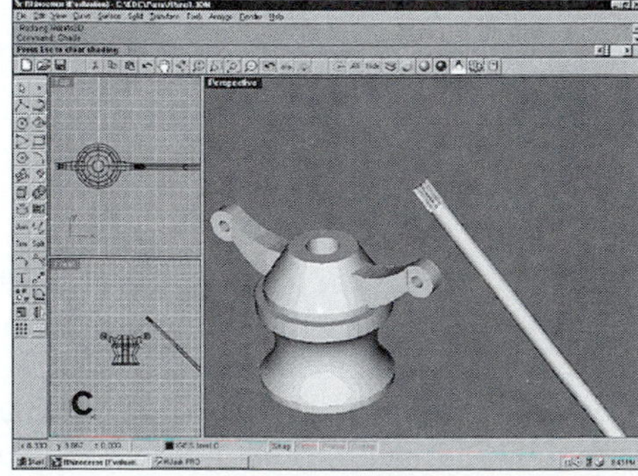

FIGURE 5.62
The two objects in (a) are two models from the same assembly that have been moved out of their assembled location. In (b), the model on the right has been scaled by a factor of two and is now half as big as it was before. In (c), the same object was rotated 45° and its vertices were transformed to different locations in the CAD database. These changes are also reflected in the top and front views on the left side of the screens.

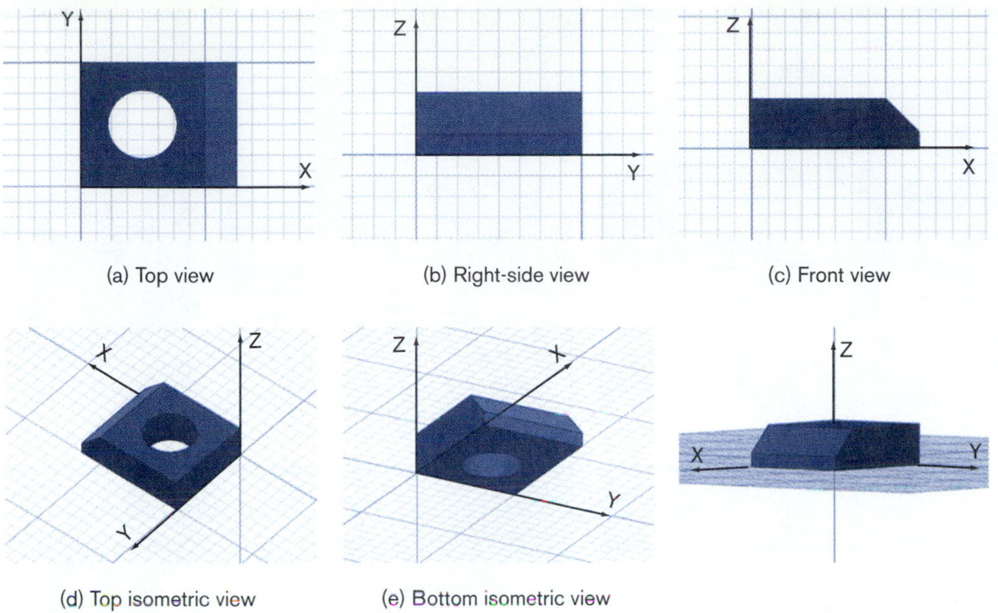

FIGURE 5.63
Note that the location of the model relative to the coordinate axes does not change in any of the different views. Changing the view does not transform the model itself.

different volume. The rotated object is situated at a different location in model space.

Viewing Transformations

A viewing transformation does not change the coordinate system or where your CAD model is stored on it, it simply changes your view of the model. The model's vertices are stored in the computer at the same coordinate locations, no matter which direction it is viewed from on the monitor. Figure 5.63 shows a 3D model and the coordinate axes used to define its location as they would appear in different views.

Although the location of your model on the coordinate system doesn't change when the view does, the CAD software does use mathematical methods to transform the matrix containing the model's vertices and sometimes other information (such as the connectivity of the vertices) to produce the new appearance of the model on the screen. For a particular viewing direction, or line of sight, a transformation matrix is calculated to represent the new appearance of the model. This viewing transformation is stored as a separate part of the model file (or a separate file) and does not affect the coordinates of your model or how it is stored.

The commands used to change the view of the model can change the view's location on the screen (sometimes called *pan*), its size (sometimes called the *zoom*), and its rotation (the relative

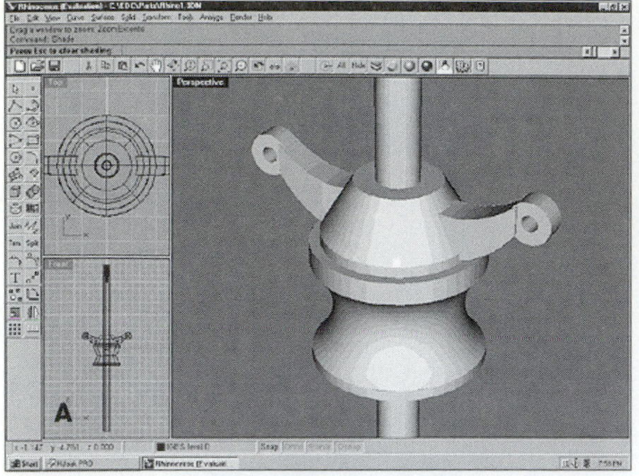

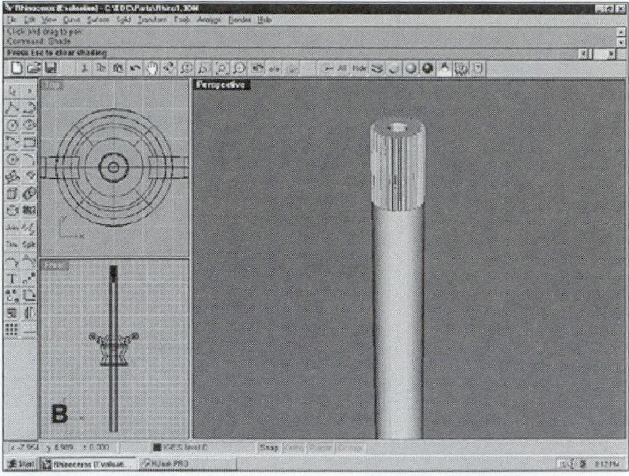

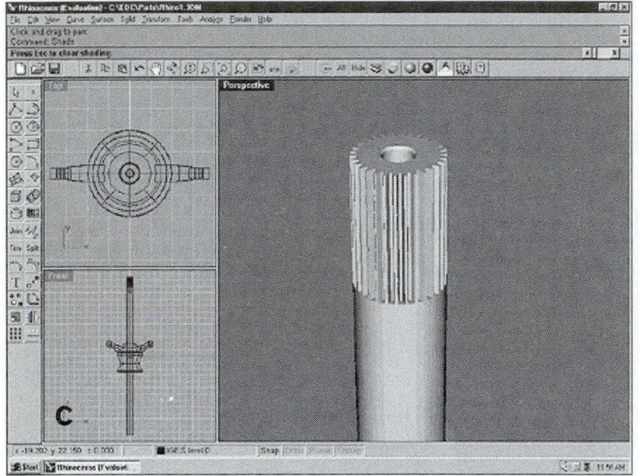

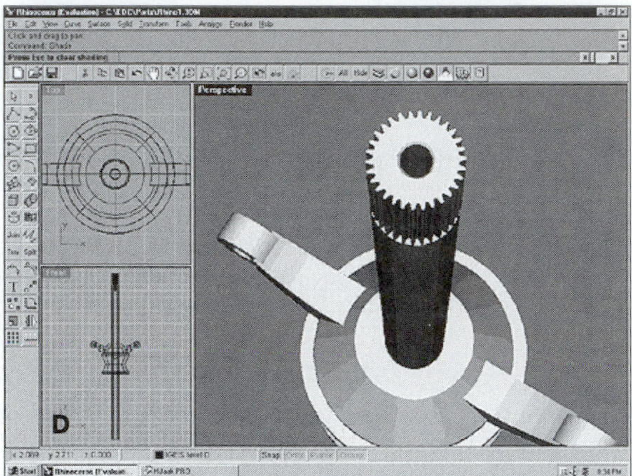

FIGURE 5.64
Panning moved the view of the objects in (a) to expose a different section of the assembly in (b). In (c), the view was enlarged to make detail more visible. In (d), the view was rotated to display the objects from a different line of sight. Note that in each case, the viewing transformation applies to all of the objects in the view and does not affect the location of the objects on the coordinate system (shown in the top and front views on the left side of the screen)

viewing direction). Unlike the geometric transformations, view transformations are applied to a viewing area of the screen, not individual objects. These commands are not the same as moving, rotating, or scaling the model in 3D space. They apply only to the viewing transformation.

Three common viewing transformations are illustrated in Figure 5.64. Panning moves the location of the view on the screen. If your monitor were a hole through which you were viewing a piece of paper, panning would be analogous to sliding the piece of paper to expose a different portion of it through the hole. Zooming enlarges or reduces the view of the objects, and operates similar to a telephoto lens on a camera. A view rotation is actually a change of viewpoint; the object appears to be rotated, but it is your point of view that is changing. The object itself remains in the same location on the coordinate system.

What is the difference between using a view control to zoom in on an object and using a command to increase the scale of the object on the coordinate system? Using viewing controls to change the

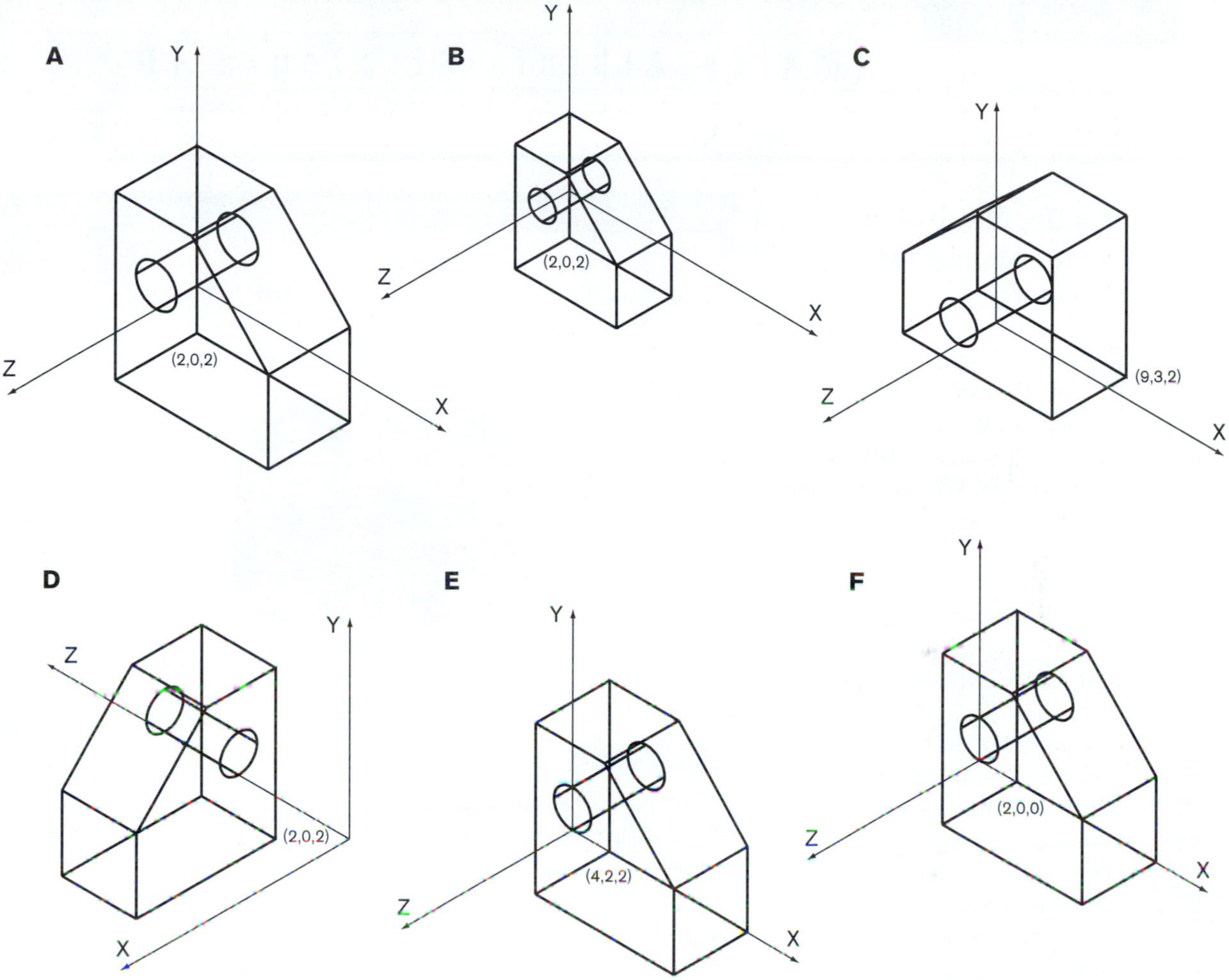

FIGURE 5.65
Three of these models are the same, but the viewing location, zoom, or rotation has changed. Three have been transformed to different locations on the coordinate system.

zoom transforms the coordinates for the viewing transformation file, changing just your view. Commands to scale the object on the coordinate system transform the actual 3D CAD database, changing the actual coordinates of the model.

Look at the six models and their coordinates in Figure 5.65. Which are views that look different because of changes in viewing controls? Which look different because the objects were rotated, moved, or scaled on the coordinate system?

You will use the basic geometric shapes and concepts outlined in this chapter to build CAD models and create accurate freehand sketches. The ability to visualize geometric entities that define an object's shape on the Cartesian coordinate system will help you create orthographic views as well as manipulate the coordinate system to build CAD models. In the next chapter, you will learn more of the CAD-based tools that can be used to create model geometry.

INDUSTRY CASE

Marty Albini, Strategix ID

The Geometry of 3D Modeling: Use the Symmetry

Strategix ID used magnets to create a clean, quiet, zero-maintenance brake for the exercise bike it designed for Park City Entertainment. When copper rings on the bike's iron flywheel spin past four rare earth magnets, they create current in a circular flow (an *eddy current*) that sets up a magnetic field. This opposing magnetic field dissipates power and slows the wheel. Moving the magnets onto and off of the copper rings varies the amount of resistance delivered. When Marty Albini modeled the plastic magnet carrier for the eddy-current brake, he started with the magnets and their behavior as the carrier moved them onto and off of the copper rings (see Fig. 5.66). "There is no one way to think about modeling a part," Albini said. "The key is to design for the use of the part and the process that will be used to manufacture it. Where you want to go with the design should drive the way you model it, and that is all part-dependent." To make the magnet carrier symmetrical, Albini started by modeling one half of it.

The magnet carrier was designed as a part in the larger flywheel assembly, parts of which were already completed. Each pair of magnets was attached to a backing bar that kept them a fixed distance apart. To begin, Albini started with the geometry he was sure of: the diameter of the magnets, the space between them, and the geometry of the conductor ring. He sketched an arc sized to form a pocket around one of the magnets so that its centerpoint would be located on the centerline of the conductor ring (see Fig. 5.67). He then sketched another similar arc but with its centerpoint positioned to match the distance between the centers of the two magnets. He connected the two arcs with parallel lines to complete

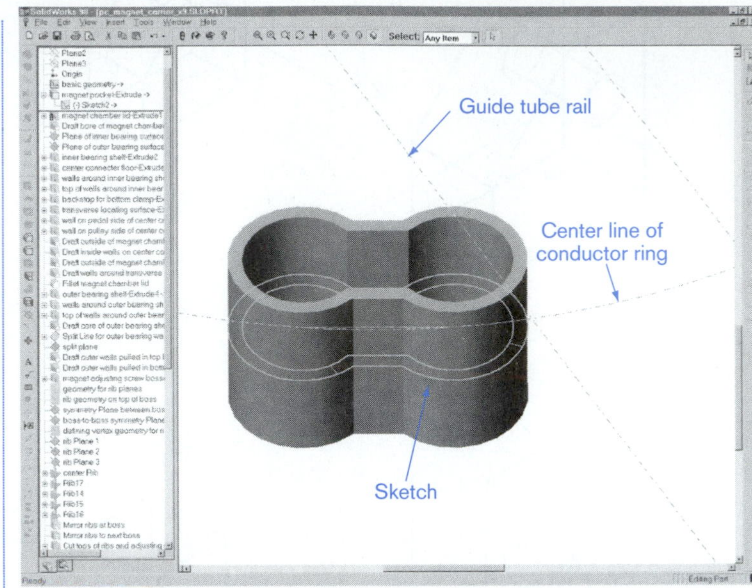

FIGURE 5.67
The magnet carrier was extruded up and down from the sketch, shown here as an outline in the middle of the extruded part. Notice that the sketch is tangent to the guide tube rail, and the centers of the arcs in the sketch are located on the centerline of the conductor ring.

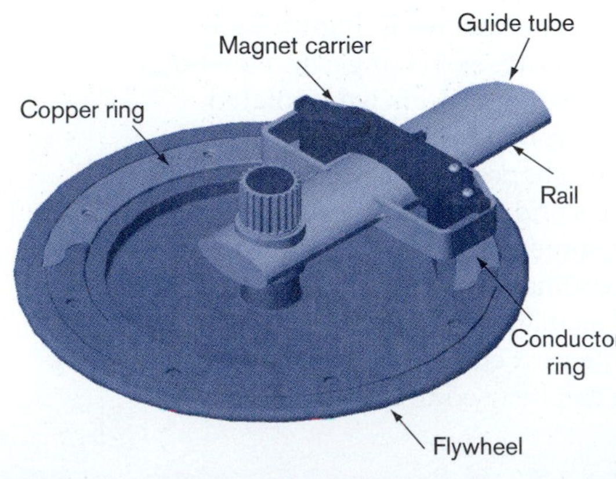

FIGURE 5.66
The magnet carrier for the brake was designed to move onto and off of the conductor ring by sliding along an elliptical guide tube, pulled by a cable attached to the small tab in the middle of the carrier.

the sketch of the inside of the carrier. This outline was offset to the outside by the thickness of the wall of the holder. (Because this is an injection-molded plastic part, a uniform wall thickness was used throughout.) One final constraint was added to position the carrier against the rail on the elliptical tube that it would slide along: the outside of the inner arc is tangent to this rail. With the sketch geometry fully defined, Albini extruded the sketch up to the top of the guide tube and down to the running clearance from the copper ring.

To add a lid to the holder, Albini used SolidWorks's Offset command to trace the outline of the holder. First he clicked on the top of the holder to make its surface the active sketch plane. This is equivalent to changing the user coordinate system in other packages: it signals to SolidWorks that points picked from the screen lie on this plane. He then selected the top edges of the holder and used the Offset command with a 0 offset to "trace" the outline as a new sketch. To form the lid, he extruded the sketch up (in the positive Z direction) the distance of the uniform wall thickness.

SolidWorks joined this lid to the magnet holder automatically and seamlessly because both features are in the same part and have surfaces that are coincident. This built-in operation is similar to a Boolean join in that the two shapes are combined to be one, but it is not a step that the user must take.

For the next feature, Albini created a "shelf" at the height of the rail that the holder will slide on. Using Offset again, he traced the outline of the holder on the sketch plane, then added parallel and perpendicular lines to sketch the outline of the bottom of the shelf. The outline was then extruded up by the wall thickness. The distance from the outside of the magnet holder to the edge of the shelf created a surface that would sit on the rail (see Fig. 5.68).

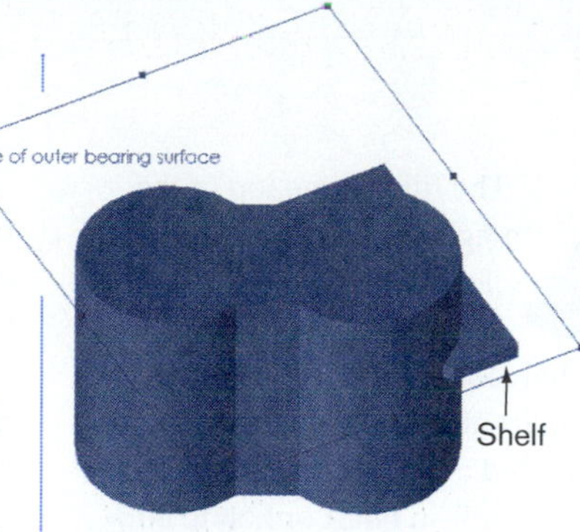

FIGURE 5.68
The surface of the rail was used as the sketch plane for the "shelf" that the magnet carrier will slide on.

Two walls were added by offsetting the edge of the shelf toward the magnet holder by the wall thickness, then offsetting the edge again by 0. Lines were added to connect the endpoints into an enclosed shape to be extruded. (In SolidWorks, an extrusion can be specified to extend in one or both directions, and to extend to a vertex, a known distance, the next surface, or the last surface encountered.) For the walls, Albini extruded them to the top surface of the magnet holder "lid."

The connecting web between the magnet holders needed to match the shape of the elliptical tube in the flywheel assembly (see Fig. 5.69). To make it, Albini sketched an ellipse on the newly created wall. Ellipse is a sketching primitive that can be specified by entering the length of the major and minor axes. Albini used the dimensions from the tube for the first ellipse sketch, then drew a second one with the same centerpoint, but with longer axes so that a gap equal to the wall thickness between them would be formed. The two ellipses were trimmed off at the bottom surface of the shelf and at the midpoint, and lines were drawn to make a closed outline. The finished sketch was extruded to the outside surface of the opposite wall.

More walls were sketched and extruded from the bottom surface of the shelf, then the wall over the connecting web was sketched and extruded down to the web.

The next step was to add the rounded edges for the top of the magnet holder. Albini issued the Fillet command and selected to

(continued on p. 194)

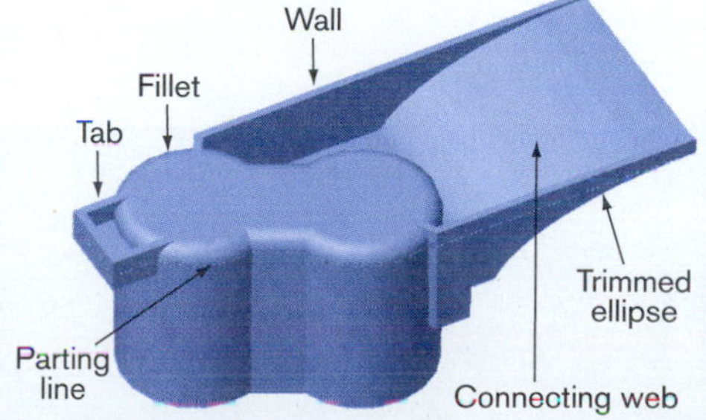

FIGURE 5.69
This view of the magnet carrier shows the elliptical shape of the connecting web and the rectangular shape of the tab. The parting line for the part, shown here as a dotted white line, is located at the edge of the fillet on the top of the magnet chambers.

INDUSTRY CASE

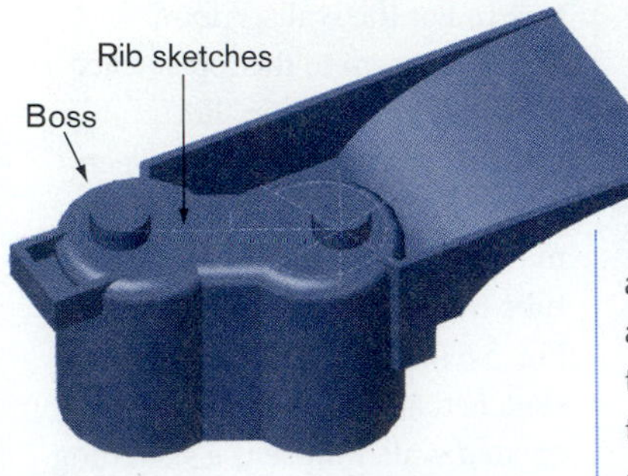

FIGURE 5.70
Sketched circles were extruded to form the bosses on the top of the magnet chamber. The dotted lines shown here on the top of the chamber pass through the centerpoint of the bosses and were used to locate the center rib and radial ribs.

round all of the edges of the top surface at once. As it created the fillet, SolidWorks maintained the relationship between the wall surfaces that intersected the top edge of the holder and extended them to the new location of the edge.

Next, Albini created a tab at the end of the part that would rest on the plastic collar in the assembly that went all the way around the magnet carrier. He first extruded a rectangular shape up from the top of the collar to form the "floor" of the tab. The walls of the tab required two additional extrusions.

The fillet at the top of the magnet holder provided the location for the parting line—the line where the two halves of the mold would come apart and release the part. Albini added a parting plane and used the built-in Draft option to add taper to the part so it would come out of the mold. After selecting all of the surfaces below the parting plane, he specified a draft angle, and SolidWorks adjusted all of the surfaces. This feature of SolidWorks makes it easy to add the draft angle after a part is finished. When draft is added, the geometry of the part becomes more complex and harder to work with. A cylinder with draft added becomes a truncated cone, for example, and the angles at which its edges intersect other edges vary along its length.

The next step was to add the bosses at the top of the magnet chambers that would support the bolts controlling the depth of the magnets. As it was a design goal to make the top of the chamber as stiff as possible to limit flex caused by the attraction of the magnets to the flywheel, the bosses were placed as far apart as possible and ribs were added for rigidity. The bosses were sketched as circles on the top surface of the magnet holder with their centers concentric to the holes in the bar connecting the magnets below. Both bosses were extruded up in the same operation.

Ribs in SolidWorks are built-in features. To create a rib, you simply draw a line and specify a width, and SolidWorks will create the rib and end it at the first surface it encounters. To create the

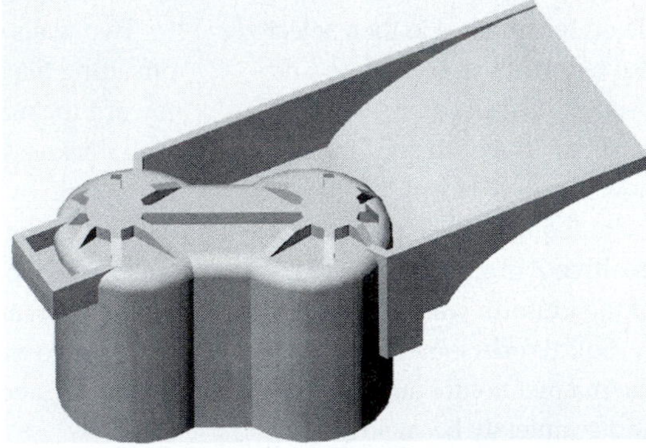

FIGURE 5.71
This view of the magnet carrier shows the symmetry of the ribs and the shape that resulted from "slicing off" the top of the bosses after the ribs were formed.

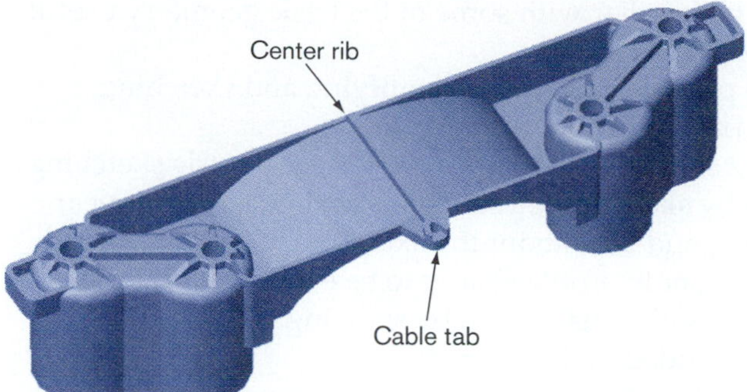

FIGURE 5.72
Circles concentric to the bosses were extruded to form the holes shown here in the finished magnet carrier.

center rib, Albini sketched a line on the plane at the top of the bosses and specified a width (ribs on a plastic part are usually two-thirds of the thickness of the walls). The rib was formed down to the top surface of the holder lid.

For the ribs around the bosses, Albini did as Obi Wan Kenobi might have advised: "Use the symmetry, Luke." He sketched the lines for ribs radially from the centerpoints of the bosses (see Fig. 5.70). To create the ribs, he created four of them on one boss, then mirrored them once to complete the set for one boss, then mirrored all of the ribs from one boss to the other boss. Once all the ribs were formed, he cut the tops off the ribs and bosses to achieve the shape shown in Figure 5.71. The result was a stiffer rib and a shape that could not be achieved with a single rib operation.

To complete the part, circles were drawn concentric to the bosses and extruded to form holes that go through the part (see Fig. 5.72). Draft was added to the ribs and walls to make the part release from the mold easily. Fillets were added to round all the edges, reducing stresses and eliminating hot spots in the mold. Then the part was mirrored to create the other half. The center rib and tab for attaching the cable were added and more edges filleted. Draft was added to the inside of the holder, and the part was complete.

(This case study is provided as a courtesy by the owner of the intellectual property rights, Park City Entertainment. All rights reserved.)

Marty Albini is senior mechanical engineer at Bozeman, Montana's Strategix ID, a firm that does industrial design for the fitness and medical industries. Albini graduated with a B.S. in Applied Mechanics from the University of California at San Diego and was hired as a development engineer in Hewlett Packard's research and development lab for plotters and printers. After returning to Montana, Albini invested in his own 3D design software and system and did freelance design work before joining Strategix ID, where he uses his CAD skills for projects such as the exercise bike described here.

Skills Summary

The skills that you develop from material in this chapter will help you in creating 2D CAD drawings and 3D CAD models for parts. By now, you should be familiar with some of the basic geometry useful in creating CAD drawings.

Exercises 1–6 provide practice in identifying and sketching primitive shapes joined by Boolean operations.

Exercises 7–12 show profile shapes. Practice isometric sketching and visualization by sketching these shapes as they would appear revolved 360°, 180°, and 270° about the axis shown.

Exercises 13–16 show profile shapes to be extruded. Practice isometric sketching and visualization by showing them as they would appear if extruded in the Z direction.

Exercises 17–19 provide practice with converting and interpreting different coordinate formats used to describe point locations.

EXERCISES

1. Sketch some objects that you use or would design that have right-hand and left-hand parts, such as a pair of in-line skates or side-mounted stereo speakers for a boom box.
2. In solid modeling, simple 3D shapes are often used to create more complex objects. These are called *primitives*. Using an isometric grid, draw seven primitives.
3. What is a Boolean operation? Define two Boolean operations by sketching an example of each in isometric view.
4. Consider primitives and Boolean operations that could be used to create a "rough" model of each of the items shown below. Using the photos as underlays, sketch primitives that could be used to create items a–d.

   a. Handlebar-mount gun rack

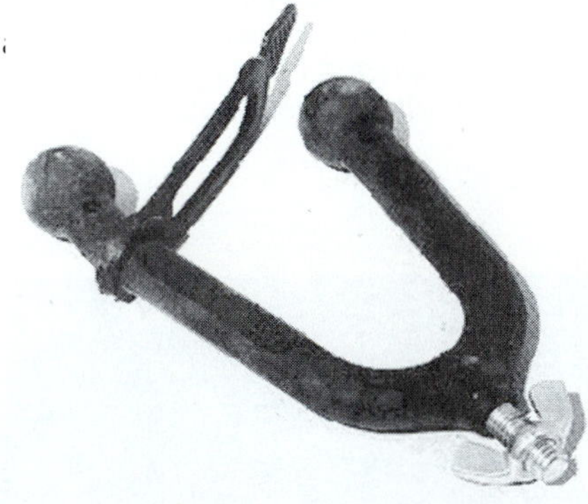

b. ACME Corporation reduction gears

c. Ashcroft Model 1305D deadweight pressure tester

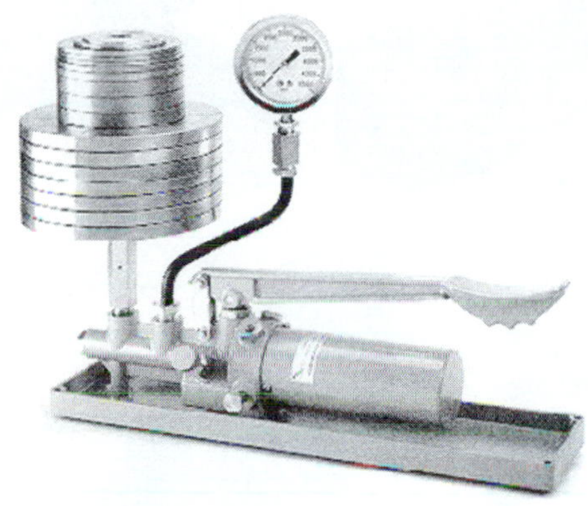

d. Davis Instruments solar-powered digital thermometer

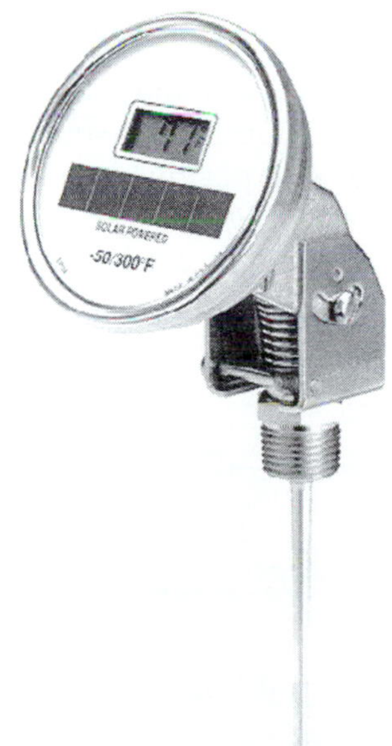

5. Use nothing but solid primitives to create a model of a steam locomotive. Sketch the shapes and note the Boolean operations that would be used to union, difference, or intersect them, or create the model using Boolean operations with your modeling software. Use at least one box, sphere, cylinder, cone, torus, wedge, and pyramid in your design.
6. Identify the solid primitives and Boolean operations you could use to create the objects below.

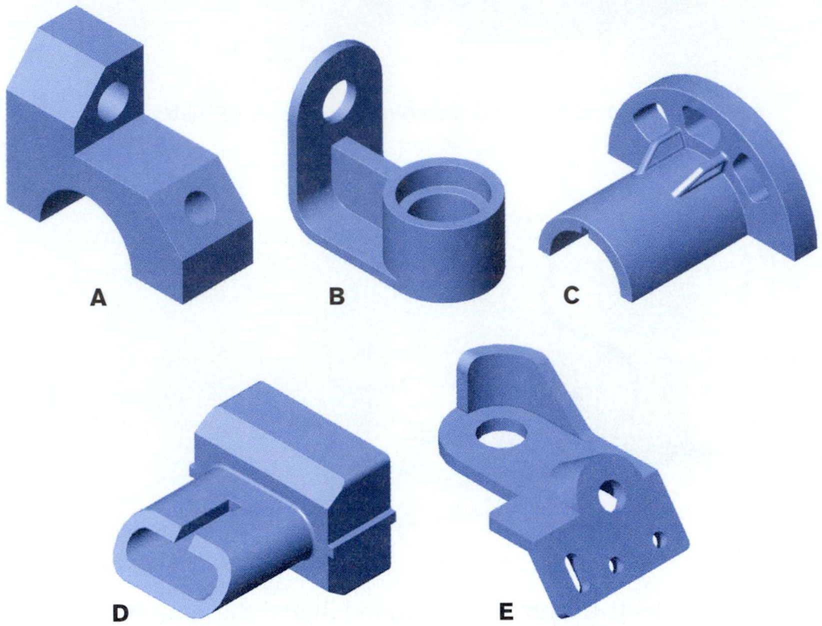

Exercises 7–12. Use an isometric grid to help sketch the solids formed by revolving the following shapes about the axis shown. Coordinates are defined by the x-y-z icon, with positive X to the right, positive Y up, and positive Z out of the page.

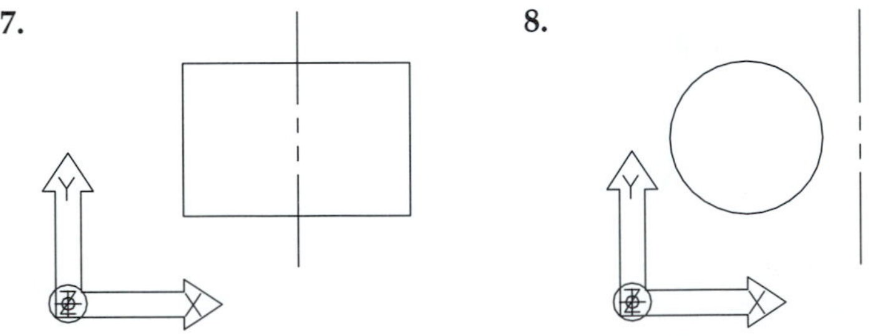

9.

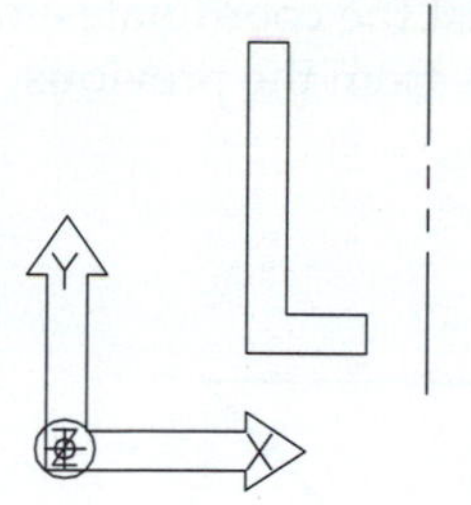

10.

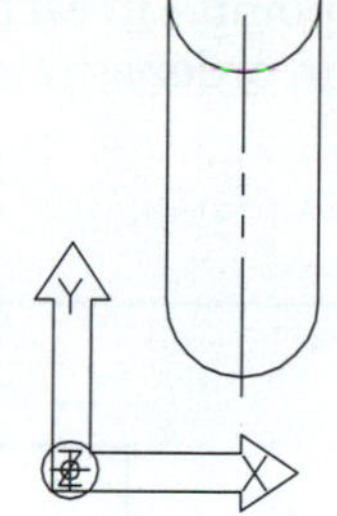

11.

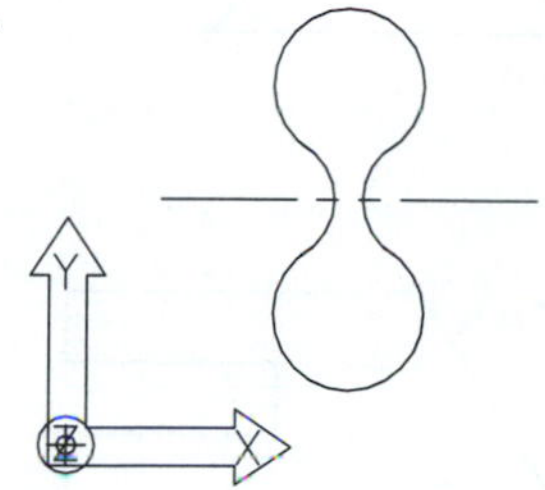

12.

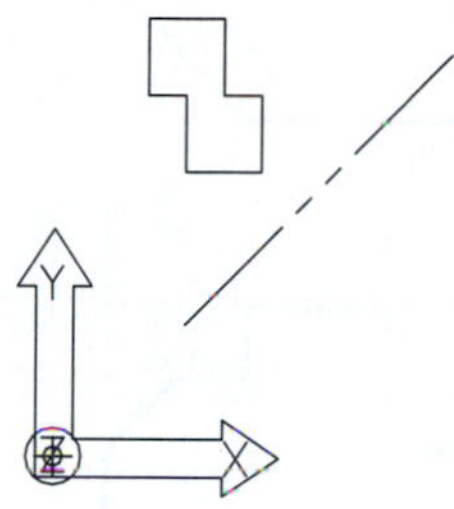

Exercises 13–16. Use an isometric grid to help sketch the solids formed by extruding the following shapes along the axis specified. Coordinates are defined by the x-y-z icon, with positive X to the right, positive Y up, and positive Z out of the page.

13. Extrude 6 inches in the positive Z direction.

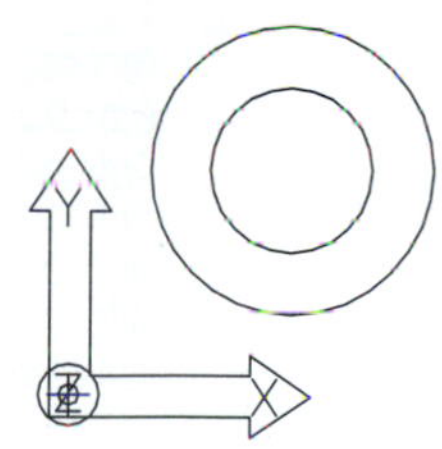

14. Extrude 4 inches in the positive Z direction.

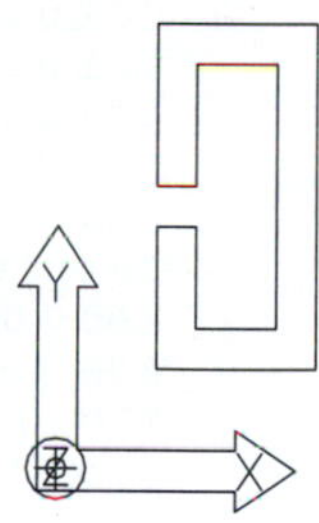

15. Extrude 6 inches in the positive Z direction.

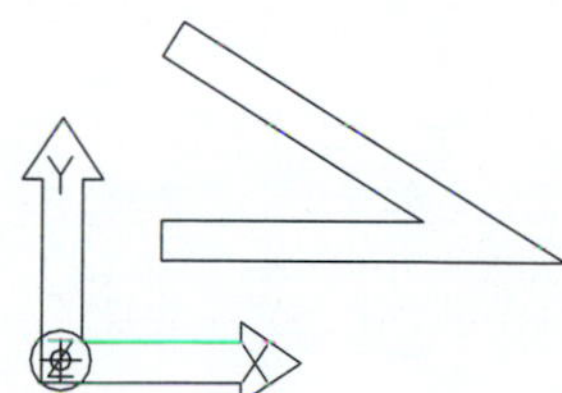

16. Extrude 4 inches in the positive Z direction.

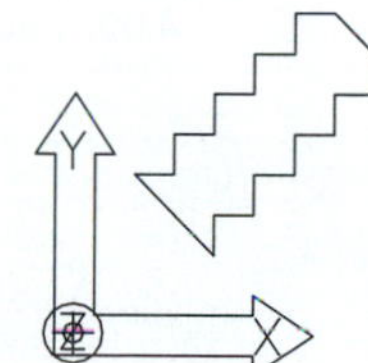

17. Starting at point A in each figure below, list the coordinates for each point in order as relative coordinates from the previous point.

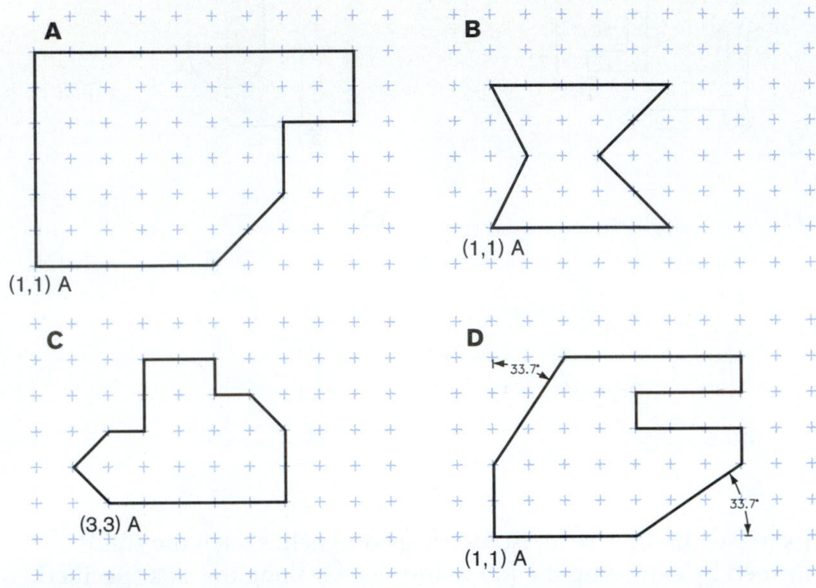

18. Plot the coordinates in each list below on grid paper. Each point represents the endpoint of a line from the previous point, unless otherwise indicated. Relative coordinates are preceded by @.

A
X, Y
1.00, 1.00
4.00, 1.00
4.00, 2.00
6.00, 2.00
6.00, 1.00
8.00, 1.00
8.00, 4.00
5.00, 4.00
4.00, 5.00
1.00, 5.00
1.00, 1.00

B
0.00, 0.00
3.00, 0.00
4.00, 1.00
5.00, 0.00
6.00, 1.00
7.00, 0.00
8.00, 1.00
9.00, 0.00
10.00, 1.00
10.00, 3.00
9.00, 4.00
8.00, 3.00
7.00, 4.00
6.00, 3.00
5.00, 4.00
4.00, 3.00

C
0,0
@2<0
@3<30
@3<-30
@2<0
@4<90
0,4
@4<-90

D
2,2
@-1<0
@3<90
@4<-30
@3<30
@1<0
@3.24<230
@4<180

19. Using the information provided on the drawing, determine the coordinates you would use (absolute, relative, or polar) and the order in which you would enter them to create the figures shown below.

A

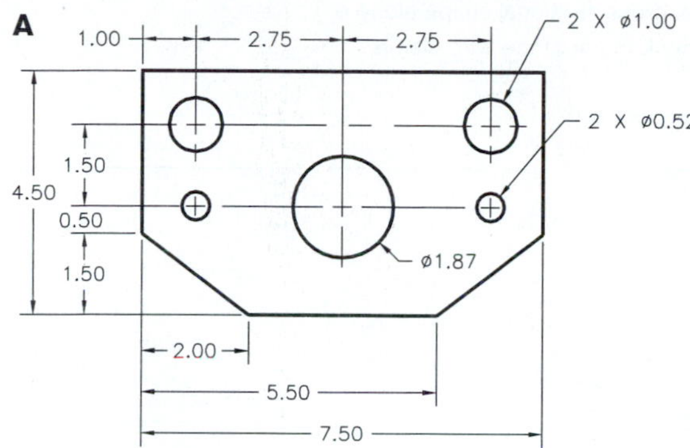

B

C

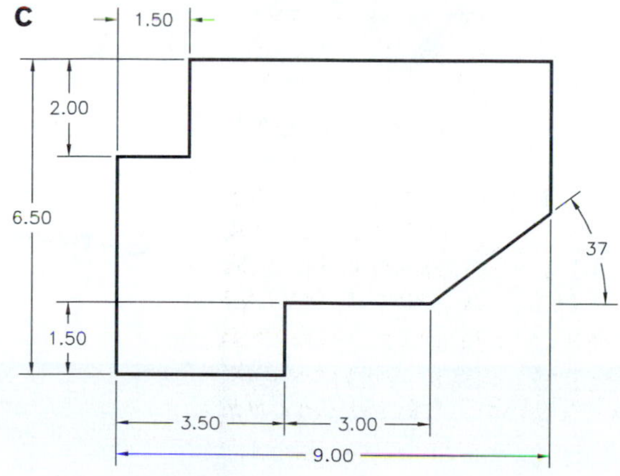

DESIGNER'S NOTEBOOK

Solid modeling was Albert Brown's choice for refining the design of the parts shown in this sketch. The coaxial cable was formed by revolution; the cast arm was extruded and the holes formed with a Boolean operation. The wire clamp was created by sweeping its cross-sectional shape along a curved path. The resulting solid model allowed the designer to check the fit of the wire clamp. (Courtesy of Albert W. Brown, Jr.)

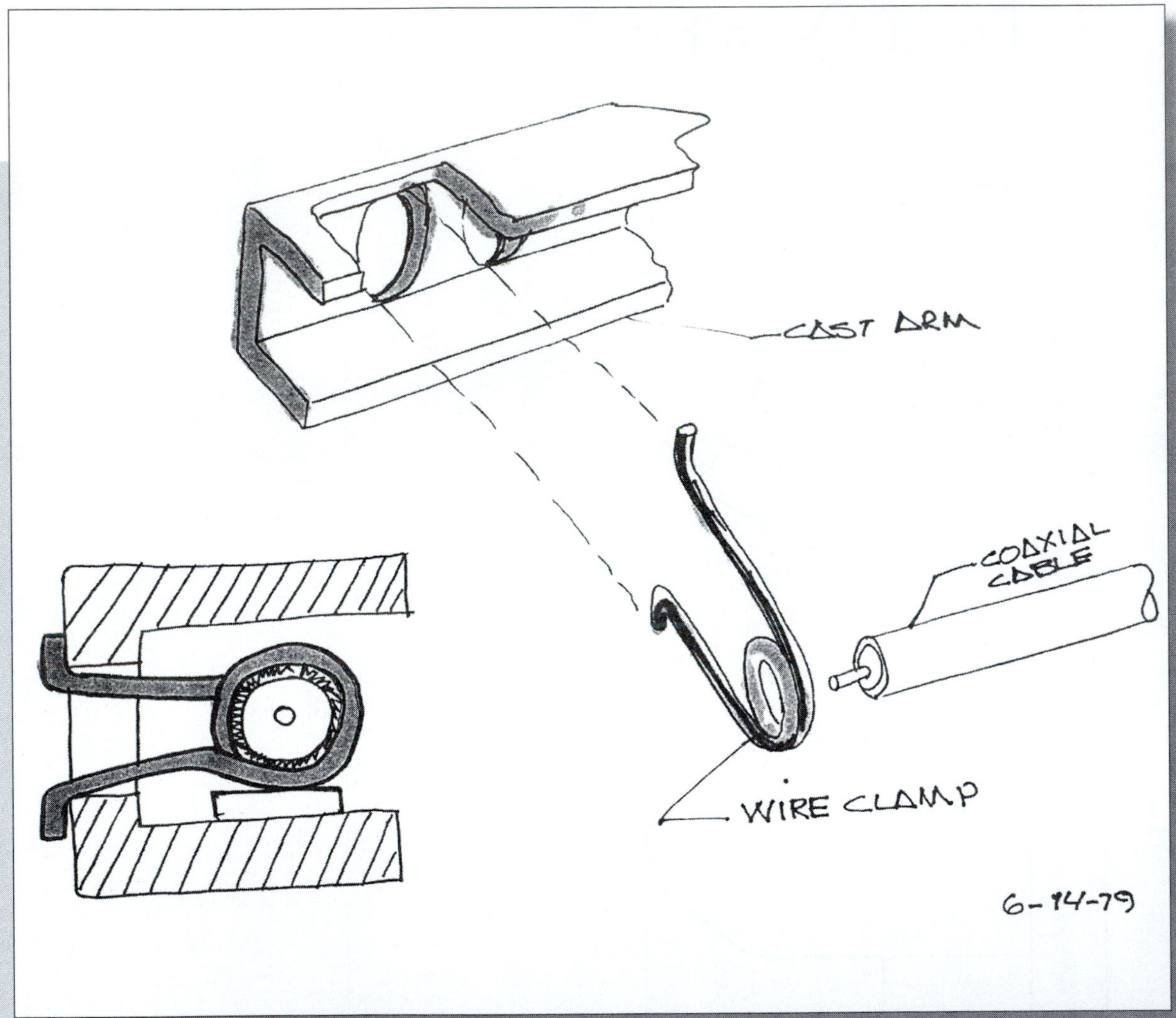

CHAPTER 6

Modeling for Refinement

Objectives

When you have completed this chapter you will be able to

- describe the modeling methods available to represent your design;
- list a set of qualities models have that can help you select among them;
- describe how models can be used in refining designs;
- select which model to use to solve a particular problem;
- select which type of model to use at various stages in the design process;
- describe which modeling method contains the most information about a design.

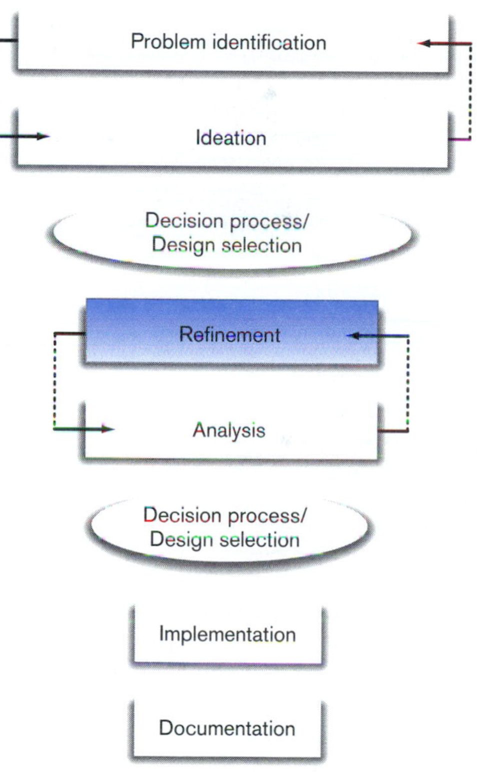

During the ideation phase of the design process, graphics tools are selected for their ability to capture a design idea and to help team members visualize possible solutions to the design problem. Because the goal during this phase is to generate and explore a wide range of options, hand sketches and conceptual representations of the design are preferable to more detailed drawings that take longer to create and you might be more reluctant to modify. Remember, you cannot model what you have not planned and thought about. Freehand sketches are still the best way to get your initial ideas down on paper.

When you move into the refinement stage, however, the criteria for the best means of describing your design change. During this middle stage of the design process, illustrated in Figure 6.1, you need to refine the design concept to include the detail necessary to fully define it (for manufacture), as well as to test it against the constraints established in the design specifications. At the same time, you need to preserve your ability to easily modify the design. During refinement, these factors are the most important in determining how best to represent your design.

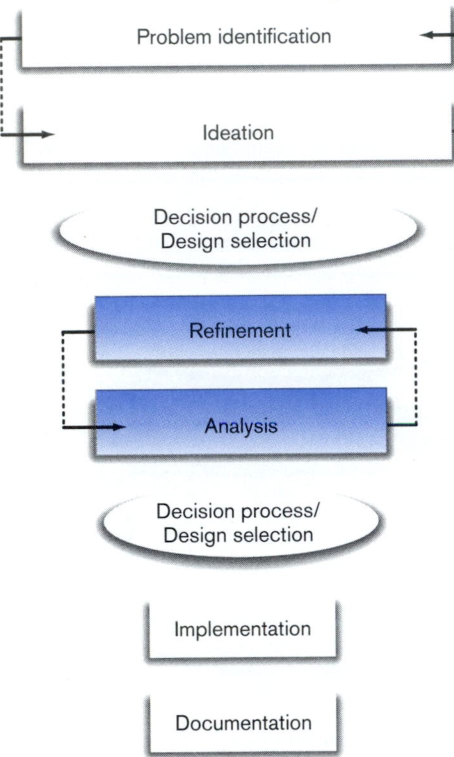

FIGURE 6.1
In the refinement and analysis loop, the details of the design are worked out, tested, adjusted, and tested again until the design is fully defined and meets the criteria established. Models in this phase need to represent the design accurately enough to be used for testing as well as to convey the details of the design into the next phase of the process, where it will be implemented.

Refining the Design Idea Through Modeling

Figure 6.1 depicts the transition from ideation and visualization to refinement and analysis as a narrowing of the process. From the several solutions generated during ideation, only a subset of those solutions are selected for further consideration. The process for selecting a design (or designs) will vary according to the situation, as you learned in Chapter 2. In each case, however, the process determines whether money for developing the design further will be committed. When a design concept is selected, it then moves to the next stage of the process: refinement.

Remember that the process of refining the design is iterative. Refinement is part of the second iterative loop in the design process: refinement and analysis. This loop illustrates how the design is continually refined until a final solution is realized. In other words, the refined design is tested, the test results suggest modifications to the design, and once the design is changed, it needs further analysis. The dynamics of this interplay of testing and refinement have been a powerful force in the development of 3D CAD modeling software.

Today, creating a CAD model is an important step in the refinement process because the software tools make it possible to incorporate a huge amount of information about the design in a format that remains easy to modify. The accuracy and completeness of the CAD model make it possible for you and others to use it to test and evaluate the design. By making changes easy—and in some cases, automating the update process—CAD modeling encourages refinement and can eliminate some barriers to a thorough review of the design.

Other aspects of today's CAD modeling software encourage refinement by making it easier to get feedback and test data about a design. Accurate part descriptions make it possible to test how well parts will fit together while they are still being refined. Realistically shaded models can be used to determine customer acceptance and styling issues early in the design process. Many analysis programs can import 3D CAD files and perform analysis on the model geometry, saving a considerable amount of time. Some analysis packages even suggest optimizations that can be made directly to the CAD model. Animation and kinematic analysis software can use CAD data to determine whether the design will function properly before any parts are actually built.

Within this trend to greater integration of modeling and analysis, however, are many different methods that can be used to create CAD models. Each modeling system has its own strengths and weaknesses for capturing design information and effectively communicating during the refinement process. Consider the three sketches shown in Figure 6.2. Computer-generated models can be used to model and refine the design for each of these items, but each may be better suited to a different kind of model, and each model type may require different modeling software.

In this chapter, you will learn about different modeling methods, how they represent and store design information, and the advantages and disadvantages of each. Armed with this information, you will be able to choose the method that will be the most effective for a particular design problem.

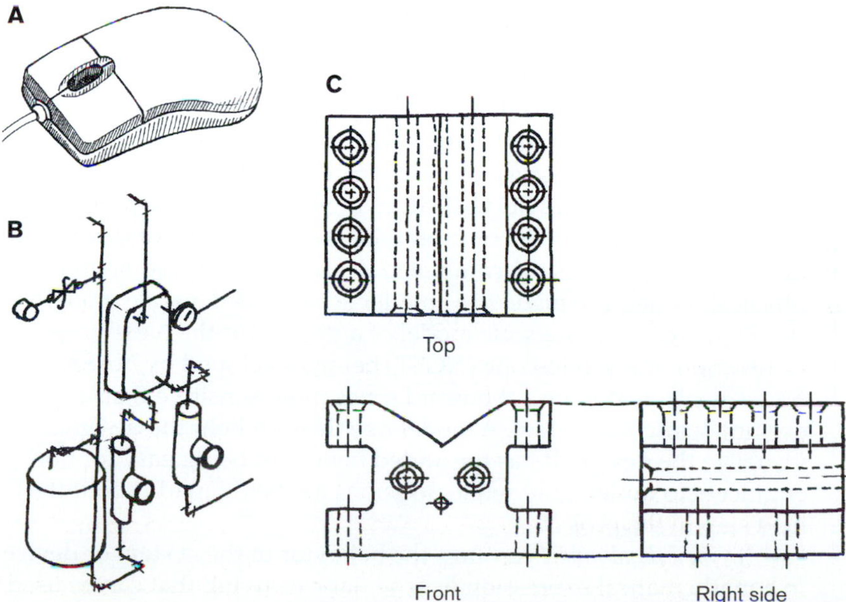

FIGURE 6.2
The same modeling software may be used to model (a) the ergonomic mouse, (b) the piping system, and (c) the V-block, but different modeling methods and packages will vary in their suitability for each of the three.

What Is a Model?

In general, a *model* is a representation of a system, device, or theory that allows you to predict its behavior. The dictionary defines a model as:

1. A small object, usually built to scale, that represents in detail another, often larger object.
2. (a) A preliminary work or construction that serves as a plan from which a final product is to be made: *a clay model ready for casting.* (b) Such a work or construction used in testing or perfecting a final product: *a test model of a solar-powered vehicle.*
3. A schematic description of a system, theory, or phenomenon that accounts for its known or inferred properties and may be used for further study of its characteristics: *a model of generative grammar; a model of an atom; an economic model.*

The dictionary definition helps further define three key qualities that are shared to varying degrees by all models used in engineering design:

- They represent in detail another object, the product, or system to be produced.
- They are a plan from which a final product will be made.
- They are used for studying and testing the design to provide an approximation of how it will behave.

FIGURE 6.3
The scale model shown in (a) represents a design submitted by Ball Aerospace to NASA for the Next Generation Space Telescope (NGST). The mock-up was the largest and most complex scale model ever created by Ball Aerospace. Created at 1:10 scale, it represents a telescope that stands over 9 feet tall with a primary mirror of 8 m in diameter when fully deployed. Because the primary mirror is so large, it must be able to fold in on itself in order to fit inside a launch vehicle, as shown in (b). This scale model demonstrated the deployment by using seventeen computer-controlled motors. Because of the improbability of servicing this telescope in space, this design is also partially functional even if the mirrors should fail to be fully deployed. (Courtesy of Ball Aerospace & Technologies Corp.)

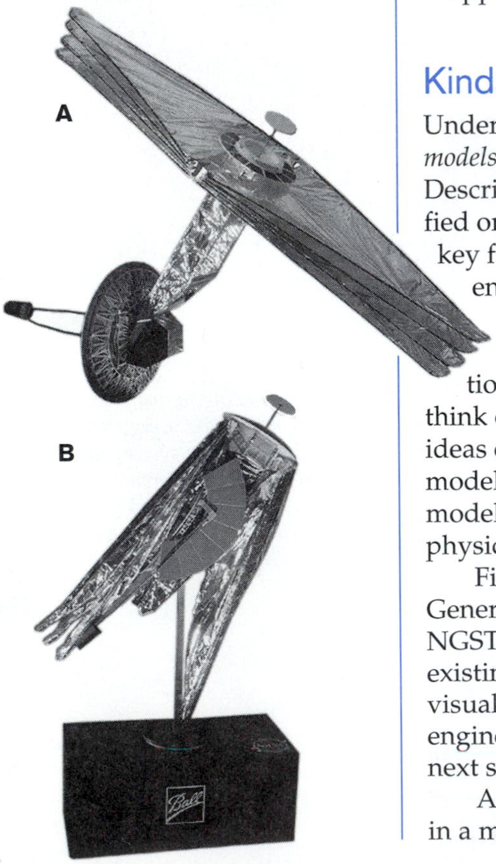

A

B

Kinds of Models

Under the broad umbrella term *model* are several types. *Descriptive models* represent a system or device in either words or pictures. Descriptive models sometimes use representations that are simplified or analogous to something that is more easily understood. The key function of a descriptive model is to *describe,* or provide enough detail to convey an image of the final product.

You can think of a set of written specifications for a design as a verbal descriptive model. If all of the specifications are followed, the system will perform correctly. You can think of sketching as a type of descriptive model of your design ideas on paper. 2D and 3D CAD drawings are also descriptive models. A physical model or prototype is another type of descriptive model that represents the completed design, although sometimes physical models are made to a smaller scale (called a scale model).

Figure 6.3 shows a scale model of a design for the Next Generation Space Telescope (NGST) being developed by NASA. NGST is planned to be a thousand times more sensitive than any existing telescope system. A model like this can help the designers visualize the design. It has the added benefit of being easy for non-engineers to understand while decisions are being made about the next step in the project.

An *analytical model* captures the behavior of the system or device in a mathematical expression or schematic drawing that can be used

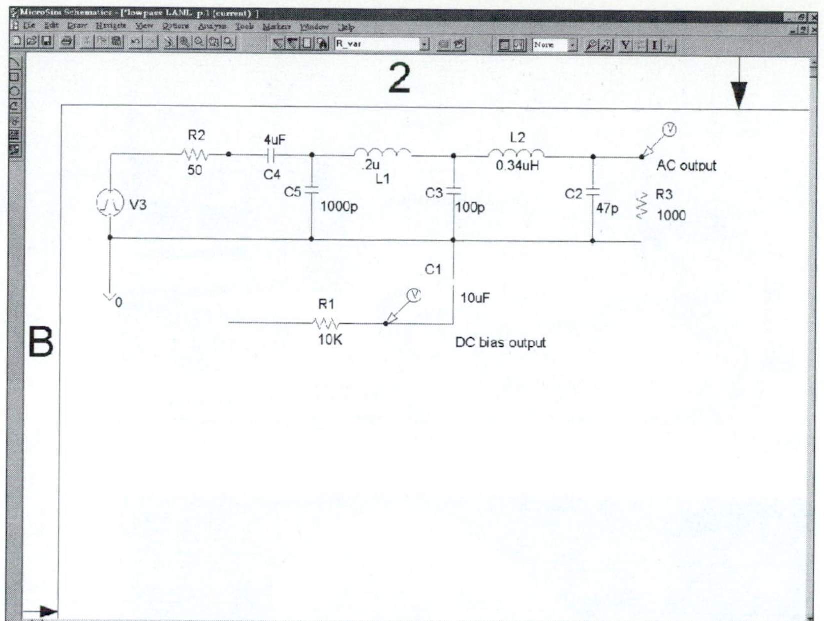

FIGURE 6.4
This analytical model created in MicroSim, a circuit modeling software package, can be used to test the operation of a circuit in lieu of building the circuit in a lab or calculating the operation manually. (Courtesy of Chris Yakymyshyn, Montana State University.)

to predict future behavior. The electrical circuit model shown in Figure 6.4 is an example of an analytical model. Each component of the circuit and its properties is represented in the model by equations and relations that simulate the way the component would behave in a real circuit. The electrical engineer designing this circuit can change values to make predictions about what will happen in a real circuit that is wired the same way as the model. The mathematics built into the model of the circuit could be done manually to evaluate the design, but like a CAD model, the computer-based model makes it easier, less expensive, and faster to test the design.

Part of creating an effective analytical model is determining which aspects of the system's behavior to model. In the circuit model, some information, such as interference between some components, is left out because it is too complex to represent effectively. A finite element model, such as that used to generate the stress plot shown in Figure 6.5, simplifies the CAD model in a similar way. The *finite element analysis (FEA)* model breaks the model into smaller elements; by reducing a complicated system to a series of smaller systems, the stresses can be more easily solved. Understanding and using analytical models effectively require knowing how the model differs from the actual system so you can interpret the results correctly.

In the design process, you will use different types of models where they are appropriate. During the ideation phase of the design, sketching is probably the best modeling technique to use. As you begin to refine the design, you will use both descriptive and analytical models to represent the design more accurately and to provide insight into its behavior.

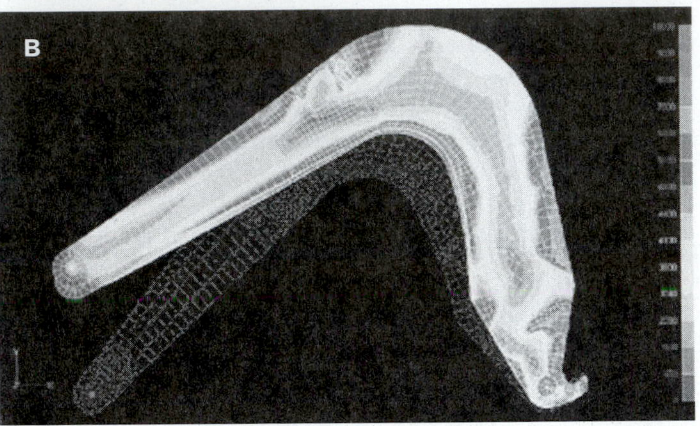

FIGURE 6.5
Calculating the stresses on the gooseneck shown in (b) are vital to the design of the heavy-duty equipment it is part of [shown in (a)]. The finite element software uses smaller "elements," indicated by the segments of the mesh covering the model, for the stress analysis. (Courtesy of Smith Equipment, USA.)

A 3D CAD model combines qualities of descriptive and analytical models. Because a 3D CAD model accurately depicts the geometry of a device, it can fully describe its shape, size, and appearance as a physical or scale model would. Additional information about the final product, such as the materials it will be made from, can also be added to the model description stored in the computer database. Figure 6.6 shows a CAD model of NASA's Space Infrared Telescope Facility (SIRTF) to be launched in 2001.

A 3D CAD model can also be used as an analytical model would be. That is, the 3D model of the device can be used to study its characteristics (see Fig. 6.7). As more sophisticated software is available to analyze, animate, and predict the behavior of the design under various physical conditions, the 3D CAD model becomes a more perfect representation of the final object, and better suited for testing and study.

Model Qualities

Each of the different modeling methods has its strengths and weaknesses. In general, models can have qualities that make them more or less useful for certain purposes and phases in the design process.

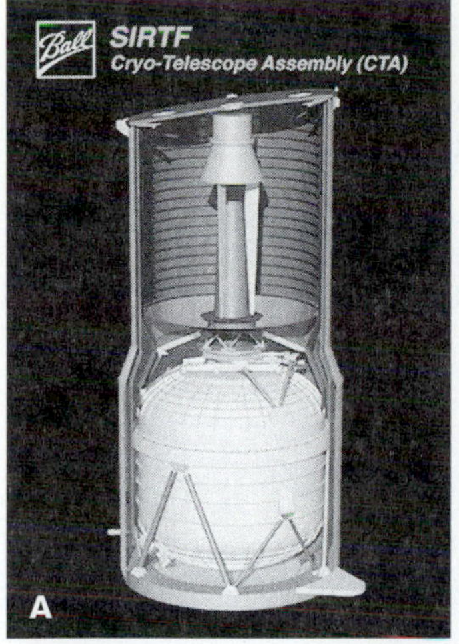

FIGURE 6.6
This rendered view of the preliminary CAD model of the SIRTF assembly, shown in (a), is difficult to distinguish from the photograph of its 1:10 scale model, shown in (b) next to a 1:10 model of the Hubble Space Telescope. (Courtesy of Ball Aerospace and Technologies Corporation.)

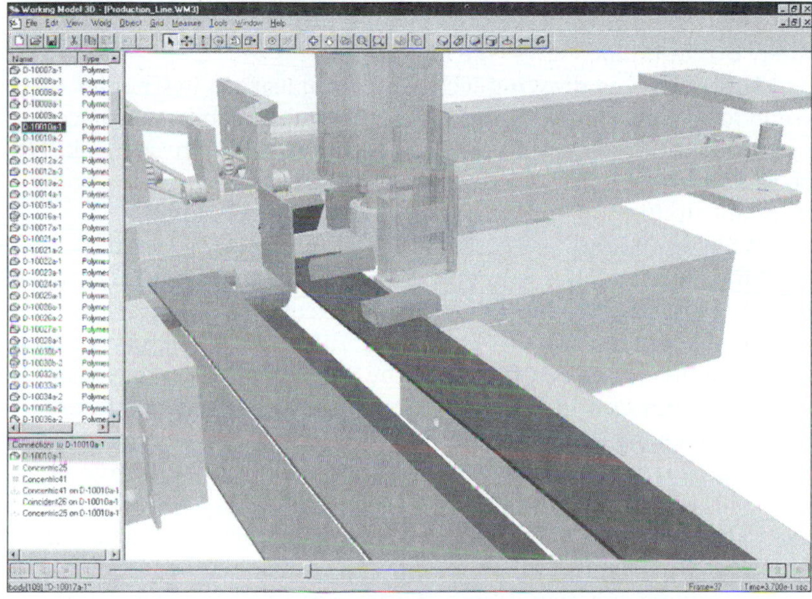

FIGURE 6.7
The 3D model of this packaging assembly line conveyor was used to create a simulation of the process in Working Model® that revealed how the system would behave when the belts were in motion. (Courtesy of Walt Walczak, AutoBotics, Inc., Minneapolis, MN.)

Because design is an iterative process, being able to change the model easily can be very important. Also, there can be a significant investment in equipment and effort in creating CAD models, so it is important for them to be useful for a variety of purposes. What are the qualities that can make models more or less useful?

Good models are

- visual (presenting information graphically);
- understandable (containing detail in a format that is suited to the audience);
- flexible (able to be updated and modified quickly and easily);
- cost-effective (providing benefits equal to or greater than the cost of creating the model);
- measurable (able to be used to extract size and shape information);

- accurate (providing information with enough precision to be useful);
- robust (containing enough information to depict all necessary aspects of the system and reflect the design intent).

There are many types of models you might use when refining the design of a system or product. In the next section, we will assess each one according to the qualities identified above.

2D Models

Paper Drawings

2D sketches and multiview drawings created with pencil and paper are visual representations of the design. They vary in their robustness; all of the information needed to define the object can be contained in multiview paper drawings, although it may require many orthographic views (see Fig. 6.8).

The multiview drawing techniques you learned in Chapter 4 were designed to improve the robustness, measurability, and accuracy of paper drawings. These same techniques require some skill to be understood, however, and make multiview drawings less understandable to the non-engineer than a 3D model would be.

While the equipment costs associated with paper drawings are minimal, they can take as long or longer to create than CAD drawings. In addition, changes usually involve considerable erasing and redrawing, making them difficult to modify as the design is refined. For example, it is difficult to move the views around on the sheet to make room for new information. Because they are so much more difficult to change, the labor costs associated with paper drawings outweigh the equipment savings, so they are less cost effective in most situations than CAD-based drawings.

Paper drawings are considered accurate if they vary by less than plus or minus one-fortieth of the drawing scale. In other words, a paper drawing of a map drawn at a scale of 1 inch = 400 feet has good accuracy if you can measure from it to plus or minus 10 feet. This makes paper drawings not particularly measurable or accurate—which is why proper dimensioning technique was developed. Proper dimensioning overcomes this limitation and provides measurements to the desired accuracy on the drawing. It is not a good practice to make measurements from paper drawings.

Paper drawings can be very effective for quickly communicating the design of small parts for manufacturing. A properly dimensioned sketch can quickly convey all of the information needed to make a part that is needed only one time (see Fig. 6.9). Also, some manufacturing facilities are not able to read electronic files; these require paper drawings to be sent to them.

Drawings created with paper and pencil are very important during the ideation stage of design and for simple design tasks in which the need to revise and analyze the model is less demanding. But other CAD modeling methods provide more options for analysis

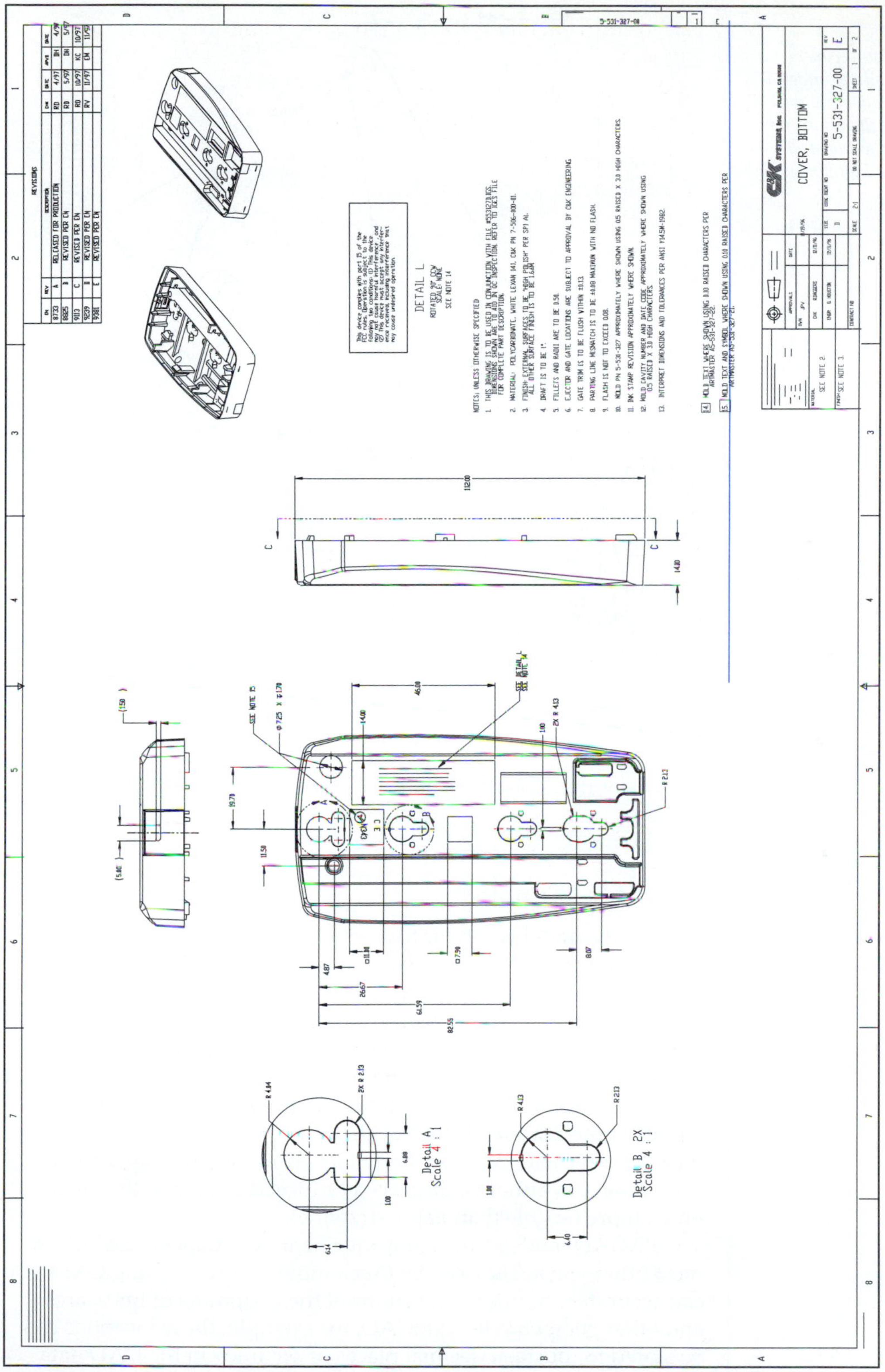

FIGURE 6.8

This multiview drawing includes front, top, and right-side dimensioned views, additional views that fully describe the part, plus notes that add information about its manufacture. (Courtesy of C&K Systems.)

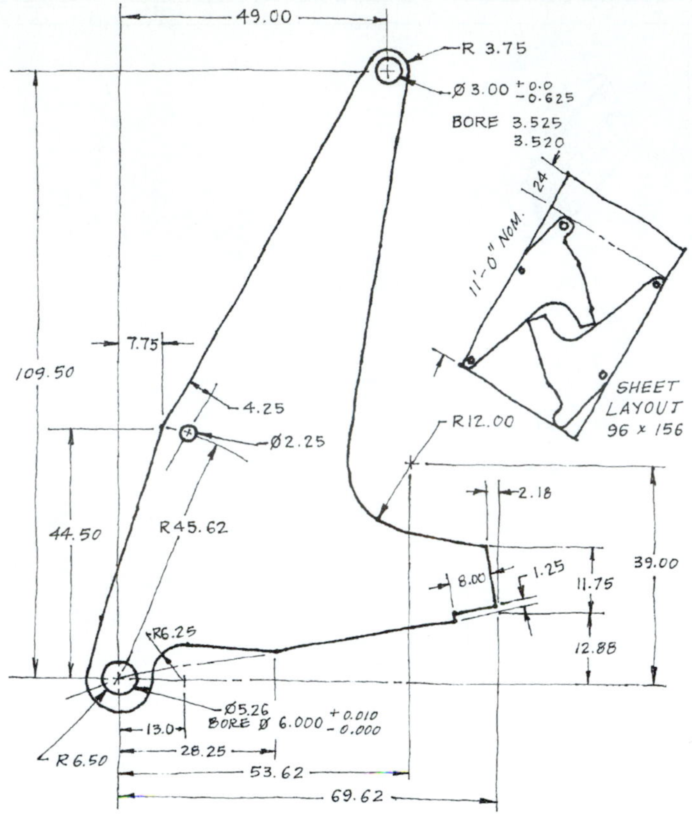

FIGURE 6.9
This fully dimensioned paper drawing contains all the information needed to manufacture this part, which will be flame cut from a 96-by-156-inch sheet of steel. (Courtesy of Smith Equipment, USA.)

and easier iteration through the refinement loop in the design process.

2D CAD Models

A 2D CAD model is similar to a paper drawing, but it is created using a computer. 2D CAD models share the visual characteristics of paper drawings but are much easier to change. Most CAD systems have a large variety of editing tools. CAD owes its original popularity to this ability to quickly edit and reuse drawing geometry in different formats. Standard symbols can be added and changed, and 2D CAD files can quickly be plotted to different scales. Other capabilities of the 2D model, such as the ability to separate different types of information onto different levels in the drawing (called *layers*) that can either be displayed or turned off, make the model much more flexible than paper drawings.

2D CAD models are an improvement over paper drawings in most other ways. They can be much more accurate. Using CAD, you can accurately define the locations of the endpoints of lines, arcs, and other geometry. In AutoCAD, for example, these locations can be stored to fourteen decimal places of accuracy in the CAD database. The database can be queried and the information returned as accurately as you originally created it. Of course, the old adage Garbage In, Garbage Out applies, meaning that a CAD drawing is only as accurate as you make it. If your endpoints do not connect, or you locate distances on the screen by eye instead of entering them precisely, you will not have accurate results.

CAD models are more measurable than paper drawings for another reason. The CAD model is usually created to represent the object full size, unlike paper drawings, which often depict the object to a smaller or larger scale. Because features are represented using their actual sizes, you can make measurements and calculations from 2D CAD models. For example, if you have a 2D CAD layout showing where equipment will be located, you can make accurate measurements from the CAD drawing database. Also, because you can "snap" to locations on objects, you can determine sizes and distances that may or may not be dimensioned on a paper drawing. If, for example, you need a clearance of 10 feet from the center of a tank to the location of another piece of equipment, you can determine whether the clearance is provided in the design by measuring from the CAD file.

The increased accuracy and measurability make 2D CAD models more robust than paper drawings. In addition, nontext information can be saved in a database and linked to the drawing information. Huge amounts of information can be stored in a 2D CAD file, or in linked spreadsheets or databases, for later retrieval. The 2D database is limited, however, in its ability to represent information that depends on a 3D definition (such as the volume inside the object), so it is not generally useful for determining masses and other physical properties of the object.

The strengths and weaknesses of the multiview projection used to fully define the object in a paper drawing are shared by 2D CAD models. To create them and read them you need to be able to interpret the 2D views to "see" the 3D object. Your ability to make measurements also depends on a good understanding of orthographic projection. In order to measure from a 2D CAD model, you need to show distances and angles true length in a view. Orthographic projection and descriptive geometry are needed to create these views. *Descriptive geometry* is the study of producing views of an object that show true lengths, shape, angles, and other information about an engineering design. Having mastered these subjects, you can use auxiliary views created with a 2D CAD system along with the traditional methods of descriptive geometry to solve many engineering problems.

The flexibility and accuracy offered by the 2D CAD database make 2D CAD systems a cost-effective tool in a wide variety of businesses. For many civil engineering projects, the difficulty of capturing all of the 3D information needed to model irregular surfaces, such as terrain, may not be worth the benefits of working in 3D. For projects such as highway design, civil mapping, electrical distribution, and building systems, 2D CAD models may provide enough information and be created more quickly from the information at hand (see Fig. 6.10). The advantages of using 2D CAD are that areas and perimeters can be calculated accurately, drawings can be revised quickly, and they are more accurate than paper drawings.

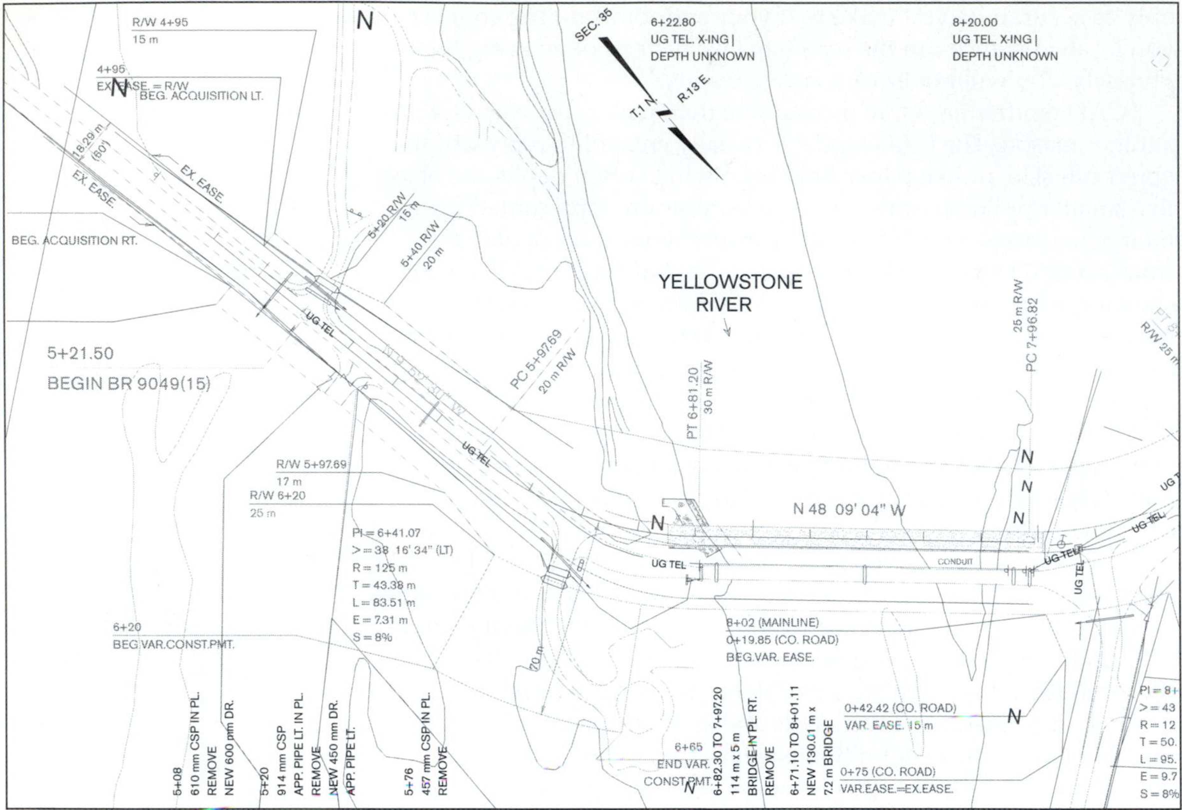

FIGURE 6.10
Large-scale projects such as this highway plan are often modeled in 2D CAD. Most 2D CAD systems associate dimension values with the entities they describe, so dimensions can be updated if the drawing changes. (Courtesy of Montana Department of Transportation, MSU Design Section, Bozeman, MT.)

3D Models

Despite the flexibility and accuracy built into 2D CAD models, all 2D models must be interpreted to visualize the 3D object. In order to convey the design to individuals unskilled in engineering drawing techniques—or to evaluate properties of the design that are undefined in 2D representations—3D models are used.

Physical Models

Physical models provide an excellent visual reference that is understandable to non-engineers. Physical models are sometimes called *prototypes,* especially when they are made full size or used to validate a nearly final design for production. People unskilled in the use of engineering drawings can interact with them and get a good feeling for what the design will look like and how it will function. We are used to experiencing things through touch and appearance. We are also used to being able to change our view of things by walking around them. Many problems with designs are discovered and corrected when a physical prototype is made.

The robustness and cost effectiveness of physical models are often linked. A simple model made of clay or cardboard may be enough to accomplish the goals for the model, but the cost of attain-

FIGURE 6.11
Rapid prototypes are physical models used for design verification. (Courtesy of DTM Corporation.)

ing other goals, such as determining the fit of many parts in a large-scale assembly or producing a model realistic enough for market testing, may be prohibitive until very late in the process. For example, the full-size prototype of the BMW 850I cost more than $1 million to produce. The amount of information and detail built into the model adds to its cost.

The accuracy of a physical prototype also drives up its cost. Many objects are designed to be mass produced so that the cost of individual parts is reduced. When each part in the model must be produced one at a time, the cost of the prototype can be many times greater than the manufacturing cost of the final product. The more the model must match the final product in terms of materials used and final appearance, the greater the cost can be. A less realistic prototype can be less costly when a highly accurate physical model is not required. Today's rapid prototyping systems (which you will learn more about in Chapter 10) are popular because they offer a quick and less expensive means of generating a physical model of smaller parts without the cost of machining or forming parts one at a time (see Fig. 6.11).

Physical models are a very good visual representation of the design, but they are not always very measurable. If they are not made from the materials that will be selected for the design, their weight and other characteristics will not match the final product. Sometimes, due to the size of the project, the physical model must be made to a smaller scale than the final design. The accuracy of the final part also may not be possible with the materials and processes used for the prototype.

Probably the least attractive feature of a physical prototype is its lack of flexibility. Once a physical prototype has been created, making changes to it can be expensive, difficult, and time-consuming. Because of this, full-size physical models are not usually used until fairly late in the design process—when the likelihood of major design changes is less. This can limit the usefulness of the model even though it provides important feedback about items that are not working well. When the information comes late in the design process, it can be too expensive to return to a much earlier stage of the design to pursue a different approach. Only critical problems may be fixed, and solutions to problems found late in the process are generally constrained to cause the least amount of redesign while fixing the problem. When a prototype is created late in the process, there may not be time or resources to create yet another physical prototype to show the solutions to the design problems, so new problems introduced by the change may not be seen until actual parts are produced.

Despite the trade-offs between robustness and cost, even very costly physical models have been a cost-effective way for companies to avoid much more costly errors in manufacturing.

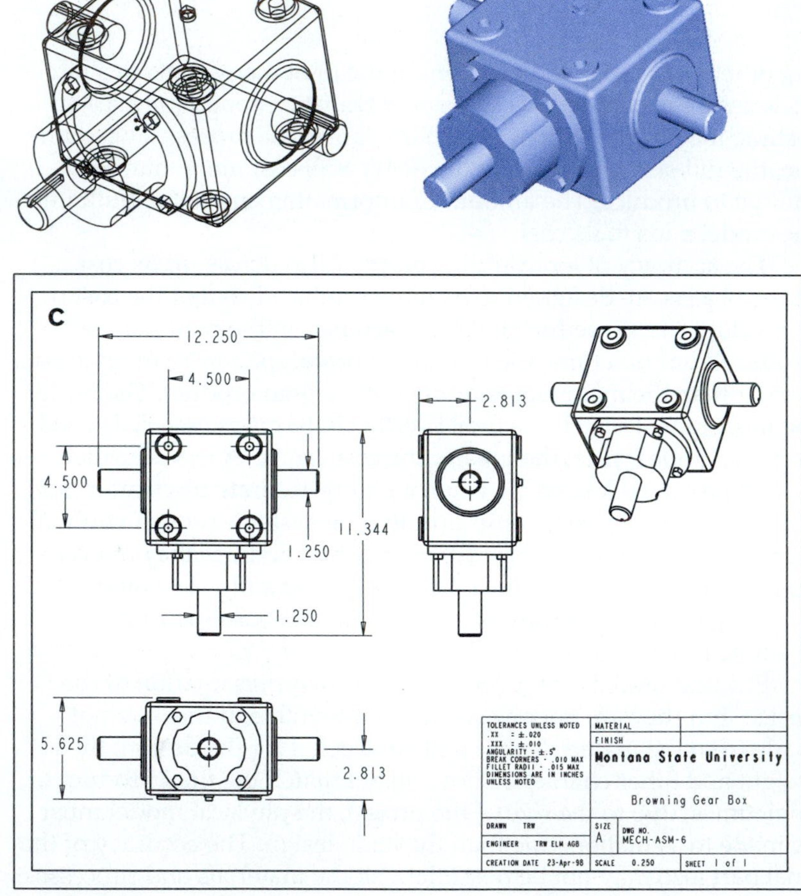

FIGURE 6.12

The 3D model can be displayed from any viewing direction; it can be an outline (or wireframe), as shown in (a); it can be shaded and rendered to produce the realistic view shown in (b); and it can generate accurate 2D views for multiview drawings.

Computer-Generated 3D Models

A 3D CAD model offers all the benefits of a 2D model, plus many benefits of a physical model. As a visual representation, 3D CAD models offer options for different audiences: standard 2D multiview drawings can be generated from the model, as can realistically shaded and rendered views. Because a 3D CAD model accurately depicts the geometry of the device, it can eliminate the need to create a physical model—or make it easy to create one from the data stored in the model. The many options for viewing the model make it understandable to a wide range of individuals who might be involved with the design's refinement (see Fig. 6.12).

Virtual Reality

Virtual reality is the term used to refer to interacting with a 3D computer-generated model as if it were real—the model simulates the way the user would interact with a real device or system. Using a virtual reality display, users are *immersed* in the model so that they can move around (and sometimes through) it and see it from different points of view. The headset display shown in Figure 6.13 uses two displays set about 3 inches apart (the typical distance between a person's eyes). Each display shows a view of the object as it would be seen from the eye looking at it, which creates a stereoscopic view similar to the one sent to the brain by our two eyes when we look at the world around us. Many of these types of headsets also provide input to control the viewpoint by moving your head, so that as you turn to look around, a new view is created corresponding to the direction you turn your head.

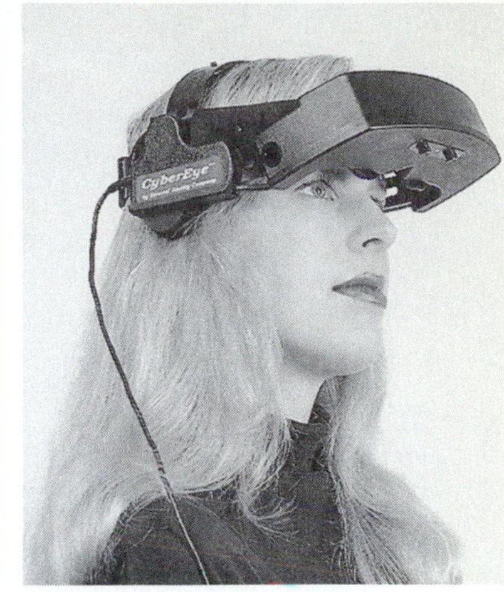

FIGURE 6.13
The CyberEye CE-200W is one example of a head-mounted display that allows the wearer to view images created by the computer as if the user were in the 3D space, or virtual reality, itself. (Courtesy of General Reality.)

Some types of 3D interfaces, such as a 3D mouse or controller, let the user interact with the items that are described in 3D in the model. Some systems contain a large amount of information about the objects stored in the 3D database that can be used to enhance the illusion of reality. Gloves such as the one shown in Figure 6.14 can use data in the model to provide physical feedback to the user when an object is encountered. If a virtual object is squeezed, feedback to the glove provides the feeling of the resistance a solid object would give when squeezed. Some systems may even interpret how much force would crush the object and provide this sensation back to the user.

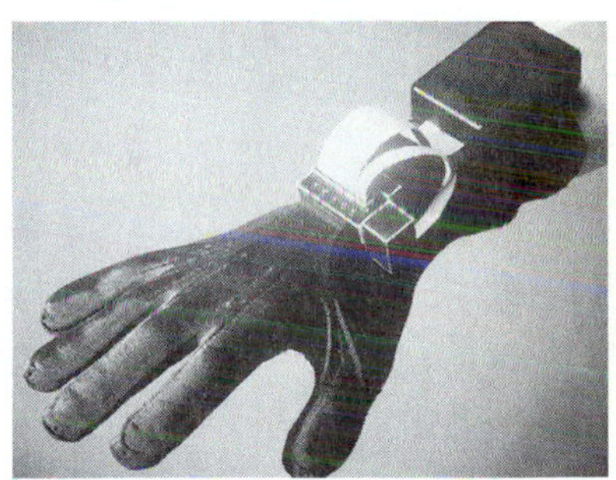

FIGURE 6.14
Input controllers, such as the 5th Glove shown here, provide feedback to the user through the glove and interpret the user's hand motions as input to the computer. (Courtesy of General Reality.)

Some 3D CAD systems can represent the object realistically enough for users, designers, and manufacturers to get the same type of information they would have previously gotten from creating a physical prototype. The term *virtual prototype* describes these types of uses of the CAD database.

3D CAD models are at least as robust as 2D or physical models; those that contain information about the volume inside a 3D shape are more robust. As more sophisticated software is available to analyze, animate, and predict the behavior of the design under various physical conditions, the 3D CAD model becomes a better representation of the final object—and better suited for testing and study.

Like 2D CAD models, 3D computer-generated models offer a high degree of accuracy and measurability and a higher degree of flexibility. Each new generation of CAD software automates more common tasks, allowing the designer to focus more on the design and less on the mechanics of changing the model. This flexibility allows the CAD model to start as a preliminary work that is refined and tested until it is ready to be used as a plan for the final product.

Its ability to be used throughout the process, serve in lieu of a physical model, and be reused and modified indefinitely make 3D CAD models very cost effective for some purposes, especially in designing mechanical assemblies. But not all 3D CAD models are the same. You should be familiar with different types of 3D modeling systems so you can select the methods and software most suitable for your design.

Types of 3D Models

There are several methods that can be used to generate 3D models using CAD. Consider the ergonomic mouse, the piping layout, and the mechanical part shown in the beginning of the chapter. Each of these items has characteristics that may be best modeled using a different 3D CAD approach. At the same time, each of the major 3D modeling methods—wireframe modeling, surface modeling, solid modeling, and parametric modeling—has its advantages and disadvantages. When you understand them, you will be better able to select the method that suits your design purpose.

Many CAD systems incorporate all of these modeling methods into one single software package. Other software packages may have the ability to perform only one or two of these methods. Be sure to select a package that supports the modeling method you intend to use.

Wireframe Modeling

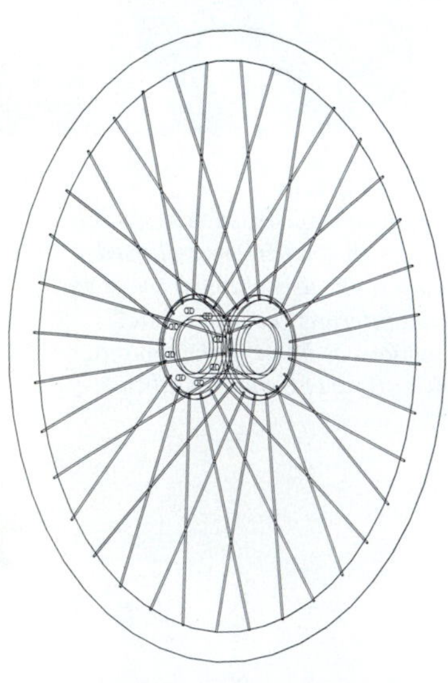

FIGURE 6.15
A wireframe model represents edges and contours of an object using lines, circles, and arcs oriented in 3D space.

Wireframe modeling represents the edges and contours of an object using lines, circles, and arcs oriented in 3D space. Endpoints of the lines, centers, and radii of the circles and pertinent information about arcs and other types of drawing entities are stored in the drawing database. The modeling method gets its name from the appearance of the model, which resembles a sculpture made of wires, as shown in Figure 6.15.

You create a wireframe model in much the same way that you create a 2D CAD drawing. Each edge, representing where surfaces on the object intersect, is drawn in 3D space using simple geometric tools built into the software, such as 3D line or arc. The X, Y, and Z coordinate locations of the endpoints are stored in the drawing database along with the type of entity. Figure 6.16(a) shows a wireframe model of a box with a hole. Figure 6.16(b) shows a printout of the drawing geometry exported as an *initial graphic exchange specification (IGES)* file. Notice the X, Y, and Z coordinates stored for the drawing entities.

A

B

```
124,-1.,0.,0.,0.,0.,0.,1.,0.,0.,1.,0.,0.;                 1P0000001
100,2.,-11.,4.,-11.,5.,-12.,4.;                           3P0000002
110,12.,2.,4.,12.,2.,0.;                                  5P0000003
110,11.,2.,5.,1.,2.,5.;                                   7P0000004
110,0.,2.,0.,12.,2.,0.;                                   9P0000005
110,1.,2.,5.,0.,2.,4.5;                                  11P0000006
110,0.,2.,4.5,0.,2.,0.;                                  13P0000007
100,2.,-6.,3.,-4.5,3.,-4.5,3.;                           15P0000008
124,1.,0.,0.,0.,0.,0.,-1.,0.,0.,1.,0.,0.;                17P0000009
100,0.,11.,4.,12.,4.,11.,5.;                             19P0000010
110,1.,0.,5.,11.,0.,5.;                                  21P0000011
110,12.,0.,4.,12.,0.,0.;                                 23P0000012
110,0.,0.,4.5,1.,0.,5.;                                  25P0000013
110,12.,0.,0.,0.,0.,0.;                                  27P0000014
110,0.,0.,4.5,0.,0.,0.;                                  29P0000015
100,0.,6.,3.,7.5,3.,7.5,3.;                              31P0000016
110,12.,2.,0.,12.,0.,0.;                                 33P0000017
110,12.,0.,0.,0.,0.,0.;                                  35P0000018
110,0.,2.,0.,12.,2.,0.;                                  37P0000019
110,0.,0.,0.,0.,2.,0.;                                   39P0000020
110,11.,2.,5.,11.,0.,5.;                                 41P0000021
110,11.,2.,5.,1.,2.,5.;                                  43P0000022
110,1.,0.,5.,11.,0.,5.;                                  45P0000023
```


FIGURE 6.16
The entities in the wireframe model shown in (a) are described in terms of the coordinates of their endpoints in the CAD database. When the CAD file is converted to IGES, a standard format that can be read by other systems, the resulting file lists the entities and their locations, as shown in the file segment in (b). Lines in the IGES file starting with 110 describe line entities. (See Appendix 49 for more about the IGES format.)

Wireframe modeling has the advantage that it is quick to generate on the computer's display and does not require particularly expensive or powerful hardware or software. Its disadvantage is that it doesn't store all of the information about an object, such as the appearance of its surfaces or the volume contained inside. Because wireframe models do not contain surface information, they cannot be shaded to look like a realistic object. And because wireframe models do not contain any information about the object's volume, mass properties cannot be analyzed directly from the model. Some objects, such as those with complex surfaces, can also be hard to represent accurately using wireframe modeling.

3D wireframe models can be beneficial to quickly show some types of designs, such as process control, piping, sheet metal parts, mechanical linkages, and layout drawings showing critical dimensions that use relatively simple geometric shapes. Interferences and clearance distances can be difficult to visualize in 2D drawings of piping or distribution systems, and it can be critical that they are not overlooked in the design. It may not be cost effective or useful to build a more detailed solid model of the design when a wireframe model is sufficient.

To quickly check clearances for the design of an electrical substation, for example, the engineer may create a wireframe model of the system and its conductors. High-voltage electrical conductors that are too close to other equipment may produce an electrical arc that jumps to the other conductive equipment, causing a short. The distances from one conductor to the next in any direction can easily be checked with a 3D wireframe model.

Another example where wireframe modeling may be used very effectively is in process control design. In designing these systems, it may not be necessary to model the pipe itself, as standard sizes and fittings will be used. The length, branching, and angles of the piping

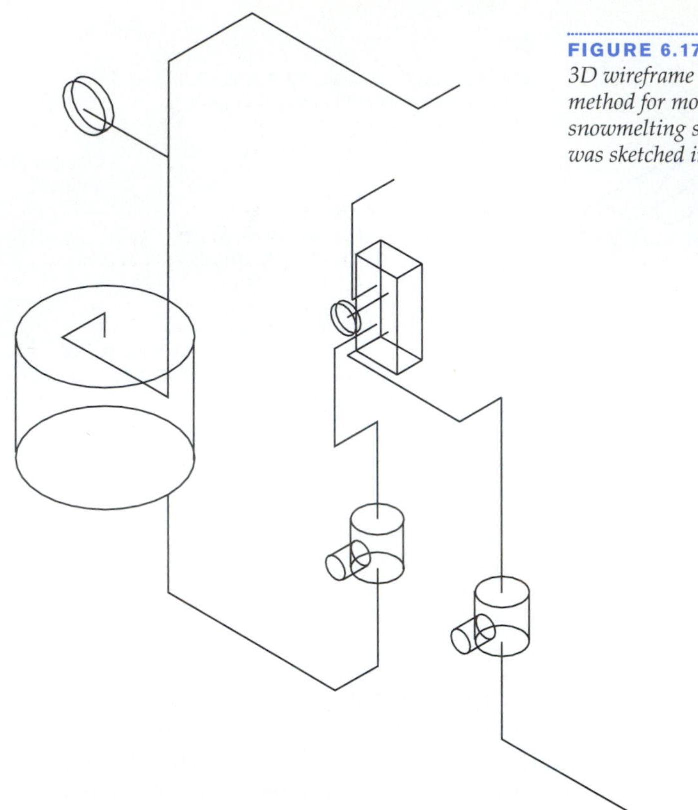

FIGURE 6.17
3D wireframe is a suitable method for modeling the snowmelting system that was sketched in Chapter 3.

in the system need to be evaluated, however, against other aspects of the 3D structure. The 3D wireframe can be used to fit the layout around other 3D equipment, which can be difficult to visualize in 2D. The centerline of the piping can be represented in a 3D wireframe and used to route and show piping in 3D, as shown in Figure 6.17. This can eliminate instances in which clearances are acceptable in two dimensions, but not in a third. In a system with moving parts, however, or where more complex shapes need to be fit, you may need to use a different modeling method to assess interferences.

How do wireframe models fit the characteristics discussed previously for their effectiveness in the design process? Wireframe models are more visual in many ways than a 2D model because they are three-dimensional and can be viewed from any direction. The model can be used to produce all of the standard views as well as any others needed to aid in visualizing the object.

Because wireframe models don't include surfaces that can be shaded, however, they are not as realistic looking as models that can be. Without visible surfaces to show the faces of the object, the viewer must mentally add the surfaces to interpret the 3D shape. Because you can see through the model, some shapes cannot be represented unambiguously, and it may be difficult to see from a single view which areas are holes and which are surfaces. The model in Figure 6.18 shows an ambiguous wireframe model. Without rotating or checking coordinate locations, you could visualize the hole on the top or the bottom.

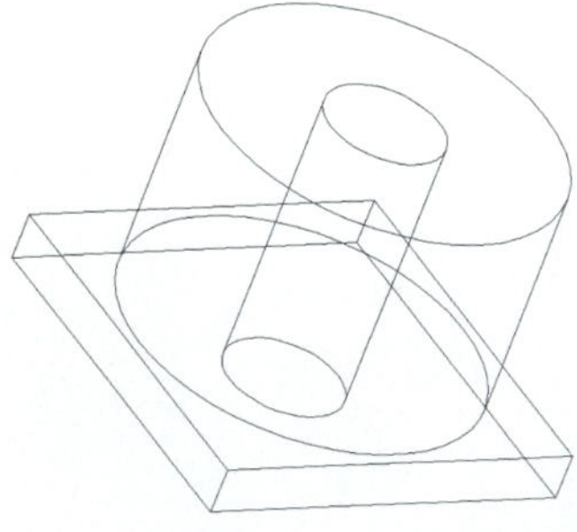

FIGURE 6.18
Are you looking at the model from below or from the top? Does the hole go through the bottom surface?

Wireframe Modeler Versus Wireframe Display

Wireframe modeling gets its name from the look of the model on the computer screen, but it is not the *look* of the model that determines whether it is a wireframe model. Wireframe modeling was the first 3D modeling method to be developed. Its depiction of edges and contours grew out of 2D modeling practices. The term *wireframe* is also used to describe an economical way of displaying a model on the computer screen, in which just the edges and contours are shown. Representing objects on the screen using wireframe display is simpler and faster than showing shaded views. The computing power needed to generate and display a model has always been a constraint in the design of CAD modeling systems, and wireframe representation was developed when the kinds of multimedia display common on home computers today were available on only the most powerful mainframe computers. As computer hardware evolved, more computing-intensive modeling techniques became feasible. More information can be stored with the model, and more complex displays, such as shaded views, can be generated from the data. But there is still a trade-off between complexity and speed; more complex and detailed information means more processor time.

CAD modelers have always addressed this trade-off by representing the model data in two ways: first, as it is stored in the database (the coordinate locations for each entity in the model, for example) and second, as it will be displayed on the screen. The economy of wireframe display makes it a good choice for working with computer models—whatever the underlying modeling method is. Surface, solid, and parametric modelers all offer a wireframe display and a range of tools for controlling that display. Each of them, however, creates a database with more information than can be found in that created by a wireframe modeler, which stores only vertex and edge information.

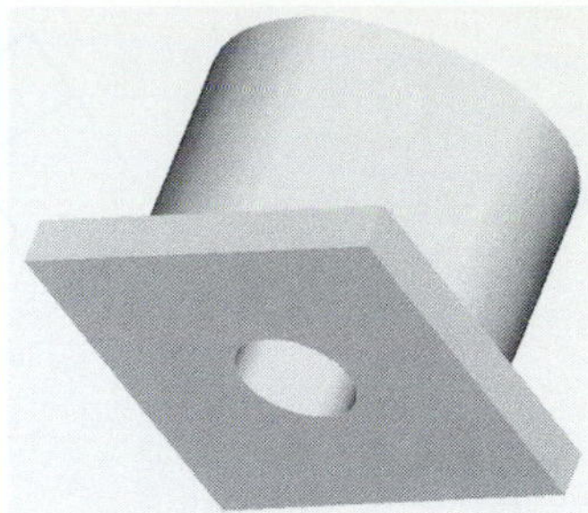

FIGURE 6.19
Adding faces to a wireframe model can remove the ambiguity about some features.

Some wireframe modelers have added surface modeling features that help overcome the difficulty in visualizing wireframe models. Using the viewpoint information and the geometry stored in the database, the modeler may allow you to add "faces" to the model; these can be used to shade it or remove lines that would be hidden in the view. Figure 6.19 shows a shaded view of the wireframe model in Figure 6.18. Each wireframe modeling package may offer a different approach to generating more realistic-looking displays, but in each case, it will be adding capabilities not found in a true wireframe modeler.

3D wireframe models are more robust than 2D drawings because they can store all of the types of information that can be stored in a 2D CAD drawing as well as information about the 3D shape and distances in the third dimension. Remember that before you can make accurate measurements of oblique edges and surfaces in a 2D drawing, you must first create an auxiliary view in which

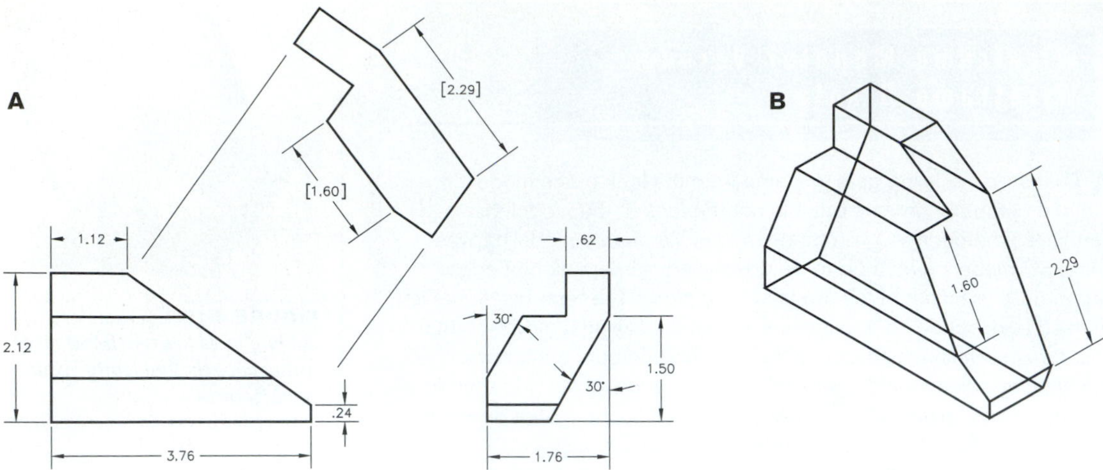

FIGURE 6.20
A 2D CAD package can display the length of a line, but only after orthographic projection techniques have been employed to ensure that the line is shown true size. The auxiliary view in (a) was added to the drawing by projecting points from the front and side view of the object. AutoCAD's List command was then used to display the length of the line in the auxiliary view, where it appears true size. The length of the line in the AutoCAD wireframe 3D model in (b) can be displayed from any view by selecting the line in the model and listing its dimensions.

that feature is shown true size. In a 3D wireframe drawing, that information is already contained within the 3D wireframe model and can be measured directly. For example, the length of edge A in Figure 6.20 can be ascertained from the 2D model created in AutoCAD, but only after a 2D auxiliary view is projected from the principal views of the object. The 3D wireframe model of the same part can be used to list the length of the edge directly.

A well-executed wireframe model can provide accurate drawing distances and clearances in 3D that can be measured from the model. Locations of geometry in the digital database are stored to several decimal places of accuracy. Because only 3D lines, circles, arcs, and other similar entities make up the database, information such as surface area and volume is not contained in wireframe models. Some models can also be ambiguous as to which part of the object is solid and which is a void or a hole. Measurements can be made only between the entities that are stored in the model, which are mainly part edges, centerlines, and other similar features. The lack of volume and surface information in a wireframe model can make it not very useful for export to machining operations, for example.

Wireframe models are as flexible as any computer-generated model. The endpoints of the lines, arcs, circles, and other drawing entities that make up a wireframe model can quickly be edited to make changes. Because each vertex, line, or circle, arc, or entity making up the model definition is stored in the database, each one can be edited individually. However, this can be more time-consuming than some other modeling methods, which group the information in ways that make it easier to edit.

Because 3D wireframe modeling software and hardware is relatively inexpensive (compared with more expensive solid modeling and parametric solid modeling software, which also requires more expensive hardware to run), it can be very cost effective. 3D wireframe drawings, although simple in many ways, can be easy to create, are handled by many low-cost software packages, and can provide a good tool for modeling geometrically simple shapes in 3D space.

Surface Models

Surface modeling defines the model by defining its surfaces. You can think of a surface model as similar to an empty box. The outer surfaces of the part are defined (although unlike an actual box, these surfaces do not have any material thickness). Because the surfaces are defined, they can be shaded to provide a realistic appearance for the design.

Surface modelers have traditionally been used more by industrial designers than mechanical engineers. Industrial designers, who create the outside envelope for a device, need modeling tools that will convey a realistic image of the final product. Software tools developed for their use focused first on the challenge of modeling the exterior of the product, not the mechanical workings within. Today, many surface modelers offer specialized tools for lighting and rendering a model, allowing designers to create photorealistic images that cannot be distinguished from the actual product.

At Timex, for example, the industrial design department used Alias Wavefront's Alias Studio in the design of the Timex TurnAndPull alarm. The product team was concerned that the layered inner rings on the watch would look too deep or too busy. A fully rendered surface model of the watch was realistic enough for buyers and others on the team to see exactly how the watch would look when manufactured (see Fig. 6.21). The model not only allowed Timex to proceed with the design as planned, but it was eventually used to feature the watch in the Timex catalog long before an actual product photo could have been available.

FIGURE 6.21
This photorealistic image of the Timex TurnAndPull alarm was created in Alias Studio, a surface modeling and rendering package. Because photoquality images can be generated from its CAD models, Timex rarely uses photos to promote products that have been modeled and rendered in this way. (Courtesy of Dana Rockel, Timex Corporation.)

The role of the 3D CAD database as the source of a complete product description and the concurrent engineering environment are contributing to the development of modeling software that satisfies the aesthetic needs of industrial designers as well as the functional information needed by engineers. Understanding surface modeling techniques will help you assess the kinds of features that are available in your modeler—and whether a dedicated surface modeler is required.

Surface Information in the Database

Most surface modeling software store a list of the vertices of a part and how they connect to form edges in the CAD database. The surfaces between them are generated mathematically from the lines, curves, and points used to define them. Some surface modelers store additional information to indicate which is the inside and which is the outside of the surface. This is often done by storing a vector (directional line) perpendicular to the outside of the surface. This allows the model to be shaded and rendered more easily.

Storing the definitions of the surfaces in the CAD digital database is termed *boundary representation (BREP)*, meaning that the information contained in the database represents the external

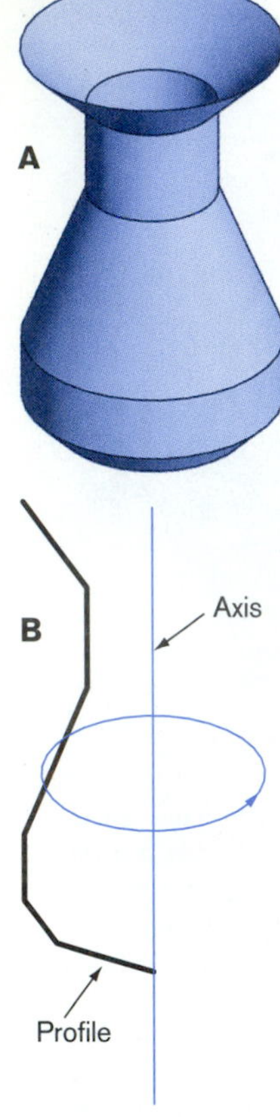

FIGURE 6.22
The surface model in (a) was created by revolving the profile about the axis shown in (b).

boundaries of the surfaces making up the 3D model. Remember that these surfaces have no thickness.

The basic methods used to create surface models are

- extrusion and revolution;
- meshes;
- spline approximations.

Extruded and Revolved Surfaces

Surface modelers create surfaces using the same basic concepts of extrusion and revolution that you saw in Chapter 5. You can define a surface by defining a 2D shape or profile and the path along which it is to be revolved or extruded.

Figure 6.22 shows a surface model created by revolution, and the profile and axis of revolution used to generate it.

Regular geometric entities can be revolved or extruded to create surface primitives such as cones, cylinders, and planes. Like the solid primitives you saw in the last chapter, surface primitives are sometimes built into surface modeling packages to be used as building blocks for more complex shapes.

Meshes

Mesh surfaces are defined by a matrix of vertices. The 3D location of each vertex is stored in the CAD database, and each group of vertices is used to define a flat plane surface. Figure 6.23 shows a mesh surface and a list of some of its vertices.

Some mesh surfaces may be referred to as *triangulated irregular networks (TINs)* because they connect sets of three vertices with triangular faces to serve as the surface model. A mesh surface can be useful for modeling uneven surfaces, such as terrain, where a completely smooth surface is not necessary. A sufficiently large matrix will result in correspondingly smaller triangles that can better approximate a smooth surface. Mesh surfaces may be detailed enough for some purposes, but may not approximate smooth surfaces well enough for realistic models. Making the mesh more refined will produce a

FIGURE 6.23
A mesh surface is composed of a series of planar surfaces defined by a matrix of vertices, as shown in (c). The wireframe view of the mesh in (a) appears more like a surface in the rendered view shown in (b).

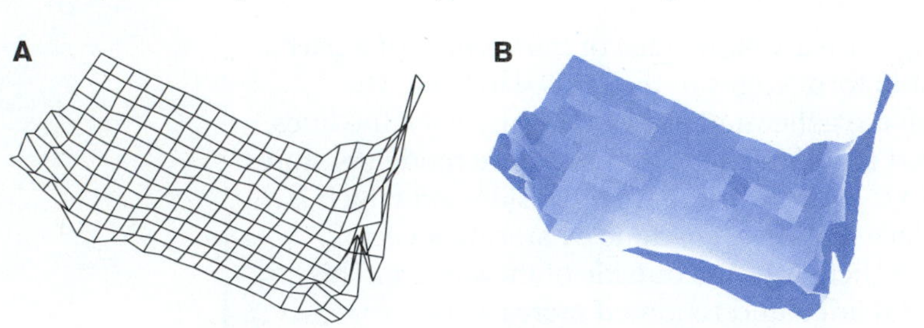

C
```
481,5066,4485[
482,5066,4483[
483,5066,4485[
484,5066,4490[
485,5066,4500[
486,5066,4510[
487,5066,4510[
488,5066,4505[
489,5066,4500[
490,5066,4495[
491,5066,4500[
492,5066,4520[
493,5066,4530[
```

smoother surface, but the size of the CAD file will increase dramatically and result in slower performance when using the software. Some modeling packages have other capabilities that allow you to define a smoothed representation of the surface instead of the somewhat bumpy appearance of a strictly mesh surface.

NURBS-Based Surfaces

The mathematics of non-uniform rational B-spline (NURBS) curves that you learned about in Chapter 5 underlie the third method used to create surfaces in today's surface modeling systems. A NURBS surface is defined by a set of vertices in 3D space that are used to mathematically define a smooth surface. The mathematics used to define the surface employs two parameters, u and v, to generate functions that blend the shape of the surface from one defining curve to another. Like a B-spline, the surface can be edited by editing these vertices, or control points, individually.

Rational curves and surfaces have the advantage that they can be used to generate not only free-form curves, but also analytical forms such as arcs, lines, cylinders, and planes. This is an advantage for surface modelers that use NURBS techniques, as the database does not need to accommodate different techniques for describing different kinds of surfaces. AutoCAD's AutoSurf, for example, uses NURBS surfaces exclusively, whether the surface was created using a surface primitive, mesh, extrusion, or revolution.

Spline curves can be used as input for revolved and extruded surfaces, but can also be lofted or swept. *Lofting* is the term used to describe a surface fit to a series of curves that do not cross each other. Sweeping creates a surface by sweeping a curve or cross section along one or more "paths." In both cases, the surface blends from the shape of one curve to the next (see Fig. 6.24).

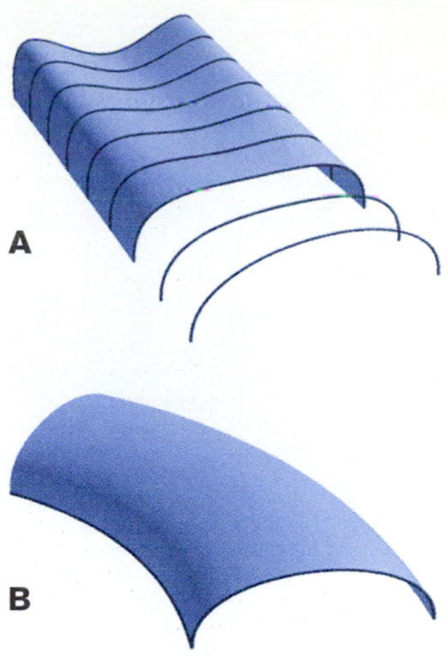

FIGURE 6.24
A lofted surface, as in (a), blends a series of curves that do not intersect into a smooth surface. A swept surface, shown in (b), sweeps a curve along a curved path and blends the influence of both into a smooth surface model.

Lofting was used to build surface models for the organic shapes in the human heart embryo models created at Oregon Health Sciences University. In order to study hearts that are only 0.8 mm across, Kent Thornburg and Jeffrey Pentecost created models that allow them to generate 5-cm-wide physical models (see Fig. 6.25). Using cross sections of embryos from the Carnegie Collection of Human Embryos at the Armed Forces Institute of Pathology, the team members first made digital photomicrographs of each slide. Using the computer, they traced the outlines of heart tissues at each level. The result was a series of closed contours upon which they could place interior and exterior surfaces to define the heart and its inner cavities. By modeling embryonic hearts at different stages, the researchers were able to use the models to examine the way different tissue layers expand and move as the heart develops.

NURBS surfaces can also be created by meshing curves that run perpendicular to each other, as illustrated in Figure 6.26.

Reverse Engineering

Mesh data, or a set of data points similar to those stored as vertices in a mesh, is what is generated by most reverse engineering packages that trace a physical object or model to generate data for a surface model. Points on the object are captured as vertices, then

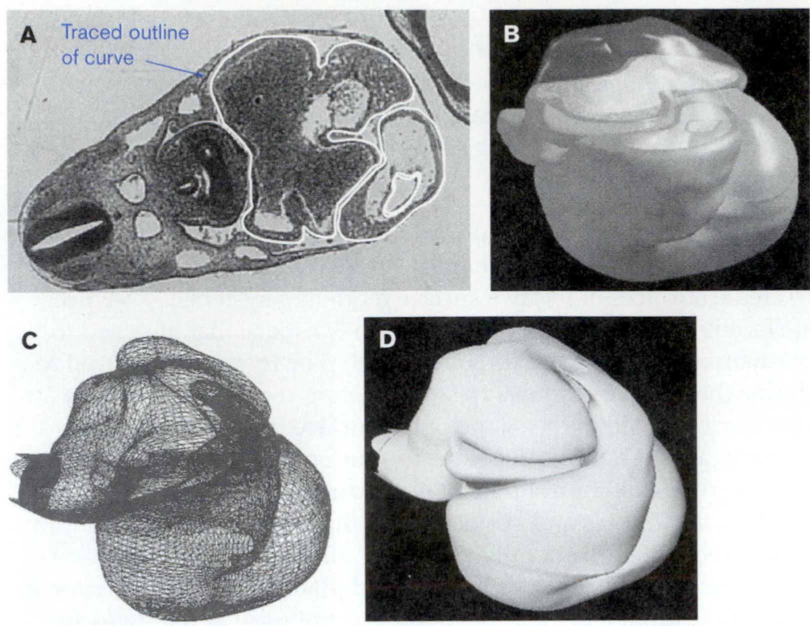

FIGURE 6.25
The embryonic heart models started from NURBS curves derived from the cross sections of heart tissue, shown in (a). Lofting combined the cross sections into a surface model of the heart, as illustrated by the cutaway rendered view of the model in (b). The resultant model, shown in a wireframe view in (c), also generated source files for a stereolithography machine so that a physical model, shown in (d), could be made from the CAD model and used to study blood flow in the developing heart. (Images courtesy of Dr. Jeffrey O. Pentecost, Director, Visible Embryo Heart Project, and Dr. Kent Thornburg, Director, Congenital Heart Research Center, at Oregon Health Sciences University, with the cooperation of Alias Wavefront and Silicon Graphics, Inc. See Science, *Nov. 21, 1997.)*

translated into a digital surface representation. Reverse engineering can be an easy way to capture the surface definition of an existing model, and it can allow designers to use traditional surface sculpting methods and port them to the digital database.

The conceptual design of the Romulus Predator supercar, for example, was first modeled in clay at one-quarter scale, as many automobile bodies are. A cloud of data points was digitized and imported to the ICEM Surf surface modeler (see Fig. 6.27). Rather than generate a mesh surface, however, spline curves were fit to the data points and NURBS surfaces created between them to model the

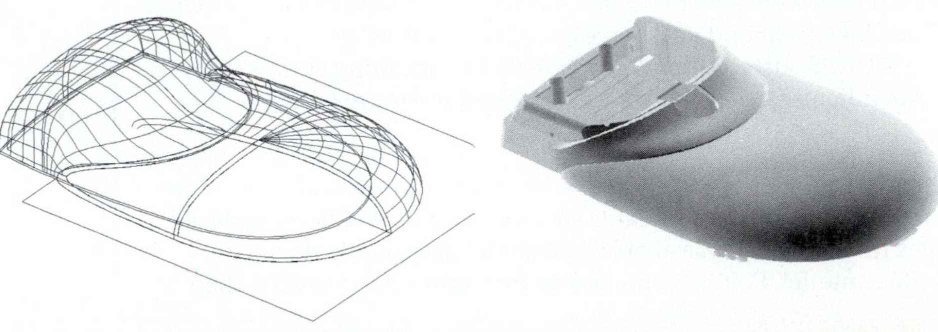

FIGURE 6.26
When the spline curves shown here are used to generate a NURBS surface, the functions defining each curve are blended. (Courtesy of Robert Mesaros.)

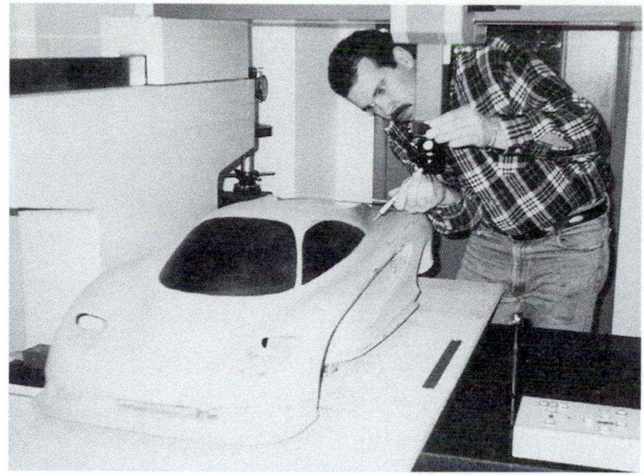

FIGURE 6.27
A coordinate measuring system, such as the one shown here digitizing the clay model of the Romulus Predator supercar, can capture hundreds of thousands of data points that can be used to define surfaces in a modeling package. Seventeen hundred data points were captured by digitizing one-half of the clay model of the Predator. By digitizing the data and mirroring it in ICEM Surf, side-to-side symmetry was ensured in the digital model. The "better" side of the hand-built model was selected for digitizing. (Courtesy of M&L Auto Specialists, Inc.)

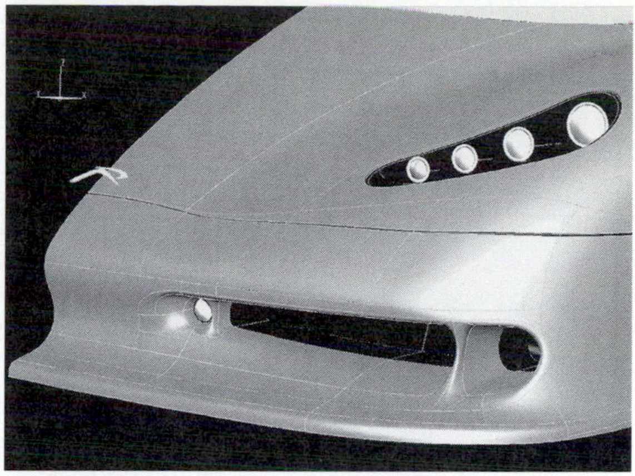

FIGURE 6.28
The white lines on the ICEM Surf surface model of the Romulus Predator indicate the individual surface patches that make up the model. Areas of the car with drastic curvature changes are divided into smaller patches, which keeps the mathematical order of the patches lower and reduces the overall complexity of the surface. (Courtesy of M&L Auto Specialists, Inc.)

body design. This model was then used to further refine the design, instead of the multiple clay models that would have been created to visualize each major change in the design.

Complex Surfaces/Combining Surfaces

To create a surface model, you do not create the entire surface at once—just patches that are combined into a continuous model. Just as curves are made of individual segments that are smoothed into a continuous curve, surfaces are made of entities referred to as *patches* (see Fig. 6.28). Like a spline curve, a patch can be *interpolated,* or approximated.

A *Coon's patch* is a simple interpolated surface that is bounded by four curves. Mathematical methods interpolate the points on the four boundary curves, highlighted in Figure 6.29, to determine the vertices of the resulting patch.

Surface patches are joined by blending the edges of the patches. Areas created by blending, or the creation of fillets, corners, or offsets, are called *derived surfaces*. They are defined by mathematical methods that use the edges of the patches being combined to create a smooth joint.

Some surface patches must be trimmed before they can be joined. A circular patch, for example, may first be defined as a rectangular patch, then trimmed to a circle before being blended with other surface patches.

New modeling techniques are being developed to overcome some of the shortcomings of surfaces that are built up of many patches.

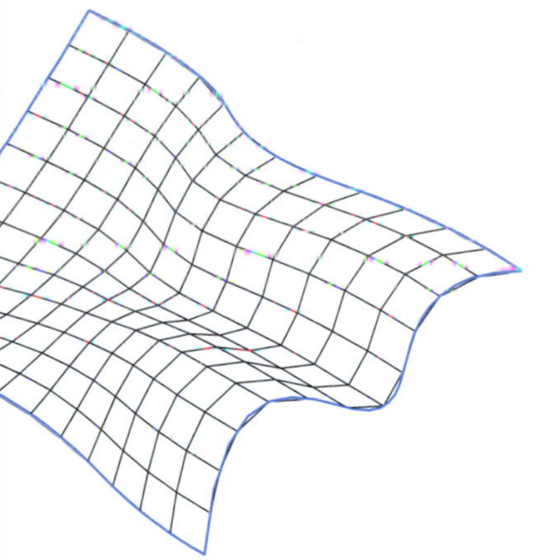

FIGURE 6.29
The lines between the boundary curves in this Coon's patch surface represent the shape of the surface and illustrate the influence of the boundary curves on the interpolated surface between them.

Some surface modeling systems provide the use of Boolean operations, which you learned about in Chapter 5, while others do not. Systems that do not provide Boolean operations or good tools for trimming surfaces can be difficult to use to create a feature such as a round hole through a flat surface because you must define the exact shape of the surface, including the hole, by specifying its edges.

Editing Surfaces

Once a surface has been defined, editing it may depend on the method used by the surface modeler to create and store the surface. Usually the information describing the locations of vertices is editable, and each vertex can be changed. If the CAD surface modeling software retains the original 2D profile that was extruded or revolved to produce the surface, it can be used to facilitate changes to these surfaces. If not, it can be difficult or impossible to efficiently edit these surfaces. Trimming can also affect how easy or possible it is to modify a surface.

For NURBS surfaces, the control points used to define them allow the surfaces to be edited as a spline curve would be. Each control point can be edited to produce local changes in the surface model. (Editing Bezier surfaces, by the same token, produces global changes to the surface.) Grabbing and relocating vertices is a highly intuitive way to edit a surface model that contributes to its usefulness in refinement. For example, the NURBS surfaces used to define the shape of the Predator's exterior made it easy to use the surface model to explore design options in several areas. The roof scoop on the clay model wasn't visible enough, but the control points in the surface model could be adjusted in real time by the designer until the scoop looked right on the screen. The term *tweaking* is used to describe editing a model by adjusting control points individually to see the result.

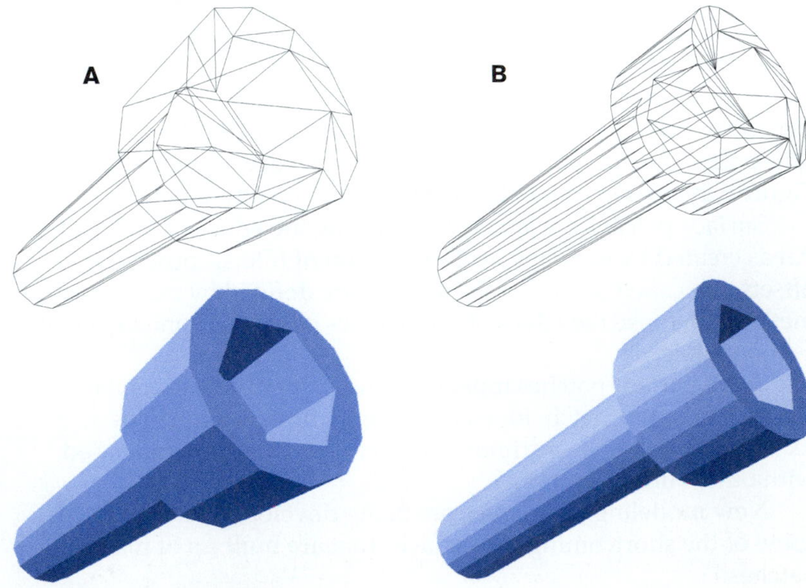

FIGURE 6.30
Mesh surfaces with more vertices can more accurately represent a curved or irregular surface, but they also require more computing power. The mesh in (a) has fewer vertices and larger facets than the mesh used in (b). The finer mesh in (b) more accurately represents the cylindrical shape of the bolt. (Courtesy of Immersive Design.)

Surface Model Accuracy

Surface models can be stored with greater and lesser degrees of accuracy depending on the amount of effort and information stored. Compare the meshes shown in Figure 6.30. The mesh at left contains fewer vertices than the mesh shown at the right. While the surface shown at the right is more accurate, the additional information stored in the model makes its file size larger. This in turn requires more powerful hardware to store and interact with the model. Depending on your purpose for the model, the smaller number of vertices may be enough to suit your purpose. Surface models to be used for computer-aided manufacturing generally require a high degree of accuracy. Sometimes the trade-off between speed and accuracy can be resolved by setting the surface display to a less accurate faceted representation and storing a smoothed or more highly defined surface definition in the database. As with wireframe modelers, it is important to distinguish between the display of the model and the surface definition stored in the database.

Spline information used to create surfaces also can be represented with varying accuracy. Splines stored using more vertices provide you with more control points that you can use to move the shape of the surface. As with meshes, however, this can increase the file size and cause poorer system performance. In addition, if you want to model a surface smoothly, having fewer control points may allow you to model the surface with more fluid curves. Remember, too, that you can often control how closely the spline is pulled toward the vertex. The tolerance of the spline can add to its accuracy as well as its size in the database (see Fig.6.31).

Whether the modeler is storing a smoothed or faceted representation of the surface will also determine how accurate the model is

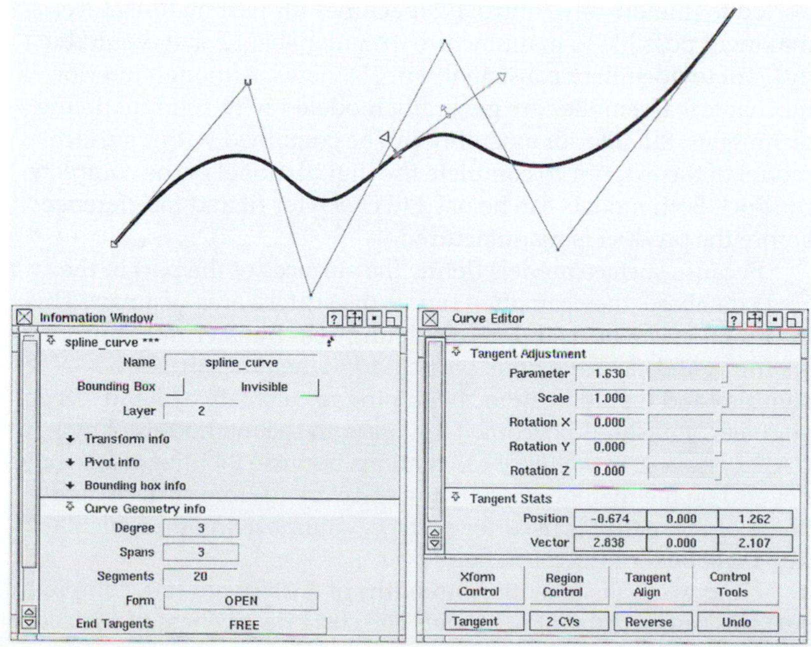

FIGURE 6.31
Most surface modelers give you a great deal of control over the points used to define spline curves. This dialog box allows the user to adjust spline properties.

for analytical purposes. A mesh surface used for area calculations may not be as accurate as one created using a smoothing algorithm such as NURBS or Bezier.

Using Surface Models

The visual information stored in a surface model suits it to very different uses than a wireframe model does. Often surface models store much wireframe modeling information about a model, but their primary strength is in the degree to which they represent the appearance of the surfaces (and can convey these complex definitions to computer-aided manufacturing). Customers purchase products not only on their function, but also on their styling. Industrial designers must make sure that a product will look good enough that people will want to buy it. Realistically shaded models can be used with potential customers to determine their reactions to the appearance of various designs. You can use lighting and apply different materials in most surface modeling software to create very realistic looking results. Many consumer products often start with the surface model, then engineer the interior parts to fit the shape of the styled exterior. When surface models can be used in place of a physical prototype—or the product itself for promotional purposes—the savings add to their cost effectiveness.

Because the relative locations of surfaces from a particular direction of sight can be calculated, surface modeling systems can automatically remove back edges (or represent them as hidden lines). Surface definitions remove the ambiguity inherent in some wireframe models and allow you to see holes and front surfaces by hiding the nonvisible parts of the model. Various surface modeling software may perform more or less well in this category depending on the specific algorithms (software programming routines) used.

The complex surfaces defined by a surface model can be exported to numerically controlled machines for part manufacture, making it possible to manufacture irregular shapes that would be difficult to document consistently in 2D views. Although interior mechanical assemblies are generally modeled with solid modeling techniques, the interior assembly can be combined with a surface model of the exterior to complete the digital model of the complete product. Both models can be used to check for fit and interference before the product is manufactured.

Because surface models define the surfaces of the part in the CAD database, they can often report the surface area of a part. This information can be useful in calculating heat transfer rates, for example, and can save time, particularly when the surface is complex and would be time-consuming to calculate by hand. As an engineer, you should become familiar with the method used by your CAD system to make these calculations because their accuracy may depend on the method used by the software to store surface data. Some of the methods used for surface, volume, and other calculations will be discussed in Chapter 10.

Despite the visualization strengths of surface models, complex surfaces can be difficult to model. The cost effectiveness of surface

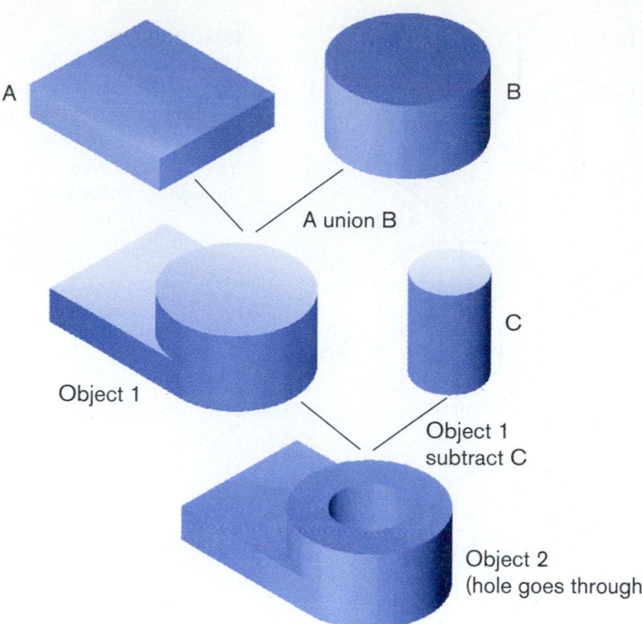

FIGURE 6.32
The model tree in a CSG system stores the primitives in the order in which they were used and records the operations used to create the model.

modeling depends on the difficulty of the surface, the accuracy required, and the purpose for which the model will be used. Many modeling software systems offer a combination of wireframe, surface, and solid modeling capabilities, making it possible to weigh the benefits against the difficulties and the time required to create each type of model.

Solid Models

Solid models go beyond surface models to store the information about the volume contained inside the object. They store the vertex and edge information of the 3D wireframe modeler, the surface definitions of the surface modeler, plus volume information included in neither. The qualities of solid models more closely approximate those of physical models than any other modeling method, making them especially useful for defining, testing, and refining the designs they represent.

Solid modeling software can be categorized according to the way in which data is stored in the CAD database: boundary representation (BREP), *constructive solid geometry (CSG)*, or a hybrid of the two.

CSG systems are so named because they allow you to subtract, join, and intersect primitive geometric shapes to create more complex solid models. These primitives are stored in the CAD database along with the operations used to form the final model. Because the primitives are still available, they can be used to make changes later. CSG systems usually use the metaphor of a model tree storing the information needed to create the model. Figure 6.32 shows a part and the model tree used to create it.

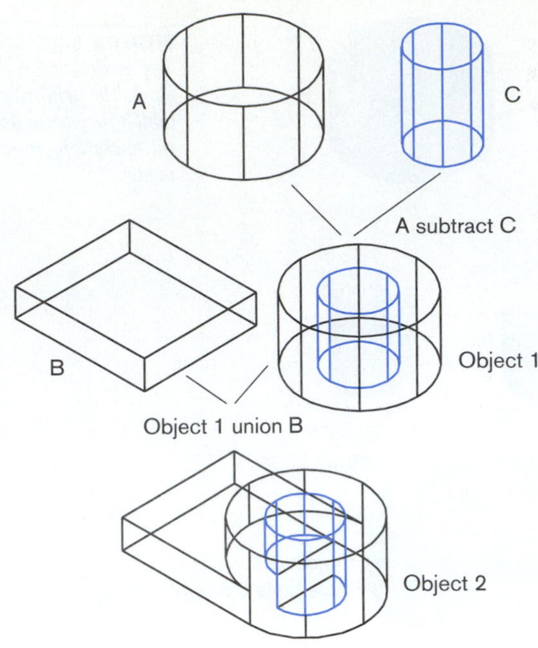

FIGURE 6.33
Unioning A and B after the through hole is created by subtracting C subverts the design intent—the hole no longer goes through the entire part.

Typical solid primitives used by many CAD systems are the box, cylinder, cone, wedge, sphere, and torus that were introduced in Chapter 5. The order in which these shapes are added and subtracted can affect how flexible the resulting model will be. For example, if part A in Figure 6.33 is unioned to part B before hole C is subtracted, the resulting part will have a hole through the entire assembly. However, if hole C is subtracted from part B before A and B are unioned, then the hole will no longer extend through part A.

CSG Method and Boolean Operations

When the CSG method is used for solid modeling, the model must be "rebuilt" by the Boolean operations used to create the model and stored in the model tree whenever you want to display a shaded view of the model or one with hidden lines removed. Both of these operations depend on the surface boundaries of the model. For example, if you decide to change the size of the hole in the model shown in Figure 6.33, the software must calculate new surface boundaries before you can shade the model again. Since the boundary definitions are not stored in the database, as they are for BREP systems, they must be calculated each time you want to show a shaded view of model surfaces. This takes time and processing power. Sometimes this results in a decision not to spend the time to show a shaded representation of the model, which translates into a lesser ability to use shading to help you interpret model views.

BREP Method and Boolean Operations

BREP systems get their name from the fact that they store only the boundaries of the solid object in the CAD database (along with some indication of which side of the surface is inside the object). Like surface modeling systems, BREP-based systems can make it more difficult to change the model once it has been stored.

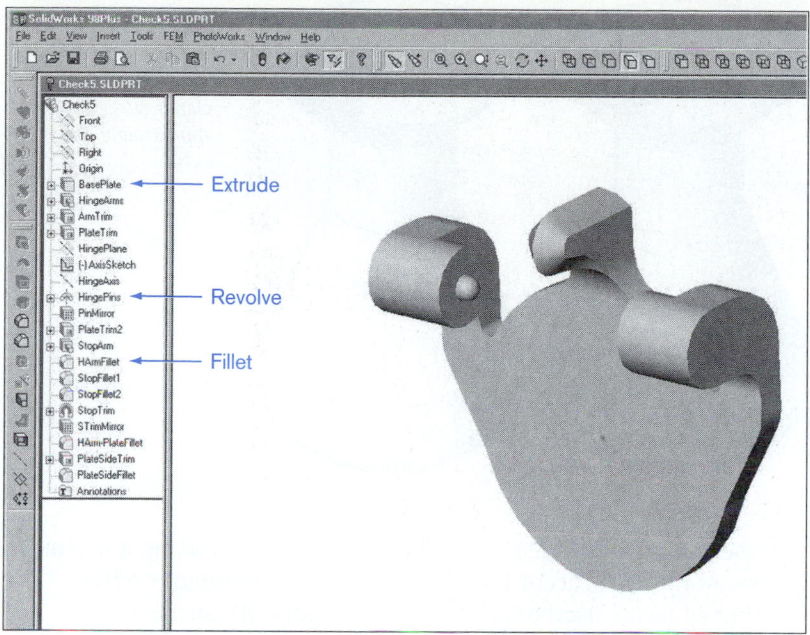

FIGURE 6.34
The operations used to create this model—e.g., extrusion, revolution—are stored in the history of the model and can be edited later as needed. The model tree on the left can be used to return to a previous stage of the model for editing purposes. (Courtesy of Brad Wright, Salient Technologies, Inc.)

Many BREP systems also provide Boolean operators, allowing you to add, subtract, and intersect volumes with the ones stored in the database to create more complex shapes, even though this is traditionally associated with CSG systems. However, once a feature has been subtracted, joined, or intersected, the model that results is stored by its boundaries in the CAD database, not in a model tree that stores the shapes and operations used to define them. This means that the primitive shapes used to create the model art are not available to separate or change when you want to edit the model.

Since BREP surfaces are always defined, however, you can shade the model or remove hidden lines in the display whenever you want, without going through any intermediate steps.

Hybrid Systems

Many systems are a hybrid of CSG and BREP modeling. Hybrid systems may store the model both with BREP and CSG methods so you have access to the primitives and Boolean operations that created the solid as well as to the totally defined boundary representation.

One way in which hybrids make both representations available is to use CSG as the primary representation of the model in the database. A secondary, BREP, representation is calculated and stored for display purposes. Both are updated whenever changes are made. While this eliminates the trade-offs involved in choosing one or the other, it can result in a much larger file size.

Solid Model Accuracy

Like surface models, the accuracy of 3D solid models depends on whether they use faceted or accurate representation to store the model geometry in the database. Some systems use planar shapes to represent smoothly contoured surfaces. This is called a *faceted*

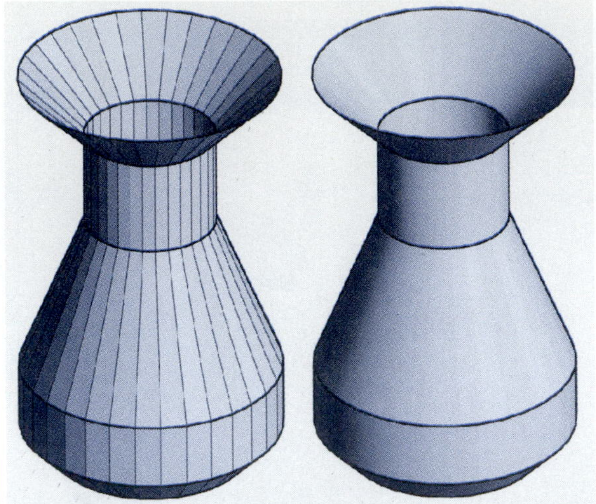

FIGURE 6.35
The faceted representation on the left uses planar surfaces to approximate smooth curves.

representation (see Fig. 6.35). If the model is stored using a faceted representation, the accuracy of the model is dependent on the number of facets used to approximate the smooth surface.

Tessellation lines are lines (or a grid of lines) used to indicate surfaces in a wireframe view, and may or may not reflect the accuracy of the surface (see Fig. 6.36). As with transformations (covered in Chapter 5), it is important to distinguish when tessellation lines represent the accuracy of the *view* of the model, and when they represent the accuracy of the model stored in the database. If the software uses an accurate representation, the actual geometry of the model is stored in the model database; the tessellations are used simply to represent it on the screen in the wireframe view. The number of tessellations in the display can be increased at any time to create a smoother appearance, but they do not increase the accuracy of the model.

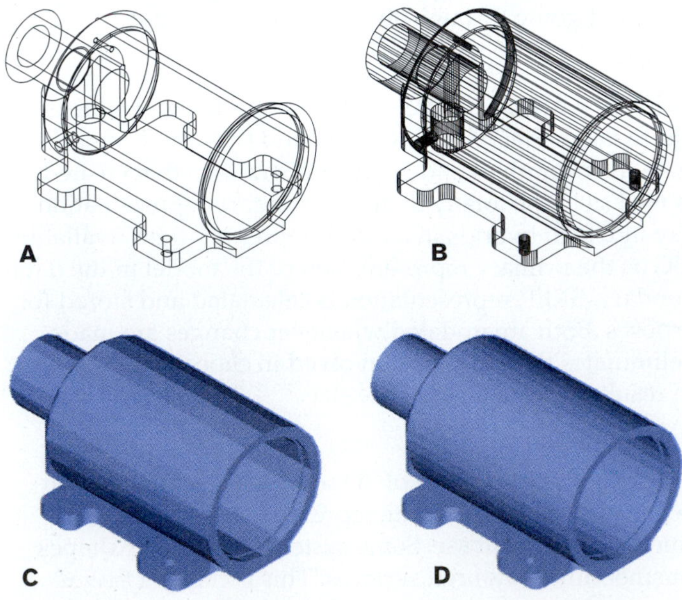

FIGURE 6.36
Four tessellation lines per surface (AutoCAD Release 14's default setting) are used for the wireframe view of the solid model shown in (a). This setting was changed to 30 lines per surface and the view regenerated to achieve the more realistic view in (b). For shaded views, another variable controls the number of facets used to depict a surface. The setting used for the shaded view in (d) is four times the default setting used for the view in (c). Both variables, however, control only the view of the model, not the accuracy of the database itself.

If the software uses a faceted representation, the model itself is stored using facets, and the tessellation lines on the screen reflect the accuracy of the model. You typically cannot increase the number of facets (indicated by the tessellation lines) used in the representation after a model has been created. To improve the accuracy—and increase the number of tessellation lines on screen—you would need to change the settings in the software to increase the number of facets per surface, then re-create the model.

Volumes and surface areas (and other calculations dependent on them) can vary in accuracy depending on the number of facets a model was created with. It is important for you to understand the methods your modeling software uses so you can determine how reliable the information is that you acquire from the model. It is always a good idea to compare hand calculations to the volume and surface areas listed by the software for your models. (You will learn more about doing this in Chapter 10.)

Modeling Kernel

Another consideration in choosing a solid modeler is the *modeling kernel,* or engine, used by the software to perform basic functions such as Boolean operations, blending, mass properties, exporting of geometry, and chamfering and filleting. Although every CAD package operates slightly differently, some of them share the same modeling kernel. ACIS and Parasolid are two kernels used in many popular programs.

The differences between modeling kernels are a result of their development history. The ACIS kernel was developed first for regular solid geometry that was well suited to computer-aided manufacturing applications. The Parasolid kernel offered more options for the blends and curves needed for consumer product modeling. Today, the differences between the kernels are disappearing rapidly as each adds capabilities lacking in the other. The ACIS kernel, for example, now supports lofting operations and modeling of helical shapes that the Parasolid kernel offered first.

Not every modeling package uses a third-party kernel such as ACIS or Parasolid. Pro/Engineer and CATIA are two that use their company's own, proprietary modeling engine.

The differences between the kernels themselves are less important than understanding which one your modeler uses and how that will affect the translation of data from one package to another. CAD software that uses the same modeling kernel can often maintain all of the information or intelligence built into the model when it is translated to different software using the same kernel. CAD files translated between software packages that use different modeling kernels may contain less information once they are translated. Features created with certain operations, such as surface blending, or tangent sweeping, may not translate correctly if the operation is unsupported in the other application. Downstream applications that use data from the model, such as analysis packages or computer-aided manufacturing tools, are particularly sensitive to data translation issues because they often interact directly with the kernel.

As new data formats are developed, such as the STEP standard for solid modeling data, the kernel used by the solid modeling package may become less important. In the meantime, being aware of the kernel used by other companies and individuals with whom you need to share data can help you choose a modeling package that will minimize data loss through translation.

Using 3D Solid Models

Many of the examples used so far in this text have illustrated how an accurate representation of a 3D object can provide valuable engineering information. In terms of visual understandability, measurability, and accuracy, solid models share the strengths of wireframe and surface models. Like surface models, their ability to replace physical models and control computer-aided manufacturing methods adds to their cost effectiveness as a modeling method.

In addition, 3D solid models are particularly suited for use with analysis packages. They contain all of the information about an object's volume, and volume is important to many engineering calculations. Mass properties, centroid, moments of inertia, and weight can be calculated from the solid model itself as often as needed during the refinement process.

Solid models also provide information needed for other analyses. Finite element analysis (FEA) methods break up a complex object into smaller shapes to make stress, strain, and heat transfer easier to calculate. Because the solid model defines the entire object, FEA software can frequently use this information to automatically generate an FEA mesh for a part. This can allow you to include this type of analysis earlier in the design process, incorporate changes, and test the new model with FEA analysis again. Some FEA and solid modeling packages are integrated enough that FEA optimization code can export the new model directly, and the designer can then work from that model if desired.

Parametric Models/Intelligent Models

Parametric modeling is a special kind of solid modeling that is becoming a popular choice for certain types of parts because of its ability to reflect the design process. Easier to update than other modeling methods, simple parametric models can be designed early on and, as additional information becomes available and the design evolves, much of the work of updating the model and drawings associated with it can be handled by the software. Like a CSG solid modeling package, the parametric model stores the steps used to create the model and makes it possible to revisit them to edit the model. Relationships defined between parts of the model are maintained by the software when changes are made. In addition, drawings created from the parametric model are linked to the model itself such that changes in the model will update the drawing, and changes in the drawing will update the model. To make the best use of parametric modeling, the design intent and an object's fit with other parts in the overall design should be considered early on in the design process. While this calls for an initial investment of time, the advantages of better planning up front may be one of the largest rewards of parametric modeling. Parametric modeling was used to help assemble the individual parts into the assembly model shown in Figure 6.37.

You will learn more about parametric modeling in the next chapter.

Choosing the Right Modeling Method

When using modeling in the design process, it is important to choose the method that makes sense for your product, the stage in the design process, and the cost. Table 6.1 recaps the modeling methods covered in this chapter according to the model qualities identified early in the chapter.

Key considerations are the time required to model the part and the purposes it will be used for. A simpler method that provides all the needed information is more cost effective than one that takes longer to model. On the other hand, time invested in a complete digital model of the product can pay for itself if it is used to generate visuals that help shape the design, reduce manufacturing difficulties, foster concurrent engineering—or promote the product later.

The accuracy of the modeling method is another key consideration. What kinds of analysis must be or could be completed with a model? What level of accuracy would be required to interface to computer-aided manufacturing programs?

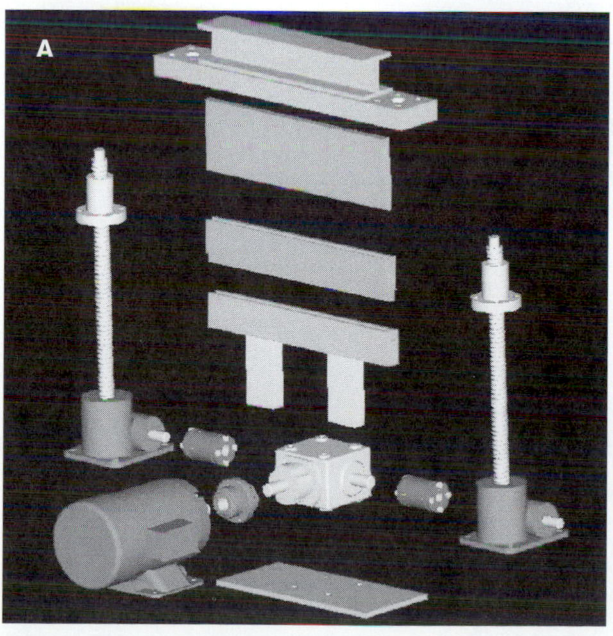

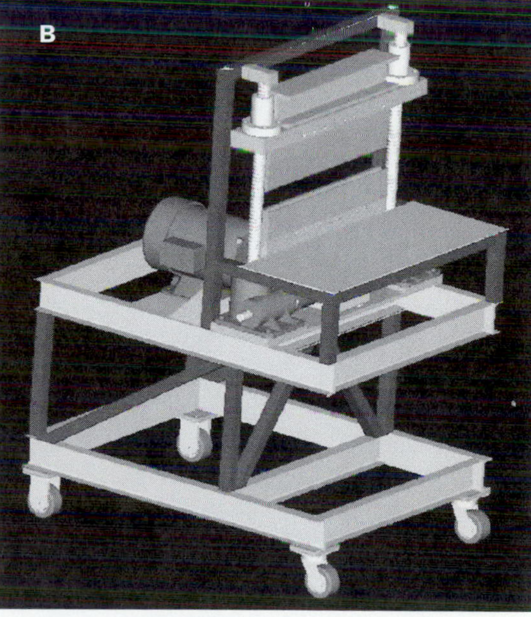

FIGURE 6.37
The parts that make up this brake press were modeled individually as solids, then combined into a single assembly model to represent the finished product. The exploded view in (a) shows the individual parts clearly; the assembled view in (b) shows how the parts will fit together.

TABLE 6.1

Characteristics of Modeling Methods

Characteristic	Paper Drawings	2D Wire	Physical Model
Visual	2D views require interpretation and may be less understandable to a non-engineer.	2D views require interpretation and may be less understandable to a non-engineer.	Similar to actual object. Can be seen from all angles. May be smaller scale, unrealistic material.
Understandable	May require experience interpreting multiview drawing technique to be understood.	May require experience interpreting multiview drawing technique and descriptive geometry to be understood.	Can interact in a way similar to real object to aid in understanding. May be different scale, material than real object.
Flexible	Changes require erasure; difficult to accommodate large changes without redrawing.	Editing capabilities and layers make models more flexible than paper drawings.	Can be modified to a degree, but substantial changes may require a new model.
Cost effective	Low equipment costs, but man-hours may be greater than other methods. Cost effective for visualization and for documenting simple parts for manufacture. May be necessary if electronic formats cannot be read.	Useful for projects where the information gained from a 3D model outweighs the cost of capturing it, as in civil mapping and electrical circuit and distribution design.	Varies with the information contained in the model (where more information is usually more expensive) and the cost of manufacturing errors that could be prevented by the model.
Measurable	Not suited to be measured from.	Because objects are drawn full-size, measurements may be taken from the model; knowledge of descriptive geometry required to derive information not shown in standard views.	Depends on how closely the model matches the materials and processes to be used to create the actual product.
Accurate	Requires proper dimensions to reflect design accurately.	Offers a high degree of accuracy, when orthographic drawing techniques are understood.	Depends on how closely the model matches the materials and processes to be used to create the actual product.
Robust	May require multiple views to fully define the shape of an object.	More robust than paper drawings and can include links to large amounts of data; lacks information about volume found in 3D models.	Can include as much information as the actual product.
Application	Effective for sketches and quickly conveying the design of simple parts for manufacturing.	Useful for largely 2D information such as maps, layout drawings, and electrical circuits.	Good for testing fit of parts, interaction with other products and people, and analyzing aspects that cannot be simulated.

3D Wireframe	Surface	3D Solid
Can be viewed from any direction and used to create standard 2D views; lack of surfaces may make views ambiguous.	Usually includes lighting and background options that can be used to create photo-realistic images of the model.	Shaded and rendered views present a realistic view of the object; offers automated 2D view generation; with most equipment only 2D view is available on monitor, part must be rotated to picture clearly.
Requires viewer to mentally add surfaces to see the 3D shape.	Shaded views are easily understood by the viewer.	Shaded views are easily understood by the viewer.
Editable, but changing individual objects more time-consuming than in other methods.	Each vertex is editable, but ease of changes depends on method used to create and store the surface.	CSG or hybrid models offer the most flexibility; can edit at the level of the solid object instead of individual geometric entity.
Requires less computing power and software generally lower priced; good for modeling geometrically simple shapes.	Most cost effective for modeling irregular surfaces that must be conveyed to manufacture with a high degree of accuracy; photo-realistic displays can offset modeling costs by adding value in marketing.	Completeness of the solid model cost effective when model can be used for multiple purposes, such as testing, presentation, and CAM.
Full-size entities are measurable without auxiliary views; can only measure between wireframe entities, however.	Can be used to calculate surface area.	Can be used to measure not just size, but also weight, mass, and other physical properties.
Offers a high degree of accuracy for the entities represented.	Depends on method used to store model data; smoothed surfaces generally more accurate than mesh surfaces.	Accurate representations are more accurate than faceted; high degree of accuracy possible.
Model stores only vertices and edges; lacks surface or volume information; may be ambiguous as to voids or holes.	Fully defines the surfaces of a part; lacks volume information.	Includes volume as well as surfaces and vertices; can eliminate need for physical model.
Good for applications where relationships in 3D space must be modeled, simple geometric shapes are sufficient for the objects to be modeled, and realistic views are not required.	Good for designs that include irregular or free-flowing surfaces and for those where the appearance of the product is a critical design criterion.	Good for applications where fit between parts and other engineering properties need to be evaluated before manufacture.

As you have seen, 3D modeling packages that are strong in surface modeling are needed to model the geometry of smoothly contoured features such as the ergonomic mouse shown earlier in the chapter. Piping systems and other structures of simple geometric shapes can sometimes be modeled very effectively using 3D wireframe modeling. Many inexpensive CAD packages support this type of modeling. To create a model that can be used to check parts for fit and interference; calculate the weight of a final assembly from the many individual parts; generate a rendered view of the assembly that can be used in manufacturing, marketing, service and repair; and drive the computer-aided machinery that will create the parts themselves, however, a solid modeling package is required. The same assembly model created using parametric modeling methods would have the added advantage of updating automatically when key dimensions change.

Whether you choose your own modeling package or learn to use the tools provided by your employer, you should be aware of the strengths and limitations of the software so you can use it most effectively. Table 6.2 shows a comparison of the types of modeling features available in some popular CAD packages that were considered medium priced ($5,000–$10,000 per seat) in early 1999.

TABLE 6.2

3D Modeling Software	Company Name	Wireframe	Surface	Solid Modeling Engine	Parametric
Alias Studio	Alias/Wavefront	yes	yes	n/a	no
AutoCAD Release 14	Autodesk	yes	yes	ACIS	no
CADKEY 97	Baystate Technologies	yes	no	ACIS	no
CATIA	Dassault Systems	yes	yes	Proprietary	yes
I-DEAS	SDRC	yes	yes	Proprietary	yes
MicroStation Modeler	Bentley Systems	yes	yes	Parasolid	yes
Mechanical Desktop	Autodesk	yes	yes	ACIS	yes
Pro/Engineer	Parametric Technology Corp.	yes	yes	Proprietary	yes
Solid Edge	EDS Unigraphics Solutions	yes	no	Parasolid	yes
SolidWorks 98	SolidWorks Corporation	yes	yes	Parasolid	yes
Ironcad	Visionary Design System	no	no	ACIS	yes
Unigraphics	EDS Unigraphics Solutions	yes	yes	Parasolid	yes

INDUSTRY CASE

Mark Gerisch, M&L Auto Specialists, Inc.

3D Modeling Methods: Not Just a Pretty Picture

When Mark Gerisch decided to make the Romulus Predator a true supercar—an exciting, aerodynamically sleek vehicle with a top speed in excess of 200 mph—he knew he needed to update the traditional methods he had been using to build replicas of classic cars. His vision of a totally digital design process caught the attention of Hewlett-Packard and made the Predator the center of a partnership that epitomized and validated the concept of "virtual product development." "I like to see how other people do business. I enjoy watching mistakes and triumphs and what caused both, then try to evaluate what is the best method and how I can use that in my business," said Gerisch of his interest in CAD/CAM systems. "The Predator was an opportunity to validate our designs in a digital format that could then be transferred over the Net or to subproviders. The car hooked people in as a metaphor for the future of design. Today's students are going to be working side by side with people in South America, Israel, and the Far East, and they're all going to be linked into the same model and work with it through a 3D viewer."

Gerisch's ultimate goal was to use 3D modeling to completely visualize the design of the vehicle and all of its parts before building a physical prototype, then to use this expertise to perfect and build prototypes for auto manufacturers. In the course of realizing his vision, Gerisch and the Predator team used a variety of models and software tools from ICEM Technologies, Unigraphics, Engineering Animation Inc., the MacNeal-Schwendler Corporation, Adobe, Rapidline Industries, and NewTek, all running on hardware provided by Hewlett-Packard.

The Predator started as a collaboration between Gerisch and Ron Sims, a GM stylist, that resulted in a clay model of the Predator body. Wind tunnel testing showed the design to be aerodynamically viable, so the work of refining the shape of the car began (see Fig. 6.38). Instead of building multiple physical models to visualize changes, the clay model was digitized and transformed into a surface model using ICEM Surf.

To digitize the model, the team used a coordinate measuring machine to capture 1700 points in 3D space that defined the surface of half of the vehicle. In ICEM Surf, these data points were mirrored over to define the other half of the vehicle, then used to generate a series of spline curves. Surface patches were then created with the splines to define the entire surface of the car (see Fig. 6.39). Because they had a model to start with, digitizing its form made it easy to capture the shape in the surface model. An alternative approach would have been to scan in photos or sketches of front, side, top, and back views of the car, then orient these sketches in 3D space on a set of coordinate planes as you would interlock pieces of paper to make a 3D model. By situating the side view along the centerline and perpendicular to

(continued on page 242)

FIGURE 6.38
The original clay model for the Predator was used to evaluate the aerodynamics of its lines in a wind tunnel at the University of Michigan. The model generated almost 2000 pounds of downforce at 200 mph.

INDUSTRY CASE

FIGURE 6.39
The surface model of the Predator was made up of more than one hundred surface patches, joined together into a single smooth surface model by the ICEM Surf software. The thin lines on the body in this figure illustrate the way the model is broken into the smaller sections, or patches.

the front view, the sketches would provide a profile that could be used to construct splines, as was done with the digitized data. This would require more modeling effort to capture the actual shape of the car, but not as much as starting with a blank screen and modeling the surfaces from scratch.

The surface model made it possible to make changes to the shape of the body and see the results immediately. The design of the roof scoop and the shape of the roof of the car were two areas that developed significantly from the original model of the car. Although dissatisfied with them, Gerisch wasn't able to say immediately how they should change. By pulling on the surface model and seeing how it changed the overall shape of the roof, he was able to see different options without building another physical model. Because the lines of the car are so important, even small changes to the design have to be evaluated for the way they change the reflective surfaces. The surface modeling software not only handled the changes by keeping the transitions between patches smooth, it also produced photorealistic images on screen that made it possible to see the way the exterior, or class A, surfaces would reflect light. "But the data is not just a pretty picture," Gerisch is quick to point out. "The surface is a math-driven object on the screen. All of the raw data has been transformed into data that can be used to drive a numerically controlled machine to cut parts for manufacture." In fact, the model was used to cut the "formers" used to shape the body of the car and to generate code for CNC milling machines that would shape the parts that needed to fit against the interior, or class B, surfaces of the body of the vehicle. (See Fig. 6.40.)

The surface data was also translated and imported into Unigraphics Solutions's solid modeling software, where the rest of the components were modeled and assembled. Because both systems use NURBS surfaces, the fully defined surface was incorporated into the solid assembly without data loss. This model became the digital representation of the car, with all its components modeled, even though not every model was 100% accurate. Gerisch used standard components whenever possible in the Predator. For these components, models were created to represent the dimensions and mass of the parts that would be important to evaluating their fit and clearance in the car (see Fig. 6.41).

The translators available with the Unigraphics software made it possible to use data from the model with Engineering Animation's software as well as with the analysis software from the MacNeal-Schwendler Corporation. EAI's software allowed the designers to see realistic views of the complete assembly and to fly through the assembly in real time (see Fig. 6.42).

FIGURE 6.40
This wooden shaping buck was cut by numerically controlled milling machines from the data stored in the digital model of the car. It was used to create molds for the first physical prototype of the car's body. The NC data will be used to mill the molds themselves for production of the Predator.

EAI's VisMockup was used to check for fit and interference. This aspect of the project was particularly appealing to Gerisch, since it meant the difference between building a physical prototype that would work and one that wouldn't. "Do we need the computer system to actually build vehicles? No, because it's people who make the vehicles. But the computer system allows you to first validate different iterations of an idea easily and affordably. Second, it allows you to get in there and run a kinematic program on an idea to make sure that it is actually going to function. When you build the car, there might still be a few small mistakes, but nothing as horrendous as doors that open up and whack into the side of the vehicle, or glass that doesn't fit, or seats that won't fit into the cockpit as they should."

The Unigraphics model was similarly imported into MSC/PATRAN and MSC/NASTRAN for structural analysis and testing. These analysis packages helped the team remove weight without loss of structural strength and calculate stresses and vibration.

Although the CAD model was the hub of the Predator project, Gerisch points out that being able to move data from one package to another is the key to virtual prototyping. "Don't expect one package to do it all. The competitive edge for CAD software companies is giving you the ability to bring in your files through translators without loss of data." Understanding the right tool for the job and being open to new approaches is Gerisch's advice to anyone seeking to benefit from new design technologies—as he has from the modeling tools brought to bear on the Predator.

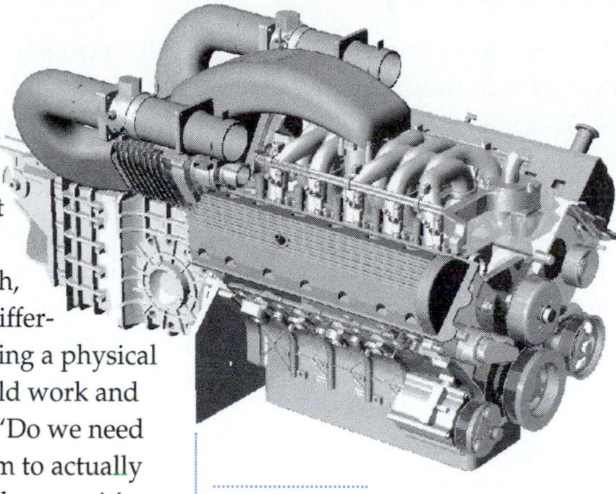

FIGURE 6.41
All of the components of the Predator were modeled as solids and integrated into a Unigraphics assembly model, whether they were new components or off-the-shelf components that needed to be accounted for in the assembly of the complete vehicle. A dual supercharger assembly was devised to generate over 500 hp from the 6.8L Ford Triton V10 engine to enable the Predator to reach speeds over 200 mph.

(All figures courtesy of M&L Auto Specialists, Inc., designers of the Romulus Predator.)

Mark Gerisch is owner and president of M&L Auto Specialists in Two Rivers, Wisconsin. The company has evolved from an exotic autobody restoration shop into a high-tech custom automotive prototyping company. The transition to 3D modeling made through the Romulus Predator has enabled the firm to take CAD files from major automakers as the foundation for the prototypes they build. Gerisch was also the driving force behind Constant Velocity, a new firm that won the SEMA "Best Engineered New Product" award in 1997 for a totally new approach to an independent suspension system—an all-inclusive unit that requires no cutting, welding, or drilling, but bolts into the vehicle as a single unit.

FIGURE 6.42
This view of the Predator was generated by EAI Visproducts visualization software using the CAD files from Unigraphics. The body of the car is rendered as if it were translucent material to reveal the interior structure of the car.

Skills Summary

Now that you have completed this chapter you should be able to list the qualities that models have that make them useful in the design process. You should be able to describe four CAD modeling methods that you might use to refine your design ideas. Not all modeling methods capture the same amount of information about the parts or device. You should be able to select the best method to use for a particular design modeling problem. Exercises 1–4 allow you to test your understanding of the modeling process. Exercises 5 and 6 ask you to apply a particular CAD modeling process to a design problem. Exercises 7–48 provide practice with your modeling package.

EXERCISES

1. You are a member of a team designing an automatic door-opening device so wheelchair users can enter and exit buildings on campus. (Reference: Chapter 2, Exercise 5.) Which of the modeling methods available to your design team are appropriate in each of the following situations:
   - **a.** Preliminary design review with your immediate engineering (technical) supervisor.
   - **b.** Discussions of installation details with the campus architectural staff.
   - **c.** Meeting with the Disabled Users Committee, a group of nontechnical advocates for accessibility by the disabled.
   - **d.** Meeting with machine shop technicians who will fabricate your prototype.
2. Which modeling method contains the most information about a design? Identify a situation in which this is not the best model to use.
3. Recent developments in *Global Positioning System* (GPS) technology have resulted in small, relatively inexpensive, handheld GPS receivers such as Garmin Corporation's Model GPSIII shown at left. The antenna of the device, which connects to the back of the case with a coaxial connector, must be afforded a clear view of the sky for the system to "lock on" to satellites in the GPS constellation. The antenna's coaxial connector (type BNC) also permits the use of different antenna types or relocation of the antenna with respect to the handheld display module.

   A need exists to design a mount to locate the antenna at a remote location, such as on top of a backpack or clipped to a hat, in order to protect the electronics and display from damage while providing a clear sky view for the antenna. What types of models could be used to develop this remote antenna mount? At what stage of the process would each type of model be appropriate?

(Figure reprinted with permission of Garmin Corporation.)

4. For the situations described below, choose a modeling method and explain why you would use it.
   a. Designing a prosthetic limb.
   b. Routing a piping system for a building.
   c. Designing a one of a kind machined part.
   d. Verifying clearances for an electrical substation located near an airfield.
   e. Mapping an oil reservoir.

5. Create a 2D CAD layout of the antenna mount introduced in Exercise 3. Use the approximate dimensions shown in the orthographic projection drawing (below) to define the shape of the mount. List some of the design constraints that you face as you consider alternative designs.

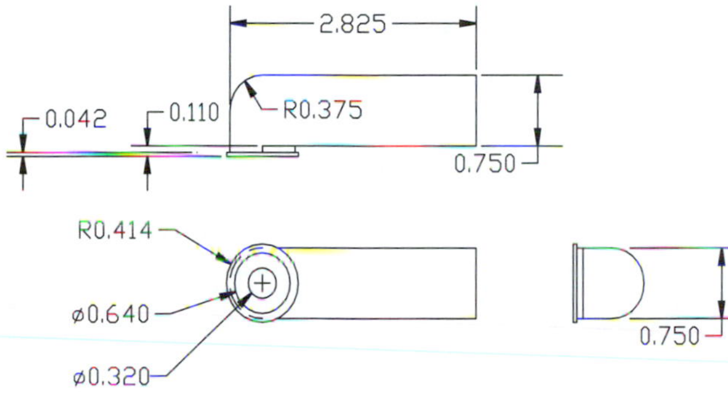

(Figure reprinted with permission of Garmin Corporation.)

6. Using a CAD system, create a solid model of the antenna shown in Exercise 5.

Exercises 7–10: Use your modeling package to create solid, surface, or 3D wireframe models of the items in Chapter 5, Exercise 4 a–d.

Exercises 11–38: Use your modeling package to create solid, surface, or 3D wireframe models of the objects depicted in Chapter 4, Exercises 7–34. Use the units indicated to model the objects full size.

Exercises 39–44: Use a solid or surface modeling package to revolve the shapes in Chapter 5, Exercises 5–10 about the axes shown.

Exercises 45–48: Use a solid or surface modeling package to extrude the shapes in Chapter 5, Exercises 11–14.

DESIGNER'S NOTEBOOK

This sketch from Albert Brown's design notebook shows the relationship between the centerline of the lead screw and the center of gravity of the stage. The critical dimensions shown in the sketch need to be captured in the parametric model in order to reflect the design intent for this device. (Courtesy of Albert W. Brown, Jr., Affymax Research Institute.)

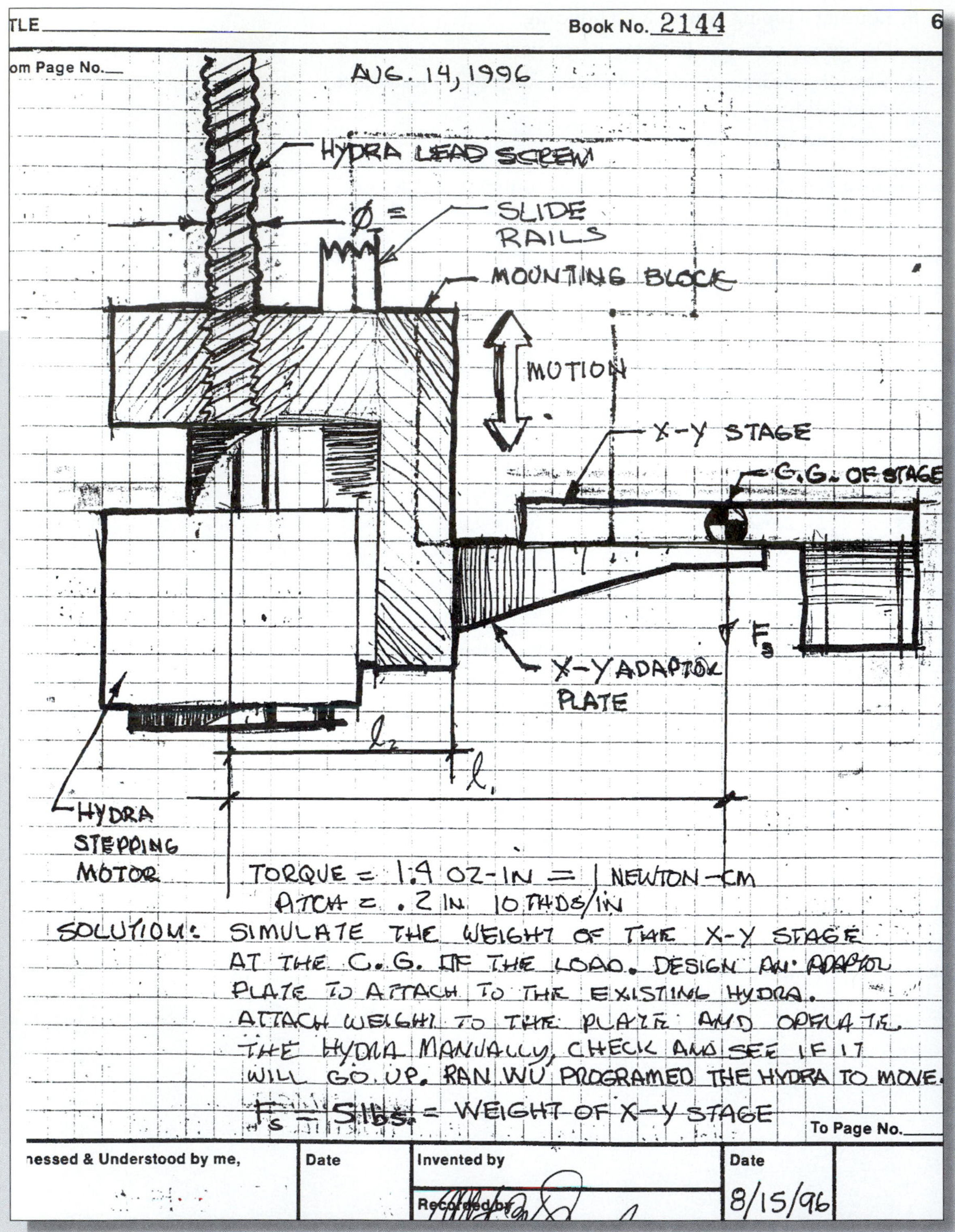

CHAPTER 7

Parametric Modeling and Design

Objectives

When you have completed this chapter you will be able to

- describe how a parametric modeler differs from a solid modeler;
- define design intent and how design intent is reflected in a parametric model;
- define the basic functions of a parametric modeler;
- identify features and the role of a base feature;
- identify constraint relationships and parametric dimensions.

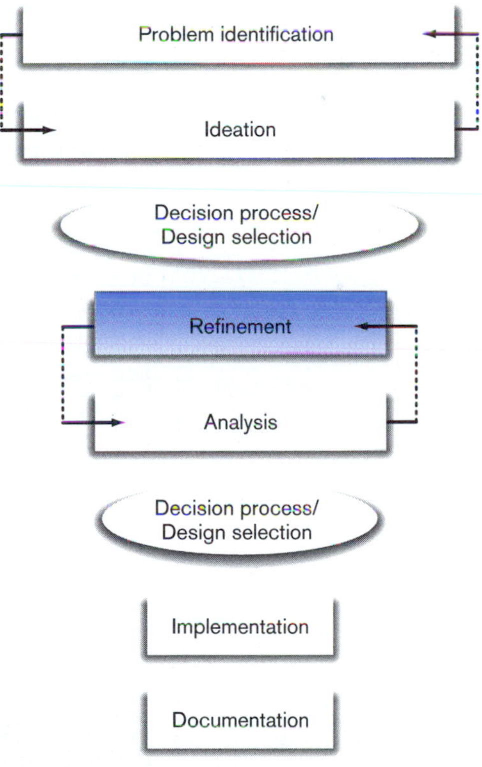

In Chapter 6 you learned how various types of 3D models can be used to refine your engineering designs and how CAD modeling software differs in the kinds of information it captures in the design database. In general, 3D models provide more information than 2D models, and 3D solid models provide more information than 3D wireframe or surface models. Parametric solid models add even more to the design database: intelligence.

In a *parametric model*, the different parts of the model "know" how they relate to one another. Their dimensions and geometric properties are defined in relation to each other, through *parameters*. When one element changes, so do any other elements that are defined in terms of that element. Through the relationships built into the model, the designer uses the ability of the software to automate tasks associated with refining the model.

Advantages of Parametric Modeling

Parametric modeling has several advantages for engineering design. One of these is that models can be updated by changing the sizes and relationships that define the model as the design changes. In traditional solid modeling methods, changing the size of one feature may require several others to be changed, or the model may need to be totally re-created. Because of the iterative nature of the design process, a design is repeatedly modified as it is refined. To be useful, models need to reflect changes that occur throughout the design process. With parametric modeling, you can update models to reflect changes in the design more quickly than with other methods. This responsiveness can result in more cycles during the refinement stage (or more cycles in less time) and, ultimately, a better design.

Nep'tune Sea Technology, Ltd. made the decision to switch from a 2D design method to a parametric 3D modeler, SolidWorks, so it could get its products to market faster while preserving the iterations needed for complex, accurate designs. Nep'tune is a Finnish engineering company that specializes in the design of vehicles for subsea applications—passenger and research submarines. To meet increasingly rigorous design, performance, and safety standards, the company found that 2D methods made it hard to complete the desired number of design iterations and still meet product development deadlines. Its first project using the parametric 3D modeler, a floating bridge for ferries, took only eighteen weeks to produce (see Fig. 7.1).

The company enjoyed two other benefits of parametric modeling. First, the ease with which parametric models can be updated also makes it possible to create families of designs. Nep'tune's floating bridge allows cars to pass from the ferry to land and vice versa during loading and unloading. Because each landing site is slightly different, the main dimensions of a given bridge vary slightly from one site to another. The bridge manufacturer that hired Nep'tune had several designs previously created for specific landing sites. To

FIGURE 7.1
Nep'tune's parametric model of the floating bridge for ferries was modified to create a family of designs for the variations in ferry landings. (Courtesy of Nep'tune Sea Technology, Ltd.)

eliminate confusion in manufacturing, the customer wanted a design that could be updated for new bridges. Nep'tune proposed to design one bridge, then produce the others by changing the length and width factors. The parametric model of the bridge it designed made it easy to change the pertinent dimensions and have the software update the related parts to a new size. The investment in the bridge model was used to refine the original bridge design, then used again as a starting point for variations of differing sizes in the product family.

Nep'tune also enjoyed the modeling software's built-in capability to analyze mass properties. The weight and volume data for a floating bridge need to be evaluated during design so that the resulting bridge floats at the correct level. Calculating the weight of a floating bridge design would have taken a week or two using Nep'tune's previous methods, but the built-in capabilities of the solid modeling software made it possible to monitor this data throughout the design process (see Fig. 7.2). The ease with which the model could be modified made it possible to analyze the model, make changes, and analyze the model again. The responsiveness of the parametric model makes it possible for analysis to occur earlier in the design process, when it can be used to optimize the design.

While the ability to easily update the model is a clear advantage, parametric modeling can also improve designs by focusing the modeler on the design intent for the product. In order to create a parametric model that will update successfully when changes are made, the model—specifically, the relationships that build intelligence into the model—must reflect the design intent for the system or device. The extra attention and upfront planning required to capture design intent in the model may make designers consider the function and purpose of their designs more carefully, which in turn results in better designs.

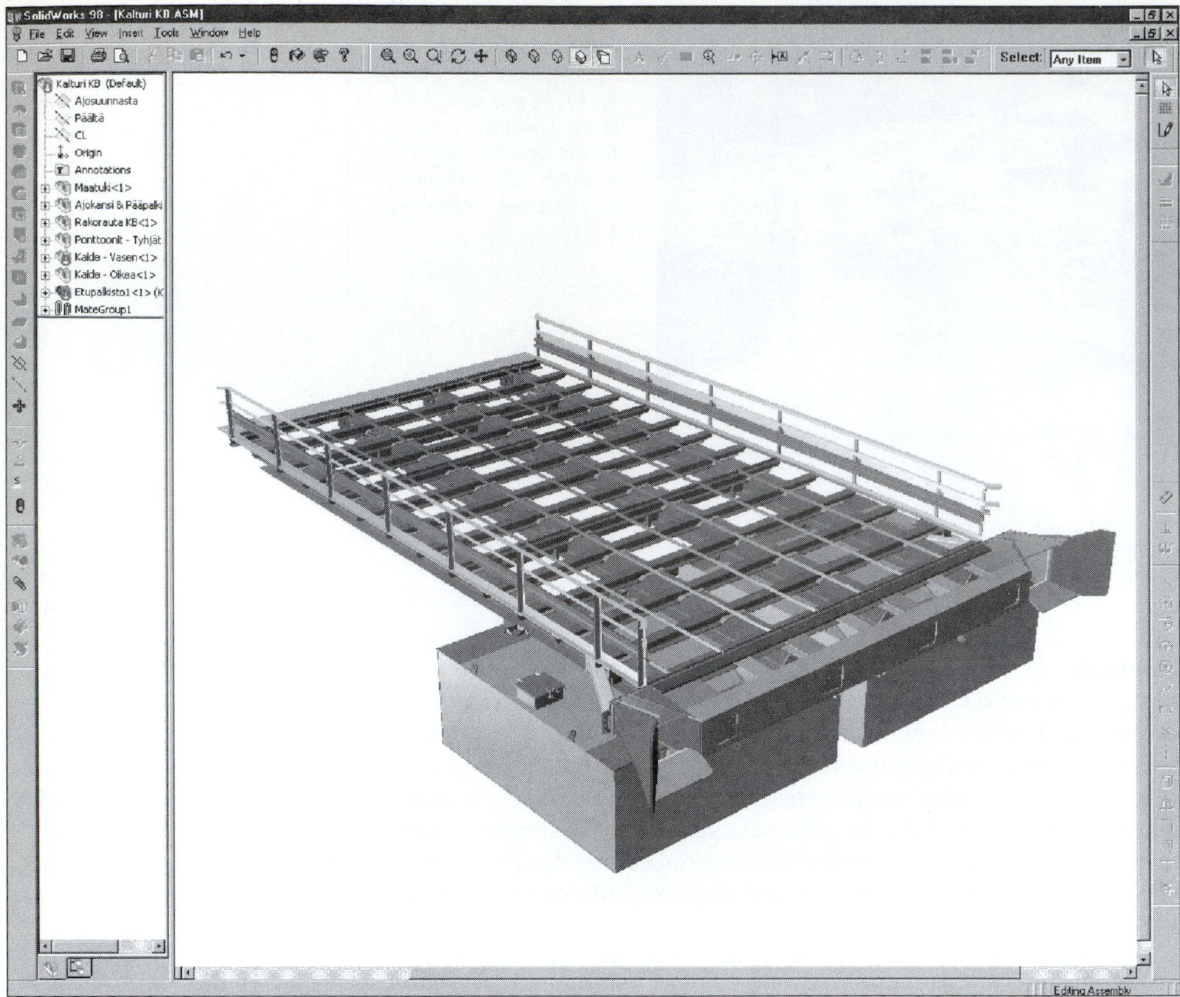

FIGURE 7.2
Nep'tune's bridge model, created in SolidWorks, involved more than seventeen hundred parts. (Courtesy of Nep'tune Sea Technology, Ltd.)

In this chapter, you will learn how to create parametric solid models that reflect the design intent, and will maintain your design intent when you want to change or update the model.

Parameters Control the Geometry

In a parametric model, an object's features are defined by the sizes and relationships that the designer builds into the model. These parameters are stored in the model and used to generate the part. Two basic kinds of parameters are used to drive the model geometry. *Size parameters* are the dimension values stored with the model. *Geometric parameters,* or constraints, define and maintain the geometric properties of an object, such as tangency, verticality, and so on.

A parameter is a named quantity whose value can change depending on the circumstances of its application. Because the value of a parameter can change, it is similar to a variable. Like a variable, a parameter can be used to define other parameters. If the width of a part is always twice the length, the width parameter would be defined as "2 * length," where *length* is the name of the length dimension.

Unlike a variable, however, a parameter is never abstract: it always has a value assigned to it. In the example above, the parameter length would be defined by a value, such as 3.5. This value would then determine the value stored in the width parameter, in this case, 7. Depending on the circumstances, the length parameter might be changed to hold the value 5. The relationships defined in the model would then cause the width to automatically update to the new value of "2 * length," or 10.

Figure 7.3 shows a part that has been modeled in Pro/Engineer, a 3D parametric design package. The parameters driving the model's geometry are indicated on the drawing. When the parameter value for the length of the part was changed, the part updated automatically to the part shown in part (b) of the figure.

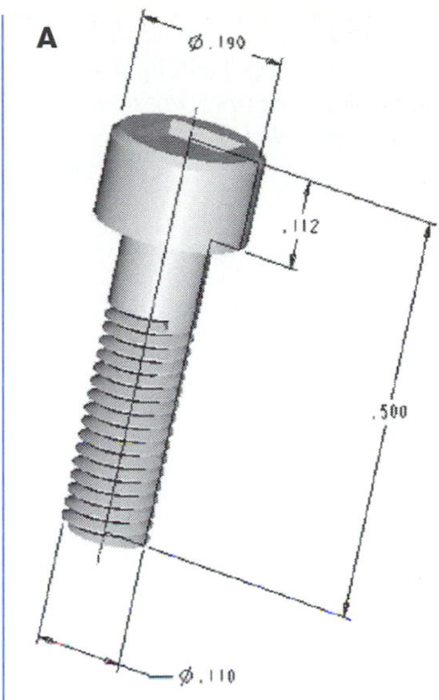

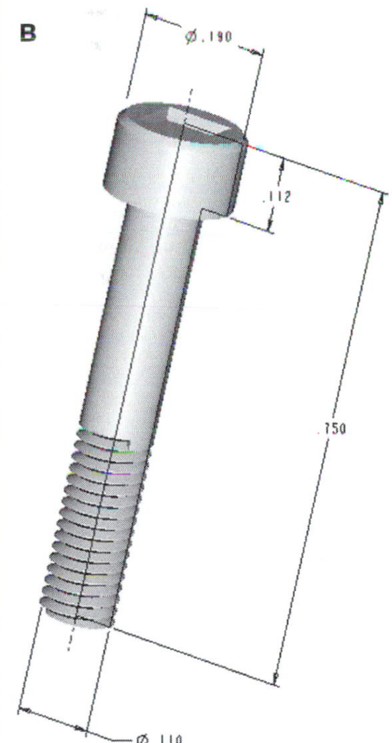

FIGURE 7.3
When the length dimension for the bolt in (a) was lengthened to .750, the part updated to reflect the new value for the parameter. Notice that the threaded length remained the same and the bolt head remained attached after the update. The parametric modeler maintained the size and geometric relationships in the model when the length dimension changed.

Feature-Based Modeling

Parametric modeling is also called *feature-based modeling* because its models are combinations of features. You may use commands to add, subtract, and intersect features to form the definition of the solid parametric model as you would with other solid modelers that allow Boolean operations. In this way, creating a parametric model is similar to creating any other solid model. However, in parametric modeling the resulting part is not just the stored CSG or BREP definition of the solid, but is derived from the dimensions and constraints that define the geometry of the features. Each time a change is made, the part is re-created from these definitions. Individual features and their relationships to one another make up the parametric model.

A *feature* is the basic unit of a parametric solid model. Each feature has properties that define it. When you create a feature, you specify the geometric constraints that apply to it, then specify the size parameters. The modeler stores these properties and uses them to generate the feature. If an element of the feature, or a related part of the model, changes, the modeling software regenerates the feature in accordance with the defining properties assigned to it. For example, an edge that is defined to be tangent to an arc will move to preserve the tangency constraint if the size of the arc is changed.

Some features, such as holes and fillets, are predefined features of parametric modelers with additional properties that can be maintained as a model updates. If a hole is defined as through (it goes all the way through the part feature), for example, and the depth of the surface it is located in increases, the hole will be extended so it remains a through hole.

When you look at the design sketch for the fixed height shaft support shown in Figure 7.4(a), can you identify the features that make up its shape? Figure 7.4(b) shows the major cylinder, hole, plate, end plate, slot, and rounds that form the model of the shaft support.

Being able to easily update models to reflect changes in the design is a key ability of parametric modelers. Compare this with

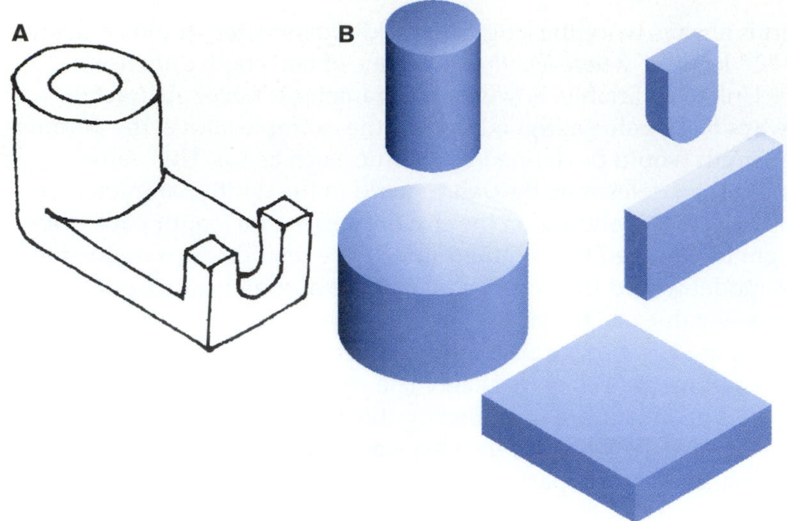

FIGURE 7.4
The features that make up the support shown in (a) are shown individually in (b).

what you learned about other solid modeling techniques in Chapter 6. If you're using a nonparametric solid modeling program to change the size of a feature, you may have to re-create the feature to that size. This can result in considerable time and effort, especially for interrelated features.

Consider the hole in the shaft support. To make it smaller in a BREP solid modeler, you could union a new solid to fill in the hole, then re-create the hole by subtracting a smaller cylinder. You can perform operations only on what is stored in the database. In a BREP model, the surface boundaries, edges, and vertices are stored to define the model.

In a CSG modeler, changing the size of the hole is a bit easier because the primitive you used to create the hole is still stored in the database. You can return to the primitive, change its size, then update the model.

To change the size of a parametrically modeled solid feature, however, you just type in the new value for the hole diameter, then issue the command to regenerate or update the model. The database stores the dimensions and constraints used to generate the model geometry and uses these to re-create the feature each time the model updates. For this reason, many parametric modeling programs are referred to as *history-based modeling programs*. To change a model, you return to the stage at which you added a certain feature, then change the parameters you specified (see Fig. 7.5 and Fig. 7.6).

In one way, parametric modeling software is similar to solid modeling software that uses CSG techniques: in both you can go back and change the size of primitive features. But parametric modeling software goes beyond this to capture the relationships between features. When one feature is changed, any other features that are related to that feature will change with it. Using the shaft support example again, the relationship between the major cylinder and the hole could be defined so the hole's diameter would always be 1 inch smaller than the cylinder's diameter. A change to the size of the cylinder would cause the hole's diameter to change, too, as

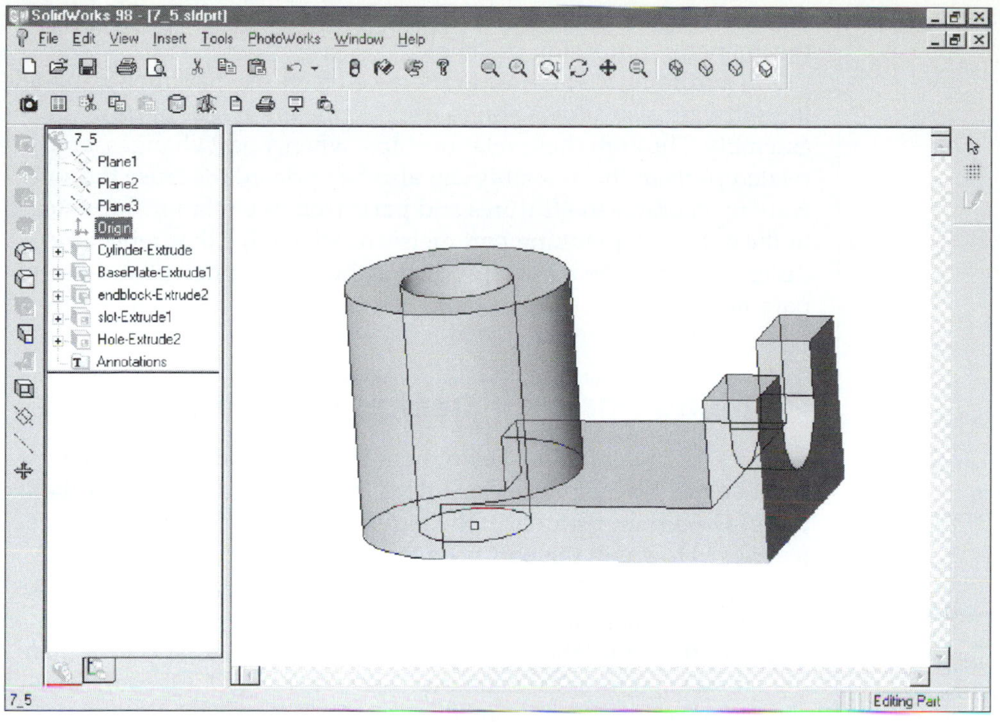

FIGURE 7.5
This illustration of the shaft support model shows how the part history is displayed in SolidWorks. The window at the left side of the screen lists the features in the order in which they were created. This window can be used to select any feature for editing. Notice that the icon before the feature in the history list indicates the operation used to create the feature. Three different extrusions were used here; the cylinder was extruded, the next two features were extruded and unioned to an existing feature, and the last two were extruded and subtracted from an existing feature.

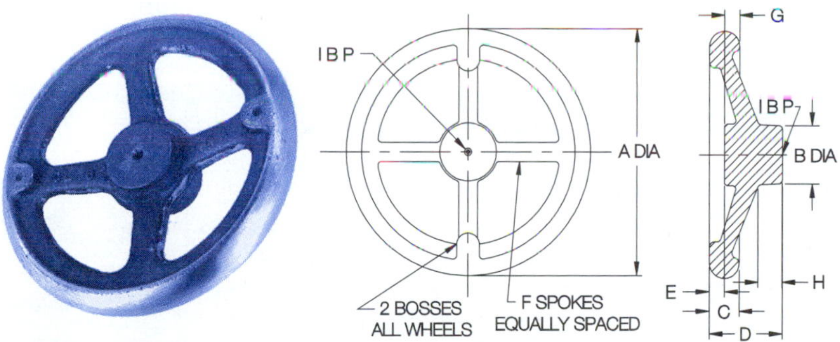

FIGURE 7.6
A parametric model is a good choice for catalog parts that may be ordered in a range of sizes. The dimensions of the hand wheel shown here can be changed to fit each of the sizes listed in the table and the model updated to the new size. (Courtesy of Carr Lane Manufacturing Co. and Earl E. Walker.)

ANGLED SPOKES

<table>
<tr><th>PART NO.
CAST IRON</th><th>A
DIA</th><th>B
DIA</th><th>C</th><th>D</th><th>E</th><th>F</th><th>G</th><th>H</th></tr>
<tr><td>CL-4-HWAF</td><td>4</td><td>1-1/4</td><td>5/8</td><td>1-11/16</td><td>1/8</td><td rowspan="4">4</td><td>9/32</td><td rowspan="2">3/4</td></tr>
<tr><td>CL-5-HWAF</td><td>5</td><td>1-1/2</td><td rowspan="2">3/4</td><td>1-7/8</td><td>3/8</td><td>5/16</td></tr>
<tr><td>CL-6-HWAF</td><td>6</td><td>1-5/8</td><td>2</td><td>13/32</td><td rowspan="2">3/8</td><td rowspan="2">13/16</td></tr>
<tr><td>CL-8-HWAF</td><td>8</td><td>1-7/8</td><td rowspan="4">1</td><td>2-7/16</td><td>1/2</td></tr>
<tr><td>CL-10-HWAF</td><td>10</td><td>2-1/4</td><td>3-1/32</td><td>11/16</td><td rowspan="3">8</td><td rowspan="2">7/16</td><td>1-1/16</td></tr>
<tr><td>CL-12-HWAF</td><td>12</td><td rowspan="2">3</td><td>4-3/32</td><td>1-1/16</td><td>1-3/4</td></tr>
<tr><td>CL-14-HWAF</td><td>14</td><td>5-1/8</td><td>1-7/32</td><td>9/16</td><td>2-3/16</td></tr>
</table>

the modeling software will preserve the size relationship defined in the diameter dimension parameter.

In addition to relationships between features, parametric modeling software also allows you to use parameters across parts in an assembly. Through these relationships, when a part changes, any related parts in the assembly can also be updated. Because the software regenerates the features and parts from the relationships stored in the database, planning parametric relationships that reflect the design intent of the part or product is the key to efficient and useful parametric models.

Planning Parts for Design Flexibility

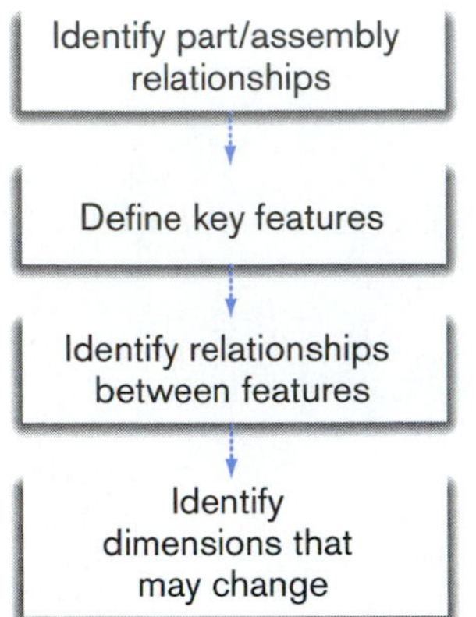

FIGURE 7.7
Four key considerations when planning a parametric model.

Design intent refers to the key dimensions and relationships that must be met by the part. Where does the part need to fit into other parts? Do certain features of the part need to be aligned with other features? How will the part need to change as the design evolves? By thinking about these relationships, you can organize the features you create so that the overall structure of the part and its relationships are the foundation for later features. Four key aspects of planning for design intent are summarized in Figure 7.7. Of course, you can't know everything about a design before you begin, and there will be changes that will require more work than others. But starting your model with the geometric relationships in the design in mind will allow you to benefit from the power of parametric modeling. To see how a parametric model can be created to reflect your design intent for a part, you should first understand some basic things about how most parametric modeling software works.

Parametric modeling software in many ways parallels the design process. To start creating a part, you first create a 2D sketch of the key shape of a feature, such as that shown in Figure 7.8.

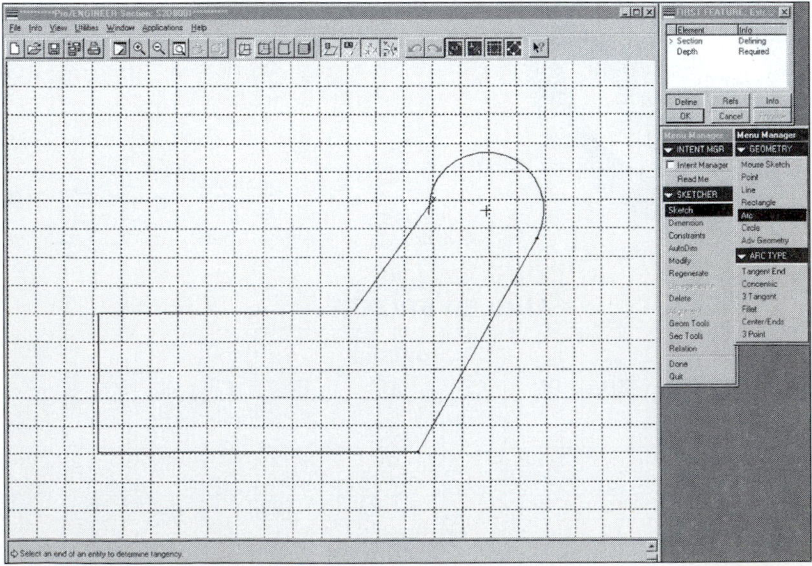

FIGURE 7.8
Parametric modeling software refers to the 2D shape of a feature as a sketch. Like a hand-drawn sketch, it captures the basic geometry of the feature as it would appear in a 2D view.

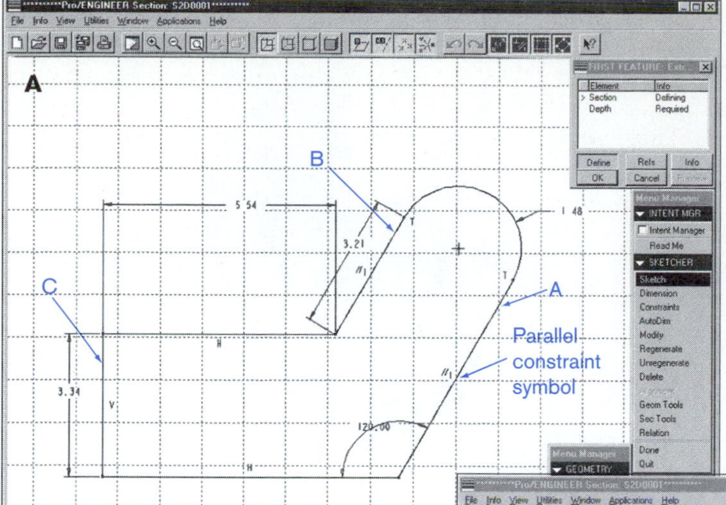

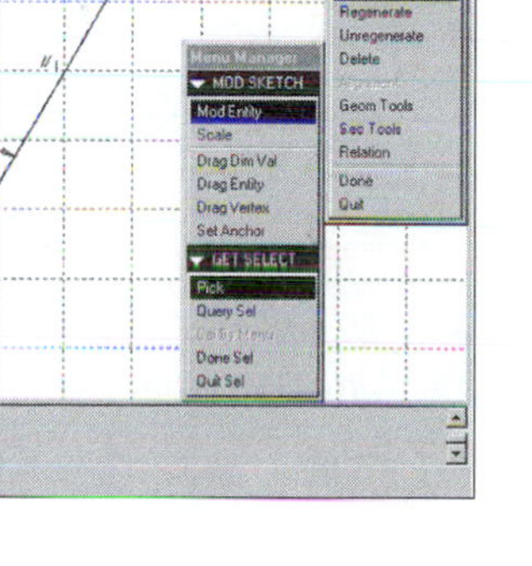

FIGURE 7.9

(a) This view of the sketch shows the constraints that have been applied by the software and by the designer. Lines A and B have been interpreted to be parallel, as indicated by the parallel constraint symbol. The length of line C was sketched so that the software interpreted the dimensional constraint to be 3.34. The designer changed this dimension to 3.5 (the desired length), and the length of the line in the sketch changed to reflect the new parameter as shown in (b).

The sketch may be rough in appearance because the software will apply constraints to define the relationships between the simple 2D geometric elements of your sketch. This process is similar to the way one of your engineering colleagues interprets your hand-drawn sketches. If lines appear to be perpendicular or nearly so, your colleague assumes that you mean them to be perpendicular, without you having to mark the angular dimension between the lines. In a similar way, parametric modelers apply constraints to your rough sketch, so that lines that appear nearly perpendicular, parallel, vertical, horizontal, concentric, or colinear will have a constraint relationship added to them so that they function that way in the model. This process is called "solving" the sketch in some modeling software, as the software is interpreting and assigning dimensions and constraints to generate a 2D "profile" of the feature. Before you generate the actual feature, you can review and change the dimensions and geometric constraints so they are what you intended. Figure 7.9 shows the same geometry once the sketch constraints have been added.

To generate a 3D feature, you then select a command to extrude, revolve, sweep, or blend the profile geometry. Figure 7.10 shows the profile in Figure 7.9 as it would appear after being (a) extruded, (b) swept, (c) revolved to create different types of features from the 2D

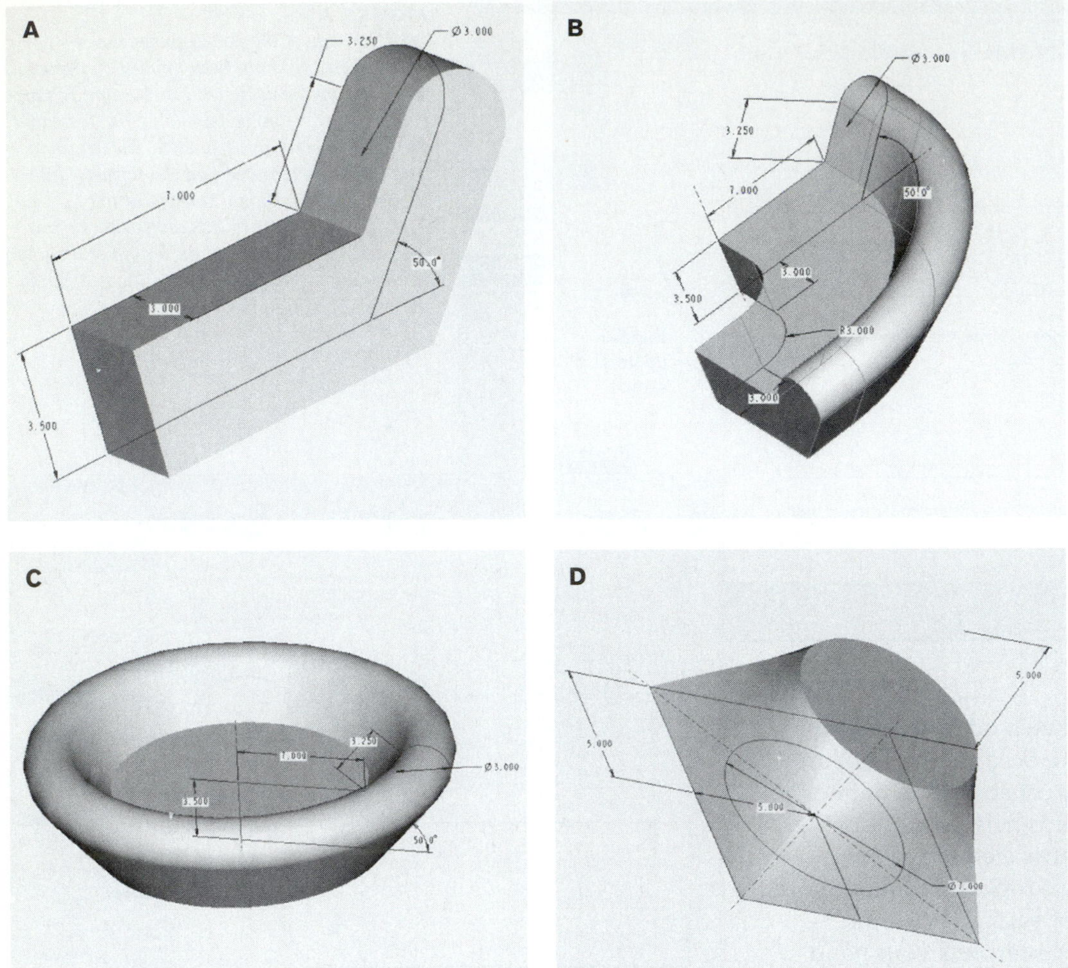

FIGURE 7.10
The sketch from Figure 7.9 can be used to create the features shown in a, b, and c. The feature in (a) was extruded 3 inches. The feature in (b) was swept along a curved path. The feature in (c) was revolved 360° about an axis. The object shown in (d) was created by sketching a square section and a round section which were blended to create the complex shape shown here.

sketch. Figure 7.10(d) shows a feature created by blending two different sketched shapes.

Sketch Constraints

When the parametric modeling software interprets, or solves, your sketch, it is defining the geometric parameters, or constraints, that it will hold constant. To do so, it evaluates your sketch against a set of rules stored in the software. A line, for example, that is drawn within 3° or 4° of being horizontal will be constrained to remain horizontal. Your sketch will be altered so the line you drew is, in fact, horizontal. The software will apply the constraints it needs to exactly define the sketch.

Parametric modeling programs have a command that allows you to show the constraints that are applied to the sketch geometry so you can review them. Table 7.1 illustrates some of the constraints that might be added to the sketch geometry and shows the symbol used to indicate them on screen in Pro/Engineer. Other modelers

TABLE 7.1

Sketch Constraints

Constraint	Rule	Symbol
Horizontal	Lines that are close to horizontal are constrained to be horizontal (usually you can control the number of degrees for this assumption, typically set to 3° or 4°).	H
Vertical	Lines that are close to vertical are constrained to be vertical (usually within 3° or 4°).	V
Equal length	Line segments that appear equal length are constrained to be equal.	L_1, L_2, etc.
Perpendicular	Lines that appear close to 90° apart are constrained to be perpendicular.	⊥
Parallel	Lines that appear close to parallel are constrained to be parallel.	$//_1$
Colinear	Lines that nearly overlap along the same line are assumed to be colinear.	no symbol displayed
Connected	Endpoints of lines that lie close together are assumed to connect.	dot located at intersection
Equal coordinates	Endpoints and centers of arcs or circles that appear to be aligned horizontally or vertically will be constrained to have equal X or equal Y coordinates.	small thick dashes between the points
Tangent entities	Entities that appear nearly tangent are constrained to be tangent.	T
Concentric	Arcs or circles that appear nearly concentric are concentric.	dot at shared center point
Equal radii	Arcs or circles that appear to have similar radii are constrained to have equal radii.	R
Points coincide	Points sketched to nearly coincide with another entity are assumed to coincide.	no symbol displayed

use different symbols and means of displaying the constraints. These constraints may apply to other lines in the sketch or to geometry in an existing feature.

Not all parametric modeling programs apply all of these constraints, nor do they apply them in the same fashion. You should become familiar with the constraints applied by the software you use and determine their effect on the sketch. You also may be able to set the threshold values for various constraints. For example, instead of having lines at 3° be constrained as horizontal lines, you may wish to increase this value to 5°.

Most parametric modeling programs also let you remove or override a constraint that is applied, or specify that the exact geometry is drawn. Some, such as Autodesk's Mechanical Desktop, will also let you manually add a constraint that you want applied. Others, such as versions of Pro/Engineer prior to 19, will not let you

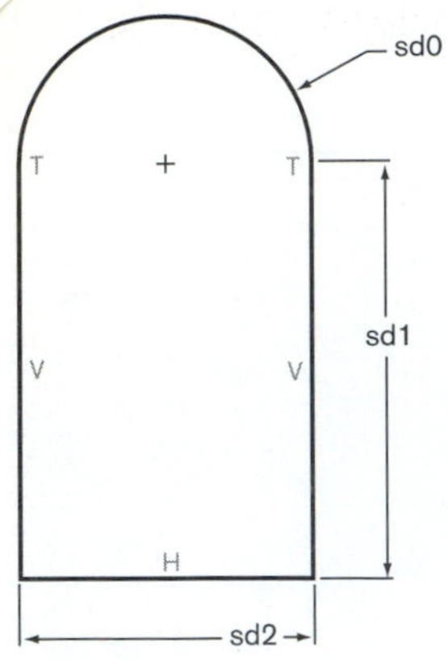

FIGURE 7.11
This sketch is overconstrained.

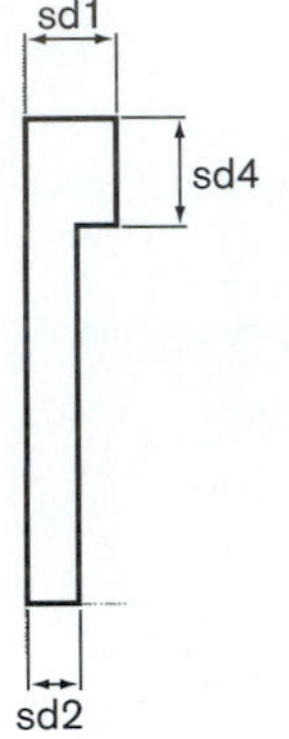

FIGURE 7.12
This sketch is underconstrained.

manually add a constraint. When you cannot add a constraint manually, you can often dimension the sketch to produce the desired geometry. For example, if two lines are too far off in the sketch to be constrained as perpendicular by the software, you could add an angular dimension of 90° between them. With the dimensions driving the geometry to the condition you want, the software may apply the perpendicular constraint. Once the constraint is applied, you can often delete the dimension you added.

It is a good practice generally to let geometric constraints determine much of your sketch geometry, then dimension the sizes and relationships that cannot be determined from constraints. You can use a parametric modeler as you would a solid modeler, using dimensions values to define all of the features, but this would not build the intelligence into the model that allows it to automate much of the work of updating the model.

Overconstrained and Underconstrained Sketches

Before the parametric modeler will create a feature from the sketch, the geometric relationships in the sketch must be unambiguously defined through the constraint relationships and dimensions provided. An overconstrained or overdimensioned sketch is one that has too many things controlling its geometry. An underconstrained or underdimensioned sketch is one that has too few dimensions or constraints and is therefore not fully defined.

The sketch shown in Figure 7.11 is overconstrained. Because the top arc is constrained to be tangent to both vertical lines and its radius is dimensioned, the width of the sketch is defined. The horizontal dimension shown at the bottom defines this width again. If both the arc radius dimension and the overall width dimension were allowed in the model and one of them changed, it would be difficult or impossible for the part to update correctly. If the overall width changed and the tangent constraint or arc radius did not, the sketch geometry would be impossible.

The part shown in Figure 7.12 is underdimensioned and therefore underconstrained. It lacks a height dimension controlling its vertical size. Without this dimension the sketch is not fully defined. Some parametric modeling software will create a 3D feature from an underdimensioned sketch by making assumptions for the dimensions that are not given. This can be helpful when you are trying out some ideas, but generally it is best to plan your model so that it reflects your design intent. Since a dimension's value can easily be changed, you can guess a value and change it later, but the dimension chosen to define the sketch is very important.

You should make sure that the dimensions you choose provide the proper relationships between the sketched geometry and other existing features. It is very important which surfaces you choose to create the dimensions between. You should not let the rules of your software package determine these dimensions, as it is unlikely that they will reflect your design intent.

When you are dimensioning the sketch to create a feature, you should place the dimensions so you can view them clearly, but you

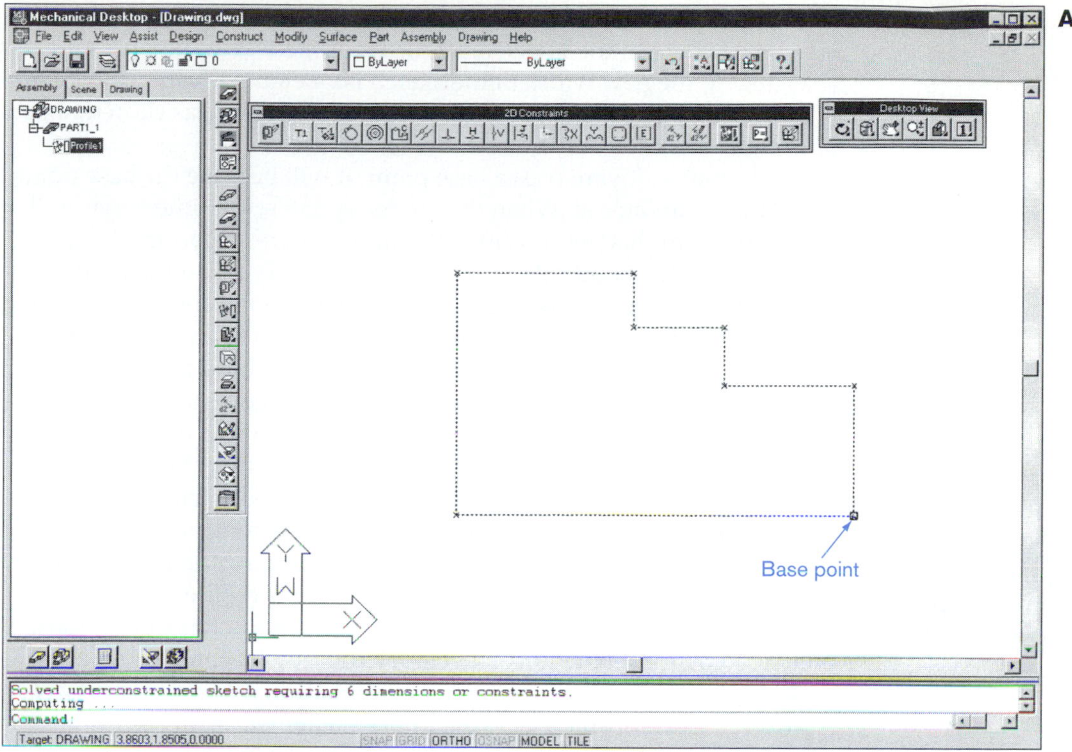

do not need to be overly concerned about following all of the dimensioning practices for placing dimensions in a drawing. Later, when you go into drawing mode to create a drawing from your model, you will be able to clean up the appearance of the dimensions. You you will be able to move dimensions, change their placement, and arrange them so that they follow the standard practices. However, the practices for dimensioning were developed to help make dimensions easy to read. Placing dimensions outside the object outline and keeping the dimensions a reasonable distance from one another will also make the dimensions in your model easier to read.

FIGURE 7.13
The base point fixes the sketch—and the model—on the coordinate system and serves as a fixed point for the software when the model is updated.

Defining your drawing geometry with dimensions that reflect your design intent is important. If you do a good job of selecting the dimensions that will control the drawing geometry, you should have little cleanup work to do later when you are creating drawing views from the model geometry.

Setting the Base Point

Some parametric modeling software (like Autodesk's Mechanical Desktop) assume a *base point* in the sketch for the first feature you create. The base point is fixed on the coordinate system; the other sketch geometry is located on the coordinate system based on its relationship to this base point. The resulting feature will stay fixed on the coordinate system at this base point. When dimension values change, they will change relative to that point. There is only one base point per model.

Consider the sketch in Figure 7.13(a). The base point is indicated by the square in the lower right corner. If the width dimension is changed, the new width of the sketch is calculated from the base point. The rest of the sketch geometry is reoriented, as shown in part (b) of the figure.

If your software uses a base point, it will become the base point for the entire model. When the model updates, all dimensions will update from this point. Whether your software supports a base point or not, a good rule of thumb is to identify a fixed point on your model that you would use as a starting point for measuring the part for inspection purposes. Use this point as the base point, or locate your dimensions from it as if it were the base point.

Locating most of your dimensions from a common edge or major surface can be effective for two reasons. First, it helps you create dimensions that will be useful for inspecting the part after manufacture; the dimension values in your model will measure from the same point that an inspector will use to measure the finished part. Second, it helps you anticipate how the model will update. When a value is changed, the features on the model will update (move) relative to the edge referenced in their dimensions. When you locate dimensions relative to other features you may have difficulty predicting how the model will update because you have to add the effect of changes to the intermediate features.

FIGURE 7.14
These two sketches of the same part have very different relations built into the dimensions for the sketch. The sketch in (a) relates all dimensions to the left edge of the part. The sketch in (b) has a chain of relationships built into the dimensions.

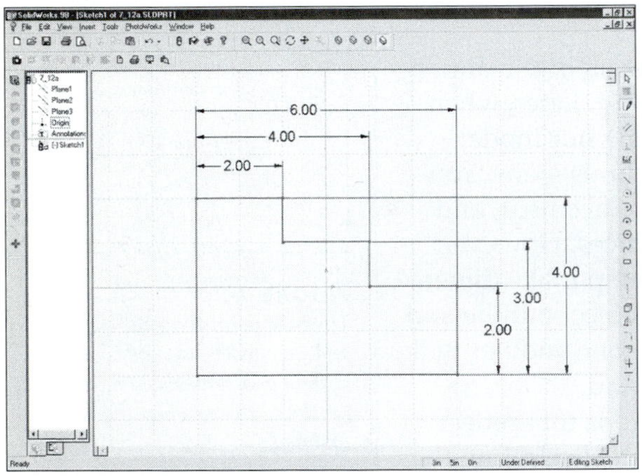

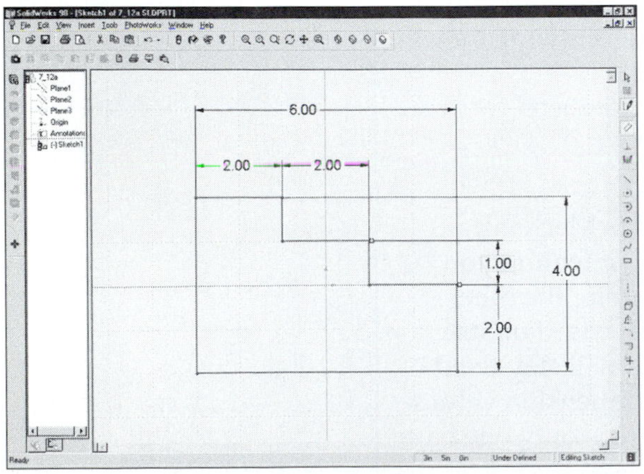

More important than any rule of thumb governing dimension selection or placement, however, is to consider thoroughly the parts you are designing and to make the drawing geometry, constraints, and dimensions reflect the design intent for the part. In some cases, dimensioning a part from another feature is required to reflect the design intent. Figure 7.14 shows two sketches that are of the same geometry but dimensioned differently. When the 2.00 height dimension is changed to 2.50, how will each of these drawings update?

In the first case, shown in (a), the height of the 2.00 "step" will change to 2.50, narrowing the gap between it and the 3.00 step. In (b), the height of the 2.00 step will change to 2.50, and the height of the next step will update to 3.50 units so it remains 1.00 unit above it. In this case, the narrow gap will occur between the 3.50 and the 4.00 step. Either is correct to the extent that it reflects the intended design.

The Base Feature

One thing that can help you to create useful relationships in a parametric model is to start out your model with a good base feature. The *base feature* is the first feature you create. The other features are created using relationships that locate them to the base feature. Subsequent features update based on their relationships to the base feature or one another. You should generally create a fairly large, significant feature as the base feature.

Figure 7.15 shows the cylinder that could be created as the base feature for the shaft support and the other features that will relate to it. The major cylinder would make a good base feature because it is a significant feature of this part, and the other features make sense related to it. That is, the design intent of the part centers on the cylinder. Its size and location will determine the size and shape of other features. The width of the base plate will depend on the stability needed for the size of the cylinder. The hole for the shaft must be centered inside the cylinder and sized to be no more than 1 inch smaller in diameter (so at least 0.5 inch material remains). Using the cylinder as the base feature also makes sense from the standpoint of the part's role within a larger assembly, as its size could depend on the shaft being designed in a different part. By building your model around the major cylinder, you make it possible to update all features automatically if the cylinder size changes.

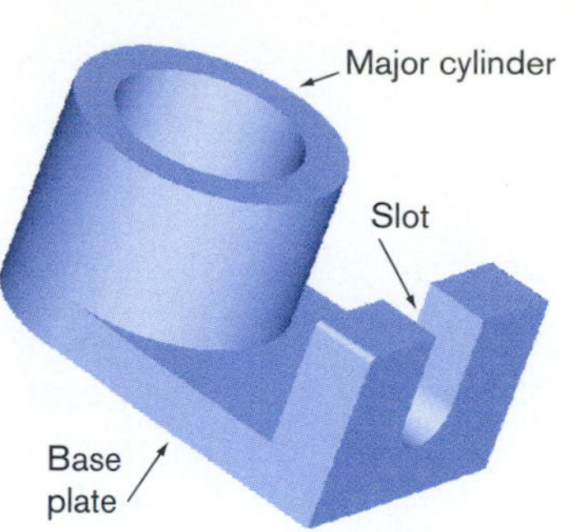

FIGURE 7.15 *The major cylinder for the shaft support could be a good base feature.*

Adding Features to the Model

When you add features to the model, the way you constrain the sketch determines how it relates to the other features. Just as the base point served to locate the base feature, at least one of the constraints applied to subsequent sketches must tell the software how the new feature relates to an existing feature.

To create the second feature, you can use a dimension to locate it relative to an existing edge, or use a geometric constraint to align one of the sketch entities to the existing geometry. That is, the center point of a hole could be dimensioned to be a certain distance from the end of an existing feature, or it could be constrained to be concentric to a cylinder or arc on an existing feature. (Pro/Engineer has a sketch command called Use Edge that allows you to use an edge on an existing feature as a line in your sketch. This command creates the line in the sketch and constrains the sketched line to be colinear with the existing geometry in a single step.)

The way you dimension, constrain, or align your second feature determines how it updates with respect to the first feature. Use the things you know about the design to add features. Where are the fixed points on the model, and how does the feature you are adding relate to them? What are the conditions that the new feature needs to satisfy at all times? If existing features were to change, how would the feature you are adding need to change? In some cases, a simple dimension from an existing edge may be all that is needed to add the feature. In other cases, you will want to build in geometric or size relationships needed in the design. The basic process of parametric part modeling is illustrated in Figure 7.16. Starting with the base

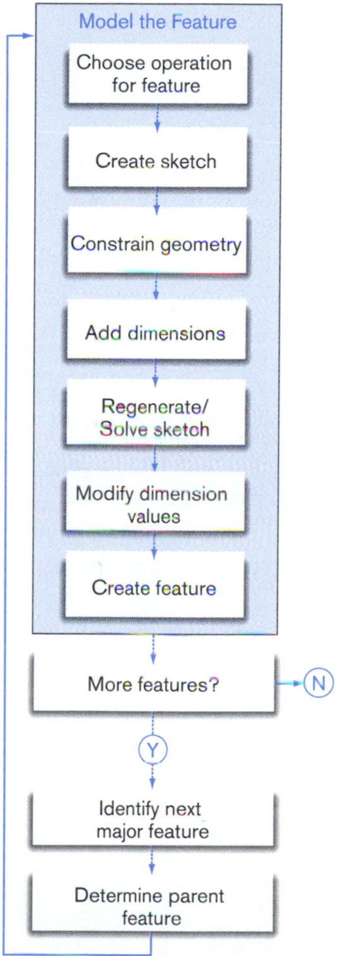

FIGURE 7.16 *This flowchart of parametric part modeling illustrates why it is called feature-based modeling. Each feature, starting with the base feature, is modeled in turn using the same basic process. Notice that geometric constraints generally precede dimensional constraints in the process.*

feature, you build a model by adding features and relating them to existing features. For each feature, the same basic process applies.

Parent-Child Relationships

Parametric modeling software stores information about each feature and its relationship to its parent feature. The term *parent-child relationship* is used to describe how one feature is derived from another feature. This relationship is defined by the geometric constraints applied to the sketch as well as its parametric dimensions. Planning ahead when creating a parent-child relationship is important if you want to select good base features and know what your model will do when you change parameters later.

The base feature is the parent feature for the next features created from it. Just as real children depend on their parents, child features are dependent on their parent features. If a parent feature moves, the child feature moves in relation to it. For example, when the hole feature is added to the part, its location will be defined to be concentric to the major cylinder feature. When the cylinder's location changes, the geometric constraint driving the hole feature will cause it to move to remain concentric to the cylinder feature. If the hole feature is located relative to the base plate or to some other feature and the cylinder was moved, the hole would not move with it.

If you delete a feature, you will also delete its child features unless you redefine them so that they are related to a different parent feature. This is another reason for selecting a good base feature; you don't want the first feature you create to be something that might be deleted later in the design. Not all features will have child features, but all features other than the base feature must be related to some parent feature.

One way to visualize the parent-child relationships as you plan your model is to build a tree diagram. The relationships in the Check5 model in Figure 7.17 are illustrated in the Parent/Child Relationships window. The feature named ArmTrim has HingePins and PlateTrim as its children, along with the sketches that formed their cross-sectional shapes. These features may or may not have children (or "grandchildren") of their own. The parent features for ArmTrim are the HingeArms feature, Baseplate feature, the Origin of the coordinate system, the Front datum plane, and Sketch3.

FIGURE 7.17
A tree diagram showing the dependencies between features can help you plan your parametric model. Notice how the list of parent and child features from this SolidWorks model shows the dependencies between features.

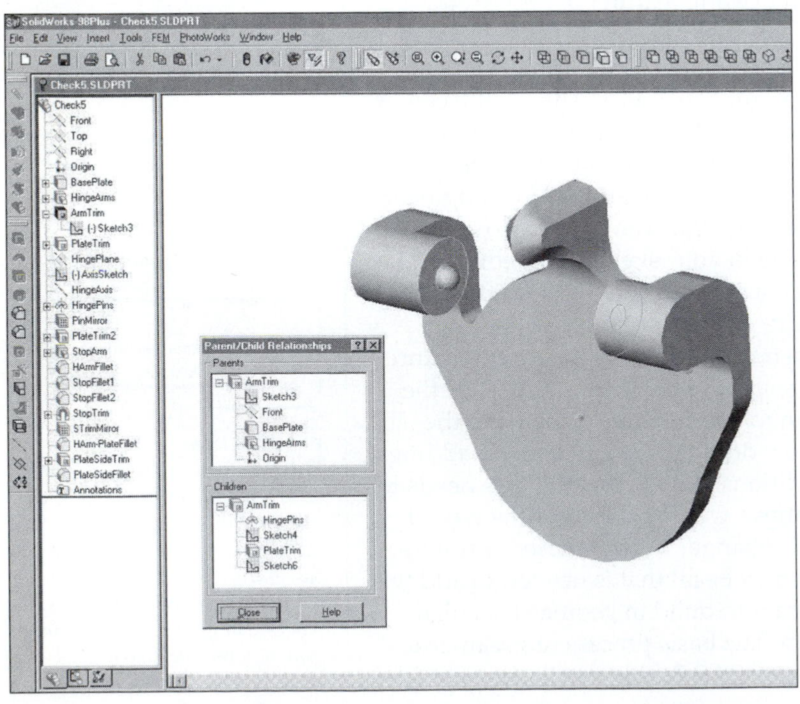

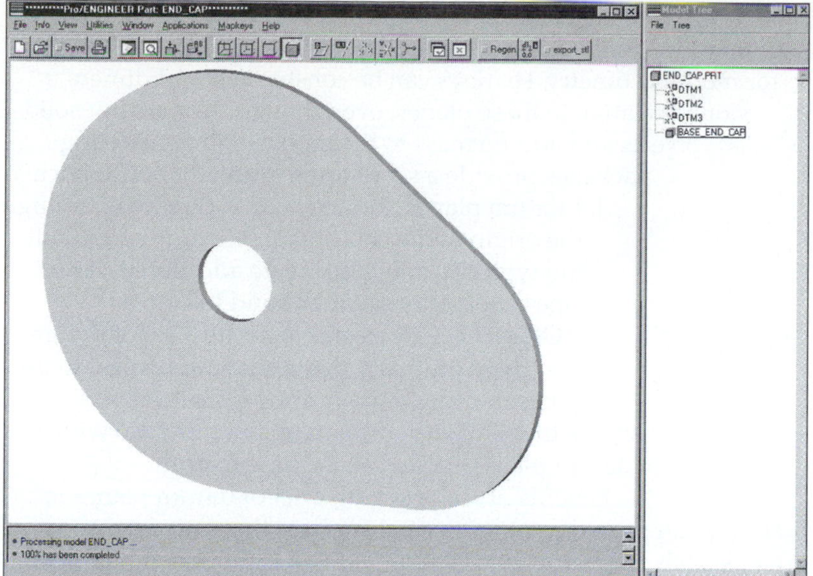

FIGURE 7.18
The base feature is the first feature created.

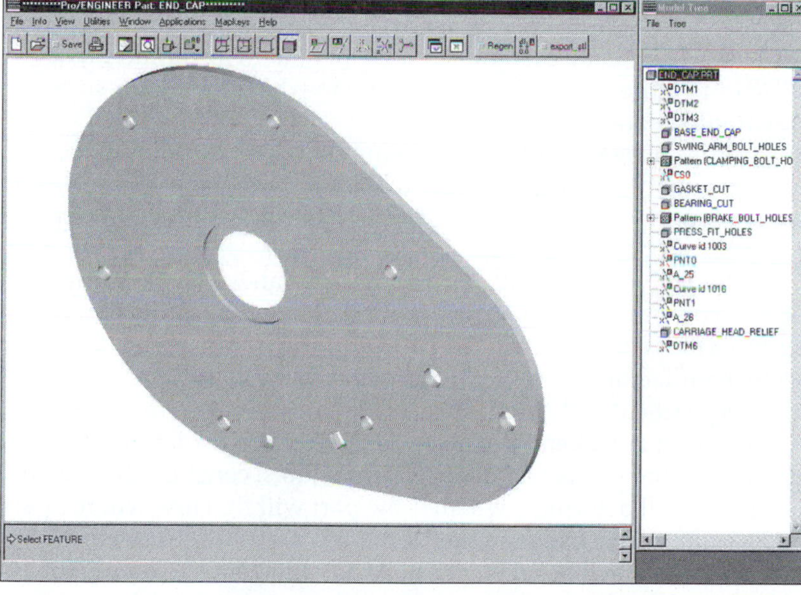

FIGURE 7.19
Subsequent features are children of previous features.

Some parametric modelers, such as Pro/Engineer, allow you to "play back" the model to see how it is being generated by the underlying parameters. The sequences shown in Figure 7.18 and Figure 7.19 illustrate how the features used to create the parts are ordered in relationships that reflect the design intent for the parts.

Datum Planes and Surfaces

In engineering drawing, a *datum* is a theoretically exact point, axis, or plane derived from the true geometric counterpart of an actual part. A datum provides a reference surface, axis, or point used for inspecting the manufactured part. Theoretically exact planes used in

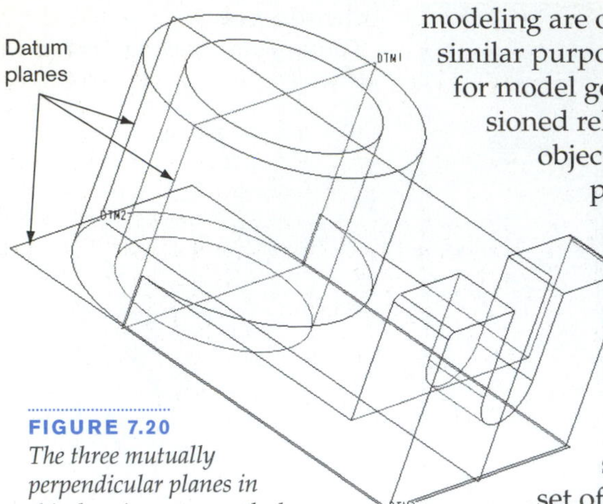

FIGURE 7.20
The three mutually perpendicular planes in this drawing serve as the base feature for the model.

modeling are often called *datum planes* because they are used for similar purposes. They are planes that can be used as a reference for model geometry. Features can be constrained and dimensioned relative to these planes, even though they are not solid objects and add no mass to the model. Some modeling packages provide a set of three mutually perpendicular datum planes, located so that they pass through the origin of the coordinate system, as a default. You typically may also create additional datum planes at other orientations and locations.

Often it is very useful to create a set of datum planes (three mutually perpendicular planes) to use as the base feature in your drawing. Figure 7.20 shows the shaft support part again, this time with a set of datum planes created as the base feature.

One of the benefits of starting with a set of datum planes is that they can provide a set of normal surfaces that you can use to orient sketch geometry. This is particularly helpful when there are no planar surfaces on the part that you can use to easily build subsequent features. With a cylinder as the base feature for the shaft support, for example, there is no flat surface to which the end of the base plate can be constrained to be parallel. The datum plane DTM1 in Figure 7.20 provides a surface to which the plate can be constrained parallel.

Datum planes also provide a common reference point for parts in an assembly. When parts share a set of datum planes, it is easier to align the part with other parts when you bring them together in an assembly model.

Whether you start with a set of datum planes as a base feature or not, you should consider the surfaces that will serve as datum surfaces on your part. Parts are often inspected to see whether they were manufactured acceptably, based on a similar set of three mutually perpendicular, theoretically exact planes. Measurements of the finished part are taken from surfaces or edges coincident with one of the datum planes. As you define your features, consider the dimensions that will determine whether the part will fit. How will the part be measured after manufacture? If there is a surface from which measurements will be taken, it can be advantageous to identify it as a datum surface in your part. Unlike a datum plane, a *datum surface* is an actual surface on the part. It is created in the same way as any surface, but you will use it as a reference location to inspect other features (see Figure 7.21).

Like the base point in a sketch, a datum plane can serve as a reference from which the other features build. If you define a set of datum planes in your model, your part's datum surface, identified for use in inspecting the finished part, may be located on one of the planes. If you do not use a set of datum planes as a base feature, you can locate new features from existing surfaces on the base feature. You can also use these datum surfaces on existing features to align the part with others in an assembly. Having datum surfaces defined in your part can be helpful when you are defining tolerances for

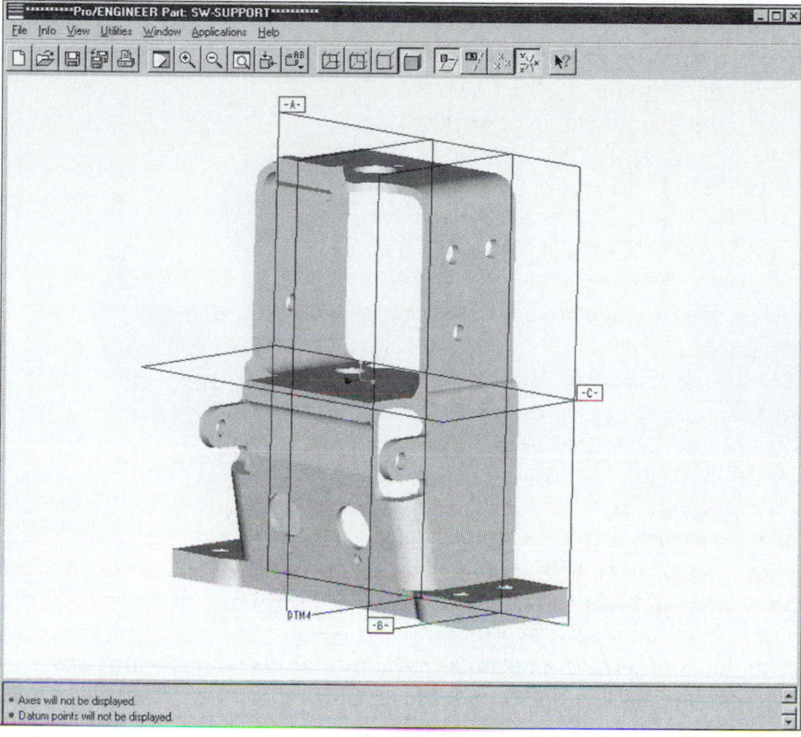

FIGURE 7.21
Surface C on this fixture is a datum surface or reference location that will be used to inspect the location of the holes seen on the right side of the part. (Courtesy of Mark Perkins.)

various features. Defining your model using the dimensions that will be used to inspect the finished part will make it easier to determine the impact of variations from the stated dimension that occur during manufacture. (You will learn more about tolerances in Chapter 12.)

Datum planes and datum surfaces can be added and identified at any time, but considering them as you plan your model will help you use them more effectively in capturing design intent in your model.

Parametric Dimensions

The dimensions in a parametrically modeled drawing have values that can be changed to update the model. Each dimension has two components: its name and its numerical value. Naming the dimensions allows them to be used in equations or relations to define other parts of the model geometry. The numerical value may be entered as a number in the definition of the dimension, or it may be derived from an equation. Figure 7.22(a) shows the shaft support part with its dimension values displayed; Figure 7.22(b) shows the same part with the dimensions displaying their parameter names. The parametric modeling software allows you to switch between numeric and parametric display for the dimensions in your model. Some modelers also allow you to switch to a third kind of display that reveals the formulas used to relate dimensions to each other.

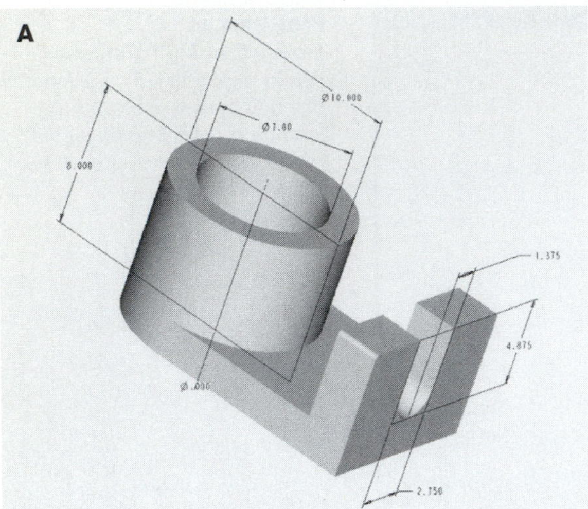

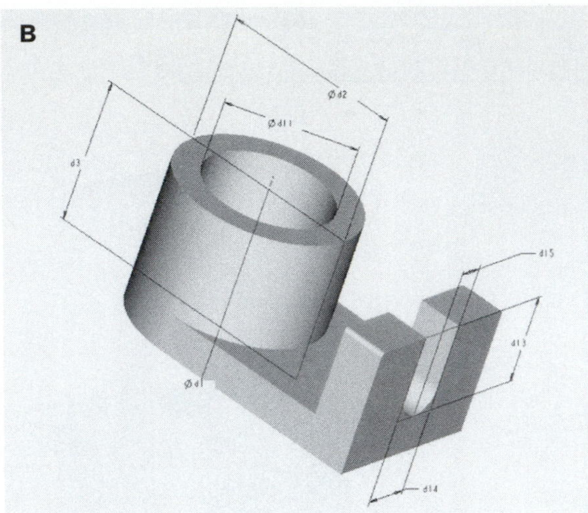

FIGURE 7.22
These figures illustrate two ways that dimensions in the parametric model can be displayed. (a) The numerical display shows the current value for the dimensions. (b) Showing the parameter name can help you locate dimension names to be used in the dimension for another entity.

Formulas in Dimensions

Like the geometric constraints in the model, the size parameters create relationships between features of the part. For example, if you wanted to maintain a constant material thickness around the outside of the central hole, you could enter the equation d1 = d2 + 1.00 for the diameter of the outer cylinder. By doing so, no matter what value is entered for d2, the driven dimension (d1) will be 1.00 overall more than d2.

Parametric dimensions should reflect your design intent. Planning them to do the work for you is the key to building design flexibility into your model. Consider adding four holes to a rectangular base plate. The location of the four holes can be specified in several ways, but they should be centered 0.75 inch from the edge of the plate. If the width of the plate changes, the hole location should also. By defining the location of the holes in terms of the plate's width and the datum plane that passes through the cylinder's centerline, the hole's vertical locations will stay 0.75 inch less than half the base plate's width away from the center of the base plate. The equation for the hole location would be h_location = height/2–.75, where height is the vertical dimension and h_location is given from the center of the part.

Consider the dimension formulas used to define features in the parts shown in Figure 7.23 to see how they reflect the underlying relationships in the model.

Equations in parametric dimensions generally use a syntax and operators similar to those used in a spreadsheet or other programming notation. (In fact, some modelers make it easy to import dimension values from and export them to an outside application. Complex formulas can be used to calculate sizes in the spreadsheet or program (as in Figure 7.24), with the resulting values imported back into the modeling package. You will learn more about this in Chapter 10.) Each modeling package will stipulate the notation and

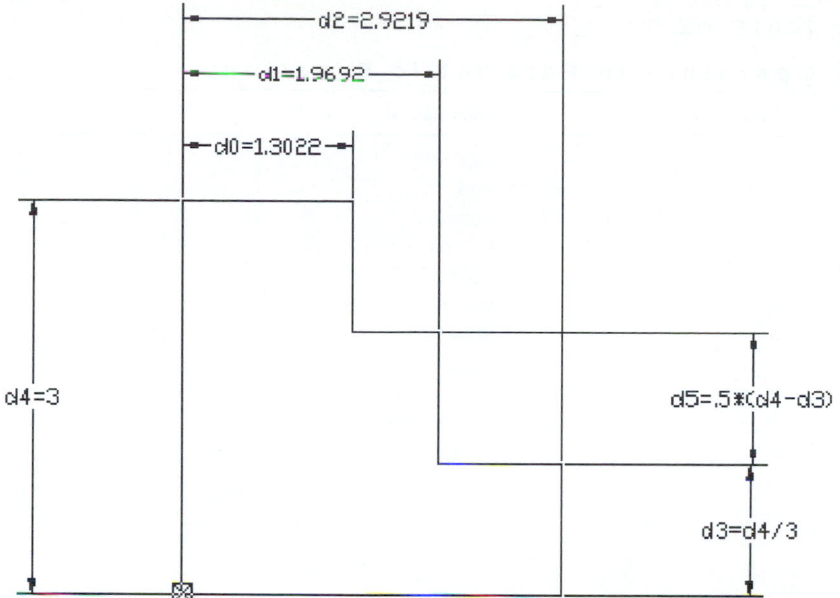

FIGURE 7.23
Can you determine the dimension values for these features based on the formulas shown?

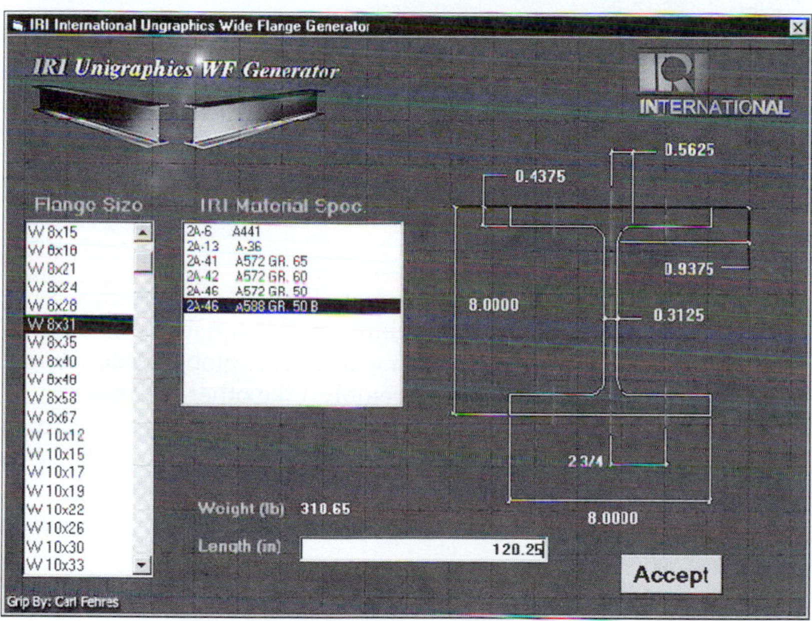

FIGURE 7.24
By naming dimension parameters in their standard parts, IRI International can quickly resize the parts by importing parameter values for different sizes. The screen shown here is from a program that will generate the dimensions for different sizes of a wide flange, a structural part used in oil rigs. The part was modeled parametrically in Unigraphics, so it could be easily resized. Options for resizing the flange are not unlimited, however; the flange size is constrained by the shape and size of industry standard stock material that it will be manufactured from. The wide flange generator program uses a table of industry standard shapes from the AISC Structural Handbook to calculate the dimensions of the flange that could be manufactured from the stock at the desired width and length. These dimensions are then written to a file using Unigraphic's GRIP programming language; the file can be read by the parametric modeler and the dimensions used to update the part to the new size. (Courtesy of Carl Fehres, IRI International.)

syntax needed for equations in parametric dimensions. Table 7.2 lists the operators that can be used in Autodesk's Mechanical Desktop.

As you create parametric dimensions, it's important to keep track of the relationships you create. Most software allows you to name and document your parametric dimensions so you (and your colleagues) can interpret parameter names more readily. The software will assign a name to a dimension by default, but you may want to give key dimensions more recognizable names so they are easier to interpret in your model.

TABLE 7.2

Operators in Parametric Equations

Operator	Description
^	Exponent
+	Add
–	Subtract
*	Multiply
/	Divide
%	Modulus (remainder)
sqrt	Square root
log	Logarithm
ln	Natural logarithm
floor	Rounds down to nearest integer
ceil	Rounds up to nearest integer
sin	Sine
cos	Cosine
tan	Tangent
asin	Arcsin (sin^{-1})
acos	Arccos (cos^{-1})
atan	Arctang (tan^{-1})
sinh	Hyperbolic sine
cosh	Hyperbolic cosine
tanh	Hyperbolic tangent
pi	3.141593
e	Base number for natural logarithm or 2.718282
exp(x)	e^x

Parameters that are common to more than one part in an assembly are called *global parameters*. To eliminate redundancy, most software programs allow you to stipulate and name global parameters that can be used throughout the assembly. Like other parameters, this is a powerful means of ensuring that parts work together and that critical dimensions are updated in all parts of an assembly. With global parameters, it is even more important to document them. You will learn more about the role of parameters in intelligent assemblies in Chapter 8.

Editing the Model

One of the biggest advantages of parametric modeling is that you can edit the features after they are created. If you have several parts open at a time in your software, you may have to identify which is the active part, then select the feature you want to edit. Once you select the feature you want to edit, the dimensions for the feature are displayed in a way that lets you change their values. Alternatively, you can also return to the original sketch that you used to create the feature and modify the sketch. Figure 7.25 shows the shaft support drawing with a change being made to the overall length of the part.

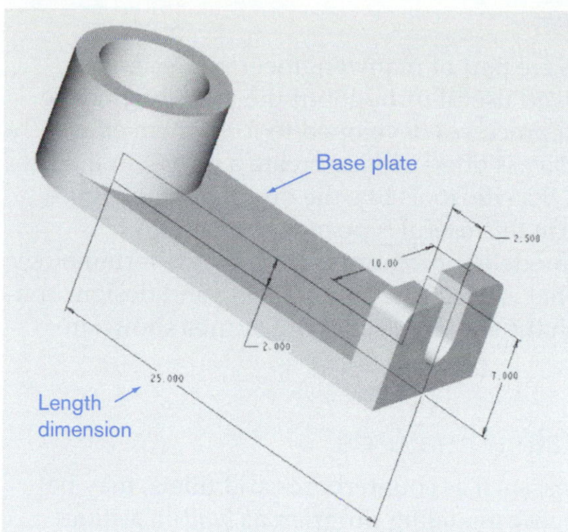

FIGURE 7.25
The length of the shaft support has been changed by editing the dimension value and updating the model.

Depending on how the model was created you may have more or less success in updating the model. For example, the base plate that connects to the cylinder (indicated in the figure) can be created many different ways. Each way may have advantages or disadvantages for how the part updates.

The plate could be generated by sketching the shape of the bottom surface on a sketch plane aligned with the bottom surface of the cylinder and extruding it up a specified height, as shown in Figure 7.26(a). It also could be sketched on the midplane and extruded to both sides, as shown in Figure 7.26(b). Finally, a rectangular sketch could be drawn on an offset plane parallel to the indicated datum plane and extruded to the cylinder feature, as shown in Figure 7.26(c). Each of these will work about the same when you want to change the length of the part. However, the third option would not be effective in making the plate wider than the cylinder. When a feature is extruded to an existing feature by using Pro/Engineer's To Next Surface option, for example, the new feature cannot be updated so that it is wider than the surface it is extruded to. The boundary or end of the surface is limited by the next surface. When there is no portion of the surface for the feature to end at, it would continue infinitely and therefore is undefined.

FIGURE 7.26
(a) Extruding upwards, (b) extruding to both sides of a center plane, (c) extruding to the cylinder from an offset plane. All of these methods could be used to generate the plate feature for the shaft support, but they vary in the ways they can be updated.

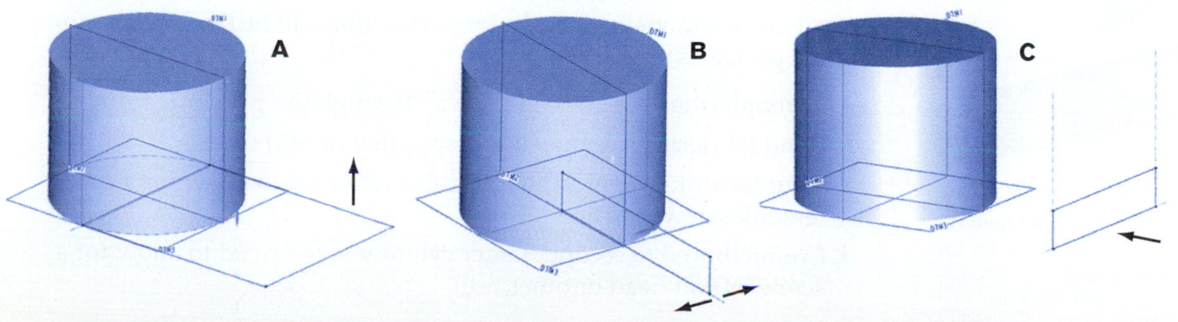

Standard Features

Certain useful features are part of many engineering designs. Features that have proved useful throughout the years often have specific manufacturing processes developed to form them efficiently. For example, a *counterbore* is often used to create a recess for a bolt head or other fastener. Specific tools like the one shown in Figure 7.27 make it easy to form this useful type of feature.

Many parametric modeling programs use the same terminology for standard features that is used in manufacturing and design. You should be familiar with the standard types of features shown in Table 7.3.

FIGURE 7.27
This counterbore tool used on a milling machine forms a depression around the top of a hole. The center part of the tool fits inside the hole to help locate the counterbore concentrically.

Working With Built-in Features

Many standard features, such as counterbores and fillets, may be available in the parametric modeling program as *built-in features*. These features can be placed with a dedicated command on surfaces that are already defined in the part, in the same way a solid primitive might be added to (or subtracted from) a model. Some features are solids, and others are negative solids; that is, they subtract a solid from the model. A hole, for example, is a negative solid; it subtracts a cylinder-shaped solid from the model. The features have properties built into them that the modeling software will preserve, just as it preserves the geometric and size relationships you set.

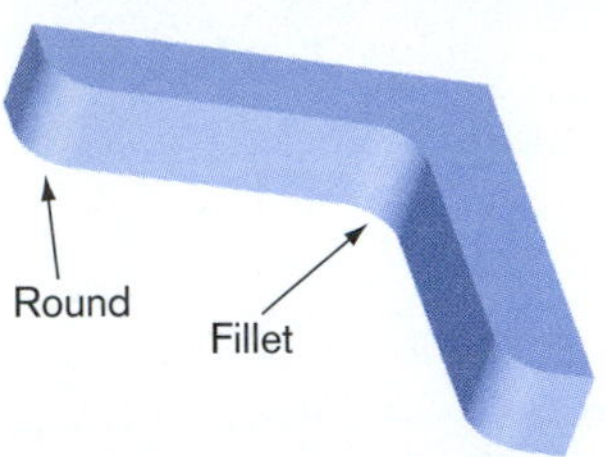

FIGURE 7.28
Fillet and rounds

A fillet is a common built-in feature that rounds the edge formed by two surfaces. A fillet can be created by selecting the edges to be filleted (several can be selected at once) and specifying the radius of the fillet. Rounds are another type of standard feature created in the same way. When the surfaces move or are lengthened, the fillet or round adjusts automatically.

Figure 7.28 shows a typical fillet and a typical round with a uniform radius.

Sophisticated parametric modelers allow you to create uniform radius fillets and rounds, chained fillets and rounds that blend around a corner of a surface, constant chord fillets and rounds, and variable radius fillets (shown in Fig. 7.29) using different methods. (The constant chord fillet is used in situations similar to that in (b) where a constant radius fillet would result in thicker material on the uphill and downhill side of the blend. The constant chord distance creates a fillet with a uniform amount of added material.)

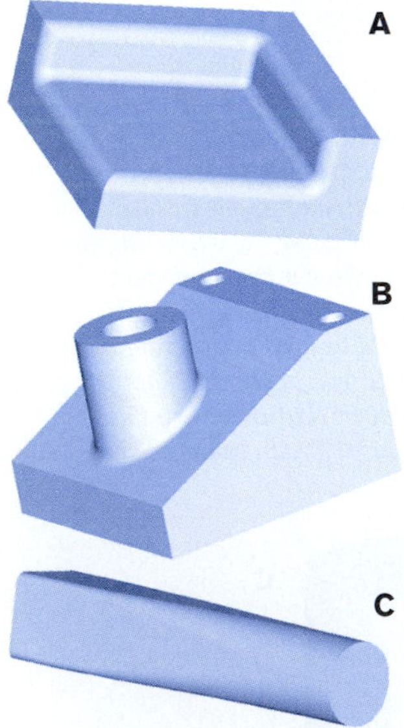

FIGURE 7.29
(a) Chained fillets and rounds; (b) constant chord fillet; (c) variable radius round

Some built-in features make it easy to create standard features in conjunction with other features. An example would be the options available for modeling a hole. Properties that can be set for holes in some packages are

- Through (the hole goes all the way through the part or feature)
- Blind (material is removed to a specified depth)
- Countersunk (a conical shape is also removed to allow for a countersunk screwhead)
- Counterbored (a second, larger cylinder is removed to allow for a screw or bolt head or other use)

TABLE 7.3

Terminology for Standard Features

Standard Feature	Definition	Illustration
Fillet	A rounded interior blend between two surfaces; some uses are to strengthen joining surfaces or to allow a part to be removed from a mold.	
Round	A rounded exterior blend between two surfaces; used to make edges easier to handle, improve strength of castings, and make parts easier to remove from mold.	
Counterbore	A cylindrical recess around a hole, usually to receive a bolt head or nut.	
Countersink	A conical-shaped recess around a hole, often used to receive a tapered screw.	
Spotface	A shallow recess like a counterbore, used to provide a good bearing surface for a fastener.	
Boss	A short protrusion beyond the normal surface of a part, often used to provide a strong bearing surface.	
Lug	A flat or rounded tab protruding from a surface, usually to provide a method for attachment.	
Flange	A flattened collar or rim around a cylindrical part to allow for attachment.	
Chamfer	A small angled surface, used on cylinders to make them easier to start into a hole, or on a plate to make it easier to handle.	
Neck	A small groove cut around the diameter of a cylinder, often where it changes diameter.	

(continues)

TABLE 7.3 (cont.)

Standard Feature	Definition	Illustration
Keyway/Keyseat	A shaped depression cut along the axis of a cylinder or hub (shown) to receive a key, used to attach hubs, gears, and other parts to a cylinder so they won't turn on it.	
Knurl	A pattern formed on a surface to provide for better gripping or surface area for attachment. Often used on knobs and tool handles.	
Bushing	A hollow cylinder that is often used as a protective sleeve or guide, or as a bearing.	

The Hole command may also allow you to choose the specialized constraints for locating holes, such as concentric, placed by edge, and on point, as illustrated in Figure 7.30.

Complex Shapes

Complex shapes can often be made using special surface modeling commands that create surfaces that cannot be made easily with regular extrusion, revolution, sweeping, and blending. These surfaces can then be added to the model as any feature would be (see Fig. 7.31). Because the complex surface can slow down model regeneration, an approximation of the surface may be used as a placeholder until later in the refinement process.

Parametric Modeling Modes

Most parametric modelers have three main modes that link together to provide a full range of functionality in creating and documenting your designs: *part mode,* which allows the creation of individually manufactured parts of a single material; *assembly mode,* which allows you to link multiple parts together into assemblies or *sub-assemblies;* and *drawing mode,* which allows you to create orthographic, section, and pictorial drawings to provide the final documentation drawings for the design.

Assembly Mode

The assembly capabilities of a parametric modeling program enable many companies to use the single design database to improve the collaborative environment for a cross-functional design team,

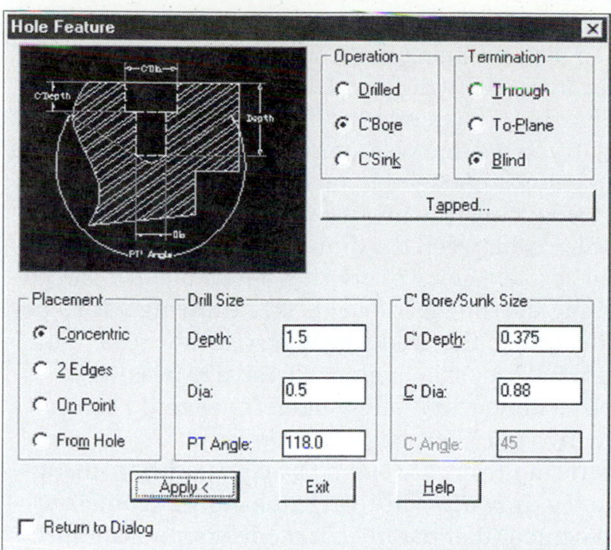

FIGURE 7.30 *Placement methods for holes from Mechanical Desktop.*

making concurrent engineering feasible. By planning the critical relationships to capture the design intent of the assembly—just as one would for each individual part—the parametric assembly can be used to coordinate efforts of several different individuals. The entire assembly and each part can be stored in the design database for members of the design and manufacturing team to view, measure, or change (if authorized), while the parts are being designed.

The assembly mode can be used to plan a framework for the model that makes it possible to use global parameters and build fully intelligent assemblies. For example, if several parts will assemble onto a shaft, the shaft diameter may be made a global parameter. The hole sizes on the parts that fit onto the shaft may be created by adding or subtracting a value from the global parameter for the shaft diameter. If the shaft diameter is changed, all of the parts that fit with it may automatically update. Using global parameters can be an effective way to coordinate the design effort for a team. You will learn more about using assemblies to refine your design in Chapter 8.

A

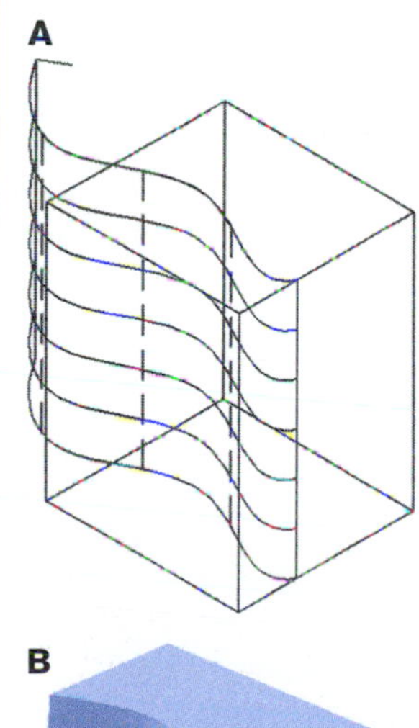

B

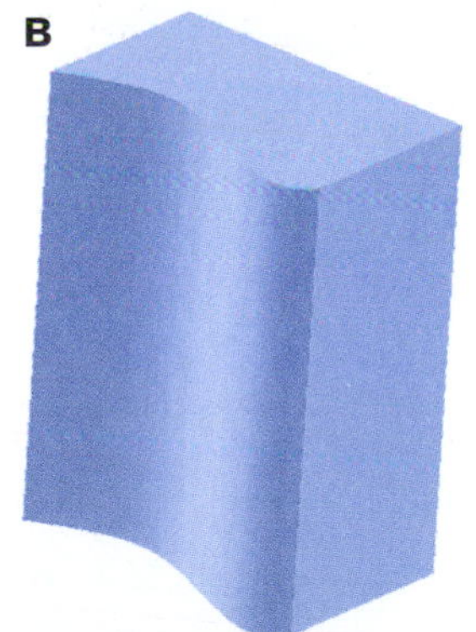

FIGURE 7.31 *This surface patch was created using the surface modeling commands built into Autodesk Mechanical Desktop, then added to the solid model as a feature.*

Drawing Mode

An advantage of all solid modeling software is that drawings can be generated from the solid model. Drawing mode in parametric modeling software adds a significant dimension to this advantage: *associativity.*

Bidirectional Associativity

Dimensions are said to be *associative* if they change when the model changes. When a set of drawing views have been created from the model and a change is made to the model, the drawing views will automatically update because they are associated with the actual model. Dimensions in most CAD packages are driven by the geometry of the model. The dimensioning process can be automated to a degree because the software will report the dimension value for a selected entity from the data in the CAD database. If a line is drawn to be 4.235 units long, 4.235 is the default value for its length dimension. If the length of the line changes, the length dimension can be updated automatically to show the new length of the line. This is a one-way process, however; changing the dimension value has no effect on the geometry. If the line was supposed to be 4.23 units long and the dimension value is changed to this value, the line will remain 4.235 units long; only the dimension text will change. In addition, the association between the dimension and the line entity will be broken. Just as a text entry overwrites a formula in a spreadsheet cell, overwriting the dimension value on a drawing with a new value overwrites the link to the model geometry. If the line length changes again, the new dimension value will have to be entered manually. Associative dimensions allow model changes to be used to update the drawing views, but not the reverse.

Many parametric modelers are *bidirectionally associative:* changing a dimension in the drawing will change the model geometry, and vice versa. Remember that parametric modelers store and use the dimensions and constraints you enter to generate the model. When dimensions are shown in a drawing, the default values are the dimensions used to create the model. Whether these dimensions are changed in part mode or drawing mode, the new values are stored in the database and the model and all associated drawings will update to reflect the new dimension values. For this reason, one can say that bidirectional associativity allows drawing changes to update the model and model changes to update the drawings.

Nonparametric Dimensions

In general, the dimensions used to create the model are also the ones needed to dimension the part on a drawing. Most parametric modelers also offer the opportunity to add two kinds of nonparametric dimensions in drawing mode. These dimensions may be added to make the drawing views meet drawing standards, or to show the size of a feature on the drawing differently from the way it appears on the model.

A *reference dimension* (also called a driven dimension) is an associative dimension, but not one used to create the parametric model. It is sometimes necessary to add a dimension to a drawing that was not needed to create the model. A reference dimension is a one-way link from the database; it cannot be used to change the model, but it will update if the model changes.

A *cosmetic dimension* is a dimension that has no link to the database; it is simply a text label. To dimension a feature at a size different from what is in the model, you must use a cosmetic dimension or otherwise break the link to the database value for the feature.

In general, you should not make much use of these nonparametric dimensions. A well-created parametric model contains all the dimensional relationships needed. If you find that you are making lots of use of nonparametric dimensions to make the drawing look good or change information, you should reconsider how you are creating the model in the first place.

You will learn more about drawing modes and creating drawings from models in Chapter 12.

INDUSTRY CASE

Brandon Larocque, Barrett Technology

Parametric Modeling: Capturing Design Intent

When William Townsend was a Ph.D. student at MIT, he designed a four-axis robot arm that simulates the motion of the human arm. This human-scale dexterous arm works with the subsequently developed Barrett wrist and robot hand and is now being produced and manufactured by Townsend's engineering firm, Barrett Technology. To prepare the design for larger-scale production, the original 2D drawings were converted to parametric solid models. Converting the original 2D drawings to parametric solid models enhanced Barrett Technology's ability to quickly make design changes and evaluate the results. By redesigning the robot arm the company was able to make it the lightest one on the market and save 50% over previous manufacturing costs.

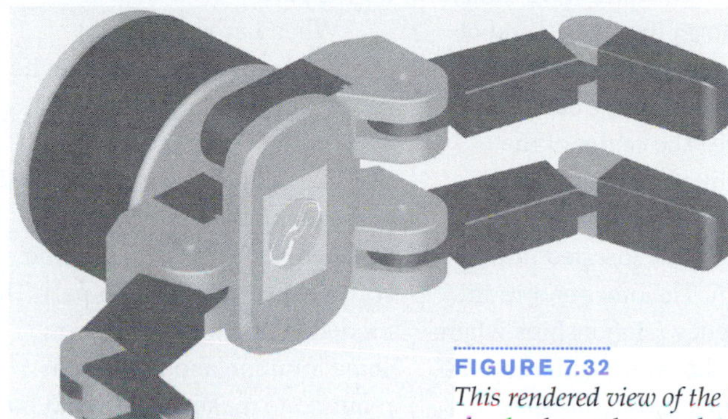

FIGURE 7.32
This rendered view of the BarrettHand clearly shows the top plate in relation to the rest of the hand.

To benefit from parametric modeling, it was important to design the parts with the design goals in mind. "When we have co-op students working here with us, I always remind them not to model blindly," said Brandon Larocque, project manager at Barrett Technology. "You need to keep your design intent in mind. Even though the software is a powerful tool, if you tie up a feature or overconstrain it, you are limiting the amount of flexibility you have in changing it easily later."

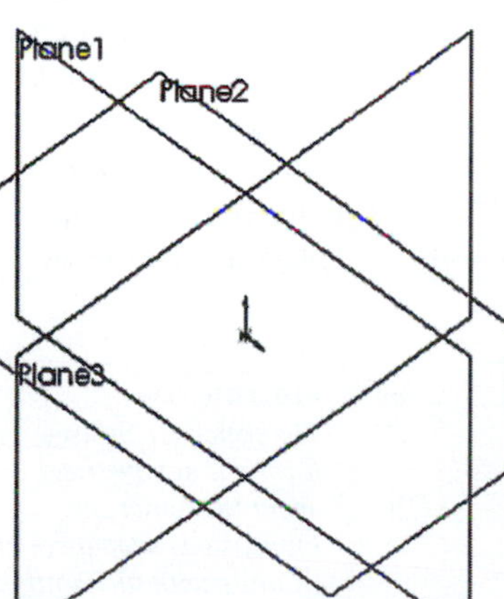

FIGURE 7.33
The model of the top plate began with SolidWorks's default sketch planes. The axis at the intersection of the planes indicates the origin.

The top plate of the Barrett Hand is a good example of how Larocque approaches parametric design (see Fig. 7.32). The part has a total of fourteen features and took about thirty minutes to model in SolidWorks. When a new part is created, SolidWorks provides a default set of three orthogonal planes that intersect at the origin (see Fig. 7.33). The user can add more planes, but SolidWorks allows you to sketch on any surface or plane it has created. If you sketch a square on plane 1 and extrude it, you can use any of the cube's surfaces as sketch planes. Larocque estimates that 80% to 90% of the parts he models don't need additional sketch planes.

The top plate of the hand began as a symmetrical sketch built around a centerline. Half of the shape was built, then mirrored

(continued on p. 276)

INDUSTRY CASE

across the centerline (see Fig. 7.34). The parametric relations in the mirrored half of the part were set up so they would be the same as those in the original half. To create the sketch, three radii were dimensioned to form the curved end of the plate: one in each corner and a larger one to form the curved end of the plate. The center of the largest radius is constrained to always lie on a horizontal construction line inserted into the base sketch. The other end-radii have tangency relationships where they meet the large radius and the straight lines that define the other edges of the plate. These tangency constraints were automatically created by SolidWorks when the radii were sketched in place. The three radii were constrained to be mutually tangent so the tool path generated by the model would be a single smooth curve through the three radii. The only other dimensions needed to define the part were those for the overall length of the part and the rectangular shape at the top of the part. All distances (dimensional constraints) were defined from the center of the part (which also coincided with the origin in SolidWorks). The sketch was then extruded to become the base feature for the part.

When Larocque thinks about selecting a base feature, he chooses one that will support additional features and reflect the way the part may be modified and eventually built. "I think modeling in 3D is similar to the way a machinist shapes a part. The modeling operations give you some intuition about the steps required to make the part and how many different setups a part will require. If you can make all of your features from the same surface of a part, chances are the machinist will be able to make them from a single setup as well. If you can form a part with a single revolution, you can probably make the part on a lathe in a single setup. Each time you add a sketching plane that is not normal or parallel to the default planes, you can assume the part may require an additional setup in order to machine it. Reducing a part's design from four setups to three can reduce the machining costs by 25%."

In the case of the top plate, all of its features except the bevel and the counterbores were created from the same side of the part.

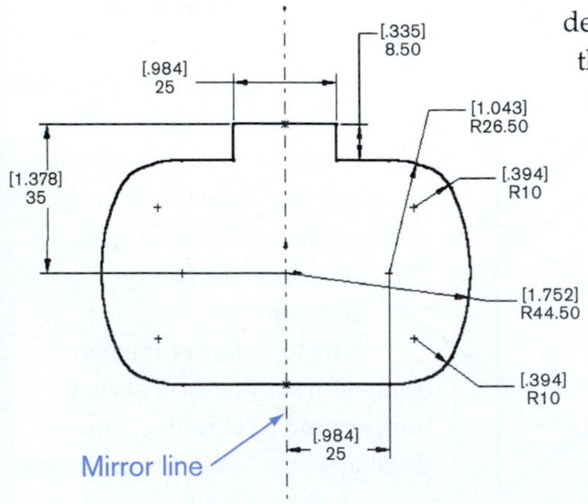

FIGURE 7.34 *The sketch for the base feature is symmetrical about the centerline. Dimensions shown are in SI units with inch units as alternates in brackets.*

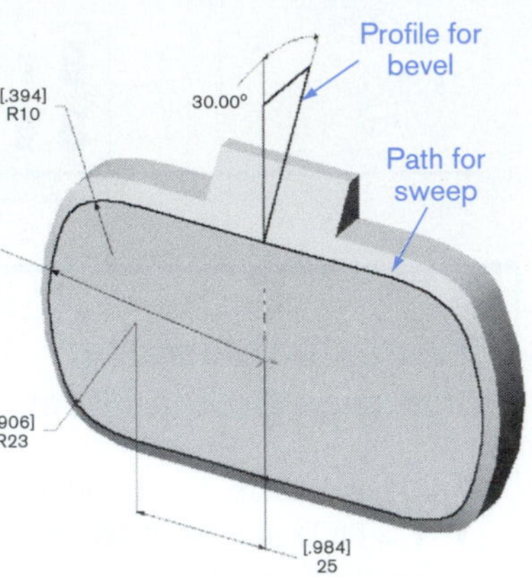

FIGURE 7.35 *The triangular shape of the bevel is shown here as a sketch on the mirror plane. The sketch was swept along the contour on the back of the plate, shown highlighted here.*

Because the plate is symmetrical, its center point also served as the design center of the part. The plate is designed to mate with two bearings set a specified distance apart. By dimensioning the part about a point at the intersection of the mirror line of the plate and a line through the center points of the two bearing bores, any change in the distance between the bearings requires few changes to update the part.

The next step was to create the bevel around the outside of the base feature. Using the mirror plane (across which the first sketch was mirrored) as the sketching plane, Larocque sketched a triangular shape with a 30° angle (see Fig. 7.35). This sketch was the profile for the bevel. A contour was sketched on the back surface

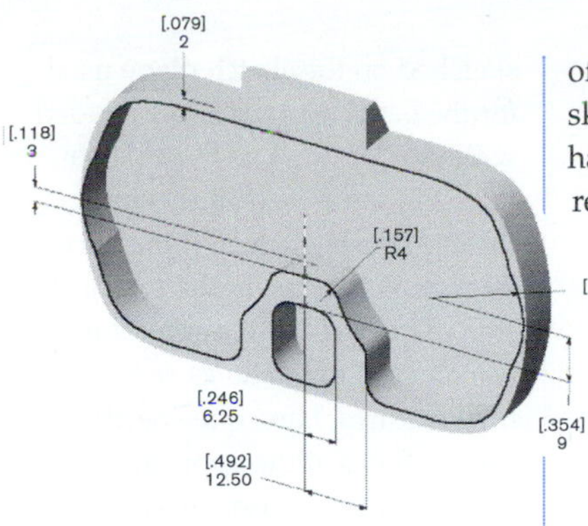

FIGURE 7.36
The sketch for hollowing out the plate included two distinct outlines that were extruded as a single feature to hollow out the part.

of the plate to serve as the path for the triangular profile. After selecting both sketches, he selected a sweeping operation to create the feature. SolidWorks prompted him to identify the profile (the triangle), the path (the contour sketch), and the sweep operation (a cut). The profile was dragged normal to the sweep path to cut the bevel on the plate. A machinist cutting this feature into actual metal would similarly use a 30° tapered mill to sweep out the same path defined in the model.

The next feature hollowed out the part. Sketching on the surface of the part that lies on the original sketch plane, Larocque created half of the shape of the area to be removed, dimensioned it from the centerline, and mirrored it to create the complete outline. The sketch was then extruded to the appropriate depth to be cut from the base feature (see Fig. 7.36).

The fourth feature was a semicircular cut-out to provide clearance for a motor pinion. This circle was sketched on the original sketching plane, its diameter dimensioned, and its center point was constrained to the center of the part (which corresponds to the distance from the centers of the bearings). Using a constraint to position the feature on the part's centerline guaranteed that the feature would stay centered even if the overall size of the part changed. To form the feature, Larocque specified a blind extrusion to a specified depth (see Fig. 7.37).

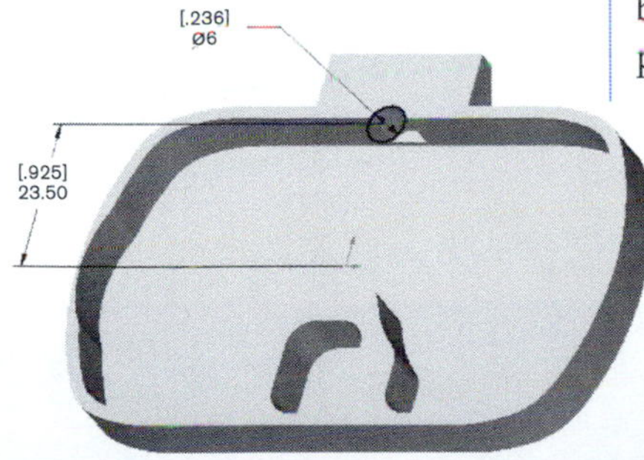

FIGURE 7.37
The sketch for the motor pinion cut-out was constrained to the center of the part.

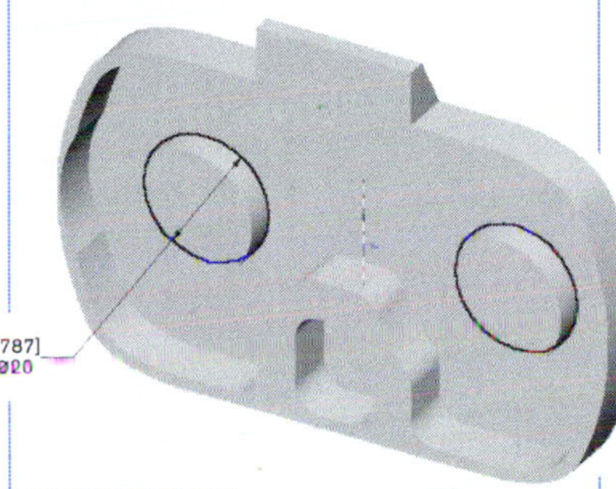

FIGURE 7.38
The sketch for the bosses was drawn on the new surface created by the hollowing operation.

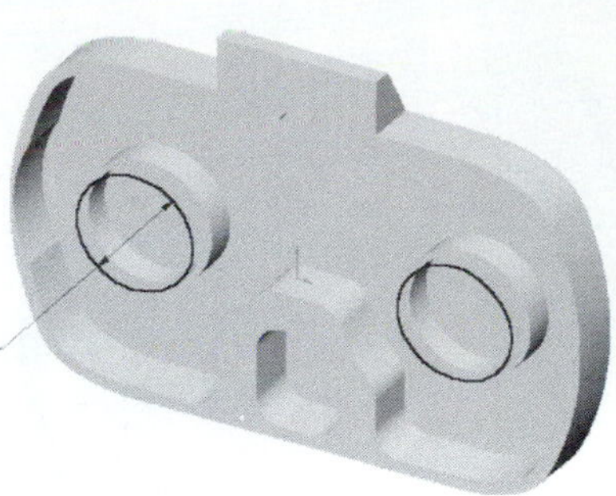

FIGURE 7.39
The bearing bores were sketched on top of the bosses and extruded down to the desired depth.

With the part hollowed out, the new interior surface was used as the sketch plane for the bosses that would be the bearing bores (see Fig. 7.38). The bosses could have been formed on the initial sketching plane in the same operation that hollowed out the part, but their height would have been constrained by the depth of the part. If the depth were to change, the height of the bosses would too. By using the new surface, the relation of the bosses to the inside surface of the plate would remain constant, even if the depth of the surface changed.

To create the bosses, Larocque created symmetric 20-mm circles with center points constrained to the horizontal centerline of the part. The circles were extruded, and two new concentric circles were sketched on the tops of the new features (see Fig. 7.39). From

(continued on p. 278)

INDUSTRY CASE

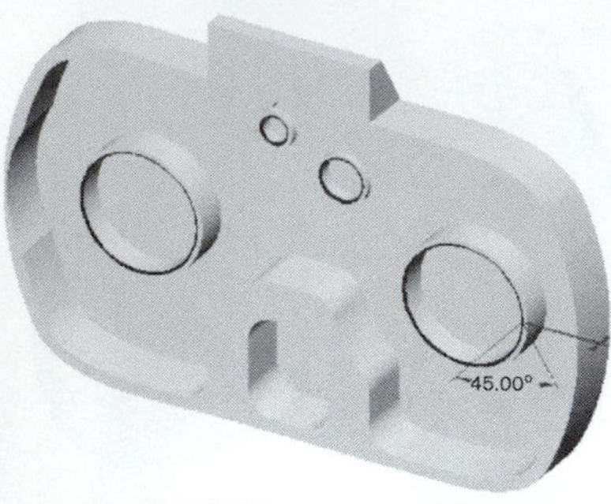

FIGURE 7.40
The chamfers were added to all four bearing bores at once.

the tops of the bosses, Larocque extruded cuts back toward the bottom. Because the boss is to serve as a bearing bore, the depth of the hole in the middle needs to correspond to the width of the bearing, not the height of the boss. By extruding it from the top of the boss, the hole will remain at the same depth even if the boss becomes taller or shorter. The same process was repeated to make the two smaller bosses.

The next feature was a chamfer on the inside edges of the bearing bores. SolidWorks allowed Larocque to multiselect all of the bosses at once, then insert a *chamfer* (see Fig. 7.40). A chamfer is a built-in feature with its own dialog box for setting its angle and depth. The single operation created the four chamfers, all to the same specifications. They comprise a single feature that cannot be decoupled into individual chamfers. Independent control of each chamfer would require creating four separate features.

Another built-in feature, a fillet, was added next. Larocque selected the two corners where the rectangular protrusion intersected the body of the part and added both fillets with the same operation.

The H-shaped pocket at the top of the plate was sketched as a symmetric feature next. This feature is an alignment pocket that should move with the bearing bores. Its location and radii were dimensioned from the center of the part (to coincide with the centers of the two 20-mm bosses), and its length constrained to the centerline of the part (see Fig. 7.41). "Most of the features after the base feature are dimensioned from the bearing bores because they are the defining characteristic of the part. I set up a relationship between the bearing bores and the alignment pocket because the alignment pocket determines where the bearings will fall."

The final features were the six holes and the counterbores to accommodate the fasteners to go into them. The first hole was given a diameter dimension, then the other holes were constrained to be the same size as the first (see Fig. 7.42). All six were sketched on the sketch plane used for the bevel contour and extruded with the "through all" setting so they will always go all the way through the part, even if the part thickness changes. From the same side of the part, the counterbores were added to the holes as the final features. Larocque specified counterbores at the appropriate depth that are concentric with the hole (see Fig. 7.43).

When Larocque picked the center point of the existing hole as the location for the counterbore, SolidWorks's AutoRelations feature interpreted his action as a concentric constraint. This feature can be toggled off to make it easier to place an object near another without snapping to a constrained relationship. SolidWorks makes it

FIGURE 7.41
The alignment pocket sketch was dimensioned from the center of the part.

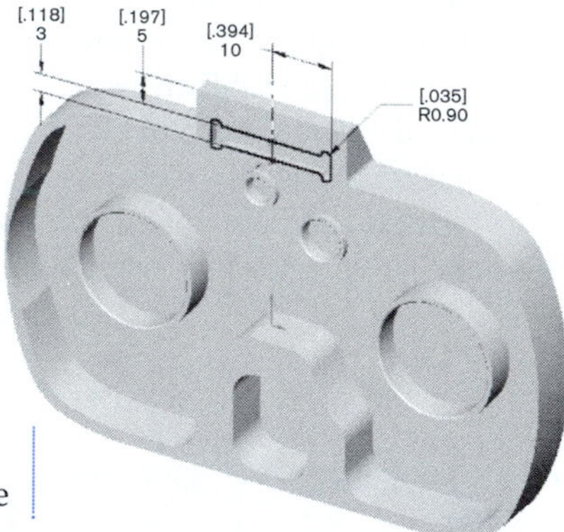

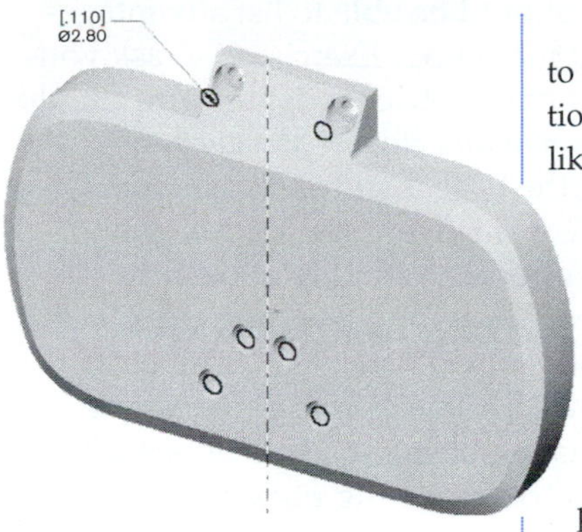

FIGURE 7.42
Only one of the holes displays a dimension here; the other five were constrained to have the same dimension as the first.

easy to review and modify any existing constraints. A feature can be selected in the 3D view of the model or in the Feature Manager (a chronological list of features on the left-hand side of the screen), and Display or Delete Relations will display all of the relations for the part so they can be edited. When two features are selected, Add Relations displays a list of all the possible relations those two geometric entities could have. For two circles, for example, the only possible constraints would be concentric or co-radial.

When asked about the degree to which he uses parametric relations, Larocque explained, "We like to keep the model somewhat segmented so we have more control over the updates, but we make sure to add relations where they make sense. For example, constraining symmetrical parts to a centerline as we did with the top plate of the hand can save enormous amounts of time when the part can update automatically. This aspect of the model was instrumental in our ability to make the BarrettHand lighter and less costly to manufacture."

FIGURE 7.43
The counterbores were the last features added to complete the plate model.

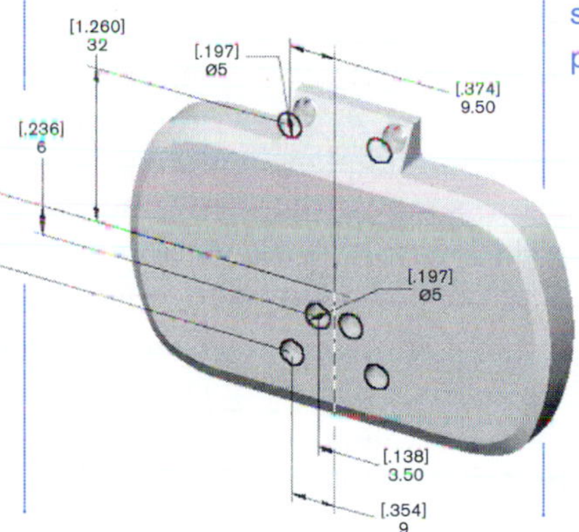

(Designs courtesy of Barrett Technology, Inc., Cambridge, MA.)

Brandon Larocque is project manager for Barrett Technology in Cambridge, Massachusetts. Trained as a mechanical engineer at Massachusetts Institute of Technology, Larocque began designing with pencil and paper before working with several 2D CAD packages, including AutoCad, and finally moving onto SolidWorks, a 3D parametric modeling and design automation package. Larocque now uses various solution partner programs for SolidWorks such as Cosmos FEA and Immersive Design's Interactive Product Animator, which let him simulate assemblies and create product animations.

Skills Summary

Having finished this chapter, you should be able to list advantages of parametric modeling over solid modeling. Exercises 1–3 ask you to answer general questions about parametric modeling. You should be able to correctly use the term *parent and child relationships* as applied to part features. Selecting the base feature of the model is important as it is the first parent feature. Exercises 4–6 ask you to demonstrate your understanding of parent-child relationships and base features. Associativity between drawing views and the 3D model is an important characteristic of sophisticated parametric modeling programs. Exercises 7 and 8 allow you to demonstrate your understanding of bidirectional associativity. Exercises 9–12 provide practice with your parametric modeling package.

EXERCISES

1. Parametric, or "feature-based" modeling uses defined relationships between model elements to control various features. What two basic types of parameters are used to control features?
2. Why are the linear, angular, or other physical dimensions of a parametric model feature given a name? Give an example where this practice might be useful.
3. An acronym sometimes used to describe the process of preparing for the creation of a parametric model is TAP, representing the steps Think, Analyze, and Plan. Consider what it means to think, analyze, and plan a drawing or model and compare the TAP process for 2D, 3D wireframe, 3D solid, and parametric modeling techniques. Why are these steps especially important prior to beginning the parametric model?
4. Using a parametric modeling package such as Pro/Engineer or Autodesk Mechanical Desktop, complete a parametric model of the GPS Antenna from Exercise 6-4 (the 2D orthographic projection shown below). Identify which parts you would design first and what you would create as the base feature. Consider the importance of relationships between the various elements of the part: coaxial connector, length, width, height, and shape.

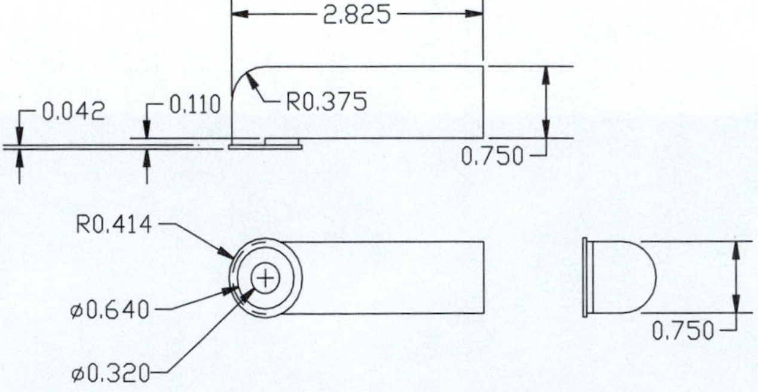

5. What is meant by "parent-child" relationship in parametric modeling?
6. For each of the parts to be modeled below, determine the feature you would use as the base feature and explain why you chose it as the base feature.

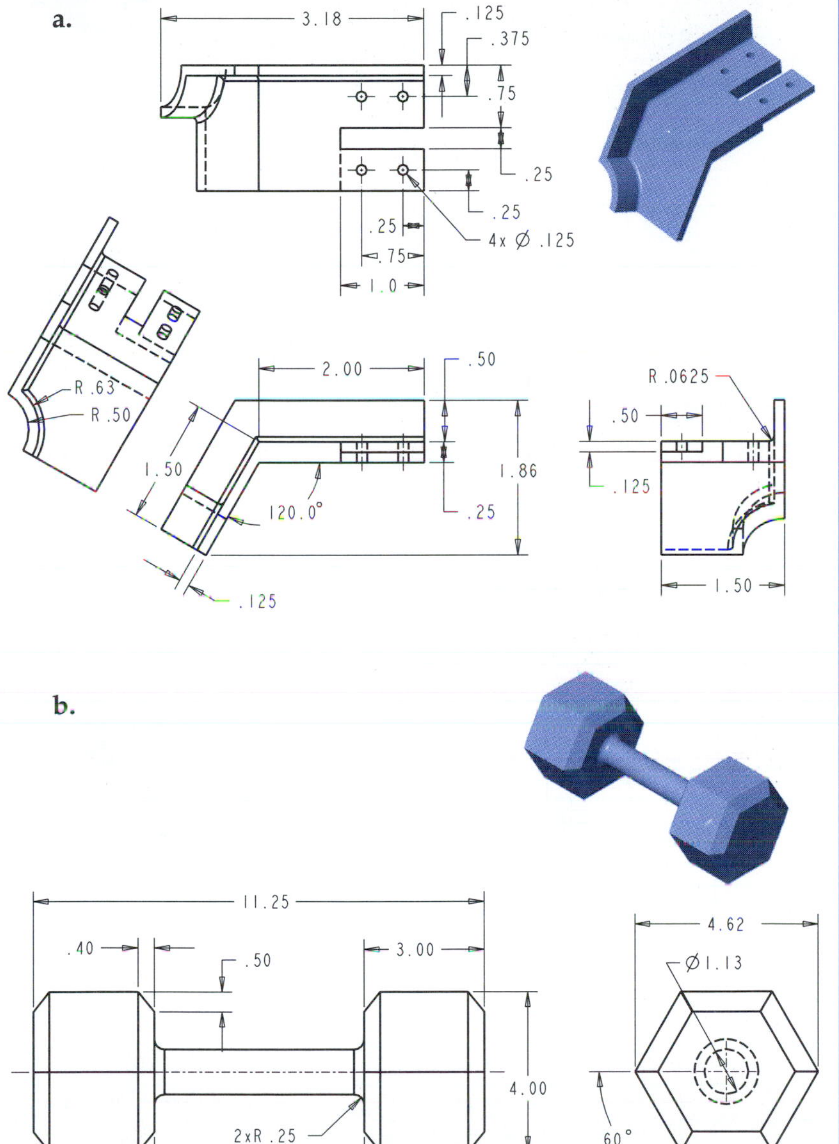

c.

d.

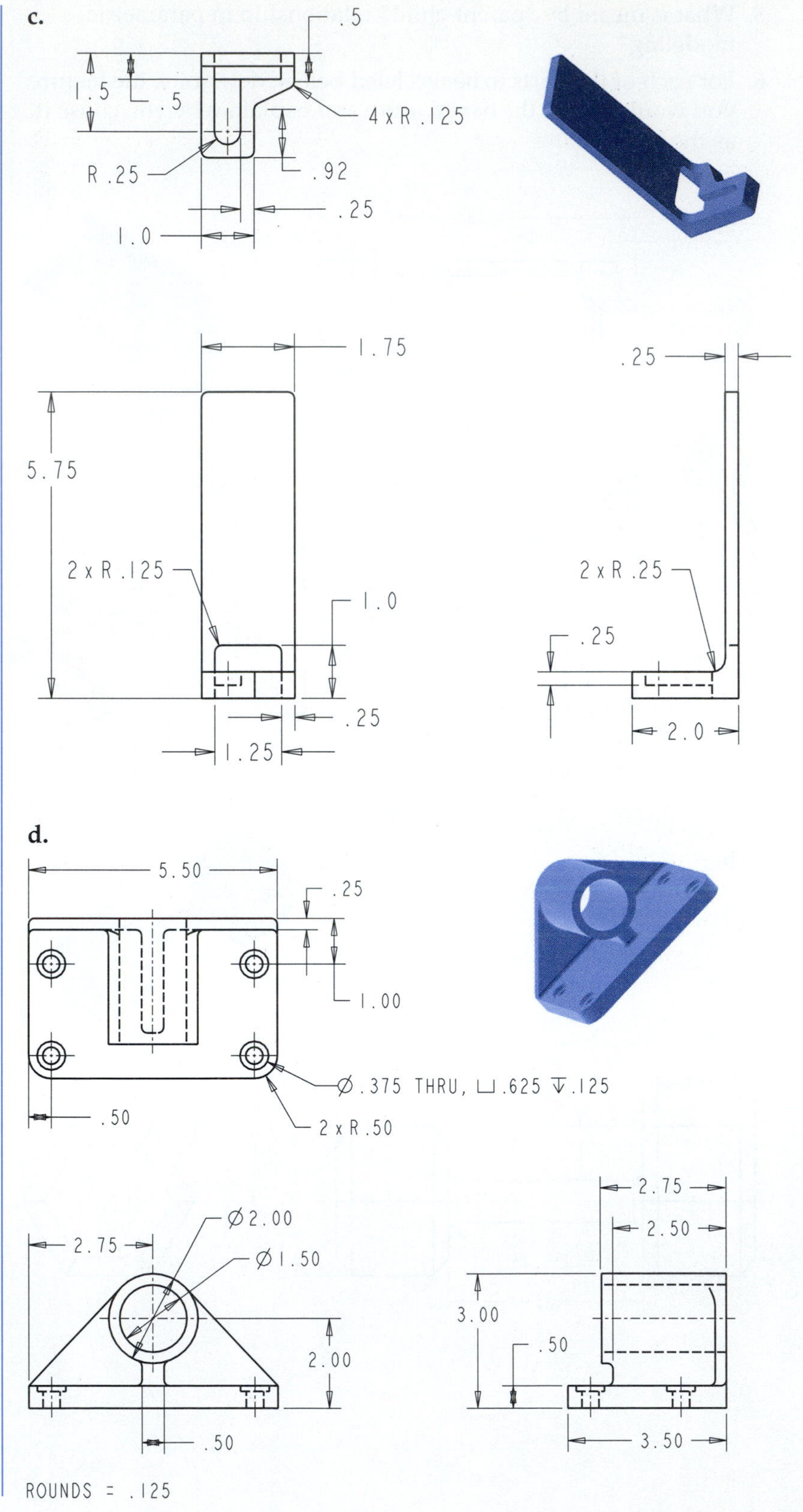
.5
1.5
.5
4 x R .125
R .25
.92
.25
1.0
1.75
.25
5.75
2 x R .125
2 x R .25
1.0
.25
.25
1.25
2.0
5.50
.25
1.00
Ø .375 THRU, ⌴ .625 ↧ .125
.50
2 x R .50
2.75
Ø 2.00
2.75
2.50
Ø 1.50
3.00
.50
2.00
.50
3.50
ROUNDS = .125

e.

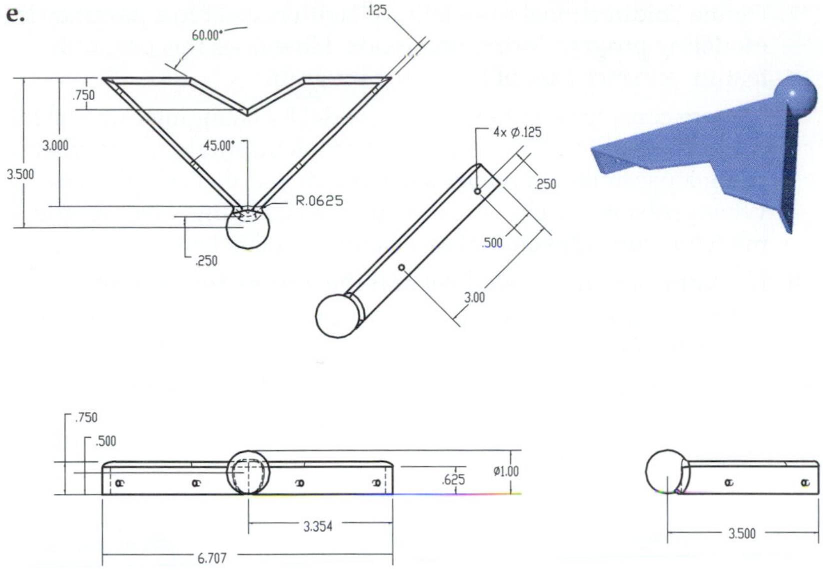

f.

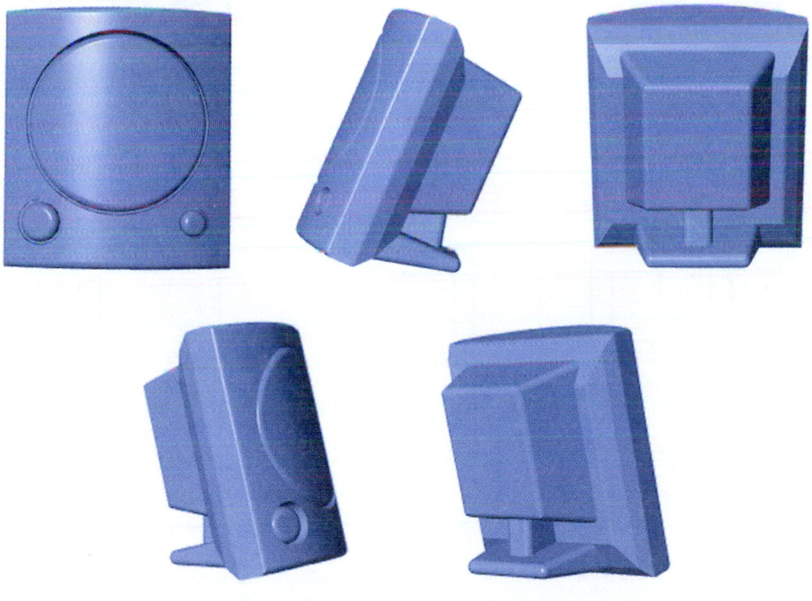

7. Define "bidirectional associativity" with respect to a parametric modeling program's drawing mode. How does this program feature enhance ease of use of the program?

8. You are considering modifying a model by changing dimensions in a drawing view. What potential problems might occur when using a parametric modeler *without* bidirectional associativity? What problems might you anticipate when using a parametric modeler *with* bidirectional associativity in this task?

9. Use your parametric modeler and the information in the Industry Case from Barrett Technology to create a model of the top plate of the BarrettHand.

10. Create a parametric model of the first part in each set below, save the part, then edit the part to create the other configurations in each set. Save all part files.

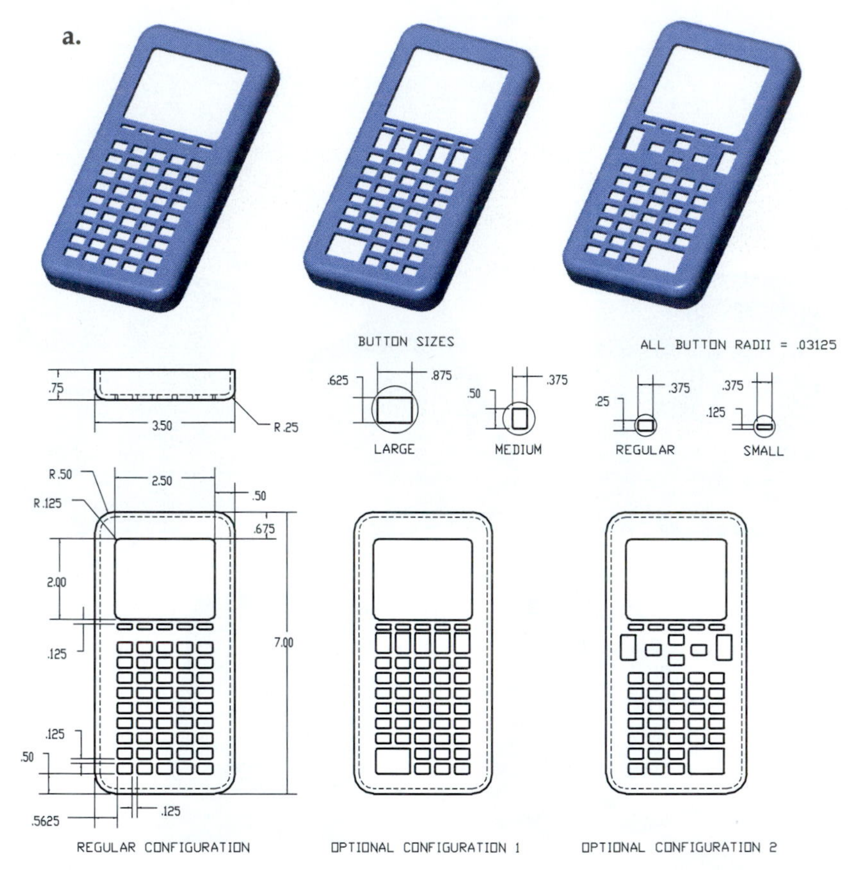

NOTE: CALCULATOR BODY THICKNESS = .125

b.

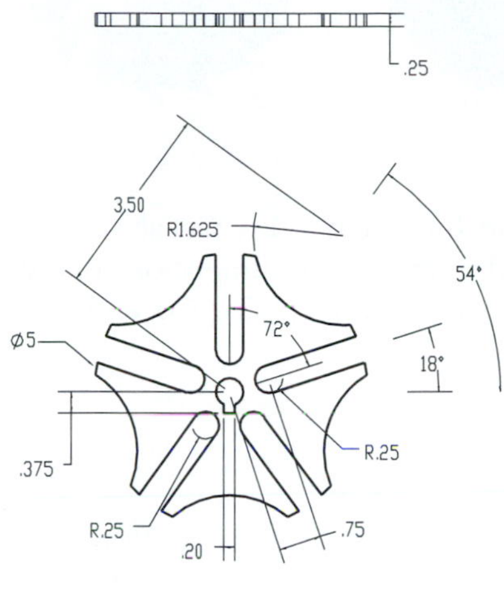

GENEVA GEAR CONFIGURATION # 1

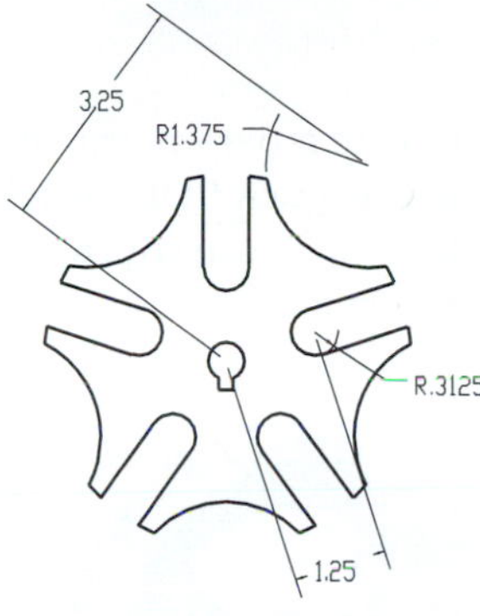

GENEVA GEAR CONFIGURATION # 2

c.

BASE FEATURE W/ REGULAR RIB

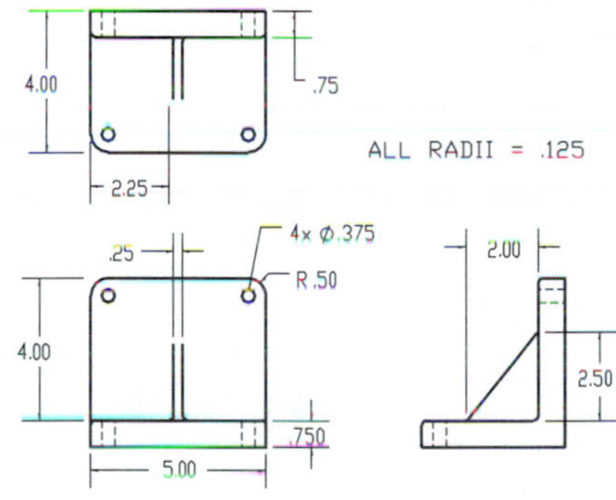

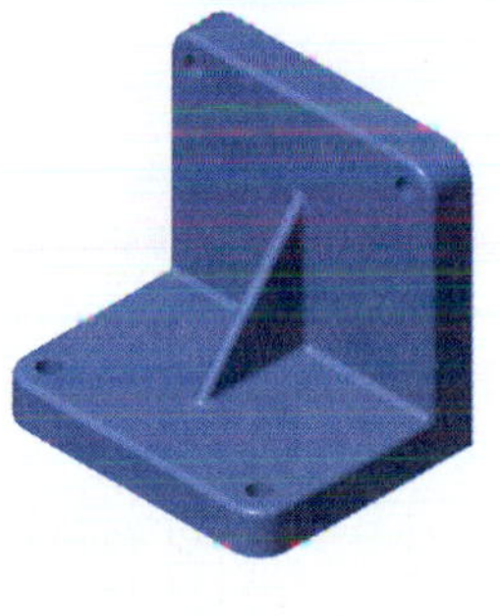

STRENGTHENED RIB # 1

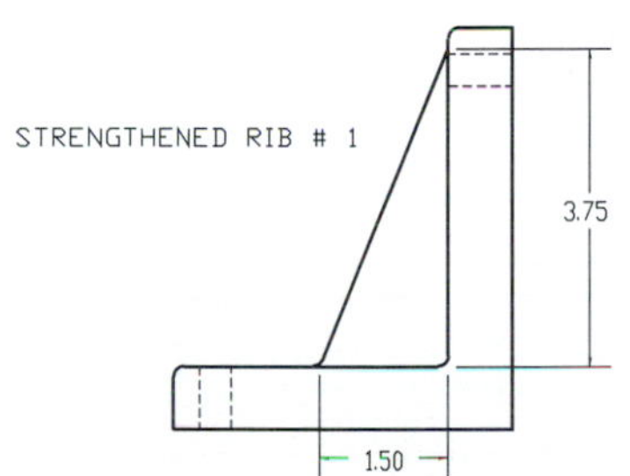

STRENGTHENED RIB # 2

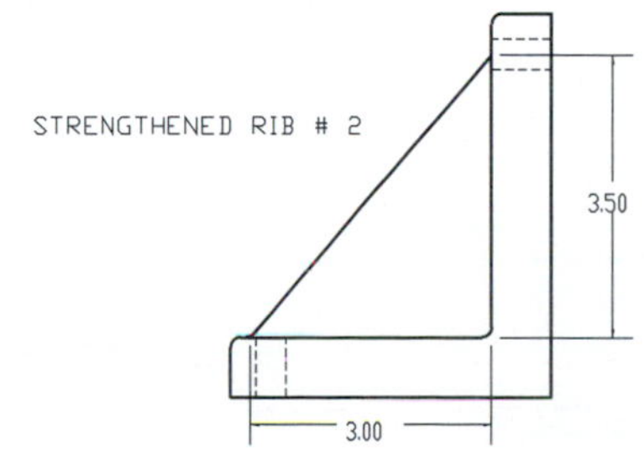

NOTE: RIB THICKNESS REMAINS .25

d.

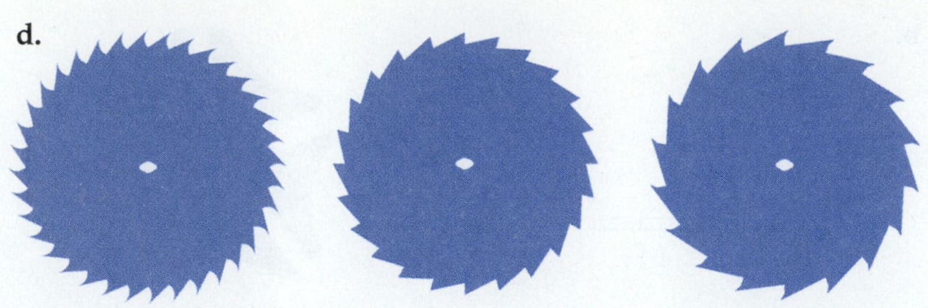

11. Create a parametric model of the 4-spoke hand wheel shown below such that it can be resized to match the dimensions in the table. (Figure courtesy of Carr Lane Manufacturing Co. and Earl E. Walker.)

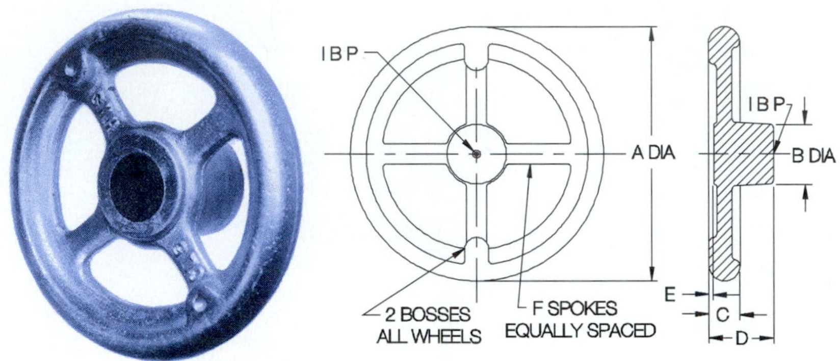

STRAIGHT SPOKES

PART NO. CAST IRON	A DIA	B DIA	C	D	E	F
CL-4-HWSF	4	1-1/4	5/8	1-11/16	1/8	4
CL-5-HWSF	5	1-1/2	3/4			
CL-6-HWSF	6	1-5/8		1-13/16		
CL-8-HWSF	8	1-7/8	1	2-1/16	5/32	
CL-10-HWSF	10	2-1/4		2-1/2		8
CL-12-HWSF	12	3		3-1/4	1/8	
CL-14-HWSF	14			3-3/4	3/32	

12. Create a parametric model of the swing washer shown below such that it can be resized to match the dimensions in the table. Capture size relationships between features in the parametric dimensions where possible. (Figure courtesy of Carr Lane Manufacturing Co. and Earl E. Walker.)

1018 STEEL, CARBURIZED-HARDENED, BLACK OXIDE FINISH

Permanently attached C washer that swings out of the way for clear loading. Can be reversed to swing into position either clockwise or counterclockwise. Shoulder Screw furnished.

A RADIUS — IBP — E DIA — C — B RADIUS — D — G DIA — IBP — H THD.

SCW / P / E R / 1SCW

USA

PART NO.	STUD SIZE	A	B	C	D	E DIA	G DIA	H	SHOULDER SCREW (FURNISHED)
CL-1-SCW	3/8 or M10	3/4	3/8	1	1/4	13/32	3/8	5/16-18	CL-24-SS
CL-2-SCW	1/2 or M12	1	1/2	1-1/4	3/8	17/32			CL-2-SS
CL-3-SCW	5/8 or M16	1-1/8	9/16	1-1/2	3/8	21/32	1/2	3/8-16	CL-3-SS
CL-4-SCW	3/4 or M20	1-1/4	5/8	1-3/4	1/2	13/16			CL-4-SS

METRIC

PART NO.	STUD SIZE	A	B	C	D	E DIA	G DIA	H	SHOULDER SCREW (FURNISHED)
CLM-1-SCW	M10 or 3/8	3/4	3/8	1	.236	13/32	10mm	M8	CLM-1006-SS
CLM-2-SCW	M12 or 1/2	1	1/2	1-1/4	.375	17/32			CLM-1010-SS
CLM-3-SCW	M16 or 5/8	1-1/8	9/16	1-1/2	.375	21/32	13mm	M10	CLM-1310-SS
CLM-4-SCW	M20 or 3/4	1-1/4	5/8	1-3/4	.472	13/16			CLM-1312-SS

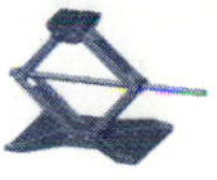

DESIGNER'S NOTEBOOK

Albert Brown's sketches for this Mini Drop-Jet Reservoir define some of the important relationships for features in the design. Notice the equations in the lower right defining the cap diameter equal to the body diameter (as shown in the revised sketch at the bottom left). When the parts are modeled parametrically this relationship can be used to build intelligence into the model. Then if the body diameter changes, the cap diameter can update to the correct size automatically. (Courtesy of Albert W. Brown, Jr., Affymax Research Institute.)

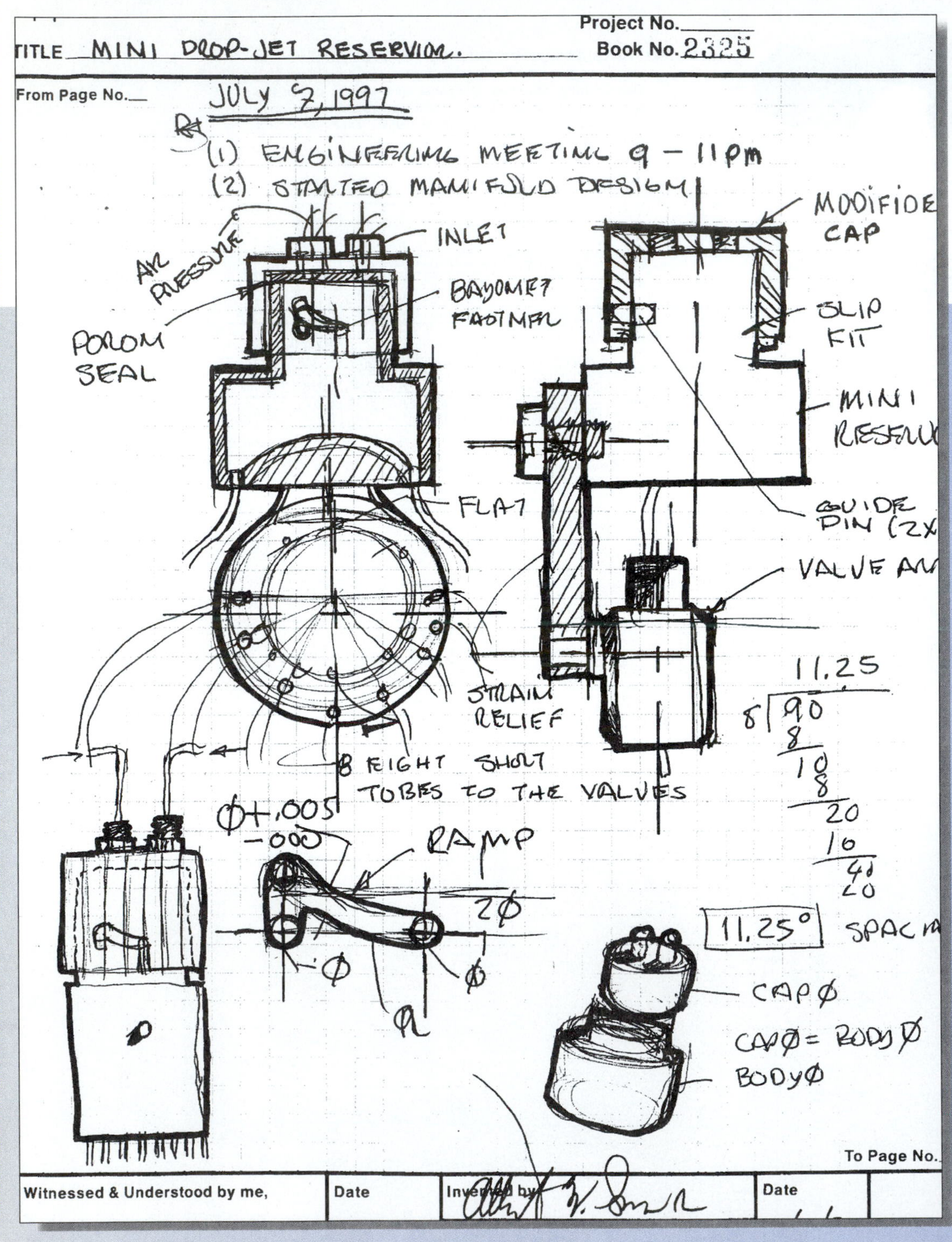

CHAPTER

8

Modeling for Manufacture and Assembly

Objectives

When you have completed this chapter you will be able to

- distinguish between static and dynamic assemblies;
- explain how parts can be added to an assembly;
- describe the role of intelligent assemblies in top-down design;
- identify how standard and static parts are used in assembly models;
- describe issues in modeling fasteners and springs;
- use an assembly model to test for fit and interference;
- describe manufacturing considerations to be reflected in a model.

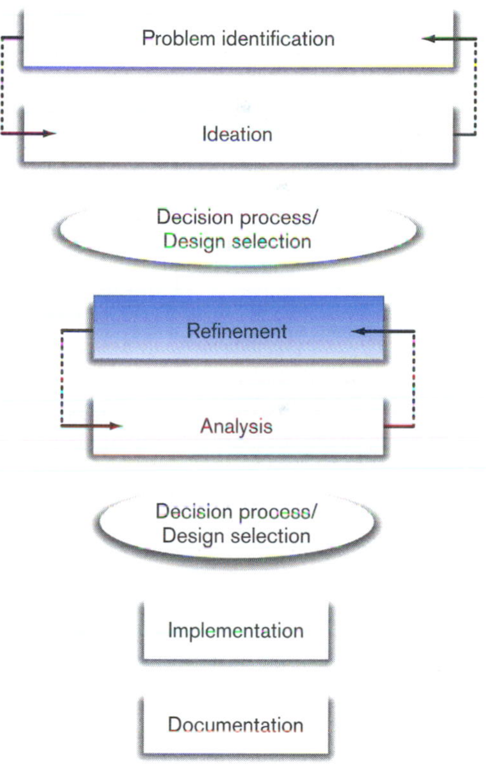

When you use 3D modeling for design, you are building a 3D database of the design. This database is stored electronically on a computer system, either on your local hard drive or on a network. If parts are modeled accurately, you will be able to use the database to interface directly with computer-aided manufacturing systems to create molds, dies, machined parts, and other fabrication elements.

Modeling accurate parts that reflect the design intent is the foundation of a good 3D database. But today's modeling software does more to help you ensure that your design will be ready for manufacture. You can use the information in the database to see how parts will fit together in the final product. Individual parts can be checked to see if they interfere with others in the assembly. Variations in manufacturing can be anticipated and evaluated in the model to see whether they will affect the function of the design—and whether parts will be rejected because they cannot be assembled. Finally, but perhaps most important, considering the design intent of the entire assembly up front can make it easier to work with others concurrently and to produce models of complex designs more effectively.

Combining Parts in an Assembly Model

Your electronic database should contain all of the individual part files created for the design. The database may be organized by levels within the design. At the highest level, all of the parts are shown in the final assembly. (Fig. 8.1 shows the many individual parts in a coffee brewer assembly.) Within a large assembly, subgroups may be broken out into subassemblies. *Subassemblies* are groups of parts that

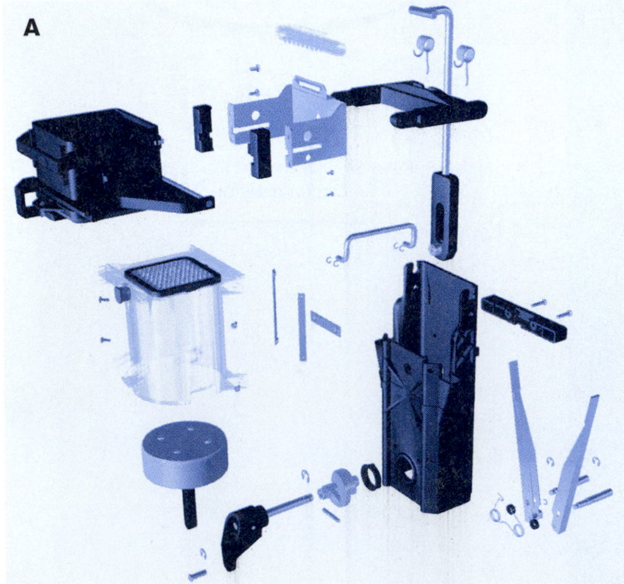

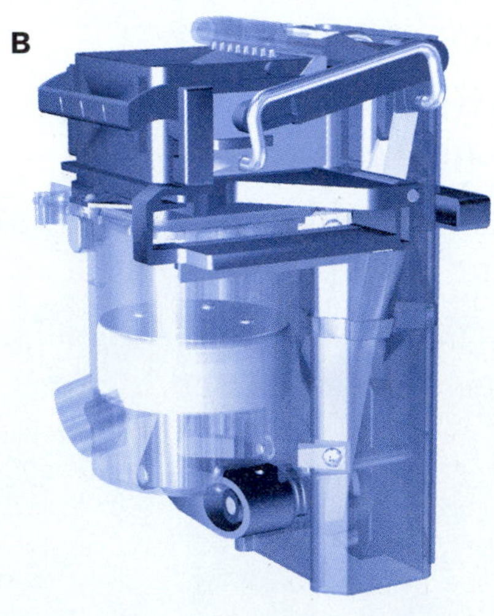

FIGURE 8.1
The exploded view of the Zuma coffee brewer in (a) shows the individual parts that make up the assembly, shown assembled in (b). Each part is stored in its own part file and linked through external references to the assembly model. (Courtesy of VKI Technologies, Inc.)

fit together to create one functional unit, often one that can be pre-assembled in some fashion to fit into the assembly as a unit. Breaking a design into levels can make it easier to divide tasks among different groups working on the same project, and it can help in presenting large, complex assemblies.

You can combine individual parts into assemblies using your 3D CAD software in different ways. One method is to insert *copies* of your solid parts into a single assembly model. Assemblies that are created by copying 3D solid parts like these are *static assemblies*; that is, they don't change to reflect alterations in the individual parts. Because there is no link established between the part files and the assembly, if the individual parts are changed, they must be recopied or re-inserted into the assembly drawing to update the appearance of the assembly. The two drawings are completely separate, and changes made to either one will not affect the other. This is in contrast to a *dynamic assembly*, in which parts are imported into the assembly through a linking process that allows the software to update the assembly when individual part files are modified.

External References

External references (as used in AutoCAD and Mechanical Desktop) are one way of creating a dynamic assembly. An *external reference* (called an XREF in AutoCAD) doesn't copy the part file into the assembly; it creates a link to the part file that imports a viewable representation of the part into the assembly file. If the part file changes, the assembly will automatically show any changes that have been made to the part. When you are using externally referenced parts in an assembly, you must have the externally referenced files available in order to view them. The data displayed in the assembly is stored in the externally referenced files and imported through the link each time you open the assembly. If the part files are not available, you cannot view them in the assembly drawing.

Figure 8.2 illustrates the difference between assemblies created by copying parts from one file to another, and those created using external references. When part file A is copied and placed into the assembly file, changes to the original part file do not change the assembly file. When part file A is externally referenced by the assembly file, as in Figure 8.2(b), changing part file A also changes the assembly file. An advantage of external references is that they do not copy the contents of the part file to the assembly model; this helps to keep the size of the assembly file from getting overly large.

It is important to consider the links between assembly and part files when you are managing your work. Consider, for example, a part that is linked to two different dynamic assemblies. If you make a change to this linked part, it will change in both assemblies. If that is your intention, it can be a great time saver. If you are modifying the part to change it in only one of the assemblies, you will get unwanted results the next time you open the second assembly drawing. And if you delete a drawing that is referenced by other drawings, the referenced information will not be available.

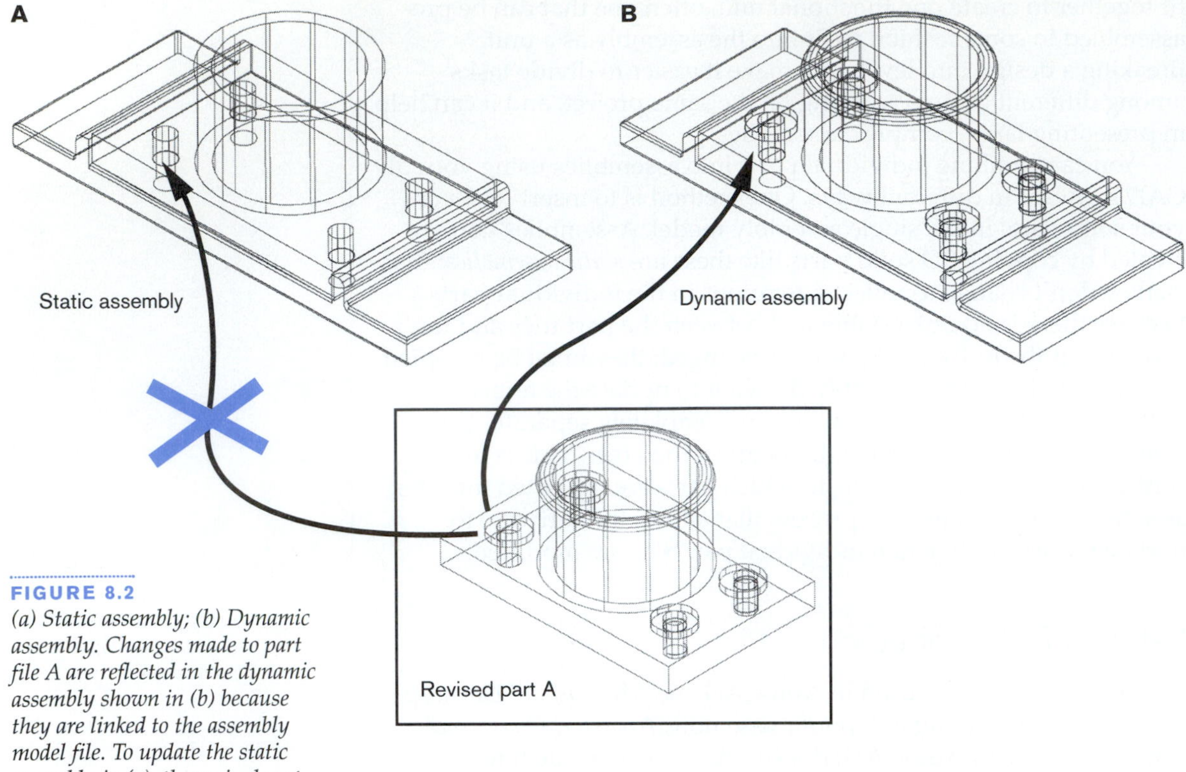

FIGURE 8.2
(a) Static assembly; (b) Dynamic assembly. Changes made to part file A are reflected in the dynamic assembly shown in (b) because they are linked to the assembly model file. To update the static assembly in (a), the revised part would have to be copied into the assembly file again.

Parametric Assemblies

Parametric assemblies are similar to assemblies that are created using externally referenced parts. When you create a parametric assembly, the part file is linked, not copied, to the assembly drawing. Like assemblies with external references, parametric assembly files stored on your computer system may rely on subassembly and part files to be available in order for you to open them for viewing or making changes. If you lose the file for a part, it will not show up in the assembly.

The bidirectional associativity built into many parametric modelers adds an additional dimension to parametric assemblies. Remember that this type of associativity lets you alter a model by making changes to a drawing view of the model. In the same way, some parametric modeling software allows you to make changes to parts in the assembly model that automatically update and store the changes in the individual part files. This two-way updating capability is illustrated in Figure 8.3.

Parametric assemblies also differ in the parent-child relationships that are built into the model. The first part that is added to the assembly may act as the parent part for other parts you add to the assembly, just as the base feature is the parent for any other features in the part model. Remember that these parent-child relationships in a parametric model are critical. If the parent is moved, the children

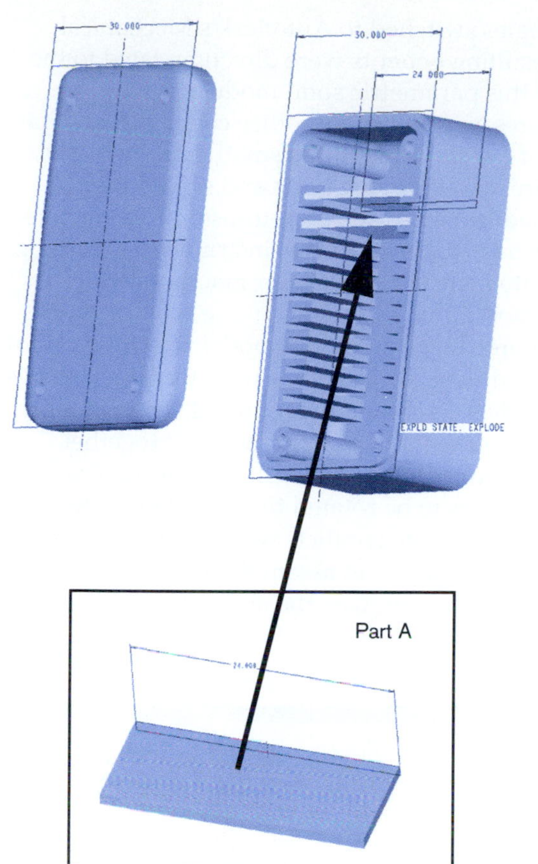

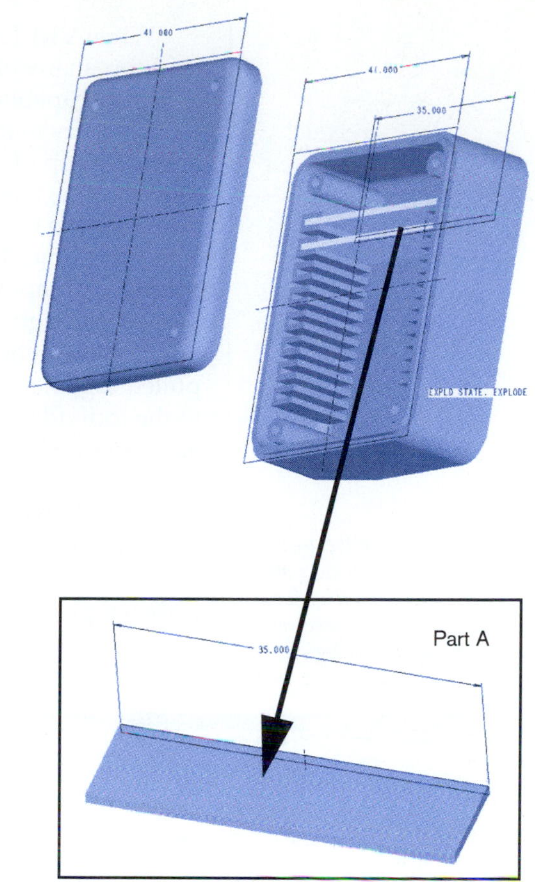

FIGURE 8.3
The nature of the links between parametric part and assembly files is analogous to the bidirectional associativity between model and drawing. This figure illustrates how a change made to part A in the parametric assembly updates the part file, and vice versa. The link between the part file in a dynamic assembly using external references is one-way only; you cannot make changes in the assembly that will be stored with the part.

move along with it. This can be a great advantage: if one of the parts in the assembly moves or changes, the child parts move to update their positions relative to it. It also follows, however, that if you delete a part that is the parent of some other part in the assembly, then the child part's location will not be defined. As the modeling software tries to generate the assembly from the parts and relationships stored in the database, undefined parts could cause the file to fail to open or the software to crash. One way to prevent this is to insert a standard empty file as the first part in your assembly so that all other parts can be defined as children of this blank part that you will never change.

Using Dynamic Assemblies

Dynamic assemblies provide an advantage to design teams that share a network of files. The assembly model can help coordinate individual part design with the overall product effort. Most systems will let you dynamically update the appearance of a part if someone is actively changing it while you are viewing the assembly. This keeps the assembly current and ensures that the part you are designing will fit with the other parts in the assembly.

When VKI Technologies switched to Autodesk's Mechanical Desktop, many of the resulting benefits were directly related to the assembly capabilities of this parametric solid modeler. VKI Technologies manufactures and distributes office coffee makers that brew one cup of coffee at a time. While these small machines may seem simple, they require many moving parts and sophisticated electronics to be contained in a compact, easy-to-use, and attractive coffee machine. The Mechanical Desktop was instrumental in fitting the many components into increasingly smaller machines while reducing design time almost 50%.

To design the Zuma brewer, an assembly model was created that pulled together sixty-two different parts by using external references to the individual part models, as shown in Figure 8.4. The assembly allowed the designer to be sure that the components fit together compactly yet didn't interfere with one another. The Mechanical Desktop allowed moving parts to be rotated through their range of motion in the assembly to eliminate conflicts with other parts, too.

One of the biggest advantages of an assembly model is its role as a virtual prototype. For the Zuma brewer, the ability to visualize

FIGURE 8.4
The individual parts in the assembly model of the Zuma brewer are listed in the window on the left side of the Mechanical Desktop screen. (Courtesy of VKI Technologies, Inc.)

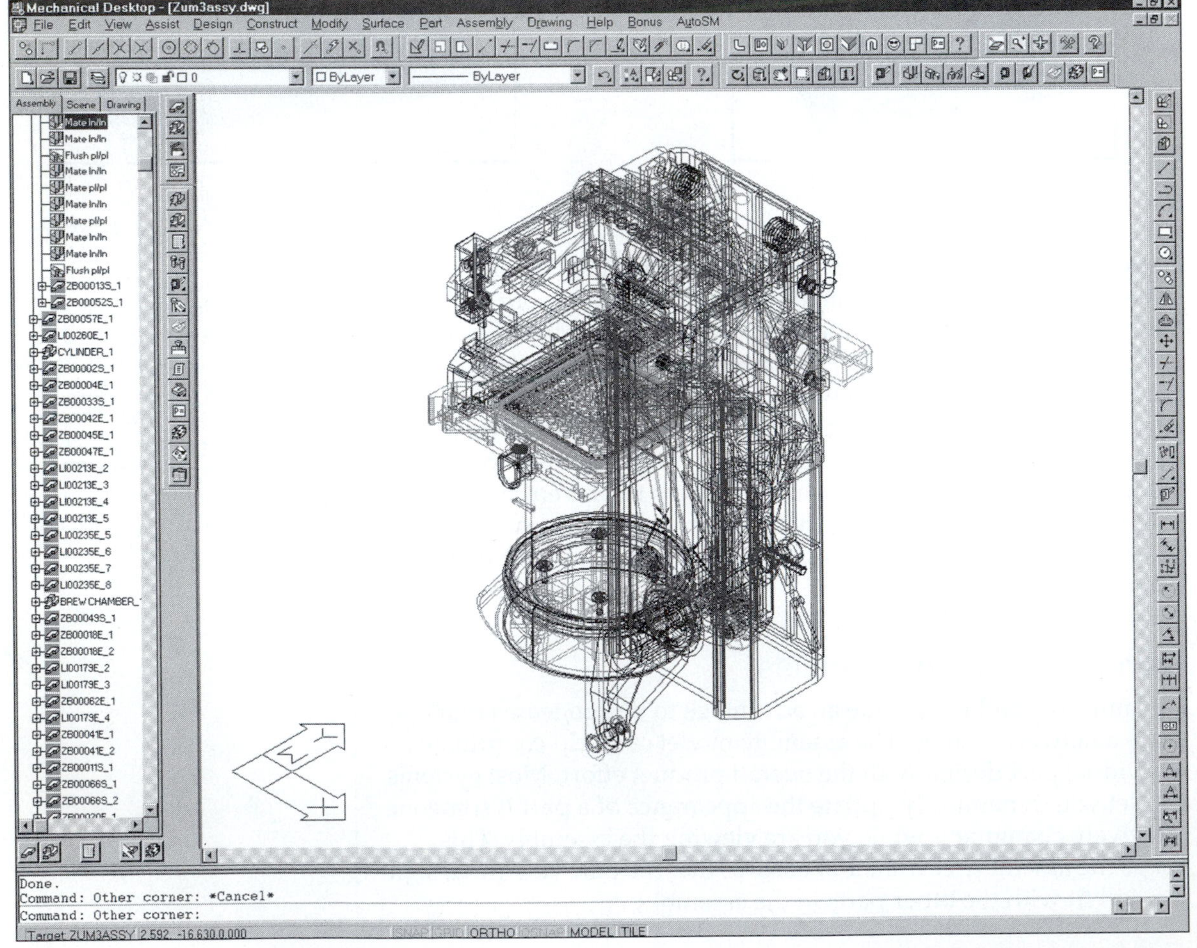

parts in 3D and evaluate them for fit with other parts eliminated the need to create physical models of the parts in many cases. When the brewer design was complete, the model was used to generate molds for the fifteen plastic parts in the brewer. When the real molded parts were available, they fit together well enough to make a functional brewer from the first set of molded parts. With 2D methods, each part would have to be machined or created on the shop floor to fully visualize it in 3D and to ensure its fit with another component. With many parts and a limited amount of time, it was not uncommon for problems of fit and interference to be discovered after manufacture, not before. The 3D assembly model created in Mechanical Desktop eliminated much of the time and cost of checking the design for manufacturing and assembly issues.

Managing Assembly Files

The links between files that make dynamic assemblies possible also underscore the importance of managing your files in a way so that you don't delete needed files. You should keep your files organized on disk by using directory structures and naming conventions that eliminate the need to continually hunt for parts of a project. The extra time it takes to name and store your file in the appropriate directory can save hours of searching for the file at a later date.

File management is especially important when you are working in a design team that's using a networked computer system. You must take responsibility to manage your files so that they are easy to find for everyone. Without good communication and an organized system for storing files, it is not unusual for one engineer to spend considerable time modifying a part only to have the latest modification not be used in the assembly because it was stored in the wrong directory.

It can also happen that two different people will modify the same part, undoing each other's changes. If you work as part of a team, it should be clear who has the responsibility for making changes to different parts. One person should have "ownership" and the ability to make changes to the part file. If a change in another part requires a change to a part you are responsible for, you should be notified of the change to the other part so you can update your part accordingly. Some systems control changes to part files by allowing other team members to work with the file in read-only mode. Everyone can see the part, and perhaps even indicate changes in the drawing, but the changes will not actually be made to the part until the owner of the part file OKs them. This can prevent different engineers from concurrently changing the same part.

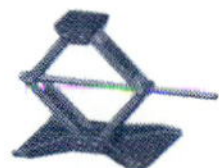

Your modeling software may have features and commands that affect how files are linked or externally referenced and the implications for others who may access the files. As you learn to create assembly models, you should be aware of these options and how your work group chooses to use them. Some companies invest in drawing management systems that help keep the design database organized. You will learn more about drawing management in the next chapter.

Assembly Constraints

How you add parts to an assembly model is another key difference between solid modeling software and parametric modeling software. To add a copy of a part or an external reference, you locate the geometry in the assembly using the same tools you use when creating the part model. In most solid modeling software, you specify a base point for the part, then tell the software where to locate that base point when the part is added to the assembly. Establishing a meaningful base point when you save the part file can make it easier to add parts to an assembly. Using the center point of a washer as a base point, for example, can make it easy to insert the part in alignment with the centerline of a hole in another part.

With parametric modeling software, you use assembly constraints to create relationships between parts. Remember that the first part you add to the assembly becomes the parent part. As you add parts, you specify the geometric constraints that reflect the design intent for the assembly. Mating parts have features that should fit together in an assembly. When you create an assembly using parametric modeling, you can use the assembly constraints available to align mating parts to capture your design intent in the assembly. For example, if you want two holes to line up, you can use an assembly constraint to align them. If a part changes, it will still be oriented in the assembly so that the holes align. Each software package will offer a similar set of constraint options, so you should become familiar with those available to you. Table 8.1 lists some of the common Pro/Engineer assembly constraints and their definitions.

Just as feature relationships are important to the way you create a model, assembly relationships can make your assembly model work for you. As you add parts, use constraints that orient the new part using relationships that will persist in the assembly.

You can create a subassembly in much the same way you create an assembly: by making an assembly of the subassembly components. This subassembly can be added to the main assembly in the same way you add a part. Organizing the model so that it comes together as it will on the assembly line can be useful in visualizing assembly difficulties. If a group of components are likely to be changed or replaced, linking all the subparts to a main component can make it easy to substitute an alternative design for that group of parts.

Planning ahead is essential to creating assemblies efficiently and getting the most out of them, as you will see in the next section.

Intelligent Assemblies

So far, we've talked about assemblies as a means of pulling all the parts of your design together. To work efficiently and concurrently, it is important that each designer be clear on the design intent so that problems of fit are minimized at later stages. Parametric modeling software also allows you to start with an assembly framework that can be used to define the design intent or parameters for individual parts and to help coordinate the work of different team members.

TABLE 8.1

Assembly Constraints

Name	Definition	Illustration
Mate	Mates two planar surfaces together.	Mate Mate Mate
Mate Offset	Mates two surfaces together so they have an offset between them.	Mate Mate Mate offset Offset
Align	Aligns two surfaces, datum points, vertices, or curve ends to be coplanar; also aligns revolved surfaces or axes to be coaxial.	Align Align Align
Align Offset	Aligns two planar surfaces with an offset between them.	Align Mate Align offset Offset
Insert	Inserts a "male" revolved surface into a "female" revolved surface, aligning the axes.	Insert Insert Mate
Orient	Orients two planar surfaces to be parallel and facing in the same direction.	Orient Insert Mate
Coordinate System	Places a component into an assembly by aligning its coordinate system with an assembly coordinate system.	Coord sys Assembly coordinate system

Using this method, parts are designed so they link to a skeleton framework in the assembly. A *skeleton* is a 3D drawing that defines major relationships in the assembly using lines, arcs, curves, and points. By creating the framework for each part up front, all parts do not have to be finished before they can be assembled. Parts can be assembled onto the skeleton at any stage of completion. By allowing the assembly to evolve as the parts are designed and refined, each designer can see the parts the others are creating—or at least the critical relationships between parts—by looking at the assembly.

Top-down design is a term used to designate starting the design by thinking about the function of the entire system, breaking that down into subassemblies or component groups based on their major functions, and, finally, defining each part that must be manufactured and assembled to create the design. *Bottom-up design* usually refers to a process that starts at the part level, sizing individual components and building the design up from them. This is typically the approach when components are standardized parts. The term *middle-out design* has been used to refer to a combination of these two methods, where some major standardized parts are used in the assembly to begin the design. The new components are designed to fit with these parts and function in the overall assembly.

A good example of middle-out design is the Romulus Predator, the "supercar" featured in Chapter 6, which combines racing capability with everyday street driving. Like most exotic cars, the Predator will never be mass-produced, so design and production costs must be recouped from a relatively small manufacturing run. To remain competitive, however, the design team made the decision to use a number of commercially available components, such as brakes, steering column, steering wheel, and engine, to keep development costs down. Even though the parts were to be purchased off the shelf, they were modeled in 3D to be part of the digital assembly for the car. Custom parts could then be designed to work with them as if they were part of the original design. The result is a selling price of only $150,000 for the Predator (shown in Fig. 8.5)—in a market in which cars sell for $250,000 to $1 million.

Layout Drawings

Traditionally, top-down design has been facilitated by layout drawings. Accurate layout drawings were created on the board to show the relationships between major functional items in the design. Layout drawings are especially useful when the product or system being designed has to fit with existing equipment. By defining the critical distances for fits with the existing equipment in the layout drawing, the design team has a parameter for the window of operation for new equipment.

FIGURE 8.5
The Predator's use of off-the-shelf components is an example of middle-out design. The high-performance engine and other components were selected and assembled to work within the supercar aerodynamics of the body design. This view of the Predator was taken from the surface model created in ICEM Surf. (Courtesy of M&L Auto Specialists, Inc.)

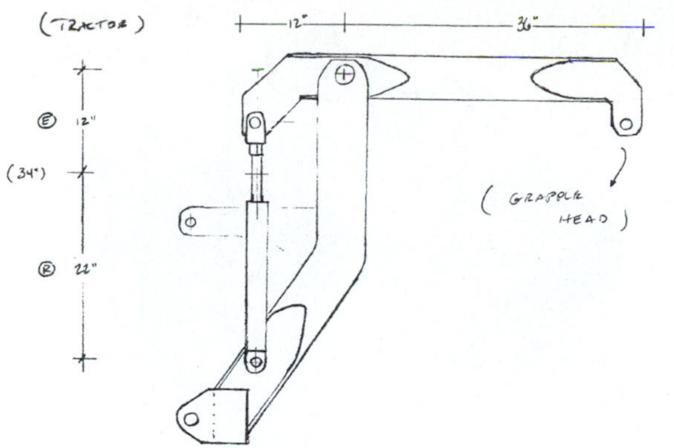

FIGURE 8.6
Layout drawing (Courtesy of Implemax Equipment Co., Inc.)

A typical layout drawing is shown in Figure 8.6. Note that the drawing is not complete in every detail, but shows the major centerlines, sizes, and relationships between parts in the assembly.

Skeleton Models

A wireframe skeleton in a 3D model serves the same purpose as a layout drawing. It is a framework on which the individual components can be located. Figure 8.7 shows a skeleton model used to control the location and position of parts in the assembly model of a clamp.

To create a skeleton model, first define the critical dimensions in your assembly. What dimensions are fixed—due to a physical space requirement, sizing requirement, or some other relationship? What dimensions are likely to change later in the design process? How will each part relate to the others in the assembly? Begin your skeleton as a part drawing made up of a parametric framework of 3D planes, lines, curves, and points that identify the basic relationships between parts in the assembly.

Figure 8.8 illustrates a skeleton model created for a laser printer designed and manufactured by Tektronix, Inc. Laser printers, like most "mechantronic" devices, include a large number of mechanical and electronic components organized into subassemblies. The printer division creates skeleton models of each subassembly, which can then be combined into a top-level assembly.

The skeleton drawing defines the interfaces where parts must come together. For the laser printer, which has a lot of moving parts, the skeleton can also define the limits within which moving parts must operate. By establishing this framework up front, each designer can upload a part to the assembly at any time to see how it will operate within the constraints established by the framework. Figure 8.8(b) shows how the subassembly might look with just a few parts attached to it. The completed assembly model is shown in Figure 8.9.

The actual part drawings can be assembled directly onto this skeleton framework. Because the skeleton is made up of lines, planes, points, and other entities that do not have volume, using a skeleton will not change the mass property analysis for the assembly.

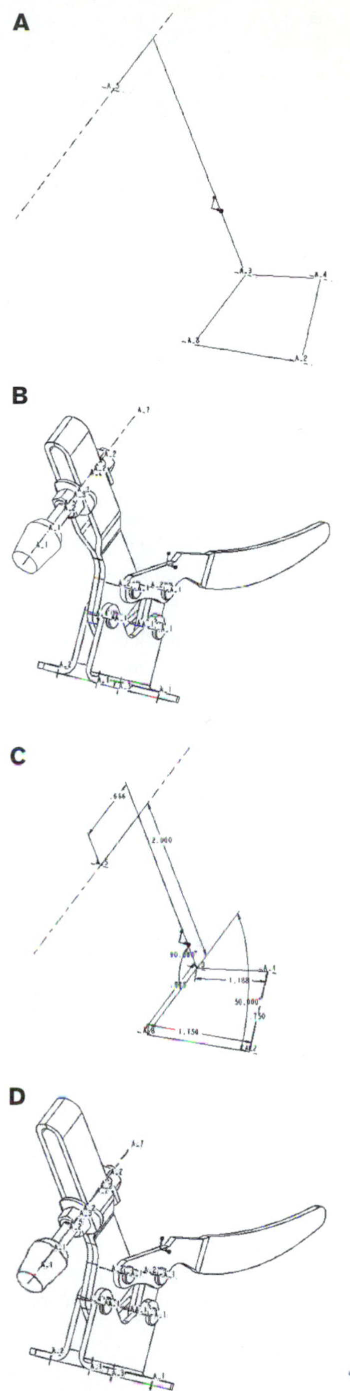

FIGURE 8.7
(a) Skeleton model for the clamp assembly; (b) Parts assembled onto the skeleton; (c) Changing the dimensions of the skeleton controls the positions in the assembly; (d) Resulting change in the assembly. (Courtesy of Mark Perkins.)

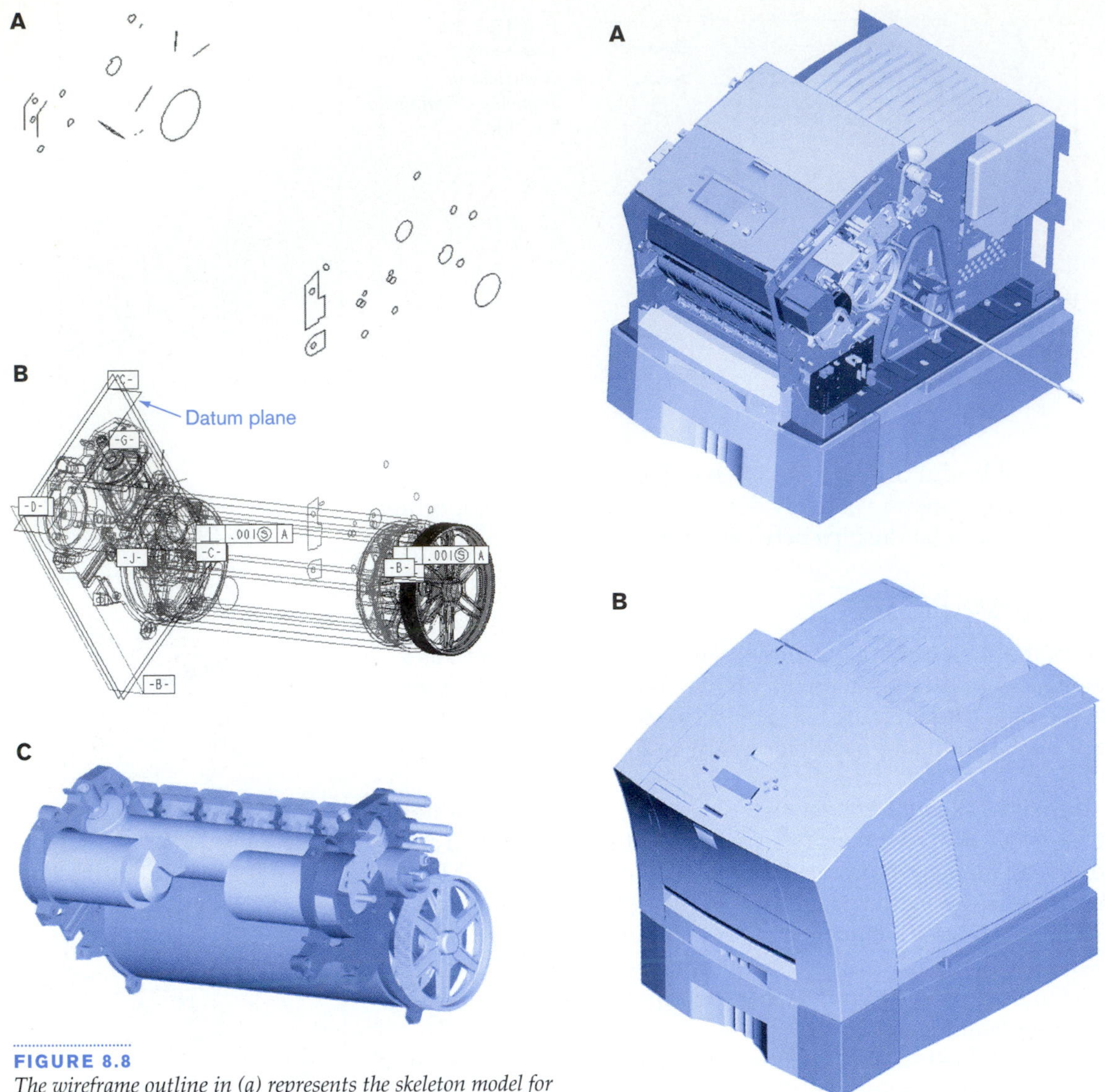

FIGURE 8.8
The wireframe outline in (a) represents the skeleton model for a subassembly of a laser printer. The skeleton establishes a framework in 3D space and defines relationships between parts or groups of parts. Each of the elements in the skeleton locates a centerpoint or an edge on a set of datum planes and axes, shown in (b). As parts are completed, they can be added to each subassembly of the printer in the proper location, as shown in (b). The completed subassembly is shown in (c). (Courtesy of the Color Printing and Imaging Division of Tektronix, Inc.)

FIGURE 8.9
The completed subassemblies for the printer are combined into a complete assembly model of the printer. The printer assembly is shown in (a) with the covers removed and in (b) as the finished printer will appear. (Courtesy of the Color Printing and Imaging Division of Tektronix, Inc.)

Because the dimensions between the planes, lines, points, and other entities making up the skeleton are parametric, they can be changed as needed as the design evolves to preserve the interfaces between parts. Changing the skeleton's dimensions also allows you to check the clearances and interferences between parts as the device is operated to different positions.

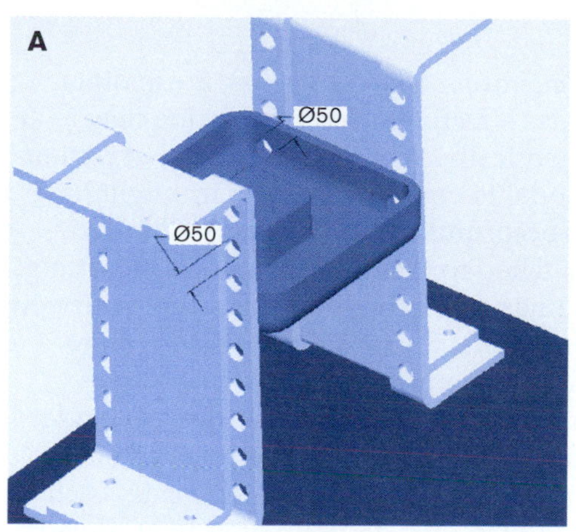

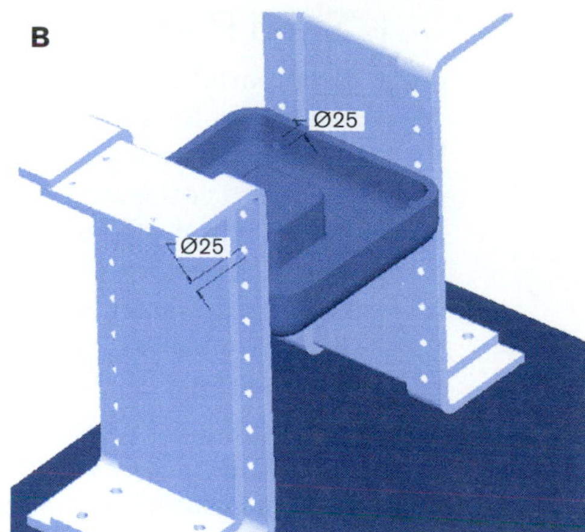

FIGURE 8.10
The size of the holes in the brackets and tray in (a) are both defined by the same dimension: a size parameter called Hole_Size that has the value .50. This global parameter is used for all holes in the assembly that need to fit a round head machine screw of diameter .50. If a different fastener is selected, all of the holes can be changed automatically by changing the global dimension parameter. In (b), the global parameter has been changed to .25.

Global Parameters

As you define the skeleton framework, you may also define *global parameters* for the assembly. You could create parts directly from the skeleton so that the parts inherit the parameters used to define the skeleton, but global parameters may be needed to capture other aspects of the design intent.

Remember that global parameters are parameters that are the same across multiple parts. You can use global parameters to control the size of a feature on one part so that mating features on other parts update automatically. Consider the fit between the mating parts shown in Figure 8.10. The holes on the mounting brackets and tray must align and must be roughly the same size so that a machine screw will fit through both parts. By creating a global parameter for the size of the hole, called in this case Hole_size, you can change the size of the hole feature on both the brackets and the tray at once.

Global parameters can be defined for any dimension to be shared by parts in the assembly. You may define a global parameter from the skeleton itself to drive critical distances on part features.

Global parameters also make it possible to build relationships into the assembly as you would for an individual part. The drilling rig in Figure 8.11 contains more than five thousand parts but was designed to be sized easily to fit different drilling situations. Changes can be made at the assembly level (the top level) to change the width from any corner, the overall height, and the height of any bay. Equations in the dimension parameters for individual parts reference global parameters that, when changed, cause all parts in the assembly to update to the new size.

FIGURE 8.11
The basic style of this drilling rig can be generated in different sizes from the same parametric model. IRI International has a set of models for different derrick styles. For a new rig, the model of the basic structure can be reshaped in just a few days so work can begin on the design of custom parts for the rig. (Rendered model view courtesy of Wayne Fehres and IRI International.)

Seed Parts

Seed parts, also called *templates* or *prototype drawings,* are another technique for assembling parts effectively and for starting new drawings quickly and systematically. *Seed parts* are files that contain elements you want every model to contain. A seed part might have a set of datum planes defined according to the company standard, named view orientations matched to the datum planes, unit settings, coordinate systems, layer names, and other items. Starting your new parts from a seed part saves you the time it would take to create these settings and elements in a new file.

Seed parts can help you assemble parts by providing a standard orientation. For example, when you assemble two parts, you may want a common datum surface to face the same direction on each part. Using a seed part, where standard datum planes have already been set up and named to make them easy to identify, can help you quickly insert your part in the correct orientation with respect to the rest of the assembly.

Seed parts with consistent layer names can eliminate confusion in assembly drawings, too. In most software, each time you add a part to an assembly, all of its layers are also added. By starting parts with the same basic set of layers, you can keep the assembly organized and make it easy for other users to identify on which layer a feature would typically be shown.

The seed part file can also aid in the creation of standardized drawing views from your models. If your company has a standard *title block,* tolerance block, notes, and set of required views, seed parts or prototype drawings can help produce drawings that all have the same high-quality appearance even when created by different users. Even if it is not a company standard, setting up a file with this information eliminates having to re-create it each time. Having standard notes that you can edit or delete if not appropriate also can remind you to include important information on the drawing.

Each modeling package comes with default settings for the software and for new document files. Any settings that are stored with the drawing or model file can be included in a seed part. If you find that you are consistently changing a setting with every new drawing, the setting might be a candidate for inclusion into a seed part or template drawing.

Parametric Drawing Elements

Another way to add borders, title blocks, and other drawing elements to your drawing is to link them parametrically as you would a part file. This allows you to create a title block and reuse it for all of your drawings. This title block can even be set up to prompt you for the engineer's name, approval date, material information, tolerances, and other information as parameters. It is important to remember that if you use parametric drawing features (or external references) for title blocks and other drawing information, deleting the original title block from the directory where it resides removes it from all of the drawings where it was used.

Building a Useful Database: Bringing It All Together

In order to fully enjoy the benefits of a digital assembly model, your design database should include all of the parts designed for the assembly. Most parametric modeling software allows you to add and import non-parametric parts to the parametric assembly so they can be referenced during the design refinement process.

Standard Parts

Using standard parts (or stock parts that are readily available) can be cost effective for two reasons: First, they do not have to be designed. Standard parts are purchased from a supplier, and stock parts have already been designed for another project. Second, their bulk manufacture may be less expensive than manufacturing the quantity needed for a particular project. Some companies maintain a library of stock parts on the network so they can be reused. Like standard parts, stock parts can be manufactured in quantity for a lower per-piece price if they are shared by more products. A stock parts database can be useful even if the manufacturing quantities never get very high, as the company will save design time for every part that is reused.

If a stock part is not available within the company, you may find that a part that will meet your need is available from a supplier. One good place to check for stock parts is the World Wide Web. Often suppliers make solid or parametric models of their parts available free on the Web. It has been estimated that engineers spend 20% of their time redrawing standard parts to show how they will fit with the parts being designed for the system or product. If you can find the model already available from the supplier you can spend your time designing the new components you need to get the job done. A number of Web sites that are useful in engineering design are listed in the appendices at the back of this book. Many of these sites provide their parts in standard 2D and 3D formats on the Web or on disk.

The Thomas Register recently purchased Autodesk's library of standard manufacturer's parts (called PartSpec). Thomas Register now makes these manufacturer's parts available to qualified mechanical engineers for free. Some of the parts are represented by 2D CAD files, some by 3D solid models, and others by parametric models. The interface makes it easy to locate and view parts from different manufacturers, as shown in Figure 8.12.

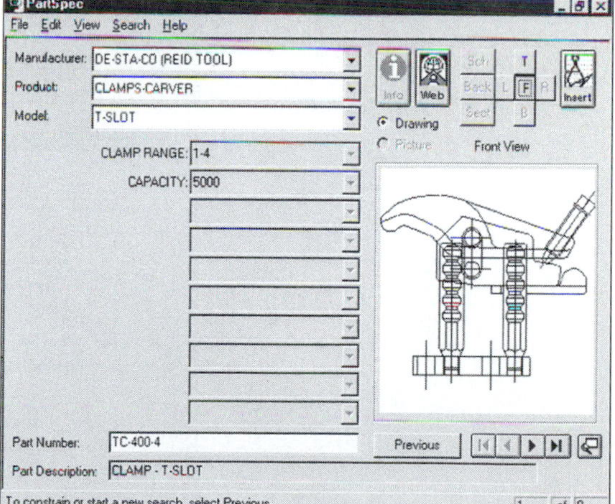

FIGURE 8.12
The PartSpec interface makes it easy to locate parts by specifying the materials and sizes needed. Once a part is located, the drawing of the part may be inserted into the file where it is needed.

Common parts are also available in libraries that ship with the Pro/Engineer software from Parametric Technologies, or from third-party vendors that make stock part libraries. These parts have the advantage of being parametrically modeled so their sizes can be changed. One such company is CIMlogic, which makes the CIMlogic toolbox of standard parts, such as helicoil inserts, fasteners, and common beams, which are compatible with AutoCAD, Mechanical Desktop, and SolidWorks software. Another is InPart's DesignSuite, a Web-based database of parametric parts that customizes models of standard parts to fit your assembly upon request (see Fig. 8.13).

Adding Static Parts to a Parametric Database

Some parts that you get from an outside vendor or import from a different modeling system may not contain the parametric information you need to edit them. These parts are sometimes called *static parts* because they were not created with parametric dimensions that allow them to be updated. They contain just the solid or surface

A

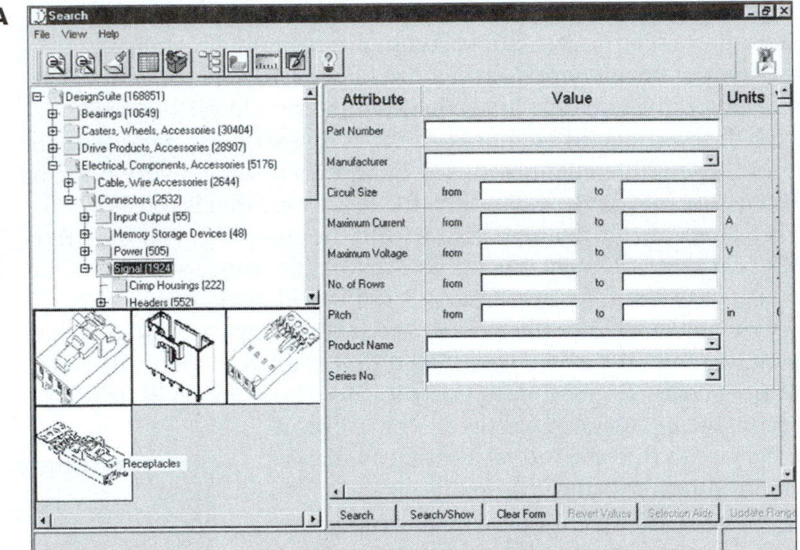

B

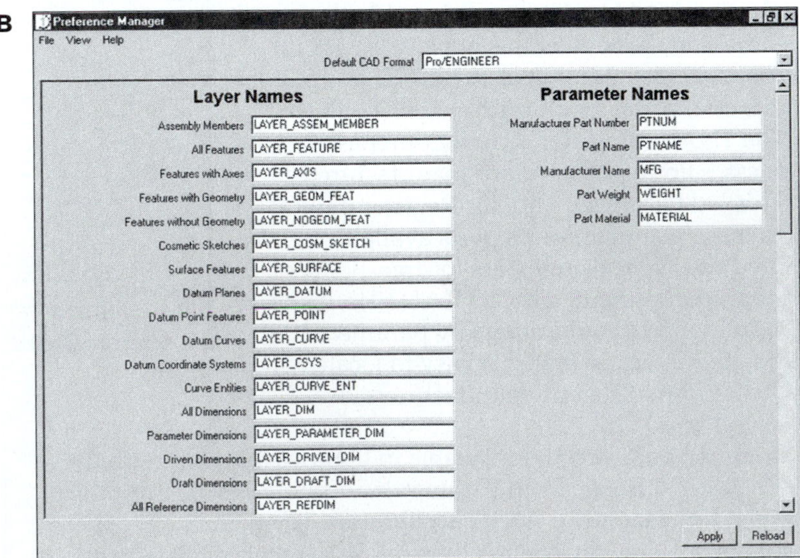

FIGURE 8.13
The DesignSuite database organizes parts by type, as shown in the menu at the left side of the screen in (a), and lets you search for standard parts by the kinds of size and spacing parameters typical of the part. Before downloading the file, you may use the screen in (b) to choose a CAD format and customize the model to use the same layer and parameter names as other models in your assembly. (Courtesy of Parametric Technology Corporation.)

information from the software they were created in. Ordinarily you will do little or no modification to stock parts, so the inability to edit the files is no disadvantage. You can call them into your assembly drawing and *add* new parametric features to these parts, but you cannot delete or resize the existing features easily.

Threads and Fasteners

Fasteners are a common kind of standard part added to assembly drawings. It can be desirable to show the fasteners in the assembly so the person doing the assembly can look at the drawing to determine where the fasteners are located. But standard fasteners are usually purchased and do not need to be modeled. In Chapter 2, you were introduced to some thread terminology in the context of sketching fasteners. To select standard fasteners, you should be familiar with the terms used to describe different varieties and how they are represented in a drawing.

Thread Forms

The shape of the thread is called the *thread form*. Most thread is based on the sharp-V thread form, which has 60° angles for its roots and crests. This thread form is still used for some purposes, such as brass pipe fittings.

Other thread forms are common in different applications. These forms are illustrated in Figure 8.14.

Unified thread is based on a sharp-V thread form, but the root is rounded and the crest may be rounded or flattened. This is the form that was agreed upon by the United States, Great Britain, and Canada during World War II. The Allies found it hard to find replacement parts because they did not have an agreed upon standard, so they adopted the unified thread form in 1948 to promote interchangeability of parts.

Metric thread is similar to the unified thread, but it has a shallower depth of thread. This is the agreed-upon standard for international fasteners today.

Square, acme, standard worm, and *buttress* thread are all thread forms that are used to transmit power.

Knuckle thread is rolled from sheet metal or cast from aluminum or steel. It is used on bottle tops and lightbulbs and for various other purposes.

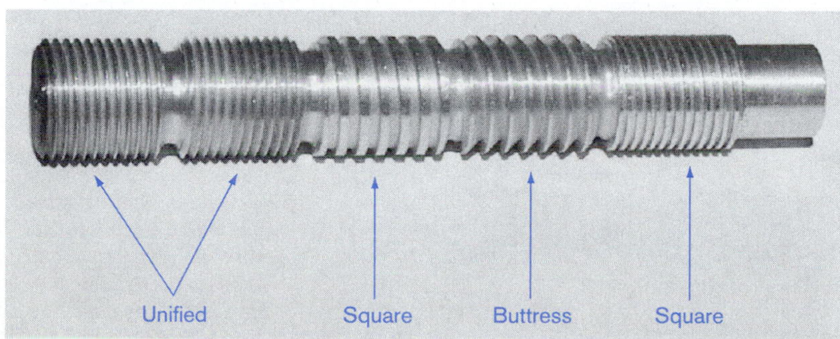

FIGURE 8.14
This photo contrasts different thread forms, all manufactured on the same shaft. Of course this part would not be useful in design, but it illustrates right- and left-hand unified thread, as well as square and buttress thread forms. Note how the profile of the thread shows its form most clearly.

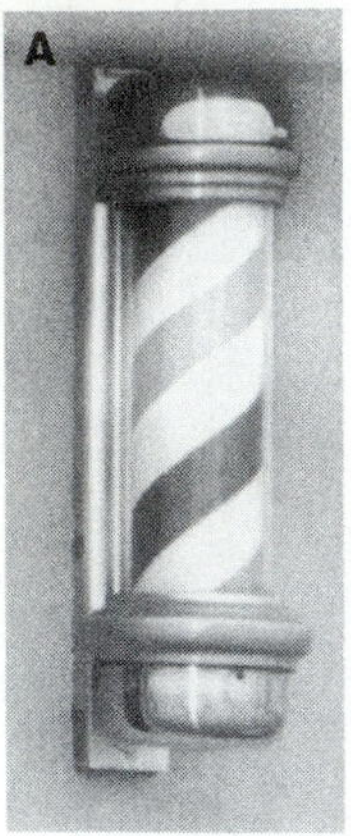

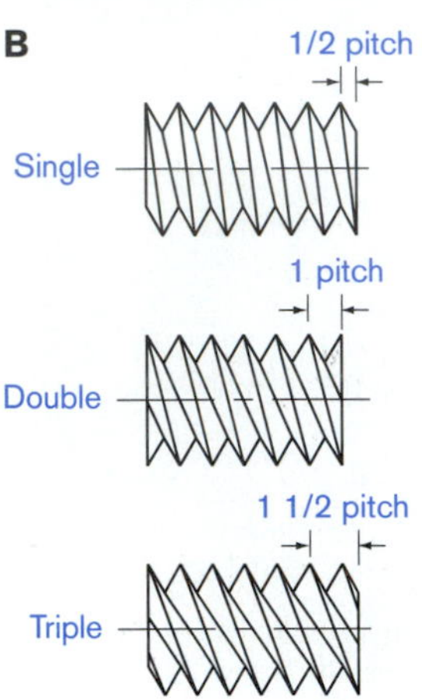

FIGURE 8.15
Like the two-color striping on the barber pole, double and triple thread have two or three helixes running parallel around the shaft.

Double and Triple Thread

Double and triple thread are used for adjusting screws, to provide for quick opening and closing of valves, and for quick assembly of parts that do not need to withstand large forces. Typically, thread is single; single thread has one single helix formed on the shaft or inside a hole. Double thread has two helixes that parallel each other along the shaft or inside the hole; triple thread has three helixes paralleling each other. Double thread advances two pitch distances in one complete turn; triple thread advances three.

Figure 8.15 shows examples of double and triple thread. Notice that you can determine whether it is double or triple thread by looking at the number of threads ending at the end of the bolt. Double and triple thread are always noted as such on drawings. If not designated, thread is understood to be single thread.

Right- and Left-Hand Thread

Most thread advances when turned toward the right or in a clockwise direction. Left-hand thread advances when turned in a counterclockwise direction—toward the left. It is labeled LH on the drawing. Thread that is not designated as left-handed with the note LH is always considered to be right-hand thread. A shaft with right- and left-hand thread at opposite ends will advance in both directions when the shaft turns—for example, as a turnbuckle does. Figure 8.16 and Figure 8.17 show examples of right- and left-hand thread.

Thread is not typically modeled because this complex helical surface adds greatly to the complexity and file size of the drawing.

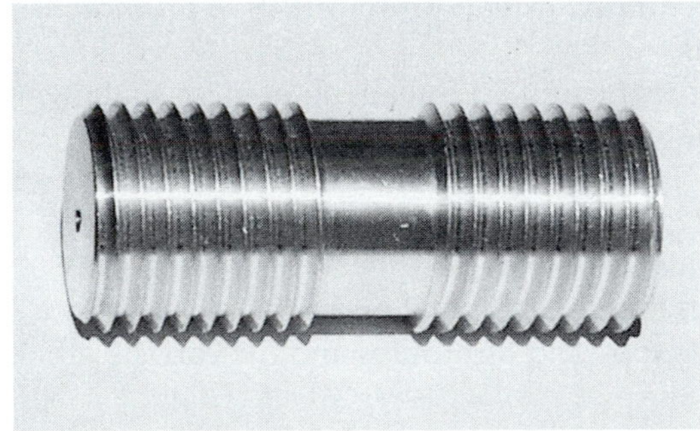

FIGURE 8.16
The thread on the right end of this shaft is right-hand thread. It advances when turned toward the right. The left-hand end uses left-hand thread, which advances when turned toward the left.

FIGURE 8.17
This bottom bracket is used to attach the cranks on a bicycle to the frame. One end is threaded with right-hand thread and the other side with left-hand thread so the pedal motion won't unscrew the bottom-bracket bearing cups.

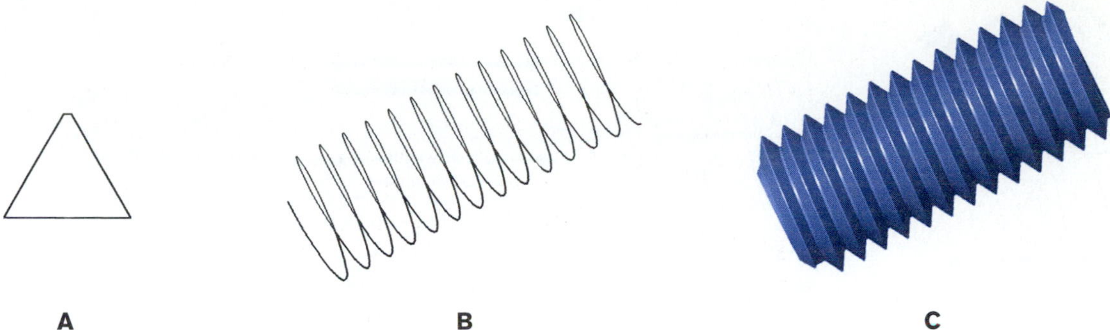

FIGURE 8.18
(a) Cross-sectional shape of thread; (b) Helical path; (c) Thread model formed by sweeping cross-sectional shape along a helical path.

Many 3D modeling packages allow you to model thread accurately, but it is better to add this type of detail when you are finished modeling and have worked out the major details of the design. Sometimes the major and minor diameter of the thread are represented so that they can be used in checking clearances and interferences. If not, the nominal size (the general size used to identify the fastener) of the threaded hole or shaft is modeled. A comprehensive list of fasteners and their proportions is included in Appendices 18–33.

For non-standard fasteners, you may want to model the thread accurately. To model thread, draw the cross-sectional shape of the thread as shown in Figure 8.18(a) and sweep the cross-section along a helical path as shown in Figure 8.18(b) to form the shape of thread as shown in Figure 8.18(c). Some modeling software allows you to define the helical path curve using an equation. Other software provides a helical sweep command that allows you to enter the number of revolutions and diameter parameters, which can be changed later. Surface modeling techniques can also be used to model thread, but it can be difficult to edit later.

Springs

A spring designed for the purpose of being compressed is called a *compression spring*. *Extension springs* are designed to be extended. Both are usually formed in the shape of a helical coil that can be compressed or extended and still return to its original shape.

The properties that determine the behavior of a spring are its material, thickness, major diameter, length, number and direction of the helical coils, and type of ends. To select a spring for a particular purpose, or to specify a standard part in your drawings, you need to specify these properties.

Springs may be made of materials such as hard drawn steel, spring steel, stainless steel, or brass. The thickness of the spring may also be referred to as the wire size; it is the width of the unwound coil. Springs usually have a round or square cross-sectional shape but can have other shapes, such as elliptical or rectangular. The *major diameter* is the outside diameter of the spring's coil. The *free length* of a spring is the length of the spring when it is not compressed or extended. For an extension spring, you would specify the coil length

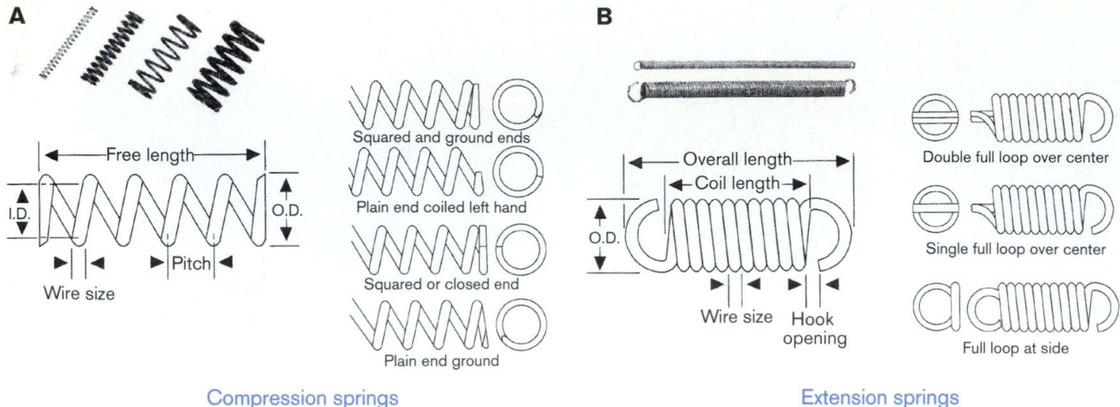

FIGURE 8.19
(a) Compression springs; (b) Extension springs. Compression and extension springs differ in how they are specified and how their ends may be finished. Basic terminology and common end types are illustrated here. (Courtesy of Reid Tool Supply Company.)

FIGURE 8.20
Springs are sometimes represented as a single 3D curve in a 3D model when the complexity of the shape outweighs the benefits of modeling it.

and the *overall length,* which is the length of the coil plus the hooks at each end used to attach the spring. The *pitch* of the spring is the distance between the coils, just as pitch is the distance between the crests on a threaded item. Like thread, the helical coil of a spring can be wound right- or left-handed. Finally, the ends of the spring can be finished in several ways. Figure 8.19 illustrates several of these options.

Modeling Springs

When you are using 3D modeling, it is uncommon to model springs accurately, unless it is critical to show them in the assembly. They are usually stock parts that are specified and purchased rather than designed individually. If you must show springs, they may sometimes be modeled as a single 3D curve, as shown in Figure 8.20, instead of a solid, or surface model. Because of their shape, springs are complex objects that greatly increase the file size of your drawings. You should model them accurately only when there is a design or production benefit to be gained from doing so.

Fillets and Rounds

Just as you do not model thread in the drawing unless it is essential, minor fillets and rounds are usually best left until late in the design process. These features make the object more complex and increase the file size, usually without adding much necessary information to the design. One way to define fillets or rounds is to specify a radius and select two surfaces. Each time a surface changes, the fillet or round is recalculated for the new location. If the angle of the surface changes (such as when draft angle is added to a part), another element is added to the calculation needed to regenerate the feature. For this reason, fillets and rounds may not update properly when values change and the model is regenerated.

To document your design and convey it accurately to manufacturing, you may want to show fillets and rounds in the final version of the parts that are released to manufacturing, but you should generally wait until just before that time to add them.

In some cases, it may not be necessary to add these features to your model. Often a note on the printed drawing that accompanies the CAD file stating "round all sharp edges to R3" (or some other value for the radius of the round) may be perfectly acceptable. The smaller file and sometimes cleaner file that results when you do not show the minor fillets and rounds may even work better for generating the numerically controlled (NC) machine code needed to manufacture the part.

FIGURE 8.21
The inside pocket was formed by cutting away material with a rotating endmill. The rounds in the inside corners can be smaller or larger, depending on the cutting tool used, but they cannot be square using this process.

In other cases, it may be difficult to manufacture the part inexpensively without fillets and rounds on the part, even if they did not exist in the model. Certain manufacturing processes, such as milling, use rotating cutting tools that cannot make a square corner. These processes create small rounds on the part that correspond to the size and type of the cutter used. Figure 8.21 shows a rounded inside corner formed by a milling machine cutter. The radius of the round is equal to one-half of the diameter of the cutter. Figure 8.22 shows some milling machine cutters.

Whether to include small fillets and rounds in the model is an issue that you should discuss with the person responsible for manufacturing the parts. If you are unsure, you may want to include them.

FIGURE 8.22
An endmill is a cutting tool for removing metal and forming an edge. The endmill can have a square or ball end, to form a square or rounded corner, at the base of an inside pocket or trough. (Courtesy of Reid Tool Supply Company.)

Using Your 3D Model to Determine Fits

When parts are accurately modeled in 3D and organized into an assembly, you can use the assembly to perform fit and interference checking on the models. Doing a thorough study of how the parts will fit once assembled can save time and money. After tooling has already been produced to manufacture parts, changes to the design become very expensive.

Because the parts in the assembly can be shaded and viewed from any direction on the computer monitor, you can visually inspect the fits and clearances. You can also make measurements and list dimension values to compare parts to one another. Most solid and parametric modeling software also provides a command for checking the interference between two parts.

Interference Checking

Interference is the name for the amount of overlap one part has with another. When you use a command to check the interference between two parts in an assembly, the solid modeling software will report that the solids do not interfere, or will indicate the amount of overlap—sometimes by creating a new solid to represent the overlap.

Because current parametric modeling software makes it easy to assemble parts early in the design process, interference checking tools are one way to facilitate more effective concurrent engineering. For example, Boeing's designers are responsible for checking their work in the digital pre-assembly model of the aircraft to identify places where the system or component they are designing will interfere with work in progress in other areas of the design. Boeing's proprietary visualization tools and assembly database work together to pinpoint problems with the assembly during refinement. Each designer then takes steps to resolve the interference by making contact with the appropriate designers for the systems involved. Interference checking with an assembly model created in Mechanical Desktop, Pro/Engineer, or another solid modeling package can serve the same role in design situations that involve fewer individuals than those typically found at Boeing.

Tolerances

The manufactured parts for your design must fit together when assembled, but no part can be manufactured exactly. Some manufacturing processes are very accurate, but there is always some slight variation between the actual manufactured parts and the dimensioned drawings or 3D models. *Tolerance* is a statement of the total amount a dimension on the manufactured part may vary from the specified value in the drawing or the model.

There are a number of ways to state tolerance values for dimensions; you will learn more about them in Chapter 12. Figure 8.23 shows the shaft support drawing you saw in Chapter 7 dimensioned with limit tolerances for the size of the central hole. A *limit tolerance* specifies the upper and lower allowable value for the dimension when measured on the actual part.

Manufacturing processes have certain ranges of accuracy. Requiring a high tolerance—one with very little allowable deviation—can limit the choice of manufacturing process to those that can be expected to meet that level of accuracy. Specifying very small tolerance values can increase the cost of the finished piece. Toler-

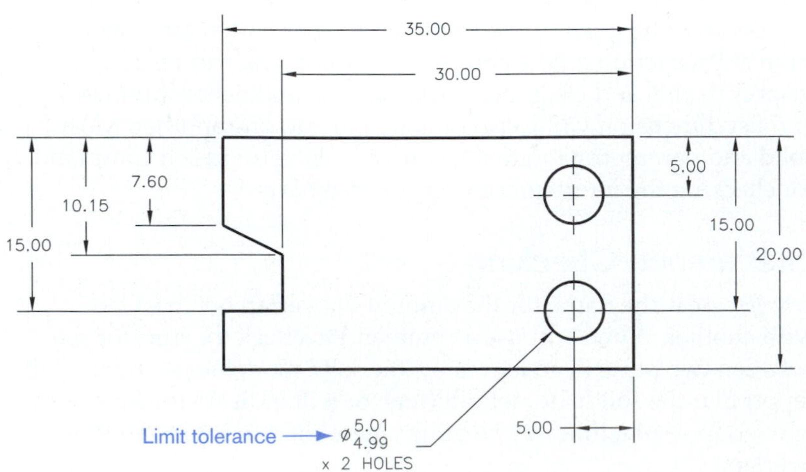

FIGURE 8.23
The limit tolerance shown here indicates that the size of the hole in the manufactured shaft support may be as small as 4.99 and as large as 5.01 and still be an acceptable part.

ances in general should allow as much variation as possible without affecting the functionality of the design.

Material selection also plays a role in the accuracy of the finished parts. Parts can be manufactured more accurately from some materials than others. In addition to determining the tolerance range for a part based on its function in the assembly, you should talk to personnel in manufacturing or the vendors who will make the parts. You want to ensure that you fully understand the issues in manufacturing the part to the tolerance you have specified and the effect that it will have on the piece price for the finished parts.

Pro/Engineer and some other solid modeling software allow you to specify a tolerance range for a dimension and then apply that to the feature at either the upper or lower end of the dimension range. By doing so, you can run a check on how the parts will fit at the minimum and maximum clearance values. The options available for doing so are illustrated in Figure 8.24.

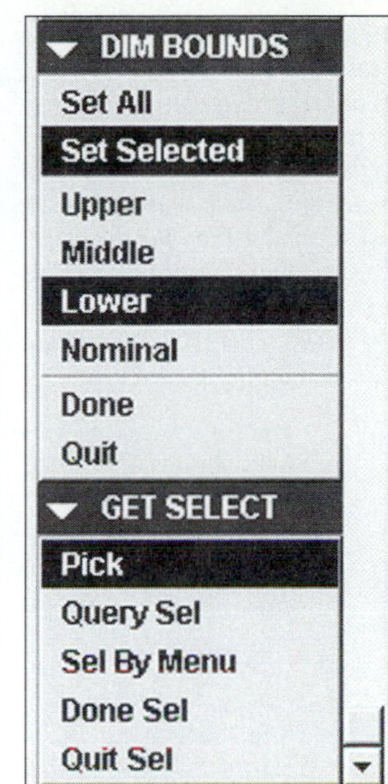

FIGURE 8.24
To evaluate interferences in the context of the allowable tolerances in Pro/Engineer, you would use this menu to specify which end of the range to evaluate.

Even if your software does not provide a special function for checking fits at the lower and upper limits of the tolerance range, you can systematically change the dimensions for mating features to represent the minimum clearance, then inspect the model to see how the parts fit. Then you can change the values to represent the maximum clearances and inspect the fits. This ability to inspect clearances between mating parts is one of the benefits of using solid modeling to create your parts.

Fit Between Mating Parts

Tolerances for individual features are determined by their fit requirements with mating parts in the assembly. As you might expect, the maximum clearance between two parts (or loosest fit) occurs when the external member is at its largest size and the internal member is at its smallest size. The minimum clearance (or tightest fit) happens when the external member is at its smallest size and the internal member is at its largest size. *Allowance* is the term for the minimum clearance.

There are four main categories of fits between mating parts: clearance fits, interference fits, transition fits, and line fits.

1. A *clearance fit* requires a space or clearance between the internal and external member.
2. An *interference fit* requires that the internal member is always larger than the external member. The parts must be forced together or the outer part heated to enlarge it enough to fit the internal part before it shrinks back to size. The tolerances for this kind of fit, often called a *force fit,* are illustrated in Figure 8.25.
3. A *transition fit* indicates that either a clearance or interference fit is acceptable. That is, some parts will fit with a clearance and others will have to be force fit.
4. A *line fit* indicates that the ranges for the dimensions ensure that either a clearance or a surface contact will result between the parts.

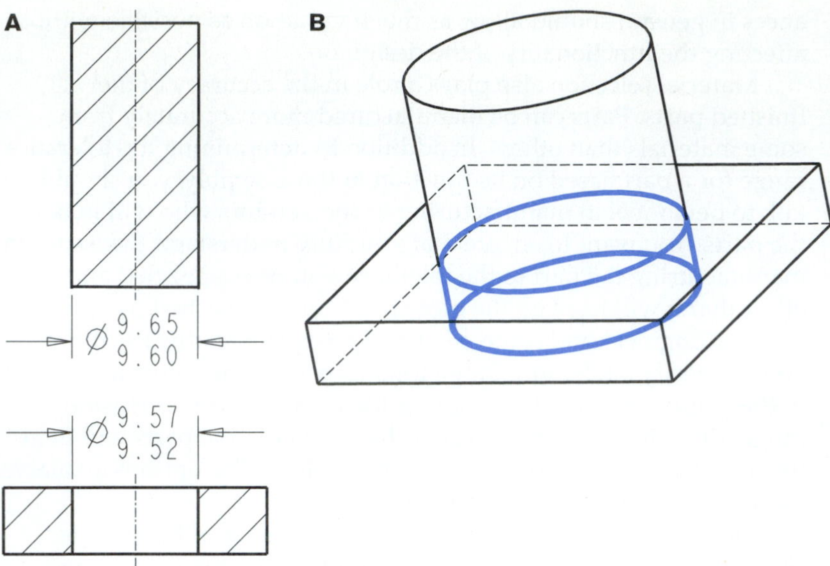

FIGURE 8.25
The tolerances for the hole and shaft in (a) specify that the mating parts will interfere, resulting in a force fit between the two. When the diameter of the hole is at its largest, it is still smaller than the diameter of the shaft that will fit into it. This interference is shown by the solid model in (b), where the interference is indicated by the shaded area.

You will learn more about these fits and the values recommended for them in Chapter 12.

Nominal Size and Basic Size

Two terms you should understand in connection with tolerance and fit are *nominal size* and *basic size*. These terms are often confusing because in general use they mean nearly the same thing. However, these are two different terms when used correctly in engineering graphics. ANSI states that *nominal size* is "the designation used for purposes of general identification." The nominal size is often expressed in common fractions, such as 1/2 inch.

You might think of nominal size as similar to the way lumber is typically sized. A two-by-four board is the general designation (nominal size), however, the actual size of a two-by-four is approximately 1-1/2 by 3-1/2 inches. The general designation of a hole as 1/2 inch in diameter is useful early in the design before accurate tolerance studies have been undertaken, but it does not always mean that the final specified size will be exactly 0.5000 inch. Typically, features may be created at their nominal size early in the design. As tolerance studies are undertaken, these values may change to reflect the more accurate sizes required for the parts to function properly in the assembly.

The *basic size* is the theoretical starting point from which the tolerance is assigned based on your design requirements. A basic dimension is the term for the theoretically exact dimensions used in geometric dimensioning and tolerancing (GD&T) to define an envelope from which tolerance variations are established. You will learn more about using GD&T in Chapter 12.

Think back to the example of the two-by-four. If you were manufacturing two-by-fours, the basic size for a theoretically exact two-by-four would be 1.50 by 3.50 inches. However, no two-by-fours would meet this exact specification, so a tolerance range is specified around that perfect size to describe allowable manufacturing variation.

Two systems for calculating a tolerance based on the desired type of fit and the basic size of the feature are the *basic hole system* and the *basic shaft system*. The basic hole system is the most common because standard machine tools, such as drills and reamers, are often used to create the hole to the specified size. Then, the shaft is machined down to the specified tolerance so that it fits inside the hole providing the desired type of fit.

In the basic hole system, the lower limit or minimum hole size is used as the basic size for the hole feature. Designing around the hole size as the basic size, you can use standard fit tables to look up the tolerance values that you can use to produce a desired type of fit. (Refer to Appendices 12–17 for a set of standard fit tables.)

The basic shaft system uses the maximum size of the shaft as the basic size to determine the tolerance. This is a less common method, but it is sometimes used when many features are to fit on a shaft made of a material that is hard to machine without changing its desired properties. In the basic shaft system, the holes in mating parts are toleranced to fit with the shaft size, which is used as the basic size from which the tolerance is determined. See Figure 8.26 for examples of designs that might use each method.

Tolerances per Manufacturing Process

As the engineer, you are responsible for fully defining exactly what you want manufactured, and for prescribing a scheme by which the final manufactured parts can be inspected to determine whether they are acceptable. The term *as per manufacturing process* means that the typical tolerance for the specified manufacturing process is acceptable for the part. You can use this designation when specific fits do not matter, such as on the exterior of a part where there is no fit with a mating part, or when you have modeled the part accurately and

FIGURE 8.26
(a) The basic hole system is the most commonly used method for determining tolerances because holes are often formed with standard sized tools. To use it, start with the minimum size of the hole and determine the tolerances for the shafts using fit tables.
(b) When a shaft is available in standard sizes and is difficult to machine, or in an assembly where many parts must fit on a single shaft, the basic shaft system may be preferred for determining its tolerances.

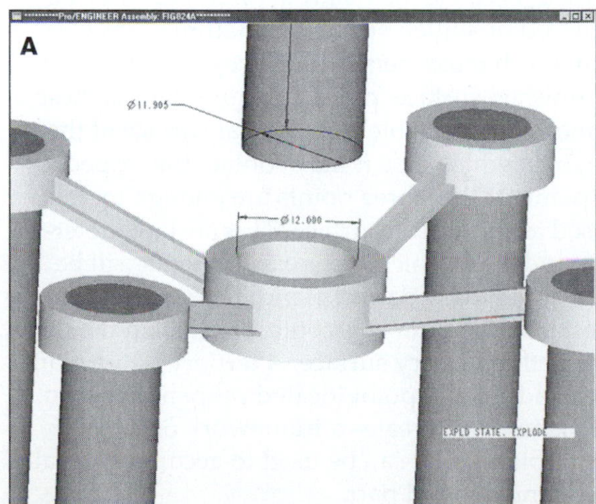

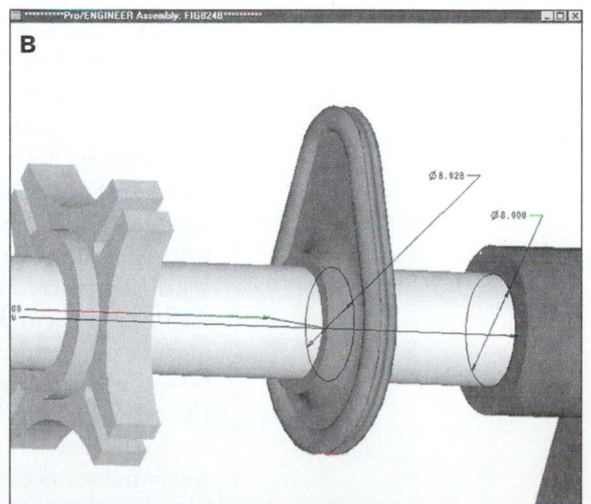

NC manufacturing processes will be used to make the part, with an acceptable allowance. When you specify tolerances as per manufacturing process, you give up the ability to later reject parts that do not fit in your assembly, unless you have provided documented requirements for the critical dimensions. If the fit between mating parts is important in the function of the system or device you are defining, then you should specify the allowable tolerance range for the dimension. You should also perform a fit study to determine whether the assembly will function at the size ranges you have specified.

Measurement and Inspection

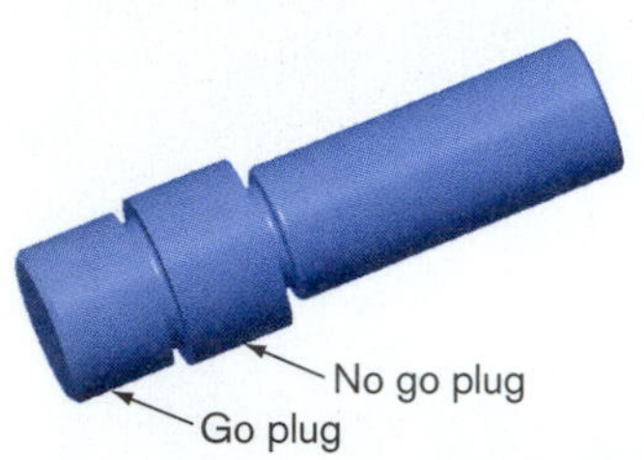

FIGURE 8.27
A go–no go gage

During and after manufacture, parts are inspected to determine if they meet the specified design requirements. Measurement and inspection can be performed a number of different ways, depending on the requirements for the accuracy and fit of the parts. Parts can be measured with a machinist's scale, calipers, micrometer, or other graduated measuring devices (ones that have markings that indicate the measured size).

If it is not necessary to measure the size of the features, but only to determine whether parts will fit, gaging can be used. One of the most common types of gage is a *go–no go plug gage*. This type of fixed gage is typically used for checking hole sizes. One end of the gage is machined to the size of the minimum cylinder that the hole should accept; the other end is machined to be just larger than the maximum size that should fit an acceptable hole. A part is acceptable if the small end of the plug gage fits into the hole and the slightly oversized cylinder at the other end does not fit into the hole. Figure 8.27 shows an example of a go–no go gage.

To inspect parts accurately, measurement and inspection devices are generally designed with tolerances that are one-tenth that of the tolerance being measured on the part. Their operation also depends on measurements taken consistently from one part to the next. To establish a starting point for accurate measurement, datum surfaces are defined and specified. Because the manufactured part may be rough or have only curved or angled surfaces, a datum surface provides a plane from which measurements can begin. A part may rest on an accurately finished surface (often a granite slab, as these don't rust and are dimensionally stable). The actual surface of the part may be very rough, in which case it may contact the inspection surface at only three points. These three points are enough to define a plane that can be used as the datum surface. This surface, determined by three points and from which the measurements will be made, is referred to as the primary datum plane. The secondary datum plane can be established from two contact points on the next surface perpendicular to the primary surface. A tertiary datum can be established from one additional point located perpendicular to the other two datum planes. This creates a framework of three mutually perpendicular planes that can be used to accurately locate measurements on the manufactured part.

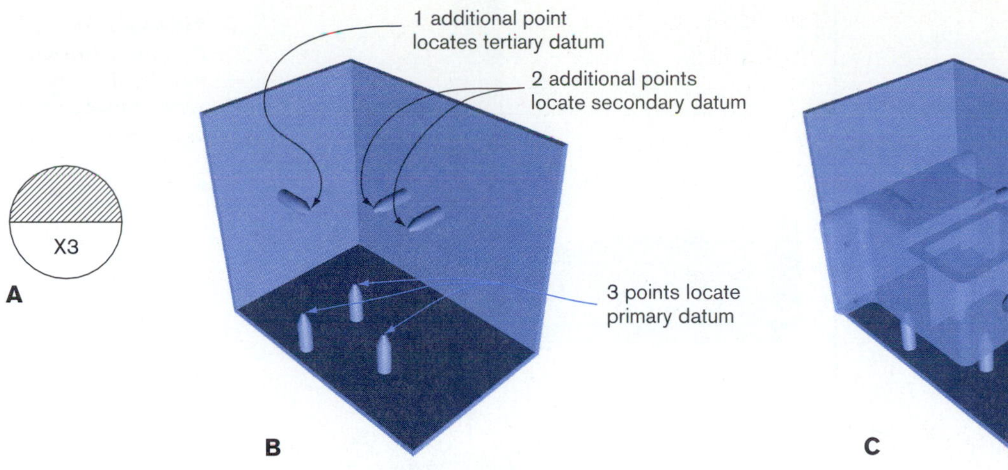

FIGURE 8.28

(a) Targets, about 12–15 mm in size, are added to the drawing to specify the principal locating points where a fixture will support the part.

(b) The primary datum surface is located by three points. The secondary datum surface is located perpendicular to primary datum (two points and the primary plane define this plane). The tertiary datum is located perpendicular to other datum surfaces (one point and the two existing datum planes define this plane).

(c) Part resting on fixture establishing primary datum surfaces.

Recall in the last chapter that datum planes can be established in your model from which to locate principal dimensions. Features on the part that will be used for measurement may be located from these planes to facilitate using the model to evaluate the effect of manufacturing deviations on fit and function. Remember that the solid model is geometrically perfect, but a manufactured part will only approximate the geometric shape. You must specify the degree of accuracy that is required for proper function of the part. Even a "flat" surface on the part may not contact the datum surface at every point on its surface. For this reason, and especially for rough or irregular parts, datum targets (illustrated in Figure 8.28 (a)) may be identified on the drawing to locate the control points on the part that will determine the datum surfaces for inspection. These points, called principal locating points, allow the inspection team to determine the theoretical planes from which measurements are to be taken.

In addition to inspecting the locations of features with respect to datum surfaces, various other geometric properties of the part may need to be inspected. It may be important that the overall flatness of a surface does not vary more than a certain amount (e.g., .002 inch) from that of a perfectly flat surface, or perhaps two surfaces must be within some tolerance range of being perfectly concentric in order for the part to work. The tolerances for these geometric properties can be specified on the drawing using special symbols for *geometric dimensioning and tolerancing (GD&T)*. Being able to visualize important relationships of how parts fit together in a 3D assembly can aid you in determining key geometric features and the effect that manufacturing variations will have on the fit between parts.

Modern inspection equipment can produce very accurate measurements of the finished part. Inspection can be contact or non-contact inspection. The use of gages, probes, and other measurement devices that touch part surfaces is termed *contact inspection*. *Non-contact inspection* includes various optical methods of inspection, laser triangulation, ultrasound, and other techniques. Coordinate measuring machines (CMMs) can provide accurate measurement and inspection using both contact and non-contact methods. Figure

FIGURE 8.29
Coordinate measuring machine (Used by permission of Brown & Sharpe.)

8.29 shows a coordinate measuring machine. 3D point locations on the inspected part as measured by a coordinate measuring machine can be compared to determine geometric properties of the part. For example, if 3 points are used to establish a theoretical plane on the surface, other points can be compared to those to determine the variation in flatness for the surface. Some CMM software can compare the measured part to data from the CAD file to depict the variation from the perfect geometry.

Temperature and material can also affect the accuracy of the part. If it is not stated otherwise, measurements are intended to be made at 68° Fahrenheit (20° Celsius).

You will learn more about tolerance and inspection in Chapter 12.

Accessibility Checking

Your assembly model can also be used to be sure that your design can be assembled. Whether the assembly will be done by robots or humans, parts and fasteners must be accessible. Many devices and systems also need to be repaired or upgraded later, requiring not just access during the assembly phase, but during repairs or upgrades. You can use your 3D design database to help check for fits and clearances in assemblies and subassemblies. You can also use it to check whether people will be able to reach a part to remove or repair it after the system is assembled.

Ergonomic analysis software can help to determine accessibility for assembly and repair. *Ergonomics* studies the ability of humans to use a system. Many ergonomic analysis packages have 3D models of human beings (anthropomorphic data) that you can position inside the designed system to see how people will fit. Sophisticated

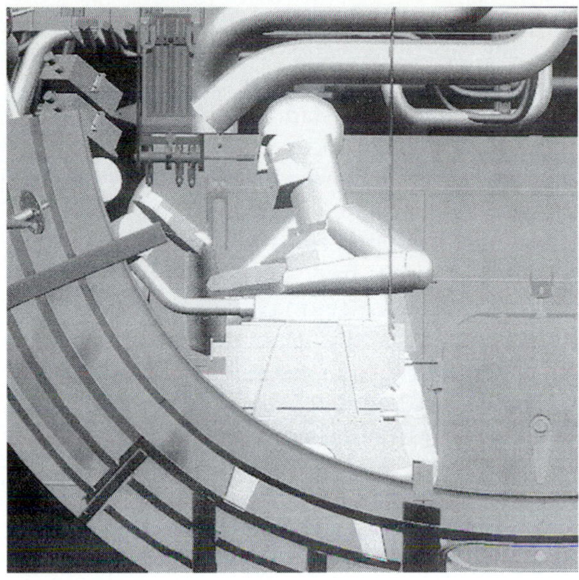

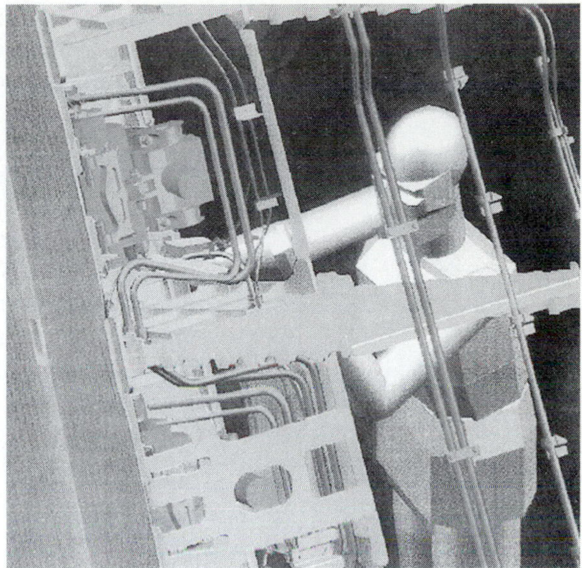

FIGURE 8.30
Planning access to an airplane for service and repair is an important part of its design. This digital mannequin was imported into the CATIA software used by Boeing to evaluate the accessibility of its parts. (Courtesy of The Boeing Company.)

ergonomic analysis systems allow you to choose various ranges of human sizes. This can help you see how your design will work when used by an average-sized woman, a tall man, or a child. Some systems even have demographic populations built in so you can evaluate how well a design will work for the average French woman or the average American man.

Even if robots will assemble the components, people will almost always perform repairs. Evaluating repair issues can be even more difficult than assembly issues, because the assembly can proceed in such a way that interior parts are fit together before their covers obscure access and visibility. You can use ergonomic analysis or human models to check and see if parts can be reached to repair them, as shown in Figure 8.30.

Usability is another ergonomic consideration: operator visibility, the complexity of controls, and the ability of the operator to correctly map the control to its function may be critical to the operation of the device you are designing.

Without software tools, you can still use the assembly model to consider how the parts will be assembled and whether they are accessible. By assembling the model in the order in which the parts will be assembled in the factory, you can visualize the location of the components as they are being built. Measurements from the model can be used to determine whether a tool can be applied as needed, and whether a distance is greater than the reach of the average person.

Design for Manufacturing

To do a good job designing any part, you should understand the manufacturing processes that are involved in producing your design. Even though you may not need to understand specifically how to operate the equipment to produce the parts yourself, you should understand the basic capabilities and limitations of various production methods, as well as the range of tolerances available from that process. You should also understand material selection and the role it plays in manufacturing processes and tolerances. The more you know about manufacturing processes, the better you will be able to design low-cost parts that function as you need them to. Chapter 11 discusses manufacturing processes in more detail.

In this section you will learn some basic information about the impact that three manufacturing processes may have on the design of individual parts. Keep in mind that this is not intended to be a thorough presentation of everything you need to know about a particular process or all processes. You will learn more about manufacturing in your engineering courses and as you gain job experience.

Design for manufacturability is a key phrase that implies the importance of designing parts that can be manufactured inexpensively. In order for companies to compete in today's world market, they must design and manufacture parts that are functional and inexpensive. Sometimes seemingly small decisions and variations in the design can add to the cost of the part. Even a small amount per part can be important when thousands or millions of the part will be produced. During the process of designing a part, you should communicate with the people responsible for seeing it manufactured and work together to produce the best design possible. The design and modeling issues presented here for three common low-cost production methods will alert you to issues that you should consider and investigate as you model parts. The processes used to manufacture parts may change over the lifetime of production for a part. As a device becomes more and more accepted by consumers and production volumes increase due to demand, it may become cost effective to produce it in new ways.

Modeling Injection-Molded Plastic Parts

Many plastic parts are formed by injecting plastic into a mold. Injection-molded plastics are popular for low-cost parts for several reasons: the material is attractive, noncorrosive, colorful, lightweight, and can provide any number of surface textures and graphics. Molded plastic parts allow for good design control overall, high-volume production, and relatively low cost per part.

Most molds have two halves. The side of the mold in which the plastic is injected is called the *active side* of the mold. After the plastic is injected and allowed to cool, the two halves of the mold separate and the part (or multiple parts) is removed. Figure 8.31 shows an illustration of an injection molding machine with the two mold halves separated.

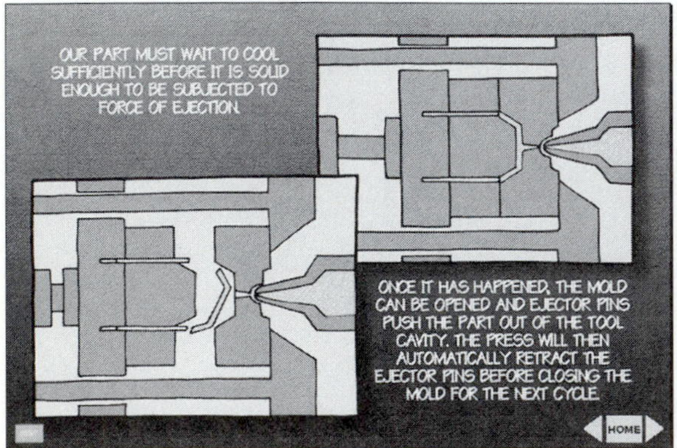

FIGURE 8.31
Injection molds have two halves that separate to release the plastic part. This image is taken from a cartoon education series on plastics manufacturing sponsored by GE Plastics. Visit their Web site at www.ge.com/plastics to learn more about plastics manufacturing. (Courtesy of GE Plastics.)

Injection-molded plastic parts have certain characteristics you may wish to take into consideration when designing for them. Molded parts in general must include *draft*, or taper, that allows the part to be removed from the mold. Plastic shrinks as it cools, and the shrinkage plus the draft allows the parts to be removed from the mold into a collection bin. If the shape of the part requires it, ejection pins may be necessary to force the part out of the mold. Usually the mold designer will determine the necessary cooling lines, ejection pins, and other features, but you should be aware of the general process. Ejector pins often may leave a slight mark on the finished part. If these marks would negatively affect the cosmetics of your design, you may need to change the shape of the part so that it removes from the mold easily and does not require ejector pins. Alternatively, you may want to identify important cosmetic surfaces on the part when you send drawings or files to the manufacturer so the mold maker can use this information in designing the mold.

The *parting line* is the line on the part where the two mold halves come together. If you look at common molded plastic parts, such as the one shown in Figure 8.32, you may be able to see a fine line in the plastic where the two halves of the mold came together. You may also be able to see marks where ejector pins pushed the part out of the mold.

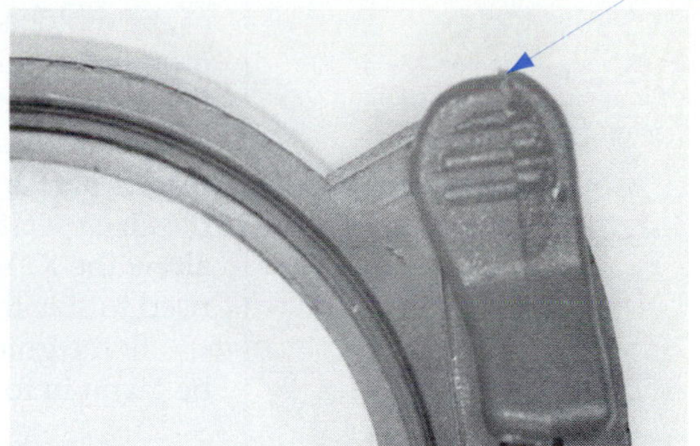

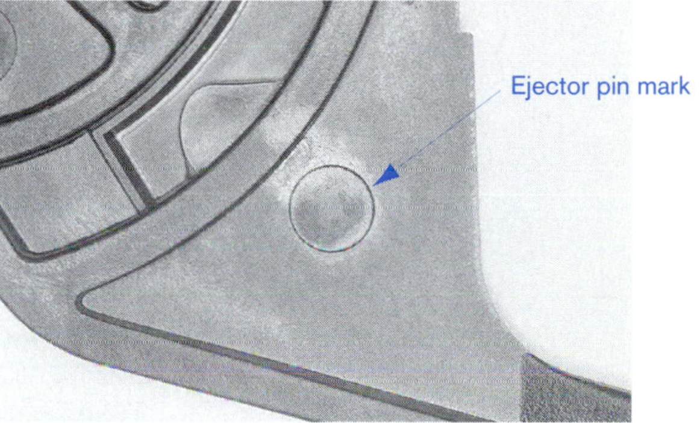

FIGURE 8.32
Injection molded parts may have a visible line where the mold halves joined, or marks from ejector pins.

When you are modeling parts that will be created using injection molding, consider how the part will remove from the mold. Although there are some molding methods that allow you to make interior holes (such as slides), in general the part cannot have a shape that will not allow it to be removed from the mold. Surfaces that slope back into the part would be impossible to remove from the mold. Consider the part shape in Figure 8.33; the lip of the bowl makes it impossible to remove from the mold.

As you design molded parts, you should plan for the parting line and design it so that both sides will remove from the mold. You may want to add a construction plane to your CAD model, then project its edge onto the part as shown in Figure 8.34 in order to represent the parting line in your model. You then add draft to the surfaces of the part on each side of the parting line.

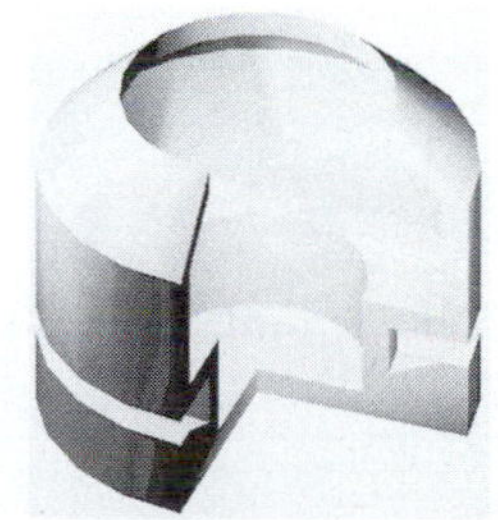

FIGURE 8.33
Can you see why this part would not remove from the mold?

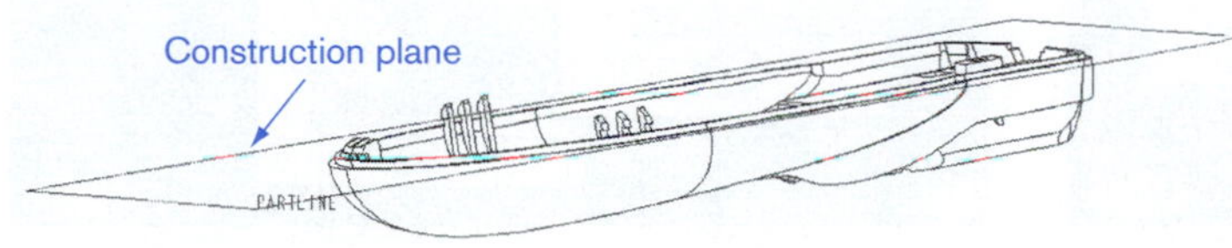

FIGURE 8.34
The construction plane in this part identifies the location of the parting line for the mold. The draft angle is added to the part on both sides of the parting line. (Courtesy of Rob Mesaros, Strategix ID.)

The amount of draft is decided by the size of the part, the finish of the mold, and the type of material and its shrink rate, among other factors. The Society of the Plastics Industry publishes a guideline to help you determine the exact amount of draft you should add to the part. Usually, the mold designer will determine the shrinkage for the part and size the mold cavity so that the final parts will be to the size specified. The shrinkage for the part is not always uniform along the X, Y, and Z axes of the part. Many CAD packages can be used to size the part for shrinkage as well as adding the draft.

Some guidelines for designing injection molded plastic parts can be given in four general rules:

1. *Try to maintain a constant wall thickness (see Fig. 8.35).* Uniform wall thickness is important when designing plastic parts. The thickness of the plastic affects the rate at which it cools. Uniform wall thickness helps prevent sink marks in the plastic and keeps the part flat and uniform once molded. If wall thicknesses must change, make the change gradually instead of abruptly.
2. *Round all inside and outside corners (see Fig. 8.36).* Rounding the corners of the part makes it remove from the mold easily. Rounded corners also improve uniform mold filling and help relieve stress concentrations in the mold.
3. *Use the maximum allowable draft angle, and never less than 1° per side (see Fig. 8.37).* Draft allows the parts to remove from the mold. When parts are designed without enough draft, a vacuum can be created between the part and the side of the mold during removal. In extreme cases, this can damage the mold. When molded parts will have a surface texture, the draft angle must be increased to allow the part to be removed without damaging the texture.
4. *Projections should be not more than 70% of the normal wall thickness (see Fig. 8.38).* For best results, projections should not be more than two and a half to four times the normal wall thickness in length.

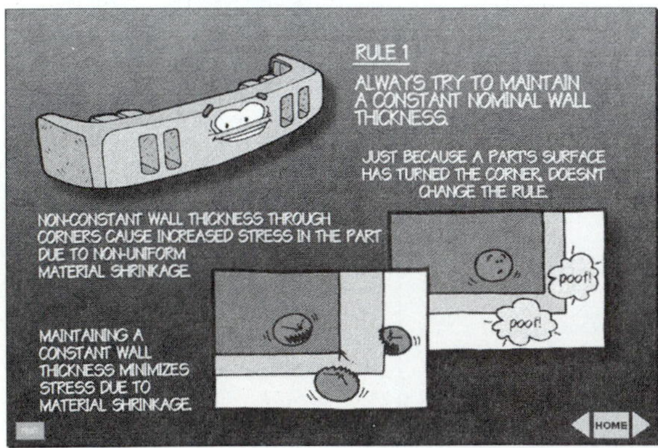

FIGURE 8.35
Always try to maintain a constant wall thickness on injection-molded parts. (Courtesy of GE Plastics.)

FIGURE 8.36
Round all inside and outside corners to improve release from the mold and relieve mold stress concentrations. (Courtesy of GE Plastics.)

FIGURE 8.37
Use the maximum draft angle. (Courtesy of GE Plastics.)

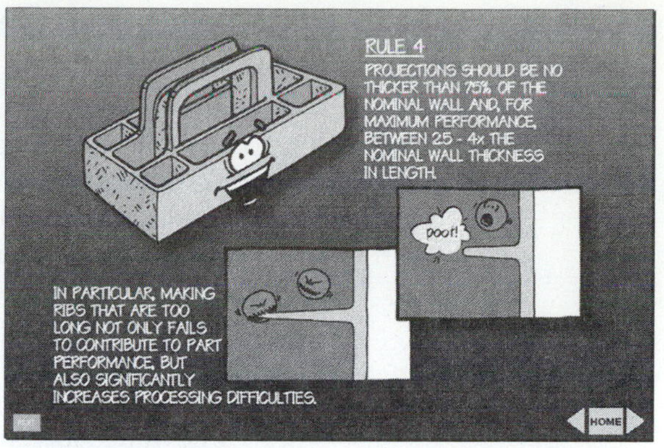

FIGURE 8.38
Size projections so they are no more than 70% of the wall thickness and no longer than four times the wall thickness. (Courtesy of GE Plastics.)

Cast Parts

Casting is the process of forming a part by pouring molten metal into a hollow mold. Many of the design issues for cast parts are similar to those for plastic parts. For example, surfaces must have draft and rounded corners to make it possible to remove the part from the mold. The Aluminum Association publishes *Standard for Aluminum Sand and Permanent Mold Castings,* which is a useful guide to design requirements for these types of parts.

Modeling Machined Parts

Machining processes remove material to produce the part shape desired. That is, a rotating cutting tool shaves away material to form the shape of the part. This type of process was the first to be computer controlled and is a popular way to form metal parts.

Many machined parts start from common stock shapes such as round, square, plate, hex, block, and bar stock steel that come in standard sizes. When it is possible to use a standard size in your design, it may lower the cost to produce the part.

As you learned earlier, certain types of features, such as a perfectly square interior corner, may be difficult to produce with this process. If you are planning to have your part manufactured using NC machining, you should try to avoid these types of features unless they are needed. (You can make perfectly square interior corners using the electrodischarge machining process, however.) Fillets and rounds that result from NC machining methods provide additional strength and smooth the corners so they are not sharp when the part is handled.

Figure 8.39 shows a solid model and the NC-machined part created from the model.

A

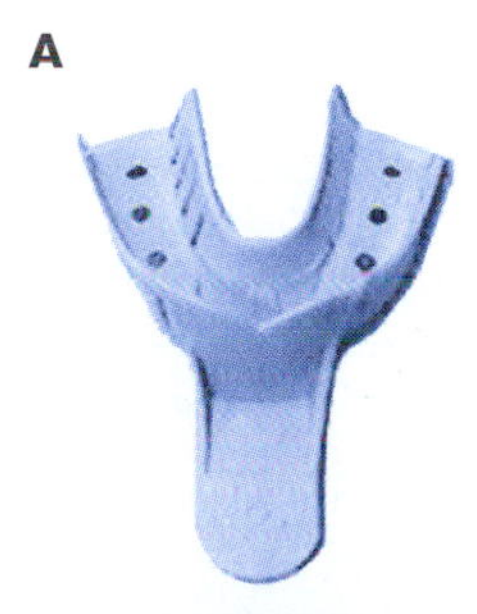

B

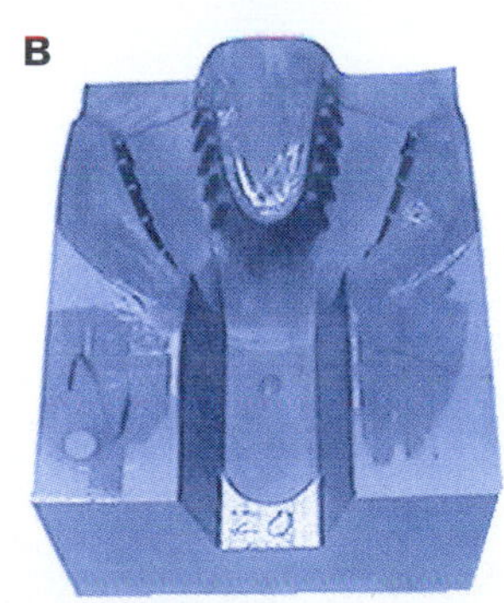

FIGURE 8.39
The 3D model for the mouthguard in (a) was used to create the program for the milling machine that created the mold cavity part shown in (b). (Courtesy of Advanced Coordinate Technologies.)

Modeling Sheet Metal Parts

Sheet metal parts, such as the one shown in Figure 8.40, are laid out as a flat pattern that is then bent into shape.

Sheet metal parts are joined together by welding, soldering, or using hems. Figure 8.41 shows standard sheet metal hems.

When modeling sheet metal parts, you may want to start by modeling them in 3D so you can visualize how they will fit with mating parts. Eventually the flat pattern for the part will have to be laid out in order for the part to be manufactured. Each surface on the part must be shown true size in the flat pattern. Additional material for hems or overlaps to be welded must be added and the bend allowance factored in. The bend allowance is determined by how much the metal compresses and stretches when it is bent. Its value depends on the thickness and material of the metal. You will learn more about the bend allowance in Chapter 11.

A key consideration in designing sheet metal parts is the fact that the shape cannot overlap itself when it is laid out flat.

A

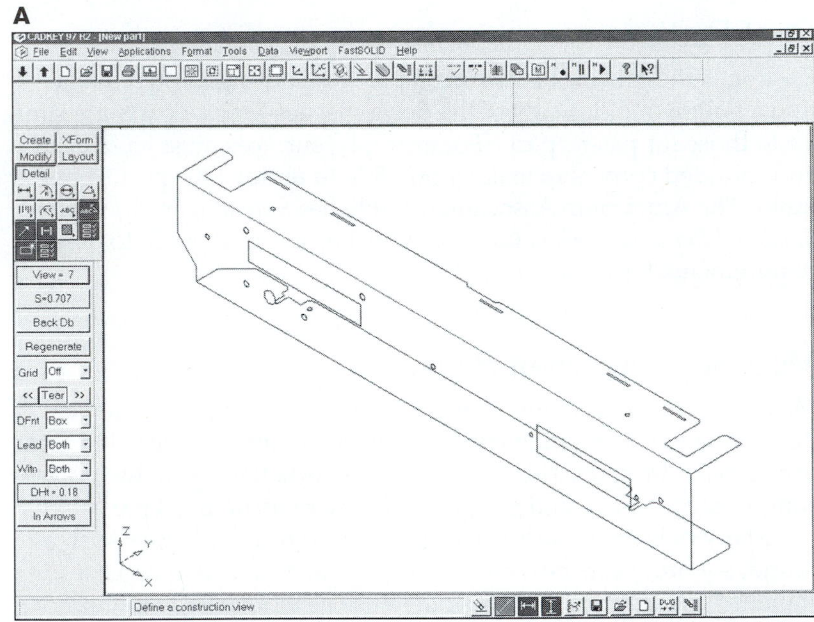

B

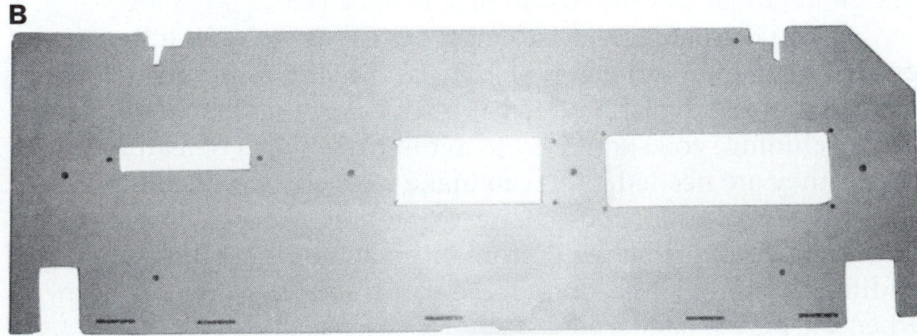

FIGURE 8.40
The 3D part in (a) will be formed by cutting the pattern from a piece of sheet metal (b) and bending it into shape. (Courtesy of J.E. Soares, Inc.)

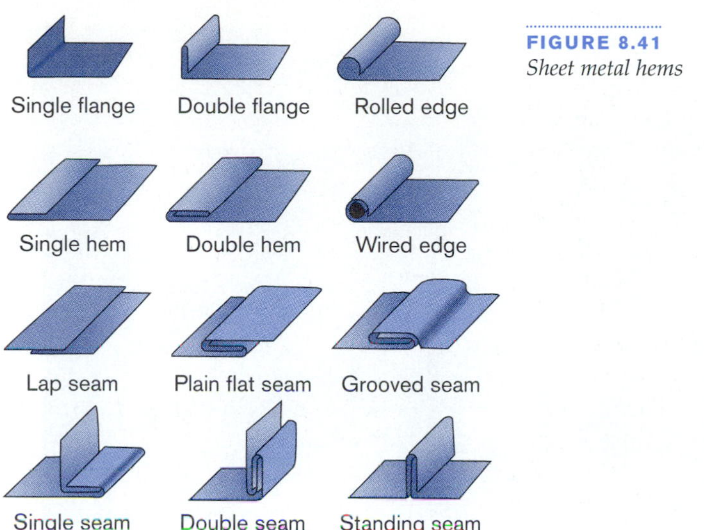

FIGURE 8.41
Sheet metal hems

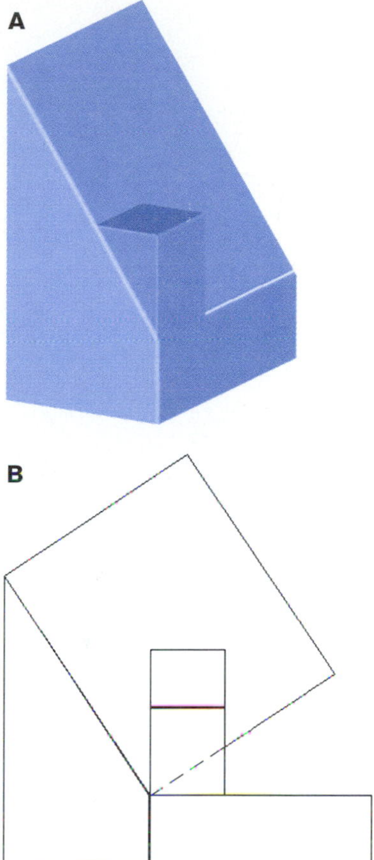

FIGURE 8.42
(a) This 3D shape cannot be made of a single piece of sheet metal. (b) The flat pattern overlaps itself.

Figure 8.42 is an example of a part that cannot be manufactured as designed from a single piece of sheet metal because its features will overlap when laid out flat.

Other sheet metal parts may be difficult to manufacture because of the way they must be bent. The brake press in Figure 8.43 is commonly used to form sheet metal parts by bending the metal along a straight line. Without special tooling, some parts may be impossible to bend without deforming other sections of the part.

Another consideration in sheet metal part design is arranging the flat pattern so that the parts nest together to waste the least amount of material. Software packages that can aid in the development of the flat pattern from a 3D wireframe, solid, or surface model oftentimes will help with pattern nesting to reduce waste material.

FIGURE 8.43
Brake press

FIGURE 8.44
Automated punch press

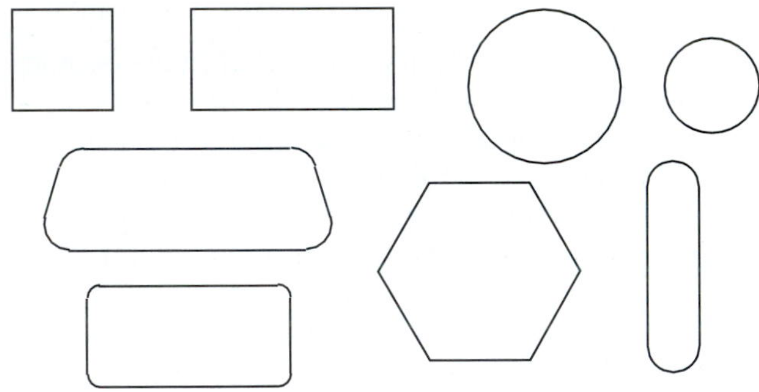

FIGURE 8.45
These 2D outlines are a library of shapes that correspond to standard punches.

Holes in sheet metal parts are often created by a punch, such as the automated punch press shown in Figure 8.44. Standard punches can quickly create openings in the sheet metal part; Figure 8.45 shows a library of shapes corresponding to standard punches. To help in designing sheet metal parts, you may want to keep a library of 2D outlines of standard punch shapes. You can then use these standard shapes to quickly create 3D model geometry by importing the 2D section into your file and extruding and subtracting the feature to create the opening.

In this chapter you have seen how the design database can serve as a proving ground for the digital assembly. By creating a comprehensive database of the design, you have an electronic representation of the design that can be used for testing, analysis, and documentation. In the next chapter, you will learn more of the ways the electronic database can be used to organize your design work and facilitate concurrent engineering.

INDUSTRY CASE

Stan McLean, VKI Technologies

Modeling Sheet Metal Parts: Think Flat

A key design challenge for the office coffee brewers manufactured by VKI Technologies is how to build more functions into increasingly smaller spaces. When Stan McLean designed the Zuma coffee brewer, he used Autodesk's Mechanical Desktop to model the brewer's sixty-two parts into a compact design that brews one cup of coffee at a time (see Fig. 8.46). The Zuma brewer, like all of the VKI vending machines, uses a number of sheet metal parts—brackets and three- or four-sided boxes—to hold the various components together. His approach to modeling sheet metals parts reflects their role in the brewer and is well suited to the design challenge of optimizing the fit of the various components.

"A vending machine is essentially a sheet metal box. Once I have the rough dimension of the machine, I start placing components into 3D model space within the outline of the machine. Our vending machines are built around a series of standard parts that we purchase, so we've created a model library of motors, pumps, valves, and fasteners. I place these into the assembly, then start moving them around to see how they will fit."

To do this, McLean usually has two files open at the same time: the layout (or assembly) model (as shown in Fig. 8.47), and the individual part model he is working on. By using external references to link individual parts into the assembly, McLean can update the assembly each time he changes an individual part. "I'm always loading the parts and checking the fit, then going to the part file and making modifications, then reloading the part to see if it is fitting the way I want it to."

With the individual components in place, McLean can begin modeling the sheet metal parts needed to combine them. For the Zuma brewer, he used AutoSM, add-on software that runs inside the Mechanical Desktop, to automate some of the sheet metal modeling functions. "A sheet metal part is a series of flanges or a box with various bends at various angles. I could create a rectangle in Mechanical Desktop, dimension it, extrude it to the gauge of the sheet metal I'm using, then change the user coordinate system as needed and repeat the process in another plane to create the flanges. I could

(continued, page 326)

A

B

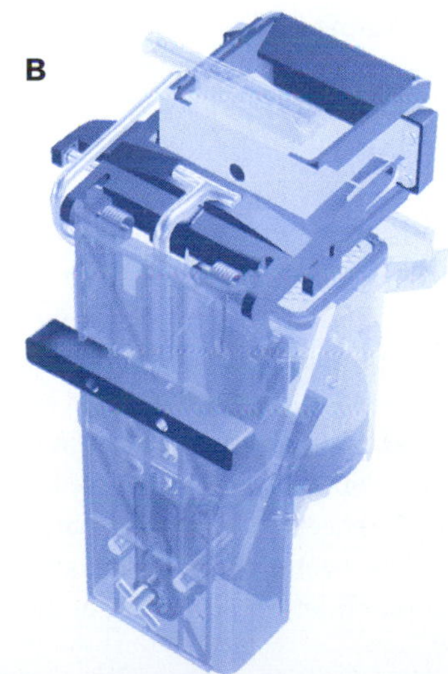

C

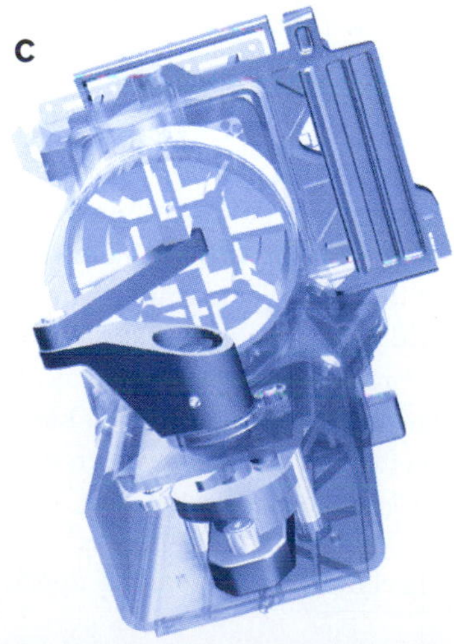

FIGURE 8.46
These three views of the Zuma brewer rendered some parts to be transparent so the internal structure would be more visible. The three-quarter rear view in (b) shows the sheet metal bracket in Figure 8.48. It holds the plastic part at the top through which water is added to the brewer.

INDUSTRY CASE

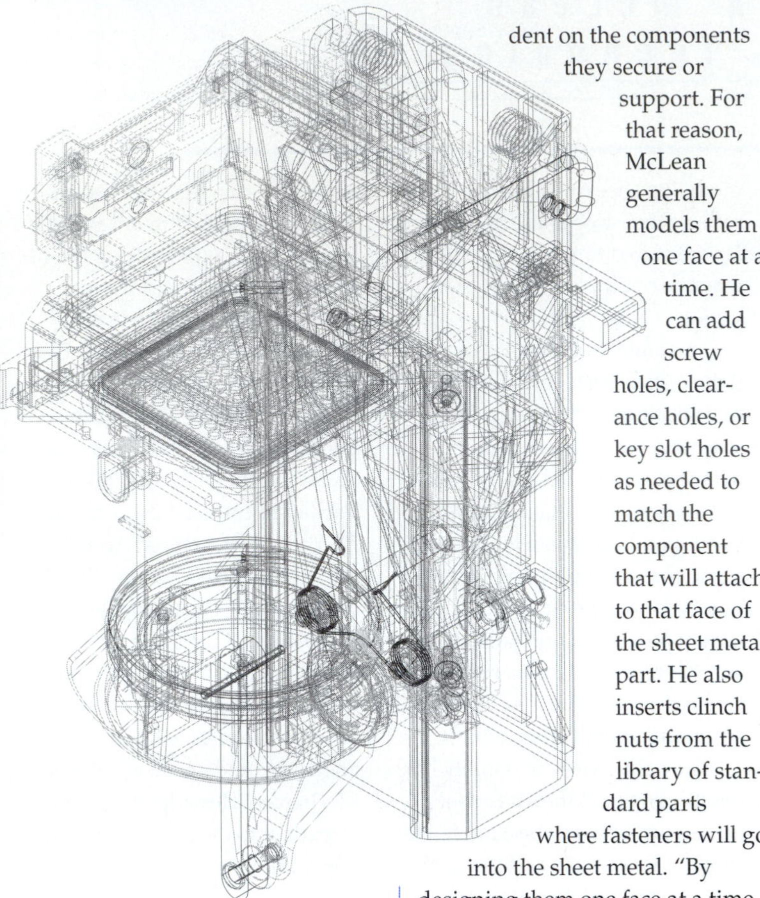

FIGURE 8.47
This wireframe view illustrates the compactness and complexity of the Zuma assembly. McLean kept this assembly model of the brewer open as he worked to refine the design of individual parts and how they fit together.

then fillet the corners to add the bend radius. AutoSM simplifies the process by automating many of the steps. For example, when I add a flange, it automatically adds the radius at the bend and calculates clearance at the corner based on the parameters I've entered for that gauge sheet metal."

The design of the brewer's sheet metal parts is largely dependent on the components they secure or support. For that reason, McLean generally models them one face at a time. He can add screw holes, clearance holes, or key slot holes as needed to match the component that will attach to that face of the sheet metal part. He also inserts clinch nuts from the library of standard parts where fasteners will go into the sheet metal. "By designing them one face at a time, I can adjust the faces as component locations change until everything is where it is supposed to be."

AutoSM offers another means of creating sheet metal parts via its "Clad a solid" command. To create a sheet metal part, McLean could create a solid box with the dimensions of the part, add holes and slots as needed, then issue the Clad command. AutoSM would first prompt him to select the faces of the solid box to be retained as sheet metal faces. Any holes in the box would be retained on the faces of the sheet metal part. Corner reliefs, overlapping corners, and corner fillets could then be added to the sheet metal enclosure created by the Clad command. AutoSM would then transform the box into a solid sheet metal part. Although the command is quick, McLean finds it less useful for designing parts such as his, in which hole locations and flange dimensions must evolve with the overall design of the assembly.

AutoSM uses parameters entered by McLean for the bend allowance, also called the bend setback, and the gauge of the metal in use. Although standards exist for these values, VKI Technologies does all of its own sheet metal manufacturing and has developed accurate values for the dies and materials used there. For 20-gauge satin coat porous sheet metal to be bent with a 1/32-inch nose die, the machine shop uses a bend allowance of 0.065. For 22-gauge metal, which is slightly thinner, the bend allowance would be 0.054. For 18-gauge, it would be 0.085.

Once the parts have been designed, McLean uses Mechanical Desktop to create multiview, fully dimensioned 2D drawings of the 3D sheet metal part and a bill of materials for any clinch nuts it includes. He then uses AutoSM's Unfold command to create the flat pattern for the part (see Fig. 8.48). AutoSM prompts him for a reference face for the part, then uses the parameters for the bend allowance to create the flat. For an L-shaped folded part with a 2-inch and 3-inch flange (their outside dimensions) to be made of 20-gauge metal, the overall length of the flat would be 2 + 3 – 0.065, or 4.935 inches.

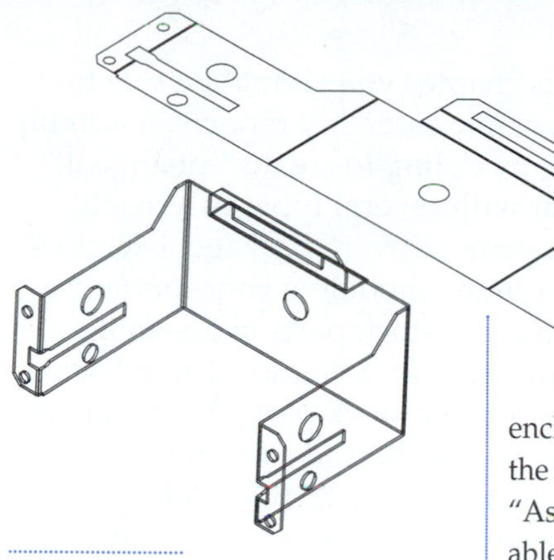

FIGURE 8.48
This sheet metal bracket appears here in its 3D folded form and as the flat pattern generated by AutoSM that will be used to cut the sheet metal.

AutoSM then creates a "block" (a group of drawing objects) of the flat pattern that McLean can insert into the multiview drawing of the part. Because the sheet metal parts will be cut with a numerically controlled machine, McLean also saves the flat pattern as a separate drawing file (using the Wblock command in Mechanical Desktop) that can be exported to C3CAM, the software VKI uses to generate the NC code for the part. The result is a complete documentation drawing of the folded part and an electronic file of the flat shape that can be used to cut the shape from the sheet metal.

McLean wrote his own program for nesting the flat patterns most efficiently on a sheet of metal. Commercial programs exist for optimizing this arrangement, but most are geared to the challenge of "jobbing," where several different parts are combined on a single sheet. VKI Technologies doesn't mix parts on a sheet, but generally manufactures a large quantity of the same part in each run.

Although AutoSM speeds up the modeling of sheet metal parts, it does not prevent the creation of impossible parts. "Because you're creating this bent-up enclosure, you have to think about the flat all the time," says McLean. "As you build it, you have to be able to visualize the flat part at the same time because you can create a solid or sheet metal enclosure with flanges that, when unfolded, will overlap each other. You also have to know how the metal is going to be bent. If you have a part with several bends, and you do them in the wrong order, you can get to a point where you can't complete the bend because the part will either hit the machine or hit the die."

McLean's manufacturing background helps him "think flat" and allows him to optimize his designs in ways that others can't. "Sometimes I look at a flat where the gap between two flanges is about 0.35, and I know that the punch operator will want to use a 0.2 punch. I might go back to the model and make some modifications so that the gap in the flat pattern will be exactly 0.2 so the punch operator won't have to punch up and down both sides to make the part."

Short of manufacturing experience, however, McLean's advice for sheet metal parts is to "keep it simple." "Instead of welding three parts together, is it better to make a single, more complicated part? This is a trade-off with no simple answer. The question is, Where are you actually saving? If you get to the point where a part has ten to fifteen bends in it, you're probably better off going to two parts and welding them together."

All illustrations courtesy of VKI Technologies, Inc.

Stan McLean is senior designer at Montreal, Canada's VKI Technologies, a manufacturer and distributor of coffee vending machines. McLean's model of the Zuma coffee brewer in Mechanical Desktop 1.2 won the 1997 Design Your World contest for digital design. McLean's background in manufacturing gave him experience with an NC punch press and the opportunity to write code for numerically controlled machines before he moved into engineering in 1989. He has always modeled his designs in 3D, starting with the 3D tools available in wireframe modeling software and adopting new tools for assembly and parametric modeling as they were developed.

Skills Summary

Now that you have completed this chapter you should be able to distinguish between static and dynamic assembly models. Assembly constraints are used in parametric modeling to create "intelligent" assemblies. You should be familiar with several types of assembly constraints you may encounter in your software package. Exercises 1–6 test your understanding of assembly modeling concepts and constraints. Often it is desirable to show stock parts in assemblies. Sometimes these parts can be acquired from a vendor. Thread, standard fasteners, and springs are among the stock parts that you may not want to model as they can be easily indicated by a standard note. Exercises 7–9 ask you to evaluate trade-offs in modeling stock parts. Fit between mating parts in an assembly is an important design consideration. You should be able to list four general categories of fit between mating parts and be familiar with ways that a CAD database can be used to check fit. Exercise 10 addresses the use of global parameters with mating parts. You should be able to list three manufacturing processes that are commonly used to manufacture parts and some of the considerations that may arise when modeling a part that will be manufactured using one of these methods. Exercise 11 asks you to consider issues in modeling for a particular manufacturing process.

EXERCISES

1. Care must be taken in parametric assemblies with regard to the parent-child relationship of component parts. What problems might occur, and how could you prevent drawing management problems when deleting or changing component parts in an assembly?
2. Define the differences between *static assemblies* and *dynamic assemblies.* Which method would offer advantages to design teams who access a common database of files through a network?
3. Many of the recent generation of computer modeling programs have capabilities for solids modeling and parametric modeling. Research the capabilities of the CAD program, solid modeler, or parametric modeler at your school or work place. What nomenclature, commands, and procedures are used by the program to define capabilities in
   a. external references
   b. global parameters
   c. interference checking
   d. dynamic assemblies

4. A parametric modeler is used to create a gear cluster assembly by inserting modeled bicycle gear cogs and spacers onto a splined hub. Axial and radial alignment of the elements is critical to maintain an accurate model. Define a sensible base point to use when adding individual cogs (gears) to the assembly.

5. Drawing templates, prototype drawings, and seed parts are one means to help assemble parts effectively. What are three advantages of using these file types when beginning a new model?

6. As directed by your instructor, create a dimensioned sketch or skeleton model that defines the assembly relations for the devices that follow:

   a. Vise grip

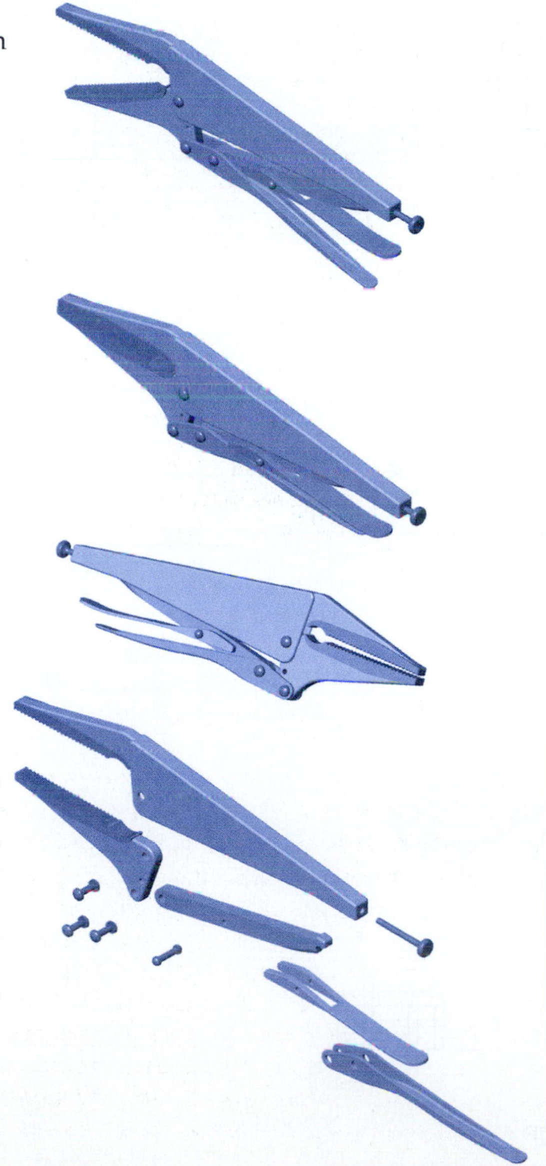

b. Jack

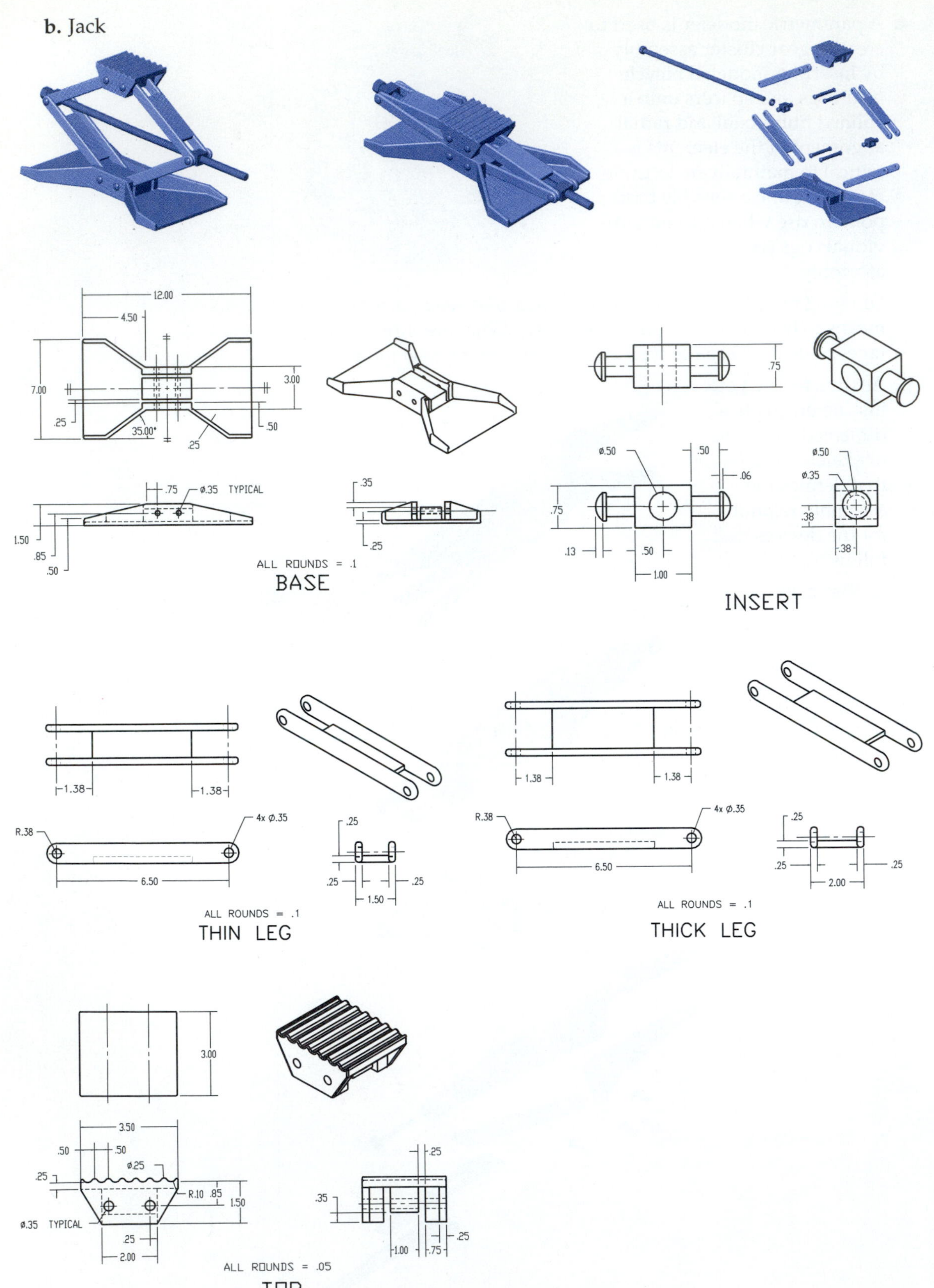
12.00
4.50
7.00
3.00
.25
35.00°
.25
.50
.75
ø.35 TYPICAL
1.50
.85
.50
.35
.25
ALL ROUNDS = .1
BASE
.75
ø.50
.50
.06
.75
.13
.50
1.00
ø.50
ø.35
.38
.38
INSERT
1.38
1.38
R.38
4x ø.35
6.50
.25
.25
.25
1.50
ALL ROUNDS = .1
THIN LEG
1.38
1.38
R.38
4x ø.35
6.50
.25
.25
.25
2.00
ALL ROUNDS = .1
THICK LEG
3.00
3.50
.50
.50
ø.25
.25
R.10
.85
1.50
ø.35 TYPICAL
.25
2.00
.25
.35
.25
1.00
.75
ALL ROUNDS = .05
TOP

b. *(continued)*

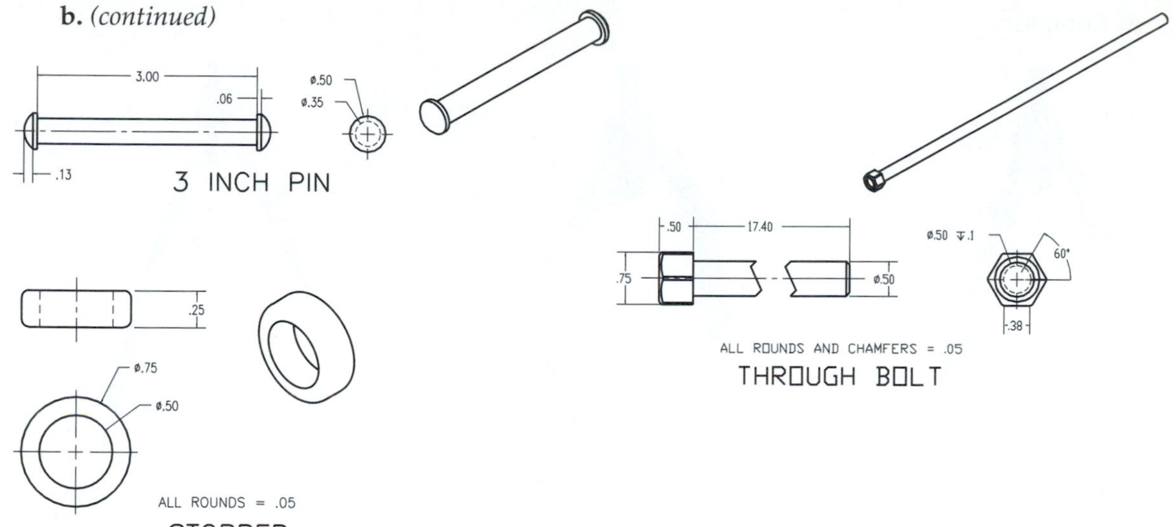

c. Remote control

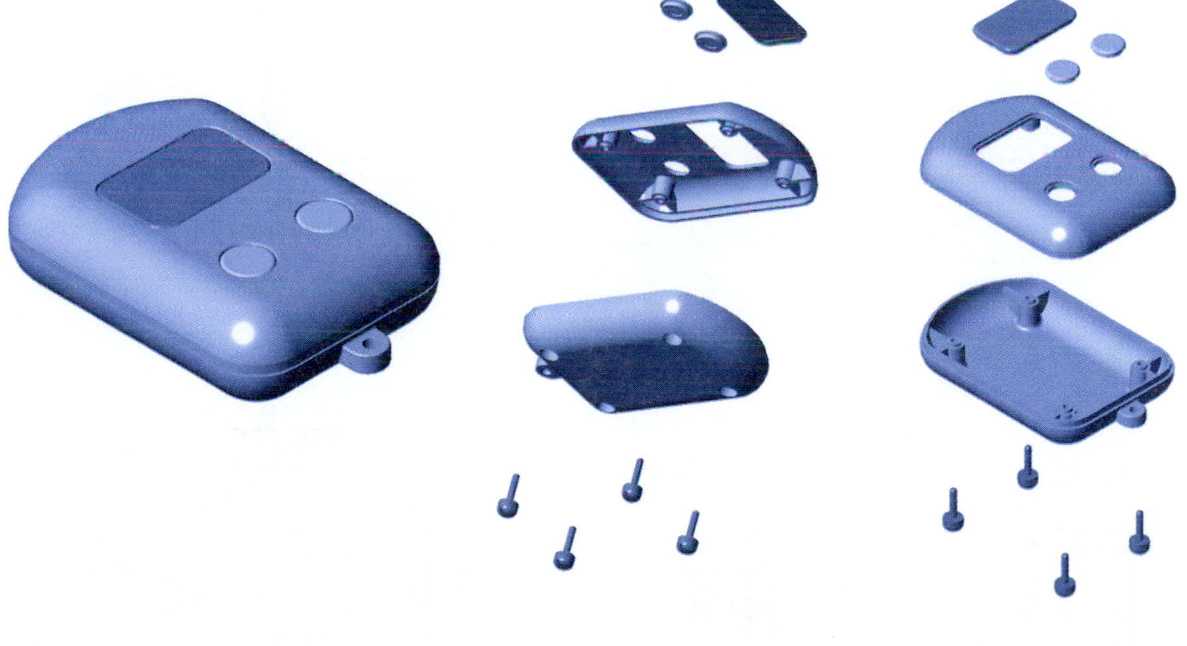

d. Compass

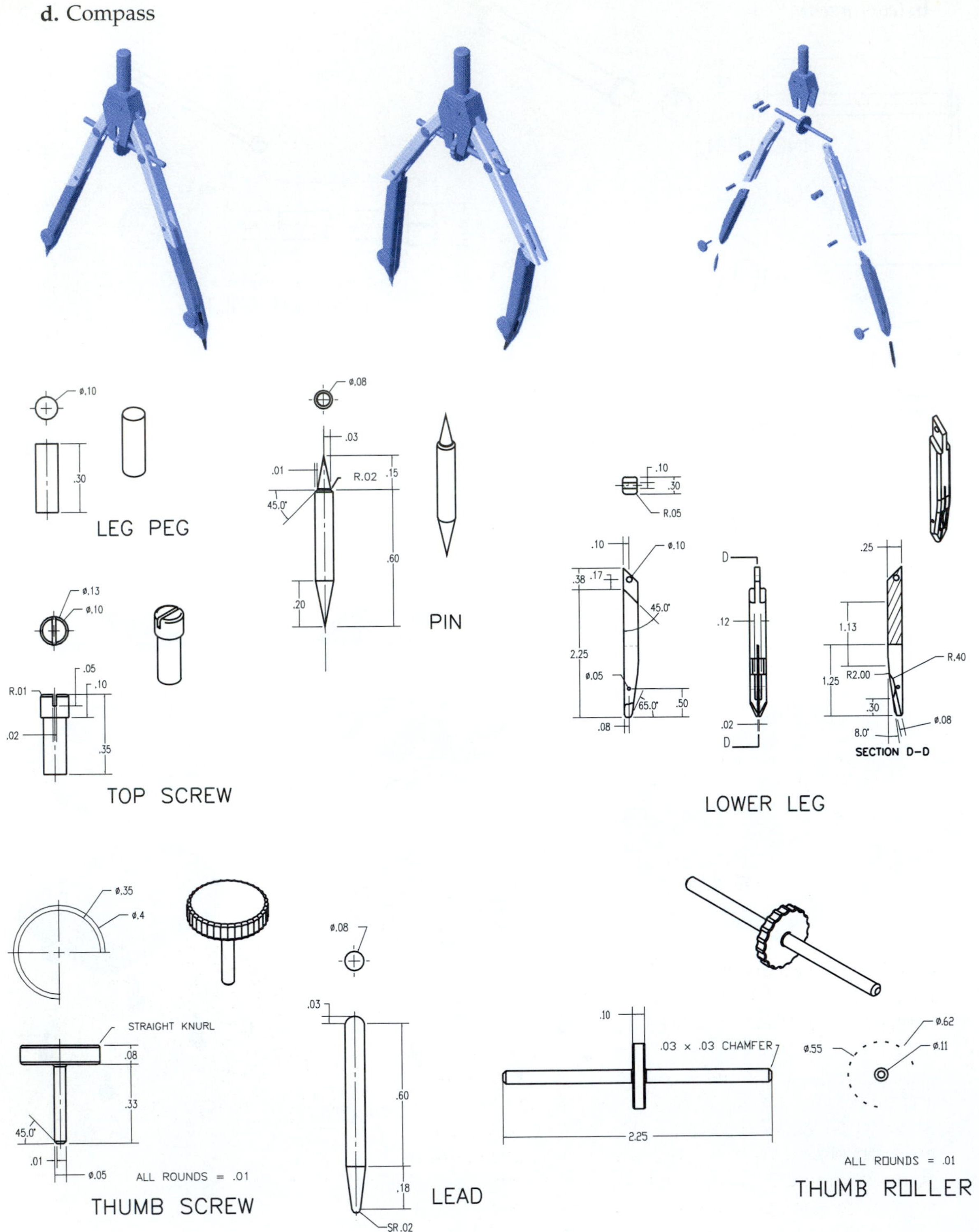

ø.10
.30
LEG PEG
ø.08
.03
.01
R.02
.15
45.0°
.60
.20
PIN
ø.13
ø.10
.05
R.01
.10
.02
.35
TOP SCREW
.10
.30
R.05
.10
ø.10
.38
.17
45.0°
2.25
ø.05
65.0°
.50
.08
D
.12
.02
.25
1.13
R.40
1.25
R2.00
.30
ø.08
8.0°
SECTION D-D
LOWER LEG
ø.35
ø.4
STRAIGHT KNURL
.08
.33
45.0°
.01
ø.05
ALL ROUNDS = .01
THUMB SCREW
ø.08
.03
.60
.18
SR.02
LEAD
.10
.03 x .03 CHAMFER
2.25
ø.62
ø.55
ø.11
ALL ROUNDS = .01
THUMB ROLLER

d. *(continued)*

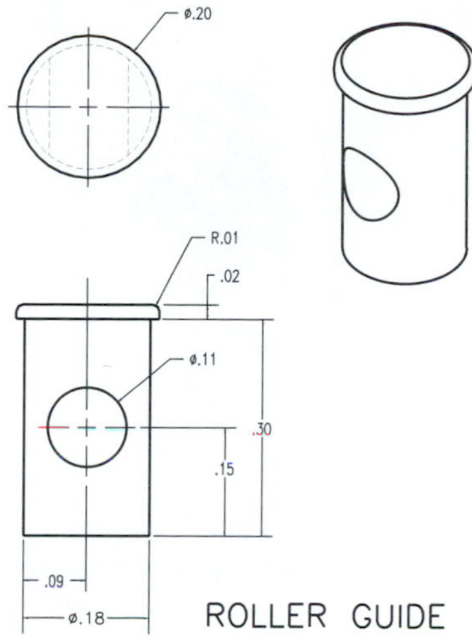

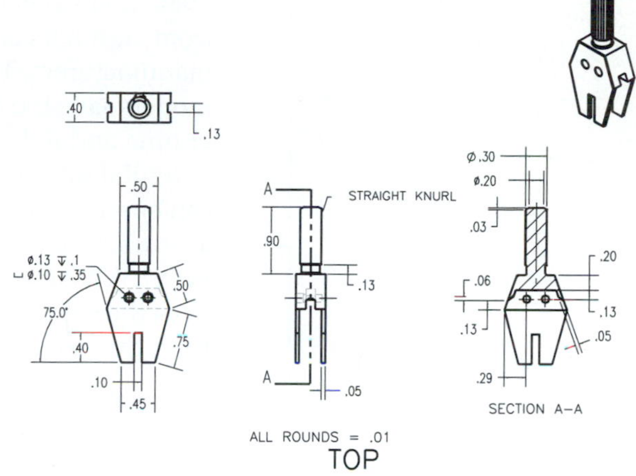

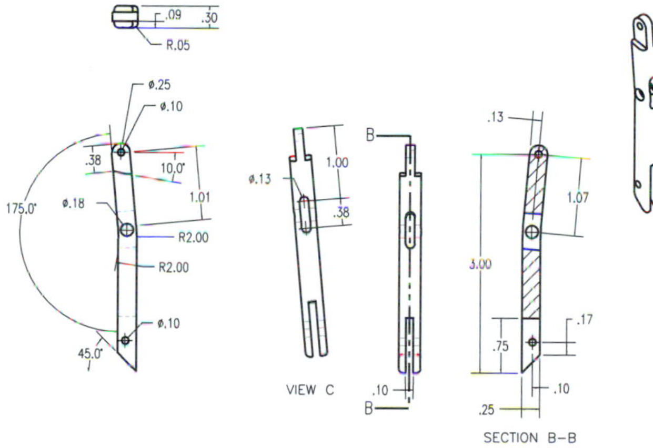

7. An entry-level engineer is put on the job of creating a detailed Pro/Engineer assembly drawing of a trailer-mounted pump and motor assembly. He models major components such as the diesel motor, the pump, and trailer frame as well as all the small parts such as wheels, tires, nuts, bolts, and lights. When the job is complete (after ten solid weeks of work) and all the drawing elements are correct, he presents his work to his supervisors. What reaction would be expected from engineering management? Was this an efficient use of resources? Do you have suggestions that would streamline the process?

8. While 3D modeling can describe part and assembly configuration in accurate detail, many parts or features are rarely included in a model until very late in the process if included at all. List four parts or features that fall into this category and give reasons for omitting detail.

9. Component models of detail parts or assemblies can be obtained from suppliers and manufacturers. These models save the designer time and reduce the potential for error, but can be difficult to use in a parametric model. Why? What advantage does the supplier gain from this service?

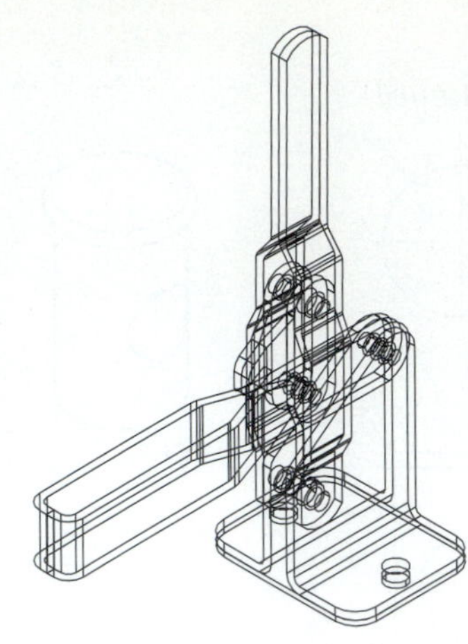

10. Parametric assemblies often have *global parameters* defined to assist in developing mating features of different parts. Identify which features of the two (mating) parts described in the following drawings would be candidates for shared parameters. (Note the different plotted scale of the parts!)

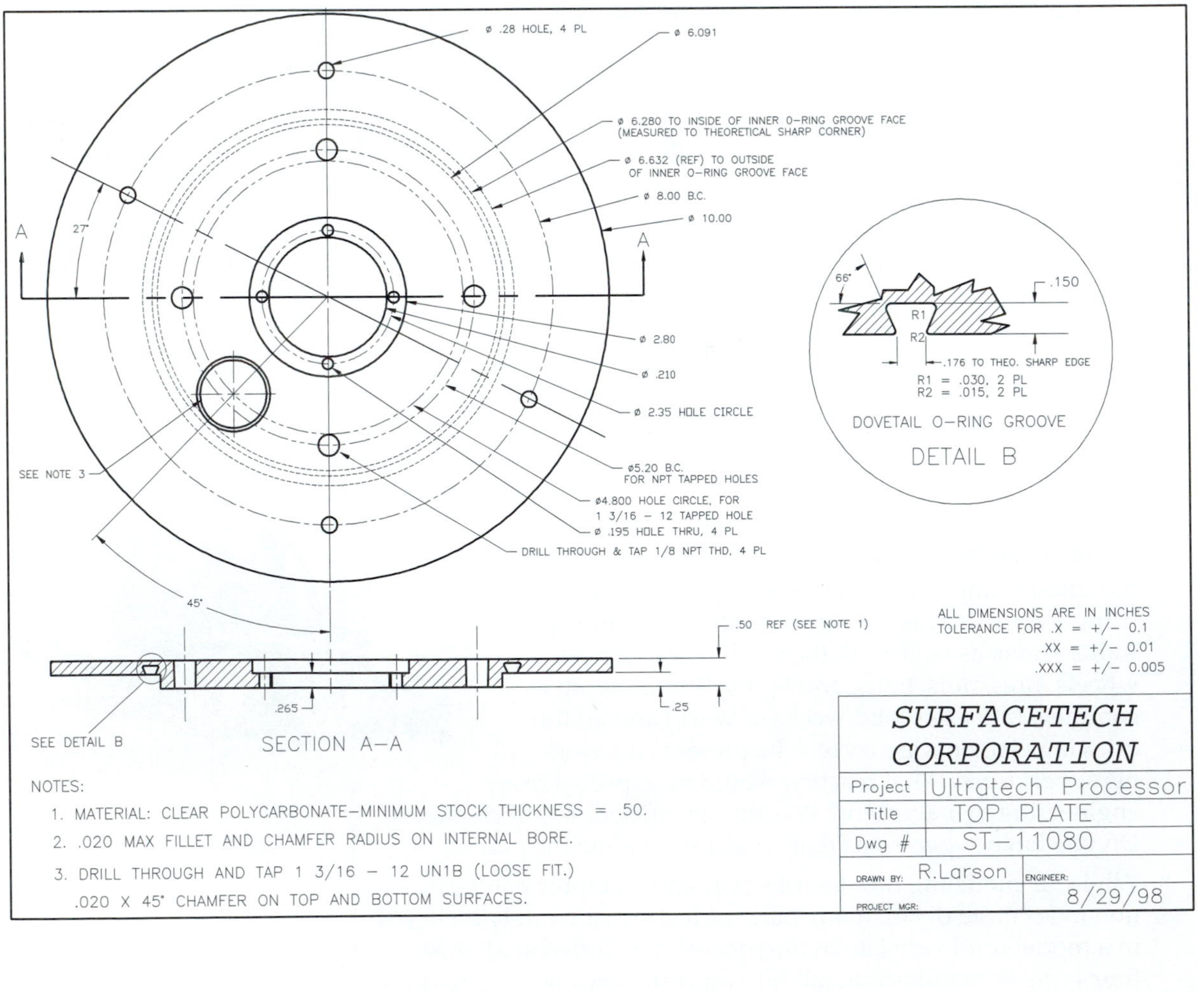

ø .28 HOLE, 4 PL
ø 6.091
ø 6.280 TO INSIDE OF INNER O-RING GROOVE FACE
(MEASURED TO THEORETICAL SHARP CORNER)
ø 6.632 (REF) TO OUTSIDE
OF INNER O-RING GROOVE FACE
ø 8.00 B.C.
ø 10.00
A
27°
A
ø 2.80
ø .210
ø 2.35 HOLE CIRCLE
ø5.20 B.C.
FOR NPT TAPPED HOLES
ø4.800 HOLE CIRCLE, FOR
1 3/16 – 12 TAPPED HOLE
ø .195 HOLE THRU, 4 PL
DRILL THROUGH & TAP 1/8 NPT THD, 4 PL
SEE NOTE 3
45°

66°
.150
R1
R2
.176 TO THEO. SHARP EDGE
R1 = .030, 2 PL
R2 = .015, 2 PL
DOVETAIL O-RING GROOVE
DETAIL B

.50 REF (SEE NOTE 1)
.265
.25
SEE DETAIL B
SECTION A–A

ALL DIMENSIONS ARE IN INCHES
TOLERANCE FOR .X = +/– 0.1
.XX = +/– 0.01
.XXX = +/– 0.005

NOTES:

1. MATERIAL: CLEAR POLYCARBONATE–MINIMUM STOCK THICKNESS = .50.
2. .020 MAX FILLET AND CHAMFER RADIUS ON INTERNAL BORE.
3. DRILL THROUGH AND TAP 1 3/16 – 12 UN1B (LOOSE FIT.)
.020 X 45° CHAMFER ON TOP AND BOTTOM SURFACES.

SURFACETECH CORPORATION			
Project	Ultratech Processor		
Title	TOP PLATE		
Dwg #	ST–11080		
DRAWN BY:	R.Larson	ENGINEER:	
PROJECT MGR:			8/29/98

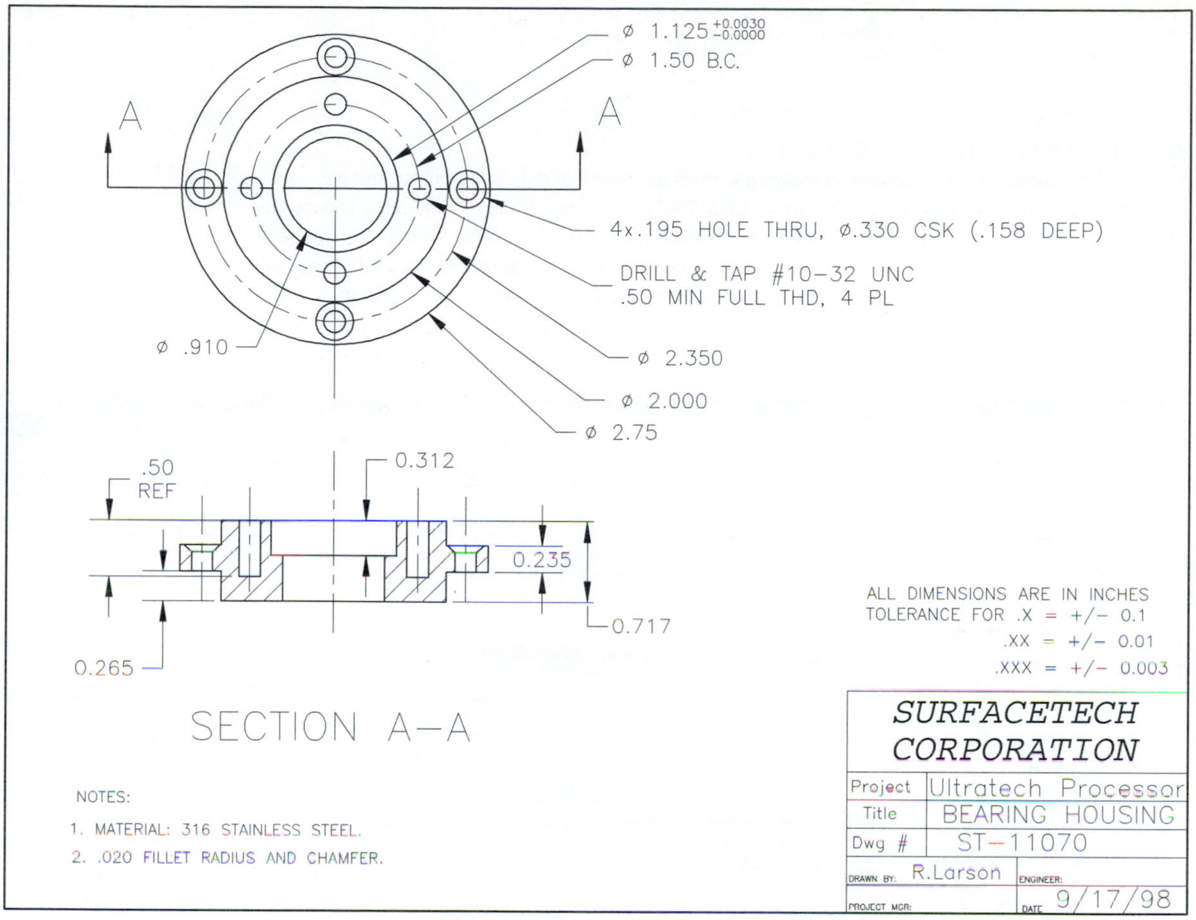

Ø 1.125 +0.0030 −0.0000
Ø 1.50 B.C.
A A
4x.195 HOLE THRU, Ø.330 CSK (.158 DEEP)
DRILL & TAP #10−32 UNC
.50 MIN FULL THD, 4 PL
Ø .910
Ø 2.350
Ø 2.000
Ø 2.75
.50 REF
0.312
0.235
0.717
0.265

SECTION A−A

NOTES:

1. MATERIAL: 316 STAINLESS STEEL.
2. .020 FILLET RADIUS AND CHAMFER.

ALL DIMENSIONS ARE IN INCHES
TOLERANCE FOR .X = +/− 0.1
.XX = +/− 0.01
.XXX = +/− 0.003

SURFACETECH CORPORATION	
Project	Ultratech Processor
Title	BEARING HOUSING
Dwg #	ST−11070
DRAWN BY: R.Larson	ENGINEER:
PROJECT MGR:	DATE 9/17/98

11. A designer must anticipate and understand the manufacturing processes that will be used to fabricate the created components. Describe differences in the model for a part designed for polyethylene (plastic) injection molding versus one for machined aluminum. Can you anticipate unique model requirements for a part created using hand lay-up resin-transfer molded composite (a process similar to that used for fiberglass)?

DESIGNER'S NOTEBOOK

Drawing and document management is evolving to meet the needs of data stored electronically, but also to facilitate collaboration around electronic data. Tri-Cycle Product Design created the secure Web site shown here using software that makes it easy to publish models on the Web, where they can be viewed and commented on by the client or used by engineers who are in a different location. These new software tools seek to address many of the document control issues in an electronic working environment. (Courtesy of Mark Wheeler, Tri-Cycle Product Design, Inc., with permission from Ovation Products Corporation.)

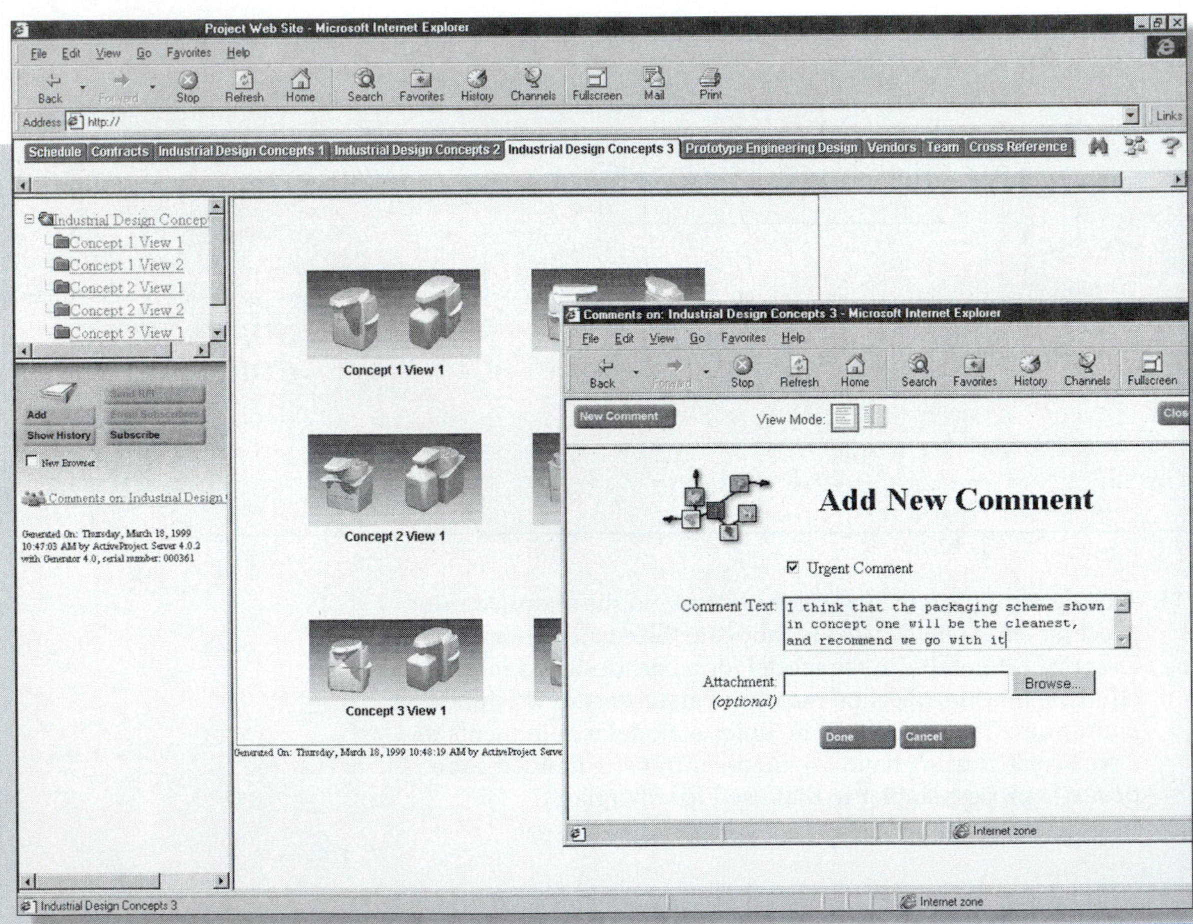

CHAPTER 9

Drawing Control and Data Management

Objectives

When you have completed this chapter, you will be able to

- present the rationale for controlling engineering documents;
- describe the role of the drawing or CAD file in engineering documentation;
- list the basic steps in approving and controlling CAD data;
- organize your files and directories in a consistent manner;
- practice good file handling and backup procedures;
- describe the basic principles of product data management.

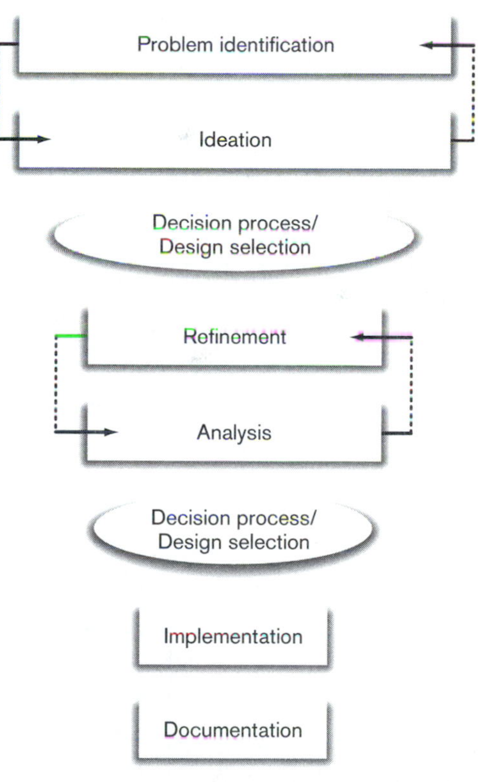

Engineering drawings, whether hand-drawn or computer-generated, have been an important part of the engineering design process for some time. The drawing's role in documenting engineering design was developed to satisfy the need for control in several areas. Legally, designs must be documented—for the life of the product at least, but in many cases permanently. Drawing standards and practices have been developed to satisfy legal requirements for design documentation. Drawings also serve as a contract with the manufacturer that will produce the design. To control how drawings are released for manufacture, companies and industries have developed procedures to eliminate misunderstandings and costly mistakes. When decisions are made to change a design, engineering change orders document these changes so they become a part of the permanent record of the design. Regulatory bodies, standards organizations, and case law have all contributed to the rules for retaining and producing documentation for your engineering designs. You should be familiar with the requirements for designs in the industry in which you work.

At the same time, you need to be prepared to address issues of document management that relate to the use of computer-aided design tools and the practice of concurrent engineering. Concurrent engineering depends on the ability of a team of individuals to work together on interrelated tasks and with the design database. The team's access to current and accurate information is crucial to its ability to work simultaneously on different aspects of the same project. In the last chapter, you learned how the design database can be used to streamline the design process by bringing the team effort together. Managing the flow of design database information to and among members of the team is a contemporary aspect of drawing control that is being addressed through new software tools and company processes.

When Boeing made the decision to go to 100% digital design of its Boeing 777 aircraft, the company modified not only how it archived its designs, but also how the digital data would be used among team members. The ideas behind the proprietary systems it developed presaged today's commercial systems that use a central database to store and serve data to a wide range of other software tools. Boeing's system is based on two large databases: The DIRRECT database (for Digital Information Repository for Release Exchange, Control Tracking) archives all finished designs and their associated models. Each individual drawing is located in the database by a unique multipart number that can be used to identify major characteristics of the part (such as the product it was designed for, creator division, and so on) and that can be used to search and sort the information. In the past, Boeing's archives were 100% drawings on mylar—a preferred form for archival drawings that satisfies all legal requirements. Today the archive has been expanded to include 3D solid models of the parts as well. The DIRRECT database serves not only to archive part designs, but also to track any changes made to them. When a part is "checked out" of the database, the user can see whether anyone else is currently working on the model so two people are not modifying the same part at the same time.

The second database addresses some of the *work flow* issues surrounding concurrent engineering; that is, how documents are shared and passed among users. EPIC (for Electronic Pre-Assembly Interaction on CATIA) gathers and stores all 3D models of the aircraft's design as they are being designed. EPIC provides the most current version of the parts to the virtual model of the aircraft so that it can be used for interference checking—much as an assembly model would be. It is also used to create fly-throughs of the computer mock-up of the aircraft and its systems and to serve parts for finite element analysis. Given the scale of the projects Boeing teams work on, however, the database also controls ownership of the information in the database and tracks interferences detected in the model to ensure that they are resolved. One of the barriers to success in concurrent engineering is the amount of time engineers spend sharing information among teams and informing others of changes made. Through their data management tools, Boeing has defined an approach that lets the engineers "check" their work and initiate contact with teams only as needed. This approach frees the engineering teams to work independently and concurrently and lets the computer do more of the data management.

Whether you work at a large or small company, you need to be aware of the issues in document management and the tools being developed to save you time and facilitate your involvement in team-based concurrent engineering.

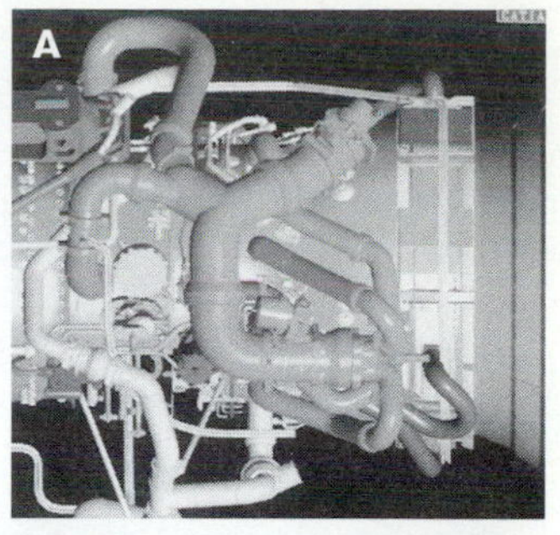

FIGURE 9.1
The EPIC database serves as the model of the aircraft shown in (a). Large screen display in (b) facilitates checking for interferences visually. (Courtesy of The Boeing Company.)

Requirements for Engineering Documentation

Traditionally, paper drawings were used throughout the design process as well as for documentation. Practices for approving and releasing drawings, making drawing changes, and distributing and storing drawings were developed over time to fit the use of "hard copy" as a medium. With the advent of design databases stored on computer systems, these practices had to be adapted to fit the "soft copy" represented by an electronic file. By failing to adapt their practices from paper drawings to electronic files, some companies have not documented their designs properly or retained their engineering design records in a systematic and permanent way.

The information contained in the 3D design database that you will use either to produce paper drawings or to send as files for NC machining, mold design, or other manufacturing processes is an important part of the engineering design record. You can think of released drawings as a contract between you and the company that will build the structure or system, or the manufacturer that will produce the part or device. It is important to be able to document what was actually provided in this drawing.

Companies need to manage these records so they can produce the documentation for various purposes, such as defending themselves against product safety liability and patent infringement lawsuits. In the event of legal issues, you must be able to produce the document as it was provided and in a method that is admissible in a court of law. Because electronically stored files can be altered, they may not be considered an acceptable method for documenting engineering designs. In addition, most legal requests are for copies of all versions of the design, and all copies stored and used within the company. It can be costly and embarrassing if they are not well organized and correct, or if there are multiple versions in use that vary from one to the next.

Legal standards for how long engineering drawings need to be retained vary from state to state and industry to industry. For example, nuclear power plant drawings are kept for years, but the drawings for a medical device may be kept only a year or two after the product is obsolete (and no longer being manufactured or sold). In most cases, industry standards groups weigh the risk of record destruction in the context of product and public safety and make recommendations for its members. American Records Management Association (ARMA) standards for records retention are another source of information about engineering drawing retention. A company—or an individual consultant—needs to understand these standards to determine how long to retain engineering drawing records. Even if the legal standard is less stringent, a court may rule that a firm should meet the industry's common practice to avoid a finding of negligence.

Industries that are regulated by the Food and Drug Administration (FDA) should also be aware of its Guidelines for Electronic Records and Signatures, a ruling that was finalized in early 1998. Regulatory agencies such as the FDA consider engineering drawings "specifications" or "documents" and have clarified their record-keeping guidelines to include electronic forms such as graphics files. Companies that are undertaking records management guidelines should be aware of the most current rulings and efforts by industry standards and regulatory groups to encompass electronic media in their recommendations.

In addition to legal requirements for documenting engineering designs, effective storage and retrieval of engineering design documentation can make a difference in a company's ability to succeed in today's world marketplace. The effective use of a 3D design database can provide many benefits outside of just reduced drafting time or a shortened product development cycle.

Understanding the process that paper drawings went through for approval, release, and storage can help you understand good practices for the approval, release, and storage of your electronic CAD data.

Drawing Approval and Release

In the days of creating paper drawings on a drawing board, a designer would give sketches or drawings to the drafter, who would use triangles, T squares, templates, and other hand tools to prepare neat, legible drawings. The designer and drafter would work on a drawing for some period of time until it was determined to be complete. The title block on the drawing would be used to document the change from a draft to a finished drawing. The drafter would sign and date the *drawn by* block, perhaps a checker would sign off

GOLDEN VALLEY ELECTRIC ASSOCIATION, INC.

Fairbanks, Alaska

FIGURE 9.2
Title block with approval signatures (Courtesy of Golden Valley Electric Association.)

and date the *checked by* block, and the engineer would approve the drawing for release by signing and dating the *approved by* block. A supervising engineer might also sign off to approve the drawing. Then the drawing or set of drawings and contract would be released to manufacturing or to the contractor to be produced. Copies of the approved drawings would be circulated to various departments within the company as required, and one set of the printed drawings and contract would be stored for the permanent record. Figure 9.2 shows a title block that would be used to gather approval signatures on the paper drawing.

The initial drawing was usually stored in a fireproof drawing vault so it could be retrieved later if changes or additional prints were needed. Mylar, a waterproof polyester film more durable than paper, was often used as a medium for documentation drawings, but other media, such as microfilm and microfiche, were also used to store the drawing. The drawing number oftentimes indicated to some extent the project, type of drawing, and even where the drawing was stored to help in its retrieval. Figure 9.3 shows a drawing number system that indicates the type of drawing and location.

Drawing Numbering System

Categories

C Cover
C1 Title Cover
C2 Drawing Index-Cover

E Electrical
E1 General Electrical Arrangement
E2 One Line Electrical
E3 Three Line Electrical
E4 AC Relaying Schematics
E5 CD Relaying Schematics
E6 S&C Circuit Switcher Electrical
E7 Transformer Electrical
E8 Recloser Electrical
E9 Panelboard Electrical
E10 RTU Electrical
E11 Building Electrical
E12 Miscellaneous Electrical

EQ Equipment (Layouts, Bill of Mat'ls, Outlines, etc.)
EQ1 S&C Circuit Switcher-Equipment
EQ2 Transformer Equipment
EQ3 Recloser Equipment
EQ4 Panelboard Equipment
EQ5 Miscellaneous Equipment

F Foundation
F1 Foundation Arrangements & Details
F2 Fencing & Grounding Details
F3 Conduit & Cable Layout Foundation

M Mechanical
M1 HVAC Mechanical
M2 Boiler Mechanical
M3 Waste & General Water Mechanical
M4 Steam & Steam Heat Mechanical
M5 Mechanical Standards
M6 Piping Mechanical
M7 Drains & Vents Mechanical
M8 Chemical Mechanical
M9 Fuel Oil Mechanical
M10 Fire Protection Mechanical
M11 Control Diagrams Mechanical
M12 Lube Oil (Turbine) Mechanical
M13 Steam Heating Mechanical

S Site
S1 Property Plat Site
S2 Site Plans
S3 Grading Plans & Sections Site

ST Structural
ST1 Specifications & Erection Drawings
ST2 Plan Arrangement Structural
ST3 Elevations Structural
ST4 Structural Details
ST5 Structural Bill of Materials

FIGURE 9.3
The drawing number used by the Golden Valley Electronic Association consists of three parts and is used as the actual computer file name. The first three letters are the three-letter designation for the substation or generation plant (e.g., GHS for Goldhill Substation) followed by a backslash (\). The second part is the designation for the drawing category (e.g., E2 for One Line Electrical) followed by a hyphen (-). The third part is the drawing number within the category. For example, GHS\E2-1 would be the first one line electrical drawing from Goldhill Substation. (Courtesy of Golden Valley Electric Association.)

P

FORM 686 REVISED 10-28-68

SHEET 1 OF 1 Nº H10547

ENGINEERING CHANGE DATE: 6-9-98	PART NO.	HE0004-AV-01-01			DRAFT.	WTG
	ON HAND				CK'D.	JMN
WARNING CHANGE DATE: 6-9-98	ON ORD.				APPR.	FB
	RES.				REL. BY	FB

COPIES TO: IND. ENG. X PROD. CON. X INSP. X SERVICE OTHER

ENG. APPR.	
REQ. BY DEPT. - NAME	
PART NO.	HE0004-AV-01-01 REV. A
CHANGE	CORRECTED BOM ITEMS 19,20,34,35 & 36
REASON	DRAWING CORRECTION
ACTION REQUIRED — PROD. CON. ON HAND	PAMPA- BUILD
ON ORD.	PEMEX
EFF. ON UNIT	20794-95-96
TOOLING	
SERVICE	

FIGURE 9.4
This engineering change order is a word processing document that is completed for each change to a released drawing. (Courtesy of IRI International.)

Change Orders

In the imperfect real world, released drawings often require some type of correction during the process of constructing or manufacturing the product or system. After a drawing has been released, however, it could not be changed unless an *engineering change order* (ECO), shown in Figure 9.4, was approved. An ECO, also called an engineering change notification (ECN), details the nature of the change in a separate document. After the ECO was approved, the drawing was revised and a revision noted on the drawing.

Revision Block

After revision, the drawing was not simply replaced. A dated *revision block*, shown in Figure 9.5, was added to the drawing. A revision block describes briefly what the change was and may also indicate the number of the engineering change order (which contains

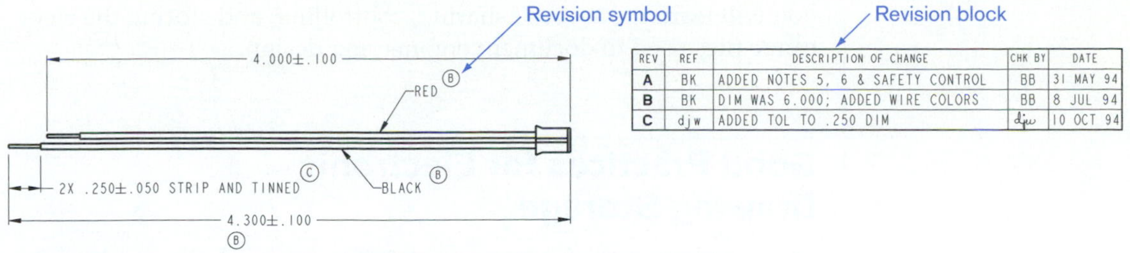

FIGURE 9.5

The revision block shown here identifies the change made to the drawing, but the change is also marked with a revision symbol to make it easy to locate. (Copyright 1998 Tektronix, Inc. All rights reserved. Reproduced by permission.)

more information about the change). The drawing was then approved again. A small number (circled or sometimes contained in a triangle) was added to call attention to the revision on the drawing. Some companies also included an easily visible revision number near the drawing number in the title block. Annotating the drawing number helps ensure that two people discussing the print from two different locations can verify that each is looking at the same revision of the drawing.

Once a revision had been made and approved, the drawing was then printed again and a copy stored for the permanent record. The new prints were circulated to all who received the previous print so they could update their files. Some companies required that old prints be collected to eliminate the possibility that some departments might continue to them.

A Drawing as a Snapshot in Time

Each of the paper drawings archived and referenced in the process described above served to document a design at a particular point in time. Some companies continue to print and store a paper copy of the CAD file as their permanent record to provide this documentation. If the electronic CAD file is updated and no longer matches the drawing, the paper copy acts to preserve the design information. In this kind of system, the same approval and archival practices just described are applied to the paper drawings generated from the CAD database. This is a perfectly acceptable practice if you can retrieve these paper drawings later as needed.

Today, however, practices are being developed to allow the same level of control around electronic files that store the design data. New quality standards in some industries allow for electronic documents to serve as the permanent record if they are properly controlled. For example, the process of approval described for paper drawings could be applied to CAD drawings that are "frozen" at each approval. Instead of storing the paper version, however, the company would store the electronic file.

However, adapting the paper process just described to electronic files covers only a portion of the design process and product information that a company needs to control. In the rest of this chapter, you will learn more about sharing, controlling, and storing the electronic files used to document engineering design.

Good Practices for Electronic Drawing Storage

Organized practices for storage, approval, retrieval, file naming, and tracking revision history for electronic CAD data are even more important than they are for paper-based systems. You've already learned a lot about the many benefits that motivate companies to invest in a 3D CAD database, such as improved communication between functional areas, computer integrated manufacturing, shorter design cycle time, cost savings, better access to information, and improved visualization capabilities. If the CAD files and documentation are not well organized, however, many of these benefits may not be realized.

Chances are, you will work in a company that has developed and articulated its standards of data management, and you will be expected to adhere to them. The company may even have invested in data management software, such as that used at Boeing. In either case, much of the responsibility for managing data will be yours. Understanding the issues in personal file management will help you devise your own system, if you must, and help you appreciate the pros and cons of various approaches.

Storing Electronic Files

The advent of personal computers on each designer's desktop may have contributed to the difficulty of managing electronic data. Each designer may organize the files he or she is working on differently, or keep multiple copies of a file in different directories. When others need to view or edit the file, it may be hard to be sure which is the current version. Without a thorough approval process for release of drawings, the engineer may neglect to track and store the revision history. Even when previous revisions of the drawing are stored electronically, they may not be useful because they don't satisfy the requirements for a static snapshot of the design at the time of release.

Many companies run into difficulties with their CAD data because they start out small and don't implement an organized system for managing the files. By the time they realize that they require better organization, they have thousands of poorly organized files and many CAD users with poor file storage habits. It is a very important part of your job as an engineer to manage the engineering design records that you produce.

Organized Directory Structures

Using an organized directory structure enables you to retrieve your CAD files and other electronically stored engineering data. Think of a directory on the disk as a kind of file folder. You would not put all of your paperwork loose in your file cabinet; neither should you scatter files over your hard disk.

Most people create a directory structure in a way that makes sense for the types of projects they do. A project-based directory structure, such as the one shown in Figure 9.6, allows you to store all files related to a particular project in the same directory.

Within each project, you may want to have subdirectories for different parts or different kinds of data associated with the project. A good rule of thumb is to think about situations when you (or others) will want to retrieve the data. Will you think of it in terms of the project? (Remember that a key to working with assemblies is the availability of the individual part files associated with it.) Or will you think of it in terms of another characteristic? Your CAD file directories may be project based, while other files may be organized differently. Making the directories work as you do will help you find what you need easily.

It is not a good idea to have a single "work" directory where you store all of your files, even if you intend to move them to another project-based directory later. This chore may not get done, or you may not remember which version of the file is current. Nor is it a good practice to store your work files in the directory that contains the application software. (When you install software upgrades, you may accidentally overwrite or delete your work files.) Developing good habits that help you organize your files as you create them will eliminate confusion and save time later.

If you are using a networked computer system, you can store company documents on the network in a directory that others in the company can access. Because CAD files require large amounts of storage space, you need to manage the space on the hard disks and drives available to you. Keeping copies of all files on your personal system should be weighed against the frequency with which you need those files, the time involved in retrieving the files as needed, and the likelihood that the copy of the file on your system will still be current the next time you need it.

If you work for a company that has implemented a system to control their electronic data, you may be required to keep all of your CAD files in a workspace allotted to you on the network server. This

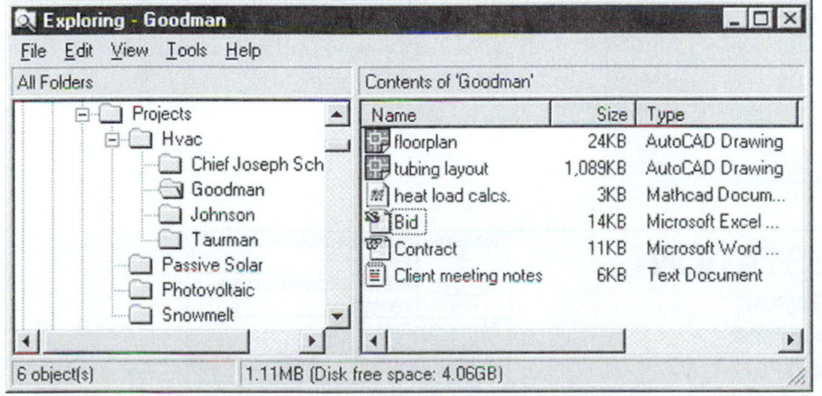

FIGURE 9.6
The top level of this directory structure is divided by project type. Each subfolder includes the CAD drawings for the individual project, plus bid documents, calculations, and letters to the client relating to that particular project.

makes it possible for the company to control (and regularly back up) this data, and it removes the burden of doing so from each individual employee.

File Naming Conventions

Organizing your files is as important as naming them systematically. If a companywide policy for naming files does not exist, you should develop one for your own files to make them easy to find. File names serve a purpose very similar to the title block in a drawing. Without opening the document or studying the model inside, you should be able to identify key information. Multipart file names may be used to indicate, for example, the part, the project it is part of, who created it, and a version number.

If you use numbers for your file names and assign them in a systematic way, users familiar with the system can search for kinds of documents, or for files by name. For example, you may have a sort of Dewey decimal system for types of drawings or documents your company typically produces, so that numbers starting with 100 are electrical schematics, 200s are assemblies, and so forth. The next set of numbers can further identify the type of file by another set of characteristics, such as project group numbers. A systematic alphanumeric code may work in a similar fashion.

A systematic system can make it easy to organize files and drawings, but the drawing name/number need not be the same as the CAD file name. To keep file names short, a different system may be used than the one to identify a drawing. In this case, or if drawing names are assigned in a fairly random way, a company may maintain an organized database where users can look to find the particular drawing or file name. Sometimes drawing names are standardized only when a part is approved and added to company archive. If so, the numerical system used for the CAD file name may also be used for the paper drawing name. Figure 9.7 shows two naming systems you might use for CAD files.

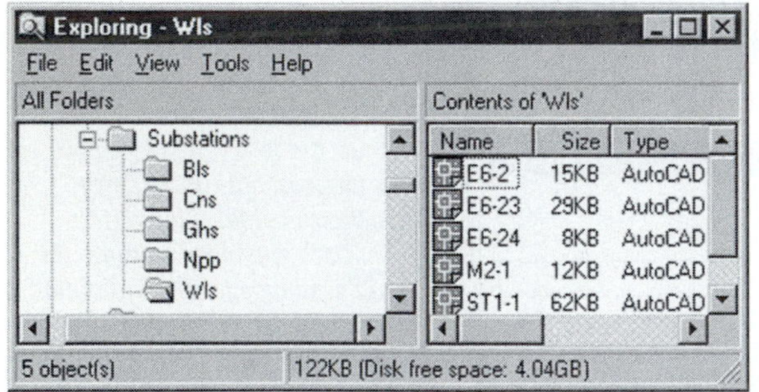

A

B

FIGURE 9.7
Naming CAD files systematically can help you retrieve them. (a) Using the drawing name or number as the CAD file name can make it easy to retrieve the file. (b) Using a file name that codes the project name as the first characters can make it easy to sort the files by project.

Additional data may be added to the file name over time. The revision number can be added when revision drawings are approved. Once a company has file retention guidelines in place, file names can also incorporate a tag that indicates when it may be removed from long-term storage. This kind of tag is generally added by the group that manages the company's archives, but it demonstrates the role of the file name in encapsulating information and aiding in file management.

Whatever system a company uses is definitely better than having each employee name drawings as he or she pleases.

Drawing Standards

Like file naming standards, drawing standards can help you work more productively and can contribute to the usefulness of the drawings as company records. Engineering drawings are the property of the company, not the individual engineer. As such, their usefulness should not depend on a single individual being able to locate or interpret them.

Company standards may be in place to introduce consistency in the way drawings are constructed. Standardizing the layers in a drawing, for example, can make them more navigable. Architectural engineering firms use layers to organize the many systems, such as plumbing and electrical, that are part of building design. Each of these systems will be assigned a set of layers and layer names so that each drawing can be manipulated in the same way. Notes and drawing text are commonly stored on a separate layer across drawings so this information can be turned on and off as needed when working with the model. Because CAD software carries layer information along with parts, standardizing layer names can prevent unnecessary confusion when files are combined.

The colors used for layers and different drawing elements can also be standardized, as shown in Table 9.1. Center lines may be red, for example. Because color is used to change pen widths on a pen plotter, this kind of standardization can facilitate printing the drawing with the appropriate line weights.

Many other aspects of creating a drawing can be standardized. The borders required for each different sheet size, the information contained in the title block (and how it is to appear), and the fonts and letter sizes to be used for different items are frequently spelled out in a company's drawing standards. The text of certain notes—such as manufacturing standards or safety control notices—are often standardized so that legally appropriate and consistent information is provided in all cases. In addition, libraries of symbols and stock parts may be provided to help you work more efficiently and produce engineering drawings that can easily be understood and reused by others.

The guidelines for drawing standards can be codified and stored with drawing archives as a navigational aid in the future. In some cases, the drafting group will help ensure that drawings meet company standards; in others, it is the responsibility of the designer

TABLE 9.1

Standard Base Layers

Line Type	Pen Number	Line Thickness (AutoCAD)
Center	1	.003
Section	2	.005
Dim, Format	3	.005
Object	4	.009
Hatch	5	.003
Bubbles	6	.004
Material, Symbols, Text	7	.005
Phantom	8	.003
Hidden	9	.003

Note: *The layers used for certain drawing elements have been standardized company wide. Each element in the first column is placed on a separate layer in the drawing. Linetype, color, and line thickness for the various elements are set by layer in the prototype files used to start all drawings. (Courtesy of IRI International.)*

to check the company's published standards. The standards may be enforced by the records management group responsible for archiving design documents, which will refuse to accept drawings that are not prepared according to standards. While it may be easy to rename a file after it is created to make it consistent with company standards, you should address drawing standards by starting new drawings from a prototype file or seed part that provides the company's common framework.

Permission and Ownership

File use in a networked environment adds a dimension of ownership to file management. *Ownership* is a term used to refer to the security systems built into network operating systems that allow the system to restrict access to data on the network. Even personal computer operating systems allow some form of file protection to prevent files from being erased, but networked systems allow for more sophisticated password-based security for files and directories.

To control access to data over a network, the network administrator gives different kinds of permission to different users. *Permission* defines a user's level of access to a file. For example, some users may have read-only permission. They can read or view a file, but cannot save changes to it. Write permission allows a user to create and change files. Change permission allows a user to read and change files or folders, but not to delete them. The permission can be set globally, or for each set of files to provide the level of access needed.

Using the operating system's built-in file protection is a low-cost way to require approval for edits to electronic files, as shown in Figure 9.8. You can discuss security with your network administrator

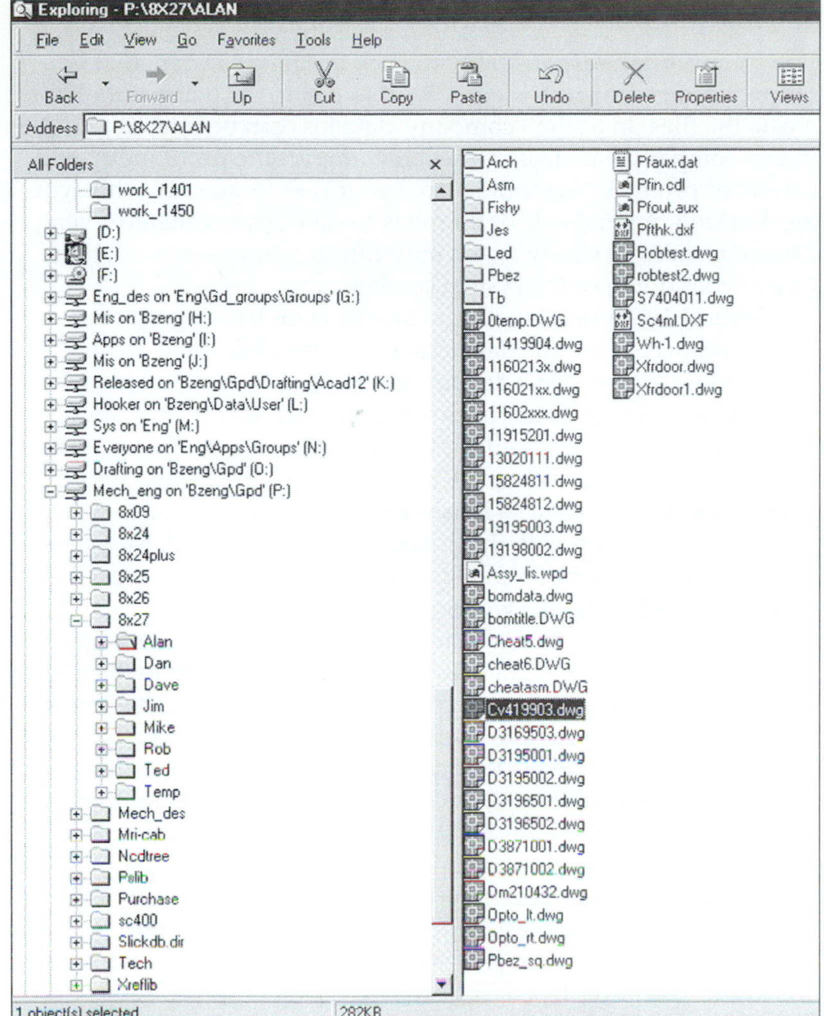

FIGURE 9.8
This screen capture shows how files and directories are organized on the drives and servers accessible to engineers. Each designer has a folder (directory) inside the project. All of the designers can see each other's work, but can only save work into their own folders. The project manager has permission to access files in all of the folders for the project. (Courtesy of Alan Hooker, Powerhouse Technologies.)

and use it to allow others to view your files but not change them. This can be effective when others need to remain current with your work.

In a concurrent, team-based environment, you may need a more elaborate system to regulate the use of shared files. Operating system permissions are generally allocated by user for groups of files in a directory. To change permission based on the file type requires a different kind of software. For example, a file that has been approved and released may have its new status reflected in its file name, and all users may have read-only access to the file from that point forward. The network itself may be configured into work groups that allow easy access to files among team members until the project's control passes to another group, say, from engineering to manufacturing. You will learn more about file management for concurrent engineering later in this chapter.

Backing Up Drawing Files

It is an important responsibility to back up the CAD data that you create. Losing your files is equivalent to losing the hours it took to create the files. In a small company, data loss can be catastrophic, but in any company data loss can severely impact the profitability of a project or product. You should save your files frequently while you are working. A good rule of thumb is to save approximately every fifteen minutes so you will lose only fifteen minutes of work if the power goes down or the system crashes.

Several different methods are available for backing up your data. Simply copying the data to a removable disk is one approach that works well for single documents, but it is too time-consuming for routine backups. Many personal computer operating systems, hard disk drives, and tape drives come with software that makes it easy to keep current backups of an entire hard disk. These utilities generally compress data so it uses less storage space, and they often can be set to back up only those files that have changed. Many also allow you to set the backup to happen while you are not working so you don't need to devote working time to the process.

For CAD data, you may want to keep previous versions as well as the current version of files. To do this, add a number to the file name that indicates the version number: for example, Plate-1.prt. Each time you want to save a version for future reference, save the file as a name with the next number in the sequence: Plate-2.prt. When you back up the data, the older file will remain, and the newer version will be stored with it.

If you work for a large company with a networked computer system, it will usually be the network administrator's responsibility to ensure that a workable backup system is in place. Larger companies usually back up changed files nightly and do full backups daily, weekly, or monthly, depending on their needs. A secure copy of the data should be stored off-site so that in the event of a disaster, the backup files can be recovered. There are many horror stories of companies whose hard drives and backup copies were destroyed in an earthquake. Because backups are frequently made on magnetic media, you also need to consider the conditions under which the backups are being stored. One company stored its backup tapes on low shelves in a climate-controlled vault that was vacuumed nightly. Unfortunately, the magnetic field from the vacuum's motor eventually erased information from the lower rows of tapes.

Storage Media

New kinds of storage media for electronic files are being developed every day, and many of them provide a greater level of security than magnetic media. As mentioned before, a changeable (or uncontrolled) electronic file is generally not acceptable for the permanent record for engineering design documentation. Electronic data written to *write once read many (WORM)* storage media not only produces an unchangeable permanent record, but results in storage that is more stable than magnetic media. Recordable CD-ROM storage is an attractive option for both of these reasons. Many WORM CD-ROM systems can store up to 800 MB per disk at a cost of less than $1 per disk This is a very inexpensive way to produce archival storage of your electronic CAD data. Figure 9.9 shows an optical drive.

Other storage media offer varying combinations of stability, shelf life, speed, and capacity. Electronic CAD files can require sizable amounts of hard disk space. The choice of media for backing up files on a daily or weekly basis and that used for long-term storage should make best use of the options available. For backup media, the data transfer rate from the system to the backup and the cost per megabyte of storage may be more important than the shelf life of the medium (as it will be replaced so often and must accommodate large amounts of data). For archival media, on the other hand, stability and shelf life may make a more expensive medium a better choice in the long run, and even outweigh the fact that it may take longer to write files to the storage media.

FIGURE 9.9
Optical drives, writable CD-ROMs, and WORM drives use high-capacity media suitable for storing large files such as CAD data. (Courtesy of Pinnacle Micro, Inc.)

Using the 3D Design Database in Concurrent Engineering

So far in this chapter, we've focused on the methods used to preserve engineering data stored in graphics files. But effective use of the digital design database goes beyond thinking of it as documentation needed to transfer manufacturing information to the shop floor, or to be stored as the engineering design record for the product. To fully realize the role of the design database in concurrent engineering, document management and drawing control become enterprisewide concepts that mirror and support the goals of the company in getting high-quality products to market more quickly and at less expense.

Concurrent engineering teams are composed of members from many diverse groups. Individuals within the company at the same location interact regularly with others in remote locations as well as external team members, such as multinational manufacturing suppliers and contractors. Team members may represent engineering design, marketing, manufacturing, engineering, the shop floor, procurement, service and support, and clients, each interacting with the 3D design database in different ways. The model in Figure 9.10 illustrates the many tasks associated with concurrent engineering and how they draw and provide information to the design database.

The sharing implied by this model poses unique challenges for electronic file management and control. Companies that wish to succeed at concurrent engineering need ways for a diverse and geographically distributed group of individuals to work with and contribute to a common database of information. By providing access to information that doesn't depend on time zones, telephone accessibility, or geographical location, the design database can eliminate time spent in making connections to get data, and can free up time for the team to work together with the information. Making it easier to find needed information can also contribute to better

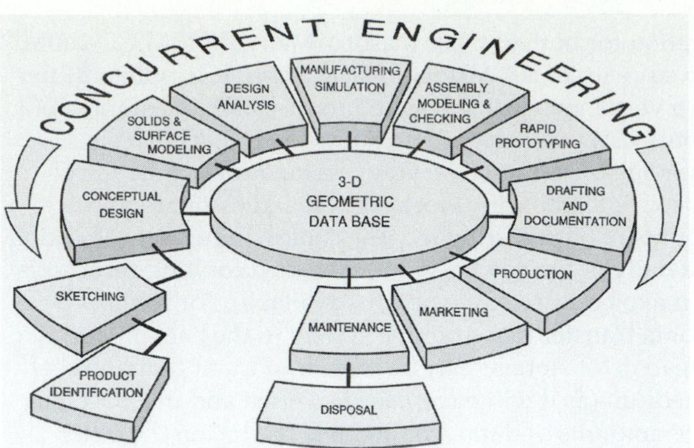

FIGURE 9.10
This model of concurrent engineering reflects a broad view of the design database that integrates information for and from all groups that participate in making a product. (Reprinted with permission from Davor Juricic and Ronald Barr. Copyright 1997 by the American Society for Engineering Education, www.asee.org.)

decision making. Finally, controlling the accuracy of the information used by various team members can eliminate wasted effort. To realize all of these benefits, software tools and methodologies are being developed to create databases that not only provide document control, but also facilitate access to complete information about a product.

Quality Management

A key motivation for many companies that undertake systematic data management is quality certification. Managing engineering documents and electronic data is one step in a quality management system. Many companies in today's marketplace are striving to become ISO 9000 certified as an indication of their quality manufacturing processes.

ISO 9000

The *International Organization for Standardization* (which is always abbreviated ISO regardless of its translation) is a worldwide federation of national standards bodies from some one hundred countries. ISO publishes many standards, most of which are specific to nuts, bolts, and standard parts as you saw in the last chapter. ISO 9000 is different. It is a generic management system standard designed to help companies that manufacture their own products ensure a consistent level of quality in design, production, installation, and service. When the European Community agreed to unify trade regulations, they adopted ISO 9000 as the standard to be met by companies that wished to manufacture products to be sold in the EU. ISO 9000 registration means that the company has identified and documented the processes that affect the quality of the product or service it provides.

ISO 9000 encompasses procedures and documentation across an organization, but two areas set out the criteria for the practices that are used to create, approve, store, retrieve, and revise engineering drawings: design control as well as document and data control.

To fulfill the design control requirement, a company must show that all of its designs are being controlled to meet quality requirements spelled out in the standard. The standard envisions a process that includes clear design criteria, good communication among team members, testing and verification of designs, and well-defined approval processes. The standard is very similar to the design process that you have been learning about in this text and that is practiced by many of the companies profiled in the examples and cases. The following checklist was prepared by the ISO 9000/QS-9000 Support Group for companies to use in evaluating how close they are to meeting the design control standard:

1. Do you have a written design development procedure that your company follows religiously?
2. Are you using a project management-type approach to design development?
3. Are properly trained, educated, and experienced staff assigned to the design teams? Are these teams properly supported?
4. Is a product brief updated and transmitted regularly among interested parties?
5. Are the relationships among departments and teams spelled out in some form of document (usually an organizational chart)?
6. Are product requirements documented and reviewed for clarity and consistency?
7. Are design outputs published as calculations, requirements, and prints, and are these verified by an independent source?
8. Do the design outputs meet regulatory, safety, health, environmental, and other external requirements?
9. Are acceptance criteria also part of the design output?
10. Are the acceptance criteria and other product requirements tested or otherwise verified?
11. Is a documented system of design verification used with qualified personnel to assure that design outputs meet customer requirements?
12. Do you have a written procedure for the modification of an existing design that requires formal approvals?

This checklist suggests a broad range of documents in addition to CAD data that supports quality assurance. The ISO 9000 document and data control standards further define the documents a company must control. These include:

- drawings
- specifications
- blueprints
- inspection instructions
- test procedures
- work instructions
- operation sheets
- quality manuals
- operational procedures
- quality assurance procedures

The document control standard specifically allows for documents to be stored digitally but calls for clear procedures for storing and revising documents that apply the same principles of control as you saw in the paper drawing control process at the beginning of this chapter. Another checklist from the ISO 9000/QS-9000 Support Group identifies some of the qualities of this kind of document control:

- Document has central approval authority.
- There is a master list of documents.
- The right documents are in the right place.
- There are no obsolete documents in circulation.
- The nature of any change shall be explained.
- Documents are re-issued after a specified number of revisions.
- Documents are signed, dated, and numbered.
- Time-sensitive documents are removed when out of date.
- Approval is needed to revise, remove, or copy a controlled document.
- All corrections are signed and dated.
- A separate procedure is used to initiate corrections.

A survey of companies that went through this registration procedure found that 89% indicated "greater operational efficiency" and 48% stated they were "more profitable after their registration." In addition to these benefits, successful registration provides small businesses with the opportunity to bid for more work and use their quality certification as a credential.

ISO 9000 does not specify an exact standard to which registered companies must conform; rather, it requires the company to develop, implement, and certify its own quality process. ISO 9000 criteria, then, are another consideration when developing a data management system.

Product Data Management

One way to achieve the goals of data storage and to expand the use of the database in a concurrent environment is to develop or invest in a new kind of software tool called a product data management system. *Product data management (PDM)* refers to computer systems that provide organized management, storage, and retrieval of not just engineering data, but all documents related to managing a product.

Data management systems can be implemented at different levels in the organization. Many organizations start out with a system at the *work group level*. For example, CAD files may be controlled and shared with a system that serves just the engineering and manufacturing group. These people have similar needs for information and work with many of the same documents, so the system needed is relatively simple. At the *enterprise level*, the data management system would include the CAD files as well as many other documents, and would

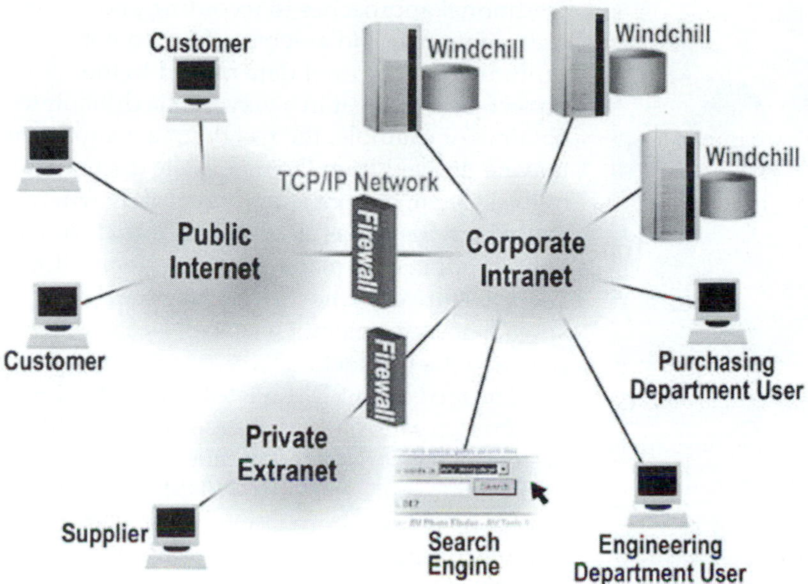

FIGURE 9.11
PTC's Windchill is an enterprise-wide product data management system designed to be extended to serve data outside the organization. This illustration shows how company users would search for and access data over the internal network (or intranet). Data from the database would be provided to external users over the Internet (for public data) or via a secure, password-protected site (to share non-public information with suppliers and other partners in remote locations). (Courtesy of Parametric Technology Corporation.)

serve them to users across the company. (See Fig. 9.11.) As you might guess, an enterprise-level PDM has many more issues of access and control than a work group-level system.

But the ability of different team members to use the database according to their needs—for financial or manufacturing data, for example—can make them more productive and creative members of the team. When Chrysler commissioned R/Greenberg Associates to help prepare a commercial for the 1998 Dodge Intrepid to air during the Winter Olympics, the CAD database was the star of the show. The Dodge Intrepid is the first Chrysler vehicle to be completely designed, assembled, and then tested digitally before a physical model was made. When the advertising agency BBDO Detroit was introduced to the car through a virtual model driven by the CAD data, it proposed to highlight the advanced technology used to develop the car in the advertising. It created a virtual tour of the car that displayed the complexity involved in car design. Instead of simulating the vehicle for the commercial, however, the agency decided to work with the engineering data. Having access to the data added accuracy and detail to the commercial and the ad campaign, depicted in Figure 9.12, that would not have been possible if the agency had created the models from scratch—and it was the access to the data that inspired the creative concept in the first place.

Organized by Product

The PDM system attempts to manage all of the product-related information—specifications, part information, NC programs, CAD drawings, models, spreadsheets, test results, electronic images, and paper documents—throughout the life cycle of the product. This

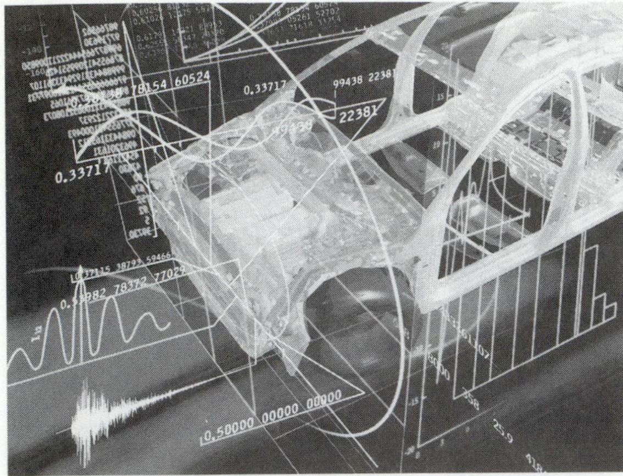

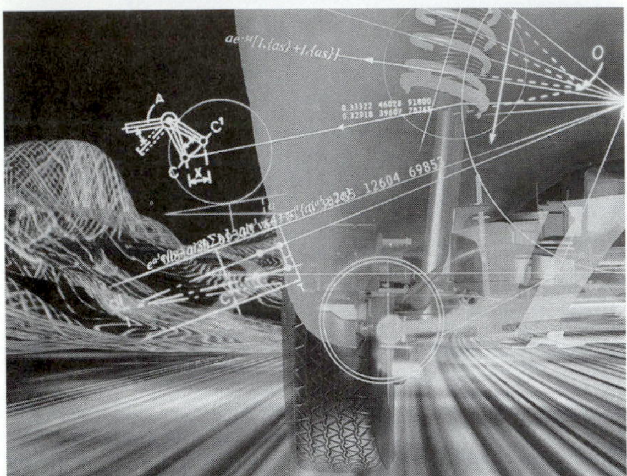

FIGURE 9.12
The inspiration for the advertising introducing the Dodge Intrepid came from the easy access to the CAD data for the digitally designed vehicle. (Courtesy of the Dodge Division of the Chrysler Corporation.)

includes release to manufacturing forms, change orders, and approval documentation. Traditional approaches to recording and preserving part and assembly files do not capture other pieces of data related to the project, or capture it in a way that is difficult to obtain. For example, the weight of a component may be derived from the model file document in the system, but not stored separately where it could be retrieved as an attribute of that component. Product data management systems allow users to store documents—be they CAD files, spreadsheets, or memos—as well as attribute data about a product.

The *product* in PDM relates to how the information is organized. A database, like your file system, can be organized in many ways. The information managed by the PDM system has traditionally been stored by functional area: manufacturing has the programs used to control the manufacturing machinery, marketing has the written specifications for the final product, and purchasing may have the final bill of materials. To get one of these pieces of information, the user needs to know how it is stored and by whom. Each may be stored in a different database. A PDM system uses the product as the organizing principle. To find a piece of information, the user can start with the product it relates to. Figure 9.13 shows a screen from SDRC's data management system, Metaphase.

The organizing principle distinguishes a PDM system from an electronic data management system (EDMS), another term applied to computer management of engineering data. EDMS refers to a system for storage and retrieval of electronic data, such as engineering CAD files, but it is not necessarily organized along product lines.

A Relational Database

All data management systems use some kind of database to store the data they control. Because product data management systems are organized around a product, they generally use a database with *relational* capabilities to allow the system to serve information in many different ways.

Some data management systems use a simpler kind of database to organize data, often referred to as flat files. Flat-file databases are analogous to a filing cabinet. Each record is like a card in the cabinet. If each record is a document, then the attributes of the document can be stored as searchable fields on the card. For example, a CAD file

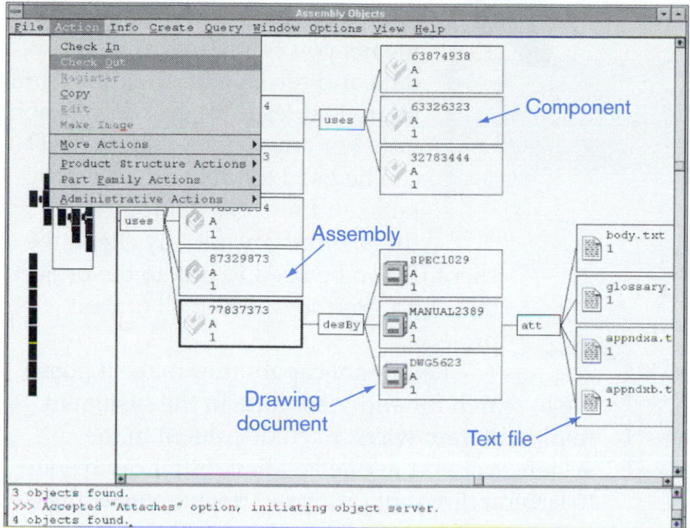

FIGURE 9.13
SDRC's Metaphase manages many different kinds of documents in addition to CAD data. This screen shows one way the document type is indicated to the user. (Courtesy of Metaphase/ SDRC.)

might have fields that reflect the project it belongs to, the type of part it is, the material it is made from, the engineer who created it, its revision number and the number of the engineering change order associated with it, and so on. Any data associated with that document that is defined as a field on the card can be used to search the database. The database is called a flat file because you can present a picture of the relationships in the database in 2D, as information on a series of cards as shown in Figure 9.14.

A drawback of a flat database is the difficulty of showing multiple relationships. A drawing of a standard part, for example, should be stored only once in the database, although it is a part of many different projects. If the project name for a part is a single field, which project should it be associated with? If more fields are added to accommodate multiple project names, each of these fields would have to be searched anytime the user wanted to find a list of parts by project.

A relational database overcomes this limitation by breaking data into tables of data that are related to each other through a common field. Figure 9.15 shows how data might be organized in a relational database. A client table could assign a unique client ID to each client and contain fields for all of the information about a client. A project table would assign a unique project ID to every project and include other fields of information related to the project, such as the client ID. Both tables share the client ID field, making it possible to link the

FIGURE 9.14
The structure of a flat file is like an index card. The top-level structure—the document—is the card, and any data pertaining to the document is a field on the card. If the top level of the database were a project, then the fields associated with it would be the individual documents that relate to the project.

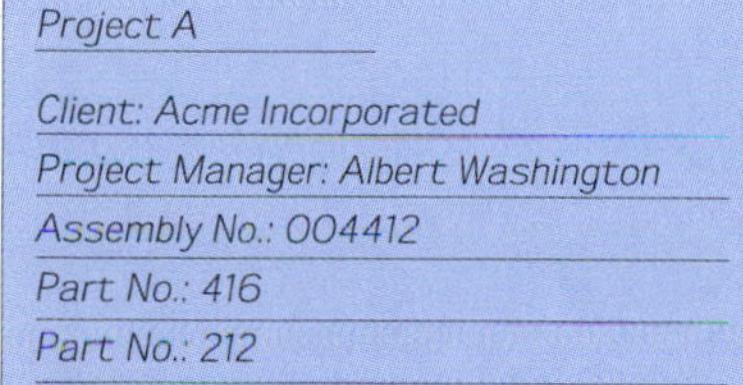
Project A
Client: Acme Incorporated
Project Manager: Albert Washington
Assembly No.: 004412
Part No.: 416
Part No.: 212

Project B
Client
Project Manager
Assembly No.
Part No.
Part No.

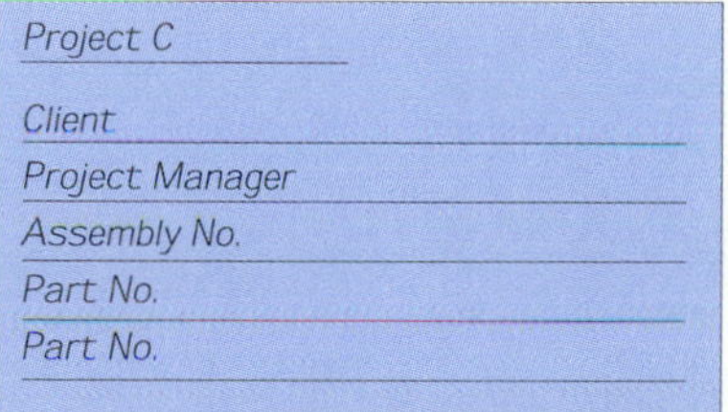
Project C
Client
Project Manager
Assembly No.
Part No.
Part No.

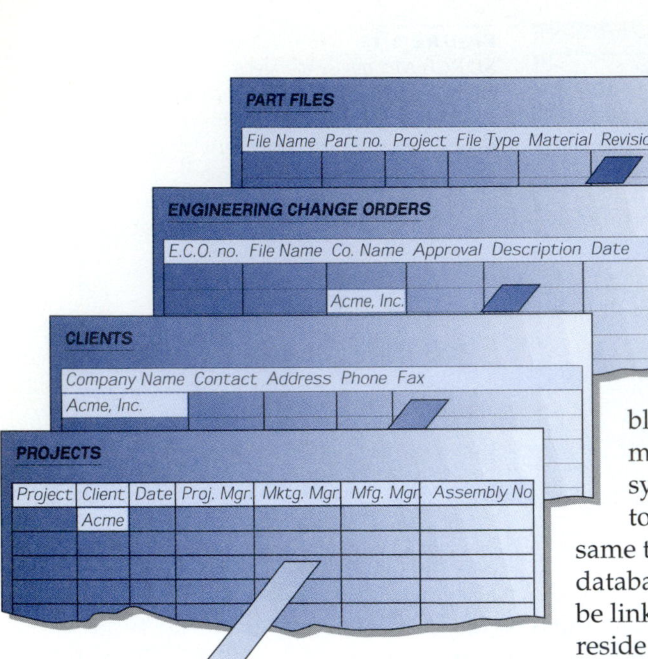

FIGURE 9.15
Each entity in a relational database is stored in a table and linked to other entities by the fields they have in common.

data between the two tables. That means you can search any table of data and retrieve information from related tables through the common link. For any project, the client ID can be used to link to the client table and retrieve data stored there. Similarly, for any client, the client ID can be used to link to the project table and produce a list of all of their projects.

Relational capabilities make it possible to search for and view data in the system in many different ways. Each document in the system is stored in one location, but it can belong to several different groups of documents at the same time through the relationships defined in the database. A table of standard parts, for example, could be linked by ID number to any number of projects, yet reside in the database in a single location.

The relational capability makes it possible to query the system in a variety of ways. A product-based structure, for example, would relate every document to a product manufactured by the company. Documents could be stored in separate tables by document type, such as drawing, 3D model, technical publication, or spreadsheet file. Each kind of file could be stored with attributes relevant to its type, such as title block information for a drawing, or publication date for a technical manual. The relations in the database make it possible to search for data in almost any combination, as shown in Figure 9.16. A user could locate, for example, all of the sheet metal parts fabricated by a particular vendor for a particular customer, or all stainless steel fasteners less than 10-mm long.

FIGURE 9.16
This view of the database illustrates the kinds of attributes that data may have in a product data management system. (Courtesy of Metaphase/SDRC.)

Assembly Objects

File Action Info Create Query Window Options View Help

Item Identifier	Class	Rev	Seq	Nomenclature	Standard Part?	Make/Buy	Proje
87932983	Assembly	A	1	Pulley, Motor Assemb	False	Make	8373
uses							
87623783	Assembly	A	1	Pulley, Motor Assemb	True	Make	8373
uses							
78230394	Assembly	A	1	Clamp Assembly	False	Make	8373
uses							
63874938	Component	A	1	Clamp	True	Buy	8373
63326323	Component	A	1	Pin	True	Buy	8373
32783444	Component	A	1	Catch	True	Buy	8373
78238233	Assembly	A	1	Rim Assembly	False	Make	8373
76538234	Assembly	A	1	Hub Assembly	False	Make	8373
87329873	Assembly	A	1	Motor	False	Make	8373
77837373	Component	A	1	Pulley	True	Buy	8373
desBy							
SPEC1029	Document	A	1	---	---	---	8373
MANUAL2389	Document	A	1	---	---	---	8373
att							
body.txt	Text File	---	1	---	---	---	8373
glossary.t	Text File	---	1	---	---	---	8373
appndxa.tx	Text File	---	1	---	---	---	8373
appndxb.tx	Text File	---	1	---	---	---	8373
DWG5623	Document	A	1	---	---	---	8373
SC923739	Component	A	1	Chromate Coating	True	Buy	8373
SC923383	Component	A	1	Zinc Plating	True	Buy	8373
useBy							
87932983	Assembly	A	1	Pulley, Motor Assemb	False	Make	8373
84723874	Assembly	A	1	Pulley, Motor Assemb	False	Make	8373
84723875	Assembly	A	1	Pulley, Motor Assemb	False	Make	8373

4 objects found.
>>> Accepted "Is Used By Assemblies" option, initiating object server.
4 objects found.

Managing Work Flow

Storage and retrieval of data is the key aspect of a product data management system, but its contribution to concurrent engineering is its ability to control access to and ensure accuracy of shared data. *Work flow* is the term used to describe how information moves through the system's checkpoints and out and back to team members.

The key to any electronic document control system is that a master record of all product data is stored securely on a networked computer system. The system provides access to the documents according to the user's defined level of permission. To work with a document, a user must check it out, as one would check out a book from a library. Changes that are made to the electronic file must be approved before the document is returned to the system. (See Fig. 9.17.) The earlier version of the document is stored as it appeared before the revision in order to provide a design history. In effect, PDM systems provide the same types of controls that were previously in place in manual drawing systems.

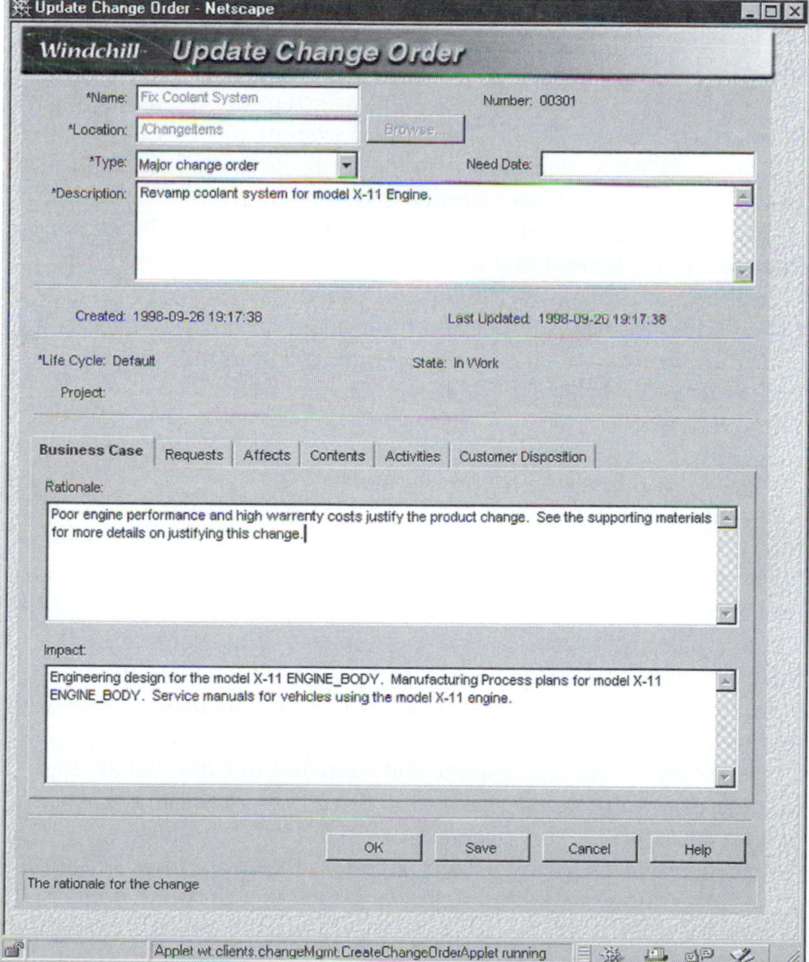

FIGURE 9.17
This electronic form captures change order information for the Windchill PDM. Notice that the user accesses the form through a Web browser. Fields on the form correspond to fields in the database that allow the information to be searched and presented in a variety of ways. (Courtesy of Parametric Technology Corporation.)

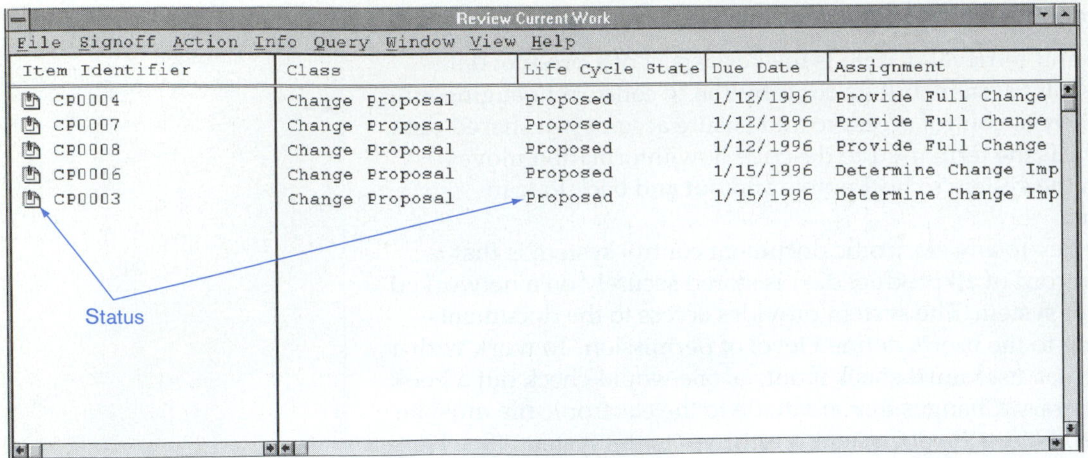

FIGURE 9.18
The status of each document in this sample PDM system is indicated by an icon next to the item identifier. (Courtesy of Metaphase/SDRC.)

How it provides these controls is a key difference, however. Most allow for sophisticated levels of access that fit the organization and how they work. Only certain individuals will have the authority to create or change documents for a particular project. In a concurrent environment, the approval model with checkpoints corresponding to drawing approval may not be enough. Coding drawings according to their status—submitted, checked, approved, released—may offer more flexibility in how files are shared. Figure 9.18 shows file status information stored in a PDM system.

Many PDM systems also provide for tools that allow input from members who do not have permission to change a model, but may "redline" a document (mark possible changes that are flagged and that can be implemented if they are approved by someone with permission to change the file) or add "sticky notes" that can be reviewed by the file's owner and other team members. This capability can improve team communication and interaction without adding meeting time or paperwork.

Archiving Work History

The controls built into work flow systems in the product data management system function similarly when data is to be archived for long-term storage. PDM systems generally offer more information about the project history than a simple archive of files would. The system might capture time spent in each phase of the project, the number of revisions made at each stage, which team members were involved, and other information that may be used to measure and improve productivity.

The system can also expand the snapshot in time that an engineering drawing represents. At any given point in time, the system can produce not just the drawing or CAD model, but also a complete set of data for a project at that moment in time. This kind of archive can make it possible to backtrack to a problem point in the design or revisit a key decision point and pursue an alternative line with the design from that point.

Individual Productivity

One of the benefits that companies expect to get from an enterprise-wide product data management system is improved productivity that translates to reduced time-to-market for their products. Rockwell Corporation reported that implementing a PDM system reduced the time it took its project managers to cost a product from four to five days to four to five minutes because all of the data was easily available. Similarly, design engineers can spend up to 25% to 30% of their time looking for information, retrieving information, waiting for copies of drawings, and archiving new data. An effective PDM system can eliminate this unproductive time because the designer knows where and how to look for released designs and other data. By providing easier access, the system can also promote re-use of existing designs and reduce the time spent "reinventing the wheel."

Many of the issues of file management can also be automated or aided by PDM software. By establishing work groups in the system that are linked to projects, the system can be designed to notify members of the group whenever a file is updated or changed. Designers who have downloaded copies of files can work with them until notified that they need to download a later version. Other PDM and file management systems, especially those implemented over distance networks, seek to reduce download times by comparing file dates and versions on the server and the receiving machine. For any set of files, the system will download only those that are more current than those already available. Graphical preview capabilities built into many systems also make it easier to identify files, especially large CAD files, before downloading them.

Interface Tools

The underlying database is the foundation for the PDM system, but a wide range of software tools reduce the effort involved in keeping the database up to date and gathering information from it. The interface between the user and the system recalls a variety of metaphors. A common one for early systems was the library card. Others employ the familiarity of the Windows Explorer or a Web browser to reduce the time it takes to learn how to use the system. (See Fig. 9.19.)

Graphical displays simplify the effort of locating CAD data and designing workflows (see Fig. 9.20). Retrieving data from a directory inside another directory that is inside another directory on a different server may require a significant amount of navigation and result in a very complex path when the file needs to be saved under a different name. The PDM system provides an interface to the data that handles the navigation and file retrieval. This can be especially important with parametric assembly files that require all of their parts to move with it. The PDM system can be configured to make the files available on an as-needed basis to your workspace.

Most product management systems also include a viewer. This utility is a part of the PDM interface that allows you to preview files or view them without retrieving and opening them.

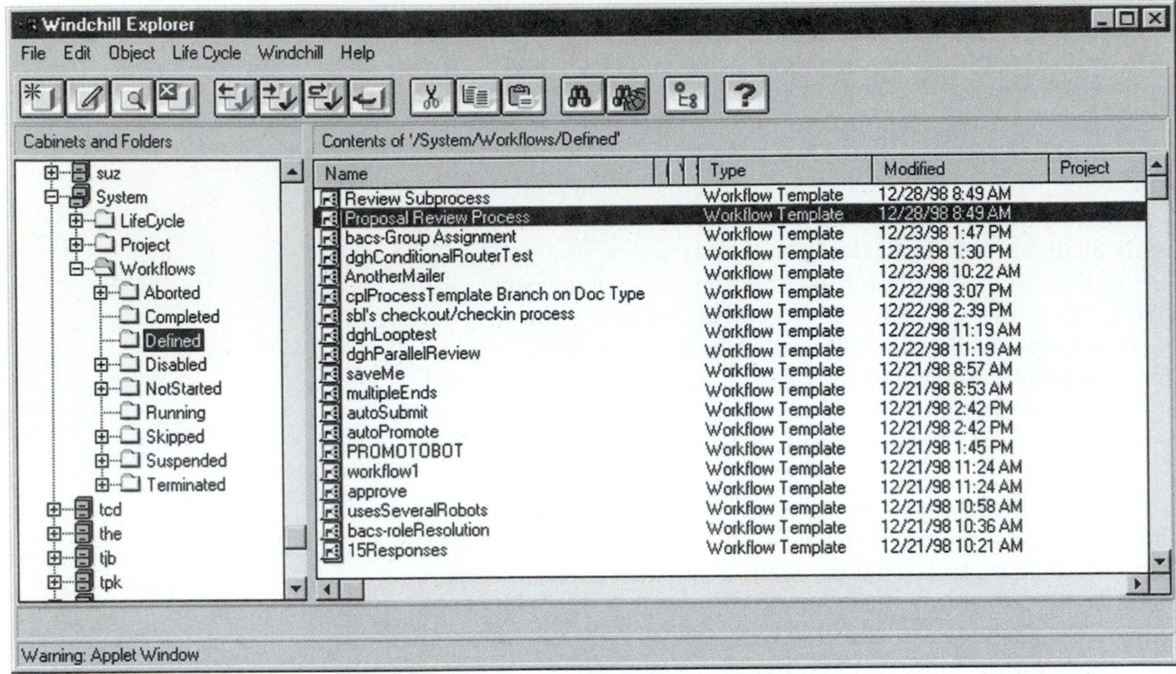

FIGURE 9.19

The relationships among elements on this screen from the Windchill PDM require no explanation to someone familiar with the Windows Explorer interface. (Courtesy of Parametric Technology Corporation.)

As data management systems become more prevalent, expect to see more functions built into them. Today, some systems include messaging tools that allow you to send notices or instant messages to others on a product team. Others organize and store project e-mail around thematic threads that team members can read and review as needed. Automated file management tools, file transfer routines, and data query tools all may be built into the user interface to capture and disseminate data with the least amount of effort on the part of the user.

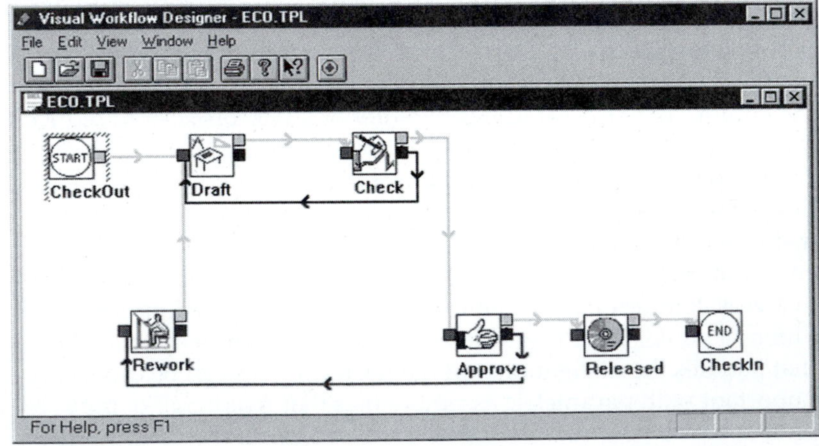

FIGURE 9.20

This graphical representation of the approval process for a CAD drawing is used to design the flow of documents from one stage to the next in AutoEDMS. (Courtesy of ACS Software, Inc., California, USA.)

Data Management and the Web

Implementing a product data management system, or even a simpler document database, is a significant investment in capturing information, defining the use of various documents, and specifying the hardware and software needed. For smaller companies, the investment in a full-blown PDM system may not be necessary, but the strategy of organizing and retrieving engineering design data is still applicable on a smaller scale.

Smaller companies or those that must interact regularly with suppliers and partners in distant locations can use an Internet-based virtual private network as the means to distribute access to project data. Tri-Cycle Design is a small engineering firm that uses the ActiveProject system (shown in the Designer's Notebook for this chapter) to enable them to bring in consulting engineers to work on specific projects (see Fig. 9.21). By creating a secure site on the World Wide Web, network users can employ local Internet connections to

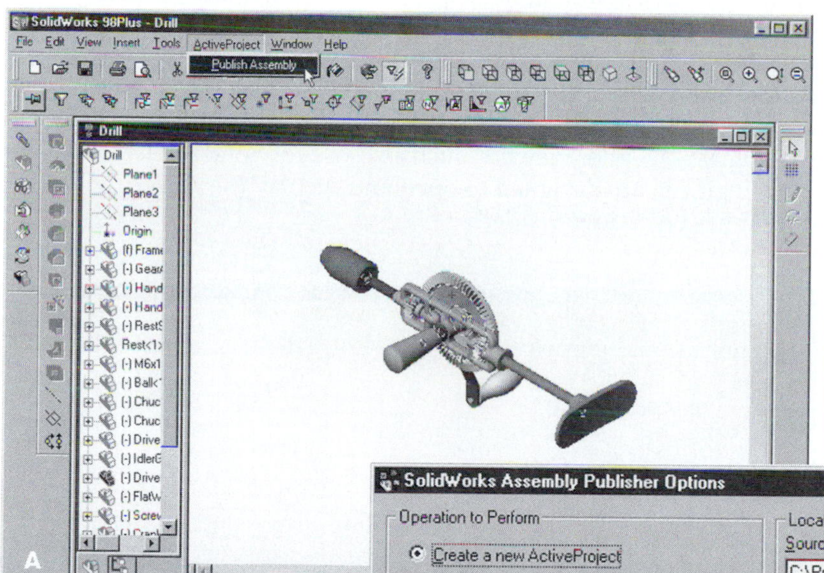

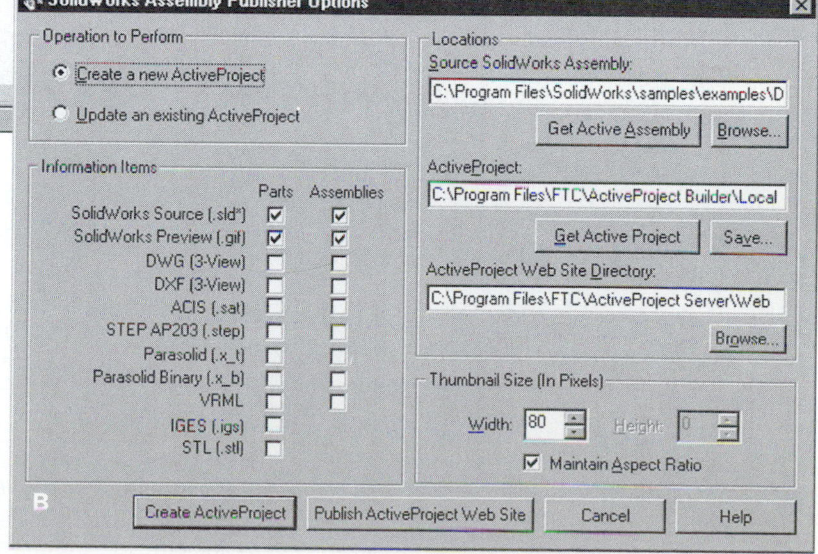

FIGURE 9.21
ActiveProject is an application that creates secure Web sites and makes it easy to export files from a CAD package. In (a), ActiveProject has been installed as an add-in to SolidWorks so that the user can publish a part or assembly without leaving the modeling environment. The dialog box in (b) offers a wide range of file format options for publishing the data. (Courtesy of FrameTech Corporation.)

gain access to the project data. Many of the same messaging and document management functions provided by a PDM system can be provided through browser and Internet-based tools.

Database servers being developed to automate product Web sites can also be adapted to provide access to the most current information and files. HTML documents can link to any kind of document type and make that available for downloading or viewing (if the user has the right kind of viewer installed). In addition, the graphical interface on the Web has spawned a full range of tools for previewing graphical information that can be used to navigate through and select from CAD information.

CAD Files on the Web

One of the most useful developments for publishing CAD data on the Web is a viewer. A viewer is a small application that allows you to open and view CAD drawings without having the application package itself. These viewers can operate as a plug-in for your Web browser, or as a stand-alone helper application. Some viewers, such as Dr. DWG, NetView, and Vdraft Internet Tools, allow you to view any *.dwg* and *.dxf* files created by AutoCAD. Other viewers, such as Autodesk's Whip! plug-in, let you view *.dwf* files, compressed drawing files that must be exported in that format from AutoCAD (see Fig. 9.22). Other viewers, such as those for SolidWorks and Pro/Engineer, are designed to open and display the native CAD file after it has been downloaded.

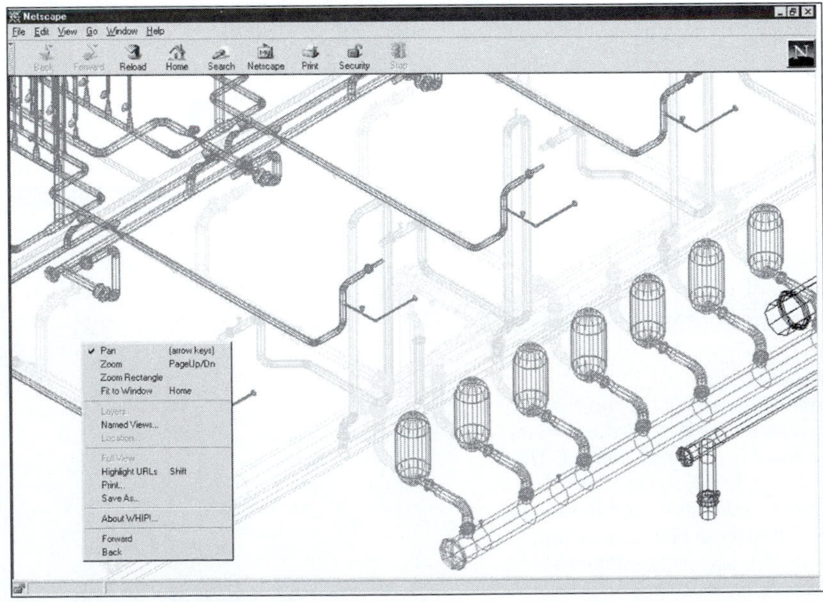

FIGURE 9.22
AutoCAD drawings stored as .dwf files, a format developed by Autodesk for publishing data on the Web, can be displayed in a Web browser. The format allows links to be embedded in the model that will take the user to a URL as any Web link would.

FIGURE 9.23
AutoCAD Release 14 provides a toolbar of Internet tools that allow you to open a URL, insert a URL, save to a URL, and more.

All viewers offer some ability to zoom, pan, and print the model. Without the cost of printing or mailing, a viewer makes it possible to make your CAD data viewable to a wider audience—including customers and non-engineers who want to preview a design but may not be able to interpret a 2D drawing.

CAD packages continue to add features that make it easier to interact with team members over the Web and publish CAD models on the Web. Internet file operations built into AutoCAD, Pro/Engineer, and others make it possible to go to a server on the Internet and retrieve a file without leaving the CAD package, as shown in Figure 9.23. Files that are exported for publication on the Web can now include hyperlinks to other CAD documents, so a user who wants more information about a particular component can click on that component to link to another CAD file or a related document with information about the file. With the ability to redline or annotate these files that some viewers offer, the CAD document itself can be used to carry communication among team members.

The graphical nature of the Web and the rapid pace at which tools are being developed for it make it an ideal environment for sharing data among team members and for publishing CAD data. By preserving the interactivity of the model and the complexity of the data, yet making the model available to individuals who do not have a CAD workstation on their desktops, Web viewers overcome many of the limitations of paper drawings for non-engineers. Data generated for the Web and served over the Web should be taken into account when planning for data management systems of the future.

INDUSTRY CASE

Carl Fehres, IRI International

Managing Engineering Drawings: A Library Metaphor

When the IRI International Houston Engineering Office decided to purchase drawing control software, it asked Carl Fehres to implement the system and devise the most efficient way for drawings to be routed and released to manufacturing. IRI International was already ISO 9000 compliant, but wanted to more efficiently control how drawings were handled so data would not be overwritten. "A drawing control system is implemented for three basic reasons," Fehres explains. "First, it's a way for people to collaborate and safely exchange information. The system helps to communicate to others when you are working on a drawing so that someone else doesn't try to work on the same drawing at the same time. The second reason is to make it possible to search for historical data. Reusing past work can prevent re-engineering a product and save the company time and money. A well implemented electronic document control system enables users to quickly find either a match for their design requirements or a good starting point for a new design. You don't want designers—especially new ones—to have to reinvent the wheel. When a new engineer starts here, the easiest way to get them up to speed is to give them a drawing similar to what they are designing and point out how they might modify it. Third, you need a system that will archive drawings from past jobs for permanent storage. All drawings need to be filed and stored as a record of what has happened on a project."

Fehres emphasizes that there is no one way to handle drawing control; the system recently implemented was designed to support the way the Houston office works (see Fig. 9.24). Typically, when a project begins, the lead engineer creates a layout drawing of the project, then assigns sections of the project to junior designers on the team. When a designer feels his or her work is done, the lead engineer reviews the AutoCAD drawing that the designer has prepared and informally approves it. At that point, the drawing goes to a drawing checker. "The drawing checker makes sure that the design intent is captured by the part and that the drawing is suited for manufacturing," Fehres explains. "The checker looks for mistakes in dimensioning practices, errors in the drawing itself (such as an entity drawn as 3 inches but labelled as 2.5 inches), fit (that Part A fits inside Part B if it is supposed to), and finish. It is a very meticulous, time-consuming job." If errors are found, they are marked on the drawing, which is returned to the designer, where it is corrected and checked again. When a drawing is OK'd by the checker, it then goes to the project's lead engineer for formal approval and signature. The vice president of engineering and the drafting manager must also sign off on the drawing before it is considered approved and ready to be released for manufacture.

Because IRI engineers use Unigraphics 3D parametric modeling software as well as AutoCAD, the drawing

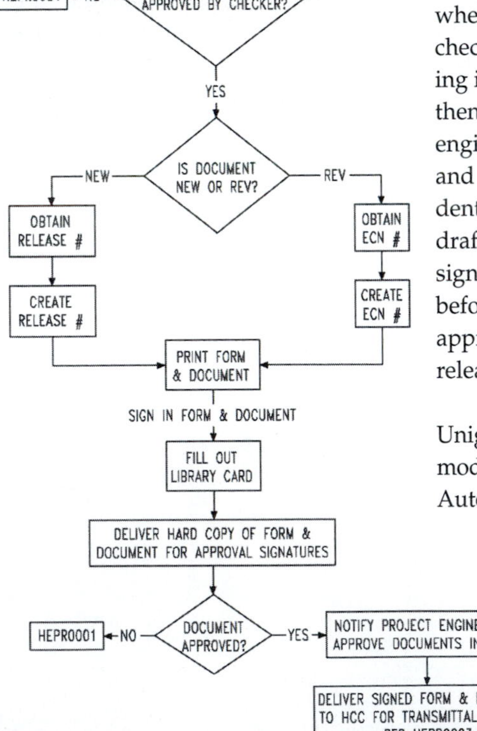

FIGURE 9.24 *This flowchart depicts the workflow and checkpoints in the drawing approval process.*

control system is built around the 2D drawings needed for manufacturing and documentation—whatever their source. IRI purchased software called Network File Manager—NFM3—to serve as the database for the drawing control system. NFM3 is tailored to AutoCAD document handling, but it also handles other document types, such as Microsoft Word and Excel files, that are associated with a project.

The NFM3 software uses a library metaphor to identify documents and control access to them, as shown in Figure 9.25. Each project is a "library" from which documents can be checked in or out. Two levels of access, user and check, are built into the system to allow access but prevent more than one user from working on a document at the same time. "Say user A wants to work on a document. When he or she checks it out, the system transfers a copy of the document from the server to his or her workspace (which is also on the Windows NT network server) and locks down the original copy so no one else can work on it. Other people can *copy* the original drawing, but no one else can check it out. When user A is finished with the drawing, he or she checks it back in and it can then be checked out by any other user with the right permission." The system displays the name of the individual who has a given file at any time, so others can find a file that is checked out if they need to.

In addition to controlling access, the system changes the status of the document to keep track of approvals and revisions.

(continued, page 368)

FIGURE 9.25

The drawing control system lets any user view the database as a list of libraries and their subdirectories. In the listing on the left, the documents are organized into subdirectories by drawing type. .DWG directories store AutoCAD drawing files, .DOC and .XLS directories store Word and Excel files respectively. Other document types, such as the graphics formats .TIF and .PS, can also be stored in the system. In the listing at right, a one-line listing shows the file name, the status of a drawing (Status), and the name of the person (Owner) to whom a file is checked out.

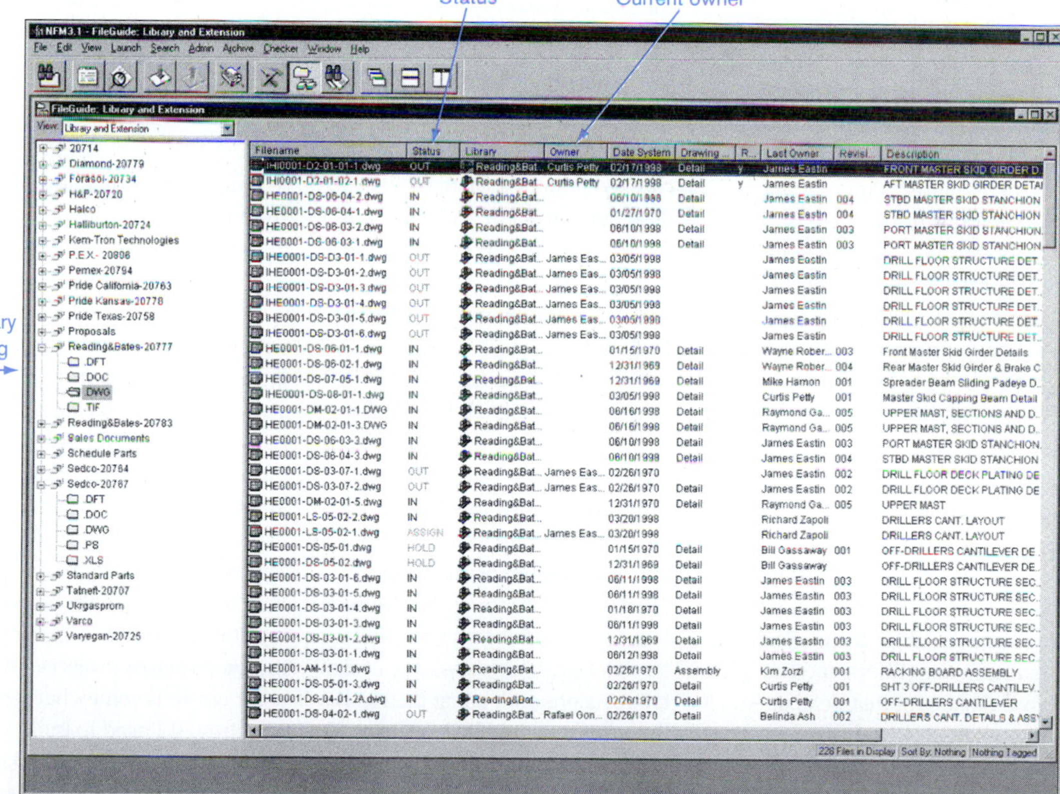

INDUSTRY CASE

Library Card: 1 Houston Engineering

1 Houston Engineering | Close | Help

Drawing Info | Revisions | Document Info | Description, Notes

Filename	Library	Status	Revision #	Size
HE0001-DM-02-03-6.dwg	Reading&Bates-20777	TAG/carlf	002	

Drawing Number	Drawing Assembly	MFG #	Drawing Type	IRI Part Number
HE0001-DM-02-03	HE0001-DM-02	20777	Detail	

Drawing Name: R&B MAST / JACKET DETAILS

Last Owner: Kim Zorzi

Owner:

Released	Released By	Release Form Number	Customer
	NS	R20777-R31	Reading & Bates

FIGURE 9.26
Each library has four tabs: Drawing Info, Revisions, Document Info, and Description Notes, that contain fields of information about the file. The Drawing Info tab, shown here, includes several fields that are linked to the title block on the AutoCAD drawing. This data need not be entered, but is picked up from the drawing automatically when it is checked into the system. Note the Release Form Number field that identifies the release to manufacturing document associated with this drawing. Information encoded in the drawing number, such as drawing type, is also displayed here as a database field.

When a designer is ready to have a drawing checked, she checks it into the NFM3 system, where it acquires *Hold* status. The drawing checker and designer sign the file out and in as corrections are made. When the lead designer approves it, he also has access rights that allow him to submit the drawing into the official library. At that point, its status changes to *In*.

Before a drawing goes to manufacturing, the lead engineer must prepare a release document. The release form is a Microsoft Word document that lists the part drawings to be released for build. The signed document is transmitted as hard copy to manufacturing, stored as hard copy in engineering, and added as an electronic file to the project library.

When an approved drawing needs to be changed, a similar process is used for the engineering change notice (ECN), another Word document that details the changes made to a drawing after it has been released. Like the original drawing, the revised drawing is checked into the system by the designer and is put on Hold status until a drawing checker approves it. When the lead engineer approves it and changes its status to In, however, she also selects an option in NFM3 called Create Revision. This signals the database to put the new document at the top level and store the original document in a subdirectory below it.

"We keep up to ten revisions of a drawing," explains Fehres. "Each time a drawing gets revised, it bumps the older versions down a level. Users can access any of the ten revisions and, using the viewer that comes with NFM3, see the differences graphically. The viewer overlays two documents, using red for one and green for the other, and black to show lines that are the same."

Like the release form, the ECN is transmitted as hard copy (along with the revised drawing) to manufacturing as well as stored electronically with the project files. All project drawings, revisions, release forms, ECNs, and other documents in the library are also stored as paper copies in a "stick" file, which serves as the archive for the project.

Another strength of the drawing control system is locating drawings across projects. "A lot of what we do is somewhat repetitive in nature. If I need to build a deadline anchor, for example, I know we've built thirty of them before.

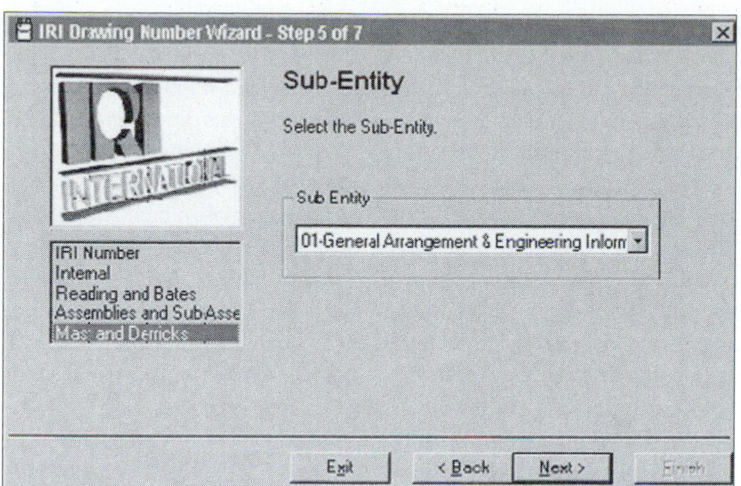

FIGURE 9.27
Visual Basic was used to create a drawing number generator that builds the number from the data entered by the lead engineer. Some of the prompts for project information are shown here.

This one is going to be different, but I still want to look at what we did before. I can go into the drawing control system, search on the keyword *deadline*, and it will bring up all the drawings that were issued with that description," says Fehres. "The database has some powerful search tools. You can search on any number of things in any of the database fields."

The database fields are organized into four different tabs on the drawing's library card (see Fig. 9.26). Users can search on fields that relate to the document itself, its status, revisions, and release data. The library card also serves to cross reference a drawing to related documents, such as its release form and ECNs.

Once a drawing is located, the name tells the user a great deal about what it contains. IRI wrote a drawing name generator program that lead engineers use when they are parceling out parts of a project to designers (see Fig. 9.27). The program, written in Visual Basic, queries the user about the drawing being created, then encodes the information into a drawing number. A drawing number, such as HE0001-DM-02-03, tells the user that the drawing belongs to the first job (0001) done by Houston Engineering (HE). The next two letters identify the drawing type and its major entity. The type of drawing (in this case, D for detail), can be detail, assembly, piping, layout, and so on. Choices for major entities are:

- **B** Blocks and references
- **C** Common rig components
- **D** Drilling machinery
- **E** Electrical
- **F** Fluids or mud systems
- **H** Hydraulics
- **M** Masts and derricks
- **P** Piping
- **S** Substructures and drill floors
- **T** Trailerized and rig equipment
- **U** HVAC
- **V** Standard architectural items
- **W** Standard individual components

Each major entity also has sub-entities that further describe the drawing. In this case, the major entity is Masts and derricks, for which sub-entities include: General arrangement drawing, Mast stem, Derrick stem, Guide tracks, Guide dollies, Shoes and shims, Shim poles and catline booms, Pipe clamps, Crown and watertable beams, Racking boards and belly boards, and Casing boards. The 02 in this drawing number indicates a mast stem sub-entity. The last two numbers uniquely identify this part from other similar parts with the same classification.

Although the system works well, Fehres can foresee another growth step in the near future—to a relational database instead of the flat database structure of NFM3. "The current system creates a new library for every job that we get. If we produce a deadline anchor for Job A, it will be stored in that directory. If Job D wants to use the same deadline anchor, we have no way to make that file appear in the Job D directory without storing another copy of the same file in the system. Right now, we use a cross-reference note, but it would be nice to have a hyperlink in the Job D library that would call up the drawing in the Job A library. Better still would be a hyperlink to a certain revision of that anchor, but there is no way to do that with a flat database."

(continued, page 370)

INDUSTRY CASE

Farther in the future would be a move to a system that managed the 3D model files as this system does the 2D drawings. "There are some fundamental differences in the way a 3D parametric modeling system and a 2D drafting system work. Each parametric drawing file is linked to the model files depicted in it. Because of this, a drawing file created from a 3D modeling system should be stored with all corresponding model data. In addition, a 3D assembly model is linked to each of the individual part files in it. When a model, assembly, or drawing is being worked on, it and all corresponding linked data should be checked out of the document handling system as well. The size of the assemblies we use (ten thousand parts and above) and the current state of computer technology prevents us from efficiently using such an integrated system now." Because the overhead involved in opening such large assemblies and having the network control the access to individual part drawings in real time is prohibitive, Fehres's office has developed a more manual system for checking out model data with drawings. Today, designers who work in 3D archive a drawing and all linked model data in one compressed file in NFM3. "Because not everyone is able to view Unigraphics drawings electronically, we plot all Unigraphics drawings to postscript format and use a postscript viewing application for printing and on-line viewing. We store this postscript file in NFM3 as the official drawing file." When a drawing is exported to postscript, the parametric associativity between the drawing and the 3D model is lost. By archiving the Unigraphics drawings with the Unigraphics model, design engineers can revise the model when they revise the drawing, then export the drawing again to be re-entered into the drawing control system.

All illustrations are used with the permission of Carl Fehres and IRI International.

Carl Fehres is mechanical engineer for IRI International in Houston, Texas, an engineering and manufacturing firm that produces oil drilling tools, drilling structures, mobile rigs, and oil field machinery for the petroleum industry. Fehres earned a master's degree in Mechanical Engineering from Montana State University, then worked as a stress analyst at Atec, Inc., where he used Cosmos/M to build an award-winning FEA model. As a designer for Dreco, he learned to use Unigraphics and developed a system for efficiently modeling structural designs. Since joining IRI International in 1997, he has developed more programs in Unigraphics to enhance design productivity.

Skills Summary

Now that you have completed this chapter, you should understand issues in managing engineering drawings and product data. The exercises presented here allow you to describe how engineering documents are controlled and the role of drawings and CAD data in engineering documentation. You should be able to list the basic steps in approving and controlling CAD data. You should also be able to describe various systems for producing engineering drawing numbers. Good file handling and backup procedures require organization and a systematic approach. You should be able to describe how to use folders to organize files and directories in a consistent manner. You should be familiar with terminology related to engineering documentation and be able to define commonly used acronyms.

EXERCISES

1. Develop a file naming convention and directory structure that would work for the documents related to your course work. Consider all of the courses you are likely to take over your college years and how you might want to access documents from them. What different kinds of documents would you want to include? Anticipate making the documents available over a network to your prospective employer and create a naming structure that would convey meaningful information about the document file. Draw a flowchart of the directory structure you would use to organize the files. Show the steps needed to access a particular type of document.
2. A three-person company designs, fabricates, and sells a line of biological laboratory equipment. All drawings are produced using a desktop CAD system. The company sells products to about twelve labs, with each user needing small custom features on the base product line. Contrast its document control needs with that of a company such as Boeing, which must keep track of commercial aircraft design details. What differences and similarities exist between the drawing control requirements of the two companies?
3. Why might you wish to store electronic files in a different directory than your application software?
4. Provide the source words and a one- or two-sentence definition of the following common engineering and computer acronyms (some research or investigation may be required):

PDM	FEA	ISO	ECO	ECN	VRML
CAD	EDMS	WORM	ASTM	HTML	.DXF

5. What is the key to a successful product data management system?

6. It is common to begin a project with a "drawing tree" or "drawing breakdown structure" (DBS) to identify the relationship between detail components, subassemblies, and assemblies. The drawing tree below shows the drawing hierarchy for a toy squirt gun. Despite the simple nature of toys, they are often produced by the millions; it is important to have a sensible means to number or identify computerized drawing files for this project. Design a convention for file naming and numbering for the toy described by the following DBS:

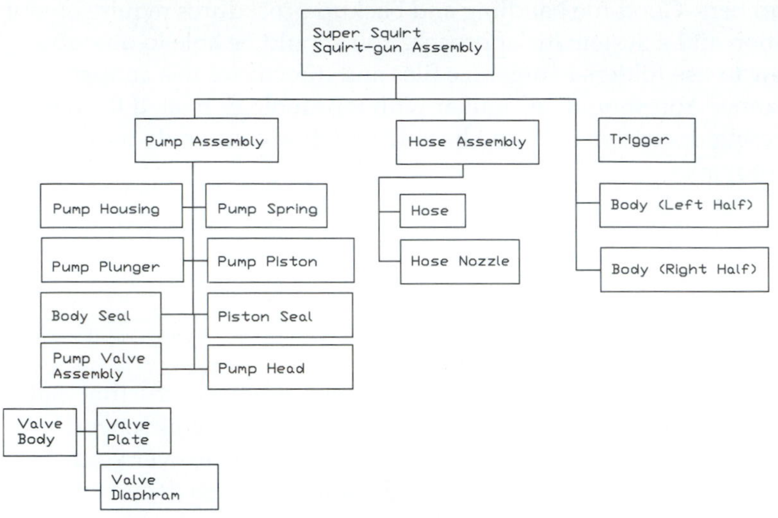

7. The Boeing Company uses database management programs to control how designers access component models in the digital database. How have the Boeing systems streamlined the necessary interaction between engineering teams from different physical locations who wish to modify the same model? (Ref: chapter discussion.)

8. A block diagram of drawing release process steps for a "conventional" paper drawing has been started for you on the right. Complete this "flowchart" and then create a block diagram of typical steps in the drawing release process for a computerized part in a digital database.

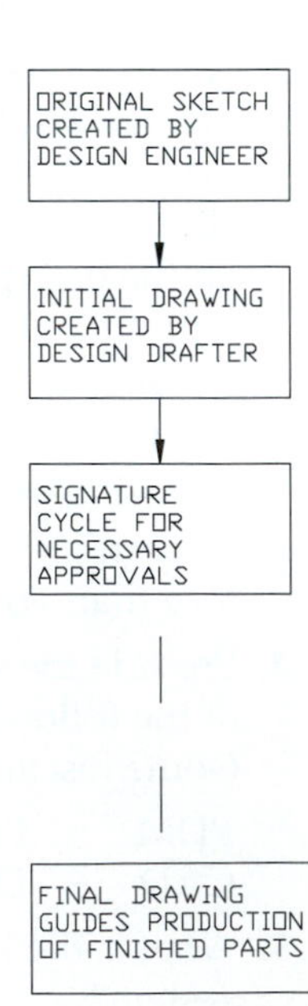

9. A designer creates two original mating part drawings on a CAD system, and stores them digitally in a specific folder on her PC. The drawings are printed and the prints are issued to the machine shop to guide part fabrication. After about 100 parts of the expected 300-part "run" have been produced, a change is requested by the shop to permit a larger bolt size to be used for assembly. The bolt size request affects both mating parts, so that interchangeability between the old and

new parts will be lost. The designer complies and issues new drawings.

a. How should these new drawings be stored? Distributed? Labeled?

b. What procedure would you recommend to identify both drawings and parts produced?

c. Can you anticipate problems that might arise if good practices are not followed?

10. A common practice on multi-layer drawing databases is to create a "layer map" directly on the drawing describing contents of each drawing layer and the conventions used for things like color, line weight, etc. Why might this be important? Who would use the information? Explain.

11. ISO 9000 requirements for document and quality control are listed in a 12-item checklist in this chapter. State the logical reason for each of these 12 items, including potential consequences of ignoring each of the recommendations.

12. Visit the company Web site for your CAD package and download the viewer software (if available). What kinds of files can the viewer open? What viewing and annotation options are built into the viewer?

13. Visit the ENOVIA Web site (www.enovia.com) and read the white papers on IBM's ENOVIA product management system.

DESIGNER'S NOTEBOOK

Even the mass of the paint is calculated into the mass properties for space vehicles. For the SIRTF space telescope, the low emissivity, silicon-grit based coating was to be 24 mils thick. In the calculations shown here, the surface area data from the model was used to calculate the mass of the various "paints," one of which turned out to be almost 9 kg. This additional mass resulted in the use of a different coating to keep the mass of the telescope within the design constraints for the project. (Courtesy of Scott Adams, Ball Aerospace & Technologies Corp.)

TELESCOPE PAINT

15 DEC 97

29

$$V_{AMES\ B.B.} = 5.35\,m^2 \cdot 6.0\times10^{-4}\,m = 0.00321\,m^3$$

$$m_{AMES\ B.B.} = \left(1300\,kg/m^3 \cdot 0.00321\,m^3 \cdot 0.8\right) + \left(3200\,kg/m^3 \cdot 0.00321\,m^3 \cdot 0.2\right)$$

$$= 3.3384\,kg + 2.0544\,kg = 5.3928\,kg$$

$$V_{AMES\ M.T.\&A.T.} = 1.19\,m^2 \cdot 6.0\times10^{-4}\,m = 0.000714\,m^3$$

$$m_{AMES\ M.T.\&A.T.} = \left(1300\,kg/m^3 \cdot 0.000714\,m^3 \cdot 0.8\right) + \left(3200\,kg/m^3 \cdot 0.000714\,m^3 \cdot 0.2\right)$$

$$= 0.74256\,kg + 0.45696\,kg = 1.1995\,kg$$

$$V_{DESOTO\ B.B.} = 5.35\,m^2 \cdot 2.032E\text{-}4\,m = 0.001087\,m^3$$

$$m_{DESOTO\ B.B.} = \left(1300\,kg/m^3 \cdot 0.001087\,m^3 \cdot 0.8\right) + \left(3200\,kg/m^3 \cdot 0.001087\,m^3 \cdot 0.2\right)$$

$$= 1.1306\,kg + 0.6958\,kg = 1.8264\,kg$$

$$V_{DESOTO\ M.T.\&A.T.} = 1.19\,m^2 \cdot 2.032E\text{-}4\,m = 0.000242\,m^3$$

$$m_{DESOTO\ M.T.\&A.T.} = \left(1300\,kg/m^3 \cdot 0.000242\,m^3 \cdot 0.8\right) + \left(3200\,kg/m^3 \cdot 0.000242\,m^3 \cdot 0.2\right)$$

$$= 0.2515\,kg + 0.1548\,kg = 0.4062\,kg$$

OUTER SHELL PAINT DUST COVER

STANDARD DESOTO BLACK $\rho = 1300\,kg/m^3$

DUST COVER -2 SIDES

$$A_{D.C.} = 1.56883\,m^2$$

$$V_{D.C.} = 1.56883\,m^2\left(2.032E\text{-}4\,m\right) = 0.000319\,m^3$$

(8 mils)

OUTER SHELL-BOTTOM

$$A_{O.S.B.} = 8.48328\,m^2/4 = 2.12082\,m^2$$

$$V_{O.S.B.} = 2.12082\,m^2\left(2.032E\text{-}4\,m\right) = 0.000431\,m^3$$

OUTER SHELL-MIDDLE

$$A_{O.S.M.} = 4.41144\,m^2/2 = 2.20572\,m^2$$

$$V_{O.S.M.} = 2.20572\,m^2\left(2.032E\text{-}4\,m\right) = 0.000448\,m^3$$

CHAPTER 10

Using the Model for Analysis and Prototyping

Objectives

When you have completed this chapter, you will be able to

- extract mass properties data from your CAD models;
- evaluate the accuracy of mass properties calculations;
- define the file formats used for exporting CAD data;
- describe how analysis data can be used to update parametric models;
- list analysis methods that can use the CAD database;
- describe how rapid prototyping systems create physical models from CAD data.

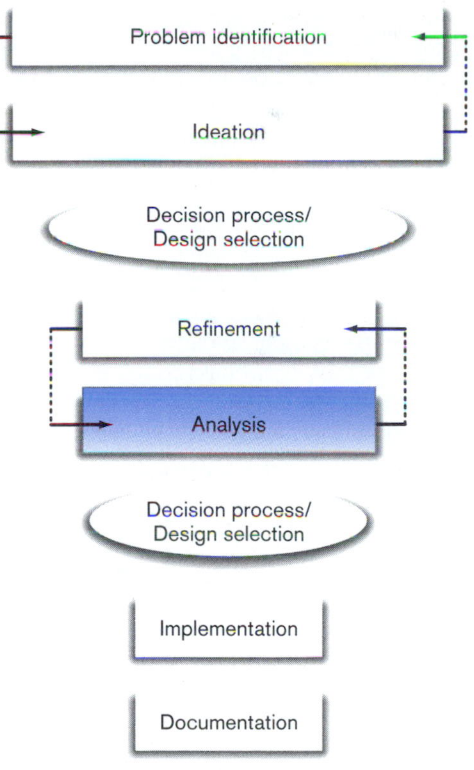

In the refinement stage of the design process, 3D modeling tools allow you to create a highly accurate digital definition of your design. But the refinement phase also forms an iterative loop with the analysis phase of design. In this loop, test results feed back into model changes, and the revised model is used for further testing. Your 3D solid model database not only defines the geometry of the design, but also provides information about the design that can be used to analyze its engineering properties.

When you accurately model a design, you can use the resulting database to determine the designed object's mass, volume, surface area, and moments of inertia. This and other information can be read directly from the modeling software or can be exported to other programs for further analysis. You can use software tools such as a spreadsheet, equation solver, motion simulator, and finite element analysis package with data from your model to check for stress concentrations; determine deflections, shear forces, bending moments, heat transfer properties, and natural frequencies; perform failure analysis and vibration analysis; and make calculations based on model data.

You can also use model data to create a prototype of the design that can be used for testing. Rapid prototyping equipment uses the 3D CAD database to create physical parts, but other "virtual prototypes" are computer-based simulations created from model data. Both kinds of prototypes allow customers and others to interact with and evaluate the design.

In this chapter, you will learn about the kinds of information that can be extracted from your models, find out how information is shared between the modeling software and different analysis tools, and see some examples of how you might use these different software tools to analyze your designs.

Factors of Safety

A *factor of safety* expresses the ratio of two quantities, such as *maximum safe speed/expected operating speed* or *failure load for a part/expected highest load in use*, that will be used as a design guideline. For example, if the failure load for a part is 250 pounds per square inch and the expected highest load in use is 125 pounds per square inch, the safety factor for the part would be 2. Factors of safety for commercial aircraft are within the range of 1.2 to 1.5. Factors of safety for elevators may be as high as 14. For some types of equipment, such as that used in public buildings, codes specify safety requirements that must be met. You should be familiar with any codes or standards that apply to your design area.

Determining Mass Properties

The size, weight, surface area, and other properties available from a 3D model are frequently part of the design constraints your design must satisfy. The surface area of a part can determine whether a part cools as quickly as it should. The volume of a molded part will determine how much material is needed to fill the mold—which in turn determines a key component of the manufacturing cost.

Consider the example of a satellite or other space vehicle. The mass of the finished product is extremely important in determining the thrust required for the vehicle to leave Earth's atmosphere. Accurate mass information can spell the difference between success and failure and, for manned space flights, the difference between life and death. At the same time, designs seek to minimize the total mass of the system, so the built-in factors of safety cannot be as high as the factors of safety used in designing bridges, for example, which can be built of heavy steel. This smaller margin for error increases the importance of accurate mass properties information.

Calculating the mass of a model as complex as a satellite by hand would be so time-consuming that it might add years to the design process. Traditionally, engineers wrote their own FORTRAN programs to perform such calculations, and even then frequently relied on approximations to make the task feasible. Today, most solid modeling and parametric modeling CAD systems have commands that allow you to generate a wide range of mass property information about the model directly from your modeling software.

The following are mass property calculations commonly available in today's solid modelers:

- volume
- surface area
- mass
- centroid or center of gravity
- moments of inertia
- radii of gyration

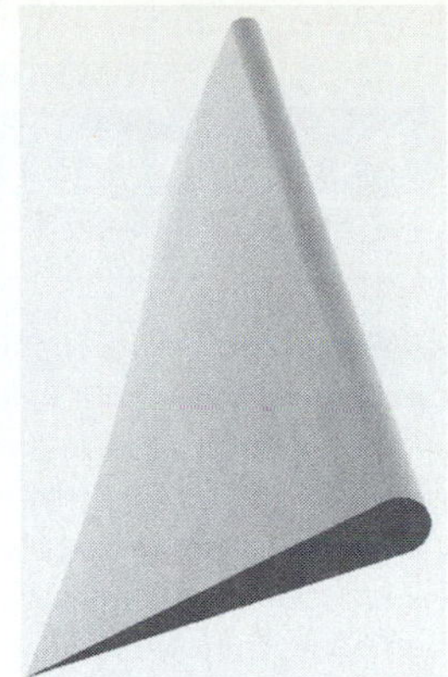

FIGURE 10.1
The volume of this turbine blade would be difficult to calculate by hand.

Acquiring mass properties information directly from the model saves time, but it also provides accurate information that would otherwise be difficult to calculate. The blade shown in Figure 10.1 was modeled in Pro/Engineer. Its volume would be difficult to calculate using hand methods. Figure 10.2 shows the mass properties for the part as generated from Pro/Engineer. Like other modeling software, Pro/Engineer can report the volume of the part—regardless of its shape—with accuracies of up to plus or minus one hundred-millionth of a percent (± 0.00000001).

```
INFORMATION WINDOW
                    MASS PROPERTIES OF THE PART FAN_BLADE

                         VOLUME =  1.1310529e+03 INCH^3
                   SURFACE AREA =  1.5128767e+03 INCH^2
                        DENSITY =  1.6400000e-01 POUND / INCH^3
                           MASS =  1.8549267e+02 POUND

        CENTER OF GRAVITY with respect to _FAN_BLADE coordinate frame:
X   Y   Z           1.1340471e+00  1.9639266e+01  3.3294231e+00  INCH

     INERTIA with respect to _FAN_BLADE coordinate frame:  (POUND * INCH^2)

INERTIA TENSOR:
Ixx Ixy Ixz         1.1011841e+05 -2.0286730e+03 -5.8218232e+02
Iyx Iyy Iyz        -2.0286730e+03  4.7037957e+03 -1.0938398e+04
Izx Izy Izz        -5.8218232e+02 -1.0938398e+04  1.0651360e+05

    INERTIA at CENTER OF GRAVITY with respect to _FAN_BLADE coordinate frame:
                                                          (POUND * INCH^2)
INERTIA TENSOR:
Ixx Ixy Ixz         3.6517547e+04  2.1025925e+03  1.1818657e+02
Iyx Iyy Iyz         2.1025925e+03  2.4090434e+03  1.1904903e+03
Izx Izy Izz         1.1818657e+02  1.1904903e+03  3.4730380e+04

                PRINCIPAL MOMENTS OF INERTIA:  (POUND * INCH^2)
I1  I2  I3          2.2369931e+03  3.4754067e+04  3.6665910e+04

        ROTATION MATRIX from _FAN_BLADE orientation to PRINCIPAL AXES:
                    -0.06105        -0.10234         0.99287
                     0.99747         0.02994         0.06442
                    -0.03632         0.99430         0.10025


Hit Space or Return to continue ('q' to quit, 'b' for previous page)
```

FIGURE 10.2
Mass properties data for the turbine blade may be generated by Pro/Engineer to a high degree of accuracy.

Mass Property Calculations

Mass Properties of a Right Cylinder

You will learn more about using mass property calculations, such as moments of inertia, in your engineering statics and dynamics, machine design, and design of structures courses. The equations and illustrations in this example illustrate these properties as they would apply to a 1-inch diameter *right cylinder* (a cylinder in which the base is perpendicular to the height), shown in Figure 10.3.

FIGURE 10.3 *For this example cylinder: π is approximately equal to 3.14; $r = 0.5$ inch, the radius of the base; $d = 1$ inch, the diameter of the circle; $l = 2$ inches, the height of the cylinder.*

Volume (v)

The *volume* of a solid is basically the amount of space it takes up; volume is measured in cubic units. For many basic geometric solids, like a cylinder, volume can be calculated using simple formulas. Complex shapes can be calculated either by breaking them down into simpler geometric solids—whose volumes are easier to calculate—and adding those volumes together, or by using calculus-based methods. CAD software usually uses approximate methods to evaluate integrals to determine the volume and mass properties of any shape.

v = volume

The volume of a 1-inch diameter cylinder, in which the 2-inch length is perpendicular to the circular base, is the area of the base times the length.

$$v = \pi \cdot r^2 \cdot l = 3.14 \cdot .5^2 \cdot 2 \text{ in} = 1.57 \text{ in}^3$$

Surface area (S_a)

The *surface area* of a solid is the square measure of its exterior surface. It is also the total area of the boundary surfaces of a solid or surface model. You can think of the surface area as the measure of the exterior surface of the object flattened out into a plane (as shown in Fig. 10.4). Once again, it is easy to calculate for simple cases, but complex for irregular shapes. CAD software usually uses approximate methods to evaluate integrals to determine the surface area of any shape.

FIGURE 10.4

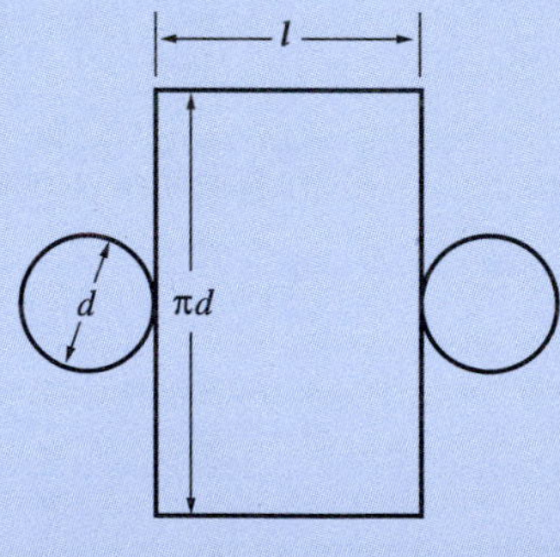

S_a = surface area

The surface area of the cylinder is the area of the rectangular shape that encloses the cylinder ($\pi d \cdot l$) plus the area of the two circular ends ($\pi r2$).

$$S_a = \pi d l + 2(\pi r2) = (3.14 \cdot 1 \cdot 2) + 2\,(3.14 \cdot .5^2) = 7.85 \text{ in}^2$$

Mass

The *mass* is roughly the measure of the resistance that an object has to changing its steady motion.

m = mass
= volume • mass density
= v • mass density

Mass Property Calculations (cont.)

Mass density

Mass density for a given material is usually found by looking up the value in an engineering materials table. This is typically information you provide in order to calculate the mass for a given object. It is very important for you to make sure the units for your model (millimeters or inches, for example) match the units for the mass density that you provide (milligrams per cubic millimeter, for example). If they do not, you must convert the volume value reported from the software or you will be off by a large factor in your calculations.

The mass of this cylinder, if it was made from copper, is its volume times the mass density of the material, which is 0.295.

m = 1.57 in^3 • 0.295 lb/in^3
= 0.46315 pounds per cubic inch

Centroid or center of gravity

A *centroid* is a point defining the geometric center of an object. The centroid and center of gravity coincide when the part is made of a uniform material in a parallel gravity field. Most CAD software reports the centroid relative to the coordinate system that is selected. Make sure that you understand how the coordinate system is oriented when interpreting this value. Approximations of double integrals are used by the CAD software to calculate centroids.

C = centroid

The centroid for the cylinder is stated in terms of the three axes of the 3D coordinate system. Given the location of the origin at one end of the cylinder, the centroid for this part is at (1,0,0) inches.

$C = C_x, C_y, C_z$
$C_x = 1/2 = 2/2 = 1$
$C_y = 0$
$C_z = 0$
$C = 1,0,0$

Moments of inertia

A *moment of inertia* is the measure of the resistance that an object has to changing its steady motion about an axis. A moment of inertia depends on the mass of the object and how that mass is distributed about the axis of interest. A moment of inertia is the second moment of mass of the object relative to the axis. Products of inertia are similar to moments of inertia, but they describe the mass distribution relative to two axes of interest. Make sure that you understand how the coordinate system is oriented when interpreting this value.

I = moment of inertia
I = second moment of mass about an axis

For this cylinder,
$I_x = \int(y^2 + z^2)dm$
$I_y = \int(x^2 + z^2)dm$
$I_z = \int(x^2 + y^2)dm$

Radii of gyration

The *radius of gyration* is the distance from the axis of interest where all the mass can be concentrated and still produce the same moment of inertia. Make sure you understand how the coordinate system is oriented when interpreting this value.

k = radius of gyration about an axis

For this cylinder,

$$k_x = \sqrt{\frac{I_x}{m}}$$

$$k_y = k_z = \sqrt{\frac{I_y}{m}}$$

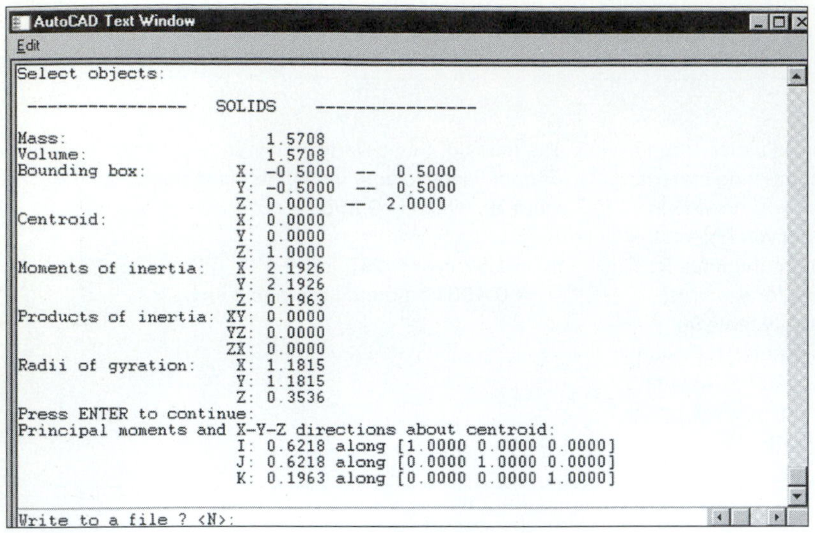

FIGURE 10.5
Mass properties for the cylinder from AutoCAD v.14.5

Understanding Mass Property Calculations

Before you make use of the mass properties generated by your modeling software, it is important to understand how the values are derived. As you learned in Chapter 6, modeling methods, algorithms, system variables, and units used by your CAD software can make a difference in the accuracy of the mass properties it calculates. For example, models that store the faceted representation of the part may produce less accurate calculations than models that store the accurate geometry.

Different modeling software uses different algorithms to calculate the volume of a solid. Some of them can have error ranges of plus or minus 20%. One algorithm used to calculate the volume of a solid simply encloses the solid in a bounding box, then breaks it up into rectangular prisms parallel to the bounding box. (A bounding box is the smallest cube that the solid will fit inside.) The lengths of the rectangular prisms are determined by the surface boundaries of the solid. The volumes of the prisms are then summed to find the volume of the solid. As you can imagine, the accuracy of this method depends greatly on the number of subdivisions that the object is broken into. The direction from which the prisms are created also affects the ability of the prisms to approximate the shape. Typically, both of these settings are variables with ranges that you can set prior to calculating the mass properties.

Other systems use an iterative process to approximate the integral that describes the volume. These often let you specify a range of accuracy for the calculations. The more accurately you require the calculations to be processed, the more time they will take.

Verifying Accuracy

You should verify the range of accuracy for the mass property results from your software. A good way to verify the results you are getting for mass properties is to model a shape in which the geometry is known (such as the 1-inch diameter cylinder used in the example above), then compare the properties calculated by hand with those given by the CAD system. Use your CAD software to model the cylinder shown in Figure 10.6: it has a diameter of 1 inch and a height of 2 inches. Calculate the volume by hand using the equations shown in the sidebar on mass properties.

$$
\begin{aligned}
v &= \pi \bullet r^2 \bullet l \\
&= 3.14 \bullet .5^2 \bullet 2 \text{ in}^3 \\
&= 1.57 \text{ in}^3
\end{aligned}
$$

Then list the same information from your solid modeling software. Is it the same? A modeling package such as AutoCAD Release 12, which uses faceted models and an approximate method to calculate volumes, will yield different results from one that stores an accurate representation, such as AutoCAD Release 14 (see Fig. 10.6).

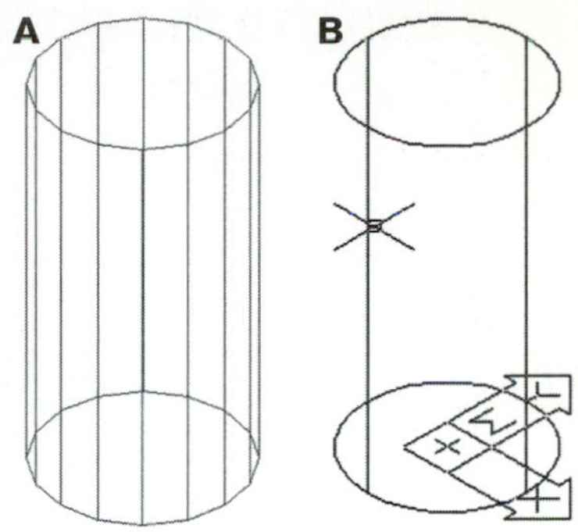

FIGURE 10.6
The faceted models stored by AutoCAD Release 12, as shown in (a), produce a less accurate volume calculation than the more accurate form of the model stored by AutoCAD Release 14, shown in (b).

Even after checking the results with a simple object, you should routinely estimate the values for your models to make sure that the computed values from your system are reasonable. If there is a significant difference between the software's report and your hand calculation, take time to determine what is producing the variation. Don't just believe the values reported by the software and base important decisions on those results. (There is sometimes a tendency to think the software values are right because they are printed neatly on the screen or page. Make sure that you understand the values and that you have an understanding of the range for their accuracy before you use them.)

Units and Assumptions

Some information reported under mass properties, such as the centroid, moments, and radii of gyrations, is related to the coordinate system in which the model is stored or to one you have selected. You must make sure that you understand the model orientation with respect to the coordinate system or you may misinterpret the numbers reported.

The type of units specified or assumed is especially important in calculating mass properties. Using mismatched types of units is a common reason mass property values may not match your expected values. The common unit systems in the United States are the *inch-pound-second (ips)* and *foot-pound-second (fps)* systems. Elsewhere, it's the *System International (SI),* which uses meters, kilograms, and seconds (mks) as the basic units, although the centimeter-gram-second (cgs) system is also sometimes used, and machine parts are often modeled using millimeters. You must make sure that you understand the units that are used in the calculation of the mass properties. If you misinterpret the units, or use different units between the model and the material properties that you enter, the value reported by the software for the mass may be wrong by a huge factor.

Materials

Many modeling packages allow you to assign material properties to the part, which will be used in calculating mass properties. This information is often stored in a separate file, called a *material file.* Once you have defined a material file, it can be stored in a library and reused, saving you the effort of having to look up and redefine the material information each time it is needed. For example, the

Pounds Mass Versus Pounds Force

Don't confuse pounds mass with pounds force when using U.S. units. An object on Earth that has a mass of 1 pound also has a weight of 1 pound. If you send that object to the moon, where the gravitational force is about one-sixth what it is on Earth, the object will still have 1 pound mass; however, its weight in pounds will be about one-sixth of what it weighed on Earth. Mass equals weight divided by gravitational acceleration ($m = W/g_c$).

material file shown in Figure 10.7 was created in Pro/Engineer and used to assign material properties to the blade part shown in Figure 10.1. The calculated mass of the blade depends on the mass density for the material that is entered in this file and the units for that mass density.

It is important when entering this information to match the units to the other units you are using. For example, if the mass density for the material is expected to be entered in slugs per cubic inch, and you enter the value using slugs per cubic *feet*, the resulting calculations for the mass will be off by a significant factor—1728. When using material files from an existing library, you must take particular care to notice the units in which the information was entered and make sure that it matches the units you are using for your model.

FIGURE 10.7
This material file was created and stored to define the material properties of 1020 steel, the steel used for the turbine blade. A template is provided by Pro/Engineer for creating these files in the appropriate format.

```
INFORMATION WINDOW
        MATERIAL  1020_STEEL

This file may be edited using available editor.
Just type on the necessary lines appropriate values
after the "=" sign. Comments are not permitted on
lines containing material properties names.

YOUNG_MODULUS                    =     3.000000E+07
POISSON_RATIO                    =     2.900000E-01
SHEAR_MODULUS                    =     1.100000E+07
MASS_DENSITY                     =     2.830000E-01
THERMAL_EXPANSION_COEFFICIENT    =     6.300000E-06
THERM_EXPANSION_REF_TEMPERATURE  =
STRUCTURAL_DAMPING_COEFFICIENT   =
STRESS_LIMIT_FOR_TENSION         =     6.500000E+04
STRESS_LIMIT_FOR_COMPRESSION     =
STRESS_LIMIT_FOR_SHEAR           =
THERMAL_CONDUCTIVITY             =     2.700000E+01
EMISSIVITY                       =     6.000000E-01
SPECIFIC_HEAT                    =     1.000000E-01
HARDNESS                         =     1.110000E+02
CONDITION                        =
INITIAL_BEND_Y_FACTOR            =     5.000000E-01
BEND_TABLE                       =
```

Specific Gravity

Specific gravity is the value that can be used to relate the density of different materials by comparing them with the density of water. The mass density of distilled water at a temperature of 20 degrees Celsius is 1 gram per cubic centimeter and its specific gravity is 1.00. Notice that specific gravity has no units. Gold has a specific gravity of 19. This essentially means that a volume of gold is 19 times as heavy as the same volume of water. The specific gravities for some other common materials are listed below:

Aluminum alloys	2.8	Stainless steel	7.8
Brass	8.6	Titanium	4.4

Some CAD software uses a density of 1.00 for all materials and reports the mass based on this. To determine the actual mass of a part, you must first determine the density for the material, then calculate the true number for the mass. When a CAD system uses a density of 1.00 to calculate the mass of your model, you can calculate the mass for a specific material by knowing its specific gravity (which can be looked up in a materials table). Knowing that the mass density of water is 1 gram per cubic centimeter, you can use the specific gravity to calculate the actual mass of a part.

If you do not check the calculations by hand and inadvertently use the density of 1.00 (as though your model was made of water), your incorrect mass value may cause a significant error.

Exporting Data from the Database

In some cases, the mass properties information provided by the solid model is enough to determine whether a key criterion is met; in other cases, the information from the model needs to be analyzed further or combined with additional information. The surface area of a model, for example, can be used with an estimate of paint per square foot to gauge the amount and cost of paint required.

Most solid and parametric modeling software allows you to export a wide range of data from the CAD database so it can be used in other applications. *Exporting* means saving the desired data in a file format that can be read by other software applications. Mass properties data is commonly exported to text files that can be read by a spreadsheet, for example. Parameter data and the geometry of the model itself can also be exported to be used in other applications. Conversely, parameter information can be imported *into* parametric modeling software from spreadsheets or other analysis tools.

To use your CAD database with other applications, you should be aware of how data is transferred from one application to the other, and of the options available to you in the package you are using.

File Formats

Each software application has its own *native file format*. A native format is designed to store any and all information created with that application in an efficient form. Each application has its own codes and characters stored with the data that are interpreted by the application when the file is opened. For this reason, file formats are identified by an extension. Microsoft Word, for example, creates .doc files. AutoCAD's native format is .dwg. If you try to open a .doc file in a different application—and do not get an error message—you may see strange characters on the screen. These are elements in the file used by Word that the other application doesn't know how to interpret.

Other file formats are standardized so they can be read by many different applications. A text-only or *ASCII* file (which often uses a .txt extension) uses nothing more than the standard 256 ASCII characters (the older ASCII standard had only 128 characters). ASCII is an acronym for American Standard Code for Information Interchange, an early standard that allows data to be shared reliably among all different kinds of computers. Each letter of the alphabet plus other common characters are coded the same way on any machine that reads and writes text files. If you've ever saved a formatted word processing document as a text file, you know that it does not save special formatting (such as bold and italic) or any graphics or equations in the file. It does, however, create a copy of the words in your document that can be read on any computer. (Text-based tools and formats enabled the growth of the Internet and the World Wide Web. The HTML language used on the Web is a text-only format that was designed to allow formatted text and graphics to be displayed from a text-based file.)

Today, there are many options for exporting data and importing data into another, target application. (The target application is the one receiving the information; the application originating or sending the data is referred to as the source.) Most software applications have filters (or translators) that allow them to write or read non-native file formats. These non-native formats may be standardized formats, or they may be the native formats of other popular packages. AutoCAD's .dwg format, for example, can be read by many applications (for example, Visio's IntelliCAD and Bentley Systems's Microstation software) that have recognized the large number of .dwg files in existence and made it easy for their users to work with these files. Conversion utilities may also be available to allow you to convert your CAD file into another format. These utilities can be useful when exporting to a standardized format.

Sometimes changing to a different format causes a loss of data—either in the new format or when you need to be able to reconvert the data back into a form readable by your CAD software. You can never increase the amount of information in a file by translating it; you can only keep it the same or decrease the level of information. You should explore the options you have for writing a file in the format used by the target application—as well as the ability of the target application to read your CAD files—before you export the data.

A

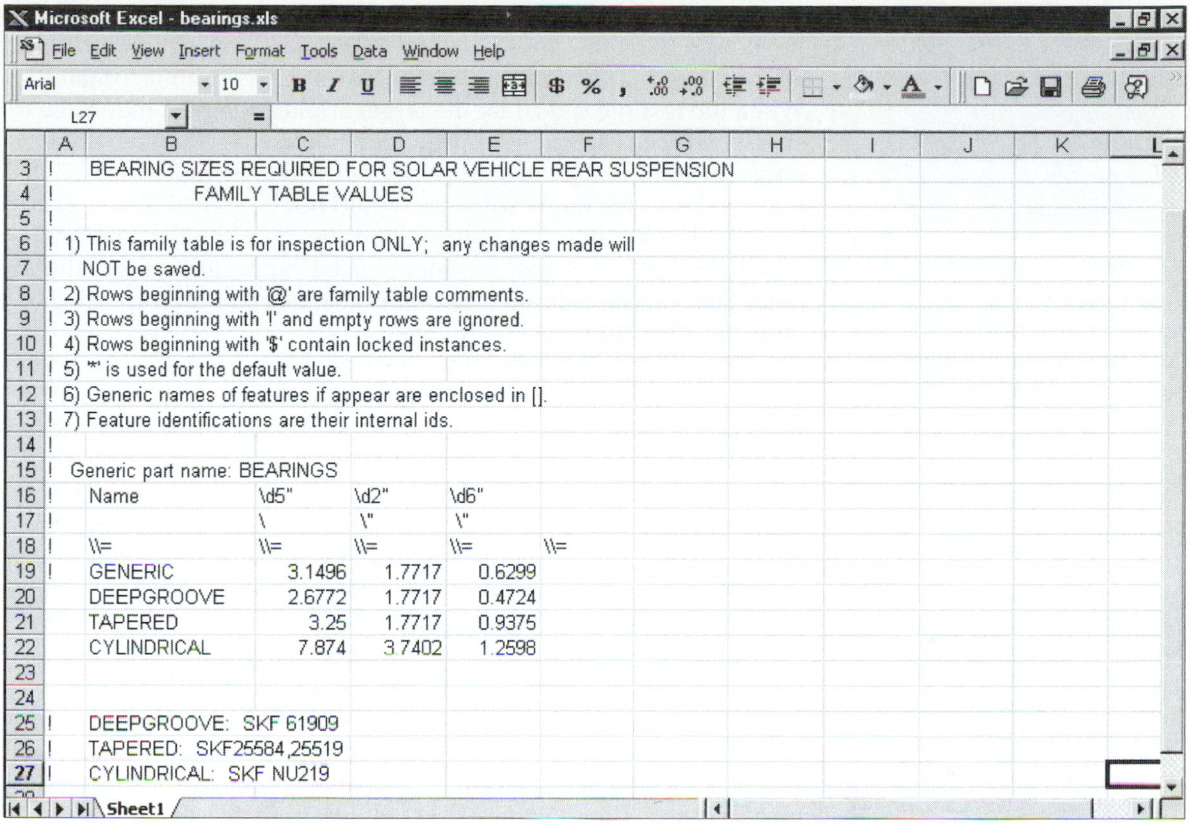

Microsoft Excel - bearings.xls

File Edit View Insert Format Tools Data Window Help

Arial 10 | L27 =

	A	B	C	D	E	F
3	!	BEARING SIZES REQUIRED FOR SOLAR VEHICLE REAR SUSPENSION				
4	!	FAMILY TABLE VALUES				
5	!					
6	!	1) This family table is for inspection ONLY; any changes made will				
7	!	NOT be saved.				
8	!	2) Rows beginning with '@' are family table comments.				
9	!	3) Rows beginning with '!' and empty rows are ignored.				
10	!	4) Rows beginning with '$' contain locked instances.				
11	!	5) '*' is used for the default value.				
12	!	6) Generic names of features if appear are enclosed in [].				
13	!	7) Feature identifications are their internal ids.				
14	!					
15	!	Generic part name: BEARINGS				
16	!	Name	\d5"	\d2"	\d6"	
17	!		\	\"	\"	
18	!	\\=	\\=	\\=	\\=	\\=
19	!	GENERIC	3.1496	1.7717	0.6299	
20		DEEPGROOVE	2.6772	1.7717	0.4724	
21		TAPERED	3.25	1.7717	0.9375	
22		CYLINDRICAL	7.874	3.7402	1.2598	
23						
24						
25	!	DEEPGROOVE: SKF 61909				
26	!	TAPERED: SKF25584,25519				
27	!	CYLINDRICAL: SKF NU219				

Sheet1

B

```
! Generic part name: BEARINGS,,,,
!,Name,"\d5""","\d2""","\d6"""
!, ,\," \""," \""",
!,\\=,\\=,\\=,\\=,\\=
!,GENERIC,3.1496,1.7717,0.6299,
 ,DEEPGROOVE,2.6772,1.7717,0.4724,
 ,TAPERED,3.25,1.7717,0.9375,
,CYLINDRICAL,7.874,3.7402,1.2598,


!,DEEPGROOVE:  SKF 61909,,,,
!,"TAPERED:  SKF25584,25519",,,,
!,CYLINDRICAL:  SKF NU219,,,,
```

C

```
!  Generic part name: BEARINGS
!  Name          \d5"     \d2"     \d6"
!                \        \"       \"
!  \\=           \\=      \\=      \\=      \\=
!  GENERIC   3.1496   1.7717   0.6299
   DEEPGROOVE   2.6772   1.7717   0.4724
   TAPERED   3.25   1.7717   0.9375
   CYLINDRICAL   7.874   3.7402   1.2598


!  DEEPGROOVE:   SKF 61909
!  TAPERED:   SKF25584,25519
!  CYLINDRICAL:   SKF NU219
```

D

```
! Generic part name: BEARINGS
! Name "\d5""" "\d2""" "\d6"""
!   \ " \""" " \"""
! \\= \\= \\= \\= \\=
! GENERIC 3.1496 1.7717 0.6299
  DEEPGROOVE 2.6772 1.7717 0.4724
  TAPERED 3.25 1.7717 0.9375
  CYLINDRICAL 7.874 3.7402 1.2598


! DEEPGROOVE:  SKF 61909
! "TAPERED:  SKF25584,25519"
! CYLINDRICAL:  SKF NU219
```

FIGURE 10.8
The parameter data in the spreadsheet in (a) were exported from Pro/Engineer. The text file used to export would look like (b) if it was exported as comma-delimited text; like (c) if it was tab-delimited; and like (d) if it was space-delimited.

Common Formats for Export

The formats you use to export data from the CAD database will depend on what you are exporting. If you are exporting mass properties data, attribute data, or dimension parameters, the export options will offer a choice of formats, most of which will be text-based. You may find some CAD packages that will export the data to a common spreadsheet format, such as Microsoft Excel's .xls format, but all of them will offer text as an option.

When you export a table of data as text, you will commonly have to choose among comma-delimited, space-delimited, or tab-delimited text (see Fig. 10.8). Tables of data are generally exported so that each record (or row of data in a table) is separated by a line break (signaling the software to go to the next line). Columns of data are separated *(delimited)* by a character—a space, comma, or tab. When the text file is read by the target application, you may have to specify which character was used to delineate the columns. If your data entries have a lot of spaces in them, space-delimited would be a bad choice, as the target software would not be able to distinguish between the spaces in the entries and those between the columns. Tab-delimited is the most commonly used export format for tabular data. The comma-delimited format is also popular for importing to and exporting from CAD software.

If you plan to export your CAD model for analysis, to generate NC tool paths, or to do animation or kinematic analysis, you will need to use some kind of *graphic exchange format.* These are standardized formats that capture graphic information in the same way that a text file captures text data. Each varies in the degree to which it captures the information in the model. The required accuracy for graphical data can also vary between graphic standards. This can be a problem because endpoints of lines that may have been considered connecting in one CAD package, where the accuracy indicates that endpoints connect if they are within 0.000001, may not connect when exported to a format that uses accuracies of 0.0000000001. The new format may not know how to interpret this difference in accuracy. The result may be that lines that previously connected no longer connect when translated. Remember, you cannot add accuracy or information to your CAD data through translation.

The Initial Graphics Exchange Specification (IGES) is a graphics format capable of exporting wireframe, surface, or solid information. This format captures the 3D information and is commonly used to export the model for computer-aided manufacturing. It was developed to export engineering drawing data, with emphasis on the 2D and 3D wireframe model information, text, dimensions, and limited surface information typical of early CAD-drawn mechanical parts. The IGES standard is controlled by the National Computer Graphics Association (NCGA). The NCGA administers the National IGES User Group (NIUG), which provides access to information on IGES. The volunteer organization IGES/PDES (Product Data Exchange using STEP) helps maintain the evolving IGES standard and publishes a specification for IGES. Because it is evolving, there are various versions of IGES in existence, with various capabilities. Because IGES has capabilities for exporting a wide variety of information, different *flavors* of it exist. You can often select options for how you want the information translated. Appendix 49 lists the types of entities and how the information for them is stored in the IGES standard. Refer to this appendix to interpret the segment from an IGES file shown in Figure 10.9.

The **STandard for the Exchange of Product model data (STEP)** is another 3D format that was agreed upon more recently by the International Standards Organization (ISO). STEP may eventually allow more information to be transferred between CAD platforms than IGES, but it is still relatively new.

Standard for the Exchange of Product Model Data

The STEP standard is being developed by the International Standards Organization to provide a format for representing and exchanging computer-readable product data. The initiative seeks to provide a neutral mechanism for describing product data throughout the life cycle of a product–not just its design, but also its manufacture, use, maintenance, and disposal. This means that the standard would not only provide a neutral file format for exchange of 3D model information, but also address additional product information such as that currently handled by a variety of database systems or a product data management (PDM) system. As the standard is being developed, U.S. industry is represented in the process by PDES (Product Data Exchange using STEP), a voluntary activity coordinated by the National Institute of Standards and Technology (www.nist.gov).

```
                                                                      S0000001
1H,,1H;,23HMASTERCAM version 6.13a,10HSLTEST.GE3,9HMASTERCAM,1H1,16,8,
                                                                      G0000001
24,8,56,,1.,1,4HINCH,1,0.01,13H980608.150243,0.00005,100.,,,8,0,;
                                                                      G0000002
     110       1       1       1       1            0         00000000D0000001
     110       0       3       1       0                             0D0000002
     110       2       1       1       1            0         00000000D0000003
     110       0       3       1       0                             0D0000004
     110       3       1       1       1            0         00000000D0000005
     110       0       3       1       0                             0D0000006
     110       4       1       1       1            0         00000000D0000007
     110       0       3       1       0                             0D0000008
     124       5       1       0       0            0         00010000D0000009
     124       0       3       1       0                             0D0000010
...
     100       0       3       2       0                             0D0000042
     110      34       1       1       1            0         00000000D0000043
     110       0       3       2       0                             0D0000044
     110      36       1       1       1            0         00000000D0000045
     110       0       3       1       0                             0D0000046
     110      37       1       1       1            0         00000000D0000047
     110       0       3       1       0                             0D0000048
     110      38       1       1       1            0         00000000D0000049
     110       0       3       2       0                             0D0000050
110,0.,-1.6974091509,-0.375,0.,-0.5,-0.375;                          1P0000001
110,0.75,0.,-0.375,0.5,0.,-0.375;                                    3P0000002
110,0.875,-1.8232830142,-0.375,0.875,-2.4167906074,-0.375;           5P0000003
110,0.75,-0.25,-0.375,0.5,-0.25,-0.375;                              7P0000004
124,1.,0.,0.,0.,0.,1.,0.,0.,0.,0.,1.,0.;                             9P0000005
100,-0.375,0.5,-0.5,0.5,0.,0.,-0.5;                                 11P0000006
100,-0.375,0.75,-0.5,1.2457224307,-0.5652630961,0.75,0.;            13P0000007
110,1.2457224307,-0.5652630961,-0.375,1.0008780126,-2.425041092,    15P0000008
-0.375;                                                             15P0000009
..
S0000001G0000002D0000050P0000039                                     T00000
```

FIGURE 10.9
The IGES standard defines and codes entities in the CAD database using only comma- and space-delimited ASCII characters arranged in a standard format. See Appendix 49 for an explanation of an IGES file's structure.

Autodesk's **Drawing Exchange Format (DXF)** is popular for exporting 2D geometry between CAD platforms and other applications that work with the 2D geometry only. Obviously, exporting a 3D part to this format will take only some 2D portion of the data. In general, don't try to use it for 3D objects.

The **STL (STereo Lithography)** format is used to export 3D geometry to rapid prototyping systems. STL format translates the surface of the object into triangular facets. The size and number of these facets determine how accurately the STL file matches the original object (see Fig. 10.10).

These formats are useful both for exporting data and for importing files that have been created on a system other than your own. Static parts, for example, may be available from a supplier in both DXF and IGES formats. To add the 3D part to your model, you would import the IGES definition. To simply document the part in a drawing made from your model, the .dxf file might be sufficient for the 2D information.

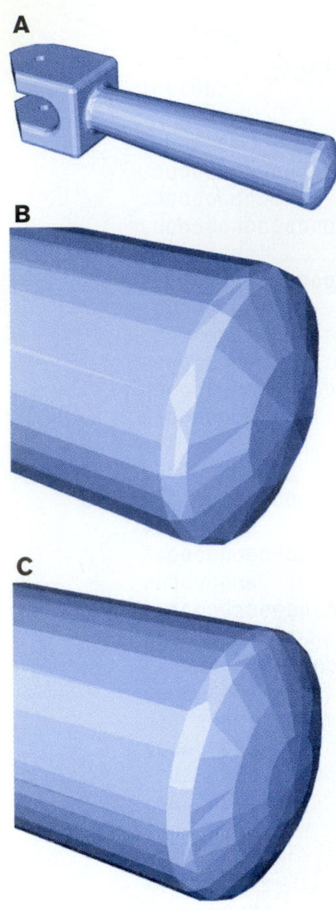

FIGURE 10.10
The handle model in (a) was modeled in Pro/Engineer, then exported in the STL format. The enlarged view in (b) shows the faceted appearance of the rounded end of the handle in this file format. This effect is reduced if the model is exported to STL with more facets that are smaller, as shown in (c).

Vector Versus Raster Data

Graphics programs are generally of two main types: *raster* or *vector.* Almost all CAD systems (including AutoCAD, Microstation, or Pro/Engineer) use vector-type data. This means that the type of information stored in the database contains the endpoints of lines, centers, and radii of arcs, and other information based on the geometry of the objects. The CAD software draws a particular entity on the screen based on this vector information. The definition of the entities allows them to be drawn, printed, or converted to any size or format. All of the previously mentioned graphical formats are used to exchange vector-type information.

Raster graphics programs (such as Photoshop or Paintshop Pro), on the other hand, store information about the discrete pixels, or dots, that make up a graphical image. Because this information is stored "dot by dot," it cannot contain useful engineering information about the individual entities in the image, such as the center of a circle. Some examples of these formats are .bmp, .pcx, .tif, and .gif. These formats contain the information about the color and intensity of the pixels that should be formed to display the image. Images produced by scanning a picture or document are raster-type images. You cannot directly scan a drawing and use it in a vector-based CAD program. First, its raster-type information must be converted to vector information using a conversion algorithm. This is often unreliable, because the connectivity of endpoints, line types, and other information can be difficult to translate reliably in every case.

Many vector-type CAD programs allow you to export and import raster-type images. For example, a shaded model may be exported to a raster format for printing or inclusion in a text document. Or a scanned image, such as a company logo, or picture of an actual part, may be imported and placed in a vector drawing, as in Figure 10.11. Raster data in the CAD drawing remains a set of pixels, however; the entities in the drawing cannot be edited individually.

Translating Data

Before selecting a particular CAD platform, you should investigate the export formats that are available and make sure that you can successfully transfer the information between the different software packages you may be using. This is not always an easy task. If you plan to use your CAD data with another package, export a test part first to be sure the conversion will suit your needs. Each time you transfer from one application, such as the CAD modeling package, to another, there is a possibility that some data will be lost. If you want to bring the data back into your CAD database from another application, this testing is even more important. Don't leave this testing to the last minute when you may be facing a critical deadline. Translating CAD data is often not a simple matter, and it may take some time to determine the best method. You may also want to change the type of CAD entities originally used to create your model if you determine that you will have difficulty translating them for a critical purpose.

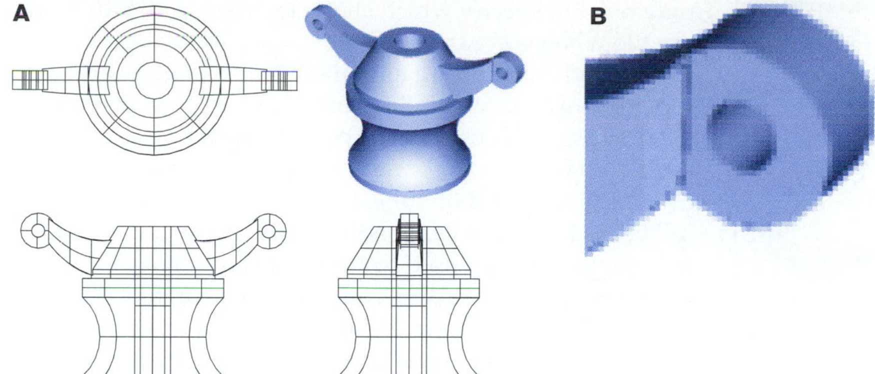

FIGURE 10.11
A rendered view of the model was exported to a raster-based file format, .jpg, then imported back into the vector-based drawing file in (a) to illustrate the part as an isometric view would. The rendered view is a set of pixels and cannot be edited. The enlarged view of the raster image in (b) shows that it is made of pixels, not editable points, lines, and arcs.

Downstream Applications

As the CAD model has evolved from a 2D drawing to a 3D database of design information, other software packages have evolved alongside it to facilitate the many kinds of engineering analysis done to evaluate product and system design. These *downstream* applications may use the data already created for the solid model so you don't have to define the model in another format. In this section, you will learn about the many kinds of analysis software that will work with data from the CAD database.

Spreadsheets

A spreadsheet is a popular application in engineering that you may already be familiar with. Once you have mastered the basic approach, creating useful spreadsheets is quick and easy. Values, text, and equations are entered into cells arranged in rows and columns. Each cell can be defined to store text and numeric data, or to store the results of an action on data stored in other cells. Most spreadsheets offer a wide range of built-in functions—from finding a simple sum to more complex mathematical functions. Advanced users can create sophisticated problem-solving tools using spreadsheets, but a spreadsheet is handy for any kind of calculation.

Spreadsheets were the first desktop application that made "what-if" analysis easy to do. Once relationships have been set up, it is easy to change values and test many different scenarios with little effort. This ability to evaluate, modify, and reevaluate is integral to the iterative process of engineering design.

The rows and columns make it easy to check the calculations done by the spreadsheet and easy to see results. Most spreadsheets also provide excellent graphics capabilities, including the ability to create graphs and charts from a range of values. Spreadsheets also automate much of the task of formatting the output in different ways.

To use information in the CAD database with a spreadsheet, you must usually export it to a file, then import the file into the spreadsheet program. When importing CAD data to a spreadsheet, you

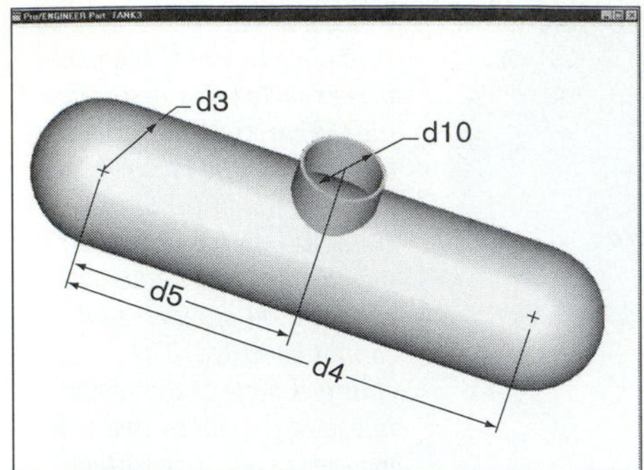

FIGURE 10.12
Tank modeled in Pro/Engineer

may need to specify which character was used to delimit the file when you exported it.

Spreadsheets are an extremely helpful engineering tool not only for calculations but also for projecting and keeping track of costs. An important aspect of an engineer's job is to design parts so they can be manufactured cost effectively and on time. For example, the tank shown in Figure 10.12 was modeled in Pro/Engineer. Its Model Info command was used to export information about the model, such as dimensions and surface area, to a file named tank3.inf (shown in Fig. 10.13). (Pro/Engineer adds this extension to indicate that it is an information file—in this case it is a text file type.) This file was then imported into a spreadsheet (tank.inf_file.xls) so the values could be used to estimate the cost of producing the tank. Another spreadsheet, tank.costs.xls, shown in Figure 10.14, was created to multiply cost estimates for weld lengths, materials, and other information by the quantities and values in the tank data. Once the calculations are complete, the same spreadsheet data can be used to produce a report, graph the costs, or recalculate the estimate based on updated cost or tank data.

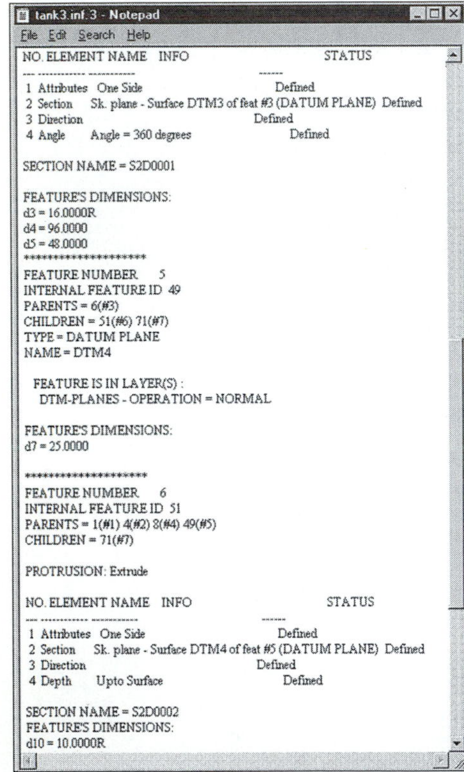


```
NO. ELEMENT NAME   INFO                                    STATUS
--- ------------ -----------                               ------
 1  Attributes   One Side                                  Defined
 2  Section      Sk. plane - Surface DTM3 of feat #3 (DATUM PLANE)  Defined
 3  Direction                                              Defined
 4  Angle        Angle = 360 degrees                       Defined

SECTION NAME = S2D0001

FEATURE'S DIMENSIONS:
d3 = 16.0000R
d4 = 96.0000
d5 = 48.0000
********************
FEATURE NUMBER    5
INTERNAL FEATURE ID  49
PARENTS = 6(#3)
CHILDREN = 51(#6) 71(#7)
TYPE = DATUM PLANE
NAME = DTM4

  FEATURE IS IN LAYER(S) :
   DTM-PLANES - OPERATION = NORMAL

FEATURE'S DIMENSIONS:
d7 = 25.0000

********************
FEATURE NUMBER    6
INTERNAL FEATURE ID  51
PARENTS = 1(#1) 4(#2) 8(#4) 49(#5)
CHILDREN = 71(#7)

PROTRUSION: Extrude

NO. ELEMENT NAME   INFO                                    STATUS
--- ------------ -----------                               ------
 1  Attributes   One Side                                  Defined
 2  Section      Sk. plane - Surface DTM4 of feat #5 (DATUM PLANE)  Defined
 3  Direction                                              Defined
 4  Depth        Upto Surface                              Defined

SECTION NAME = S2D0002
FEATURE'S DIMENSIONS:
d10 = 10.0000R
```

FIGURE 10.13
Tank parameters as exported to a text file with the extension inf

Equation Solvers

Equation-solving software, such as MathCAD and TK Solver, is used to solve simultaneous equations and to visualize the results using built-in graphing functions. Equation solvers, like spreadsheets, allow you to easily change values and run different cases, but they also allow you to run a range of values to select an optimum solution from among several solutions. Their ability to optimize a solution makes them a valuable tool.

MathCAD's interface uses a standard notation so you can type equations in the MathCAD editor the way you would ordinarily write them. MathCAD's interface is very simple to learn and operates as an electronic scratchpad. You can type in it as you might write on a piece of paper.

You export data for use in an equation solver in the same way you would for a spreadsheet. In the example in Figure 10.15, the model information for the tank modeled in Pro/Engineer was exported and loaded into MathCAD where the convective heat loss for the tank was calculated from the surface area.

TK Solver's interface is more similar to spreadsheet software. It uses a *variable sheet* to keep track of the variables, where you can have both the input variables and the outputs for variables. The *rule sheet* keeps track of the equations or rules. (It is billed as a rule-based programming language, not just an equation solver.) When you solve, the output values are shown on the variable sheet.

A

Microsoft Excel - Tank costs.xls

B5 = =0.283

	A	B	C	D	E	F	G	H
1	Tank Dimensions							
2	Tank Length (in)	128.00						
3	Tank Radius (in)	16.00		Shop Rate (/hr)	$45.00			
4	Opening Diameter (in)	20.00		Weld time (hr/ft*in)	1.00			
5	Density of steel (lb/in^3)	0.28						
6	Weld Length (in)	322.19						
7								
8								
9								
10								
11				Material Costs				
12		Thickness (in)	Cost for 8'x8' plate	Volume of plate (in³)	Volume of material used (in³)	Cost per in³	Cost per Pound	Material Cost for tank
13	ASTM A285	1/4	$738.74	2304	3418	$0.32	$1.13	$1,095.94
14		3/8	$966.04	3456	5027	$0.28	$0.99	$1,405.05
15	ASTM A516	1/4	$731.81	2304	3418	$0.32	$1.12	$1,085.66
16		3/8	$849.37	3456	5027	$0.25	$0.87	$1,235.36
17								
18		Manufacturing Costs						
19	Weld Time (hr/ft	Thickness (in)	Weld time per foot (min)	Weld cost				
20	ASTM A285	1/4	15.00	$302.06				
21		3/8	22.50	$453.09				
22	ASTM A516	1/4	15.00	$302.06				
23		3/8	22.50	$453.09				
24								

Sheet1 / Sheet2 / Sheet3

Ready NUM

B

Microsoft Excel - Tank costs.xls

A1 = Tank Dimensions

	A	B	C	D	E	F	G
1	Tank Dimensions						
2	Tank Length (in)	='Tank inf file.xls'!$B$4					
3	Tank Radius (in)	='Tank inf file.xls'!$B$4		Shop Rate (/hr)	45		
4	Opening Diameter (in)	='Tank inf file.xls'!$B$7		Weld time (hr/ft*in)	1		
5	Density of steel (lb/in^3)	=0.283					
6	Weld Length (in)	='Tank inf file.xls'!$B$4					
7							
8							
9							
10							
11					Material Costs		
12		Thickness (in)	Cost for 8'x8' plate	Volume of plate (in³)	Volume of material used (in³)	Cost per in³	Cost per Poun
13	ASTM A285	0.25	738.74	=8*8*B13*144	=(4*PI()*$B$3)+($B$2*(2*F	=C13/D13	=F13/$B$5
14		0.375	966.04	=8*8*B14*144	=(4*PI()*$B$3)+($B$2*(2*F	=C14/D14	=F14/$B$5
15	ASTM A516	0.25	731.81	=8*8*B15*144	=(4*PI()*$B$3)+($B$2*(2*F	=C15/D15	=F15/$B$5
16		0.375	849.37	=8*8*B16*144	=(4*PI()*$B$3)+($B$2*(2*F	=C16/D16	=F16/$B$5
17							
18		Manufacturing Costs					
19	Weld Time (hr/ft)	Thickness (in)	Weld time per foot (min)	Weld cost			
20	ASTM A285	0.25	=B20*$E$4*60	=($E$3*$B$6/12)*C20			
21		0.375	=B21*$E$4*60	=($E$3*$B$6/12)*C21			
22	ASTM A516	0.25	=B22*$E$4*60	=($E$3*$B$6/12)*C22			
23		0.375	=B23*$E$4*60	=($E$3*$B$6/12)*C23			
24							

Sheet1 / Sheet2 / Sheet3

Ready NUM

FIGURE 10.14

The parameters from the tank were used to calculate cost information in Microsoft Excel, as shown in (a). The formulas used in the spreadsheet are visible in the view of the same spreadsheet shown in (b). Note that the formulas in column B2-5 relate directly to data in the tank.inf_file.xls, the spreadsheet into which the tank.inf file was imported. If updated tank parameters are exported and imported into a new version of the tank.inf_file.xls spreadsheet, the calculations in tank.costs.xls will update with this new information.

A

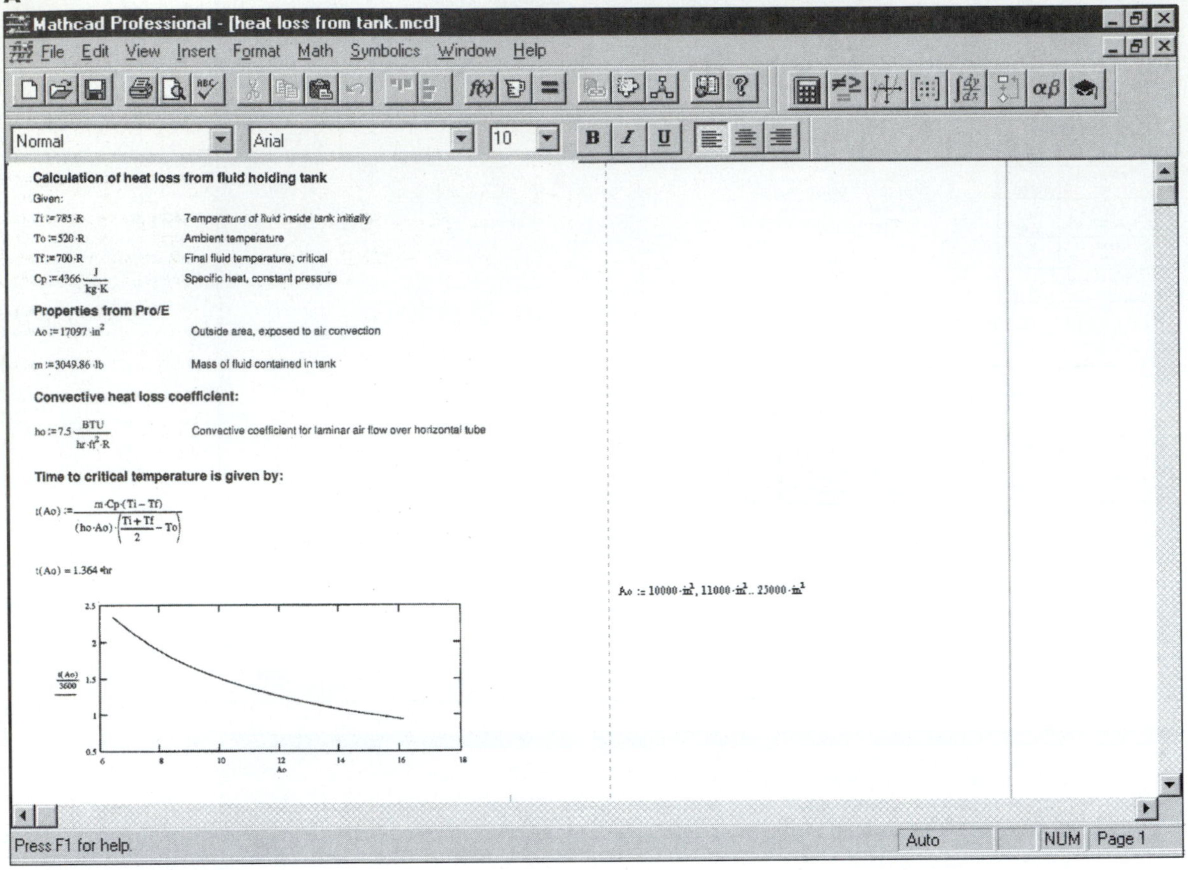

Mathcad Professional - [heat loss from tank.mcd]

File Edit View Insert Format Math Symbolics Window Help

Normal | Arial | 10

Calculation of heat loss from fluid holding tank

Given:

Ti := 785·R — Temperature of fluid inside tank initially

To := 520·R — Ambient temperature

Tf := 700·R — Final fluid temperature, critical

$Cp := 4366 \cdot \frac{J}{kg \cdot K}$ — Specific heat, constant pressure

Properties from Pro/E

$Ao := 17097 \cdot in^2$ — Outside area, exposed to air convection

m := 3049.86·lb — Mass of fluid contained in tank

Convective heat loss coefficient:

$ho := 7.5 \cdot \frac{BTU}{hr \cdot ft^2 \cdot R}$ — Convective coefficient for laminar air flow over horizontal tube

Time to critical temperature is given by:

$$t(Ao) := \frac{m \cdot Cp \cdot (Ti - Tf)}{(ho \cdot Ao) \cdot \left(\frac{Ti + Tf}{2} - To\right)}$$

t(Ao) = 1.364·hr

$Ao := 10000 \cdot in^2, 11000 \cdot in^2 .. 25000 \cdot in^2$

Press F1 for help. Auto NUM Page 1

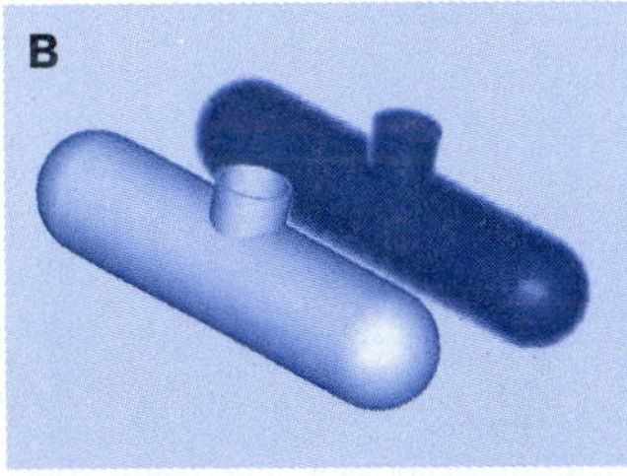

FIGURE 10.15

The assembly model of the tank in (b) can report the combined mass properties of the tank and the fluid, which are two different materials. Modeling both allows you to calculate the heat loss more accurately. The MathCAD calculations for convective heat loss shown in (a) use the combined mass properties.

One of TK Solver's biggest advantages is its ability to *backsolve* for variables regardless of the position in an equation. This makes it easy for you to test different values. For example, if you entered the equation for Newton's second law $F = \frac{ml}{t^2}$, Force = mass•length/time^2, in most programs, you would be able to solve this form only for force. In most computer programs, the equal sign is an assignment function and does not work in the way it does mathematically. In most software A = B means "assign the evaluated value of B to the variable or storage location called A." In TK Solver, you can define any of the variables listed in the equation and the remaining one is solved for. This makes it very easy for you to try different values or use a guess and check approach to selecting initial values without constantly having to rewrite the equations.

Graphs

For many years engineering graphics courses included the preparation of graphs and charts by hand methods. Comparing data visually helps you to easily grasp relationships between variables, compare results, and observe relationships between groups of data. Graphs provide a visual way to interpret data, which makes them easier for most people to understand than symbolic or numerical data.

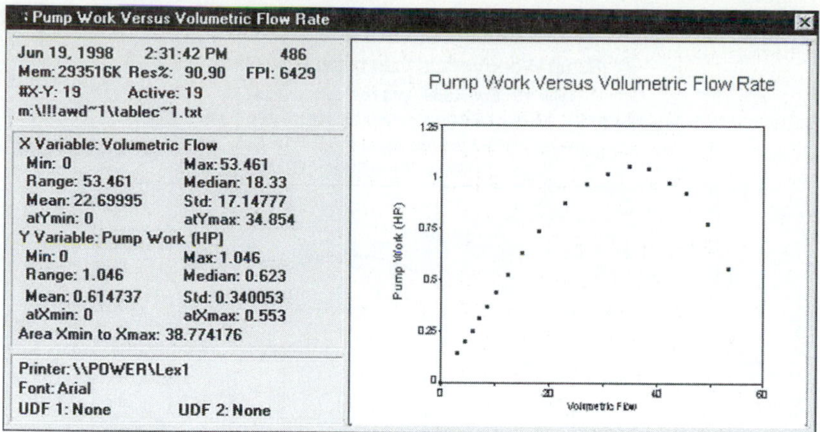

FIGURE 10.16
Empirical data for flow and horsepower was plotted using Labview software to visualize the relationship between the two.

Consider the example shown in Figure 10.16. The graph shows the relationship of empirical data (data actually measured) for a centrifugal pump. The X axis of the graph shows the volumetric flow rate, and the Y axis shows the horsepower used to drive the pump. Visually you can easily determine the point at which the efficiency of the pump falls off. This information can help you design the system for efficiency—a combination of sizing the piping for a desirable flow rate and selecting a pump that will operate near its peak efficiency for that flow rate.

Many excellent graphing packages and curve-fitting programs, such as Easy Plot, TableCurve, Sigmaplot, and Maxsyma (in addition to graphing functions built into spreadsheets), are available to aid you in creating charts and graphs used in engineering.

Curve-fitting programs let you generate an equation that fits the data from actual measurements such as the pump data above. You can then use this equation to plug in independent variable values and determine a value for the dependent variable, which you can use as design information.

TableCurve is a curve-fitting and graphing program that was used to plot the data shown in Figure 10.16. Figure 10.17 is a plot of the same data with an equation fit to it using TableCurve. When you execute the command to fit a curve to the data, TableCurve returns a list of the top three thousand equations that would fit this curve. Each equation is ranked by its R^2 value. (The R^2 value is a statistical term that reflects how accurately the equation fits the data.) The higher the R^2, the better the equation captures all of the data points. You can then select from the list of possible equations the one you consider the most likely match for the data. Figure 10.18 shows a list of the top matching equations for the empirical centrifugal pump data.

For the centrifugal pump, once you have determined an equation giving the relationship between flow rate and horsepower, you can plug in a value for flow rate and determine the horsepower that will be required to run the pump. Having this information will allow you to determine the cost of operating your system at a particular flow rate, because the cost of power to run the pump is based on the horsepower. You can also use this information to help make

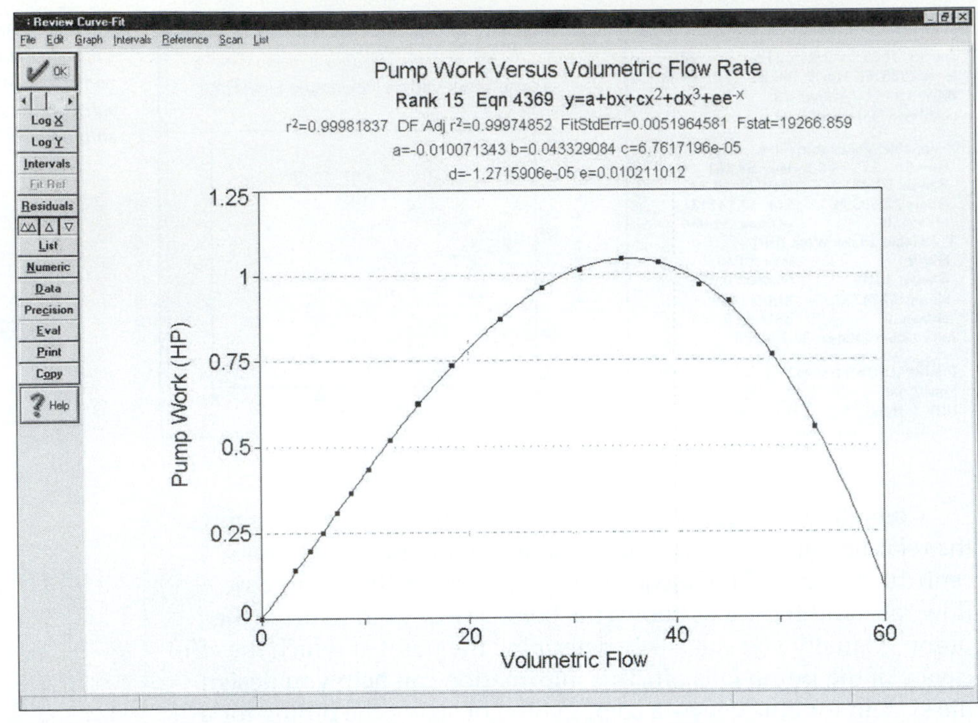

FIGURE 10.17 *TableCurve can fit a curve to the plotted empirical data for flow and horsepower.*

FIGURE 10.18 *The equations shown here were generated by TableCurve to fit the data collected for volumetric flow rate and horsepower for a specific centrifugal pump. The equations are ranked by their R^2 values, so the top ones listed fit the data most accurately.*

355 Equations [Rank, r², DOF Adj r², Fit Std Err, F-statistic, FP, Eq#, Eqn]

File Edit List Filter Sort

Rank	r^2	DOF Adj r^2	Fit Std Err	F-statistic	FP	Eq#	Eqn
2	0.9998586	0.9998042	0.0045852	2.475e+04	22	**4342**	$y=a+bx+cx^2+dx^{2.5}+ee^x$
3	0.9998576	0.9997864	0.0047754	1.825e+04	13	**7004**	$y=(a+cx+ex^2)/(1+bx+dx^2+fx^3)$
4	0.9998472	0.9997885	0.0047660	2.29e+04	21	**4689**	$y=a+bx+cx^3+de^x+ex^{0.5}$
5	0.9998349	0.9997715	0.0049537	2.12e+04	26	**4698**	$y=a+bx+cx^3+de^x+ee^{-x}$
6	0.9998251	0.9997578	0.0050999	2e+04	79	**8054**	[Beta_] $y=a((x-b+cm)/c)^{d-1}*(1-(x-b+cm)/c)^{e-1}/(m^{d-1}*n^{e-1})$ $m=(d-1)/(d+e-2)$ $n=(e-1)/(d+e-2)$
7	0.9998224	0.9997541	0.0051387	1.97e+04	22	**4220**	$y=a+bx+cx^{1.5}+dx^3+ee^x$
8	0.9998222	0.9997538	0.0051416	1.968e+04	20	**4740**	$y=a+bx+cx^3+dx^{0.5}+ee^{-x}$
9	0.9998215	0.9997323	0.0053453	1.457e+04	106	**8059**	[ADC] $y=a+(b/4)(1+erf((x-c+d/2)/(2^{0.5}e)))(1-erf((x-c-d/2)/(2^{0.5}f)))$
10	0.9998209	0.9997520	0.0051598	1.954e+04	17	**4585**	$y=a+bx+cx^{2.5}+dx^3+ex^{0.5}$
11	0.9998207	0.9997518	0.0051628	1.952e+04	16	**4224**	$y=a+bx+cx^{1.5}+dx^3+ex^{0.5}$
12	0.9998207	0.9997517	0.0051637	1.951e+04	14	**4360**	$y=a+bx+cx^2+dx^3+ex^{0.5}$
13	0.9998194	0.9997500	0.0051810	1.938e+04	21	**4233**	$y=a+bx+cx^{1.5}+dx^3+ee^{-x}$
14	0.9998193	0.9997498	0.0051835	1.936e+04	15	**4174**	$y=a+bx+cx^{1.5}+dx^2+ex^3$
15	0.9998184	0.9997485	0.0051965	1.927e+04	19	**4369**	$y=a+bx+cx^2+dx^3+ee^{-x}$
16	0.9998178	0.9997478	0.0052043	1.921e+04	22	**4594**	$y=a+bx+cx^{2.5}+dx^3+ee^{-x}$
17	0.9998173	0.9997470	0.0052117	1.915e+04	18	**4205**	$y=a+bx+cx^{1.5}+dx^{2.5}+ex^3$
18	0.9998148	0.9997619	0.0050687	2.7e+04	11	**2090**	$y=a+bx+cx^3+dx^{0.5}$
19	0.9998120	0.9997397	0.0052871	1.861e+04	16	**4341**	$y=a+bx+cx^2+dx^{2.5}+ex^3$
20	0.9998097	0.9997553	0.0051387	2.627e+04	12	**2023**	$y=a+bx+cx^{1.5}+dx^3$
21	0.9998088	0.9997352	0.0053318	1.83e+04	25	**4206**	$y=a+bx+cx^{1.5}+dx^{2.5}+ee^x$
22	0.9998064	0.9997320	0.0053648	1.808e+04	20	**4356**	$y=a+bx+cx^2+dx^3+ee^x$
23	0.9997965	0.9997384	0.0053138	2.457e+04	10	**2040**	$y=a+bx+cx^2+dx^3$
24	0.9997936	0.9997141	0.0055402	1.695e+04	23	**4581**	$y=a+bx+cx^{2.5}+dx^3+ee^x$
25	0.9997927	0.9997129	0.0055519	1.688e+04	17	**4173**	$y=a+bx+cx^{1.5}+dx^2+ex^{2.5}$
26	0.9997852	0.9997238	0.0054600	2.327e+04	13	**2071**	$y=a+bx+cx^{2.5}+dx^3$
27	0.9997813	0.9996971	0.0057026	1.6e+04	16	**4346**	$y=a+bx+cx^2+dx^{2.5}+ex^{0.5}$
28	0.9997788	0.9997156	0.0055407	2.26e+04	15	**2099**	$y=a+bx+cx^3+de^{-x}$
29	0.9997778	0.9996000	0.0064842	7070.5497	17	**7006**	$y=(a+cx+ex^2+gx^3)/(1+bx+dx^2+fx^3+hx^4)$
30	0.9997770	0.9996912	0.0057583	1.569e+04	21	**4355**	$y=a+bx+cx^2+dx^{2.5}+ee^{-x}$
31	0.9997623	0.9996944	0.0057434	2.103e+04	12	**2039**	$y=a+bx+cx^2+dx^{2.5}$
32	0.9997380	0.9996372	0.0062417	1.335e+04	19	**4210**	$y=a+bx+cx^{1.5}+dx^{2.5}+ex^{0.5}$
33	0.9997368	0.9996616	0.0060434	1.899e+04	17	**2086**	$y=a+bx+cx^3+de^x$
34	0.9997191	0.9996110	0.0064628	1.246e+04	23	**4219**	$y=a+bx+cx^{1.5}+dx^{2.5}+ee^{-x}$
35	0.9997120	0.9996013	0.0065431	1.215e+04	8	**6001**	$y=a+bx+cx^2+dx^3+ex^4$

other cost-based decisions, such as whether to spend the money for a larger-sized pipe to increase the flow rate, and the length of time needed to pay back the additional cost of the pipe through increased energy efficiency.

Finite Element Analysis

Finite element analysis (FEA) refers to the analysis method of breaking up a complex shape into discrete smaller parts *(finite elements)* for which the properties such as stress, strain, temperature distribution, fluid flow, and electric and magnetic fields can be more easily found. The collective results for these smaller elements are then linked together to determine the overall solution for the complex shape. The combination of smaller elements needed to cover a 2D or 3D shape is called a *mesh*. Many finite element analysis programs are available that can use CAD model geometry as the basis for FEA mesh.

Some FEA software is linked to or runs inside the CAD interface. For example, AutoFEA from ANSYS runs inside AutoCAD, and its commands are invoked from a menu on the AutoCAD menu bar. A similar FEA package (Pro/Mechanica) is available for Pro/Engineer. Other FEA software requires you to export the CAD model. Most FEA software will import CAD geometry from IGES, STEP, DXF, or other standard formats. COSMOS/M, for example, is FEA software that can import IGES files, which you can create in practically any CAD package.

Selecting the proper element type when forming the mesh is essential to getting accurate results from the FEA program. Element types are shapes that will be used to make the mesh. The choice of element type will depend on a number of factors, such as the material for the part, its geometric properties, and anticipated load.

A good understanding of material properties is necessary when using FEA methods. One property of a material is its *elasticity*. Only the simplest finite element codes represent materials as entirely elastic. Most have capabilities that go far beyond simple elastic behavior. It is the element shape used in the mesh that determines how the material behavior is modeled. Some elements are used strictly to represent elastic materials and small strain; some are elastic and include large strains; some are plastic; some are viscoelastic. Some elements even allow for coupled behavior: for example, temperature and stress for thermoelastic problems in which temperature affects stress and vice versa; electric current and stress for piezoelectric materials in which an electric current causes stress and vice versa. You must understand the range of material behavior likely to be encountered and select an appropriate element type in order to get valid results from FEA analysis.

Most modern finite element codes have automatic mesh generation. The quality of your FEA results also depends on the quality (or refinement) of the mesh. For example, regions where stress gradients are steep require a finer mesh to give good results. For symmetrical parts, where results will be the same on both sides of the axis of symmetry, time can be saved by cutting the model in half (or quarter) and analyzing only a portion of it. This keeps the file size and computation time down, which is particularly important when generating fine meshes.

After the mesh has been created, you use the FEA software to enter the forces. Determining how to represent the forces on the object as they are applied to the mesh is another important consideration.

FEA software will never tell you when your assumptions, element type, or mesh density is not right for the problem you are analyzing. It will produce some type of result. You must have the engineering background to interpret the validity of the result.

To illustrate how CAD data is used in FEA, consider an example in which it is used to determine suitable initial sizing for a compressor blade. The blade will be attached by mounting slots that will slide onto the hub. A model of the compressor blade is shown in Figure 10.19.

The geometry for the blade will be exported using an IGES export format to COSMOS/M FEA software. The Pro/Engineer settings for the IGES file are shown in Figure 10.20.

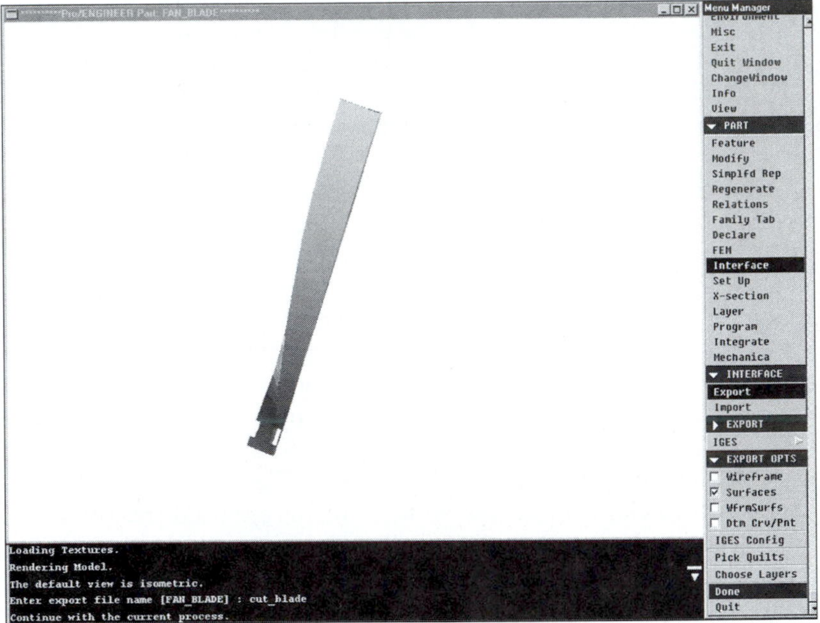

FIGURE 10.19
This shaded model shows the compressor blade with mounting slots as modeled in Pro/Engineer. The commands for exporting the surfaces to an IGES file have been selected from the menu on the right side of the screen.

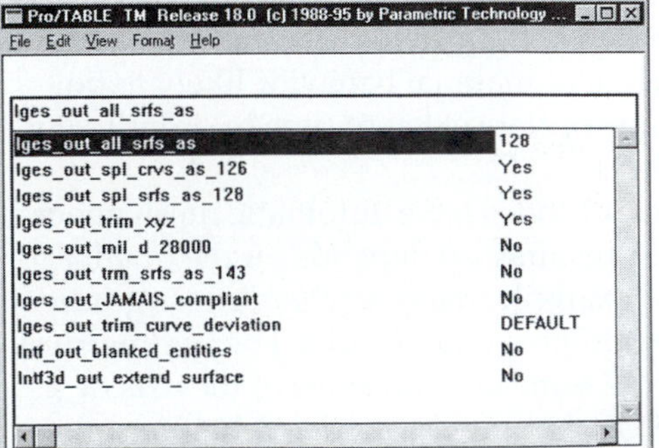
Pro/TABLE TM Release 18.0 (c) 1988-95 by Parametric Technology ...

File Edit View Format Help

Iges_out_all_srfs_as

Iges_out_all_srfs_as	128
Iges_out_spl_crvs_as_126	Yes
Iges_out_spl_srfs_as_128	Yes
Iges_out_trim_xyz	Yes
Iges_out_mil_d_28000	No
Iges_out_trm_srfs_as_143	No
Iges_out_JAMAIS_compliant	No
Iges_out_trim_curve_deviation	DEFAULT
Intf_out_blanked_entities	No
Intf3d_out_extend_surface	No

FIGURE 10.20
The table shows the IGES export configuration that Pro/Engineer software will use to export the file. Because IGES can accommodate many different aspects of the CAD file, you can often configure the CAD software to indicate what should be included in the end file.

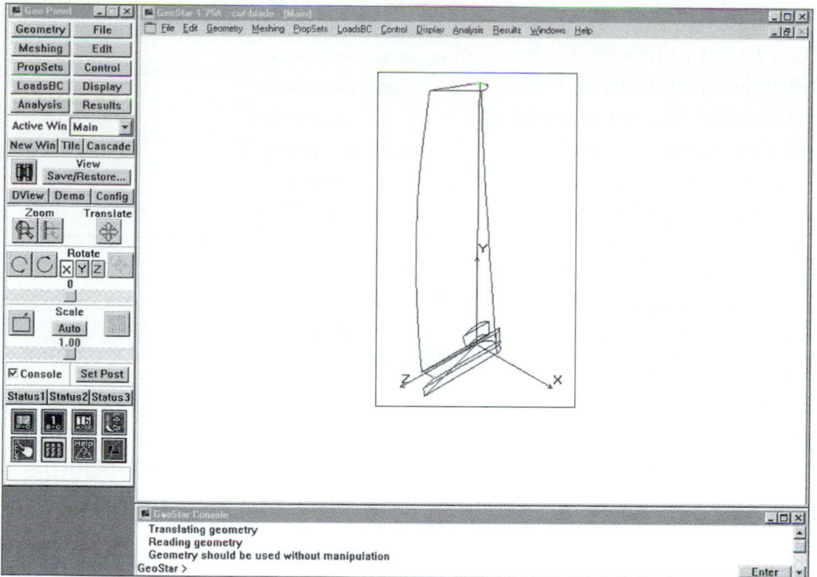

FIGURE 10.21
The blade is shown here after being imported into COSMOS/M. The coordinate system for the part is the same coordinate system that was used for the part in Pro/Engineer.

The blade will be analyzed for the pressure differences on the front and back side of the blade to determine the loads that will be needed for further structural analysis. To model the forces on this compressor blade accurately, the rotational forces for the blade would also be added. They have been left out in this analysis to make it easier for you to see what is represented in the figures.

Once the surface of the blade has been exported as an IGES file, the file is imported into COSMOS/M. Figure 10.21 shows the blade geometry once it has been imported into COSMOS/M.

The type of element that will be used to form the mesh is then selected. Figure 10.22 shows a list of the element types available. Each of these element types has various strengths and weaknesses for modeling particular features for particular material conditions. For this example, a tetrahedral element was selected.

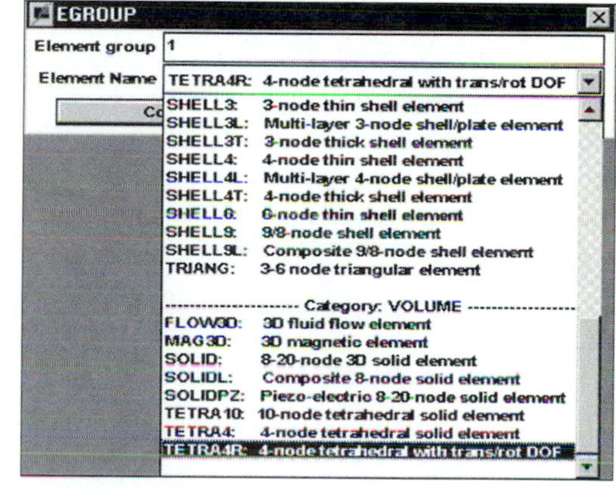

FIGURE 10.22
The COSMOS/M menu of element types

Once the element is selected, you can identify which material properties will be used for the part. In this case, titanium is chosen for the material from a list of predefined materials (see Fig. 10.23). The units are important because the material properties are specified in

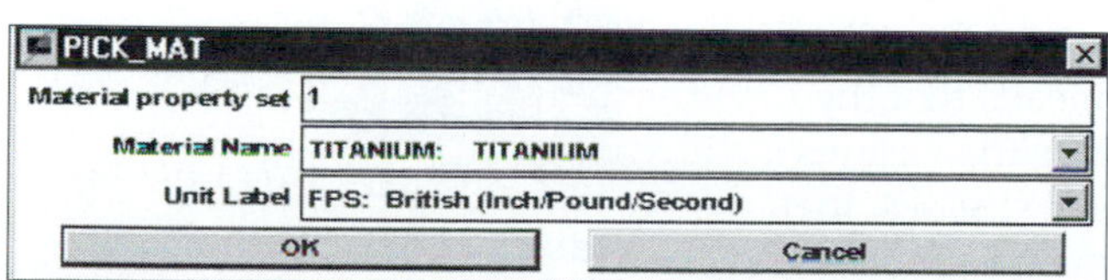

FIGURE 10.23
Titanium is selected as the material. The units must match the model units.

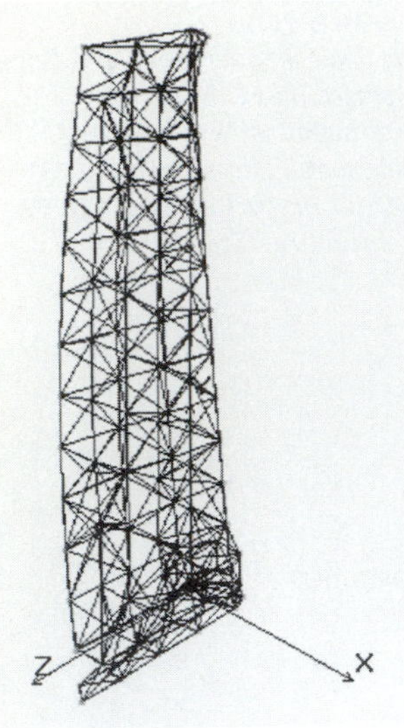

FIGURE 10.25
The mesh breaks the part into elements that can be analyzed as discrete solvable parts.

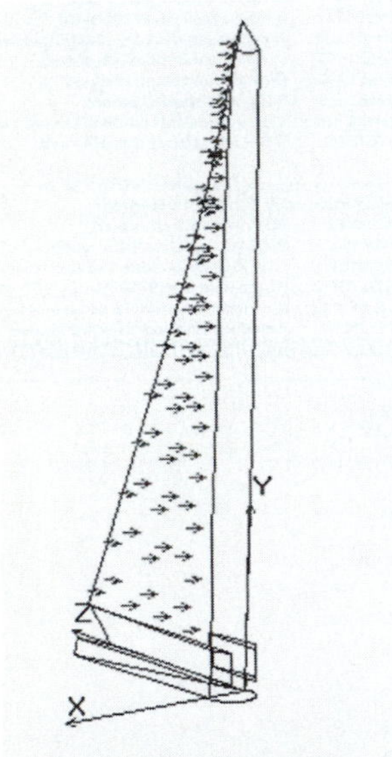

FIGURE 10.26
The pressure behind the blade is added. The small arrows at each node indicate the direction this force acts on the model geometry.

the file and must match the units for the model. Notice the designation British FPS in the file. The units are in inches, pounds, seconds, which match the units used to model the blade.

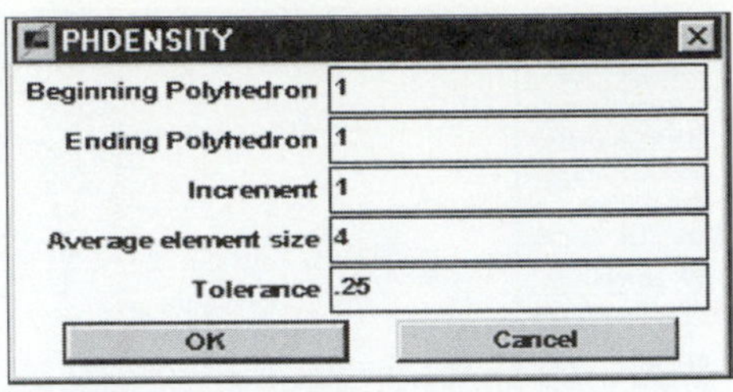

FIGURE 10.24
The mesh density (here specified as the size of the element) and tolerance are input into a dialog box. In this case the elements will be 4 inches in size.

The mesh density is an important factor for the accuracy of FEA results. Figure 10.24 shows the dialog box for setting the mesh density and tolerance. Once the mesh density has been selected, COSMOS/M's automatic mesh creation function can be used to fit a mesh to the part as shown in Figure 10.25.

Next, boundary conditions and loads are applied to the model to represent the forces acting on it. The bottom of the cut surface will be constrained to be fixed so that it doesn't move in either the X, Y, or Z directions. This will represent the fact that this surface will be used to attach the compressor blade to the hub. COSMOS lets you select an entire region (surface) to use, then constrains the nodes at each point in the selected region of the mesh automatically.

Forces on the blade are added to the model next. In this example, only the pressure behind the blade is being modeled. The arrows pointing toward the nodes of the mesh in Figure 10.26 represent the forces that have been applied.

Now the model is set up and ready to select an analysis method. Figure 10.27 shows the menu selections for the analysis methods that can be used in COSMOS/M. Von Mises effective stress analysis was selected because it analyzes the combined results of stresses in all of the different directions. (The von Mises effective stress is defined as the uniaxial tensile stress that would create the same distortion energy as is created by the combination of applied stresses.) This means that the analysis combines the various shear and normal stresses, then represents them as a single stress that you can relate back to material properties. This is useful for determining maximum loading.

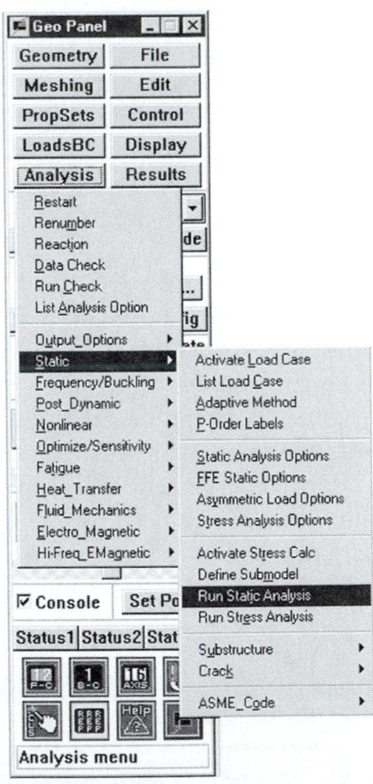

FIGURE 10.27
The Analysis menu offers a selection of methods, then a menu of options for each. Von Mises effective stress analysis was selected for the compressor blade.

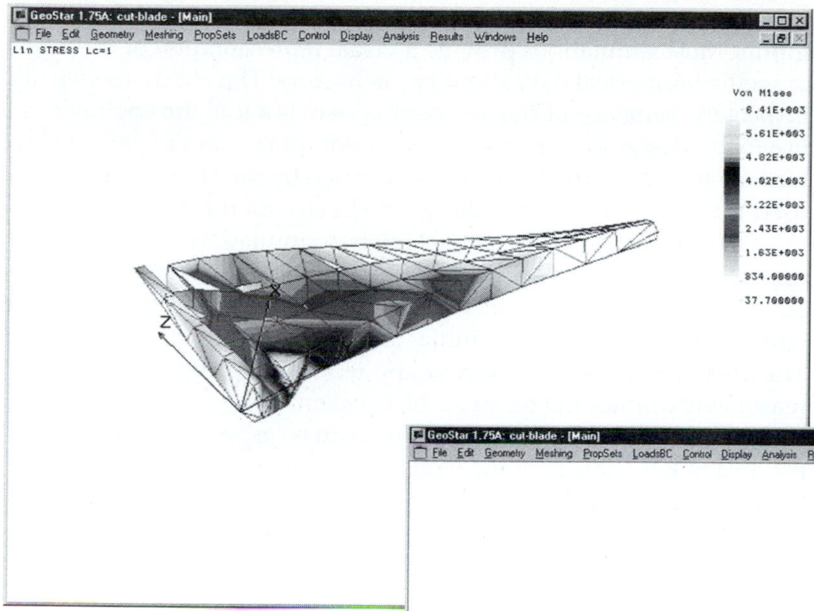

FIGURE 10.28
The von Mises plot for the compressor blade. The coarse mesh that was chosen for this part should make you skeptical of the results for this analysis. The key indicates values for the stress regions.

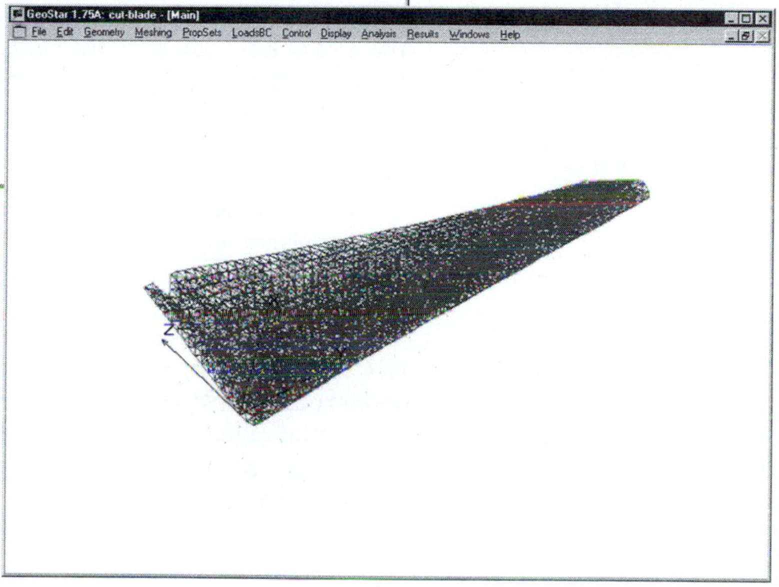

FIGURE 10.29
The compressor blade analysis using a finer mesh

The results of the analysis are shown in Figure 10.28. This plot would appear on screen in color. Hot colors (such as orange) are used to represent higher stresses. Cool colors (such as blue) represent lower stresses. Figure 10.29 shows the part remeshed using a denser mesh. The coarse mesh used above might be used for base testing, but it is unlikely to yield useful analytical results.

Useful FEA analysis combines your understanding of the FEA method, element selection, material properties, and the refinement or density of the mesh. Exporting the CAD model data makes FEA analyses less time-consuming and more readily available for use in design optimization.

Simulation Software

Simulation software allows you to simulate the function of a mechanism or mechanical system using rules that define how the parts will behave. Once you have modeled the mechanism's behaviors accurately you can set the simulation into motion to see how the model

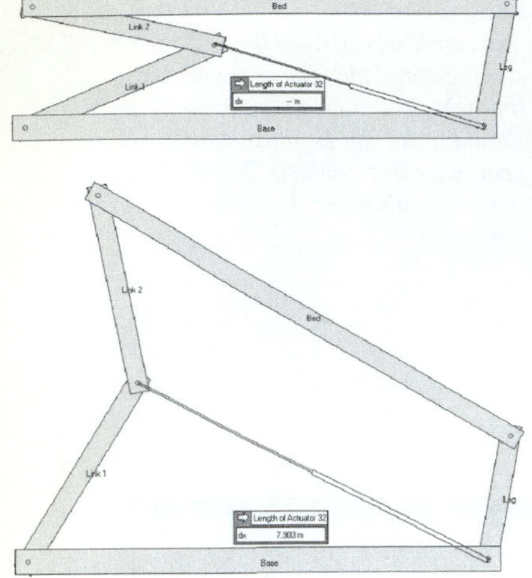

FIGURE 10.30
This four-bar linkage was modeled in the 2D version of Working Model.

will behave under different conditions or in response to different inputs. Most simulations provide a visual representation of the action as well as numerical data about key behaviors. The ability to visually inspect the behavior of the device or system is a tool the engineer can use to test designs early on—before prototyping—and to get a feeling for how they will work. The data generated by the simulation can be used to assess how well the design meets criteria for the project.

ADAMS and Working Model are two simulation software packages that allow you to see and measure a mechanical device in action. Other simulation software exists for specific purposes such as wind tunnel simulations, manufacturing cell simulations, and so on. The broad term *simulation* can be applied to any program that reasonably mimics the behavior of a system.

In mechanical design, simulations can be especially effective in the area of kinematics, or the study of motion. Although parts may be rotated through a range of motion in the CAD assembly model, simulating complex interactions among parts in an assembly or linkage, such as the four-bar linkage shown in Figure 10.30, is better done in a dedicated package such as Working Model.

In most cases, the CAD model can be exported as a .dxf file (for 2D models) or as an IGES file and then imported into Working Model. Once the model is imported, part and assembly constraints are converted to Working Model's joint types so they are constrained to the kind of action defined by the joint. Materials and their properties are assigned to each part in the assembly (either in your CAD model or in Working Model), as are motors and power sources. The appropriate forces are applied to the model, and other environmental settings (such as gravity) can be set to simulate action on the moon or in space.

The 2D version of Working Model is still used by many designers to see motion that is hard to visualize. Albert Brown, of Affymax Research Institute, used Working Model to model the way a robot gripper would move plastic trays used in drug testing. After importing the .dxf file, adding joints and applying forces, he was able to "see" how the trays would move. When the robot gripper was built, the motion was exactly what he had seen in the simulation. As design alternatives are being generated, the ability to see different shapes and materials in action can contribute to different design options.

Once the 3D model has been built, more complex analysis can be done with the 3D version of Working Model (now called Working Model Motion). The CAD model becomes a "virtual prototype" after it is imported into Working Model and joints and constraints are applied. In motorcycle design, for example, stability and safety of the design are a function of how the torque and stress induced by acceleration are handled by the vehicle. By bringing the CAD model into Working Model, the designer can test the vehicle as if it were a real prototype and take accurate measurements of torque and tension. Variations on the design can be "road-tested" as the design is optimized. The packaging system conveyor belt design shown in Figure 10.31 was modeled in SolidWorks, then brought into Working Model to predict the behavior of a complex, high-speed system.

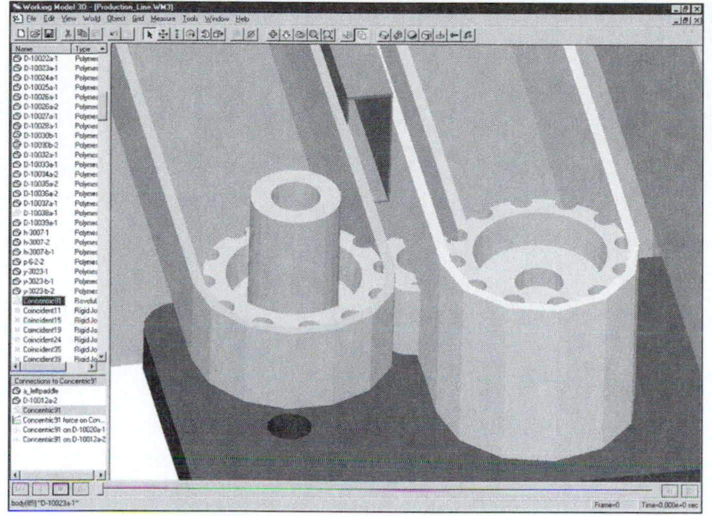

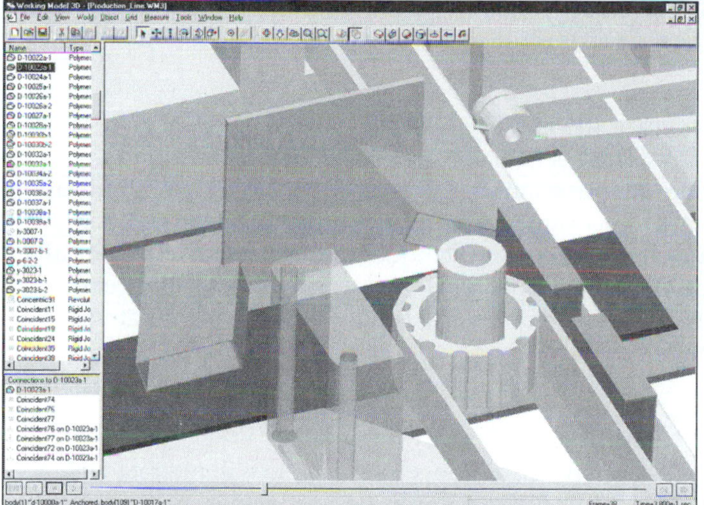

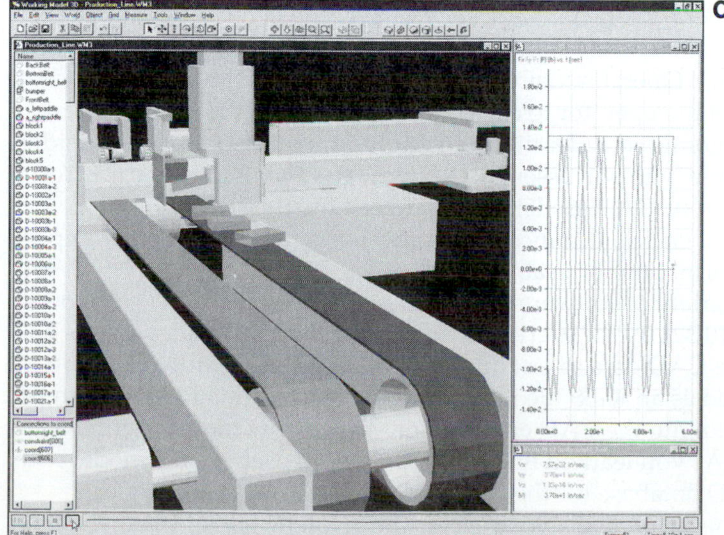

FIGURE 10.31

Walt Walczak designed this packaging system to move boxes at the rate of 800 per minute, then flip them 90 degrees so they lie flat on the belt. The boxes, 1/2 inch deep × 1-1/4 inches wide × 2 inches long, start in a vertical position, as shown in (a).

Several options for flipping the boxes were modeled and tried in Working Model; because the boxes are spaced about 3/8 inch apart, it was important to develop the right approach with the right amount of force. The final design used a rotary "pusher" to push the boxes over, shown in (b).

Working Model allows you to add meters to measure the behavior of the system, as shown in (c). The meters display results as the simulation is running and also save them to a file that can be evaluated separately. (Courtesy of Walt Walczak, Autobotics, Inc., Minneapolis, MN.)

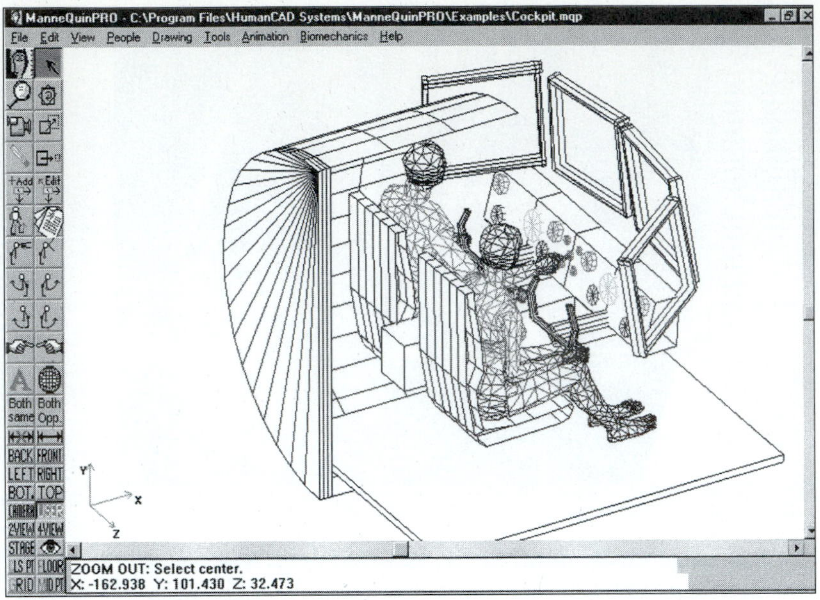

FIGURE 10.32
An airplane cockpit model can be imported into Mannequin Pro software and used to evaluate several different usability factors, including the fit for different-size humans, visibility from a seated position, and access to the control panel. (Courtesy of HumanCAD Systems, Inc., a division of BCAM International.)

For several CAD packages (such as Autodesk Mechanical Desktop, Pro/Engineer, SolidWorks, and Solid Edge), Working Model offers a more seamless interface to make design optimization even easier. Once a simulation has been created from the CAD model, any change to the CAD model can be used to automatically update the Working Model. This interaction makes it even easier to use the analysis done in Working Model to guide the iterative process of design.

Human Factors

Human factors analysis considers how people will interact with a design. As you are designing, you need to consider the qualities of the users who will use the system or product. Human populations have general characteristics, such as average height and language, that may affect their ability to use or operate the device you are designing. Operator visibility, the complexity of the controls, and the ability of the operator to correctly map a control to its function are all ergonomic considerations. In addition, you also need to consider the people who will assemble, service, and repair the product. You can use your CAD model to help determine the sizes for access openings, critical distances that affect users' reach, and other important information for usability.

A variety of software products, such as Mannequin Pro (shown in Fig. 10.32), WorkSafety, Transom Jack, and Ergoman, will work with the CAD data to test human factors. The CAD model can be imported as an IGES file into the software, or human models can be imported into the CAD environment to see how they fit with the model. As you learned in Chapter 8, Boeing used such a model with the CAD database for the 777 to determine how maintenance personnel would be able to access the airplane for replacement or repair. Being able to visually assess whether an average-size person would be able to reach the critical parts saved time and guesswork.

The Mannequin Pro software models typical humans from eleven different populations, including databases of human sizes such as 1988 Natick US Army and NASA-STD-3000. These populations provide a range of ethnic groups, size percentiles, body types, ages, cultures, and gender, which you can then fit with your CAD model to see how different categories of people might interact with the design (see Fig. 10.33). Mannequin's library includes typical predefined body and hand positions, linked for realistic human motion ranges, that you can import and export among your design software using common file formats. It even provides physically challenged mannequins that you can use to test accessibility. You can also use the software to "see" the view from the mannequin's eyes as it moves along a path that you define, as shown in Figure 10.34, and to simulate lifting, pushing, and pulling by adding forces and torque in any direction on any body part.

Ergonomic analysis using similar models and databases is an effective way to test work safety conditions for machine operators and others who will be in close proximity to moving parts as well.

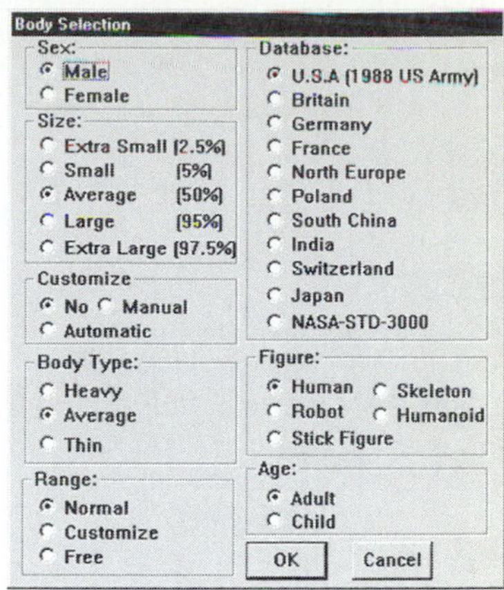

FIGURE 10.33
This dialog box from Mannequin Pro is used to select or change the characteristics of the human model being used for testing. (Courtesy of HumanCAD Systems, Inc., a division of BCAM International.)

Integrated Modeling and Design Software

The iterative nature of design refinement is fueling software tools that will allow the design to be modeled and analyzed in one software environment. Many analysis packages already provide tools that make it easier to go back and forth between the analysis package and the CAD model.

Working Model, as you have seen, can read some CAD files and update the Working Model file to mirror changes made to the CAD model. It also offers a feature called *automated constraint mapping*

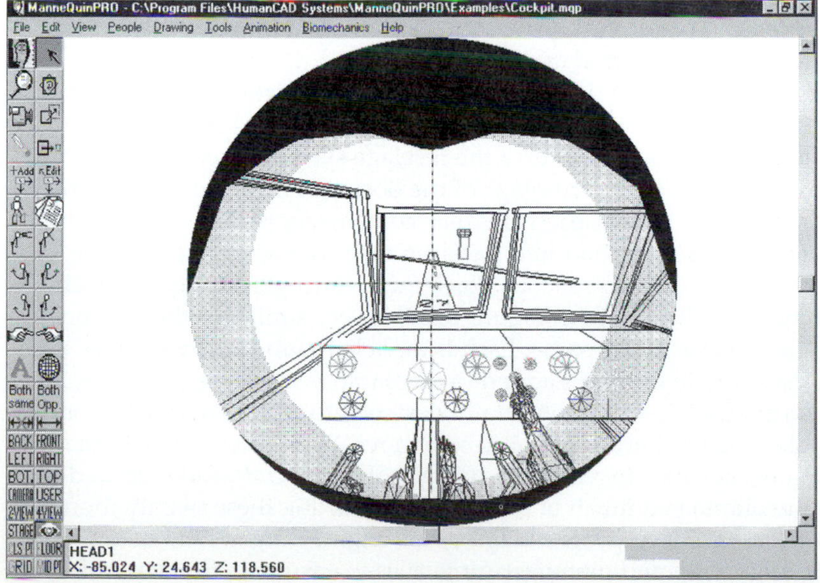

FIGURE 10.34
This view through the pilot's eyes illustrates the ability of the pilot to see the runway from the airplane cockpit in Figure 10.32. Operator visibility is an important factor in meeting safety requirements as well as functional goals. (Courtesy of HumanCAD Systems, Inc., a division of BCAM International.)

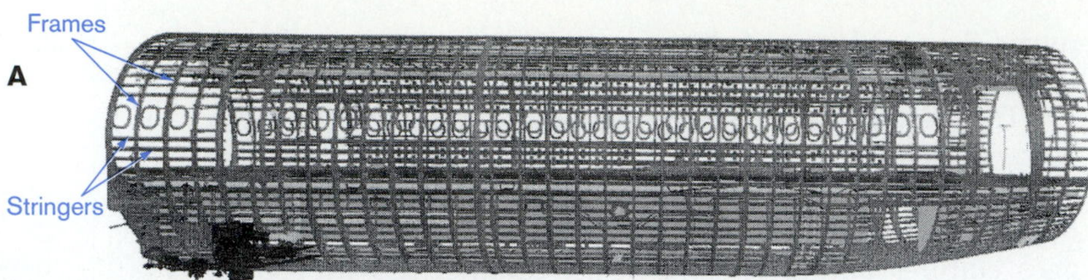

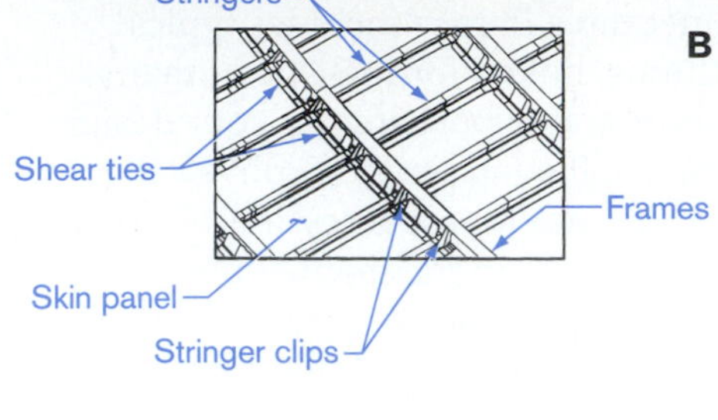

FIGURE 10.35
Design and manufacturing rules for the many similar parts in the aircraft fuselage were used to create a system that would automate much of the part modeling in CATIA, Boeing's 3D modeling system. The structural framework is visible in the assembly model of the fuselage in (a); the system of frames, stringers, stringer clips, and shear ties used to support the skin is illustrated in (b). (Courtesy of The Boeing Company.)

which interprets assembly constraints in the CAD model and translates them automatically to the appropriate joint or constraint in Working Model. Working Model also offers a seamless integration with an FEA solver so the finite element analysis can reflect loads and stresses induced when the object is not at rest.

Design rules that can be used for design optimization have also generated a closer link between parametric modelers and equation solvers. For example, DesignLink software from Universal Technical Systems, Inc. allows you to bidirectionally link the parameter names in your Mechanical Desktop or Pro/Engineer parametric model with the variable names used in TK Solver. When you bring the parameter values from your model into TK Solver, you can use the equation solver to analyze them and evaluate them in the context of design rules expressed as equations. Solving the equations with a more optimal value for one of the parameters will generate new values for others that can be brought into the parametric modeling package directly. The design can be updated to reflect the new values, and the process repeated as necessary.

Using design rules to automate the optimization and/or creation of model geometry has been successfully used at Boeing in the design of certain aspects of the fuselage system for commercial aircraft. The central fuselage of the Boeing 777, for example, is made of fifty skin panels, each of which contains more than four hundred parts. These four hundred parts are the framework to which the aircraft's "skin" is attached. Each frame, stringer, stringer clip, and shear tie, illustrated in Figure 10.35, is very similar to the next one, but not exactly the same. Each is modeled individually because each may vary in the holes added to accommodate wiring, fasteners, and so on. Each is also individually designed to weigh as little as possible. Chemical milling is used to remove all unnecessary material from each part to reduce the weight of the aircraft. Rule-based design has eliminated much of the work of modeling these twenty thousand parts. The Boeing knowledge-based engineering system uses both design rules and manufacturing rules to ensure that each part is

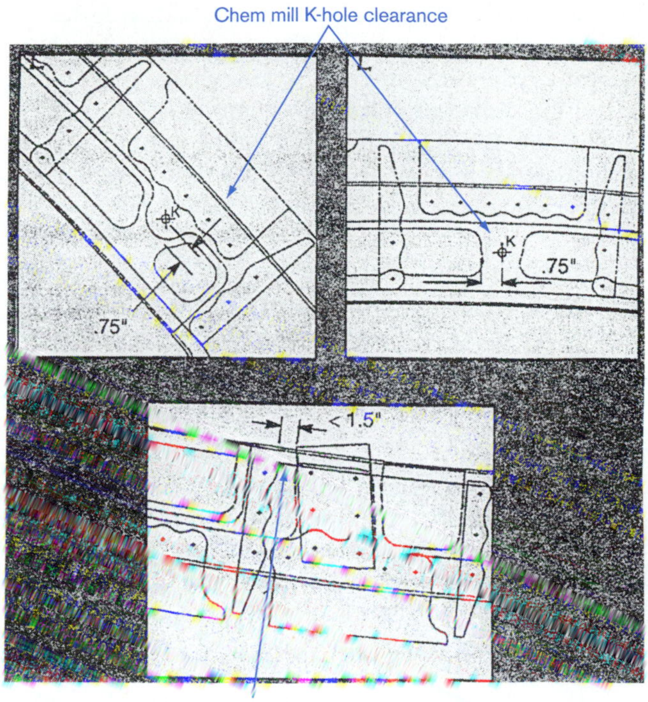

FIGURE 10.36
A design rule may prescribe the clearance needed for a part or feature. (Courtesy of The Boeing Company.)

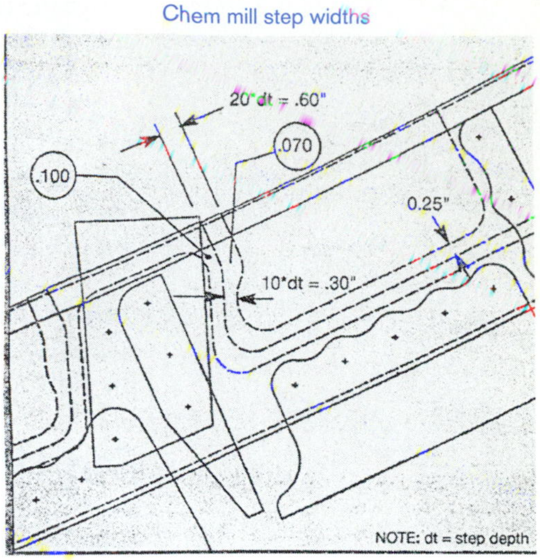

FIGURE 10.37
A manufacturing rule expresses the limits of the manufacturing process, such as chemical milling, that will be used. (Courtesy of The Boeing Company.)

structurally sound and optimized for weight (see Figures 10.36 and 10.37). For example, one design rule is that chemical milling to reduce the thickness of the part must stop 0.75 inch from a hole in the part. This ensures that the strength of the area around the attachment won't be compromised. Manufacturing rules are derived from the limits of the manufacturing process used to remove excess weight. No part may be designed that could not be milled. The parameters defining each feature of the part are evaluated against these rules both to optimize the design and to automate much of the part creation.

As you use your CAD database for analysis in downstream applications, you should familiarize yourself with the options being developed that can facilitate the exchange of information among packages and the use of analytical tools in design refinement.

Prototyping Your Design

No matter how accurate your CAD model, it is never exactly the same as the manufactured part. Traditionally, building a prototype was the best way to ensure that a design could be manufactured and that it would operate as desired when it was built. Today, the information in the 3D solid modeling database has affected the role of the physical prototype in two key ways: it has made it possible to simulate a prototype with the 3D model, and it has made it faster and cheaper to create a physical prototype than ever before.

A simulated prototype—often referred to as a virtual prototype—can serve many of the purposes of a physical model. Throughout this text, you've seen how the 3D solid model can be used to evaluate appearance, customer appeal, fit and clearance for assembled parts, mass properties, kinematics, and other characteristics of the design.

The same information in the CAD database can also be used to direct rapid prototyping processes that generate physical models relatively inexpensively. In the next section, you will learn about these processes and how they operate.

Rapid Prototyping

Because the volumes contained in a design are fully defined in a solid modeling CAD system, physical models can be created by nontraditional technologies that translate the data into a physical entity. *Rapid prototyping (RP)* systems allow the engineer to develop a prototype directly from a CAD design within minutes or hours instead of the days or weeks it might otherwise take to create a prototype part.

What is it worth to the design process to have an actual part that people can hold in their hands? As a visualization tool and a means for checking the fit with other parts, a physical model is a valuable aid in reducing the time it takes a company to develop a product idea from a sketch to a product that is available in the marketplace. Approximately 10% of all manufacturing and design shops spend over $100,000 each year for prototypes. The resulting confidence in the design and improved ability to communicate with the customer about the design in an understandable way is an important advantage of rapid prototyping. Rapid tooling processes use the same 3D CAD information to produce molds and other tooling that can reduce the time to market even further.

Rapid prototypes are especially useful for prototypes of complex molded parts. Molds for fairly simple plastic parts can cost from $20,000 to $50,000, making them prohibitively expensive to create just to check a design appearance. Using rapid prototyping, single parts can be produced in a matter of hours and used to verify the design. In addition, complex shapes can be created as easily as simpler ones.

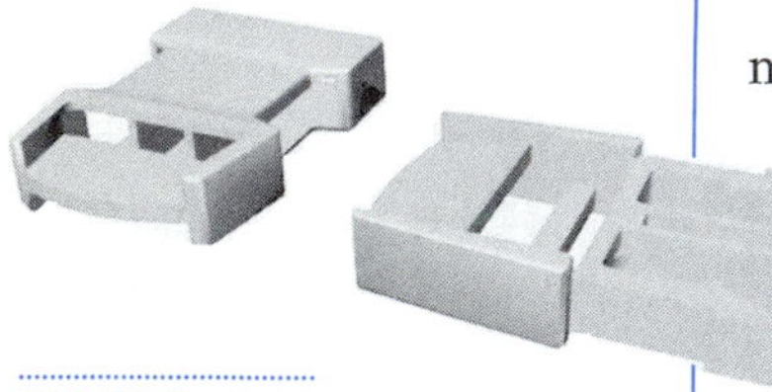

FIGURE 10.38
Complex molded shapes can be prototyped cost effectively with rapid prototyping systems.

Despite its advantages over traditional processes, rapid prototyping is not lightning fast. A part that is 2 by 3 by 1 inch may take three or more hours to create. However, this does not usually depend on the part's complexity, just the size and accuracy built into the prototype file. Parts that would ordinarily have to be molded or cast can be created in the same amount of time required for a rectangular block of about the same dimensions. The complexity of each slice does not have much effect on the time needed to create the part. While a rectangular block is easy to manufacture using traditional machining methods, the buckle shown in Figure 10.38 is a good candidate for rapid prototyping because of its complex shape.

Translating the Model

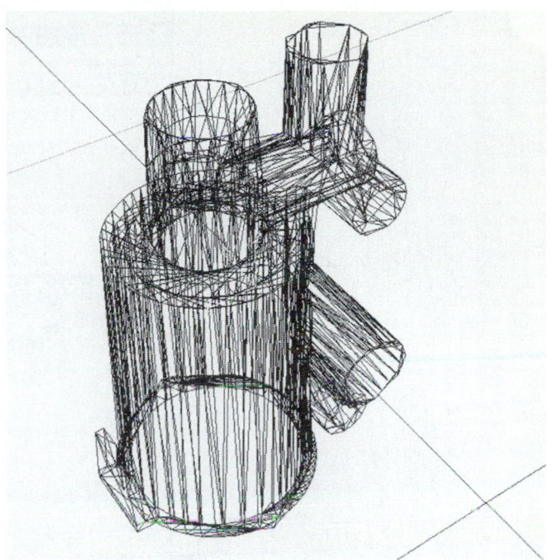

FIGURE 10.39
Triangular facets that define the boundaries of the model will be more visible on curved surfaces and when the facet size used is large.

Today's major rapid prototyping systems all work on a similar principle: they slice the CAD model into thin layers, then create the model layer by layer from a material that can be fused to the next layer until the entire part is realized.

To send a CAD file to most rapid prototyping systems, you must export a file in the STL file format. This file type was developed to export CAD data to an early rapid prototyping system. Since then, it has become the de facto standard for exporting CAD data to RP systems.

STL files define the boundaries of the CAD model using triangular facets. This format transforms any model into a standardized definition, but it has the disadvantage of generating a very large file when a realistic shape is required. You generally have the option of setting the size of the facets when you export your model. If the facets are small and the model complex, the resulting STL file will be very large. If a larger size for the triangular facets is used, however, the prototyped part will have noticeable facets on its curved surfaces, as shown in Figure 10.39.

Once the CAD file has been exported, the STL file is read by software that generates thin slices through the model that will be used to create the layers. This software is usually part of the rapid prototyping system. QuickSlice (from Stratasys) is one such software package, illustrated in Figure 10.40, which is used to generate the cross sections for their rapid prototyping system.

Once again, you have options in generating the slices that will affect the model accuracy and the time it takes to create a part. Generally speaking, the thinner the slice, the more accurate the part—however, thinner slices can also mean that it will take longer to generate the prototype part.

The accuracy of the model's surface is also limited by the material used in the process and how small the layers and features are that it can make. Figure 10.41 shows the edge of a prototype part created by depositing individual layers of material in the X-Y plane and dropping the table down in the Z plane in order to create the next layer on top of the first. This process necessarily produces a jagged edge. The size of the *jaggies*, as these are often called, is dependent on the thickness of the layer of material that is deposited. This thickness is limited by the size of the smallest particle that can be fused together. RP systems that have more than three axes of movement can reduce or eliminate the jagged appearance by filling in material on angled edge surfaces.

Current Rapid Prototyping Systems

Rapid prototyping systems vary in the types of materials used, the size of the model that can be created, and the time it takes to generate a part. If your company owns a rapid prototyping system, the choice

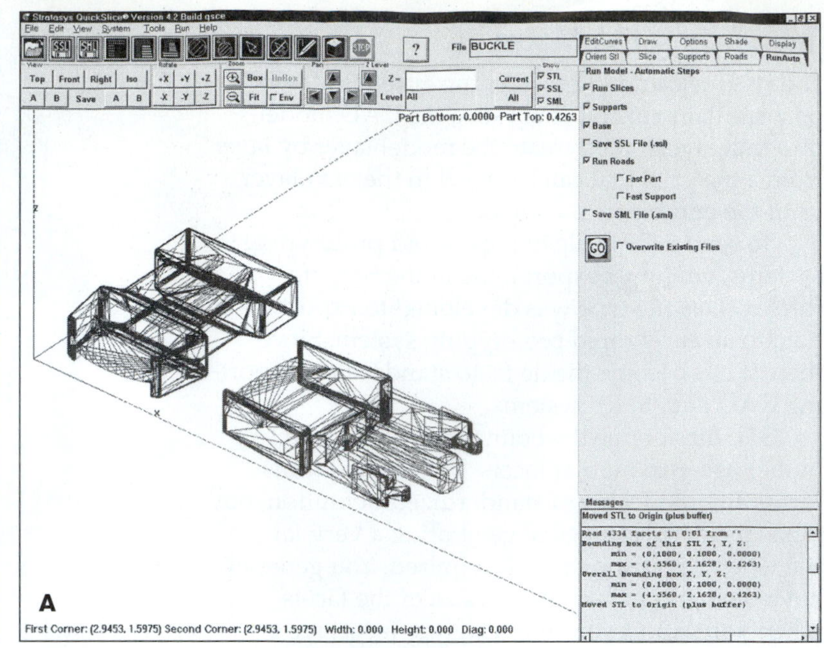

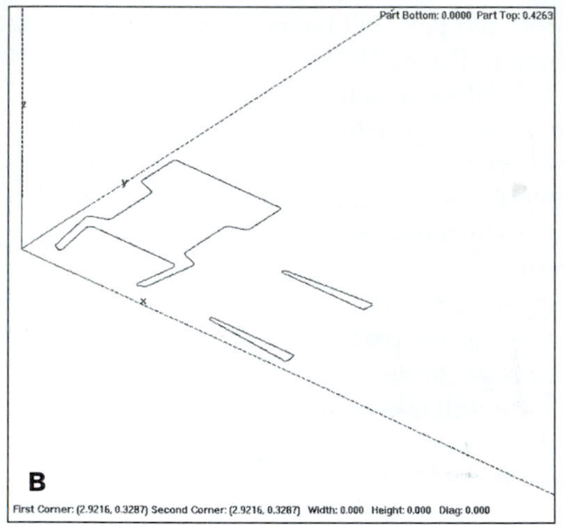

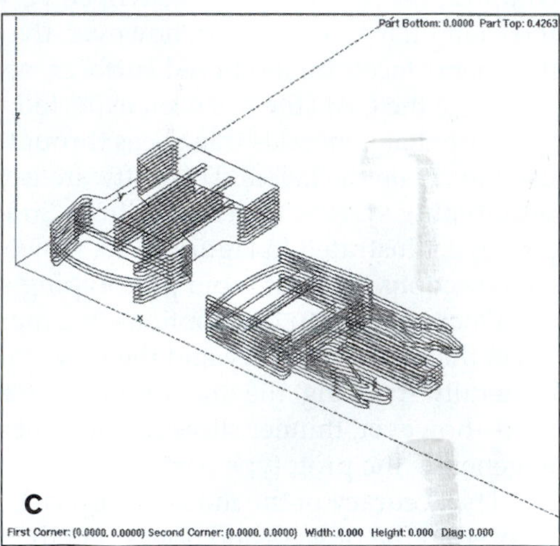

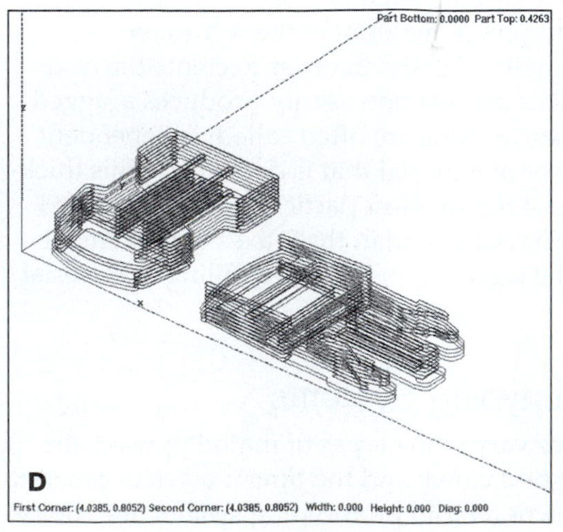

FIGURE 10.40
The model as imported into QuickSlice as an STL file is shown in (a). Slices through the model are generated at .01 intervals, as shown in (b) and (c). The software will also generate supports needed for segments of the part that do not rest on one of the lower slices; these are shown in (d) as lighter lines. These supports are not needed for all RP systems.

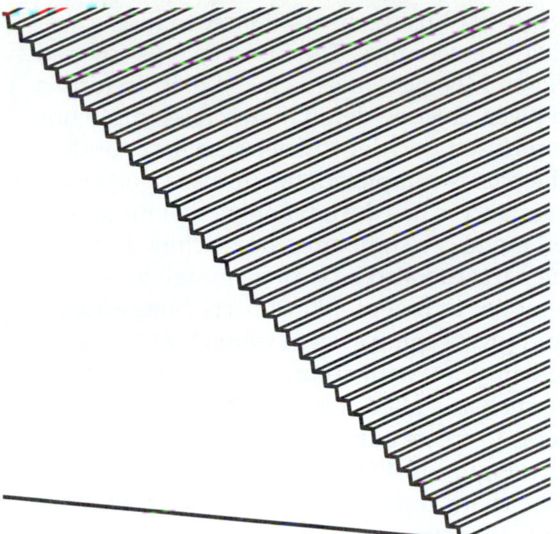

FIGURE 10.41
The process of building up the part by layers produces a jagged effect on non-normal surfaces. This effect can be minimized by thinner layers.

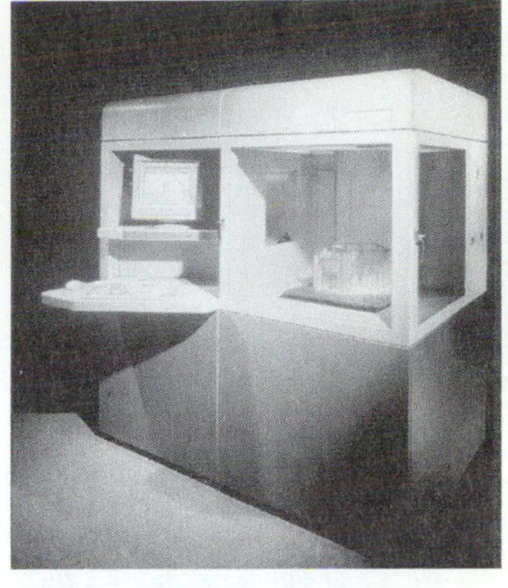

FIGURE 10.42
The SLA 5000 is a stereolithography system that uses laser-hardened resin to prototype parts. (Courtesy of 3D Systems.)

FIGURE 10.43
Cross-sectional slices are hardened in a resin pool by the SLA system. (Courtesy of 3D Systems.)

of system may be moot. If not, you should consider the design questions that the prototype needs to answer when selecting an appropriate system. The accuracy, size, durability, and time it takes to create a prototype are dependent on the process and material used.

The main categories of rapid prototyping equipment are stereolithography, selective laser sintering, fused deposition modeling, and 3D printing. Most rapid prototyping systems can create parts up to about 10 cubic inches in size. Laminate object manufacturing and topographic shell fabrication are two less common methods that allow you to create prototypes of larger parts.

Stereolithography apparatus (SLA) uses laser-hardened resins to form the model. Figure 10.42 shows an SLA system from 3D Systems Corporation. The system software controls a focused laser beam in a pool of light-sensitive polymer. The laser hardens each layer in the shape of the cross section or slice of the part. As successive layers are hardened, they are submerged slightly into the resin pool (see Fig. 10.43) and the next layer is hardened on top of them. Holes and pockets in the model are formed by uncured resin which easily pours out of the resulting part. SLA systems create durable parts that can be painted and finished to look very similar to the finished product. The range of accuracy for SLA parts can be up to ± 0.002 milliliters. Because of this accuracy the prototype parts created using SLA can have relatively smooth surface finishes. SLA is also the most established technology, since it was the first method on the market.

Solid ground curing (SGC) systems are similar to SLA systems except that they use ultraviolet light to cure an entire cross section at once in the polymer pool. A negative of the shape of the cross section is created on a glass plate using electrostatic toner (similar to a copying machine), then used to mask ultraviolet light in the shape of the cross section. With no lasers to replace, these systems are cost-effective and create accurate, durable parts.

FIGURE 10.44
Selective laser sintering fuses powdered metals instead of polymer resins to form prototype parts that can be machined. Other materials may be sintered in the Sinterstation 2500plus shown here to produce prototype parts with different characteristics. (Courtesy of DTM Corporation.)

Selective laser sintering (SLS) uses a focused laser to fuse powdered metals, plastics, or ceramics. The fused layer is covered with additional powder and the next layer is fused to it. To form a hole in the prototyped piece, the powdered material is simply not fused in that area. The unfused powder still acts as a base for the next layer, but when the part is completed, the unfused portions are simply poured out. This process has the advantage that models created from powdered metal can sometimes be machined for further refinement. The parts can also be strong enough to be used in certain types of assemblies as one-of-a-kind parts. Figure 10.44 shows an SLS-type rapid prototyping system from DTM Corporation. This system can create parts with accuracies of plus or minus 50μm. Other materials, such as a glass-filled nylon, may be used with the sintering process to create parts with varying degrees of flexibility and durability. Figure 10.45 illustrates an elastomeric material with rubber-like characteristics that makes it suitable for prototyping gaskets and athletic equipment.

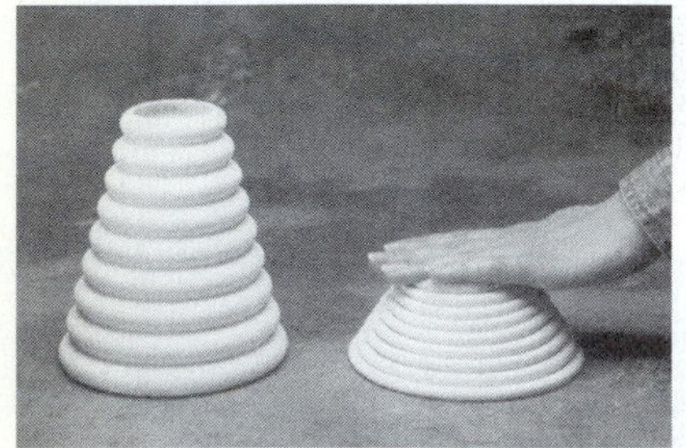

FIGURE 10.45
A new material from DuPont was used with the DTM laser sintering process to produce a rapid prototype with rubber-like qualities. (Courtesy of DTM Corporation.)

Fused deposition modeling (FDM) systems use molten plastic deposited in layers corresponding to cross sections on the part. Because the soft molten plastic cannot be deposited in thin air, to make a hole or an overhang, a second type of plastic is used to create a support structure. Because the two plastics are different materials that do not readily adhere to one another, the support structure can be separated from the actual part. (Figure 10.40(d) shows these supports being formed by the RP software.) Figure 10.46 shows an FDM system from Stratasys Corporation and some of the parts and support structures that it creates. A part that is about 3 by 2 by 1 inch takes about three hours to create.

Laminated object manufacturing (LOM) produces solid parts from sheets of material, such as paper or vinyl. LOM systems can be used to create larger prototype parts, as shown in Figures 10.47 and 10.48. Like all rapid prototyping processes, software first generates cross-sectional slices through the model. Instead of fusing the layer, however, a computer-controlled laser cuts it from the first sheet of

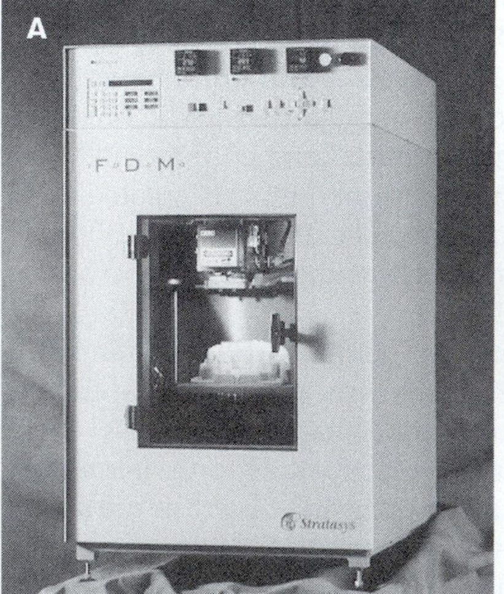

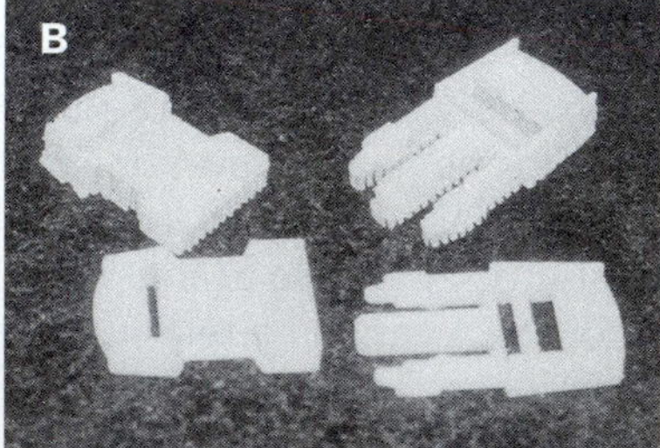

FIGURE 10.46
The fused deposition modeling system in (a) requires supports for sections of the part that are unsupported by lower slices. The buckle with and without supports is shown in (b). (Part (a) courtesy of Stratasys, Inc.)

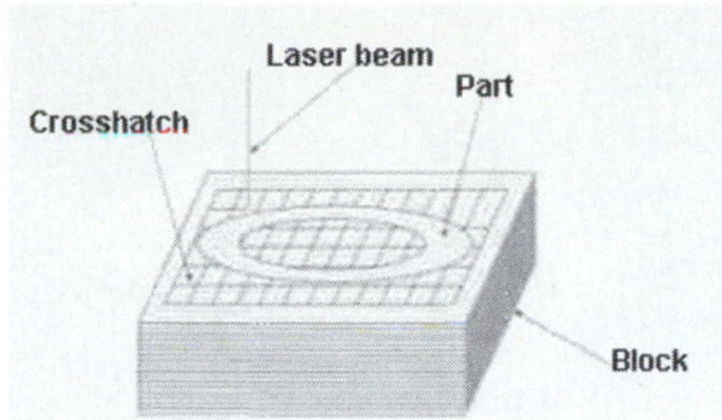

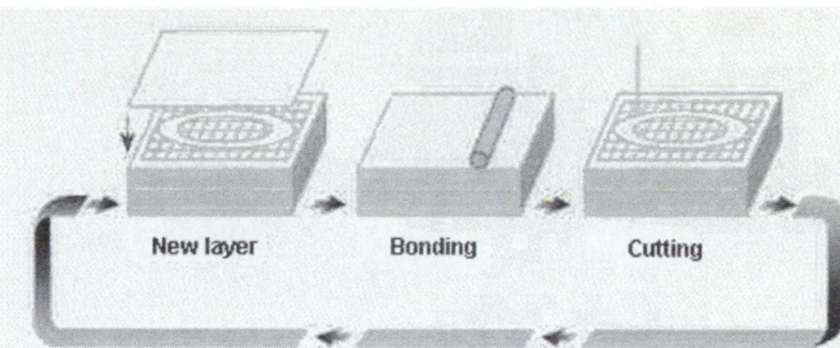

FIGURE 10.47
Laminated object manufacturing forms large-scale prototype parts by bonding sheets of material. (Courtesy of Helisys, Inc.)

material. Then a heated roller bonds the next sheet to the previous layer, and the next cross section is cut from this sheet. The material that will later be removed is cut into crosshatched shapes to make removal easier. The Helisys LOM-2030H system can create a part with a maximum size of 32 inches in length, 22 inches in width, and 20 inches in height (813 millimeters in length, 559 millimeters in width, and 508 millimeters in height) and a maximum part weight up to 450 pounds (204 kilograms).

Topographic shell fabrication (TSF) uses layers of high-quality silica sand fused together with wax to build shells that can be used to mold rapid prototypes of large-scale parts. The sand is deposited in layers and then fused together with molten wax sprayed from a computer-controlled three-axis nozzle. More sand is deposited and then the next layer is fused. The layers range from 0.05-inch to 0.15-inch thick and take about 10 minutes per square foot of model to print. Once all of the slices have been deposited, the sand/wax shell is smoothed, then lined with plaster or other material. The shell is then used as a temporary mold for creating parts of fiberglass, epoxy, foam, concrete, or other materials. This method is able to handle very large shapes up to 11 by 6 by 4 feet.

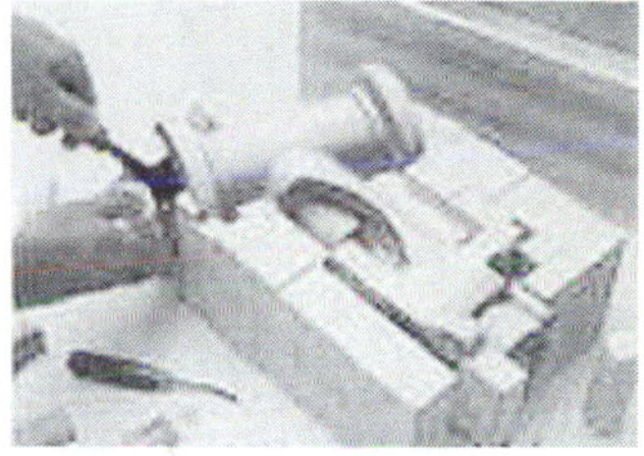

FIGURE 10.48
The large size of this part made it a candidate for LOM. The part is shown here as the cross-hatched material is being removed from the block within which the part was formed. (Courtesy of Helisys, Inc.)

3D Printing

Rapid prototyping systems referred to as 3D printing systems "print" layers of molten thermoplastic material. These low-cost machines were designed to enable the use of prototypes early and often in the design cycle. At Kodak, for example, a proposed product idea was to be discussed at a 10:30 meeting. At 8:30, the team was asked to bring a model of the idea to the meeting. Z Corporation's 3D printer, similar to that shown in Figure 10.49, was able to build a prototype from the CAD files in time for the meeting.

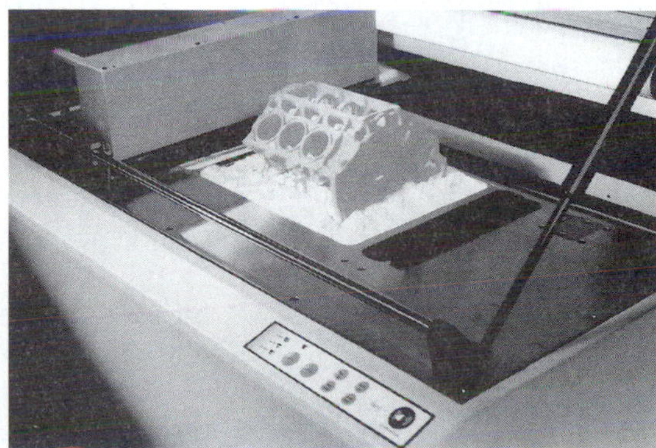

FIGURE 10.49
3D printers, such as the Zcorp Z402 shown here, have decreased the speed and cost of rapid prototypes even further. (Courtesy of Z Corporation.)

The relatively low cost 3D printing systems (currently under $50,000) can be operated safely enough that they can sit next to the regular office printer or copier. Another advantage of 3D printer systems is that those having a three-axis print head can deposit the plastic on the edges of the part in a way that creates a smooth model surface (see Fig. 10.50).

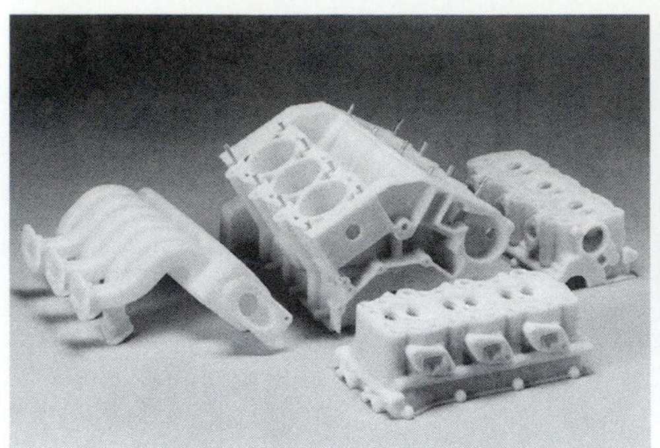
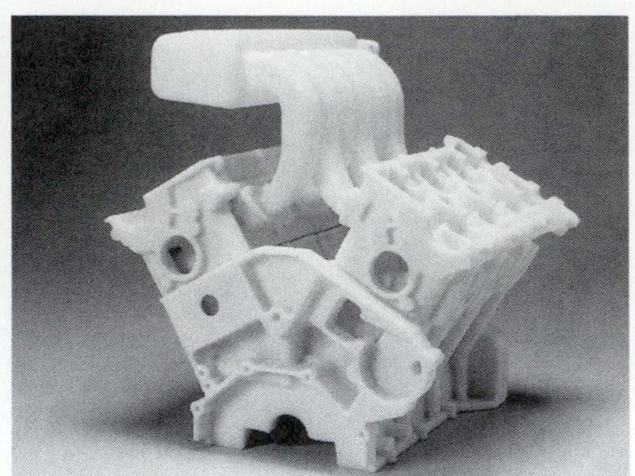

FIGURE 10.50
This engine block was formed with a 3D printing system. (Courtesy of Z Corporation.)

Rapid Tooling

Rapid prototyping systems were developed to produce parts without having to create a mold or complete intermediate steps needed to manufacture a part. *Rapid tooling* is a similar process, but one that creates the *tool* (usually a mold for molded plastic or cast metal parts) through a rapid prototyping process, not the part itself. Metal injection molds and molds for cast metal parts are often one of the most expensive and time-consuming parts of the design process. Rapid tooling processes can reduce the amount of time involved in producing these tools. The resulting rapid tool can be used to produce test products and to get products to market early.

Rapid tooling can be accomplished by several different methods. One is direct mold design, in which the tool itself is created using a selective laser sintering–type process. Another method uses rapid prototyping to produce a master part from which a silicone rubber mold is formed. That mold is then used to make other parts. A third method (a more traditional process) uses computer-controlled machining technology to create the cavity in a mold blank to quickly create the mold.

DTM corporation's RapidTool process is an example of *direct mold design*. RapidTool's SLS-type process uses powdered carbon steel pellets coated with a thermal plastic binder. Using a CAD file as input, the SLS process melts (or sinters) the powder to form a green mold shape consisting of metal particles bound by smaller areas of polymer. When the green mold is heated in a furnace, the plastic polymer burns off, leaving only the metal mold as shown in Figure 10.51. The mold is then machined to a tolerance of ±0.005 inch to eliminate any defects and can be drilled, tapped, welded, and plated like a conventional mold.

FIGURE 10.51
This mold was created by DTM's RapidTool process. The cavity side of the mold is on the left, the core on the right. (Courtesy of DTM Corporation.)

Using different materials, cores and molds for sand casting can be created directly using an SLS system, then cured (hardened) in a conventional sand casting oven. Another application is in *investment casting* (a method you will learn more about in Chapter 11). In investment casting, an original shape called a *master* is used to create the proper shaped opening in a mold. The master is typically made of a wax material so that it will melt out of the mold when molten metal is poured in. Investment casting masters can be produced using SLS processes.

Cores and Cavities

Molded parts are formed by cavities and cores. The *cavity* is the part of the mold that forms the outside shape of the object. The colloquial use of the term *mold* generally refers to the cavity. Holes that will be formed in molded parts are formed by cores. A *core* is a solid shape that fits inside the mold. It will form a hole in the cooled cast metal or molten plastic. Cores for cast metal parts are often made of packed sand. After the cast metal is cooled, the part is pounded to loosen the particles of sand from one another so they can be poured out through a small hole. This is similar to the way non-hardened resin or unfused powdered metal is poured out of holes in rapid prototypes.

Direct Shell Production Casting (DSPC) is a system licensed to Soligen Inc. which is used in directly creating molds for metal casting. It is based on 3D printing technologies developed at Massachusetts Institute of Technology. First the 3D CAD designed part is used to create a 3D digital model of the mold including the solid cores needed to produce hollow sections. In addition, the CAD file is used to modify the cavity shape by adding fillets and removing features such as small holes that will be machined in the finished casting. If necessary, multiple cavities are defined to produce a mold that forms several parts at once.

To produce the mold, the mold model is sliced into layers. A powder of ceramic material is deposited and liquid binder is "printed" onto the powder in the shape of the cross section of the mold. The process is repeated until the entire mold is printed. Then the mold is fired, producing a rigid ceramic. The unbound powder is removed from the mold to produce the cavities. The mold is then used to produce cast metal parts.

Silicon rubber molds are another way to produce rapid tooling. This process uses an accurately prototyped part that is then coated with silicon rubber to form a mold from which more parts can be molded. The MCP Vacuum Casting System, which uses this process, is capable of producing large plastic parts that weigh as much as 12 pounds and span 2 feet by 3 feet.

Rapid NC machining of mold inserts from a 3D CAD file also promotes rapid tooling. Even though this is more of a traditional process that creates the mold cavity by removing material, the NC machined cavity combined with standard mold blanks can often lead to shorter tooling times for injection molded parts. Metal spraying is another method that can be used in rapid tool production for less complex parts.

Despite the ease with which rapid tooling and rapid protoyping create physical models from CAD data, a strong understanding of traditional and current manufacturing methods will enable you to produce more cost-effective and producible parts. In the next chapter, you will learn about key manufacturing processes and how they can inform your design decisions.

INDUSTRY CASE

Scott Adams, Ball Aerospace

Mass Properties: From I-DEAS to Excel

Scott Adams uses Microsoft Excel to gather and report mass properties for the space telescopes and science instruments engineered by Ball Aerospace & Technologies Corporation. A majority of their designs will be carried into space by a launch vehicle and operate in space where the mass properties of the design can be critical to their operation. These designs must meet mass restrictions, but also be designed so that their center of gravity, moments of inertia, and other mass properties will work with the launch vehicle and the attitude determination and control systems of the spacecraft.

Adams may work with data from different CAD software packages during the concept stage, but most projects for which a contract has been awarded use I-DEAS Master Series.

Adams has done extensive accuracy evaluations for the I-DEAS software and takes care to "crank up the accuracy" for models that include revolved or swept shapes. I-DEAS uses triangular facets to define the surface generated from a 2D profile; where the swept or revolved surface meets itself at the starting point, there can be overlap that adds mass and skews the center of gravity to that side. By increasing the accuracy (more, smaller triangles to define the solid), Adams reduces this error when generating mass properties data. The default range of accuracy for the model is 1% error; Adams increases this value to 0.01% error for swept surfaces.

To begin the mass properties analysis, Adams first requests that all design engineers update the flight hardware models in the project library on the network. He then gets a reference (uneditable) copy of the top assembly and uses it to generate a mass properties report using a project-defined global coordinate system. I-DEAS can calculate, and put in a user-defined format, data for all subassemblies and their respective parts at once. This can take anywhere from a few seconds to a few hours.

I-DEAS organizes the data to first show the mass properties for the entire assembly, then for each of its parts (see Fig. 10.52). The data in this report, however, have no "intelligence." A change to or an addition of one of the underlying parts will not change the assembly

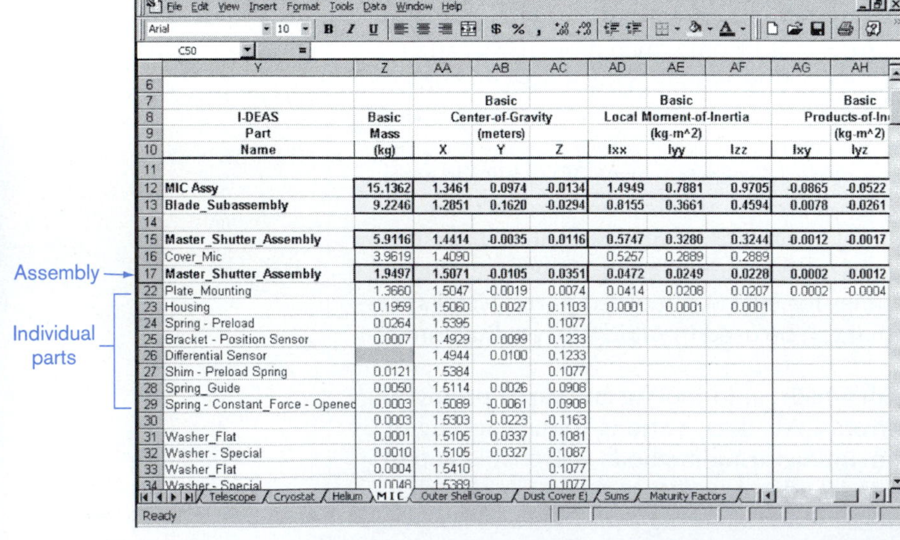

	Y	Z	AA	AB	AC	AD	AE	AF	AG	AH
6										
7				Basic			Basic			Basic
8	I-DEAS	Basic		Center-of-Gravity			Local Moment-of-Inertia			Products-of-In
9	Part	Mass		(meters)			(kg-m^2)			(kg-m^2)
10	Name	(kg)	X	Y	Z	Ixx	Iyy	Izz	Ixy	Iyz
11										
12	**MIC Assy**	**15.1362**	**1.3461**	**0.0974**	**-0.0134**	**1.4949**	**0.7881**	**0.9705**	**-0.0865**	**-0.0522**
13	**Blade_Subassembly**	**9.2246**	**1.2851**	**0.1620**	**-0.0294**	**0.8155**	**0.3661**	**0.4594**	**0.0078**	**-0.0261**
14										
15	**Master_Shutter_Assembly**	**5.9116**	**1.4414**	**-0.0035**	**0.0116**	**0.5747**	**0.3280**	**0.3244**	**-0.0012**	**-0.0017**
16	Cover_Mic	3.9619	1.4090			0.5257	0.2889	0.2889		
17	**Master_Shutter_Assembly**	**1.9497**	**1.5071**	**-0.0105**	**0.0351**	**0.0472**	**0.0249**	**0.0228**	**0.0002**	**-0.0012**
22	Plate_Mounting	1.3660	1.5047	-0.0019	0.0074	0.0414	0.0208	0.0207	0.0002	-0.0004
23	Housing	0.1959	1.5060	0.0027	0.1103	0.0001	0.0001	0.0001		
24	Spring - Preload	0.0264	1.5395		0.1077					
25	Bracket - Position Sensor	0.0007	1.4929	0.0099	0.1233					
26	Differential Sensor		1.4944	0.0100	0.1233					
27	Shim - Preload Spring	0.0121	1.5384		0.1077					
28	Spring_Guide	0.0050	1.5114	0.0026	0.0908					
29	Spring - Constant_Force - Opened	0.0003	1.5089	-0.0061	0.0908					
30		0.0003	1.5303	-0.0223	-0.1163					
31	Washer_Flat	0.0001	1.5105	0.0337	0.1081					
32	Washer - Special	0.0010	1.5105	0.0327	0.1087					
33	Washer_Flat	0.0004	1.5410		0.1077					
34	Washer - Special	0.0048	1.5389		0.1077					

FIGURE 10.52
I-DEAS generates mass properties data that can be exported to an ASCII file for printing or to a tab-delimited file for import into a spreadsheet such as Microsoft Excel. This Excel screen capture shows the raw I-DEAS data after being brought into Excel. The assembly's properties are presented at the top, followed by the mass properties for each part in the assembly.

properties (unless the Generate a Mass Properties Report command is run from I-DEAS again). To facilitate updating the data, Adams links the data to a spreadsheet where formulas have been set up for summing and combining the various parts into the summary data for the assembly.

For most projects, "hardware" such as nuts, bolts, and wire harnesses are not modeled, so their mass must be added to the data generated from the model. Paint and bonding materials are also too difficult to model, and are added by the analyst. Adams adds these components to the list of parts for the given subsystem or assembly in the spreadsheet, so they are rolled up into the overall mass properties calculations (see Fig. 10.53).

One exception to this rule was the design of a spectrometer on a telescope that rotates to follow the sun. Measurements made by this spectrometer (looking for ozone holes) depend on its scanning sunlight through the atmosphere. Because the instrument rotates, it had to be closely balanced to ensure that other instruments on the satellite, to which this instrument is attached, are not disturbed. Everything on the instrument was modeled in I-DEAS and the model was used to estimate the size of the shims needed to balance the instrument. The estimates were so close that the large shim needed 1/8 inch and the small shim needed only 1 mil of material removed to pass the requirement. 90% of the mass properties calculations were derived from the model, not from weighing actual flight hardware (parts).

Adams uses an Excel workbook to manage several spreadsheets—one for each of the subsystems—that help him build a complete analysis, update it easily, and report it in various ways. For example, the design engineers want to see data for all of the parts, but the systems engineers want only summary data at the system level (see Fig. 10.54). The customer may get summary data for the top-level assembly only.

On a monthly basis, Adams references the top assembly from the library on the network and generates new data for any parts that have changed. (The assembly model for the SIRTF, a cryogenic infrared telescope, is shown in Fig. 10.55.) I-DEAS has a "suppress" option that allows him to skip unchanged parts when generating mass properties data, so his exported data includes only those parts that need to be updated. He replaces the old values in the cells of raw data for the parts, and they are picked up by the formulas in the summary spreadsheet.

Part of the Excel workbook is a units worksheet that was created to globally convert units in the mass properties spreadsheet data. Most design engineers at Ball Aerospace work in English units, but the data Adams prepares for the customer must be in SI (metric) units. This worksheet also helps to convert data to more convenient units. For example, the values that result from moments of inertia calculations become very large when stated as lb * in^2, so the units spreadsheet converts them to slugs (1 slug = 14.6 kilograms). Part of Adams's job is ensuring that the

(continued on page 416)

FIGURE 10.53

The formulas in this sheet use the raw I-DEAS data to predict the mass properties for the final assembly. The drawing number distinguishes items designed for the project from standard parts. The material listed for each part is used to calculate the predicted mass. Formulas in the spreadsheet compile the properties data from each of the parts in the assembly and update the assembly properties at the top of the sheet.

Microsoft Excel - sirtf_98_06.xls

File Edit View Insert Format Tools Data Window Help

G79 =

SPACE INFRARED TELESCOPE FACILITY

Cryogenic Telescope Assembly

Projected Mass Properties

Scott Adams

Row	Item	Dwg #	Material	Predicted Mass (kg)	Center-of-Gravity (meters) X	Y	Z	Predicted Local Moment-of-Inertia (kg-m^2) Ixx	Iyy	Izz	Predicted Products-of-Inertia (kg-m^2) Ixy	Iyz	Ixz
11	MULTIPLE INSTRUMENT CHAMBER (MIC)												
12	**Multiple Instrument Chamber**	**545009-500**		**17.548**	**1.344**	**0.095**	**-0.013**	**1.69**	**0.89**	**1.10**	**-0.10**	**-0.06**	**0.03**
13	**Baseplate Assy, MIC**	**545270-500**		**10.861**	**1.285**	**0.156**	**-0.028**	**0.93**	**0.41**	**0.53**	**0.01**	**-0.03**	**0.00**
14													
15	**Cover/Shutter Blade Assy**	**545296-500**		**6.687**	**1.441**	**-0.004**	**0.012**	**0.65**	**0.37**	**0.37**	**0.00**	**0.00**	**0.01**
16	Cover, MIC	545297-001	Aluminum	4.477	1.409	0.000	0.000	0.59	0.33	0.33	0.00	0.00	0.00
17	**Shutter Blade Assy - Photon Shutt**	**545022-500**		**2.210**	**1.507**	**-0.011**	**0.035**	**0.05**	**0.03**	**0.03**	**0.00**	**0.00**	**0.00**
22	Plate, Mounting	545412-500	Al 6061-T651	1.544	1.505	-0.002	0.007	0.05	0.02	0.02	0.00	0.00	0.00
23	Housing	545023-500	Cres A286	0.221	1.506	0.003	0.110	0.00	0.00	0.00	0.00	0.00	0.00
24	Spring, Preload	545034-001	Cres A286	0.030	1.540	0.000	0.108	0.00	0.00	0.00	0.00	0.00	0.00
25	Bracket, Position Sensor	545040-001	Vespel	0.001	1.493	0.010	0.123	0.00	0.00	0.00	0.00	0.00	0.00
26	Differential Sensors [3]	KD5100-004	Spec	0.000	1.494	0.010	0.123	0.00	0.00	0.00	0.00	0.00	0.00
27	Shim, Preload Spring	545035-001	SS 15-5PH	0.014	1.539	0.000	0.108	0.00	0.00	0.00	0.00	0.00	0.00
28	Guide, Spring	545042-001	Al 6061-T651	0.006	1.511	0.003	0.091	0.00	0.00	0.00	0.00	0.00	0.00
29	Spring, Constant Force - Open (H4D12)	545033-001	Inconel 718	0.000	1.509	-0.006	0.091	0.00	0.00	0.00	0.00	0.00	0.00
30	Snubber	545041-001	Vespel	0.000	1.530	-0.022	-0.116	0.00	0.00	0.00	0.00	0.00	0.00
31	Washers, Flat [3]	NAS620C4	Cres A286	0.000	1.511	0.034	0.108	0.00	0.00	0.00	0.00	0.00	0.00
32	Screws, Cap Socket Head [3]	NAS1352N02-4	Cres A286	0.001	1.511	0.033	0.109	0.00	0.00	0.00	0.00	0.00	0.00
33	Washers, Flat [6]	NAS620C4L	Cres A286	0.000	1.541	0.000	0.108	0.00	0.00	0.00	0.00	0.00	0.00
34	Screws, Cap Socket Head [6]	NAS1352N04-8	Cres A286	0.005	1.539	0.000	0.108	0.00	0.00	0.00	0.00	0.00	0.00
35	Shim, Latch Assembly	545026-001	SS 302	0.001	1.525	-0.008	0.138	0.00	0.00	0.00	0.00	0.00	0.00
36	**Blade Assembly - Open**	**545026-500**		**0.271**	**1.510**	**-0.079**	**0.085**	**0.00**	**0.00**	**0.00**	**0.00**	**0.00**	**0.00**
37	Shaft, Take-up Spool	545031-001	Al 6061-T6	0.001	1.512	-0.037	0.098	0.00	0.00	0.00	0.00	0.00	0.00

CTA Summary / Structures Summary / Telescope / Cryostat / Helium / MIC / Outer Shell Group / Dust

Ready

INDUSTRY CASE

Microsoft Excel - sirtf_98_06.xls

SPACE INFRARED TELESCOPE FACILITY
Cryogenic Telescope Assembly
Projected Mass Properties - Summary with Cold Instruments
Scott Adams

Subsystems	Predicted Mass (kg)	Center-of-Gravity (meters) X	Y	Z	Predicted Local Moment-of-Inertia (kg-m^2) Ixx	Iyy	Izz	Predicted Products-of-Inertia (kg-m^2) Ixy	Iyz	Ixz
DELIVERY - DRY, STOWED	**361.714**	**1.244**	**0.007**	**0.009**	**65.46**	**219.00**	**215.07**	**0.59**	**-0.12**	**6.55**
TELESCOPE	54.405	2.021	0.000	0.006	5.76	18.04	18.01	0.00	-0.02	0.21
CRYOSTAT	118.521	0.969	-0.006	-0.006	18.14	29.74	29.46	0.62	0.38	0.81
MIC w/ INSTURMENTS	70.332	1.310	0.038	-0.014	4.06	1.79	2.01	-0.04	-0.30	0.18
OUTER SHELL GROUP	113.644	1.044	0.005	0.040	36.53	108.17	104.55	-0.28	-0.15	5.78
DUST COVER - EJECTABLE	4.812	2.947	0.000	-0.001	0.72	0.46	0.32	0.00	0.00	0.11
LAUNCH - WET, STOWED	**412.122**	**1.191**	**0.006**	**0.008**	**69.54**	**231.35**	**227.42**	**0.72**	**-0.12**	**6.72**
TELESCOPE	54.405	2.021	0.000	0.006	5.76	18.04	18.01	0.00	-0.02	0.21
CRYOSTAT	118.521	0.969	-0.006	-0.006	18.14	29.74	29.46	0.62	0.38	0.81
HELIUM	50.408	0.815	0.000	0.000	4.07	4.22	4.22	0.00	0.00	0.00
MIC w/ INSTURMENTS	70.332	1.310	0.038	-0.014	4.06	1.79	2.01	-0.04	-0.30	0.18
OUTER SHELL GROUP	113.644	1.044	0.005	0.040	36.53	108.17	104.55	-0.28	-0.15	5.78
DUST COVER - EJECTABLE	4.812	2.947	0.000	-0.001	0.72	0.46	0.32	0.00	0.00	0.11

Principle Axes and Moments of Inertia

	(kg-m^2)		Degrees	Degrees	Degrees
AXIS NO.1 MOI	231.4	located	90.35 from x-axis	2.36 from y-axis	87.66 from z-axis
AXIS NO.2 MOI	69.3	located	2.44 from x-axis	89.75 from y-axis	87.57 from z-axis
AXIS NO.3 MOI	227.7	located	92.42 from x-axis	92.35 from y-axis	3.37 from z-axis

CTA Summary / Structures Summary / Telescope / Cryostat / Helium / M I C / Outer Shell Group / Dust

Ready

FIGURE 10.54
An Excel workbook can contain many worksheets and formulas in any worksheet can reference information from all of them. The tabs at the bottom of the window represent different subsystems and summary sheets. This summary of mass properties for the entire cryogenic telescope assembly is one of the sheets in the workbook.

data generated from the models is in the correct units, that materials have been applied correctly, and that the density values for the materials applied use the value for mass density, not the one for weight density. Because the mass of an object is the same everywhere, but the weight of an object depends on the gravitational system, this is a critical difference. If weight densities were used, the mass properties values for the device would be different once it was in orbit.

A critical issue for Adams is the coordinate system referenced by the mass properties data. In order to roll up the data, subsystems and assemblies must share a global coordinate system with the other parts of the mission. For example, when Ball Aerospace designs a science instrument that will be integrated into a spacecraft that will in turn be carried into space by a launch vehicle, the mass properties of the science instrument must be rolled up into the mass properties for the spacecraft and then up to the launch vehicle. In the same way that the parts of an assembly are often defined by a skeleton drawing, the location of the various components on the global coordinate system is specified by the customer. This common set of coordinate axes allows the mass properties data to be combined

Skills Summary

Now that you have completed this chapter, you should begin practicing using your CAD models for engineering analysis. Exercises 1–6 allow you to practice your skills in extracting mass properties data from your CAD models and using them to solve simple engineering problems. Exercises 7–15 ask you to evaluate the accuracy of mass properties calculations based on results and algorithms presented in the problems. You should also be familiar with exporting CAD data to various packages and selecting the best file format for the export. Exercises 16–17 test your ability to select the appropriate export format and address its limitations. Exercises 18–19 ask you to describe how various rapid protoyping systems work to create physical models from CAD data.

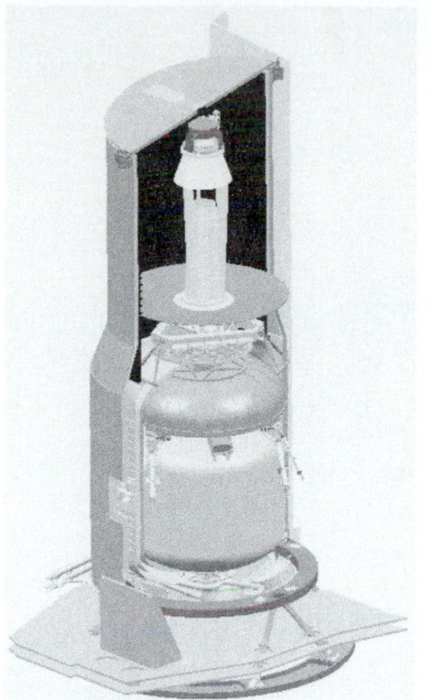

FIGURE 10.55
The Space Infrared Telescope Facility (SIRTF) model from which the mass properties data was drawn.

correctly. Adams has encouraged design engineers to create the global coordinate system as a part in the assembly and use it to define their models. This visible reminder ensures that all models relate correctly to the coordinate system defined by the launch vehicle.

Once the project is complete, Adams publishes the data and writes a report in Microsoft Word. The customer report presents summary data only; the internal report documents the contents of the complete spreadsheet. One hard copy of the report is signed by Adams and the lead systems engineer and archived; an electronic copy is published on the internal company network.

Two years ago, Ball Aerospace was using proprietary FORTRAN programs to do its mass properties calculations. Today, Adams can envision a switch to a dedicated relational database program, such as dbMASS, which would do all of the calculations he does now with the spreadsheet, but handle more of the work of updating parts from the I-DEAS model.

Although automating its process for calculating mass properties has saved Ball Aerospace a large percentage of the time it previously took to do the calculations, determining accuracy and checking the calculations are still a very important part of Adams's job.

All figures courtesy of Ball Aerospace & Technologies Corporation.

Scott Adams is now a mechanical design engineer for Ball Aerospace & Technologies Corporation in Boulder, Colorado. In his previous position as analyst, he handled mass properties analysis of the projects from NASA and other aerospace customers. Adams earned a mechanical engineering degree at Montana State University in 1994. He is familiar with several CAD packages, including I-DEAS Master Series, Pro/Engineer, and AutoCAD.

EXERCISES

Exercises 1–6: Mass Properties Evaluation. Use your solids modeler to determine mass of the following objects. Define a datum, then locate and dimension the centroid. Find the moment of inertia and radii of gyration about a defined major axis for each part.

Assume all objects are milled from 6061-T6 aluminum, with specific gravity of 0.28. Small fillets or chamfers do not significantly change your solution and can be neglected in the model, but don't neglect important machining details such as the internal radii formed by the 0.25-or 0.50-inch diameter cutters!

Dimensions shown are in inches.

1. Simple rectangular block.

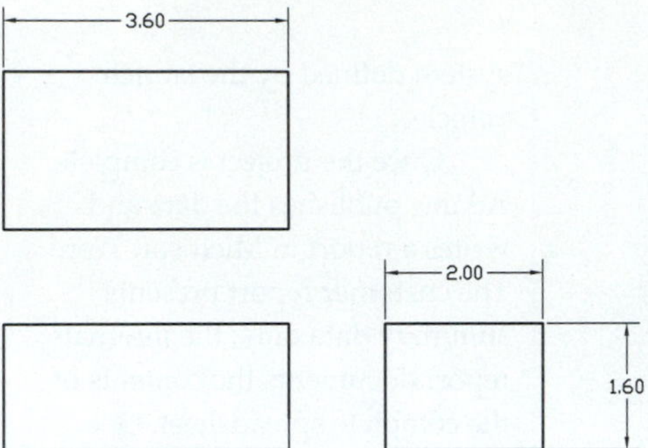

2. Rectangular block with internal recess.

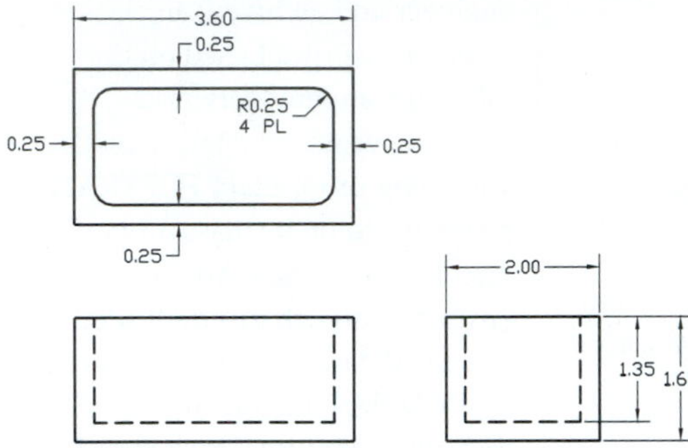

3. Rectangular block with split internal recess.

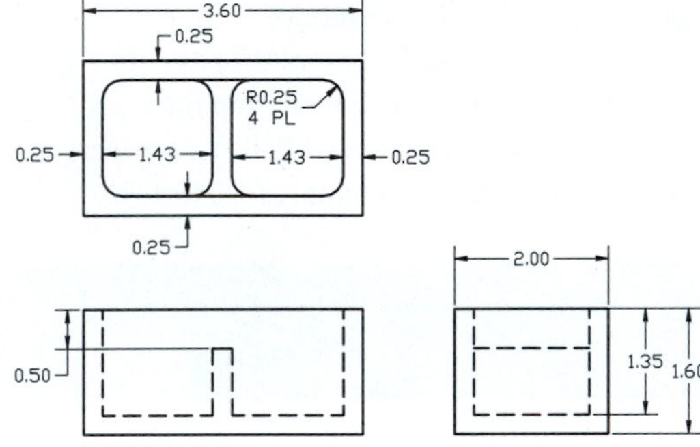

4. Asymmetrically milled block.

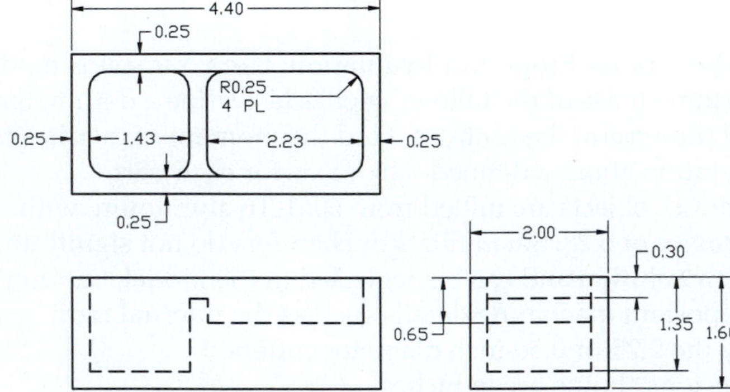

5. Slotted tube.

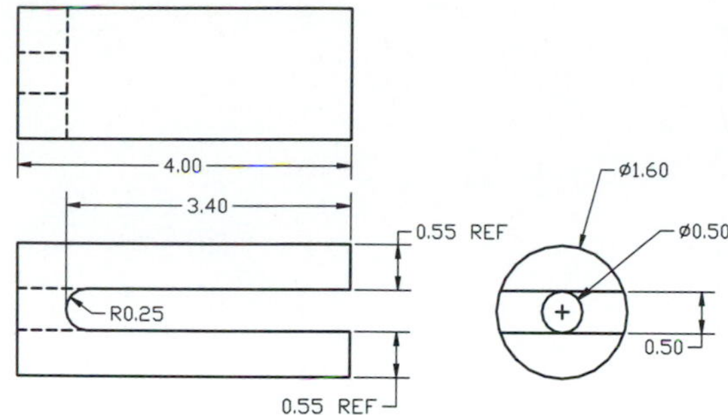

6. Slotted hexagonal hollow rod.

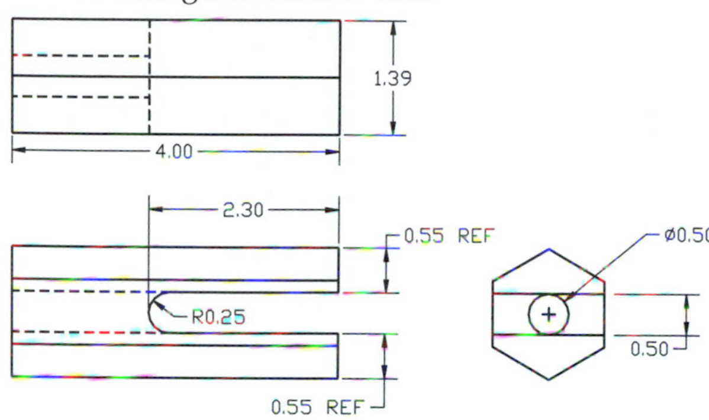

Exercises 7–10: Create a table in Excel or other spreadsheet program comparing the mass properties as calculated by your CAD modeler with hand-calculated exact values, for the following figures. Be sure to define a common datum for each figure, so that proper comparisons can be made. Include a column showing percent difference from the exact value. How does part-axis orientation affect your accuracy?

7.

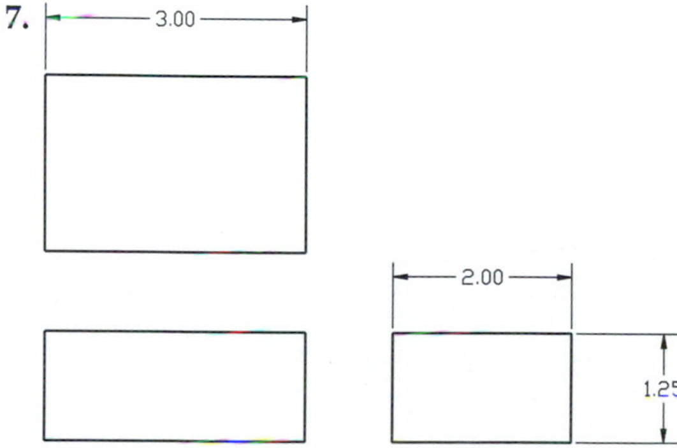

8.

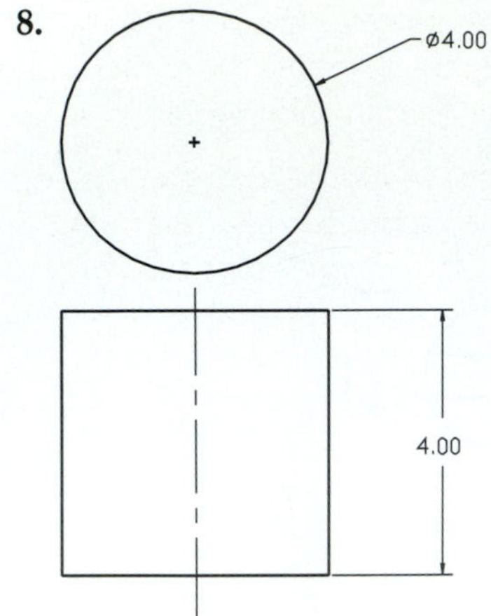

9.

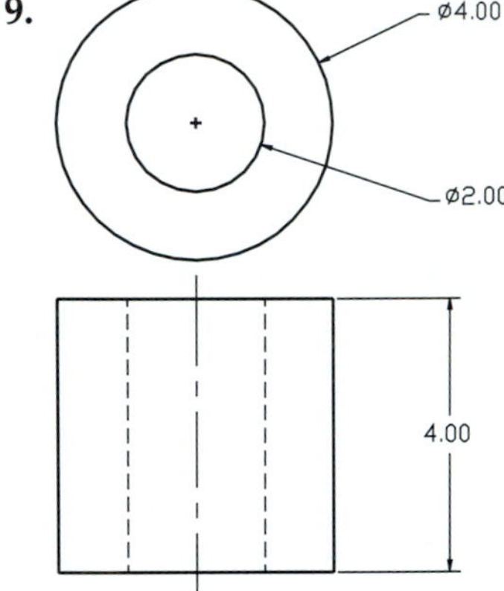

10.

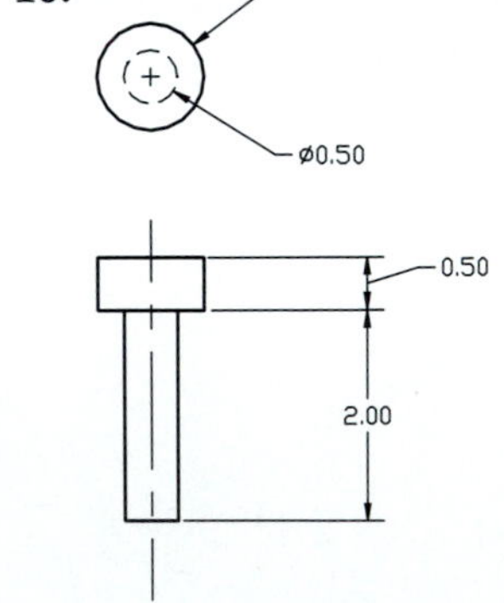

Exercises 11–14: Use the mass properties data for the models in Exercises 7–10 to create a spreadsheet that computes the material, cost, and labor needed to coat the surface area of the parts.

15. Material changes are common in the development phase of engineered parts. Consider a component part modeled first using 6061-T6 Aluminum, and then with 1020 Steel. No dimensional changes are made. Do each of the following properties of a solid change with changing material? How?

  a. Volume

  b. Mass

  c. Density

  d. Centroid

  e. Moment of inertia

  f. Radius of gyration

16. Vector data and raster data are both used in graphics programs. Which type of data format is better suited for CAD systems that have mass properties capability?

17. Define the steps involved in creating a stereolithographic rapid prototype part from a solid model. What are some of the considerations when deciding whether a CAD model is a good candidate for prototyping using stereolithography?

18. Give a brief explanation of the physical process used to create rapid prototype parts by each of the following methods. What are the advantages and disadvantages of each method?

  a. Stereolithography

  b. Fused deposition modeling

  c. Selective laser sintering

  d. 3D printing

19. How does "Rapid Tooling" differ from "Rapid Prototyping"? In what situations might either method be used to create actual working components rather than just shape-representative models?

DESIGNER'S NOTEBOOK

Notice item #4 "Met with casting vendor today" in the notes on this page from Albert Brown's notebook. Engineers need to interact knowledgeably with manufacturers who will produce the parts for their designs. Many manufacturing processes have relatively long lead times and must be coordinated before the design becomes final. Manufacturers also have valuable information about designing to accommodate their production process. (Courtesy of Albert W. Brown, Jr., Affymax Research Institute.)

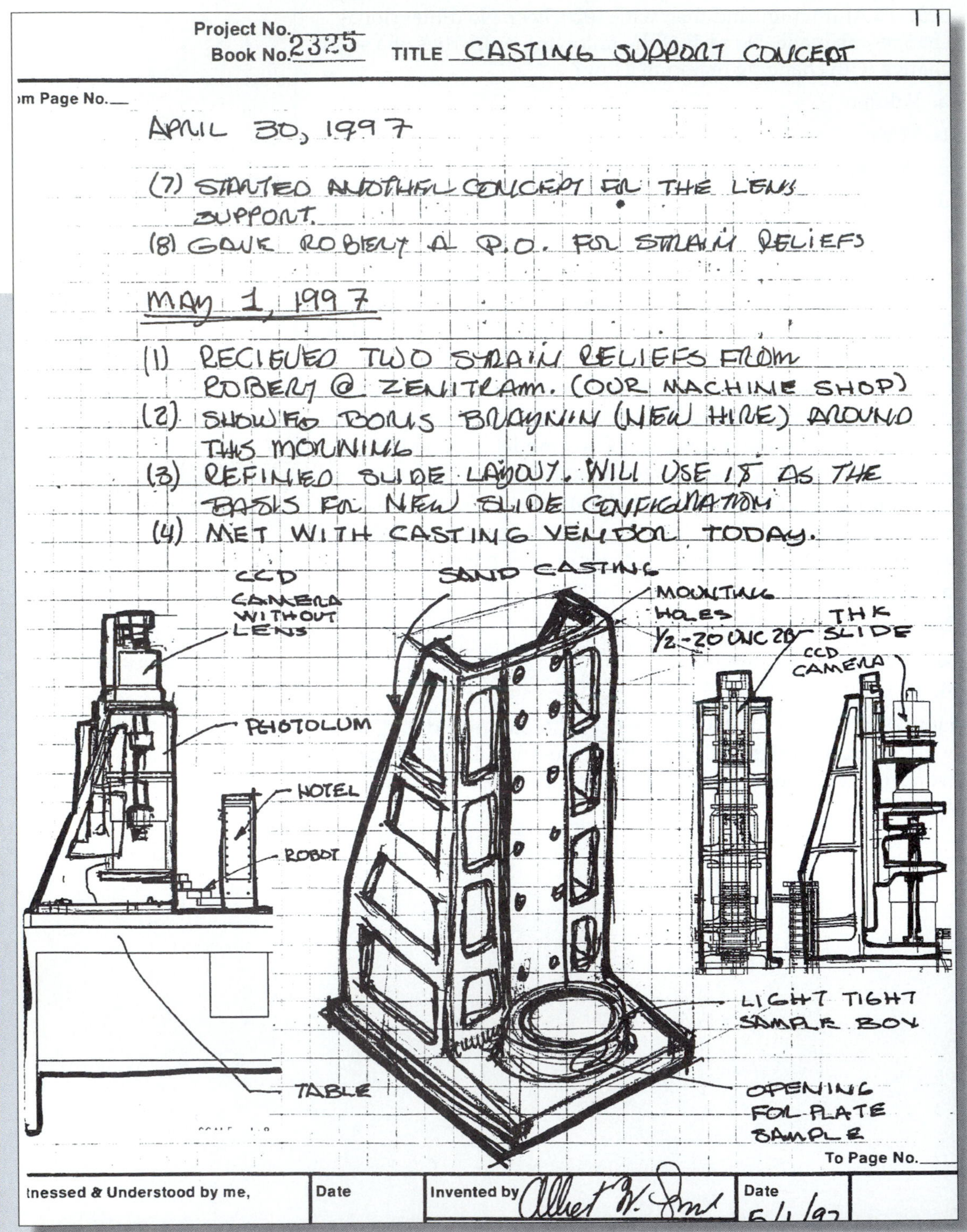

Project No.
Book No. 2325 TITLE CASTING SUPPORT CONCEPT

om Page No.

APRIL 30, 1997

(7) STARTED ANOTHER CONCEPT FOR THE LENS SUPPORT.
(8) GAVE ROBERT A P.O. FOR STRAIN RELIEFS

MAY 1, 1997

(1) RECIEVED TWO STRAIN RELIEFS FROM ROBERT @ ZENITRAM. (OUR MACHINE SHOP)
(2) SHOWED BORIS BRAYNIN (NEW HIRE) AROUND THIS MORNING
(3) REFINED SLIDE LAYOUT. WILL USE 18 AS THE BASIS FOR NEW SLIDE CONFIGURATION
(4) MET WITH CASTING VENDOR TODAY.

To Page No.

tnessed & Understood by me,	Date	Invented by Albert W. Brown	Date 5/1/97

CHAPTER 11

Implementation

Objectives

When you have completed this chapter, you will be able to

- describe the link between implementation and documentation;
- list the steps in common manufacturing processes;
- describe how CAD can be used to generate NC machine code;
- identify manufacturing-related design issues.

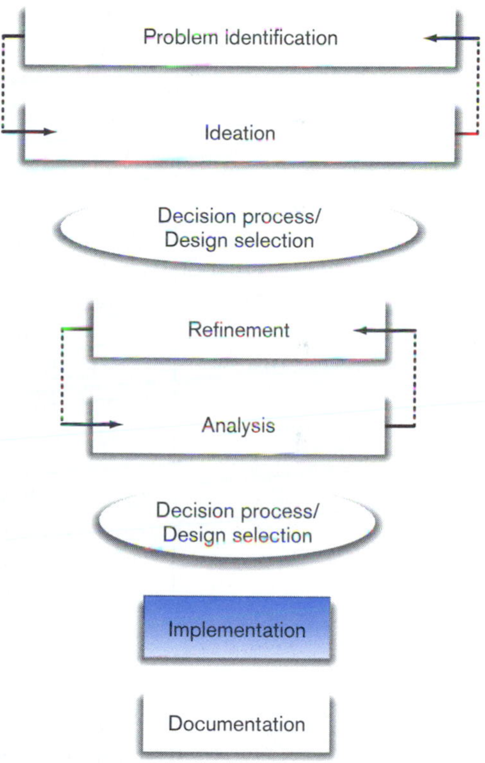

The ultimate goal of engineering design is to manufacture a product or system. The process of designing does not end with the development of a 3D CAD model: it includes considerations for the life cycle of the device—from its manufacture, assembly, and packaging, through its repair and eventual disposal. The implementation phase of the design process is the transition from sketches, drawings, or models of the design to specifications for making the commercial product or system.

An understanding of basic manufacturing processes will help you create the computer files needed to create the product directly from the CAD database and/or to provide all of the necessary information to manufacturing personnel on a documentation drawing. (You will learn more about documenting your designs in the next chapter.) During the implementation phase, decisions are made about the processes that will be used to make the parts. The choice of one process over another can require modifications to the design, and each process has characteristics that must be reflected in your drawing or model.

For example, if you decide that a part will be molded or cast, draft must be added to the surfaces so the part can be removed from the mold. You need to specify a range for the draft angle that will still allow the parts to function. Machining and cutting processes use tools of various shapes and sizes to form parts from raw material. The size and shape of your part will imply a certain combination of cutting tools that can make it harder or easier to manufacture. Using standard sizes, you may be able to save time and money without affecting the quality of your design.

Each process also has an expected range of accuracy. You must specify the *tolerance,* or the total amount of variation that an acceptable part can have from the stated dimensions. If you specify a very precise tolerance range for a part, certain types of processes may not be capable of creating the part to that specification. Conversely, understanding the expected accuracies of a process will allow you to determine whether the accuracy is acceptable for your part, or whether you need to choose a different process.

All of the decisions made during implementation are reflected in the documentation you will create for the design, but that doesn't mean you have to know everything about manufacturing. Today's concurrent engineering process emphasizes the early involvement of suppliers and manufacturing personnel who will work with you to anticipate implementation issues throughout the design process. This is in contrast to the "over the wall" process used in the past: Marketing tossed a product idea to engineering, who designed the parts without much discussion with the people who would manufacture the parts. When the design was complete, the engineers tossed it over the wall to manufacturing, who made choices about processes and costs. "Throwing it over the wall" described the lack of communication between the different groups. Problems in manufacturing the parts as designed were not discovered until late in the design process, when it is costly to make major design changes. Concurrent engineering brings manufacturing considerations into

the design process earlier to get products to market sooner and produce them cost effectively.

The advent of numerically controlled manufacturing processes and the development of plastic, ceramic, and composite materials, as well as a wide variety of metals, provide a much broader choice of materials and processing methods than were available fifty years ago. The great variety of materials and processes require more, not less, communication with those who will make the parts. Often the manufacturing specialist has a wealth of knowledge combined with practical experience about particular materials and their processing, the tolerances that can be expected from the process, lead times, and the effect of manufacturing methods and tolerances on the part cost. This information can help you select a material and specify it correctly.

Engineering colleagues are another source of implementation information. There is a wide variety in the methods used for huge steel or concrete structures and those used for miniaturized parts, with every range in between (see Fig. 11.1). Experienced designers usually have practical experience with the way products in a particular area are manufactured, and they bring this experience to bear in designing similar new products. Product research and reviewing standards in a particular area are necessary parts of the design process.

To design manufacturable products and implement the designs well, you need to understand the basics of manufacturing methods and how to communicate the information necessary for manufacture of your design. Classes in material science and manufacturing processes will provide more specific information, but a basic understanding will empower you to ask the right questions and interpret the information you receive. Because new methods and materials are continually being developed, you should plan to be a lifelong learner in the field of engineering, updating yourself on new technologies and processes as they develop. In this chapter, you will learn about the most common manufacturing processes.

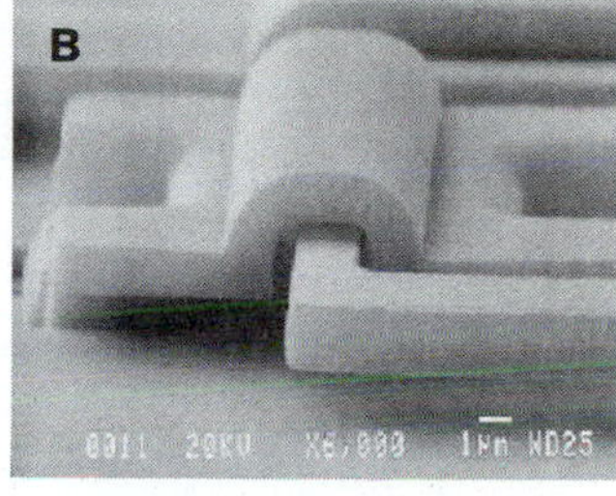

FIGURE 11.1
Manufacturing the heavy equipment trailer rated to carry 10 tons, shown in (a), uses different manufacturing methods from those used to produce the micromechanical hinge shown in (b), which is measured in microns. (Courtesy of Smith Equipment and MEMSCAP.)

Manufacturing Processes Overview

A part usually starts with raw stock or material. Figure 11.2 shows a rack of the kind of stock material you would find in a machine shop.

FIGURE 11.2
Machined parts typically start from standard stock shapes and sizes. The cost of a part is often reduced if it is sized close to a standard stock size or if portions of it can be left as rough stock.

There are three basic ways that raw material can be changed into a finished product through manufacturing:

- by removing material
- by joining or adding material
- by redistributing material

Milling, drilling, turning (on a lathe), laser cutting, chemical milling, and punching are examples of processes that remove material to form a part. Welding, soldering, plating, coating, and assembly are examples of processes that add material to form a part. Molding, pressing, forging, extruding, and spinning are examples of processes that redistribute material to form a part. These various operations are frequently used in combination.

Today's technologies make it seem possible to manufacture almost anything, but most parts are made through a combination of and variations on the basic processes listed here. The ability to envision the steps needed to make a part can help you design parts more effectively.

Manufacturing Processes

Processes That Remove Material

Milling is a type of machining process. It uses rotating cutters to remove material to form the part. The raw material is anchored to a table that moves in any X-Y-Z direction to position the material for the cutting operation. Contemporary milling machines may also tilt the material to allow for a wider range of cuts.

Turning is the term applied to machining performed with a lathe. A lathe holds the raw material in a clamp (called a chuck) and rotates it. As it rotates, a cutting blade is moved into it to remove material from its circumference. Turning is used to add features to cylindrical stock, such as a chamfer or neck on a rod.

Drilling is another machining operation used to create holes. Different drill bits are used to create holes of different shapes and sizes. Circular features such as spot faces, counterbores, and countersinks are formed by drilling.

Boring is a machining operation similar to drilling, but it uses a more rigid boring tool in place of a drill bit. Boring is used to enlarge a drilled hole to more accurately achieve a specified size, and is also used for large holes. Boring forms holes that are round and straight.

Manufacturing Processes (cont.)

Reaming is another machining operation also used to enlarge a drilled or bored hole. Reaming creates a smoother finish on the inside of the hole than drilling or boring.

Broaching is an operation used to cut keyways and square slots. The broach tool has teethlike prongs that are smaller at the tip and increase in size along the shaft of the tool. The tool moves back and forth to shave material from the edges of the square hole. As the tool is forced through the workpiece, each succeeding stroke penetrates deeper and enlarges the opening.

Shaping (performed by a shaper) moves a cutting tool back and forth as the workpiece is moved into it to gradually remove material. Shaping may be used to form keyways, dovetails, and other shapes that cannot be formed by a rotating cutter.

Punching uses a rigid tool to create holes and other voids in thin metal sheets. The tool, having the same shape as the desired opening, is pressed through the material and the material inside the tool shape is removed.

Stamping is similar to punching, but the material outside the tool shape is removed (like a cookie cutter).

Laser cutting uses a laser to cut through the workpiece. The laser beam melts and evaporates portions of the material to form the cut.

Grinding removes small amounts of material, usually metal, with an abrasive grinding wheel that rotates at high speeds. Grinding is used to smooth rough edges.

Sanding is a surface finishing operation that moves an abrasive sheet over the part.

Lapping produces a smoother surface finish than sanding by sliding a fine abrasive material on a long flat band (a lap) across the surface.

Electric discharge machining (EDM) is a newer machining process that uses an electric current to erode the workpiece, often a mold. Current passing through an electrode (often formed by traditional machining methods) cuts the workpiece metal and leaves an impression in the shape of the electrode.

Electrochemical machining dissolves and removes material by inducing chemical reactions on the surface of the workpiece.

Processes That Add Material

Welding fuses two metal parts by heating them together at the joint. A gas flame, electric arc, or laser may be used to heat the joint.

Soldering joins two metals together with solder, a soft metal melted at the joint. Solder is made of tin and lead and melted by a hot soldering iron.

Brazing joins two metals with a harder metal than solder, often brass or zinc. The filler metal is melted to form the joint.

Plating uses an electrochemical process to coat a metal with another metal, such as chrome or nickel.

Coating bonds a layer of a material to a surface.

Processes That Redistribute Material

Forging hammers or presses hot metal into the desired shape.

Casting is used to form metal parts by pouring molten metal into a mold of the desired shape, then allowing the metal to cool and harden. The mold (often sand or ceramic) is generally broken to remove the part.

Molding refers to forming parts by pouring or forcing molten material into a cavity of the desired shape. The mold (often steel) may be more refined and reusable than molds used for castings.

Extrusion forms parts by forcing material through a shaped opening.

Spinning presses a smooth tool against a rotating piece of sheet metal to form the desired shape.

Sheet Metal Bending is a process that uses force applied by grippers to thin metal sheets to bend them into the correct shape.

Machining: A Manual Example

Today's practices for creating dimensioned drawings grew out of the need to communicate information about the design to the machine shop. In the early days of manufacturing, scaled drawings were given to the machinist who would make the parts. The machinist would scale from the drawing and use the sizes determined to manufacture the part. The machining processes were not highly accurate, and variations between the measurement on the print and the part often were not a huge concern because the machinist would build the entire assembly in one shop and work with the parts to make them fit to function as desired. Parts from one machine were not interchangeable with similar parts from another machine, as they were usually made to fit in the specific machine. For parts to be interchangeable, each one must be manufactured to within a certain range or tolerance so it will fit reliably with other parts. Dimensioned and toleranced drawings grew out of the need to communicate this information.

Understanding the way a part is made by hand can help you better understand the processes underlying many of today's automated methods. Most machine shop practices grew out of methods used to make parts manually. Early automated processes, in turn, first simply automated the manually controlled machining processes. To better understand machining methods, you will first see how traditional machining methods are combined to create a manually machined part for a one-of-a-kind device—a fixture for an impact tower to be used in materials testing.

Whether the part is a one-of-a-kind part that will be machined by an operator or a mass produced product, there are many similarities in the type of information that needs to be communicated to the manufacturer. Stepping through this example of a seemingly simple part will help you understand the process that raw stock goes through to be transformed into a finished product. When you go on to create documentation drawings in Chapter 12, you will understand the type of information that needs to be communicated through your drawing. Even if you will use direct manufacturing methods to go from CAD file to finished part, many issues in identifying critical dimensions and tolerances are the same.

The example will also illustrate some of the key differences between making a model and fashioning a part from actual materials. Creating an actual part involves a number of considerations that don't arise in modeling. For example, the part must be clamped or held in some fashion while it is being machined. How the part is clamped can make portions of the part difficult to reach with the machine tool. Modern machining can overcome some of the limitations of manual methods, but it is possible to easily model a part that is very difficult to machine (or manufacture by other methods). An understanding of the general process of machining can help you design and document parts more effectively. Many of the same issues of holding and inspecting the workpiece will pertain whether the part is manufactured using contemporary or traditional machining processes.

FIGURE 11.3
The fixture shown here is sitting inside the impact tower. The tower can be adjusted to drop a weight from a range of heights to impact samples of different materials. The new fixture will be designed and manufactured to hold the rectangular material sample resting on the standard testing fixture.

Designing the Impact Tower Fixture

The process of design begins with a need. In this case, a new fixture is needed to hold a specific composite material sample for impact testing. The impact tower shown in Figure 11.3 is used to test the strength of material samples by dropping a weight onto the sample from a determined height and measuring the results. The rectangular object in the figure is sitting on a standard fixture used in materials testing. The new fixture must hold this rectangular material sample.

Critical dimensions for the new fixture are measurements taken from the impact tower (where it must fit in place) and the size of the sample to be tested. The design of the new fixture was modeled in Pro/Engineer and is represented as a shaded model in Figure 11.4.

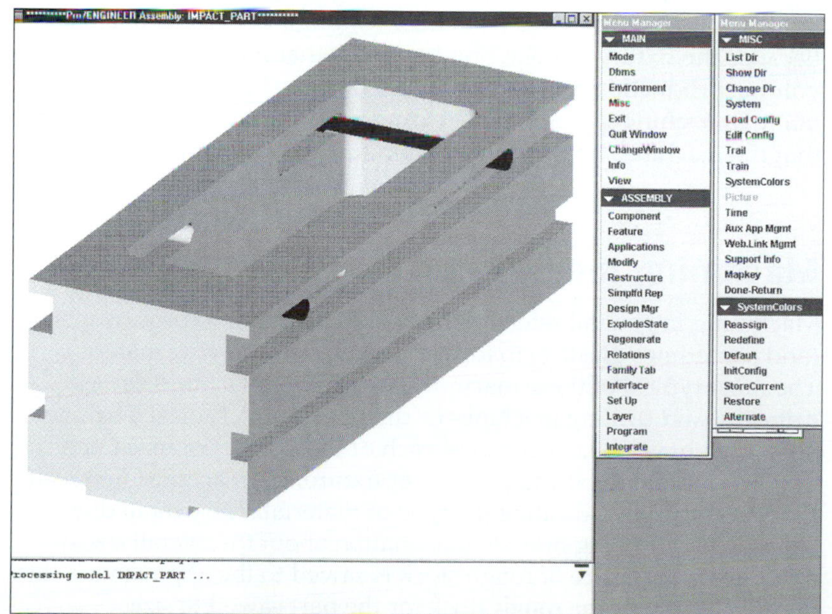

FIGURE 11.4
This shaded model shows the new fixture's design.

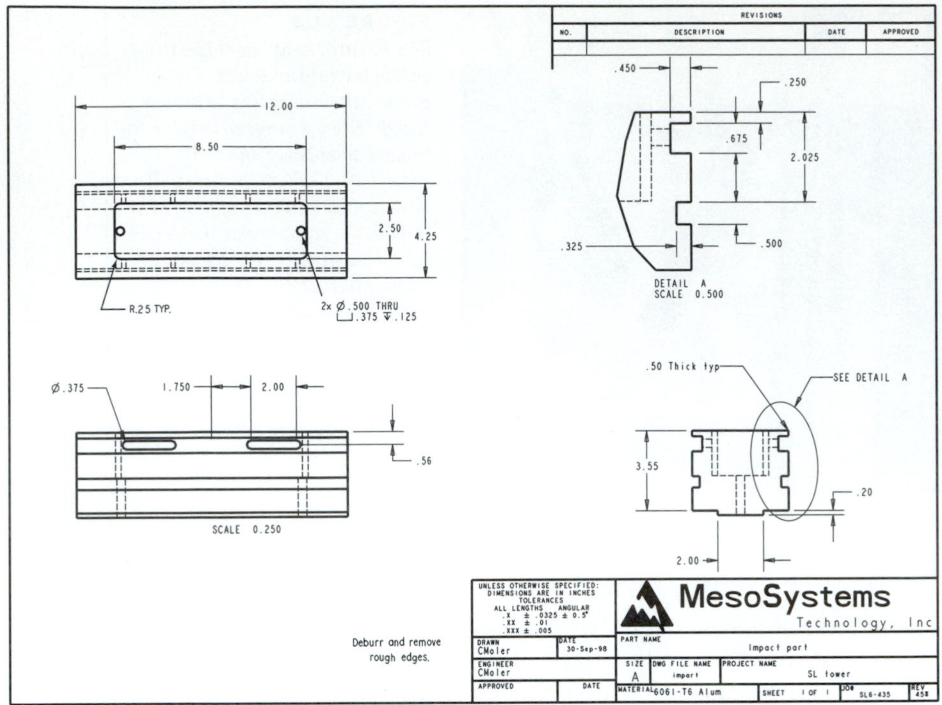

FIGURE 11.5 *Documentation drawings, such as this one of the fixture base, convey all of the necessary information for manufacturing the part.*

After a model is created, the next step is to prepare engineering drawings showing all of the information needed to manufacture this part. The drawing shows clearly the shape of the part, exactly what type of material will be used to create it, all of the necessary dimensions, any special material finishes or surface finishes, and the range of tolerance for the dimensions of an acceptable part. (You will learn more about specifying tolerance in your engineering drawings in Chapter 12.) Figure 11.5 shows the finished drawing of the impact fixture that was generated from the Pro/Engineer model database. Because the fixture is a one-of-a-kind part that will not go into volume production, the tolerances are those to be expected from manual machining methods. The drawing contains the information that the machine shop will work from to create the impact fixture.

Machining the Fixture

Machining in general refers to the use of machines to process metals (and sometimes plastics) to form a feature by removing material. The most typical of these machines are milling machines, lathes, grinders, and drilling machines or drill presses. A typical machine shop may have one to several of each of these types of machines.

To start making the impact tower fixture, the machinist first reads the drawing note indicating the type of material. The overall dimensions on the drawing provide information about the overall size of stock needed. A piece of rough stock is sawed to the approximate size. Figure 11.6 shows the rough stock for the part sawed to size.

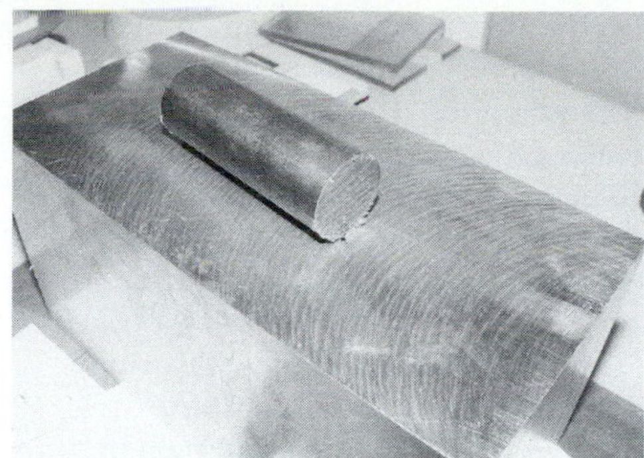

FIGURE 11.6
Rough stock is sawed to the approximate size to begin making the parts. Stock for the fixture base and one of the rollers are shown here.

FIGURE 11.7
Blue toolmaker's ink is sprayed on the surface of the part so that the features to be machined can be scribed on the part—almost as if a 2D drawing view of the part were drawn on the surface. Most modern machinery has numerical readouts and zero presets so the operator can locate these features directly without scribing lines on the part.

The rough stock is sprayed with blue toolmaker's ink so that features to be machined can be marked on its surface. Then a scribe is used to mark lines with the correct dimensions onto the rough piece as shown in Figure 11.7.

Milling Machine

A *milling machine* will be used to machine an accurate surface on the faces of the part. A milling machine removes material from the piece of rough stock, or workpiece, using a rotating cutter. Typically the part or stock is clamped to the milling machine's table and the table is moved into the cutter to remove material. The movements for the table are described by the X, Y, and Z axis directions, left to right called the X direction, forward and back called the Y direction, and up and down called the Z direction. Various types of cutters, such as ball endmills, square endmills, side cutters, angle cutters, and others, can be used in the milling machine to produce a wide variety of features. A disadvantage of milling is that there is a lot of waste material. As much as 50% of the raw stock is removed in the form of metal chips, which are typically wasted and not recycled.

The accuracy of manufacturing processes depends on many factors, such as the material being used, the tool, how well the machinery is maintained, and the operator. Most manually operated milling machines are easily capable of accuracies as small as one-thousandth of an inch, and even to half a thousandth of an inch (0.001 inch to 0.0005 inch or 0.25 to 0.10 millimeter).

For the impact tower fixture, the milling process first will produce an accurate surface on the workpiece from which the slots that will be machined next can be located. A facing mill is used because it removes material along the length of the cutter to create an accurate flat surface. Figure 11.8 shows the facing mill being used on the rough stock.

FIGURE 11.8
A facing mill as long as the part is deep is used here to machine the overall dimensions of the part to size.

In the figure, the workpiece is clamped to the movable table that feeds it into the cutter. The operator is machining the rough part down to the required overall dimensions, stopping periodically to measure the part.

FIGURE 11.9
The rough edges are removed from the part using a grinder.

After milling, the part is inspected again to ensure that the final dimensions of the part are within the stated tolerance listed on the drawing. As per a note on the drawing, all of the rough edges are removed so that the part will be easier to handle. Figure 11.9 shows the operator removing the rough edges from the part.

The part is sprayed with toolmaker's ink again so the interior features such as the slots and the pocket can be located. The slots are located by measuring from the specified finished surfaces. They are machined using an endmill like the one shown in Figure 11.10. Endmills are so called because they remove material with cutting edges that are located at the end of the tool.

An electronic measurement system is used to read the overall part dimension to within 0.001 inch (one-thousandth of an inch). Since the interior pocket is dimensioned from the center line of the part, the actual center of the part is calculated and new lines are scribed that locate the pocket and the counterbores (that will let a bolt head sit below the surface of the part). Figure 11.11 shows the measurement system being used to check the finished dimensions. Surface plates of granite or aged cast iron are frequently used as the base for making measurements because they are a very stable material and their surfaces can be made flat to a high accuracy.

FIGURE 11.10
An endmill removes material from the rotating end of the tool.

The two holes and the counterbored areas for the bolt heads will be machined first, while there is still a solid block of material. If the pocket is created first, the relatively small thickness of the wall where the counterbore will be located could cause problems. Figures 11.12 through 11.15 show steps in creating the holes, counterbores, and pocket.

Another common machine tool that could have been used to create the holes and counterbores for the impact tower is a drill press.

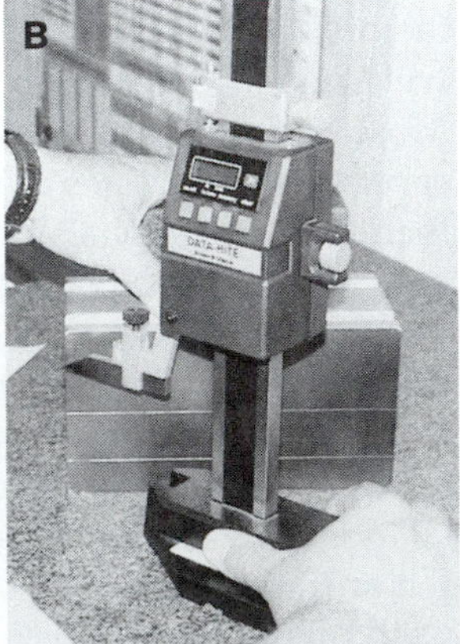

FIGURE 11.11
Digital measuring systems can aid in locating features accurately. This measuring device not only measures the part (a) but can also be used to scribe lines in the toolmaker's ink at the desired location (b).

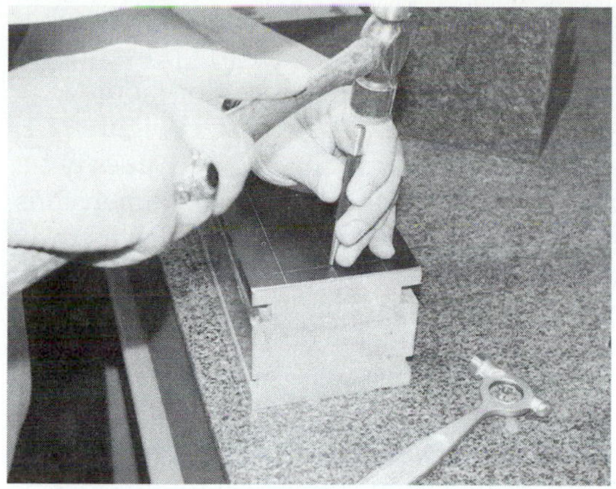

FIGURE 11.12
A center punch is used to create small depressions that will make it easier to start drilling the holes.

FIGURE 11.13
For this part, the through holes are drilled using a milling machine. This allows the holes and the pocket to be formed without having to move the part to a drill press, then return to the milling machine and clamp the part down a second time.

FIGURE 11.14
The counterbore is created next using a special tool that locates the counterbore so that it is concentric to the hole.

FIGURE 11.15
Finally, the pocket in the fixture is milled.

Drill Press

A *drill press* is one of the most commonly used machine tools. It is used to drill, tap, spot face, counterbore, countersink, ream, and bore holes. Drill presses, such as the one shown in Figure 11.16, have a rotating spindle that holds the tool. The spindle is forced down against the stationary workpiece. The tolerance for the size of features created by typical drilling operations is from several thousandths of an inch up to a couple of hundredths of an inch. The tolerance for the location can be from one or two-thousandths of an inch up to several hundredths depending on the operator of a manual machine.

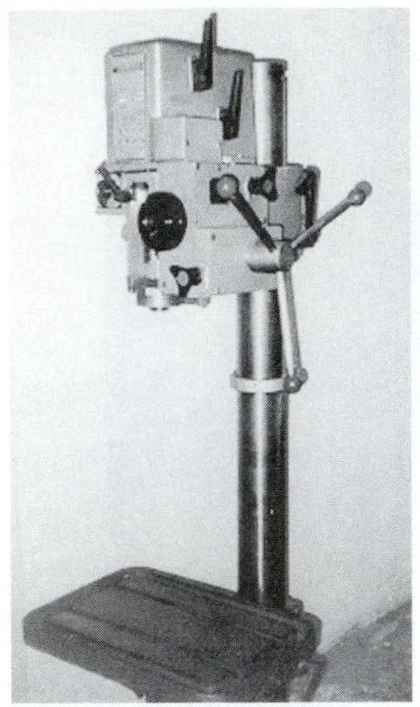

FIGURE 11.16
Drill press

The accuracy of the hole size and location depends on a number of factors, such as the condition of the drill bit and drill press, the diameter of the hole being drilled, and other factors. For more accurate holes, reamers and boring bars such as those in Figure 11.17 can be used in place of a drill bit. In general, the engineer does not need to state which machine tools will be used, only the dimensions and tolerances. The machine operator will determine which methods to use to achieve that tolerance while still providing the lowest cost part.

Lathe

Two rollers that will fit inside the pocket of the impact tower fixture and move along the slots are created from cylindrical rough stock and turned down to size using a lathe. A *lathe* holds the workpiece in a *chuck* (a kind of clamp) that rotates, spinning the piece so that a cutting tool can be forced against it to remove material. Figure 11.18 shows the operator tightening the piece of rough stock for the rollers into the chuck on a lathe, also shown in Figure 11.19.

After the rollers are formed, the finished parts are inspected to determine whether they are inside the specified tolerances that are listed on the drawing. After assembly, the fixture is ready for use in the impact tower. Figure 11.20 shows the finished parts.

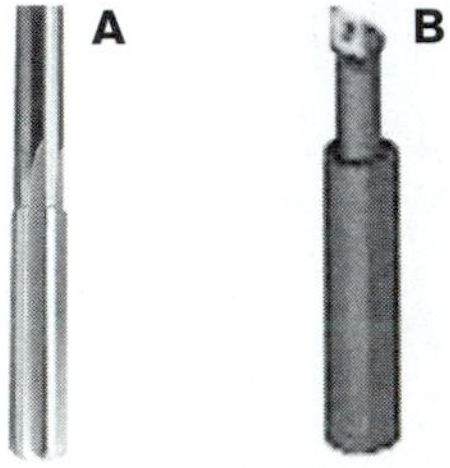

FIGURE 11.17
The reamer (a) and boring bar (b) shown here create more accurately finished holes than can be achieved with a drill bit. (Courtesy of Reid Tool Supply Company and Triad Tooling.)

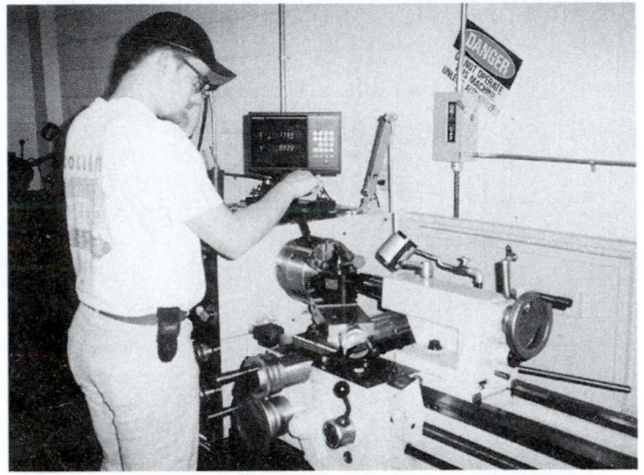

FIGURE 11.18
John Johnson tightens the chuck holding the cylindrical stock which has been cut to rough length.

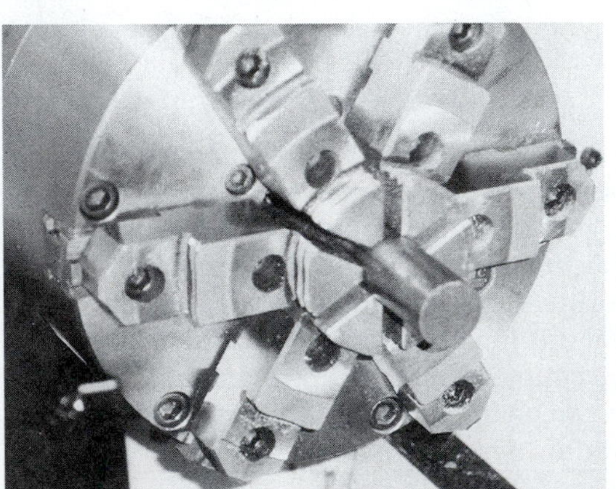

FIGURE 11.19
The chuck tightens around the workpiece to hold it in the lathe.

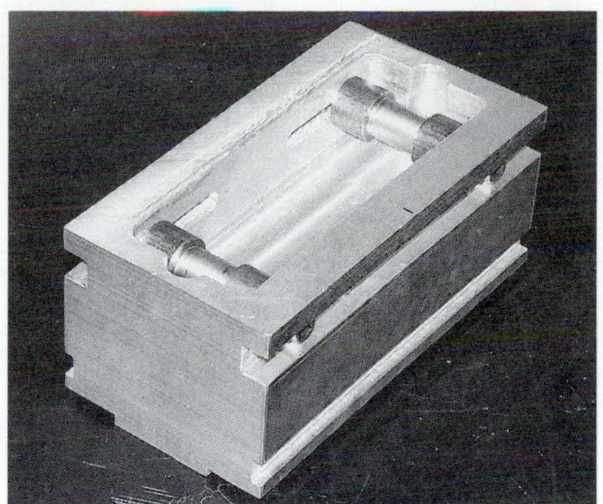

FIGURE 11.20
The finished parts for the impact tower fixture

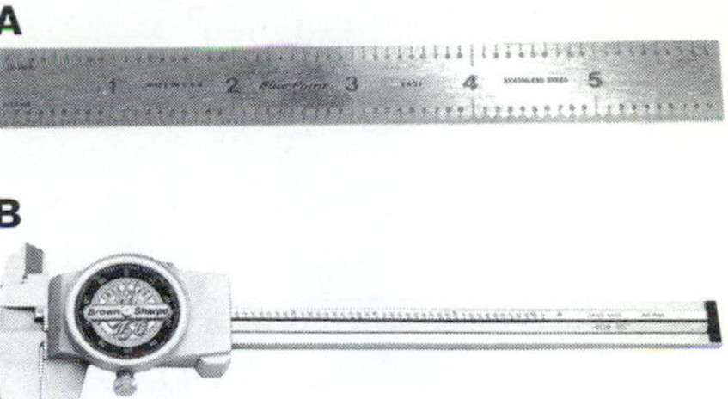

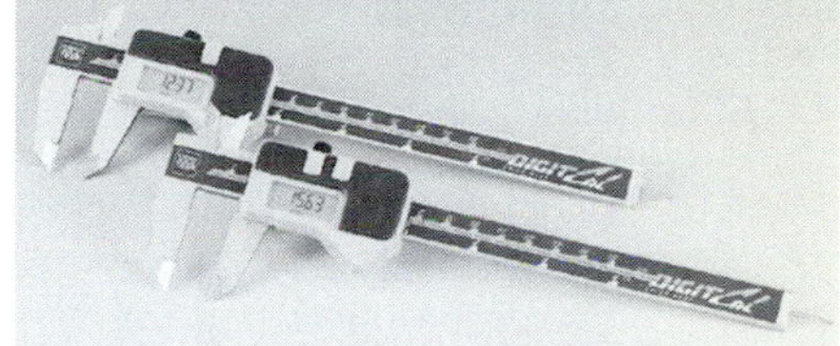

FIGURE 11.21
(a) Inch machinist scale; (b) Dial calipers; (c) Digital calipers. (b and c used by permission of Brown & Sharpe.)

Inspection

In order to measure the part to determine whether it is within the stated tolerance, a number of different tools may be used. A machinist scale is a steel ruler with its smallest divisions 0.01 inch apart. It is frequently used in the shop to measure parts to accuracies of tenths to hundredths of an inch. Similar metric scales with divisions in millimeters are also frequently used to make measurements within a range of plus or minus one millimeter. Dial calipers can be used to measure parts within 0.025 millimeter or a thousandth of an inch. Digital calipers can measure parts to within half a thousandth of an inch and can be easier for the operator to use and read. Examples of these measuring devices are shown in Figure 11.21. For even greater accuracies, a coordinate measuring machine uses physical probes to make measurements to within a few ten-thousandths of an inch. Using lasers, the measuring machine can achieve accuracies of ± 0.00001 inch. Figure 11.22 shows a coordinate measuring system and a typical probe.

Dimensions specified on a drawing are applicable at 20° Celsius (68° Fahrenheit). Compensation must be made for measurements taken at other temperatures. Keep in mind that most materials expand when heated. When measurements are made using highly

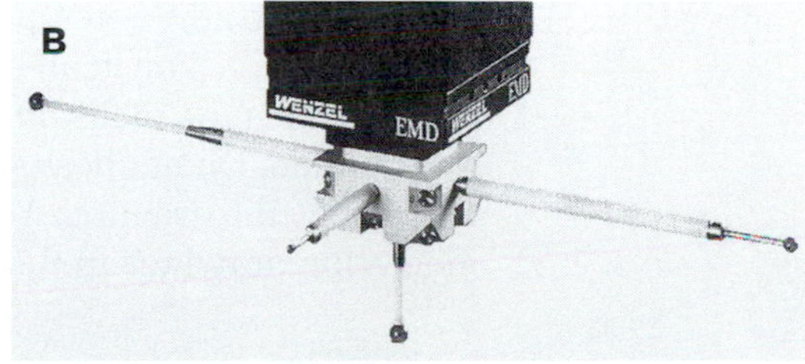

FIGURE 11.22
The coordinate measuring machine in (a) is designed to measure large-scale parts, such as this auto body. A close-up view of a probe used to gather measurement data is shown in (b). (Used by permission of (a) Brown & Sharpe, and (b) Wenzel America, Ltd., Elmsford, NY, EMD, Budd Lake, NJ.)

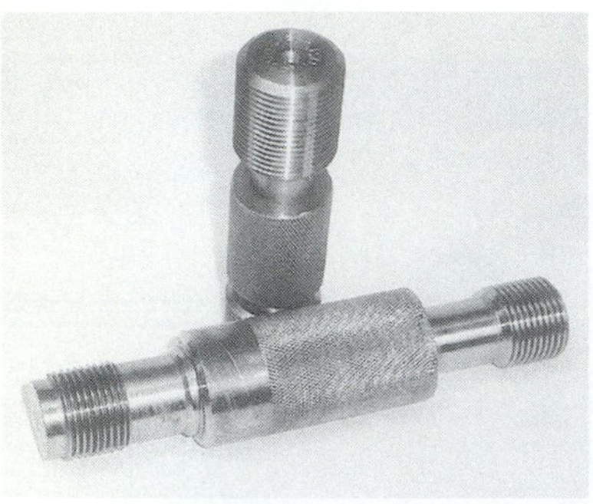

FIGURE 11.23
These fixed gauges have two threaded ends. The "no go" end is too large to fit in the properly sized threaded hole. The "go" end is the smallest acceptable thread size. The threaded hole on the part is acceptable as long as the go end can be threaded into the hole and the no go end cannot be.

accurate equipment, the temperature can have a significant effect. For example, if an operator holds an aluminum part in his or her hand on the way to the inspection station, the heat of the hand can cause a variation in the part size that will be noticeable if measuring to a ten-thousandth of an inch.

Fixed gauges are also used for inspecting. A fixed gauge is carefully manufactured to precise dimensions. If a correctly sized gauge fits inside the interior feature, then the feature must be large enough. To check that the feature is not too large, another gauge sized just slightly over the largest acceptable size should not fit into the feature. Figure 11.23 shows *go–no go* gauges for thread.

The impact tower fixture illustrates the steps in machining a part by hand, but also many aspects common to automated methods. Although much machining and manufacturing today are computer-controlled, someone or something must still hold the material to be shaped (as the workpiece is clamped to the table), the size of the features must be measured to determine whether the part is within the tolerance range specified, the order of manufacturing operations is important in getting good results (so the holes are drilled before the pocket is formed), and additional steps may be required to achieve the desired finish (such as smoothing the rough edges of the part).

The example also illustrates the role of the engineering drawing in implementation. Remember that the engineering drawing can be thought of as a contract between you and the manufacturer who will make the part. The drawing must accurately give the information needed to make the part; it should clearly define the part you want built. If a dimension is specified incorrectly and parts do not fit together, a great deal of time is wasted in starting over again. Keep in mind as you are learning to dimension drawings for manufacturing that measurements are better made from finished or machined surfaces than from rough surfaces, which do not provide a basis for accurate measurements. To design molds, dies, and other tooling effectively, the manufacturer needs to know which surfaces on the part will require additional machining, and which surfaces will be exposed to the user and should therefore be attractive (often called cosmetic surfaces). You will need to indicate this information in the drawing or in a written specification that accompanies the drawing. Remember to add notes such as "Debur all rough edges," "Break all rough edges," "Finish all over," or other information regarding the ultimate finishes, including painting, coating, heat treating, or other information. Do not hesitate to add any special notes that you think will clarify the drawing. You will learn more about producing documentation drawings in the next chapter.

Computer-Aided Manufacturing (CAM)

Many material processing methods today are automated versions of the hand machining processes used for the impact tower fixture. Many of these automated processes use *numerically controlled (NC)* machines. Numerical control describes the numeric or symbolic instructions given to the machine controller that drives (or positions) the machine. Milling machines, electric discharge machines (EDM), lathes, punch presses, plasma cutters, and laser cutters are among the types of machines that can be numerically controlled.

On manually controlled machines, the machine operator positions the machine using a handwheel. Each revolution of the handwheel corresponds to an amount of movement on the machine. Dials on the machine also can be used to indicate how far the machine will travel. In contrast, NC machines receive encoded information from the controller that tells exactly how many times the drive motor should rotate to produce the desired movement. This allows more accurate positioning and automatic hands-off control of the machine. The instruction set, stored in a digital form, allows multiple parts to be run from the same control code, with a high degree of repeatability between parts. Numerical control can also offer more flexibility and easier setup of the machine, in addition to higher tolerances.

Computer numerical control (CNC) is essentially the same thing, except that a computer is used to communicate the instructions to the machine. Using CNC, the program for the part can be edited more easily if the part must change, and the program can be tested and simulated before any parts are run. In addition, the computer control programs can adjust for tool wear and other irregularities while parts are being machined. Once a program is created and tested, many very accurate parts can be created using the same program.

Automatic tool changers, like the one shown in Figure 11.24, allow the type of tool used to do the cutting to be changed automatically so that even complex operations can be programmed and do not require an operator.

FIGURE 11.24
This automatic tool changer holds a variety of differently sized tools and allows automatic switching of tools during a CNC machining process. The double arm toolchanger rotates 90° and grabs the used tool and the new tool. It moves downward to extract the tools, rotates 180° to exchange tools, then moves up to secure them. (Courtesy of Cincinnati Milacron.)

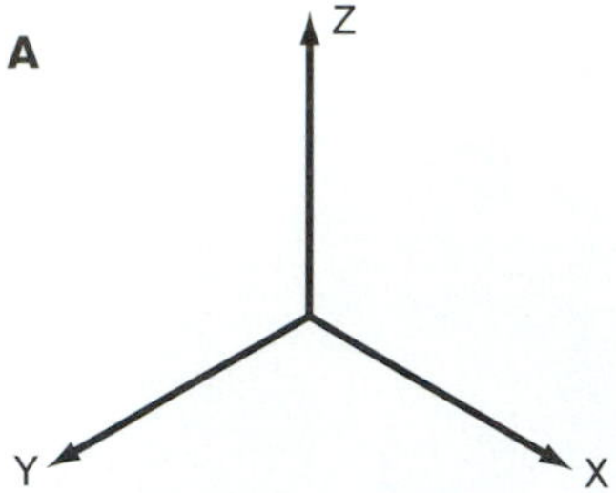

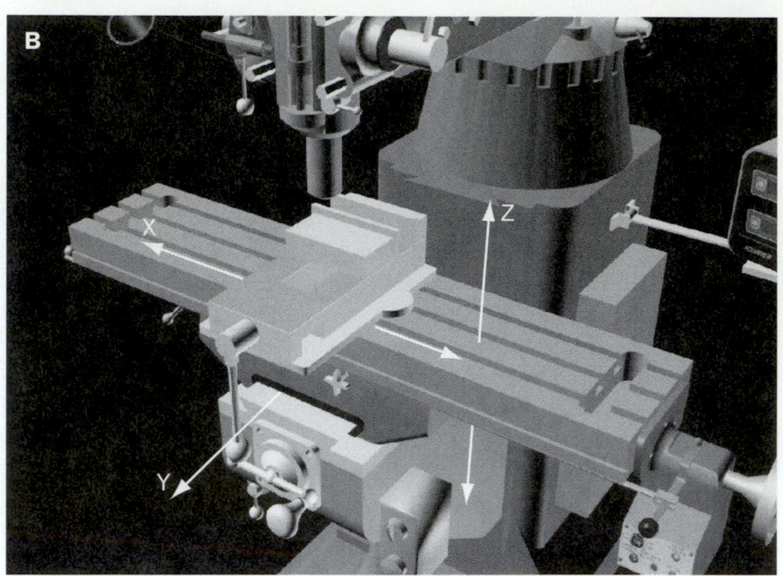

FIGURE 11.25
(a) Movement in the plus or minus direction along the X, Y, or Z axis of a coordinate system. (b) Movement in the plus or minus direction along the X, Y, or Z axis for a typical milling machine. (Courtesy of Jared Williams and Joel Troxler, Montana State University, and the NSF-sponsored Virtual Manufacturing Project.)

CNC machines can use the CAD file as the starting point for the control code needed to drive the machines. The Cartesian coordinate system that is the basis for storing your CAD model relates quite closely to the system used to control the position of the machine tools. Consider a typical milling machine. The cutter can be programmed to move back and forth (along the Y axis) or side to side (along the X axis), and the table that holds the workpiece can be programmed to move up and down (along the Z axis) to allow the machine to move in any combination of these directions (see Fig. 11.25). This provides motion in three directions, which is called *three-axis machining*. More sophisticated tools may move in additional directions by combining tilt and rotation with movements in the X, Y, and Z directions. Three-axis milling machines are very versatile, however, and through combinations of X, Y, and Z movements can create a great many different types of parts, including contoured parts and cylinders.

FIGURE 11.26
This NC-controlled oxy-acetalene cutting torch is used to cut large steel parts used in the manufacture of heavy equipment trailers such as those used in the mining industry. (Courtesy of Smith Equipment USA.)

Two-axis machines can move only in the X and Y directions. Punches and laser and plasma cutters are often two-axis machines because they are used to cut uniform thicknesses of material and do not require movement in a third direction (see Fig. 11.26). A two-and-a-half-axis machine provides programmed movement in all three directions, but has the ability to control only two axes simultaneously, e.g., X and Y, X and Z, or Y and Z.

Communicating From CAD to Machine Code

Before a CAD file can be used to machine a part, it has to be translated into control code that the machine can understand. The process of converting the CAD file translates the finished edges of the model into paths for the cutting and shaping tools that will form the actual part. The outside edges of the part are not exactly the same points that are needed to machine the part. The path for the machine tool typically defines the location of the spindle end. To form the edge of the part, the path for the tool must be offset from the edge of the part by the cutter's radius. In a similar way, the length of the tool is built into the NC code so the CNC machine knows where the tool tip is.

Machining software (or tool path software) such as Mastercam, Surfcam, or Cimatron helps you go from a CAD file to the control code that can be understood by your particular machine. Most machining software will read common file formats such as IGES, STEP, DXF, and STL. As you learned in Chapter 10, different file types provide different types of information about the CAD model. DXF, for example, is a 2D type file format, suitable for the 2D shape needed for sheet metal stamping but not for 3D machined parts. IGES, a 3D format, currently is the most typical file format used to export to machining software.

Once the CAD file has been exported, the next step is to import the file into the machining package. Figure 11.27 shows Mastercam software into which CAD geometry for the part to be machined—in this case, 2D geometry—has been imported.

The type of operation is defined in Mastercam, in this case a pocketing operation in which the material between the outside

FIGURE 11.27
The 2D CAD geometry for the clip was exported to DXF format, then imported into the Mastercam software shown here.

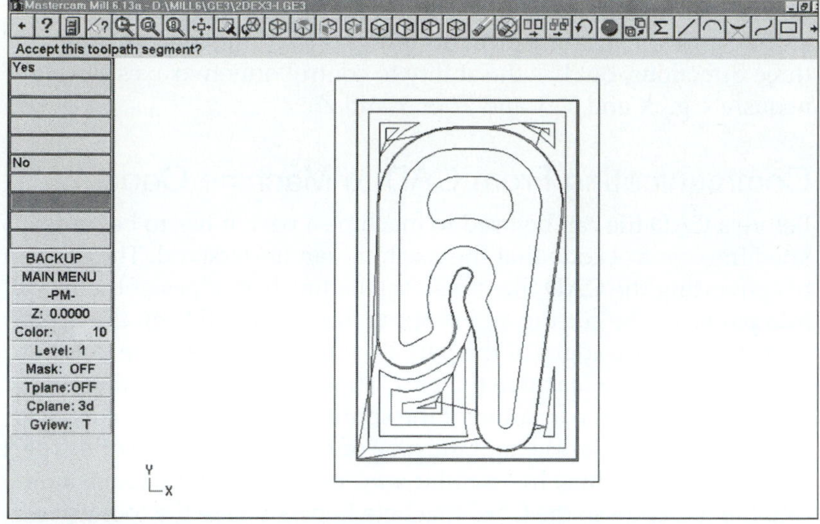

FIGURE 11.28
The tool path that will be followed to machine the part is offset from the actual part geometry by the radius of the tool.

FIGURE 11.29
Circles can be used to represent the tool's diameter in Mastercam to aid in visualizing the tool path.

rectangular shape representing the raw stock size and the interior island (the curved portion) will be removed using a 0.375-inch-diameter endmill. The Mastercam software generates the tool path that the milling machine will use to create the pocket. The tool path, shown in Figure 11.28, is the path that the center line of the spindle holding the tool will follow. Figure 11.29 shows the tool as a circular shape to make it easier to visualize the tool diameter's role in removing the material.

Figure 11.30 shows the Mastercam window where the NC portion of the pocketing parameters will be defined. A particular tool reference is selected from the library of tools that have been defined for the milling machine operation, in this case a 0.375-inch (3/8 inch) endmill. It is tool number 7 in the automatic tool changer for this machine. Other parameters, such as the feed rate (how fast the material is fed into the cutter), the plunge rate (how quickly the

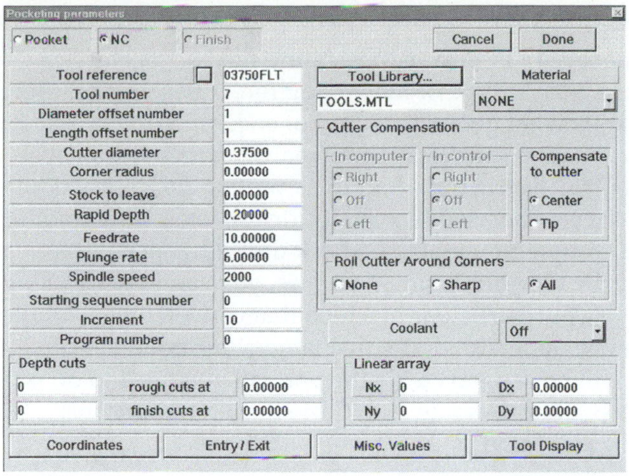

Pocketing parameters

Pocket | NC | Finish | Cancel | Done

Tool reference	03750FLT
Tool number	7
Diameter offset number	1
Length offset number	1
Cutter diameter	0.37500
Corner radius	0.00000
Stock to leave	0.00000
Rapid Depth	0.20000
Feedrate	10.00000
Plunge rate	6.00000
Spindle speed	2000
Starting sequence number	0
Increment	10
Program number	0

Tool Library... TOOLS.MTL | Material NONE

Cutter Compensation: In computer (Right, Off, Left) | In control (Right, Off, Left) | Compensate to cutter (Center, Tip)

Roll Cutter Around Corners: None | Sharp | All

Coolant: Off

Depth cuts: 0 rough cuts at 0.00000; 0 finish cuts at 0.00000

Linear array: Nx 0, Dx 0.00000; Ny 0, Dy 0.00000

Coordinates | Entry / Exit | Misc. Values | Tool Display

FIGURE 11.30
This Mastercam input screen is used to define the pocketing parameters for this machining operation.

cutter descends to the desired depth), and the rotational speed for the tool spindle, are all input using this screen.

A second pocketing operation will need to be defined to remove the material inside the large rounded end. It cannot be done with the same tool because the 0.375-inch diameter is too large. A second tool (0.25-inch diameter) will be selected for this operation. Automatic tool changer capabilities will be used to automatically replace the tool during the machining process.

Once you have created the control code for the machining process, Mastercam software allows you to test it using a simulation package. Figures 11.31 and 11.32 show still pictures of the Mastercam animated simulation. In the animation you are able to see the light-colored cylinder—representing the tool moving along the tool path—removing material to produce the shape. Notice the two different tool diameters.

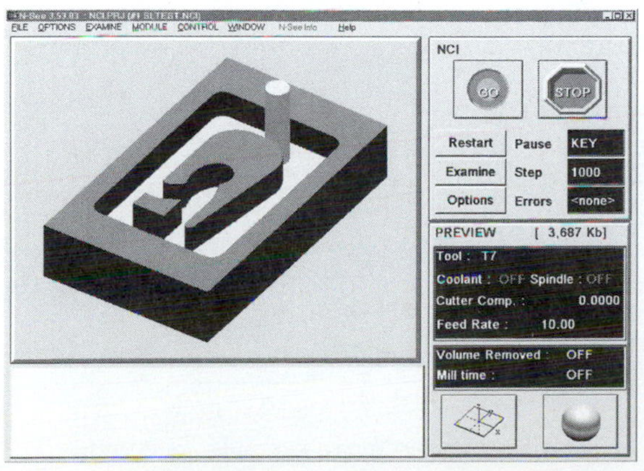

FIGURE 11.31
The Mastercam animated simulation shows the 0.375-diameter tool used for the larger pocketing operations moving along the path defined in the NC code.

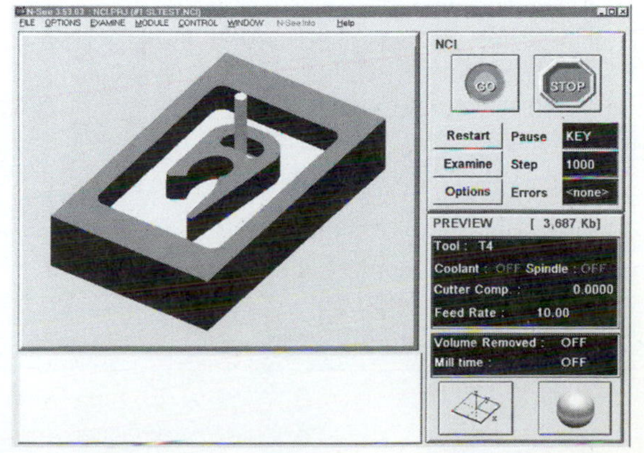

FIGURE 11.32
This shot of the animation shows the 0.25-diameter tool used to create the smaller inside pocket. The simulation shows results visually, but it will also calculate the volume removed by the operation and the approximate time required for the milling operation.

Generic Numerical Control Codes (NCI)

After all of the necessary information defining the NC machining process is input, a generic type of machining file, called an NCI file, is created. This somewhat generic method of encoding CNC machine instructions is particularly beneficial. Previous to its development, all of the NC instructions had to be written for the particular brand of machine to be used. If you decided to use a different but similar machine, you might have to go back through much of the process to create the new file. Using NCI files, you can now use a post-processor for your particular machine that converts the NCI file in a way that can be understood. If you change your mind about which machine will be used to create the part, all you need to do is rerun the post-processor to create an NC file that communicates with the particular machine.

NC Code

Generic NCI code encodes the toolpath information to create the part. A post-processor is used to go from the NCI code to the NC code that contains the specific G and M codes that can be interpreted by a particular brand of machine. An excerpt of the NC code for a specific machine is shown here next to the generic NCI code for the same part. Notice how the NC code contains the G-codes that specify machine actions, translated from the more general NCI code.

NC Code	Generic NCI Code
NC (SLTEST)	NCI (SLTEST)
N0(TIME - 14:52 DATE - 08-06-99)	1011
N10G40G49G80G90Z0	0. 0. 0. 0. 0. 0. 0. 0. 0. 0.
N20M01	1012
N30T7M6	0 0 0 0 0 0 0 0 0 0
N40G53X-0.3737Y-2.7477S2000M03	1013
N50G43H1Z0.2	0 0.375 0. 0 0 0 0 0 4 d:\mill6\mtools\03750flt.ge3
N60M9	1001
N70Z0.2	0 10 10 14 1 1 0 2500 10. 0 -0.3125 -0.3125 0.5 0. 0. 0. 0 1.
N80G1Z-0.5F6.	0
N90Y0.4306F10.	0 -0.3125 -0.3125 0.5 -2.
N100X1.5663	1
N110Y-2.7477	0 -0.3125 -0.3125 -0.5 6.
N120X-0.3737	1
N130X-0.2737Y-2.6477	0 -0.3125 1.6125 -0.5 10.
N140Y0.3306	1
N150X1.4663	0 2.8125 1.6125 -0.5 10.
N160Y-2.6477	1
N170X1.0593	0 2.8125 -0.3125 -0.5 10.
N180G3X1.1967Y-2.4508I-.1211J.2309	1
N190G1X1.4415Y-0.591	0 -0.3125 -0.3125 -0.5 10.
N200G3X0.75Y0.1975I-.6915J.091	1

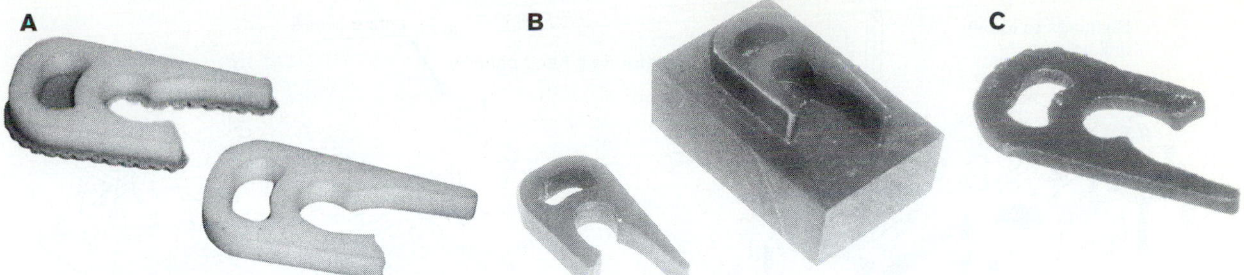

FIGURE 11.33
The same NCI code was used to send this part to (a) an FDM rapid prototyping system, (b) a milling machine, and (c) an NC plasma cutter. A different post-processor was used to generate G-codes that each machine could read.

G-Codes and M-Codes

NC machine control code is written in G-code and M-code. G-codes specify the specific motion for the CNC-controlled machine. For example, G00 X1.0000 Y2.0000 specifies a rapid linear motion (G00) to X coordinate 1.0000 and Y coordinate 2.0000. M-codes are used to specify other types of instructions, such as initiating a tool change.

Figure 11.33 shows three parts that were created on different CNC machines from the same NCI file. The file was translated by different post-processors to communicate with the different machines.

Not all CNC machines have the same capabilities. They can use different methods to interpret the coordinate information for the tool path, for example. Linear interpolation uses straight line segments for each movement of the cutter. If the desired path is circular, the circular shape is broken up into numerous straight line segments. The higher the number of segments, the more closely the end result approximates a circle. *Circular interpolation* means that the cutter can be controlled so that its movement is a true circular or arc path. Instead of requiring endpoint data for all of the numerous straight line segments needed to define a circular shape, the machine can accept the starting point of the curve, endpoint, and radius, and can be controlled to move along a circular path using this data, producing smooth circular shapes. Helical, parabolic, and cubic interpolation add more sophisticated capabilities for creating additional smooth shapes.

The number of axes of machine control, cutter size, and cutter type also determines the kinds of parts that can be created and their finished appearance. For example, on a three-axis milling machine, rounded corners are necessarily produced using a series of small steps. The movement is controlled in the X and Y directions and the plunge depth in Z, so rounds usually have very small rectangular steps. When making mold cavities or other refined features, these are polished out to make smooth blends.

Tooling

Tooling as a general term refers to any of the devices that work along with machinery to produce a part. In this sense, molds, drill bits, and milling machine cutters are all tools. In addition, specialized

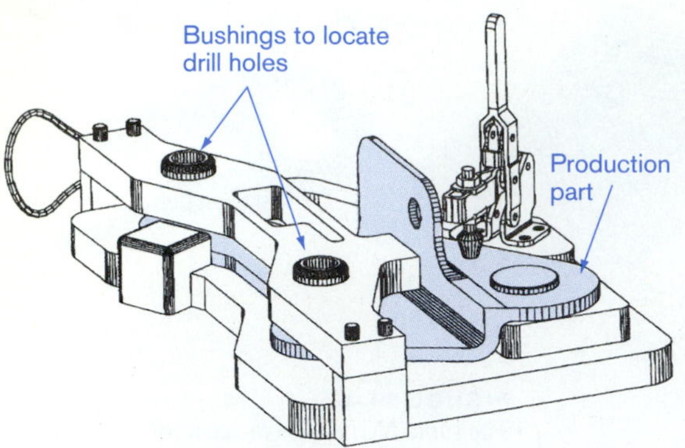

FIGURE 11.34
Jigs like the one shown here are used to guide a drill bit during manufacturing. (Courtesy of The Boeing Company.)

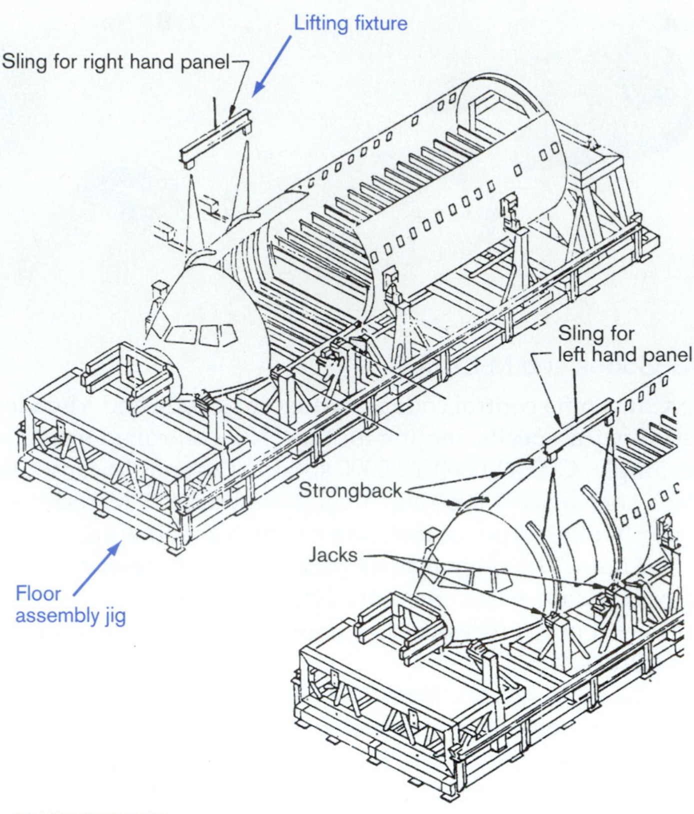

FIGURE 11.35
Assembly jigs accurately locate parts for improved assembly. For large-scale assemblies, such as the aircraft fuselage shown here, jigs position parts that would be difficult to assemble accurately without special tooling. (Courtesy of The Boeing Company.)

tools are often designed to hold or guide the workpiece for machining and assembly operations. Tool design departments devise the specialized tooling that is needed to accurately manufacture and assemble parts. The investment in this kind of tooling can eliminate the need to check location and placement on every part and make it possible for less-skilled labor to assemble parts with a high degree of accuracy. There are two main categories of tooling: *jigs* and *fixtures*.

Jigs

A *jig* is tooling that is used to guide a tool during a machining process (see Figs. 11.34 and 11.35). A jig may hold the workpiece at the same time as it is guiding the cutting tool. It may be moving or stationary.

Fixtures

A *fixture* is a device designed to attach to a machine to hold a part (see Fig. 11.36). Fixtures do not move or guide the cutting tool.

Fixtures are often used in assembly and for inspection as well as for holding work while it is being machined.

Jigs and fixtures are modeled and drawn in the same way that any parts are. They may be designed along with an assembly and shown in the assembly drawing for the part. Using the CAD model for the part and the jig or fixture, critical clearances and fits with the tooling can be checked. Tool designers must often work with very tight tolerances. A tolerance of one-tenth of the tolerance required for the actual part may be necessary for the tooling, because inaccuracies in the tooling would produce additional inaccuracy in the parts being produced. Some of the most precise machining required is that for manufacturing tooling.

FIGURE 11.36
This drill fixture is used for precision small hole drilling. (Courtesy of Reid Tool Supply Company.)

Metal Forming

Not all metal parts are formed by machining processes. Casting and forging are two of the most basic processes for forming metal shapes; both form the part by redistributing material.

Casting is the process of pouring molten metal into a hollow mold to form a metal part. Prior to rapid tooling methods, castings always started from a pattern of the part to be manufactured. A *pattern* is a model of the part made from some other material, often wood, that can be used to form the mold.

Sand Castings

Sand casting is a gravity casting method, in which the molten metal is poured into the mold and not forced in under pressure. For a sand casting, a pattern is made and a mold is created by packing sand around the pattern. The mold is prepared in two halves (called the *cope* and the *drag*) with a *sprue* (or opening) through which the molten metal will be poured, as shown in Figure 11.37. Risers allow

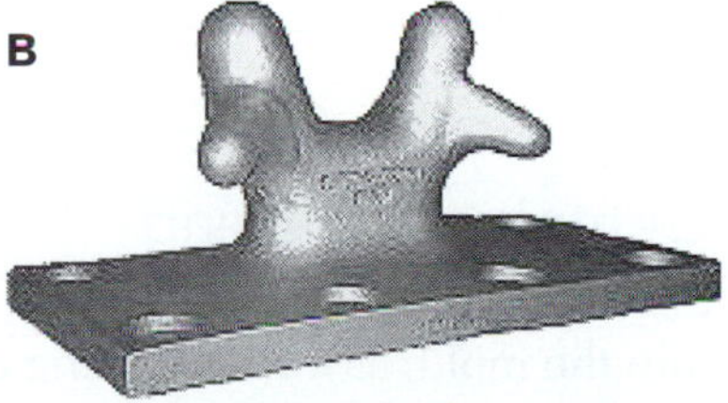

FIGURE 11.37
Molten metal is being poured into the opening of a sand casting mold in (a). The bollard in (b) is a sand cast part. Notice the pebbly surface on its unfinished surfaces.

extra metal to exit so the mold can be entirely filled, then provide extra material to keep the mold full as the metal shrinks during cooling. Vent holes allow air inside the mold to escape when the molten metal is poured in. After the mold is filled and the part cools, the two mold halves are separated and the part removed. Excess material from the sprue, risers, and runners is removed. Sand castings have a rough, pebbly appearance.

Sand casting can be used to form parts from a wide variety of metals, including ferrous metals (ones containing iron). The basic process is fairly simple; in fact, you can purchase a home sand casting kit on the Web.

Small holes (under 1 inch) in the casting are typically drilled after the casting is complete. Larger holes (1 inch and over) may be formed by cores in the casting. Cores are located inside the model to form a hollow in the casting. Cores can reduce the weight of the casting and eliminate the need to machine the hole or cavity later. Made from sand and a binding agent (or from ceramics), a core is removed from inside the part by shaking or tumbling the part to break up the core so the sand that formed it can be poured out.

Surfaces on the cast part can be machined to produce smoother flatter surfaces. When they are, extra material is allowed when the part is molded so it can be machined away. *Permanent casting* is similar to sand casting except that the mold is formed from cast iron or steel and can be reused, unlike a sand mold.

Investment Casting

Investment casting uses a wax pattern that is coated with a gypsum product to form a mold, as shown in Figure 11.38. The original wax pattern is melted out to create a cavity. The cavity can then be filled with molten metal. Investment casting is a precision casting process that can use a wide variety of metal alloys. It also has a relatively low tooling cost, so it can be used for small batches of parts. Production costs per part are somewhat higher than other casting methods, but the end part needs very little finishing. Class rings, refined jewelry, and dental prosthetics are often made using this process.

Centrifugal Casting

Centrifugal casting does not rely on gravity to force the molten metal into the mold, as a sand casting does, but on centrifugal force. The mold is rotated rapidly to force the molten metal outward into the mold cavity. True centrifugal casting methods produce a hole through the center of the part, because the material moves out from the center toward the mold. Centrifugal casting is useful for creating cylindrical shapes such as pipe and tubing. It can also use multiple cavity molds that rotate so that the hole is formed in the central sprue. In this way, all of the mold cavities are entirely filled with metal. Centrifugal casting can produce better surface finish than sand casting. Most commercial jewelry is made using this process.

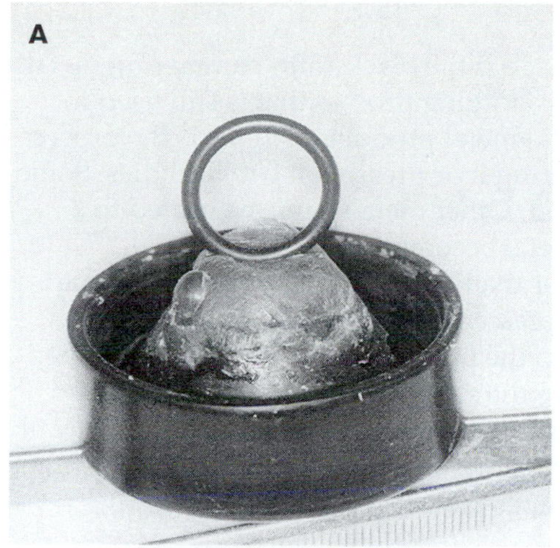

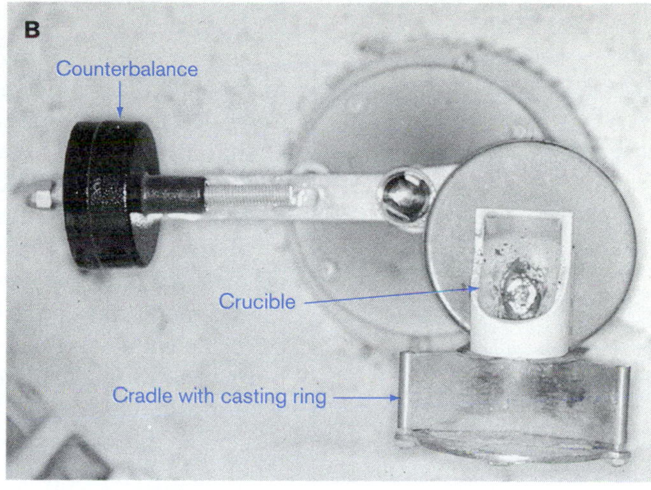

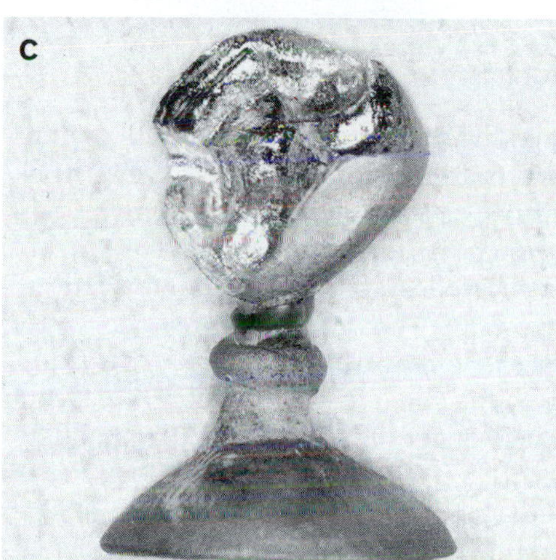

FIGURE 11.38
Investment casting uses a wax pattern attached to a sprue to create a cavity; a pattern for a ring is shown in (a). The wax pattern is then melted out and the cavity filled with molten material. A top view of a centrifugal casting machine used in this process is shown in (b). Investment casting is used for products such as the dental crown (still attached to the sprue) shown in (b). (Courtesy of Daniel Gaffney.)

Die Casting

The *die casting* process uses a permanent steel die (or mold) and forces the metal in under high pressure and temperatures. Die casting can be used to quickly create small part features with close tolerances. Die cast parts often have a high-quality surface finish and do not require secondary machining. Unlike sand casting, molds for die casting can be reused many times, and die cast parts typically have thinner walls than sand cast parts (see Fig. 11.39). However, the die has a relatively high initial cost, so die casting is not economical for small batches of parts (under two thousand).

Die casting cannot be used for ferrous metals and certain alloys. Because the molten metal is forced into the mold under pressure, there is typically some excess material (called flash) that must be trimmed from the part. This flash often occurs where the two halves of the mold come together.

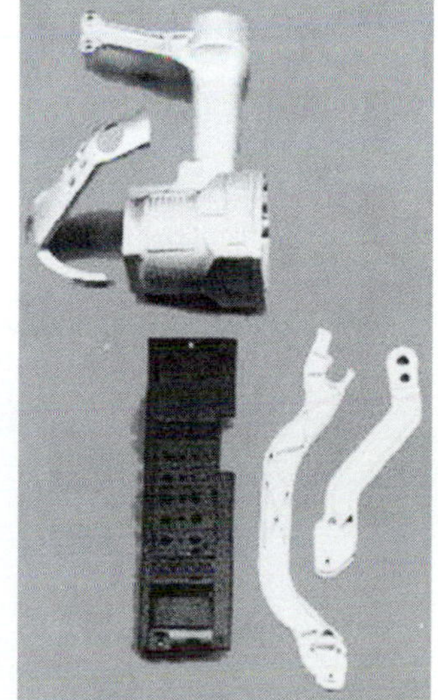

FIGURE 11.39
These die cast parts can be made with relatively thin walls.

Forging

A familiar metalworking image is the blacksmith hammering metal to shape it. Blacksmithing is a forging process that is still used for horseshoeing and ornamental metal products. *Forging* is the process of shaping metal by hammering it or pressing it between dies. Some soft metals can be cold forged. Other metals must be heated to a temperature at which the metal becomes workable.

Forged parts are stronger than a similar cast or machined part because the material is compressed and the grain structure of the forged metal stays parallel to the part contours. Cast and machined parts have weak areas where the shape changes direction. Forged parts typically have lower tolerance ranges and require thicker wall sections than cast parts. But casting methods, especially die casting, can create more complex shapes than forged parts.

Sheet Metal Manufacturing

The third major process for making parts from metal is sheet metal manufacturing. Sheet metal parts are cut from flat sheet metal stock, typically made of .125″ or thinner galvanized steel. Sheet metal parts are usually formed through punching or stamping, then bent or welded to produce more complicated shapes. The ultimate pattern for a sheet metal part must necessarily be flat. Sometimes the designer lays out the flat pattern. Sometimes the engineer provides the 3D shape or 2D drawings to the sheet metal manufacturer, who lays out the drawing. Specialized sheet metal software can make it easier to lay out the flat pattern.

For multiple parts to be created from the same sheet of metal with the least waste, the flat patterns are nested so that each part lies as closely as possible to the next one, creating the largest number of parts from the standard sheet metal stock size. This reduces waste and keeps the per piece price as low as possible. Some sheet metal software also provides pattern nesting capabilities, and specialized software can be purchased for pattern nesting.

Sheet Metal Fasteners

Self-tapping sheet metal screws are often used in assembling sheet metal parts because their one-step operation provides faster assembly than standard nuts and bolts. Threaded inserts, such as *swage nuts* or *clinch nuts* are often used to provide a threaded hole in a sheet metal part. Because sheet metal is relatively thin and made of relatively soft materials, fastener inserts work well because they can be made from stronger materials and quickly pressed into hexagonal or other shaped holes punched in the sheet metal. Threaded studs that have been welded on and other types of threaded studs are also popular for designing with sheet metal to make assembly easier. Tabs and slots that can be formed when sheet metal parts are punched are usually the least expensive way to fasten sheet metal parts.

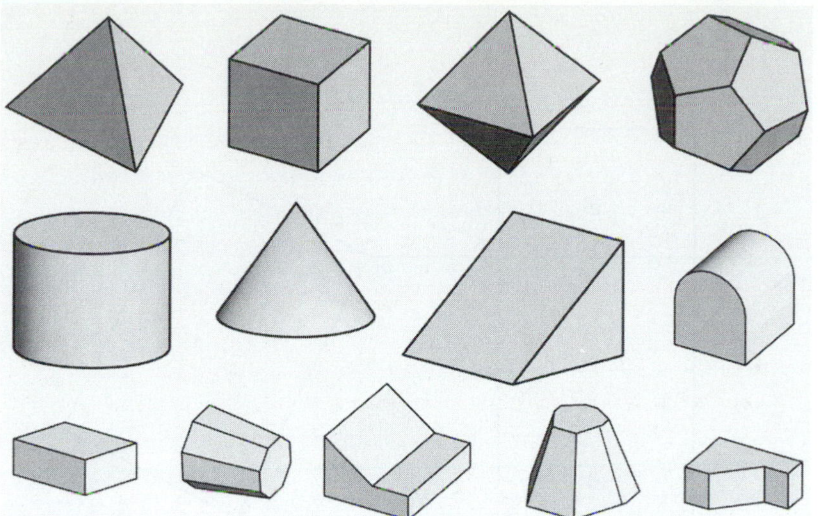

FIGURE 11.41
Each of the models shown here are developable because they contain only planar or single-curved surfaces.

FIGURE 11.40
Typical sheet metal fasteners (Courtesy of Emhart Fastening Technologies.)

Some examples of typical sheet metal fasteners are shown in Figure 11.40.

Pattern Layout

To create a shape that will be cut from a sheet of material, the shape must be *developable.* A developable shape is one that is capable of being laid out flat (see Fig. 11.41). Developments are used to create sheet metal parts as well as in fabric and package design. Developable shapes include *polyhedra* (shapes with boundaries made up of plane surfaces) and single-curved surfaces such as cylinders and cones.

Irregular surfaces and double-curved surfaces, such as spheres and ellipsoids, can be developed only by approximating portions of their surfaces as flat surfaces.

Extra material must be added to the developed shape to provide for seams and attachments. Figure 11.42 shows examples of standard hems and joints that are frequently used in sheet metal applications.

FIGURE 11.42
An assortment of developable shapes made from sheet metal.

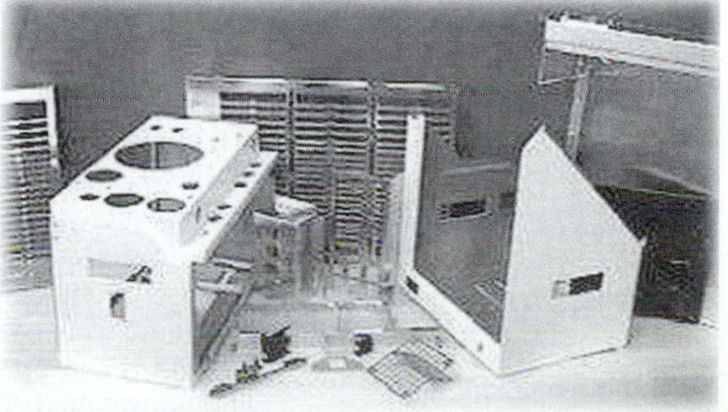

Developments

To create a development, each surface of the object is laid out true size. Special software can be used to automatically develop the flat pattern from a solid, surface, or wireframe model of the part. These packages automatically generate the true-size views of surfaces in the configuration needed for the pattern layout. Without such software, you must create true-size views of each surface. In Chapter 4,

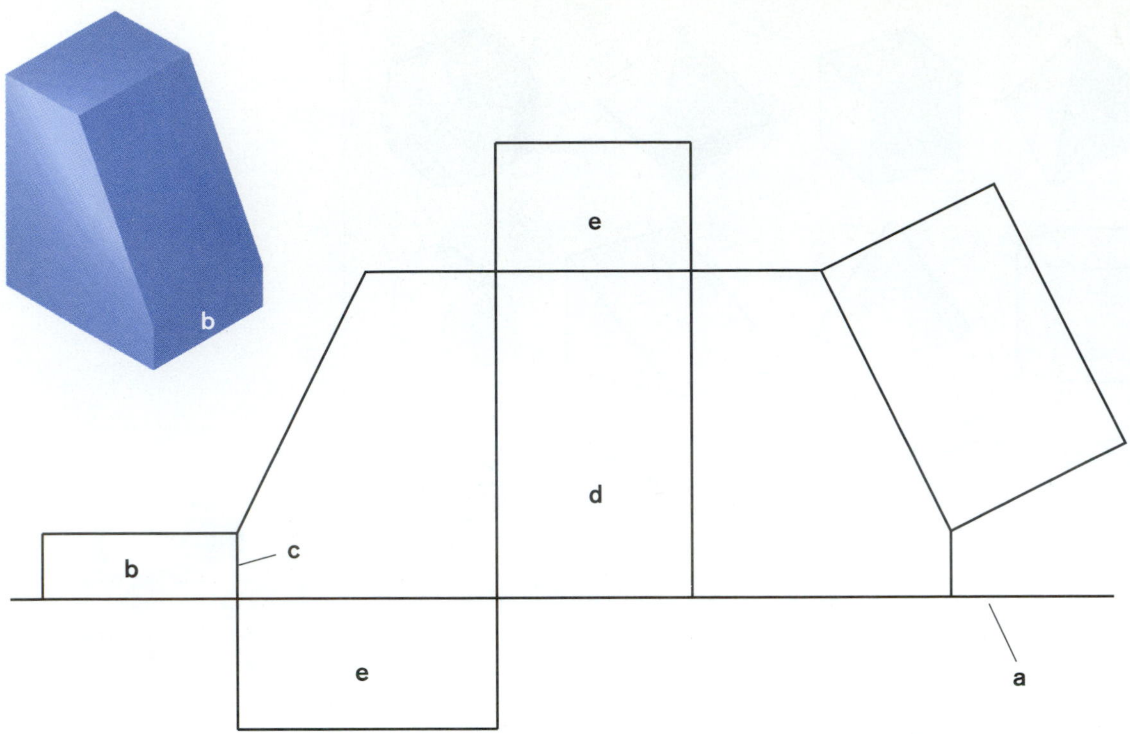

FIGURE 11.43
To produce a development for a prismatic shape: (a) Choose the stretch-out line and draw it an arbitrary length (you can make it longer or shorter if needed later). (b) Select the surface you will start the development from and draw it true size along the stretch-out line. (c) Starting with the common edge between the starting surface and the next surface, lay out the next surface true size. (d) Continue around the object until you have laid out all of the surfaces along the stretch-out line. (e) Add the top surface and bottom surface along an edge you have already laid out, where there is a common intersection, then add any other surfaces in a similar way. (f) Add tabs, seam, or hem allowances if necessary to the outside edges.

you learned how to project an auxiliary view for any surfaces that are not shown true size in the orthographic views. In Chapter 12, you will learn how to create these views using your CAD package.

Views of the object are arranged along the stretch-out line for the part. The stretch-out line is a line chosen along the object where major surfaces can be laid out in a straight row, allowing them to later be folded or bent into the shape.

To create a development, determine the true lengths of each surface edge. You can usually do this using your CAD software by listing out the value of the edge line. Be sure to read out the actual value for the edge length and not a projection on the coordinate system where it may be shown foreshortened. Figure 11.43 shows the steps needed for the development of a prismatic shape.

To develop a cylindrical shape, you unroll the cylinder along a stretch-out line equal to the length of its circumference. Figure 11.44 illustrates the steps in developing a cylindrical shape.

Bend Allowance

If you are creating a pattern that will be folded from thin paper, you do not usually need to think about adding extra material for the bends, but thicker materials and metals, such as sheet metal, cannot be bent exactly at 90°, so extra material is required to form the bend. This additional material is called a *bend allowance*. See Figure 11.45 for illustrations of bend allowance and bend allowance terminology.

Where the sheet metal is bent, the metal on the inside portion of the bend is compressed. The metal on the outside portion of the bend is stretched. The neutral axis of the bend is the area inside the bend where the metal is not being compressed or stretched. The length of the neutral axis is the amount that is added to the inside pattern for the bend allowance. For typical sheet metal, the location of the neutral axis is often considered to be .44 of the metal thickness from the inside surface of the bend.

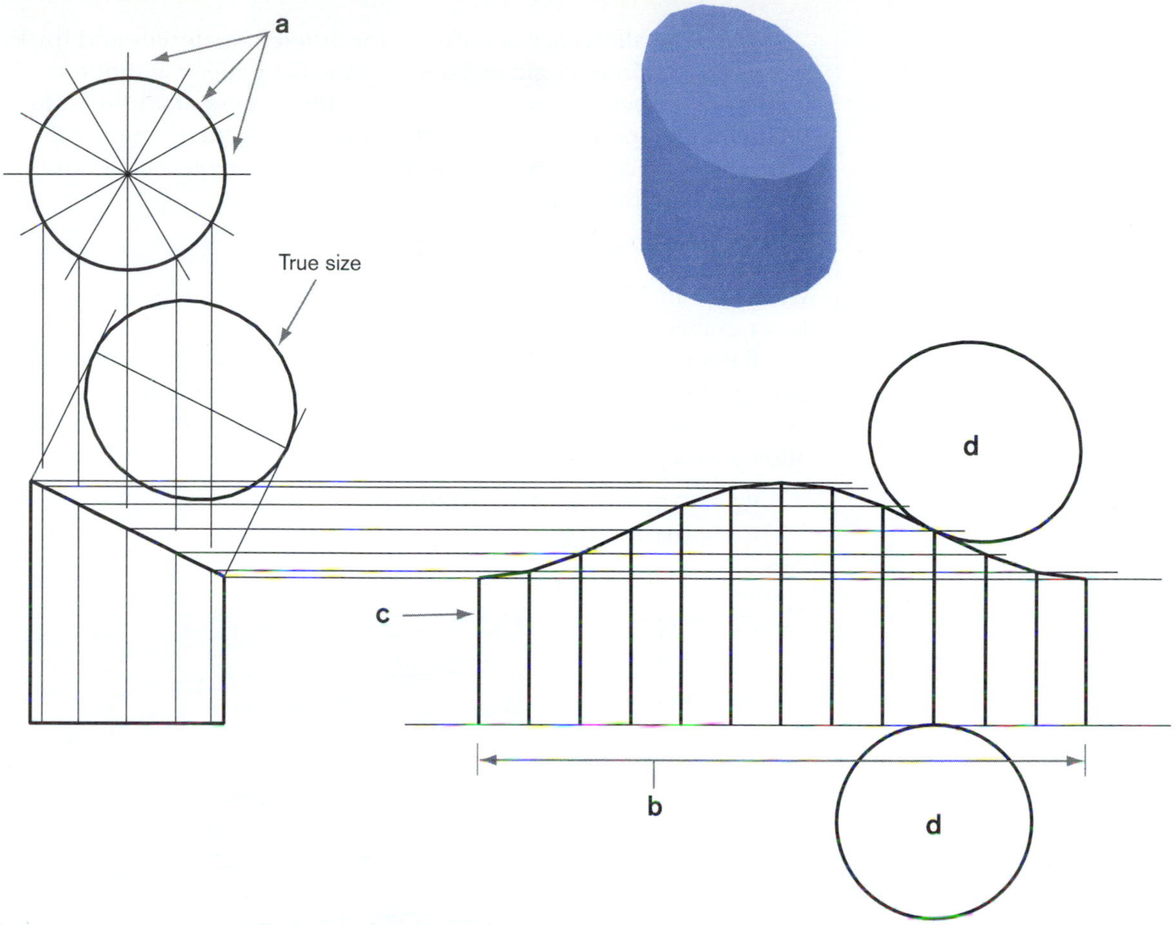

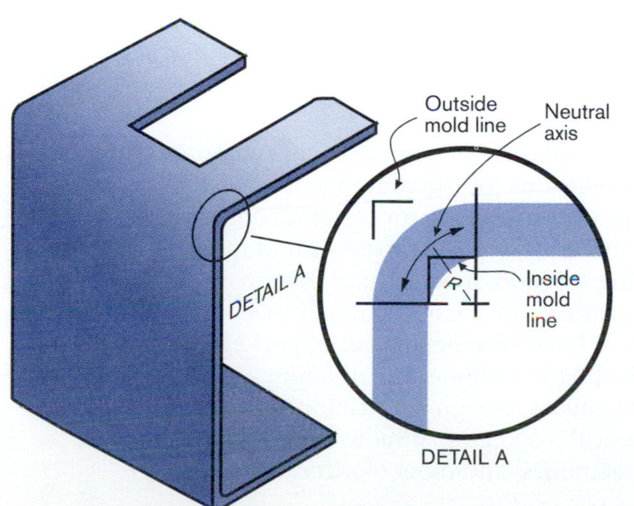

FIGURE 11.45
During bending of sheet metal parts the metal on the inside portion of the bend is compressed. The neutral axis is where the metal is neither stretched nor compressed.

FIGURE 11.44
To produce a development for a cylindrical shape cut by a plane: (a) Divide the circumference of the cylinder into a number of equal parts. (b) Lay out the circumference of the cylinder along the stretch-out line. (c) Construct the height of each element of the cylinder at the points you located. (d) Add the true-size shapes for the top and bottom surface for a closed pattern.

The bend allowance is different for different materials and thicknesses. Many times engineers develop the flat pattern without including the bend allowance and leave the process of adding it to the manufacturing personnel to determine. A standard empirical formula (one gained from experience and measurement) that you can use to approximate the bend allowance is

$$BA = (0.017453R + 0.0078T)\,N$$

where R = inside radius of the bend, T = metal thickness, and N = number of degrees of bend.

Refer to Figure 11.46 for the following example. To calculate the bend allowance for .125" thick sheet metal bent at 90° with an inside bend radius of .375", substitute the values in the equation for bend allowance as follows:

$$BA = (0.017453 \times .375 + 0.0078 \times .125)\ .675$$
$$BA = .011$$

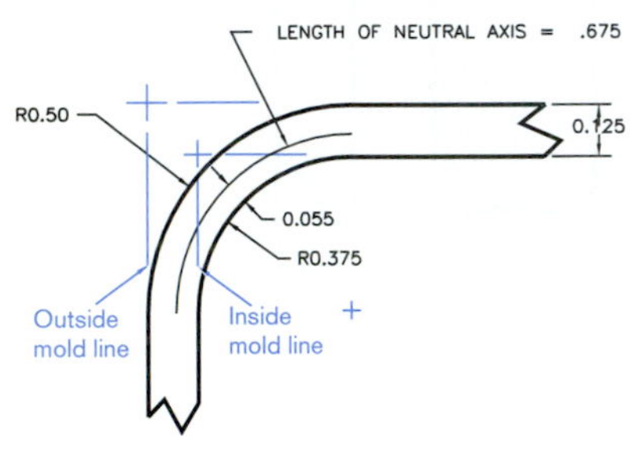

FIGURE 11.46
Use the length of the neutral axis, the bend radius, and the metal thickness to calculate the bend allowance according to the empirical formula: BA = (0.017453R + 0.0078T) N, where R = radius of the bend, T = metal thickness, and N = number of degrees of bend.

Stamping

Stamping is the process whereby sheet metal is pressed between dies to form simple parts, such as those shown in Figure 11.47. Sheet metal stamping is a very cost-effective way to produce certain parts, especially if the part can be formed in a single stamping operation.

FIGURE 11.47
Sheet metal stamping can be used to form simple parts like this. (Courtesy of Performance Stamping Co. Inc., Carpentersville, IL.)

Packaging

Packaging engineers perform many and varied tasks, including preparing specifications for new packaging components; determining critical dimensions for the component to be packaged; designing packaging and labeling for cartons, containers, and other packaging; and implementing the processes and machinery to produce packaging for food, medical, and other consumer products. Packaging engineers often have industrial or mechanical engineering backgrounds or have specific degrees in packaging engineering. (Electronics packaging is another very different area of engineering dealing with the analysis, design, fabrication, and assembly process for electromechanical packaging of radio frequency assemblies, printed circuit assemblies, and other electronic products.)

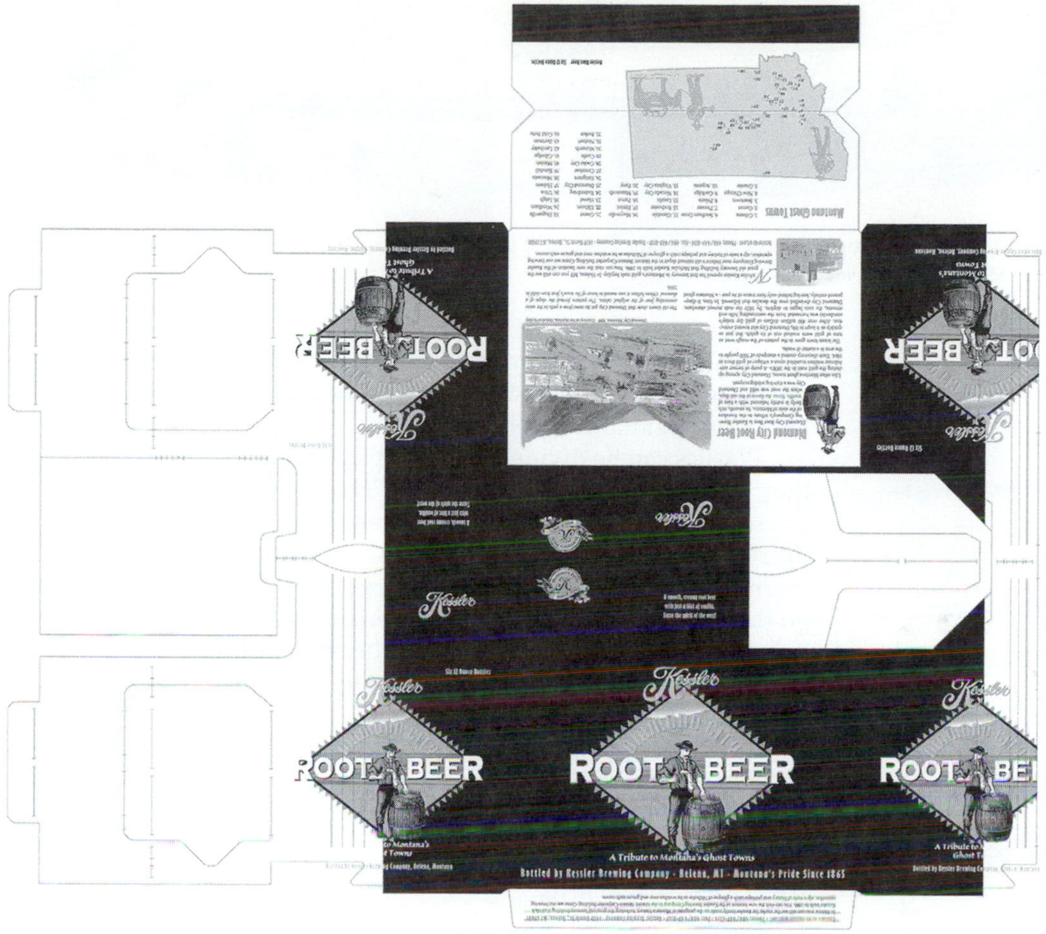

FIGURE 11.48
The flat pattern for this root beer package shows the unfolded shape that is printed, then stamp cut and folded to produce typical six-pack packaging. (Courtesy of Kessler Root Beer.)

Among the skills that packaging engineers and designers use are developing the flat pattern for cardboard packages. In addition they often design molded plastics that fit with or encase product components. The same CAD models that are used to produce the part definition can also be beneficial in designing the packaging. The model can be used to determine critical dimensions, and surfaces of the model may be offset and used to define (and create the molds for) plastic-bubble packaging. Figure 11.48 shows the flat pattern for a six-pack beverage container.

Plastics Manufacturing

Processes used to manufacture plastic parts generally redistribute material to form the part. The processes, while relatively young, offer parts with a number of advantages. They are inexpensive to manufacture in high volumes, pleasing to look at, and provide a wide range of formable shapes, colors, and material properties, such as strength and resistance to ultraviolet radiation, weathering, and

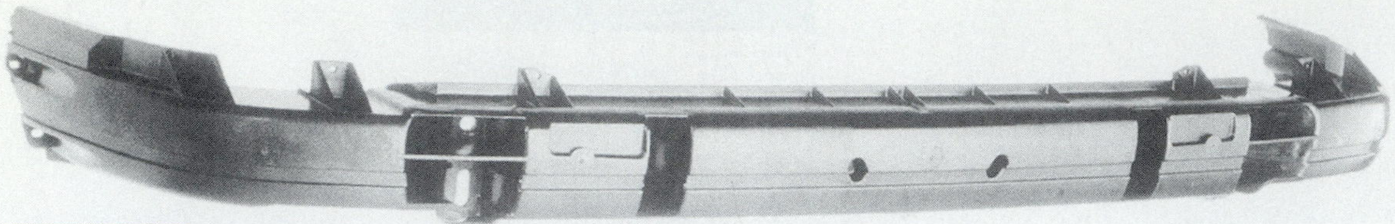

FIGURE 11.49
The 1998 Mercury Mountaineer sport/utility vehicle was the first in the world to use a thermoplastic bumper system. (Courtesy of GE Plastics.)

chemicals. Plastic parts are being used more and more in engineering design. Figure 11.49 shows the plastic bumper used on a 1998 Mercury Mountaineer vehicle.

Plastics can be divided into two basic categories based on how they react to heat during processing. *Thermoset* plastics go through a physical change from solid to liquid and back to solid when heated. *Thermoplastics* melt when heated and set when cooled. In processing thermoplastics, the mold is cooled to set the part. Thermoset plastics go through a chemical change to permanently and irreversibly become a new cross-linked plastic. Thermoset plastics will cross-link due to heat or pressure. They are usually cured in a warm mold to aid in this process.

The most common plastics processing methods are *injection molding, compression molding, blow molding, transfer molding, extrusion,* and *thermoforming*.

Injection Molding

Injection molding uses powdered or ground plastic, which is fed from a hopper into a preheating chamber. A hydraulically powered screw forces the powdered plastic into a chamber where it is melted, or plasticized. A shot of the proper amount of molten plastic is measured in the chamber and injected using high pressures through a sprue into the mold. Under slightly less pressure, plastic is added to fill the mold completely. Once the plastic in the sprue has cooled so that no more material can be forced in, the next shot of plastic is prepared in the chamber. As soon as the parts in the mold are sufficiently cooled, the mold press opens and the parts are ejected into a collection bin. The press then closes, and the next shot of melted plastic is forced into the mold. Recycled plastic (from trimming off the sprue and other waste) can be reground and fed back into the hopper (see Fig. 11.50).

Extrusion Molding

Extrusion molding heats plastic particles (including regrind and additives) in a heating chamber. A powered screw forces the heated plastic through a shaped die to produce the shape of the extruded plastic part. Plastic pipe, tubing, bulk plastics, and other uniform shapes can be created with this process. Metals, such as aluminum, can also be extruded in a similar process. See Figure 11.51 for extrusion molding machinery.

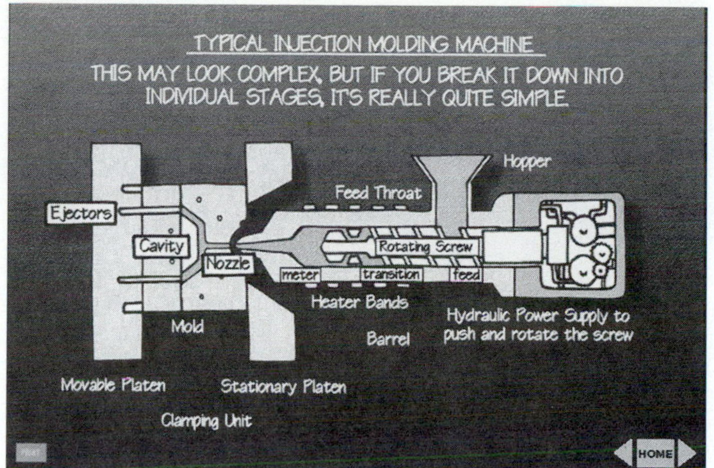

FIGURE 11.50
The injection molding process shown in cartoon form (Courtesy of GE Plastic.)

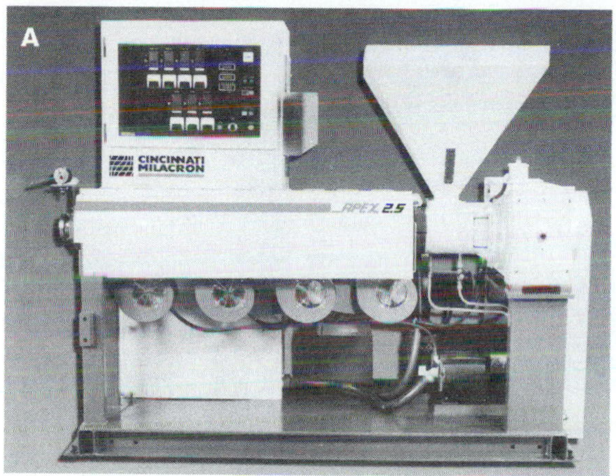

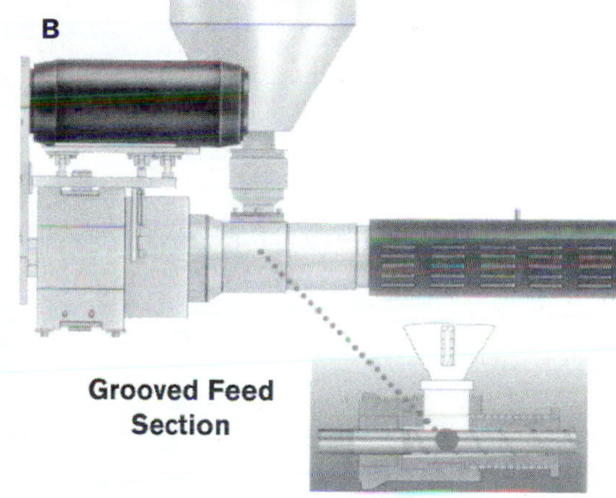

FIGURE 11.51
The extrusion molding machinery in (a) uses a feed screw mechanism, illustrated in (b), to force the heated plastic through the die. (Courtesy of Cincinnati Milacron.)

Injection and Extrusion Blow Molding

Injection blow molding is similar to injection molding in that a hollow tube of plastic shaped like a test tube, called a *parison,* is formed by injection molding methods. A mold is then closed around the parison, and compressed air is forced inside it so that it expands to fill the mold cavity. The mold cavity defines both the shape and finish of the final product (see Fig. 11.52). The part is formed when the hot plastic contacts the cold mold. Bottles, automotive fluid reservoirs, ductwork, toys, wheels, and other parts are commonly formed using blow molding. Figure 11.53 shows an injection blow molding machine.

Extrusion blow molding also uses compressed air to expand the parison into a mold, but the parison is extruded, not injection molded. A hollow tube of material is extruded to serve as the parison. When the mold closes around it, the parison is simultaneously cut and sealed before air is forced into it to expand it. Figure 11.54 shows parisons being extruded for this process. Figure 11.55 shows an extrusion blow molding machine.

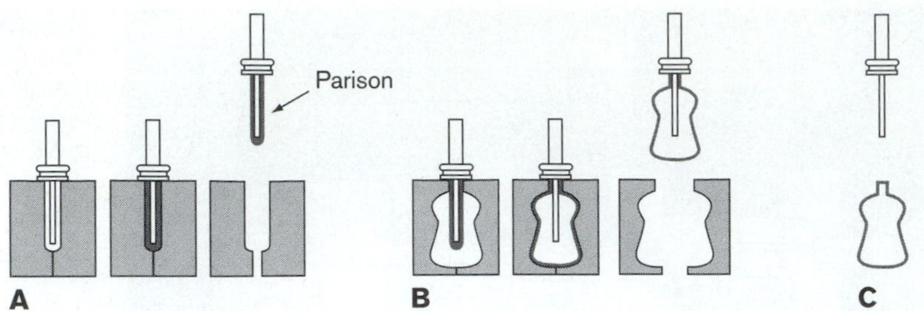

FIGURE 11.52

This three-part figure illustrates the steps in injection blow molding. In (a), the plastic is injected around a rod in a mold to form the parison. In (b), the rod and still pliable parison move to the next station where a mold in the shape of the final part (in this case, a bottle) closes around each parison. Compressed air enters the parison through the rod to "blow" it into the shape of the mold. At (c), the molded part is removed from the rod and ejected from the machine.

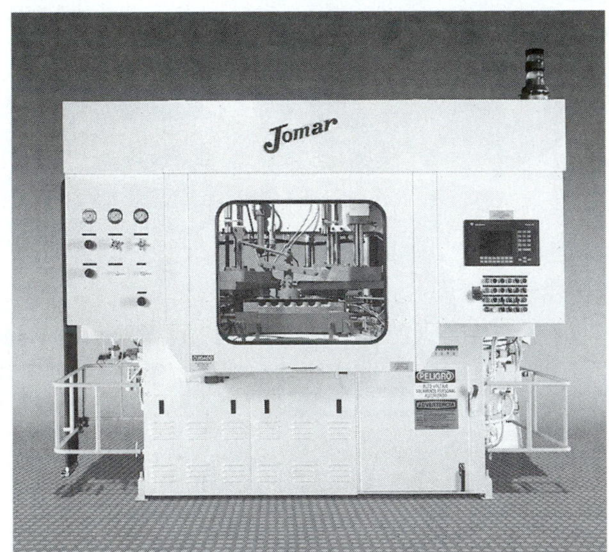

FIGURE 11.53

An injection blow molding machine (Courtesy of Jomar Corporation.)

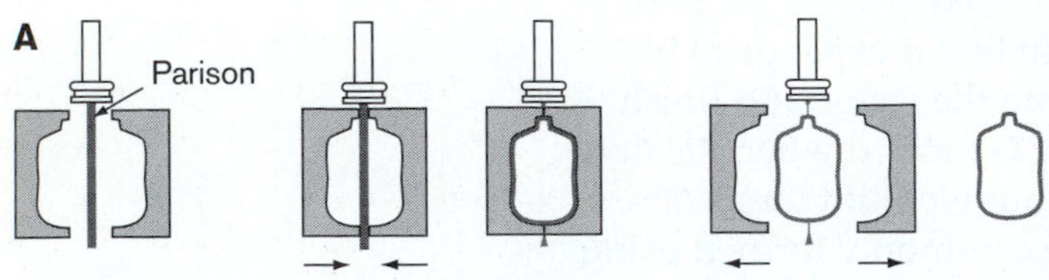

FIGURE 11.54

(a) In extrusion blow molding, the parison is extruded as a hollow tube, then the mold closes around it and seals it. Air is forced into the parison to expand it into the shape of the mold. After the mold opens, the part is trimmed and ejected. (b) These parisons are being extruded for extrusion blow molding. (Photo courtesy of Jomar Corporation.)

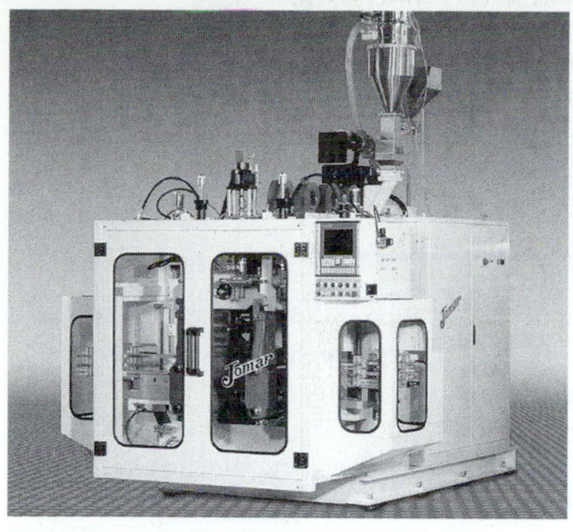

FIGURE 11.55
An extrusion blow molding machine (Courtesy of Jomar Corporation.)

Compression Molding

Compression molding uses thermoset resin which is placed into a heated mold cavity. When a plunger is depressed, a combination of pressure and heat sets the plastic. After the parts are cured they are ejected from the mold.

Vacuum Molding

Vacuum molding uses heated plastic sheets which are drawn down into a shaped cavity. Vacuum molding is used to produce products such as common plastic containers as well as large items like refrigerator liners, tubs, and camper tops, as shown in Figure 11.56.

This brief overview of manufacturing processes illustrates the many options available for producing the parts you design. Understanding basic processes will equip you to evaluate options and work with others to determine which materials and processes will achieve the criteria your design needs to satisfy. To complete your design, it must be documented with the information needed by the manufacturer. In this way, implementation is linked to the next phase of the design process—documentation, which is the subject of the next chapter.

FIGURE 11.56
This sophisticated vacuum former uses plastic up to 3/4 inch thick. It is best suited for deep draw forms such as refrigerator liners, tubs, and camper tops. (Courtesy of Plasti-Vac, Inc.)

INDUSTRY CASE

Hugo Haselhuhn, C&K Systems

Modeling for Manufacture: No Glues, No Screws

When C&K Systems designed and manufactured the MC-760 passive infrared motion detector, it used 3D solid modeling in SolidWorks to produce the parts more quickly and to eliminate miscommunication with the mold makers for the plastic parts. C&K Systems manufactures products for the security industry and operates 28 facilities in 18 countries around the world. C&K Systems designs and manufactures a full line of security products including microwave and passive infrared motion detectors, glass break detectors, wireless systems, control panels, CCTV products, smoke detectors, magnetic contacts, sounders, speakers, and monitoring software. The company designs over 100 new tooled parts each year.

C&K Systems's main engineering facility is in Folsom, California, but its manufacturing facility is in China, as is an engineering, marketing, and sales office in Hong Kong. "Most of our products—about 80% to 85%—are injection molded plastic," says Hugo Haselhuhn, director of engineering there. "We do steel cans for the control panels—sheet metal box and lid, if you will—and we also have a number of zinc die cast components that we use in our microwave sensors. Almost everything we do also has a printed circuit board inside, so the PC board designers and the mechanical engineers have to work in concert to make sure that the PC boards fit into the metal or plastic enclosures that we develop."

When Haselhuhn joined C&K Systems in 1995, mechanical engineers were using AutoCAD to develop 3D models of their design concepts. At that time, there were limitations to the amount of detail that could be shown in the model, so the 3D model was passed to the drafting team, which created 2D drawings of the final parts. As the team members created the 2D drawings, there were many places where implementation details weren't clear; for example, how the draft should be added to the part, or what the radii for curved surfaces should be. As they prepared the drawings, the drafters were in constant communication with the design engineer to determine what was intended for the part. This was a time-consuming process. In addition, a change to any dimension in the 2D drawings necessitated changing that dimension in every view where it appeared, so the drafter was continually checking and rechecking the drawings. "The typical check process was about a week for a package. One of our sensors typically has a front cover, a back cover, a lens, and maybe two or three internal plastic parts, such as a lens retainer or a light pipe for an LED. So there are five to ten plastic parts that all have to fit together perfectly (see Fig. 11.57). The checker spent about a week making sure that the dimensions on each part drawing matched associated dimensions on other part drawings so that everything would fit together. They used the same method to perform dimensional analysis on the parts

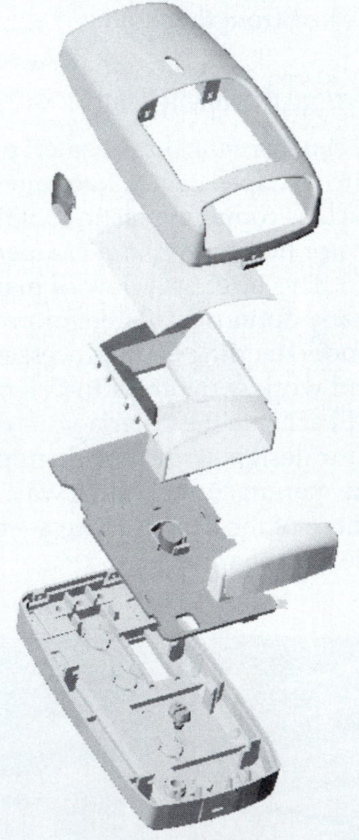

FIGURE 11. 57
The MC-760's injection-molded case and internal parts are designed to snap together to eliminate glue and screws in their assembly.

to make sure that for normal manufacturing processes the tolerances listed would satisfy the build requirements and the assembly would fit."

After the drawings were completed, they were sent to a model maker in Taiwan, who made a prototype of the assembly. "He made prototypes for us out of plastic and balsa wood, or whatever materials he could get," explains Haselhuhn. "The prototype was representative of what we wanted, but many times it was not dimensionally accurate. Making the prototype would take about three weeks."

After evaluating the prototype and making any necessary changes, the 2D drawings then went to the mold maker, who would tool the molds for the plastic parts. "Our staff in Hong Kong was our go-between with mold makers in China. They would take our drawings, get quotations from vendors, and pick the vendor based on price and delivery date. One of the problems that arose consistently was errors due to misinterpreting the drawings. Because we could get tools in about half the time and half the price that we'd be spending in the United States, we were willing to live with that shortfall. But we had to plan into our development schedules anywhere from two to four weeks to fix the tool after the first shot."

Miscommunication between engineer and drafter, and between 2D drawing and mold maker, motivated Haselhuhn to propose that the company switch to 3D parametric solid modeling. In January 1996, the decision was made to model their designs in SolidWorks. Today, the mechanical engineers create models that represent the final part, and those models are used to generate drawings and rapid prototypes and to create the tooling for the molds.

"When we were doing 2D drawings, we would typically add a note to say, 'All draft is 1° or 2° unless otherwise stated.' The draft would not be represented in the drawings. With the solid model, we make sure that all parts are an exact representation of what the tool is going to be. You can't make an approximate representation of your part in a computer file and expect the tool vendor to use your file and modify it exactly." Parting line, ejector pins, slides, and undercuts are now factored into the design of the part, but the engineer doesn't have to know it all. "We recently changed our engineering organization to focus on teams, and part of that team is our engineer in Hong Kong and the toolmaker in China. We try to get preliminary models to both our Hong Kong engineers and the toolmaker so that they can do a review of the design before it's complete. They can come back and say, 'This part's going to be too thin and won't mold properly' or 'You're going to end up with a lot of sink in this area because you haven't made that boss correctly.' In most situations, though, the engineers here have enough experience to know what to do and what not to do."

Tolerance is another element built into the models. "We have standard tolerances of about ± 0.1 millimeter for injection molded parts of handheld size. The designers understand that when two parts fit together, they have to take into account the low tolerance on the external piece and the high tolerance on the internal piece so that when they fit together, they *always* fit together. We had some difficulties when we first started modeling in this way. The design engineers would design everything line to line; that is, where the internal part and external part fit together with no interference. Because they didn't take into account tolerancing, however, the vendor could be small on one piece and large on the other—within the tolerance range—and the part might not fit together. You definitely have to understand tolerances within whatever industry you're working to make sure that parts are designed to fit within the scope of that particular manufacturing technique."

The checking process has gone from a week per package to about a day per package because the solid model is used to assemble the parts, check for interference, and evaluate the tolerances (see Fig. 11.58). "In some situations

(continued page 460)

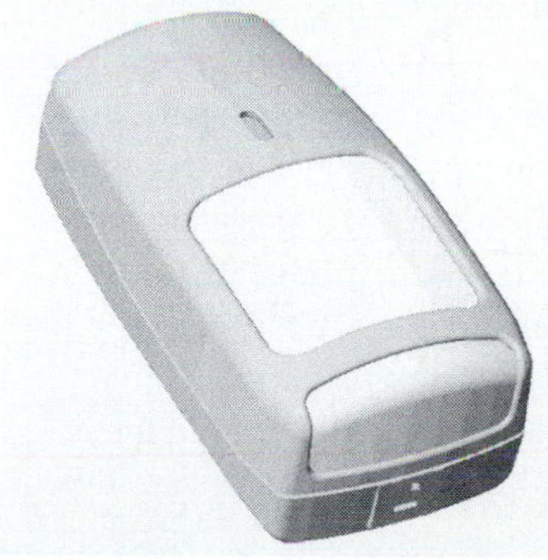

FIGURE 11.58
Assembling the individual solid parts into the assembly model shown here shortened the days needed to check the fit and interference in the model to a few hours.

INDUSTRY CASE

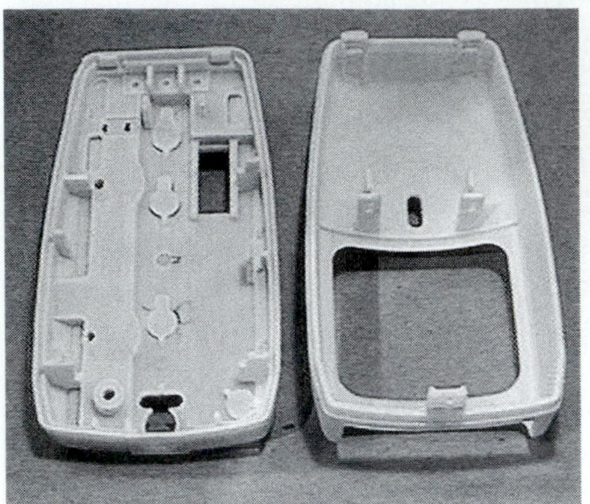

FIGURE 11.59
The two halves of the rapid prototype shown here illustrate the internal details of the MC-760's case. The tab at the bottom of the case that snaps the case together is an example of a feature that would require a slide in the core half of the mold. The tab is formed around the slide, then it is moved to the side to allow the part to eject from the mold.

where we have snap fits, we know that we want interferences. We may have a tab that is going to deflect 3 millimeters, for instance, in order to make it a snap fit, so we will check to see that there is an interference. If there's another part that should not interfere and it does, then we have to adjust it."

Prototyping has gone from three weeks to three or four days. "With SolidWorks, we can export an STL file to any of a variety of rapid prototype service bureaus." The stereo lithography method (SLA) is used to prototype parts such as a light pipe, because the liquid polymer is a translucent material. "One of our recent products was a fairly large part, maybe 14 inches long by 8 inches wide by 4 inches deep. It was a lens cover for an alarm product that fits on a wall outside of a building and has a sounder plus a strobe light. These are mainly used in the U.K. and Europe, where central station alarm monitoring isn't as prevalent as it is in the U.S. In that case, we definitely wanted something translucent for the prototype so we could see the light flashing through the lens." For other parts, selective laser sintering (SLS) is preferred because the material used (powdered nylon with fiberglass strands) results in a stronger, more flexible model. "The goal I've given my designers is no glues and no screws. This means that everything needs to snap together. The SLS prototype is flexible enough that we can assemble parts and test the snap fit" (see Fig. 11.59).

After making any changes to the model indicated by the rapid prototype, the SolidWorks file is exported in an IGES file format to go to the mold maker in China. This change in the process is perhaps the most dramatic, shaving more than the two to four weeks allowed for fixing the tool from the schedule. At first, only one of the vendors in China had CNC machining software (such as Mastercam) in its facility, but Haselhuhn's team worked with the others to get them to adopt it. "Because we went to solid modeling and SolidWorks, I had to take advantage of the shorter time to market that it would provide," explains Haselhuhn. "Sending an IGES file saves time because the toolmaker doesn't have to interpret drawings, he doesn't have to program his machine, and he doesn't have to translate the drawing information from English to Chinese so his journeyman machinist on the floor can understand it. All of the communication is electronic—through digital images."

Even the documentation drawings that go with the files are transmitted digitally to the office in Hong Kong where they are printed out. When the IGES file is generated for the final part, so are drawings that the toolmaker's quality control (QC) and the C&K Systems factory's incoming inspection team will use. These drawings are not fully dimensioned, however. "I want the vendor to make the tool only from the solid model and not from the drawings. The drawings are a representation of the part to be used for the QC process and perhaps for a dimensional check if they have questions within the model." Only the critical dimensions where parts fit together—where PC board grooves are and where parts fit into slots—are highlighted on the drawings that accompany the IGES file (see Fig. 11.60).

The sculpted shape of the wall-mounted MC-760 motion sensor is conveyed by the solid model directly to the rapid prototyping and manufacturing processes used to produce the parts. In the past, prototypes and molds were an approximation of the shape indicated on the 2D drawings, but they were never exact. Today, the mold maker uses the SolidWorks IGES file and Mastercam software (as shown in Fig. 11.61) to generate

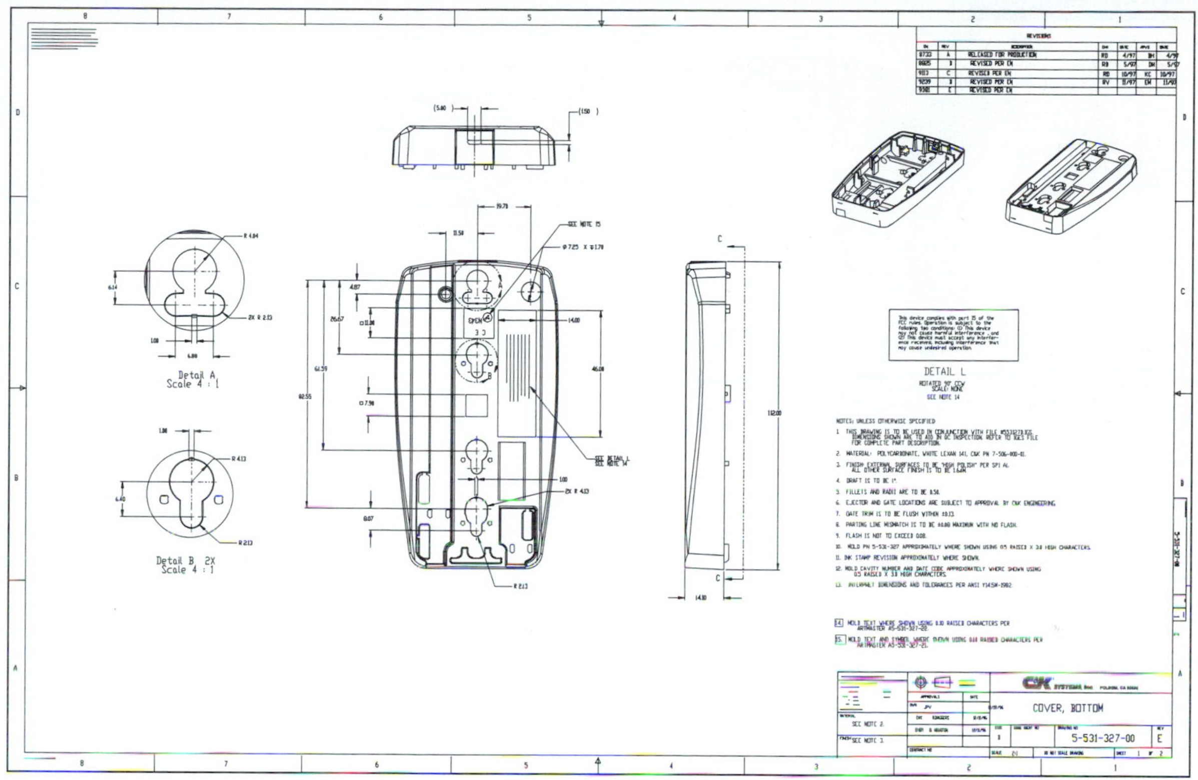

FIGURE 11.60

Documentation drawings travel with the IGES file of the model to provide a reference for the tool maker and to define the critical dimensions of acceptable parts for inspection.

the tool paths that will be used to machine the form specified in the CAD file.

Electric discharge machining (EDM) is one process used to make the steel molds used for C&K's injection-molded plastic parts. To make a mold with EDM, a pattern in the shape of the part is first machined out of copper. The copper part serves as an electrode that will be used to burn the shape of the part into the mold blank. When the copper electrode is brought close to the steel mold blank in a liquid bath, the current in the electrode burns away the metal, leaving a cavity in the shape of the part (see Fig. 11.62). "EDM allows you to have very complex shapes in the tool, because the CAD file can be used to machine the complex shape into the copper electrode," explains Haselhuhn.

"When the external half of the tool is done, which is called the cavity half, the other half of the tool—the core half—needs to reflect all the internal details of the part. Our sensors are like clam shells; the front and back snap together, and the outsides are relatively smooth, sculpted forms. Internally, the case may have ribs

(continued page 462)

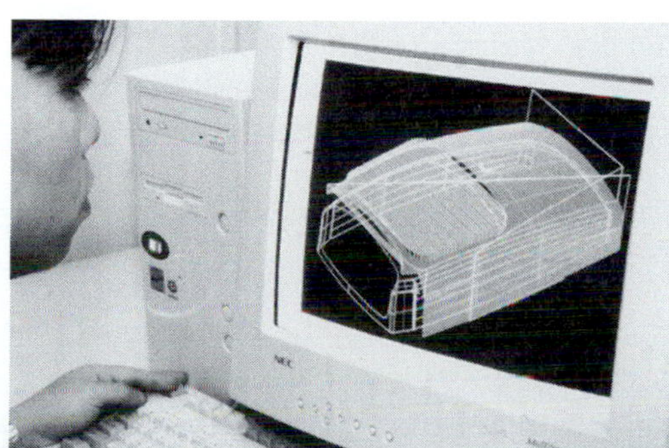

FIGURE 11.61

The engineering staff in Hong Kong works with the mold maker's staff to translate the IGES files into tool paths for NC machining. The tool paths for the top part of the case are shown here in Mastercam.

INDUSTRY CASE

FIGURE 11.62
This copper electrode in the shape of the outside of the MC-760 was used in the electric discharge machining process to form the cavity half of the mold, shown in the lower half of the photo.

for strengthening the sides, grooves for the PC boards, and tabs used for snap fitting the light pipes. To form these features the core half may have internal slides or undercuts that cannot be formed using EDM alone." Slides are parts of the mold that can be used to form features, then retracted to allow the part to be removed from the mold. To make these features, the toolmaker uses the CAD file to machine the individual pieces that represent the internal details, then welds them or assembles them on a slide to form the core of the mold.

Making the two halves of the mold requires the mold maker to edit the IGES file to select features needed for each half of the mold. Internal details may be removed when NC code for the cavity is formed. Internal features of the part that will be formed by slides may be removed so the file can be used to machine the rest of the core. The documentation drawings sent with the file serve as a fixed point for working with the file so that critical features aren't omitted or misplaced.

The transition from 2D paper drawings to electronic files that are routinely sent via e-mail and FTP has abbreviated the time to market for C&K Systems's products and has shortened the distance between team members across the Pacific Ocean. More importantly, the CAD files have improved the communication necessary for producing accurate snap-fit parts every time.

All illustrations used with the permission of C&K Systems.

Hugo Haselhuhn joined C&K Systems, in Folsom, California, in August 1995 as the Mechanical Design Manager. Haselhuhn has worked in many positions in electro-mechanical product development, as a manufacturing engineer, and now as Director of Engineering at C&K Systems. He started using CAD in 1986 and has used AutoCAD, CADKEY, and Pro/Engineer in previous positions. Haselhuhn was instrumental in making the switch from 2D to SolidWorks solid modeling at C&K Systems.

Skills Summary

Your understanding of the basic manufacturing processes presented in this chapter will help you create CAD models and dimensioned drawings that describe all of the information necessary to manufacture a part. Exercises 1–4 allow you to test your ability to identify and select from various manufacturing operations. Use what you have learned of numerically controlled machinery to answer the questions for Exercises 5 and 6. Exercises 8 and 9 check your ability to use the information in the chapter on manufacturing sheet metal parts. Exercise 10 asks you to investigate how some common items are manufactured.

EXERCISES

Partial list of parts-forming operations:

a. Centrifugal casting
b. Drilling
c. Milling
d. Sawing
e. Injection molding
f. Stamping
g. Investment casting
h. Turning
i. Broaching
j. Punching
k. Rolling
l. Compression molding
m. Oxy-acetylene gas torch cutting
n. Forging

1. Parts-forming operations involving material removal or material redistribution are sometimes subdivided into *Primary Operations* (to shape raw material into the geometry required for the finished part) and *Secondary Operations* (to finalize raw shapes to that of the finished part). Group the above operations as Primary or Secondary Operations. Many can be grouped with both categories: Which ones, and why?
2. Choose six of the above listed operations, and provide a complete description of each process: What tools are used to accomplish each? Give an example of parts formed with each method.
3. Various means of fabricating the same part can be developed. Create your own list of manufacturing steps that could be used to create each of the following components. For each part, create an alternate list with *different* manufacturing methods. (If you don't know what the part is, investigate!)

   a. Bicycle crank arm
   b. Tent stake
   c. Garden rake

d. Polyethylene pail
e. Plastic liquid detergent bottle
f. Automobile disk brake rotor
g. Ball-type trailer hitch
h. Coin (e.g., a penny)

4. Economics play a large part in the selection of manufacturing methods to be used. Often it is a matter of how many parts might be produced. Contrast the methods that might be used for creating a single prototype plastic fork (eating utensil) versus ten thousand of the same item.
5. Numerically controlled (NC) machining operations or computer-numerically-controlled (CNC) processes use digital instruction sets to define motion of the tool. Describe the two main numerical control code types.
6. What differences exist between 2-axis, 2-1/2-axis, and 3-axis milling machines? Investigate and report on the unique capabilities of 4-axis milling machines.
7. Tools, jigs, and fixtures used to assist in part manufacturing can have a great impact on the accuracy of the parts produced. Why might the tooling require a tolerance of 1/10 that of the part to be fabricated? What does this imply for the tools, jigs, and fixtures used to build the tooling that in turn is used to build a part?
8. What is meant by "bend allowance" on sheet metal parts?
9. Use the empirical formula BA = (0.017453R + 0.0078T)N (see text for description) to create a spreadsheet that defines the bend allowances for the following cases:

Material Thickness	Bend Angle Desired Degrees	Radians
10 gage = 0.1354"	45°	$\pi/2$
12 gage = 0.1046"	45°	$\pi/2$
16 gage = 0.0598"	45°	$\pi/2$
24 gage = 0.0239"	45°	$\pi/2$
10 gage = 0.1354"	90°	π
12 gage = 0.1046"	90°	π
16 gage = 0.0598"	90°	π
24 gage = 0.0239"	90°	π
10 gage = 0.1354"	135°	$3\pi/2$
12 gage = 0.1046"	135°	$3\pi/2$
16 gage = 0.0598"	135°	$3\pi/2$
24 gage = 0.0239"	135°	$3\pi/2$

10. Identify the collection of manufacturing processes needed to fabricate each of the following items. Develop a flow chart for manufacturing and assembly of the item: some research will be needed. If your assembly contains purchased parts (i.e., tires) simply call them out at the level where they are installed.
    - a. Retractable-tip ballpoint pen
    - b. Single-axle tilt-bed utility trailer
    - c. Kitchen toaster
    - d. Aluminum extension ladder

DESIGNER'S NOTEBOOK

The tolerances built into the manufacture of the X-Y bead picker bridge are noted on this sketch of an adaptor that fits onto it. The adaptor must accommodate a bridge width that can vary from 15.85 to 15.91. (Courtesy of Albert W. Brown, Jr., Affymax Research Institute.)

Project No._______

E____________________ Book No. 2144 67

Page No.___

AUG 22, 1996

(1) GOT A FAX FROM NEWPORT ENGINEER 510
ANA, FOR THE SLIDE 1-800-222-6440
(2) ADDED MIKE HOLDER TO SLIDE IN PRO-E
(3)

15.88±.03

BRIDGE OF X-Y BEAD PICKER

ADAPTOR FOR V

(4) DESIGNED AN ADAPTOR TO ATTACH THE INK-JET (8X) HEAD TO THE BEAD PICKER X-Y STAGE BASE UNIT.
(5) RAM + JOE SUCESSED IN SENDING MOVE COMMANDS TO THE HYDRA.
(6)

To Page No.___

Witnessed & Understood by me,	Date	Invented by	Date
		Recorded by	8-22-96

CHAPTER 12

Documentation Graphics

Objectives

When you have completed this chapter, you will be able to

- locate drawing standards that apply to your project;
- create drawings from 3D models to an appropriate scale;
- create a set of working drawings that fully document a design;
- add section, auxiliary, and detail views to a drawing;
- dimension a drawing and indicate tolerances;
- document surface finish;
- add notes and title blocks.

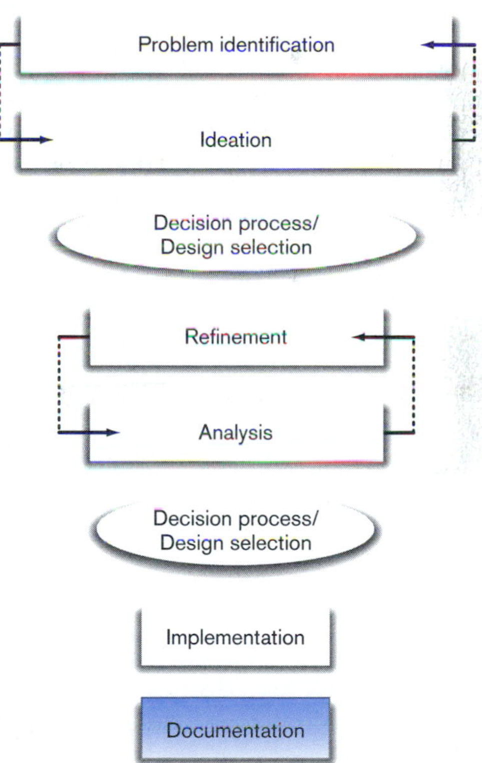

Throughout the design process, engineers use a variety of ways to document design ideas. As you have progressed through this text, you have learned how to sketch design ideas, select a modeling method, and produce CAD models defining the design—all of which are part of the design record. The documentation phase of the process, however, refers to the production of engineering drawings that unambiguously describe the part or process so it can be made. Documentation locks down a design at a point in time and can serve as a contract with a manufacturer, a legal record of the design, and a tool for communicating key aspects of the design's function.

Documentation and implementation could be considered another loop in the design process. The dimensions and tolerances specified on the documentation drawing convey the criteria that the manufacturing process must meet. How much variation from the exact measurements is acceptable? If the manufacturer can't meet the criteria, changes may be made to the design or a different process selected. In preparing good documentation drawings, the engineer has to work with information about different manufacturing processes and specify the criteria required for the parts to function properly. In this chapter, you will learn about dimensioning drawings and describing the tolerances for the manufactured part so that it can be produced as inexpensively as possible and inspected to see if it is an acceptable part.

You will also learn more about the standards that govern how to show information in a drawing. A variety of standards have been developed to establish a "language" for documentation drawings. These standards define symbols and formats that make it possible to express large amounts of information in a small space. Many of the standards were designed to facilitate the transition from design concept to manufactured part and reflect the information needed for common manufacturing processes.

As more manufacturing processes become automated, the tools used to document engineering design will continue to evolve. Engineering drawing practices were developed for design communication among people; as more information is conveyed electronically from the CAD database to the manufacturing device, drawing practices and standards will evolve to support the new processes. The role of documentation in the process is unlikely to change; it will just be performed differently, through methods that increasingly use the electronic database to convey the information.

As you create documentation drawings, you are expressing the key aspects of your design in a way that is easy for others to interpret. Today, many CAD programs have commands that "automatically" produce dimensioned drawing views. Even though these tools make it much easier to produce documentation drawings, they do not have all of the necessary intelligence to produce correctly dimensioned and toleranced drawings. You must add your understanding of good engineering drawing techniques to guide the choice of views, placement, scale, and other factors that produce a good drawing in keeping with drawing standards.

The CAD Database as Design Documentation

Some companies maintain the CAD database of the 2D or 3D model of the object, when released for manufacture, as documentation for the design. It is an important advantage of producing an accurate CAD database that you can use the model as a basis for manufacturing. Today's CAD packages offer tools for incorporating tolerances and manufacturing notes into the database itself that improve its ability to document the design. A good understanding of the type of information that is available in your CAD database combined with the knowledge of how to show critical dimensions and tolerances clearly are important in achieving the most benefit from your 3D CAD software.

To use the CAD database as design documentation, however, companies must consider the legal requirements for maintaining a permanent record of the design as discussed in Chapter 9. For some industries, a permanent record (or snapshot) of the design used for production must be maintained. A changeable record on the computer may not be considered a legally acceptable practice, or it may be acceptable only if a standard of drawing control is met.

Some companies may use the 3D model with electronic annotations stored in the file or a related database as the final documentation for the product. Other companies produce 2D original drawings from the 3D model to communicate the design for manufacture and to provide design documentation for the project. The 3D model may be stored, but the design record is the fully dimensioned 2D drawings.

Other companies use a combination of the computer files and 2D drawings to document the design. The 2D drawings are used to communicate information about critical tolerances and other information that may not be easily visible in the 3D file. The CAD file serves as the interface to automated manufacturing processes, but the drawing allows the company to call attention to those elements of the design that are critical to its function (see Fig. 12.1). Because the manufacturer may not have the same software that was used to create the CAD model, the documentation needs to use a format that can be interpreted by the manufacturer, mold maker, or others who will create or inspect the parts. This is frequently a combination of electronic files in a common 3D format (such as IGES) for the model, and 2D CAD documentation drawings (either printed or in a common 2D file format such as DXF) showing critical dimensions.

Whether the 2D drawings are printed on paper or stored electronically, correctly shown orthographic views still provide much of the basis for communicating and documenting the design. Correctly shown drawing views are also used to communicate information for user manuals and repair manuals, as well as for manufacturing and inspection.

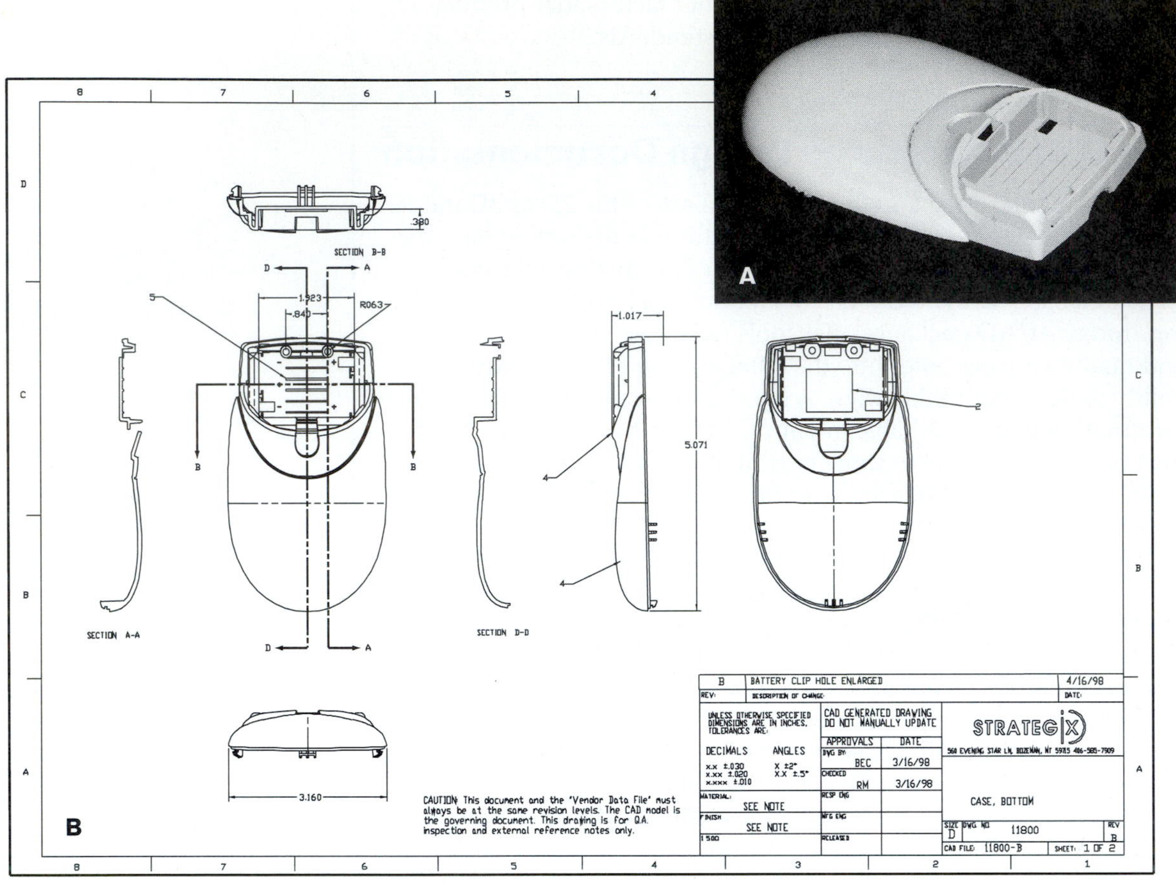

FIGURE 12.1

(a) A shaded view of the parametric 3D model for the bottom half of the case for a personal training device. (b) A dimensioned drawing for the same part with manufacturing tolerances listed in the title block. (Courtesy of Rob Mesaros, Strategix ID.)

Standards

You should be familiar with three broad classes of guidelines for preparing documentation drawings: national and international standards applicable to engineering drawings in general, discipline-specific standards applicable to a design area, and individual company drawing standards.

The American National Standards Institute (ANSI) describes a series of codes and standards that are commonly used in the United States. The American Society of Mechanical Engineers (ASME) publishes the ANSI standards, which are organized into groups and referred to by number. Appendix 1 includes a listing of the standards available from ASME and the information for ordering them. Each ANSI standard is broken down into a series of documents covering specific topics. Each of these documents is revised individually as needed. Many of the standards are safety standards or dimensional standards for interchangeable parts, in addition to drawing standards.

The International Standards Organization (ISO) publishes standards that govern international engineering standardization and the creation of metric drawings for international use. Other standards that may apply are those published by the American Society for Testing and Materials (ASTM) for materials and testing. These are very useful when you wish to specify a particular material, grade, standard stock size, or finish. The Society of Automotive Engineers (SAE) standards also specify material grades. The Industrial Fasteners Institute provides documentation standards for fasteners in addition to those provided by ANSI. The National Institute of Standards and Technology (NIST) is a governmental agency that may publish standards relevant for your design area. For federally and state funded projects, you should research government contracting regulations that may apply.

There may be other standards and practices for your particular area of interest. For example, dimensional standards in bridge construction differ greatly from those for manufacturing a laser optical device. Each design area has standards that apply to the processes used in that line of work. The American Congress on Surveying and Mapping (ACSM) provides guidelines for surveying and mapping standards. Civil engineering design and building systems have a number of standards they follow. ASTM also publishes some guidelines for building design and construction. The National Institute for Building Sciences (NIBS) publishes guides for federal construction that can be helpful for these areas. Steel fabrication standards and practices are available from the American Institute of Steel Construction. Appendix 2 lists the above organizations and their contact information.

ANSI Drawing Standards

The ANSI standards covered in the Y14, Y1, and Y32 series are pertinent to drawings and terminology, but additional information useful for representing particular types of equipment can be found in other standards as well. Table 12.1 lists some of the drawing topics covered in the Y14 standard.

TABLE 12.1

Selected Topics From the ANSI Y14 Standard

Name	Designation	Year
Decimal inch drawing sheet size and format	Y14.1	1995
Dimensioning and tolerancing	Y14.5M	1995
Line conventions and lettering	Y14.2M	1993
Metric drawing sheet size and format	Y14.1M	1995
Multiview and sectional view drawings	Y14.3M	1994
Pictorial drawing	Y14.4M-89 R94	1989
Screw thread representation	Y14.6-78 R87/93	1978

The standards that convey how engineering drawings should appear have been developed in order to make it easier for you to clearly communicate your design to manufacturers and others. Figure 12.2 shows a few of the standard symbols for piping. Details in the standards such as the sizes for text, arrowheads, and other

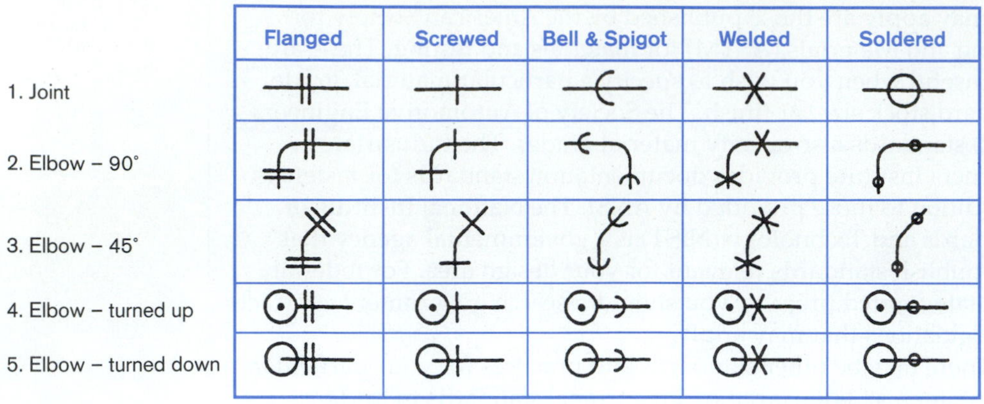

	Flanged	Screwed	Bell & Spigot	Welded	Soldered
1. Joint					
2. Elbow – 90°					
3. Elbow – 45°					
4. Elbow – turned up					
5. Elbow – turned down					

FIGURE 12.2
Standard symbols serve as a shorthand for information in the drawing. More of the ANSI standard Graphic Symbols for Pipe Fittings, Valves, and Piping are found in Appendix 45. (ANSI Y32.2.3–1949 (R1994).)

items in the drawing have been developed by groups of people who use this information in an industrial setting. Following the standards will help you produce good clear drawings that convey all of the needed information.

Working Drawings

Working drawings is the term applied to drawings that convey all of the information needed to manufacture and assemble a design. The term was coined to distinguish the drawings used to document a final design from those that might be created in the course of designing a project. As you have seen in the drawing control chapter, working drawings are frequently prepared by drafters, and they are checked carefully and approved by the designer and others before being released to production.

Working drawings include:

- an assembly drawing that shows how the parts assemble together;
- part drawings showing all of the information needed to manufacture each nonstock part;
- a parts list indicating all of the parts used in the assembly;
- specifications or written instructions of any additional details required in manufacturing and assembling the design.

Assembly Drawings

Assembly drawings show how all of the parts in an assembly fit together. A good assembly drawing shows the parts clearly enough that a worker who has the parts could put them together correctly. A single view may be enough. Figures 12.3 through 12.7 illustrate orthographic, isometric, and section views used for assembly drawings.

Section views are often used for assembly drawings because they can show the interior details of how parts fit together. To create a section view from an assembly drawing, first create an assembly

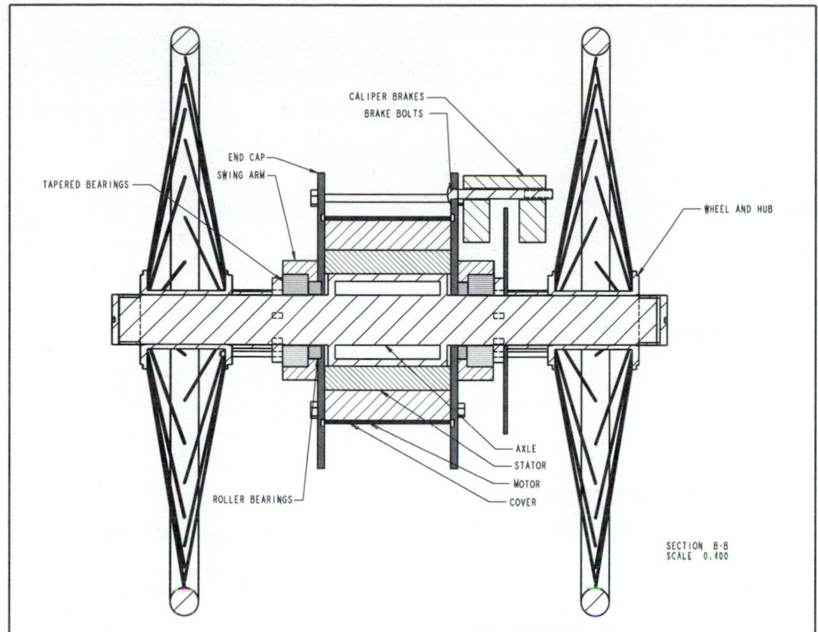

FIGURE 12.3
A section view of an assembly reveals the way parts fit together inside.

model where all of the parts are inserted in their fully assembled positions. Then use the CAD system's commands to select the cutting plane through the object (typically through the center of the object). Some CAD systems will have commands that automatically generate correctly shown hatch patterns or hatching boundaries from the section. With other CAD programs, you will have to use a command to slice the fully assembled object in two, then generate the hatching yourself. Yet other programs have sophisticated selections for the types of sections to define from the model.

For some assemblies, an exterior orthographic view as shown in Figure 12.4 or an exploded orthographic view as shown in Figure 12.5 may be sufficient to show how the individual parts assemble.

Isometric views are very useful for creating assembly drawings because they are easily interpreted. To create an assembly drawing from a CAD model, first assemble all of the parts fully together in an assembly model. Then use the fully assembled file to create an isometric view that shows the parts clearly as shown in Figure 12.6. If the fully assembled file does not show the parts clearly, then an exploded isometric view may be best as shown in Figure 12.7. To create an exploded isometric view from a CAD model, use the fully assembled parts and move them along the X, Y, or Z axis directions. Move each part far enough that it does not overlap other parts in the isometric view. Use this to generate the assembly drawing. Some CAD software, such as Pro/Engineer, has a command for viewing the model in an exploded assembly view, making it even easier to create exploded assembly drawings.

Assembly drawings do not typically show dimensions unless a critical distance needs to be maintained during assembly. For example, if two gears must be exactly 4 inches apart on a shaft, that dimension might be provided on the assembly drawing.

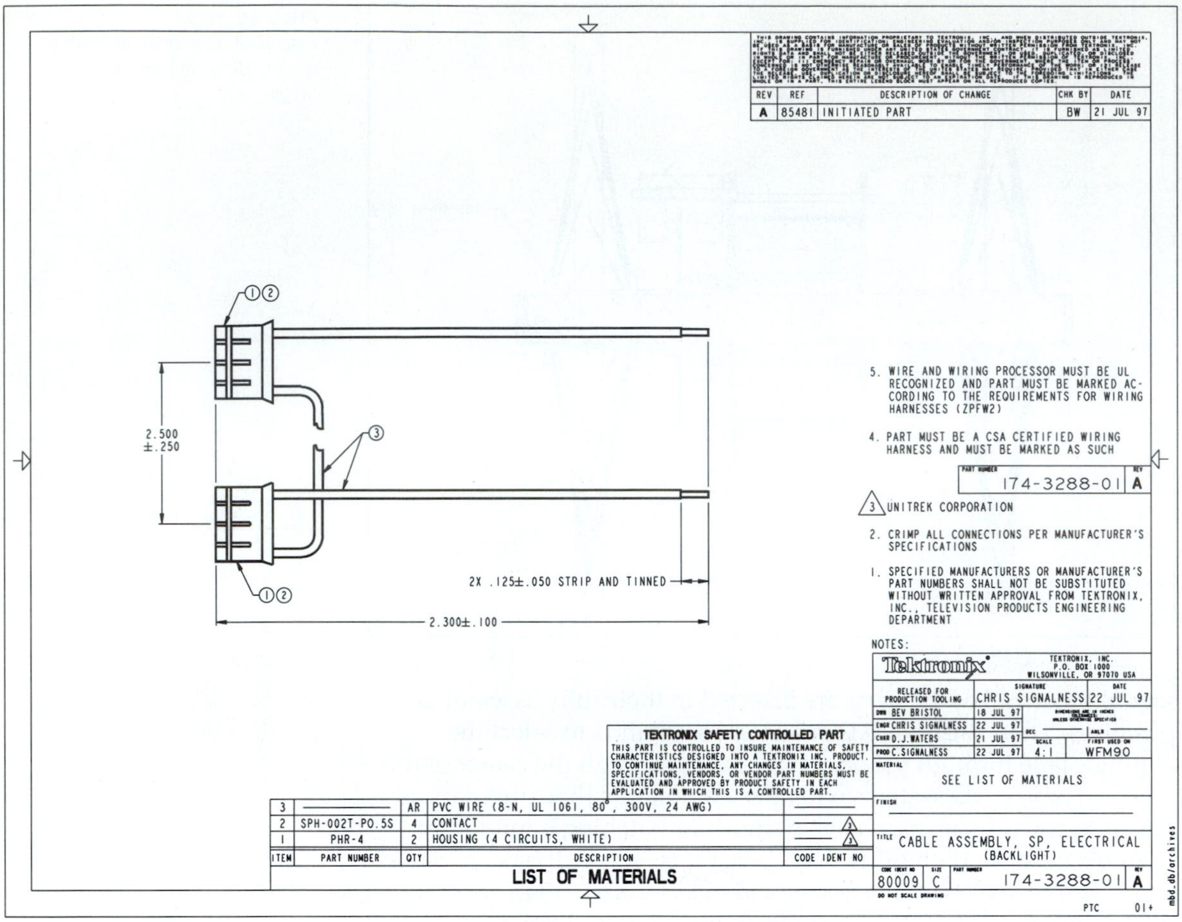

REV	REF	DESCRIPTION OF CHANGE	CHK BY	DATE
A	85481	INITIATED PART	BW	21 JUL 97

2.500 ±.250

2X .125±.050 STRIP AND TINNED

2.300±.100

NOTES:

1. SPECIFIED MANUFACTURERS OR MANUFACTURER'S PART NUMBERS SHALL NOT BE SUBSTITUTED WITHOUT WRITTEN APPROVAL FROM TEKTRONIX, INC., TELEVISION PRODUCTS ENGINEERING DEPARTMENT
2. CRIMP ALL CONNECTIONS PER MANUFACTURER'S SPECIFICATIONS
3. UNITREK CORPORATION
4. PART MUST BE A CSA CERTIFIED WIRING HARNESS AND MUST BE MARKED AS SUCH
5. WIRE AND WIRING PROCESSOR MUST BE UL RECOGNIZED AND PART MUST BE MARKED ACCORDING TO THE REQUIREMENTS FOR WIRING HARNESSES (ZPFW2)

PART NUMBER 174-3288-01 REV A

TEKTRONIX SAFETY CONTROLLED PART

THIS PART IS CONTROLLED TO INSURE MAINTENANCE OF SAFETY CHARACTERISTICS DESIGNED INTO A TEKTRONIX INC. PRODUCT. TO CONTINUE MAINTENANCE, ANY CHANGES IN MATERIALS, SPECIFICATIONS, VENDORS, OR VENDOR PART NUMBERS MUST BE EVALUATED AND APPROVED BY PRODUCT SAFETY IN EACH APPLICATION IN WHICH THIS IS A CONTROLLED PART.

ITEM	PART NUMBER	QTY	DESCRIPTION	CODE IDENT NO
3	——	AR	PVC WIRE (8-N, UL 1061, 80°, 300V, 24 AWG)	——
2	SPH-002T-PO.5S	4	CONTACT	—— 3
1	PHR-4	2	HOUSING (4 CIRCUITS, WHITE)	—— 3

LIST OF MATERIALS

Tektronix — TEKTRONIX, INC. P.O. BOX 1000 WILSONVILLE, OR 97070 USA

RELEASED FOR PRODUCTION TOOLING	SIGNATURE: CHRIS SIGNALNESS	DATE: 22 JUL 97
DWN BEV BRISTOL	18 JUL 97	DIMENSIONS ARE IN INCHES TOLERANCES: UNLESS OTHERWISE SPECIFIED
ENGR CHRIS SIGNALNESS	22 JUL 97	DEC / ANLR
CHKR D.J. WATERS	21 JUL 97	SCALE 4:1
PROD C. SIGNALNESS	22 JUL 97	FIRST USED ON WFM90

MATERIAL: SEE LIST OF MATERIALS

FINISH

TITLE: CABLE ASSEMBLY, SP, ELECTRICAL (BACKLIGHT)

CODE IDENT NO 80009 | SIZE C | PART NUMBER 174-3288-01 | REV A

DO NOT SCALE DRAWING

PTC 01+

mbd_db/archives

FIGURE 12.4
This assembly drawing uses a single orthographic view. (Copyright 1998 Tektronix, Inc. All rights reserved. Reproduced by permission.)

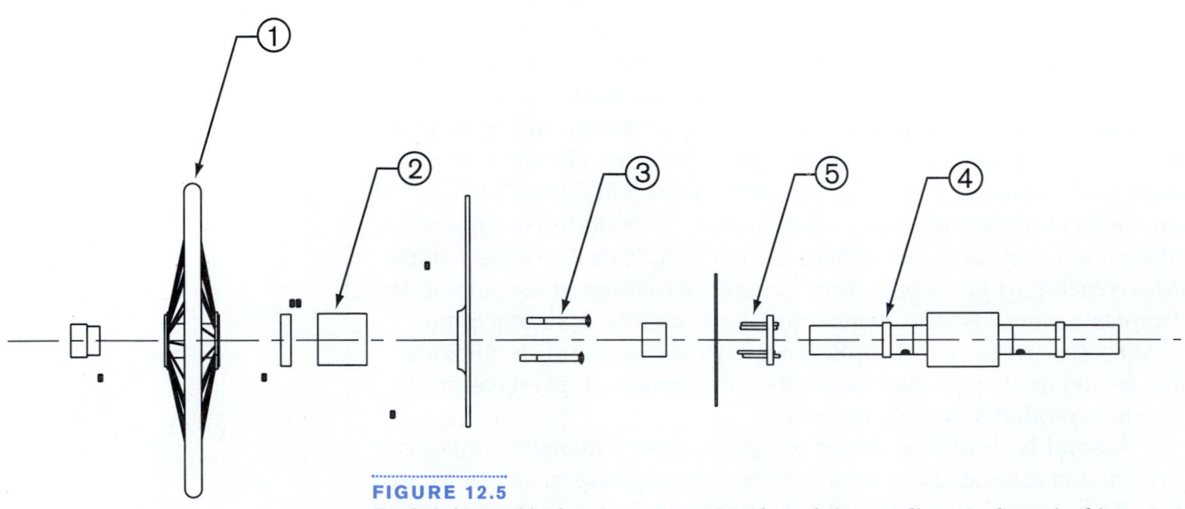

FIGURE 12.5
Exploded assembly drawings move parts along their centerlines, as shown in this drawing that uses a single orthographic view.

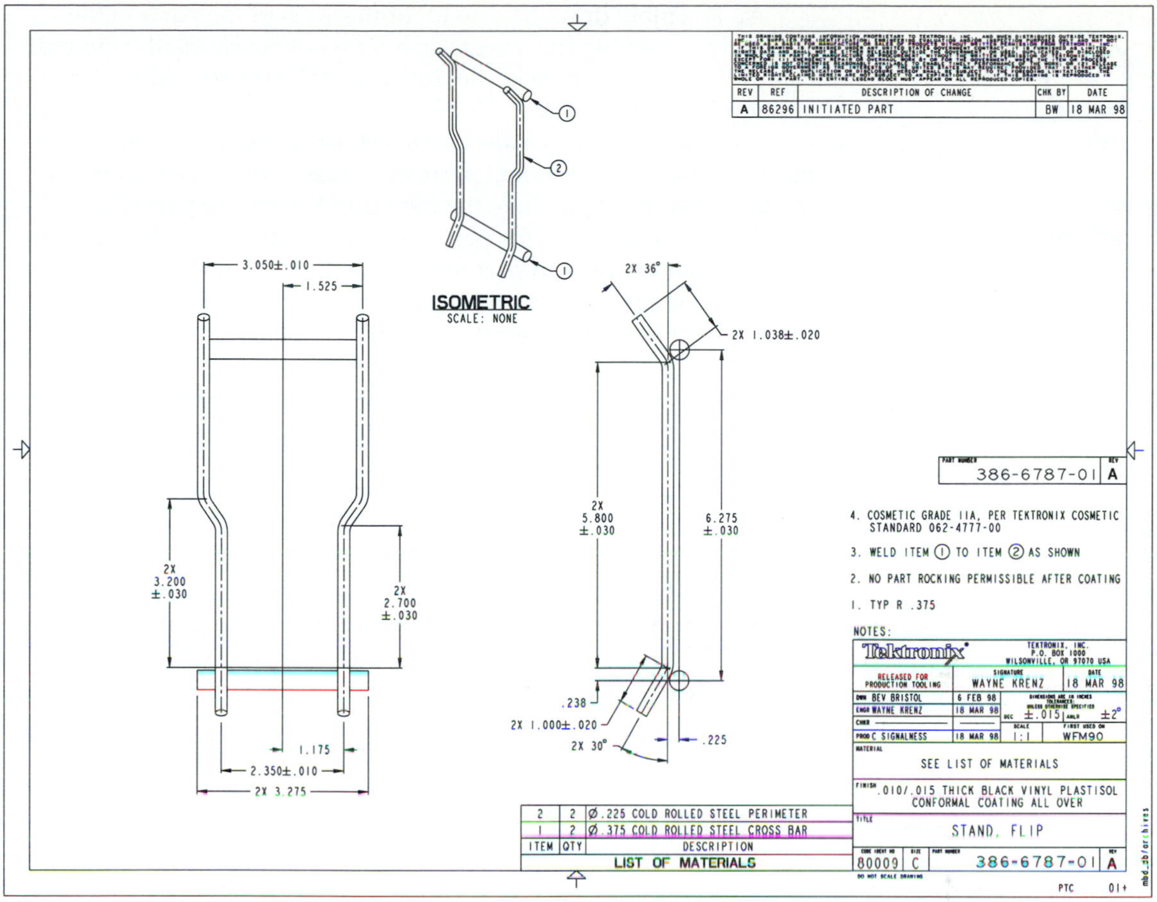

FIGURE 12.6
This assembly drawing uses a fully assembled isometric view to show how parts assemble. (Copyright 1998 Tektronix, Inc. All rights reserved. Reproduced by permission.)

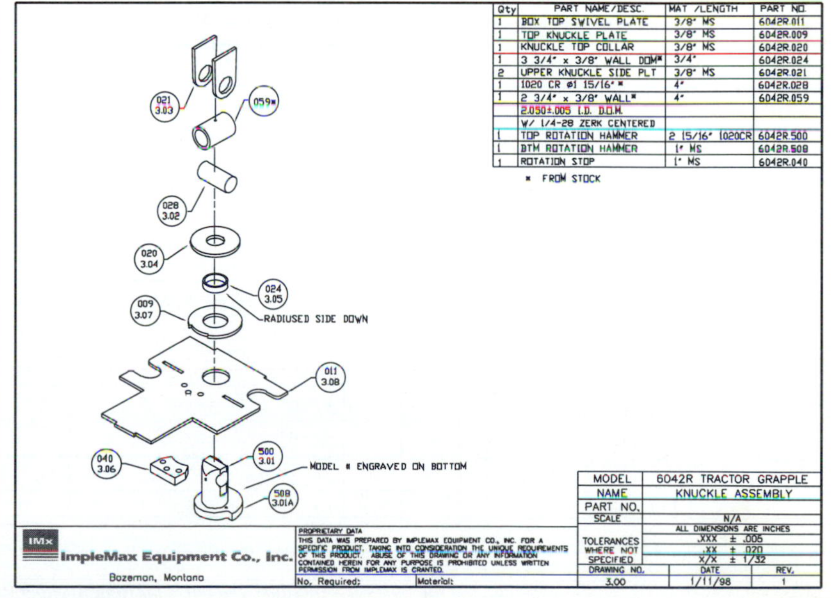

FIGURE 12.7
Exploded isometric views are frequently used for assembly drawings. (Courtesy of Implemax Equipment Co., Inc.)

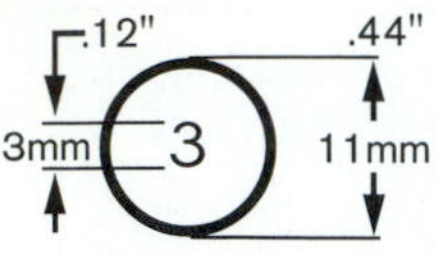

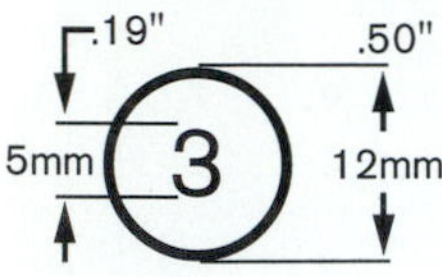

FIGURE 12.8
Dimensions and sizes for ball tags

An assembly drawing also identifies each of the parts in the assembly. Assembly drawings use *ball tags* or *bubble numbers* to indicate which part is which in the drawing. Ball tags should be drawn as shown in Figure 12.8

A parts list or *bill of materials (BOM)* provides a list of and information about the individual parts in the assembly. At a minimum, the parts list should include the item number for the part (the number in the ball tag), the part name, the material for the part, and the quantity of each part required in the assembly. Most companies also include their standard part numbers or stock numbers on the parts list to make it easy to track the purchasing and handling of parts.

Parts lists can be generated automatically by some CAD software. Assembly drawings are generally created by assembling the separate part files to form assembly models, then generating the views from it. If the information describing the part and its material has been stored with the individual part files, this information can be extracted automatically in the assembly drawing and displayed as the parts list. This is beneficial because if a part changes, the assembly drawing and parts list can be automatically updated. Figure 12.9 shows a typical parts list that uses company part numbers.

Part Drawings

Part drawings, also called *detail drawings,* are prepared for parts that will be manufactured or modified for the project. They are not necessary for standard or stock parts, so you should not waste time creating detailed part drawings for nuts, bolts, bearings, pumps, and other items that you will purchase and use as is.

Part drawings (see Fig. 12.10) usually show one part per drawing sheet, not multiple parts on one sheet, even if the part is small.

Item	QTY.:	Part Name:	Mat./Length	Part No.:
1	1	OUTER TONG SIDE PLATE	3/16" MS	6042R.007
2	1	INNER TONG SIDE PLATE	3/16" MS	6042R.008
3	1	TOP TONG PLATE	3/16" MS	6042R.012
4	1	BOTTOM TONG PLATE	3/16" MS	6042R.013
5	1	TONG HYDRAULIC BASE	1" MS	6042R.018
6	2	INTERNAL TONG SPACER	3/16" MS	6042R.031
7	1	3/16" X 1 3/16" B.S.	5"	6042R.458
8	1	3/16" X 1 11/16" B.S.	5"	6042R.033
9	1	2 3/4" x 3/8" WALL DOM	10 11/16	6042R.053
		W/ 1/4-28 ZERK		

FIGURE 12.9
This parts list includes the parts by item number, but also includes the company part number used for inventory. (Courtesy of Implemax Equipment Co., Inc.)

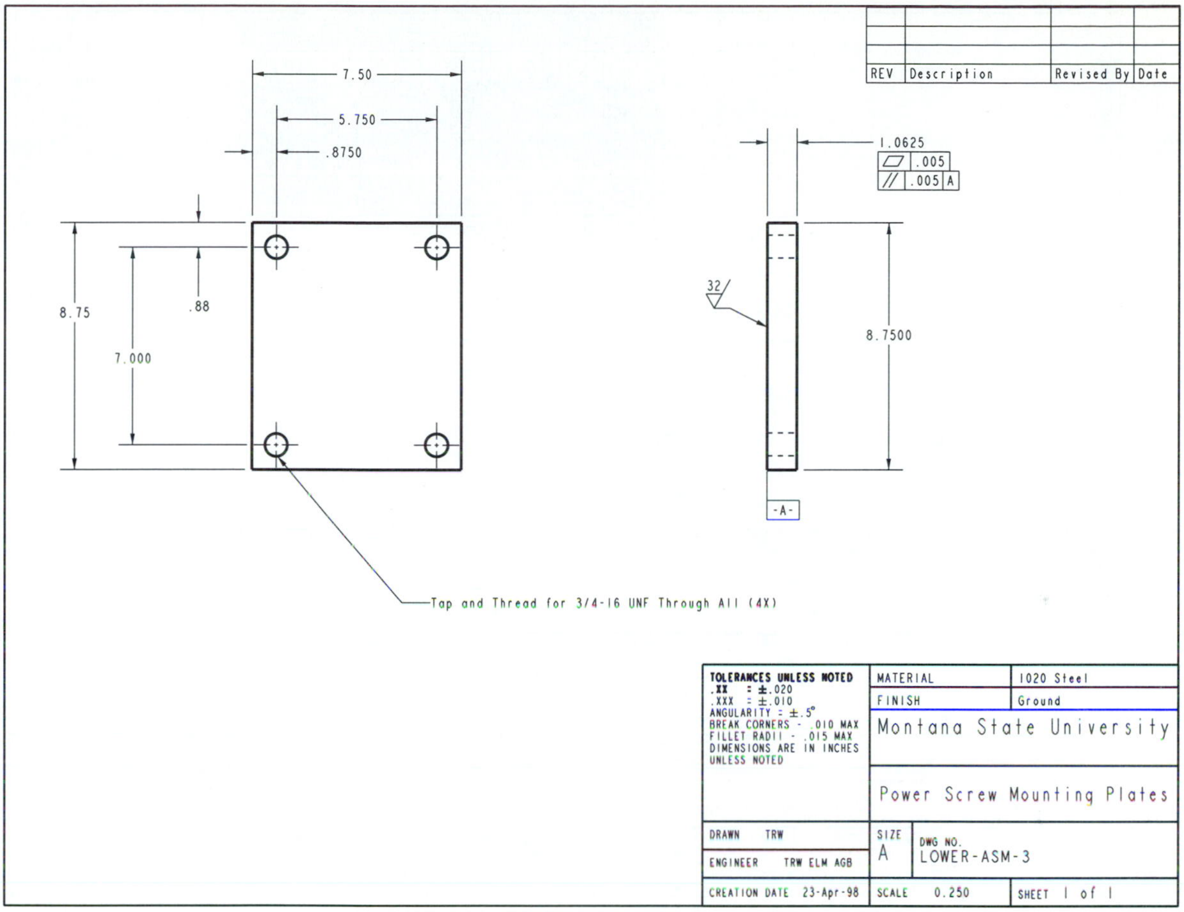

REV	Description	Revised By	Date

7.50
5.750
.8750
8.75
7.000
.88
1.0625
.005
.005 A
32
8.7500
-A-
Tap and Thread for 3/4-16 UNF Through All (4X)

TOLERANCES UNLESS NOTED .XX : ±.020 .XXX : ±.010 ANGULARITY : ±.5° BREAK CORNERS - .010 MAX FILLET RADII - .015 MAX DIMENSIONS ARE IN INCHES UNLESS NOTED	MATERIAL 1020 Steel FINISH Ground Montana State University Power Screw Mounting Plates
DRAWN TRW ENGINEER TRW ELM AGB	SIZE A DWG NO. LOWER-ASM-3
CREATION DATE 23-Apr-98	SCALE 0.250 SHEET 1 of 1

FIGURE 12.10
Part drawing

By doing so, each part number corresponds to a single drawing that has all of the information for that part. Also, if the same part is used in another assembly, the drawing can be used as is without causing confusion. For very small projects, occasionally more than one part drawing will be shown per sheet, but in general it is not a good practice.

Part drawings include:

- drawing views
- dimensions
- tolerances
- material designation
- finishes
- notes
- title and revision block

Figures 12.11 through 12.15 illustrate a set of working drawings for VKI Technologies's Zuma brewer, profiled in the case study in Chapter 8. Individual part drawings for the parts shown in the assembly drawing were prepared for 41 of the parts; 7 of the parts were standard parts that did not require part drawings. Only four of the part drawings are shown here; they illustrate detailed drawings for sheet metal, plastic, and machined parts.

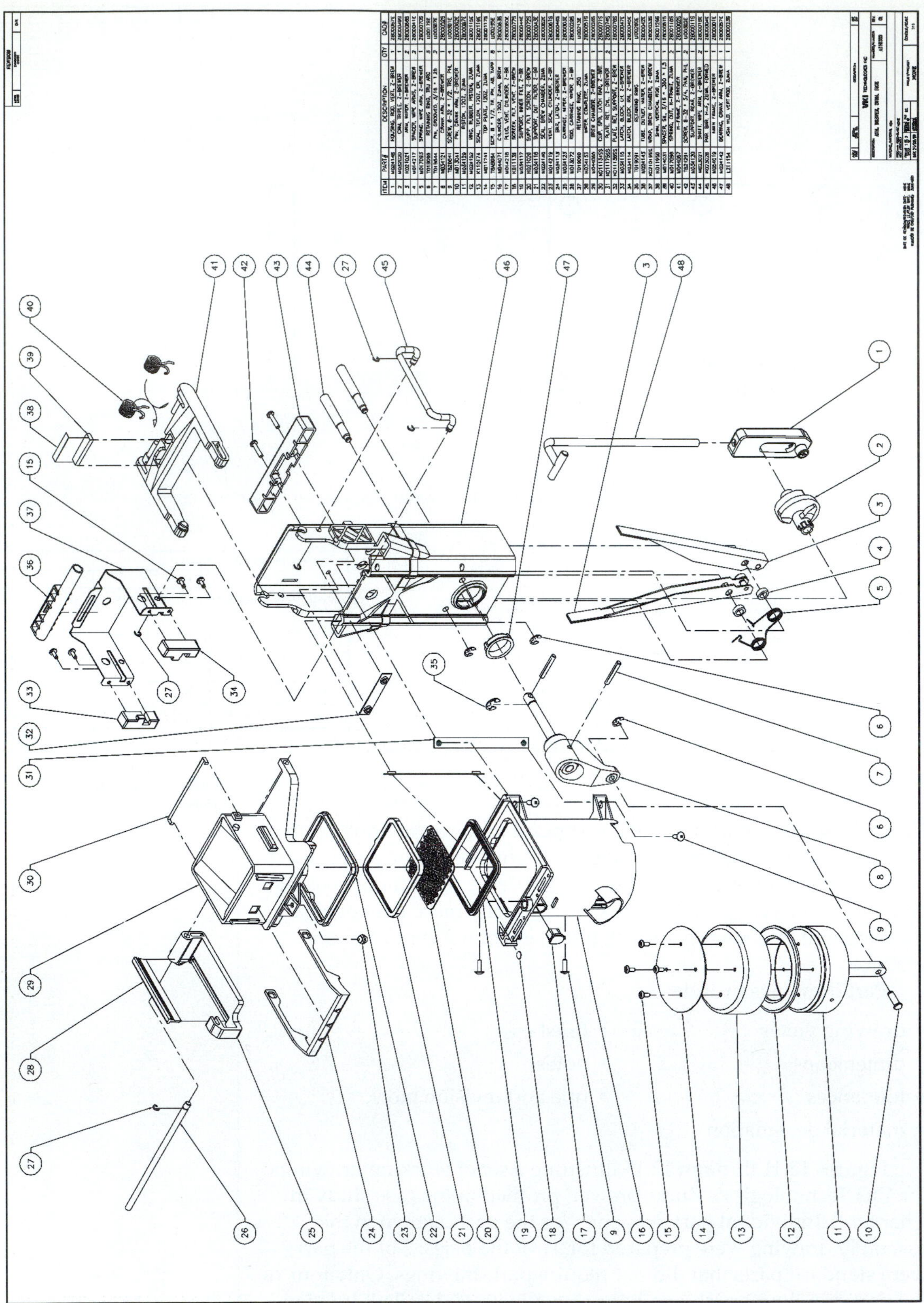

FIGURE 12.11
Assembly drawing for the Zuma Brewer (Courtesy of Stan McLean, VKI Technologies.)

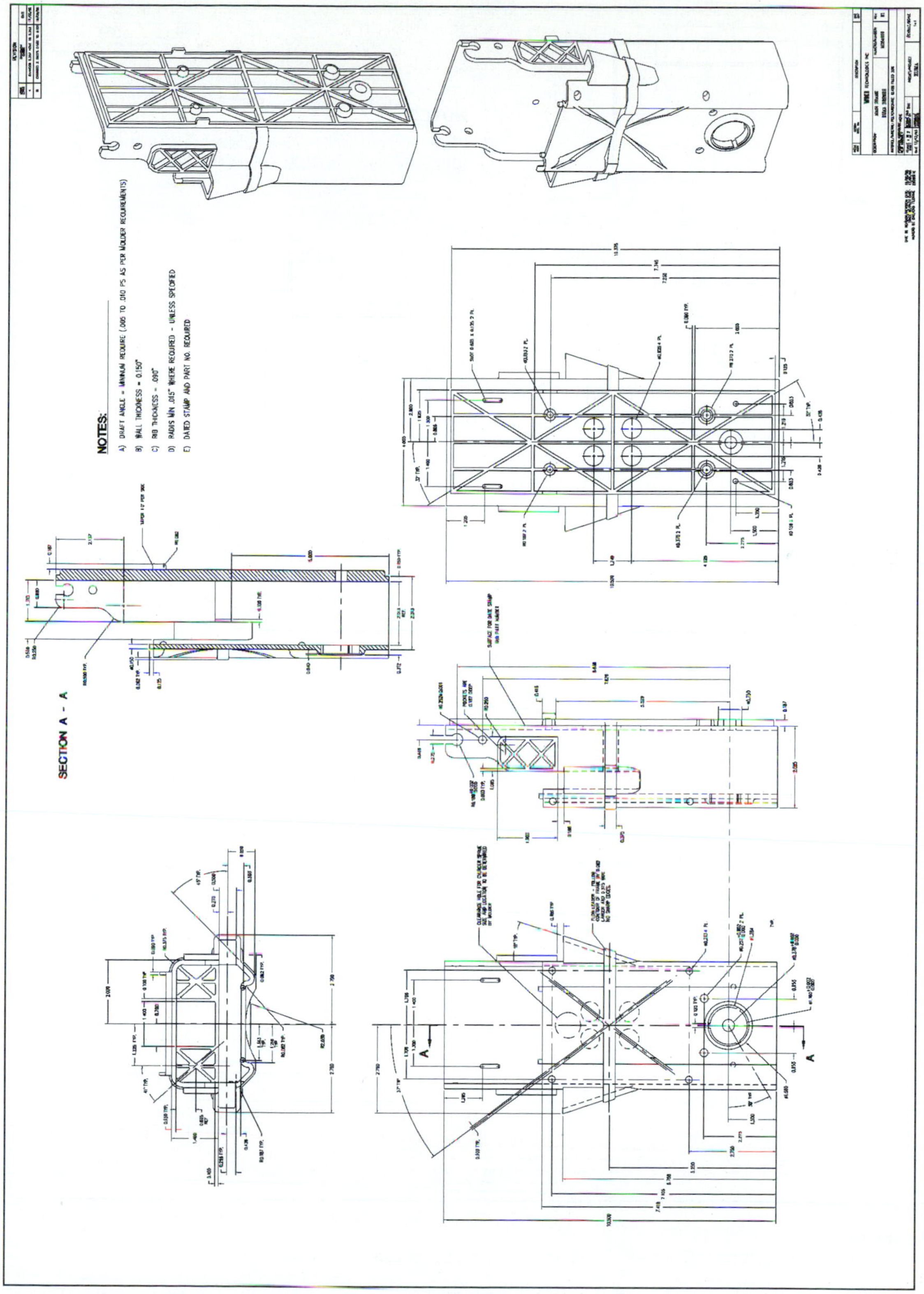

NOTES:

A) DRAFT ANGLE = MINIMUM REQUIRE (.005 TO .010 PS AS PER MOLDER REQUIREMENTS)
B) WALL THICKNESS = 0.150"
C) RIB THICKNESS = .090"
D) RADIUS MIN .015" WHERE REQUIRED - UNLESS SPECIFIED
E) DATED STAMP AND PART NO. REQUIRED

SECTION A - A

FIGURE 12.12
Part drawing for a plastic part (Courtesy of Stan McLean, VKI Technologies.)

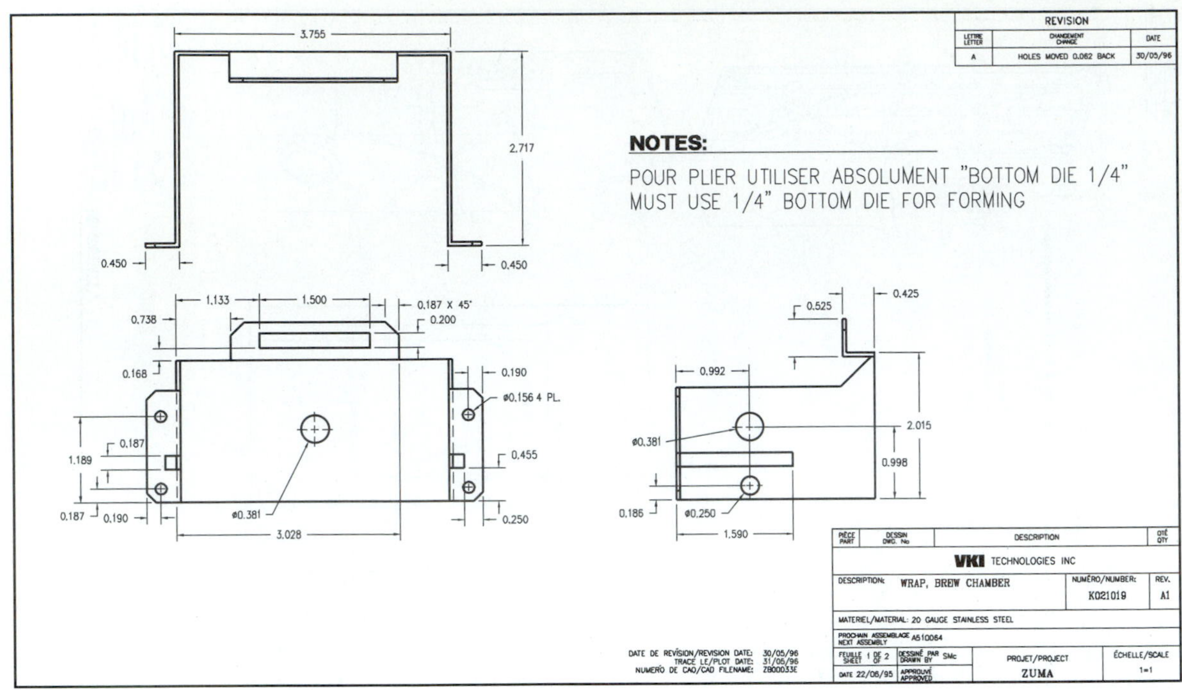

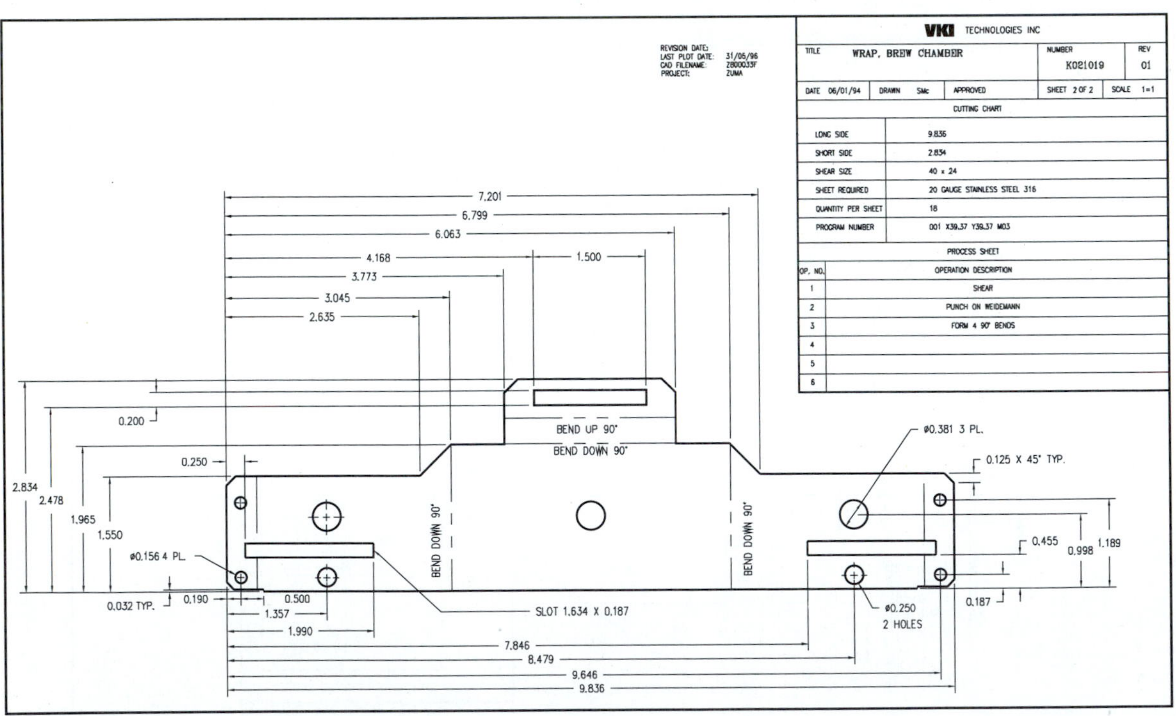

FIGURE 12.13

Part drawing for a sheet metal part (Courtesy of Stan McLean, VKI Technologies.)

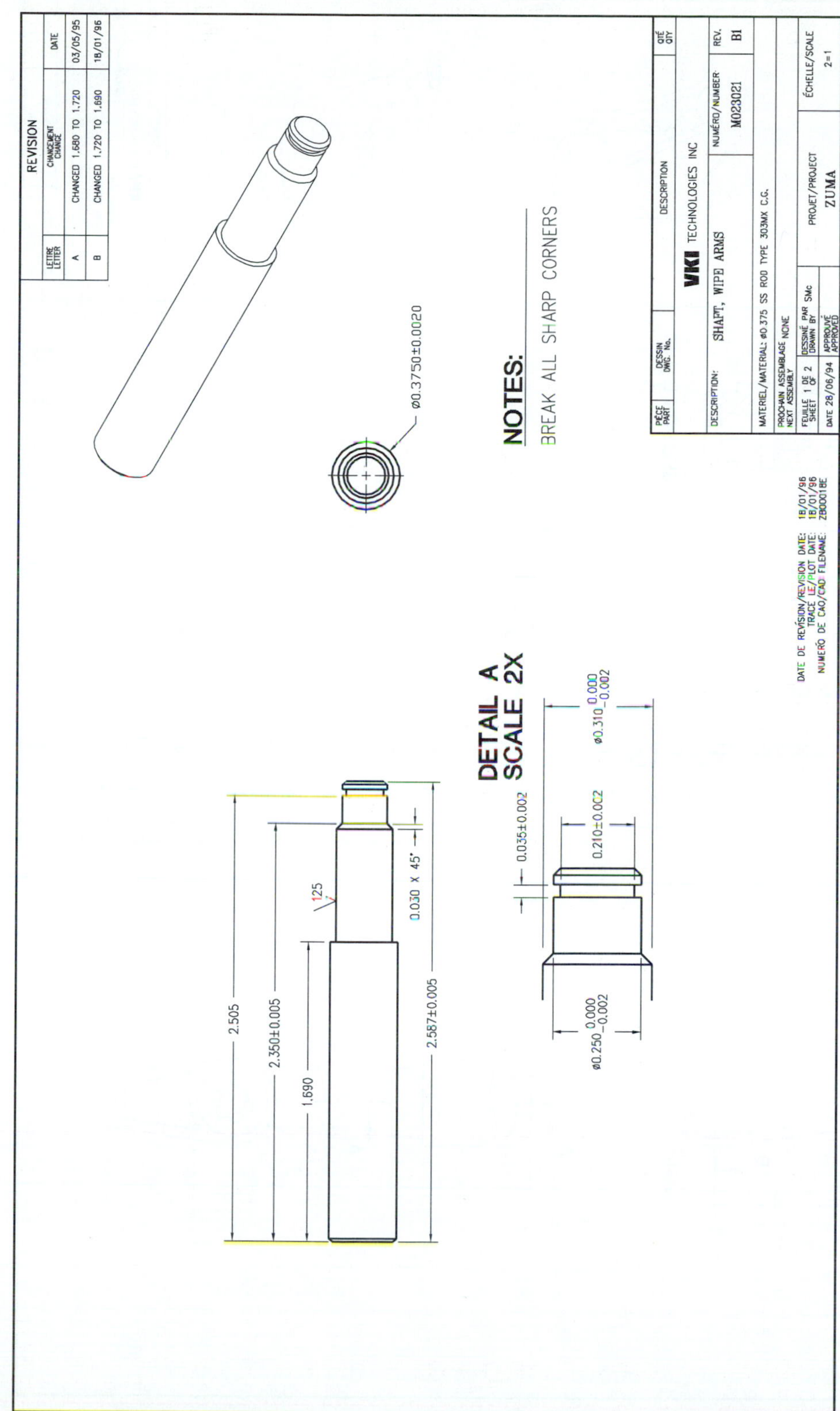

FIGURE 12.14

Part drawing for a machined part (Courtesy of Stan McLean, VKI Technologies.)

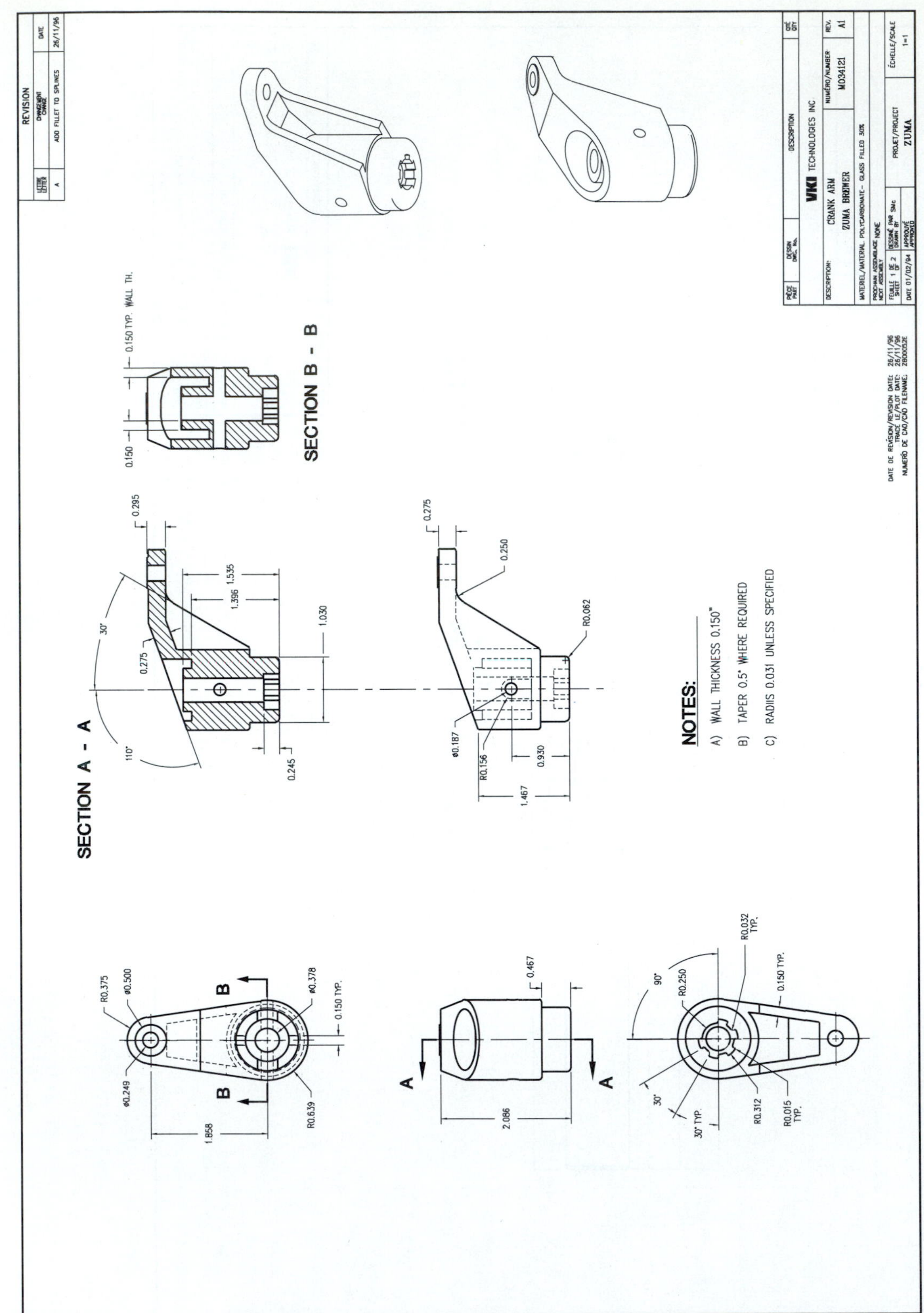

FIGURE 12.15

This part drawing includes section views to better define the interior of the part. (Courtesy of Stan McLean, VKI Technologies.)

Preparing Working Drawings

Each working drawing is prepared in a similar way:

1. Plan the views needed and determine the drawing size.
2. Determine the scale for the views.
3. Locate the views on the sheet.
4. Add the title, revision blocks, notes, and other standard elements.
5. Dimension and tolerance the views.

These basic steps will be explained in more detail in the rest of this chapter. If you keep in mind that your intent for the drawing is to communicate, you will be able to judge if your drawing is good. Ask yourself, "Does this drawing show the reader the information I am trying to convey?" If not, revise it until the information is presented clearly.

Standard Sheet Sizes

When you are developing documentation drawings, one of the first things to consider is the size of the sheet of paper the drawing will be plotted on. Many companies standardize on a single size for all plotted drawings, or on a few standard sheet sizes. This is convenient for the company, because there is no need to stock a full range of different sizes of plotter paper, but also for the manufacturer, field engineer, or anyone else who is likely to work with sets of printed drawings, because it is sometimes easy to lose smaller sheets that are included in sets of larger drawing sheets. Tables 12.2 and 12.3 describe the ANSI and ISO standards for drawing sheet sizes.

TABLE 12.2

ANSI Standard Sheet Sizes (for Inch Sheets)

Designation	Sheet size in inches
A	11 × 8.5
B	17 × 11
C	22 × 17
D	34 × 22
E	44 × 34

TABLE 12.3

ISO Standard Sheet Sizes

Designation	Metric sheet size (in mm)	English equivalent sheet (in inches)
AO	1189 × 841	44 × 34 (E)
A1	841 × 594	34 × 22 (D)
A2	594 × 420	22 × 17 (C)
A3	420 × 297	17 × 11 (B)
A4	297 × 210	11 × 8.5 (A)

If your company allows the use of more than one sheet size, you should select a size that will allow you to show the drawing views at a reasonable scale so the information in the drawing is clearly visible. For complex parts and assemblies that require many views or for very detailed parts, use large sheet sizes. For small parts or assemblies, you can use smaller sheets.

The standards for the size of drawing elements such as lettering, dimension text, gaps, and line weights are based on their plotted sizes on the drawing sheet. Although some CAD packages make it fairly easy to resize the drawing, others do not, so it is usually a good idea to decide on the sheet size before you start laying out and dimensioning the views.

To determine the sheet size needed, first consider the number of views you wish to show and whether they will fit full size on a drawing sheet. If they will not, you can experiment with the model to see how a view will appear at different reductions. This can give you an estimate of the amount of scaling that will still produce a legible view. Alternatively, you can determine the longest horizontal and vertical dimensions in the standard views and use these values to calculate the scale needed to fit them on the sheet. (You will learn about scale in the next section.)

The development of portable computers has allowed more companies to provide collections of drawings in digital format via the Internet or on CD-ROM and has contributed to printed drawings being less and less common. Nevertheless, the 2D views in the drawing to be viewed via the Internet or on CD-ROM are generally arranged as on standard sheets so they may be printed easily, with text, dimensions, and drawing details visible at the printed size. Portable computers with small screens allow you to use zoom commands to enlarge an area of interest to see the details.

Scale

The CAD database should always be created so that parts are modeled full size so the database can be used for analysis and manufacturing. When 2D views of a part are being prepared for plotting on a drawing sheet, however, the part will not necessarily fit on the paper full size. Drawing views are typically shown to a standard scale on the sheet. A *scale* is the ratio of the full size to the size of the enlarged or reduced view of the part. Common scales are used so that the reader can relate to the size of the actual object. That is, it is easier to visualize what an object that is 100 or 200 times larger than the view would look like than it is to visualize the object 157 times larger. The scale for the plotted drawing is noted in the title block. The scale note gives the relationship between plotted units to real world units.

Metric scales are indicated by separating the plotted units and the real world units with a colon. English units use an equal sign to separate the two values (although for large map scales a colon is often used). Some typical scales are listed in Table 12.4; a more complete listing of standard scales is in Appendix 3.

TABLE 12.4

Typical Scales

Scale	Note Format	Description
Metric scales	1:1	Full
	1:2	Half
	1:5	Fifth
	1:10	Tenth
	1:20	Twentieth
	1:100	Hundredth
	2:1	Double
Metric scales for civil and mapping (*for mapping the units are typically meters)*	1:100	One plotted unit equals one hundred real world units
	1:500	One plotted unit equals five hundred real world units
English scales	1 = 1	Full
	1 = 2	Half
	1 = 4	Quarter
	1 = 8	Eighth
	1 = 10	Tenth
	2 = 1	Double
English scales for civil and mapping	1 inch = 10 feet	One inch equals 10 feet
	1 inch = 20 feet	One inch equals 20 feet
	1 inch = 1000 feet	One inch equals 1000 feet
	1:24000 or 1 inch = 2000 feet	One inch equals 2000 feet
	1:62500 or 1 inch = 1 mile	One inch equals 1 mile
English scales for building construction	1 inch = 10 feet	One inch equals 10 feet
	1 inch = 20 feet	One inch equals 20 feet
	1 inch = 50 feet	One inch equals 50 feet

Developing Views From 3D Models

Many CAD packages provide commands to project 2D views automatically from the 3D model to create standard documentation drawings. Figure 12.16 shows a 3D model and the 2D drawings that were created directly from the model using Pro/Engineer software.

For many parametric modeling software packages, creating orthographic views from the model is simple. Two basic steps are required: selecting a location on the page where the orthographic view will be placed, and selecting a surface on the object that would be parallel to the orthographic viewing plane. Adding more orthographic views is even easier, as they are automatically placed in relation to the initial, front view. The software orients and aligns the views around the front view in the drawing automatically. The same basic approach also allows the easy creation of section, isometric, and auxiliary views.

Even if your CAD package does not automate the creation of orthographic drawing views to this extent, creating orthographic

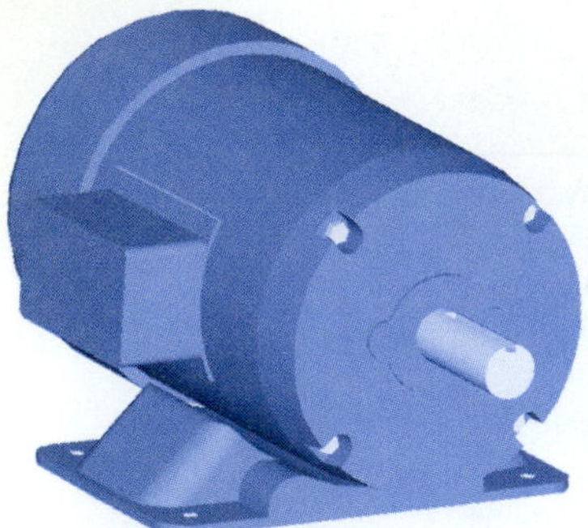

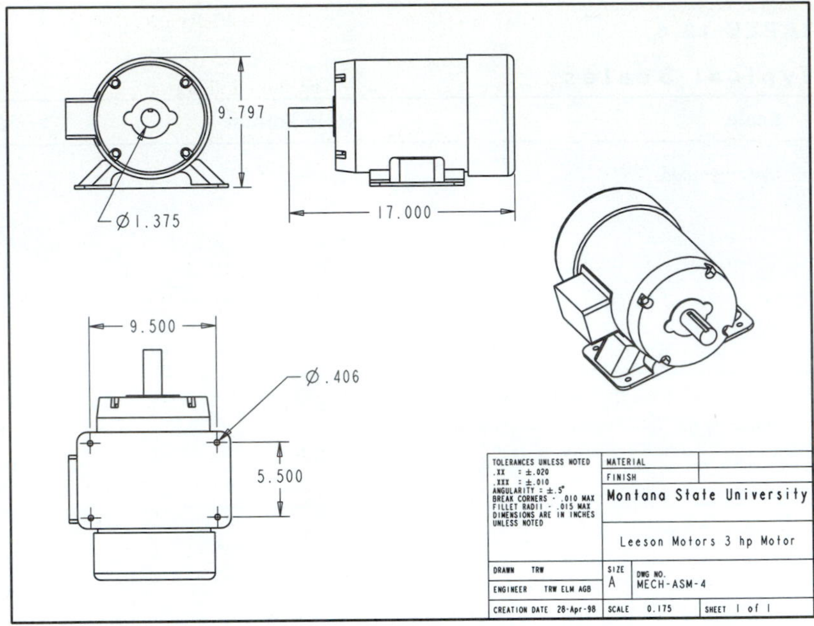

FIGURE 12.16
These drawing views were generated automatically from the 3D model of the part.

views directly from the 3D model still offers an advantage. Even engineers trained in orthographic projections of views will have difficulty visualizing complex orthographic views correctly all of the time—and the views can be time-consuming to create. Using the 3D object to develop the views can be a big help in getting the projections right.

When you are selecting which views to show, keep in mind the practices that make a good sketch:

- Show the shape of the object clearly in the front view, which is usually placed in a central location on the drawing sheet. (See Chapter 4 to review selecting the front view.)
- For objects that have a uniform thickness, only one view is necessary to fully define the part. The thickness can be defined in a note (see Fig. 12.17).
- For more complicated parts, show at least two drawing views in the standard arrangement on the sheet to make it easy to interpret the views. Remember that the reader of the drawing is not as familiar with the shape of the part as you are. Additional views can be placed on a second drawing sheet if necessary, although it is preferable to have the entire part on one sheet if it can be shown clearly. If multiple sheets are used, indicate that in the title block with a note such as "Sheet 1 of 3."
- Make the drawing clear and easy to interpret. Use different line thicknesses to make the object stand out clearly.
- Show the object at a size where its details are clear, or include an enlarged detail view showing those features to a larger scale. Figure 12.18 shows an example of an enlarged detail.

- Except in the case of enlarged details, show all views at the same scale.
- Clearly note the difference in scale for enlarged details.

Show only as many views as necessary. Do not add unnecessary views, as they tend to confuse the reader. Similarly, every hidden line need not always be shown. You can simplify views by leaving out hidden lines that do not add to the interpretation of the view. If you are going to leave out hidden lines in a view, however, it is a good idea to alert the reader by adding a note or labeling the view as a partial view.

Placing the Views

As you plan the drawing and the views to be included, allow for plenty of white space on the drawing. Leave room for dimensions between the views and separate the views from one another with white space (use 1 1/2 inches of space between dimensioned views if

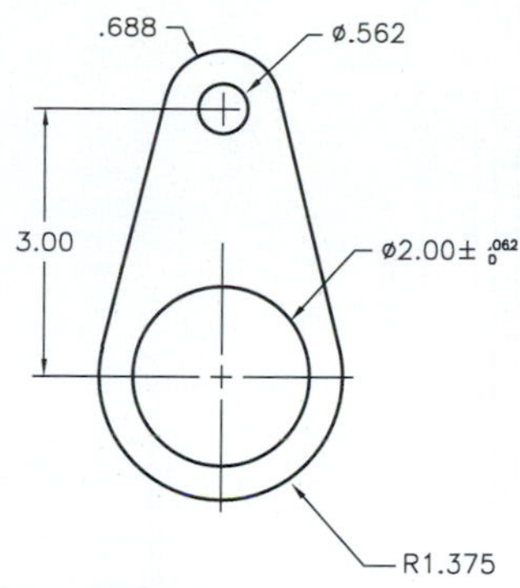

FIGURE 12.17
Save time by eliminating unnecessary views. This object is fully defined with a single view and a note. (Courtesy of Implemax Equipment Co., Inc.)

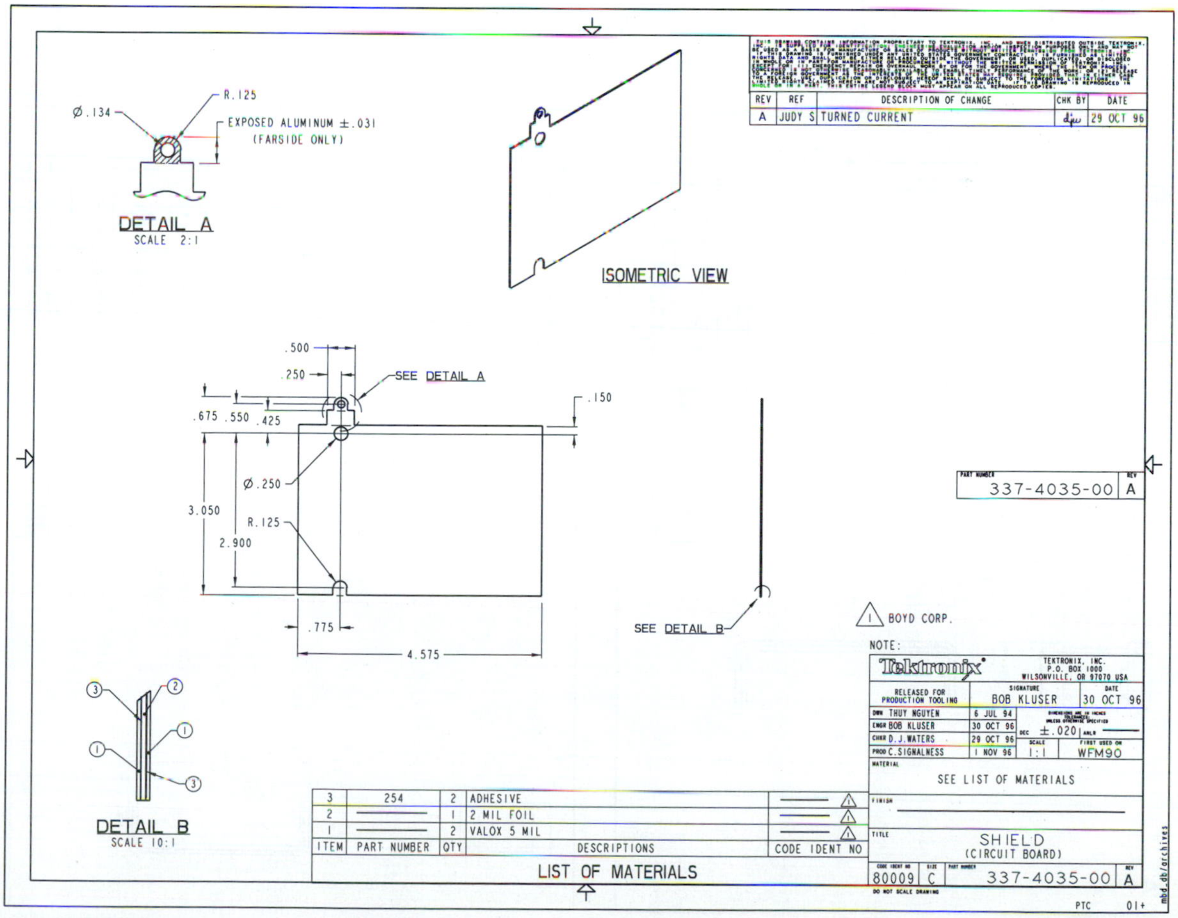

ITEM	PART NUMBER	QTY	DESCRIPTIONS	CODE IDENT NO
3	254	2	ADHESIVE	
2		1	2 MIL FOIL	
1		2	VALOX 5 MIL	

LIST OF MATERIALS

FIGURE 12.18
Enlarged detail A shows the dimensions for the exposed aluminum tab/hole more clearly. (Copyright 1998 Tektronix, Inc. All rights reserved. Reproduced by permission.)

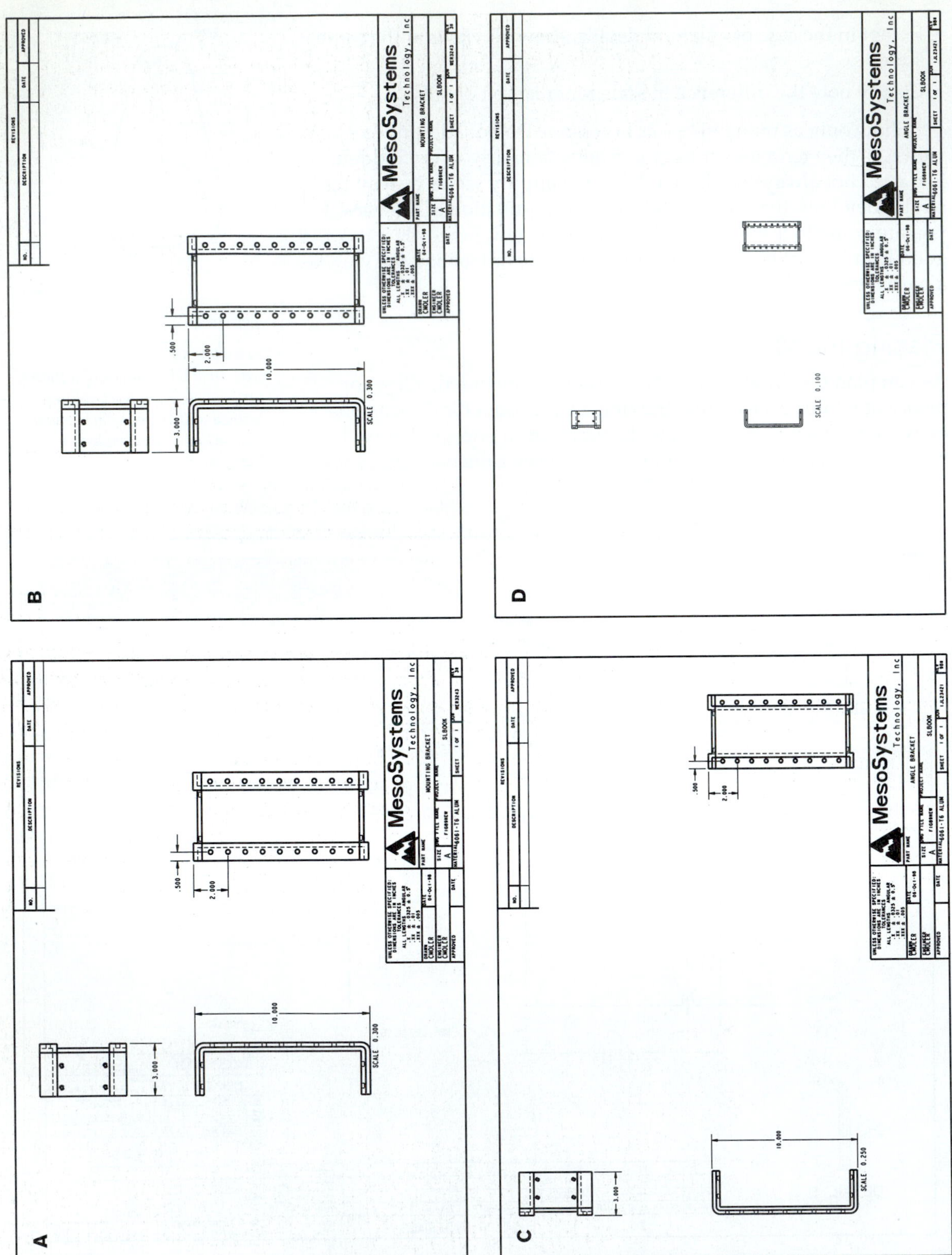

FIGURE 12.19

View spacing: (a) Good practice; (b) Too little space between views; (c) Too much space between views; (d) Views are too small to be legible.

you are unsure how much is enough). The space between drawing views does not have to be equal, yet the views should appear to be related. Don't space them too far apart. If there is not enough white space in your drawing, or if the views are too small to see the details clearly [as in Fig. 12.19(d)], consider using a larger sheet size.

Isometric Views

Isometric views can be included in detail drawings to help others interpret the drawing. Isometric views are often shown in the upper right-hand area of the drawing, as there is often room for it there. If you add an isometric view, it does not have to appear at the same scale as the other views. Often a smaller scale is just as clear and fits on the sheet better. It is not necessary to indicate the scale of the isometric view on the drawing. Figure 12.20 shows an example of an isometric view added to a detail drawing. Remember that hidden lines are not usually shown in isometric views.

If you can create an isometric view easily from a 3D model, add it to the drawing. In most packages, adding an isometric view involves using a direction of sight that will produce a view as though looking through the diagonal of a cube. In some CAD packages, such as AutoCAD, you can specify the endpoint of a vector

FIGURE 12.20
The isometric view provides an easy visual reference for the part described in the orthographic views.

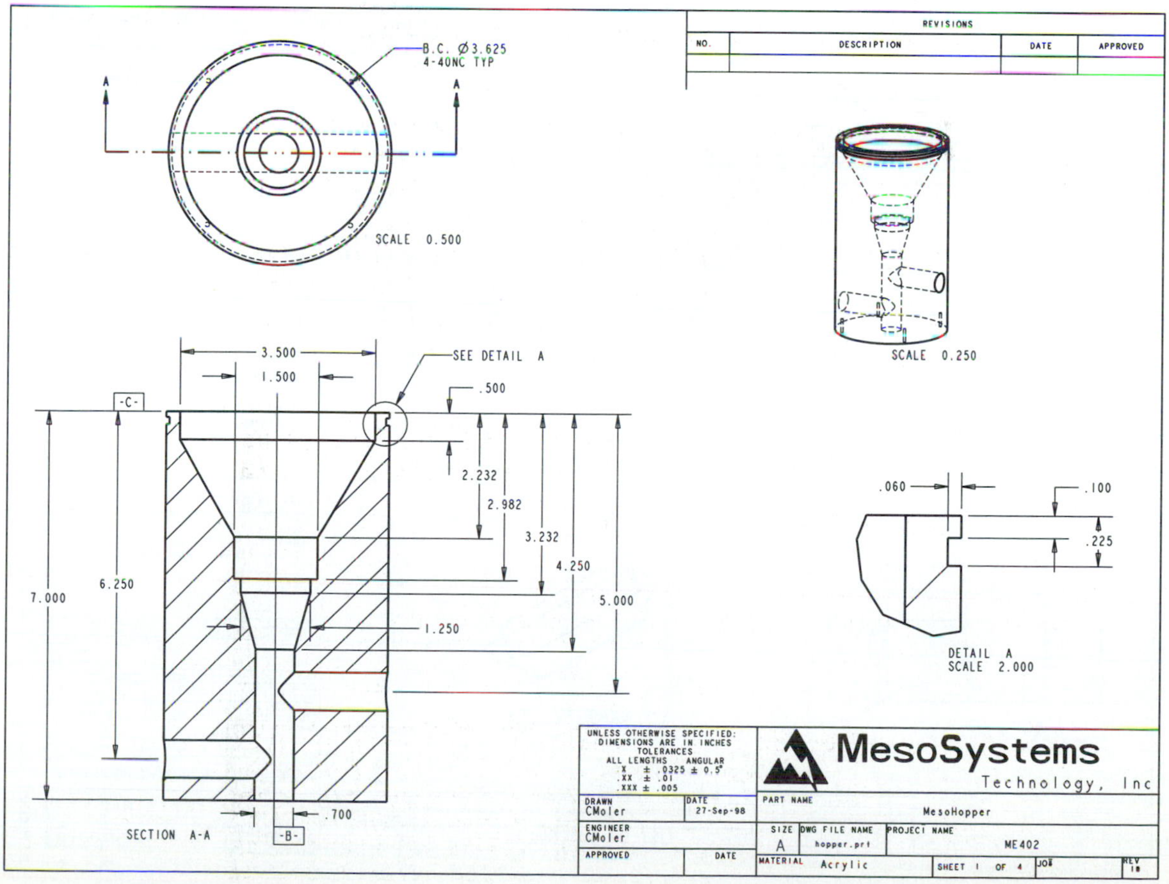

REVISIONS
NO. DESCRIPTION DATE APPROVED
B.C. Ø3.625
4-40NC TYP
A A
SCALE 0.500
SCALE 0.250
3.500
1.500
SEE DETAIL A
.500
-C-
2.232
2.982
3.232
4.250
5.000
7.000
6.250
1.250
.700
SECTION A-A
-B-
.060
.100
.225
DETAIL A
SCALE 2.000
UNLESS OTHERWISE SPECIFIED:
DIMENSIONS ARE IN INCHES
TOLERANCES
ALL LENGTHS ANGULAR
.X ± .0325 ± 0.5°
.XX ± .01
.XXX ± .005
MesoSystems Technology, Inc
DRAWN CMoler DATE 27-Sep-98
PART NAME MesoHopper
ENGINEER CMoler
SIZE A DWG FILE NAME hopper.prt PROJECT NAME ME402
APPROVED DATE
MATERIAL Acrylic SHEET 1 OF 4 JOB REV

toward the origin to produce the viewing direction. For example, a 1-unit cube with one corner at the origin would produce an isometric view when viewed as though looking through point 1,-1,1 toward the origin. Most modern CAD packages have preset isometric viewing directions for the model that you can select.

Auxiliary Views

Auxiliary views are used to create true size views of surfaces that do not appear true size in any orthographic view. A surface may be fully defined in a drawing without appearing true size, but a true size view is often necessary for pattern layouts and to show features true size for dimensioning. When sketching or using 2D CAD, an auxiliary view is projected from existing views, using the methods you learned in Chapter 4. Using 3D CAD, you can create a true-size view by aligning the viewing plane with the surface (which is the same as selecting a direction of sight for the view that is perpendicular to the surface).

Primary auxiliary views are auxiliary views that are projected directly from one of the standard orthographic views, such as the top, front, or side view. *Secondary auxiliary* views are auxiliary views that are projected from a primary auxiliary view.

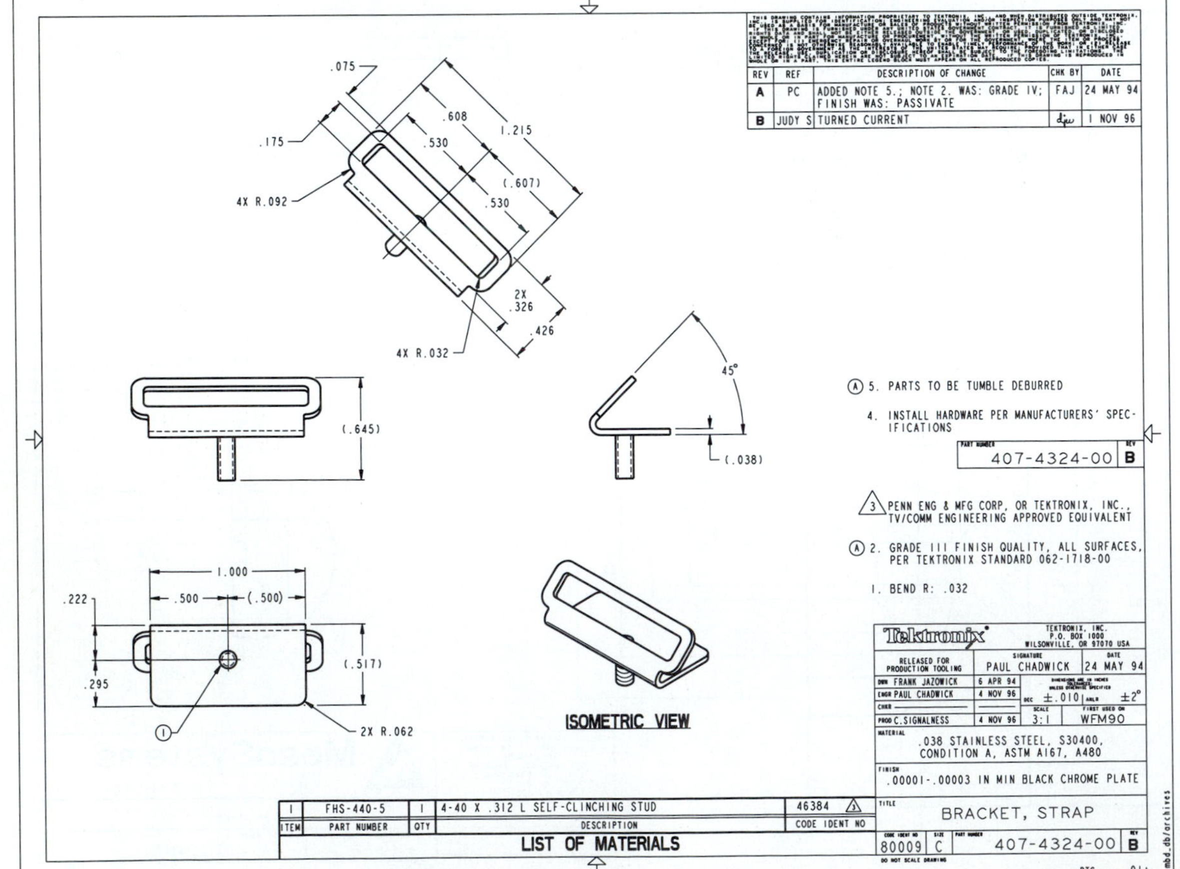

FIGURE 12.21
An auxiliary view for an inclined surface. (Copyright 1998 Tektronix, Inc. All rights reserved. Reproduced by permission.)

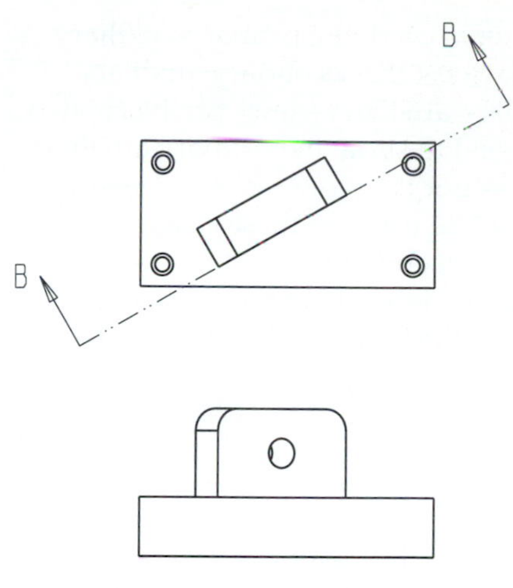

FIGURE 12.22
The viewing plane line in this figure indicates the direction of sight for the auxiliary view.

Auxiliary views should be located on the drawing in the same way that other orthographic views are related (as much as possible). In other words, they should project directly from a view shown in the drawing. The auxiliary view should be shown in its proper orientation with the view from which it is projected. Figure 12.21 shows a 2D projection of the auxiliary view for an inclined surface. For primary auxiliary views, it is usually easy to show them in direct projection.

If the auxiliary view cannot be fit on the drawing sheet in the proper projection, then a viewing plane line (representing the edge view of the auxiliary viewing plane—and similar to a cutting plane line for section views) should be used to indicate the viewing direction for the auxiliary view. The rotation of the auxiliary view should be preserved in its projected orientation even when it is placed elsewhere in the drawing. Figure 12.22 shows a viewing plane line used to indicate the direction of sight for an auxiliary view that is located away from its projected location. The arrows on the viewing plane line indicate the direction of sight.

Figure 12.23 shows the steps in creating an auxiliary view using Pro/Engineer software. Auxiliary views can be created as easily as other views by selecting the viewing direction on the part and deciding where to place the view on the drawing sheet.

FIGURE 12.23
3D solid modeling software makes it easy to create auxiliary views. Using Pro/Engineer software you (a) select the location for the view, (b) select the surface to be shown, then (c) select the scale for the view.

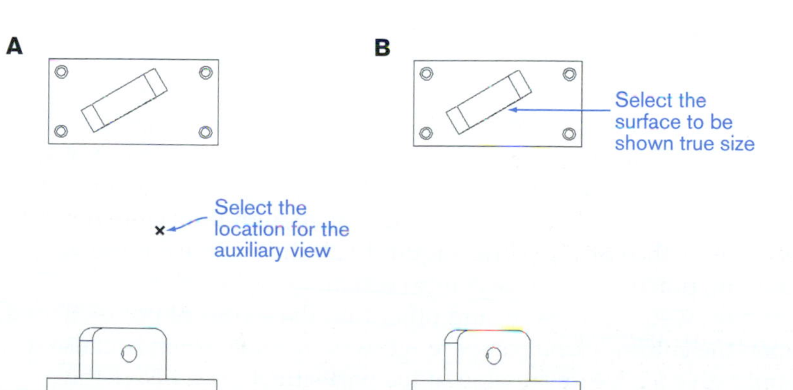

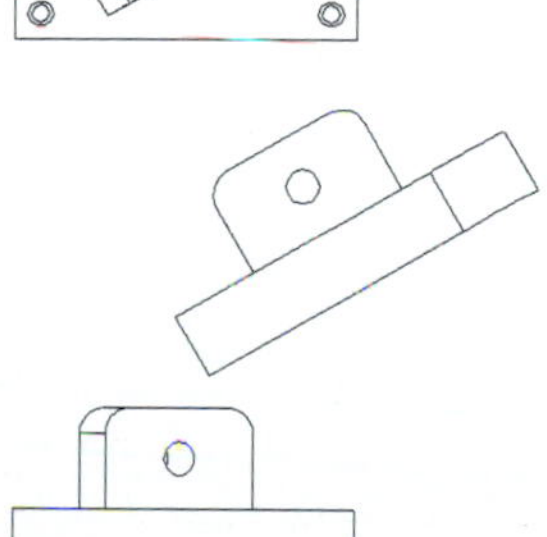

When 2D drawing practices were used, the primary auxiliary view had to be created in order to draw the secondary auxiliary view. Now with 3D CAD, secondary auxiliary views can be created and placed directly without the need to show the primary auxiliary view. However, this practice leaves out the view where a viewing plane line or direct projection could be used. In these cases, it is acceptable practice to draw a line perpendicular to the indicated surface in the standard orthographic view where the surface is shown foreshortened and label it as the viewing direction for the auxiliary view.

Section Views

Any number of section views may be included in a drawing. When many interior features are shown using hidden lines, the view may be difficult to interpret. A section view can show interior parts clearly and make the drawing easier to understand. Section views also allow you to dimension interior details where the features show clearly. Section views are useful for parts where elements overlap each other in a view, such as the keypad shown in Figure 12.24.

Each section view is related to one of the orthographic views in the drawing by a cutting plane line, as you learned in Chapter 4. Each section view is identified by the letters on the cutting plane line. Sections may either replace standard orthographic views or be located anywhere on the sheet or a separate sheet arranged in a logical order (usually alphabetically by the letters identifying the cutting plane line).

CAD modeling software can make it easy to quickly create section views. To add a section view to a drawing using Pro/Engineer software, you need only to select where to place the view on the sheet and to select a feature that can be used to define the cutting plane. Figure 12.25 shows the steps in creating a section view using Pro/Engineer software.

Hatching

Hatching, also called cross-hatching or section lining, is used to indicate the solid portions of the object that were cut through when producing the section view. Hatching should be plotted with a thin black line. The typical spacing for hatch lines is 1/16 to 1/8 inch apart. Hatching should always run the same direction on any single part.

To show different parts in a section view of an assembly, use different hatch angles and/or different hatch patterns for each part. Thin parts, such as gaskets, can be filled in solid when they are too small to show hatching effectively. Do not hatch parts that have no interior detail in assembly section views, such as solid rods, bolts, and screws. Parts like these are shown "in the round," meaning that you show them sitting in the assembly as though they were not cut through by the cutting plane. Figure 12.26 shows an example of hatching practices in an assembly section.

Ribs, webs, gear teeth, and other thin flat parts are not sectioned when the cutting plane passes lengthwise through them because it would give a false impression of the part's thickness. Leave the hatching off these features, as shown in Figure 12.27.

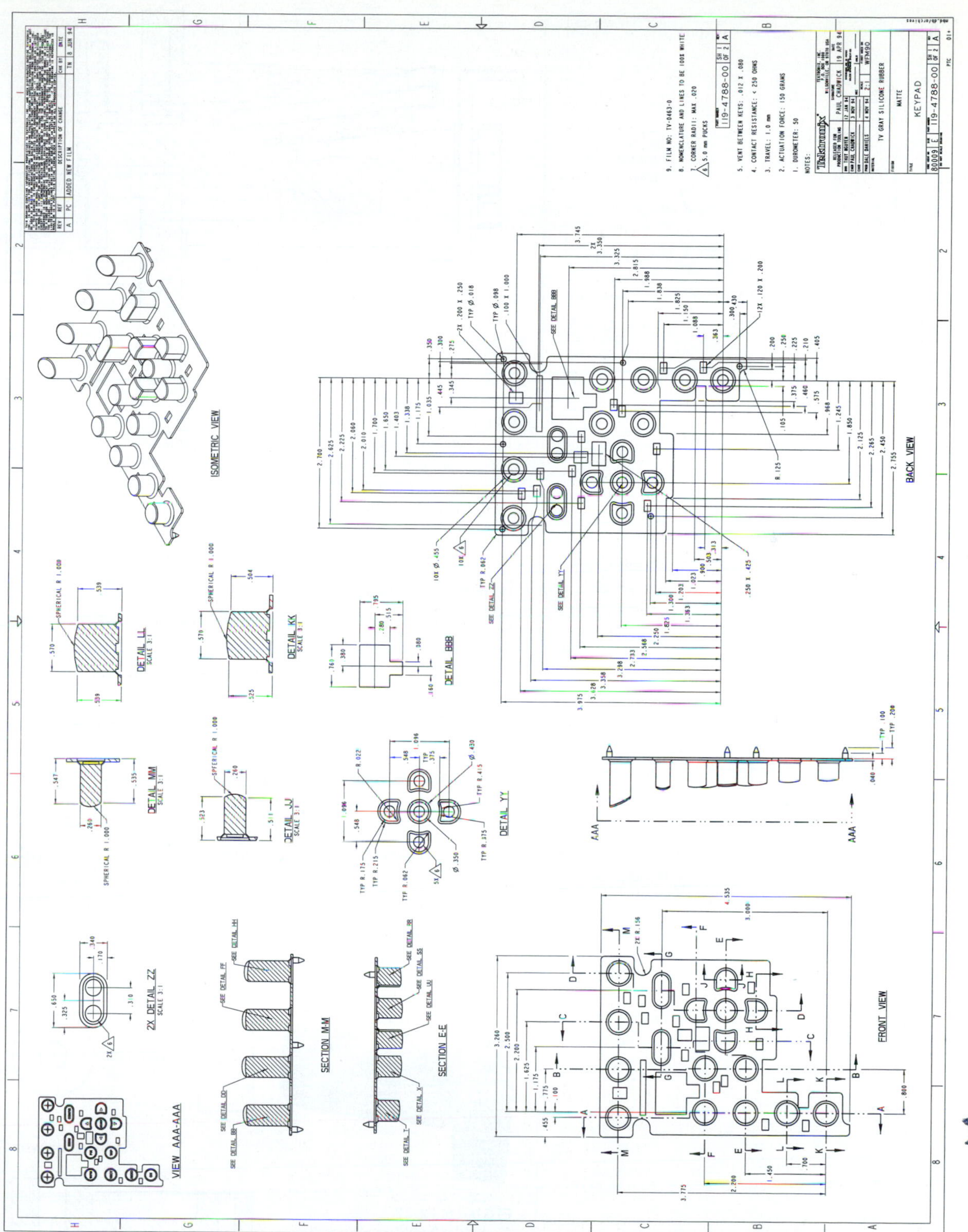

FIGURE 12.24

Section views for each of the keys fully define the keypad shown here. Note that the isometric view of the keypad makes it easy to relate the various views to the part as a whole. (Copyright 1998 Tektronix, Inc. All rights reserved. Reproduced by permission.)

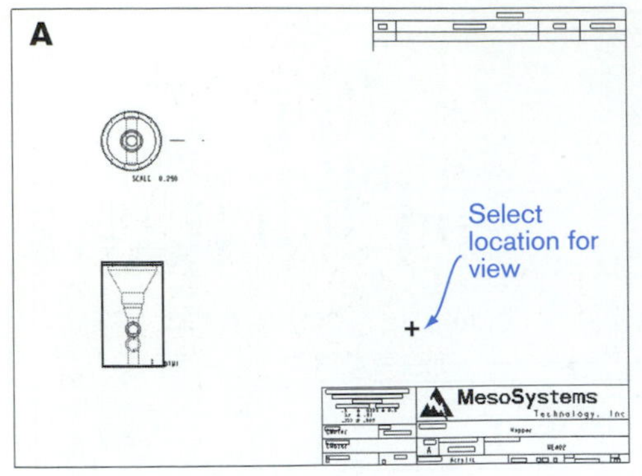

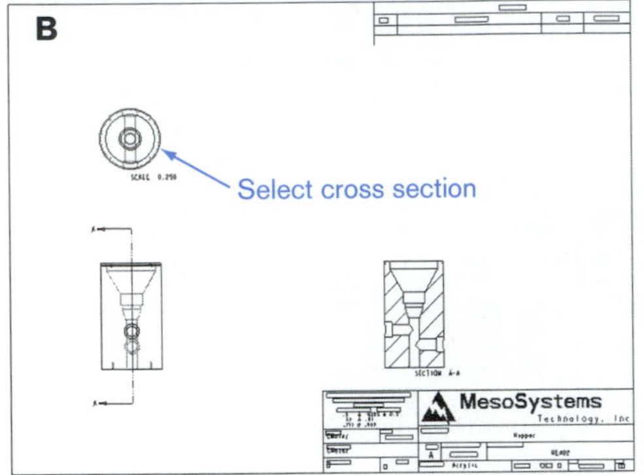

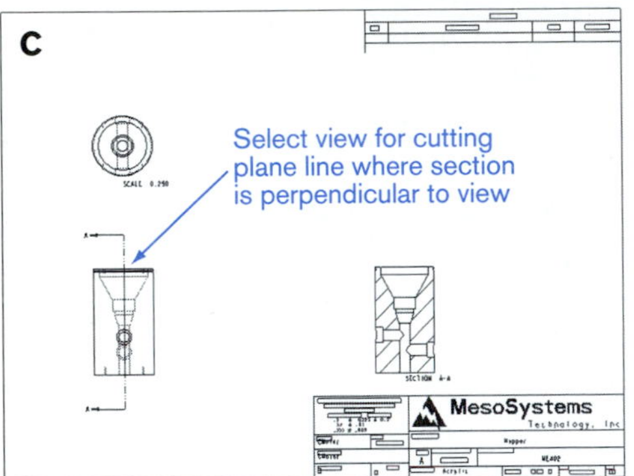

FIGURE 12.25
3D CAD software makes it easy to create section views. Using Pro/Engineer software you (a) select the location for the view, (b) select the cutting plane, then (c) select where to locate the cutting plane line.

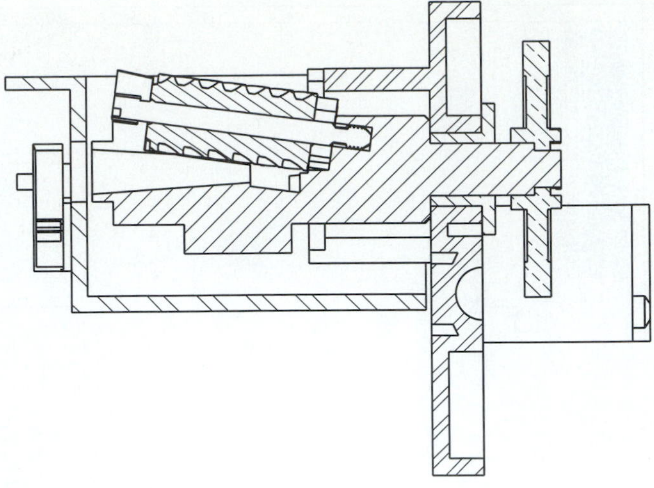

FIGURE 12.26
Use a different angle for the hatch pattern to indicate different parts in an assembly section. Plot hatching with thin, black lines. Thin, flat parts can be filled in solidly. Do not hatch parts that have no interior detail; show them "in the round."

Types of Sections

There are a number of different types of section views that can be used to show interior detail. A *full section* shows the part cut entirely through, typically along the centerplane.

Half Sections A *half section* shows the object as though a quarter of it were removed, so that half of the resulting view is shown in section and the other half shows the exterior view of the objects. Figure 12.28 shows a pictorial and orthographic half section. The cutting plane line indicates the direction of sight and the portion of the object that was removed to produce the section. Notice that the two halves of the half-section view are divided by a centerline pattern.

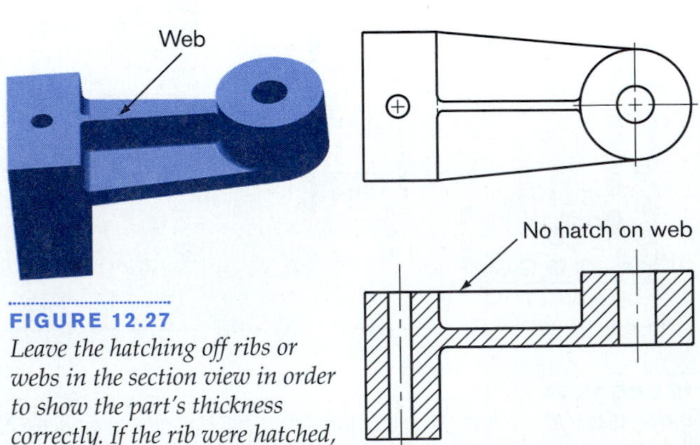

FIGURE 12.27
Leave the hatching off ribs or webs in the section view in order to show the part's thickness correctly. If the rib were hatched, the object would appear solid.

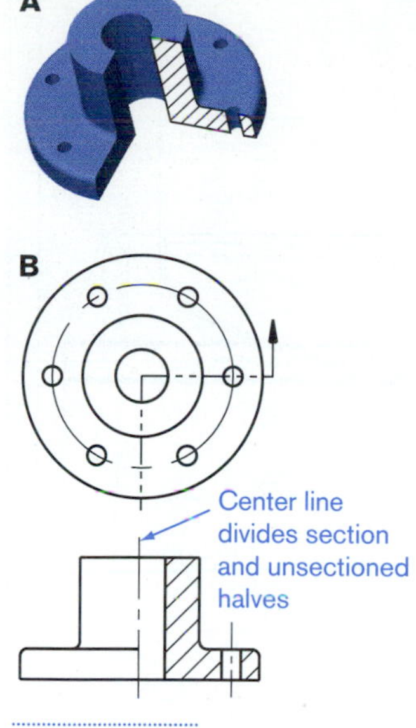

FIGURE 12.28
(a) Pictorial view of a half section; (b) This front view is shown as a half section. Notice the cutting plane line in the top view.

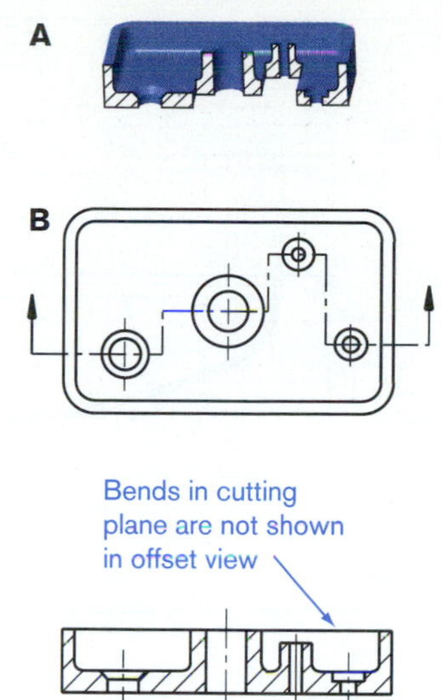

FIGURE 12.29
(a) Pictorial view of an offset section; (b) This offset section replaces the typical front view. Notice that there are no lines indicating the bends in the cutting plane line in the section view. To interpret the offset section, the cutting plane line must be shown, as it is here in the top view.

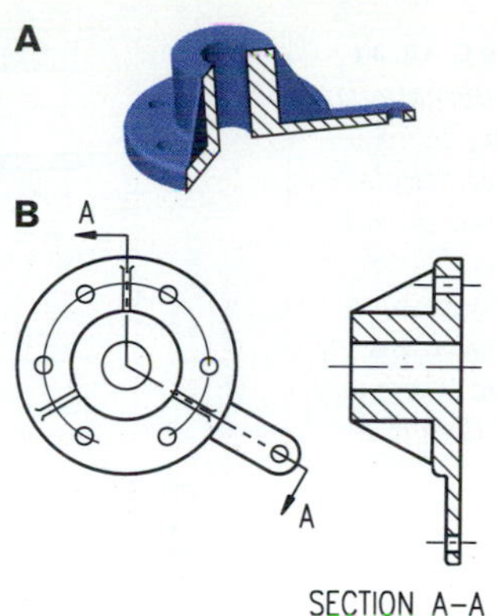

FIGURE 12.30
(a) Pictorial view of an aligned section; (b) The side view is shown as an aligned section. Notice that there are not lines indicating where the cutting plane was bent to pass through the feature.

Also, hidden lines are typically not shown in section views, including half-section views. Of course, you would use this type of view only when the part being shown in section is symmetrical.

Offset Sections *Offset sections* use a cutting plane line that is bent at 90 degree angles so that it passes through features that do not all lie in the same plane on the object. This allows you to show many features in one section view. The cutting plane line must be shown in the drawing so that you can interpret the features. Bends in the cutting plane line are not shown in the section view. Offset sections can be difficult to create with some CAD packages. Figure 12.29 shows pictorial and orthographic views using an offset section.

Aligned Sections An *aligned section* is used to pass through angled arms, holes, or other features located around a central cylindrical shape. The aligned cutting plane is bent to pass through the feature. In the section view, the cutting plane and section are rotated into a single plane, so that the angled feature shows its true size in the section view. Bends in the cutting plane line are not indicated in the section view. Many CAD packages also have difficulty producing aligned sections. Figure 12.30 shows pictorial and orthographic views of an aligned section.

Broken Out Sections A *broken out section* shows a partial view of a part or assembly's interior. A rough break line in the section view indicates which portions of the surface are broken to reveal the

FIGURE 12.31
This international drawing shows enlarged details using broken out sections. (Courtesy of Viessman Manufacturing Co., Inc./Viessman Werke GmbH & Co.)

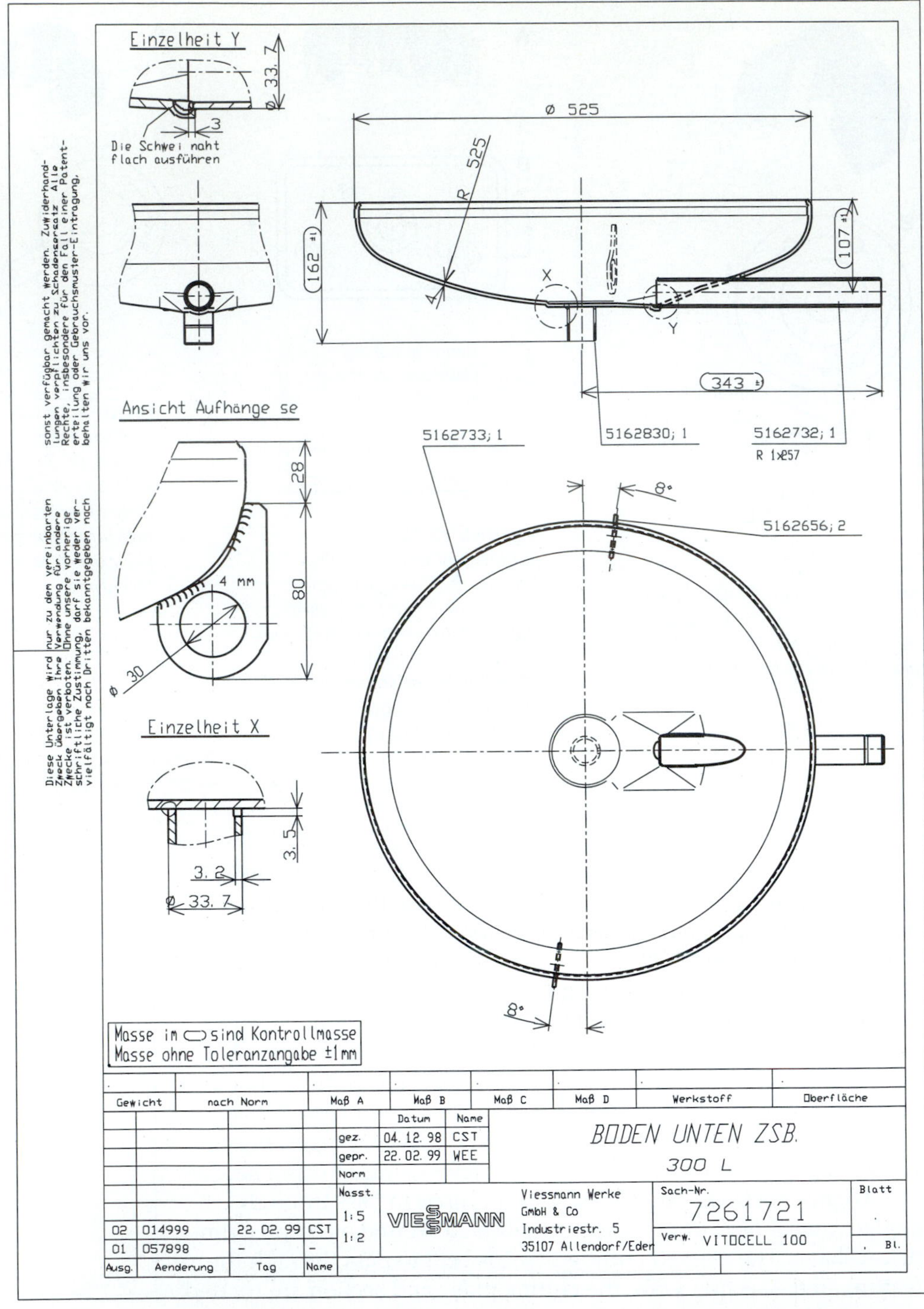

interior. In ANSI standard drawings, the rough break is a thick jagged line. Figure 12.31 shows an international drawing of the bottom base assembly for a boiler. Notice the details X and Y called out in the front view (shown here at the top of the sheet). Details X and Y (labeled Einzelheit X and Y) to the left of the sheet show portions as broken out sections to reveal details of the interior

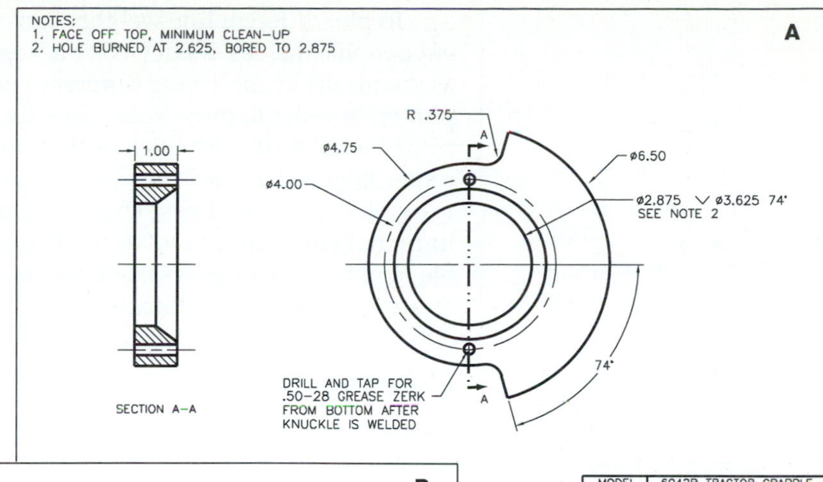

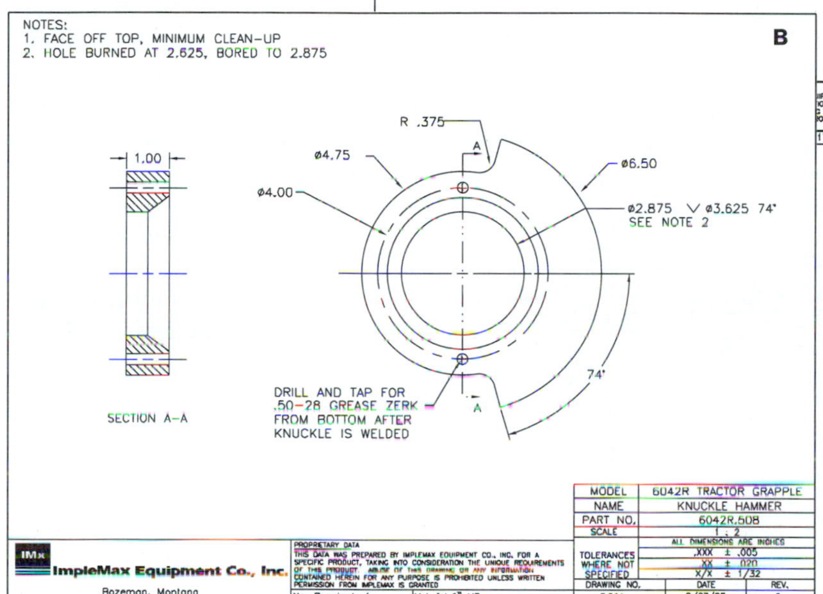

FIGURE 12.32
(a) Good use of thick and thin lines allows the object lines to stand out. (b) Having all line weights the same makes it difficult to see the object lines in this drawing. (Courtesy of Implemax Equipment Co., Inc.)

assembly. Keep in mind that most international drawings use first-angle projection. In first-angle projection, the top view is placed below the front view.

Linetypes

One of the most important things for making a readable drawing is the ability to plot the object using thick lines. The preferred weights for standard engineering linetypes, as you saw in Chapter 4, are defined in ANSI Y14.2M. Object (or visible) lines and cutting plane, viewing plane, and short break lines are plotted thick. Hidden lines, center lines, dimensions, notes, and most other lines are thin. The thin lines do not distract your eye from clearly seeing the shape of the object. Figure 12.32 shows a good example and a poor example of using different line weights.

To plot different line weights, most plotters offer several different pen thicknesses. Most CAD packages use the color of the entity as the means of specifying different pen thicknesses to the printer or plotter. In order to print lines correctly, it is important to associate color consistently with thick or thin lines. For example, make all center lines red and set the pen thickness for red lines in the drawing to be plotted as thin lines. Then remember not to use red for any lines that you want to plot as thick lines. Even printers that are capable of only black or grayscale lines still typically use the color of the entity to select the line thickness for printing.

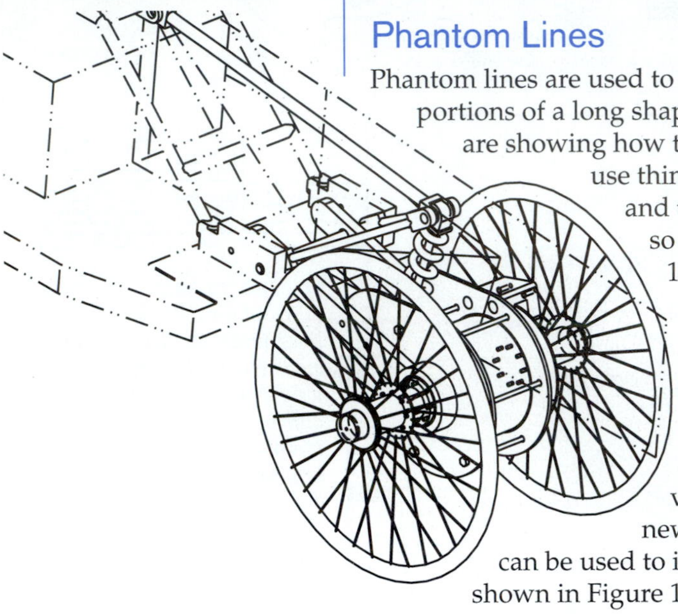

FIGURE 12.33
Phantom lines indicate existing equipment that the new part needs to work with.

Phantom Lines

Phantom lines are used to show existing equipment or to indicate portions of a long shape or structure that are not shown. If you are showing how the object will fit with existing equipment, use thin phantom lines for the existing portions and thick object lines for the portion to be built so that it will stand out clearly (see Fig. 12.33). Because you don't want to draw more than is necessary in your documentation drawings, phantom lines can allow you to show part of an object without having to draw it all. In assembly drawings, phantom lines can be used to show alternate locations for a part in a single view. For example, if an arm can swing to a new location in the assembly, phantom lines can be used to indicate its other extreme location, as shown in Figure 12.34.

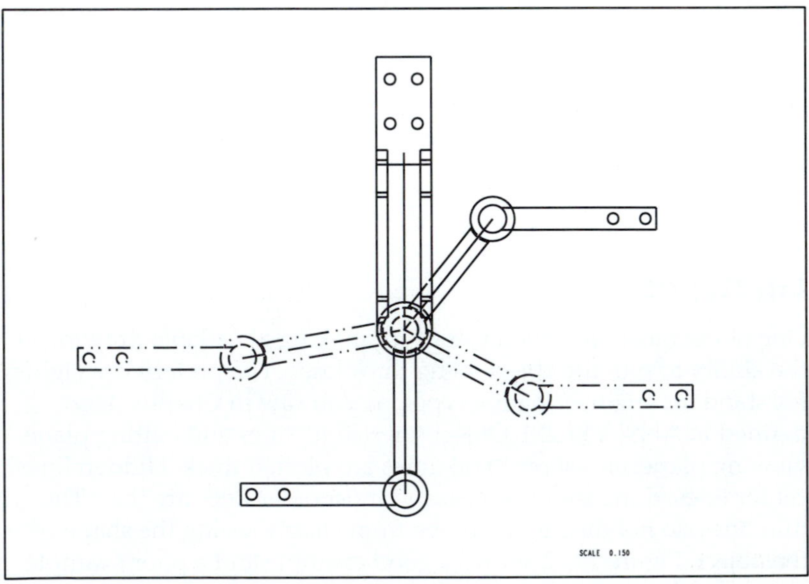

FIGURE 12.34
The phantom lines are used to show the alternate location for the angled arm in the assembly.

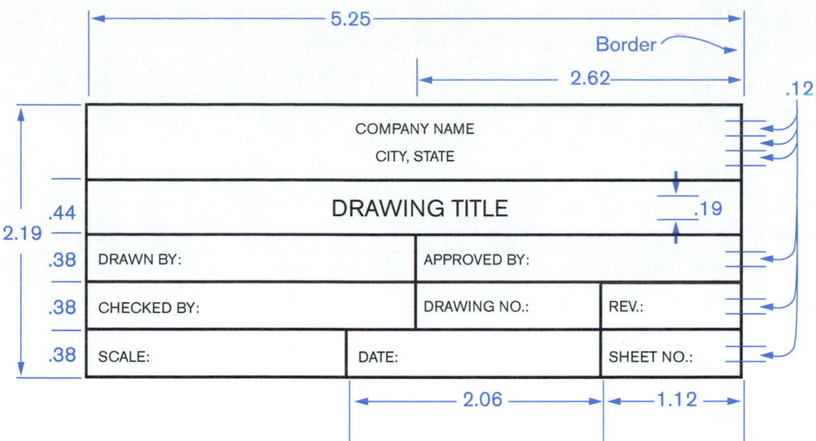

FIGURE 12.35
The standard for a title block shown here indicates the size of the block, spacing between the lines, and the size of the lettering. Some title block formats may include the revision block; others locate it separately in another location on the drawing.

Title and Revision Blocks

Every drawing needs a title block and, as revisions are made, a revision block. A *title block* organizes information needed to identify and interpret the drawing in a consistent location that is easy to read. Most companies have a standard format for title and revision blocks. Figure 12.35 illustrates the dimensions and typical information for an ANSI-preferred title block. A variety of sizes and formats of title blocks for standard sheet sizes are shown in Appendix 4.

Companies often make the standard title block available in a CAD library so it can quickly be inserted into the drawing. A variety of title blocks and borders for each drawing sheet size may be available. This eliminates re-creating the elements of the title block each time and ensures consistency among drawings. Many software packages allow you to define fields in the title block so that the information in those fields, such as part number or designer name, can be extracted into a drawing tracking system or linked to other documents.

A title block generally includes:

- drawing title, or the name of the part or assembly shown
- company name and address
- date (when the drawing is completed)
- designer's name and release signature
- checker's name and release signature
- name and release signature of the chief engineer or other company official
- scale
- drawing number
- revision number

Additional information may also be included in the title block, such as:

- material
- general tolerance

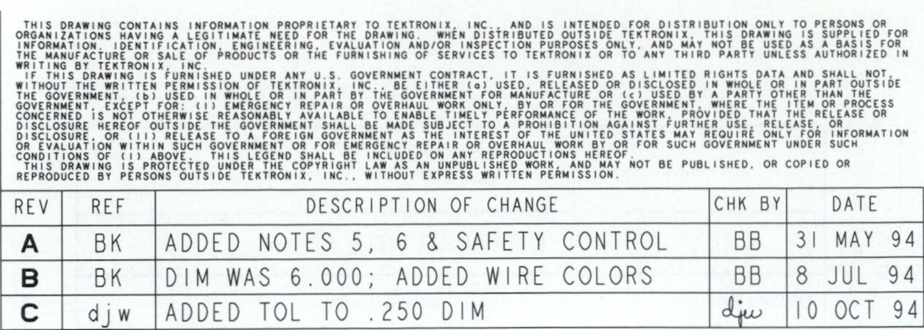

THIS DRAWING CONTAINS INFORMATION PROPRIETARY TO TEKTRONIX, INC., AND IS INTENDED FOR DISTRIBUTION ONLY TO PERSONS OR ORGANIZATIONS HAVING A LEGITIMATE NEED FOR THE DRAWING. WHEN DISTRIBUTED OUTSIDE TEKTRONIX, THIS DRAWING IS SUPPLIED FOR INFORMATION, IDENTIFICATION, ENGINEERING, EVALUATION AND/OR INSPECTION PURPOSES ONLY, AND MAY NOT BE USED AS A BASIS FOR THE MANUFACTURE OR SALE OF PRODUCTS OR THE FURNISHING OF SERVICES TO TEKTRONIX OR TO ANY THIRD PARTY UNLESS AUTHORIZED IN WRITING BY TEKTRONIX, INC.

IF THIS DRAWING IS FURNISHED UNDER ANY U.S. GOVERNMENT CONTRACT, IT IS FURNISHED AS LIMITED RIGHTS DATA AND SHALL NOT, WITHOUT THE WRITTEN PERMISSION OF TEKTRONIX, INC., BE EITHER (a) USED, RELEASED OR DISCLOSED IN WHOLE OR IN PART OUTSIDE THE GOVERNMENT, (b) USED IN WHOLE OR IN PART BY THE GOVERNMENT FOR MANUFACTURE OR (c) USED BY A PARTY OTHER THAN THE GOVERNMENT, EXCEPT FOR: (I) EMERGENCY REPAIR OR OVERHAUL WORK ONLY, BY OR FOR THE GOVERNMENT, WHERE THE ITEM OR PROCESS CONCERNED IS NOT OTHERWISE REASONABLY AVAILABLE TO ENABLE TIMELY PERFORMANCE OF THE WORK, PROVIDED THAT THE RELEASE OR DISCLOSURE HEREOF OUTSIDE THE GOVERNMENT SHALL BE MADE SUBJECT TO A PROHIBITION AGAINST FURTHER USE, RELEASE, OR DISCLOSURE, OR (II) RELEASE TO A FOREIGN GOVERNMENT AS THE INTEREST OF THE UNITED STATES MAY REQUIRE ONLY FOR INFORMATION OR EVALUATION WITHIN SUCH GOVERNMENT OR FOR EMERGENCY REPAIR OR OVERHAUL WORK BY OR FOR SUCH GOVERNMENT UNDER SUCH CONDITIONS OF (I) ABOVE. THIS LEGEND SHALL BE INCLUDED ON ANY REPRODUCTIONS HEREOF.

THIS DRAWING IS PROTECTED UNDER THE COPYRIGHT LAW AS AN UNPUBLISHED WORK, AND MAY NOT BE PUBLISHED, OR COPIED OR REPRODUCED BY PERSONS OUTSIDE TEKTRONIX, INC., WITHOUT EXPRESS WRITTEN PERMISSION.

REV	REF	DESCRIPTION OF CHANGE	CHK BY	DATE
A	BK	ADDED NOTES 5, 6 & SAFETY CONTROL	BB	31 MAY 94
B	BK	DIM WAS 6.000; ADDED WIRE COLORS	BB	8 JUL 94
C	djw	ADDED TOL TO .250 DIM	djw	10 OCT 94

FIGURE 12.36
Typical revision block (Copyright 1998 Tektronix, Inc. All rights reserved. Reproduced by permission.)

- heat treatment
- finish
- hardness
- estimated weight
- superseding drawing numbers

A *revision block* is used to document changes made to the drawing. A typical revision block is shown in Figure 12.36.

Dimensioning

Dimensions are used to indicate the size and location of features on the part. Dimensioning is very important to preparing correct engineering drawings. Once you have clearly shown the drawing views, you need to provide the dimensions needed to make the part. Even if a company will manufacture the part directly from the model, it is customary to provide a dimensioned drawing as reference. This drawing may be marked as "reference only" so the manufacturer understands that if there is a discrepancy he or she is to use the digital model database as the basis for creating the part. In other cases, the drawing may provide only critical dimensions that the manufacturer can use to be sure the digital file has been translated correctly and to resolve any discrepancies between the two. If the part will not be manufactured directly from the model, then it is even more important to provide fully and completely dimensioned drawings, as they are the sole means of communicating the design to the manufacturer.

Dimensioning and tolerancing practices are outlined in ANSI Y14.5M. In general, the standards clarify dimensioning and tolerancing practice so that a drawing cannot be interpreted in more than one way. There are several factors in good dimensioning. First is the proper use of dimensioning elements such as extension lines, gaps, text, and arrowheads. Showing them in the proper relationship and at standard sizes is key to their appearance. Second is the way the dimensions are placed around the drawing views. Following a few key rules can avoid confusion when the dimensions are read. Finally, and perhaps most important, is choosing the dimensions that define the part so they are unambiguous, tolerances can be applied consistently, and the resulting manufactured part will be what the designer intended. To do all of these well, you must achieve a balance among the various practices, rules of thumb, and a good understanding of the role the manufacturing technician will play in creating the part.

Units

Dimensioned drawings may use either System International (SI) metric standard units or U.S. customary foot/inch units (also called English units). Most companies have a standard for the type of drawing units they use. Although the United States is one of the very few countries that has not converted to the metric standard (only Burma, Liberia, and the United States have not), many U.S. companies have adopted metric units for their drawings.

Advantages of Metric Units

Many people agree that the metric system is easier to learn and convert and has the potential to be a worldwide standard for measurement. When the Metric Conversion Act was passed by the U.S. Congress in 1975, it encouraged the change to metric units as a voluntary measure. Most companies did not see any advantage to changing their work practices and material sizes, and few companies adopted metric standards for their drawings. When the Omnibus Trade and Competitiveness Act of 1988 designated the SI system as the preferred system for U.S. trade and commerce, it stipulated that federal agencies use the metric standard to the extent economically feasible. It does not require that U.S. industries, states, or other organizations convert to metric, but it may require companies to produce bid documents and plans in metric units for federally funded projects.

The nonmetric nature of many U.S. products may cause them to suffer a lack of acceptance in the world marketplace, where metric is the standard. Many U.S. companies have already switched to metric units to increase their competitiveness. General Motors is an example of one company that decided to convert to metric units relatively early. Its total conversion costs were only 1% of what it had estimated it would spend. When IBM converted its processes to metric, it also rethought its corporate practices and was able to reduce its standard fastener sizes from 30,000 down to 4000. An advantage of metric units that is often overlooked is the possibility that manufacturing workers may make fewer errors working with metric drawings because the typical unit is whole numbers of millimeters, whereas inch drawings require greater use of decimals or even fractions.

Standard Dimension Appearances

Dimensions are drawn between extension lines that relate the dimension to the feature on the part. Dimension lines, dimension values, and extension lines are all thin black lines in the drawing. This way they do not overpower the object outline and yet reproduce well and can be easily read. Typical sizes for the items that make up a dimension are given in Figure 12.37.

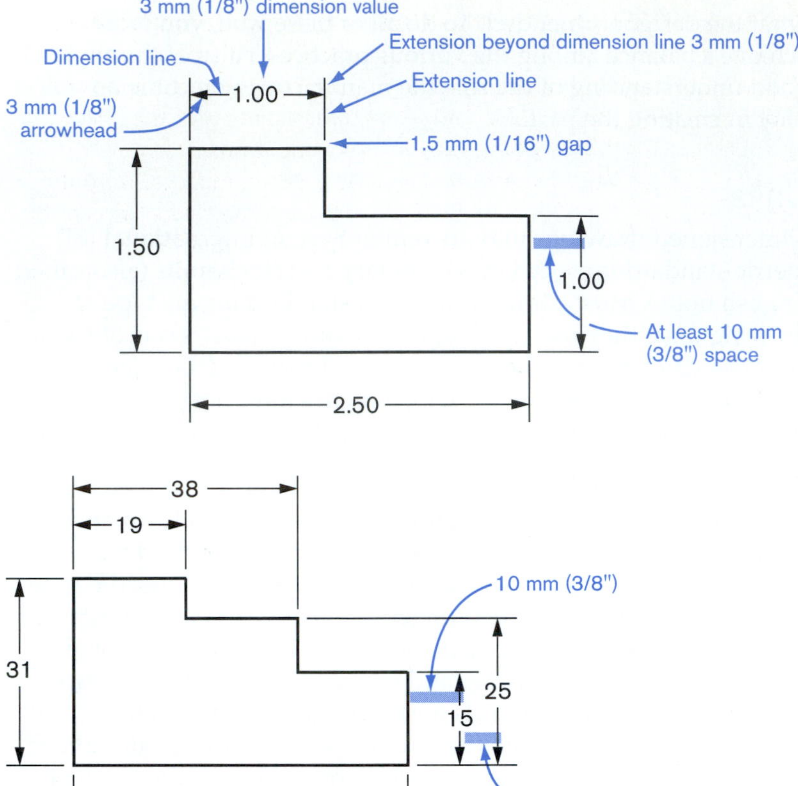

FIGURE 12.37
Typical sizes for dimension features

FIGURE 12.38
Dimension lines should be offset consistently from the object outline and from subsequent dimensions.

Dimension Lines

Dimension lines are used to show the direction and extent of the dimension. They typically end in arrowheads. The dimension line is usually broken so that the dimension value can be shown in line with the dimension line.

The closest dimension line should be spaced no less than 10 mm (0.375 inch) away from the object's outline. Subsequent rows of dimensions should be at least 6 mm (0.25 inch) away from the preceding row of dimensions, as shown in Figure 12.38.

Extension Lines

Extension lines extend from the feature that is being dimensioned to 3 mm (0.125 inch) beyond the dimension line. The dimension line is drawn perpendicular to the extension lines. Center lines can be extended beyond the edge of the part to serve as extension lines for dimensions. Extension lines, including extended center lines, are not broken when they cross the edge of the part.

Dimension Value

The *dimension value* represents the size in units between the extension lines. The dimension value specifies the size of a feature or its distance from another feature on the part.

A simple, single-stroke, gothic style of lettering should be used to create the dimension values. Most CAD systems provide a font that looks similar to the gothic letter shapes you learned to sketch in Chapter 3. AutoCAD's roman simplex shape font (romans.shx), shown for the dimension values in Figure 12.39, is one such font. Many modern CAD systems allow you to use most of the common font types available in other Windows applications, such as PostScript and True Type fonts. If you use these, make sure that other people who will open the CAD file have the same fonts available. Otherwise the software will have to use a substitute font. Substitute fonts are rarely exactly the same size and frequently cause dimensions and drawing notes to overlap other drawing elements. Do not get carried away with the use of exotic fonts. They typically make the drawing difficult to read and can cause problems for other people opening your drawings. You may also find that your ability to control the way the type appears is more limited with exotic fonts than with the fonts built into the CAD package.

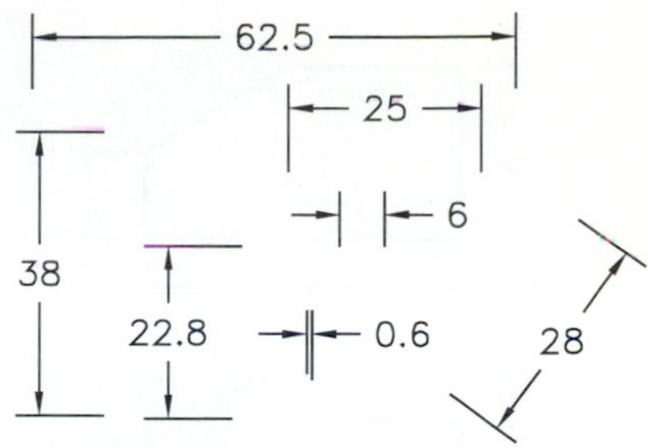

FIGURE 12.39
Dimensions in SI units are typically in whole millimeters or shown to one decimal place. Decimal values less than 1 are preceded by a 0.

Metric Dimension Values The millimeter is the base unit for dimensioning machine drawings in the SI system. Most metric dimensions are given in whole millimeters, unless the precision for the part requires more decimal places to be shown. Whole numbers are not followed by a decimal point and zeros. Dimensions less than 1 mm are written as decimals preceded by a 0. Do not include mm or other designations with the dimension values, but make sure that the type of units is clearly labeled in or near the drawing title block. Print the letters *SI* in a larger size than the standard lettering height, for example, or add a note that reads "All measurements in millimeters." Figure 12.39 illustrates the form of properly written millimeter dimensions.

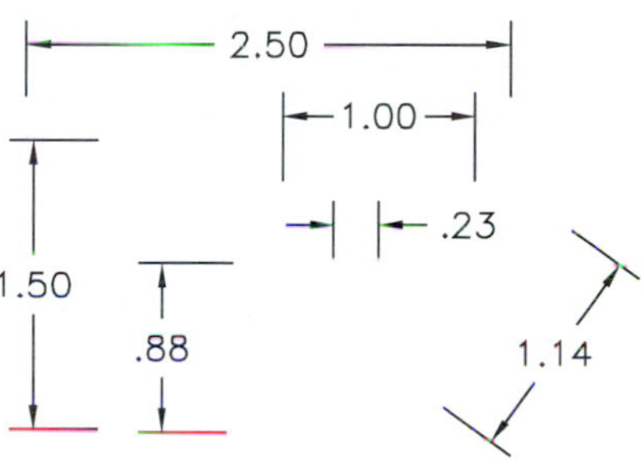

FIGURE 12.40
Inch dimensions are often shown to two decimal places as the default. Decimal values less than 1 are not preceded by a 0.

Inch Dimension Values Decimal inches, not fractions, are the base units used for inch dimensioning. Typical inch dimensions are written to two decimal places, unless the precision of the part requires more. When converting traditional fractions, such as a 5/8-inch hole diameter to decimal numbers, round the value to two decimals, or .68 inch, if that is all of the precision required. Do not be concerned that the rounded value does not specify the exact drill size. The manufacturer will determine what size drill, ream, bore, or other tool is needed to form the hole by reading the value from the drawing along with the stated tolerance. When dimensioning in inch units, whole numbers should typically be followed by a decimal point and two 0s. Values less than 1 are not preceded by a 0. Do not include inch marks (") with the dimension value, but clearly label the units in the title block as inches, or add a note near the title block stating, "All measurements in inches." Figure 12.40 illustrates the preferred form for inch dimensions.

Decimal Points in Values Make sure that the decimal points in dimension values stand out clearly when you plot or print your drawing. They should be dark, uniform, and large enough to stand out and reproduce clearly. In some CAD systems (AutoCAD, for example) you can edit the standard fonts to produce lettering variations. If your decimal points are not easily legible, you may want to edit the font, select a different font, or use a slightly thicker pen

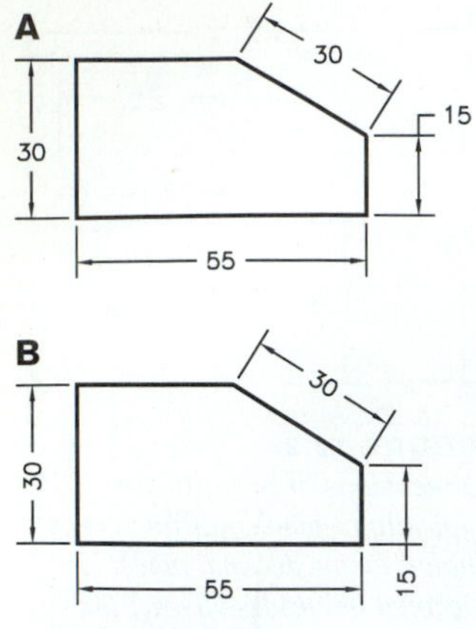

FIGURE 12.41
(a) Unidirectional dimensions are read from the bottom of the sheet. (b) Aligned dimensions are read from the bottom or right side of the sheet.

width when printing or plotting to make sure that decimals are seen. It is not an unusual error for a part or device to be built to ten times the desired size because a decimal point was not visible on a print.

Orientation

All dimension values and notes should be *unidirectional*; that is, all text is lettered horizontally from left to right so it can be read from the bottom of the sheet, as shown in Figure 12.41(a). Another acceptable, but less common, orientation is *aligned*. When dimensions are aligned, the dimension text is aligned with the dimension line so that it can read from the bottom or right of the sheet, as shown in Fig. 12.41(b). That is, horizontally aligned dimensions are read from the bottom of the sheet, and vertical dimensions from the right side of the sheet (never from the left).

Placement of Dimensions

Placing dimensions around the views requires attention to a few basic rules. Dimension lines should not cross one another. Nor should extension lines be drawn across dimension lines. Placing shorter dimensions closer to the object will help avoid crossing other dimension or extension lines with longer dimensions. These rules of thumb are often poorly observed by automatic dimensioning algorithms of modern CAD packages. The drawing in Figure 12.42(a) shows how the dimensions appear when placed automatically by Pro/Engineer parametric modeling software. Tools built into the software allow you to clean up the dimension placement so dimensions can be read clearly and lines do not cross unnecessarily. Figure 12.42(b) shows the same drawing after the dimensions have been properly placed and located around a central view.

Place dimensions where the feature is shown true size and shows its shape clearly. Because the front view typically shows the shape of the object best, most of the dimensions will be located around the front view. This also makes it easy for the manufacturer to locate the dimensions, as they are mostly grouped together around the central view.

Place dimensions off the view unless placing the dimension on the view makes the drawing easier to read. Placing dimensions off the view makes it easier to see the object lines, hidden lines, and center lines on the view that define the shape of the features. If plac-

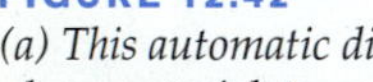

FIGURE 12.42
(a) This automatic dimension placement violates rules about dimension and extension lines crossing. (b) After repositioning, the dimensions reflect good practice.

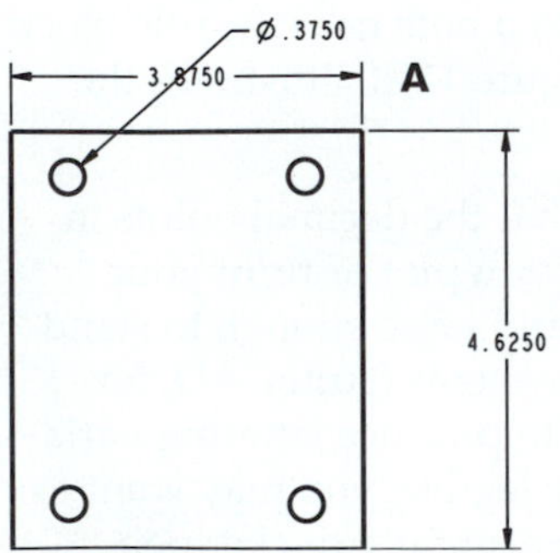

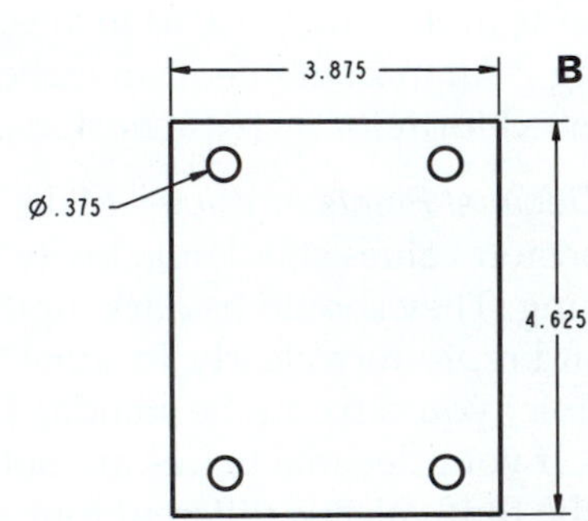

ing the dimension off the view has the opposite effect, it is better to make an exception. For example, if you are dimensioning an interior slot on a complex shape where the extension lines for the slot would have to cross many other features, or be placed in a crowded area, you may want to place the dimension on the view as shown in Figure 12.43(b).

Show each dimension only one time in the drawing and attached to one view only.

Place dimensions between views when possible to help show that the value is related to the feature as seen in both views. Remember, however, to connect it to only one of the views, as shown in Figure 12.44.

Stagger dimension values so they do not all line up in a neat row, as shown in Figure 12.45, which makes them difficult to read. Avoid using more than three rows of dimensions on one side of a view, which is also difficult to read—but it often cannot be avoided.

Choosing the Best Dimensions to Show

The dimensions you show in the drawing are the ones needed to manufacture and inspect the part to determine whether it meets the specified tolerances for its manufacture. As the engineer, you should dimension the drawing so that the manufactured part will perform its function properly in the assembled product.

Some drawings show only critical features and leave the undimensioned features to be created as manufactured from the CAD database, with the tolerances defined as per manufacturing process. Often these are exterior features, such as the outside of a plastic

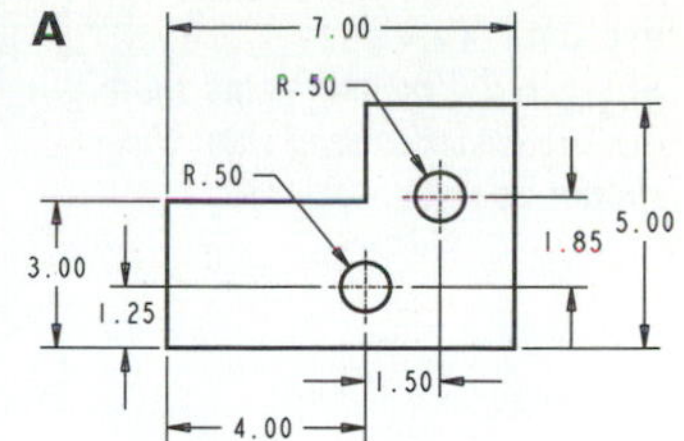

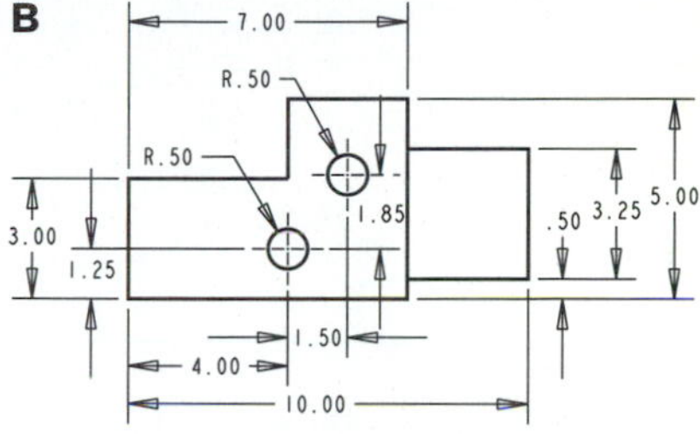

FIGURE 12.43
(a) Place dimensions off the view whenever possible. (b) A dimension may be placed on the view if the alternative would violate other rules of placement.

FIGURE 12.44
Show each dimension only once, but choose a location between views that contributes to interpreting the size or location of the feature in both views.

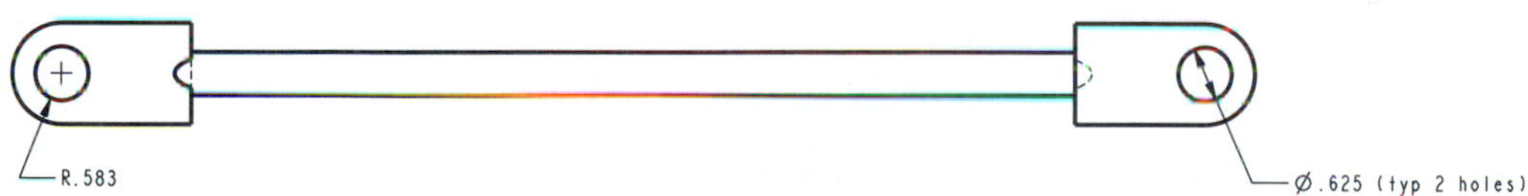

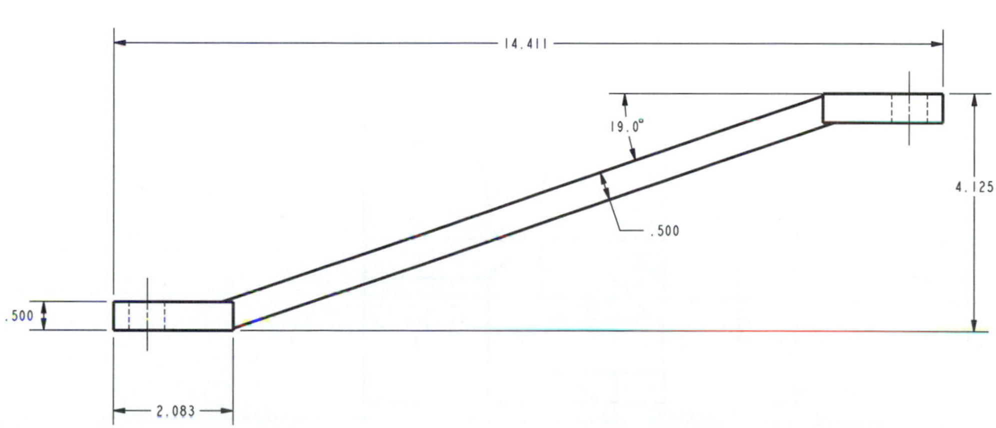

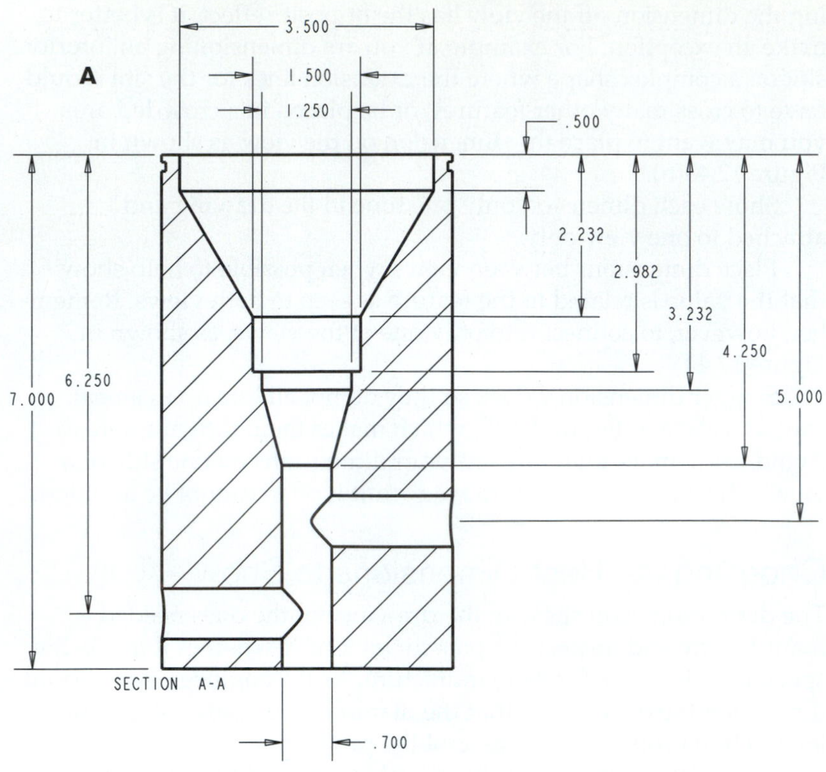

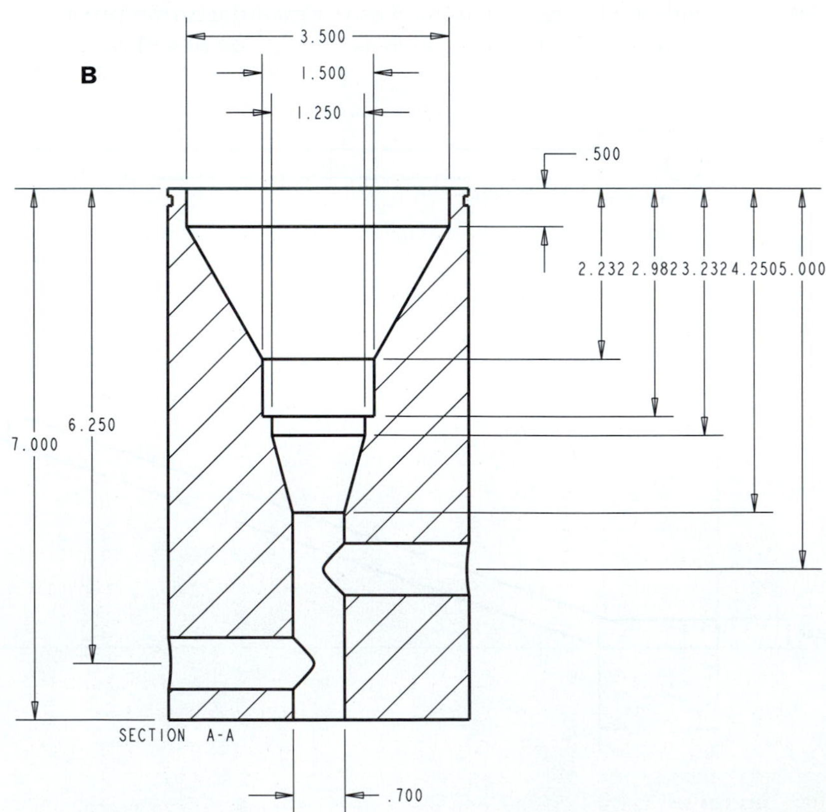

FIGURE 12.45
Staggered dimensions, as shown in (a), are easier to read than those shown in (b).

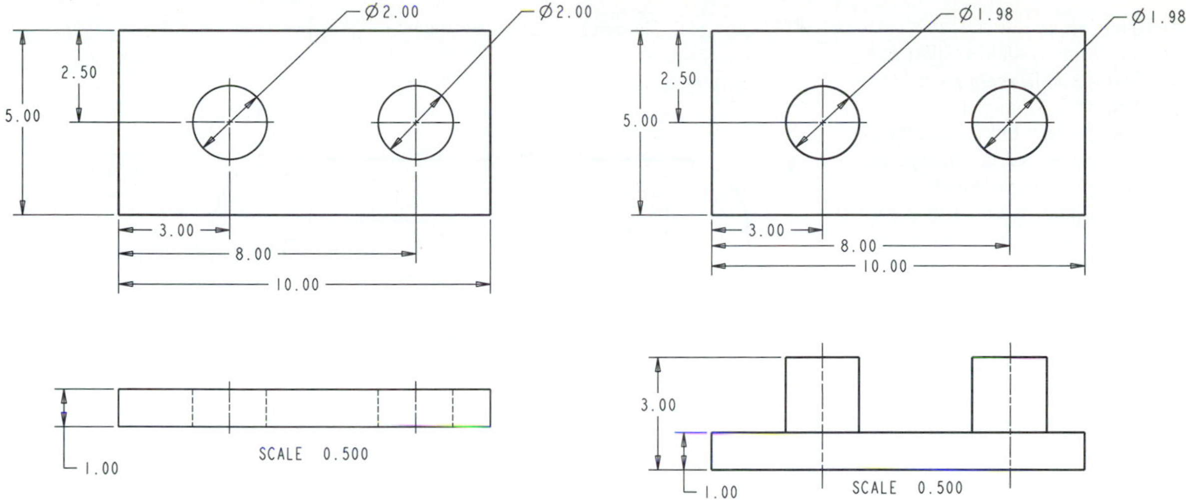

FIGURE 12.46
The two mating parts shown must fit together when manufactured. Providing their mating dimensions on the drawing helps ensure that they will fit.

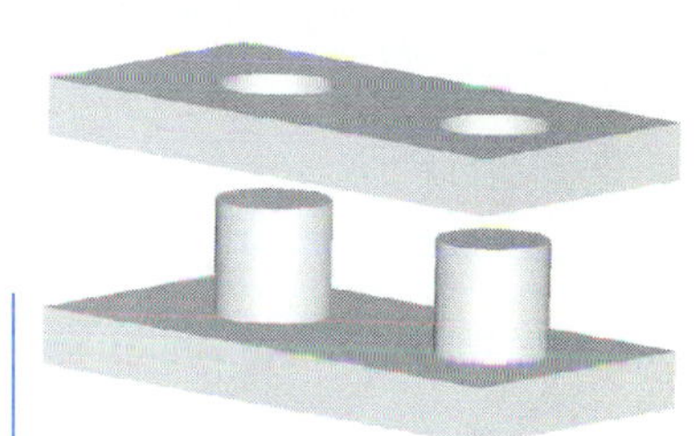

housing, that do not fit with other parts and are not critical to the function of the assembly.

Mating parts are parts that fit together in the assembly. A *mating dimension* is the term for the dimension on mating parts that allows them to fit together when assembled. You will want to provide mating dimensions, or to dimension mating features on different parts in a similar way so that the parts fit together when assembled. Figure 12.46 shows two mating parts and their dimensioned drawings. Notice that mating dimensions are shown on the parts.

Most CAD packages offer *associative dimensioning*, a feature that automatically inserts the dimension value for a selected entity when a dimension is created. This makes it easy to add dimensions to a drawing that accurately reflect the CAD database. If the model is modified, dimension values can be updated automatically to reflect the new dimension stored in the CAD database. Most packages also offer the option to hand-enter dimension values that override the value provided by the database. When dimensioning a drawing from the model, it may seem quicker to change a dimension on the drawing instead of going back and changing the model. If you do, however, the CAD database will not reflect the sizes and locations on your drawing and will not be useful for computer-controlled manufacturing.

When parametric modeling is the basis for your engineering drawing, choosing the best dimensions to show can be easy. Because parametric design encourages you to give a lot of thought up front to the relationships in the design and how to size the various features, all of the part features will already be located in relation to the existing base feature or to other features on the part. It is a common mistake when dimensioning drawings by hand to remember to give the size of a feature, but to forget to locate it on the part. If you captured the design intent in the model, the dimensions you used to

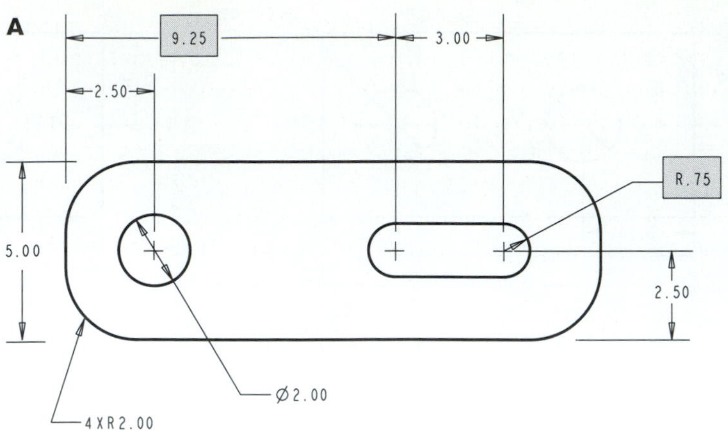

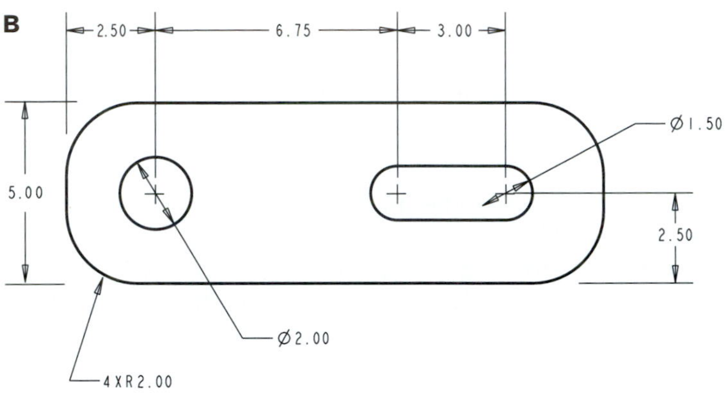

FIGURE 12.47
(a) This drawing shows a part that required the addition of placed (or nonparametric) dimensions in order for the part to be produced as the designer intended. These placed dimensions are highlighted.
(b) The parametric dimensions for the model used to create the drawing in (a) show that the slot was dimensioned from the location of the left hole. This dimension had to be overridden in the drawing so that the feature would be located from the left edge of the part. In addition, the R.75 dimension was added to replace the diameter dimension used to model the part. If the model had been created with the dimensions shown in (a), the drawing would have been generated correctly without the need for placed dimensions.

create it will generally be the same ones that you should show in the drawing. If they are not, you may want to reconsider your parametric design database and ask if it really reflects your design intent.

The *bidirectional associativity* of the parametric modeler makes it possible to change dimensions on the model or in the drawing and have the change stored in the CAD database. In addition, most parametric modeling software offers the ability to add additional dimensions to a drawing. These dimensions may update as an associative dimension will, but changing them will not change the database. This feature allows you to add dimensions to the drawing that were not used to create the model. Overuse of this option in the CAD package probably means that your parametric model does not adequately reflect your design intent. Traditionally, *reference dimensions* are used to add information to a drawing, but are identified with REF to signify that they should not be used to manufacture the part.

Examine Figure 12.47(a) and (b). Notice that the dimensioned drawing contains dimensions that were not used to create the part model shown in (b). The highlighted dimensions were added in drawing mode, but would have been generated automatically if the part had been created with the dimensions shown in (a).

When you are dimensioning a drawing, keep the manufacturing process in mind. Remember that if you can size features to be manufactured with standard tools and stock sizes, the parts will cost less. You do not need to indicate the manufacturing process that will be used to form the hole of the size you have indicated within the specified tolerance range, but you should be aware of standard sizes and the tolerances possible with various methods. You will learn more about tolerances later in this chapter. The manufacturer will be the best judge of how to make the part in a cost-effective manner.

Dimensioning Arcs and Circles

Specify the diameter of circular shapes and the radius of arcs. Remember that the dimensions you provide are for the manufacture and inspection of the part. Tools for forming holes are sized by their diameter, not radius. Also, once a hole has been drilled, it is easy to measure its diameter using calipers, or even a machinist scale, but it is not easy to measure the radius.

Diameter dimensions are preceded by the diameter symbol Ø. Radius values are preceded by the uppercase letter *R*.

Don't forget to give location dimensions for circular shapes, such as holes, or pins. Most circular shapes are located with a dimension from the center point to another feature. Many radius shapes are self-locating if the entire end of a feature is rounded or a corner is rounded.

Figure 12.48 shows examples of correct practices for dimensioning circles and arcs.

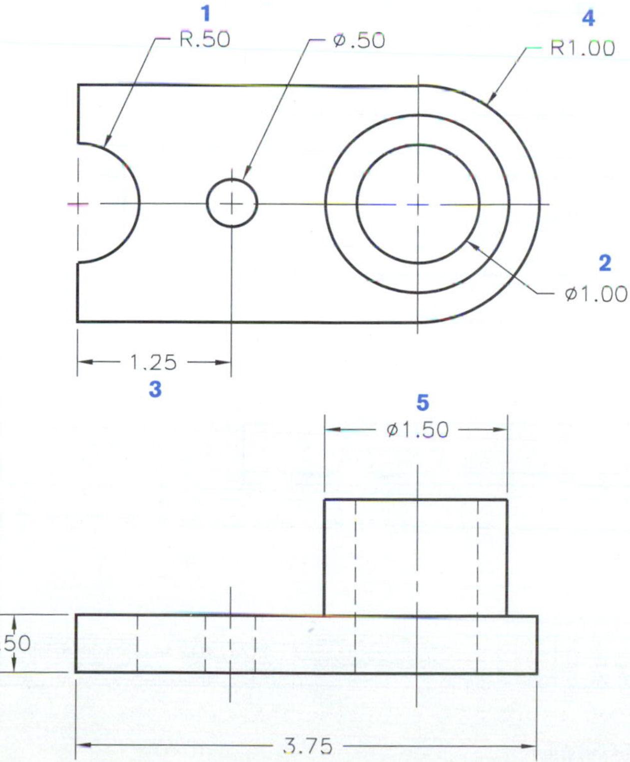

FIGURE 12.48
Dimensioning circles and arcs

1. Give radius of arcs.
2. Give diameters of circles.
3. Locate holes where you see their circular shapes.
4. Rounded ends are self locating.
5. Dimension the diameters of cylinders where they appear rectangular.

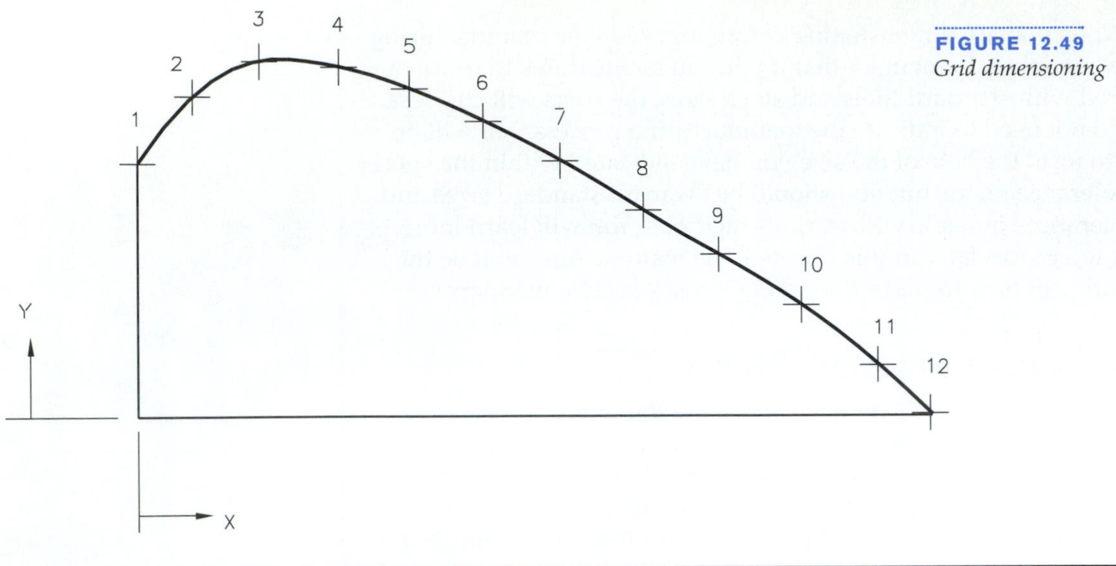

FIGURE 12.49
Grid dimensioning

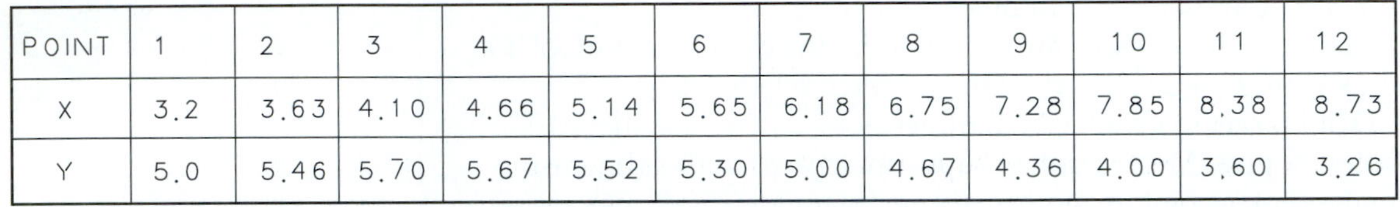

POINT	1	2	3	4	5	6	7	8	9	10	11	12
X	3.2	3.63	4.10	4.66	5.14	5.65	6.18	6.75	7.28	7.85	8.38	8.73
Y	5.0	5.46	5.70	5.67	5.52	5.30	5.00	4.67	4.36	4.00	3.60	3.26

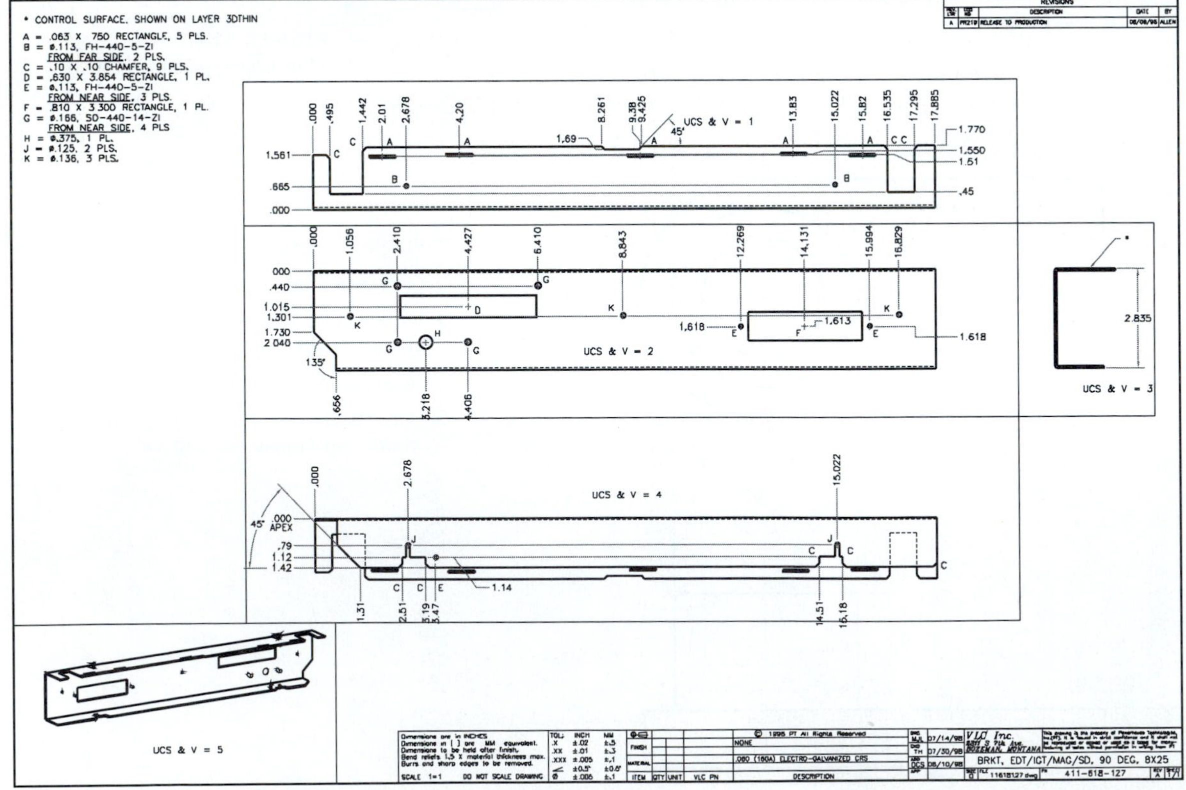

FIGURE 12.50
Coordinate dimensioning (Courtesy of Powerhouse Technologies, Inc.)

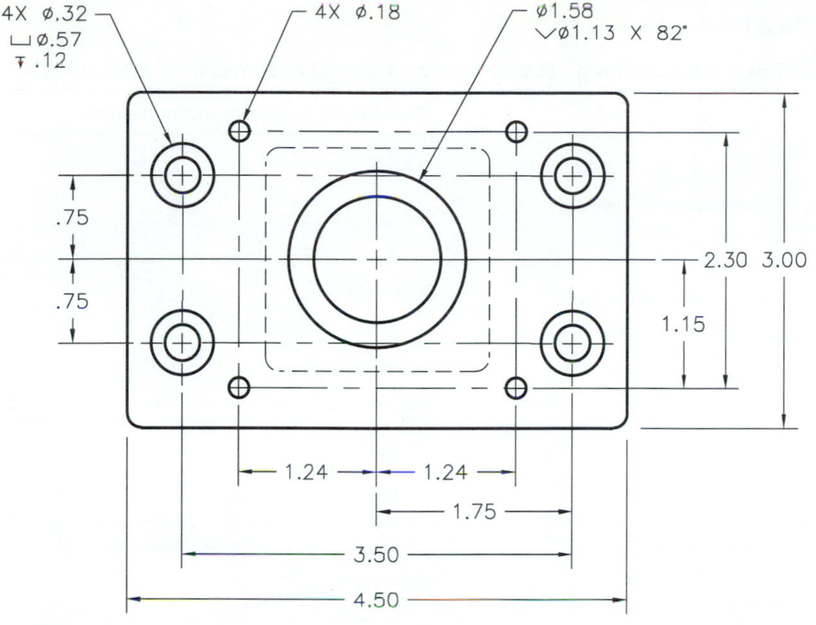

FIGURE 12.51
Several dimensioning symbols are used here to dimension common features.

Coordinate and Grid Dimensioning

Grid dimensioning may be used to specify points along an irregular curve. *Coordinate dimensioning* is often used for parts that have complex interior hole patterns. A table may be provided specifying the hole diameters and their coordinates. Figure 12.49 and Figure 12.50 show examples of grid and coordinate dimensioning.

Coordinate dimensioning often works well for parts that will be NC machined as the part typically must have a 0,0,0 location relative to the machine. The other features can be dimensioned grid fashion relative to the 0 point. Because the file is often generated directly from the CAD model, the drawing is for reference only. It is still important to provide a drawing, though, because a value changed in the exported file through a glitch in the software will not be caught by the machine; it will cut to the location specified, whether or not that is what you intended. Providing the drawing allows the manufacturer to check through the file to ensure that the dimensions are correct.

Standard Symbols for Dimensioning Common Features

Common features such as countersunk holes, counterbored holes, spotfaced holes, square holes, taper, thread, spherical radius, and spherical diameter all have standardized symbols to help define them clearly on the drawing without a large number of separate dimensions. Figure 12.51 illustrates the use of these symbols in a drawing. The standard shape and size of these symbols is described in Table 12.5.

TABLE 12.5

Dimensioning Symbols for Common Features

Feature	Symbol (h = letter/symbol height)
Counterbore or spotface	2h, h
Countersink	90°, h
Square	h, h
Conical taper	0.5 h 0.5 h, 2 h, 30°
Depth	h, 0.6 h, h, 60°
Diameter	Ø
Spherical radius	SR
Spherical diameter	SØ

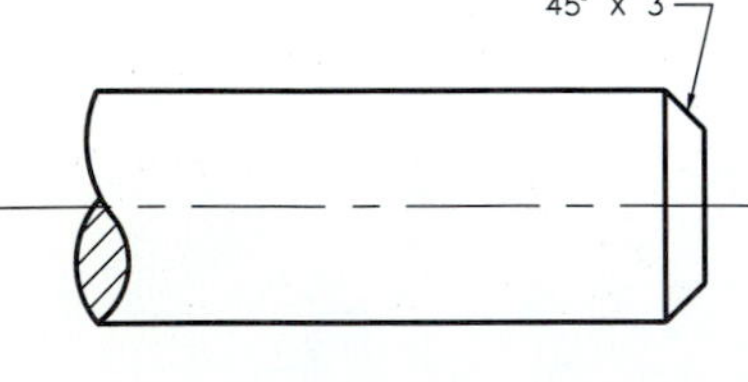

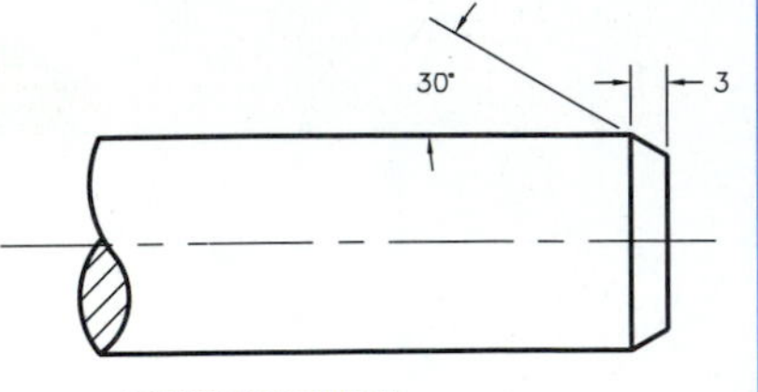

FIGURE 12.52
Methods of dimensioning chamfer

Dimensioning Chamfer

A *chamfer* is a small angled surface formed between two surfaces. Chamfers can be dimensioned by specifying a linear dimension from the surface of the part to the start of the chamfer and an angle, or by giving two linear dimensions (see Fig. 12.52). Chamfers of 45° can be given as a note.

Dimensioning Knurling

Knurling is a pattern that is formed on the outside of a part, often to provide a gripping surface for better handling, or for press fitting parts. Knurling is specified by a note listing the type of knurling pattern (for example, diamond or straight), the pitch (distance from one groove to the next), and the diameter before and after knurling (see Fig. 12.53). If the diameter after knurling is not important, it can

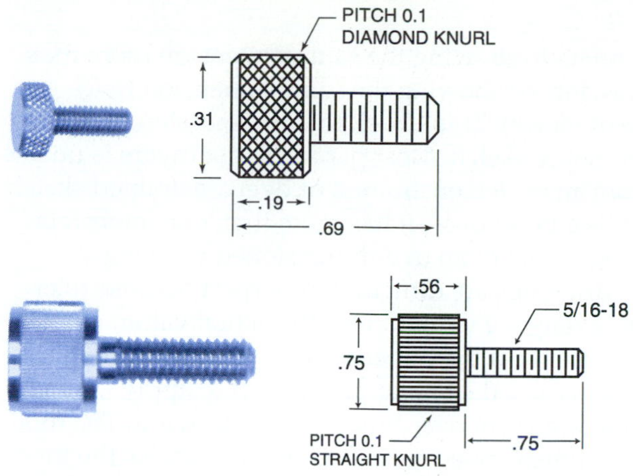

FIGURE 12.53
Dimensioning knurling (Photos courtesy of Reid Tool Supply Company.)

be omitted. If the entire feature is not knurled, dimension the length and location of the knurled portion.

Dimensioning Keyseats and Keyways

Keys are rectangular or semicircular shapes that are used to prevent parts, such as gears or wheels, from turning on a shaft. *Keyseats* and *keyways* are the slotted shapes that receive the key. (Keyseats are in the shaft; keyways are in the hub of the part on the shaft.) There are several standard key types, such as Woodruff keys, whose dimensions are provided in Appendices 37 and 38. Keyseats and keyways are dimensioned by specifying the width, depth, and location of the keyseat, as shown in Figure 12.54. If needed, the length of the keyseat is also dimensioned.

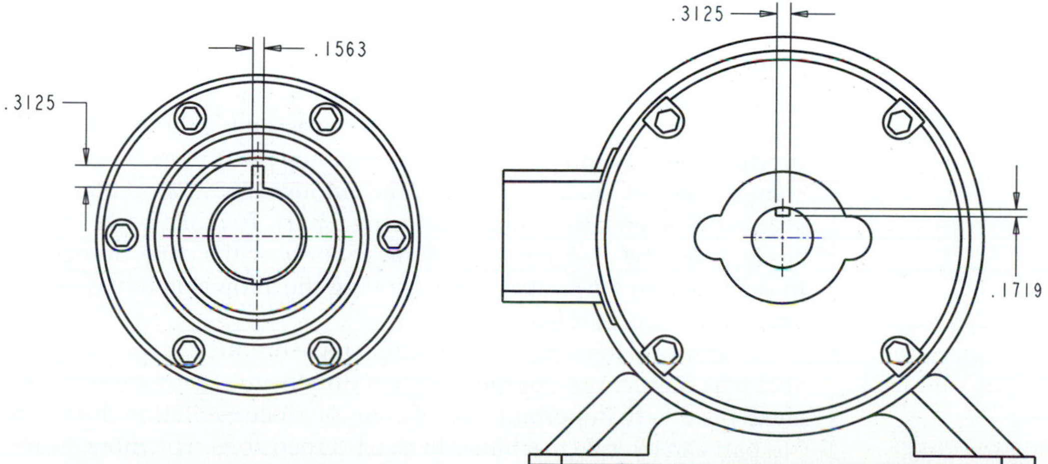

FIGURE 12.54
Dimensioning keyseats and keyways

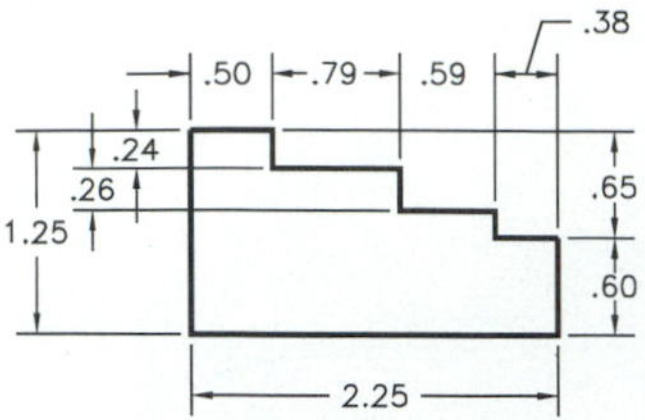

FIGURE 12.55
Overdimensioned drawing. Both the length and the height of this part are overdimensioned. Adding the maximum tolerance to each set of dimensions results in a maximum overall length of 2.27 or 2.34, and a maximum height of 1.27 or 1.29.

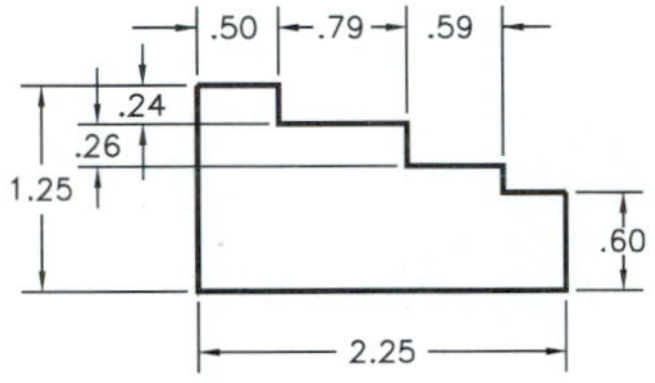

FIGURE 12.56
Fully dimensioned drawing. Removing unnecessary dimensions eliminates the ambiguity about the size of the part when tolerances are applied.

Overdimensioning

Overdimensioning refers to showing the same dimension more than one way in the drawing, or showing the same dimension twice (perhaps in different views). Think back to the discussion of overdimensioning a parametric sketch. Most parametric software is unable to create a part from an underconstrained or overconstrained sketch because the sketch is ambiguous—it has more than one interpretation. The same thing is true of an overdimensioned drawing. Overdimensioned drawings are difficult to interpret because tolerances (the allowable range of variation for the actual value) must be applied to the dimensions. If the dimension is stated in more than one way, it is not clear how the tolerance is meant to apply.

Figure 12.55 illustrates an overdimensioned drawing. The total width of the part is dimensioned, as is each feature across the top. The tolerance stated in the general note for the part is ±.02 inch, so the overall length of the part can be as much as 2.27 or as small as 2.23. Across the top of the part the overall size is dimensioned as the series .50, .79, .59, and .38. Each of these dimensions can also have a tolerance of ±.02 as stated in the note. This means that the overall width that was just dimensioned as 2.25 ±.02 is dimensioned here as 2.26 (because of the rounded two-digit numbers) and the allowed tolerance is ±.08. Clearly the machinist will not be able to interpret this drawing! The vertical dimensions are equally confusing because the height of the drawing is specified two different ways. Figure 12.56 shows the drawing with extra dimensions removed so that it can be interpreted clearly. If you are specifying the overall size of the part in a dimension, be sure to leave out one dimension in a chain of dimensions. This way it is clear how the tolerance applies. Select to leave out the dimension for the feature that you can allow to vary in size without affecting the function of the part.

Tolerancing

Most of today's goods are mass produced. High-volume production is one way of keeping the price of goods lower. Many parts are manufactured in widely separate areas across the world, then brought together for assembly. In order to assemble these mass produced goods and to repair them later, the parts must be interchangeable. Yet, there is always some amount of variation that occurs in the process of manufacturing a part. To make parts interchangeable, the designer must think about how the parts fit together in assembly and indicate on the drawing the allowable range for variation of the part features.

Tolerance is the total amount that a measurement on an acceptable part may vary from the specified dimension. Stating tolerances clearly is a very important part of creating documentation drawings. No part can ever be machined to exact dimensions. The more accurately you measure, the smaller the variations in the part size that can be identified.

REPOSITORY: DIRRECT
STATUS: RELEASE
STATUS DATE: 11-29-97
VERSION: 5
MODEL NAME(S): 146W1106M05-BD-201 FRAME INSTL/ASSY STA 1567
PLOTSHEET: 146W1106S06-BP
PLOT EXECUTION DATE: 11-29-97
CAD CODE: 5 - UPDATE BY DCN ONLY

Tolerance note

G—C PCM GRID CHECK POINT

T—H Ø.247 -.260 TOOL HOLE WHERE SHOWN. USE IS A MANUFACTURING OPTION.

UNLESS OTHERWISE SPECIFIED DIMENSIONS ARE IN INCHES
TOLERANCES:
ANGLES±2° DECIMALS±.03

SEE PL FOR LIST OF MATERIAL, USAGE AND NOTES

USED ON	DRAWN J. XXX	342-XXXX	10-22-96	BOEING CORPORATE OFFICES SEATTLE, WA 98124
777	CHECKED R. XXXX	342-XXXX		
	STRESS L. XXX		10-23-96	FRAME INSTL -
SECT NO. 46	ENGR J. XXX	342-XXXX	10-23-96	STA 1567.5,
	M. E. M. XXX	342-XXXX	10-23-96	S-14L TO S-14R
CHNG NO. -	GROUP T. XXXX		11-08-96	SIZE J / CAGE CODE 81205 / DWG NO. 146W1106
GROUP ORG BODY B-YU31	PROJ E. XXXX		11-08-96	
	BMT B. XXXX	342-XXXX	10-23-96	SCALE 1/1 / PIN 1461-1000 / SH 6

PCM (PHOTO CONTACT MASTER). FULL SIZE DIMENSIONALLY STABLE COPIES AVAILABLE FOR SHOP USE. CHECK ACCURACY BEFORE USING. TEMPERATURE & RELATIVE HUMIDITY CONDITIONS OTHER THAN THOSE NOTED MAY AFFECT ACCURACY. UNLESS OTHERWISE SPECIFIED: UNDIMENSIONED FEATURE TOLERANCE: ±.03. TEMP °F RH % DATE

DRAWING RECORDS

DWG NO. 146W1106 SHEET 6

FIGURE 12.57
General tolerance note (Courtesy of The Boeing Company.)

Specifying a tolerance states a range for the size and location of features on the part that will still allow the part to function properly in the design. Tolerances should be stated as generously as possible, as this provides a wider variety of processes that can be used to manufacture the part and keeps the price of the part lower. Specifying very high precision for parts increases the part price and should be used only if the design requires the added precision.

The primary ways to indicate tolerances in a drawing are

- a general tolerance note;
- a note providing a tolerance for a specific dimension;
- a reference on the drawing to another document that specifies the required tolerances;
- adding limit tolerances to dimensions;
- adding direct plus/minus tolerances to dimensions;
- geometric tolerances.

Many of these tolerancing methods can be used in combination with one another in the same drawing.

General Tolerance Notes

General notes are usually located in the lower right corner of the drawing sheet near the title block. Often general tolerance notes are included in the title block itself. For example, a general tolerance note might state, "ALL TOLERANCES ±1 MM UNLESS OTHERWISE NOTED. ANGLES ±1 DEGREE." This indicates that for a dimension value written as 25, for example, any measurement between 24 and 26 on the actual part would be acceptable.

Many companies have standard title blocks that they insert into CAD drawing files that contain general tolerancing standards for the type of production that is common to their industry. Figure 12.57 shows an example of a general tolerance note.

Another way general tolerances are stated is with a table on or near the title block indicating the tolerance by the number of digits used in the dimension. For example:

Digits	Tolerance
.X	±.2 inch
.XX	±.02 inch
.XXX	±.001 inch
X°	±1°

This type of table indicates that single place decimal dimensions have a tolerance of ±.2. For example, a dimension value written as 3.5 could range anywhere from 3.3 to 3.7 on the actual part and still be acceptable. A dimension written as 3.55 could range from 3.53 to 3.57 on the actual part. And a value written as 3.558 could range from 3.557 to 3.559 and be acceptable. It is uncommon to see more than three decimal places listed for inch drawings because precisions of ±.0001 are very high precision manufacturing and would be unlikely to be indicated merely by a general tolerance note.

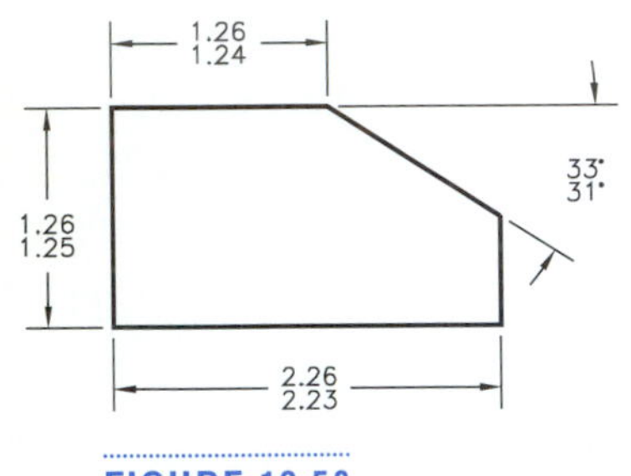

FIGURE 12.58
Limit tolerances

Limit Tolerances

Limit tolerances state the upper and lower limits for the dimension range in place of the dimension values. Figure 12.58 shows examples of limit tolerances in a drawing. The upper value is always placed above the lower value or, if the two values are written horizontally, to the left of the lower value separated by a dash, as in "32–29."

Plus/Minus Tolerances

Tolerances can also be specified by a range that is added or subtracted from the dimension value. This is often referred to as *plus/minus tolerancing*. Plus/minus tolerancing can be either bilateral or unilateral. *Bilateral* tolerances specify a range to be added and subtracted from the base value for the dimension. Bilateral tolerances can be equal or unequal. For equal bilateral tolerances you can write the dimension as the value plus or minus the single range. For example, 3.00±.01. For unequal bilateral tolerances, it is necessary to state the two separate values; for example, 3.00+.01/–.02. For unilateral tolerances, either the added or subtracted value is zero (0). When stating unilateral tolerances, you should still include the zero value and not leave it off. You can show it as just a 0 or it can be preceded by the plus (+) or minus (–) sign. Figure 12.59 shows examples of unilateral and bilateral tolerances.

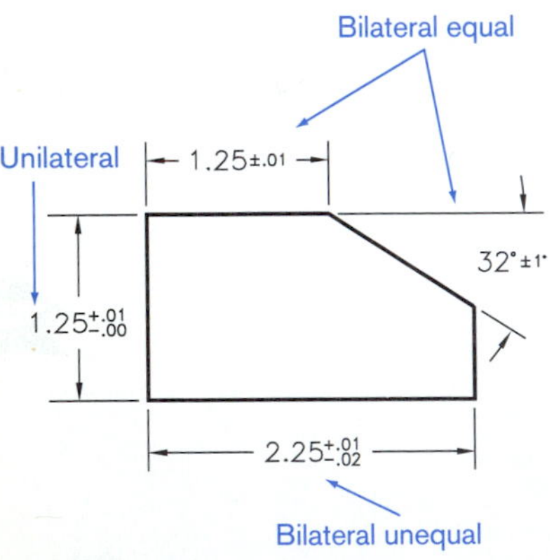

FIGURE 12.59
Unilateral and bilateral plus/minus tolerances

Tolerance Stacking

When dimensions are specified as a chain, the tolerances for the part may add up. A chained dimension uses the end of one dimension as the beginning of the next. *Tolerance stacking* refers to the way the tolerance for one dimension is added to the next dimension in the chain and so on from one feature to the next, resulting in a large variation in the location of the last feature in the chain.

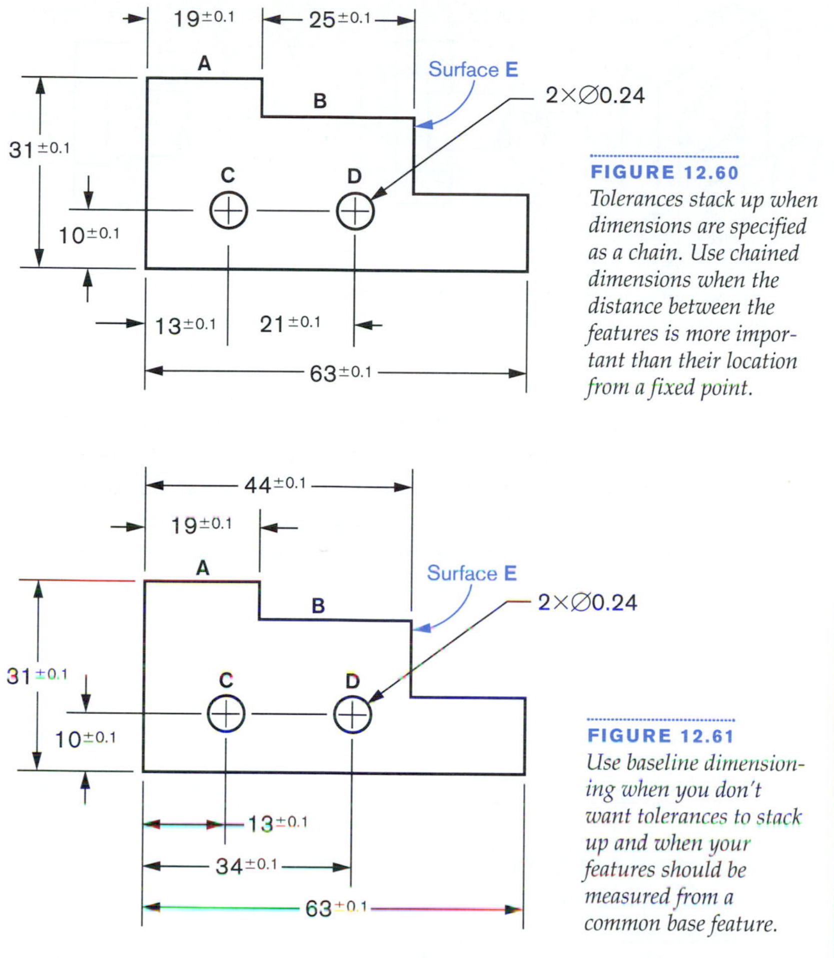

FIGURE 12.60
Tolerances stack up when dimensions are specified as a chain. Use chained dimensions when the distance between the features is more important than their location from a fixed point.

FIGURE 12.61
Use baseline dimensioning when you don't want tolerances to stack up and when your features should be measured from a common base feature.

Figure 12.60 illustrates this effect with a part where the surfaces labeled A and B and the holes labeled C and D are dimensioned chain fashion. Consider the location of surface E relative to the left-hand surface of the part. When surface A is at its maximum size of 19.1 and surface B is at its maximum size of 25.1, surface E is 44.2 away from the left end. Similarly, when the center of hole C is at its maximum distance from the left end, 13.1, the maximum distance for hole D from the left end is 13.1 + 21.1 or 34.2 from the left end of the part. Tolerance stacking is not necessarily bad, if that is the intent for the relative locations of the features. You should be aware of the effect that tolerance has on chained dimensions and specify the tolerances this way when you want the tolerance to accumulate.

Baseline Dimensioning

Baseline dimensioning locates a series of features from a common base feature. Tolerances do not stack up because dimensions are not based on other, toleranced dimensions. Figure 12.61 illustrates how the part in Figure 12.60 could be dimensioned using baseline dimensioning. Notice that surface E can now be a maximum of 44.1 from

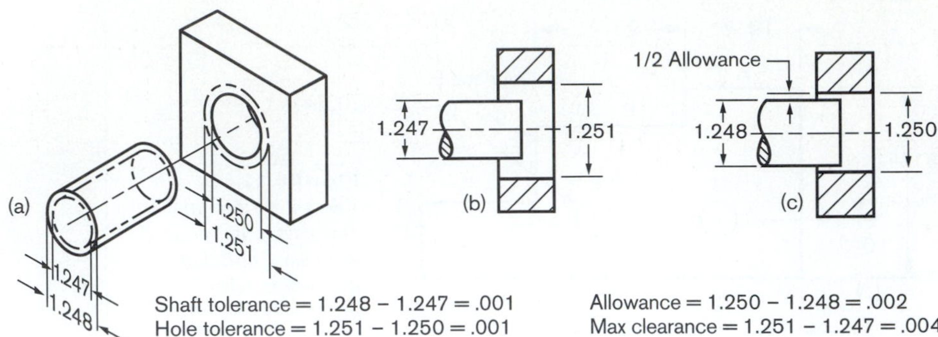

FIGURE 12.62
The dimensions and tolerances shown here produce a clearance fit because the internal member will always be smaller than the external member.

the left end surface on an acceptable part, and the location for the center of hole D is a maximum of 34.1 from the left end surface.

Baseline dimensioning can make it easy to inspect the part because features are measured from a common base feature. Dimensioning from a zero point as the base feature can also be a useful technique for dimensioning parts for NC machining.

Fit

Fit refers to how tightly or loosely mating parts must fit together when assembled. Different types of fit have been developed through years of typical machining practice. In the past, before interchangeable parts, the designer and machinist worked closely together. The machinist would measure the drawing, which did not include dimensions, and make the part as measured. A note or spoken instruction would let the machinist know that the parts should fit together so they had a sliding fit, a running fit, a force fit, or some other term. The machinist would fashion the parts until they functioned as intended. The categories used to describe types of fit still derive from these practices, but now they have been standardized to make them more consistent in use.

One way to determine the necessary tolerance for a part is by deciding on the type of fit and looking up the required tolerance for that type of fit in a standard fit table. Both U.S. customary inch units and metric unit fit tables are available in Appendices 6–15.

There are broad categories that describe the types of fit between mating parts: clearance fits, interference fits, transition fits, and line fits. In a *clearance fit,* as illustrated in Figure 12.62, the internal member is always smaller than the external member so that the parts will slide together. For an *interference fit,* the internal member is always larger than the external member so that the parts must be forced together. A *transition fit* indicates that either the internal or external member can be larger so that parts either slide together or can be forced together. A *line fit* indicates that the internal member is sized so that there is a slight clearance or so that it contacts the external member. The limit tolerances on the drawings in Figure 12.63 specify an interference and transition fit. Within these basic categories are additional designations that more specifically describe the desired fit. These fit classes are listed in Table 12.6.

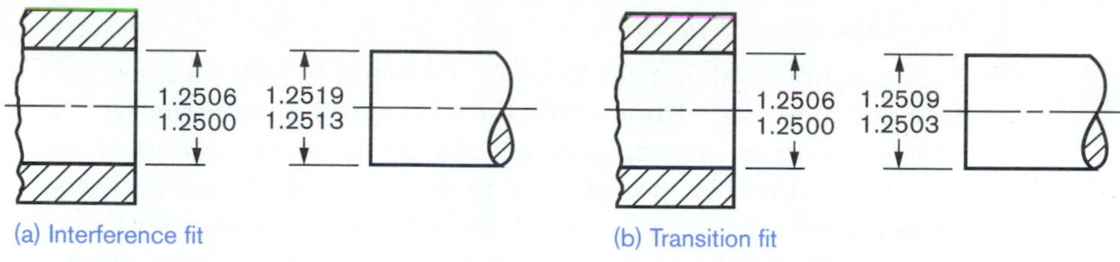

FIGURE 12.63
(a) Interference fit; (b) Transition fit

TABLE 12.6

Classes of Fit (ANSI B4.1-1967)

Category	Class	Description
Running or sliding	RC1	Close sliding fit intended for accurate location of parts that must assemble without perceptible play.
	RC2	Sliding fit intended for accurate location but with greater maximum clearance than RC1, where parts move and turn easily but are not intended to run freely, and in larger sizes may seize with small temperature changes.
	RC3	Precision running fit, the closest fit that can be expected to run freely. Intended for precision work at slow speeds and light journal pressures, but not suitable when appreciable temperature differences will be encountered.
	RC4	Close running fits intended to run freely on accurate machinery with moderate surface speeds and journal pressures, when accurate location and minimal play are desired.
	RC5, RC6	Medium running fits are intended for higher running speeds or heavy journal pressures or both.
	RC7	Free running fits are for use when accuracy is not essential, or when large temperature differences are likely.
	RC8, RC9	Loose running fits for use when wide commercial tolerances are necessary together with an allowance on the external member.
Clearance locational fit	LC1–LC11	Locational clearance fits are intended for parts that are normally stationary but can be freely assembled or disassembled. These fits run from snug parts requiring accuracy of location (LC1) through medium clearance fits for parts such as spigots, to the looser fastener fits, where freedom of assembly is of prime importance.
Transition locational fits	LT1–LT6	Transition fits are a compromise between clearance and interference fits for applications where accuracy of location is important but either a small amount of clearance or interference is permissible.
Interference locational fits	LN1–LN3	Locational interference fits are used where accuracy of location is most important and for parts requiring rigidity and alignment with no special requirements for bore pressure. These fits are not intended for parts designed to transmit frictional loads from one part to another through the tightness of the fit. For these types use force fits.
Force and shrink fits	FN1	Light drive fits require light assembly pressures and produce more or less permanent assemblies. These are suitable for thin sections, long fits, and in cast-iron external members.
	FN2	Medium drive fits are suitable for ordinary steel parts, or for shrink fits on light sections. They are about the tightest fits that can be used with high-grade cast-iron external members.
	FN3	Heavy drive fits are suitable for heavier steel parts or for shrink fits in medium sections.
	FN4, FN5	Force fits are suitable for parts that can be highly stressed, or for shrink fits in which the heavy pressing forces required are impractical.

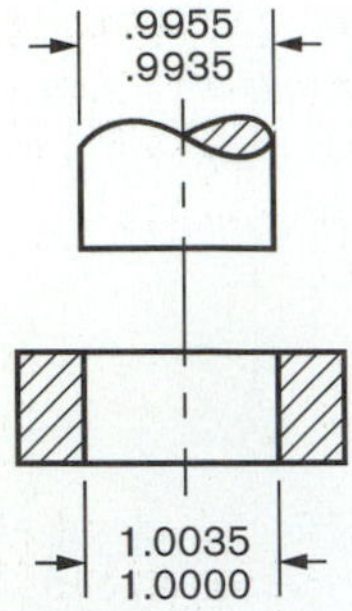

FIGURE 12.64
Calculating the tolerances for an RC8 clearance fit

The Basic Hole System

The basic hole system uses the external member (often a hole) as the basic dimension from which the toleranced dimensions to achieve the fit are determined. The reason for this is that holes are often formed with standard sized tools such as drills and reamers, while the internal shaft is turned down to size from standard stock, which can be easily machined to the required size to fit the more standard hole.

Figure 12.64 illustrates how to use fit tables based on the basic hole system. The hole to be formed is nominally a 1.00-diameter hole. The cylindrical shaft that will fit inside the hole is desired to assemble freely with a loose running fit (RC8). Based on the nominal size, look up the tolerance for the hole using the portion of the fit table reproduced in Table 12.7. For the hole, the values given are +3.5 and –0. For the shaft (the internal member) the values listed are –4.5 and –6.5. The values in the table are given in thousandths of an inch, so the limit tolerance for the hole will be 1.0035–1.0000. The limits for the shaft will be 0.9955–0.9935. The smallest actual hole size will always be larger than the largest actual shaft size. This will produce a clearance fit where the smallest actual hole size will always be larger than the largest actual shaft size. If you decide that the last digit (implying accuracies of one ten-thousandth of an inch) is not necessary in this case (because an RC8 fit is a relatively loose fit) and round the values to 1.004–1.000 for the hole and 0.996–0.994 for the shaft, it will still produce a clearance fit similar to an RC8 fit but not an exact RC8 fit.

The Basic Shaft System

The basic shaft method for calculating fits takes the size of the shaft as the basic dimension from which the toleranced dimensions are derived. The basic shaft method is not commonly used, but may be preferable when multiple parts will fit onto a common shaft size, or when the shaft is formed of a cold worked material that is difficult to machine to size. (See Figures 12.65 and 12.66 respectively for the preferred fits for the basic hole system and the preferred fits for the basic shaft system.)

Allowance

The difference between the smallest hole size and the largest shaft size is the *allowance*. Allowance is the minimum clearance or maxi-

TABLE 12.7

Portion of RC8 Fit Table

	Class RC8		
		Standard limits (in thousandths)	
Nominal size range, inches	Limits of clearance	Hole H10	Shaft c9
0.71–1.19	4.5 10.0	+3.5 –0	–4.5 –6.5

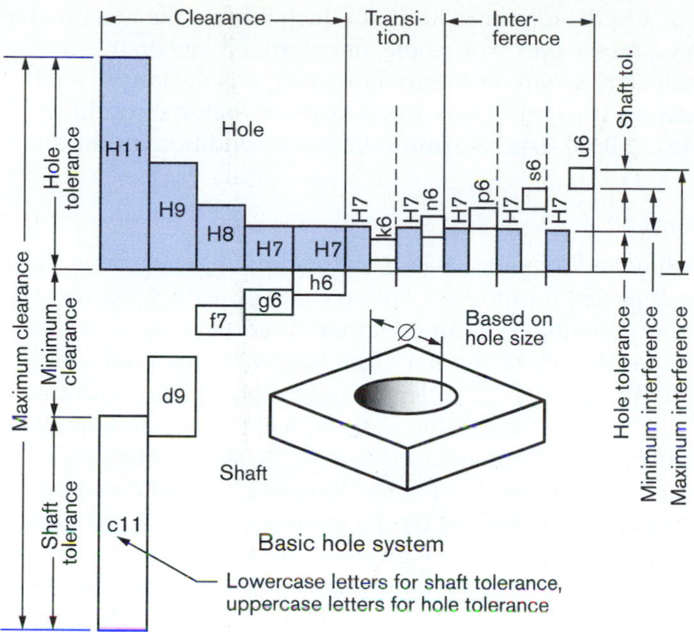

FIGURE 12.65 *The preferred fits for the basic hole system*

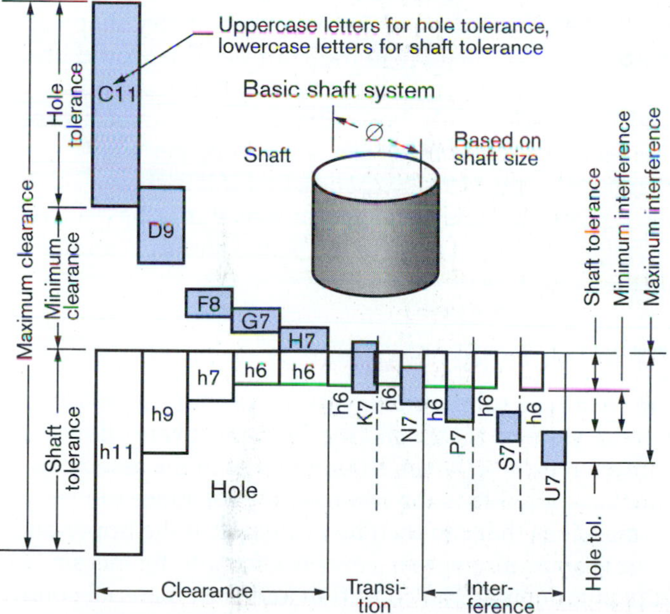

FIGURE 12.66 *The preferred fits for the basic shaft system*

mum interference that is desired between two members when they are at maximum material condition. The allowance for the parts in Figure 12.64 is .0045.

Maximum Material Condition

The term *maximum material condition* is used to eliminate confusion between the largest sizes for holes and shafts. Maximum material

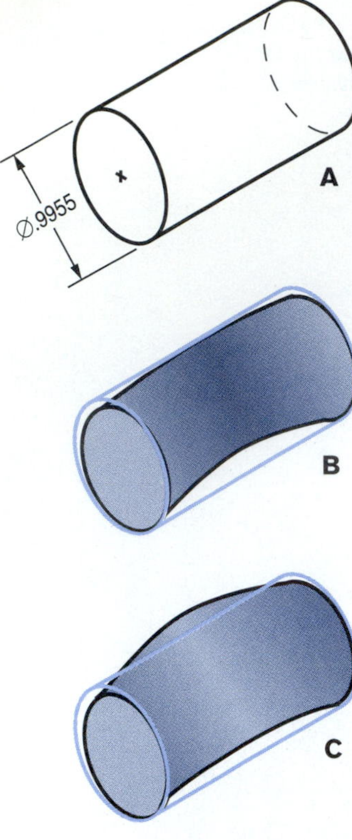

FIGURE 12.67
Perfect form boundary. (a) The perfect form for the shaft is a cylinder with the diameter of the shaft at maximum material condition. (b) The actual part shown is acceptable because, although it is bent, it does not exceed the perfect form boundary. (c) This part is not acceptable because it exceeds the perfect form boundary, even though its diameter at any point is within the stated tolerance.

condition means the tolerance limit at which there is the most material remaining on the part. For a hole, or external member, it is the smallest hole. For a shaft, or internal member, it is the largest shaft. For the example in Figure 12.64, the maximum material condition for the hole is 1.0000. The maximum material condition for the shaft is 0.9955.

Perfect Form

The dimension and tolerance stated on the drawing determine the *perfect form boundary* for the part. For an external feature like a shaft, the perfect form boundary is an imaginary boundary equal to the largest limit for the size of the shaft (the maximum material condition). For the shaft in the example, it would be a cylinder with the diameter of 0.9955. An acceptable part can have its actual diameter at any location along the shaft anywhere in the stated tolerance range, and it can be tapered or bent as long as it doesn't exceed the perfect form boundary implied by the tolerance. Figure 12.67 illustrates the perfect form boundary for the shaft. When an exterior feature is at its minimum size, the tolerance does not imply that it has to be perfectly formed, but it may not extend beyond the perfect form boundary.

A hole is at maximum material condition when it is at its smallest size limit. The perfect form boundary for an internal feature, like a hole, is an imaginary boundary of the perfect geometry shape at the smallest size limit. For the hole in the example, this would be a cylinder that is 1.00 in diameter. If the surface of the actual hole extended inside this perfect form boundary, it would not be acceptable. When the hole is at its smallest size, it must be perfectly formed (e.g., it must be perfectly perpendicular if it is shown perpendicular). When the hole is at a larger size, it may be drilled slightly at an angle, as long as none of the actual surface forming the hole is inside the perfect form boundary.

Modeling for Tolerance Studies

When you are creating a solid model, you can use the model to perform tolerance studies. Most solid modeling software allows you to check for interferences between different parts in the assembly. Some software has capabilities for you to apply tolerance ranges to parametric dimensions that can then be evaluated at the upper and lower limits for the size range. You can check for interferences when the part is at maximum material condition and least material condition (the smallest shaft and the largest hole). Commands to check interferences usually report the size of the interference. If you are intending a force fit, this will be the allowance.

The type of tolerance designation you use can make it harder or easier to model and change tolerances. Consider the use of limit tolerances in the last example. The size for the hole was 1.0035–1.0000. The nominal size for the hole is 1.0000, which is the lower limit. A machinist aiming to produce the minimum size hole (a reasonable practice since it is easy to make holes larger, but very difficult to make them smaller!), can easily see the starting dimension, making the drawing easy to interpret. If you want to change the stated tolerance, you can add all of the additional tolerance to the upper range (making the 1.0035 dimension larger) and leave the model at the lower end size. Then you can change the value (if it is a parametric dimension) to the upper range to model the tolerance easily. For the mating shaft part, its upper limit size can remain the same, and the lower limit can be changed to add additional tolerance if desired.

Equal bilateral tolerances can make drawings easy to check because the value for a dimension includes the nominal size intended for the part (again easily visible to the machinist), with equal upper and lower tolerances. However, changing the tolerance zone for a bilateral tolerance means changing the mean size for either the hole or the shaft. If you add additional plus/minus range to the size of the hole, you will have to change the starting diameter size for the shaft in order for it to still fit. In general, bilateral tolerances work well for parts where the allowable variation is equally distributed around both sides of the nominal size. Welded assemblies and loose tolerances are an example.

Metric Fit Tables and Designations

The ISO system of preferred metric limits and fits is included in the ANSI B4.2 standard. Like the basic hole system for inch designations, this system can be used for holes, cylinders, and shafts, and is also adaptable for fits between parallel surfaces including features such as keys and slots. Terms for metric fits are similar to those for decimal-inch fits.

Tolerance: the difference between the permitted minimum and maximum sizes of a part.

Basic size: the size from which limits or deviations are assigned.

Deviation: the difference between the basic size and the hole or shaft size (comparable to the tolerance in the U.S. customary inch system).

Upper deviation: the difference between the basic size and the maximum allowable size of the part, similar to the maximum tolerance in the the U.S. customary inch system.

Lower deviation: the difference between the basic size and the minimum allowable size of the part, comparable to minimum tolerance in the the U.S. customary inch system.

Fundamental deviation: the deviation closest to the basic size, comparable to the minimum allowance in the the U.S. customary inch system.

International tolerance grade (IT): is a set of tolerances that vary according to the basic size to provide a uniform level of accuracy within the grade. For example, in the dimension 50H8 (for a close-running fit), the IT grade is indicated by the number 8. There are 18 IT grades—IT01, IT0, and IT1 through IT16. Smaller grade numbers indicate smaller tolerance zones.

Tolerance zone: describes the relationship of the tolerance to basic size. In the dimension 50H8, for the close-running fit, the tolerance zone is specified by a letter and a number, H8. An uppercase

TABLE 12.8

Basic Size (in millimeters)

1st choice	2nd choice	1st choice	2nd choice	1st choice	2nd choice	1st choice	2nd choice
1	1.1	5	5.5	25	28	120	140
1.2	1.4	6	7	30	35	160	180
1.6	1.8	8	9	40	45	200	220
2	2.2	10	11	50	55	250	280
2.5	2.8	12	14	60	70	300	350
3	3.5	16	18	80	90	400	450
4	4.5	20	22	100	110	500	550

letter indicates that the tolerance applies to the hole; a lowercase letter indicates that the tolerance applies to the shaft. The number is the IT grade.

Hole-basis system of fits: uses the hole diameter (basic size) as the minimum size. A table of fits will indicate whether the hole basis or shaft basis is being presented.

Shaft-basis system of fits: uses the shaft diameter (basic size) as the maximum size.

Preferred Metric Sizes The preferred basic sizes for computing tolerances are given in Table 12.8. Choosing basic diameters from the first column yields readily available stock sizes for round, square, and hexagonal products.

Geometric Dimensioning and Tolerancing

Geometric dimensioning and tolerancing (GD&T) uses special drawing symbols to specify geometric characteristics as well as dimensional requirements on engineering drawings. GD&T is a means of more closely controlling how much a feature can deviate from the perfect geometry implied by the drawing. Using GD&T, you can specify tolerance zones that reflect the geometry of the feature you are trying to control. GD&T symbols allow you to control features such as parallelism, perpendicularity, and other geometric characteristics of the part that may otherwise be implied by the perfect form boundary for the part, but not controlled specifically. Use GD&T to control the geometry of features when it is important that they are formed accurately to ensure the fit and function of mating parts. In any particular drawing, not every feature needs to be tightly controlled. The manufacture of some parts or features can be stated with more general tolerances.

In this section, you will be introduced to the geometric characteristics defined by the GD&T standard and the symbols used to define them. ANSI/ASME Y14.5M–1994 is the current GD&T standard, and many of the illustrations in this section are taken from that standard. As you build your skill in engineering drawing, you may wish to take a separate course in GD&T to learn to apply the basic principles presented here.

TABLE 12.9

Geometric Characteristics (ANSI Y14.5M–1994)

	Tolerance	Characterstic	Symbol
For individual features	Form	Straightness	—
		Flatness	⏥
		Circularity	○
		Cylindricity	⌭
For individual or related features	Profile	Profile of a line	⌒
		Profile of a surface	⌓
For related features	Orientation	Angularity	∠
		Perpendicularity	⊥
		Parallelism	//
	Location	Position	⌖
		Concentricity	◎
		Symmetry	⌯
	Runout	Circular runout	↗
		Total runout	⌰

Geometric Characteristic Symbols

The geometric characteristic symbols can be divided into those for individual features, such as *form tolerances* (which relate to the shape of a feature) and those used for related features such as *orientation* or *location tolerances* (which control a feature in relation to another feature). Table 12.9 shows the geometric characteristics divided into these categories and the symbols used to indicate them.

The Feature Control Frame

Geometric tolerances are indicated by means of geometric characteristic symbols in a feature control frame. A *feature control frame* contains all of the symbols and modifiers necessary for specifying the geometric tolerance for a feature. It begins with the *geometric characteristic symbol,* which tells the type of geometry being controlled. The next compartment in the frame contains the geometric tolerance value for the feature, which specifies the total width of the tolerance zone. The simplest feature control frame contains just

A
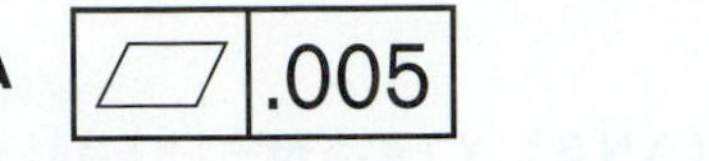

B
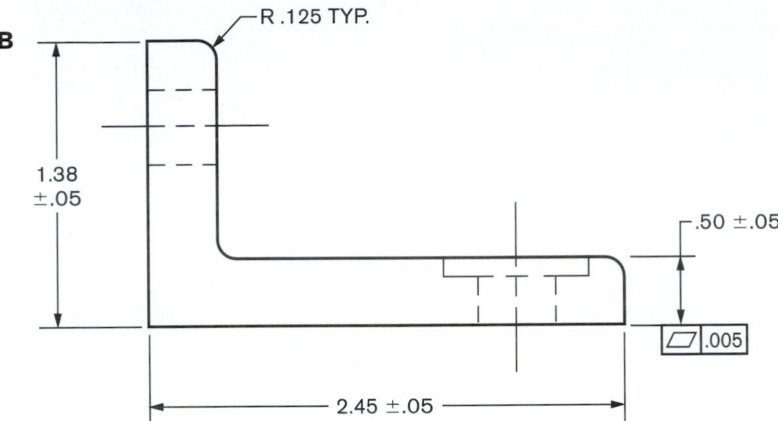

FIGURE 12.68
(a) A simple feature control frame; (b) Feature control frame attached to an extension line controlling the flatness of the bottom surface.

the geometric characteristic symbol and the tolerance, as shown in Figure 12.68. The placement of additional symbols in the feature control frame is governed by Y14.5 standard and will be illustrated throughout this section.

Feature control frames can be related to a feature in the drawing by:

- adding the frame below a note or dimension pertaining to the feature;
- a leader line from the frame to the feature;
- attaching a side or end of the frame to an extension line from the feature (if it is a plane surface);
- attaching a side or end of the symbol frame to a dimension line pertaining to a feature of size.

Form Tolerances for Individual Features

Straightness, flatness, circularity (roundness), cylindricity, and sometimes profile are form tolerances that apply to single features. Think back to the discussion of the perfect form boundary implied by a tolerance for the size of a feature. When the feature is at its maximum material condition (largest size for external features like shafts, smallest size for internal features like holes), its form must be perfect in order to fit inside the perfect form boundary. That is, a shaft would have to be perfectly straight. But if the shaft were at its smallest size, it could be tapered or bent and still meet the specified tolerance. Sometimes you may need to control the tolerance more precisely than this. Specifying the straightness, flatness, roundness, cylindricity, and even the profile tolerance can provide more control over the feature.

Figure 12.69 illustrates a form tolerance using a dimensioned and toleranced drawing of a shaft. The feature control frame indicates that any longitudinal element of the cylinder forming the shaft

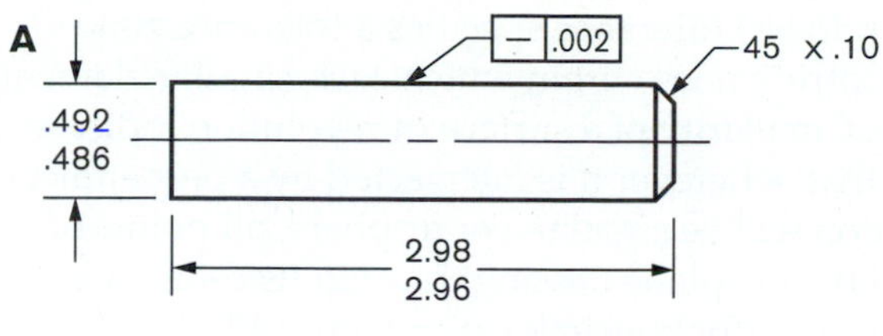

FIGURE 12.69
(a) The feature control frame indicates that any element of the shaft must be straight within .002. (b) To interpret the straightness control, imagine two straight parallel lines set apart by the tolerance amount. Any line element making up the feature being controlled must lie entirely between the perfectly straight lines. (c) An acceptably straight element fits within the tolerance zone.

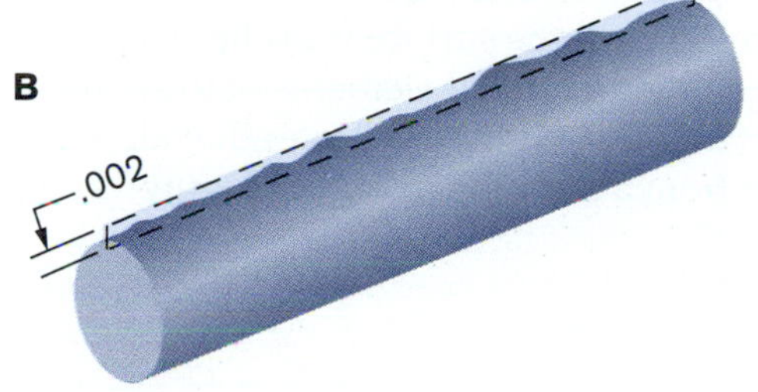

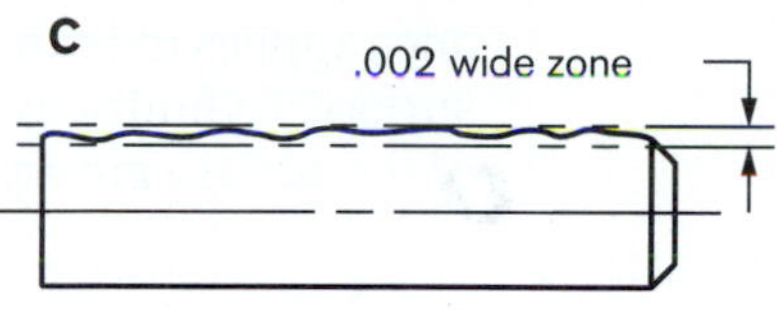

must be straight within .002 inch. The pictorial drawing in (b) shows an exaggerated view of two straight parallel lines .002 inch apart running the length of the shaft. Any line element of the shaft must fall within the zone described by the straight lines. The orthographic view in (c) shows an acceptably straight element of the cylinder as it fits in the tolerance zone.

A **straightness tolerance** specifies a tolerance zone within which an axis or all points of the indicated element must lie. Straightness indicates that the element of a surface or an axis is a straight line. (See Figure 12.70.)

A **flatness tolerance** specifies a tolerance zone defined by two parallel planes within which the surface must lie. Flatness of a surface is having all elements in one plane. (See Figure 12.71.)

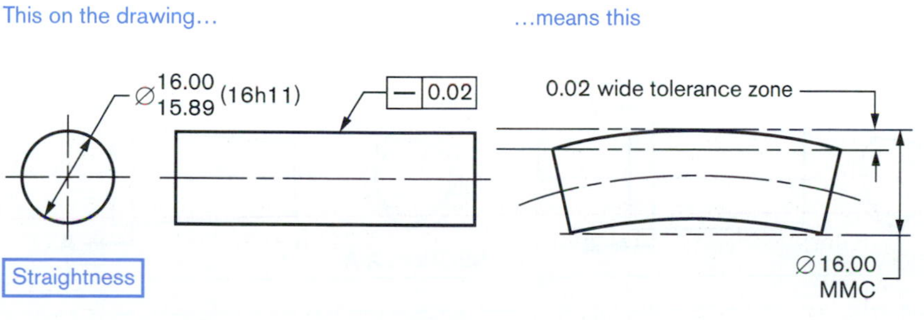

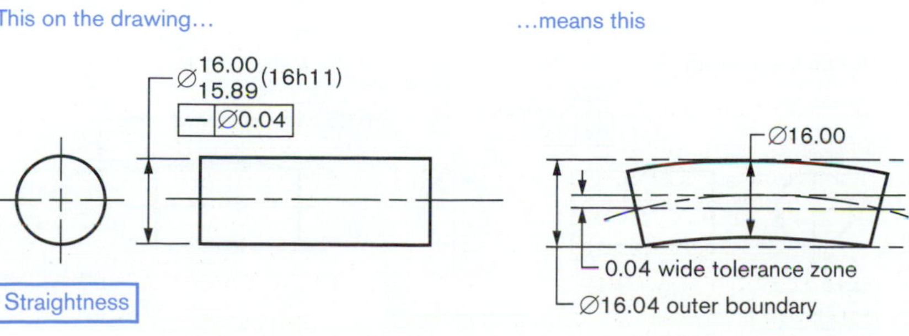

FIGURE 12.70
Straightness (ANSI Y14.5M–1994.)

A **circularity (roundness) tolerance** specifies a tolerance zone bounded by two concentric circles within which each circular element of the surface must lie. Circularity of a surface of revolution indicates for a cone or cylinder that, wherever it is intersected by a perpendicular plane, its cross section will be circular. For a sphere, all points of the surface intersected by any plane passing through its center are equidistant from that center (form a circle). (See Figure 12.72.)

A **cylindricity tolerance** specifies a tolerance zone bounded by two concentric cylinders within which the surface must lie. This tolerance applies to both circular and longitudinal elements of the entire surface. Cylindricity of a surface of revolution indicates that all points of the surface are equidistant from a common axis. (See Figure 12.73.)

FIGURE 12.71 *Flatness (ANSI Y14.5M–1994.)*

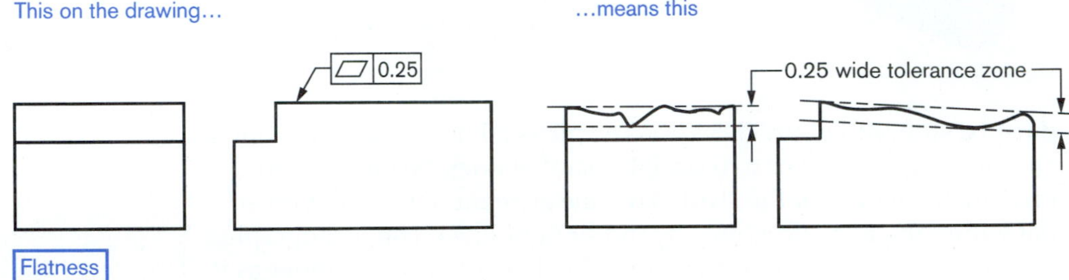

FIGURE 12.72 *Circularity (roundness) (ANSI Y14.5M–1994.)*

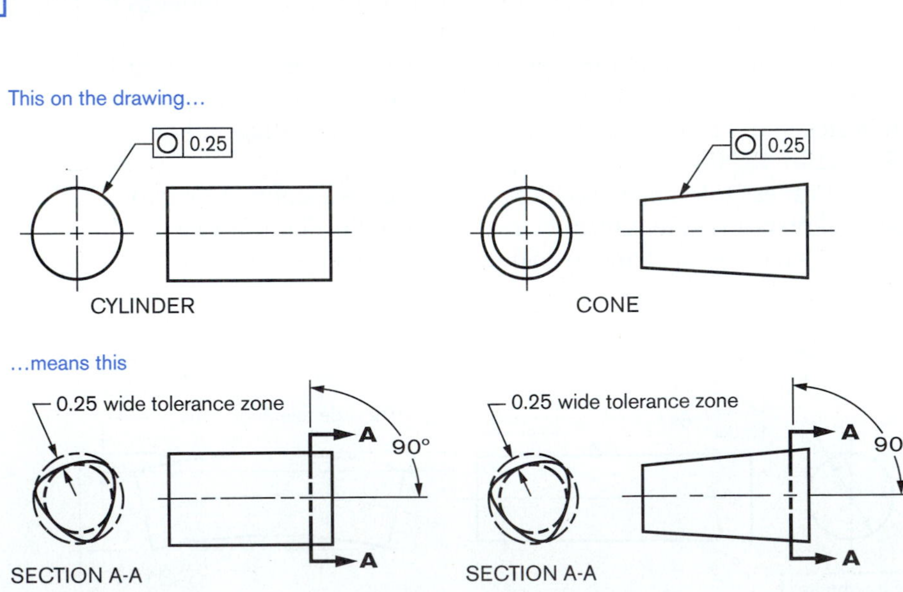

FIGURE 12.73 *Cylindricity (ANSI Y14.5M–1994.)*

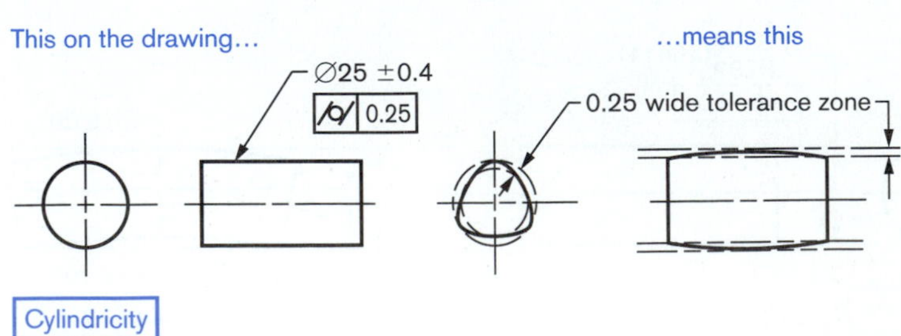

A **profile tolerance** specifies a uniform boundary or zone along the true profile within which all elements of the surface must lie. A profile is the outline of an object in a given plane (a 2D figure). Profiles are formed by taking cross sections through the figure. The resulting profile is made up of straight lines, arcs, other curved lines, and 2D elements. (See Figures 12.74–12.76.) The profile of a surface is a 3D zone that extends along the length and width (or circumference) of the indicated feature. This can also be applied to parts that have a constant cross section, to a surface, to a revolution, or to parts such as castings by profile tolerances that apply all over (through the use of the all over symbol) on the feature control frame leader.

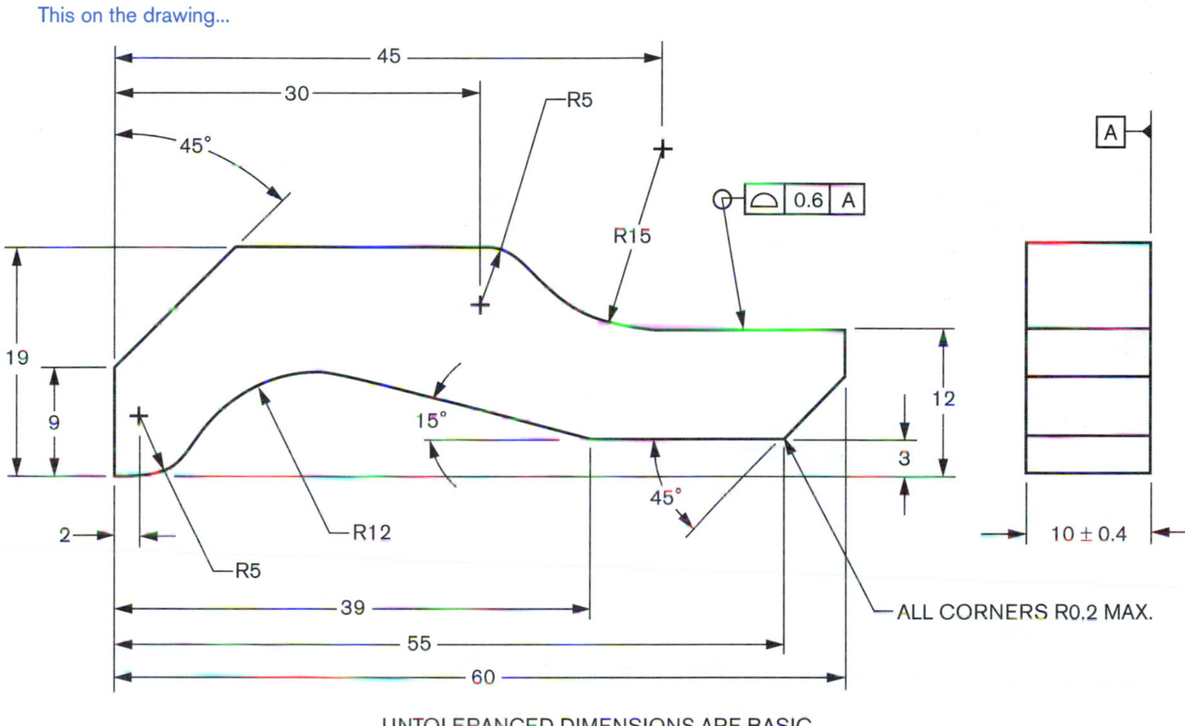

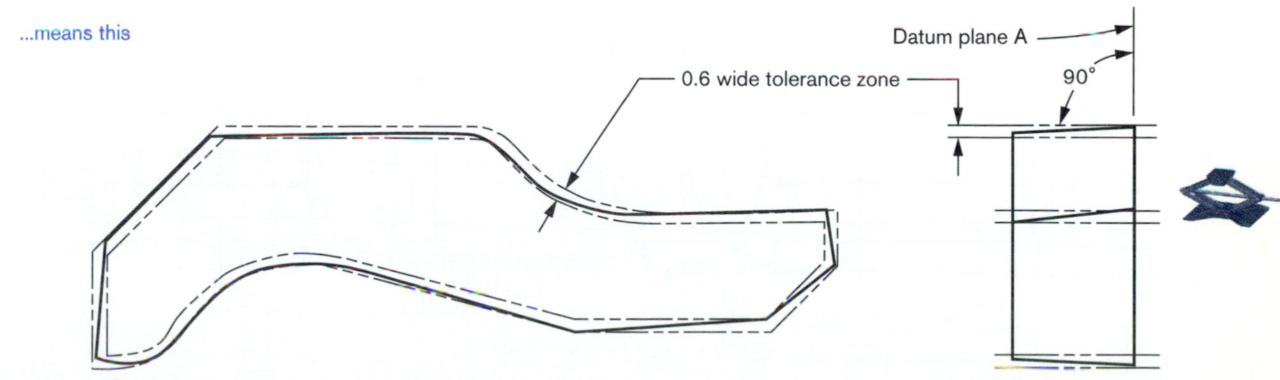

FIGURE 12.74
Profile of a surface all around (ANSI Y14.5M–1994.)

This on the drawing...

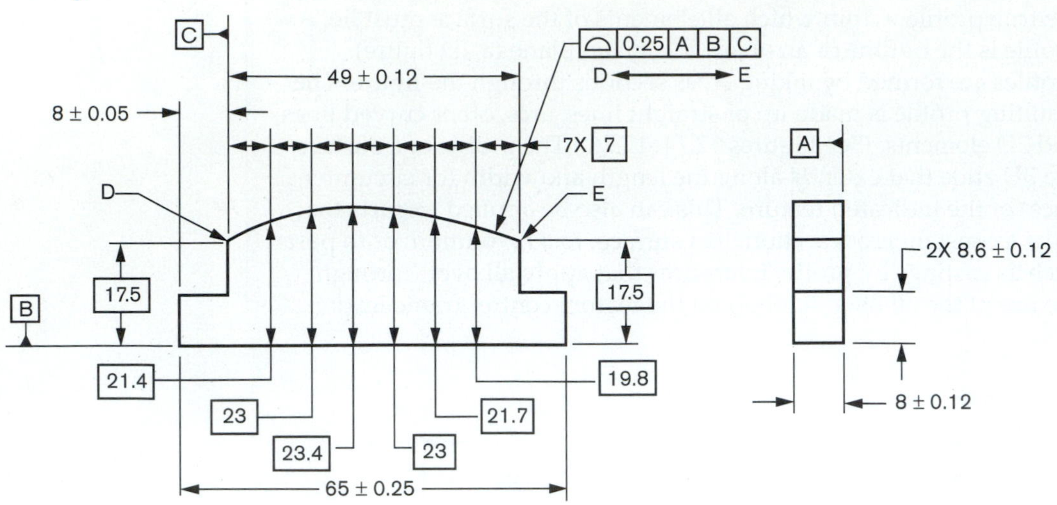

...means this

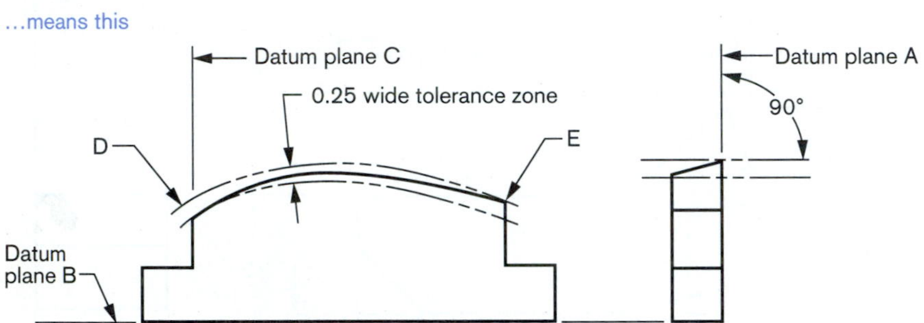

FIGURE 12.75
Profile of a surface between points

This on the drawing...

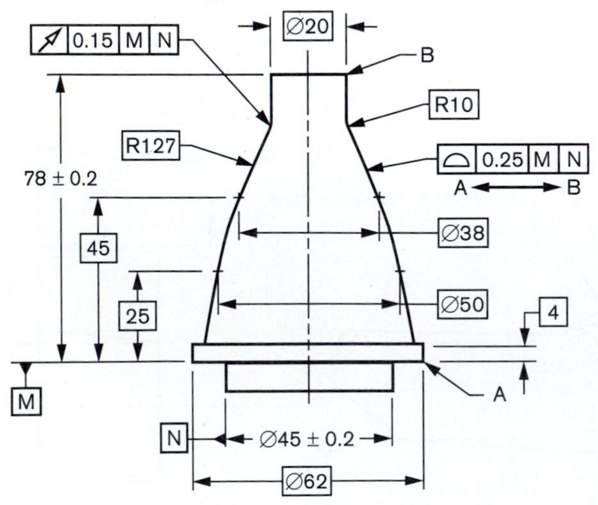

...means this

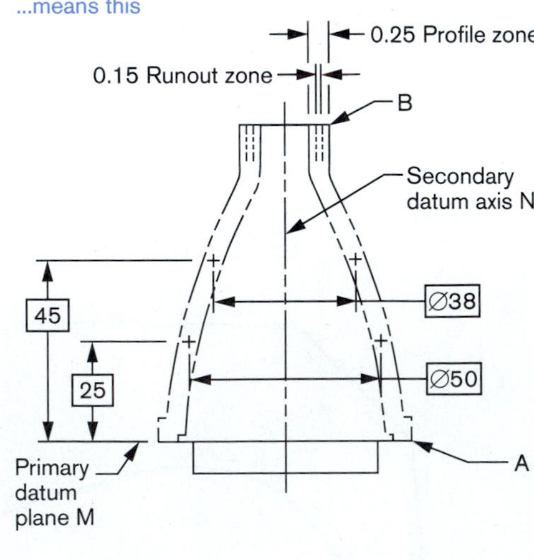

FIGURE 12.76
Profile of a surface of revolution

Datum Surfaces and Features

Datum surfaces and datum features are used as references to control other features on the part. Datum features should be actual features on the part, such as a point, line, plane, cylinder, or other geometric form that is assumed to be exact.

For example, you may wish to indicate that surface F of the short leg of the bracket shown in Figure 12.77 must be perpendicular to surface G, the bottom of the bracket, within a certain tolerance. A feature, such as the bottom surface, that is used as a starting point for a measurement is called a datum feature. It establishes a theoretically exact plane from which a measurement can be made. Figure

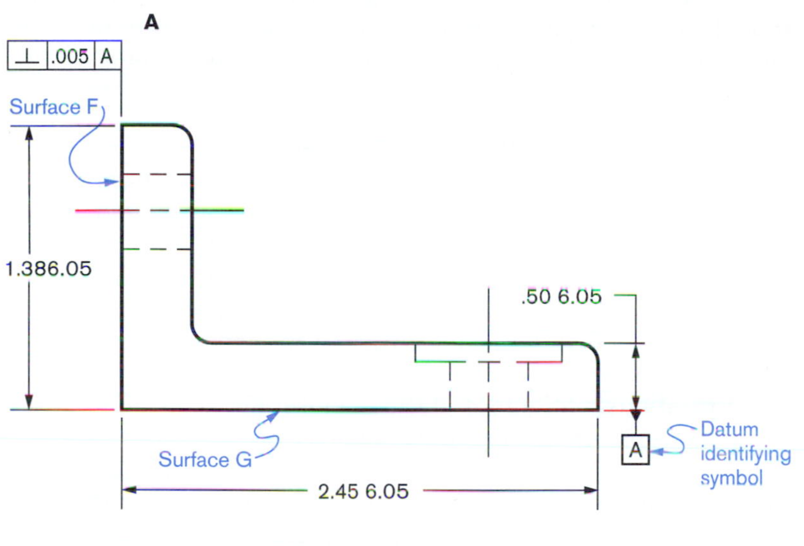

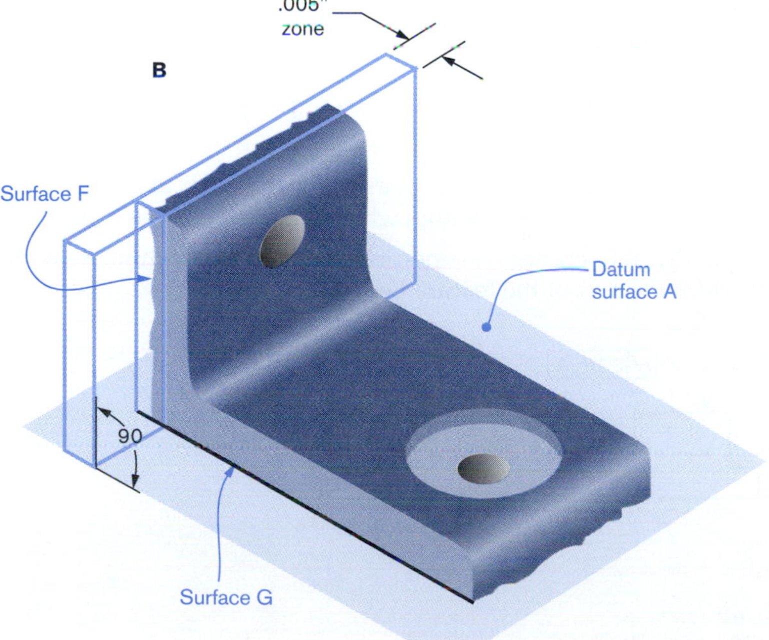

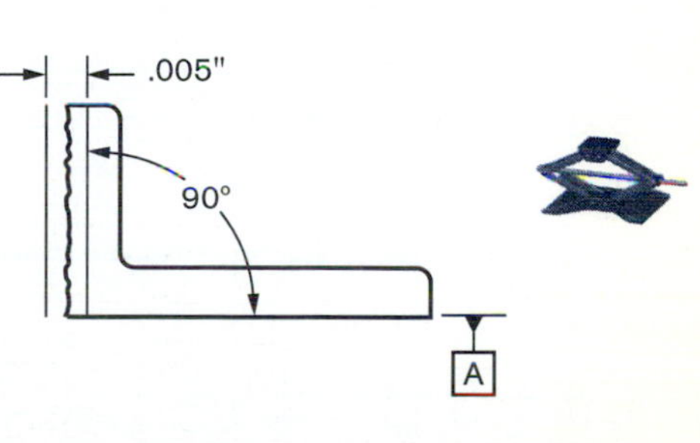

FIGURE 12.77
Surface G on the part is used to establish the theoretically exact datum A from which the perpendicularity of surface F is determined. The feature control frame indicates that surface F must be perpendicular to datum plane A within a .005-inch-wide tolerance zone.

12.77(a) shows an angled bracket with a feature control frame controlling the perpendicularity of surface F to within .005 of datum A, which is bottom surface G. The pictorial drawing in (b) illustrates datum A, a theoretically exact plane established from surface G, which will be used to measure the perpendicularity of surface F. Note that not every geometric characteristic requires a datum—only those that relate two different features to one another; for example, perpendicularity, parallelism, or concentricity.

Datum Identifiers

Datum identifiers are added to a drawing to indicate the reference feature from which a location or to which a geometric relationship may be established. A *datum-identifying symbol* is a capital letter in a square frame with a leader attaching it to the indicated feature via a triangle. The triangle may be filled or not filled. Letters of the alphabet (except *I, O,* and *Q*) are used as datum-identifying letters.

Tolerances for Related Features

Parallelism, perpendicularity, and angularity are tolerances that relate the orientation of one feature to another. Position and concentricity tolerances are geometric characteristics that relate the location of one feature to another. Runout tolerances relate how well a circular or cylindrical part rotates about its axis. Tolerances for related features require that datum references be specified in the feature control frame.

A **parallelism tolerance** specifies a tolerance zone defined by two parallel planes (see Figure 12.78) or by two lines parallel to a datum plane or axis within which the surface or axis of the feature must lie. The parallelism tolerance may also specify a cylindrical tolerance zone parallel to a datum axis within which the axis of the feature must lie. (See Figure 12.79.)

A **perpendicularity tolerance** indicates that a surface, median plane, or axis is at 90 degrees to a datum plane or axis. (See Figures 12.80 and 12.81.) A perpendicularity tolerance specifies one of the following:

- A zone defined by two parallel planes perpendicular to a datum plane, datum axis, or axis within which the surface of the feature lies.
- A cylindrical tolerance zone perpendicular to a datum plane within which the axis of the feature lies.

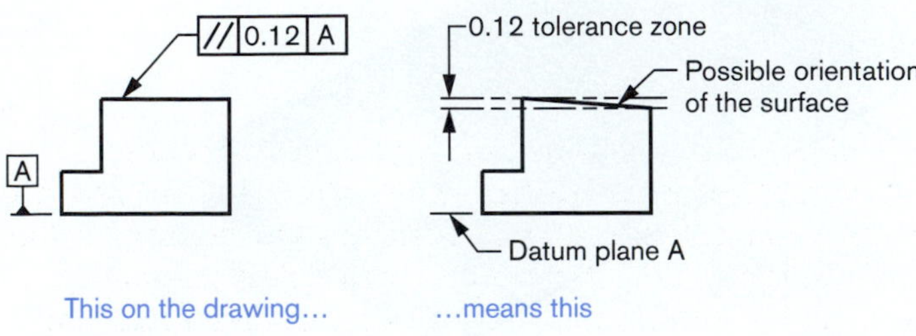

FIGURE 12.78
Parallelism for a plane surface

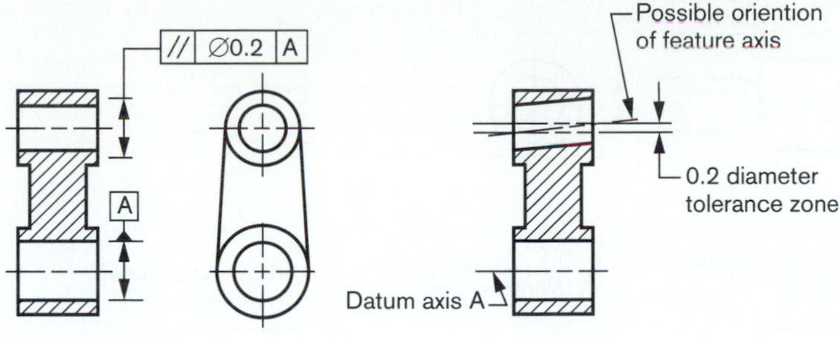

FIGURE 12.79
Parallelism for an axis

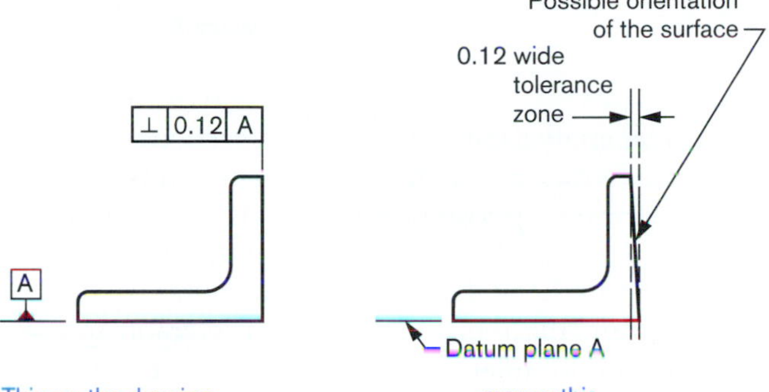

FIGURE 12.80
Perpendicularity for a plane

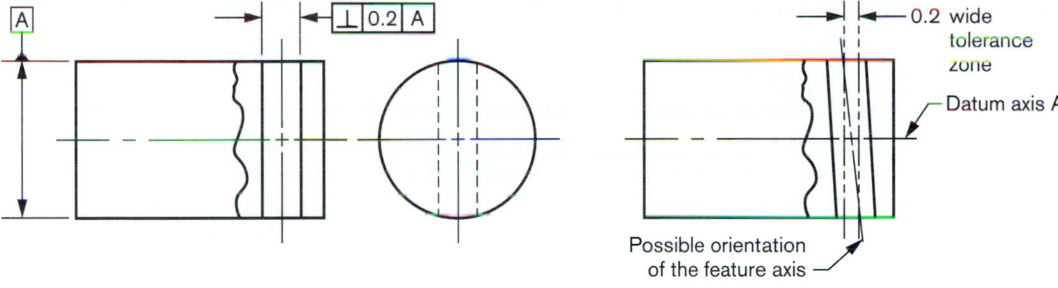

FIGURE 12.81
Perpendicularity for an axis

A **concentricity tolerance** indicates that a cylinder, cone, hex, square, or surface of revolution feature shares a common axis with a datum feature. A concentricity tolerance controls the location for the axis of the indicated feature within a cylindrical tolerance zone whose axis coincides with a datum axis. The controlled feature must have its axis entirely within the cylindrical tolerance zone established by the datum axis. To measure concentricity, the axis of the

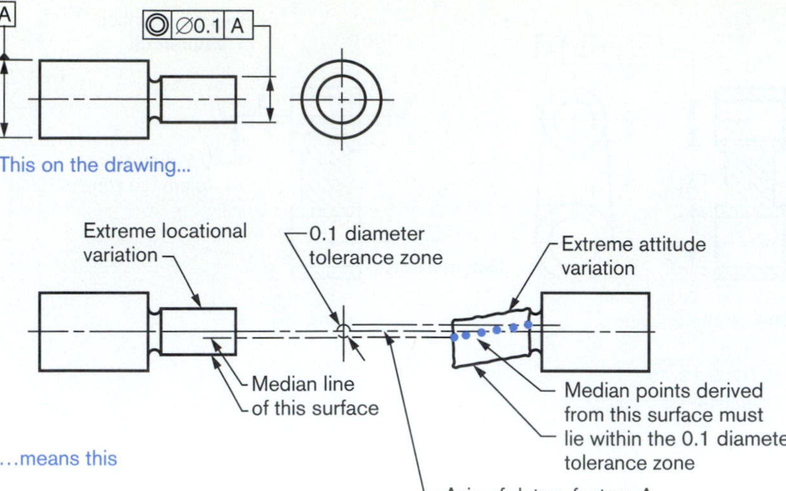

FIGURE 12.82
Concentricity

feature is established by analysis of multiple cross-sectional elements of the surface (which calculates the median points between multiple sets of diametrically opposed elements). All of the center points for the cross sections must lie in the cylindrical tolerance zone for concentricity. See Figure 12.82. Because measurements to determine concentricity are difficult to make, it is often preferable to use positional tolerance or runout tolerance instead. (You will learn about positional tolerance later in this chapter.)

Runout

Runout is another geometric characteristic that can be specified in a feature control frame. Runout is a measure of how a circular or cylindrical part rotates about its axis. When a perfectly formed wheel is rotated on its axis, the surface of the wheel does not dip above or below the reference plane. To measure runout, an indicator is mounted on a stationary surface and touching the rotating feature so it can determine how much the rotating feature moves above or below the plane. The full indicator movement (FIM) is used to measure the runout. In other words, if the indicator moved from –1 to +1, the FIM would be 2.

Two types of runout can be specified in the feature control frame. *Circular runout* positions the indicator at a single location, where the variation in the circular shape of the rotating object is measured by the indicator. (See Fig. 12.83.)

To measure *total runout,* the indicator is moved along the rotating surface and the total allowable variation of the indicator over the surface is specified.

Basic Dimensions

Basic dimensions are the theoretically exact, untoleranced dimensions that specify the perfect location, size, shape, or angle of a feature. A basic dimension is identified by enclosing the dimension

This on the drawing ....

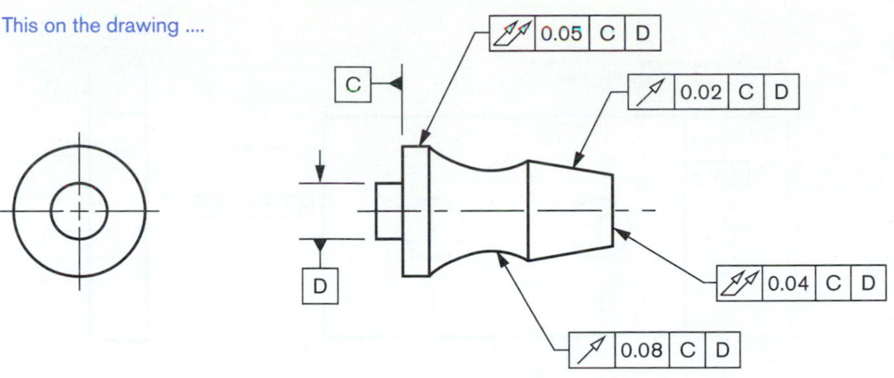

Means this ....

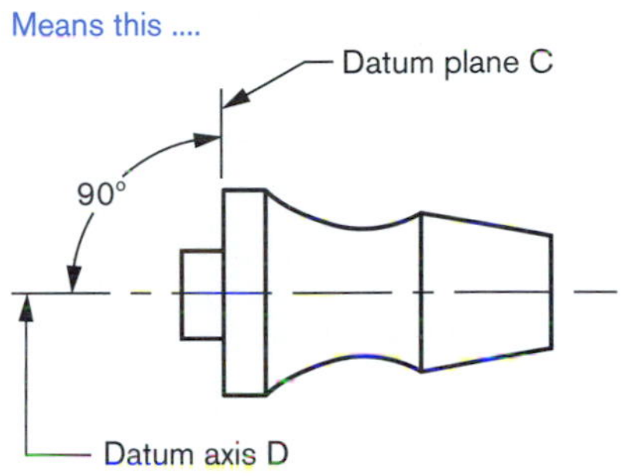

At any measuring position, each circular element (for circular runout) and each surface (for total runout) must be within specified runout tolerance when the part is mounted on datum surface C and rotated 360° about datum axis D.

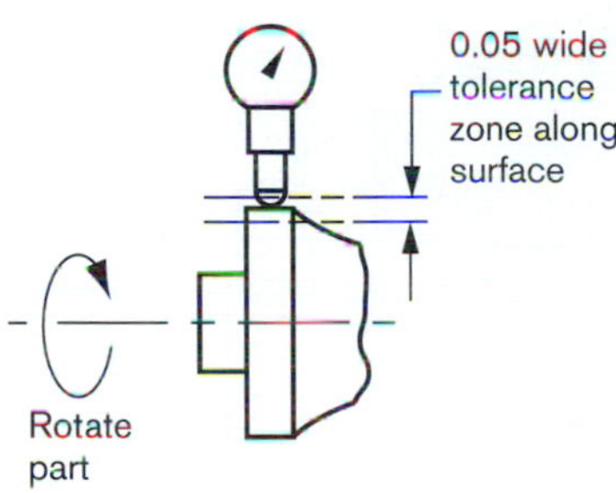

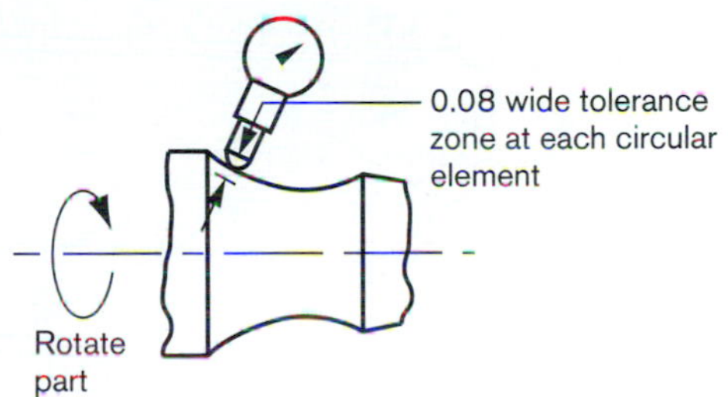

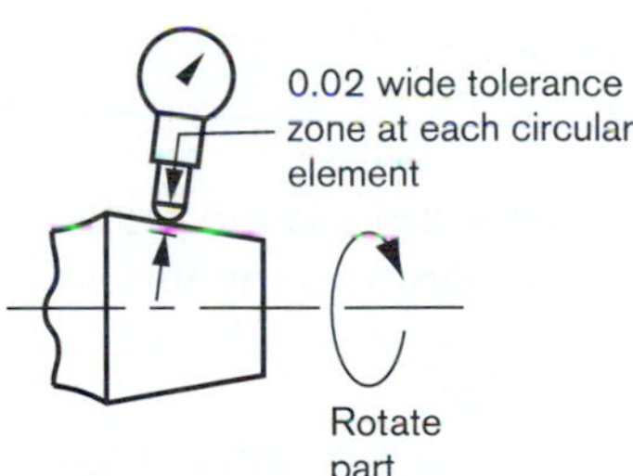

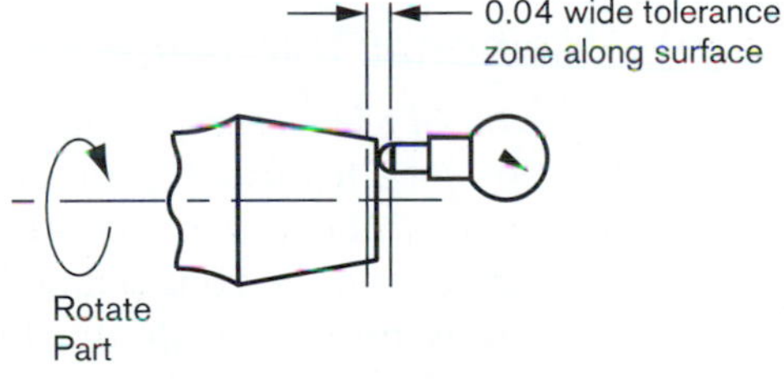

FIGURE 12.83
Circular and total runout

in a box, as shown in Figure 12.84 (a). Basic dimensions specify theoretically exact values from which variations are defined through feature control frames as well as toleranced features or notes. Sometimes it is noted on the drawing: "GENERAL TOLERANCES DO NOT APPLY TO BASIC DIMENSIONS" in order to prevent general tolerances from being applied to basic dimensions by those not skilled in interpreting GD&T symbols.

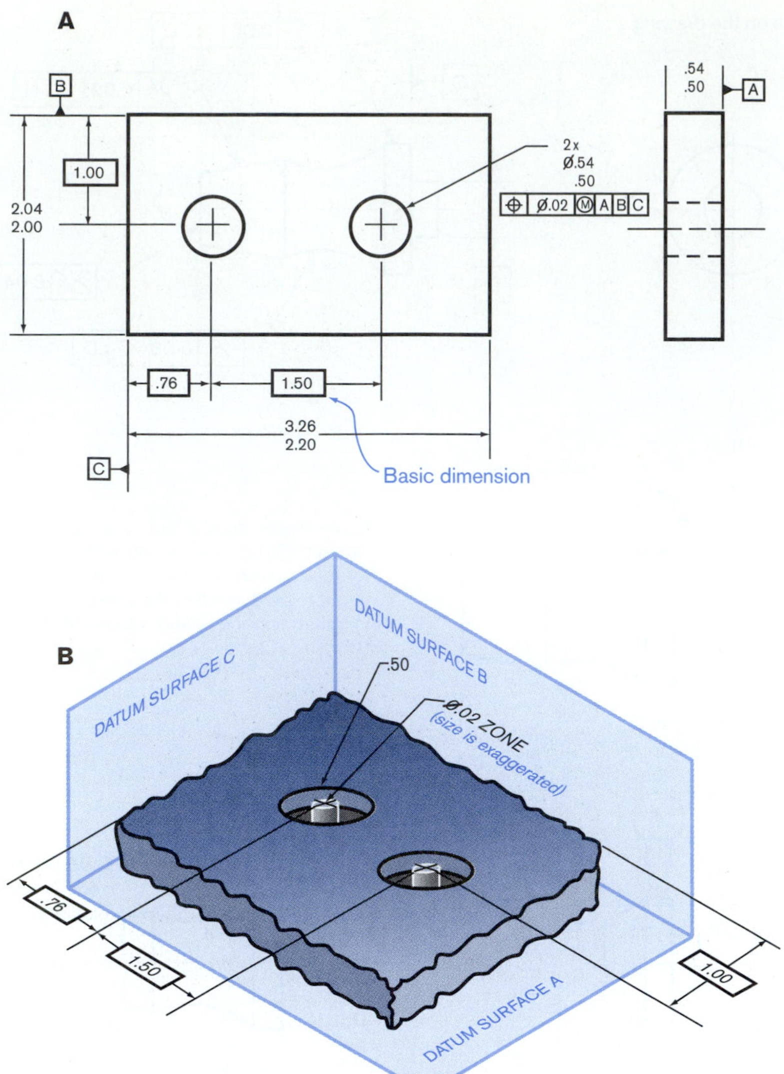

FIGURE 12.84
(a) The feature control frame in the orthographic views can be interpreted such that when the holes are at their smallest size, their positions can vary from the theoretically exact basic dimension locations within a zone that is .02 in diameter. (b) Pictorially, the small cylinders at the center of the holes represent the tolerance zone for the locations of the hole centers. The centerline of the cylinder forming the hole must remain inside this zone.

Positional Tolerance

A *positional tolerance* establishes a tolerance zone around the perfect location for a feature. A positional tolerance is identified by its characteristic symbol and linked to a size feature. The axis of the feature must be within the zone defined by the tolerance.

For example, consider the locations of the holes shown in Figure 12.84. Basic, theoretically exact, dimensions are used to locate features at true position. The location dimensions for the holes in the figure are shown as basic dimensions. The positional tolerance specification requires that the location for the center of the hole remain inside a cylindrical zone .02 in diameter. The circled M in the example is a modifier that indicates that the tolerance applies when the hole is at maximum material condition. As you learned earlier, the perfect form requirement for the hole specifies that all of the elements on the hole's surface must be on or outside a cylinder

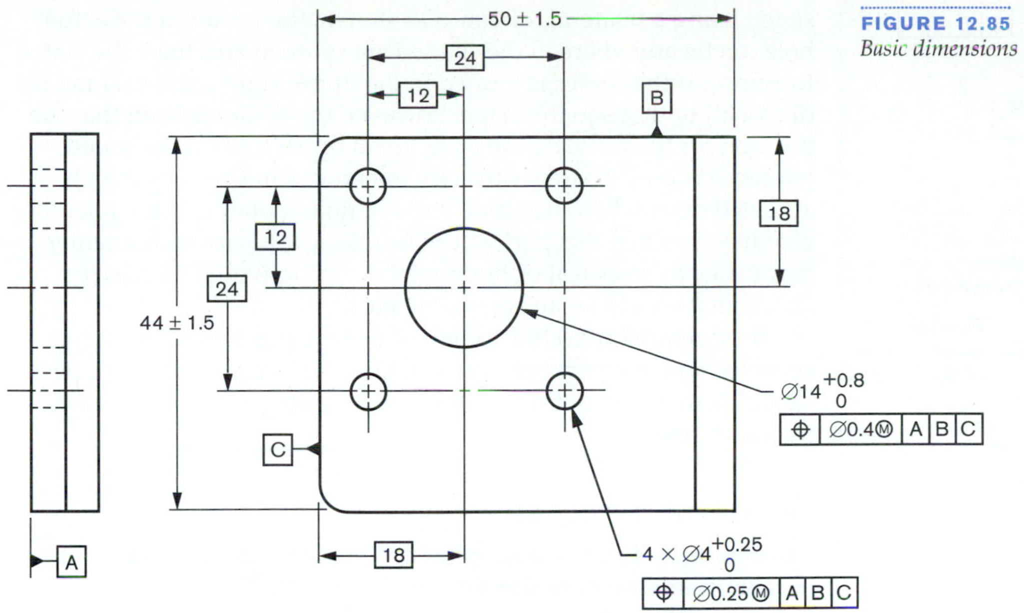

FIGURE 12.85
Basic dimensions

whose diameter is equal to the minimum diameter (maximum material condition). In this case, the hole must have its center in the .02 cylindrical zone but also may not exceed the perfect form boundary.

Figure 12.85 shows a more complex example of positional tolerance for holes.

Using GD&T symbols to specify positional tolerances overcomes the limitations of stating dimensions using only limit tolerances or plus/minus ranges. The general shape of the tolerance zone described by limit or plus/minus tolerances is typically rectangular, regardless of the shape of the feature. This rectangular tolerance zone may not fully convey the tolerance needed for certain features.

For example, Figure 12.86 shows the end of a plate with two circular holes. If the location dimension for the center of a hole is

FIGURE 12.86
Limitations of rectangular tolerance zones for some features

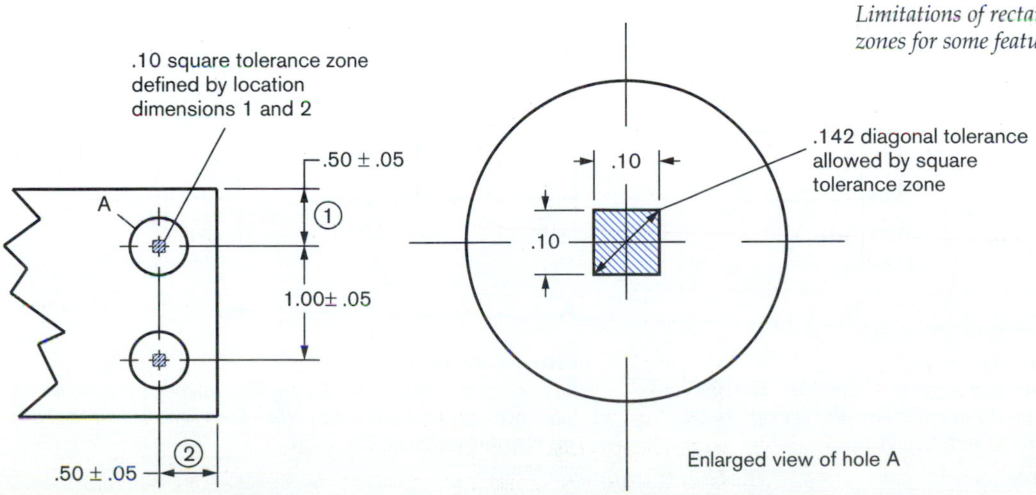

.10 square tolerance zone defined by location dimensions 1 and 2
.50 ± .05
A
①
1.00± .05
②
.50 ± .05
.10
.10
.142 diagonal tolerance allowed by square tolerance zone
Enlarged view of hole A

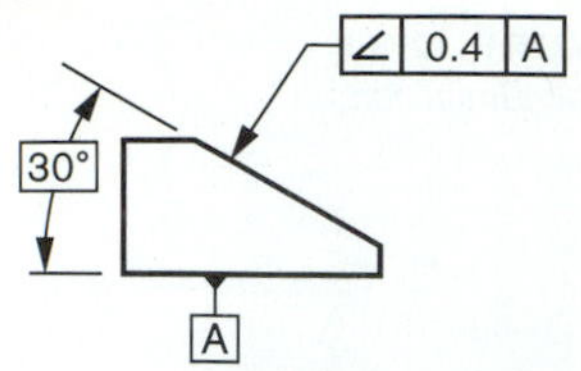

This on the drawing...

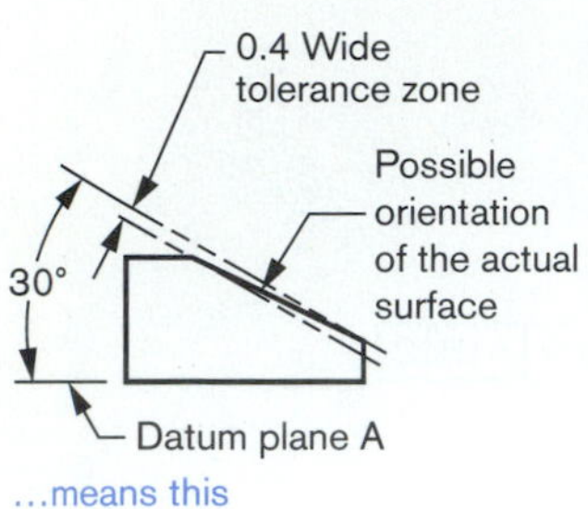

...means this

FIGURE 12.87
Angularity

stated using a bilateral tolerance as shown, the center of the actual hole can be anywhere in the shaded area shown and meet the stated tolerance. If the center is located in the upper right corner (along the diagonal) of that square shape, however, the distance from the true position for the circle's center can be off by .071 when the stated tolerance is only .05. Parts that are toleranced in this way may be accepted even when they may be truly unacceptable. If the tolerance is reduced so that this part would be rejected, a part with a center that varies in a vertical or horizontal direction would be rejected although it would be an acceptable part.

In comparison, GD&T (often called true position tolerancing) specifies how the part can vary from the perfect geometry implied on the drawing by defining either the diameter or the width of a tolerance zone.

Tolerancing Angles

An **angularity tolerance** specifies a tolerance zone defined by two parallel planes on opposite sides of the specified perfect or *basic angle* (other than 90°) from a datum plane or axis within which the surface or the axis of the feature must lie. (See Figure 12.87.)

Bilateral tolerances have traditionally been added to angles by specifying an angular dimension and a plus/minus variation for the angle (often of 1°). Using bilateral tolerances, the wedge-shaped tolerance zone increases as the distance from the vertex of the angle increases. (See Figure 12.88.)

To avoid the accumulation of tolerance farther away from the angle's vertex, make the angular measure a basic dimension as illustrated in Figure 12.89. The angle is indicated as a basic dimension, and therefore no angular tolerance is specified. The tolerance zone is defined by two parallel planes around the true location of the surface, resulting in improved angular control.

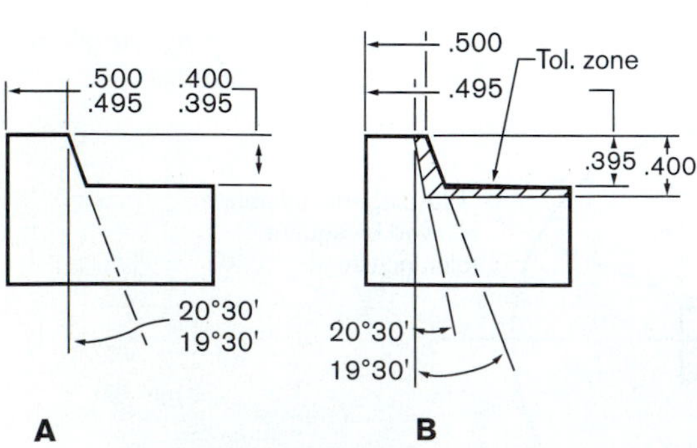

FIGURE 12.88
Bilateral tolerancing of angular dimensions. The effect of the 1° tolerance on the angular measure is magnified as the distance from the vertex increases.

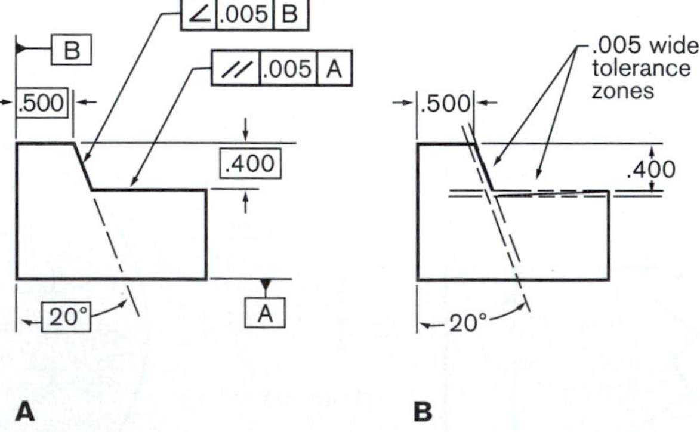

FIGURE 12.89
Basic angular tolerancing. An angular tolerance creates a tolerance zone parallel to the perfect form specified by the basic angular dimension.

Supplementary Symbols and Modifiers

Supplementary symbols or modifiers include the following:

Ø The diameter symbol often precedes the specified tolerance in a feature control symbol to describe a circular or cylindrical shaped tolerance zone. This symbol for diameter should precede the tolerance value.

Ⓜ *MMC*, or *maximum material condition*, means that a feature on the finished part contains the maximum amount of material permitted by the toleranced dimensions shown for that feature. Holes and slots are at MMC when at minimum size. Shafts, tabs, bosses, and other external features are at MMC when they are at their maximum size.

Since all features may vary in size, it is necessary to make clear on the drawing at what limit of size for the feature the characteristic applies. In all but a few cases, when a feature is not at MMC, additional positional tolerance is available without affecting function.

For example, if a tolerance is indicated to apply at maximum material condition, then the full tolerance stated is available when a hole is at its smallest size or a shaft is at its greatest size. As the actual part deviates from maximum material condition toward the minimum material condition, a bonus tolerance may be allowable. Consider trying to fit a fixed pin into a hole. As the hole size becomes larger, the pin can be off from its perfect location by a larger amount and still fit through the hole. When MMC is indicated, the *bonus tolerance* is the difference between the actual hole size measured on the feature and the maximum material condition. In this case, this extra tolerance can be added to the positional tolerance without affecting the function of the parts.

To avoid possible misinterpretation as to whether maximum material condition applies, it should be clearly stated on the drawing by adding an MMC symbol to each applicable tolerance or by suitable coverage in a document referenced on the drawing.

Ⓛ *LMC*, or *least material condition*, indicates that the feature contains the least amount of material (as opposed to MMC); LMC indicates the maximum size for an internal feature like a hole or slot and the minimum for an external feature such as a shaft or tab.

Ⓟ The projected tolerance zone symbol indicates that the tolerance zone applies beyond the boundary of the object by the distance stated to the left of the symbol. Projected tolerance zones are typically used when a threaded or press fit fastener might interfere with a mating part if it is inserted into a hole that is angled. By stating a projected tolerance zone, the true position of the center for the hole must apply beyond the edge of the part, ensuring that a fixed fastener will not be angled too much to fit through the hole in a mating part.

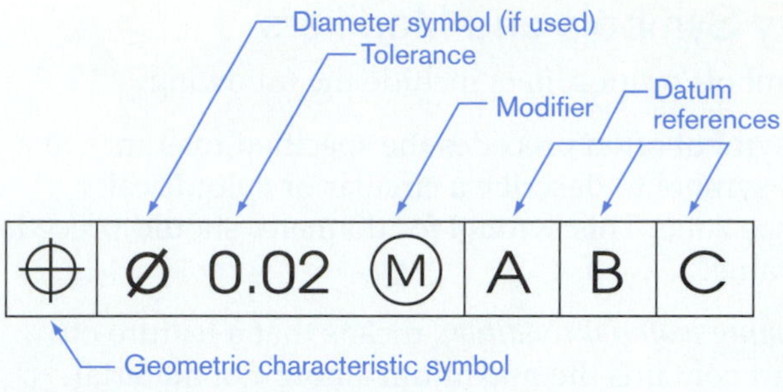

FIGURE 12.90
A feature control frame using modifiers and datum references

Do not use these symbols in local or general notes in the drawing. Instead, either write out the entire terms or use standard abbreviations such as MMC and LMC.

Regardless of Feature Size

The "regardless of feature size" (RFS) note or modifier has been used in the past to indicate that it does not matter whether features are at maximum or least material condition; the tolerance still applies as written. A circled S modifier in the feature control frame has also been used in the past to indicate RFS on drawings. ANSI Y14.5M-1994 states that unless a different modifier is used, it is understood that tolerances apply regardless of feature size; it is not necessary to indicate it on the drawing specifically.

In the feature control frame shown in Figure 12.90, the geometric characteristic symbol is that for a positional tolerance. The next portion shows the total allowable tolerance zone. The tolerance zone for the feature being located in this case is a diametral-shaped zone (indicated by the diameter symbol) 0.02 inch around the perfect position for the feature. This is followed by any modifiers, in this case a circle M, the modifier for maximum material condition, meaning that this tolerance zone applies when the actual part is at maximum material condition. The remaining boxes indicate that the measurements are to be made from datums A, B, and C. The order of the datum symbols indicate which is the primary, secondary, and tertiary datum, respectively.

Surface Control

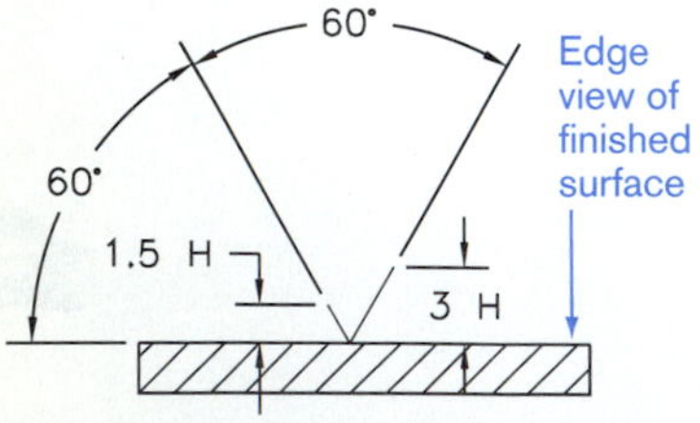

Documentation drawings also specify whether surfaces on a part are to be *finished,* or machined to a desired degree of smoothness. Surfaces on a cast or forged part are not smooth when the part is formed. Machining removes material from the surface to smooth it so it may move against other parts or assemble properly. *Finish marks* indicate which surfaces are to be machined and the quality of the finish (see Fig. 12.91). These marks are not needed for parts to be made by machining processes or from sheet metal. If all surfaces of a part are to be finished, "Finish all over" (or FAO) may be added as a note in lieu of finish marks.

60° 60° 1.5 H 3 H Edge view of finished surface

FIGURE 12.91
Surface finish marks are placed in every view where a finished surface appears on edge, even if it is shown as a hidden line. (H = letter/symbol height.)

Before a part is cast or forged, additional material is added to any surfaces that will be machined later. For example, if a slot is to be formed by a core in a casting, its surface will be rough and pebbly. If the slot needs to be smooth for the part to function properly, additional material is added to the casting so the slot can be machined smooth when the part is finished. Many times, the determination of how much material must be added to the part is left up to the manufacturer. When it is not, some companies provide two sets of drawings for parts that are to be forged or cast. One drawing shows all of the information necessary for the forging or casting, and a second drawing—for the machine shop—shows the finished dimensions for the part.

The finish marks may indicate not only that the surface is to be machined, but also special requirements for the finish. The process used to finish the part, such as milling or grinding, and the direction in which tool is applied to the surface can result in varying degrees of smoothness, called *surface texture*. Controlling the surface texture can be critical to the function of bearings and seals and to reducing the friction between moving parts. If the texture of the surface is important to its function, it should be documented in the drawing by adding additional information to the finish mark.

Finished surfaces have the following qualities:

- roughness
- waviness
- lay

The *roughness* of a surface is a measure of the irregularities in the surface's finish, as shown in Figure 12.92. Roughness is measured in micrometers (μm), which are millionths of a meter, or in microinches (μin.), which are millionths of an inch.

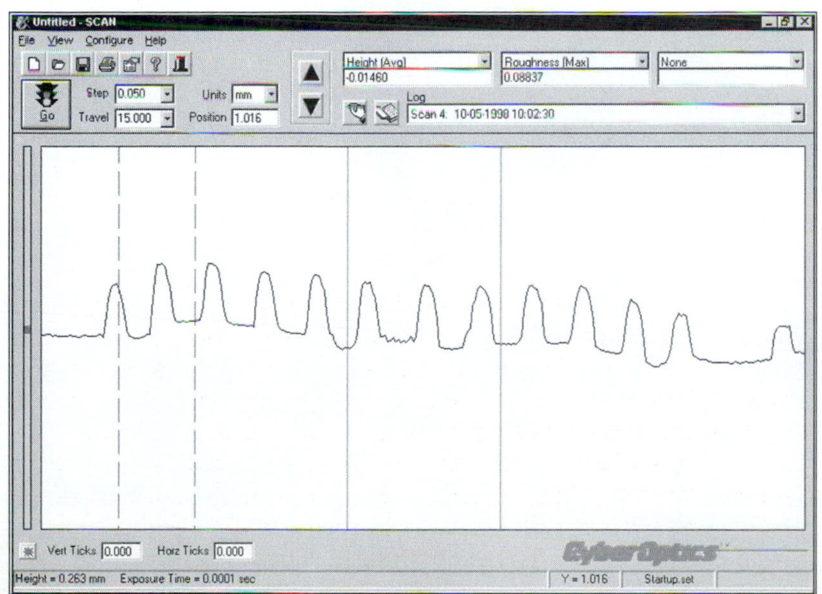

FIGURE 12.92
This plot is from CyberScan, a device used to measure surface roughness. Roughness measures the finest irregularities in the surface. (Courtesy of Advanced Coordinate Technology.)

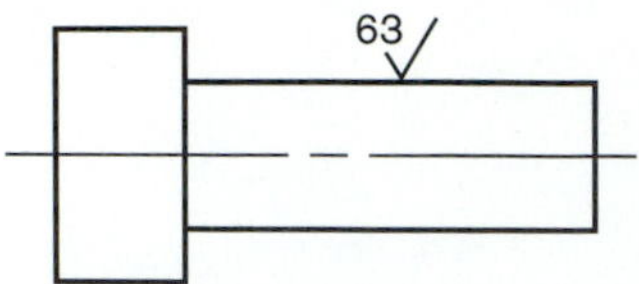

FIGURE 12.93
The finish mark indicates that an arithmetic average roughness of 63 microinches must be met for the surface of the cylinder.

A 

B 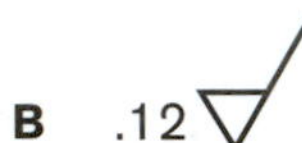

C 

FIGURE 12.94
(a) The standard finish mark indicates that the surface may need to be machined to meet requirements. (b) Machining is required. Additional material must be allowed. A number to the left of the symbol indicates the value for the additional material required. (c) Machining is prohibited. The surface must be produced by a process such as forging or casting.

Figure 12.93 shows the surface finish mark used with a value indicating the average allowable roughness height.

Figure 12.94 shows variations of the surface finish mark that are used to indicate when additional material must be added in a casting or forging because machining will be required, and when removing material by machining the surface is prohibited.

The *waviness* of a surface is the more widely spaced irregularity of the surface due to warping, machine vibrations, heat treating, or other similar factors. Waviness is measured in millimeters or inches. See Figure 12.95 for an illustration of measuring roughness and waviness.

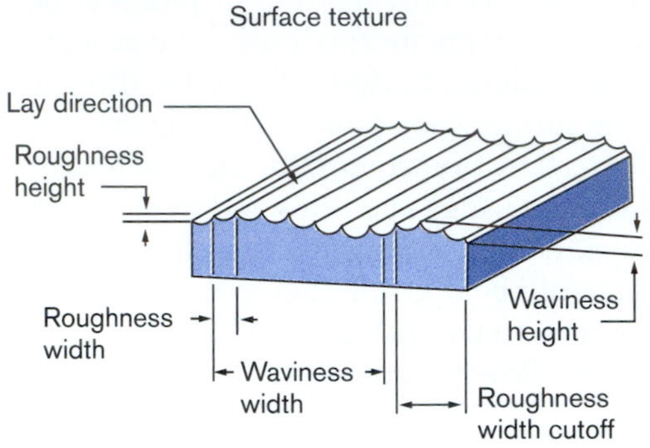

FIGURE 12.95
Roughness and waviness are defined in terms of their height (deviation from the mean plane of the surface) and width (the distance between peaks and valleys), as illustrated here. Both measures can be used to define the degree of smoothness of the surface.

The *lay* of a surface describes the direction or arrangement of the primary surface pattern, as shown in Figure 12.96. If it is important to the design of the part, the lay symbol is used to indicate the desired arrangement for the surface pattern caused by the machining process (see Figures 12.97 and 12.98). Different degrees of roughness can be expected from different manufacturing processes (see Fig. 12.99.)

FIGURE 12.96
The rectangular block shown has a visible pattern (lay) for the primary surface of the part.

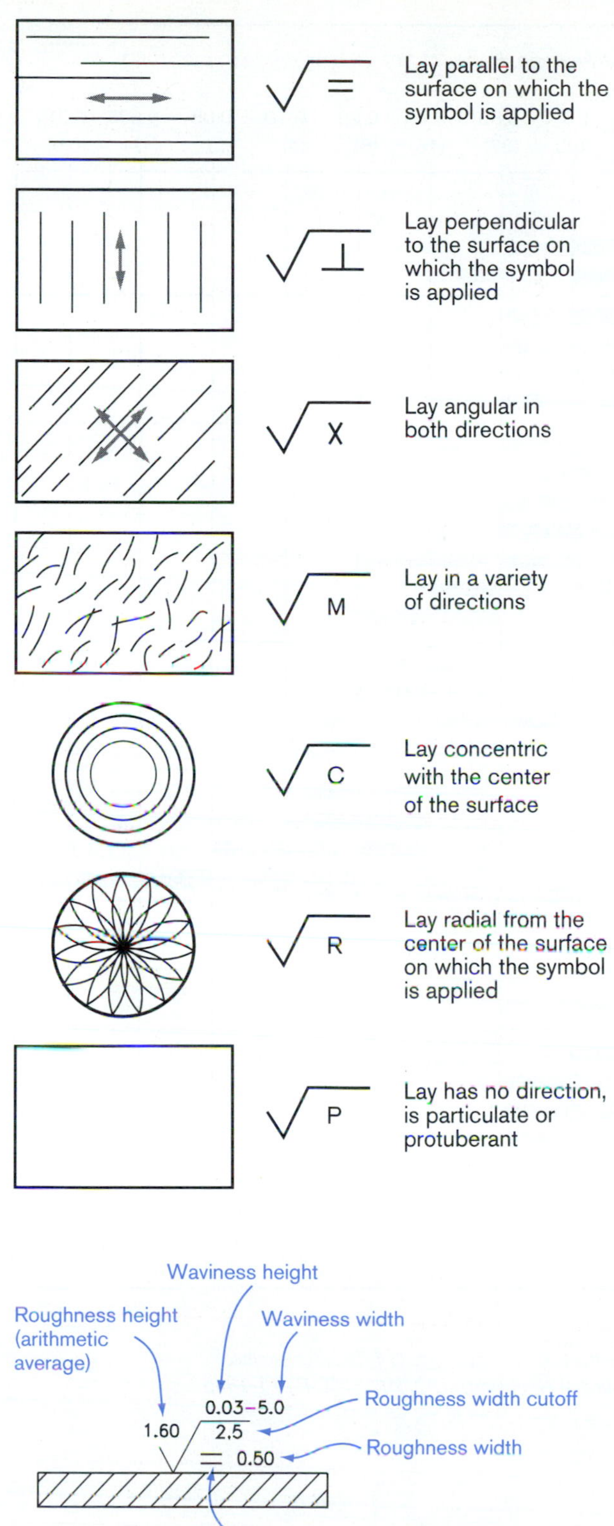

FIGURE 12.97
These symbols indicate the direction of the lay in relation to the surface or, in the case of parallel or perpendicular lay, to the edge where the symbol is located. (ANSI Y14.36-1978 (R1987).)

FIGURE 12.98
Surface finish marks specifying lay and roughness and waviness limits

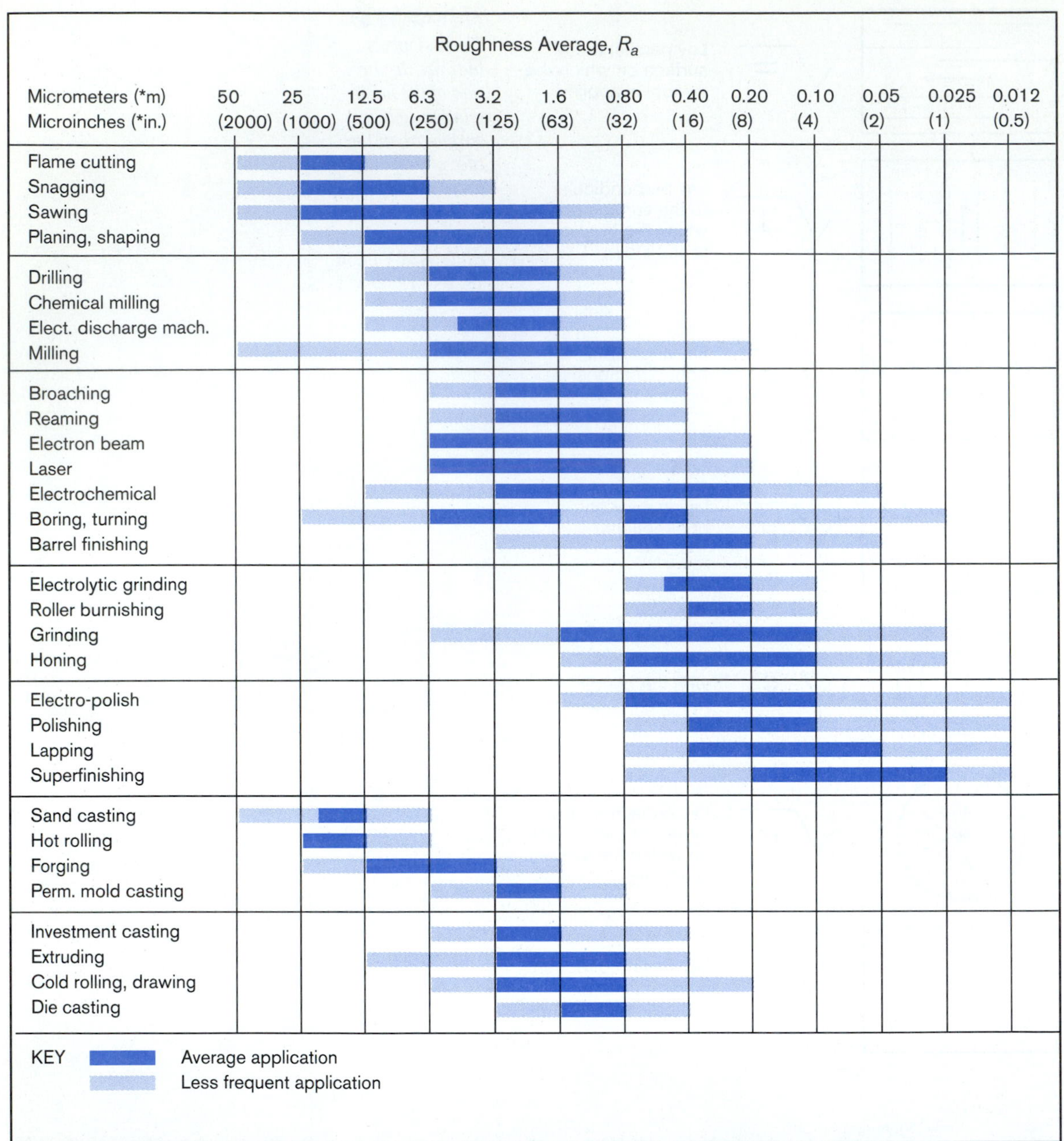

FIGURE 12.99

Different manufacturing processes produce different degrees of smoothness. The range of values for surface roughness shown here are typical of the processes listed under normal conditions. (ANSI/ASME B46.1-1985.)

Specifications

Written specifications are often used to supplement the requirements for constructing designs. For example, a written document that accompanies a set of drawings for a design might indicate special considerations for material type, strength, textures, appearances, handling, tolerances, finishes, work processes, assembly techniques, safety requirements, packaging, shipping, or any other detailed information that may be necessary to bid or construct the design. An example specification is shown in Figure 12.100. Well-written specifications help ensure that products quoted from various vendors are equal quality. Many organizations, such as ASTM, ASME, UL, among others, publish standards and specifications that can help guide you in writing the detailed specifications required to ensure that the product or design is manufactured as you intend.

16.8 Uncovered, exposed pipes shall be provided with plates at the points where they pass through floors, finished walls, and finished ceilings, including points where covering is terminated above the floor. Where necessary to cover heads of fittings, special deep escutcheons shall be provided in lieu of plates. Plates shall be not less than 0.018-inch thick. Plates on chromium plated pipe or tubing shall be brass, chromium plated with bright or polished finish. Wall and ceiling plates shall be secured with round head set screws, not with spring clips. Unless otherwise specified, plates shall be of the one-piece type. Wall and ceiling plates may be flat, hinged pattern.

16.9 Plates on exposed fixture connections shall be as hereinafter specified in the description of the fixture trim.

17 Pressure Gages

17.1 Pressure gages shall be stem mounted, shall have a 3-1/2" (80mm) dial and shall be in accordance with Federal Specification GG-G-760. The gage connection shall be provided with T-handle stop. Dial gages shall be midpoint of normal pressure. Dial shall be marked in both English and metric S.I. units (kPa).

17.2 A pressure gage shall be provided in the suction and discharge connection of each pump.

FIGURE 12.100
Written specifications

Specification documents can be very lengthy due to the scope of a project, so they are usually well organized. Each item may be identified by a number corresponding to the part or step in the project to make the specifications easier to follow.

Standardized parts that may be in an assembly are often listed in a specification. Prewritten specifications for these parts are often available from manufacturers to aid you in specifying their parts in your assembled product.

Patent Drawings

The same CAD database that is used to produce the design documentation can also be used to create the drawings that are necessary to apply for patents. Patent drawings were previously produced by hand. You may have seen some of the elaborately hand-drawn representations of devices that were required for patents year ago.

The U.S. Patent office has standards similar to the types of standards that are used to produce good engineering drawings. Drawings may be submitted on one of three sheet sizes:

U.S. size 8.5 inches by 14 inches
U.S. size 8.5 inches by 13 inches
International size 210 mm by 297 mm

No borders may be drawn on the sheets, and minimum margin sizes must be maintained. No writing or drawing lines may extend into the margin area, except for the specific identification required at the top of each sheet. All sheets within a single application must be the same size.

Patent drawings are required to be submitted in black and white. No color plots, drawings, or photographs may be used. Either instrument or CAD-produced drawings can be used. (Sketches are acceptable during the application process, but since formal drawings are required if it is accepted, you should probably start with drawings.) Drawing lines must be crisp and sharp, so CAD drawings should be plotted or laser printed on high quality paper (at least 20# bond). (Jagged dot matrix-style printed lines are not accepted.) Photocopies are acceptable, since three copies of each drawing are required to be submitted. The drawings cannot be returned, even if changes are necessary, so it is not a good idea to send your only original artwork with the patent application.

Unlike most engineering drawings, where it is preferred to show views in alignment on the sheet, patent drawings show each separate view as one figure. The figures are numbered consecutively, for example, Fig. 1A, Fig. 1B, Fig. 2. Often, specific dimensions and tolerances are not required to patent the general design or innovation. Instead exploded isometric or perspective drawings with clear reference numbers identifying the parts, very similar to assembly drawings, are preferred. Center lines should be used to show how

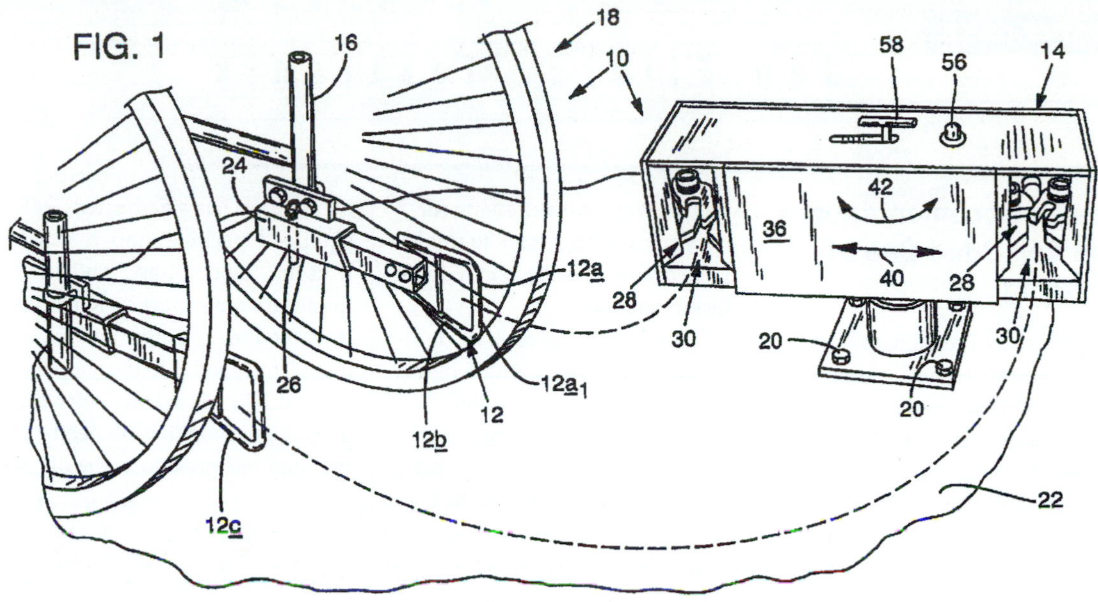

FIGURE 12.101
This patent drawing is taken from the patent for a securement system for a rollable mobility aid that was invented by David Ullman, Katherine Hunter-Zaworski, Derald Herling, and Garrett Clark. (U.S. Patent 5,344,265.)

parts in exploded views assemble. The reference number for a part or feature should remain the same throughout the set of drawings. The drawings must show every feature that is listed in the patent claims. If standardized parts (motors, CPUs, etc.) are used, they can be represented symbolically (per an appropriate standard) and do not have to be drawn in detail.

Most rules for creating good-quality engineering drawings still apply. For example, hatching should be thin, black, and uniformly spaced, running only a single direction on the part; phantom lines may be used to show a secondary position for a part. See Figure 12.101 for a patent drawing.

There is no limit to the number of drawings that may be submitted to fully document the parts, and a variety of views may be used. Border sizes, sheet sizes, how the drawings are signed, and other requirements are spelled out in a guide published by the U.S. patent office that may be ordered from the U.S. Government Printing Office. Another good reference is *Patent It Yourself*, by David Pressman (Nolo Press, 1997).

INDUSTRY CASE

Jae Ellers, Tektronix

Design Documentation: You Can Build *One* of Anything

Design documentation in the Tektronix Color Printing and Imaging Division is a pragmatic mix of parametric models and printed drawings taken from the models—both created in Pro/Engineer. The CAD database created for the part conveys the design information to the manufacturer and is used as the source of the drawing views that define the quality standards for the manufactured part. "The drawing is the legal interface between us and the company doing the manufacturing," explains Jae Ellers, a mechanical engineer. "Until that changes, we have a drawing for every part, and we have to think of the drawing as the actual design documentation."

For Ellers, the process of creating drawings begins after a model has been refined, tested, and prototyped. "There are basically two ways to do drawings: either build an original drawing and update it as the model changes until it becomes a production drawing, or use the model and sketches as needed to get a working prototype part, then generate the drawings from the part that works. I've seen people do both, but the latter is more efficient for me."

Detail drawings are not required for prototypes because the division has a dedicated, well-equipped machine shop, referred to as the "model shop," on site. Engineers can export an IGES file from the Pro/Engineer solid model and send it to the shop to be built. The Mastercam software used in the shop doesn't need all the solid information in the file, so

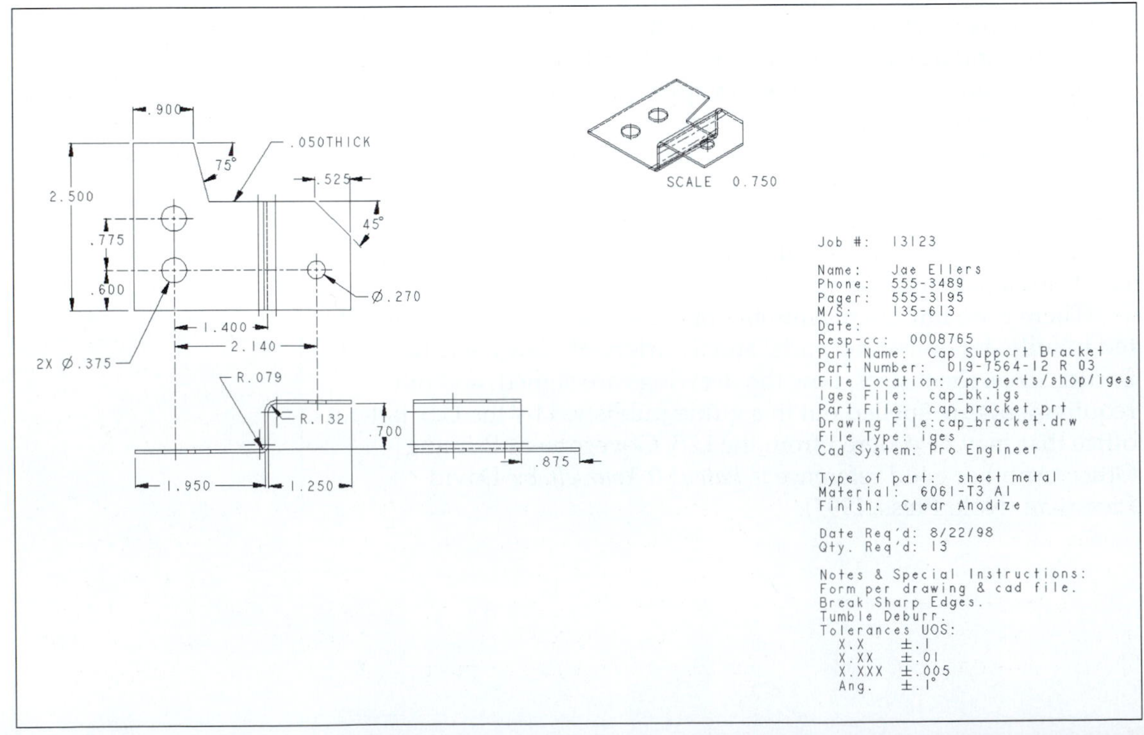

FIGURE 12.102
After modeling this sheet metal part, Ellers generated drawing views of the part and displayed the dimensions used to model it. He used a simple tolerance block, shown at the bottom of the notes column, to start the tolerancing process. These tolerances apply to dimensions according to the degree of accuracy specified by the dimension value and do not necessarily reflect the tolerances that may be achieved by a specific manufacturing process.

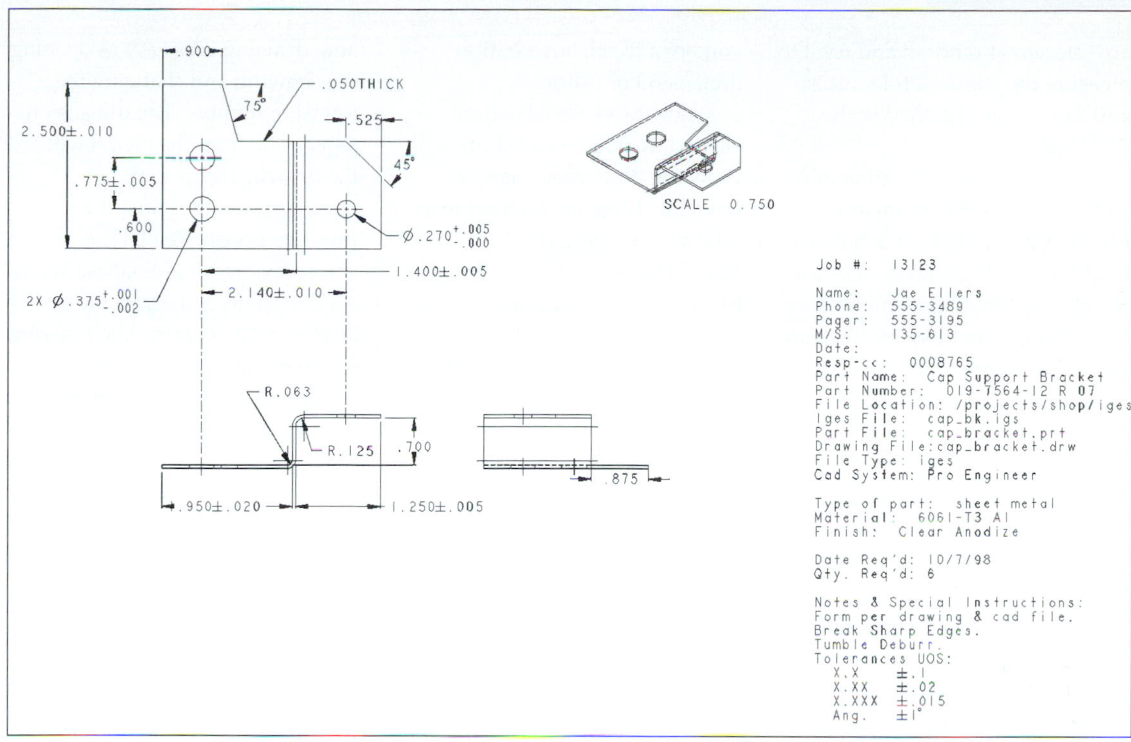

FIGURE 12.103
After conferring with model shop personnel, Ellers modified the drawing to reflect their input. First, the block tolerance was changed to reflect the larger tolerances required for sheet metal parts. Hole sizes were then toleranced individually according to the tolerances for standard punches and the fit needed for the part. Bend radii were changed to match the bends produced by standard dies (e.g., from R.132 to R.125 and from R.079 to R.063). Finally hole locations were toleranced to reflect where closer tolerances were required to ensure fit with other parts (e.g., the .775 distance between the two .375 diameter holes is toleranced more closely (±.005) than the standard tolerance in the block (±.015).

they can export just a wireframe definition for a simple part, or a surface definition for more complicated parts. An isometric view of the model with critical tolerances or other manufacturing notes sketched on it is often enough to convey what the shop needs to know to manufacture an acceptable part. Ellers points out that this is simply a time-saving approach that eliminates "spending hours generating drawings that may not be used again if the prototype doesn't work."

Once a design is ready to be produced, however, Ellers begins the drawings for production by identifying what he calls the "critical-to-function" features and dimensions and tolerances. These are the criteria that must be met for the part to operate correctly. From there, he looks at the data transfer capability of the company that will make the part. If the vendor can work with an electronic file—either the Pro/Engineer model or an IGES export from the model—the printed drawing will show only dimensions and tolerances for these critical-to-function features. "We only fully dimension a drawing if we need to—for example, if someone is still working on a system where they need to type in the dimensions for a part or if their system can't read the files we write."

Even though a vendor works with an electronic file, there is still room for human error in the process. The drawing provides documentation of the part that will be used to inspect the manufactured parts. While a part is being made, the engineers write up a quality plan (an inspection procedure) that lists the features dimensioned on the drawing. When parts come back, a statistically significant sample (in this case, approximately thirty) of the parts

(continued page 550)

INDUSTRY CASE

are selected at random and used to measure the features, tolerances, and finishes as specified in the drawing.

"Often, the exact location of a feature isn't as important as its function. A pin that fits in a hole to hold the part together is an example of a part that has some leeway. Even though we model the feature where we want it to be, ultimately it doesn't matter exactly where it comes in as long as the two line up and function as they should." The opposite is true of the critical-to-function features, where dimensions are always provided on the drawing so they can be used to determine if a part has been manufactured acceptably.

Typically, a project will have an assembly drawing at the top that documents how all the parts go together, subassembly drawings for each subassembly, then individual part drawings at the bottom. To document a single part, Ellers uses the tools built into Pro/Engineer to create drawing views and add dimensions (see Fig.12.102). "For a simple part, such as a pulley, two views are all that is needed to show the part. For a screw, a single isometric view with a note is sufficient. If you have a 440 socket-head cap screw with a cadmium plated finish, that's all you need to know for the standard part. The picture isn't even really needed. For some of the more intricate castings in the print head, however, it may take three orthographic views, an isometric view, and section views from AA to MM because there are so many different places where you need to call out a critical dimension or feature."

Once the views have been created, clicking on a feature will show the dimensions used to model it. These are referred to as "shown" dimensions. Pro/Engineer will also allow you to add dimensions, referred to as "driven" dimensions, by picking on entities and displaying their current length, radius, diameter, etc. "If you capture the design intent in the parametric model, you can show the dimensions in the drawing without having to click lines or points to create the dimensions." That's a big advantage of creating parametric models that capture your design intent correctly. Another advantage of "shown" dimensions is their bidirectional associativity. "If you make a change to a shown dimension in a drawing, the model updates. That means manufacturing guys can change the dimension on the drawing and the model will update without their having to know how to use the modeling software." Driven dimensions, on the other hand, are associative in only one direction. They will update when the model does, but changing a driven dimension on a drawing won't update the model.

The associativity between the model and the drawing makes it easy to generate new drawings for any model revisions. When a model goes to production, its status is coded as Rev A and the file is protected. To make a change, an engineer must issue an internal change notice (ICN). Once the model is changed, generating a new drawing is as easy as opening the drawing and changing the Revision number. The dimensions regenerate from the model when the drawing is opened. The current drawing is then the revised version, Rev B. The Tektronix drawing database keeps each revision of the part file associated with the appropriate drawing file so that the dimensions shown on a given drawing match the revision of the part that was requested.

Once the dimensions are shown, Ellers uses Pro/Engineer to automate their placement—to a degree. The Clean Dims command will move the dimensions around so dimension lines don't intersect other dimension lines. He still finds it necessary to make adjustments, however, to align the dimensions properly and display the dimension lines and arrows. "Cleaning up the dimensions is important for readability's sake. If you had two dimensions close to one another and, instead of having the dimension centered between the witness [extension] lines, had to locate the dimension values outside the lines, it might be difficult to interpret them. If this drawing went to someone who could not use the model but had to enter the data manually, it would be easy to get the numbers in the wrong place."

To finalize the drawings and conform to company standards, Ellers may send the drawing at this point to the in-house drafting group. Four or five draftsmen serve the approximately two hundred and fifty engineers in the division and are a resource for preparing accurate drawings with

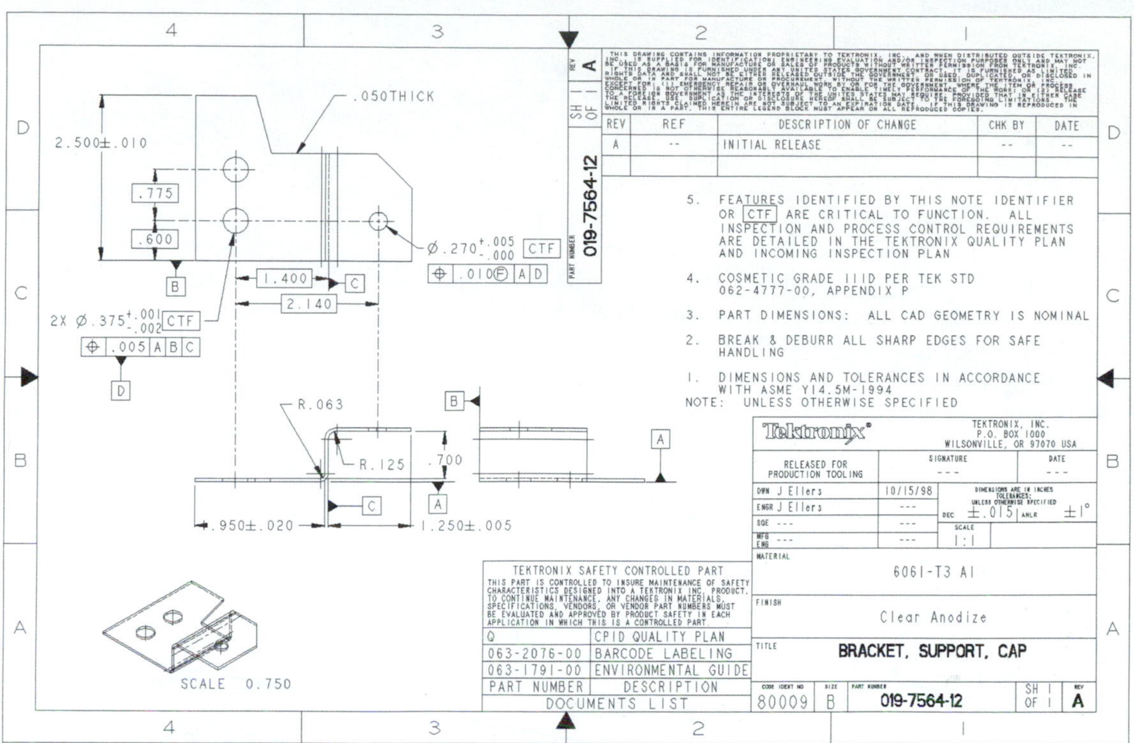

a consistent look. "Pro/E is really powerful in that you can communicate in the model the tolerances, the surface finish, and the geometric dimensioning and tolerancing. You can even associate notes with surfaces in the part that can be "shown" in the drawing. You can put all of this in the model itself, generate drawings, and show the information you feel is needed, then hand the drawing off to drafting to let them put the notes and tables in the right place, add the bill of materials, and further clean up the drawing." This approach is efficient, in Ellers's view, because it takes advantage of what each group is comfortable with. An engineer could do the entire drawing, but engineers are typically more proficient with the modeling tools in the software. A draftsman is more proficient with the software's drawing tools, but will not know enough about the design to determine which features are critical to dimension and tolerance. "The other end of the spectrum is to hand drafting a model and ask them to start the drawing. Typically, drafting would show all of the dimensions, move them around, then create enough views and section views that they can see every feature in detail. The engineer could then delete views and dimensions that were not needed."

All Tektronix drawings are dimensioned according to the geometric dimensioning and tolerancing standards as defined in the ASME Y 14.5M 1994 standard. In addition, company standards for the appearance of the title block, the size and form of the lettering, and the format for legal notices on the drawing are documented and stored in a central location, usually with the administrative assistant for a group of engineers. Keeping the standards in a central location

(continued page 552)

FIGURE 12.104
In the release drawing shown here, Ellers has applied the Tektronix format to the drawing information, adding notes, title block, and revision block that conform to company standards. Non-critical dimensions have been removed from the drawing, and critical-to-function dimensions have been identified with the letters CTF. The geometric dimensioning and tolerancing symbols have been applied to all dimensions and tolerances shown (in keeping with the company standard), with feature control frames added to all critical-to-function features. Certain dimensions, including the closely toleranced .775 distance between the holes, have been specified as basic dimensions in this drawing.

INDUSTRY CASE

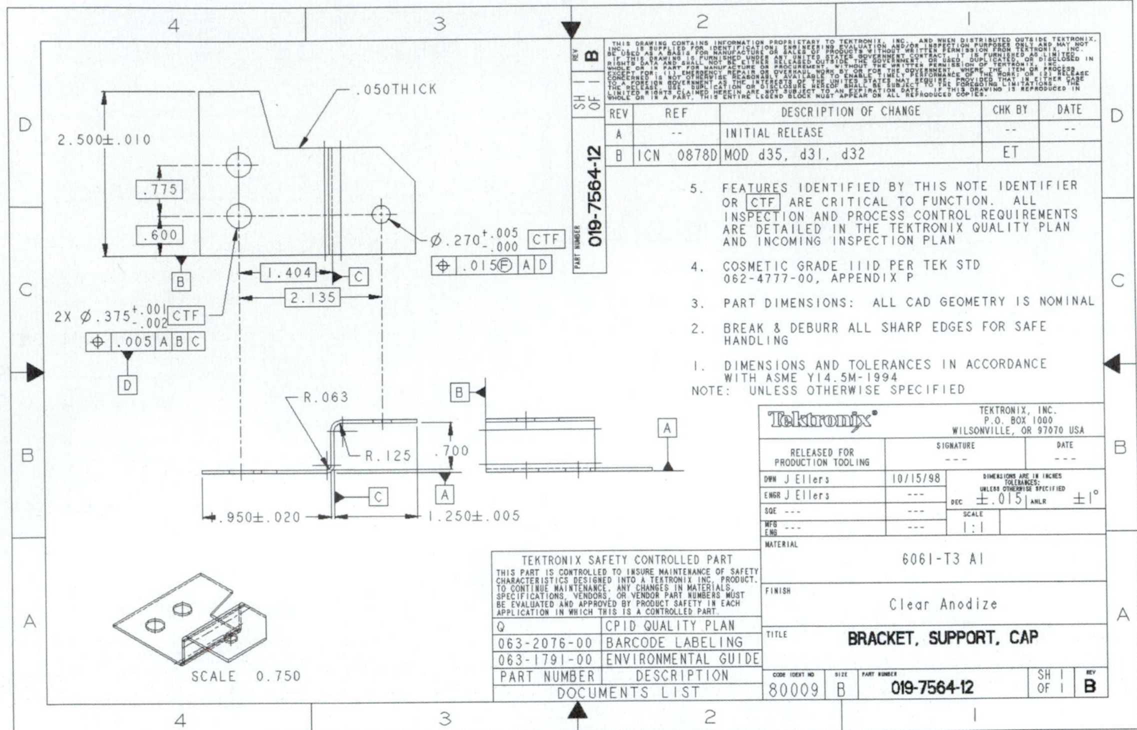

FIGURE 12.105
After the part was made, measurements taken from the actual part were used to modify the dimensions on the drawing (and in the model). Notice that the revision block lists the changed dimensions by the names used in the parametric model.

discourages individuals from relying on paper copies in their own files that may not be current. (Controlling the standards in this way also meets one of the ISO 9001 documentation requirements.)

Some of the consistency between drawings is accomplished by starting all drawings from a template stored in the file. Pro/Engineer allows you to store formats that set up the zones on the outside of the drawing and include 2D entities such as the title block that link to data associated with the model file. The engineer's name and the part number of the file are inserted where appropriate in the drawing through these links. Pro/Engineer also makes it easy to access boilerplate for notes relating to various processes. If the engineer needs to add a sheet metal note, for example, he or she can bring in a text file that can be modified for that situation. More often, though, the consistency is achieved by the human element. The drafters bring their knowledge and expertise to bear and work together to achieve a consistent look in the drawings.

A key part of preparing drawings for manufacture is specifying the tolerances—the degree to which a part can vary from the modeled size and still be acceptable. "You can always build *one* of anything," Ellers explains. "Our model shop builds prototypes of our parts to the nominal size specified in the model. A model can be built in the shop and work perfectly, but a vendor may build the part to the standard tolerances. If the engineer didn't consider the tolerance of the manufacturing process used to make the parts, the parts may not work when fitted together."

Once again, Ellers takes a very pragmatic approach to determining the tolerances he builds into his designs. First, he distinguishes between two different kinds of tolerances. Interface tolerances just depend on the desired type of fit and can be looked up in a table of fits. Manufacturing tolerances depend on the process to be used and the equipment available. For the latter, Ellers will choose a vendor, fax him or her a copy of the part design, and discuss how it might be made. "They will go through the part with you and tell you what they can do in certain places For example, they might say, this is a tooling hole, so the diameter tolerance is going to be 0.002; from the hole over to the feature, the tolerance will be 0.006, then to go around the bend, that will be 0.015 from the hole." (See Fig. 12.103.) After hearing what the options are for the part, Ellers then has three choices: to live with the accuracy the vendor can achieve and specify those tolerances on the drawing, change the design, or try a different process to achieve a higher accuracy.

The vendor is not the only resource to be used to determine tolerances, however. There are organizations for all of the major manufacturing processes—such as die casting, sheet metal, molded plastics—that publish books and offer seminars. Like fit tables, these resources list the tolerances to be used for parts of varying dimensions. "Some people develop rules of thumb that they go by—if it's a machined part, they put in ±.005; if it's a sheet metal part, they put in ±.015. To hold the manufacturer to the highest standard, however, you need to investigate the tolerances for each individual part."

To add tolerances to the Pro/Engineer model, Ellers first selects the datum planes that will be the base datum for the manufactured part. After that, he uses the Geometric Dimensioning and Tolerancing dialog box to associate a tolerance with a feature. He selects the geometric tolerance to apply, then the material condition (Maximum Material Condition, Least Material Condition, or Regardless of Feature Size (the default)) he wants to apply it at, then the datum planes and features that it references, and Pro/Engineer creates the appropriate symbol. With these conditions built into the model, they can then be shown in any drawing view as needed (see Fig. 12.104).

Once the tolerances are in place, Ellers can use Pro/Engineer to evaluate their impact on fit and function. By showing all of the features at least material condition, or maximum material condition, the part can be used to run an interference check in the assembly model. Pro/Engineer shows volumes that interfere in a highlighting color and calculates the volume of the interference for the user.

Ellers feels the biggest challenge in getting from a design of the model to a manufactured part is "unexpected interactions between real parts with real tolerances that don't work together like they did in the prototype. You can come up with clever designs that overcome tolerancing problems, but the most important thing to do is to define the functionality you need and get the interface defined. Once you've defined what you need, you can go out and find the process to support it. I've seen a part go from die cast aluminum (at $7 per piece) to injection molded plastic (at $3 per piece), then to sheet metal (at $1 per piece). There's a whole different set of design considerations for each of these three processes, so you have to define your needs first. If you do, you'll have the information you need to evaluate the trade-offs between processes and select one that will give you the tolerances you need. You can then incorporate these process capabilities into the tolerances specified in your design."

Jae Ellers is a mechanical engineer in the Tektronix Color Printing and Imaging Division in Wilsonville, Oregon. Ellers earned a B.S. in mechanical engineering at Montana State University and joined Tektronix in 1995, where he worked in the Print Engine Design Group. He is now in the Advanced Development Group, where the print heads and tooling for their production and development are designed. Ellers began working in CAD with AutoCAD and now uses Pro/Engineer for his design work at Tektronix.

Skills Summary

Now that you have completed this chapter, you are ready to create dimensioned multiview drawings from CAD models. You should also be able to add tolerances to your drawings to describe the allowable variation that an acceptable part may have from the stated dimensions. Exercises 1–5 allow you to practice your skills in interpreting orthographic views and then creating the model and a dimensioned multiview drawing from it. Exercises 6–10 check your ability to model the shape given and create properly dimensioned drawings to describe it. Exercises 11 and 12 provide questions to test your understanding of how dimension choice can influence tolerance stacking. Exercises 13–20 allow you to build your skills in dimensioning and tolerancing real objects. These problems ask to you to stretch your abilities by creating models and making reasonable design decisions for the sizes of the features, then creating dimensioned drawings from your designs.

EXERCISES

Exercises 1–5: Dimensioning. Using your CAD package, create a solid model of each of the objects shown. From your model, develop orthographic views, and fully dimension all features and locations. Place the views properly so that all views can be included on one sheet. Use size B (11" × 17") sheets for your full-sized, dimensioned drawings. Include a border and a title block on all drawings.

1. Dished washer. Assume grid squares are 0.500 inch.

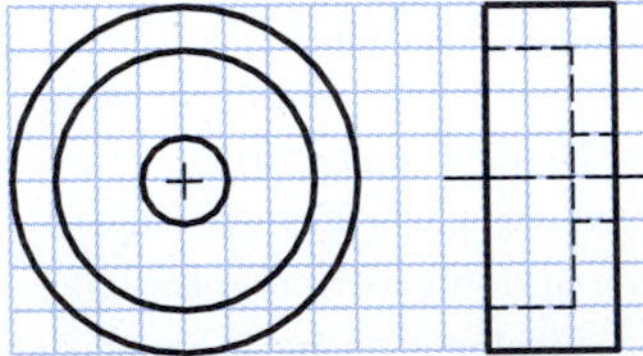

2. Offset bushing. Assume grid squares are 0.250 inch.

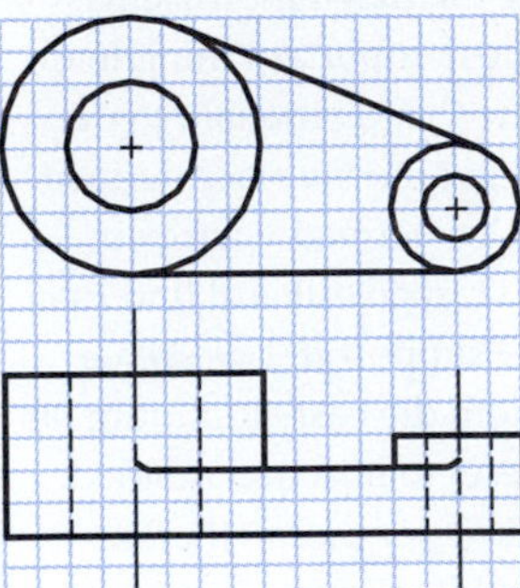

3. L-Bracket. Assume grid squares are 0.250 inch.

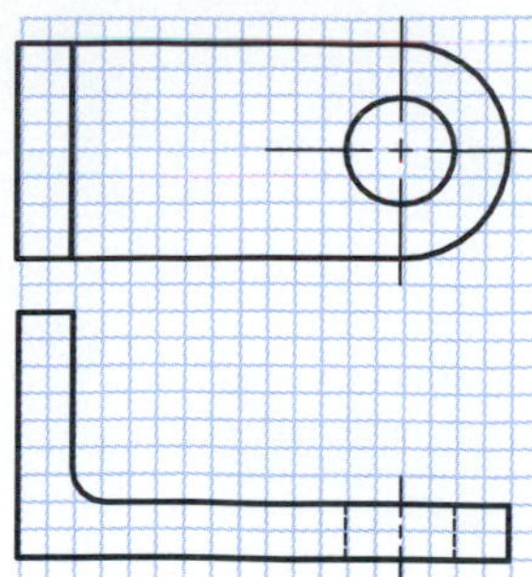

4. Bellcrank arm. Assume grid squares are 0.250 inch. Radii all 0.25 inch.

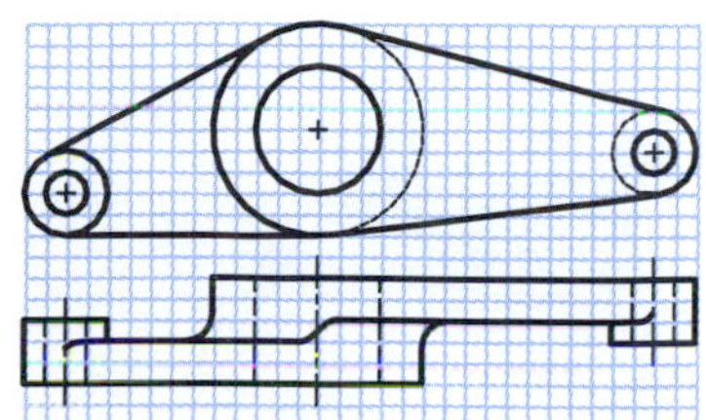

5. Shaft mounting bracket. Grid squares are 0.250 inch. Radii all 0.25 inch.

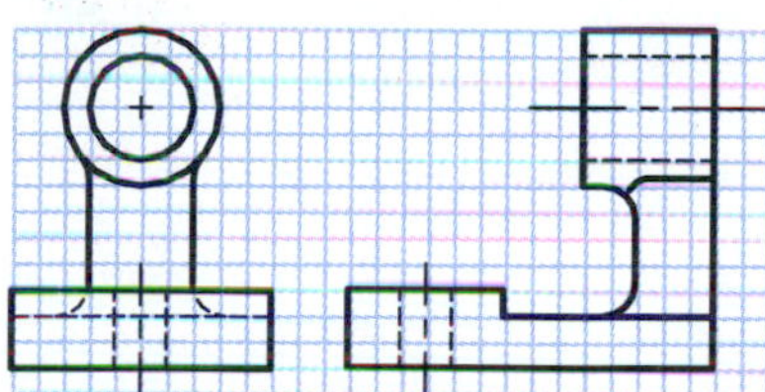

Exercises 6–10: Dimensioning. Create solid models of the objects shown below (some were already encountered in Chapter 3). Perform the following drawing operations on each completed model:

a. From your solid model, develop the orthographic projection views needed to fully define the object.

b. Place the required views properly, so that all views can be included on one sheet. Use standard-size drawings (choose size a, b, or c, as appropriate).

c. Fully dimension all part size elements, features, and locations.

d. Include tolerances as follows:

Feature	Tolerance
Through-hole diameter	+/−0.005 inch
Hole location from datum	+/−0.030 inch
Overall part dimensions	+/−0.010 inch
Fillets and chamfers	+/−0.010 inch

6. Notched block. Each grid square is 0.50 inch.

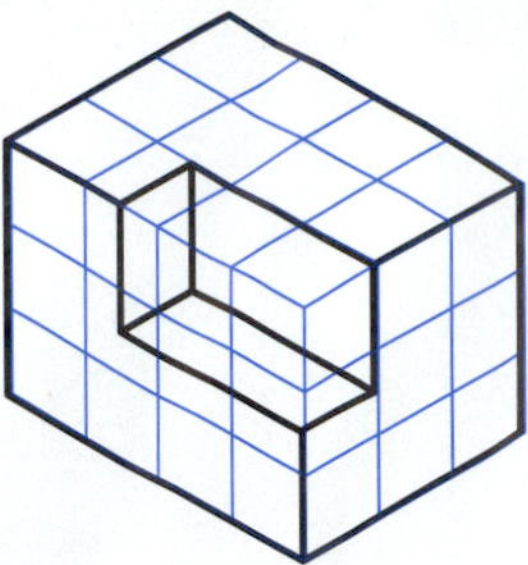

7. Single-rivet nut plate. All holes through. Each grid square is 0.1250 inch.

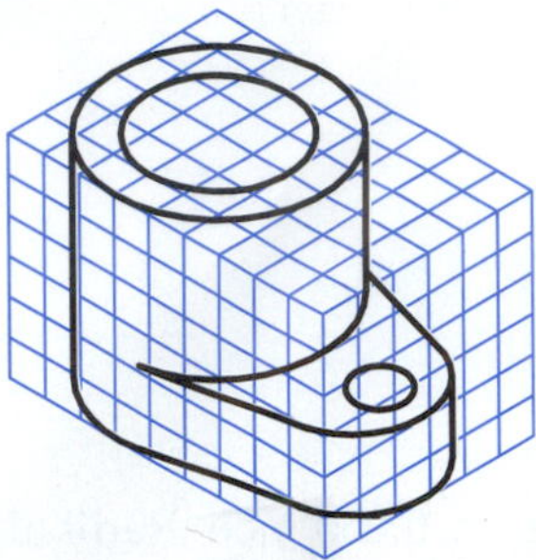

8. Saddle bracket. All holes are through. Assume that each grid square is 0.50 inch.

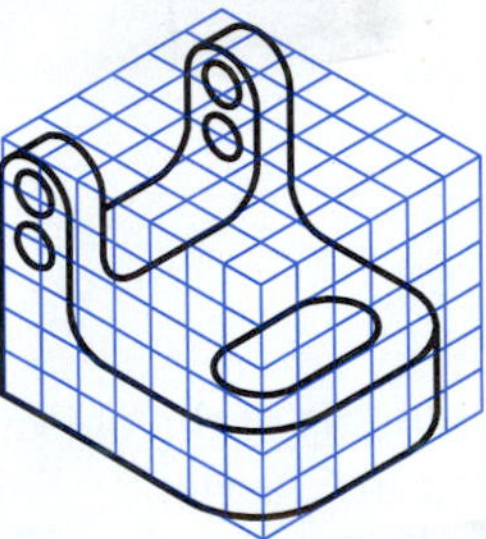

9. Keeper. Large holes pass through; small holes are 0.375 diameter, drilled and tapped to 0.50 depth. Assume grid squares are 0.50 inch. Radii are 0.50 inch.

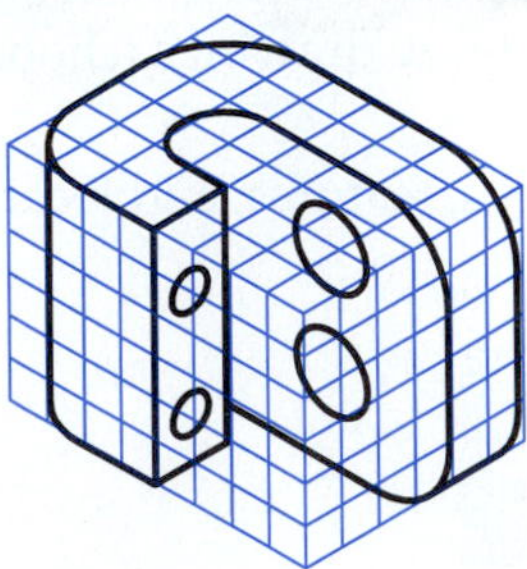

10. Rail mount shaft support. All holes pass through. Rail track on underside of support extends the length of the part. Grid squares are 0.50 inch apart.

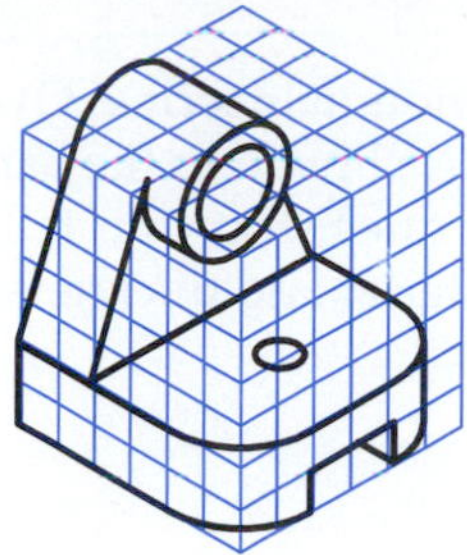

11. Explain "Tolerance Stacking." How can the selection of *chain dimensioning* versus *baseline dimensioning* affect the tolerances on a finished part?

12. Why would you want to leave out one of the chain dimensions in the following part? Re-draw the part using correct dimensioning practices. Add surface finish callouts as needed to specify roughness height of 32 microinches on each of the steps.

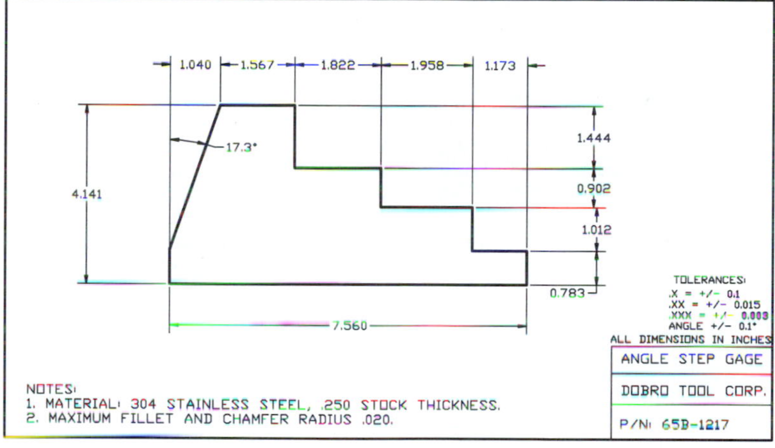

Exercises 13–18: Dimensioning and Tolerancing Applied to Real Parts. You may not know exact dimensions of existing parts, or you may have ideas for improvement—so use your knowledge to design your own parts! Here is an opportunity to apply your own creativity and make improvements where you can. Pay particular attention to materials, surface finishes, and tolerances.

For each object, perform the following operations:

a. Create a new design, including solid models of the objects shown.

b. From your solid model, develop the orthographic projection views needed to fully define the object.

c. Place the required views properly, so that all views can be included on one sheet. Use standard-size drawings (choose size a, b, or c, as appropriate).

d. Fully dimension all part size elements, features, and locations.

e. Include tolerances of all features. ANSI-Y14.5 dimensioning and tolerancing may be utilized.

f. Add surface finish callouts to fully define machining requirements.

13. Bolt hanger for use in rock climbing and mountaineering. Overall size is about 2 inches. A climbing rope is clipped to the hanger which is bolted to a rock face: This device must stop your fall, so it must be strong! Choose appropriate material and thickness.

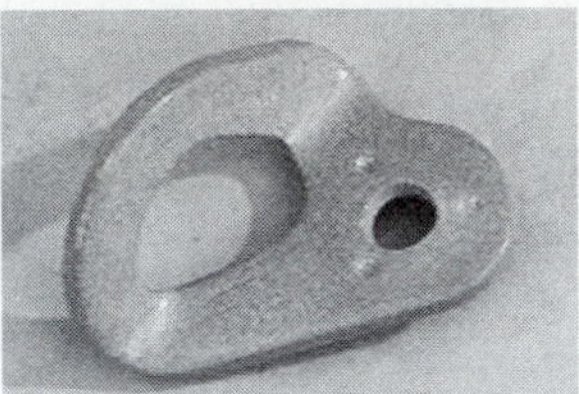

14. Climbing carabiner, to be used for general mountaineering purposes. It can clip a rope into the bolt hanger from Exercise 13. Most are made of 6061 aluminum, and have a spring-loaded gate that opens with a light squeeze.

15. A rear cargo rack is a useful addition to a mountain bike. This one is made mostly of extruded aluminum, with welded joints. It is sized to bolt to the chainstays near the rear hub, and to the seat tube near the brake mounting bosses.

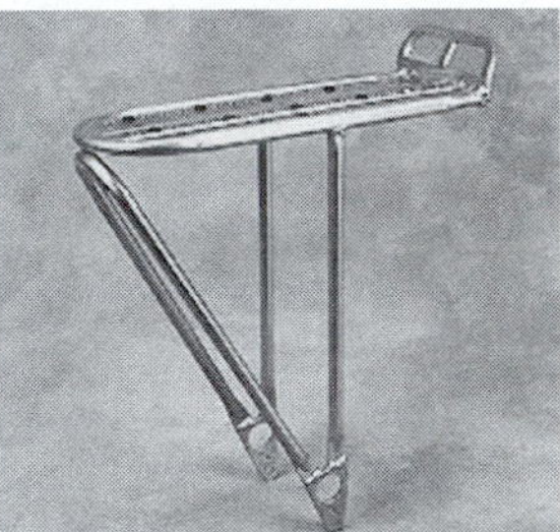

16. Campers use lightweight tongs to lift pots off a cookstove. The lighter the better! About 5" long.

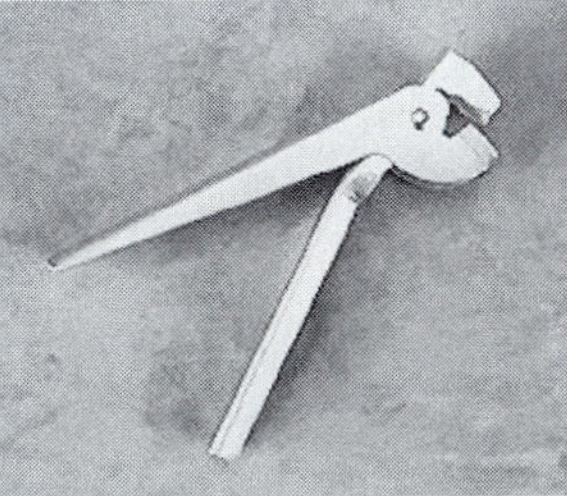

17. Socket wrench sets often include a u-joint to permit wrenching at a slight angle. This one is for a 3/8" square drive socket set.

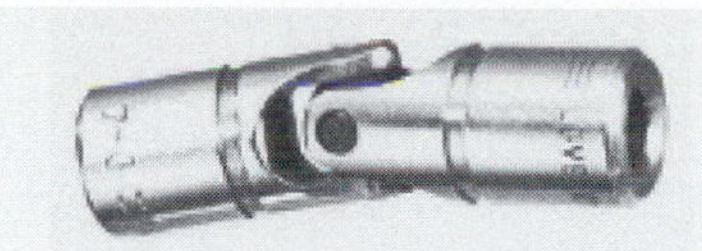

18. A folding saw for tree pruning. The 1/8" wide Stainless Steel blade includes extra sharp teeth. Blade length is around 12". Wood handle on this model, but what material will your handle use?

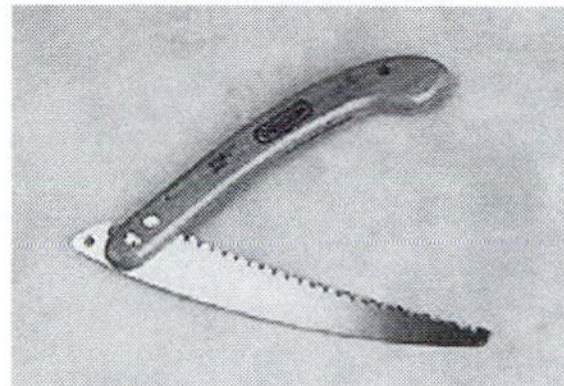

Challenge Exercises 19 & 20: Advanced Dimensioning, Tolerancing and Design. Create the model(s) and the 2-D drawings needed to fully define the parts. Create section views as needed to clarify construction details.

19. This cut-away view shows a gear-type shaft coupling used to transmit power between such elements as a motor output shaft and a pump input shaft. The hubs are fit onto the two shafts—usually with a shrink fit or interference fit. The external ring provides the connection. Precise alignment is mandatory, as is extreme strength and perfect balance. Design yours for mounting on 0.500" diameter shafts. (Photo courtesy of Lovejoy, Inc.)

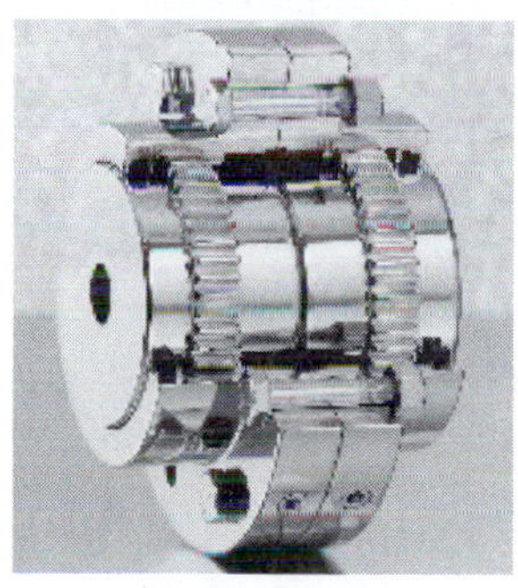

20. Air-powered actuator cylinders are used in many industries for positioning and lifting. Many sizes and configurations are available. This one uses a stainless steel body and a 1/4 inch diameter bore, with a captured air-powered piston stroke of about 3 inches. (Photo courtesy of Bimba Manufacturing Co.)

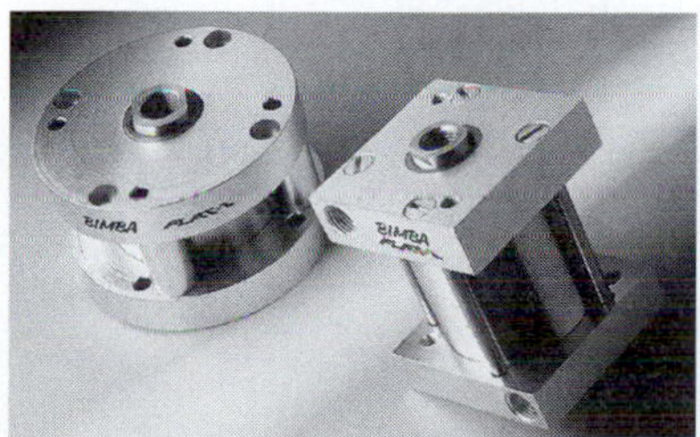

DESIGNER'S NOTEBOOK

The three concept sketches for a bar code reader shown here illustrate the range of motion planned for the design. Animation techniques are one way to effectively present design ideas that incorporate movement critical to the function of the device. (Courtesy of Albert W. Brown, Jr., Affymax Research Institute.)

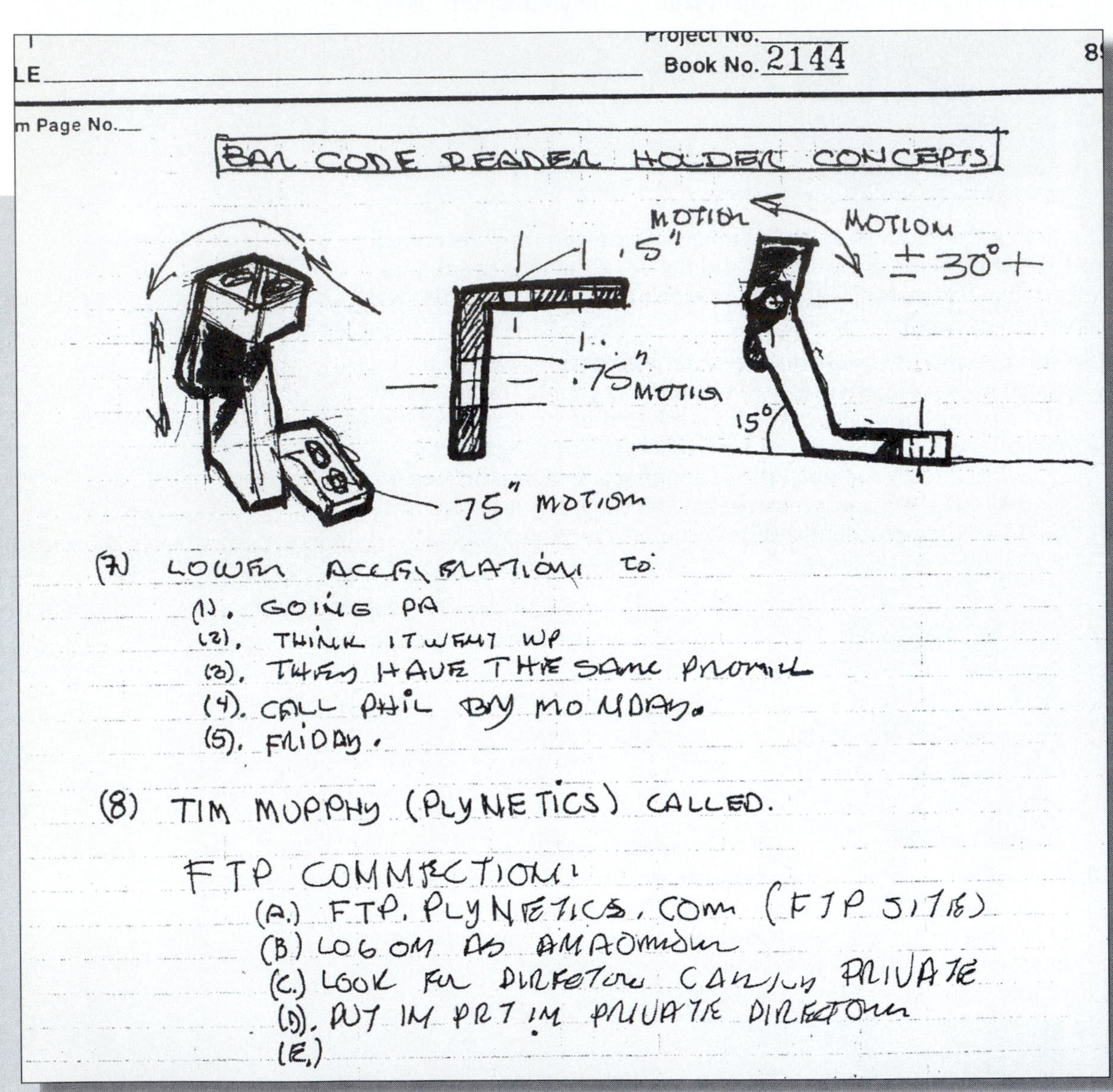

Project No.
Book No. 2144

LE

m Page No.

BAR CODE READER HOLDER CONCEPTS

MOTION MOTION ±30°+

.5"

.75" MOTION

15°

.75" MOTION

(7) LOWER ACCELERATION TO:
(1). GOING PR
(2). THINK IT WENT UP
(3). THEY HAVE THE SAME PROBLEM
(4). CALL PHIL BY MONDAY.
(5). FRIDAY.

(8) TIM MURPHY (PLYNETICS) CALLED.

FTP CONNECTION:
(A.) FTP. PLYNETICS. COM (FTP SITE)
(B) LOG ON AS ANONOMOUS
(C.) LOOK FOR DIRECTORY CALLED PRIVATE
(D). PUT IN PRT IN PRIVATE DIRECTORY
(E.)

CHAPTER

13

Animation and Presentation Graphics

Objectives

When you have completed this chapter, you will be able to

- plan a presentation of your design ideas;
- include graphics in reports and presentations;
- choose among different graphics formats;
- describe rendering processes and algorithms;
- plan and storyboard an animation;
- describe how VRML worlds are created.

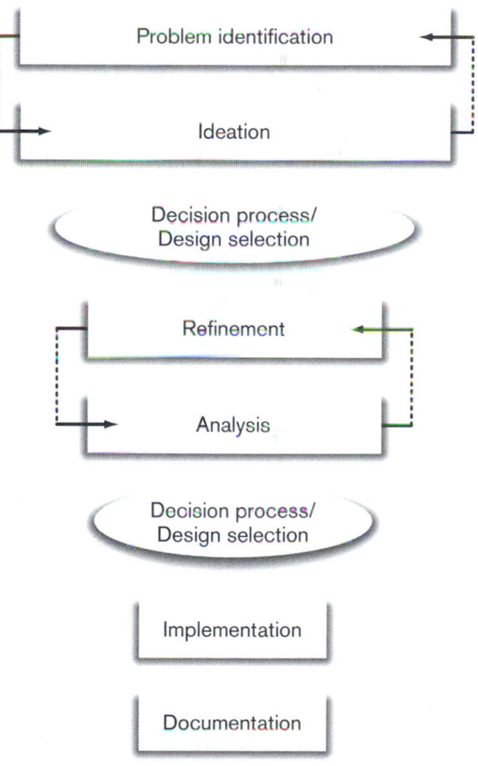

Throughout the design process, engineers are called upon to present design ideas, schedules, solutions to problems, and other information. Making an effective presentation at a design review meeting can mean the difference between implementing a concept or not. Presenting a prototype design to a customer group can yield feedback to make the design truly competitive—or result in a discontinued project. Concurrent engineering makes communicating effectively with others an even more important workplace skill. It has been estimated that employees in general spend 37% of their time in meetings. Using graphics effectively in your presentations can help you communicate your message clearly and in a way that will be remembered.

Today, the CAD database you create for engineering design can also be the foundation of your presentation graphics. A variety of software tools can use the CAD model data to produce a design presentation suited to any audience. Just as the database can be exported in a variety of formats for use in design analysis, so too can it be exported to graphics formats that can be used in printed reports, audiovisual presentations, computer-based animations, and photorealistic displays.

Presentation software, such as Microsoft PowerPoint, is one way to create presentations that combine text and graphics for visual impact. Animation software, such as Autodesk's 3D Studio and EAI's VisFly, can be used to create movies that show how your design will function—and to add sounds and backgrounds to capture attention. Rendering software, built into many modelers, can generate views of your design that may be hard to distinguish from the real thing.

When IRI International was designing the drilling platform shown in Figure 13.1, for example, the client was concerned that there be no viewing obstructions from the operating booth. The viewpoint of the operator was used to create the view shown here, and the lighting for that latitude and time of day was used to reflect how the platform would appear from the booth. The client was reassured by the image, and there were no surprises when the platform was constructed.

In this chapter, you will learn about the techniques and formats used by today's presentation graphics software so you will be able to use them in your work—and evaluate new graphical communication tools as they are developed.

Effective Presentations

To create effective presentations you need to keep your audience in mind. The listener needs to be able to understand and see the information clearly. For groups of customers or marketing personnel, you would not include the same level of technical information that you would when presenting your ideas to fellow engineers or technical staff. You may be intimately acquainted with details of the project, but you should not assume that your audience will have or want the same detailed knowledge.

FIGURE 13.1 *To create this rendered view, materials were applied to the different parts of the solid model to correspond to the appearance of the materials that it would be manufactured from. The sun in the image was added to simulate daylight in that geographical location, and background images were selected to suggest the landscape for the rig's installation. The viewpoint is that of the operator. (Courtesy of Wayne Fehres and IRI International.)*

A simple formula for structuring your presentation breaks it into three steps:

- Tell them what you're going to tell them.
- Tell them.
- Tell them what you told them.

While this may be somewhat simplified, it is a good starting point. Use a clear title and statement of purpose to introduce your topic. Letting the audience know what you will talk about provides a context that allows them to better interpret the importance of remarks in the body of your talk. The body of your presentation should include three or four reasons or arguments in support of the purpose. Use data, charts, graphs, pictures, photographs, and rendered views to support your points and aid understanding. A clear summary at the end reminds the audience of the important points and invites them to act on the information.

Planning your presentation from the viewer's perspective will help you organize text, graphics, sound, and other media to focus attention and eliminate distraction. Storyboarding is a planning technique developed for video production that works equally well with any complex presentation. A *storyboard* is a series of sketches that show the key events in an animated sequence. For a live presentation, the storyboard can help you arrange the elements that are visible to the audience at each stage of the presentation. A

FIGURE 13.2
A computer projection system such as this one from InFocus allows the user to project high-resolution graphics directly from the computer onto a screen. (Courtesy of InFocus Systems, Inc.)

contemporary presentation often integrates text, graphics, charts, sound, video, or a live demonstration with the speaker's words. Each element should contribute to the speaker's main points, not distract from them. A storyboard can be as simple as an annotated outline of the speaker's text or a series of sketches at key points in the outline. By anticipating how and when elements will appear, you can avoid gaps in the presentation, awkward transitions, and visual aids that are out of synch with your spoken text. (You will learn more about storyboarding later in this chapter.)

You may create graphics to enhance your presentation in a number of ways. You may draw or write on a flip chart, for example, use acetate transparencies with an overhead projector, or create computer-based graphics on slides to be shown with a projection system. (Figure 13.2 shows a computer projection system.) Whatever method you choose, make sure that the text and graphics you use to enhance your presentation are clearly visible and accurate, and that they add information that aids the viewer's understanding.

There are some rules of thumb for producing visual aids that are easily seen and understood. These rules address the number of elements on a single visual aid and the best colors and text sizes for visibility. Bulleted lists work well for presenting information, although using too many in a row can make the presentation dull. The following tips can help you create a good single slide or chart:

- Keep titles short.
- Communicate only one message per slide.
- Use only one thought per line.
- Limit yourself to six words per line.
- Limit each slide to six lines.
- Limit colors, fonts, and effects.

Text

A rule of thumb for presentation text is to keep the text size large and use a simply shaped font. Fonts can be categorized as *serif* or *sans serif*. A serif is the short stroke that finishes the ends of letter shapes. (This sentence is printed in a serif font.) Sans serif fonts, such as Helvetica, lack these strokes. (The heading "Text" above this paragraph is printed in a sans serif font.) Sans serif fonts enlarge better and are frequently used for projecting text in a presentation. Fonts with complex shapes can be difficult to read and may distract from your message.

Limit yourself to one or two fonts throughout the presentation. Generally, words set in a different font indicate that the information is in a different category from that set in another font. Headings are commonly set in a different font to distinguish them from the body text. If you want to add emphasis to a word, use similar styles of the same font, such as bold or italic. This sets off the text without causing the viewer to wonder what the font change signifies.

Points are a printing measurement used to size type. The size of the letters in this line of text is 10 points. Most word processing software will not offer font sizes much larger than 96 points on the font menu, but you may be able to type in sizes above 96 points to adjust the font to a larger size. Table 13.1 illustrates how large type should be to be legible to viewers at varying distances.

Using an overhead projector or projection system, the lettering is enlarged when it is projected on the screen at some distance away. Table 13.2 gives recommended lettering heights for creating slides that will be projected on an 8-by-8-foot screen.

The color of the text you use can also affect legibility. Like changing the font, changing the color of the text can be distracting if it is not clear what the relationship is between the color and the content of the message. Use one color for your headings and another for body text, adding colors only as needed to set off a certain kind of text. For example, a table of figures representing cost data in your

TABLE 13.1

Flip Charts, White Boards, Posters

Plotted or written letter height	Font size	Visible distance
6 mm / 0.25 inch	24 point	2 meters / 4 feet
12 mm / 0.50 inch	48 point	3 meters / 6 feet
20 mm / 0.75 inch	72 point	6 meters / 12 feet
25 mm / 1 inch	96 point	8 meters / 17 feet
40 mm / 1.5 inches	144 point	12 meters / 25 feet
50 mm / 2 inches	*	15 meters / 30 feet
80 mm / 3 inches	*	25 meters / 50 feet
100 mm / 4 inches	*	35 meters / 75 feet
130 mm / 5 inches	*	40 meters / 85 feet
150 mm / 6 inches	*	50 meters / 100 feet

*Font size not usually available.

TABLE 13.2

Overhead or Projection Systems

Letter height	Font size	Viewable distance
1/4 inch	24 point	8 feet
3/8 inch	36 point	20 feet
1 inch	96 point	32 feet
3 inches	*	90 feet

*Font size not usually available.

presentation may have a different color to group it and distinguish it from the other text.

In general, you want your text to stand out from the background. Use black, dark blue, or red text on a white background; or white or yellow text on a black background. Avoid using background and foreground colors of the same intensity, as they provide little contrast.

Charts and Graphs

Using graphical information can help reduce the amount of text on screen. Charts and graphs are a good way to present data because they quickly convey the relationships you want to show. Showing a table may be necessary, but if you know what you want the audience to see, consider converting the data to a graphical form that expresses the information.

Pie charts, bar charts, and line charts are some of the most common and most easily interpreted ways to present data. Many spreadsheet programs, such as Microsoft Excel or Quattro Pro, have built-in functions for creating all three of these types of charts. In most cases, you need only select the rows and columns of data in the spreadsheet and use the charting function to select the kind of graph you wish to present. Options for displaying and labeling the graph can be selected from a dialog box.

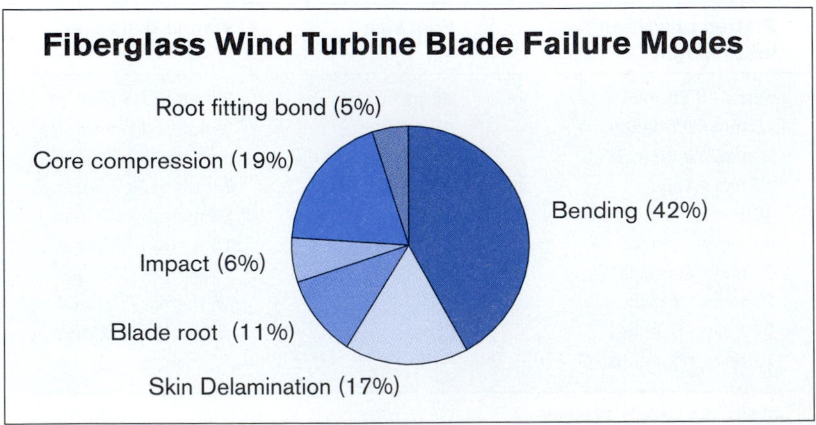

FIGURE 13.3
Pie charts show the relative proportion of categories that make up the whole.

Pie Charts

Pie charts divide a circle into areas representing parts of a whole. Use pie charts to show relative sizes of the categories that make up the whole. For example, a pie chart might effectively show how the cost of a product is divided into design, manufacturing, and assembly costs. Two pie charts with the same categories are an effective way to show differences in relative percentages. Avoid trying to show fine contrasts between data with this type of chart. Pie charts show data best when there are six or fewer categories forming the slices of the pie (see Fig. 13.3).

To fill in the areas, use contrasting colors or alternating solid and white areas for adjacent slices. If you are using patterns, place a solid or contrasting color next to a patterned area. Keep in mind that 6% to 10% of the male population is color blind and may have difficulty distinguishing shades of red and green or green and blue. Try not to use these colors next to each other even if they look different to you.

Names for the segments and the percentages of the whole that they represent are often placed outside the circle. If the general trend is obvious (say that area A is half of the pie, and B is twice that of C) you do not need to label the percentages. If some are close, adding the percentages below the names can help the viewer determine which is larger.

Bar Charts

Bar charts use vertical or horizontal bars to represent data. Bar charts let you quickly contrast discrete categories of data. They are good for showing quantities, trends, and how something has changed over time. For example, the number of defects in a series of part samples taken at regular intervals could quickly show whether the number of defects was increasing, decreasing, or remaining stable (see Fig. 13.4).

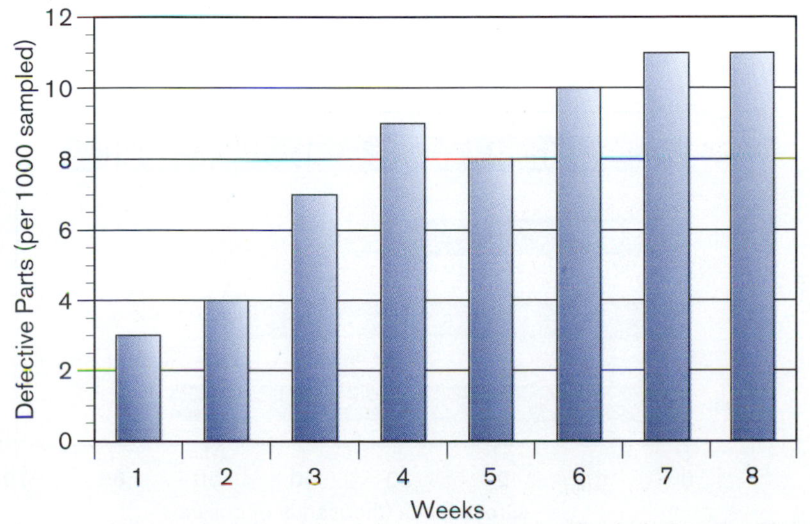

FIGURE 13.4
Bar charts contrast the value for discrete categories of data.

When creating bar charts, try to keep the number of bars representing items to eight or less. If you have a number of less significant items, consider making a group named "Other" to reduce the total number of bars. If you are graphing monthly data and must use twelve bars, for example, consider putting the information on two slides.

Fill each bar in the same data set with the same color. If you are going to contrast a second data set (such as yearly data from two separate years) use a contrasting color for the second data set. A 3D appearance can make bar charts more interesting to look at, but plain 2D bars are generally the easiest to read. Keep titles, labels, and tags reading horizontally on the slide. Vertical lettering is hard to read.

Sometimes symbols are used in place of the bar to indicate the number and liven up the chart. This works well for small units in which each symbol is clearly visible. Resist the urge to get cute or clutter the slide with too many elements (see Fig. 13.5).

Line Charts

Line charts are useful for showing relationships between two variables or to indicate how something has changed over time. Individual data points are connected by lines in a line chart, and the slope of the line (up or down) indicates the direction the data or process is trending (see Fig. 13.6). A line chart can be useful to show how performance varies in terms of energy expended, or how heat dissipation correlates to changes in surface area. When time is the variable plotted on the X axis, the chart is referred to as a *time series*.

To contrast multiple data sets, connect each set of data points with a readily visible line color or pattern. If you are marking the data points, use a different symbol for each set of data. Try to limit the number of sets of data to three or four on a single chart.

Area Charts

Area charts are similar to line charts, except they show the areas between lines filled in to represent quantities or volumes (see

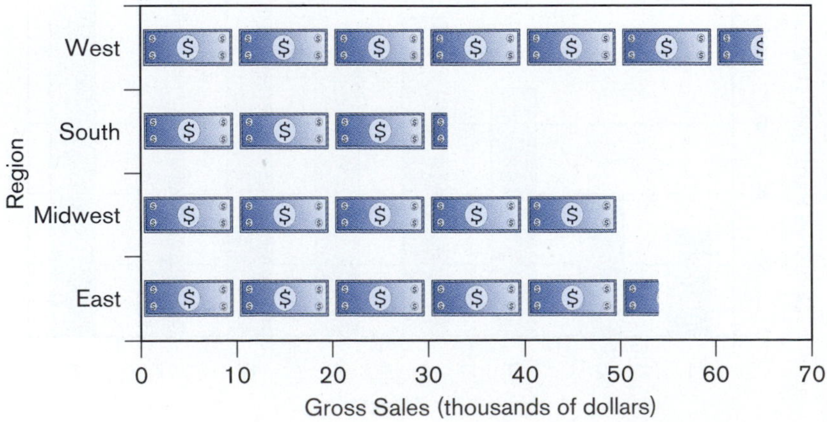

FIGURE 13.5
A stack of symbols can express the identity of a category of data, but should not be used if they detract from the readability of the chart.

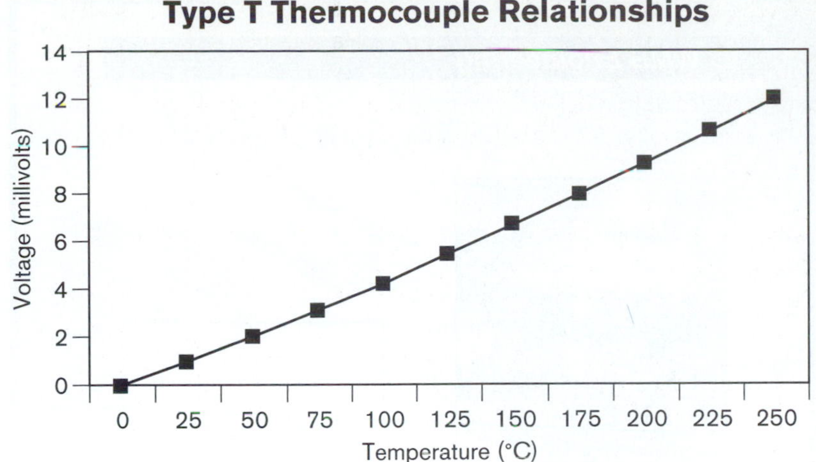

FIGURE 13.6
Line charts show the relationship between two variables or the behavior of a variable over time.

Fig. 13.7). When shading the areas, choose contrasting colors, alternating dark and light. Don't put two patterns next to each other.

Presentation Software

Presentation software, such as Microsoft PowerPoint or Aldus Persuasion, can help you create presentations that are clear, editable, and easy to present. Formatting and organizing tools allow you to quickly create a set of slides that may be printed and used as acetate transparencies with an overhead projector, printed in various sizes and formats to be used as handouts, or displayed directly from the computer via a projection device. Using the computer makes it easy to edit and refine the presentation and to modify it for different audiences.

FIGURE 13.7
Area charts emphasize the quantity or volume between the curves in the chart.

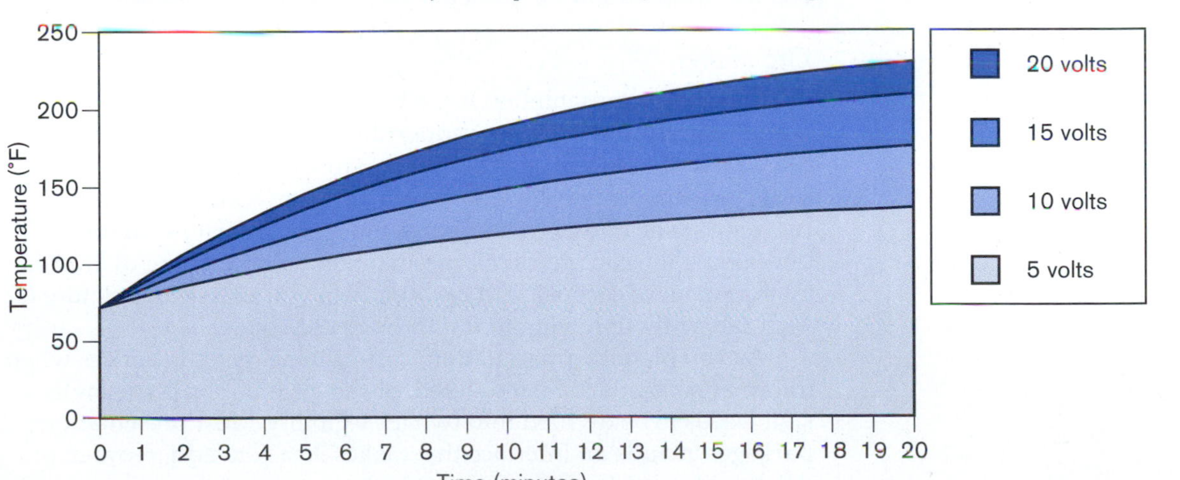

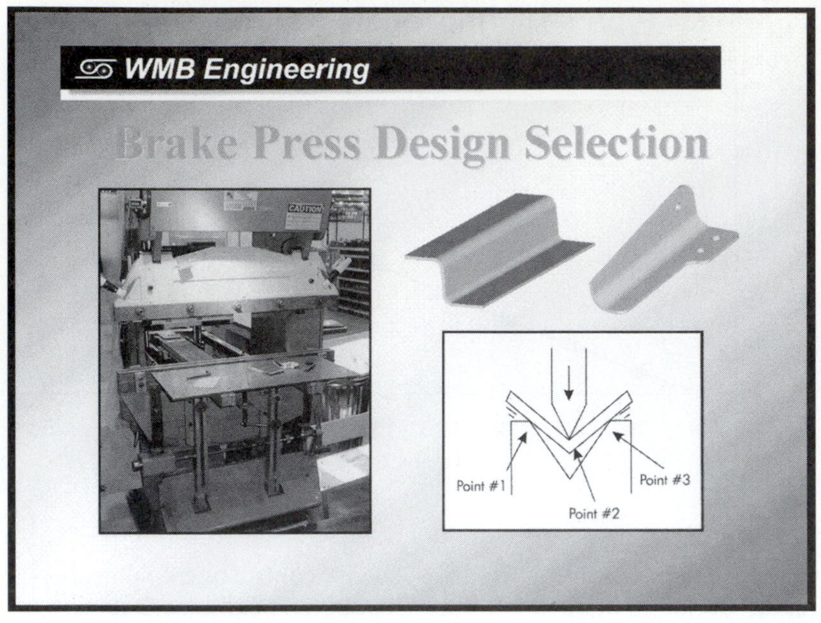

FIGURE 13.8
This opening slide for a PowerPoint presentation clearly states the topic of the talk, identifies the company, and includes images that add interest and illustrate the brake press in the title. (Courtesy of Trent Wetherbee, Ezra Merrill, and Andrew Beddoe.)

Formatting

Presentation software provides formatting tools that contribute to a consistent appearance for the slides in the presentation. Several slide layouts are available that provide predefined text sizes, colors, and background designs. After you have chosen a layout you may immediately begin entering text into the prescribed areas on the slide.

Many options are available for creating your own layout or customizing one of the predefined ones. In all cases, you can define information on a master slide that will appear on each slide in the presentation, such as the heading on the slide in Figure 13.8, or a date at the bottom of the slide. Each successive slide will have the same format until you change it. If you subsequently decide to change some of the elements on the master slide, they will change on all of the slides in the presentation.

Organizing

Outlining your presentation is the best way to organize it so that it flows logically and presents supporting information as needed. Most presentation software allows you to view your presentation as slides or in an outline view and to go back and forth between the two views. The outline view shown in Figure 13.9 is useful for developing the content of the presentation without being distracted by the layout of each slide. It can also help you see the big picture of your talk without having to flip through each slide.

As you plan the presentation, there will be many instances when information needs to be reordered, or the information on a single slide needs to be divided into two for legibility. Most presentation packages include an interface that makes it easy to add, reorder, or delete slides, as shown in Figure 13.10.

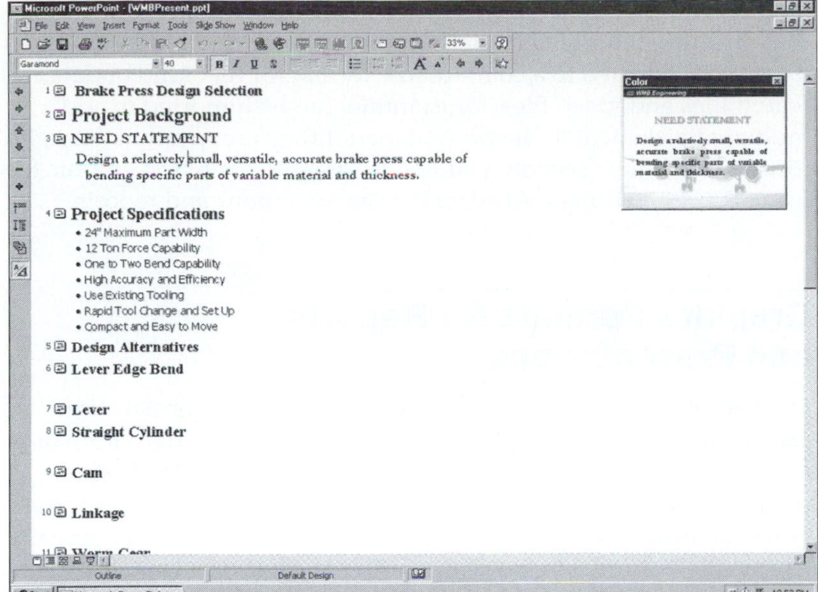

FIGURE 13.9
The brake press presentation is shown here in PowerPoint's outline view.

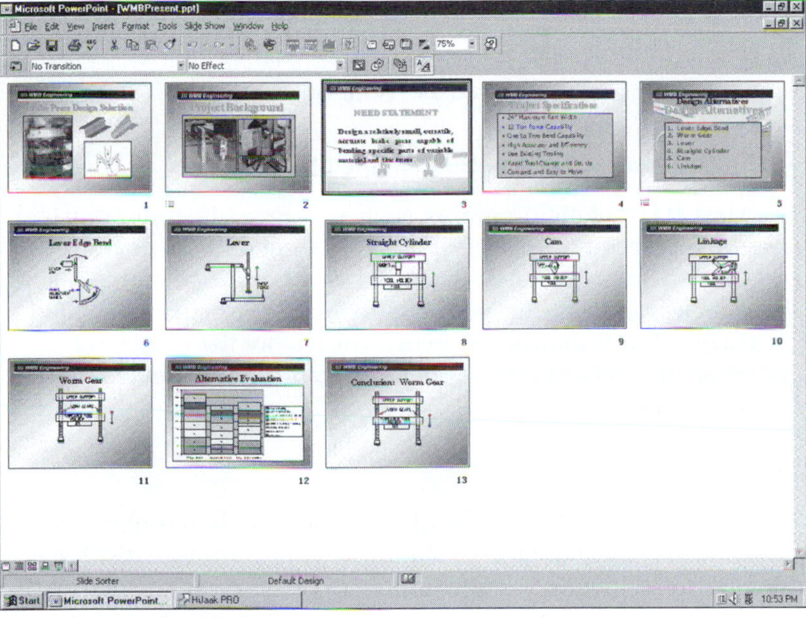

FIGURE 13.10
PowerPoint's slide sorter lets you easily reorder the slides in your talk. Once you have developed a presentation, you can easily edit it for presentation to different audiences or to add material.

Importing Graphics

The opening slide shown in Figure 13.8 combined shaded representations of parametric models, a photograph, and 2D drawings with text on the same slide. Most presentation software allows you to incorporate a wide variety of graphics, charts, and text styles into the slides in the presentation.

Because presentation software was designed to work with productivity tools such as spreadsheets, word processors, and databases, data from these applications is easy to add and format in the presentation. Each presentation package varies somewhat in the

kinds of other files that may be added to the presentation. Many files, including most 3D models, cannot be imported unless they have been exported to a common file format for that kind of data. Sound files and video files, for example, can be imported to add a multimedia element to the presentation if they are saved in the right format. In the next section, you will learn more about the graphic file formats used to bring CAD data into presentations and reports.

Graphics Formats for Reports and Presentations

In Chapter 10 you learned about the many standard formats used for exchanging CAD data among applications. These standard formats were designed to preserve the 2D or 3D definitions of the CAD model so that it could be transferred to another application. Standard graphics formats serve the same purpose, and, like the CAD formats, vary in the information they convey. To get good results with your CAD images, you need to be aware of the strengths and limitations of the formats you may use to export or render the image.

Three key concepts affect the results you can expect from the image you export:

- whether the image is stored as vector or raster data;
- the resolution of the file;
- the color information captured in the file.

Most 2D drawing programs, such as Adobe Illustrator or Corel Draw, use a vector format for storing drawing data, as do most CAD programs. This format defines the drawing entities as endpoints of lines, circles, and arcs that can be individually selected and edited. This is in contrast to raster formats, which store the image as a collection of pixels, or dots. All rendered and shaded images are by definition raster images.

Resolution

Resolution refers to the number of pixels used to define an image. When you speak about the resolution of your computer display, you are referring to the maximum number of pixels used to make up the screen image. A resolution of 1280 × 1024 means that the screen displays 1280 pixels across and 1024 pixels down. For printed images, resolution is referred to in terms of dots per inch, or *dpi*. If you look closely at a photograph in a newspaper, book, TV, or computer screen, or from your printer, you will see that the image is composed of individual dots. When you view them from a distance they blend together so that they are not noticeable. A typical laser printer capable of 300 dpi forms images on the paper using 300 dots for each printed inch. A full-screen image that is 1280 pixels wide would produce on that printer an image that was 1280/300, or 4.27 inches wide. If the printer were capable of 600 dots per inch, the screen would print at 2.13 inches wide.

More pixels, or dots, per inch produce a better image, but the number of pixels needed to produce a high-quality image also depends on whether the image will be viewed on screen or as a printed image. High-quality printed material routinely uses 2400 or more dots per inch, but computer-based images can look extremely good at much lower resolutions. Images on the World Wide Web are rarely more than 72 or 96 pixels per inch.

Determining the resolution needed for an image file is not an issue for vector-based images. Vector-based images can be scaled to print or display at any resolution without loss of quality. For raster images, however, the number of pixels used to define the image when it is exported to the file is fixed. Enlarging the file will always result in some loss of image quality. If you will be printing the image to a printer capable of 300 dots per inch, you can multiply the size of the image by 300 to determine approximately how many pixels you would require for the printed picture to look optimal. If you are going to post the image to the Web, or import it into a presentation package where it will be viewed on a computer screen, you can multiply by a smaller number, such as 96.

Color Depth

The number of colors in the file is the third consideration for graphic images. Your computer monitor has a defined capability to display colors that is dependent on your *video card*. Your video card may be capable of true color (32-bit), high color (24-bit), or 256 colors (8-bit). You should determine the capabilities of your system and keep them in mind when saving or rendering your images to a file. Like resolution, greater color depth results in a larger file size. If you are going to display the image on a system capable of only 256 colors, there is no reason to include more color information in the file than you will be able to display. Likewise, if your printer is capable of a range of only 256 colors (which is common), even images that display with more color information on your screen may not print with the same appearance.

Most files to be used for computer presentations are created at no more than 256 colors (8-bit). Broadcast quality images are generally 24-bit color, which uses 8 bits for each of the three color channels in the RGB image (red, green, and blue). A fourth channel of data, referred to as *Alpha*, is included in 32-bit color. This channel, which uses another 8 bits, is used to assign transparency to the pixels in the image. The Alpha channel makes it possible for rendering packages to specify entities to be transparent.

Graphic File Formats

Most CAD packages offer a number of different file formats for saving images to files. Each differs in its ability to capture the image and reproduce it in different media. Before you save an image to a file, you should consider whether it will be printed or displayed on the screen, the kinds of file formats that the target application will accept, the color information in your file, and the color information

File Formats for 2D Images

.EPS Encapsulated PostScript
PostScript is a graphics format that uses a vector format to store drawing information. It has the advantage of being scalable without loss of image quality. To include 2D drawing views or portions of wireframe displays in a printed report, this format may produce better line quality in a smaller file size than a raster format. Not all systems will write to encapsulated PostScript; many write PostScript files with a .PS extension.

.TGA Targa
Targa files are a popular format for rendered CAD images because they can store files at any resolution and color depth. Targa files can be read by several animation packages, such as Autodesk's 3D Studio. Targa files may be compressed when they are created to reduce the file size somewhat.

.TIF Tagged Image Format
TIFF files are frequently used in desktop publishing and are a common format for capturing images from the screen display. Most presentation packages will import TIFF files.

.BMP Windows Bitmap
The Windows Bitmap format is common to Windows applications and suitable for use with presentation software. The format is less popular for rendered images or those that require 32-bit color.

.JPG JPEG
JPEG is a standard file format established by the Joint Photographic Experts Group. JPEG files allow you to control the degree to which the data is compressed and are popular for posting on the Web. Some image quality can be lost as compression is increased, however. JPEG does not support the Alpha channel.

.GIF Graphics Interchange Format
GIF files are limited to 8-bit color and were developed to enjoy extremely high file compression and portability among personal computers. GIF files are a popular format for images to be posted on the Web. Most GIF files store images at 72 or 96 dpi.

that can be shown in the target application. Several of the most common formats used for importing graphics to print or display in a presentation are described in the box below.

Object Linking and Embedding

Many Windows applications offer another way to add images (and data) from one application to another. *Object linking and embedding* (OLE) refers to the use of the Copy and Paste functions common to Windows applications to embed or link data. To embed data from a Windows-based CAD package, for example, you could select an entity or group of entities, choose Copy, then switch to another application, such as Microsoft Word, and choose Paste to add the selected entities to the Word document. If your CAD package supports this function, this is an easy way to add images and data to reports.

To link an object, you select and copy the entities in the same way, then switch to the target application. In that document, however, you would choose Paste Link to insert the object as a link to the original document. Object linking is similar to adding an external reference. When the original document changes, the target document can be updated automatically. Object linking is useful for reports and presentations that include information from a spreadsheet that is likely to change. By linking the information instead of copying it, it will change as the original document does.

Rendering

Solid and surface models can be shaded to produce realistic views that help you communicate about your design, but these images typically cannot be imported into presentations or, often, printed without being exported as a rendered file. A rendered file is one that stores a single 2D view of the solid with the colors, materials, lighting, viewing direction and sometimes backgrounds you have added to the model.

How the object appears when it's rendered depends on the lighting, materials, and resolution of the screen or file, and the rendering algorithm used. Figure 13.11 shows an example of a solid model and the rendered view produced from it.

Lighting

Lighting is very important to the appearance of rendered views. The placement of lights and the amount of light striking various surfaces of the object are a key aspect of its appearance when rendered.

Most rendering software has options for the types of lighting that you can model: omnidirectional, parallel, or ambient, although not all packages refer to them by these names. AutoCAD's rendering

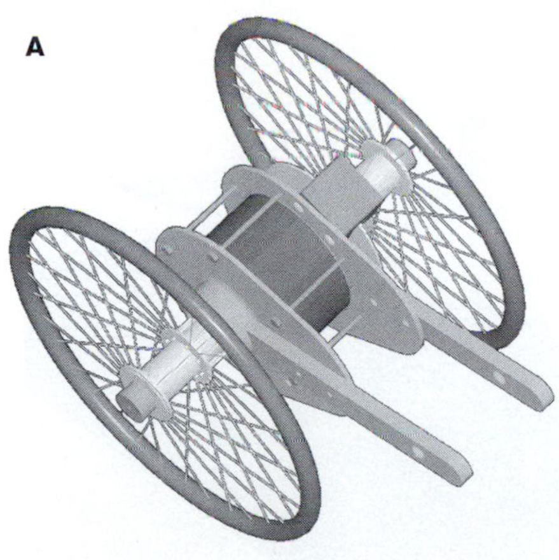

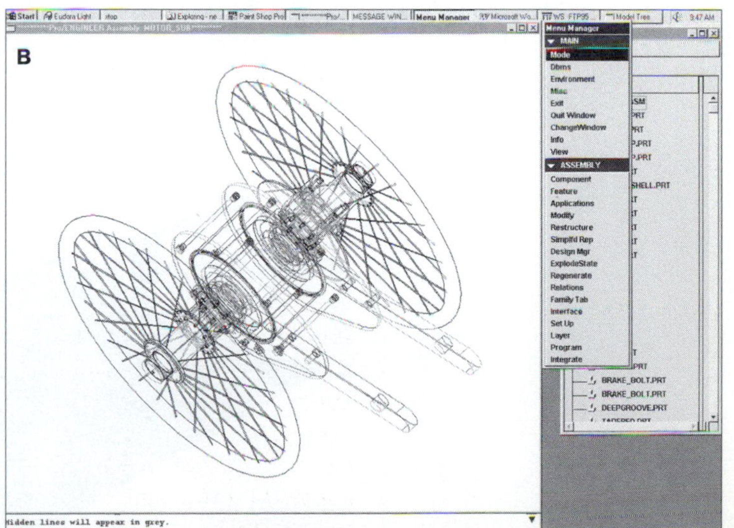

FIGURE 13.11
The rendered view in (a) was generated from the solid model shown in (b).

package, for example, refers to an omnidirectional light such as sunlight as a "distant light."

Omnidirectional lights cast their light in all directions. Two types of lighting that might be modeled as omnidirectional lights are sunlight and point lights. A *point light* is like a bare lightbulb. It shines in all directions from its location, but objects that are farther away from it receive less light than closer objects do. The sun, on the other hand, shines equally brightly on all objects that are not blocked by another object. These two kinds of omnidirectional lights are illustrated in Figure 13.12.

FIGURE 13.12
(a) Sunlight and (b) bare lightbulbs are modeled as omnidirectional lights.

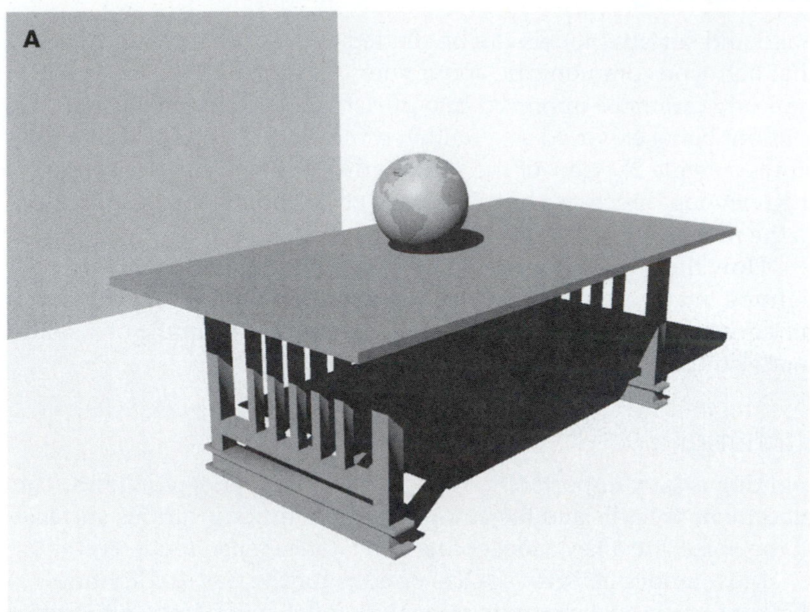

Ambient light is similar to room lighting. It strikes equally on all surfaces of the object. This produces a very flat lighting effect, as shown in Figure 13.13.

A *parallel light* shines in a certain direction, as shown in Figure 13.14. Spotlights are an example of parallel lights. To create a spotlight effect, the light shines from the source toward some target. None of the light shines in other directions. *Hotspot* and *falloff* are two terms that refer to the appearance of a spotlight. The *hotspot* is the bright spot at the center of the spotlight created by the direct light from the bulb. The *falloff* is the circle of light reflected from the

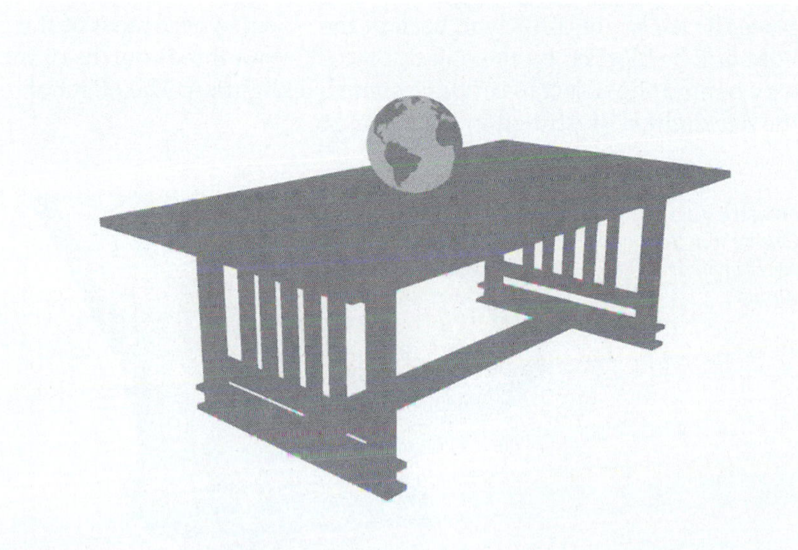

FIGURE 13.13
Ambient light produces a flat lighting effect.

FIGURE 13.14
A parallel light shines from a source toward a target.

back of the spotlight-type fixture. It is less intense. When using AutoCAD's rendering, you can adjust the size of the hotspot and falloff for spot-type lights.

Placing Lights

To place lights for a rendered view, think about how a studio would set up lights for photographing a subject. You might start with one main light that illuminates most of the object from the front. A rendered view of an object with just one light near the front of the object is shown in Figure 13.15.

With just one light source things tend to look flat. Adding a back light with its intensity set to 25% to 30% of the main light will highlight the back edges of the object so that it doesn't look so flat. Don't place the backlight directly in back of the object, where most of the light will be blocked by the object; place it about three-fourths of the way behind the object to produce some backlighting. The effect of the backlight is illustrated in Figure 13.16.

FIGURE 13.15
One light in front of the object produces a flat effect.

FIGURE 13.16
Backlighting is not as intense as the light in front of the object.

FIGURE 13.17
This rendering uses a combination of lights to illuminate the objects in the scene.

Pure white-colored light rarely occurs naturally. To make your renderings look more realistic, adjust the color of the light. Using a color of light that is a little toward blue will simulate outdoor lighting. Using light that is colored toward yellow looks like incandescent lighting, such as that from a lamp. This gives a warmer appearance to a subject.

Spotlights can be used to call attention to an important feature or to set a mood. Figure 13.17 shows a scene with a good combination of different types of lighting.

Materials

Applying materials to objects to be rendered gives them the appearance of the real object. Materials can simulate the color and surface texture of the object and allow the model to reflect light like the real thing. Materials for rendering are similar to paint, except that they can be transparent, shiny, or even self-illuminating (which looks like a neon light). These materials are unlike materials used in engineering because they do not have mass density, modulus of elasticity, or other engineering properties. They have properties that affect only how the material appears. These properties include *reflectivity* (or shininess), *refractivity* (or the degree to which light is bent), transparency/opacity, color, pattern, and bumpiness.

Most rendering packages include a library of materials that are predefined to look like different kinds of steel, glass, stone, and so on. For these the properties have been preset. New materials can be defined by editing materials in the library or starting from scratch to define each of the properties that an object's appearance should have. The dialog box for editing materials in 3D Studio is shown in Figure 13.18.

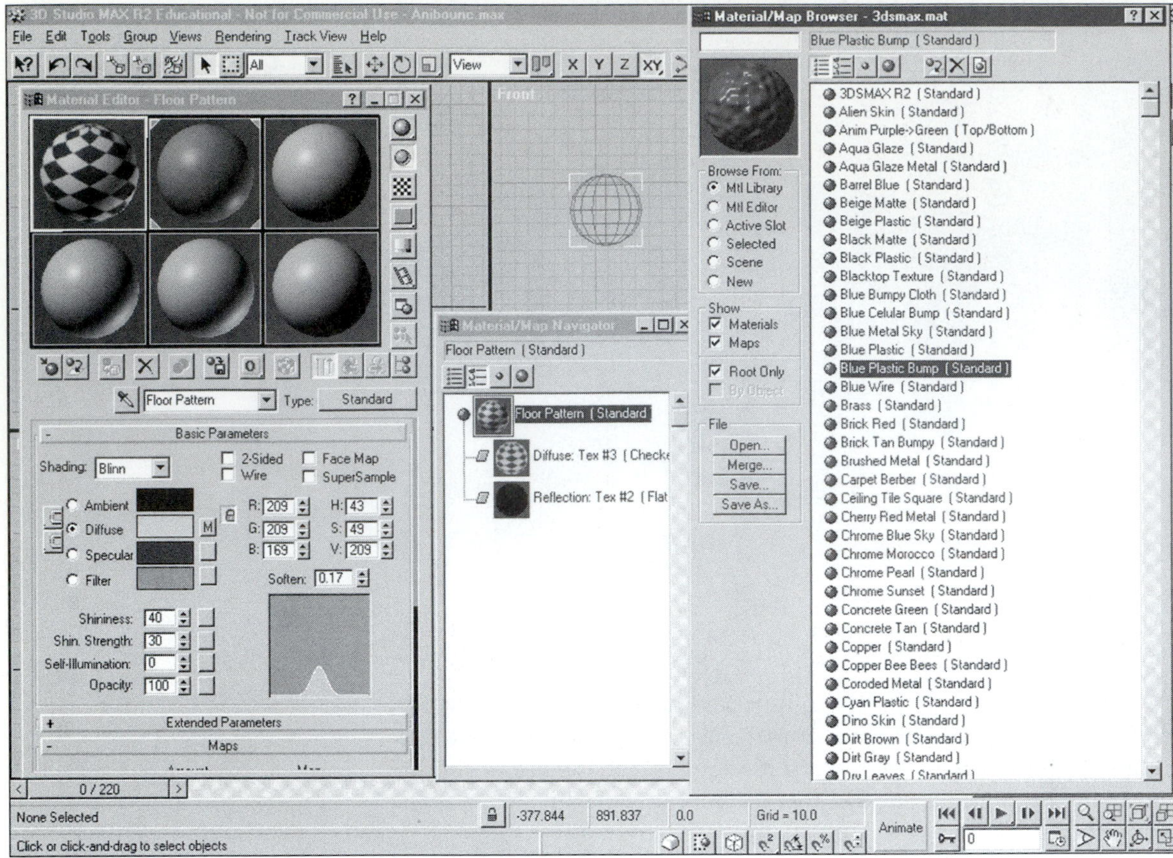

FIGURE 13.18
The Materials Editor screen from 3D Studio Max includes a sample window at the top that lets you preview materials as you change the settings.

Rendering Options

Views of the model can be rendered to the screen or to a file where the image can be saved. Rendered views can be saved only to one of the raster formats, also called a *raster graphic* file or *pixel graphic* file, such as TIFF, JPEG, BMP, and GIF. Essentially, raster graphics store the color and intensity value for each dot or pixel making up the picture.

Choosing Resolution and Color Depth

When you render the file, you must specify a resolution. As you have already learned, resolution refers to the number of pixels per inch on the screen or in a printed file. In most rendering software, you are able to select varying numbers of pixels for the final output image. If you are rendering to your screen, you won't be able to choose a resolution higher than your screen can display, but you may choose a lower one. Figure 13.19 illustrates the difference in quality between rendered images. When rendering to a file, a higher number of pixels produces a better image but also takes more time to render. If you will be printing the images to a printer capable of 300 dots per inch, you can multiply the desired size of the image by 300 to determine how many pixels you need for an optimal image.

FIGURE 13.19
A resolution of 640 x 480 was specified for the rendered image in (a). The same part was rendered at 4096 x 3072 to produce the image in (b).

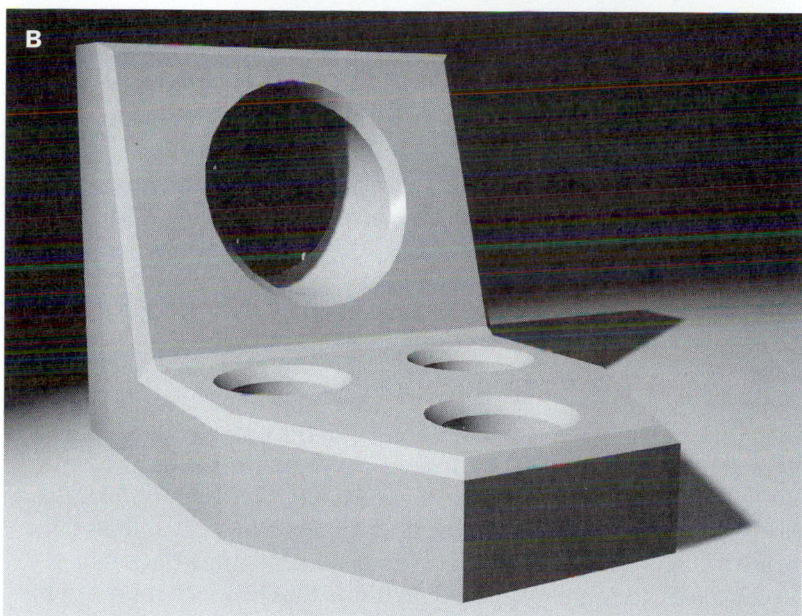

You should apply the same rules of thumb to determining the color information to be included in the file. When rendering to the screen, the number of colors your computer system can represent is the maximum number for the rendering. For posting to the Web, using no more than 256 colors will allow the image to be viewed on the greatest number of systems. (And the .GIF format can store no more than 256 colors.) If you are going to print the file, you should consider the color information that your printer can print. If you are using transparent materials in the file, you should be sure to specify 32-bit color.

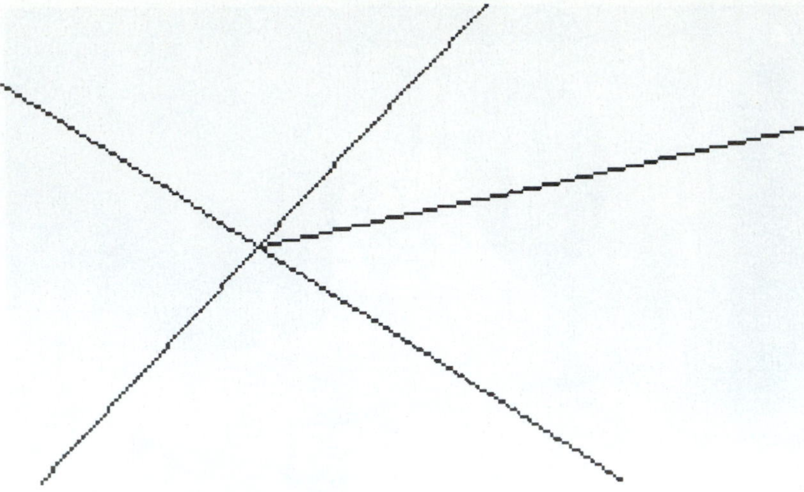

FIGURE 13.20
At low resolutions, angled lines in raster images can look jagged.

Antialiasing

Another setting that can be turned on or off before rendering is antialiasing. Each pixel on the screen or dot on the printed sheet is a discrete unit that can be on or off. To create an angled line, a number of dots are drawn on the screen or printed. Certain angled lines will appear very jagged (see Fig. 13.20)—almost like a small set of stairs—at lower resolutions due to the size and orientation of the pixels forming them. *Antialiasing* reduces this effect by shading the adjacent pixels. Because antialiasing adds to the time needed to render an image, many rendering packages allow you to turn it off until you are ready to render the final image.

Rendering Algorithms

Each time you render an image, you have the option of selecting the algorithm to be used. There are several popular algorithms for rendering. The most common are *flat shading, Gouraud shading, Phong shading, ray tracing,* and *radiosity*. Each of these has advantages and disadvantages. The simpler algorithms take less time but often produce less accurate results. Algorithms that produce more sophisticated results often take a long time.

Flat Shading

The flat shading algorithm for rendering calculates a different color for each flat polygon of the object. This looks great when the object is comprised of flat surfaces, but it does not work as well for rounded shapes and complex contours. This algorithm is very quick to produce the shaded result, which can be an advantage for quick checking in the initial stages of your work. The object is often broken up into triangular facets or polygons to define the different surfaces for shading. Using flat shading, each triangular facet is shaded entirely one color. This sometimes creates a sharp line between two

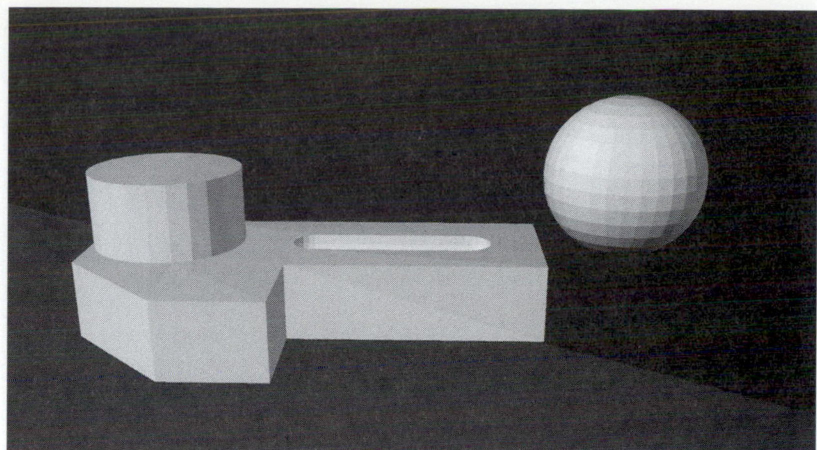

FIGURE 13.21
Flat shading

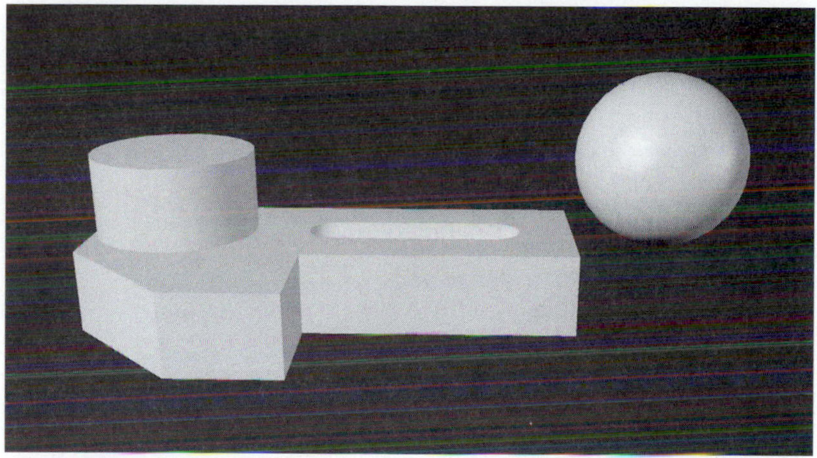

FIGURE 13.22
Gouraud shading

areas on the object where no distinction exists on the object. Flat shading also does not cast accurate shadows on surfaces. Figure 13.21 shows an example of flat shading.

Gouraud Shading

Gouraud shading is named for the algorithm's developer. This shading algorithm is more realistic than flat shading. Gouraud shading calculates a color for each vertex on the object depending on how much light is striking it, then blends each set of vertices to determine the color for the pixels making up a flat surface (see Fig. 13.22). The colors of the vertices at the ends of an edge are averaged, and that color is used to show the edge. Gouraud-shaded objects do not have the incorrect sharp edges of flat-shaded objects, but this algorithm still does not correctly show shadows.

Phong Shading

Phong shading, also named for its inventor, is more realistic than Gouraud shading. It calculates the color for each pixel individually.

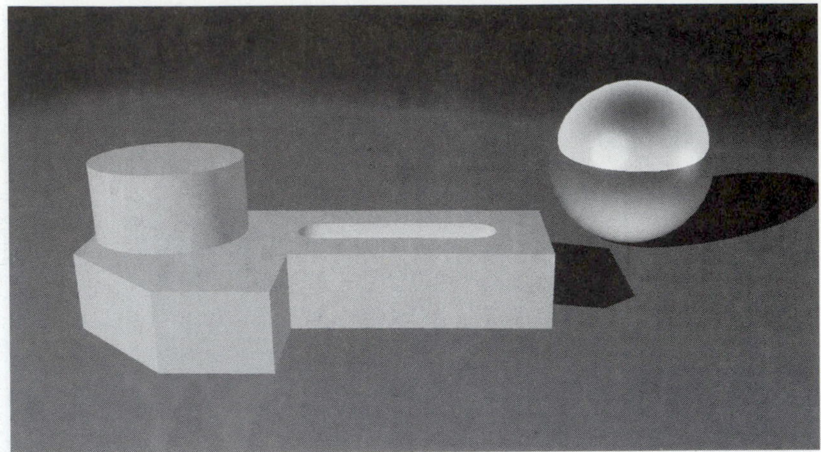

FIGURE 13.23
Phong shading

FIGURE 13.24
Ray tracing

Phong shading takes longer than either Gouraud shading or flat shading because of the additional complexity of its rendering calculations. Using Phong shading allows reflections and shadows to be rendered. Figure 13.23 shows an example of Phong shading.

Ray Tracing

Ray tracing algorithms produce sophisticated reflections, refractions, and complex shadows. Ray tracing generates the appearance of the object by tracing the path of the light to its source. Light rays are reflected from shiny materials and are bent as they pass through materials that refract light. Ray tracing calculates this effect by tracing the path of light rays. Ray tracing requires many calculations, making it one of the slowest rendering methods. Its sophistication also produces the best results. In ray traced rendered objects (shown in Fig. 13.24), highlights and shadows appear correctly. For many objects, however, Phong shading may look nearly as good, because without reflected or refracted light there is little difference.

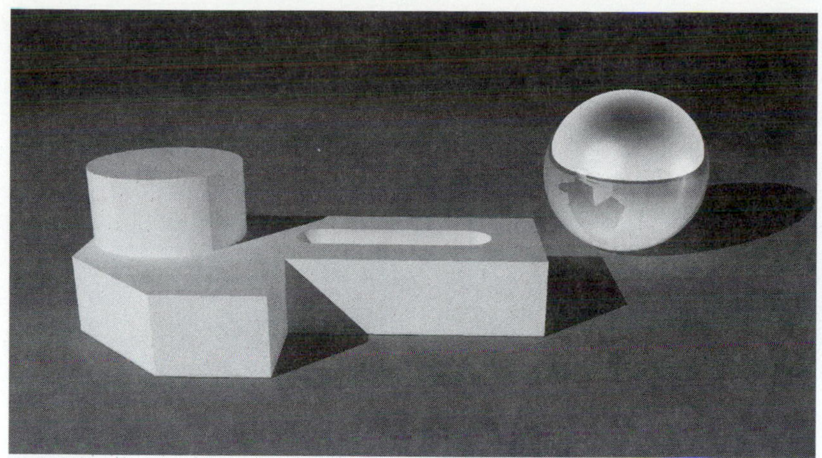

FIGURE 13.25
Radiosity

Radiosity

Radiosity rendering algorithms divide surfaces into small areas, or patches. Each patch is assumed to be evenly lighted across its entire surface. Equations that describe the interaction of the illumination from one patch to its neighboring patches are used to determine the rendered appearance. Using radiosity (see Fig. 13.25), the light energy reflected in a given direction is independent of the direction from which the light energy arrives at that patch. This means that instead of tracing the incoming direction of the light to the location like ray tracing algorithms do, radiosity algorithms need only keep track of the total incident light energy on each patch. Very sophisticated renderings, with a lot of reflected color and light, can be created using radiosity algorithms.

When you understand the basics of how these algorithms work, you can select the most appropriate one for your rendering task. If shadows do not show correctly or you are using a transparent material, you may want to choose a ray tracing algorithm. In general, you should start with the least sophisticated methods for quickly checking the orientation of the objects, materials, and lights in the view. Once you are satisfied with the scene, you can select the appropriate algorithm for the finished appearance.

Animation

Animation software lets you add motion to 3D models to show the movement of parts in an assembly; check interferences; fly through a design site to visualize a building or system before it is constructed; or train workers how to install, assemble, or repair a product.

The technique used to create animations is similar to that for a flip book. To create the appearance of smooth motion, many individual views are rendered so that the scene in each is slightly changed from the previous one (see Fig. 13.26). When these scenes are played back rapidly, the objects appear to move. This effect varies by individual, but for most people, the eye starts to blend the individual scenes into a single moving image when they appear at a rate of at least 10 per second. A single rendered view in an animation is often

FIGURE 13.26
Each frame in the animation is slightly different from the previous one. More frames allow for smaller differences between them and a smoother motion in the animation.

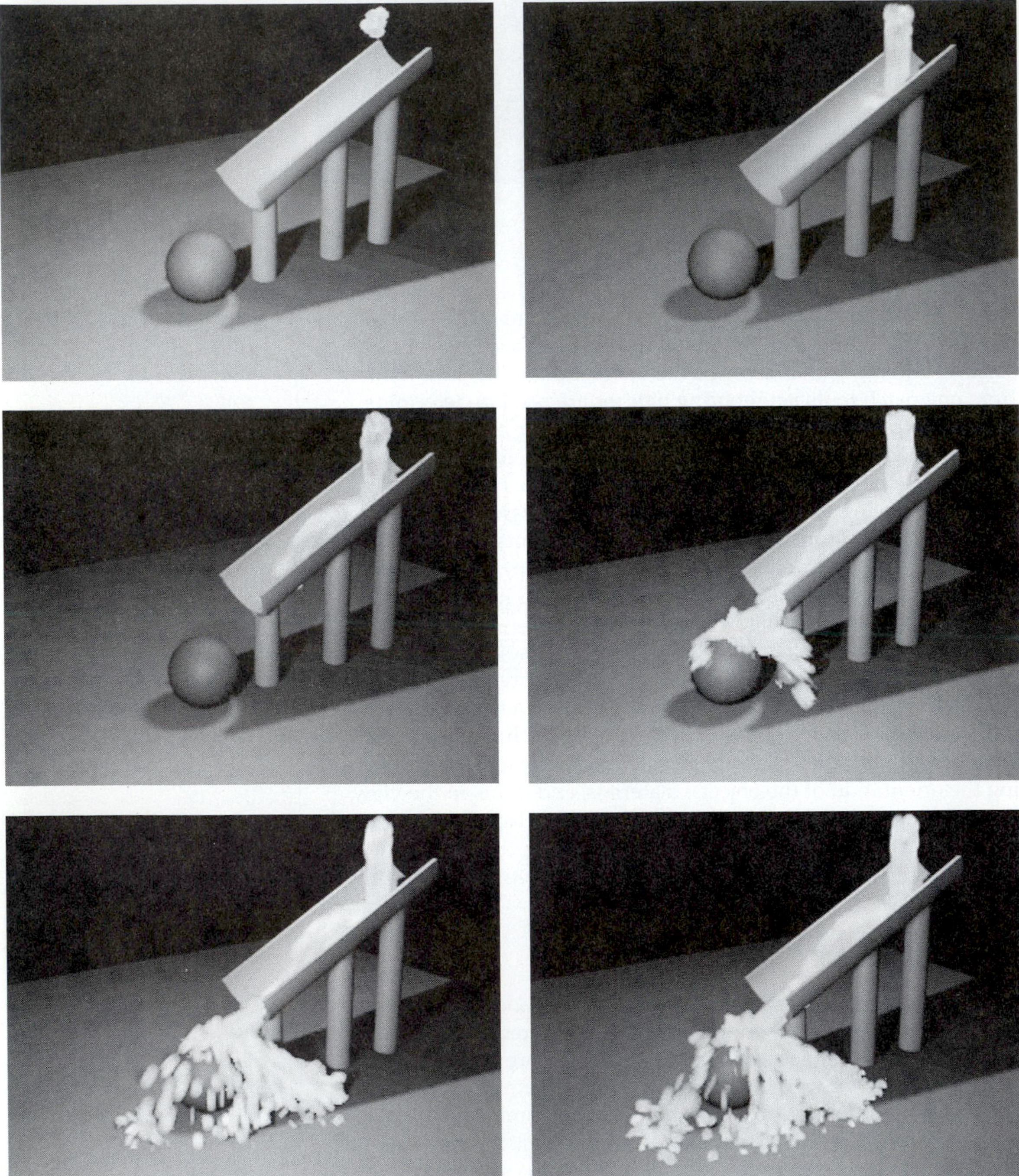

referred to as a *frame*. For the motion to appear smooth, the frames need to be played back at about 30 frames per second. Older cartoon animation sometimes played at 12 frames per second, but this is what gave a jerky appearance to the motion. Today's cartoons are generally 24 frames per second, motion pictures are 24 frames per second, and U.S. television is 30 frames per second.

The task of rendering an animation is computationally intense. The resolution, color depth, complexity of the objects, and number of frames in the animation all are factors in how long it can take to render an animation. For a full-length 3D animated movie such as Disney's *Toy Story*, this can mean 800,000 hours on a networked series of 117 workstations. It has been estimated that if *Toy Story* were rendered on a single processor, it would have taken 43 years to render. Even a short, 30-second animation typically contains nine hundred frames. Each individual frame may take up to several minutes to render. For a 30-second animation in which each frame takes 2 minutes to render, the rendering time would be 30 hours on a typical Pentium computer. If changes are made, the animation must be rendered again.

Planning Your Animation

Planning is an important part of creating and rendering a good animation. Just like any project, you should consider the need for the project, its purpose, the audience, the desired length, and how it will be viewed. If your objectives for the animation are clear, you can produce an animation that will meet your needs. Without planning, you may find that you spend a lot of time playing around with the software, but not meeting the project needs in the end. A storyboard is a good way to plan the project.

Storyboarding

A *storyboard* uses pictures to show the general appearance of key events in the animation. Sketches in a storyboard can be very rough as long as you can get the idea across. Planning your animation ahead will save time in the long run because you will spend less time redoing scenes. When contracting for animation to be performed by another company, the storyboard can be construed as a contract document, in the same way that engineering drawings are. If you agree to the storyboard, the animator will know what to plan for. If you are creating an animation yourself, the storyboard will help you know what to plan for, and you will get better results than if you just dive in.

You do not need any special supplies to create a storyboard, although commercially printed ones are available. You can sketch your ideas for the animation on a tablet, loose paper, or 3-by-5-inch cards. Cards are useful for creating storyboards because you can easily rearrange the order of key scenes. Also, the small format of the card will help prevent you from getting too detailed in creating the storyboard. When storyboarding your animation, keep your

FIGURE 13.27
These frames from a storyboard illustrate key scenes and the elements that will be on screen at that moment.

audience in mind, your goal for creating the presentation, the length of the final project, and how certain parts or features will best be shown. Figure 13.27 shows frames from a storyboard.

Importing Geometry

Various file formats can be used to export to animation software. Most 3D animation software, such as Autodesk's 3D Studio Max, uses surface models of the object. For example, 3D Studio Max will import AutoCAD surface and solid models, DXF files, and STL files. Because animations primarily present the external appearance of shapes and not their engineering properties, they do not need the volumetric information.

Before you begin an animation, check the software you will be using to see what types of CAD geometry it can import. You should be able to use a common file format so you won't have to re-create the objects to add animation to them. If not, 2D information about the model can be imported into the animation package and used as a starting point for creating the geometry.

Once an object has been imported, it can be modified further, if desired, and materials can be applied to control the appearance of the object when the animation is rendered. Materials may be applied from a library or created with a materials editor as you would with rendering software.

Keyframing Methods

Keyframe animation software lets you select various transformations, such as position, rotation, and scale, to apply to the object in a particular frame, giving it the name *keyframe animation*. A *key* is basically a transformation applied to an object or set of objects. The software calculates the steps between the keyframes for you. If you were going to animate how the parts in your design assemble together, for example, you would start by specifying where the objects are located in frame 1. In frame 1, the parts might be spread out from one another (as if they are lying on a work table). By frame 50, you may want to have the first two major parts in their assembled positions. To do this, you would select frame 50 and add a key for position by moving the two objects so they are in their assembled position. You do not need to change the locations of the objects in the frames between 2 and 49. The animation software will calculate these intermediate positions from the starting and ending locations for each part, as shown in Figure 13.28. This process is called *tweening*.

Keyframes are logical items to document in your storyboard for the animation. Think about which frames will have key motions in them and sketch only those frames. They are the ones you will have to set up in the animation.

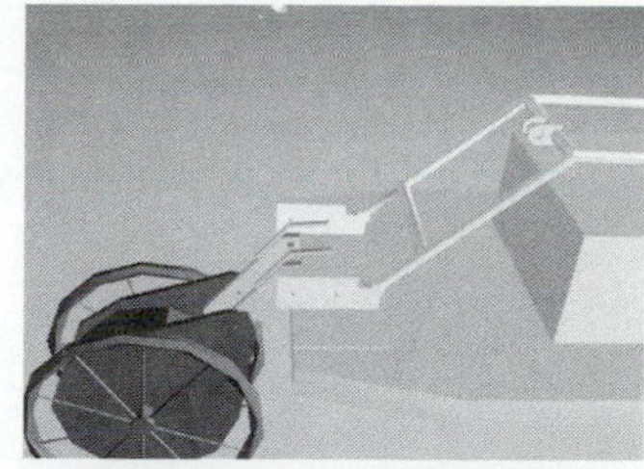

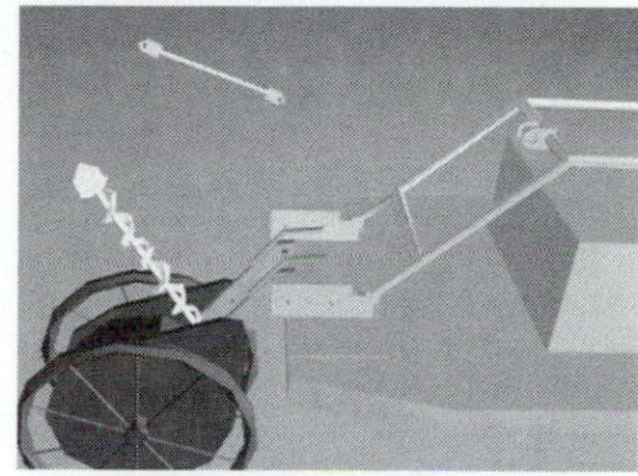

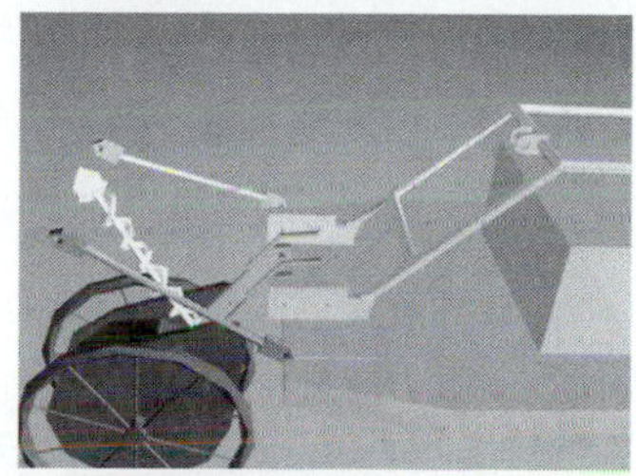

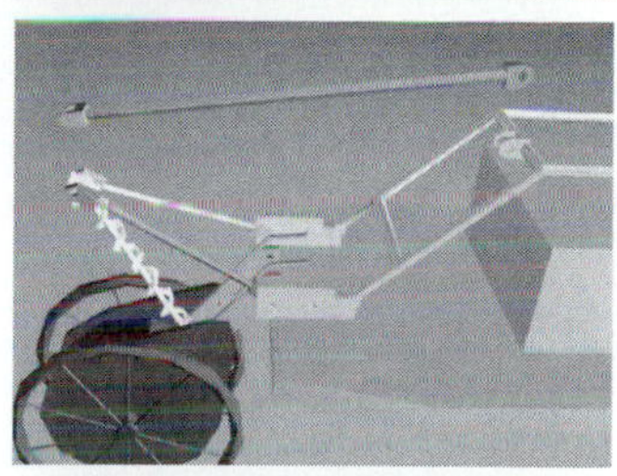

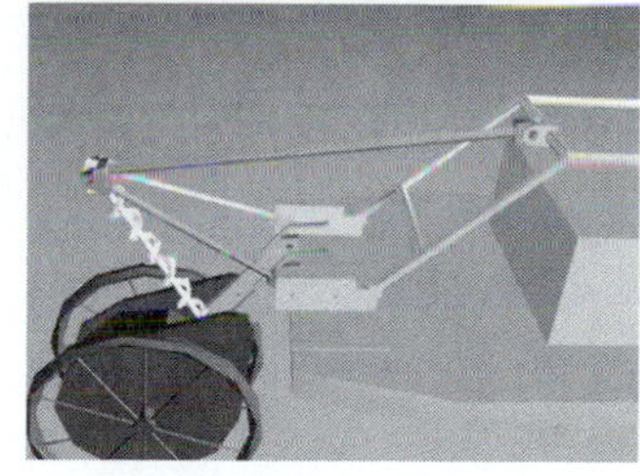

FIGURE 13.28
Tweening calculated where each part should be located in the frames between each one shown here.

Transformations

Location is not the only transformation that can be specified at a key. The mesh objects themselves can be modified so their appearance changes. Without changing the number of vertices in the object, the location of the vertices can be changed in a keyframe, and the tweening process will fill in the incremental changes in the frames between them. This process of transforming an object from one shape to another, shown in Figure 13.29, is called *morphing*. Objects can also be rotated, scaled, squashed, and hidden at a keyframe.

Mesh objects are not the only items that can be transformed at a keyframe. Lights are placed in the animation scene as they would be in a rendered view. Different types of light may be added, and each one can have its own values for intensity, color, and so on. Changing the lighting during the animation can draw attention to different parts of the objects, and can light the action as stage lighting does in a live production.

The point of view for the animation is determined by the location of the camera, which is placed in the scene as a light or object would be. Cameras, too, can move. Moving the camera simulates the

FIGURE 13.29
The shape of the mesh in frame (a) morphs in the succeeding frames to the ending shape in (b).

way a camera operator would move to capture the action and contributes to the sense of the animation as a movie. For a fly-through animation, the camera moves through the objects to capture the vantage point of an observer moving through the scene. This camera movement gives the viewer the impression of being in the scene.

Inverse Kinematics

Inverse kinematics is a technique that lets you constrain joints between different objects in the animation so they move only in ways that the finished product would be able to move. Inverse kinematics is a useful feature for animating machinery realistically. (Of course, it's also great for character animation.) Once you have constrained a joint to have a certain number of allowable degrees of rotation in various axes, you can animate the object by moving one of the objects. Any other object that is linked to it will display the proper behavior for the joint. See Figure 13.30 for an example of inverse kinematics.

Rendering the Animation

As you define each keyframe, you can play the animation and make adjustments to refine the motion and transformations, adding or deleting frames as needed to slow or speed the motion. When finished, a rendering algorithm is selected and the animation is rendered to a file.

A number of different animation file formats exist that allow the file to be played back by dedicated viewers as well as other media players for video files. .FLI and .FLC files are standard animation formats, but animations can be saved as .AVI files, MPEG, and Quicktime movies as well, which are popular animation formats for personal computers.

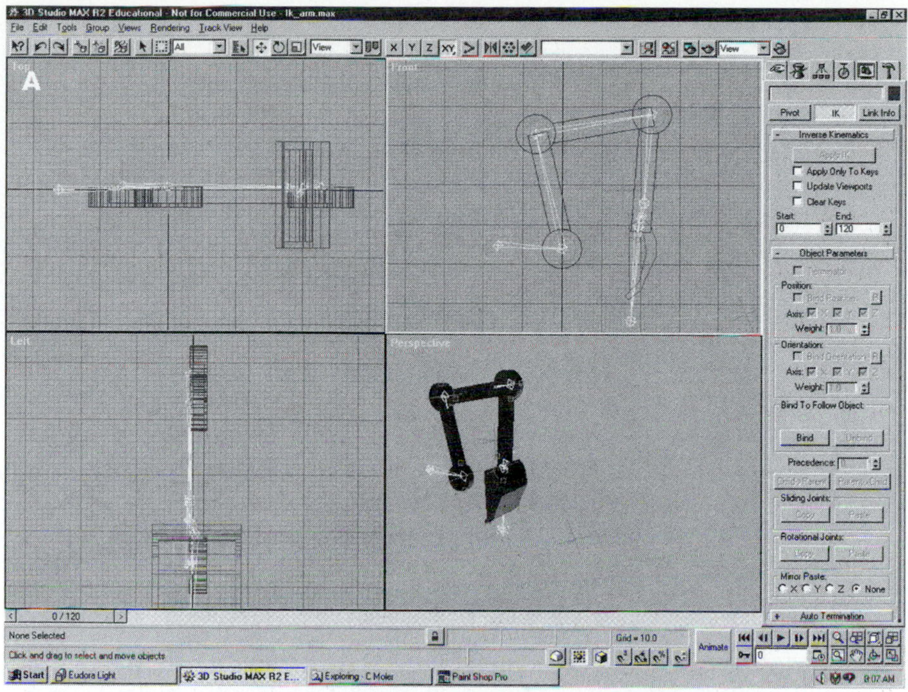

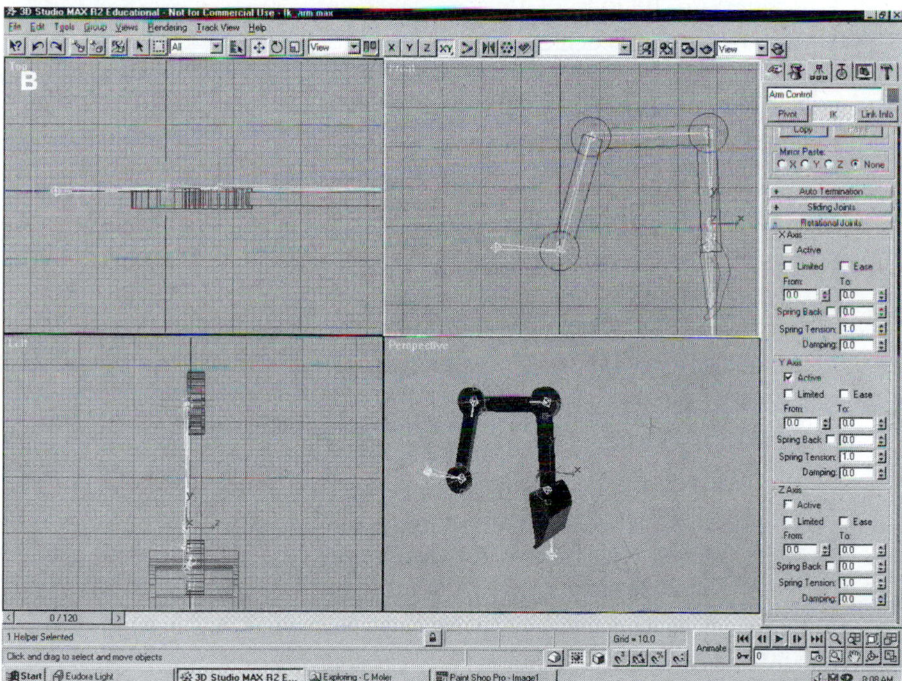

FIGURE 13.30

Inverse kinematics was used to constrain the movement of parts in this excavator arm. When the bucket is moved, the other parts of the arm move to follow it. Just as your finger, hand, and arm move to follow when someone pulls on your fingertip, the object parameters for the parts in the excavator arm can be set (with the dialog box shown in (a)) so they move when the bucket does. The dialog box options shown in (b) are an example of those used to constrain the motion of a joint. The joint where the shovel attaches can be constrained so that the shovel cannot move past the point shown in (b); outward movement past that point causes the rest of the arm to move.

VRML

Another format used to publish animated 3D information on the Web is *virtual reality modeling language (VRML).* VRML is used to create and animate 3D environments that the viewer can move around in. Using a VRML plug-in with a Web browser, the VRML file can be viewed and "experienced" over the Web, as shown in Figure 13.31. Unlike animation packages, VRML is used to create virtual 3D settings, called "worlds," that allow the users to control what they see, just as if they were in the model and could turn to view it from any angle.

VRML is a formatting language like HTML (hypertext markup language) that was developed to create virtual worlds on the Web using nothing more than text-based commands. Its syntax is more complicated than HTML, however, as it must define the 3D objects as well as their appearance and behavior. The VRML standard defines the content and format of the world file, which has a .wrl suffix.

Having first been introduced in 1994, VRML is still a very young language. VRML 1.0 offered very little animation capability. In the racetrack model shown in Figure 13.31, Bruce Olson could only use the VRML 1.0 commands to make the 3D models move. To do so, he used 3D Studio to model the race cars and a 20-mile-long rod for each of the cars. Each car was attached to the rod, but 10 feet above it. He then exported the 3D models from 3D Studio so they could be used by the VRML program he would write. Using the VRML Spin command, he programmed each of the rod/car combinations to spin at a certain rate. When he created the ground above the rod, the illusion of the cars moving on the track was complete. The same mechanism was used to make the boats on the lake move. With VRML 2.0, much more complex animations became possible. VRML objects can move from one location to the other and interpolator objects fill in the frames between locations to create a smooth motion, much as an animation package would tween them. With VRML 97, the language standard was first recognized by ISO as an international standard

Because VRML models are displayed in a 3D environment, the viewer can move around and see the scene from any vantage point. General navigational controls are provided by the browser plug-in, as shown in Figure 13.32. In addition, the programmer can preset viewpoints for the viewer. Top, front, and side views, for example, can be preset to simulate orthographic views. In a large world, such as the racetrack model above, preset viewpoints might take the viewer to different parts of the world, such as the viewing stands or the finish line.

Many CAD platforms allow you to export models in a format that can be published as a VRML document on the Web. VRML files use nodes to define objects and fields to define their attributes. Like a 3D solid model, nodes can have children, making it possible to build up complex objects from simpler ones. The Shape node contains information about the geometry of the model and its

FIGURE 13.31
The cars and boats in this 25-mile park were created in 3D Studio and exported to be used as images in the .wrl file. The VRML 1.0 "world" is shown here as it appears in Netscape Navigator with the Live 3D plug-in. Notice the navigation commands along the bottom of the screen: walk, spin, look, slide, and point. Lamp and view are additional controls for viewing the virtual world. (Courtesy of Dynojet Research, Inc. Models and images by Bruce A. Olson.)

FIGURE 13.32
A mouse click activates this VRML animation of a robot that is spot welding a car. The file is displayed in the Cosmo viewer for VRML 2.0, which offers a standard control panel that allows the user to zoom, pan, and rotate the model view. A second set of controls allows the viewer to move in the 3D space, tilt, and slide. The menu at the bottom left offers a choice of viewpoints set by the creator. (VRML model created by Qiming Wang and reprinted courtesy of the National Institute of Standards and Technology.)

appearance. Complex polygons can be imported from other modeling packages, or four primitive shapes (box, cylinder, sphere, and cone) can be combined into more complex models. As in an animation package, attributes are applied to 3D geometry to control its appearance. The Material attributes control the color, shininess, reflective color, and transparency of an object. Texture files—raster images of patterns and details—can be imported and painted on an object. Like an animation package, VRML allows you to add lights, sounds, and backgrounds to create a realistic appearance for the model.

Once an object is imported or created, the Transform node makes it possible to move (translate), scale, and rotate objects. Animations are created by changing the value of any field over time.

FIGURE 13.33
This sample of VRML code illustrates the nature of the language. As an object-definition language, its syntax is similar to object-oriented programs such as C. The header at the top of this file indicates that it is written in VRML version 2.0. The first node is a Background node, which sets the color of the sky in the background. The Sound node defines the attributes of the .wav file to be played by the file. The next two nodes define two preset Viewpoints, "Start of the Road" and "Ramp over the Road." The Inline node at the bottom of the excerpt is used to call another VRML world file, making it possible to build up complex worlds from smaller VRML world files.

```
#VRML V2.0 utf8
WorldInfo {
 info []
 title "Road created with Bashir Research Tools"
}

Background {
 skyColor [0.4 0.4 1.0 ]
}

NavigationInfo {
 avatarSize [ 0.55, 3.6, 1.75 ]
 speed 4.0
}

Group{
 children[

 Sound {
  intensity .4
  maxBack 100000
  maxFront 100000
  minBack 10000
  minFront 10000
  source
  AudioClip {
   description "A noise"
   loop TRUE
   startTime 1
   url "sound.wav"
  }
 }

 Viewpoint {
  position 4 20 4
  orientation 0 1 0 -1.56
  description "Start of the Road"
  }
 Viewpoint {
  position 548 32 -155
  orientation 0 1 0 -0.9
  jump FALSE
  description "Ramp over the Road"
  }
 Inline {
   url "loop.wrl.gz"
```

To animate motion, the translation field for an object (or group of objects) would be set to change over a period of time (analogous to a number of frames in an animation). Any attribute, such as color or size, can be animated, as can the attributes of lights and sounds.

After a simulated environment has been created, it can be used to visualize and communicate design in a variety of ways. For example, Bashir Research creates 3D models of highway systems that can be used both to visualize highway sight lines and to simulate traffic flows. The company first builds CAD models of the civil engineering data for the highway, then imports the models as the foundation for a 3D environment that incorporates the engineering data. The model can be used with an animated viewpoint to simulate a driver's sightlines on the highway, or can be used with other VRML tools and simulation packages to simulate traffic flows at different volumes.

Each new release of the standard can be expected to extend the capabilities of the language. Animations in VRML 2.0, for example, are event-driven. That is, they are not just preset sequences, but can be triggered by an external event such as a mouse-click, contact with other objects in the world, or proximity to the cursor. This user interactivity makes it possible to create complex worlds that respond to the viewer's input. The next version of the language, VRML 3.0, promises to allow many users to interact with the same world at the same time, adding to the illusion of a virtual reality that can be occupied.

Many resources exist on the Web for learning more about VRML programming. Because most 3D modeling packages will export a VRML version of the model, it is possible to create animated sequences without a dedicated animation package. VRML worlds are one way to add animated sequences of your models to presentations, or to an HTML page on your Web site. World-building packages, like HTML editors, can be used to facilitate creating the VRML worlds without learning all of the formatting rules, but nothing more than a text editor is required to create a VRML world.

Today, VRML applications on the Web are limited by the computing power required to make a good simulation. Because the browser must render the model in real time to create a different view each time the user changes location, a very complex model may take too much time to download to be practical. The popularity of the Web and the speed with which applications are being developed for it, however, suggest that this limitation will not last for long.

INDUSTRY CASE

Robert Reisinger, Mountain Cycle

Animating Your Models: If a Picture Is Worth a Thousand Words . . .

"We are a visual culture," explained Robert Reisinger, CEO, founder, and lead design engineer for Mountain Cycle®, "so why not use the data in our solid models in a variety of formats to disseminate information?"

Inspired by motorcycle design and his own experience as a motocross racer, Robert Reisinger earned an engineering degree and founded Mountain Cycle to produce an innovative line of high-performance bicycles. Instead of the tubular structures commonly used for bicycle frames, Reisinger's designs use "monocoque" frames with free-flowing lines formed of sheet aluminum. (See Figure 13.34.) Other innovations include disc brakes and suspension systems for both front and rear wheels.

From the outset, Mountain Cycle has been a 100% digital company. Reisinger's early work experience included CNC programming, which helped him to capitalize on the benefits of 100% first-fit rates through the use of CAD/CAM with his first products, the Suspenders™ suspension fork and ProStop™ hydraulic disc brake. These were the bicycle industry's first inverted front suspension fork system and hydraulic disc brake, which were followed by Reisinger's first frameset, the San Andreas™. The San Andreas, Suspenders, and ProStop were designed in CadKey and the files were exported to Mastercam to generate the CNC tool paths for the parts. For his eighth new design, the ShockWave™, Reisinger made the switch to solid modeling and SolidWorks. All of the ShockWave parts were modeled and assembled in SolidWorks, then exported to Mastercam.

Having invested in the solid models, Reisinger was inspired to invest in another piece of software designed to work with SolidWorks: Interactive Product Animator (or IPA) from Immersive Design. The ease with which animations could be created from the SolidWorks models, the popularity of visual information on the Web and CD-ROM, and Reisinger's insight into human nature all contributed to his vision of how a twenty-first-century company would disseminate information: "If a picture is worth a thousand words, an animation is worth a lifetime of stories. A picture *is* worth a thousand words,

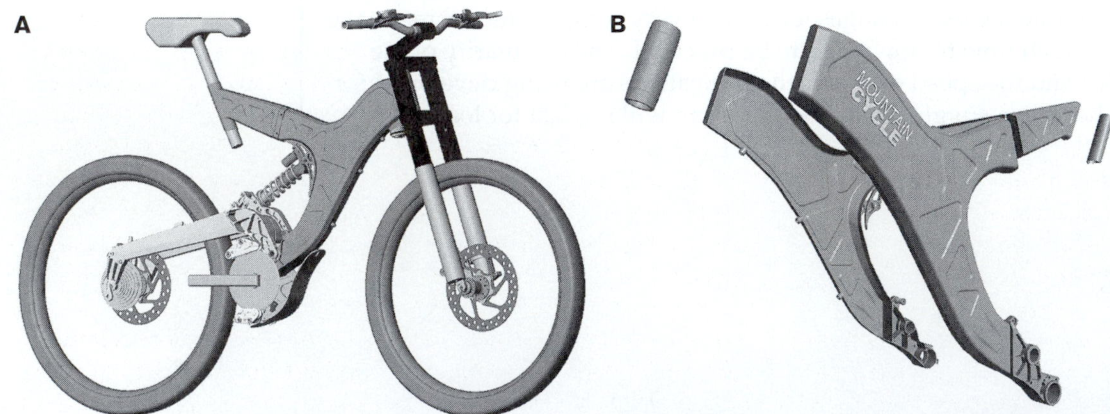

FIGURE 13.34
The ShockWave's distinctive look derives from its welded aluminum frame. Each section of the frame is molded in two parts that are welded together; (b) shows the two halves of the main frame coming together in the first step of the asembly animation. Fixtures are used in the shop to ensure that the frame sections are fitted and welded accurately.

ShockWave™ CABLE ROUTING

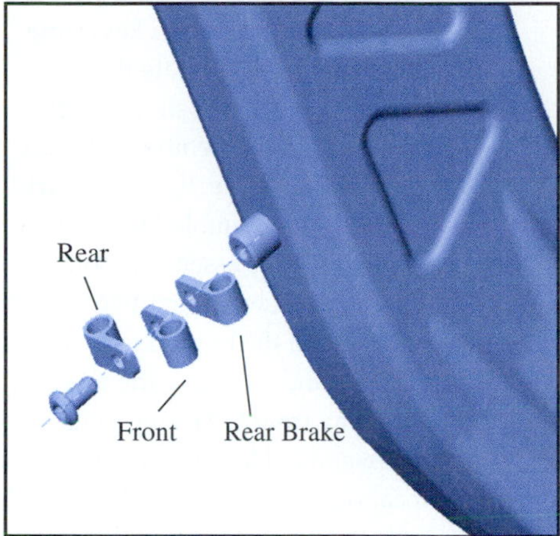

Figure 1.1

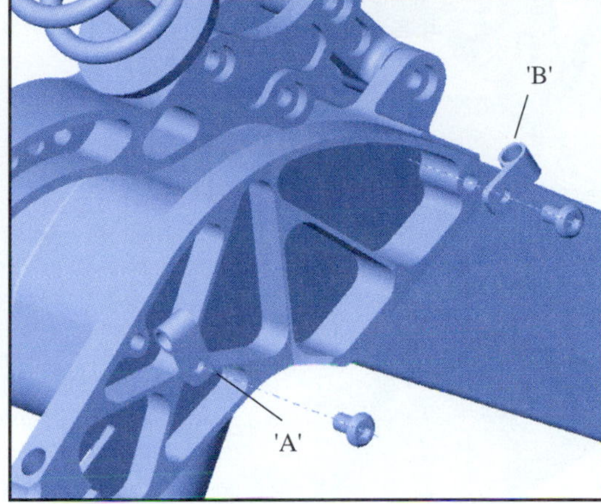

Figure 1.2

FIGURE 13.35
The steps in routing the brake and derailleur cables for the ShockWave were illustrated with these exploded views in the printed Owner's Manual. Callouts and written instructions explain the pictures and how the parts are to be assembled.

but watching something move into place gets that *sound of learning*—it's the slap on the forehead that says 'Oh, I see how that goes.' Animation attracts the mind's eye. We are a culture that likes to watch things—getting someone to flip through pages of a manual or a set-up sheet is *much* harder."

Reisinger's vision of the future included animated information that is disseminated externally—to customers and dealers—as well as internally, to illustrate part information in the database and to document assembly procedures.

Mountain Cycle first tackled external communication. "When I buy a new product, I want to understand the *intent* of the product—how the designer intended the user to interact with it or what a design was supposed to do. This desire burns in many people, but maybe not enough to read through a jargon-filled manual that leaves them saying 'What does that mean?' Our marketing goal is to be able to show the ShockWave's full intent by illustrating its full range of motion, then show the accessory items and how they go on and off the bike, and finally show its maintenance characteristics and how to fully maintain the bike."

To that end, they set about creating an animated user guide for the ShockWave. The printed Owner's Manual was fifteen pages of text and pictures. "How many jokes have been written about the hieroglyphic instructions for a child's bike?" asked Reisinger. "It's not easy sometimes to interpret the pictures." The ShockWave's brake line routing, for example, involves six small parts that must be assembled in a specified order. An exploded view of the assembly, shown in Figure 13.35, served to illustrate this process in the printed guide. The animated guide would not only illustrate the order of operations, but also show the assembly at each stage as parts are added.

Reisinger's team, in partnership with Immersive Design, first animated the ShockWave's complete assembly sequence. Greg Smith, founder and CEO of Immersive Design, explained how they began: "The current manual served as the storyboard for the animation; we didn't need more than the order of the steps documented there. Because IPA lets you create animation segments from longer animations, we captured the full process, knowing we could then break it down into smaller 15-second animations for each step in the process." The complete animation sequence was played in the booth at a trade show where it attracted a great deal of attention and confirmed the appeal of an animated manual.

IPA is a stand-alone animation package, but it can also be used as

(continued page 598)

INDUSTRY CASE

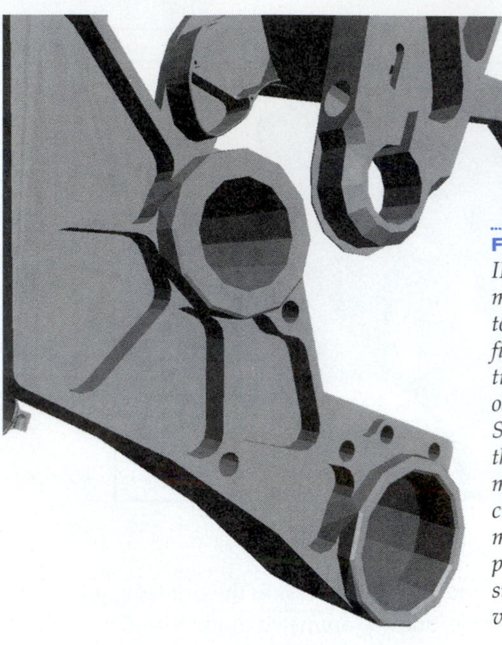

FIGURE 13.36
IPA converts the solid models to meshed surfaces to reduce the size of the files and speed up operations with the model objects. This view of the ShockWave frame reveals the facets used to approximate the flowing lines and curved surfaces in the model. The rendering process can smooth the surface so the facets are not visible in the animation.

an "add-on" to SolidWorks, Pro/Engineer, Solid Edge, and I-DEAS. The term "add-on" refers to the way the product adds capability to the solid modeling package. The Mountain Cycle team installed IPA into their SolidWorks environment, so it could be launched from the SolidWorks menu. In this configuration, IPA used the solids database created by the modeling package and eliminated the need to export the solids into a format that would be usable for animation purposes. It also allowed associativity between the objects in the animation and the solid model so that the animation sequence could be updated automatically whenever changes were made to the solid model.

Although it may appear that IPA is working with the solid model, in fact a data conversion does take place behind the scenes. "Like most animation packages, IPA requires only a fraction of the data in a solid model," explained Greg Smith. "Material properties and data required to define the volume of a solid are not needed for the 2D images in an animation. To reduce the size of the files—and make the animations more transportable—the solid geometry is converted to polygonal surface models of the objects in the assembly. A 35 MB assembly file might translate to a 1.5 MB file in IPA." Like the surfaces exported for stereolithography, the surfaces used by IPA are faceted or tessellated surfaces only, as shown in Figure 13.36. These faceted surfaces can be rendered to varying degrees of smoothness in the animation.

To create the animation, the Mountain Cycle team had to create motion paths for the objects in the assembly. A motion path specifies the locations where a part will start, move to, and end. Each point on the motion path is a keyframe for that object. To create the motion path for the shock shaft and spring subassembly in Figure 13.37, for example, the team started from the assembled location of the parts in the assembly, point X. This location was identified as one endpoint of the motion path. After clicking and dragging the assembly away from the frame to its disassembled location, the new location was defined as the second endpoint, or the keyframe for location Y. This created a motion path from X to Y that is stored with the object. Like any keyframe animator, IPA uses tweening to locate the assembly in the frames between the keyframes.

This example illustrates one aspect of IPA that made it easy to animate the ShockWave assembly: IPA retains the hierarchy of the parts and subassemblies in the SolidWorks assembly. (See Figure 13.38.) As in SolidWorks, the parts in IPA are related by the assembly constraints reflected in the model tree. Because parent-child relationships in the assembly model are preserved, it is easy to animate a group of parts in one step; moving a parent part causes all of its children to move with it. The single motion path from X to Y can move the shaft and spring assembly.

For each motion path, attributes of the object other than position can also be animated, such as color, orientation (rotation), texture, or transparency. Transparency can be used to reveal the interior structure of a part, or to cause a part to fade into view. For

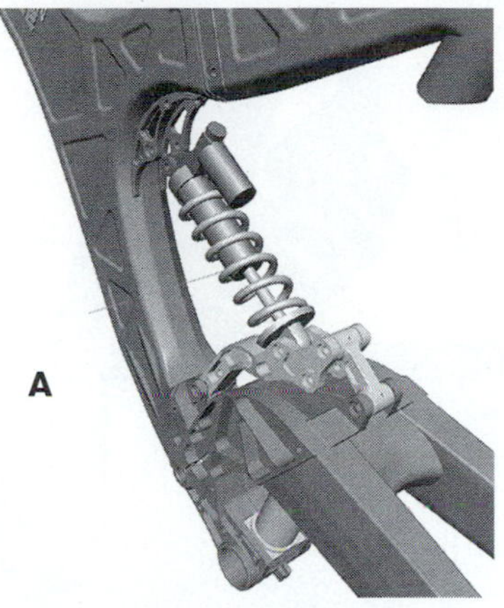

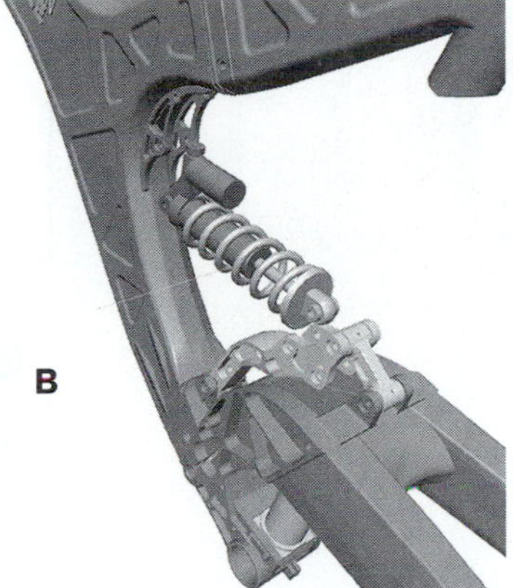

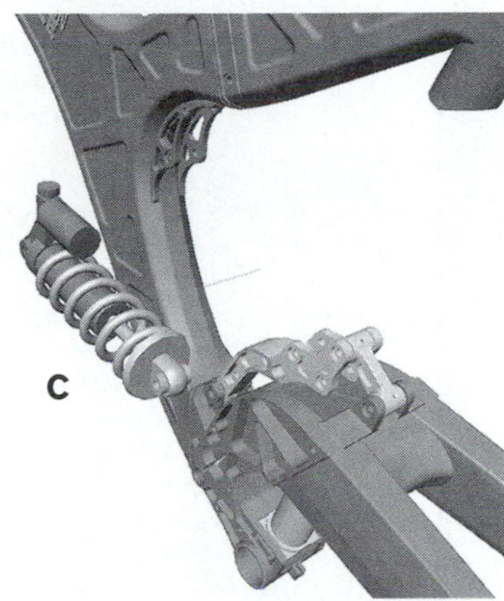

FIGURE 13.37
The first keyframe for the shock shaft and spring subassembly was its assembled position, shown in (a). The assembly was rotated up to a second keyframe, shown in (b), then out to its disassembled location, shown in (c). The motion between these frames was added by IPA to create a smooth transition. Each of the parts in the subassembly also has motion paths that move them out of the subassembly to their fully disassembled position.

the parts in the ShockWave assembly, the transparency for the objects in the first keyframe of the motion sequence was set to transparent. In the last keyframe, the setting was opaque. This caused the part to "appear," then fade to opacity as the part moved into place in the assembly. Any object's attributes can be animated by changing their value in a keyframe of the animation. Lights, backgrounds, and views also have attributes that can be animated in the same way.

The speed of the move (and fade) is determined by the number of frames between the keyframes (and ultimately, the number of

(continued page 600)

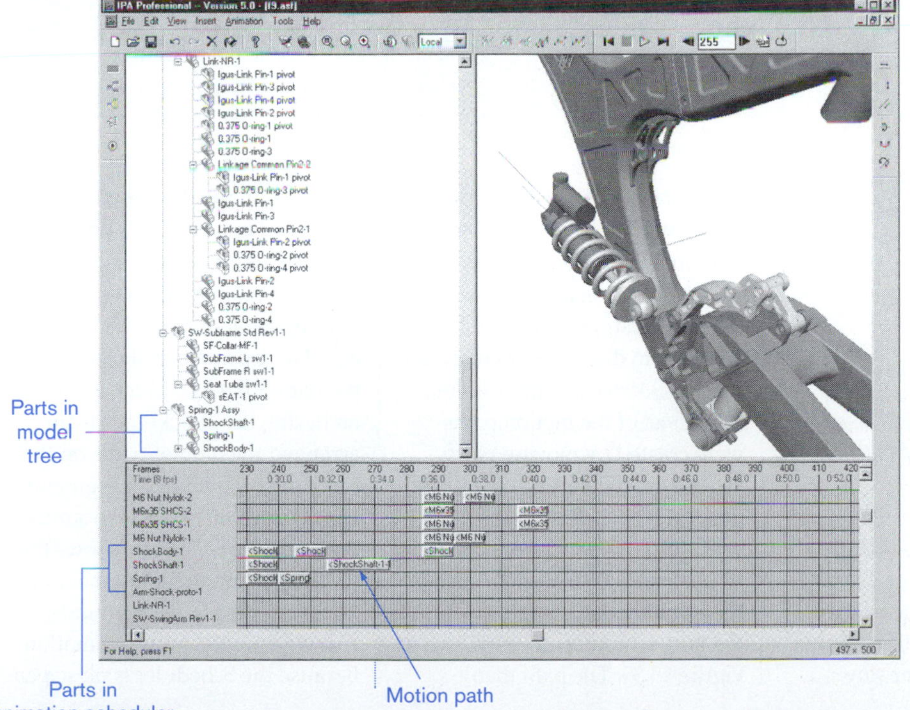

FIGURE 13.38
The relationship of the shock body, shaft, and spring are evident from this view of the animation in IPA. The tree model on the left hand side of the screen shows that the Shock Shaft, Spring, and ShockBody are children of the Spring-1 subassembly. The motion paths for the subassembly and the individual parts are arranged in the Scheduler at the bottom of the screen in order of their "appearance" in the animation.

INDUSTRY CASE

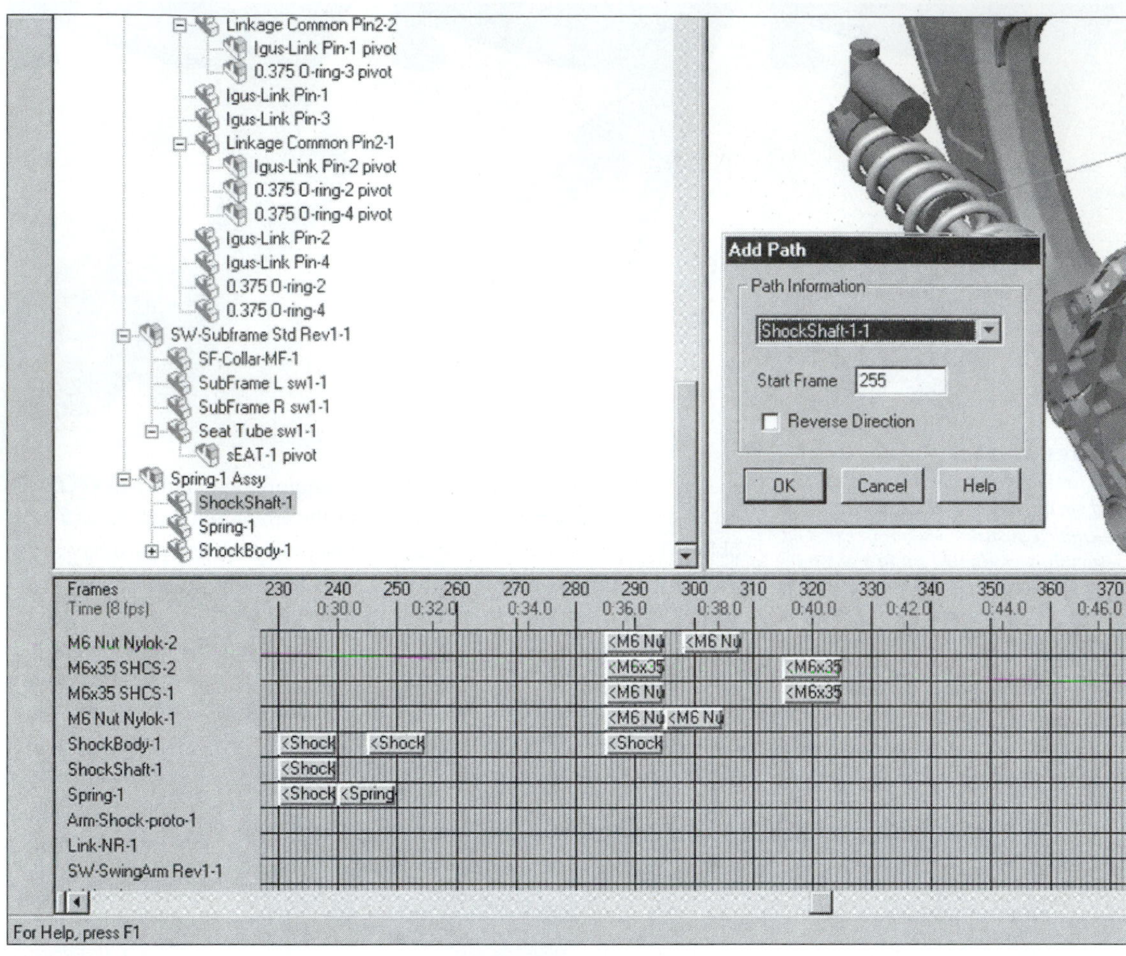

FIGURE 13.39

The Scheduler window combines the motion paths for various objects with lights and backgrounds to orchestrate the animation. After a motion path is created, the dialog box shown here allows the user to name it, specify its start frame in the animation, and set it to run in reverse. The arrows on the motion path bars in the Scheduler indicate whether the motion path is to run forward or backward in the animation.

frames per second at which the animation will run). The number of frames is an attribute of the motion path, and can be edited at any time to speed up or slow down the object's motion. Motion paths can also be reversed, so the motion path defined for the shock shaft and spring assembly can be used to illustrate disassembly, or reversed to illustrate assembly. For the ShockWave assembly animation, most of the motion paths were created by moving parts away from their assembled position, then reversed.

As motion paths were created, IPA prompted the team to specify the frame in the animation where the path would begin, as shown in Figure 13.39. The path then appeared in the IPA Scheduler. The Scheduler is similar to a timeline, with each unit equal to a frame in the animation. A motion path is a line segment, or bar, with length equal to the number of frames in the path. Like a conductor with an orchestra, the ShockWave team arranged the motion paths on the timeline to signal when segments would start and stop. Like a musical score, the Scheduler shows the layering of many different objects (including lights, backgrounds, and viewpoints) in the animation. Because the Scheduler is on screen

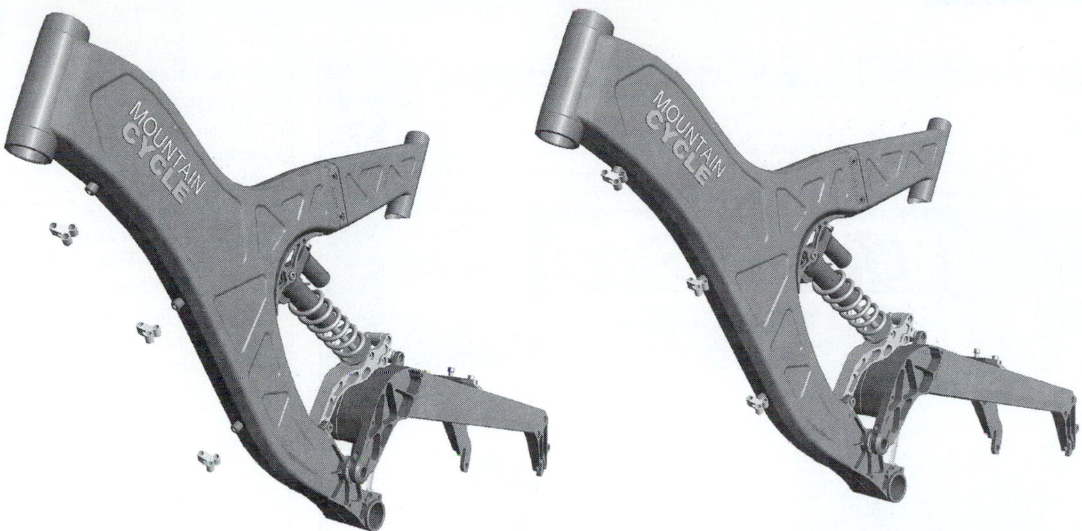

FIGURE 13.40
Because paths have names, they can be duplicated and reused for objects that need to follow the same motion path. The three adel clamps shown here move along the same axis to their assembled positions, so the motion path for one was duplicated for the second and third.

at the same time as the animation and the model tree, the ShockWave team found it easy to play the animation, adjust a motion path, and play the animation again until the desired result was achieved.

Simply setting the motion paths is not enough to create a realistic animation, however. For the guide, the animation team used the assembly sequence to capture the motion and sequence of steps in the animation. Then they studied the ShockWave's appearance to identify attributes that would add realism to the images. They matched the color and shininess of the bike, added decals that appear on the finished bike (as on the main frame in Figure 13.40), and added texture maps to simulate the material used for the seat covering. These attributes can be set for each object before motion paths are created, or may be added or changed at any time.

The completed user guide was rendered to .avi files, a Windows animation format, but it could also have been saved as a VRML or VRML2 file. Smaller sets of steps can also be saved as animated .gif files that will run in any browser. Mountain Cycle wants the guide to be available over the Web to any customer (see Figure 13.41), but will also provide it on CD-ROM to customers (and dealers) who may not have a fast connection to the Internet.

With the animated guide complete, Reisinger plans to use IPA to add visual information to their internal workflows next. He envisions a time when work orders will be linked to animated information that will reduce errors in manufacturing. "When a work order for five ShockWaves in extra-small size is approved by production and transmitted to the shop, the welding area needs to determine which fixture to use for welding the frame and which settings to use for extra-small frames. Some of our fixtures have close to twelve different configurations for different sizes and models. With IPA, we can show them how to manipulate, rotate, disassemble, and put back together the fixture in the correct orientation. When the order comes through, the welder can click to see an animation of the appropriate fixture settings." He plans to use the same information for training purposes to get new employees up to speed more quickly, then expand it to illustrate the entire product database.

"Video-based information makes it easier for people to learn more about our products on the fly. Someone in purchasing who sees the drawing of a part on the network may want to know where it fits, but doesn't have the time to

(continued page 602)

INDUSTRY CASE

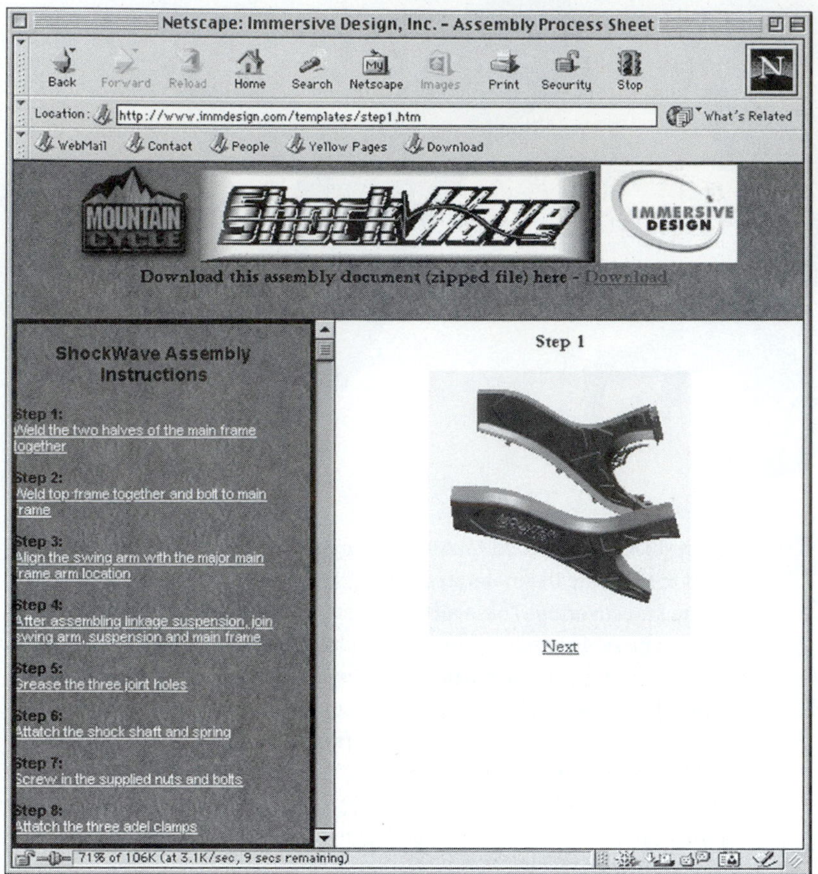

FIGURE 13.41
This Web page shows how the steps in the assembly sequence could be accessed with a common Web browser. The steps in the frame on the left are hyperlinks that correspond to steps in the assembly sequence. Clicking a link loads the animated sequence in the right-hand frame of the page.

gather that knowledge. An illustration that allows the viewer to see it move into place in the assembly makes it easier for everyone to learn more with little effort—and it gets that sound of learning again, 'Oh, that's where that goes.' "

The ShockWave images are provided courtesy of Mountain Cycle, a (d.b.a.) of Reisinger Engineering, Inc., design engineers Robert Reisinger and George Dubois. All other illustrations are used with permission from Immersive Design.

Robert Reisinger is founder, president, and CEO of Mountain Cycle in San Luis Obispo, California. Reisinger graduated with a degree in Manufacturing Engineering Technology from California Polytechnic State University, San Luis Obispo, where he subsequently taught robotics and CNC programming classes. Reisinger had previously worked for Kawasaki Motors, where he helped to design a line of motocross racing bikes. Reisinger's innovative designs and technological advancements have earned his company widespread recognition—including the addition of the San Andreas's frame to the San Francisco Museum of Modern Art's collection.

Skills Summary

Now that you have completed this chapter, you should be able to create four different types of charts and graphs that can be used effectively in presentations. Exercises 1–4 let you practice developing charts and graphs from sample data to make your point graphically. You should also be able to size text so that it is visable in a presentation. Exercises 5–7 ask you to explore different graphics file formats. Exercises 8–9 allow you to demonstrate your understanding of different rendering algorithms and how they affect the appearance of a rendered model. Exercises 10–13 suggest ways for you to incorporate your knowledge of animation into your engineering presentations.

EXERCISES

1. Create a line chart of the following pressure data. Label your chart axes with appropriate variables, and develop a meaningful title. Perform a linear curve-fit of the data and include the fitting equation on the graph. (Most graphing software supports at least linear curve fits.) Format your chart so that it would be useful as a transparency for overhead projection, or for incorporation in a computer-assisted presentation.

Distance from fluid inlet (inches)	Gauge pressure (psi)
0.00	24.2
5.00	22.2
10.00	21.0
15.00	19.1
20.00	17.7
25.00	16.7
30.00	15.0
35.00	14.1
40.00	13.5
45.00	11.5
50.00	9.4

2. Create a spreadsheet based on the information in the following "project budget" pie chart, if the total project budget was $46,700.

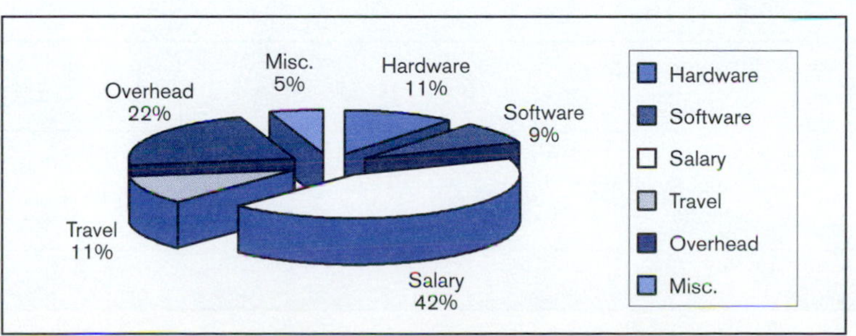

3. Create a pie chart from the following spreadsheet data. Include title and legend. Create a bar chart with the same data. Which chart do you feel better conveys the data?

Wraparound Redwood Deck – Budget

Redwood Decking	4x4 Supports & Corner Posts	Weatherproof Screws	Ready-mix Concrete	Galvanized Brackets	Construction Adhesive	Deck Stain	Total Estimate
$1,000.00	$200.00	$80.00	$140.00	$50.00	$50.00	$250.00	$1,770.00

4. Object linking and embedding (OLE) is supported by many presentation and word processing programs. Explain the difference between embedding and linking. Under what circumstances might either method be preferable?
5. Investigate the graphics exporting capabilities of the modeling software you use. In a spreadsheet, list the available file formats that are supported for exporting data, versus the presentation programs available to you that can utilize each of the file types. Keep this spreadsheet as a handy reference for future presentation work.
6. Choose a CAD model and export a rendered view as a TIF, JPG, GIF, TGA, and EPS file. Examine the resulting file sizes, then use a presentation package such as PowerPoint to create a slide of each image with a heading that identifies it by type.
7. Cut and paste an image from your CAD package to PowerPoint or Word to see the impact of the clipboard's translation of the image to .BMP format. Try this with a wireframe view as well as a rendered view and compare the results. When would this be a useful technique?
8. How does choice of the rendering resolution affect the process of rendering a solid model?
9. Various rendering algorithms are available in solid modeling CAD packages. Reproduce the following table and finish filling it in, using brief explanations of how each of the listed

Rendering Algorithm	How is shading determined?	Reflection and shadowing?	Relative time and memory required?	Other pertinent comments?
Flat Shading	Common colors are calculated for each flat surface or facet			
Gouraud Shading				
Phong Shading	reflections & shadows	Effectively displays		
Ray Tracing		One of the slowest methods	Computationally intensive:	
Radiosity				

algorithms works, recognizing strong points and limitations. A few of the blocks have been filled in for you.

10. Architects and engineers must often convince investors of the desirability of their designs before funding can be obtained. Choose a model you have created, then describe how you would incorporate a "fly-through" animation clip to present your design to an audience of potential investors. "Storyboard" your presentation with 6–10 sketches of the steps of your plan.

11. Keyframing and transforming animation software can save time and effort in creating animated presentations. Storyboard the steps to be used in animating the following cases with use of a keyframing and transformation-capable software package:

a. "Flying" animation of a bird.

b. Images of a juggler's "bowling pin" juggling act.

c. A bicycle rider pedaling along.

d. A fly-through of a proposed kitchen remodel.

e. A logger cutting down a tree.

f. A car crashing into a fallen tree.

12. The use of inverse kinematics techniques applies boundary conditions to models so that motion is limited to realistic ranges. What limits to motion might be included in animations of:

a. A bio-mechanical model of the human knee joint.

b. A pair of channel lock pliers.

c. A three-hole paper punch.

13. Storyboard a presentation of one of your design or research projects first using only slides and still pictures, then using animated sequences, sounds, and/or VRML images. Select one of the presentations to implement using available software tools.

Appendix Contents

1 ANSI Codes and Standards 608
2 Organizations That Publish Drawing Standards 610
3 Abbreviations for Use on Drawings and in Text 614
4 Standard Sheet Sizes and Title Blocks 618
5 Standard Scales 621
6 American Standard Running and Sliding Fits (hole basis) 622
7 American Standard Clearance Locational Fits (hole basis) 624
8 American Standard Transition Locational Fits (hole basis) 626
9 American Standard Interference Locational Fits (hole basis) 627
10 American Standard Force and Shrink Fits (hole basis) 628
11 International Tolerance Grades 629
12 Preferred Hole Basis Clearance Fits–Cylindrical Fits 630
13 Preferred Hole Basis Transition and Interference Fits–Cylindrical Fits 632
14 Preferred Shaft Basis Clearance Fits–Cylindrical Fits 634
15 Preferred Shaft Basis Transition and Interference Fits–Cylindrical Fits 636
16 Hole Sizes for Nonpreferred Diameters (millimeters) 638
17 Shaft Sizes for Nonpreferred Diameters (millimeters) 640
18 Screw Threads: American National and Unified (inches) 642
19 Screw Threads: American National and Unified (inches) 643
20 Screw Threads: American National and Unified (metric) 644
21 Square and Acme Threads 645
22 American Standard Taper Pipe Threads (NPT) 646
23 Square Bolts (inches) 647
24 Square Nuts (inches) 648
25 Hexagon Head Bolts (inches) 649
26 Hex Nuts and Jam Nuts 650
27 Round Head Cap Screws 651
28 Flat Head Cap Screws 652
29 Filister Head Cap Screws 653
30 Flat Socket Head Cap Screws 654
31 Socket Head Cap Screws 655
32 Round Head Machine Screws 656
33 Set Screws 657
34 Twist Drill Sizes 658
35 Cotter Pins: American National Standard 660
36 Straight Pins 661
37 Woodruff Keys 662
38 Standard Keys and Keyseats 663
39 Plain Washers (inches) 664
40 Metric Washers (millimeters) 665
41 Regular Helical Spring Lock Washers (inches and millimeters) 666
42 American National Standard 125-lb Cast Iron Screwed Fittings (inches) 667
43 American National Standard 250-lb Cast Iron Flanged Fittings (inches) 668
44 Welding Symbols 669
45 Piping Symbols, ANSI Standard 672
46 Heating, Ventilation, and Ductwork Symbols, ANSI Standard 673
47 Electronics Symbols, ANSI Standard 676
48 Geometric Dimensioning and Tolerancing Symbols 677
49 IGES Standard 679
50 Useful Formulas for Geometric Entities 682

APPENDIX 1

ANSI Codes and Standards

Name	Designation
Authorized Inspection	QAI
Automotive Lifting Devices	PALD
Boiler and Pressure Vessel Code-1995	
Chains	B29
Compressors	B19
Controls	CSD
Conveyors	B20
Cranes and Hoists	B30, HST
Dimensions	B4, B32, B36
Drawings and Terminology	Y1, Y14, Y32
Elevators and Escalators	A17, QEI
Fasteners	FAP, B18, B27
Flow Management	MFC
Gauge Blanks	B47
Gauges	B40
High Pressure Systems	HPS
Industrial Trucks	B56
Keys	B17
Machine Guarding	B15
Manlifts	A90
Measurement	B89, MC88
Metric System	SI
Nuclear	AG, B16, N278, N509, N510, N626, NOG, NQA, OM, QME
Offshore	SPPE
Operator Qualification & Certification	QHO, QRO
Pallets	MH1
Performance Test Codes	PTC
Piping	A13, B31
Plumbing	A112
Pressure Vessels	PVHO
Pumps	B73
Reinforced Thermostat Plastic Corrosion Resistant Equipment	RTP
Screw Threads	B1
Steel Stacks	STS
Storage Tanks	B96
Surface Quality	B46
Tools	B5, B94, B107
Turbines	B133
Valves, Fittings, Flanges, Gaskets	B16

ANSI Y14 Standards

Name	Designation	Year
Associated Lists	Y14.34M	1997
Castings and Forgings	Y14.8M	1997
Certification of Geometric Dimensioning and Tolerancing Professionals	Y14.5.2	1996
Chassis Frames–Passenger Car and Light Truck Ground Vehicle Practices	Y14.32.1M	1995
Decimal Inch Drawing Sheet Size and Format	Y14.1	1995
Dimensioning and Tolerancing	Y14.5M	1995
Dimensioning and Tolerancing with Mathematical Definition Package	Y14.5M & Y14.5.1M	
Gear and Spline Drawing Standards–Part 2 Bevel and Hypoid Gears	Y14.7.2-78 R84/94	1978
Gear Drawing Standards–Part 1: For Spur, Helical, Double Helical, and Rack	Y14.7.1-71 R88/93	1971
Line Conventions and Lettering	Y14.2M	1993
Mathematical Definition of Dimensioning and Tolerancing Principles	Y14.5.1M	1995
Mechanical Spring Representation	Y14.13M-81 R87/92	1981
Metric Drawing Sheet Size and Format	Y14.1M	1995
Multiview and Sectional View Drawings	Y14.3M	1994
Optical Parts	Y14.18M-86 R93	1987
Pictorial Drawing	Y14.4M-89 R94	1989
Screw Thread Representation	Y14.6-78 R87/93	1978
Surface Texture Symbols	Y14.36M	1996
Technical Report–A Structural Language Format for Basic Shape Description	Y14	1989
Types and Applications of Engineering Drawings	Y14.24M-89 R96	1991

ANSI Y32 Standards

Name	Designation	Year
Graphic Symbols for Fluid Power Diagrams	Y32.10-67 R74/87/94	1967
Graphic Symbols for Heating, Ventilating, and Air Conditioning	Y32.2.4-49 R84/93	1998
Graphics Symbols for Heat-Power Apparatus	Y32.2.6-50 R84/93	1998
Graphic Symbols for Plumbing Fixtures for Diagrams used in Architecture & Building Construction	Y32.4-77 R87/94	1977
Graphic Symbols for Process Flow Diagrams in Petroleum and Chemical Industries	Y32.11-61 R85/93	1998
Graphics Symbols for Railroad Maps and Profiles	Y32.7-72 R79/87/94	1972
Redesignation of A32.2.3-1949 R 1953 Graphic Symbols for Pipe Fittings, Valves, and Piping	Y32.2.3-49 R88/94	1967
Symbols for Mechanical and Accoustical Elements as used in Schematic Diagrams	Y32.18-72 R78/85/93	1998

APPENDIX 2

Organizations That Publish Drawing Standards

AAMA	American Architectural Manufacturers Association	AAMA 1827 Walden Office Square Suite 104 Schaumberg, IL 60173-4628	T 847.303.5664	www.aamanet.org
ACSM	American Congress on Surveying and Mapping	ACSM 5410 Grosvenor Lane Suite 100 Bethesda, MD 20814-2122	T 301.493.0200	www.survmap.org
AGA	American Gas Association	AGA 400 N. Capitol Street, NW Washington, DC 20001	T 202.824.7000	www.aga.org
AGMA	American Gear Manufacturers Association	AGMA 1500 King Street Suite 201 Alexandria, VA 22314-2730	T 703.684.0211	www.agma.org
AHAM	Association of Home Appliance Manufacturers	AHAM 20 N. Wacker Drive Suite 1231 Chicago, IL 60606	T 312.984.5800	www.aham.org
AISC	American Institute of Steel Construction	AISC One E. Wacker Drive Suite 3100 Chicago, IL 60601-2001	T 312.670.5414	www.aisc.org
AMCA	Air Movement and Control Association International	AMCA 30 W. University Drive Arlington Heights, IL 60004-1893	T 847.394.0150	www.amca.org
ANSI	American National Standards Institute	ANSI 11 W. 42nd Street New York, NY 10036	T 212.642.4900	web.ansi.org
ARI	Air-conditioning & Refrigeration Institute	ARI 4301 N. Fairfax Drive Suite 425 Arlington, VA 22203	T 703.524.8800	www.ari.org

ARL	Applied Research Laboratories	ARL S.A. En Vallaire Ouest C Case postale CH-1024 Ecublens SWITZERLAND	T 41.21.694.71.11	www.arl.ch
ASAE	American Society of Agricultural Engineers	ASAE 2950 Niles Road St. Joseph, MI 49085-9659	T 616.429.0300	www.asae.org
ASHRAE	American Society of Heating, Refrigeration, and Air Conditioning Engineers	ASHRAE 1791 Tullie Circle, NE Atlanta, GA 30329	T 404.636.8400	www.ashrae.org
ASME	American Society of Mechanical Engineers	ASME 22 Law Drive Fairfield, NJ 07004	T 800.843.2763 T 973.882.1167	www.asme.org
ASSE	American Society of Sanitary Engineering	ASSE 28901 Clemens Road Suite 100 Westlake, OH 44145	T 440.835.3040	www.asse-plumbing.org
ASTM	American Society for Testing and Materials	ASTM 100 Barr Harbor Drive West Conshohocken, PA 19428-2959	T 215.299.5585	www.astm.org
CBM	Certified Ballast Manufacturers	CBM 355 Lexington Avenue, 17th floor New York, NY 10017-6603	T 212.661.4261	www.certbal.org
CEEA	Canadian Earth Energy Association	CEEA 130 Slater Street, Suite 1050 Ottawa, ON K1P 6E2 CANADA	T 613.230.2332	www.earthenergy.org
CGA	Canadian Gas Association	CGA 243 Consumers Road, Suite 1200 North York, ON M2J 5E3 CANADA	T 416.498.1994	www.cga.ca
CSA	Canadian Standards Association	CSA International 178 Rexdale Blvd. Etobicoke, ON M9W 1R3 CANADA	T 416.747.4000	www.csa.ca
DEMKO	Danmarks Elektriske Materielkontrol	DEMKO Lyskaer 8 2730 Herlev DENMARK	T 45.44.85.65.65	www.demko.dk

DOT	Department of Transportation	US DOT 400 7th Street, SW Washington, DC 20590	T 202.366.9191	www.dot.gov
ETL	ETL Testing Laboratories	ETL Testing Laboratories Intertek Testing Services 3933 US Route 11 Cortland, NY 13045	T 800.967.5352	www.etl.com
FM	Factory Mutual	Factory Mutual 1151 Boston-Providence Tpke P.O. Box 9102 Norwood, MA 02062-9102	T 781.762.4300	www.factorymutual.com
GAMA	Gas Appliance Manufacturers Association	GAMA 1901 N. Moore Street Suite 1100 Arlington, VA 22209	T 703.525.9565	www.gamanet.org
HRAI	Heating, Refrigeration, Air Conditioning Institute of Canada	HRAI 5045 Orbitor Drive, Building 11 Suite 300 Mississauga, ON L4W 4Y4 CANADA	T 800.267.2231	www.hrai.ca
IAPMO	International Association of Plumbing and Mechanical Officials	IAPMO 20001 E. Walnut Drive, S Walnut, CA 91750-2825	T 909.595.8449	www.iapmo.org
IEEE	Institute of Electrical and Electronics Engineers, Inc.	IEEE-USA 1828 L St. NW, Suite 1202 Washington, DC	T 800-678-4333 T 212-419-7900 F 732-981-9667	www.ieee.org
IFI	Industrial Fasteners Institute	IFI 1105 East Ohio Building 171 E. 9th Street Cleveland, OH 44114	T 216.241.1482	www.industrial-fasteners.org
ISO	International Standards Organization	ISO 1, rue de Varembé Case postale 56 CH-1211 Genève 20 SWITZERLAND	T 41.22.749.01.11	www.iso.ch
NEMA	National Electrical Manufacturers Association	NEMA 1300 N. 17th Street Suite 1847 Rosslyn, VA 22209	T 703.841.3200	www.nema.org
NFPA	National Fire Protection Association	NFPA 1 Batterymarch Park Quincy, MA 02269-9101	T 617.770.3000	www.nfpa.org

NIBS	National Institute of Building Sciences	NIBS 1090 Vermont Avenue, NW, Suite 700 Washington, DC 20005-4905	T 202.289.7800	www.nibs.org
NIST	National Institute of Standards and Technology	NIST Public and Business Affairs Administration Building, Room A-903 Gaithersburg, MD 20899-0001	T 301.975.NIST	www.nist.gov
NSF	National Sanitation Foundation International	NSF International 3475 Plymouth Road Ann Arbor, MI 48105	T 313.769.8010	www.nsf.org
OPEI	Outdoor Power Equipment Institute	OPEI 341 S. Patrick Street Alexandria, VA 22314	T 703.549.7600	www.mow.org
OSHA	Occupational Safety and Health Administration	OSHA Public Affairs Office, Room 3647 200 Constitution Avenue Washington, DC 20210	T 202.693.1999	www.osha.gov
SAE	Society of Automotive Engineers	SAE World Headquarters 400 Commonwealth Drive Warrendale, PA 15096-0001	T 724.776.4841	www.sae.org
SEV	Schweiz Elektrotechnischen Vereins	SEV Luppmenstrasse 1 CH-8320 Fehraltorf SWITZERLAND	T 41.1.956.11.11	www.sev.ch
SRCC	Solar Rating and Certification Corporation	SRCC c/o FSEC 1679 Clearlake Road Cocoa, FL 32922-5703	T 407.638.1537	www.solar-rating.org
SSPMA	Sump and Sewage Pump Manufacturers Association	SSPMA P.O. Box 647 Northbrook, IL 60065-9233	T 708.559.9233	
UL	Underwriters Laboratories	UL 333 Pfingsten Road Northbrook, IL 60062-2096	T 847.272.8800	www.ul.com
USDA	United States Department of Agriculture	USDA 14th & Independence Avenue, SW Washington, DC 20250	T 202.720.2791	www.usda.gov
VDE	Verband Deutscher Elektrotechniker	VDE Stresemannallee 15 60596 Frankfurt am Main GERMANY	T 069.6308.0	www.vde.de

APPENDIX 

Abbreviations for Use on Drawings and in Text

Selected from ANSI Y1.1, 1989

A

absolute	ABS
accelerate	ACCEL
accessory	ACCESS.
account	ACCT
accumulate	ACCUM
actual	ACT.
adapter	ADPT
addendum	ADD.
addition	ADD.
adjust	ADJ
advance	ADV
after	AFT.
aggregate	AGGR
air condition	AIR COND
airplane	APL
allowance	ALLOW
alloy	ALY
alteration	ALT
alternate	ALT
alternating current	AC
altitude	ALT
aluminum	AL
American National Standard	AMER NATL STD
American wire gage	AWG
amount	AMT
ampere	AMP
amplifier	AMPL
anneal	ANL
antenna	ANT.
apartment	APT.
apparatus	APP
appendix	APPX
approved	APPD
approximate	APPROX
arc weld	ARC/W
area	A
armature	ARM.
armor plate	ARM-PL
army navy	AN
arrange	ARR.
artificial	ART.
asbestos	ASB
asphalt	ASPH
assemble	ASSEM
assembly	ASSY
assistant	ASST
associate	ASSOC
association	ASSN
atomic	AT
audible	AUD
audio frequency	AF
authorized	AUTH
automatic	AUTO
auto-transformer	AUTO TR
auxiliary	AUX
avenue	AVE
average	AVG
aviation	AVI
azimuth	AZ

B

Babbitt	BAB
back feed	BF
back pressure	BP
back to back	B to B
backface	BF
balance	BAL
ball bearing	BB
barometer	BAR
base line	BL
base place	BP
bearing	BRG
bench mark	BM
bending moment	M
bent	BT
bessemer	BESS
between	BET.
between centers	BC
between perpendiculars	BP
bevel	BEV
bill of material	B/M
Birmingham wire gage	BWG
blank	BLK
block	BLK
blueprint	BP
board	BD
boiler	BLR
boiler feed	BF
boiler horsepower	BHP
boiling point	BP
bolt circle	BC
both faces	BF
both sides	BS
both ways	BW
bottom	BOT
bottom chord	BC
bottom face	BF
bracket	BS
brake	BW
brake horsepower	BHP
brass	BRS
brazing	BRZG
break	BRK
Brinell hardness	BH
British Standard	BR STD
British thermal unit	BTU
broach	BRO
bronze	BRZ
Brown & Sharpe (wire gage, same as AWG)	B&S
building	BLDG
bulkhead	BHD
burnish	BNH
bushing	BUSH.
button	BUT.

C

cabinet	CAB.
calculate	CALC
calibrate	CAL
cap screw	CAP SCR

capacity	CAP
carburetor	CARB
carburize	CARB
carriage	CRG
case harden	CH
cast iron	CI
cast steel	CS
casting	CSTG
castle nut	CAS NUT
catalogue	CAT.
cement	CEM
center	CTR
center line	CL
center of gravity	CG
center of pressure	CP
center to center	C to C
centering	CTR
chamfer	CHAM
change	CHG
channel	CHAN
check	CHK
check valve	CV
chord	CHD
circle	CIR
circular	CIR
circular pitch	CP
circumference	CIRC
clear	CLR
clearance	CL
clockwise	CW
coated	CTD
cold drawn	CD
cold-drawn steel	CDS
cold finish	CF
cold punched	CP
cold rolled	CR
cold-rolled steel	CRS
combination	COMB.
combustion	COMB
commercial	COML
company	CO
complete	COMPL
compress	COMP
concentric	CONC
concrete	CONC
condition	COND
connect	CONN
constant	CONST
construction	CONST
contact	CONT
continue	CONT
copper	COP.
corner	COR
corporation	CORP
correct	CORR
corrugate	CORR
cotter	COT
counter	CTR
counterbore	CBORE
counter clockwise	CCW
counterdrill	CDRILL
counterpunch	CPUNCH
countersink	CSK
coupling	CPLG
cover	COV
cross section	XSECT
cubic	CU
cubic foot	CU FT
cubic inch	CU IN.
current	CUR
customer	CUST
cyanide	CYN

D

decimal	DEC
dedendum	DED
deflect	DEFL
degree	(°) DEG
density	D
department	DEPT
design	DSGN
detail	DET
develop	DEV
diagonal	DIAG
diagram	DIAG
diameter	DIA
diametral pitch	DP
dimension	DIM.
discharge	DISCH
distance	DIST
division	DIV
double	DBL
dovetail	DVTL
dowel	DWL
down	DN
dozen	DOZ
drafting	DFTG
drawing	DWG
drill or drill rod	DR
drive	DR
drive fit	DF
drop	D
drop forge	DF
duplicate	DUP

E

each	EA
east	E
eccentric	ECC
effective	EFF
elbow	ELL
electric	ELEC
elementary	ELEM
elevate	ELEV
elevation	EL
engine	ENG
engineer	ENGR
engineering	ENGRG
entrance	ENT
equal	EQ
equation	EQ
equipment	EQUIP
equivalent	EQUIV
estimate	EST
exchange	EXCH
exhaust	EXH
existing	EXIST.
exterior	EXT
extra heavy	X HVY
extra strong	X STR
extrude	EXTR

F

fabricate	FAB
face to face	F to F
Fahrenheit	F
far side	FS
federal	FED.
feed	FD
feet	(') FT
figure	FIG.
fillet	FIL
fillister	FIL
finish	FIN.
finish all over	FAO
flange	FLG
flat	F
flat head	FH
floor	FL
fluid	FL
focus	FOC
foot	(') FT
force	F
forged steel	FST
forging	FORG
forward	FWD
foundry	FDRY
frequency	FREQ
front	FR
furnish	FURN

G

Gage or gauge	GA
gallon	GAL
galvanize	GALV

galvanized iron	GI
galvanized steel	GS
gasket	GSKT
general	GEN
glass	GL
government	GOVT
governor	GOV
grade	GR
graduation	GRAD
graphite	GPH
grind	GRD
groove	GRV
ground	GRD

H

half-round	1/2RD
handle	HDL
hanger	HGR
hard	H
harden	HDN
hardware	HDW
head	HD
headless	HDLS
heat	HT
heat-treat	HT TR
heavy	HVY
hexagon	HEX
high-pressure	HP
high-speed	HS
horizontal	HOR
horsepower	HP
hot rolled	HR
hot-rolled steel	HRS
hour	HT
housing	HSG
hydraulic	HYD

I

illustrate	ILLUS
inboard	INBD
inch	(") IN.
inches per second	IPS
inclosure	INCL
include	INCL
inside diameter	ID
instrument	INST
interior	INT
internal	INT
intersect	INT
iron	I
irregular	IREG

J

joint	JT
joint army-navy	JAN
journal	JNL
junction	JCT

K

key	K
keyseat	KST
Keyway	KWY

L

laboratory	LAB
laminate	LAM
lateral	LAT
left	L
left hand	LH
length	LG
length over all	LOA
letter	LTR
light	LT
line	L
locate	LOC
logarithm	LOG.
long	LG
lubricate	LUB
lumber	LBR

M

machine	MACH
machine steel	MS
maintenance	MAINT
malleable	MALL
malleable iron	MI
manual	MAN.
manufacture	MFT
manufactured	MFD
manufacturing	MFG
material	MATL
maximum	MAX
mechanical	MECH
mechanism	MECH
median	MED
metal	MET.
meter	M
miles	MI
miles per hour	MPH
millimeter	MM
minimum	MIN
minute	(') MIN
miscellaneous	MISC
month	MO
Morse taper	MOR T
motor	MOT
mounted	MTD
mounting	MTG
multiple	MULT
music wire gage	MWG

N

national	NATL
natural	NAT
near face	NF
near side	NS
negative	NEG
neutral	NEUT
nominal	NOM
normal	NOR
north	NB
not to scale	NTS
number	NO.

O

obsolete	OBS
octagon	OCT
office	OFF.
on center	OC
opposite	OPP
optical	OPT
original	ORIG
outlet	OUT.
outside diameter	OD
outside face	OF
outside radius	OR
overall	OA

P

pack	PK
packing	PKG
page	P
paragraph	PAR.
part	PT
patent	PAT.
pattern	PATT
permanent	PERM
perpendicular	PERP
piece	PC
piece mark	PC MK
pint	PT
pitch	P
pitch circle	PC
pitch diameter	PD
plastic	PLSTC
plate	PL
plumbing	PLMB
point	PT
point of curve	PC
point of intersection	PI
point of tangent	PT
polish	POL
position	POS
potential	POT.
pound	LB
pounds per square inch	PSI

power	PWR
prefabricated	PREFAB
preferred	PFD
prepare	PREP
pressure	PRESS.
process	PROC
production	PROD
profile	PF
propeller	PROP
publication	PUB
push button	PB

Q

quadrant	QUAD
quality	QUAL
quarter	QTR

R

radial	RAD
radius	R
railroad	RR
ream	RM
received	RECD
record	REC
rectangle	RECT
reduce	RED.
reference line	REF L
reinforce	REINF
release	REL
relief	REL
remove	REM
require	REQ
required	REQD
return	RET.
reverse	REV
revolution	REV
revolutions per minute	RPM
right	R
right hand	RH
rivet	RIV
Rockwell hardness	RH
roller bearing	RB
room	RM
root diameter	RD
root mean square	RMS
rough	RGH
round	RD

S

schedule	SCH
schematic	SCHEM
scleroscope hardness	SH
screw	SCR
second	SEC
section	SECT
semi-steel	SS
separate	SEP
set screw	SS
shaft	SFT
sheet	SH
shoulder	SHLD
side	S
single	S
sketch	SK
sleeve	SLV
slide	SL
slotted	SLOT.
small	SM
socket	SOC
space	SP
special	SPL
specific	SP
spot faced	SF
spring	SPG
square	SQ
standard	STD
station	STA
stationary	STA
steel	STL
stock	STK
straight	STR
street	ST
structural	STR
substitute	SUB
summary	SUM.
support	SUP.
surface	SUR
symbol	SYM
system	SYS

T

tangent	TAN.
taper	TPR
technical	TECH
template	TEMP
tension	TENS
terminal	TERM.
thick	THK
thousand	M
thread	THD
threads per inch	TPI
through	THRU
time	T
tolerance	TOL
tongue & groove	T & G
tool steel	TS
tooth	T
total	TOT
transfer	TRANS
typical	TYP

U

ultimate	ULT
unit	U
universal	UNIV

V

vacuum	VAC
valve	V
variable	VAR
versus	VS
vertical	VERT
volt	V
volume	VOL

W

wall	W
washer	WASH.
watt	W
week	WK
weight	WT
west	W
width	W
wood	WD
Woodruff	WDF
working point	WP
working pressure	WP
wrought	WRT
wrought iron	WI

Y

yard	YD
year	YR

APPENDIX 4

Standard Sheet Sizes and Title Blocks

ANSI Standard Sheet Sizes (for Inch Sheets)

Designation	Sheet size in inches
A	11 x 8.5
B	17 x 11
C	22 x 17
D	34 x 22
E	44 x 34

ISO Standard Sheet Sizes

Designation	Metric sheet size (in mm)	English equivalent sheet (in inches)
AO	1189 x 841	44 x 34 (E)
A1	841 x 594	34 x 22 (D)
A2	594 x 420	22 x 17 (C)
A3	420 x 297	17 x 11 (B)
A4	297 x 210	11 x 8.5 (A)

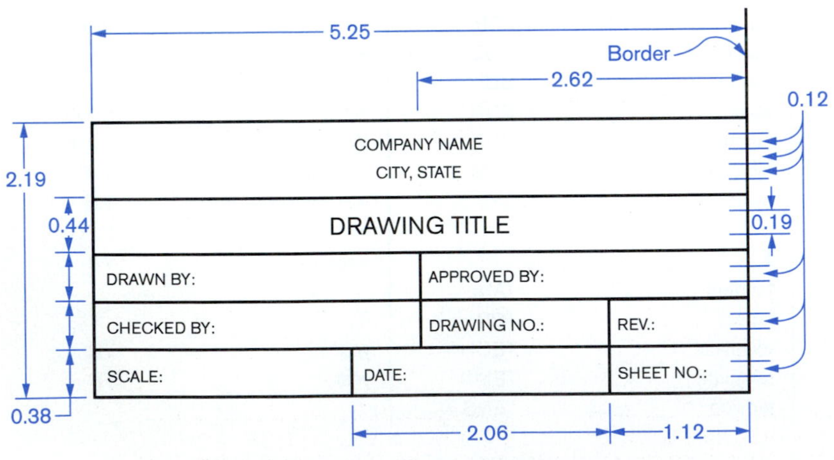

FIGURE A.1
Typical Title Block for Size A Sheet

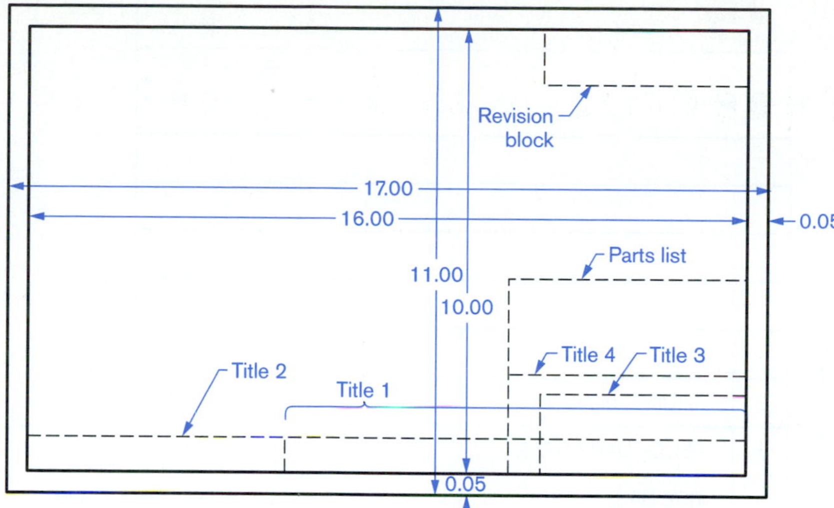

FIGURE A.2
Size B Sheet (17 x 11 inches)

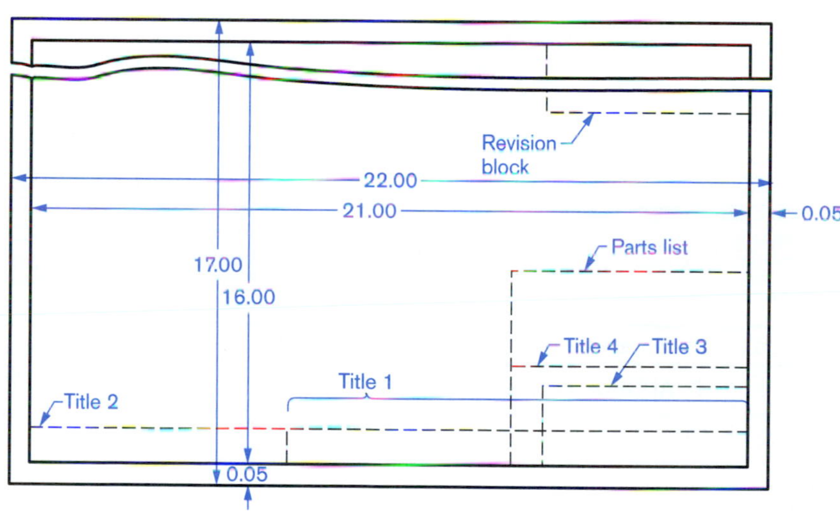

FIGURE A.3
Size C Sheet (22 x 17 inches)

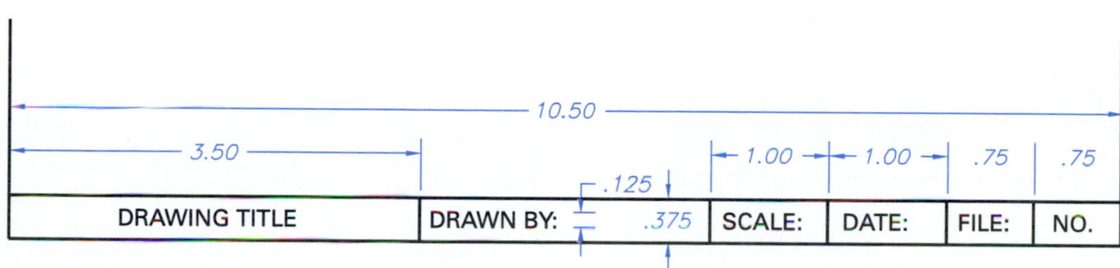

FIGURE A.4
Title Strip for Size A Sheet

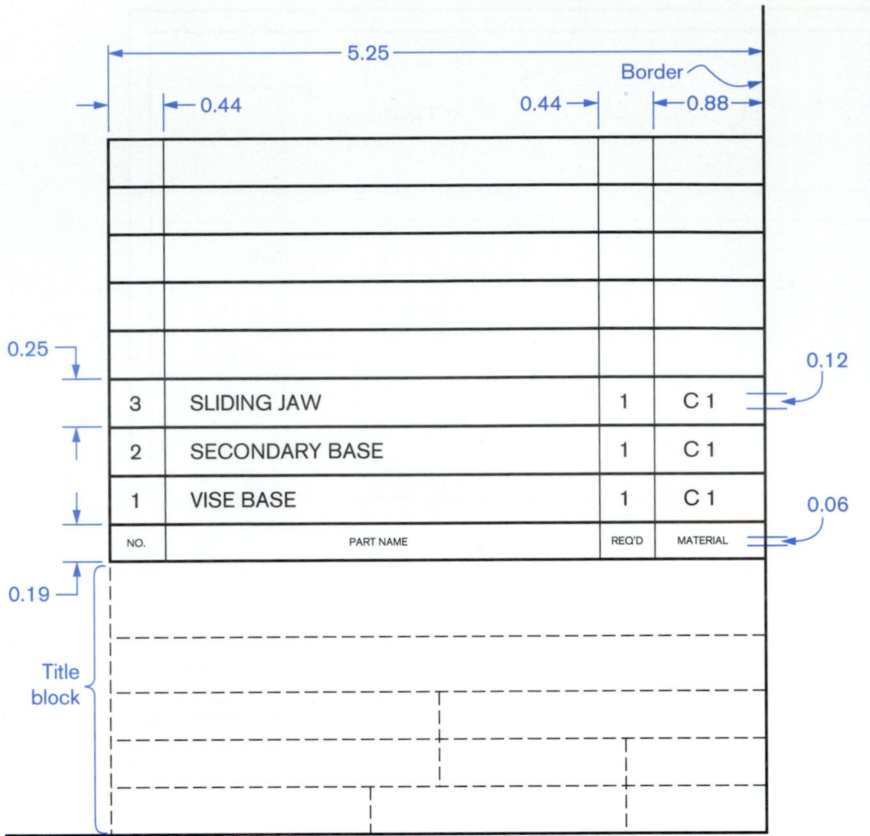

FIGURE A.5
Typical Parts List or Material List

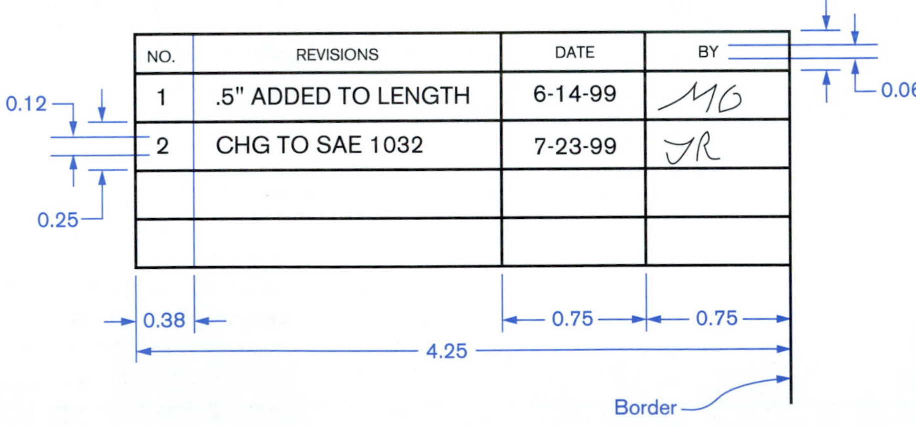

FIGURE A.6
Typical Revision Block

APPENDIX 5

Standard Scales

Metric Scales	
1:1	Full
1:2	Half
1:5	Fifth
1:10	Tenth
1:20	Twentieth
1:100	Hundredth
2:1	Double

Metric Scales for Civil and Mapping *(for mapping the units are typically meters)*	
1:100	One plotted unit equals one hundred real world units
1:200	One plotted unit equals two hundred real world units
1:500	One plotted unit equals five hundred real world units
1:1000	One plotted unit equals one thousand real world units
1:2000	One plotted unit equals two thousand real world units
1:5000	One plotted unit equals five thousand real world units
1:10000	One plotted unit equals ten thousand real world units

English Scales	
1 = 1	Full
1 = 2	Half
1 = 4	Quarter
1 = 8	Eighth
1 = 10	Tenth
2 = 1	Double

English Scales for Civil and Mapping	
1" = 10'	One inch equals 10 feet
1" = 20'	One inch equals 20 feet
1" = 50'	One inch equals 50 feet
1" = 100'	One inch equals 100 feet
1" = 200'	One inch equals 200 feet
1" = 400'	One inch equals 400 feet
1" = 500'	One inch equals 500 feet
1" = 1000'	One inch equals 1000 feet
1:24000 or 1" = 2000'	One inch equals 2000 feet
1:62500 or 1" = 1 mile	One inch equals 1 mile

English Scales for Building Construction	
1" = 10'	One inch equals 10 feet
1" = 20'	One inch equals 20 feet
1" = 50'	One inch equals 50 feet
1" = 100'	One inch equals 100 feet
1" = 200'	One inch equals 200 feet
1" = 400'	One inch equals 400 feet
1" = 500'	One inch equals 500 feet
1" = 1000'	One inch equals 1000 feet

APPENDIX 

American Standard Running and Sliding Fits (hole basis)

Limits are in thousandths of an inch.
Limits for hole and shaft are applied algebraically to the basic size to obtain the limits of size for the parts.
Data in bold face are in accordance with ABC agreements.
Symbols H5, g5, etc., are Hole and Shaft designations used in ABC System.

Nominal Size Range Inches		Class RC1			Class RC 2			Class RC 3			Class RC4		
		Limits of Clearance	Standard Limits		Limits of Clearance	Standard Limits		Limits of Clearance	Standard Limits		Limits of Clearance	Standard Limits	
Over	To		Hole H5	Shaft g4		Hole H6	Shaft g5		Hole H7	Shaft f6		Hole H8	Shaft f7
0	−0.12	0.1	+0.2	−0.1	0.1	+0.25	−0.1	0.3	+0.4	−0.3	0.3	+0.6	−0.3
		0.45	0	−0.25	0.55	0	−0.3	0.95	0	−0.55	1.3	0	−0.7
0.12	−0.24	0.15	+0.2	−0.15	0.15	+0.3	−0.15	0.4	+0.5	−0.4	0.4	+0.7	−0.4
		0.5	0	−0.3	0.65	0	−0.35	1.12	0	−0.7	1.6	0	−0.9
0.24	−0.40	0.2	0.25	−0.2	0.2	+0.4	−0.2	0.5	+0.6	−0.5	0.5	+0.9	−0.5
		0.6	0	−0.35	0.85	0	−0.45	1.5	0	−0.9	2.0	0	−1.1
0.40	−0.71	0.25	+0.3	−0.25	0.25	+0.4	−0.25	0.6	+0.7	−0.6	0.6	+1.0	−0.6
		0.75	0	−0.45	0.95	0	−0.55	1.7	0	−1.0	2.3	0	−1.3
0.71	−1.19	0.3	+0.4	−0.3	0.3	+0.5	−0.3	0.8	+0.8	−0.8	0.8	+1.2	−0.8
		0.95	0	−0.55	1.2	0	−0.7	2.1	0	−1.3	2.8	0	−1.6
1.19	−1.97	0.4	+0.4	−0.4	0.4	+0.6	−0.4	1.0	+1.0	−1.0	1.0	+1.6	−1.0
		1.1	0	−0.7	1.4	0	−0.8	2.6	0	−1.6	3.6	0	−2.0
1.97	−3.15	0.4	+0.5	−0.4	0.4	+0.7	−0.4	1.2	+1.2	−1.2	1.2	+1.8	−1.2
		1.2	0	−0.7	1.6	0	−0.9	3.1	0	−1.9	4.2	0	−2.4
3.15	−4.73	0.5	+0.6	−0.5	0.5	+0.9	−0.5	1.4	+1.4	−1.4	1.4	+2.2	−1.4
		1.5	0	−0.9	2.0	0	−1.1	3.7	0	−2.3	5.0	0	−2.8
4.73	−7.09	0.6	+0.7	−0.6	0.6	+1.0	−0.6	1.6	+1.6	−1.6	1.6	+2.5	−1.6
		1.8	0	−1.1	2.3	0	−1.3	4.2	0	−2.6	5.7	0	−3.2
7.09	−9.85	0.6	+0.8	−0.6	0.6	+1.2	−0.6	2.0	+1.8	−2.0	2.0	+2.8	−2.0
		2.0	0	−1.2	2.6	0	−1.4	5.0	0	−3.2	6.6	0	−3.8
9.85	−12.41	0.8	+0.9	−0.8	0.8	+1.2	−0.8	2.5	+2.0	−2.5	2.5	+3.0	−2.5
		2.3	0	−1.4	2.9	0	−1.7	5.7	0	−3.7	7.5	0	−4.5
12.41	−15.75	1.0	+1.0	−1.0	1.0	+1.4	−1.0	3.0	+	−3.0	3.0	+3.5	−3.0
		2.7	0	−1.7	3.4	0	−2.0	6.6	0	−4.4	8.7	0	−5.2
15.75	−19.69	1.2	+1.0	−1.2	1.2	+1.6	−1.2	4.0	+1.6	−4.0	4.0	+4.0	−4.0
		3.0	0	−2.0	3.8	0	−2.2	8.1	0	−5.6	10.5	0	−6.5
19.69	−30.09	1.6	+1.2	−1.6	1.6	+2.0	−1.6	5.0	+3.0	−5.0	5.0	+5.0	−5.0
		3.7	0	−2.5	4.8	0	−2.8	10.0	0	−7.0	13.0	0	−8.0
30.09	−41.49	2.0	+1.6	−2.0	2.0	+2.5	−2.0	6.0	+4.0	−6.0	6.0	+6.0	−6.0
		4.6	0	−3.0	6.1	0	−3.6	12.5	0	−8.5	16.0	0	−10.0
41.49	−56.19	2.5	+2.0	−2.5	2.5	+3.0	−2.5	8.0	+5.0	−8.0	8.0	+8.0	−8.0
		5.7	0	−3.7	7.5	0	−4.5	16.0	0	−11.0	21.0	0	−13.0
56.19	−76.39	3.0	+2.5	−3.0	3.0	+4.0	−3.0	10.0	+6.0	−10.0	10.0	+10.0	−10.0
		7.1	0	−4.6	9.5	0	−5.5	20.0	0	−14.0	26.0	0	−16.0
76.39	−100.9	4.0	+3.0	−4.0	4.0	+5.0	−4.0	12.0	+8.0	−12.0	12.0	+12.0	−12.0
		9.0	0	−6.0	12.0	0	−7.0	25.0	0	−17.0	32.0	0	−20.0
100.9	−131.9	5.0	+4.0	−5.0	5.0	+6.0	−5.0	16.0	+10.0	−16.0	16.0	+16.0	−16.0
		11.5	0	−7.5	15.0	0	−9.0	32.0	0	−22.0	36.0	0	−26.0
131.9	−171.9	6.0	+5.0	−6.0	6.0	+8.0	−6.0	18.0	+8.0	−18.0	18.0	+20.0	−18.0
		14.0	0	−9.0	19.0	0	−11.0	38.0	0	−26.0	50.0	0	−30.0
171.9	−200	8.0	+6.0	−8.0	8.0	+10.0	−8.0	22.0	+16.0	−22.0	22.0	+25.0	−22.0
		18.0	0	−12.0	22.0	0	−12.0	48.0	0	−32.0	63.0	0	−38.0

Class RC 5			Class RC 6			Class RC 7			Class RC 8			Class RC 9			Nominal Size Range Inches	
Limits of Clearance	Standard Limits		Limits of Clearance	Standard Limits		Limits of Clearance	Sandard Limits		Limits of Clearance	Standard Limits		Limits of Clearance	Standard Limits			
	Hole H8	Shaft e7		Hole H9	Shaft e8		Hole H9	Shaft d8		Hole H10	Shaft c9		Hole H11	Shaft	Over	To
0.6	+0.6	−0.6	0.6	+1.0	−0.6	1.0	+1.0	−1.0	2.5	+1.6	−2.5	4.0	+2.5	−4.0	0	0.12
1.6	−0	−1.0	2.2	−0	−1.2	2.6	0	−1.6	5.1	0	−3.5	8.1	0	−5.6		
0.8	+0.7	−0.8	0.8	+1.2	−0.8	1.2	+1.2	−1.2	2.8	+1.8	−2.8	4.5	+3.0	−4.5	0.12	0.24
2.0	−0	−1.3	2.7	−0	−1.5	3.1	0	−1.9	5.8	0	−4.0	9.0	0	−6.0		
1.0	+0.9	−1.0	1.0	+1.4	−1.0	1.6	+1.4	−1.6	3.0	+2.2	−3.0	5.0	+3.5	−5.0	0.24	0.40
2.5	−0	−1.16	3.3	−0	−1.9	3.9	0	−2.5	6.6	0	−4.4	10.7	0	−7.2		
1.2	+1.0	−1.2	1.2	+1.6	−1.2	2.0	+1.6	−2.0	3.5	+2.8	−3.5	6.0	+4.0	−6.0	0.40	0.71
2.9	−0	−1.9	3.8	−0	−2.2	4.6	0	−3.0	7.9	0	−5.1	12.8	−0	−8.8		
1.6	+1.2	−1.6	1.6	+2.0	−1.6	2.5	+2.0	−2.5	4.5	+3.5	−4.5	7.0	+5.0	−7.0	0.71	1.19
3.6	−0	−2.4	4.8	−0	−2.8	5.7	0	−3.7	10.0	0	−6.5	15.5	0	−10.5		
2.0	+1.6	−2.0	2.0	+2.5	−2.0	3.0	+2.5	−3.0	5.0	+4.0	−5.0	8.0	+6.0	−8.0	1.19	1.97
4.6	−0	−3.0	6.1	−0	−3.6	7.1	0	−4.6	11.5	0	−7.5	18.0	0	−12.0		
2.5	+1.8	−2.5	2.5	+3.0	−2.5	4.0	+3.0	−4.0	6.0	+4.5	−6.0	9.0	+7.0	−9.0	1.97	3.15
5.5	−0	−3.7	7.3	−0	−4.3	8.8	0	−5.8	13.5	0	−9.0	20.5	0	−13.5		
3.0	+2.2	−3.0	3.0	+3.5	−3.0	5.0	+3.5	−5.0	7.0	+5.0	−7.0	10.0	+9.0	−10.0	3.15	4.73
6.6	−0	−4.4	8.7	−0	−5.2	10.7	0	−7.2	15.5	0	−10.5	24.0	0	−15.0		
3.5	+2.5	−3.5	3.5	+4.0	−3.5	6.0	+4.0	−6.0	8.0	+6.0	−8.0	12.0	+10.0	−12.0	4.73	7.09
7.6	−0	−5.1	10.0	−0	−6.0	12.5	0	−8.5	18.0	0	−12.0	28.0	0	−18.0		
4.0	+2.8	−4.0	4.0	+4.5	−4.0	7.0	+4.5	−7.0	10.0	+7.0	−10.0	15.0	+12.0	−15.0	7.09	9.85
8.6	−0	−5.8	11.3	0	−6.8	14.3	0	−9.8	21.5	0	−14.5	34.0	0	−22.0		
5.0	+3.0	−5.0	5.0	+5.0	−5.0	8.0	+5.0	−8.0	12.0	+8.0	−12.0	18.0	+12.0	−18.0	9.85	12.41
10.0	0	−7.0	13.0	0	−8.0	16.0	0	−11.0	25.0	0	−17.0	38.0	0	−26.0		
6.0	+3.5	−6.0	6.0	+6.0	−6.0	10.0	+6.0	−10.0	14.0	+9.0	−14.0	22.0	+14.0	−22.0	12.41	15.75
11.7	0	−8.2	15.5	0	−9.5	19.5	0	13.5	29.0	0	−20.0	45.0	0	−31.0		
8.0	+4.0	−8.0	8.0	+6.0	−8.0	12.0	+6.0	−12.0	16.0	+10.0	−16.0	25.0	+16.0	−25.0	15.75	19.69
14.5	0	−10.5	18.0	0	−12.0	22.0	0	−16.0	32.0	0	−22.0	51.0	0	−35.0		
10.0	+5.0	−10.0	10.0	+8.0	−10.0	16.0	+8.0	−16.0	20.0	+12.0	−20.0	30.0	+20.0	−30.0	19.69	30.09
18.0	0	−13.0	23.0	0	−15.0	29.0	0	−21.0	40.0	0	−28.0	62.0	0	−42.0		
12.0	+6.0	−12.0	12.0	+10.0	−12.0	20.0	+10.0	−20.0	25.0	+16.0	−25.0	40.0	+25.0	−40.0	30.09	41.49
22.0	0	−16.0	28.0	0	−18.0	36.0	0	−26.0	51.0	0	−35.0	81.0	0	−56.0		
16.0	+8.0	−16.0	16.0	+12.0	−16.0	25.0	+12.0	−25.0	30.0	+20.0	−30.0	50.0	+30.0	−50.0	41.49	56.19
29.0	0	−21.0	36.0	0	−24.0	45.0	0	−33.0	62.0	0	−42.0	100	0	−70.0		
20.0	+10.0	−20.0	20.0	+16.0	−20.0	30.0	+16.0	−30.0	40.0	+25.0	−40.0	60.0	+40.0	−60.0	56.19	76.39
36.0	0	−26.0	46.0	0	−30.0	56.0	0	−40.0	81.0	0	−56.0	125	0	−85.0		
25.0	+12.0	−25.0	25.0	+20.0	−25.0	40.0	+20.0	−40.0	50.0	+30.0	−50.0	80.0	+50.0	−80.0	76.39	100.9
45.0	0	−33.0	57.0	0	−37.0	72.0	0	−52.0	100	0	−70.0	160	0	−110		
30.0	+16.0	−30.0	30.0	+35.0	−30.0	50.0	+25.0	−50.0	60.0	+40.0	−60.0	100	+60.0	−100	100.9	131.9
56.0	0	−40.0	71.0	0	−46.0	91.0	0	−66.0	125	0	−85.0	200	0	−140		
35.0	+20.0	−35.0	35.0	+30.0	−35.0	60.0	+30.0	−60.0	80.0	+50.0	−80.0	130	+80.0	−130	131.9	171.9
57.0	0	−47.0	85.0	0	−55.0	110.0	0	−80.0	160	0	−110	260	0	−180		
45.0	+25.0	−45.0	45.0	+40.0	−45.0	80.0	+40.0	−80.0	100	+60.0	−100	150	+100	−150	171.9	200
86.0	0	−61.0	110.0	0	−70.0	145.0	0	−105.0	200	0	−140	310	0	−210		

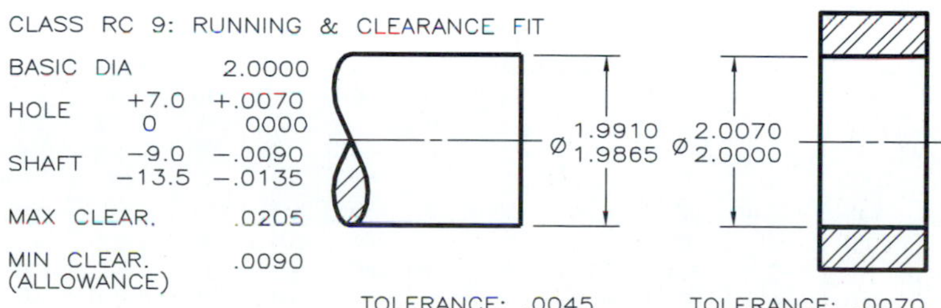

APPENDIX 7

American Standard Clearance Locational Fits (hole basis)

Limits are in thousandths of an inch.
Limits for hole and shaft are applied algebraically to the basic size to obtain the limits of size for the parts.
Data in bold face are in accordance with ABC agreements.
Symbols H9, f8, etc., are Hole and Shaft designations used in ABC System.

Nominal Size Range Inches		Class LC 1			Class LC2			Class LC 3			Class LC 4			Class LC 5		
		Limits of Clearance	Standard Limits		Limits of Clearance	Standard Limits		Limits of Clearance	Standard Limits		Limits of Clearance	Standard Limits		Limits of Clearance	Standard Limits	
Over	To		Hole H6	Shaft h5		Hole H7	Shaft h6		Hole H8	Shaft h7		Hole H10	Shaft h9		Hole H7	Shaft g6
0	−0.12	0	+0.25	+0	0	+0.4	+0	0	+0.6	+0	0	+1.6	+0	0.1	+0.4	−0.1
		0.45	−0	−0.2	0.65	−0	−0.25	1	−0	−0.4	2.6	−0	−1.0	0.75	−0	−0.35
0.12	−0.24	0	+0.3	+0	0	+0.5	+0	0	+0.7	+0	0	+1.8	+0	0.15	+0.5	−0.15
		0.5	−0	−0.2	0.8	−0	−0.3	1.2	−0	−0.5	3.0	−0	−1.2	0.95	−0	−0.45
0.24	−0.40	0	+0.4	+0	0	+0.6	+0	0	+0.9	+0	0	+2.2	+0	0.2	+0.6	−0.2
		0.65	−0	−0.25	1.0	−0	−0.4	1.5	−0	−0.6	3.6	−0	−1.4	1.2	−0	−0.6
0.40	−0.71	0	+0.4	+0	0	+0.7	+0	0	+1.0	+0	0	+2.8	+0	0.25	+0.7	−0.25
		0.7	−0	−0.3	1.1	−0	−0.4	1.7	−0	−0.7	4.4	−0	−1.6	1.35	−0	−0.65
0.71	−1.19	0	+0.5	+0	0	+0.8	+0	0	+1.2	+0	0	+3.5	+0	0.3	+0.8	−0.3
		0.9	−0	−0.4	1.3	−0	−0.5	2	−0	−0.8	5.5	−0	−2.0	1.6	−0	−0.8
1.19	−1.97	0	+0.6	+0	0	+1.0	+0	0	+1.6	+0	0	+4.0	+0	0.4	+1.0	−0.4
		1.0	−0	−0.4	1.6	−0	−0.6	2.6	−0	−1	6.5	−0	−2.5	2.0	−0	−1.0
1.97	−3.15	0	+0.7	+0	0	+1.2	+0	0	+1.8	+0	0	+4.5	+0	0.4	+1.2	−0.4
		1.2	−0	−0.5	1.9	−0	−0.7	3	−0	−1.2	7.5	−0	−3	2.3	−0	−1.1
3.15	−4.73	0	+0.9	+0	0	+1.4	+0	0	+2.2	+0	0	+5.0	+0	0.5	+1.4	−0.5
		1.5	−0	−0.6	2.3	−0	−0.9	3.6	−0	−1.4	8.5	−0	−3.5	2.8	−0	−1.4
4.73	−7.09	0	+1.0	+0	0	+1.6	+0	0	+2.5	+0	0	+6.0	+0	0.6	+1.6	−0.6
		1.7	−0	−0.7	2.6	−0	−1.0	4.1	−0	−1.6	10	−0	−4	3.2	−0	1.6
7.09	−9.85	0	+1.2	+0	0	+1.8	+0	0	+2.8	+0	0	+7.0	+0	0.6	+1.8	−0.6
		2.0	−0	−0.8	3.0	−0	−1.2	4.6	−0	−1.8	11.5	−0	−4.5	3.6	−0	−1.8
9.85	−12.41	0	+1.2	+0	0	+2.0	+0	0	+3.0	+0	0	8.0	+0	0.7	+2.0	−0.7
		2.1	−0	−0.9	3.2	−0	−1.2	5	−0	−2.0	13	−0	−5	3.9	−0	−1.9
12.41	−15.75	0	+1.4	+0	0	+2.2	+0	0	+3.5	+0	0	+9.0	+0	0.7	+2.2	−0.7
		2.4	−0	−1.0	3.6	−0	−1.4	5.7	−0	−2.2	15	−0	−6	4.3	−0	−2.1
15.75	−19.69	0	+1.6	+0	0	+2.5	+0	0	+4	+0	0	+10.0	+0	0.8	+2.5	−0.8
		2.6	−0	−1.0	4.1	−0	−1.6	6.5	−0	−2.5	16	−0	−6	4.9	−0	−2.4
19.69	−30.09	0	+2.0	+0	0	+3	+0	0	+5	+0	0	+12.0	+0	0.9	+3.0	−0.9
		3.2	−0	−1.2	5.0	−0	−2	8	−0	−3	20	−0	−8	5.9	−0	−2.9
30.09	−41.49	0	+2.5	+0	0	+4	+0	0	+6	+0	0	+16.0	+0	1.0	+4.0	−1.0
		4.1	−0	−1.6	6.5	−0	−2.5	10	−0	−4	26	−0	−10	7.5	−0	−3.5
41.49	−56.19	0	+3.0	+0	0	+5	+0	0	+8	+0	0	+20.0	+0	1.2	+5.0	−1.2
		5.0	−0	−2.0	8.0	−0	−3	13	−0	−5	32	−0	−12	9.2	−0	−4.2
56.19	−76.39	0	+4.0	+0	0	+6	+0	0	+10	+0	0	+25.0	+0	1.2	+6.0	−1.2
		6.5	−0	−2.5	10	−0	−4	16	−0	−6	41	−0	−16	11.2	−0	−5.2
76.39	−100.9	0	+5.0	+0	0	+8	+0	0	+12	+0	0	+30.0	+0	1.4	+8.0	−1.4
		8.0	−0	−3.0	13	−0	−5	20	−0	−8	50	−0	−20	14.4	−0	−6.4
100.9	−131.9	0	+6.0	+0	0	+10	+0	0	+16	+0	0	+40.0	+0	1.6	+10.0	−1.6
		10.0	−0	−4.0	16	−0	−6	26	−0	−10	65	−0	−25	17.6	−0	−7.6
131.9	−171.9	0	+8.0	+0	0	+12	+0	0	+20	+0	0	+50.0	+0	1.8	+12.0	−1.8
		13.0	−0	−5.0	20	−0	−8	32	−0	−12	8	−0	−30	21.8	−0	−9.8
171.9	−200	0	+10.0	+0	0	+16	+0	0	+25	+0	0	+60.0	+0	1.8	+16.0	−1.8
		16.0	−0	−6.0	26	−0	−10	41	−0	−16	100	−0	−40	27.8	−0	−11.8

Class LC 6			Class LC 7			Class LC 8			Class LC 9			Class LC 10			Class LC 11			Nominal Size Range Inches	
Limits of Clearance	Standard Limits		Limits of Clearance	Standard Limits		Limits of Clearance	Standard Limits		Limits of Clearance	Standard Limits		Limits of Clearance	Standard Limits		Limits of Clearance	Standard Limits			
	Hole H9	Shaft f8		Hole H10	Shaft e9		Hole H10	Shaft d9		Hole H11	Shaft c10		Hole H12	Shaft		Hole H13	Shaft	Over	To
0.3	+1.0	−0.3	0.6	+1.6	−0.6	1.0	+0.6	−1.0	2.5	+2.5	−2.5	4	+4	−4	5	+6	−5	0	0.12
1.9	0	−0.9	3.2	0	−1.6	3.6	−0	−2.0	6.6	−0	−4.1	12	−0	−8	17	−0	−11		
0.4	+1.2	−0.4	0.8	+1.8	−0.8	1.2	+1.8	−1.2	2.8	+3.0	−2.8	4.5	+5	−4.5	6	+7	−6	0.12	0.24
2.3	0	−1.1	3.8	0	−2.0	4.2	−0	−2.4	7.6	−0	−4.6	14.5	−0	−9.5	20	−0	−13		
0.5	+1.4	−0.5	1.0	+2.2	−1.0	1.6	+2.2	−1.6	3.0	+3.5	−3.0	5	+6	−5	7	+9	−7	0.24	0.40
2.8	0	−1.4	4.6	0	−2.4	5.2	−0	−3.0	8.7	−0	−5.2	17	−0	−11	25	−0	−16		
0.6	+1.6	−0.6	1.2	+2.8	−1.2	2.0	+2.8	−2.0	3.5	+4.0	−3.5	6	+7	−6	8	+10	−8	0.40	0.71
3.2	0	−1.6	5.6	0	−2.8	6.4	−0	−3.6	10.3	−0	−6.3	20	−0	−13	28	−0	−18		
0.8	+2.0	−0.8	1.6	+3.5	−1.6	2.5	+3.5	−2.5	4.5	+5.0	−4.5	7	+8	−7	10	+12	−10	0.71	1.19
4.0	0	−2.0	7.1	0	−3.6	8.0	−0	−4.5	13.0	−0	−8.0	23	−0	−15	34	−0	−22		
1.0	+2.5	−1.0	2.0	+4.0	−2.0	3.0	+4.0	−3.0	5	+6	−5	8	+10	−8	12	+16	−12	1.19	1.97
5.1	0	−2.6	8.5	0	−4.5	9.5	−0	−5.5	15	−0	−9	28	−0	−18	44	−0	−28		
1.2	+3.0	−1.2	2.5	+4.5	−2.5	4.0	+4.5	−4.0	6	+7	−6	10	+12	−10	14	+18	−14	1.97	3.15
6.0	0	−3.0	10.0	0	−5.5	11.5	−0	−7.0	17.5	−0	−10.5	34	−0	−22	50	−0	−32		
1.4	+3.5	−1.4	3.0	+5.0	−3.0	5.0	+5.0	−5.0	7	+9	−7	11	+14	−11	16	+22	−16	3.15	4.73
7.1	0	−3.6	11.5	0	−6.5	13.5	−0	−8.5	21	−0	−12	39	−0	−25	60	−0	−38		
1.6	+4.0	−1.6	3.5	+6.0	−3.5	6	+6	−6	8	+10	−8	12	+16	−12	18	+25	−18	4.73	7.09
8.1	0	−4.1	13.5	0	−7.5	16	−0	−10	24	−0	−14	44	−0	−28	68	−0	−43		
2.0	+4.5	−2.0	4.0	+7.0	−4.0	7	+7	−7	10	+12	−10	16	+18	−16	22	+28	−22	7.09	9.85
9.3	0	−4.8	15.5	0	−8.5	18.5	−0	−11.5	29	−0	−17	52	−0	−34	78	−0	−50		
2.2	+5.0	−2.2	4.5	+8.0	−4.5	7	+8	−7	12	+12	−12	20	+20	−20	28	+30	−28	9.85	12.41
10.2	0	−5.2	17.5	0	−9.5	20	−0	−12	32	−0	−20	60	−0	−40	88	−0	−58		
2.5	+6.0	−2.5	5.0	+9.0	−5	8	+9	−8	14	+14	−14	22	+22	−22	30	+35	−30	12.41	15.75
12.0	0	−6.0	20.0	0	−11	23	−0	−14	37	−0	−23	66	−0	−44	100	−0	−65		
2.8	+6.0	−2.8	5.0	+10.0	−5	9	+10	−9	16	+16	−16	25	+25	−25	35	+40	−35	15.75	19.69
12.8	0	−6.8	21.0	0	−11	25	−0	−15	42	−0	−26	75	−0	−50	115	−0	−75		
3.0	+8.0	−3.0	6.0	+12.0	−6	10	+12	−10	18	+20	−18	28	+30	−28	40	+50	−40	19.69	30.09
16.0	0	−8.0	26.0	−0	−14	30	−0	−18	50	−0	−30	88	−0	−58	140	−0	−90		
3.5	+10.0	−3.5	7.0	+16.0	−7	12	+16	−12	20	+25	−20	30	+40	−30	45	+60	−45	30.09	41.49
19.5	0	−9.5	33.0	−0	−17	38	−0	−22	61	−0	−36	110	−0	−70	165	−0	−105		
4.0	+12.0	−4.0	8.0	+20.0	−8	14	+20	−14	25	+30	−25	40	+50	−40	60	+80	−60	41.49	56.19
24.0	0	−12.0	40.0	−0	−20	46	−0	−26	75	−0	−45	140	−0	−90	220	−0	−140		
4.5	+16.0	−4.5	9.0	+25.0	−9	16	+25	−16	30	+40	−30	50	+60	−50	70	+100	−70	56.19	76.39
30.5	0	−14.5	50.0	−0	−25	57	−0	−32	95	−0	−55	170	−0	110	270	−0	−170		
5.0	+20.0	−5	10.0	+30.0	−10	18	+30	−18	35	+50	−35	50	+80	−50	80	+125	−80	76.39	100.9
37.0	0	−17	60.0	−0	−30	68	−0	−38	115	−0	−65	210	−0	−130	330	−0	−205		
6.0	+25.0	−6	12.0	+40.0	−12	20	+40	−20	40	+60	−40	60	+100	−60	90	+160	−90	100.9	131.9
47.0	0	−22	67.0	−0	−27	85	−0	−45	140	−0	−80	260	−0	−160	410	−0	−250		
7.0	+30.0	−7	14.0	+50.0	−14	25	+50	−25	50	+80	−50	80	+125	−80	100	+200	−100	131.9	171.9
57.0	0	−27	94.0	−0	−44	105	−0	−55	180	−0	−100	330	−0	−205	500	−0	−300		
7.0	+40.0	−7	14.0	+60.0	−14	25	+60	−25	50	+100	−50	90	+160	−90	125	+250	−125	171.9	200
72.0	0	−32	114.0	−0	−54	125	−0	−65	210	−0	−110	410	−0	−250	625	−0	−375		

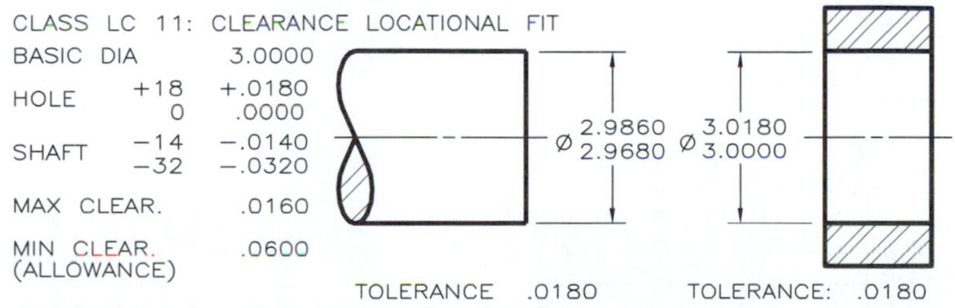

APPENDIX

American Standard Transition Locational Fits (hole basis)

Limits are in thousandths of an inch.
Limits for hole and shaft are applied algebraically to the basic size to obtain the limits of size for the mating parts.
Data in bold face are in accordance with ABC agreements.
Fit represents the maximum interference (minus values) and the maximum clearance (plus values).
Symbols H7, js6, etc., are Hole and Shaft designations used in ABC System.

Nominal Size Range Inches		Class LT 1			Class LT 2			Class LT 3			Class LT 4			Class LT 5			Class LT 6		
		Fit	Standard Limits		Fit	Standard Limits		Fit	Standard Limits		Fit	Standard Limits		Fit	Standard Limits		Fit	Standard Limits	
Over	To		Hole H7	Shaft js6		Hole H8	Shaft js7		Hole H7	Shaft k6		Hole H8	Shaft k7		Hole H7	Shaft n6		Hole H7	Shaft n7
0	− 0.12	−0.10	+0.4	+0.10	−0.2	+0.6	+0.2							−0.5	+0.4	+0.5	−0.65	+0.4	−0.65
		+0.50	−0	−0.10	+0.8	−0	−0.2							+0.15	−0	+0.25	+0.15	−0	+0.25
0.12	− 0.24	−0.15	+0.5	+0.15	−0.25	+0.7	+0.25							−0.6	+0.5	+0.6	−0.8	+0.5	+0.8
		+0.65	−0	−0.15	+0.95	−0	−0.25							+0.2	−0	+0.3	+0.2	−0	+0.3
0.24	− 0.40	−0.2	+0.6	+0.2	−0.3	+0.9	+0.3	−0.5	+0.6	+0.5	−0.7	+0.9	+0.7	−0.8	+0.6	+0.8	−1.0	+0.6	+1.0
		+0.8	−0	−0.2	+1.2	−0	−0.3	+0.5	−0	+0.1	+0.8	−0	+0.1	+0.2	−0	+0.4	+0.2	−0	+0.4
0.40	− 0.71	−0.2	+0.7	+0.2	−0.35	+1.0	+0.35	−0.5	+0.7	+0.5	−0.8	+1.0	+0.8	−0.9	+0.7	+0.9	−1.2	+0.7	+1.2
		+0.9	−0	−0.2	+1.35	−0	−0.35	+0.6	−0	+0.1	+0.9	−0	+0.1	+0.2	−0	+0.5	+0.2	−0	+0.5
0.71	− 1.19	−0.25	+0.8	+0.25	−0.4	+1.2	+0.4	−0.6	+0.8	+0.6	−0.9	+1.2	+0.9	−1.1	+0.8	+1.1	−1.4	+0.8	+1.4
		+1.05	−0	−0.25	+1.6	−0	−0.4	+0.7	−0	+0.1	+1.1	−0	+0.1	+0.2	−0	+0.6	+0.2	−0	+0.6
1.19	− 1.97	−0.3	+1.0	+0.3	−0.5	+1.6	+0.5	−0.7	+1.0	+0.7	−1.1	+1.6	+1.1	−1.3	+1.0	+1.3	−1.7	+1.0	+1.7
		+1.3	−0	−0.3	+2.1	−0	−0.5	+0.9	−0	+0.1	+1.5	−0	+0.1	+0.3	−0	+0.7	+0.3	−0	+0.7
1.97	− 3.15	−0.3	+1.2	+0.3	−0.6	+1.8	+0.6	−0.8	+1.2	+0.8	−1.3	+1.8	+1.3	−1.5	+1.2	+1.5	−2.0	+1.2	+2.0
		+1.5	−0	−0.3	+2.4	−0	−0.6	+1.1	−0	+0.1	+1.7	−0	+0.1	+0.4	−0	+0.8	+0.4	−0	+0.8
3.15	− 4.73	−0.4	+1.4	+0.4	−0.7	+2.2	+0.7	−1.0	+1.4	+1.0	−1.5	+2.2	+1.5	−1.9	+1.4	+1.9	−2.4	+1.4	+2.4
		+1.8	−0	−0.4	+2.9	−0	−0.7	+1.3	−0	+0.1	+2.1	−0	+0.1	+0.4	−0	+1.0	+0.4	−0	+1.0
4.73	− 7.09	−0.5	+1.6	+0.5	−0.8	+2.5	+0.8	−1.1	+1.6	+1.1	−1.7	+2.5	+1.7	−2.2	+1.6	+2.2	−2.8	+1.6	+2.8
		+2.1	−0	−0.5	+3.3	−0	−0.8	+1.5	−0	+0.1	+2.4	−0	+0.1	+0.4	−0	+1.2	+0.4	−0	+1.2
7.09	− 9.85	−0.6	+1.8	+0.6	−0.9	+2.8	+0.9	−1.4	+1.8	+1.4	−2.0	+2.8	+2.0	−2.6	+1.8	+2.6	−3.2	+1.8	+3.2
		+2.4	−0	−0.6	+3.7	−0	−0.9	+1.6	−0	+0.2	+2.6	−0	+0.2	+0.4	−0	+1.4	+0.4	−0	+1.4
9.85	−12.41	−0.6	+2.0	+0.6	−1.0	+3.0	+1.0	−1.4	+2.0	+1.4	−2.2	+3.0	+2.2	−2.6	+2.0	+2.6	−3.4	+2.0	+3.4
		+2.6	−0	−0.6	+4.0	−0	−1.0	+1.8	−0	+0.2	+2.8	−0	+0.2	+0.6	−0	+1.4	+0.6	−0	+1.4
12.41	−15.75	−0.7	+2.2	+0.7	−1.0	+3.5	+1.0	−1.6	+2.2	+1.6	−2.4	+3.5	+2.4	−3.0	+2.2	+3.0	−3.8	+2.2	+3.8
		+2.9	−0	−0.7	+4.5	−0	−1.0	+2.0	−0	+0.2	+3.3	−0	+0.2	+0.6	−0	+1.6	+0.6	−0	+1.6
15.75	−19.69	−0.8	+2.5	+0.8	−1.2	+4.0	+1.2	−1.8	+2.5	+1.8	−2.7	+4.0	+2.7	−3.4	+2.5	+3.4	−4.3	+2.5	+4.3
		+3.3	−0	−0.8	+5.2	−0	−1.2	+2.3	−0	+0.2	+3.8	−0	+0.2	+0.7	−0	+1.8	+0.7	−0	+1.8

APPENDIX 9

American Standard Interference Locational Fits (hole basis)

Limits are in thousandths of an inch.
Limits for hole and shaft are applied algebraically to the basic size to obtain the limits of size for the parts.
Data in bold face are in accordance with ABC agreements.
Symbols H7, p6, etc., are Hole and Shaft designations used in ABC System.

Nominal Size Range Inches		Class LN 1			Class LN 2			Class LN 3		
		Limits of Interference	Standard Limits		Limits of Interference	Standard Limits		Limits of Interference	Standard Limits	
Over	To		Hole H6	Shaft n5		Hole H7	Shaft p6		Hole H7	Shaft r6
0	− 0.12	0	+0.25	+0.45	0	+0.4	+0.65	0.1	+0.4	+0.75
		0.45	−0	+0.25	0.65	−0	+0.4	0.75	−0	+0.5
0.12	− 0.24	0	+0.3	+0.5	0	+0.5	+0.8	0.1	+0.5	+0.9
		0.5	−0	+0.3	0.8	−0	+0.5	0.9	0	+0.6
0.24	− 0.40	0	+0.4	+0.65	0	+0.6	+1.0	0.2	+0.6	+1.2
		0.65	−0	+0.4	1.0	−0	+0.6	1.2	−0	+0.8
0.40	− 0.71	0	+0.4	+0.8	0	+0.7	+1.1	0.3	+0.7	+1.4
		0.8	−0	+0.4	1.1	−0	+0.7	1.4	−0	+1.0
0.71	− 1.19	0	+0.5	+1.0	0	+0.8	+1.3	0.4	+0.8	+1.7
		1.0	−0	+0.5	1.3	−0	+0.8	1.7	−0	+1.2
1.19	− 1.97	0	+0.6	+1.1	0	+1.0	+1.6	0.4	+1.0	+2.0
		1.1	−0	+0.6	1.6	−0	+1.0	2.0	−0	+1.4
1.97	− 3.15	0.1	+0.7	+1.3	0.2	+1.2	+2.1	0.4	+1.2	+2.3
		1.3	−0	+0.7	2.1	−0	+1.4	2.3	−0	+1.6
3.15	− 4.73	0.1	+0.9	+1.6	0.2	+1.4	+2.5	0.6	+1.4	+2.9
		1.6	−0	+1.0	2.5	−0	+1.6	2.9	−0	+2.0
4.73	− 7.09	0.2	+1.0	+1.9	0.2	+1.6	+2.8	0.9	+1.6	+3.5
		1.9	−0	+1.2	2.8	−0	+1.8	3.5	−0	+2.5
7.09	− 9.85	0.2	+1.2	+2.2	0.2	+1.8	+3.2	1.2	+1.8	+4.2
		2.2	−0	+1.4	3.2	−0	+2.0	4.2	−0	+3.0
9.85	−12.41	0.2	+1.2	+2.3	0.2	+2.0	+3.4	1.5	+2.0	+4.7
		2.3	−0	+1.4	3.4	−0	+2.2	4.7	−0	+3.5
12.41	−15.75	0.2	+1.4	+2.6	0.3	+2.2	+3.9	2.3	+2.2	+5.9
		2.6	−0	+1.6	3.9	−0	+2.5	5.9	−0	+4.5
15.75	−19.69	0.2	+1.6	+2.8	0.3	+2.5	+4.4	2.5	+2.5	+6.6
		2.8	−0	+1.8	4.4	−0	+2.8	6.6	−0	+5.0
19.69	−30.09		+2.0		0.5	+3	+5.5	4	+3	+9
			−0		5.5	−0	+3.5	9	−0	+7
30.09	−41.49		+2.5		0.5	+4	+7.0	5	+4	+11.5
			−0		7.0	−0	+4.5	11.5	−0	+9
41.49	−56.19		+3.0		1	+5	+9	7	+5	+15
			−0		9	−0	+6	15	−0	+12
56.19	−76.39		+4.0		1	+6	+11	10	+6	+20
			−0		11	−0	+7	20	−0	+16
76.39	−100.9		+5.0		1	+8	+14	12	+8	+25
			−0		14	−0	+9	25	−0	+20
100.9	−131.9		+6.0		2	+10	+18	15	+10	+31
			−0		18	−0	+12	31	−0	+25
131.9	−171.9		+8.0		4	+12	+24	18	+12	+38
			−0		24	−0	+16	38	−0	+30
171.9	− 200		+10.0		4	+16	+30	24	+16	+50
			−0		30	−0	+20	50	−0	+40

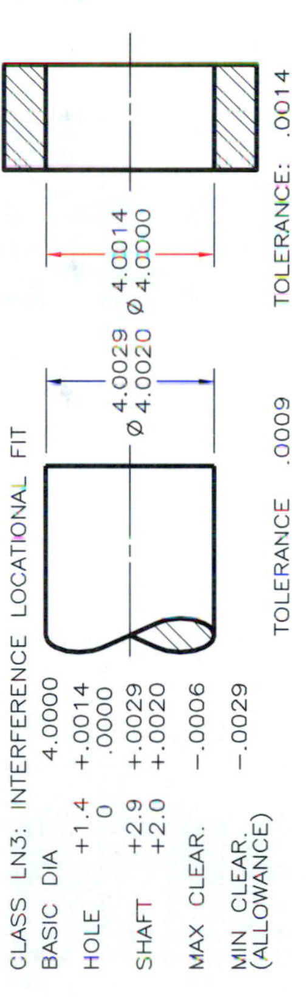

APPENDIX 10

American Standard Force and Shrink Fits (hole basis)

Limits are in thousandths of an inch.
Limits for hole and shaft are applied algebraically to the basic size to obtain the limits of size for the parts.
Data in bold face are in accordance with ABC agreements.
Symbols H7, s6, etc., are Hole and Shaft designations used in ABC System.

Nominal Size Range Inches		Class FN 1			Class FN 2			Class FN 3			Class FN 4			Class FN 5		
		Limits of Interference	Standard Limits		Limits of Interference	Standard Limits		Limits of Interference	Standard Limits		Limits of Interference	Standard Limits		Limits of Interference	Standard Limits	
Over	To		Hole H6	Shaft		Hole H7	Shaft s6		Hole H7	Shaft t6		Hole H7	Shaft u6		Hole H8	Shaft x7
0	– 0.12	0.05	+0.25	+0.5	0.2	+0.4	+0.85				0.3	+0.4	+0.95	0.3	+0.6	+1.3
		0.5	−0	+0.3	0.85	−0	+0.6				0.95	−0	+0.7	1.3	−0	+0.9
0.12	– 0.24	0.1	+0.3	+0.6	0.2	+0.5	+1.0				0.4	+0.5	+1.2	0.5	+0.7	+1.7
		0.6	−0	+0.4	1.0	−0	+0.7				1.2	−0	+0.9	1.7	−0	+1.2
0.24	– 0.40	0.1	+0.4	+0.75	0.4	+0.6	+1.4				0.6	+0.6	+1.6	0.5	+0.9	+2.0
		0.75	−0	+0.5	1.4	−0	+1.0				1.6	−0	+1.2	2.0	−0	+1.4
0.40	– 0.56	0.1	−0.4	+0.8	0.5	+0.7	+1.6				0.7	+0.7	+1.8	0.6	+1.0	+2.3
		0.8	−0	+0.5	1.6	−0	+1.2				1.8	−0	+1.4	2.3	−0	+1.6
0.56	– 0.71	0.2	+0.4	+0.9	0.5	+0.7	+1.6				0.7	+0.7	+1.8	0.8	+1.0	+2.5
		0.9	−0	+0.6	1.6	−0	+1.2				1.8	−0	+1.4	2.5	−0	+1.8
0.71	– 0.95	0.2	+0.5	+1.1	0.6	+0.8	+1.9				0.8	+0.8	+2.1	1.0	+1.2	+3.0
		1.1	−0	+0.7	1.9	−0	+1.4				2.1	−0	+1.6	3.0	−0	+2.2
0.95	– 1.19	0.3	+0.5	+1.2	0.6	+0.8	+1.9	0.8	+0.8	+2.1	1.0	+0.8	+2.3	1.3	+1.2	+3.3
		1.2	−0	+0.8	1.9	−0	+1.4	2.1	−0	+1.6	2.3	−0	+1.8	3.3	−0	+2.5
1.19	– 1.58	0.3	+0.6	+1.3	0.8	+1.0	+2.4	1.0	+1.0	+2.6	1.5	+1.0	+3.1	1.4	+1.6	+4.0
		1.3	−0	+0.9	2.4	−0	+1.8	2.6	−0	+2.0	3.1	−0	+2.5	4.0	−0	+3.0
1.58	– 1.97	0.4	+0.6	+1.4	0.8	+1.0	+2.4	1.2	+1.0	+2.8	1.8	+1.0	+3.4	2.4	+1.6	+5.0
		1.4	−0	+1.0	2.4	−0	+1.8	2.8	−0	+2.2	3.4	−0	+2.8	5.0	−0	+4.0
1.97	– 2.56	0.6	+0.7	+1.8	0.8	+1.2	+2.7	1.3	+1.2	+3.2	2.3	+1.2	+4.2	3.2	+1.8	+6.2
		1.8	−0	+1.3	2.7	−0	+2.0	3.2	−0	+2.5	4.2	−0	+3.5	6.2	−0	+5.0
2.56	– 3.15	0.7	+0.7	+1.9	1.0	+1.2	+2.9	1.8	+1.2	+3.7	2.8	+1.2	+4.7	4.2	+1.8	+7.2
		1.9	−0	+1.4	2.9	−0	+2.2	3.7	−0	+3.0	4.7	−0	+4.0	7.2	−0	+6.0
3.15	– 3.94	0.9	+0.9	+2.4	1.4	+1.4	+3.7	2.1	+1.4	+4.4	3.6	+1.4	+5.9	4.8	+2.2	+8.4
		2.4	−0	+1.8	3.7	−0	+2.8	4.4	−0	+3.5	5.9	−0	+5.0	8.4	−0	+7.0
3.94	– 4.73	1.1	+0.9	+2.6	1.6	+1.4	+3.9	2.6	+1.4	+4.9	4.6	+1.4	+6.9	5.8	+2.2	+9.4
		2.6	−0	+2.0	3.9	−0	+3.0	4.9	−0	+4.0	6.9	−0	+6.0	9.4	−0	+8.0
4.73	– 5.52	1.2	+1.0	+2.9	1.9	+1.6	+4.5	3.4	+1.6	+6.0	5.4	+1.6	+8.0	7.5	+2.5	+11.6
		2.9	−0	+2.2	4.5	−0	+3.5	6.0	−0	+5.0	8.0	−0	+7.0	11.6	−0	+10.0
5.52	– 6.30	1.5	+1.0	+3.2	2.4	+1.6	+5.0	3.4	+1.6	+6.0	5.4	+1.6	+8.0	9.5	+2.5	+13.6
		3.2	−0	+2.5	5.0	−0	+4.0	6.0	−0	+5.0	8.0	−0	+7.0	13.6	−0	+12.0
6.30	– 7.09	1.8	+1.0	+3.5	2.9	+1.6	+5.5	4.4	+1.6	+7.0	6.4	+1.6	+9.0	9.5	+2.5	+13.6
		3.5	−0	+2.8	5.5	−0	+4.5	7.0	−0	+6.0	9.0	−0	+8.0	13.6	−0	+12.0
7.09	– 7.88	1.8	+1.2	+3.8	3.2	+1.8	+6.2	5.2	+1.8	+8.2	7.2	+1.8	+10.2	11.2	+2.8	+15.8
		3.8	−0	+3.0	6.2	−0	+5.0	8.2	−0	+7.0	10.2	−0	+9.0	15.8	−0	+14.0
7.88	– 8.86	2.3	+1.2	+4.3	3.2	+1.8	+6.2	5.2	+1.8	+8.2	8.2	+1.8	+11.2	13.2	_2.8	+17.8
		4.3	−0	+3.5	6.2	−0	+5.0	8.2	−0	+7.0	11.2	−0	+10.0	17.8	−0	+16.0
8.86	– 9.85	2.3	+1.2	+4.3	4.2	+1.8	+7.2	6.2	+1.8	+9.2	10.2	+1.8	+13.2	13.2	+2.8	+17.8
		4.3	−0	+3.5	7.2	−0	+6.0	9.2	−0	+8.0	13.2	−0	+12.0	17.8	−0	+16.0
9.85	– 11.03	2.8	+1.2	+4.9	4.0	+2.0	+7.2	7.0	+2.0	+10.2	10.0	+2.0	+13.2	15.0	+3.0	+20.0
		4.9	−0	+4.0	7.2	−0	+6.0	10.2	−0	+9.0	13.2	−0	+12.0	20.0	−0	+18.0
11.03	– 12.41	2.8	+1.2	+4.9	5.0	+2.0	+8.2	7.0	+2.0	+10.2	12.0	+2.0	+15.2	17.0	+3.0	+22.0
		4.9	−0	+4.0	8.2	−0	+7.0	10.2	−0	+9.0	15.2	−0	+14.0	22.0	−0	+20.0
12.41	– 13.98	3.1	+1.4	+5.5	5.8	+2.2	+9.4	7.8	+2.2	+11.4	13.8	+2.2	+17.4	18.5	+3.5	+24.2
		5.5	−0	+4.5	9.4	−0	+8.0	11.4	−0	+10.0	17.4	−0	+16.0	24.2	+0	+22.0
13.98	– 15.75	3.6	+1.4	+6.1	5.8	+2.2	+9.4	9.8	+2.2	+13.4	15.8	+2.2	+19.4	21.5	+3.5	+27.2
		6.1	−0	+5.0	9.4	−0	+8.0	13.4	−0	+12.0	19.4	−0	+18.0	27.2	−0	+25.0
15.75	– 17.72	4.4	+1.6	+7.0	6.5	+2.5	+10.6	9.5	+2.5	+13.6	17.5	+2.5	+21.6	24.0	+4.0	+30.5
		7.0	−0	+6.0	10.6	−0	+9.0	13.6	−0	+12.0	21.6	−0	+20.0	30.5	−0	+28.0
17.72	– 19.69	4.4	+1.6	+7.0	7.5	+2.5	+11.6	11.5	+2.5	+15.6	19.5	+2.5	+23.6	26.0	+4.0	+32.5
		7.0	−0	+6.0	11.6	−0	+10.0	15.6	−0	+14.0	23.6	−0	+22.0	32.5	−0	+30.0

APPENDIX 11

International Tolerance Grades

Dimensions are in mm.

Basic sizes		Tolerance grades[1]																	
Over	Up to and including	IT01	IT0	IT1	IT2	IT3	IT4	IT5	IT6	IT7	IT8	IT9	IT10	IT11	IT12	IT13	IT14	IT15	IT16
0	3	0.0003	0.0005	0.0008	0.0012	0.002	0.003	0.004	0.006	0.010	0.014	0.025	0.040	0.060	0.100	0.140	0.250	0.400	0.600
3	6	0.0004	0.0006	0.001	0.0015	0.0025	0.004	0.005	0.008	0.012	0.018	0.030	0.048	0.075	0.120	0.180	0.300	0.480	0.750
6	10	0.0004	0.0006	0.001	0.0015	0.0025	0.004	0.006	0.009	0.015	0.022	0.036	0.058	0.090	0.150	0.220	0.360	0.580	0.900
10	18	0.0005	0.0008	0.0012	0.002	0.003	0.005	0.008	0.011	0.018	0.027	0.043	0.070	0.110	0.180	0.270	0.430	0.700	1.100
18	30	0.0006	0.001	0.0015	0.0025	0.004	0.006	0.009	0.013	0.021	0.033	0.052	0.084	0.130	0.210	0.330	0.520	0.840	1.300
30	50	0.0006	0.001	0.0015	0.0025	0.004	0.007	0.011	0.016	0.025	0.039	0.062	0.100	0.160	0.250	0.390	0.620	1.000	1.600
50	80	0.0008	0.0012	0.002	0.003	0.005	0.008	0.013	0.019	0.030	0.046	0.074	0.120	0.190	0.300	0.460	0.740	1.200	1.900
80	120	0.001	0.0015	0.0025	0.004	0.006	0.010	0.015	0.022	0.035	0.054	0.087	0.140	0.220	0.350	0.540	0.870	1.400	2.200
120	180	0.0012	0.002	0.0036	0.005	0.008	0.012	0.018	0.025	0.040	0.063	0.100	0.160	0.250	0.400	0.630	1.000	1.600	2.500
180	250	0.002	0.003	0.0045	0.007	0.010	0.014	0.020	0.029	0.046	0.072	0.115	0.185	0.290	0.460	0.720	1.150	1.850	2.900
250	315	0.0025	0.004	0.006	0.008	0.012	0.016	0.023	0.032	0.052	0.081	0.130	0.210	0.320	0.520	0.810	1.300	2.100	3.200
315	400	0.003	0.005	0.007	0.009	0.013	0.018	0.025	0.036	0.057	0.089	0.140	0.230	0.360	0.570	0.890	1.400	2.300	3.600
400	500	0.004	0.006	0.008	0.010	0.015	0.020	0.027	0.040	0.063	0.097	0.156	0.250	0.400	0.630	0.970	1.550	2.500	4.000
500	630	0.0045	0.006	0.009	0.011	0.016	0.022	0.030	0.044	0.070	0.110	0.175	0.280	0.440	0.700	1.100	1.750	2.800	4.400
630	800	0.005	0.007	0.010	0.013	0.018	0.025	0.035	0.050	0.080	0.125	0.200	0.320	0.500	0.800	1.250	2.000	3.200	5.000
800	1000	0.0055	0.008	0.011	0.015	0.021	0.029	0.040	0.056	0.090	0.140	0.230	0.360	0.560	0.900	1.400	2.300	3.600	5.600
1000	1250	0.0065	0.009	0.013	0.018	0.024	0.034	0.046	0.066	0.105	0.165	0.260	0.420	0.660	1.050	1.650	2.600	4.200	6.600
1250	1600	0.008	0.011	0.015	0.021	0.029	0.040	0.054	0.078	0.125	0.195	0.310	0.500	0.780	1.250	1.950	3.100	5.000	7.800
1600	2000	0.009	0.013	0.018	0.025	0.035	0.048	0.065	0.092	0.150	0.230	0.370	0.600	0.920	1.500	2.300	3.700	6.000	9.200
2000	2500	0.011	0.015	0.022	0.030	0.041	0.057	0.077	0.110	0.175	0.280	0.440	0.700	1.100	1.750	2.800	4.400	7.000	11.000
2500	3150	0.013	0.018	0.026	0.036	0.050	0.069	0.093	0.135	0.210	0.330	0.540	0.860	1.350	2.100	3.300	5.400	8.600	13.500

[1]IT Values for tolerance grades larger than IT16 can be calculated by using the following formulas: IT17 = IT12 × 10; IT18 = IT13 × 10; etc.

APPENDIX

12

Preferred Hole Basis Clearance Fits–Cylindrical Fits

Dimensions are in mm.

BASIC SIZE		LOOSE RUNNING Hole H11	LOOSE RUNNING Shaft c11	LOOSE RUNNING Fit	FREE RUNNING Hole H9	FREE RUNNING Shaft d9	FREE RUNNING Fit	CLOSE RUNNING Hole H8	CLOSE RUNNING Shaft f7	CLOSE RUNNING Fit	SLIDING Hole H7	SLIDING Shaft g6	SLIDING Fit	LOCATIONAL CLEARANCE Hole H7	LOCATIONAL CLEARANCE Shaft h6	LOCATIONAL CLEARANCE Fit
1	MAX	1.060	0.940	0.180	1.025	0.980	0.070	1.014	0.994	0.030	1.010	0.998	0.018	1.010	1.000	0.016
	MIN	1.000	0.880	0.060	1.000	0.955	0.020	1.000	0.984	0.006	1.000	0.992	0.002	1.000	0.994	0.000
1.2	MAX	1.260	1.140	0.180	1.225	1.180	0.070	1.214	1.194	0.030	1.210	1.198	0.018	1.210	1.200	0.016
	MIN	1.200	1.080	0.060	1.200	1.155	0.020	1.200	1.184	0.006	1.200	1.192	0.002	1.200	1.194	0.000
1.6	MAX	1.660	1.540	0.180	1.625	1.580	0.070	1.614	1.594	0.030	1.610	1.598	0.018	1.610	1.600	0.016
	MIN	1.600	1.480	0.060	1.600	1.555	0.020	1.600	1.584	0.006	1.600	1.592	0.002	1.600	1.594	0.000
2	MAX	2.060	1.940	0.180	2.025	1.980	0.070	2.014	1.994	0.030	2.010	1.998	0.018	2.010	2.000	0.016
	MIN	2.000	1.880	0.060	2.000	1.955	0.020	2.000	1.984	0.006	2.000	1.992	0.002	2.000	1.994	0.000
2.5	MAX	2.560	2.440	0.180	2.525	0.480	0.070	2.514	2.494	0.030	2.510	2.498	0.018	2.510	2.500	0.016
	MIN	2.500	2.380	0.060	2.500	2.455	0.020	2.500	2.484	0.006	2.500	2.492	0.002	2.500	2.494	0.000
3	MAX	3.060	2.940	0.180	3.025	2.980	0.070	3.014	2.994	0.030	3.010	2.998	0.018	3.010	3.000	0.016
	MIN	3.000	2.880	0.060	3.000	2.955	0.020	3.000	2.984	0.006	3.000	2.992	0.002	3.000	2.994	0.000
4	MAX	4.075	3.930	0.220	4.030	3.970	0.090	4.018	3.990	0.040	4.012	3.996	0.024	4.012	4.000	0.020
	MIN	4.000	3.855	0.070	4.000	3.940	0.030	4.000	3.978	0.010	4.000	3.988	0.004	4.000	3.992	0.000
5	MAX	5.075	4.930	0.220	5.030	4.970	0.090	5.018	4.990	0.040	5.012	4.996	0.024	5.012	5.000	0.020
	MIN	5.000	4.855	0.070	5.000	4.940	0.030	5.000	4.978	0.010	5.000	4.988	0.004	5.000	4.992	0.000
6	MAX	6.075	5.930	0.220	6.030	5.970	0.090	6.018	5.990	0.040	6.012	5.996	0.024	6.012	6.000	0.020
	MIN	6.000	5.855	0.070	6.000	5.940	0.030	6.000	5.978	0.010	6.000	5.988	0.004	6.000	5.992	0.000
8	MAX	8.090	7.920	0.260	8.036	7.960	0.112	8.022	7.987	0.050	8.015	7.995	0.029	8.015	8.000	0.024
	MIN	8.000	7.830	0.080	8.000	7.924	0.040	8.000	7.972	0.013	8.000	7.986	0.005	8.000	7.991	0.000
10	MAX	10.090	9.920	0.260	10.036	9.960	0.112	10.022	9.987	0.050	10.015	9.995	0.029	10.015	10.000	0.024
	MIN	10.000	9.830	0.080	10.000	9.924	0.040	10.000	9.972	0.013	10.000	9.986	0.005	10.000	9.991	0.000
12	MAX	12.110	11.905	0.315	12.043	11.950	0.136	12.027	11.984	0.061	12.018	11.994	0.035	12.018	12.000	0.029
	MIN	12.000	11.795	0.095	12.000	11.907	0.050	12.000	11.966	0.016	12.000	11.983	0.006	12.000	11.989	0.000
16	MAX	16.110	15.905	0.315	16.043	15.950	0.136	16.027	15.984	0.061	16.018	15.994	0.035	16.018	16.000	0.029
	MIN	16.000	15.795	0.095	16.000	15.907	0.050	16.000	15.966	0.016	16.000	15.983	0.006	16.000	15.989	0.000
20	MAX	20.130	19.890	0.370	20.052	19.935	0.169	20.033	19.980	0.074	20.021	19.993	0.041	20.021	20.000	0.034
	MIN	20.000	19.760	0.110	20.000	19.883	0.065	20.000	19.959	0.020	20.000	19.980	0.007	20.000	19.987	0.000
25	MAX	25.130	24.890	0.370	25.052	24.935	0.169	25.033	24.980	0.074	25.021	24.993	0.041	25.021	25.000	0.034
	MIN	25.000	24.760	0.110	25.000	24.883	0.065	25.000	24.959	0.020	25.000	24.980	0.007	25.000	24.987	0.000
30	MAX	30.130	29.890	0.370	30.052	29.935	0.169	30.033	29.980	0.074	30.021	29.993	0.041	30.021	30.000	0.034
	MIN	30.000	29.760	0.110	30.000	29.883	0.065	30.000	29.959	0.020	30.000	29.980	0.007	30.000	29.987	0.000

BASIC SIZE		LOOSE RUNNING Hole H11	LOOSE RUNNING Shaft c11	LOOSE RUNNING Fit	FREE RUNNING Hole H9	FREE RUNNING Shaft d9	FREE RUNNING Fit	CLOSE RUNNING Hole H8	CLOSE RUNNING Shaft f7	CLOSE RUNNING Fit	SLIDING Hole H7	SLIDING Shaft g6	SLIDING Fit	LOCATIONAL CLEARANCE Hole H7	LOCATIONAL CLEARANCE Shaft h6	LOCATIONAL CLEARANCE Fit
40	MAX	40.160	39.880	0.440	40.062	39.920	0.204	40.039	39.975	0.089	40.025	39.991	0.050	40.025	40.000	0.041
	MIN	40.000	39.720	0.120	40.000	39.858	0.080	40.000	39.950	0.025	40.000	39.975	0.009	40.000	39.984	0.000
50	MAX	50.160	49.870	0.450	50.062	49.920	0.204	50.039	49.975	0.089	50.025	49.991	0.050	50.025	50.000	0.041
	MIN	50.000	49.710	0.130	50.000	49.858	0.080	50.000	49.950	0.025	50.000	49.975	0.009	50.000	49.984	0.000
60	MAX	60.190	59.860	0.520	60.074	59.900	0.248	60.046	59.970	0.106	60.030	59.990	0.059	60.030	60.000	0.049
	MIN	60.000	59.670	0.140	60.000	59.826	0.100	60.000	59.940	0.030	60.000	59.971	0.010	60.000	59.981	0.000
80	MAX	80.190	79.850	0.530	80.074	70.900	0.248	80.046	79.970	0.106	80.030	79.990	0.059	80.030	80.000	0.049
	MIN	80.000	79.660	0.150	80.000	79.826	0.100	80.000	79.940	0.030	80.000	79.971	0.010	80.000	79.981	0.000
100	MAX	100.220	99.830	0.610	100.087	99.880	0.294	100.054	99.964	0.125	100.035	99.988	0.069	100.035	100.000	0.057
	MIN	100.000	99.610	0.170	100.000	99.793	0.120	100.000	99.929	0.036	100.000	99.966	0.012	100.000	99.978	0.000
120	MAX	120.220	119.820	0.620	120.087	119.880	0.294	120.054	119.964	0.125	120.035	119.988	0.069	120.035	120.000	0.057
	MIN	120.000	119.600	0.180	120.000	119.793	0.120	120.000	119.929	0.036	120.000	119.966	0.012	120.000	119.978	0.000
160	MAX	160.250	159.790	0.710	160.100	159.855	0.345	160.063	159.957	0.146	160.040	159.986	0.079	160.040	160.000	0.065
	MIN	160.000	159.540	0.210	160.000	159.755	0.145	160.000	159.917	0.043	160.000	159.961	0.014	160.000	159.975	0.000
200	MAX	200.290	199.760	0.820	200.115	199.830	0.400	200.072	199.950	0.168	200.046	199.985	0.090	200.046	200.000	0.075
	MIN	200.000	199.470	0.240	200.000	199.715	0.170	200.000	199.904	0.050	200.000	199.956	0.105	200.000	199.971	0.000
250	MAX	250.290	249.720	0.860	250.115	249.830	0.400	250.072	249.950	0.168	250.046	249.985	0.090	250.046	250.000	0.075
	MIN	250.000	249.430	0.280	250.000	249.715	0.170	250.000	249.904	0.050	250.000	249.956	0.015	250.000	249.971	0.000
300	MAX	300.320	299.670	0.970	300.130	299.810	0.450	300.081	299.944	0.189	300.052	299.983	0.101	300.052	300.000	0.084
	MIN	300.000	299.350	0.330	300.000	299.680	0.190	300.000	299.892	0.056	300.000	299.951	0.017	300.000	299.968	0.000
400	MAX	400.360	399.600	1.120	400.140	399.790	0.490	400.089	399.938	0.208	400.057	399.982	0.111	400.057	400.000	0.093
	MIN	400.000	399.240	0.400	400.000	399.650	0.210	400.000	399.881	0.062	400.000	399.946	0.018	400.000	399.964	0.000
500	MAX	500.400	499.520	1.280	500.155	499.770	0.540	500.097	499.932	0.228	500.063	499.980	0.123	500.063	500.000	0.103
	MIN	500.000	499.120	0.480	500.000	499.615	0.230	500.000	499.869	0.068	500.000	499.940	0.020	500.000	499.960	0.000

METRIC H11/c11: LOOSE RUNNING FIT

BASIC DIA	40 mm
HOLE	40.160 40.000
SHAFT	39.880 39.720
MAX CLEAR.	0.440
MIN CLEAR. (ALLOWANCE)	0.120

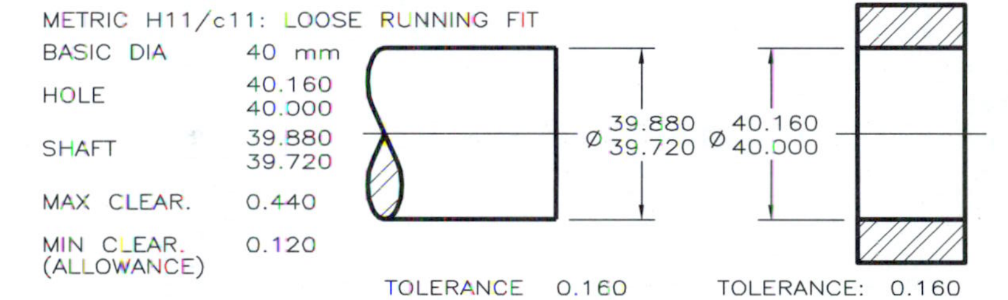

Preferred Hole Basis Transition and Interference Fits–Cylindrical Fits

Dimensions are in mm.

		LOCATIONAL TRANSN.			LOCATIONAL TRANSN.			LOCATIONAL INTERF.			MEDIUM DRIVE			FORCE		
BASIC SIZE		Hole H7	Shaft k6	Fit	Hole H7	Shaft n6	Fit	Hole H7	Shaft p6	Fit	Hole H7	Shaft s6	Fit	Hole H7	Shaft u6	Fit
1	MAX	1.010	1.006	0.010	1.010	1.010	0.006	1.010	1.012	0.004	1.010	1.020	−0.004	1.010	1.024	−0.008
	MIN	1.000	1.000	−0.006	1.000	1.004	−0.010	1.000	1.006	−0.012	1.000	1.014	−0.020	1.000	1.018	−0.024
1.2	MAX	1.210	1.206	0.010	1.210	1.210	0.006	1.210	1.212	0.004	1.210	1.220	−0.004	1.210	1.224	−0.008
	MIN	1.200	1.200	−0.006	1.200	1.204	−0.010	1.200	1.206	−0.012	1.200	1.214	−0.020	1.200	1.218	−0.024
1.6	MAX	1.610	1.606	0.010	1.610	1.610	0.006	1.610	1.612	0.004	1.610	1.620	−0.004	1.610	1.624	−0.008
	MIN	1.600	1.600	−0.006	1.600	1.604	−0.010	1.600	1.606	−0.012	1.600	1.614	−0.020	1.600	1.618	−0.024
2	MAX	2.010	2.006	0.010	2.010	2.010	0.006	2.010	2.010	0.004	2.010	2.020	−0.004	2.010	2.024	−0.008
	MIN	2.000	2.000	−0.006	2.000	2.004	−0.010	2.000	2.006	−0.012	2.000	2.014	−0.020	2.000	2.018	−0.024
2.5	MAX	2.510	2.506	0.010	2.510	2.510	0.006	2.510	2.512	0.004	2.510	2.520	−0.004	2.510	2.524	−0.008
	MIN	2.500	2.500	−0.006	2.500	2.504	−0.010	2.500	2.506	−0.012	2.500	2.514	−0.020	2.500	2.518	−0.024
3	MAX	3.010	3.006	0.010	3.010	3.010	0.006	3.010	3.012	0.004	3.010	3.020	−0.004	3.010	3.024	−0.008
	MIN	3.000	3.000	−0.006	3.000	3.004	−0.010	3.000	3.006	−0.012	3.000	3.014	−0.020	3.000	3.018	−0.024
4	MAX	4.012	4.009	0.011	4.012	4.016	0.004	4.012	4.020	0.000	4.012	4.027	−0.007	4.012	4.031	−0.011
	MIN	4.000	4.001	−0.009	4.000	4.008	−0.016	4.000	4.012	−0.020	4.000	4.019	−0.027	4.000	4.023	−0.031
5	MAX	5.012	5.009	0.011	5.012	5.016	0.004	5.012	5.020	0.000	5.012	5.027	−0.007	5.012	5.031	−0.011
	MIN	5.000	5.001	−0.009	5.000	5.008	−0.016	5.000	5.012	−0.020	5.000	5.019	−0.027	5.000	5.023	−0.031
6	MAX	6.012	6.009	0.011	6.012	6.016	0.004	6.012	6.020	0.000	6.012	6.027	−0.007	6.012	6.031	−0.011
	MIN	6.000	6.001	−0.009	6.000	6.008	−0.016	6.000	6.012	−0.020	6.000	6.019	−0.027	6.000	6.023	−0.031
8	MAX	8.015	8.010	0.014	8.015	8.019	0.005	8.015	8.024	0.000	8.015	8.032	−0.008	8.015	8.037	−0.013
	MIN	8.000	8.001	−0.010	8.000	8.010	−0.019	8.000	8.015	−0.024	8.000	8.023	−0.032	8.000	8.028	−0.037
10	MAX	10.015	10.010	0.014	10.015	10.019	0.005	10.015	10.024	0.000	10.015	10.032	−0.008	10.015	10.037	−0.013
	MIN	10.000	10.001	−0.010	10.000	10.010	−0.019	10.000	10.015	−0.024	10.000	10.023	−0.032	10.000	10.028	−0.037
12	MAX	12.018	12.012	0.017	12.018	12.023	0.006	12.018	12.029	0.000	12.018	12.039	−0.010	12.018	12.044	−0.015
	MIN	12.000	12.001	−0.012	12.000	12.012	−0.023	12.000	12.018	−0.029	12.000	12.028	−0.039	12.000	12.033	−0.044
16	MAX	16.018	16.012	0.017	16.018	16.023	0.006	16.018	16.029	0.000	16.018	16.019	−0.010	16.018	16.044	−0.015
	MIN	16.000	16.001	−0.012	16.000	16.012	−0.023	16.000	16.018	−0.029	16.000	16.028	−0.039	16.000	16.033	−0.044
20	MAX	20.021	20.015	0.019	20.021	20.028	0.006	20.021	20.035	−0.001	20.021	20.048	−0.014	20.021	20.054	−0.020
	MIN	20.000	20.002	−0.015	20.000	20.015	−0.028	20.000	20.022	−0.035	20.000	20.035	−0.048	20.000	20.041	−0.054
25	MAX	25.021	25.015	0.019	25.021	25.028	0.006	25.021	25.035	−0.001	25.021	25.048	−0.014	25.021	25.061	−0.027
	MIN	25.000	25.002	−0.015	25.000	25.015	−0.028	25.000	25.022	−0.035	25.000	25.035	−0.048	25.000	25.048	−0.061
30	MAX	30.021	30.015	0.019	30.021	30.028	0.006	30.021	30.035	−0.001	30.021	30.048	−0.014	30.021	30.061	−0.027
	MIN	30.000	30.002	−0.015	30.000	30.015	−0.028	30.000	30.022	−0.035	30.000	30.035	−0.048	30.000	30.048	−0.061

	LOCATIONAL TRANSN.			LOCATIONAL TRANSN.			LOCATIONAL INTERF.			MEDIUM DRIVE			FORCE		
BASIC SIZE	Hole H7	Shaft k6	Fit	Hole H7	Shaft n6	Fit	Hole H7	Shaft p6	Fit	Hole H7	Shaft s6	Fit	Hole H7	Shaft u6	Fit
0 MAX	40.025	40.018	0.023	40.025	40.033	0.008	40.025	40.042	−0.001	40.025	40.059	−0.018	40.025	40.076	−0.035
MIN	40.000	40.002	−0.018	40.000	40.017	−0.033	40.000	40.026	−0.042	40.000	40.043	−0.059	40.000	40.060	−0.076
50 MAX	50.025	50.018	0.023	50.025	50.033	0.008	50.025	50.042	−0.001	50.025	50.059	−0.018	50.025	50.086	−0.045
MIN	50.000	50.002	−0.018	50.000	50.017	−0.033	50.000	50.026	−0.042	50.000	50.043	−0.059	50.000	50.070	−0.086
60 MAX	60.030	60.021	0.028	60.030	60.039	0.010	60.030	60.051	−0.002	60.030	60.072	−0.023	60.030	60.106	−0.057
MIN	60.000	60.002	−0.021	60.000	60.020	−0.039	60.000	60.032	−0.051	60.000	60.053	−0.072	60.000	60.087	−0.106
80 MAX	80.030	80.021	0.028	80.030	80.039	0.010	80.030	80.051	−0.002	80.030	80.078	−0.029	80.030	80.121	−0.072
MIN	80.000	80.002	−0.021	80.000	80.020	−0.039	80.000	80.032	−0.051	80.000	80.059	−0.078	80.000	80.102	−0.121
100 MAX	100.035	100.025	0.032	100.035	100.045	0.012	100.035	100.059	−0.002	100.035	100.093	−0.036	100.035	100.146	−0.089
MIN	100.000	100.003	−0.025	100.000	100.023	−0.045	100.000	100.037	−0.059	100.000	100.071	−0.093	100.000	100.124	−0.146
120 MAX	120.035	120.025	0.032	120.035	120.045	0.012	120.035	120.059	−0.002	120.035	120.101	−0.044	120.035	120.166	−0.109
MIN	120.000	120.003	−0.025	120.000	120.023	−0.045	120.000	120.037	−0.059	120.000	120.079	−0.101	120.000	120.144	−0.166
160 MAX	160.040	160.028	0.037	160.040	160.052	0.013	160.040	160.068	−0.003	160.040	160.125	−0.060	160.040	160.215	−0.150
MIN	160.000	160.003	−0.028	160.000	160.027	−0.052	160.000	160.043	−0.068	160.000	160.100	−0.125	160.000	160.190	−0.215
200 MAX	200.046	200.033	0.042	200.046	200.060	0.015	200.046	200.079	−0.004	200.046	200.151	−0.076	200.046	200.265	−0.190
MIN	200.000	200.004	−0.033	200.000	200.031	−0.060	200.000	200.050	−0.079	200.000	200.122	−0.151	200.000	200.236	−0.265
250 MAX	250.046	250.033	0.042	250.046	250.060	0.015	250.046	250.079	−0.004	250.046	250.169	−0.094	250.046	250.313	−0.238
MIN	250.000	250.004	−0.033	250.000	250.031	−0.060	250.000	250.050	−0.079	250.000	250.140	−0.169	250.000	250.284	−0.313
300 MAX	300.052	300.036	0.048	300.052	300.066	0.018	300.052	300.088	−0.004	300.052	300.202	−0.118	300.052	300.382	−0.298
MIN	300.000	300.004	−0.036	300.000	300.034	−0.066	300.000	300.056	−0.088	300.000	300.170	−0.202	300.000	300.350	−0.382
400 MAX	400.057	400.040	0.053	400.057	400.073	0.020	400.057	400.098	−0.005	400.057	400.244	−0.151	400.057	400.471	−0.378
MIN	400.000	400.004	−0.040	400.000	400.037	−0.073	400.000	400.062	−0.098	400.000	400.208	−0.244	400.000	400.435	−0.471
500 MAX	500.063	500.045	0.058	500.063	500.080	0.023	500.063	500.108	−0.005	500.063	500.292	−0.189	500.063	500.580	−0.477
MIN	500.000	500.005	−0.045	500.000	500.040	−0.080	500.000	500.068	−0.108	500.000	500.252	−0.292	500.000	500.540	−0.580

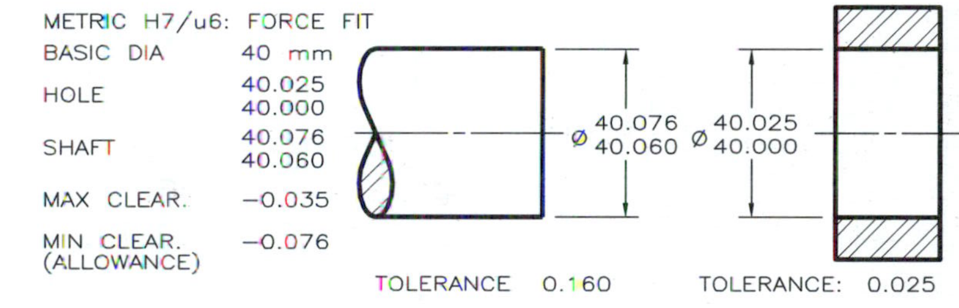

APPENDIX

14

Preferred Shaft Basis Clearance Fits–Cylindrical Fits

Dimensions are in mm.

BASIC SIZE		LOOSE RUNNING Hole C11	LOOSE RUNNING Shaft h11	LOOSE RUNNING Fit	FREE RUNNING Hole D9	FREE RUNNING Shaft h9	FREE RUNNING Fit	CLOSE RUNNING Hole F8	CLOSE RUNNING Shaft h7	CLOSE RUNNING Fit	SLIDING Hole G7	SLIDING Shaft h6	SLIDING Fit	LOCATIONAL CLEARANCE Hole H7	LOCATIONAL CLEARANCE Shaft h6	LOCATIONAL CLEARANCE Fit
1	MAX	1.120	1.000	0.180	1.045	1.000	0.070	1.020	1.000	0.030	1.012	1.000	0.018	1.010	1.000	0.016
	MIN	1.060	0.940	0.060	1.020	0.975	0.020	1.006	0.990	0.006	1.002	0.994	0.002	1.000	0.994	0.000
1.2	MAX	1.320	1.200	0.180	1.245	1.200	0.070	1.220	1.200	0.030	1.212	1.200	0.018	1.210	1.200	0.016
	MIN	1.260	1.140	0.060	1.220	1.175	0.020	1.206	1.190	0.006	1.202	1.194	0.002	1.200	1.194	0.000
1.6	MAX	1.720	1.600	0.180	1.656	1.600	0.070	1.620	1.600	0.030	1.612	1.600	0.018	1.610	1.600	0.016
	MIN	1.660	1.540	0.060	1.620	1.575	0.020	1.606	1.590	0.006	1.602	1.595	0.002	1.600	1.594	0.000
2	MAX	2.120	2.000	0.180	2.045	2.000	0.070	2.020	2.000	0.030	2.012	2.000	0.018	2.010	2.000	0.016
	MIN	2.060	1.940	0.060	2.020	1.975	0.020	2.006	1.990	0.006	2.002	1.994	0.002	2.000	1.994	0.000
2.5	MAX	2.620	2.500	0.180	2.545	2.500	0.070	2.520	2.500	0.030	2.512	2.500	0.018	2.510	2.500	0.016
	MIN	2.560	2.440	0.060	2.520	2.475	0.020	2.506	2.490	0.006	2.502	2.494	0.002	2.500	2.494	0.000
3	MAX	3.120	3.000	0.180	3.045	3.000	0.070	3.020	3.000	0.030	3.012	3.000	0.018	3.010	3.000	0.016
	MIN	3.060	2.940	0.060	3.020	2.975	0.020	3.006	2.990	0.006	3.002	2.994	0.002	3.000	2.994	0.000
4	MAX	4.145	4.000	0.220	4.060	4.000	0.090	4.028	4.000	0.040	4.016	4.000	0.024	4.012	4.000	0.020
	MIN	4.070	3.925	0.070	4.030	3.970	0.030	4.010	3.988	0.010	4.004	3.992	0.004	4.000	3.992	0.000
5	MAX	5.145	5.000	0.220	5.060	5.000	0.090	5.028	5.000	0.040	5.016	5.000	0.024	5.012	5.000	0.020
	MIN	5.070	4.925	0.070	5.030	4.970	0.030	5.010	4.988	0.010	5.004	4.992	0.004	5.000	4.992	0.000
6	MAX	6.145	6.000	0.220	6.060	6.000	0.090	6.028	6.000	0.040	6.016	6.000	0.024	6.012	6.000	0.020
	MIN	6.070	5.925	0.070	6.030	5.970	0.030	6.010	5.988	0.010	6.004	5.992	0.004	6.000	5.992	0.000
8	MAX	8.170	8.000	0.260	8.076	8.000	0.112	8.035	8.000	0.050	8.020	8.000	0.029	8.015	8.000	0.024
	MIN	8.080	7.910	0.080	8.040	7.964	0.040	8.013	7.985	0.013	8.005	7.991	0.005	8.000	7.991	0.000
10	MAX	10.170	10.000	0.260	10.076	10.000	0.112	10.035	10.000	0.050	10.020	10.000	0.029	10.015	10.000	0.024
	MIN	10.080	9.910	0.080	10.040	9.964	0.040	10.013	9.985	0.013	10.005	9.991	0.005	10.000	9.991	0.000
12	MAX	12.205	12.000	0.315	12.093	12.000	0.136	12.043	12.000	0.061	12.024	12.000	0.035	12.018	12.000	0.029
	MIN	12.095	11.890	0.095	12.050	11.957	0.050	12.016	11.982	0.016	12.006	11.989	0.006	12.000	11.989	0.000
16	MAX	16.205	16.000	0.315	16.093	16.000	0.136	16.043	16.000	0.061	16.024	16.000	0.035	16.018	16.000	0.029
	MIN	16.095	15.890	0.095	16.050	15.957	0.050	16.016	15.982	0.016	16.006	15.989	0.006	16.000	15.989	0.000
20	MAX	20.240	20.000	0.370	20.117	20.000	0.169	20.053	20.000	0.074	20.028	20.000	0.041	20.021	20.000	0.034
	MIN	20.110	19.870	0.110	20.065	19.948	0.065	20.020	19.979	0.020	20.007	19.987	0.007	20.000	19.987	0.000
25	MAX	25.240	25.000	0.370	25.117	25.000	0.169	25.053	25.000	0.074	25.028	25.000	0.041	25.021	25.000	0.034
	MIN	25.110	24.870	0.110	25.065	24.948	0.065	25.020	24.979	0.020	25.007	24.987	0.007	25.000	24.987	0.000
30	MAX	30.240	30.000	0.370	30.117	30.000	0.169	30.053	30.000	0.074	30.028	30.000	0.041	30.021	30.000	0.034
	MIN	30.110	29.870	0.110	30.065	29.948	0.065	30.020	29.979	0.020	30.007	29.987	0.007	30.000	29.987	0.000

BASIC SIZE		LOOSE RUNNING Hole C11	LOOSE RUNNING Shaft h11	LOOSE RUNNING Fit	FREE RUNNING Hole D9	FREE RUNNING Shaft h9	FREE RUNNING Fit	CLOSE RUNNING Hole F8	CLOSE RUNNING Shaft h7	CLOSE RUNNING Fit	SLIDING Hole G7	SLIDING Shaft h6	SLIDING Fit	LOCATIONAL CLEARANCE Hole H7	LOCATIONAL CLEARANCE Shaft h6	LOCATIONAL CLEARANCE Fit
40	MAX	40.280	40.000	0.440	40.142	40.000	0.204	40.064	40.000	0.089	40.034	40.000	0.050	40.025	40.000	0.041
	MIN	40.120	39.840	0.120	40.080	39.938	0.080	40.025	39.975	0.025	40.009	39.984	0.009	40.000	39.984	0.000
50	MAX	50.290	50.000	0.450	50.142	50.000	0.204	50.064	50.000	0.089	50.034	50.000	0.050	50.025	50.000	0.041
	MIN	50.130	49.840	0.130	50.080	49.938	0.080	50.025	49.975	0.025	50.009	49.984	0.009	50.000	49.984	0.000
60	MAX	60.330	60.000	0.520	60.174	60.000	0.248	60.076	60.000	0.106	60.040	60.000	0.059	60.030	60.000	0.049
	MIN	60.140	59.810	0.140	60.100	59.926	0.100	60.030	59.970	0.030	60.010	59.981	0.010	60.000	59.981	0.000
80	MAX	80.340	80.000	0.530	80.174	80.000	0.248	80.076	80.000	0.106	80.040	80.000	0.059	80.030	80.000	0.049
	MIN	80.150	79.810	0.150	80.100	79.926	0.100	80.030	79.970	0.030	80.010	79.981	0.010	80.000	79.981	0.000
100	MAX	100.390	100.000	0.610	100.207	100.000	0.294	100.090	100.000	0.125	100.047	100.000	0.069	100.035	100.000	0.057
	MIN	100.170	99.780	0.170	100.120	99.913	0.120	100.036	99.965	0.036	100.012	99.979	0.012	100.000	99.979	0.000
120	MAX	120.400	120.000	0.620	120.207	120.000	0.294	120.090	120.000	0.125	120.047	120.000	0.069	120.035	120.000	0.057
	MIN	120.180	119.780	0.180	120.120	119.913	0.120	120.036	119.965	0.036	120.012	119.978	0.012	120.000	119.978	0.000
160	MAX	160.460	160.000	0.710	160.245	160.000	0.345	160.106	160.000	0.146	160.054	160.000	0.079	160.040	160.000	0.065
	MIN	160.210	159.750	0.210	160.145	159.900	0.145	160.043	159.960	0.043	160.014	159.975	0.014	160.000	159.975	0.000
200	MAX	200.530	200.000	0.820	200.285	200.000	0.400	200.122	200.000	0.168	200.061	200.000	0.090	200.046	200.000	0.075
	MIN	200.240	199.710	0.240	200.170	199.885	0.170	200.050	199.954	0.050	200.015	199.971	0.015	200.000	199.971	0.000
250	MAX	250.570	250.000	0.860	250.285	250.000	0.400	250.122	250.000	0.168	250.061	250.000	0.090	250.046	250.000	0.075
	MIN	250.280	249.710	0.280	250.170	249.885	0.170	250.050	249.954	0.050	250.015	249.971	0.015	250.000	249.971	0.000
300	MAX	300.650	300.000	0.970	300.320	300.000	0.450	300.137	300.000	0.189	300.069	300.000	0.101	300.052	300.000	0.084
	MIN	300.330	299.680	0.330	300.190	299.870	0.190	300.056	299.948	0.056	300.017	299.968	0.017	300.000	299.968	0.000
400	MAX	400.760	400.000	1.120	400.350	400.000	0.490	400.151	400.000	0.208	400.075	400.000	0.111	400.057	400.000	0.983
	MIN	400.400	399.640	0.400	400.210	399.860	0.210	400.062	399.943	0.062	400.018	399.964	0.018	400.000	399.964	0.000
500	MAX	500.880	500.000	1.280	500.385	500.000	0.540	500.165	500.000	0.228	500.083	500.000	0.123	500.063	500.000	0.103
	MIN	500.480	499.600	0.480	500.230	499.845	0.230	500.068	499.937	0.068	500.020	499.960	0.020	500.000	499.960	0.000

METRIC C11/h11: LOOSE RUNNING FIT

BASIC DIA	40 mm
HOLE	40.280 40.120
SHAFT	40.000 39.840
MAX CLEAR.	0.440
MIN CLEAR. (ALLOWANCE)	0.120

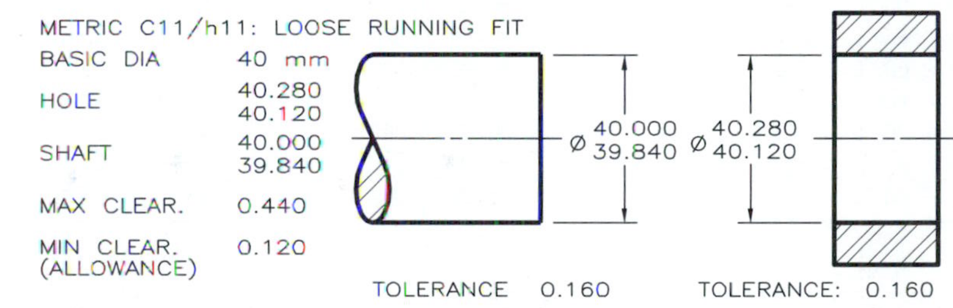

APPENDIX 15

Preferred Shaft Basis Transition and Interference Fits–Cylindrical Fits

Dimensions are in mm.

BASIC SIZE		LOCATIONAL TRANSN.			LOCATIONAL TRANSN.			LOCATIONAL INTERF.			MEDIUM DRIVE			FORCE		
		Hole K7	Shaft h6	Fit	Hole N7	Shaft h6	Fit	Hole P7	Shaft h6	Fit	Hole S7	Shaft h6	Fit	Hole U7	Shaft h6	Fit
1	MAX	1.000	1.000	0.006	0.996	1.000	0.002	0.994	1.000	0.000	0.986	1.000	−0.008	0.982	1.000	−0.012
	MIN	0.990	0.994	−0.010	0.986	0.994	−0.014	0.984	0.994	−0.016	0.976	0.994	−0.024	0.972	0.994	−0.028
1.2	MAX	1.200	1.200	0.006	1.196	1.200	0.002	1.194	1.200	0.000	1.186	1.200	−0.008	1.182	1.200	−0.012
	MIN	1.190	1.194	−0.010	1.186	1.194	−0.014	1.184	1.194	−0.016	1.176	1.194	−0.024	1.172	1.194	−0.028
1.6	MAX	1.600	1.600	0.006	1.596	1.600	0.002	1.594	1.600	0.000	1.586	1.600	−0.008	1.582	1.600	−0.012
	MIN	1.590	1.594	−0.010	1.586	1.594	−0.014	1.584	1.594	−0.016	1.576	1.594	−0.024	1.572	1.594	−0.028
2	MAX	2.000	2.000	0.006	1.996	2.000	0.002	1.994	2.000	0.008	1.986	2.000	−0.008	1.982	2.000	−0.012
	MIN	1.990	1.994	−0.010	1.986	1.994	−0.014	1.984	1.994	−0.016	1.976	1.994	−0.024	1.972	1.994	−0.028
2.5	MAX	2.500	2.500	0.006	2.496	2.500	0.002	2.494	2.500	0.000	2.486	2.500	−0.008	2.482	2.500	−0.012
	MIN	2.490	2.494	−0.010	2.486	2.494	−0.014	2.484	2.494	−0.016	2.476	2.494	−0.024	2.472	2.494	−0.028
3	MAX	3.000	3.000	0.006	2.996	3.000	0.002	2.994	3.000	0.000	2.986	3.000	−0.008	2.982	3.000	−0.012
	MIN	2.990	2.994	−.010	2.986	2.994	−0.014	2.984	2.994	−0.016	2.976	2.994	−0.024	2.972	2.994	−0.028
4	MAX	4.003	4.000	0.011	3.996	4.000	0.004	3.992	4.000	0.000	3.985	4.000	−0.007	3.981	4.000	−0.011
	MIN	3.991	3.992	−0.009	3.984	3.992	−0.016	3.980	3.992	−0.020	3.973	3.992	−0.027	3.969	3.992	−0.031
5	MAX	5.003	5.000	0.011	4.996	5.000	0.004	4.992	5.000	0.000	4.985	5.000	−0.007	4.981	5.000	−0.011
	MIN	4.991	4.992	−0.009	4.984	4.992	−0.016	4.980	4.992	−0.020	4.973	4.992	−0.027	4.969	4.992	−0.031
6	MAX	6.003	6.000	0.011	5.996	6.000	0.004	5.992	6.000	0.000	5.985	6.000	−0.007	5.981	6.000	−0.011
	MIN	5.991	5.992	−0.009	5.984	5.992	−0.016	5.980	5.992	−0.020	5.973	5.992	−0.027	5.969	5.992	−0.031
8	MAX	8.005	8.000	0.014	7.986	8.000	0.005	7.991	8.000	0.000	7.983	8.000	−0.008	7.978	8.000	−0.013
	MIN	7.990	7.991	−0.010	7.981	7.991	−0.019	7.976	7.991	−0.024	7.968	7.991	−0.032	7.963	7.991	−0.037
10	MAX	10.005	10.000	0.014	9.996	10.000	0.005	9.991	10.000	0.0000	9.983	10.000	−0.008	9.978	10.000	−0.013
	MIN	9.990	9.991	−0.010	9.981	9.991	−0.019	9.976	9.991	−0.024	9.968	9.991	−0.032	9.963	9.991	−0.037
12	MAX	12.006	12.000	0.017	11.995	12.000	0.006	11.989	12.000	0.000	11.979	12.000	−0.010	11.974	12.000	−0.015
	MIN	11.988	11.989	−0.012	11.977	11.989	−0.023	11.971	11.989	−0.029	11.961	11.989	−0.039	11.956	11.989	−0.044
16	MAX	16.006	16.000	0.017	15.995	16.000	0.006	15.989	16.000	0.000	15.979	16.000	−0.010	15.974	16.000	−0.015
	MIN	15.988	15.989	−0.012	15.977	15.989	−0.023	15.971	15.989	−0.029	15.961	15.989	−0.039	15.956	15.989	−0.044
20	MAX	20.006	20.000	0.019	19.993	20.000	0.006	19.986	20.000	−0.001	19.973	20.000	−0.014	19.967	20.000	−0.020
	MIN	19.985	19.987	−0.015	19.972	19.987	−0.028	19.965	19.987	−0.035	19.952	19.987	−0.048	19.946	19.987	−0.054
25	MAX	25.006	25.000	0.019	24.993	25.000	0.006	24.986	25.000	−0.001	24.973	25.000	−0.014	24.960	25.000	−0.027
	MIN	24.985	24.987	−0.015	24.972	24.987	−0.028	24.965	24.987	−0.035	24.952	24.987	−0.048	24.939	24.987	−0.061
30	MAX	30.006	30.000	0.019	29.993	30.000	0.006	29.986	30.000	−0.001	29.973	30.000	−0.014	29.960	30.000	−0.027
	MIN	29.985	29.987	−0.015	29.972	29.987	−0.028	29.965	29.987	−0.035	29.952	29.987	−0.048	29.939	29.987	−0.061

Dimensions are in mm.

BASIC SIZE		LOCATIONAL TRANSN.			LOCATIONAL TRANSN.			LOCATIONAL INTERF.			MEDIUM DRIVE			FORCE		
		Hole K7	Shaft h6	Fit	Hole N7	Shaft h6	Fit	Hole P7	Shaft h6	Fit	Hole S7	Shaft h6	Fit	Hole U7	Shaft h6	Fit
40	MAX	40.007	40.000	0.023	39.992	40.000	0.008	39.983	40.000	−0.001	39.966	40.000	−0.018	39.949	40.000	−0.035
	MIN	39.982	39.984	−0.018	39.967	39.984	−0.033	39.958	39.984	−0.042	39.941	39.984	−0.059	39.924	39.984	−0.076
50	MAX	50.007	50.000	0.023	49.992	50.000	0.008	49.983	50.000	−0.001	49.966	50.000	−0.018	49.939	50.000	−0.045
	MIN	49.982	49.984	−0.018	49.967	49.984	−0.033	49.958	49.984	−0.042	49.941	49.984	−0.059	49.914	49.984	−0.086
60	MAX	60.009	60.000	0.028	59.991	60.000	0.010	59.979	60.000	−0.002	59.958	60.000	−0.023	59.924	60.000	−0.057
	MIN	59.979	59.981	−0.021	59.961	59.981	−0.039	59.949	59.981	−0.051	59.928	59.981	−0.072	59.894	59.981	−0.106
80	MAX	80.009	80.000	0.028	79.991	80.000	0.010	79.979	80.000	−0.002	79.952	80.000	−0.029	79.909	80.000	−0.072
	MIN	79.979	79.981	−0.021	79.961	79.981	−0.039	79.949	79.981	−0.051	79.922	79.981	−0.078	79.879	79.981	−0.121
100	MAX	100.010	100.000	0.032	99.990	100.000	0.012	99.976	100.000	−0.002	99.942	100.000	−0.036	99.889	100.000	−0.089
	MIN	99.975	99.978	−0.025	99.955	99.978	−0.045	99.941	99.978	−0.059	99.907	99.978	−0.093	99.854	99.978	−0.146
120	MAX	120.010	120.000	0.032	119.990	120.000	0.012	119.976	120.000	−0.002	119.934	120.000	−0.044	119.869	120.000	−0.109
	MIN	119.975	119.978	−0.025	119.955	119.978	−0.045	119.941	119.978	−0.059	119.899	119.978	−0.101	119.834	119.978	−0.166
160	MAX	160.012	160.000	0.037	159.988	160.000	0.013	159.972	160.000	−0.003	159.915	160.000	−0.060	159.825	160.000	−0.150
	MIN	159.972	159.975	−0.028	159.948	159.975	−0.052	159.932	159.975	−0.068	159.875	159.975	−0.125	159.785	159.975	−0.215
200	MAX	200.013	200.000	0.042	199.986	200.000	0.015	199.967	200.000	−0.004	199.895	200.000	−0.076	199.781	200.000	−0.190
	MIN	199.967	199.971	−0.033	199.940	199.971	−0.060	199.921	199.971	−0.079	199.849	199.971	−0.151	199.735	199.971	−0.265
250	MAX	250.013	250.000	0.042	249.986	250.000	0.015	249.967	250.000	−0.004	249.877	250.000	−0.094	249.733	250.000	−0.238
	MIN	249.967	249.971	−0.033	249.940	249.971	−0.060	249.921	249.971	−0.079	249.831	249.971	−0.169	249.687	249.971	−0.313
300	MAX	300.016	300.000	0.048	299.986	300.000	0.018	299.964	300.000	−0.004	299.850	300.000	−0.188	299.670	300.000	−0.298
	MIN	299.964	299.968	−0.036	299.934	299.968	−0.066	299.912	299.968	−0.088	299.798	299.968	−0.202	299.618	299.968	−0.382
400	MAX	400.017	400.000	0.053	399.984	400.000	0.020	399.959	400.000	−0.005	399.813	400.000	−0.151	399.586	400.000	−0.378
	MIN	399.960	399.964	−0.040	399.927	399.964	−0.073	399.902	399.964	−0.08	399.756	399.964	−0.244	399.529	399.964	−0.471
500	MAX	500.018	500.000	0.058	499.983	500.000	0.023	499.955	500.000	−0.005	499.771	500.000	−0.189	499.483	500.000	−0.477
	MIN	499.955	499.960	−0.045	499.920	499.960	−0.080	499.892	499.960	−0.1808	499.708	499.960	−0.292	499.420	499.960	−0.580

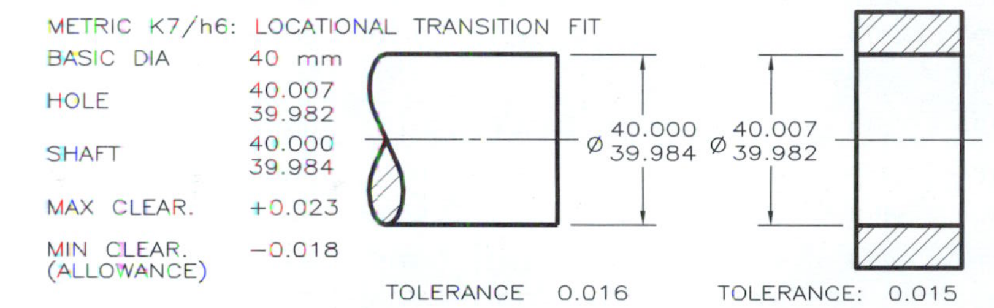

APPENDIX 16

Hole Sizes for Nonpreferred Diameters (millimeters)

Basic Size		C11	D9	F8	G7	H7	H8	H9	H11	K7	N7	P7	S7	U7
OVER	0	+0.120	+0.045	+0.020	+0.012	+0.010	+0.014	+0.025	+0.060	0.000	−0.004	−0.006	−0.014	−0.018
TO	3	+0.060	+0.020	+0.006	+0.002	0.000	0.000	0.000	0.000	−0.010	−0.014	−0.016	−0.024	−0.028
OVER	3	+0.145	+0.060	+0.028	+0.016	+0.012	+0.018	+0.030	+0.075	+0.003	−0.004	−0.008	−0.015	−0.019
TO	6	+0.070	+0.030	+0.010	+0.004	0.000	0.000	0.000	0.000	−0.009	−0.016	−0.020	−0.027	−0.031
OVER	6	+0.170	+0.076	+0.035	+0.020	+0.015	+0.022	+0.036	+0.090	+0.005	−0.004	−0.009	−0.017	−0.022
TO	10	+0.080	+0.040	+0.013	+0.005	0.000	0.000	0.000	0.000	−0.010	−0.019	−0.024	−0.032	−0.037
OVER	10	+0.205	+0.093	+0.043	+0.024	+0.018	+0.027	+0.043	+0.110	+0.006	−0.005	−0.011	−0.021	−0.026
TO	14	+0.095	+0.050	+0.016	+0.006	0.000	0.000	0.000	0.000	−0.012	−0.023	−0.029	−0.039	−0.044
OVER	14	+0.205	+0.093	+0.043	+0.024	+0.018	+0.027	+0.043	+0.110	+0.006	−0.005	−0.011	−0.021	−0.026
TO	18	+0.095	+0.050	+0.016	+0.006	0.000	0.000	0.000	0.000	−0.012	−0.023	−0.029	−0.039	−0.044
OVER	18	+0.240	+0.117	+0.053	+0.028	+0.021	+0.033	+0.052	+0.130	+0.006	−0.007	−0.014	−0.027	−0.033
TO	24	+0.110	+0.065	+0.020	+0.007	0.000	0.000	0.000	0.000	−0.015	−0.028	−0.035	−0.048	−0.054
OVER	24	+0.240	+0.117	+0.053	+0.028	+0.021	+0.033	+0.052	+0.130	+0.006	−0.007	−0.014	−0.027	−0.040
TO	30	+0.110	+0.065	+0.020	+0.007	0.000	0.000	0.000	0.000	−0.015	−0.028	−0.035	−0.048	−0.061
OVER	30	+0.280	+0.142	+0.064	+0.034	+0.025	+0.039	+0.062	+0.160	+0.007	−0.008	−0.017	−0.034	−0.051
TO	40	+0.120	+0.080	+0.025	+0.009	0.000	0.000	0.000	0.000	−0.018	−0.033	−0.042	−0.059	−0.076
OVER	40	+0.290	+0.142	+0.064	+0.034	+0.025	+0.039	+0.062	+0.160	+0.007	−0.008	−0.017	−0.034	−0.061
TO	50	+0.130	+0.080	+0.025	+0.009	0.000	0.000	0.000	0.000	−0.018	−0.033	−0.042	−0.059	−0.086
OVER	50	+0.330	+0.174	+0.076	+0.040	+0.030	+0.046	+0.074	+0.190	+0.009	−0.009	−0.021	−0.042	−0.076
TO	65	+0.140	+0.100	+0.030	+0.010	0.000	0.000	0.000	0.000	−0.021	−0.039	−0.051	−0.072	−0.106
OVER	65	+0.340	+0.174	+0.076	+0.040	+0.030	+0.046	+0.074	+0.190	+0.009	−0.009	−0.021	−0.048	−0.091
TO	80	+0.150	+0.100	+0.030	+0.010	0.000	0.000	0.000	0.000	−0.021	−0.039	−0.051	−0.078	−0.121
OVER	80	+0.390	+0.207	+0.090	+0.047	+0.035	+0.054	+0.087	+0.220	+0.010	−0.010	−0.024	−0.058	−0.111
TO	100	+0.170	+0.120	+0.036	+0.012	0.000	0.000	0.000	0.000	−0.025	−0.045	−0.059	−0.093	−0.146

Basic Size		c11	d9	f8	g7	h7	h8	h9	h11	k7	n7	p7	s7	u7
OVER	100	+0.400	+0.207	+0.090	+0.047	+0.035	+0.054	+0.087	+0.220	+0.010	−0.010	−0.024	−0.066	−0.131
TO	120	+0.180	+0.120	+0.036	+0.012	0.000	0.000	0.000	0.000	−0.025	−0.045	−0.059	−0.101	−0.166
OVER	120	+0.450	+0.245	+0.106	+0.054	+0.040	+0.063	+0.100	+0.250	+0.012	−0.012	−0.028	−0.077	−0.155
TO	140	+0.200	+0.145	+0.043	+0.014	0.000	0.000	0.000	0.000	−0.028	−0.052	−0.068	−0.117	−0.195
OVER	140	+0.460	+0.245	+0.106	+0.054	+0.040	+0.063	+0.100	+0.250	+0.012	−0.012	−0.028	−0.085	−0.175
TO	160	+0.210	+0.145	+0.043	+0.014	0.000	0.000	0.000	0.000	−0.028	−0.052	−0.068	−0.125	−0.215
OVER	160	+0.480	+0.245	+0.106	+0.054	+0.040	+0.063	+0.100	+0.250	+0.012	−0.012	−0.028	−0.093	−0.195
TO	180	+0.230	+0.145	+0.043	+0.014	0.000	0.000	0.000	0.000	−0.028	−0.052	−0.068	−0.133	−0.235
OVER	180	+0.530	+0.285	+0.122	+0.061	+0.046	+0.072	+0.115	+0.290	−0.013	−0.014	−0.033	−0.105	−0.219
TO	200	+0.240	+0.170	+0.050	+0.015	0.000	0.000	0.000	0.000	−0.033	−0.060	−0.079	−0.151	−0.265
OVER	200	+0.550	+0.285	+0.122	+0.061	+0.046	+0.072	+0.115	+0.290	+0.013	−0.014	−0.033	−0.113	−0.241
TO	225	+0.260	+0.170	+0.050	+0.015	0.000	0.000	0.000	0.000	−0.033	−0.060	−0.079	−0.159	−0.287
OVER	225	+0.570	+0.285	+0.122	+0.061	+0.046	+0.072	+0.115	+0.290	+0.013	−0.014	−0.033	−0.123	−0.267
TO	250	+0.280	+0.170	+0.050	+0.015	0.000	0.000	0.000	0.000	−0.033	−0.060	−0.079	−0.169	−0.313
OVER	250	+0.620	+0.320	+0.137	+0.069	+0.052	+0.081	+0.130	+0.320	+0.016	−0.014	−0.036	−0.138	−0.295
TO	280	+0.300	+0.190	+0.056	+0.017	0.000	0.000	0.000	0.000	−0.036	−0.066	−0.088	−0.190	−0.347
OVER	280	+0.650	+0.320	+0.137	+0.069	+0.052	+0.081	+0.130	+0.320	+0.016	−0.014	−0.036	−0.150	−0.330
TO	315	+0.330	+0.190	+0.056	0.017	0.000	0.000	0.000	0.000	−0.036	−0.066	−0.088	−0.202	−0.382
OVER	315	+0.720	+0.350	+0.151	+0.075	+0.057	+0.089	+0.140	+0.360	+0.017	−0.016	−0.041	−0.169	−0.369
TO	355	+0.360	+0.210	+0.062	+0.018	0.000	0.000	0.000	0.000	−0.040	−0.073	−0.058	−0.226	−0.426
OVER	355	+0.760	+0.350	+0.151	+0.075	+0.057	+0.089	+0.140	+0.360	+0.017	−0.016	−0.041	−0.187	−0.414
TO	400	+0.400	+0.210	+0.062	+0.018	0.000	0.000	0.000	0.000	−0.040	−0.073	−0.058	−0.244	−0.471
OVER	400	+0.840	+0.385	+0.165	+0.083	+0.063	+0.097	+0.155	+0.400	+0.018	−0.017	−0.045	−0.209	−0.467
TO	450	+0.440	+0.230	+0.068	+0.020	0.000	0.000	0.000	0.000	−0.045	−0.080	−0.108	−0.272	−0.530
OVER	450	+0.880	+0.385	+0.165	+0.083	+0.063	+0.097	+0.155	+0.400	+0.018	−0.017	−0.045	−0.229	−0.517
TO	500	+0.480	+0.230	+0.068	+0.020	0.000	0.000	0.000	0.000	−0.045	−0.080	−0.108	−0.292	−0.580

APPENDIX 17

Shaft Sizes for Nonpreferred Diameters (millimeters)

Basic Size		c11	d9	f7	g6	h6	h7	h9	h11	k6	n6	p6	s6	u6
OVER	0	−0.060	−0.020	−0.006	−0.002	0.000	0.000	0.000	0.000	+0.006	+0.010	+0.012	+0.020	+0.024
TO	3	−0.120	−0.045	−0.016	−0.008	−0.006	−0.010	−0.025	−0.060	0.000	+0.004	+0.006	+0.014	+0.018
OVER	3	−0.070	−0.030	−0.010	−0.004	0.000	0.000	0.000	0.000	+0.009	+0.016	+0.020	+0.027	+0.031
TO	6	−0.145	−0.060	−0.022	−0.012	−0.008	−0.012	−0.030	−0.075	+0.001	+0.008	+0.012	+0.019	+0.023
OVER	6	−0.080	−0.040	−0.013	−0.005	0.000	0.000	0.000	0.000	+0.010	+0.019	+0.024	+0.032	+0.037
TO	10	−0.170	−0.076	−0.028	−0.014	−0.009	−0.015	−0.036	−0.090	+0.001	+0.010	+0.024	+0.023	+0.028
OVER	10	−0.095	−0.050	−0.016	−0.006	0.000	0.000	0.000	0.000	+0.012	+0.023	+0.029	+0.039	+0.044
TO	14	−0.205	−0.093	−0.034	−0.017	−0.011	−0.018	−0.043	−0.110	+0.001	+0.012	+0.018	+0.028	+0.033
OVER	14	−0.095	−0.050	−0.016	−0.006	0.000	0.000	0.000	0.000	+0.012	+0.023	+0.029	+0.039	+0.044
TO	18	−0.205	−0.093	−0.034	−0.017	−0.011	−0.018	−0.043	−0.110	+0.001	+0.012	+0.018	+0.028	+0.033
OVER	18	−0.110	−0.065	−0.020	−0.007	0.000	0.000	0.000	0.000	+0.015	+0.028	+0.035	+0.048	+0.054
TO	24	−0.240	−0.117	−0.041	−0.020	−0.013	−0.021	−0.052	−0.130	+0.002	+0.015	+0.022	+0.035	+0.041
OVER	24	−0.110	−0.065	−0.020	−0.007	0.000	0.000	0.000	0.000	+0.015	+0.028	+0.035	+0.048	+0.061
TO	30	−0.240	−0.117	−0.041	−0.020	−0.013	−0.021	−0.052	−0.130	+0.002	+0.015	+0.022	+0.035	+0.048
OVER	30	−0.120	−0.080	−0.025	−0.009	0.000	0.000	0.000	0.000	+0.018	+0.033	+0.042	+0.059	+0.076
TO	40	−0.280	−0.142	−0.050	−0.025	−0.016	−0.025	−0.062	−0.160	+0.002	+0.017	+0.026	+0.043	+0.060
OVER	40	−0.130	−0.080	−0.025	−0.009	0.000	0.000	0.000	0.000	+0.018	+0.033	+0.042	+0.059	+0.086
TO	50	−0.290	−0.142	−0.050	−0.025	−0.016	−0.025	−0.062	−0.160	+0.002	+0.017	+0.026	+0.043	+0.070
OVER	50	−0.140	−0.100	−0.030	−0.010	0.000	0.000	0.000	0.000	+0.021	+0.039	+0.051	+0.072	+0.106
TO	65	−0.330	−0.174	−0.060	−0.029	−0.019	−0.030	−0.074	−0.190	+0.002	+0.020	−0.032	+0.053	+0.087
OVER	65	−0.150	−0.100	−0.030	−0.010	0.000	0.000	0.000	0.000	+0.021	+0.039	+0.051	+0.078	+0.121
TO	80	−0.340	−0.174	−0.060	−0.029	−0.019	−0.030	−0.074	−0.190	+0.002	+0.020	+0.032	+0.059	+0.102
OVER	80	−0.170	−0.120	−0.036	−0.012	0.000	0.000	0.000	0.000	+0.025	+0.045	+0.059	+0.093	+0.146
TO	100	−0.390	−0.207	−0.071	−0.034	−0.022	−0.035	−0.087	−0.220	+0.003	+0.023	+0.037	+0.071	+0.124

Basic Size		c11	d9	f7	g6	h6	h7	h9	h11	k6	n6	p6	s6	u6
OVER	100	−0.180	−0.120	−0.036	−0.012	0.000	0.000	0.000	0.000	+0.025	+0.045	+0.059	+0.101	+0.166
TO	120	−0.400	−0.207	−0.071	−0.034	−0.022	−0.035	−0.087	−0.220	+0.003	+0.023	+0.037	+0.079	+0.144
OVER	120	−0.200	−0.145	−0.043	−0.014	0.000	0.000	0.000	0.000	+0.028	+0.052	+0.068	+0.117	+0.195
TO	140	−0.450	−0.245	−0.083	−0.039	−0.025	−0.040	−0.100	−0.250	+0.003	+0.027	+0.043	+0.092	+0.170
OVER	140	−0.210	−0.145	−0.043	−0.014	0.000	0.000	0.000	0.000	+0.028	+0.052	+0.068	+0.125	+0.215
TO	160	−0.460	−0.245	−0.083	−0.039	−0.025	−0.040	−0.100	−0.250	+0.003	+0.027	+0.043	+0.100	+0.190
OVER	160	−0.230	−0.145	−0.043	−0.014	0.000	0.000	0.000	0.000	+0.028	+0.052	+0.068	+0.133	+0.235
TO	180	−0.480	−0.245	−0.083	−0.039	−0.025	−0.040	−0.100	−0.250	+0.003	+0.027	+0.043	+0.108	+0.210
OVER	180	−0.240	−0.170	−0.050	−0.015	0.000	0.000	0.000	0.000	+0.033	+0.060	+0.079	+0.151	+0.265
TO	200	−0.530	−0.285	−0.096	−0.044	−0.029	−0.046	−0.115	−0.290	+0.004	+0.031	+0.050	+0.122	+0.236
OVER	200	−0.260	−0.170	−0.050	−0.015	0.000	0.000	0.000	0.000	+0.033	+0.060	+0.079	+0.159	+0.287
TO	225	−0.550	−0.285	−0.096	−0.044	−0.029	−0.046	−0.115	−0.290	+0.004	+0.031	+0.050	+0.130	+0.258
OVER	225	−0.280	−0.170	−0.050	−0.015	0.000	0.000	0.000	0.000	+0.033	+0.060	+0.079	+0.169	+0.313
TO	250	−0.570	−0.285	−0.096	−0.044	−0.029	−0.046	−0.115	−0.290	+0.004	+0.031	+0.050	+0.140	+0.284
OVER	250	−0.300	−0.190	−0.056	−0.017	0.000	0.000	0.000	0.000	+0.036	+0.066	+0.088	+0.190	+0.347
TO	280	−0.620	−0.320	−0.108	−0.049	−0.032	−0.052	−0.130	−0.320	+0.004	+0.034	+0.056	+0.158	+0.315
OVER	280	−0.330	−0.190	−0.056	−0.017	0.000	0.000	0.000	0.000	+0.036	+0.066	+0.088	+0.202	+0.382
TO	315	−0.650	−0.320	−0.108	−0.049	−0.032	−0.052	−0.130	−0.320	+0.004	+0.034	+0.056	+0.170	+0.350
OVER	315	−0.360	−0.210	−0.062	−0.018	0.000	0.000	0.000	0.000	+0.040	+0.073	+0.098	+0.226	+0.426
TO	355	−0.720	−0.350	−0.119	−0.054	−0.036	−0.057	−0.140	−0.360	+0.004	+0.037	+0.062	+0.190	+0.390
OVER	355	−0.400	−0.210	−0.062	−0.018	0.000	0.000	0.000	0.000	+0.040	+0.073	+0.098	+0.244	+0.471
TO	400	−0.760	−0.350	−0.119	−0.054	−0.036	−0.057	−0.140	−0.360	+0.004	+0.037	+0.062	+0.208	+0.435
OVER	400	−0.440	−0.230	−0.068	−0.020	0.000	0.000	0.000	0.000	+0.045	+0.080	+0.108	+0.272	+0.530
TO	450	−0.840	−0.385	−0.131	−0.060	0.040	0.063	−0.155	0.400	+0.005	+0.040	+0.068	+0.232	+0.490
OVER	450	−0.480	−0.230	−0.068	−0.020	0.000	0.000	0.000	0.000	+0.045	+0.080	+0.108	+0.292	+0.580
TO	500	−0.880	−0.385	−0.131	−0.060	−0.040	−0.063	−0.155	−0.400	+0.005	+0.040	+0.068	+0.252	+0.540

APPENDIX 

Screw Threads: American National and Unified (inches)

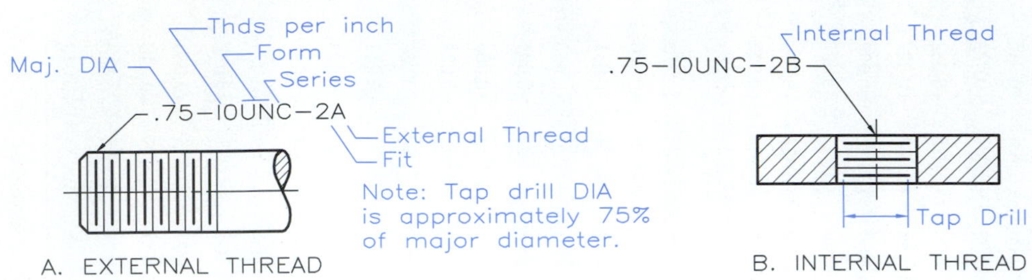

Nominal Diameter	Basic Diameter	Coarse NC & UNC		Fine NF & UNF		Extra Fine NEF/UNEF	
		Thds per in.	Tap Drill DIA	Thds per in.	Tap Drill DIA	Thds per in.	Tap Drill DIA
0	.060			80	.0469		
1	.073	64	No. 53	72	No. 53		
2	.086	56	No. 50	64	No. 50		
3	.099	48	No. 47	56	No. 45		
4	.112	40	No. 43	48	No. 42		
5	.125	40	No. 38	44	No. 37		
6	.138	32	No. 36	40	No. 33		
8	.164	32	No. 29	36	No. 29		
10	.190	24	No. 25	32	No. 21		
12	.216	24	No. 16	28	No. 14	32	No. 13
1/4	.250	20	No. 7	28	No. 3	32	.2189
5/16	.3125	18	F	24	I	32	.2813
3/8	.375	16	.3125	24	Q	32	.3438
7/16	.4375	14	U	20	.3906	28	.4062
1/2	.500	13	.4219	20	.4531	28	.4688
9/16	.5625	12	.4844	18	.5156	24	.5156
5/8	.625	11	.5313	18	.5781	24	.5781
11/16	.6875	...	...	...	...	24	.6406
3/4	.750	10	.6563	16	.6875	20	.7031
13/16	.8125	...	...	...	...	20	.7656
7/8	.875	9	.7656	14	.8125	20	.8281
15/16	.9375	...	...	...	...	20	.8906

Nominal Diameter	Basic Diameter	Coarse NC & UNC		Fine NF & UNF		Extra Fine NEF/UNEF	
		Thds per in.	Tap Drill DIA	Thds per in.	Tap Drill DIA	Thds per in.	Tap Drill DIA
1	1.000	8	.875	12	.922	20	.953
1-1/16	1.063	...	...	...	...	18	1.000
1-1/8	1.125	7	.904	12	1.046	18	1.070
1-3/16	1.188	...	...	...	...	18	1.141
1-1/4	1.250	7	1.109	12	1.172	18	1.188
1-5/16	1.313	...	...	...	...	18	1.266
1-3/8	1.375	6	1.219	12	1.297	18	1.313
1-7/16	1.438	...	...	...	...	18	1.375
1-1/2	1.500	6	1.344	12	1.422	18	1.438
1-9/16	1.563	...	...	...	...	18	1.500
1-5/8	1.625	...	...	...	...	18	1.563
1-11/16	1.688	...	...	...	...	18	1.625
1-3/4	1.750	5	1.563	...	...	...	...
2	2.000	4.5	1.781	...	...	...	...
2-1/4	2.250	4.5	2.031	...	...	...	...
2-1/2	2.500	4	2.250	...	...	...	...
2-3/4	2.750	4	2.500	...	...	...	...
3	3.000	4	2.750	...	...	...	...
3-1/4	3.250	4	...	...	...	...	...
3-1/2	3.500	4	...	...	...	...	...
3-3/4	3.750	4	...	...	...	...	...
4	4.000	4	...	...	...	...	...

APPENDIX 19

Screw Threads: American National and Unified (inches)

Constant-Pitch Threads

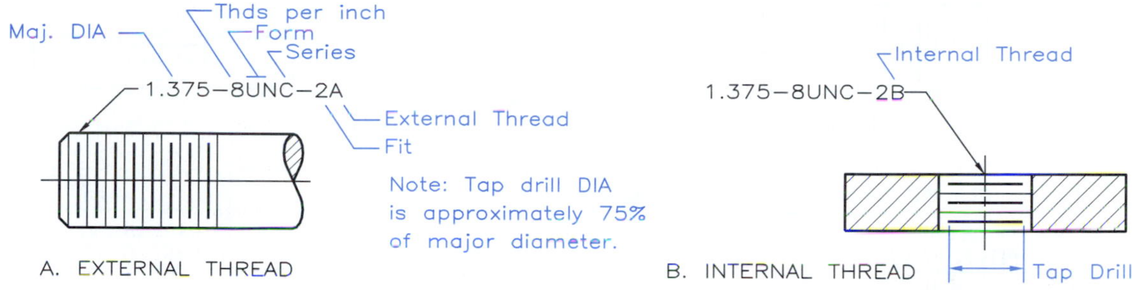

Nominal Diameter	8 Pitch 8N & 8UN		12 Pitch 12N & 12UN		16 Pitch 16N & 16UN	
	Thds per in.	Tap Drill DIA	Thds per in.	Tap Drill DIA	Thds per in.	Tap Drill DIA
.500	...	...	12	.422	...	...
.563	...	...	12	.484	...	...
.625	...	...	12	.547	...	...
.688	...	...	12	.609	...	...
.750	...	...	12	.672	16	.688
.813	...	...	12	.734	16	.750
.875	...	...	12	.797	16	.813
.934	...	...	12	.859	16	.875
1.000	8	.875	12	.922	16	.938
1.063	...	...	12	.984	16	1.000
1.125	8	1.000	12	1.047	16	1.063
1.188	...	...	12	1.109	16	1.125
1.250	8	1.125	12	1.172	16	1.188
1.313	...	...	12	1.234	16	1.250
1.375	8	1.250	12	1.297	16	1.313
1.434	...	...	12	1.359	16	1.375
1.500	8	1.375	12	1.422	16	1.438
1.563	...	...	...	...	16	1.500
1.625	8	1.500	12	1.547	16	1.563
1.688	...	...	...	...	16	1.625
1.750	8	1.625	12	1.672	16	1.688
1.813	...	...	...	...	16	1.750
1.875	8	1.750	12	1.797	16	1.813
1.934	...	...	...	...	16	1.875
2.000	8	1.875	12	1.922	16	1.938

Nominal Diameter	8 Pitch 8N & 8UN		12 Pitch 12N & 12UN		16 Pitch 16N & 16UN	
	Thds per in.	Tap Drill DIA	Thds per in.	Tap Drill DIA	Thds per in.	Tap Drill DIA
2.063	...	...	...	...	16	2.000
2.125	...	...	12	2.047	16	2.063
2.188	...	...	...	...	16	2.125
2.250	8	2.125	12	2.172	16	2.188
2.313	...	...	...	...	16	2.250
2.375	...	...	12	2.297	16	2.313
2.438	...	...	...	...	16	2.375
2.500	8	2.375	12	2.422	16	2.438
2.625	...	...	12	2.547	16	2.563
2.750	8	2.625	12	2.717	16	2.688
2.875	...	...	12	...	16	...
3.000	8	2.875	12	...	16	...
3.125	...	...	12	...	16	...
3.250	8	...	12	...	16	...
3.375	...	...	12	...	16	...
3.500	8	...	12	...	16	...
3.625	...	...	12	...	16	...
3.750	8	...	12	...	16	...
3.875	...	...	12	...	16	...
4.000	8	...	12	...	16	...
4.250	8	...	12	...	16	...
4.500	8	...	12	...	16	...
4.750	8	...	12	...	16	...
5.000	8	...	12	...	16	...
5.250	8	...	12	...	16	...

APPENDIX 

Screw Threads: American National and Unified (metric)

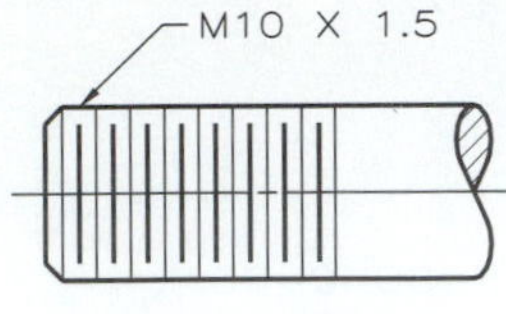

A. EXTERNAL THREAD

Note: Tap drill DIA is approximately 75% of major diameter.

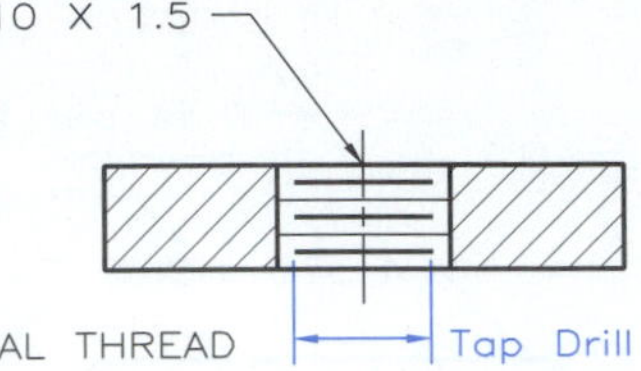

B. INTERNAL THREAD

COARSE		FINE	
MAJ. DIA & THD PITCH	TAP DRILL	MAJ. DIA & THD PITCH	TAP DRILL
M1.6 × 0.35	1.25		
M1.8 × 0.35	1.45		
M2 × 0.4	1.6		
M2.2 × 0.45	1.75		
M2.5 × 0.45	2.05		
M3 × 0.5	2.5		
M3.5 × 0.6	2.9		
M4 × 0.7	3.3		
M4.5 × 0.75	3.75		
M5 × 0.8	4.2		
M6 × 1	5.0		
M7 × 1	6.0		
M8 × 1.25	6.8	M8 × 1	7.0
M9 × 1.25	7.75		
M10 × 1.5	8.5	M10 × 1.25	8.75
M11 × 1.5	9.5		
M12 × 1.75	10.3	M12 × 1.25	10.5
M14 × 2	12.0	M14 × 1.5	12.5
M16 × 2	14.0	M16 × 1.5	14.5
M18 × 2.5	15.5	M18 × 1.5	16.5

COARSE		FINE	
MAJ. DIA & THD PITCH	TAP DRILL	MAJ. DIA & THD PITCH	TAP DRILL
M20 × 2.5	17.5	M20 × 1.5	18.5
M22 × 2.5	19.5	M22 × 1.5	20.5
M24 × 3	21.0	M24 × 2	22.0
M27 × 3	24.0	M27 × 2	25.0
M30 × 3.5	26.5	M30 × 2	28.0
M33 × 3.5	29.5	M33 × 2	31.0
M36 × 4	32.0	M36 × 3	33.0
M39 × 4	35.0	M39 × 3	36.0
M42 × 4.5	37.5	M42 × 3	39.0
M45 × 4.5	40.5	M45 × 3	42.0
M48 × 5	43.0	M48 × 3	45.0
M52 × 5	47.0	M52 × 3	49.0
M56 × 5.5	50.5	M56 × 4	52.0
M60 × 5.5	54.5	M60 × 4	56.0
M64 × 6	58.0	M64 × 4	60.0
M68 × 6	62.0	M68 × 4	64.0
M72 × 6	66.0	M72 × 4	68.0
M80 × 6	74.0	M80 × 4	76.0
M90 × 6	84.0	M90 × 4	86.0
M100 × 6	94.0	M100 × 4	96.0

APPENDIX 21

Square and Acme Threads

2.00–2.5 SQUARE

Typical thread note

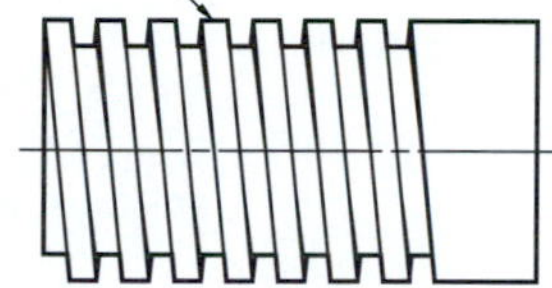

Dimensions are in inches

Size	Size	Thds per inch	Size	Size	Thds per inch	Size	Size	Thds per inch
3/8	.375	12	1-1/8	1.125	4	3	3.000	1-1/2
7/16	.438	10	1-1/4	1.250	4	3-1/4	3.125	1-1/2
1/2	.500	10	1-1/2	1.500	3	3-1/2	3.500	1-1/3
9/16	.563	8	1-3/4	1.750	2-1/2	3-3/4	3.750	1-1/3
5/8	.625	8	2	2.000	2-1/2	4	4.000	1-1/3
3/4	.75	6	2-1/4	2.250	2	4-1/4	4.250	1-1/3
7/8	.875	5	2-1/2	2.500	2	4-1/2	4.500	1
1	1.000	5	2-3/4	2.750	2	Larger		1

APPENDIX 22

American Standard Taper Pipe Threads (NPT)

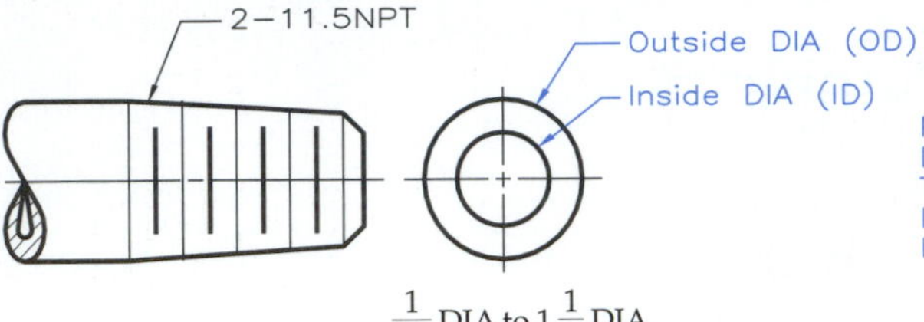

$\frac{1}{16}$ DIA to $1\frac{1}{4}$ DIA

Dimensions in inches

Nominal ID	$\frac{1}{16}$	$\frac{1}{8}$	$\frac{1}{4}$	$\frac{3}{8}$	$\frac{1}{2}$	$\frac{3}{4}$	1	$1\frac{1}{4}$
Outside DIA	0.313	0.405	0.540	0.675	0.840	1.050	1.315	1.660
Thds/Inch	27	27	18	18	14	14	$11\frac{1}{2}$	$11\frac{1}{2}$

$1\frac{1}{2}$ DIA to 6 DIA

Nominal ID	$1\frac{1}{2}$	2	$2\frac{1}{2}$	3	$3\frac{1}{2}$	4	5	6
Outside DIA	1.900	2.375	2.875	3.500	4.000	4.500	5.563	6.625
Thds/Inch	$11\frac{1}{2}$	$11\frac{1}{2}$	8	8	8	8	8	8

8 DIA to 24 DIA

Nominal ID	8	10	12	14 OD	16 OD	18 OD	20 OD	24 OD
Outside DIA	8.625	10.750	12.750	14.000	16.000	18.000	20.000	24.000
Thds/Inch	8	8	8	8	8	8	8	8

Square Bolts (inches)

DIA	E Max.	F Max.	G Avg.	H Max.	R Max.
1/4	.250	.375	.530	.188	.031
5/16	.313	.500	.707	.220	.031
3/8	.375	.563	.795	.268	.031
7/16	.438	.625	.884	.316	.031
1/2	.500	.750	1.061	.348	.031
5/8	.625	.938	1.326	.444	.062
3/4	.750	1.125	1.591	.524	.062
7/8	.875	1.313	1.856	.620	.062
1	1.000	1.500	2.121	.684	.093
1-1/8	1.125	1.688	2.386	.780	.093
1-1/4	1.250	1.875	2.652	.876	.093
1-3/8	1.375	2.625	2.917	.940	.093
1-1/2	1.500	2.250	3.182	1.036	.093

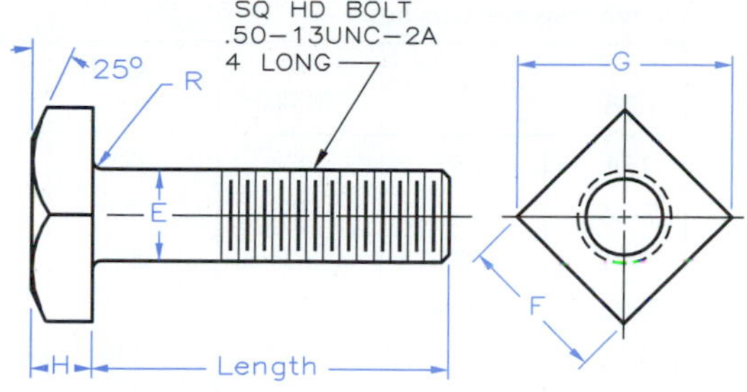

STANDARD COMMERCIAL LENGTHS

DIAMETER	.50	.75	1.00	1.25	1.50	1.75	2.00	2.50	3.00	3.50	4.00	4.50	5.00	5.50	6.00	6.50	7.00	8.00 *PLUS
.25	●	●	●	●	●	●	●											
.31		●	●	●	●	●	●	●	●		●							
.375			●	●	●	●	●	●	●	●	●							
.500			●	●	●	●	●	●	●	●	●	●	●	●	●	●	●	*14
.625					●	●	●	●	●	●	●	●	●	●	●	●	●	*12
.750							●	●	●	●	●	●	●	●	●	●	●	*13
.875										●	●	●	●	●				
1.00															●	●	●	*12
1.25															●	●	●	●

*14 MEANS THAT LENGTHS ARE AVAILABLE AT 1 INCH INCREMENTS UP TO 10 INCHES.

APPENDIX

24

Square Nuts (inches)

Dimensions are in inches.

DIA	DIA	F Max.	G Avg.	H Max.
1/4	.250	.438	.619	.235
5/16	.313	.563	.795	.283
3/8	.375	.625	.884	.346
7/16	.438	.750	1.061	.394
1/2	.500	.813	1.149	.458
5/8	.625	1.000	1.414	.569
3/4	.750	1.125	1.591	.680
7/8	.875	1.313	1.856	.792
1	1.000	1.500	2.121	.903
1-1/8	1.125	1.688	2.386	1.030
1-1/4	1.250	1.875	2.652	1.126
1-3/8	1.375	1.063	2.917	1.237
1-1/2	1.500	2.250	3.182	1.348

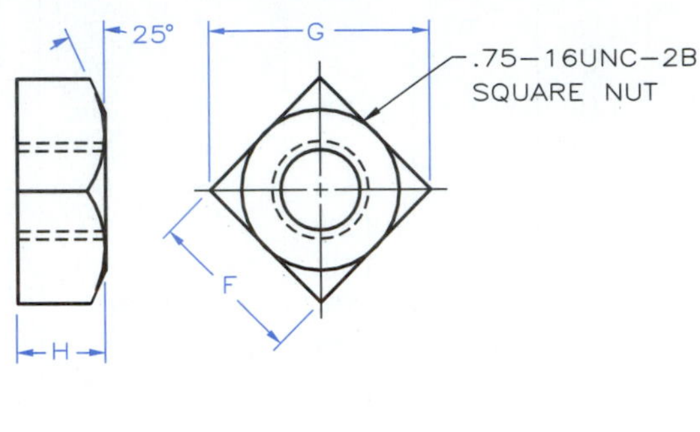

APPENDIX **25**

Hexagon Head Bolts (inches)

Dimensions are in inches.

DIA	E Max.	F Max.	G Avg.	H Max.	R Max.
1/4	.250	.438	.505	.163	.025
5/16	.313	.500	.577	.211	.025
3/8	.375	.563	.650	.243	.025
7/16	.438	.625	.722	.291	.025
1/2	.500	.750	.866	.323	.025
9/16	.563	.812	.938	.371	.045
5/8	.625	.938	1.083	.403	.045
3/4	.750	1.125	1.299	.483	.045
7/8	.875	1.313	1.516	.563	.065
1	1.000	1.500	1.732	.627	.095
1-1/8	1.125	1.688	1.949	.718	.095
1-1/4	1.250	1.875	2.165	.813	.095
1-3/8	1.375	2.063	2.382	.878	.095
1-1/2	1.500	2.250	2.598	.974	.095
1-3/4	1.750	2.625	3.031	1.134	.095
2	2.000	3.000	3.464	1.263	.095
2-1/4	2.250	3.375	3.897	1.423	.095
2-1/2	2.500	3.750	4.330	1.583	.095
2-3/4	2.750	4.125	4.763	1.744	.095
3	3.000	4.500	5.196	1.935	.095

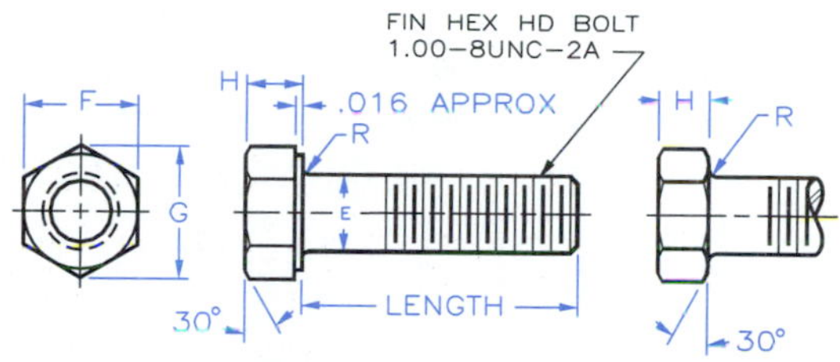

STANDARD COMMERCIAL LENGTHS

DIAMETER	.50	.75	1.00	1.25	1.50	1.75	2.00	2.50	3.00	3.50	4.00	4.50	5.00	5.50	6.00	6.50	7.00	8.00 *PLUS
.25	●	●	●	●	●	●	●	●	●	●	●	●	●	●	●			
.31	●	●	●	●	●	●	●	●	●	●	●	●	●	●	●			
.375		●	●	●	●	●	●	●	●	●	●	●	●	●	●		●	●
.500			●	●	●	●	●	●	●	●	●	●	●	●	●	●	●	
.625			●	●	●	●	●	●	●	●	●	●	●	●	●			
.750			●		●	●	●	●	●	●	●	●	●	●	●	●	●	
.875					●		●	●	●	●	●	●	●	●			●	
1.00							●	●	●		●		●		●			
1.25										●	●	●	●	●	●	●	●	*10
1.38									●	●	●		●		●		●	
1.50									●	●	●		●		●		●	*12
1.75													●	●	●		●	*9.5
2.00													●	●	●	●	●	*12

*10 MEANS THAT LENGTHS ARE AVAILABLE AT 1 INCH INCREMENTS UP TO 10 INCHES.

APPENDIX 26

Hex Nuts and Jam Nuts

MAJOR DIA		F Max.	G Avg.	H1 Max.	H2 Max.
1/4	.250	.438	.505	.226	.163
5/16	.313	.500	.577	.273	.195
3/8	.375	.563	.650	.337	.227
7/16	.438	.688	.794	.385	.260
1/2	.500	.750	.866	.448	.323
9/16	.563	.875	1.010	.496	.324
5/8	.625	.938	1.083	.559	.387
3/4	.750	1.125	1.299	.665	.446
7/8	.875	1.313	1.516	.776	.510
1	1.000	1.500	1.732	.887	.575
1-1/8	1.125	1.688	1.949	.899	.639
1-1/4	1.250	1.875	2.165	1.094	.751
1-3/8	1.375	2.063	2.382	1.206	.815
1-1/2	1.500	2.250	2.589	1.317	.880

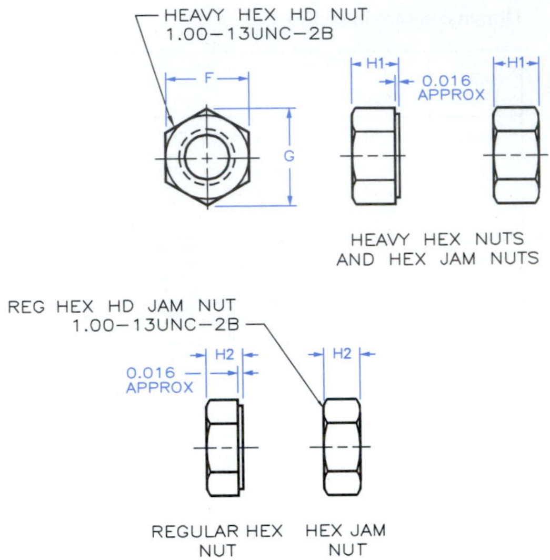

APPENDIX 27

Round Head Cap Screws

Dimensions are in inches.

DIA	D Max.	A Max.	H Avg.	J Max.	T Max.
1/4	.250	.437	.191	.075	.117
5/16	.313	.562	.245	.084	.151
3/8	.375	.625	.273	.094	.168
7/16	.438	.750	.328	.094	.202
1/2	.500	.812	.354	.106	.218
9/16	.563	.937	.409	.118	.252
5/8	.625	1.000	.437	.133	.270
3/4	.750	1.250	.546	.149	.338

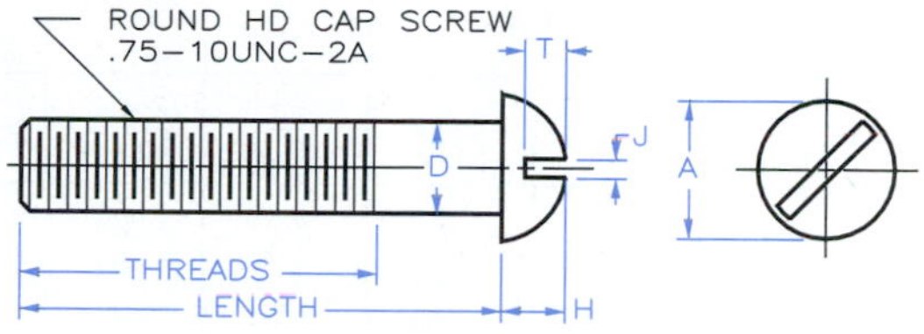

STANDARD COMMERCIAL LENGTHS

DIAMETER	.50	.75	1.00	1.25	1.50	1.75	2.00	2.50	3.00	3.50	4.00
.25	●	●	●	●	●	●	●				
.31		●	●	●	●	●	●	●	●	●	●
.375			●	●	●	●	●	●	●	●	●
.500			●	●	●	●	●	●	●	●	●
.625					●	●	●	●	●	●	●
.750							●	●	●	●	●

OTHER LENGTHS AND DIAMETERS ARE AVAILABLE, BUT THESE ARE THE MORE STANDARD ONES.

APPENDIX 28

Flat Head Cap Screws

Dimensions are in inches.

DIA	D Max.	A Max.	H Avg.	J Max.	T Max.
1/4	.250	.500	.140	.075	.068
5/16	.313	.625	.177	.084	.086
3/8	.375	.750	.210	.094	.103
7/16	.438	.813	.210	.094	.103
1/2	.500	.875	.210	.106	.103
9/16	.563	1.000	.244	.118	.120
5/8	.625	1.125	.281	.133	.137
3/4	.750	1.375	.352	.149	.171
7/8	.875	1.625	.423	.167	.206
1	1.000	1.875	.494	.188	.240
1-1/8	1.125	2.062	.529	.196	.257
1-1/4	1.250	2.312	.600	.211	.291
1-3/8	1.375	2.562	.665	.226	.326
1-1/2	1.500	2.812	.742	.258	.360

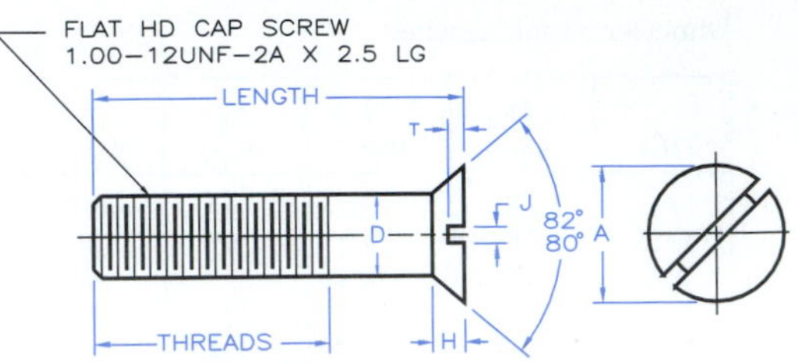

STANDARD COMMERCIAL LENGTHS

DIAMETER	.50	.75	1.00	1.25	1.50	1.75	2.00	2.50	3.00	3.50	4.00
.25	●	●	●	●	●	●	●				
.31		●	●	●	●	●	●	●	●	●	●
.375			●	●	●	●	●	●	●	●	●
.500			●	●	●	●	●	●	●	●	●
.625					●	●	●	●	●	●	●
.750							●	●	●	●	●
.875								●	●	●	●
1.00									●	●	●
1.50									●	●	●

OTHER LENGTHS AND DIAMETERS ARE AVAILABLE, BUT THESE ARE THE MORE STANDARD ONES.

APPENDIX

29

Filister Head Cap Screws

Dimensions are in inches.

DIA	D Max.	A Max.	H Avg.	J Max.	T Max.
1/4	.250	.375	.172	.075	.097
5/16	.313	.437	.203	.084	.115
3/8	.375	.562	.250	.094	.142
7/16	.438	.625	.297	.094	.168
1/2	.500	.750	.328	.106	.193
9/16	.563	.812	.375	.118	.213
5/8	.625	.875	.422	.133	.239
3/4	.750	1.000	.500	.149	.283
7/8	.875	1.125	.594	.167	.334
1	1.000	1.312	.656	.188	.371

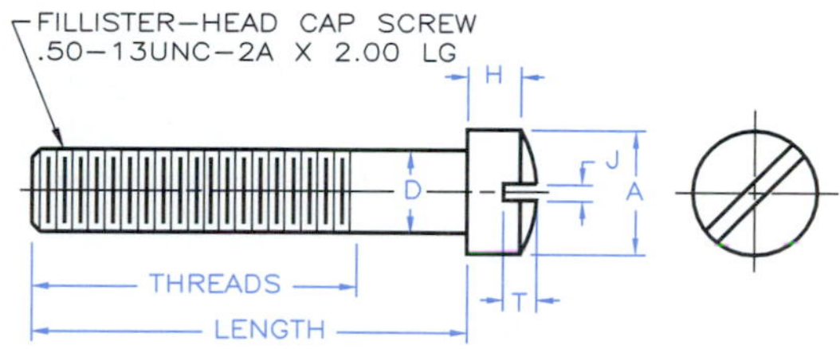

STANDARD LENGTHS

DIAMETER	.50	.75	1.00	1.25	1.50	1.75	2.00	2.50
.25	●	●	●	●	●	●	●	●
.313	●		●	●			●	
.375		●	●	●		●	●	
.500		●	●				●	●
.625			●		●		●	●
.750			●		●		●	●
.875				●	●		●	●
1.00				●	●	●	●	●

APPENDIX 30

Flat Socket Head Cap Screws

Diameter mm	inches	Pitch	A	Ang.	W
M3	.118	.5	6	90	2
M4	.157	.7	8	90	2.5
M5	.197	.8	10	90	3
M6	.236	1	12	90	4
M8	.315	1.25	16	90	5
M10	.394	1.5	20	90	6
M12	.472	1.75	24	90	8
M14	.551	2	27	90	10
M16	.630	2	30	90	10
M20	.787	2.5	36	90	12

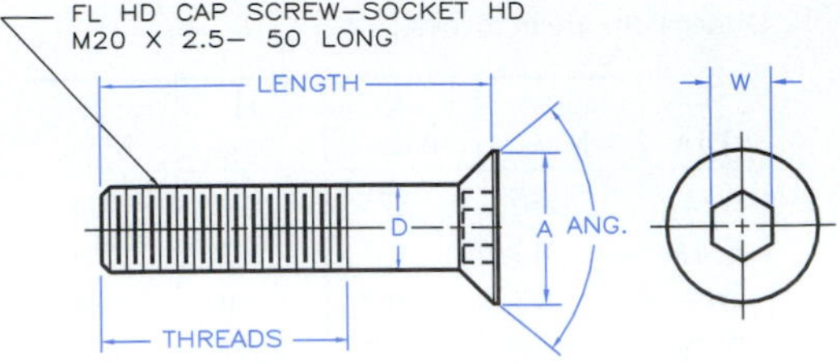

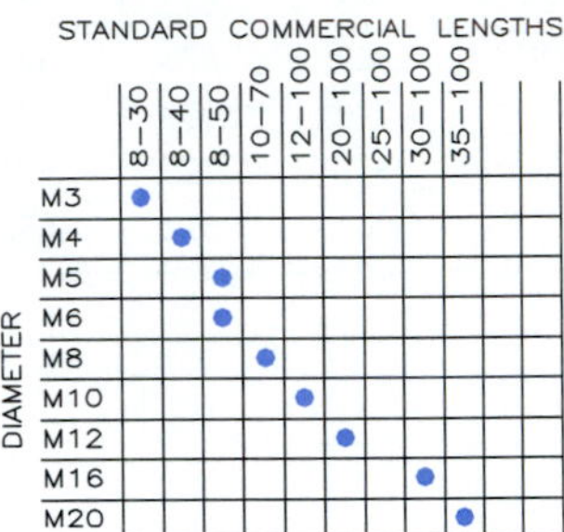

DIA 8–16: LENGTHS AT INTERVALS OF 2 MM
DIA 20–100: LENGTHS AT INTERVALS OF 5 MM

Reprinted from ANSI/ASME B18.3–1986 (R1995) and ANSI/ASME B18.3.5M–1986 (R1993).

APPENDIX 31

Socket Head Cap Screws

Diameter mm	Diameter inches	Pitch	A	H	W
M3	.118	.5	6	3	2
M4	.157	.7	8	4	3
M5	.187	.8	10	5	4
M6	.236	1	12	6	6
M8	.315	1.25	16	8	6
M10	.394	1.5	20	10	8
M12	.472	1.75	24	12	10
M14	.551	2	27	14	12
M16	.630	2	30	16	14
M20	.787	2.5	36	20	17

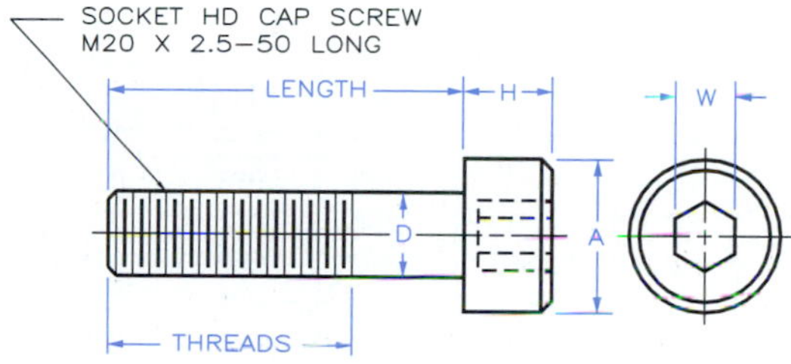

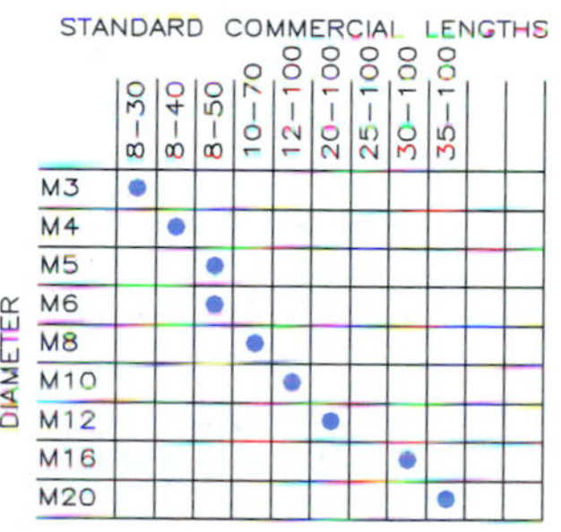

DIA 8–16: LENGTHS AT INTERVALS OF 2 MM
DIA 20–100: LENGTHS AT INTERVALS OF 5 MM

APPENDIX 32

Round Head Machine Screws

Dimensions are in inches.

DIA	D Max.	A Max.	H Avg.	J Max.	T Max.
0	.060	.113	.053	.023	.039
1	.073	.138	.061	.026	.044
2	.086	.162	.069	.031	.048
3	.099	.187	.078	.035	.053
4	.112	.211	.086	.039	.058
5	.125	.236	.095	.043	.063
6	.138	.260	.103	.048	.068
8	.164	.309	.120	.054	.077
10	.190	.359	.137	.060	.087
12	.216	.408	.153	.067	.096
1/4	.250	.472	.175	.075	.109
5/16	.313	.590	.216	.084	.132
3/8	.375	.708	.256	.094	.155
7/16	.438	.750	.328	.094	.196
1/2	.500	.813	.355	.106	.211
9/16	.563	.938	.410	.118	.242
5/8	.625	1.000	.438	.133	.258
3/4	.750	1.250	.547	.149	.320

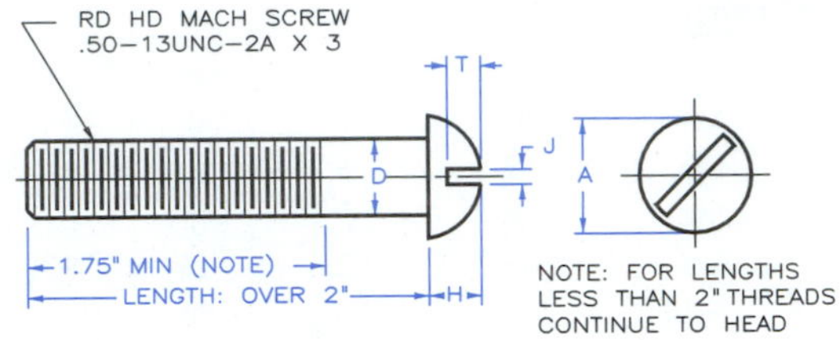

STANDARD LENGTHS

DIAMETER	.25	.50	.75	1.00	1.25	1.50	1.75	2.00	2.50	3.00
0	●	●								
1	●	●								
2	●	●	●	●						
4	●	●	●	●	●	●				
6	●	●	●	●	●	●	●	●	●	
.125	●	●	●	●	●	●	●	●	●	
.250	●	●	●	●	●	●	●	●	●	
.375		●	●	●	●	●	●	●	●	
.500		●	●	●	●	●	●	●	●	●
.625		●	●	●	●	●	●	●	●	●
.750		●	●	●	●	●	●	●	●	●

OTHER LENGTHS AND DIAMETERS ARE AVAILABLE; THESE ARE THE MORE STANDARD ONES.

APPENDIX 33

Set Screws

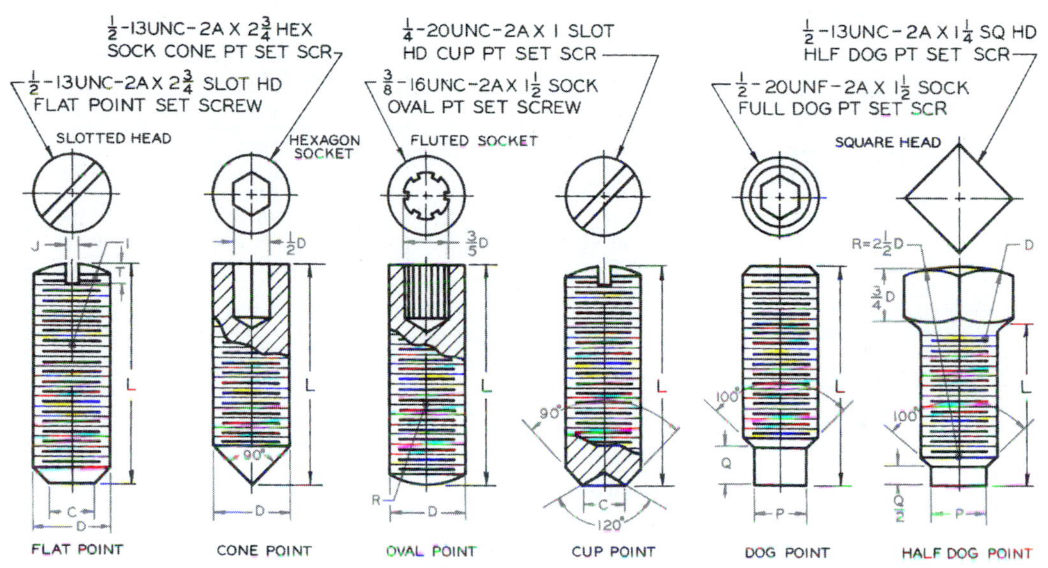

Dimensions for the set screws shown above (dimensions in inches)

D	I	J	T	R	C		P		Q	q
Nominal Size	Radius of Headless Crown	Width of Slot	Depth of Slot	Oval Point Radius	Diameter of Cup and Flat Points		Diameter of Dog Point		Length of Dog Point	
					Max	Min	Max	Min	Full	Half
5 0.125	0.125	0.023	0.031	0.094	0.067	0.057	0.083	0.078	0.060	0.030
6 0.138	0.138	0.025	0.035	0.109	0.047	0.064	0.092	0.087	0.070	0.035
8 0.164	0.164	0.029	0.041	0.125	0.087	0.076	0.109	0.103	0.080	0.040
10 0.190	0.190	0.032	0.048	0.141	0.102	0.088	0.127	0.120	0.090	0.045
12 0.216	0.216	0.036	0.054	0.156	0.115	0.101	0.144	0.137	0.110	0.055
$\frac{1}{4}$ 0.250	0.250	0.045	0.063	0.188	0.132	0.118	0.156	0.149	0.125	0.063
$\frac{5}{16}$ 0.3125	0.313	0.051	0.076	0.234	0.172	0.156	0.203	0.195	0.156	0.078
$\frac{3}{8}$ 0.375	0.375	0.064	0.094	0.281	0.212	0.194	0.250	0.241	0.188	0.094
$\frac{7}{16}$ 0.4375	0.438	0.072	0.190	0.328	0.252	0.232	0.297	0.287	0.219	0.109
$\frac{1}{2}$ 0.500	0.500	0.081	0.125	0.375	0.291	0.270	0.344	0.344	0.250	0.125
$\frac{9}{16}$ 0.5625	0.563	0.091	0.141	0.422	0.332	0.309	0.391	0.379	0.281	0.140
$\frac{5}{8}$ 0.625	0.625	0.102	0.156	0.469	0.371	0.347	0.469	0.456	0.313	0.156
$\frac{3}{4}$ 0.750	0.750	0.129	0.188	0.563	0.450	0.425	0.563	0.549	0.375	0.188

APPENDIX 34

Twist Drill Sizes

Letter Size Drills

Size	Drill Diameter inches	mm	Size	Drill Diameter inches	mm	Size	Drill Diameter inches	mm	Size	Drill Diameter inches	mm
A	0.234	5.944	H	0.266	6.756	O	0.316	8.026	V	0.377	9.576
B	0.238	6.045	I	0.272	6.909	P	0.323	8.204	W	0.386	9.804
C	0.242	6.147	J	0.277	7.036	Q	0.332	8.433	X	0.397	10.084
D	0.246	6.248	K	0.281	7.137	R	0.339	8.611	Y	0.404	10.262
E	0.250	6.350	L	0.290	7.366	S	0.348	8.839	Z	0.413	10.490
F	0.257	6.528	M	0.295	7.493	T	0.358	9.093			
G	0.261	6.629	N	0.302	7.601	U	0.368	9.347			

Source: Courtesy of General Motors Corporation.

Number Size Drills

Size	Drill Diameter inches	mm	Size	Drill Diameter inches	mm	Size	Drill Diameter inches	mm	Size	Drill Diameter inches	mm
1	0.2280	5.7912	21	0.1590	4.0386	41	0.0960	2.4384	61	0.0390	0.9906
2	0.2210	5.6134	22	0.1570	3.9878	42	0.0935	2.3622	62	0.0380	0.9652
3	0.2130	5.4102	23	0.1540	3.9116	43	0.0890	2.2606	63	0.0370	0.9398
4	0.2090	5.3086	24	0.1520	3.8608	44	0.0860	2.1844	64	0.0360	0.9144
5	0.2055	5.2197	25	0.1495	3.7973	45	0.0820	2.0828	65	0.0350	0.8890
6	0.2040	5.1816	26	0.1470	3.7338	46	0.0810	2.0574	66	0.0330	0.8382
7	0.2010	5.1054	27	0.1440	3.6576	47	0.0785	19.812	67	0.0320	0.8128
8	0.1990	5.0800	28	0.1405	3.5560	48	0.0760	1.9304	68	0.0310	0.7874
9	0.1960	4.9784	29	0.1360	3.4544	49	0.0730	1.8542	69	0.0292	0.7417
10	0.1935	4.9149	30	0.1285	3.2639	50	0.0700	1.7780	70	0.0280	0.7112
11	0.1910	4.8514	31	0.1200	3.0480	51	0.0670	1.7018	71	0.0260	0.6604
12	0.1890	4.8006	32	0.1160	2.9464	52	0.0635	1.6129	72	0.0250	0.6350
13	0.1850	4.6990	33	0.1130	2.8702	53	0.0595	1.5113	73	0.0240	0.6096
14	0.1820	4.6228	34	0.1110	2.8194	54	0.0550	1.3970	74	0.0225	0.5715
15	0.1800	4.5720	35	0.1100	2.7940	55	0.0520	1.3208	75	0.0210	0.5334
16	0.1770	4.4958	36	0.1065	0.7051	56	0.0465	1.1684	76	0.0200	0.5080
17	0.1730	4.3942	37	0.1040	2.6416	57	0.0430	1.0922	77	0.0180	0.4572
18	0.1695	4.3053	38	0.1015	2.5781	58	0.0420	1.0668	78	0.0160	0.4064
19	0.1660	4.2164	39	0.0995	2.5273	59	0.0410	1.0414	79	0.0145	0.3638
20	0.1610	4.0894	40	0.0980	2.4892	60	0.0400	1.0160	80	0.0135	0.3428

Metric Size Drills

Decimal-inch equivalents are for reference only.

Drill Diameter		Drill Diameter		Drill Diameter		Drill Diameter		Drill Diameter		Drill Diameter		Drill Diameter	
mm	in.	mm	in.	mm	in.	mm	in.	mm	in.	mm	in.	mm	in.
.40	.0157	1.03	.0406	2.20	.0866	5.00	.1969	10.00	.3937	21.50	.8465	48.00	1.8898
.42	.0165	1.05	.0413	2.30	.0906	5.20	.2047	10.30	.4055	22.00	.8661	50.00	1.9685
.45	.0177	1.08	.0425	2.40	.0945	5.30	.2087	10.50	.4134	23.00	.9055	51.50	2.0276
.48	.0189	1.10	.0433	2.50	.0984	5.40	.2126	10.80	.4252	24.00	.9449	53.00	2.0866
.50	.0197	1.15	.0453	2.60	.1024	5.60	.2205	11.00	.4331	25.00	.9843	54.00	2.1260
.52	.0205	1.20	.0472	2.70	.1063	5.80	.2283	11.50	.4528	26.00	1.0236	56.00	2.2047
.55	.0217	1.25	.0492	2.80	.1102	6.00	.2362	12.00	.4724	27.00	1.0630	58.00	2.2835
.58	.0228	1.30	.0512	2.90	.1142	6.20	.2441	12.50	.4921	28.00	1.1024	60.00	2.3622
.60	.0236	1.35	.0531	3.00	.1181	6.30	.2480	13.00	.5118	29.00	1.1417		
.62	.0244	1.40	.0551	3.10	.1220	6.50	.2559	13.50	.5315	30.00	1.1811		
.65	.0256	1.45	.0571	3.20	.1260	6.70	.2638	14.00	.5512	31.00	1.2205		
.68	.0268	1.50	.0591	3.30	.1299	6.80	.2677	14.50	.5709	32.00	1.2598		
.70	.0276	1.55	.0610	3.40	.1339	6.90	.2717	15.00	.5906	33.00	1.2992		
.72	.0283	1.60	.0630	3.50	.1378	7.10	.2795	15.50	.6102	34.00	1.3386		
.75	.0295	1.65	.0650	3.60	.1417	7.30	.2874	16.00	.6299	35.00	1.3780		
.78	.0307	1.70	.0669	3.70	.1457	7.50	.2953	16.50	.6496	36.00	1.4173		
.80	.0315	1.75	.0689	3.80	.1496	7.80	.3071	17.00	.6693	37.00	1.4567		
.82	.0323	1.80	.0709	3.90	.1535	8.00	.3150	17.50	.6890	38.00	1.4961		
.85	.0335	1.85	.0728	4.00	.1575	8.20	.3228	18.00	.7087	39.00	1.5354		
.88	.0346	1.90	.0748	4.10	.1614	8.50	.3346	18.50	.7283	40.00	1.5748		
.90	.0354	1.95	.0768	4.20	.1654	8.80	.3465	19.00	.7480	41.00	1.6142		
.92	.0362	2.00	.0787	4.40	.1732	9.00	.3543	19.50	.7677	42.00	1.6535		
.95	.0374	2.05	.0807	4.50	.1772	9.20	.3622	20.00	.7874	43.50	1.7126		
.98	.0386	2.10	.0827	4.60	.1811	9.50	.3740	20.50	.0871	45.00	1.7717		
1.00	.0394	2.15	.0846	4.80	.1890	9.80	.3858	21.00	.8268	46.50	1.8307		

APPENDIX 35

Cotter Pins: American National Standard

Nominal Diameter	Maximum DIA A	Minimum DIA B	Hole Size
0.031	0.032	0.063	0.047
0.047	0.048	0.094	0.063
0.062	0.060	0.125	0.078
0.078	0.076	0.156	0.094
0.094	0.090	0.188	0.109
0.109	0.104	0.219	0.125
0.125	0.120	0.250	0.141
0.141	0.176	0.281	0.156
0.156	0.207	0.313	0.172
0.188	0.176	0.375	0.203
0.219	0.207	0.438	0.234
0.250	0.225	0.500	0.266
0.312	0.280	0.625	0.313
0.375	0.335	0.750	0.375
0.438	0.406	0.875	0.438
0.500	0.473	1.000	0.500
0.625	0.598	1.250	0.625
0.750	0.723	1.500	0.750

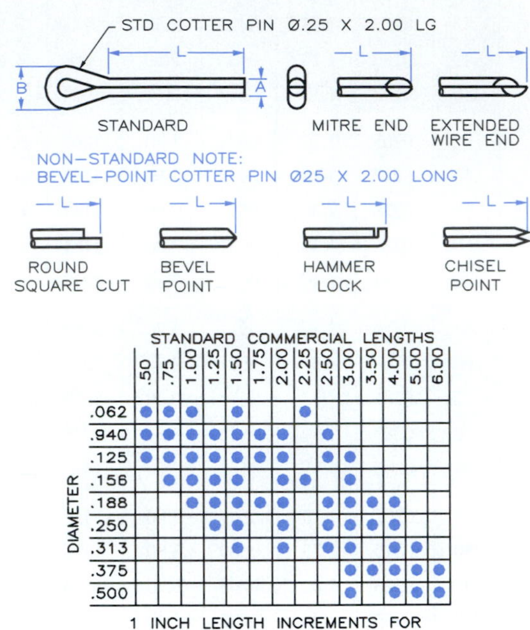

APPENDIX 36

Straight Pins

Nominal DIA	Diameter A Max	Diameter A Min	Chamfer B
0.062	0.0625	0.0605	0.015
0.094	0.0937	0.0917	0.015
0.109	0.1094	0.1074	0.015
0.125	0.1250	0.1230	0.015
0.156	0.1562	0.1542	0.015
0.188	0.1875	0.1855	0.015
0.219	0.2187	0.2167	0.015
0.250	0.2500	0.2480	0.015
0.312	0.3125	0.3095	0.015
0.375	0.3750	0.3720	0.030
0.438	0.4345	0.4345	0.030
0.500	0.4970	0.4970	0.030

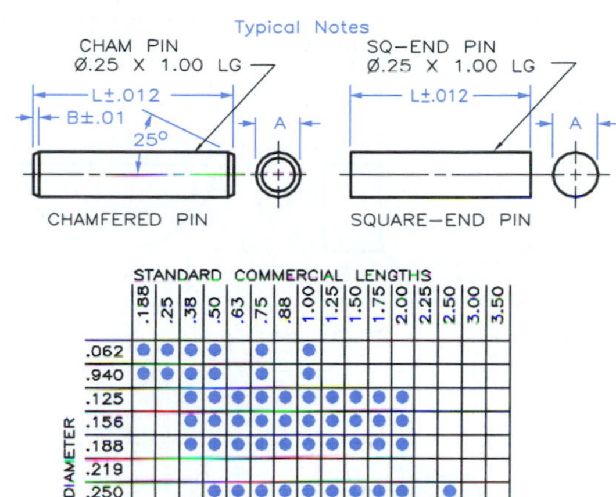

APPENDIX 37

Woodruff Keys

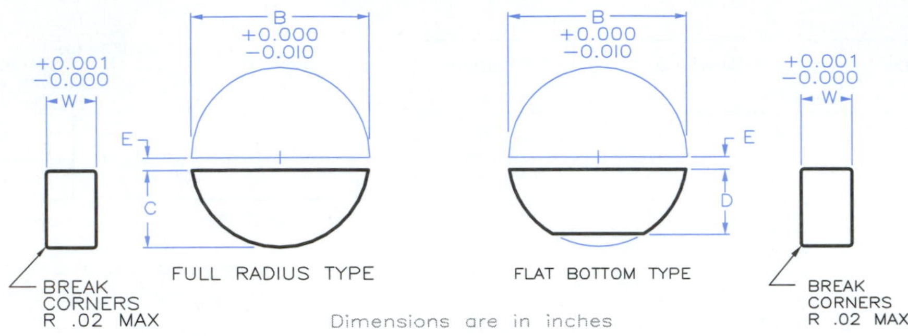

Key No.	W × B	C Max.	D Max.	E
204	1/16 × 1/2	.203	.194	.047
304	3/32 × 1/2	.203	.194	.047
404	1/8 × 1/2	.203	.194	.047
305	3/32 × 5/8	.250	.240	.063
405	1/8 × 5/8	.250	.240	.063
505	5/32 × 5/8	.250	.240	.063
406	1/8 × 3/4	.313	.303	.063
506	5/32 × 3/4	.313	.303	.063
606	3/16 × 3/4	.313	.303	.063
507	5/32 × 7/8	.375	.365	.063
607	3/16 × 7/8	.375	.365	.063
807	1/4 × 7/8	.375	.365	.063
608	3/16 × 1	.438	.428	.063
609	3/16 × 1-1/8	.484	.475	.078

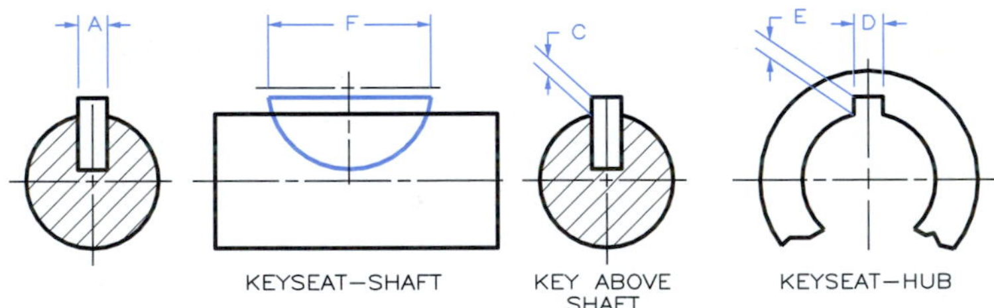

Key No.	A Min.	C +.005 −.000	F	D +.005 −.000	E +.005 −.000
204	.0615	.0312	.500	.0635	.0372
304	.0928	.0469	.500	.0948	.0529
404	.1240	.0625	.500	.1260	.0685
305	.0928	.0625	.625	.0948	.0529
405	.1240	.0469	.625	.1260	.0685
505	.1553	.0625	.625	.1573	.0841
406	.1240	.0781	.750	.1260	.0685
506	.1553	.0781	.750	.1573	.0841
606	.1863	.0937	.750	.1885	.0997
507	.1553	.0781	.875	.1573	.0841
607	.1863	.0937	.875	.1885	.0997
807	.2487	.1250	.875	.2510	.1310
608	.1863	.3393	1.000	.1885	.0997
609	.1863	.3853	1.125	.1885	.0997

Key sizes vs. Shaft sizes

Shaft DIA	to .375	to .500	to .750	to 1.313	to 1.188	to 1.448	to 1.750	to 2.125	to 2.500
Key Nos.	204	304 305	404 405 406	505 506 507	606 607 608 609	807 808 809	810 811 812	1011 1012	1211 1212

APPENDIX 38

Standard Keys and Keyseats

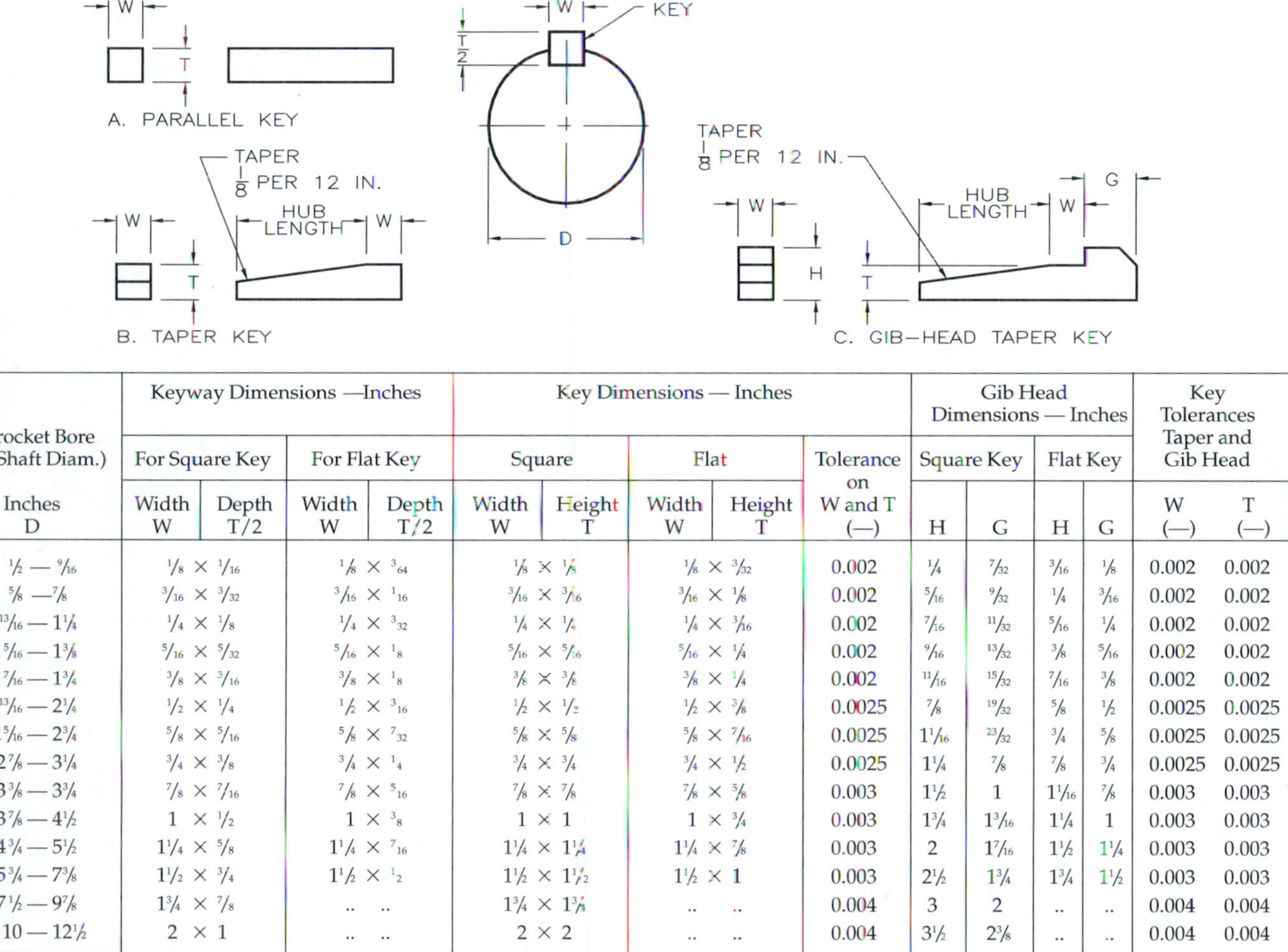

Sprocket Bore (= Shaft Diam.)	Keyway Dimensions —Inches		Key Dimensions — Inches				Gib Head Dimensions — Inches				Key Tolerances Taper and Gib Head	
	For Square Key	For Flat Key	Square	Flat	Tolerance on W and T		Square Key		Flat Key			
Inches D	Width W × Depth T/2	Width W × Depth T/2	Width W × Height T	Width W × Height T	(—)		H	G	H	G	W (—)	T (—)
1/2 — 9/16	1/8 × 1/16	1/8 × 3/64	1/8 × 1/8	1/8 × 3/32	0.002		1/4	7/32	3/16	1/8	0.002	0.002
5/8 — 7/8	3/16 × 3/32	3/16 × 1/16	3/16 × 3/16	3/16 × 1/8	0.002		5/16	9/32	1/4	3/16	0.002	0.002
13/16 — 1 1/4	1/4 × 1/8	1/4 × 3/32	1/4 × 1/4	1/4 × 3/16	0.002		7/16	11/32	5/16	1/4	0.002	0.002
1 5/16 — 1 3/8	5/16 × 5/32	5/16 × 1/8	5/16 × 5/16	5/16 × 1/4	0.002		9/16	13/32	3/8	5/16	0.002	0.002
1 7/16 — 1 3/4	3/8 × 3/16	3/8 × 1/8	3/8 × 3/8	3/8 × 1/4	0.002		11/16	15/32	7/16	3/8	0.002	0.002
1 13/16 — 2 1/4	1/2 × 1/4	1/2 × 3/16	1/2 × 1/2	1/2 × 3/8	0.0025		7/8	19/32	5/8	1/2	0.0025	0.0025
2 5/16 — 2 3/4	5/8 × 5/16	5/8 × 7/32	5/8 × 5/8	5/8 × 7/16	0.0025		1 1/16	23/32	3/4	5/8	0.0025	0.0025
2 7/8 — 3 1/4	3/4 × 3/8	3/4 × 1/4	3/4 × 3/4	3/4 × 1/2	0.0025		1 1/4	7/8	7/8	3/4	0.0025	0.0025
3 3/8 — 3 3/4	7/8 × 7/16	7/8 × 5/16	7/8 × 7/8	7/8 × 5/8	0.003		1 1/2	1	1 1/16	7/8	0.003	0.003
3 7/8 — 4 1/2	1 × 1/2	1 × 3/8	1 × 1	1 × 3/4	0.003		1 3/4	1 3/16	1 1/4	1	0.003	0.003
4 3/4 — 5 1/2	1 1/4 × 5/8	1 1/4 × 7/16	1 1/4 × 1 1/4	1 1/4 × 7/8	0.003		2	1 7/16	1 1/2	1 1/4	0.003	0.003
5 3/4 — 7 3/8	1 1/2 × 3/4	1 1/2 × 1/2	1 1/2 × 1 1/2	1 1/2 × 1	0.003		2 1/2	1 3/4	1 3/4	1 1/2	0.003	0.003
7 1/2 — 9 7/8	1 3/4 × 7/8	.. ..	1 3/4 × 1 3/4	.. ..	0.004		3	2	..	..	0.004	0.004
10 — 12 1/2	2 × 1	.. ..	2 × 2	.. ..	0.004		3 1/2	2 3/8	..	..	0.004	0.004

Standard Keyway Tolerances: Straight Keyway —Width (W) +.005 −.000 Depth (T/2) + .010 −.000

Taper Keyway —Width (W) + .005 −.000 Depth (T/2) + .000 −.010

Plain Washers (inches)

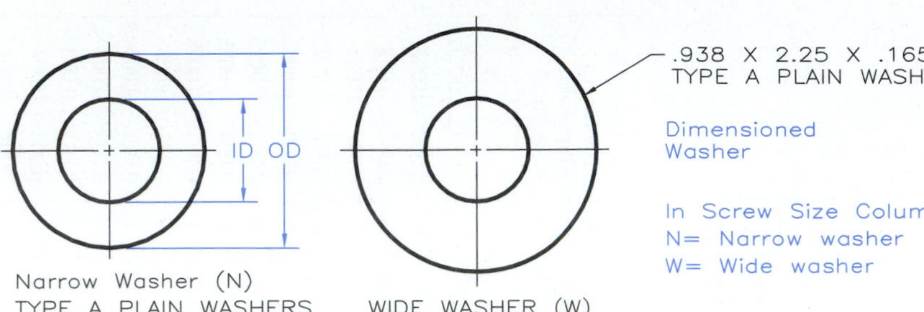

SCREW SIZE	ID SIZE	OD SIZE	THICK-NESS
0.138	0.156	0.375	0.049
0.164	0.188	0.438	0.049
0.190	0.219	0.500	0.049
0.188	0.250	0.562	0.049
0.216	0.250	0.562	0.065
0.250 N	0.281	0.625	0.065
0.250 W	0.312	0.734	0.065
0.312 N	0.344	0.688	0.065
0.312 W	0.375	0.875	0.083
0.375 N	0.406	0.812	0.065
0.375 W	0.438	1.000	0.083
0.438 N	0.469	0.922	0.065
0.438 W	0.500	1.250	0.083
0.500 N	0.531	1.062	0.095
0.500 W	0.562	1.375	0.109
0.562 N	0.594	1.156	0.095
0.562 W	0.594	1.469	0.190
0.625 N	0.625	1.312	0.095
0.625 N	0.625	1.750	0.134
0.750 W	0.812	1.469	0.134
0.750 W	0.812	2.000	0.148

SCREW SIZE	ID SIZE	OD SIZE	THICK-NESS
0.875 N	0.938	1.750	0.134
0.875 W	0.938	2.250	0.165
1.000 N	1.062	2.000	0.134
1.000 W	1.062	2.500	0.165
1.125 N	1.250	2.250	0.134
1.125 W	1.250	2.750	0.165
1.250 N	1.375	2.500	0.165
1.250 W	1.375	3.000	0.165
1.375 N	1.500	2.750	0.165
1.375 W	1.500	3.250	0.180
1.500 N	1.625	3.000	0.165
1.500 W	1.625	3.500	0.180
1.625	1.750	3.750	0.180
1.750	1.875	4.000	0.180
1.875	2.000	4.250	0.180
2.000	2.125	4.500	0.180
2.250	2.375	4.750	0.220
2.500	2.625	5.000	0.238
2.750	2.875	5.250	0.259
3.000	3.125	5.500	0.284

APPENDIX 40

Metric Washers (millimeters)

Flat Washers

SCREW SIZE	ID SIZE	OD SIZE	THICK-NESS
3	3.2	9	0.8
4	4.3	12	1
5	5.3	15	1.5
6	6.4	18	1.5
8	8.4	25	2
10	10.5	30	2.5
12	13	40	3
14	15	45	3
16	17	50	3
18	19	56	4
20	21	60	4

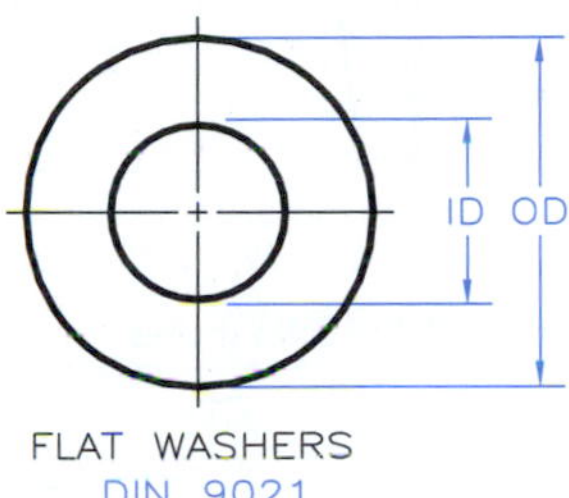

Wrought Washers

SCREW SIZE	ID SIZE	OD SIZE	THICK-NESS
2.6	2.8	5.5	0.5
3	3.2	6	0.5
4	4.3	8	0.5
5	5.3	10	1.0
6	6.4	11	1.5
8	8.4	15	1.5
10	10.5	18	1.5
12	13	20	2.0
14	15	25	2.0
16	17	27	2.0
18	19	30	2.5
20	21	33	2.5

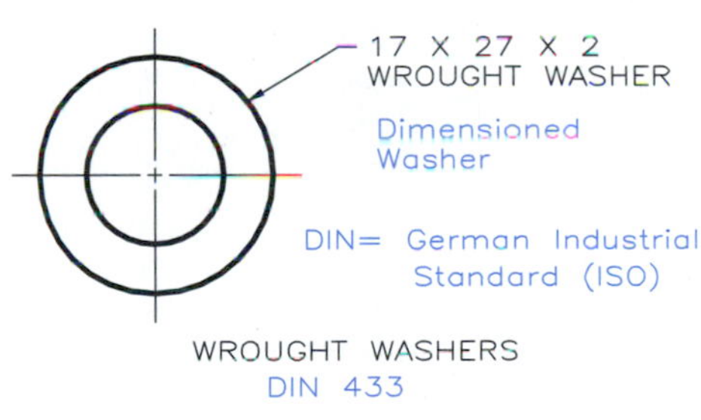

APPENDIX 41

Regular Helical Spring Lock Washers (inches and millimeters)

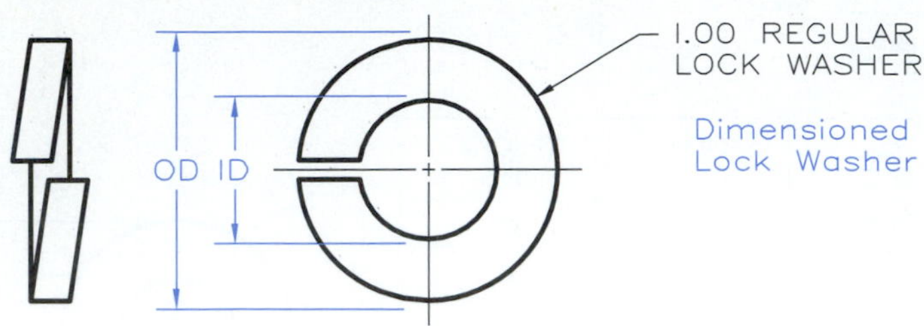

LOCK WASHERS inches

SCREW SIZE	ID SIZE	OD SIZE	THICK-NESS
0.164	0.168	0.175	0.040
0.190	0.194	0.202	0.047
0.216	0.221	0.229	0.056
0.250	0.255	0.263	0.062
0.312	0.318	0.328	0.078
0.375	0.382	0.393	0.094
0.438	0.446	0.459	0.109
0.500	0.509	0.523	0.125
0.562	0.572	0.587	0.141
0.625	0.636	0.653	0.156
0.688	0.700	0.718	0.172
0.750	0.763	0.783	0.188
0.812	0.826	1.367	0.203
0.875	0.890	1.464	0.219
0.938	0.954	1.560	0.234
1.000	1.017	1.661	0.250
1.062	1.080	1.756	0.266
1.125	1.144	1.853	0.281
1.188	1.208	1.950	0.297
1.250	1.271	2.045	0.312
1.312	1.334	2.141	0.328
1.375	1.398	2.239	0.344
1.438	1.462	2.334	0.359
1.500	1.525	2.430	0.375

METRIC LOCK WASHERS DIN 127 (Millimeters)

SCREW SIZE	ID SIZE	OD SIZE	THICK-NESS
4	4.1	7.1	0.9
5	5.1	8.7	1.2
6	6.1	11.1	1.6
8	8.2	12.1	1.6
10	10.2	14.2	2
12	12.1	17.2	2.2
14	14.2	20.2	2.5
16	16.2	23.2	3
18	18.2	26.2	3.5
20	20.2	28.2	3.5
22	22.5	34.5	4
24	24.5	38.5	5
27	27.5	41.5	5
30	30.5	46.5	6
33	33.5	53.5	6
36	36.5	56.5	6
39	39.5	59.5	6
42	42.5	66.5	7
45	45.5	69.5	7
48	49	73	7

APPENDIX

42 American National Standard 125-lb Cast Iron Screwed Fittings (inches)

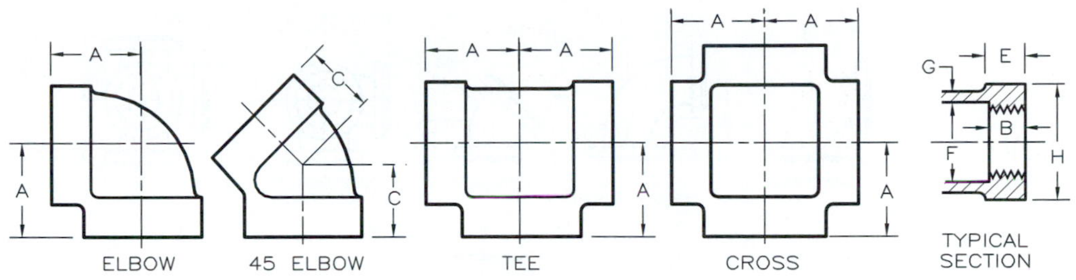

Nominal Pipe Size	A	C	B	E	F		G	H
			Min	Min	Min	Max	Min	Min
1/4	0.81	0.73	0.32	0.38	0.540	0.584	0.110	0.93
3/8	0.95	0.80	0.36	0.44	0.675	0.719	0.120	1.12
1/2	1.12	0.88	0.43	0.50	0.840	0.897	0.130	1.34
3/4	1.31	0.98	0.50	0.56	1.050	1.107	0.155	1.63
1	1.50	1.12	0.58	0.62	1.315	1.385	0.170	1.95
1 1/4	1.75	1.29	0.67	0.69	1.660	1.730	0.185	2.39
1 1/2	1.94	1.43	0.70	0.75	1.900	1.970	0.200	2.68
2	2.25	1.68	0.75	0.84	2.375	2.445	0.220	3.28
2 1/2	2.70	1.95	0.92	0.94	2.875	2.975	0.240	3.86
3	3.08	2.17	0.98	1.00	3.500	3.600	0.260	4.62
3 1/2	3.42	2.39	1.03	1.06	4.000	4.100	0.280	5.20
4	3.79	2.61	1.08	1.12	4.500	4.600	0.310	5.79
5	4.50	3.05	1.18	1.18	5.563	5.663	0.380	7.05
6	5.13	3.46	1.28	1.28	6.625	0.725	0.430	8.28
8	6.56	4.28	1.47	1.47	8.625	8.725	0.550	10.63
10	8.08	5.16	1.68	1.68	10.750	10.850	0.690	13.12
12	9.50	5.97	1.88	1.88	12.750	12.850	0.800	15.47
14 O.D.	10.40		2.00	2.00	14.000	14.100	0.880	16.94
16 O.D.	11.82		2.20	2.20	16.000	16.100	1.000	19.30

APPENDIX 43

American National Standard 250-lb Cast Iron Flanged Fittings (inches)

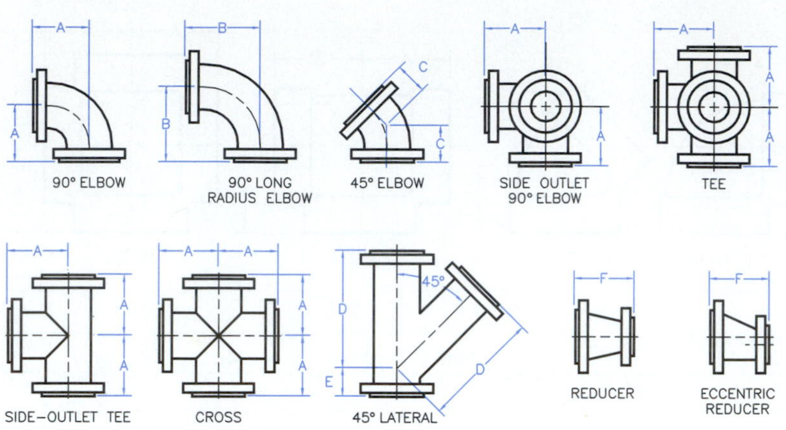

Nominal Pipe Size	Flanges: Dia of Flange	Flanges: Thickness of Flange (Min)	Flanges: Dia of Raised Face	Fittings: Inside Dia of Fittings (Min)	Fittings: Wall Thickness	Straight: Center to Face 90 Deg Elbow Tees, Crosses and True "Y" A	Straight: Center to Face 90 Deg Long Radius Elbow B	Straight: Center to Face 45 Deg Elbow C	Straight: Center to Face Lateral D	Straight: Short Center to Face True "Y" and Lateral E	Straight: Face to Face Reducer F
1	4 7/8	11/16	2 11/16	1	7/16	4	5	2	6 1/2	2	....
1 1/4	5 1/4	3/4	3 1/16	1 1/4	7/16	4 1/4	5 1/2	2 1/2	7 1/4	2 1/4	....
1 1/2	6 1/8	13/16	3 9/16	1 1/2	7/16	4 1/2	6	2 3/4	8 1/2	2 1/2	....
2	6 1/2	7/8	4 3/16	2	7/16	5	6 1/2	3	9	2 1/2	5
2 1/2	7 1/2	1	4 15/16	2 1/2	1/2	5 1/2	7	3 1/2	10 1/2	2 1/2	5 1/2
3	8 1/4	1 1/8	5 11/16	3	9/16	6	7 3/4	3 1/2	11	3	6
3 1/2	9	1 3/16	6 5/16	3 1/2	9/16	6 1/2	8 1/2	4	12 1/2	3	6 1/2
4	10	1 1/4	6 15/16	4	5/8	7	9	4 1/2	13 1/2	3	7
5	11	1 3/8	8 5/16	5	11/16	8	10 1/4	5	15	3 1/2	8
6	12 1/2	1 7/16	9 11/16	6	3/4	8 1/2	11 1/2	5 1/2	17 1/2	4	9
8	15	1 5/8	11 15/16	8	13/16	10	14	6	20 1/2	5	11
10	17 1/2	1 7/8	14 1/16	10	15/16	11 1/2	16 1/2	7	24	5 1/2	12
12	20 1/2	2	16 7/16	12	1	13	19	8	27 1/2	6	14
14	23	2 1/8	18 15/16	13 1/4	1 1/8	15	21 1/2	8 1/2	31	6 1/2	16
16	25 1/2	2 1/4	21 1/36	15 1/4	1 1/4	16 1/2	24	9 1/2	34 1/2	7 1/2	18
18	28	2 3/8	23 5/16	17	1 3/8	18	26 1/2	10	37 1/2	8	19
20	30 1/2	2 1/2	25 9/16	19	1 1/2	19 1/2	29	10 1/2	40 1/2	8 1/2	20
24	36	2 3/4	30 5/16	23	1 5/8	22 1/2	34	12	47 1/2	10	24
30	43	3	37 3/16	29	2	27 1/2	41 1/2	15	....	....	30

APPENDIX 44

Welding Symbols

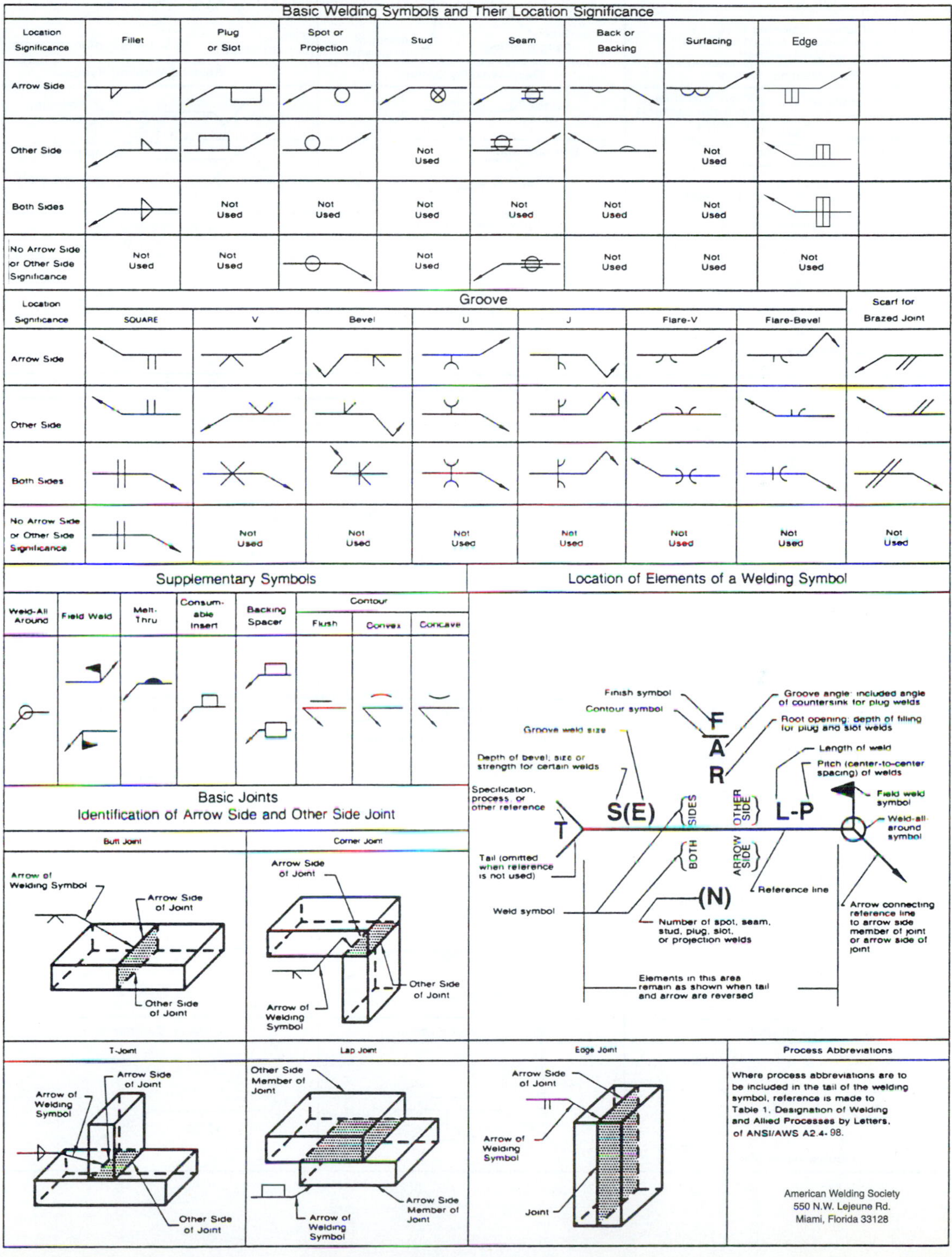

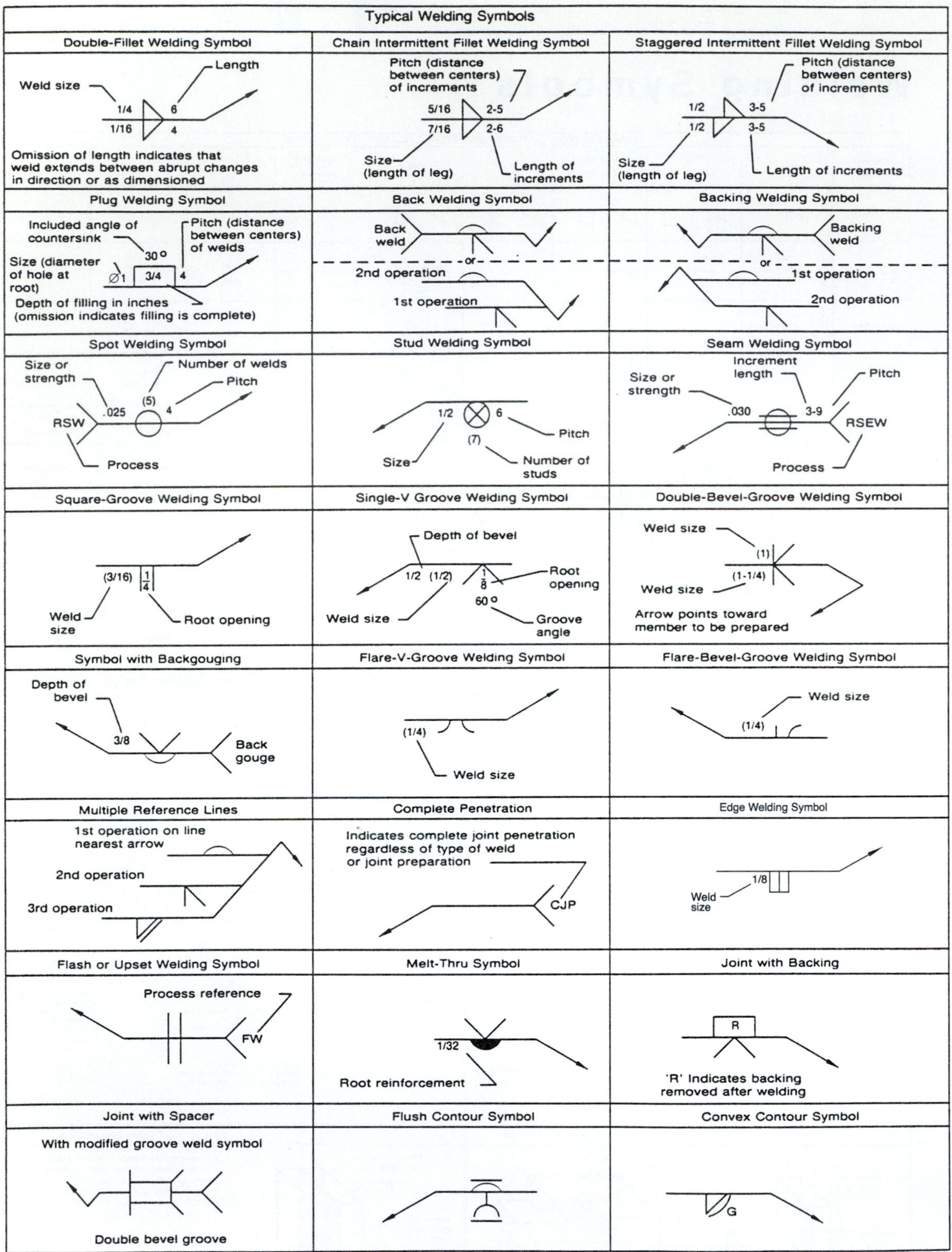

Typical Welding Symbols
Double-Fillet Welding Symbol
Weld size
Length
1/4
6
1/16
4
Omission of length indicates that weld extends between abrupt changes in direction or as dimensioned
Chain Intermittent Fillet Welding Symbol
Pitch (distance between centers) of increments
5/16
2-5
7/16
2-6
Size (length of leg)
Length of increments
Staggered Intermittent Fillet Welding Symbol
Pitch (distance between centers) of increments
1/2
3-5
1/2
3-5
Size (length of leg)
Length of increments
Plug Welding Symbol
Included angle of countersink
30°
Pitch (distance between centers) of welds
Size (diameter of hole at root)
Ø1
3/4
4
Depth of filling in inches (omission indicates filling is complete)
Back Welding Symbol
Back weld
or
2nd operation
1st operation
Backing Welding Symbol
Backing weld
or
1st operation
2nd operation
Spot Welding Symbol
Size or strength
Number of welds
Pitch
(5)
.025
4
RSW
Process
Stud Welding Symbol
1/2
6
Pitch
(7)
Size
Number of studs
Seam Welding Symbol
Increment length
Pitch
Size or strength
.030
3-9
RSEW
Process
Square-Groove Welding Symbol
(3/16)
1/4
Weld size
Root opening
Single-V Groove Welding Symbol
Depth of bevel
1/2 (1/2)
1/8
Root opening
60°
Weld size
Groove angle
Double-Bevel-Groove Welding Symbol
Weld size
(1)
(1-1/4)
Weld size
Arrow points toward member to be prepared
Symbol with Backgouging
Depth of bevel
3/8
Back gouge
Flare-V-Groove Welding Symbol
(1/4)
Weld size
Flare-Bevel-Groove Welding Symbol
Weld size
(1/4)
Multiple Reference Lines
1st operation on line nearest arrow
2nd operation
3rd operation
Complete Penetration
Indicates complete joint penetration regardless of type of weld or joint preparation
CJP
Edge Welding Symbol
1/8
Weld size
Flash or Upset Welding Symbol
Process reference
FW
Melt-Thru Symbol
1/32
Root reinforcement
Joint with Backing
R
'R' Indicates backing removed after welding
Joint with Spacer
With modified groove weld symbol
Double bevel groove
Flush Contour Symbol
Convex Contour Symbol
G

MASTER CHART OF WELDING AND ALLIED PROCESSES

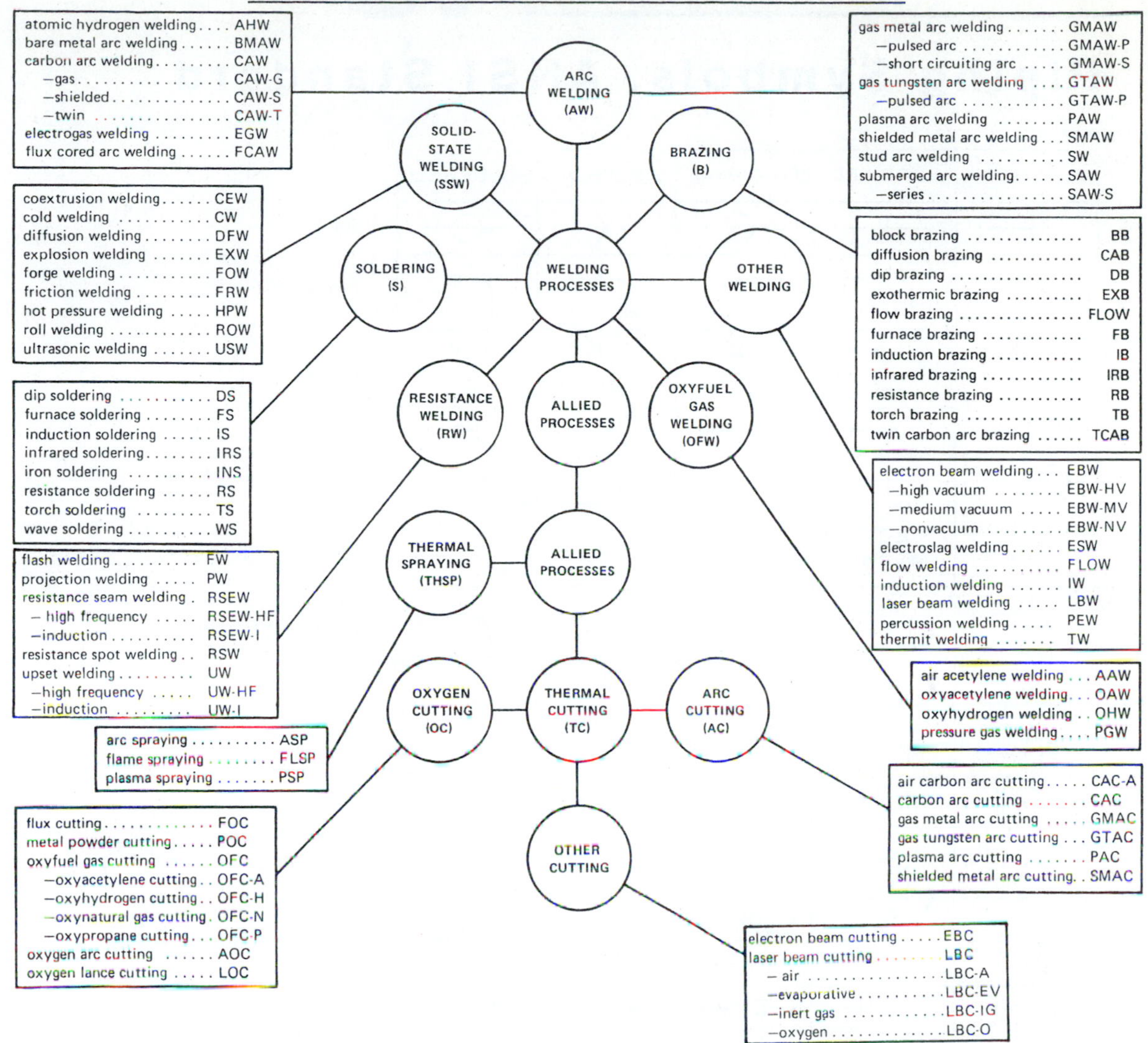

APPENDIX 

Piping Symbols, ANSI Standard

FLANGED	SCREWED	BELL & SPIGOT	WELDED	SOLDERED	
					Joint
					Elbow–90°
					Elbow–45°
					Elbow–Turned Up
					Elbow–Turned Down
					Elbow–Long radius
2 4	2 4			2 4	Reducing Elbow
					Tee
					Tee–Outlet Up
					Tee–Outlet Down
					Side Outlet Tee–Outlet Up
					Cross
					Reducer–Concentric
					Reducer–Eccentric
					Lateral
					Gate Valve–Elev.
					Globe Valve–Elev.
					Check Valve
					Stop Cock
					Safety Valve
					Expansion Joint
					Union
					Sleeve
		6 4			Bushing

APPENDIX 46

Heating, Ventilation, and Ductwork Symbols, ANSI Standard

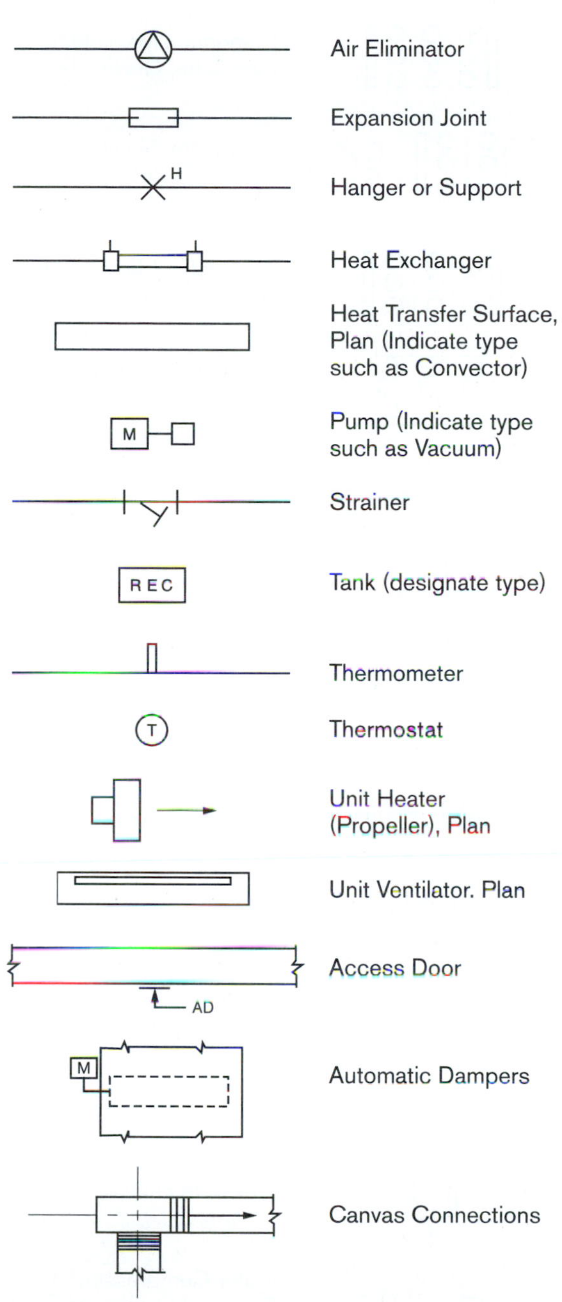

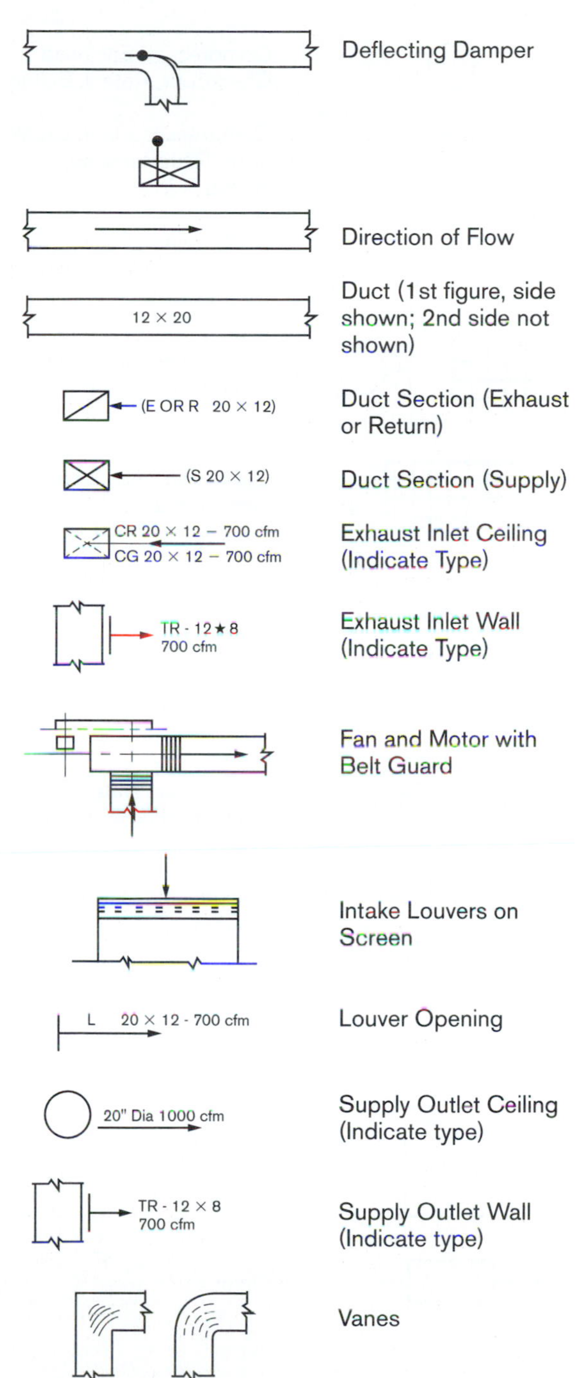

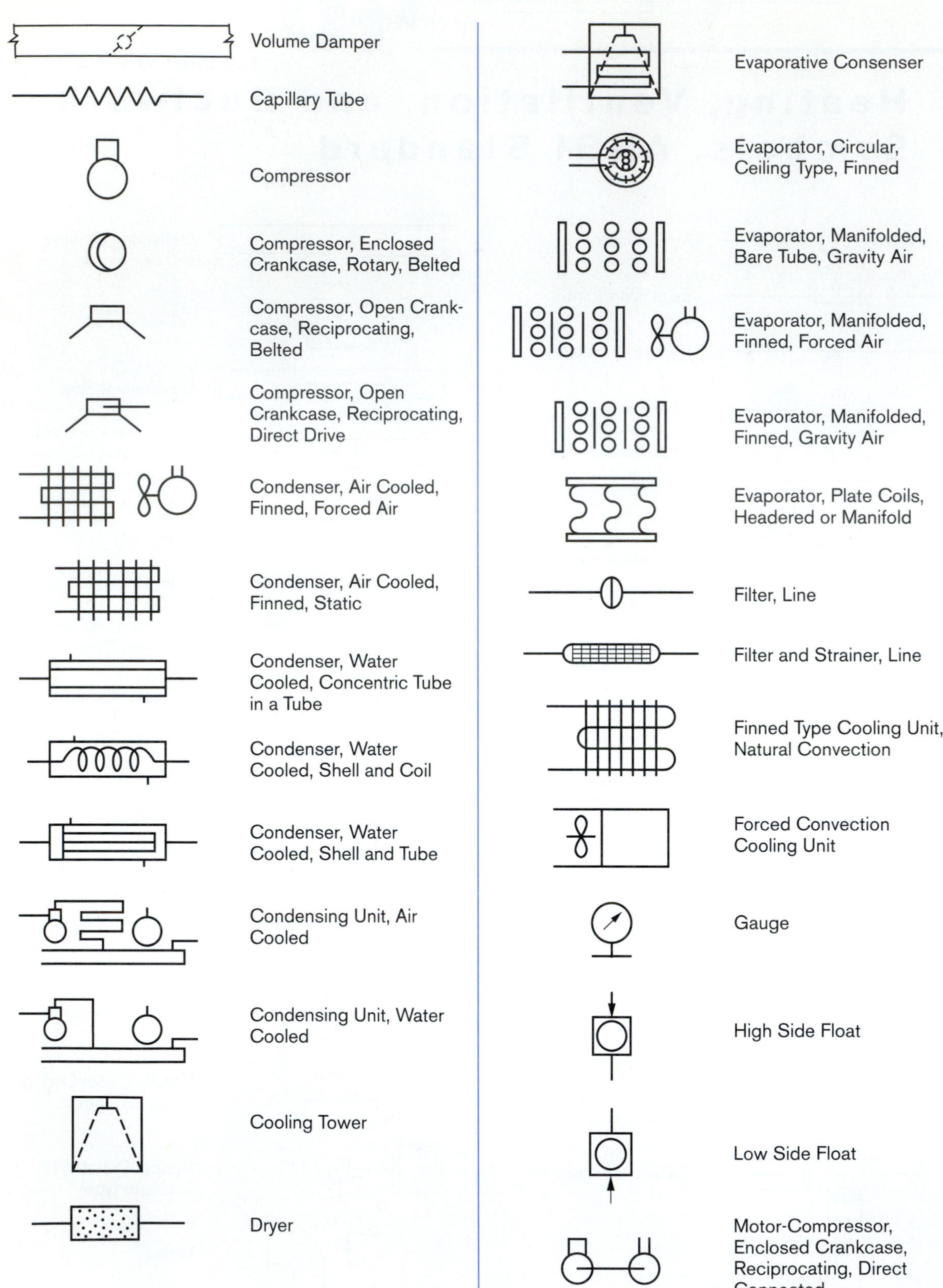
Volume Damper
Capillary Tube
Compressor
Compressor, Enclosed Crankcase, Rotary, Belted
Compressor, Open Crank-case, Reciprocating, Belted
Compressor, Open Crankcase, Reciprocating, Direct Drive
Condenser, Air Cooled, Finned, Forced Air
Condenser, Air Cooled, Finned, Static
Condenser, Water Cooled, Concentric Tube in a Tube
Condenser, Water Cooled, Shell and Coil
Condenser, Water Cooled, Shell and Tube
Condensing Unit, Air Cooled
Condensing Unit, Water Cooled
Cooling Tower
Dryer
Evaporative Consenser
Evaporator, Circular, Ceiling Type, Finned
Evaporator, Manifolded, Bare Tube, Gravity Air
Evaporator, Manifolded, Finned, Forced Air
Evaporator, Manifolded, Finned, Gravity Air
Evaporator, Plate Coils, Headered or Manifold
Filter, Line
Filter and Strainer, Line
Finned Type Cooling Unit, Natural Convection
Forced Convection Cooling Unit
Gauge
High Side Float
Low Side Float
Motor-Compressor, Enclosed Crankcase, Reciprocating, Direct Connected

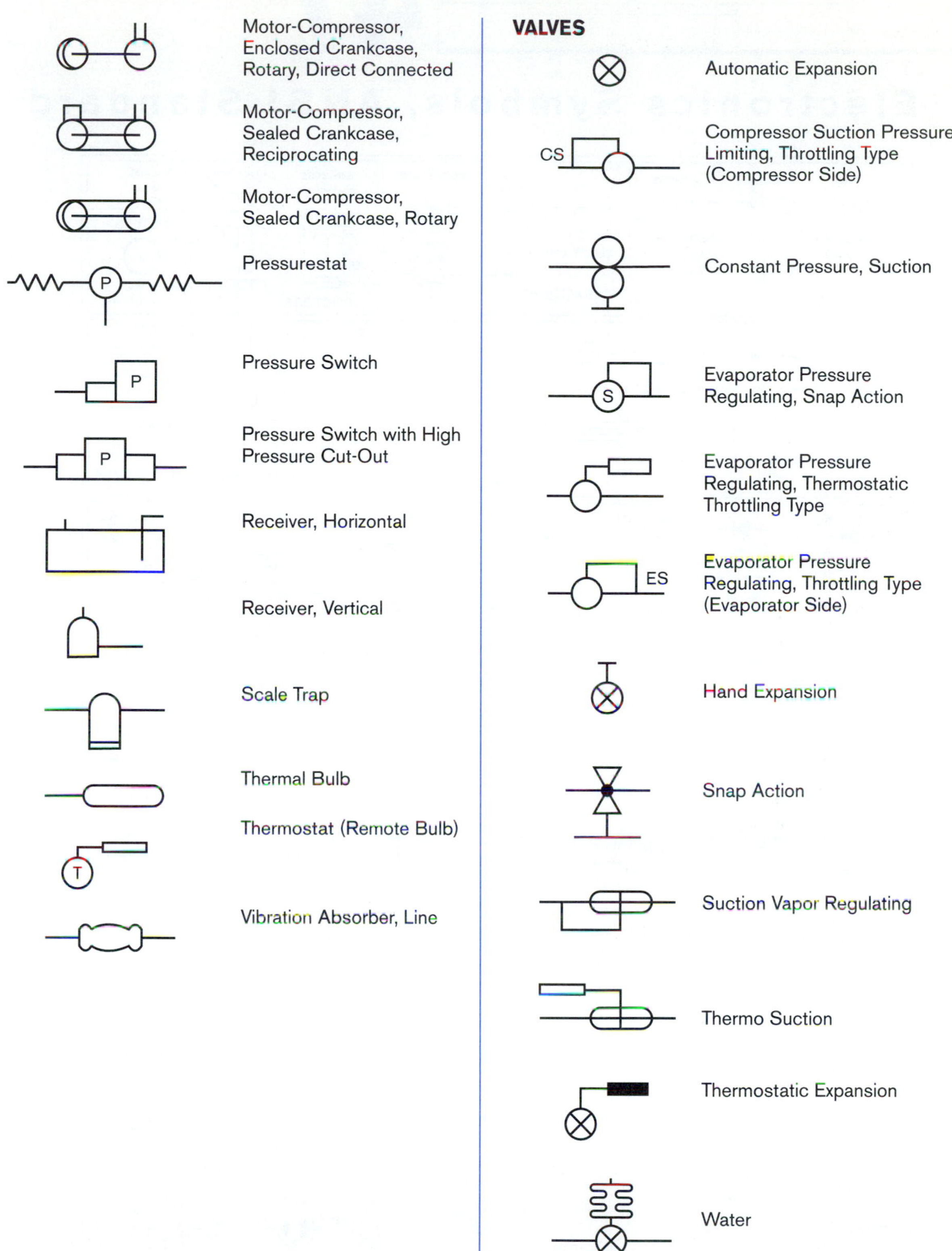
Motor-Compressor, Enclosed Crankcase, Rotary, Direct Connected
Motor-Compressor, Sealed Crankcase, Reciprocating
Motor-Compressor, Sealed Crankcase, Rotary
P
Pressurestat
P
Pressure Switch
P
Pressure Switch with High Pressure Cut-Out
Receiver, Horizontal
Receiver, Vertical
Scale Trap
Thermal Bulb
T
Thermostat (Remote Bulb)
Vibration Absorber, Line
VALVES
Automatic Expansion
CS
Compressor Suction Pressure Limiting, Throttling Type (Compressor Side)
Constant Pressure, Suction
S
Evaporator Pressure Regulating, Snap Action
Evaporator Pressure Regulating, Thermostatic Throttling Type
ES
Evaporator Pressure Regulating, Throttling Type (Evaporator Side)
Hand Expansion
Snap Action
Suction Vapor Regulating
Thermo Suction
Thermostatic Expansion
Water

APPENDIX 

Electronics Symbols, ANSI Standard

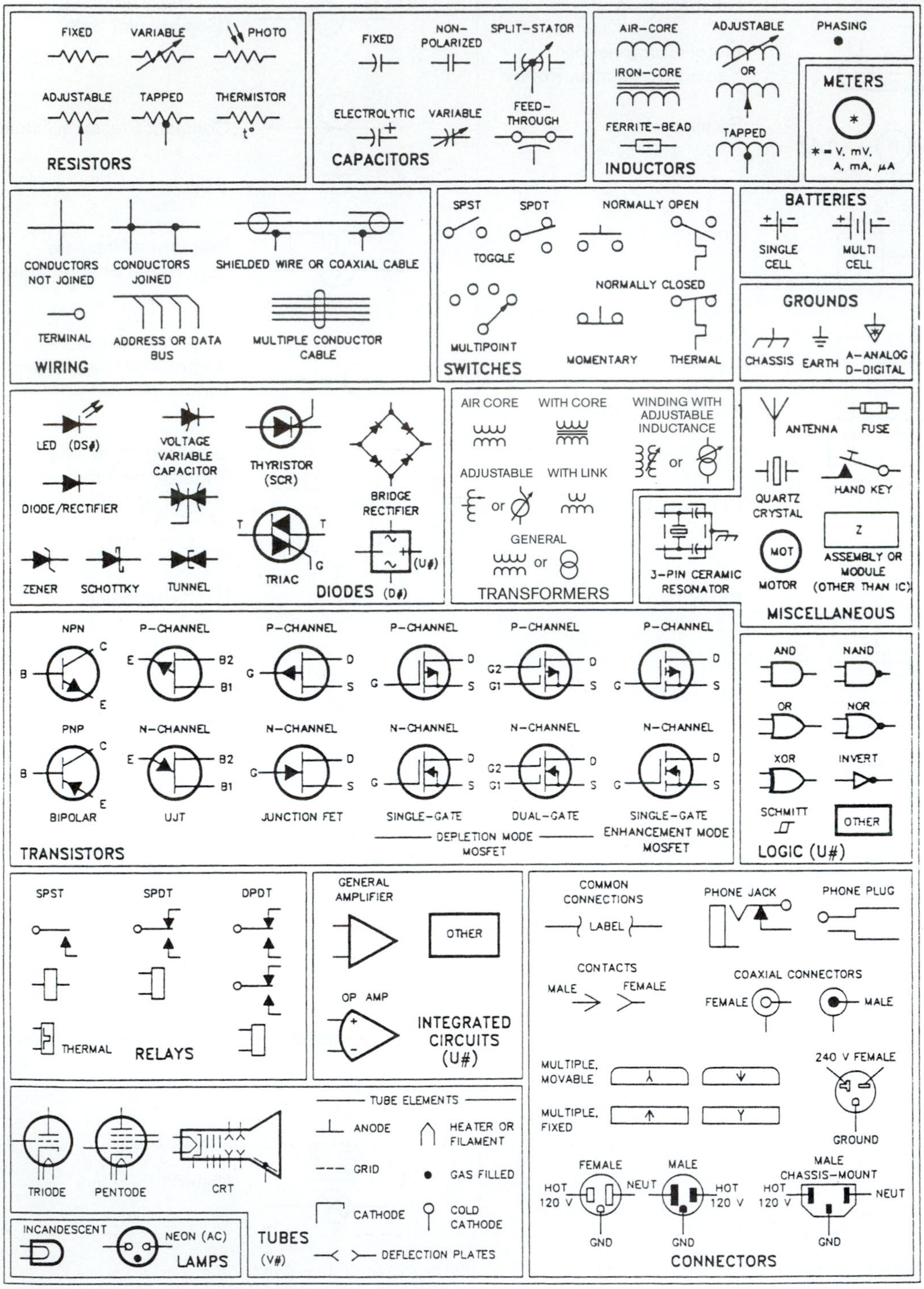

RESISTORS
FIXED · VARIABLE · PHOTO · ADJUSTABLE · TAPPED · THERMISTOR (t°)

CAPACITORS
FIXED · NON-POLARIZED · SPLIT-STATOR · ELECTROLYTIC · VARIABLE · FEED-THROUGH

INDUCTORS
AIR-CORE · IRON-CORE · FERRITE-BEAD · ADJUSTABLE OR · TAPPED · PHASING

METERS
* = V, mV, A, mA, µA

WIRING
CONDUCTORS NOT JOINED · CONDUCTORS JOINED · SHIELDED WIRE OR COAXIAL CABLE · TERMINAL · ADDRESS OR DATA BUS · MULTIPLE CONDUCTOR CABLE

SWITCHES
SPST · SPDT · TOGGLE · NORMALLY OPEN · NORMALLY CLOSED · MULTIPOINT · MOMENTARY · THERMAL

BATTERIES
SINGLE CELL · MULTI CELL

GROUNDS
CHASSIS · EARTH · A-ANALOG D-DIGITAL

DIODES (D#)
LED (DS#) · VOLTAGE VARIABLE CAPACITOR · THYRISTOR (SCR) · BRIDGE RECTIFIER · DIODE/RECTIFIER · TRIAC (T, T, G) · (U#) · ZENER · SCHOTTKY · TUNNEL

TRANSFORMERS
AIR CORE · WITH CORE · WINDING WITH ADJUSTABLE INDUCTANCE (or) · ADJUSTABLE (or) · WITH LINK · GENERAL (or)

MISCELLANEOUS
ANTENNA · FUSE · QUARTZ CRYSTAL · HAND KEY · 3-PIN CERAMIC RESONATOR · MOTOR (MOT) · ASSEMBLY OR MODULE (OTHER THAN IC) (Z)

TRANSISTORS
NPN (B, C, E) · PNP (B, C, E) · BIPOLAR
P-CHANNEL (E, B2, B1) · N-CHANNEL (E, B2, B1) · UJT
P-CHANNEL (G, D, S) · N-CHANNEL (G, D, S) · JUNCTION FET
P-CHANNEL (G, D, S) · N-CHANNEL (G, D, S) · SINGLE-GATE
P-CHANNEL (G2, G1, D, S) · N-CHANNEL (G2, G1, D, S) · DUAL-GATE
DEPLETION MODE MOSFET
P-CHANNEL (G, D, S) · N-CHANNEL (G, D, S) · SINGLE-GATE ENHANCEMENT MODE MOSFET

LOGIC (U#)
AND · NAND · OR · NOR · XOR · INVERT · SCHMITT · OTHER

RELAYS
SPST · SPDT · DPDT · THERMAL

INTEGRATED CIRCUITS (U#)
GENERAL AMPLIFIER · OTHER · OP AMP (+, −)

CONNECTORS
COMMON CONNECTIONS (LABEL) · PHONE JACK · PHONE PLUG
CONTACTS: MALE · FEMALE
COAXIAL CONNECTORS: FEMALE · MALE
MULTIPLE, MOVABLE · MULTIPLE, FIXED
240 V FEMALE (GROUND)
FEMALE (HOT 120 V, NEUT, GND) · MALE (HOT 120 V, GND) · MALE CHASSIS-MOUNT (HOT 120 V, NEUT, GND)

TUBES (V#)
TRIODE · PENTODE · CRT
TUBE ELEMENTS: ANODE · HEATER OR FILAMENT · GRID · GAS FILLED · CATHODE · COLD CATHODE · DEFLECTION PLATES

LAMPS
INCANDESCENT · NEON (AC)

APPENDIX 48

Geometric Dimensioning and Tolerancing Symbols

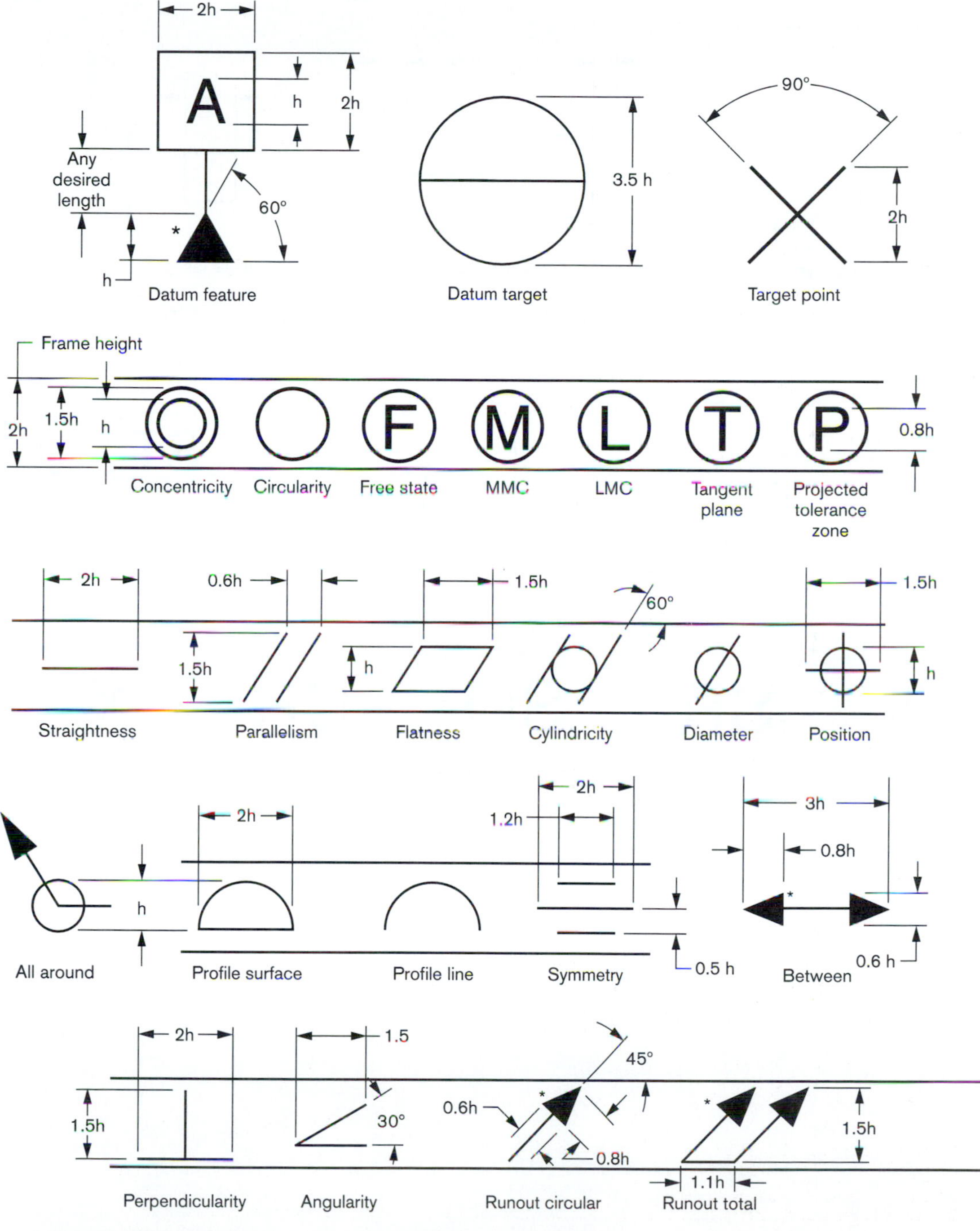

* May be filled or not filled

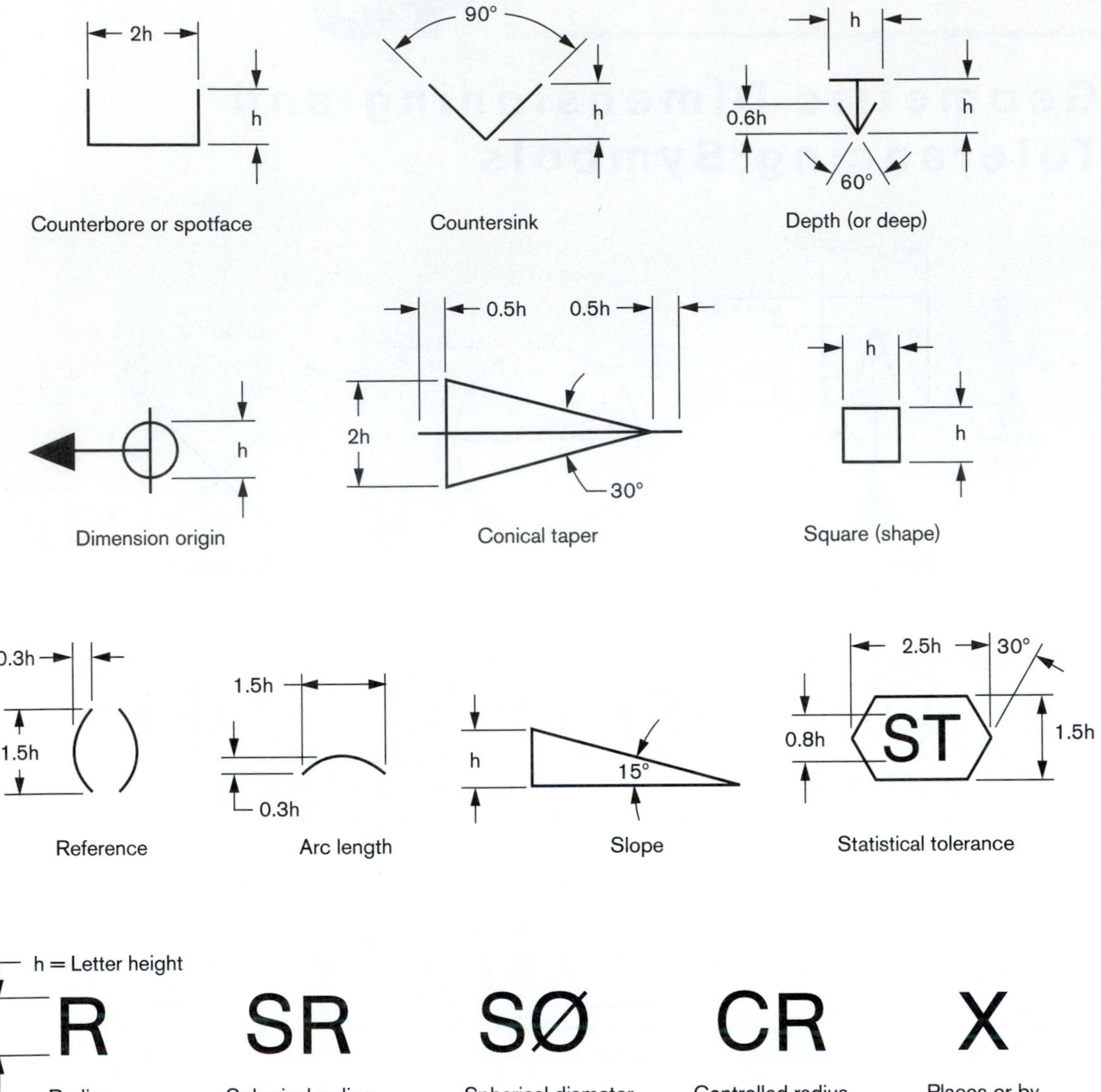
2h
h
Counterbore or spotface
90°
h
Countersink
h
0.6h
h
60°
Depth (or deep)
0.5h
0.5h
2h
30°
h
Dimension origin
Conical taper
h
h
Square (shape)
0.3h
1.5h
Reference
1.5h
0.3h
Arc length
h
15°
Slope
2.5h
30°
0.8h
ST
1.5h
Statistical tolerance
h = Letter height
R
Radius
SR
Spherical radius
SØ
Spherical diameter
CR
Controlled radius
X
Places or by

APPENDIX 49

IGES Standard

The IGES standard is a specification for the structure and syntax of an electronic comma-delimited file in ASCII, compressed ASCII, or binary format. Each 80-character line of the file is a record that ends with a semi-colon. Each record (line) is subdivided into fields by commas. Each record line has an identifier in columns 73–80. The first character of the identifier indicates the file section (start = S, global = G, and so on), and the remainder is an integer number starting with one at the beginning of each section. This number is used by IGES for cross-referencing between sections.

File Organization

Each IGES file is divided into five sections:

Start

The Start Section is set up by the person creating the IGES file and includes information about the originating system and other data that may be useful to the person receiving the file.

Global

The Global Section is a set of 24 fields that spell out the parameters needed to translate the file:

- the delimiter characters (fields 1 and 2)
- the sender's identifier (3)
- filename (4)
- ID of the software producing the file (5)
- version of the IGES processor (6)
- precision of integer, floating-point, and double precision numbers (7 to 11)
- receiver's identifier (12)
- model space scale (13)
- units (14)
- name of the units (15)
- maximum number of line thickness (16)
- maximum line thickness (17)
- time file generated (18)
- smallest distance (19)

- largest coordinate value (20)
- person and organization creating the file (21 and 22)
- IGES version (23)
- drafting standard (24)

Directory

The Directory Section is generated by the IGES pre-processor and contains a two-line entry for each entity in the file. Each entry specified the entity type and sub-type and pointers to the entity data in the next section.

Parameter Data

The Parameter Data Section specifies the data specific to each entity—coordinate values, annotation text, number of spline data points, etc. The first parameter in each entry identifies the entity type. Each entry has a pointer in columns 66–72 to the directory entry for the entity.

Termination

The Termination Section marks the end of the Data file. It includes subtotals of records that are used to ensure that the file is complete and not damaged in transmission.

IGES Data Types

The listing below illustrates the kinds of entities supported by IGES and the code used to identify the entity in the directory and parameter data sections.

Geometric Entities

100	Circular arc
102	Composite curve
104	Conic arc
106	Copious data
108	Plane
110	Line
112	Parametric spline curve
114	Parametric spline surface
116	Point
118	Ruled surface
120	Surface of revolution
122	Tabulated cylinder
124	Transformation matrix
125	Flash
126	Rational B-spline curve
128	Rational B-spline surface

130	Offset curve
132	Connect point
134	Node
136	Finite element
138	Nodal displacement and rotation
140	Offset surface
142	Curve on a parametric surface
144	Trimmed parametric surface

Annotation Entities

202	Angular dimension
206	Diameter dimension
208	Flag note
210	General label
212	General note
214	Leader (arrow)
216	Linear dimension
218	Ordinate dimension
220	Point dimension
222	Radius dimension
228	General symbol
230	Sectioned area

Structure Entities

302	Associativity definition
304	Line font definition
306	Macro definition
308	Subfigure definition
310	Text font definition
312	Text display template
314	Color definition
320	Network subfigure definition
402	Associativity instance
404	Drawing
406	Property
408	Singular subfigure instance
410	View
412	Rectangular array subfigure instance
414	Circular array subfigure instance
416	External reference
418	Node load/constraint
420	Network subfigure instance
600–699	Macro instance

APPENDIX 50

Useful Formulas for Geometric Entities

Formulas for Circles

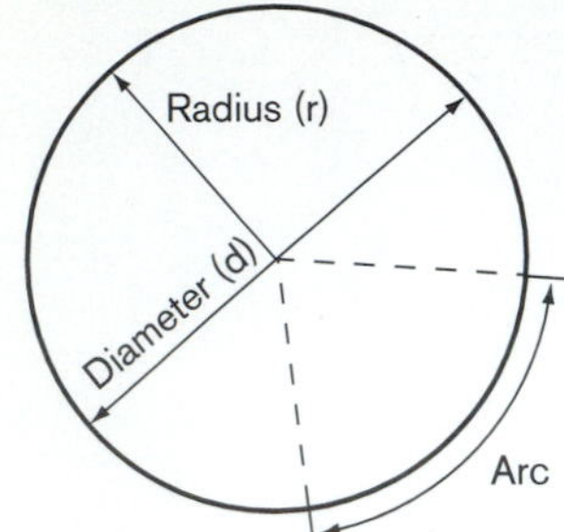

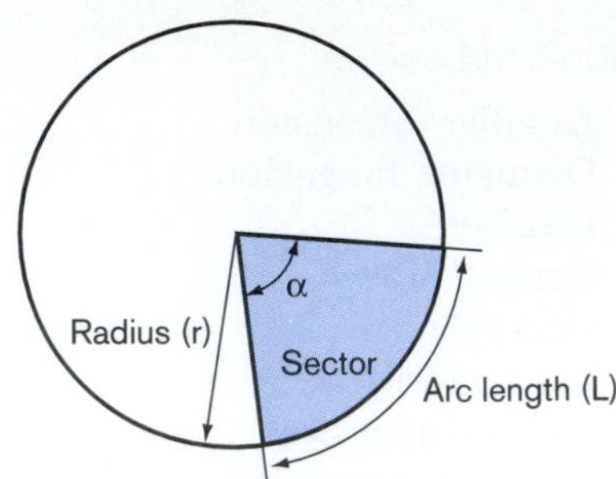

Circle

Area	$A = \pi r^2$
	$A = 3.141r^2$
	$A = 0.7854d^2$
Radius	$r = d/2$
Diameter	$d = 2r$
Circumference	$C = 2\pi r$
	$C = \pi d$
	$C = 3.141d$

(Note: 22/7 and 3.141 are different approximations for π.)

Sector of a Circle

Area	$A = \frac{3.141r^2\alpha}{360}$
Arc (length)	$L = \frac{2\pi r}{360}\alpha$
	$L = 0.01745r\alpha$
Angle	$\alpha = \frac{L}{0.01745r}$
Radius	$r = \frac{L}{0.01745\alpha}$

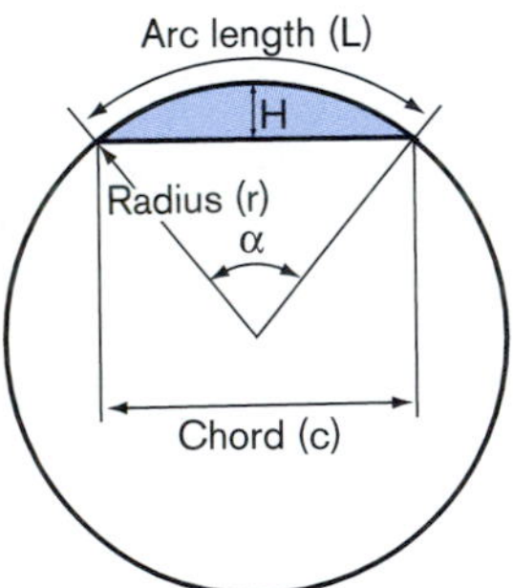

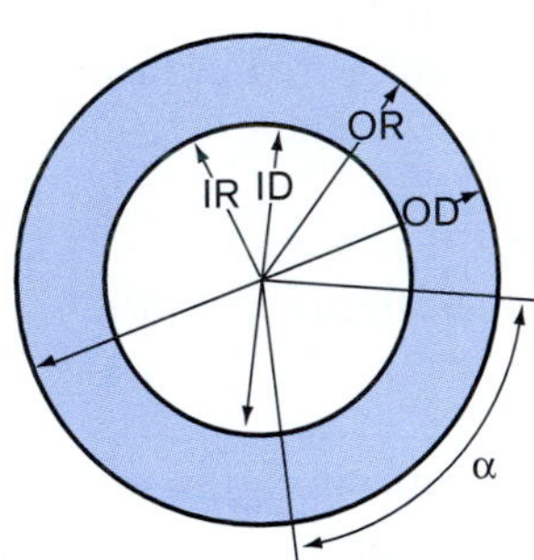

Segment of a Circle

Area	$A = \frac{1}{2}(r \bullet L - c(r - h))$
Arc (length)	$L = 0.01745r\alpha$
Angle	$\alpha = \frac{57.296L}{r}$
Height	$H = r - \frac{1}{2}\sqrt{4r^2 - c^2}$
Chord	$c = 2r \sin\alpha$

Circular Ring

Ring area	$A = 0.7854\ (OD^2 - ID^2)$
Ring sector area	$a = 0.00873\ \alpha(OR^2 - IR^2)$
	$a = 0.00218\ \alpha(OD^2 - ID^2)$

OD = outside diameter
ID = inside diameter
α = ring sector angle
OR = outside radius
IR = inside radius

Formulas for Triangles

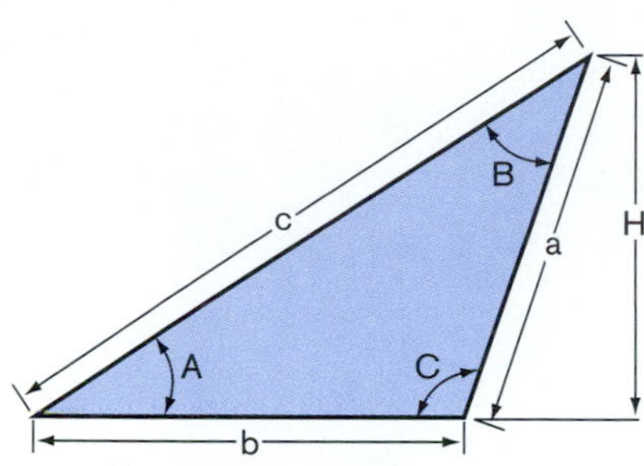

Obtuse angle triangle

Any Triangle

Area $A=1/2bH=\dfrac{Hb}{2}$

$A=\sqrt{S(S-a)(S-b)(S-c)}$

$S=1/2\ (a+b+c)$

Perimeter $P=a+b+c$

Height $H=\dfrac{2}{b}\sqrt{S(S-a)(S-b)(S-c)}$

Sum of angles $180°=A+B+C$

Acute angle triangle

Equilateral Triangle

Area $A=a^2\dfrac{\sqrt{3}}{4}=0.433a^2$

$A=0.577H^2$

$A=\dfrac{a^2}{2}$ or $\dfrac{aH}{2}$

Perimeter $P=3a$

Height $H=\dfrac{a}{2}\sqrt{3}=0.866a$

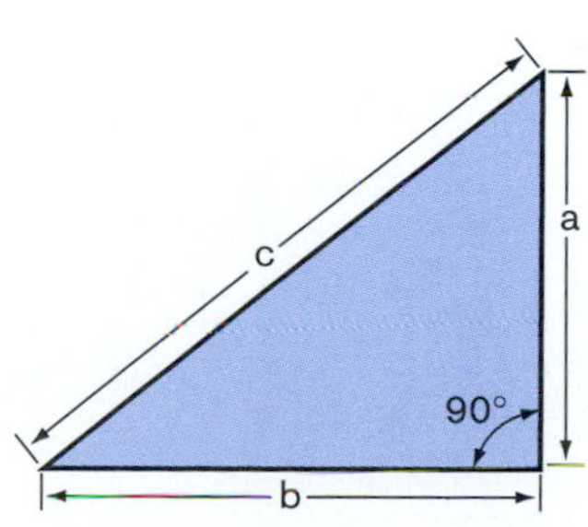

Right Triangle

Area $A=\dfrac{ba}{2}$

Perimeter $P=a+b+c$

Height $a=\sqrt{c^2-b^2}$

Base $b=\sqrt{c^2-a^2}$

Hypotenuse $c=\sqrt{a^2-b^2}$

Formulas for Four-Sided Polygons

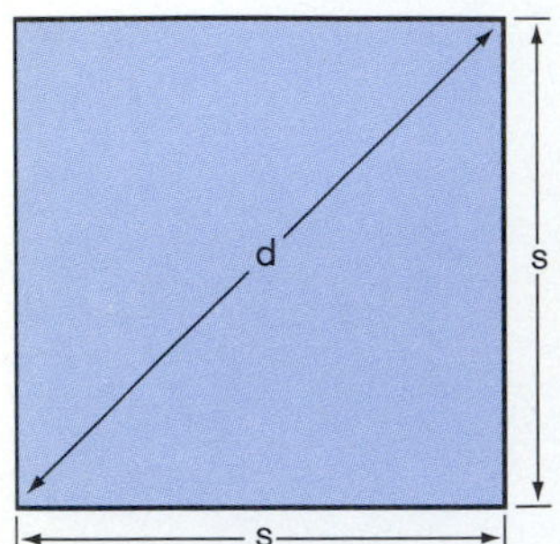

Square

Area $A = s^2$

$A = 0.5d^2$

Side $s = 0.707d$

Diagonal $d = 1.414s$

Perimeter $P = 4s$

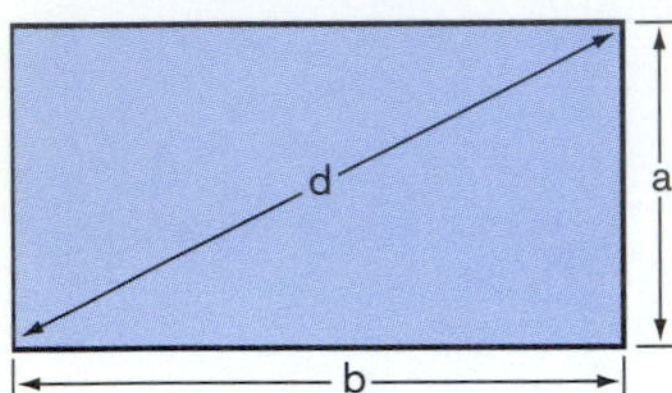

Rectangle

Area $A = ab$

Side a $a = \sqrt{d^2 - b^2}$

Side b $b = \sqrt{d^2 - a^2}$

Diagonal $d = \sqrt{a^2 - b^2}$

Perimeter $P = 2\,(a + b)$

Parallelogram

Area $A = Hb$

Height $H = A/b$

Base $b = A/H$

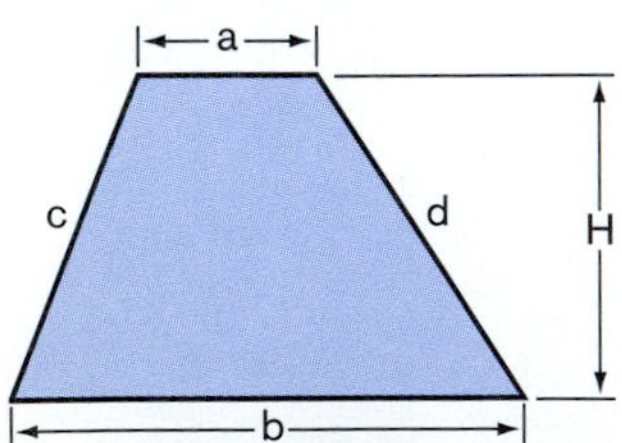

Trapezoid

Area $A = 1/2\,(a + b) \cdot H$

Perimeter $P = a + b + c + d$

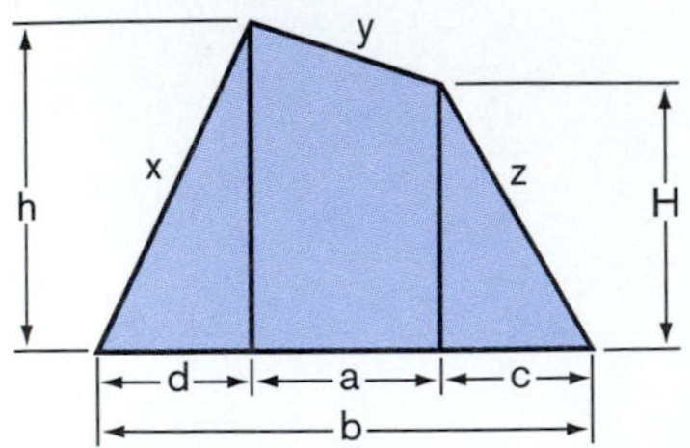

Trapezium

Area $A = \dfrac{a(H+h)+cH+dh}{2}$

Area Another method is to divide the area into two triangles, find the area of each, and add the areas together.

Perimeter $P = b + x + y + z$

Formulas for Ellipses and Parabolas

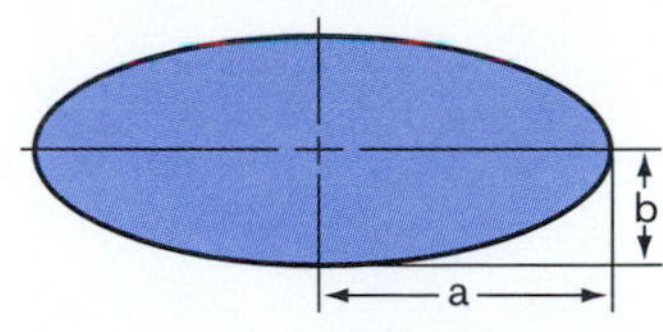

Ellipse

Area $A = \pi ab$

$A = 3.142ab$

Perimeter $P = 6.283 \bullet \frac{\sqrt{a^2 + b^2}}{2}$

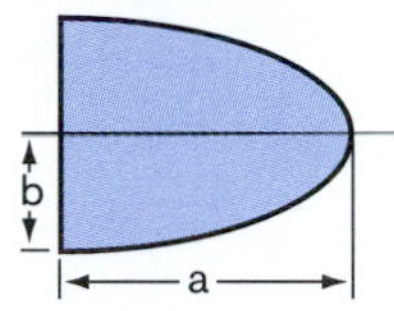

Parabola

Area $A = 2/3\ ab$

Formulas for Regular Polygons

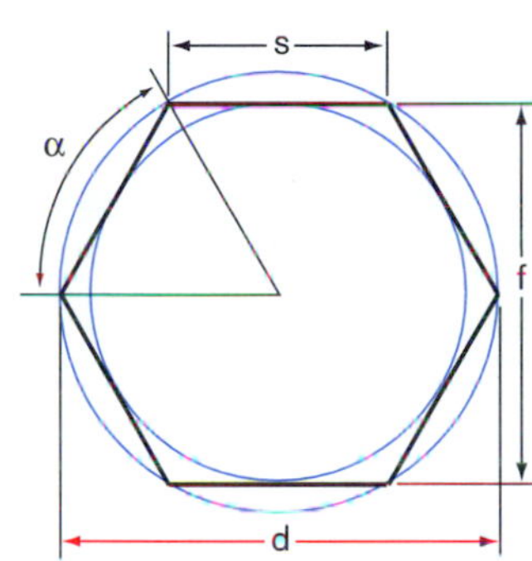

Multisided

Area $A = n \frac{s \bullet \frac{1}{2}f}{2}$

n = number of sides

Side $s = 2\sqrt{\frac{1}{2}d^2 - \frac{1}{2}f^2}$

Flats f = distance across flats; diameter of inscribed circle

Diagonal d = diameter of circumscribed circle

Perimeter P = sum of the sides

Angle $\alpha = 360/n$

Hexagon

Area	$A = 0.866f^2$
	$A = 0.650d^2$
	$A = 2.598s^2$
Side	$s = 0.577f$
	$s = 0.5d$
Flats	$f = 1.732s$
	$f = 0.866d$
Diagonal	$d = 2s$
	$d = 1.155f$
Perimeter	$P = 6s$
Angle	$\alpha = 60°$

Octagon

Area	$A = 0.828f^2$
	$A = 0.707d^2$
	$A = 4.828s^2$
Side	$s = 0.414f$
	$s = 0.383d$
Flats	$f = 2.414s$
	$f = 0.924d$
Diagonal	$d = 2.613s$
	$d = 1.083f$
Perimeter	$P = 8s$
Angle	$\alpha = 45°$

Formulas for 3-Dimensional Shapes

Cube

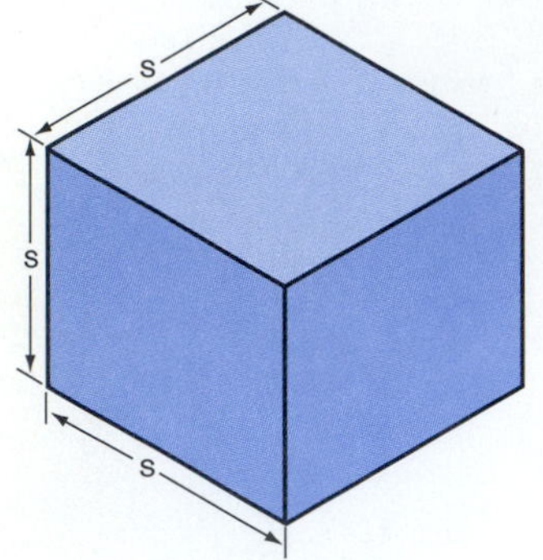

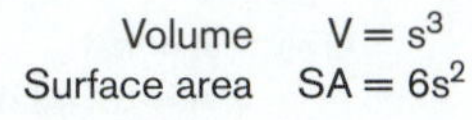

Volume $V = s^3$
Surface area $SA = 6s^2$

Side $s = \sqrt[3]{V}$

Rectangular Prism

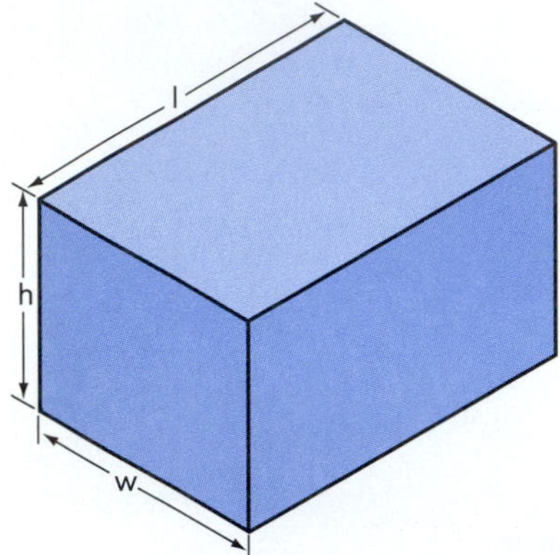

Volume $V = lwh$
Surface area $SA = 2(lw + lh + wh)$
Length $l = V/hw$
Width $w = V/lh$
Height $h = V/lw$

Cone (Right Circular)

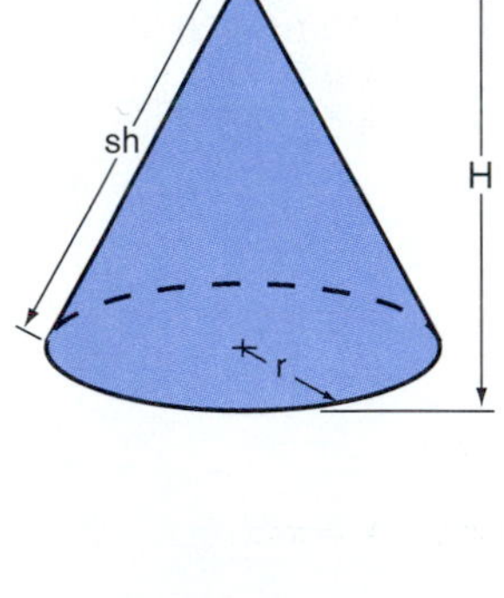

Volume $V = 1/3\ AH$*
A = area of base
$V = 1/3 \pi r^2 H$
r = radius of base

Slant height $sh = \sqrt{r^2 + H^2}$

Surface area $SA = (1/2 \text{ perimeter of base} \cdot sh) + \pi r^2$
$SA = \pi r(sh) + \pi r^2$
Lateral surface area $LSA = \pi r(sh)$

*Note: True for any cone or pyramid

Pyramid

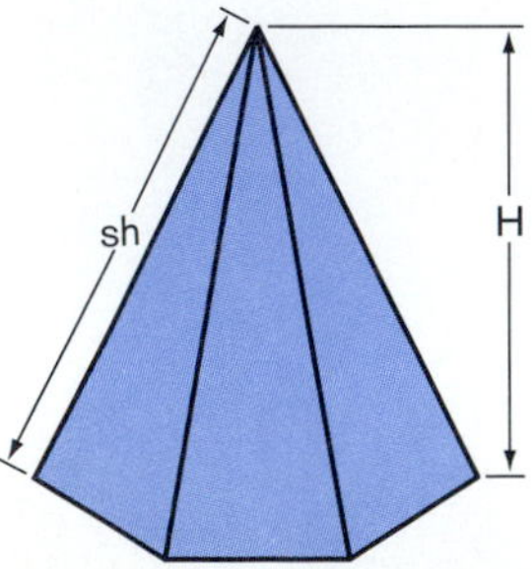

Volume $V = 1/3\ AH$
A = area of base
Surface area $SA = (1/2 \text{ perimeter of base} \cdot sh) + A$

Slant height $sh = \sqrt{r^2 + h^2}$

r = radius of circle circumscribed around base

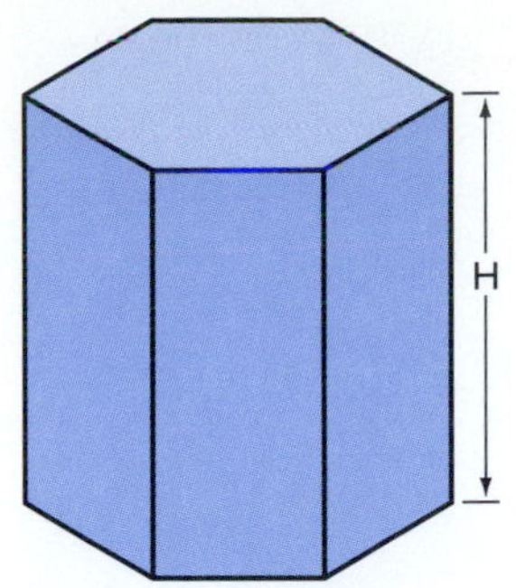

Prism

Volume $V = AH^*$

A = area of base (see Multisided Polygon)

Surface area SA = (area of each panel) + $2A$

*Note: True for any prism or cylinder with parallel bases.

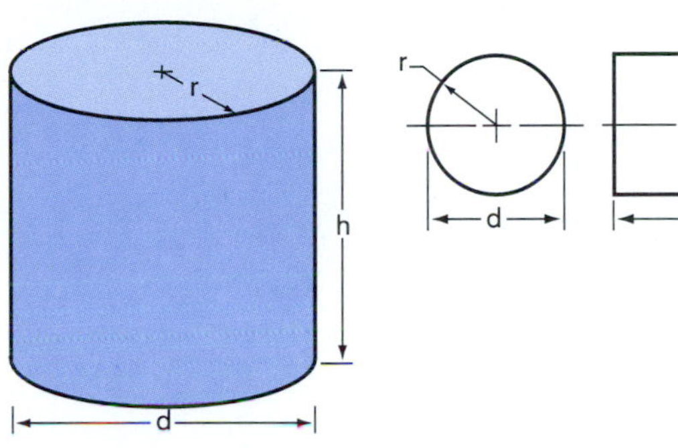

Cylinder (Right Circular)

Volume $V = Ah$

$V = \pi r^2 h$

$V = 0.7854d^2 h$

Surface area $SA = \pi dh + 2\pi r^2$

$SA = 2\pi rh + 2\pi r^2$

$SA = 6.283rh + 6.283r^2$

Lateral surface area $LSA = 2\pi rh$

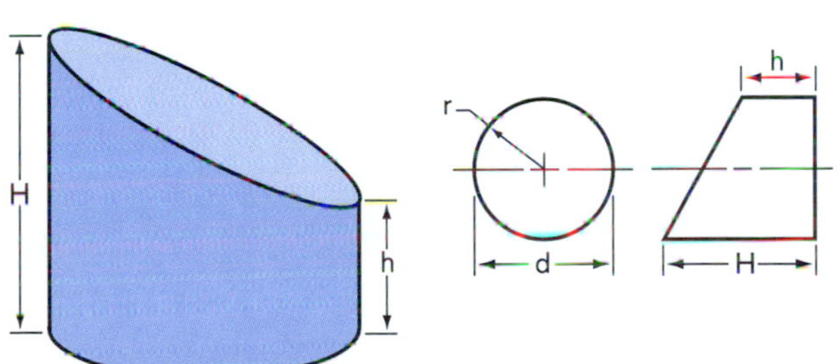

Frustrum of a Cylinder

Volume $V = \pi r^2 \dfrac{H + h}{2}$

$V = 1.5708r^2(H+h)$

$V = 0.3927d^2(H+h)$

Lateral surface area $LSA = \pi r(H+h)$

$LSA = 1.5708d(H+h)$

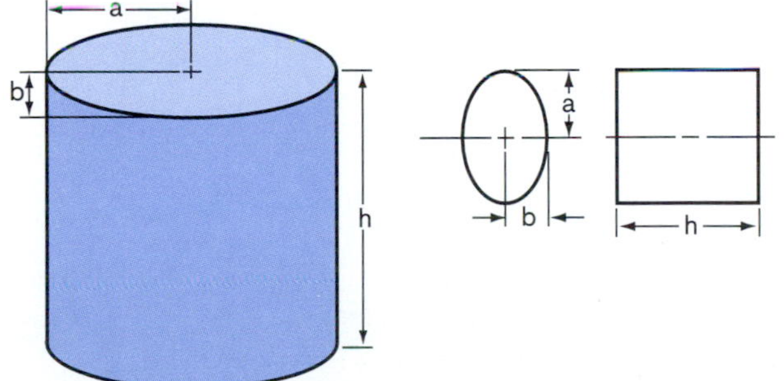

Elliptical Cylinder

Volume $V = \pi abh$

Lateral surface area $LSA = \pi h\sqrt{a^2 + b^2}$

GLOSSARY

∅ Indicates that the value is a diameter.

3D printing Lower-cost rapid prototyping process that "prints" layers of molten thermoplastic material.

absolute coordinates The exact location of a specific point in terms of X, Y, and Z from the fixed point of origin.

acme thread A thread form used to transmit power.

adjacent views Two orthographic views that share a common dimension and are located so that the dimension is aligned in both views.

aesthetic constraints Design guidelines that express what is known about taste and appeal in a certain market.

algorithm A computational procedure, as in the process used to solve a problem or render an image.

aligned Describes dimension text that runs in the same direction as the dimension line.

aligned section A section view where the cutting plane passes through angled arms, holes, or other features located around a central cylindrical shape and the section view is rotated into a single plane that shows features in the section true size.

allowance The minimum clearance or maximum interference that is desired between two members when they are at maximum material condition.

ambient The overall amount of light that exists in the environment of a rendered scene.

ambient light Light that strikes all surfaces equally, similar to room lighting.

analytical model A model that captures the behavior of the system or device in a mathematical expression or schematic drawing that can be used to predict future behavior.

angle The space between two lines (or planes) that diverge from a single point (or line).

animation A sequence of still images that give the illusion of motion when viewed in rapid succession.

antialiasing A process that reduces the jagged edges in raster images by shading adjacent pixels.

apparent intersection An intersection in a 2D view that is not a true intersection in 3D space.

approximation An inexact result that is suitable for a given purpose.

as per manufacturing process A note signifying that the typical tolerance for the specified manufacturing process is acceptable for the part.

assembly drawing A drawing that shows how all the parts in an assembly fit together.

assembly mode The operating status of a CAD package that allows you to link multiple parts together into assemblies.

associative Describing an entity that is linked to a second entity and will change when the second one does.

associative dimensions Dimensions inserted as a group of drawing objects related to geometric entities in the drawing. If the size of the geometric entity changes, the dimension values will update accordingly.

auxiliary view An orthographic view of an object using a direction of sight other than one of the six basic views (front, top, right-side, rear, bottom, left-side); used to show a surface that is not parallel to the true size of any of the principal viewing planes.

B-spline A blended piece-wise polynomial curve passing near a set of control points.

ball tag A circled number identifying each part shown in an assembly drawing. Also called a **bubble number**.

base feature The first feature created in a parametric model and the one from which other features are defined.

base point The point on a parametric sketch that remains fixed to the coordinate system when the sketch is solved.

baseline dimensioning Locating a series of features from a common base feature.

basic angle An angular dimension stated as a basic dimension.

basic dimensions The theoretically exact, untoleranced dimensions that specify the perfect location, size, shape, or angle of a feature.

basic hole system A tolerancing method that uses the lower limit or minimum size of the hole as the basic size.

basic shaft system A tolerancing method that uses the upper limit or maximum size of the shaft as the basic size.

basic size The theoretical size from which tolerance is assigned.

bend allowance An amount added to the inside dimensions of a sheet metal part to add material needed to form the bend to the flat pattern for the part.

Bezier curve A polynomial curve defined by a set of control points where the first and last control points are on the curve. This was the first CAD method that used spline approximation to create flowing curves.

bicubic surface A sculptured plane, mathematically describing a sculptured surface between 3D curves.

bidirectional associativity Describes the link between the drawing and the model in a parametric modeler. If a change is made to the model, the drawing is automatically updated to reflect the change; if a change is made to the parametric dimensions in the drawing, both the drawing and the model update automatically.

bilateral tolerance Tolerance that specifies a base value for a dimension and the allowable range of deviation from that dimension, both plus and minus.

bill of materials (BOM) A list of and information about the individual parts in an assembly.

block diagram A schematic drawing that shows the relationships in a system, using simple shapes for the entities.

BOM Bill of materials.

Boolean operation Finding the union (addition), difference (subtraction), or intersection (common area) of two or more sets, such as the data defining a CAD solid.

boss A short protrusion beyond the normal surface of a part, often used to provide a strong bearing surface.

bottom-up design A design process that starts at the part level, sizing individual components and building the design up from them.

boundary representation (BREP) An approach to creating a CAD database that stores the external boundaries of the surfaces making up the 3D model.

bounding box The smallest cube that a solid will fit inside.

brainstorming A method for generating many ideas.

BREP Boundary representation.

bump map Converts color intensity or gray scale information to height, giving the appearance that features are raised above the surface, similar to embossed letters.

bushing A hollow cylinder that is often used as a protective sleeve or guide or as a bearing.

buttress thread A thread form used to transmit power.

cabinet projection An oblique projection in which the depth of the object is shown half size.

calipers An instrument with two hinged legs used to measure internal and external diameters.

Cartesian coordinate system A rectangular coordinate system created by three mutually perpendicular coordinate axes, commonly labeled X, Y, and Z.

casting The process of pouring molten metal into a hollow mold to form a metal part.

CATIA A parametric modeling package from Dassault.

cavalier projection An oblique projection in which the depth of the object is shown full size.

cavity The side of the mold that forms the outside shape of the object.

center line A linetype used to indicate the axis of symmetry for a part or feature, the symmetrical alignment of a pattern of holes, and the path of motion for moving parts in an assembly.

centroid A point defining the geometric center of an object. The centroid and center of gravity coincide when the part is made of a uniform material in a parallel gravity field.

chained (continued) dimensioning A dimensioning method that uses the last extension line of the previous dimension as the beginning of the next.

chamfer A small angled surface formed between two surfaces.

chord length The straight line distance between the start point and the endpoint of an arc.

chuck A clamp that holds a tool or piece of material, as in a lathe or drill.

circular runout A measure of the variation in the circular shape of an object as it is rotated about a datum axis.

circular view A view of a cylinder in which it appears as a circle.

circumscribed Drawn around the outside of a base circle.

clearance fit A tolerance for mating parts in which the smallest external member is larger than the largest internal member, resulting in a space between them.

CNC Computer numerical control.

colinear Lying on the same straight line.

color depth The number of colors and bits used to define the color in an image.

composite A material made up of two or more distinct substances.

compression spring A spring designed to be compressed and returned to its original shape.

computer numerical control (CNC) Using a computer to direct machine processes.

construction lines Thin lines that serve as guides while sketching or drawing.

constructive solid geometry (CSG) An approach to creating a CAD database that stores the solids and operations used to create a 3D model.

contact inspection Evaluating the acceptability of a part by means of a measurement device such as a gage or probe that touches the part or feature.

contour Outline of an object's shape.

Coons patch A bicubic surface interpolated between four edges.

coordinate dimensioning Used for parts that have complex interior hole patterns.

coordinate values Used to identify the location of a point, using the Cartesian coordinate system: X represents the horizontal position on the X axis and Y represents the vertical position on the Y axis. In a 3D drawing, Z represents the depth position on the Z axis.

core A shape that fits in a mold to form a hole in a molded part. Also, the side of the mold that forms the interior features of the molded part.

cosmetic dimensions Nonassociative dimensions that are added to a drawing view, not derived from the model.

counterbore A cylindrical recess around a hole, usually to receive a bolt head or nut.

countersink A conical-shaped recess around a hole, often used to receive a tapered screw.

crest The top surface of a thread where the two sides join.

CSG Constructive solid geometry.

curve fitting Generating an equation that describes a set of empirical data.

cutting plane The plane that defines the cut through an object that reveals the surfaces shown in a section view.

cutting plane line A linetype used to define the location on an object where a sectional view was taken.

cylindrical coordinates The location of a point given as a radius, angle from the X axis, and distance along the Z axis.

datum plane A geometric reference point for parametric dimensons.

datum surface A theoretically exact geometric reference used to establish the tolerance zone for a feature.

decision matrix A tool for systematically ranking alternatives according to a set of criteria.

default coordinate system The coordinate system used by a CAD modeling system to define and locate geometric entities in the CAD database.

delimited Separated by a specified character, such as a space, comma, or tab.

delta angle The included angular value from the start point to the endpoint.

derived surfaces Surfaces created with mathematical methods that use the edges of the patches being combined to create a smooth joint.

descriptive geometry Mathematical techniques used to accurately describe 3D geometry in 2D.

descriptive model A model that represents a system or device in either words or pictures.

design constraint A limit on the range of options that are acceptable.

design intent The purpose or function of a feature in a part or of a part in an assembly.

design rule A relationship that must be preserved or a constraint that must be met to achieve the design intent.

detail drawing A part drawing.

development A flat pattern for a 3D shape that may be folded into the shape.

die casting A molding process that forces metal into a permanent steel die (or mold) under high pressure and temperatures.

dihedral angle The true angle between two planes.

dimension An entity used to describe the size and location of a feature on a part so that it can be manufactured.

dimension line A line drawn between extension lines with an arrowhead at each end. It indicates how the stated dimension relates to the feature on the object.

dimension value The value of the dimension being described (how long, how far across, etc.). The dimension value is placed near the midpoint of the dimension line.

dimension variables Features of dimensions that can be altered by the user in a CAD package.

documentation drawing A drawing that serves as a contract with a manufacturer, a legal record of the design, and a tool for communicating key aspects of the design's performance.

dots per inch (dpi) The number of pixels, or dots, in each printed inch of an image.

downstream applications Application software that uses CAD data so the model does not have to be created in that application.

dpi Dots per inch.

draft The taper on a molded part that makes it possible to easily remove the part from the mold.

drawing mode The operating status of a CAD package that allows you to create 2D drawings from the model geometry.

.DWG AutoCAD's native file format.

.DXF Drawing Interchange Files, popular for exporting 2D CAD geometry.

dynamic assembly An assembly model in which parts are linked to their individual part files so the software can update the assembly when individual part files are modified.

ECN Engineering change notification.

ECO Engineering change order.

EDM Electric discharge machining.

EDMS Electronic data management system.

electric discharge machining (EDM) A process that forms metal by cutting through it with electric current discharged through a tooled electrode.

electronic data management system (EDMS) A system for storage and retrieval of electronic data, such as engineering CAD files.

endmill A rotating tool that removes metal or other material with cutting surfaces locating at the end of the tool.

engineering change notification (ECN) A written description of a change to an approved documentation drawing or model used to notify manufacturing and document the change.

engineering change order (ECO) A written description of a change to an approved documentation drawing or model used to notify manufacturing and document the change.

English units U.S. customary foot/inch units.

equation solver Software used to solve simultaneous equations and work with symbolic mathematical expressions.

ergonomics The study of human interaction with a product or system.

exploded Describes an assembly in which parts are moved out of position along an axis so that each individual part is visible.

export To save data from one application in a file format that can be read by other software applications.

extension line A line that relates a dimension to the feature or entity it refers to.

extension spring A spring designed to be extended and returned to its original shape.

external reference A link from a part in an assembly model or drawing to the part's individual file (called an XREF in AutoCAD).

extrusion A manufacturing process that forces material through a shaped opening. Also a modeling process that creates a 3D shape by defining a closed 2D shape and a length.

extrusion blow molding Similar to injection blow molding except that the parison is extruded through a die. When the mold closes, the parison is simultaneously cut and sealed, then expanded into the shape of the mold by compressed air.

face A planar surface on an object bounded by edges where it intersects other surfaces.

facets Planar shapes used to represent smoothly contoured surfaces.

factor of safety The ratio of the maximum value to the expected value for an operating characteristic, such as maximum safe speed/expected operating speed, that is used as a design guideline.

falloff The angle of the cone of light from a spotlight.

FDM Fused deposition modeling.

FEA Finite element analysis.

feature Any definable aspect of an object—a hole, a chamfer, etc. Also the basic unit of a parametric solid model.

feature control frame A boxed format for geometric dimensioning and tolerancing symbols and values.

feature-based modeling Parametric modeling.

file extension The part of a file name that is composed of a period followed by one to three characters, and that helps identify the file.

fillet A rounded interior blend between two surfaces. Some uses are to strengthen joining surfaces or to allow a part to be removed from a mold.

finish marks Symbols and values that indicate which surfaces are to be machined and the quality of the desired finish.

finished Machined to a desired degree of smoothness.

finite element analysis (FEA) An analysis method that breaks up a complex shape into discrete smaller parts (finite elements) for which the properties such as stress, strain, temperature distribution, fluid flow, and electric and magnetic fields can be solved.

first-angle projection The arrangement of orthographic views used in Europe, Asia, and several countries other than the United States and Canada.

fit The definition of how tightly or loosely mating parts must fit together when assembled.

fixture A device that holds a part or work piece in position during manufacture or assembly. Fixtures do not move or guide the cutting tool.

flange A flattened collar or rim around a cylindrical part to allow for attachment.

focus A point used to define certain geometric entities. Also one of two points used to define an ellipse. Plural is **foci**.

force fit A tolerance for mating parts in which the internal member is always larger than the external member.

foreshortened Shown smaller than true size.

forging The process of shaping metal by hammering it or pressing it between dies.

form tolerances Tolerances that relate to the shape of a feature.

fractional units Units that express lengths less than 1 as fractions, for example 15 1/4.

frame A single rendered view in an animation sequence.

free length The length of a spring when it is not compressed or extended.

freebody diagram Schematic drawing that uses vectors, masses, and simplified objects to depict the forces in a mechanical system or device.

full section A section view that shows the part cut entirely through, typically along the centerplane.

fundamental deviation The deviation closest to the basic size (the minimum allowance).

fused deposition modeling (FDM) A rapid prototyping system that deposits molten plastic in layers corresponding to cross sections of the part.

G-codes The portions of CNC control code that specify the specific motion of the machine. For example, G00 X1.0000 Y2.0000 specifies a rapid linear motion (G00) to X coordinate 1.0000 and Y coordinate 2.0000.

GD&T Geometric dimensioning and tolerancing.

geometric dimensioning and tolerancing (GD&T) Specifying how a part can vary from the perfect geometry implied on the drawing by defining either the diameter or the width of a tolerance zone.

geometric parameters Constraints that define the geometric properties of a feature, such as tangency, verticality, and so on.

global parameters Parameters that are common to more than one part in an assembly.

grid dimensioning Used to specify points along an irregular curve.

half section A view that shows exterior and interior detail in the same view, as if a quarter of the object were cut away; one half of the resulting view is a section view and the other half shows the exterior of the object.

hard copy Printed or plotted material (as opposed to electronic files).

hatch A series of parallel thin lines drawn on the diagonal to indicate a surface created by a cut.

hidden line A linetype that represents an edge that is not directly visible because it is behind or beneath another surface.

history-based modeling programs CAD programs that store CSG data.

HTML HyperText Markup Language; a text-only formatting language readable by Web browsers that is used to control the display of electronic information transmitted over the Internet.

human factors analysis The study of how humans will interact with a design.

IGES Initial graphics exchange specification.

implementation The transition from sketches, drawings, or models of the design to specifications for making the product or system.

import To open and use a file created by a different application than the one being used.

inclined Perpendicular to one of the orthographic viewing planes, but not true size in any view.

Initial Graphics Exchange Specification (IGES) A graphics format capable of exporting 3D wireframe, surface, or solid information; commonly used to export the model for computer-aided manufacturing.

injection molding A molding process in which molten plastic is injected into a two-sided mold. The mold is opened to release the part.

inscribed Drawn inside a base circle.

inspection The act of examining a part against its specifications.

iterative Repetitive.

interface The portion of a system or device used to interact with or control it. A keypad, computer screen, and control panel are all interfaces.

interference The amount of overlap one part has with another when assembled.

interference fit A tolerance for mating parts in which the internal member is always larger than the external member.

interpolate To determine a value or location between two known values or locations.

intersection The point at which two lines or surfaces meet, or the area shared by overlapping solids.

intranet A private network that links servers in different locations, much as the Internet does.

inverse kinematics A means of constraining motion in an animation to match the way that the finished product would be able to move.

investment casting A molding process that uses a wax pattern of the part to form the mold.

irregular surfaces Surfaces that cannot be defined by simple geometric entities such as lines and arcs.

ISO International Organization for Standardization. Its headquarters are in Geneva, Switzerland.

ISO 9000 A generic management system standard designed to help companies that manufacture their own products to ensure a consistent level of quality in design, production, installation, and service.

isometric sketch A parallel projection in which the object is rotated 45° and tipped away from the viewing place.

jig Tooling used to guide a cutting tool.

key A transformation applied to an object or set of objects in an animation frame. Also a rectangular or semicircular shape used to prevent parts, such as gears or wheels, from turning on a shaft.

keyseat A slot in a shaft to receive a key. (Keyseats are in the shaft; **keyways** are in the hub of the part on the shaft.)

keyway A slot in a hub or material around a shaft that receives a key.

kinematics The study of motion.

knuckle thread Thread rolled from sheet metal or cast. Used on bottle tops and lightbulbs.

knurling A pattern of depressions formed on a surface to provide for better gripping.

lathe A machine tool that removes material by turning the work piece as a cutting tool is forced against it.

lay The direction or arrangement of the primary surface pattern.

layer A method of separating drawing objects so they can be viewed individually or stacked like transparent acetates, allowing all layers to show. Used to set color and linetype properties for groups of objects.

lead The distance thread advances when turned one complete turn. (Single thread advances one pitch distance in one complete turn.)

leader A line from a note or radial dimension that ends in an arrowhead pointing at the edge of the feature (or a dot if the feature is inside the object's outline).

least material condition (LMC) The tolerance limit where the part is made of the least amount of material. For example, the maximum size for a hole and the minimum size for a shaft.

limit tolerance A tolerance that states the maximum and minimum allowable dimension, not the basic dimension value. Limit tolerance specifies the upper and lower allowable value for the dimension when measured on the actual part.

limiting element The outer edge of a curved surface.

line fit A tolerance that ensures either a clearance fit or surface contact between the parts, but no interference.

LMC Least material condition.

local coordinate system A coordinate system defined by the user to assist in creating CAD geometry.

lug A flat or rounded tab protruding from a surface, usually to provide a method for attachment.

M-codes The portions of CNC control code that specify instructions other than the motion of the tool, such as tool changes.

machining The use of machines to process metals (and sometimes plastics) to form a feature by removing material.

machinist scale A steel ruler with its smallest divisions 0.01 inch apart.

major axis The long axis of symmetry across an ellipse.

major diameter The outside diameter of a spring's coil. Also the largest diameter of a threaded hole or shaft.

manufacturing rule A design constraint that expresses the limits of the manufacturing process to be used.

mass properties Data concerning the real-world object being modeled, such as its mass, volume, and moments of inertia.

mating parts Parts that fit together in an assembly.

maximum material condition (MMC) The tolerance limit where the most material remains on the part. For example, the minimum size for a hole and the maximum size for a shaft.

mesh A set of smaller elements used to cover or model a 2D or 3D shape.

metric thread The standard for international fasteners. Similar to unified thread, but with a shallower thread depth.

minor axis The short axis of symmetry across an ellipse.

minor diameter The smallest diameter of a threaded hole or shaft. Also the inside diameter of a spring's coil.

mirror line A line that defines the angle and distance at which a reversed image of a selected object will be created.

mirrored Created by reversing a copy of another object across a mirror line or plane.

MMC Maximum material condition.

model A representation of a system, device, or theory that allows you to predict its behavior.

modeling kernel Engine used by the CAD software to perform basic functions such as Boolean operations, blending, mass properties, exporting of geometry, and chamfering and filleting.

moment of inertia The measure of the resistance that an object has to changing its steady motion about an axis.

morphing The process of transforming an object from one shape to another.

mylar A material more durable than paper used for documentation drawings.

neck A small groove cut around the diameter of a cylinder, often where it changes diameter.

nominal size The designation used for purposes of general identification. Often expressed in common fractions, such as 1/2 inch.

non-contact inspection Evaluating the acceptability using optical devices, ultrasound, or other measurement devices that do not touch the part or feature.

normal Parallel to one of the three principal orthographic viewing planes where it shows true size.

NURBS curve A nonuniform rational B-spline curve. Also a B-spline defined by weighted control points or a cubic spline curve.

oblique Not parallel or perpendicular to any of the standard views.

oblique sketch An oblique projection that shows the front surface true size but varies in the foreshortening of the depth of the object.

offset The distance from an existing object at which a new object will be created.

offset section A section view that uses a cutting plane line that is bent at 90 degree angles so that it passes through features that do not all lie in the same plane on the object.

omnidirectional light Light that shines in all directions.

optimization The process of finding the most effective or favorable value or condition.

orthographic projection A method for transferring 3D shapes to 2D paper that shows surfaces and edges parallel to the viewing plane true size.

overdimensioning Showing the same dimension more than one way in the drawing, or showing it twice (perhaps in different views).

pan A view transformation that shifts the viewing area to show a different portion of the model or drawing.

parameter A named quantity whose value can change depending on the circumstances of its application.

parametric assembly A dynamic assembly model that also allows you to use global parameters and define relationships between parts.

parametric modeling A modeling method that uses parameters to define the size and geometry of features and to create relationships between features. Changing a parameter value updates all related features of the model at once.

parent-child relationship The way in which one feature is related to another feature through the geometric constraints and parametric dimensions applied to the sketch.

part drawings Individual drawings prepared for each part that will be manufactured or modified for the project.

part mode The operating status of a CAD package that allows the creation of individual parts made of a single material.

parting line A line on a molded part where the two mold halves come together.

patch A section of a surface model.

pattern A model used to form a mold, or the flat shape to be cut to form a sheet metal part.

permission A user's level of access to an electronic file on a networked system.

perspective sketch A sketch in which portions of the object that are further from the viewer appear smaller and lines recede into the distance.

phantom lines A linetype used to show existing equipment or to indicate portions of a long shape or structure that are not shown.

pictorial sketch A sketch that shows an object's height, width, and depth in a single view.

pitch The distance from a point on a thread to the corresponding point on the next thread or the distance between the coils of a spring.

pixel The smallest unit of display on a computer screen. A dot in a raster image.

placed dimensions Nonparametric dimensions added to a drawing. Placed dimensions may be associative, but are not bidirectionally associative.

plan view The top view or view looking straight down the Z axis toward the X-Y plane.

plus/minus tolerance A tolerance specified by a range that is added or subtracted from the dimension value.

points A printing measurement used to size type.

polar coordinates The location of a point as given by an angle and a distance.

positional tolerance A tolerance that controls the location of a feature.

presentation software A computer application that organizes and formats information to be displayed like slides or acetate transparencies in a presentation.

primary auxiliary view An auxiliary view projected directly from one of the standard orthographic views, such as the top, front, or side view.

primitive A simple 3D shape, such as a box, cone, sphere, or torus, that can be combined to make more complex shapes.

product data management system A system that organizes and stores all data related to managing a product, including engineering data.

profile A 2D shape that can be used to generate a 3D solid.

project To transfer information from one view of an object to another by aligning the views and using projection lines.

projection line A horizontal or vertical line that can be used to locate entities in an adjacent view. Also the set of all points that share a given coordinate value in both views.

prototype A full-size model used to validate a nearly final design for production. Prototypes may be physical models or solid models.

prototype drawing A drawing saved with certain settings that can be used repeatedly as the basis for starting new drawings.

rapid prototyping (RP) A process that forms a physical model from CAD data using nontraditional machinery and materials.

rapid tooling The use of a rapid prototyping-like process to create the tool, usually a mold, used to manufacture a part.

raster Describes programs or graphics files that store information about the discrete pixels, or dots, that make up a graphical image.

ray tracing A rendering algorithm that generates the appearance of the object by tracing the path of the light to its source.

reference dimensions Dimensions added to a drawing but not used to create the parametric model. They may be associative, but changes to them will not change the model.

reference surface A surface shown on edge in two orthographic views that can be used to locate points from one view to the other.

reflection The degree to which a surface bounces back light.

refraction The degree to which an object changes the angle of light passing through it.

related views Views that are adjacent to the same view.

relative coordinates The location of a point as given by the distance from the last point specified.

render To calculate a single, shaded view from a 3D model to which colors, materials, and lighting have been added.

rendering algorithm The computational method used to produce a rendered view from the information stored in the solid model.

resin A translucent, viscous, polymer substance, natural or synthetic, used to form plastics.

resolution The number of pixels used to define an image.

revision block A brief listing of revisions made to a drawing since it was initially released to manufacture.

revolution Creating a 3D solid or surface by revolving a 2D shape about an axis.

root The bottom surface of a thread where the two sides join.

roughness A measure of the irregularities in a surface's finish.

round A rounded exterior blend between two surfaces.

RP Rapid prototyping.

sans serif Describing fonts, such as Helvetica, that lack serifs.

scale The ratio of the full size of the object to its size in an enlarged or reduced view.

schematic Using symbols to represent components, constructs, or relationships in a drawing.

secondary auxiliary view An auxiliary view projected from a primary auxiliary view.

section views A special kind of orthographic view used to show an object's interior detail.

seed parts Files that contain elements you want every part model to contain.

selective laser sintering (SLS) A rapid prototyping system that uses a laser to fuse powdered metals together.

serif The short stroke that finishes the ends of letter shapes.

shading Creating a view of the model that hides hidden lines and applies flat colors to visible surfaces.

sheet size The size of the paper used to print or plot a drawing.

SI System International.

simulation A model of the function of a mechanism or mechanical system that can be used to predict its behavior.

size parameters Dimension values used to define a feature that are stored with the parametric model.

SLA Stereolithography apparatus.

SLS Selective laser sintering.

soft copy An electronic copy of a document.

solid modeling A type of 3D modeling that represents the volume of an object, not just its lines and surfaces. This allows for analysis of the object's mass properties.

specific gravity A value used to relate the density of different materials to the density of water.

specification A measurable statement of the objectives the design should achieve.

spherical coordinates The location of a point given as a radius, the angle from the X axis, and the angle from the X-Y plane.

spline A complex or irregular curve.

spotface A shallow recess like a counterbore, used to provide a good bearing surface for a fastener.

spotlight A light source used to highlight certain key areas and cast light only toward the selected target.

sprue The opening in a mold through which the material enters.

square thread A thread form used to transmit power.

stamping A process whereby sheet metal is pressed between dies to form simple parts.

standard parts Parts to be purchased, not manufactured with the rest of the assembly.

standard worm thread A thread form used to transmit power.

static assembly An assembly that does not update when individual part files change.

static part A nonparametric part that cannot be automatically updated.

STEP Standard for the Exchange of Product Model Data, a relatively new 3D format from ISO for transferring CAD data.

stereolithography apparatus (SLA) A rapid prototyping process that uses laser-hardened resins to form the model.

stippling Using dots to add shading to a sketch.

STL Stereo Lithography file format used to export 3D surface geometry to rapid prototyping systems. STL translates the surface of the object into triangular facets.

stock parts Parts manufactured in quantity and kept in inventory for use in several different products.

storyboard Pictures that depict the sequence and composition of key events in the animation.

subassembly A group of parts that fit together to create one functional unit.

surface modeling A type of 3D modeling that defines only surfaces and edges. It resembles an empty shell in the shape of the object.

System International (SI) The metric system of units and measures.

system variables Settings that control the default operation of a CAD package.

tap To cut internal threads.

template A drawing with certain settings that can be used repeatedly as the basis for starting new drawings. Also a prototype drawing.

tessellation lines Lines used to indicate surfaces in a wireframe view.

thermoplastics Plastics that melt when heated and set when cooled. In processing thermoplastics, the mold is cooled to set the part.

thermoset plastics Plastics that change from solid to liquid and back to solid when heated to become a new cross-linked plastic.

third-angle projection The arrangement of orthographic views used in the United States and Canada.

thread angle The angle between the sides of the thread.

thread axis The center line of the screw where it appears rectangular.

thread depth The distance from the crest to the root of the thread measured perpendicular to the thread axis.

thread form The shape of the thread.

TIN Triangulated irregular networks.

tolerance The total amount that a measurement on an acceptable part may vary from the specified dimension. Also the difference between the permitted minimum and maximum sizes of a part.

tooling Any of the devices that work along with machinery to produce a part, such as jigs, fixtures, molds.

top-down design Design that starts with the function of the entire system, then subsystems, and finally, defining each part that must be manufactured and assembled to create the design.

total runout A measure of the cumulative variation measured over the entire surface as it is rotated about a datum axis.

transformations Operations on the solid model that move it, scale it, or rotate it on the coordinate system.

transition fit A tolerance that indicates that either a clearance or interference fit is acceptable.

translation Moving the model to a new location on the coordinate system.

triangulated irregular networks (TIN) A system of triangular shapes used to define a surface.

tweaking Editing a surface model by moving individual control points.

unidirectional Using a single orientation for all text on a drawing. Letters are arranged horizontally from left to right to be read from the bottom of the sheet.

unified thread A standard adopted during World War II to make it easier to find replacement parts from allied countries.

unilateral tolerance A tolerance where the upper or lower deviation is 0.

update To regenerate the model or drawing using any new dimension values or changed parameters.

URL Universal Resource Locator; the address format used to locate information on the World Wide Web. Every document on the Web has a unique URL.

user coordinate systems A coordinate system defined by the user to assist in creating CAD geometry.

vacuum molding A molding process that draws heated plastic sheets down into a shaped cavity.

vanishing point The location where the projectors converge in a perspective drawing.

vector A directional line. Also, a way of storing a graphic image as a set of mathematical formulas.

vector data Data that defines geometric entities by means of points and vectors (that define direction and length). CAD software draws a particular entity on the screen based on this vector information.

vertex The location where three or more plane surfaces intersect to form a point.

virtual prototype A 3D solid model used in lieu of a physical prototype to test and validate a design before manufacture.

VRML Virtual Reality Modeling Language, a text-based language used to create a 3D model that the viewer can move around in.

wireframe modeling A modeling method that represents the edges and contours of an object using lines, circles, and arcs oriented in 3D space.

working drawings Drawings that convey all of the information needed to manufacture and assemble a design.

WORM Write once read many.

write once read many (WORM) A storage media that cannot be overwritten.

zoom To change the magnification of the view of the model on screen.

INDEX

A

Abbreviations, 614–617
Absolute coordinates, 169
ACIS kernel, 235
Acme threads, 305
ActiveProject system, 363
Adams, Scott, 414, 417
ADAMS software, 400
Adjacent views, 125–126
Adobe Illustrator, 572
Aesthetic constraints, 41–42
Affymax Research Institute, 34
AISC Structural Handbook, 267
Albini, Marty, 192, 195
Aldus Persuasion, 569
Algorithms, rendering, 582–585
Alias Studio, 223
Aligned sections, 495
Allowance, 311, 520–521
Alpha channel, 572
Aluminum Association, 321
Ambient light, 577
American Congress on Surveying and Mapping (ACSM), 471
American Institute of Chemical Engineers (AICE), 46
American Institute of Steel Construction, 471
American National Standards Institute (ANSI)
  cast iron flanged fittings, 668
  cast iron screwed fittings, 667
  codes and standards, 608–609
  description of, 64
  dimensioning and tolerancing practices and, 500
  electronics symbols, 67, 68, 676
  heating, ventilation, and ductwork symbols, 673–675
  piping symbols, 472, 672
  standard sheet sizes, 483
  transition locational fits, 626
American Records Management Association (ARMA), 340
American Society for Testing and Materials (ASTM), 46, 471
American Society of Civil Engineers (ASCE), 46
American Society of Heating, Refrigeration, and Air Conditioning Engineers (ASHRAE), 46
American Society of Mechanical Engineers (ASME)
  ASME Boiler and Pressure Vessel Code, an American National Standard, 42
  as information source, 46
  standards published by, 470
American Standard
  clearance locational fits, 624–625
  force and shrink fits, 628
  interference locational fits, 627
  running and sliding fits, 622–623
  taper pipe threads, 646
American Standard Code for Information Interchange (ASCII), 384
Analytical models, 206–208
Angles
  dihedral, 148
  included, 174
  true size of, 88
Angularity tolerance, 538
Animation
  function of, 585–587
  importing geometry and, 588–589
  inverse kinematics and, 590, 591
  keyframe, 589
  at Mountain Cycle, 596–602
  planning for, 587
  rendering, 590
  software for, 242, 585–591
  storyboarding and, 587–588
  transformations and, 589–590
  VRML and, 592–595
ANSI/ASME Y14.5M–1994, 524, 525
ANSYS, 395
Antialiasing, 582
Apparent intersection, 179
Approved by block, 341
Arcs
  description of, 174
  dimensions for, 509
  formulas for, 175
Area charts, 568–569
ASCII files, 384
ASME Boiler and Pressure Vessel Code, an American National Standard (American Society of Mechanical Engineers), 42
Assembly, exploded, 473–475
Assembly drawings
  description of, 102–103, 472
  elements of, 473, 476
  isometric views of, 473, 475
  orthographic views of, 473, 474
  section views of, 472–473

Assembly models
  building database for, 303–309
  capabilities of, 272–273
  combining parts in, 290–291
  constraints in, 296, 297
  design for manufacturing and, 317–327
  dynamic, 293–295
  external references and, 291
  global parameters and, 301
  intelligent, 296–302
  interference checking in, 309–317
  layout drawings and, 298–299
  managing files and, 295
  parametric, 292–293, 302
  seed parts and, 302
  skeleton models and, 299–300
Associative dimensions, 273, 507
AutoCAD
  AutoFEA, 395
  AutoSurf, 225
  external references in, 291
  file format and, 384
  Internet file operations and, 365
  List command, 222
  object snap feature and, 170–171
  relative coordinate entry in, 171–172
  Release 12, 381
  Release 14, 381
  rendering capabilities, 575–576, 578
  viewing .dwg and .dxf files created by, 364
Autodesk
  assembly capabilities of, 294
  AutoSM, 325–327
  base points and, 259
  constraints and, 257
  Drawing Exchange Format, 387
  operators in, 267
  PartSpec, 303
  3D Studio Max, 562, 588
Automatic tool changes, 437
Auxiliary views
  description of, 148–149
  primary and secondary, 490
  from 3D models, 490–492
.AVI files, 590

B

Backlighting, 578
Ball Aerospace & Technologies Corporation, 414
Ball tags, 476
Bar charts, 566–568
Barrett Hand, 275–278
Barrett Technology, 275
Base feature, 260–261
Base point, 259–260
Bashir Research, 595
Basic hole system, 313, 520
Basic shaft system, 313, 520, 521
Basic size, 312–313, 523
BBDO Detroit, 355
Bend allowance, 450–452
Bezier, Paul, 177
Bezier curves, 177
Bidirectional associativity
  description of, 274
  in parametric modelers, 292, 293, 508
Bill of materials (BOM), 476
Blind spots, 62
Block diagrams, 64, 66–67
.BMP, 574, 580
Boeing
  concurrent engineering process and, 27
  cost of design changes at, 26
  interference checking and, 310
  knowledge-based engineering system of, 404–405
  777, 4–5, 9, 31, 338, 402, 404
  teamwork at, 32
  visualization tools at, 30, 31
Bolt heads, 71–72
Bolts
  orthographic views of, 145
  schematic drawing of, 71
  square, 647
Bonus tolerance, 539
Boole, Charles, 181
Boolean operations
  BREP method and, 232–233
  CSG method and, 232
  making complex shapes with, 181–182
  surface modeling and, 228
Boring, 426
Boss, 271
Bottom-up design, 298
Boundary representation (BREP)
  description of, 223–224
  solid models and, 231–233
Brackets, 85
Brainstorming, 48–50
Brazing, 427
Broaching, 427
Broken out sections, 495–497
Brown, Albert, 2, 38, 61, 82, 104–107, 114, 164, 202, 246, 288, 400
B-splines, 177–178
Bubble numbers, 476
Built-in features, 270
Bushing, 271
Butress threads, 305

C

Cabinet projection, 94
CAD
  arcs and, 174
  benefits of using, 204–205
  Cartesian coordinate system and2D, 168
  circles and, 173
  communicating to machine code, 439–441
  coordinates for 3D, 167–171
  description of, 27–28
  dimensioning and, 502–505
  drawing geometry and, 166–167
  ellipses and, 174–175
  extruding forms and, 184
  geometric relationships and, 178–180
  irregular surfaces and, 186
  isometric views and, 473
  left-hand rule for, 168
  parts list and, 476
  planes and, 173
  points and, 171–172
  presentation graphics and, 562, 572–574 (*See also* Presentations)
  recognizing symmetry and, 182–184
  revolving forms and, 185
  right-hand rule for, 167–168
  secondary auxiliary views using, 492
  section views and, 473
  solid primitives and, 180–182
  specifying location and, 169–171
  spline curves and, 176–178
  transformations and, 187–192
  use of, 117
  user coordinate systems and, 186–187
CAD data. *See also* Data management
  backing up, 350
  concurrent engineering and, 27–28, 30, 32
  as design documentation, 469
  drawing standards and, 347–348
  example of managing, 366–370
  file formats and, 384
  formats for export and, 383, 385–387, 572
  permission and ownership and, 348–349
  product data management and, 354–362
  scale and, 484
  storage media for, 350–351
  storage of, 343–347
  translating, 388
  vector and raster data and, 388
  World Wide Web and, 364–365
Cartesian coordinate system
  isometric axes and, 88
  orthographic views and, 120–121
  reference surfaces and, 126–127
  3D CAD software and, 167–171, 187
Casting
  centrifugal, 446
  description of, 321, 427
  die, 447
  investment, 446, 447
  permanent, 446
  sand, 445–446
CATIA, 235
Cavalier projection, 94
CD-ROM
  drawings in digital format via, 484
  storage on, 350–351
Centerlines, 144–145
Center of gravity, 379
Centimeter-gram-second (cgs) system, 381
Centrifugal casting, 446
Centroid, 379
Chadwick, Paul, 16, 19
Challenger space shuttle, 41
Chamfers, 271, 512
Charts, 566–569
Checked by block, 341
Chrysler, 355
Chunk, 434
Cimatron, 439
CIMlogic, 304
Circles
  in CAD systems, 173, 174
  dimensions for, 509
  formulas for, 175
Circularity tolerance, 528
Circular runout, 534
Civil engineering design, 471
C&K Systems, 458, 460–462
Clay models, 142
Clearance fit, 311, 518, 519
Client IDs, 357–358
Clinch nuts, 448
Coating, 427
Coaxial cable clamp, 38, 114
Color depth, 573, 580–581
Communication
  function of spoken and written, 32
  technological advances in, 44
  using graphics for, 7–8, 55
  of visual ideas, 55–56
Competition reviews, 45, 48
Compression molding, 457
Compression springs, 307
Computer-aided design tools, 4–5, 338
Computer-aided manufacturing (CAM)
  communicating from CAD to machine code and, 439–442
  G-codes and M-codes and, 443
  NC codes and, 442
  NCI codes and, 442
  overview of, 436–439

Computer-integrated manufacturing (CIM), 25
Computer numerical control (CNC)
  CAD files and, 438
  description of, 437
  machine capabilities and, 443
  machine instructions and, 442
Computer projection system, 564, 565
Computer software. *See also* Parametric modeling software; *specific software*
  animation, 242, 585, 588–591
  to develop flat patterns, 449, 450
  engineering uses for, 4–5
  equation-solving, 390, 392
  ergonomic analysis, 316–317
  integrated modeling and design, 403–405
  machining, 439
  modeling features in various, 240
  presentation, 562, 569–572
  rendering, 575–585
  schematic drawing, 64
  simulation, 399–402
  solid modeling, 242, 243, 383
Computer software engineers, 29
Concentricity tolerance, 533–534
Concurrent engineering
  CAD database and, 27–28, 30, 32
  computer-integrated manufacturing and, 25
  concurrent design process and, 25–26
  design teams and, 28–30
  function of, 23–24, 424–425
  model of, 352
  teamwork and, 30–32
  3D design database in, 351–354
Cones, 62, 132, 133
Configuration design, 22–23
Constant chord fillets, 270
Construction lines, 63–64
Consumer surveys, 45, 47, 48
Contact inspection, 315
Contour sketches, 58, 63–64
Coon's patch, 227
Coordinate dimensioning, 510, 511
Coordinate measuring machines (CMMs), 315–316
Coordinates
  absolute, 169
  cylindrical, 170
  polar, 170
  relative, 169
  spherical, 170
Coordinate systems. *See also* Cartesian coordinate system
  user, 186–187
Cope, 445
Copy and Paste function, 574–575
Corel Draw, 572
Cosmetic dimensions, 274
COSMOS/M, 395–398
Cotter pins: American National Standard, 660
Counterbores, 270, 271
CSG (constructive solid geometry)
  description of, 231–232
  parametric modeling and, 251, 252
Cubes, 87
Cubic splines, 176
Curve-fitting programs, 393
Curves
  projecting, 137
  spline, 176–178
Cutting plane, 146, 147
CyberEye CE-200W, 217
Cylinders
  description of, 132
  mass properties of right, 378
  sketches of, 133
Cylindrical coordinates, 170
Cylindricity tolerance, 528

D

Data management. *See also* CAD data
  design database and, 351–354
  drawing control system and, 366–370
  overview of, 338
  product, 354–362
  requirements for, 339–344
  storage guidelines and, 343–351
  World Wide Web and, 363–365
Datum, 263–264
Datum features, 531–532
Datum identifiers, 532
Datum planes
  description of, 264
  primary and secondary, 314, 315
Datum surfaces, 264–265, 531–532
Decimal points, 503–504
Decision matrix, 50–52
Deferred judgment, 48
de Mestral, Georg, 44
Derived surfaces, 227
Descartes, Rene, 167
Descriptive geometry, 213
Descriptive models, 216
Design constraints
  aesthetic, 41–42
  description of, 6
  economic, 42
  functional, 41
  legal, 42–43
  time-related, 43
Designer's notebooks
  examples of, 2, 38, 82, 114, 164, 202, 246, 288, 336, 374, 422, 466, 560
  guidelines for, 34

Design for manufacturability, 318
Design intent, 254, 290
DesignLink software, 404
Design process
  building visual ability and, 57–63
  construction lines and sketching aids and, 63–64
  description of, 5
  ideation and, 44–56, 74–77
  model of, 9–11
  problem identification in, 40–44
  role of graphics in, 5–9 (*See also* Graphics)
  schematic drawings and, 64–73
  sketching and visualization in, 103, 108 (*See also* Sketches)
  stages in, 6
  tasks in, 21–22
Design proposals, 52–53
Design reviews
  function of, 53–54
  meetings for, 51–52
Designs
  bottom-up, 298
  middle-out, 298
  reading existing, 11–12
  selection of, 14–15, 18–19
  top-down, 298
DesignSuite, 304
Detail drawings. *See* Part drawings
Detailed representations, 70
Development plans, 53
Deviations, 523
Dial calipers, 435
Diameter, 173, 539
Die casting, 447
Difference, as Boolean operation, 181, 182
Digital calipers, 435
Dihedral angle, 148
Dimension lines, 502
Dimensions, basic, 534–535
Dimensions/dimensioning. *See also* Geometric dimensioning and tolerancing (GD&T)
  for arcs and circles, 509
  associative, 273
  baseline, 517–518
  choice of best, 505–509
  of common features, 511–514
  coordinate, 510, 511
  cosmetic, 274
  function of, 500
  GD&T and, 313
  grid, 510, 511
  in multiview sketches, 138–139
  nonparametric, 274
  parametric, 265–268
  placement of, 504–505
  reference, 274
  standard appearances and, 501–504
  units for, 501
Dimension values
  decimal points in, 503–504
  description of, 502–503
  inch, 503
  metric, 503
  orientation of, 504
Directory structure, 344–346
DIRRECT database (Boeing), 338
Documentation. *See also* Data management
  case study of, 20, 21
  function of, 468
  technical writing for, 29
Documentation drawings. *See also* Drawings; Sketches
  CAD database and, 469
  description of, 8–9, 468–469
  dimensioning and, 500–514
  examples of, 21, 430
  function of, 338, 468–469
  geometric dimensioning and tolerancing and, 524–540 (*See also* Geometric dimensioning and tolerancing (GD&T))
  legal issues and, 340
  linetypes and, 497, 498
  process of creating, 548–553
  scale for, 484, 485
  sheet sizes for, 483–484
  specifications and, 545–546
  standards and, 368, 470–472
  surface control and, 540–544
  title and revision blocks and, 499–500
  tolerancing and, 514–524
  views from 3D models and, 485–497
  working drawings and, 472–483
Double-curved surfaces, 85
Double threads, 306
Downstream applications
  description of, 389
  equation solvers and, 390, 392
  finite element analysis and, 395–399
  graphs and, 392–395
  human factors and, 402–403
  integrated modeling and design software and, 403–405
  simulation software and, 399–402
  spreadsheets and, 389–391
dpi (dots per inch), 572
Draft (molded parts), 319, 320
Drag, 445
Drawing Exchange Format (DXF) (Autodesk), 387
Drawing number system, 341
Drawings. *See also* Documentation drawings; Schematic drawings; Sketches
  abbreviations for use on, 614–617

Drawings *(cont.)*
 approval and release of, 340–341
 assembly, 102–103, 472–476
 change orders on, 342
 characteristics of, 210–212, 238
 function of, 338
 isometric, 86–94, 473
 layout, 298–299
 overdimensioned, 514
 parametric modeling and, 254–259
 part, 476–482
 patent, 546–547
 as point in time, 343–344
 revision block on, 343
 working, 472–483
Drawing standards
 ANSI, 471–472
 data management and, 347–348
 organizations publishing, 610–613
Drawn by block, 340–341
Dr.DWG, 364
Drilling, 426
Drill press, 433–434
Drop shadows, 66–67
.dwf files, 364, 384
.dwg files, 364
.dxf files, 364, 588
Dynamic assemblies, 291

E

Easy Plot, 393
Economic constraints, 42
Edges, 85, 132
Editing, surface, 228
Einstein, Albert, 54
Electrical engineers, 29
Electric discharge machining (EDM), 427, 461, 462
Electrochemical machining, 427
Electronic packaging, 452
Electronic schematics, 67–68
Electronics symbols, 676
Ellers, Jae, 548–553
Ellipses
 in CAD system, 175
 description of, 174–175
 isometric, 91–92
 perimeter of, 176
Endmills, 309
Engineering Animation software, 242
Engineering change notification (ECN), 269, 342, 368
Engineering change order (ECO), 342
Engineering design. *See* Design
Engineering notebooks, 58
Engineering release, 53
Engineers, 29
English scales, 484
EPIC database (Boeing), 339
.EPS Encapsulated PostScript, 574
Equations, 266–268
Equation-solving software, 390, 392
Ergoman, 402
Ergonomic analysis
 description of, 41, 316
 function of, 403
 products applying, 43
 software for, 316–317
Euler's formula, 86–87
Excel (Microsoft), 414–416, 566
Exploded assembly drawings, 473–475
Exporting data
 CAD database and, 383
 common formats for, 385–387, 572
 file formats and, 384
 vector and raster data and, 388
Extension lines, 502
Extension springs, 307
External references, 291, 292
Extrusion, 184, 224, 427
Extrusion blow molding, 455
Extrusion molding, 454

F

Faces, 85, 86
Faceted representation, 233–234
Factors of safety, 376
Falloff, 577–578
Fasteners
 in assembly models, 305, 307
 sheet metal, 448, 449
 sketches of, 69
Feature-based modeling, 251–254. *See also* Parametric models
Feature control frame, 525–527, 536
Fehres, Carl, 366, 370
5th Glove, 217
File formats. *See also* Graphics formats
 color depth and, 573
 description of, 384
 graphics, 572–574
 resolution and, 572–573
 for 2D images, 574
Files
 assembly models and management of, 295
 backing up drawing, 350
 naming systems for, 346–347
 permission and ownership issues and, 348–349
 rendered, 575
 storage of, 340, 344–346, 350–351

Filister head cap screws, 653
Fillets
  in assembly models, 308–309
  as built-in feature, 270
  description of, 138, 271
Finite element analysis (FEA), 207, 395–399
First-angle projection, 124–125
Fits
  classes of, 518, 519
  clearance locational, 624–625
  elements of, 518
  force, 519
  force and shrink, 628
  hole-basis system of, 524
  interference, 311
  interference locational, 519, 627
  line, 311, 518
  between mating parts, 311–312, 518
  nominal size and basic size and, 312–313
  running, 519, 622–623
  shaft-basis system of, 524
  shrink, 519
  sliding, 519, 622–623
  transition, 311, 518
Fixed gauges, 436
Fixtures, 444–445
Flange, 271
Flash, 447
Flat-file databases, 356–357
Flat head cap screws, 652
Flatness tolerance, 527
Flat shading algorithm, 582–583
Flat socket head cap screws, 654
.FLC files, 590
.FLI files, 590
Fonts, 565
Food and Drug Administration (FDA), 340
Foot-pound-second (fps) system, 381
Force fit, 311, 519
Foreshortening, 87, 88
Forging, 427, 428
Form tolerances, 525–527
Frames, 587
Freebody diagrams, 64, 65
Full indicator movement (FIM), 534
Functional design constraints, 41
Fundamental deviation, 523

G

G-codes, 443
Generic NCI codes, 442
Geometric characteristics symbol, 525–526
Geometric dimensioning and tolerancing (GD&T). *See also* Dimensions; Tolerance
  basic dimensions and, 313, 534–535
  datum surfaces and datum features and, 531–532
  feature control frame and, 525–526, 536
  form tolerances for individual features and, 526–530
  function of, 524
  geometric characteristics symbols and, 525
  positional tolerance and, 536–538
  runout and, 534
  supplementary symbols and modifiers and, 539–540
  symbols for, 315, 677–678
  tolerances for related features and, 532–534
  tolerancing angles and, 538
Geometric entities, 682–687
Geometric parameters, 250–251
Geometric relationships, 178–180
Geometry
  descriptive, 213
  drawing, 166–167
  importing, 588–589
  of projection, 117
  of solid objects, 85–86
  of 3D modeling, 192–195
Gerisch, Mark, 241–243
.GIF format, 574, 580, 581
Glass box, 123
Global parameters, 268, 301
Golden Valley Electronic Association, 341
Go-no go plug gage, 314
Gothic lettering, 99
Gouraud shading algorithm, 583
Graphic exchange format, 386
Graphics
  communication using, 7–8, 55
  documentation using, 8–9
  for presentations, 562, 564, 572–574 (*See also* Presentations)
  visualization using, 6–7
Graphics formats. *See also* File formats
  color depth and, 573
  file formats and, 573–574
  resolution and, 572–573
Graphics tools case study (Tektronix)
  analysis in, 17–18
  design selection in, 14–15, 18–19
  documentation in, 20, 21
  ideation in, 12–14, 26
  implementation in, 19–20
  problem identification in, 11–12
  refinement in, 15–17
Graphs, 392–395, 566
Gravity, 379, 383
Grid dimensioning, 510, 511
Grid paper
  isometric, 89, 90
  sketching on, 57

Grinding, 427
Gripper alignment tool, 164
GRIP programming language, 267
Guidelines for Electronic Records and Signatures (Food and Drug Administration), 340

H

Hadamard, Jacques, 54
Half sections, 494–495
Haselhuhn, Hugo, 458–462
Hatching, 146, 492, 494
Head cap screws
  filister, 653
  flat, 652
  flat socket, 654
  round, 651
  socket, 655
Heating, ventilation, and ductwork symbols, 673–675
Helvetica, 565
Hems, sheet metal, 322, 323
Hexagon head bolts, 649
Hex head bolts, 145
Hex nuts and jam nuts, 650
Hidden linetypes, 144
History-based modeling programs, 252
Hole-basis system of fits, 524
Hole sizes, 638–639
Horizon line, 96
Hotspot, 577
Houlihan, John, 74–77
HTML (hypertext markup language), 384, 592
Human factors analysis, 402–403

I

ICEM Surf software, 241, 242
ICEM Surf surface modeler, 226–227
I-DEAS software, 414–417
Ideation
  brainstorming and, 48–50
  competition reviews and, 48
  consumer surveys and, 45, 47, 48
  decision process and, 50–51
  function of, 44
  listing and, 50
  literature searches and, 45–46, 48
  product/design reviews and, 51–54
  role of visualization in, 54–56
  sketching for, 74–77
  in Tektronix case study, 12–14, 26
IGES/PDES, 386
IGES standard, 386, 679–681
Impact tower fixture
  design of, 429–430
  drill press and, 433–434
  inspection and, 435–436
  lathe and, 434
  milling process for, 431–433
Implementation
  computer-aided manufacturing and, 436–443
  machining and, 430–426
  manufacturing processes and, 425–430
  metal forming and, 445–448
  modeling for manufacturing and, 458–462
  overview of, 424–425
  plastics manufacturing and, 453–457
  sheet metal manufacturing and, 448–453
  tooling and, 443–445
Inch dimension values, 503
Inch-pound-second (ips) system, 381
Inclined surfaces, 131, 132
Individual brainstorming, 49–50
Industrial design engineers, 29
Industrial Designers Society of America (IDSA), 46
Industrial engineers, 29
Industrial Fasteners Institute, 471
Infrared motion detector, 458–462
Initial graphics exchange specification (IGES), 218, 219, 386
Injection blow molding, 455, 456
Injection-molded plastics, 318–320
Injection molding, 454
InParts's DesignSuite, 304
Inspection tools, 314–315, 435–436
Institute of Electrical and Electronics Engineers (IEEE), 46
Integrated modeling and design software, 403–405
IntelliCAD, 384
Interference checking, 309–310
Interference fit, 311
Interference locational fit, 519
International Organization for Standardization (ISO)
  function of, 64, 352
  ISO 9000, 352
  standard sheet sizes, 483
  standards published by, 470
  STEP, 386
International tolerance grade (IT), 523
Internet
  checking for stock parts on, 303
  data management and, 363–365
  drawings in digital format via, 484
  images on, 573
  locating information on, 45, 46, 48
  manufacturing information on, 24
  VRML applications on, 595
Interpolated splines, 176
Intersecting lines, 178–179
Intersection
  as Boolean operation, 181, 182
  sketches of, 134–137
Inverse kinematics, 590, 591

Investment casting, 446, 447
IRI International, 267, 301, 342, 366, 562
Irregular surfaces, 186
ISO 9000, 352–354, 366
Isometric drawings
 description of, 87–89, 473
 example of, 86
 grid paper for, 89, 90
 orientation in, 89–90
 of piping layout, 92–94
 without using grid paper, 90–91
Isometric ellipses, 91–92
Isometric lines, 88
Isometric views
 of assembly drawings, 473, 475
 of cubes, 87
 from 3D models, 489–490

J

Jigs, 444, 445
JPEG, 574, 580

K

Kettner, Scott, 13
Keyframe animation, 589
Keys, 662, 663
Keyseats
 description of, 271
 dimensioning, 513
 standard, 663
Keyways, 271, 513
Knuckle threads, 305
Knurling, 271, 512–513

L

Lapping, 427
Larocque, Brandon, 275–279
Laser cutting, 427
Laternal motion gripper, 82
Lathe, 434
Lay, of surface, 542, 543
Layers, 212
Layout drawings, 298–299
Leader, 101
Least material condition (LMC), 539
Left-hand rule, 168
Left-hand threads, 306–307
Legal issues
 design constraints due to, 42–43
 electronically stored files as, 340
 retention of sketches and, 340
Leonardo da Vinci, 96
Lettering
 guidelines for, 99–100
 of notes on sketches, 101–102
 size standards for, 100, 101
Lighting, in rendered views, 575–579
Limiting element, 130
Limit tolerance, 310, 516
Line charts, 566, 568
Line fit, 311, 518
Lines
 in CAD systems, 172
 isometric, 88
 in multiview drawings, 130
 nonisometric, 91
 parallel, 178
 perpendicular, 178
 phantom, 498
 projection, 125–126
 true size, 148, 149
Linetypes
 centerline, 144–145
 elements of, 497–498
 hidden, 144
 interpreting, 143
 list of standard, 143
Listing, 50
Literature searches, 45–46
Locational clearance fit, 519
Location tolerance, 525
Lofting, 225
Love, Sidney F., 49
Lower deviation, 523
Lug, 271

M

Machining
 description of, 428, 430
 electric discharge, 427
 electrochemical, 427
 impact tower fixture, 429–436
MacNeal-Schwendler Corporation software, 242
Magnet carrier, 192–195
Mannequin Pro, 402, 403
Manufacturing
 case parts and, 321
 computer-aided, 436–443
 computer-integrated, 25
 design for, 317–318
 information on Web on, 24
 modeling for, 458–462
 modeling injection-molded plastic parts and, 318–320
 modeling machined parts and, 321
 modeling sheet metal parts and, 322–327
 plastics, 453–457
 sheet metal, 448–453
Manufacturing engineers, 24

Manufacturing processes
  machining as, 428–430
  overview of, 425–426
  that add material, 427
  that redistribute material, 427
  that remove material, 426–427
  tolerance for, 313–314
Manufacturing test engineers, 29
Marketing surveys. *See* Consumer surveys
Markus, Karen, 151, 156
Mass, 378
Mass density, 379
Mass properties
  acquiring information for, 376–377
  calculations for, 377–379
  material properties and, 381–383
  modeling software and, 380
  of right cylinder, 378
  units and assumptions of, 381
  using I-DEAS and Excel to find, 414–417
  verifying accuracy of, 380–381
Mastercam, 439–441
Material files, 381–382
Materials, rendering, 579, 580
MathCAD, 390, 392
Mating dimension, 507
Mating parts
  description of, 507
  fit between, 311–312, 518
Maximum material condition, 521–522
Maximum material condition (MMC), 539
Maxsyma, 393
McLean, Stan, 325–327
M-codes, 443
Measurement devices, 314–315
Mechanical Desktop
  AutoSM, 325–327
  external references in, 291
Mechanical engineers, 29
Mesh surfaces, 224–225
Metal forming
  centrifugal casting and, 446
  die casting and, 447
  forging and, 448
  investment casting and, 446, 447
  sand castings for, 445–446
Metric Conversion Act of 1975, 501
Metric dimension values, 503
Metric fit tables, 523–524
Metric system, 484, 485, 501
Metric threads, 305
Metric washers, 665
Microelectromechanical structures (MEMS), 151–156
MicroSim, 207
Microsoft Excel, 414–416, 566
Microsoft PowerPoint, 562, 569–571
Microsoft Word, 384
Microstation (Bentley Systems), 384
Middle-out design, 298
Milling, 426, 431
Milling machines, 431–433
Mini Drop-Jet Reservoir, 288
Mirrored shapes, 183–184
Modeling kernel, 235–236
Models/modeling. *See also* Assembly models; Parametric models; 3D models; 2D models
  analysis of, 17–18
  characteristics of various, 238–239
  choosing best method for, 237–240
  description of, 206
  function of, 19
  generation of, 16–17
  for manufacturing, 317–327
  qualities of good, 208–210
  refining design through, 204–205
  springs in, 308
  tolerance and, 359, 522–524
  types of, 206–208
Molding
  compression, 457
  description of, 427
  extrusion, 454
  extrusion blow, 455, 456
  injection, 454
  injection blow, 455, 456
  vacuum, 457
Molds
  construction of, 318–319
  rapid prototypes of, 406
Moment of inertia, 379
Monge, Gaspard, 117
Morphing, 589, 590
Mountain Cycle, 596–602
Moving entities, 188
MPEG, 590
MSC/NASTRAN, 243
MSC/PATRAN, 243
Muller, George, 49
Multiview sketches
  auxiliary views in, 148–149
  Cartesian coordinate system and, 120–121
  cylinders, spheres, and cones in, 132–133
  description of, 116–117, 210, 211
  dimensions in, 138–139
  first-angle and third-angle projection and, 124–125
  front view for, 128–129
  hex head bolts and nuts in, 145
  interpreting linetypes in, 143–145
  intersections and tangencies in, 134–138
  lines and points in, 130–131
  microelectromechanical structures and, 150–156
  normal, inclined, and oblique edges in, 132

normal, inclined, and oblique surfaces in, 131–132
orthographic projection and, 117–118
projection lines and, 125–126
reference surfaces and, 126–127
scale in, 138
section views in, 146–147
spacing between views in, 120
standard arrangement of views in, 118–119
systems of projection and, 122–123
visualizing 3D objects in, 140–142

N

National Computer Graphics Association (NCGA), 386
National IGES User Group (NIUG), 386
National Institute for Building Sciences (NIBS), 471
National Institute of Standards and Technology (NIST), 386, 471
Native file format, 384
NC codes, 442
NCI codes, 442, 443
Neck, 271
Needs statement, 11
Nep'tune Sea Technology, Ltd., 248–250
NetView, 364
Network File Manager (NFM3), 367, 370
Next Generation Space Telescope (NGST), 206
Nodes, 592–594
Nominal size, 312–313
Noncontact inspection, 315
Nonisometric lines, 91
Nonparametric dimensions, 274
Normal surfaces, 131, 132
Notes, on sketches, 101–102
Numerical control (NC), 436–437
Numerical control (NC) codes, 442
NURBS surfaces, 225, 226, 228

O

Object linking and embedding (OLE), 574–575
Oblique sketches
description of, 94–95, 122
example of, 86
Oblique surfaces, 131, 132
Offset sections, 495
Olson, Bruce, 592
Omnibus Trade and Competitiveness Act of 1988, 501
Omnidirectional lights, 576
Optical illusions, 58–59
Orientation tolerance, 525
Original design, 23
Orthographic views
of assembly drawings, 473, 474
of bolts, 145
Cartesian coordinate system and, 120–121
description of, 117–118, 146–147
of glass box, 123
of hex head bolts, 145
parametric modeling software to create, 485–486
Osborne, Alex, 48, 49
Overdimensioning, 514
Ownership issues, 348–349

P

Packaging engineers, 452–453
Panning, 190
Parallelism tolerance, 532, 533
Parallel light, 577
Parallel lines, 178
Parametric modeling software
capabilities of, 296–297, 310, 383
creating orthographic views using, 485–486
double and triple threads and, 306
fillets and rounds and, 308–309
right- and left-hand threads and, 306–307
springs and, 307–308
standard parts and, 303–304
static parts and, 304–305
threads and fasteners and, 305
Parametric models
adding features to, 261–262
advantages of, 248–250
assembly mode and, 272–273, 292–293
base features and, 260–261
capturing design intent with, 275–279
datum planes and surfaces and, 263–265
description of, 16, 236–237, 248
design flexibility and, 254–260
dimensions in, 265–268, 507–508
drawing mode and, 273–274
editing and, 268–272
as feature-based modeling, 251–254
parent-child relationships and, 262–263
sketches for, 246
Parametric part modeling, 261–262
Parametrics, 250–251
Parasolid kernel, 235
Parent-child relationship, 262–263
Parison, 455
Part drawings
description of, 476–477
elements of, 477
examples of, 478–482
Parting line, 319
Part reviews, 53
Parts list, 476
PartSpec, 303
Patches, 227
Patent drawings, 546–547

Patents, 45
PDES (Product Data Exchange using STEP), 386
Pencils, 57
Pentecost, Jeffrey, 225
Perception, visual, 61, 62
Perfect form boundary, 522
Permanent casting, 446
Permission, 348–349
Perpendicularity, 532, 533
Perpendicular lines, 178
Perspective sketches
  description of, 95–97
  example of, 86
  guidelines for creating, 97–99
  uses for, 98
Phantom lines, 498
Phong shading algorithm, 583–584
Physical models
  changes in, 215–216
  characteristics of, 238
  elements of, 214–215
Pictorial sketches
  assembly, 102–103
  benefits and drawbacks of, 116
  description of, 86–87
  design process and, 103, 108
  guidelines for, 104–107
  isometric, 87–93
  lettering on, 99–102
  oblique, 94–95
  perspective, 95–99
  solid objects and, 85–86
Pie charts, 566, 567
Piping layout, 92–94
Piping symbols, 472, 672
Pixel graphic file, 580
Pixels, 572, 573, 580
Plain washers, 664
Planar objects, 85
Planes, 173
Plastic parts, 318–320
Plastics manufacturing
  compression molding and, 457
  description of, 453–454
  extrusion blow molding and, 455, 456
  extrusion molding and, 454
  injection blow molding and, 455, 456
  injection molding and, 454
  vacuum molding and, 457
Plating, 427
Plynetics, 20
Point light, 576
Points
  in CAD systems, 171–172
  in multiview drawings, 130–131
  type size, 565
Polar coordinates, 170
Polyhedra, 85, 86, 449
Positional tolerance, 536–538
PostScript, 574
Pounds force, 382
Pounds mass, 382
PowerPoint (Microsoft), 562, 569–571
Prebstell, Bob, 15
Preferred hole basis clearance fits—cylindrical fits, 630–631
Preferred hole basis transition and interference fits—cylindrical fits, 633–634
Preferred shaft basis clearance fits—cylindrical fits, 634–635
Preferred shaft basis transition and interference fits—cylindrical fits, 636–637
Presentations
  charts and graphs for, 566–569
  elements of effective, 562–564
  function of, 562
  graphics formats for reports and, 572–574
  object linking and embedding and, 574–575
  rendering and, 575–585
  software for, 562, 569–572
  text for, 565–566
Primary auxiliary views, 490
Primitives, solid, 180–182
Problem definition, 40–41
Product data management (PDM)
  archiving work history and, 360
  description of, 354–355
  individual productivity and, 361
  interface tools and, 361–362
  managed work flow and, 359–360
  organized by product, 355–356
  relational database and, 356–358
Product development, 15–17
Product proposals, 52
Product release, 53
Product reviews, 51–53
Products, 25, 355–356
Pro/Engineer
  assembly constraints, 296
  auxiliary views with, 491
  description of, 4
  display of CAD files and, 364
  Internet file operations and, 365
  mass property information and, 377
  material files, 382
  Model Info command, 390
  modeling engine and, 235
  parts availability and, 303
  section views with, 492–494
  tolerance range and, 311
  Use Edge, 261
  use of, 16, 18, 19, 251, 256, 257, 548, 550, 552, 553

Professional organizations. *See also specific standards*
  drawing standards published by, 610–613
  Web sites for, 46
Profile tolerance, 529, 530
Program managers, 29
Projected tolerance zones, 539
Projection
  first-angle and third-angle, 124–125
  system of, 122–123
Projection lines, 125–126
Projectors, 97
Proportion estimates, 59–60
Prototype drawings, 302
Prototypes
  changes in, 215–216
  description of, 214
  development of, 405–406
  rapid, 215, 376, 406–413, 460
  virtual, 217, 406
Punch, 324
Punching, 427
Punch press, 324

Q

Quality engineers, 29
Quality management, 352–354
Quattro Pro, 566
Quicktime, 590

R

Radiosity rendering algorithm, 585
Radius, 173
Radius of gyration, 379
Rapid prototypes
  description of, 215
  equipment for, 376
  function of, 406
  production of, 407, 408, 460
Rapid prototyping systems
  rapid tooling and, 412–413
  3D printing, 411
  types of, 407, 409–411
Raster graphics, 388, 580
Ray tracing algorithm, 584
Reaming, 427
Redesign, 23
Reference dimensions, 274, 508
Reference surfaces, 126–127
Reflectivity, 579
Refractivity, 579
Regular helical spring lock washers, 666
Reisinger, Robert, 596, 597, 601, 602
Related views, 126
Relational databases, 356–358
Relative coordinates, 169
Rendering
  algorithms for, 582–585
  animation, 590
  antialiasing and, 582
  description of, 575
  lighting and, 575–579
  materials and, 579
  resolution and color depth and, 580–581
Resolution
  choices for, 580–581
  description of, 572–573
Retina, 62
Reverse engineering, 225–227
Revision blocks, 342–343, 500
Revolution, 185, 224
R/Greenberg Associated, 355
Right-hand rule, 167–168
Right-hand threads, 306–307
Risers, 445–446
Rockwell Corporation, 361
Rods, 62
Romulus Predator supercar
  conceptual design of, 226–228
  as middle-out design, 298
  modeling methods used for, 241–243
Rotating entities, 188
Roughness, surface, 541, 544
Round head cap screws, 651
Round head machine screws, 656
Rounds
  in assembly models, 308–309
  description of, 138, 271
Rule-based programming language, 390
Running fit, 519
Runout, 534
R^2 value, 393

S

Safety engineers, 29
Sand casting, 445–446
Sanding, 427
Sans serif fonts, 565
Scales
  CAD database and, 484
  description of, 138
  list of standard, 485, 621
Scaling entities, 188
Schematic drawings. *See also* Drawings; Sketches
  block diagram, 64, 66–67
  description of, 64–66
  detailed representation, 70
  electronic, 67–68
  simplified representation, 70
  of springs, 72, 73

Schematic drawings *(cont.)*
    thread and fastener, 68–70
    thread notes and, 71–72
    thread representation, 70
Screws, head cap, 651–655
Screw threads: American National and Unified, 642–644
Secondary auxiliary views, 490, 491
Section views
    of assembly drawings, 472–473
    CAD and, 473
    description of, 492
    example of, 493
    hatching and, 492, 494
    in multiview sketches, 146–147
    orthographic, 146–147, 472–473
    types of, 494–497
Seed parts, 302
Selection design, 22
Serif fonts, 565
Serifs, 99
Set screws, 657
Shading, 60
Shaft sizes, 640–641
Shaft support model, 252, 253
Shaping, 427
Sheet metal
    bending and, 427, 450–452
    fasteners and, 448, 449
    modeling, 322–327
    packaging and, 452–453
    pattern layout and, 449–452
    stamping and, 452
Sheet sizes (paper), 483–484, 618
ShockWave, 596–602
Shrinkage, 320
Shrink fit, 519
Sigmaplot, 393
Simplified representations, 70
Simulation software, 399–402
Single-curved surfaces, 85
SIRTF space telescope, 374
Size parameters, 250
Skeleton models, 299–300
Skeletons, 297, 299
Sketches. *See also* Documentation drawings; Drawings
    assembly, 102–103
    communication through, 55
    constraints applied to, 256–258
    construction lines and, 63–64
    contour, 58, 63–64
    design process and, 103, 108
    estimating proportions and, 59–60
    function of, 6, 7
    generation of, 12–14
    guidelines for, 58, 104–107, 486–487
    importance of, 84
    isometric, 87–93 (*See also* Isometric drawings)
    lettering in, 99–102
    multiview, 116–156 (*See also* Multiview sketches)
    oblique, 86, 94–95, 122
    optical illusions and, 58–59
    overconstrained and underconstrained, 258–259
    perspective, 87, 95–99
    pictorial, 86–108 (*See also* Pictorial sketches)
    positive and negative spaces in, 60–61
    practice and, 58
    schematic, 64–77 (*See also* Schematic drawings)
    shading in, 60
    of solid objects, 85–86
    tools for, 57–58
    viewpoint in, 60
    visual perception and, 61, 63
Sliding fit, 519
Smith, Greg, 597
Society of Automotive Engineers (SAE), 471
Society of Manufacturing Engineers (SME), 46
Society of the Plastics Industry, 320
Socket head cap screws, 655
Software. *See* Computer software; *specific software*
Solid modeling software, 242, 243, 383. *See also* Computer software
Solid models
    accuracy of, 233–235
    BREP method and Boolean operations and, 232–233
    CSG method and Boolean operations and, 232
    description of, 231–232
    hybrid systems and, 233
    modeling kernels and, 235–236
    3D, 236
Solid objects, 85–86
Solid primitives, 180–182
SolidWorks
    display of CAD files and, 364
    Interactive Product Animator, 596–601
    parametric modeling and, 248, 250, 253, 275–278
    passive infrared motion detector and, 458–462
Space, positive and negative, 60–61
Space Infrared Telescope Facility (SIRTF), 208, 209, 415, 417
Specifications
    description of, 43
    for documentation drawings, 545–546
    initial graphics exchange, 386
Specific gravity, 383
Spheres, 132, 133
Spherical coordinates, 170
Spinning, 427
Spline curves
    description of, 176–178
    surface models and, 225, 226

Spotface, 271
Spreadsheets
  description of, 389–391
  for presentations, 566
Springs
  in assembly models, 307–308
  compression, 307
  description of, 72
  extension, 307
  sketches of, 72–73
Sprints, 308
Sprue, 445
Square, 305
Square and acme threads, 645
Square bolts, 647
Square nuts, 648
Stamping, 427, 452
Standard for Aluminum Sand and Permanent Mold Casting (Aluminum Association), 321
Standard keys and keyseats, 663
Standard parts, 303–304
Standards. *See also specific standards*
  ANSI, 64, 67, 68, 471, 626, 667, 668, 672–675
  drawing, 347–348, 471–472, 610–613
  types of, 470–471
Standard scales, 485, 621
Standard worm threads, 305
Static assemblies, 291
Static parts, 304–305
STEP standard, 386
STL files, 387, 588
Storyboards, 564, 587–588
Straightness tolerance, 527
Straight pins, 661
Strategix ID, 192
Subassemblies
  description of, 290–291
  skeleton models of, 299, 300
Surface area, 378
Surface control, 540–544
Surface finish marks, 542, 543
Surface models
  accuracy of, 229–230
  characteristics of, 239
  complex, 227–228, 230
  description of, 223
  editing and, 228–229
  extrusion and revolution and, 224
  meshes and, 224–225
  reverse engineering and, 225–227
  spline approximations and, 225
  storing information in database and, 223–224
  use of, 230–231, 240
Surface texture, 541
Surfcam, 439
Swage nuts, 448
Sweeping, 184
Symbols
  ANSI electronics, 67, 68, 676
  ANSI heating, ventilation, and ductwork, 673–675
  ANSI piping, 472, 672
  AWS welding, 669–671
  for dimensioning common features, 511–514
  for direction of lay, 543
  geometric dimensioning and tolerancing, 315, 677–678
Symmetry
  description of, 182–183
  parting line, 183–184
System International (SI) standard units, 381, 501, 503

T

TableCurve, 393, 394
Tangencies, 134, 137
Tangent, 179
Teams
  benefits of, 28
  CAD database and, 30–32
  concurrent engineering, 351–352
  members of, 29
  structure of, 30
Technical writers, 29
Tektronix, Inc.
  design documentation at, 548–553
  organization of teams in, 29, 30
  overview of, 12
  skeleton model for laser printer at, 299, 300
  WFM 90 project, 11–21, 26
Templates, 302
Tessellation lines, 234
.TGA Targa, 574
Thermoplastics, 454
Thermoset, 454
Third-angle projection, 124–125
Thomas Register, 46, 303
Thornburg, Kent, 225
Thread forms, 305–306
Thread notes, 71
Threads
  acme, 305, 645
  in assembly models, 305
  description of, 68
  double and triple, 306
  right- and left-hand, 306–307
  schematic representations of, 70
  screw, 642–644
  sketches of, 69–70
  square, 645
  taper pipe, 646
  uses for, 69

Three-axis machines, 438
3D models
  advantages of, 27, 56, 248
  CAD, 218–237 (*See also* CAD)
  computer-generated, 217–218
  computer programs for, 16, 18, 117
  constructing physical, 142
  as design documentation, 469
  to determine fit, 309–317
  developing views from, 485–497
  geometry of, 192–195
  interpreting views by removing portions of, 141
  parametric, 236–237, 248–280 (*See also* Parametric models)
  physical, 214–216, 238
  solid, 231–236, 239
  steps involved in, 241–243
  surface, 223–231
  understanding, 28
  visualizing, 140–141
  wireframe, 218–223, 239
3D Studio (Autodesk), 562, 588
TIFF, 580
.TIF Tagged Image Format, 574
Time-related issues, 43
Time series, 568
Timex TurnAndPull alarm, 74–77, 223
Title blocks
  description of, 302
  function of, 340, 499
  illustrations of, 341, 618–620
  information included in, 499–500
  tolerance notes in, 515
TK Solver, 390, 392, 404
Tolerance. *See also* Geometric dimensioning and tolerancing (GD&T)
  angularity, 538
  baseline dimensioning and, 517–518
  basic dimensions and, 534–535
  circularity, 528
  concentricity, 533–534
  cylindricity, 528
  datum surfaces and features and, 531–532
  description of, 310–311, 514–515
  expected range of, 424
  fit and, 311, 312, 518–522
  flatness, 527
  form, 525–530
  general notes for, 515–516
  geometric, 525
  international grades for, 629
  limit, 310, 516
  modeling and, 359, 522–524
  nominal size and basic size and, 312–313
  parallelism, 532, 533
  per manufacturing process, 313–314
  perpendicularity, 532, 533
  plus/minus, 516
  positional, 536–538
  profile, 529, 530
  for related features, 532–534
  standards for, 500
  straightness, 527
  symbols and modifiers for, 539–540
  X-Y bead picker bridge and, 466
Tolerance stacking, 516–517
Tolerance zones
  description of, 523–524, 537, 538
  projected, 539
Tooling
  description of, 443–444
  fixtures and, 444–445
  jigs and, 444
  rapid, 412–413
Toolmaker's ink, 431, 432
Top-down design, 298
Total runout, 534
Townsend, William, 275
Toy Story (Disney), 587
Transformations
  description of, 187–188
  geometric, 188–189
  viewing, 189–191
Transition fit, 311, 518
Translation, 188
Transom Jack, 402
Triangulated irregular networks (TINs), 224
Tri-Cycle Product Design, Inc., 336, 363
Triple threads, 306
Turning, 426
Tweaking, 228
Tweening, 589
Twist drill sizes, 658–659
Two-axis machines, 438–439
2D models
  CAD, 168, 212–213
  characteristics of wire, 238
  as design documentation, 469
  paper drawings as, 210–212
  wireframe models *vs.*, 218, 220–222

U

Unidirectional values, 504
Unified threads, 305
Unigraphics Solutions, 242, 243
Union, as Boolean operation, 181, 182
Upper deviation, 523
U.S. Patent Office, 45, 546
User coordinate systems, 186–187

V

Vacuum molding, 457
Vanishing point, 96, 97
Vdraft Internet Tools, 364
Vector-based images, 573
Vector graphics programs, 388
Velcro, 44
Vertices, 85, 86
Video cards, 573
Viewing transformations, 189–191
View rotation, 190
Views from 3D models
  auxiliary, 490–492
  description of, 486–487
  isometric, 489–490
  placement of views and, 487–489
  section, 492–497
Virtual prototypes, 217, 406
Virtual reality, 217–218
VisFly (EAI), 562
Vision, 62
Visual communication. *See* Graphics; Sketches
Visualization, 6–7, 54–56
VKI Technologies, 294, 325, 477
Volume, 378
VRML (virtual reality modeling language), 592–595

W

Walczak, Walt, 401
Warped surfaces, 85
Washers
  metric, 665
  plain, 664
  regular helical spring lock, 666
Waviness, surface, 542
Weighted decision matrix, 50–52
Welding, 427
Welding symbols (AWS), 669–671
Windchill (PTC), 355, 359
Wireframe models
  advantages of, 219–220
  characteristics of, 239
  description of, 218
  disadvantages of, 220–222
  software and hardware for, 222–223
Wireframe skeleton, 299, 300
Woodruff keys, 662
Word (Microsoft), 384
Workflow, 359–362
Work history, 360
Working drawings
  assembly drawings as, 472–476
  description of, 472
  part drawings as, 476–482
  preparation of, 483
Working Model
  automated constraint mapping and, 403–404
  description of, 18
  illustration of, 209
  use of, 400–403
Working Model Motion, 400
WorkSafety, 402
World Wide Web. *See* Internet
WORM CD-ROM storage, 351
Write once read many (WORM), 350

X

X direction, 431, 438

Y

Y direction, 431, 438

Z

Z direction, 431, 438
Zooming, 190
Zuma coffee brewer, 290, 294–295, 325–327